MODELING
AND
ANALYTICAL
METHODS IN
TRIBOLOGY

CRC SERIES: MODERN MECHANICS AND MATHEMATICS

Series Editors: David Gao and Ray W. Ogden

PUBLISHED TITLES

BEYOND PERTURBATION: INTRODUCTION TO THE HOMOTOPY ANALYSIS METHOD
by Shijun Liao

Classical and Generalized Models of Elastic Rods
by Dorin Ieşan

CONTINUUM MECHANICS AND PLASTICITY
by Han-Chin Wu

HYBRID AND INCOMPATIBLE FINITE ELEMENT METHODS
by Theodore H.H. Pian and Chang-Chun Wu

INTRODUCTION TO ASYMPTOTIC METHODS
by Jan Awrejcewicz and Vadim A. Krysko

MECHANICS OF ELASTIC COMPOSITES
by Nicolaie Dan Cristescu, Eduard-Marius Craciun, and Eugen Soós

MODELING AND ANALYTICAL METHODS IN TRIBOLOGY
by Ilya I. Kudish and Michael J. Covitch

MICROSTRUCTURAL RANDOMNESS IN MECHANICS OF MATERIALS
by Martin Ostoja-Starzewski

CRC SERIES: MODERN MECHANICS AND MATHEMATICS

MODELING AND ANALYTICAL METHODS IN TRIBOLOGY

Ilya I. Kudish

Kettering University

Flint, Michigan

Michael J. Covitch

Lubrizol Corporation

Wickliffe, Ohio

CRC Press
Taylor & Francis Group
Boca Raton London New York

CRC Press is an imprint of the
Taylor & Francis Group, an **informa** business
A CHAPMAN & HALL BOOK

CRC Press
Taylor & Francis Group
6000 Broken Sound Parkway NW, Suite 300
Boca Raton, FL 33487-2742

First issued in paperback 2019

© 2010 by Taylor and Francis Group, LLC
CRC Press is an imprint of Taylor & Francis Group, an Informa business

No claim to original U.S. Government works

ISBN-13: 978-1-4200-8701-7 (hbk)
ISBN-13: 978-0-367-38379-4 (pbk)

Library of Congress Cataloging-in-Publication Data

Kudish, Ilya I.
    Modeling and analytical methods in tribology / authors, Ilya I. Kudish, Michael Judah Covitch.
        p. cm. -- (CRC series, modern mechanics and mathematics)
    "A CRC title."
    Includes bibliographical references and index.
    ISBN 978-1-4200-8701-7 (hardcover : alk. paper)
    1. Tribology--Mathematics. 2. Lubrication and lubricants--Mathematical models. 3. Friction--Mathemantical models. I. Covitch, Michael Judah. II. Title. III. Series.

TJ1075.K78 2010
621.8'9--dc22   2010021421

Visit the Taylor & Francis Web site at
http://www.taylorandfrancis.com

and the CRC Press Web site at
http://www.crcpress.com

Contents

1 Basics of Asymptotic Expansions and Methods   **1**

  1.1   Introduction   1

  1.2   Ordering, Order Sequences, and Asymptotic Expansions . . .   1

  1.3   Asymptotic Sequences and Expansions   3

  1.4   Asymptotic Methods   4

  1.5   Exercises and Problems   6

2 Contact Problems for Coated and Rough Surfaces   **9**

  2.1   Introduction   9

  2.2   Some Classic Results for Smooth Elastic Solids 10

       2.2.1   Formulas and Results for Elastic Half-Space 10

       2.2.2   Formulas and Results for Elastic Half-Plane 11

  2.3   Spatial Rough Contacts Modeled by Nonlinear Coating . . . 13

       2.3.1   Problem Formulation 13

       2.3.2   Existence and Uniqueness of the Problem Solution . . 15

       2.3.3   Solution Properties 18

       2.3.4   Problems with Fixed Contact Region 22

       2.3.5   Problems for Rough and Smooth Surfaces 24

       2.3.6   A Different Approach to Problems (2.35)-(2.37) 26

  2.4   Asymptotic Analysis of Plane Rough Contacts 28

       2.4.1   Plane Contact Problems with Fixed Boundaries 30

       2.4.2   Plane Contact Problems with Free Boundaries 34

       2.4.3   Padé Approximations 38

       2.4.4   Analysis of Plane Rough Contacts with Fixed and Free

              Boundaries 39

  2.5   Numerical Methods and Results for Rough Contacts 45

       2.5.1   Numerical Methods for Problems in the Original For-

              mulations 45

       2.5.2   Numerical Methods for Problems in the Asymptotic

              Formulations 51

       2.5.3   Numerical Results for Contact Problems for Rough

              Surfaces 53

  2.6   Analysis of Axially Symmetric Rough Contacts 61

       2.6.1   Axially Symmetric Rough Contacts with Fixed

              Boundaries 66

      2.6.2   Axially Symmetric Rough Contacts with Free Boundaries 68

      2.6.3   Numerical Results for Axially Symmetric Rough Contacts 71

 2.7   An Example of an Application to Roller Bearings 73

 2.8   Closure   74

 2.9   Exercises and Problems 76

3 Contact Problems with Friction   **81**

 3.1   Introduction 81

 3.2   Plane Frictional Contacts with Fixed Boundaries 84

      3.2.1   Application of Regular Perturbations. Contact with Fixed Boundaries 84

      3.2.2   Application of Matched Asymptotic Expansions. Contact with Fixed Boundaries 85

 3.3   Plane Frictional Contacts with Free Boundaries 88

      3.3.1   Application of Regular Perturbations. Contact with Free Boundaries 88

      3.3.2   Application of Matched Asymptotic Expansions. Contact with Free Boundaries 90

 3.4   Plane Frictional Rough Contacts Modeled by Nonlinear Coating   92

      3.4.1   Existence and Uniqueness of the Problem Solution . . 94

      3.4.2   Qualitative Properties of the Solution 97

 3.5   Asymptotic and Numerical Analysis for Large Roughness . . 99

      3.5.1   Asymptotic Analysis for Large Roughness 100

      3.5.2   Numerical Solutions for Rough Contacts with Friction 101

 3.6   Closure   105

 3.7   Exercises and Problems   106

4 Rheology of Lubricating Oils   **109**

 4.1   Introduction 109

 4.2   Rheology Relationships for Lubricating Oils 110

      4.2.1   Definition of Viscosity 110

      4.2.2   The Effect of Temperature on Viscosity 111

      4.2.3   The Effect of Shear Rate on Viscosity 113

      4.2.4   The Effect of Pressure on Viscosity 115

      4.2.5   The Effect of Suspended Contaminants on Viscosity . 116

      4.2.6   Density   119

      4.2.7   Thermal Conductivity 120

 4.3   Polymer Thickening and Shear Stability 121

      4.3.1   Molecular Weight 121

      4.3.2   Dilute Solution Rheology 121

      4.3.3   Shear Stability 122

 4.4   Closure   124

4.5	Exercises and Problems	125

5 Properties of Multi-grade Lubricating Oils   129

5.1	Introduction	129
5.2	Multi-grade Lubricating Oils	130
	5.2.1 Engine Startability at Low Temperature	130
	5.2.2 Engine Oil Pumpability at Low Temperature	131
	5.2.3 Bearing Protection	132
	5.2.4 Engine Oil Viscosity Classification SAE J300	132
	5.2.5 Automotive Gear Oil Viscosity Classification SAE J306	134
5.3	Viscosity Modifiers	135
	5.3.1 Olefin Copolymer (OCP) Viscosity Modifiers	135
	5.3.2 Polyalkylmethacrylate (PMA) Viscosity Modifiers . .	137
	5.3.3 Styrene-Alkylmaleate Ester (SME) Copolymers Viscosity Modifiers	138
	5.3.4 Hydrogenated Styrene Diene (HSD) Copolymer Viscosity Modifiers	139
	5.3.5 Star Copolymer Viscosity Modifiers	141
	5.3.6 Lubricant Applications	142
5.4	Closure	145
5.5	Exercises and Problems	146

6 Degradation of Linear Polymers   151

6.1	Introduction	151
6.2	Kinetic Equation for Degrading Linear Polymers	152
6.3	Probability of Scission of Linear Polymer Molecules	156
6.4	Conditional Probability of Scission for Linear Polymers . . .	159
6.5	Lubricant Viscosity and Polymeric Molecules	161
6.6	Some Properties of the Kinetic Equation	162
6.7	A Limiting Case of the Kinetic Equation	170
6.8	Numerical Method for the Kinetic Equation	171
6.9	Numerical Solutions of the Kinetic Equation	175
6.10	Closure	182
6.11	Exercises and Problems	183

7 Degradation of Star Polymers   187

7.1	Introduction	187
7.2	System of Kinetic Equations for Star Polymers	187
7.3	Probabilities of Scission	196
7.4	Forming Star Polymeric Molecules	197
7.5	Approximation of Star Polymer Initial Distribution	200
7.6	Lubricant Viscosity and Polymer Distribution	204
7.7	Some Properties of the System of Kinetic Equations	205
7.8	Numerical Method for Kinetic Equations	220

7.9	Numerical Results for Lubricants with Star Polymers	223
7.10	Closure	230
7.11	Exercises and Problems	231

8 Review of Data on Contact Fatigue   **233**

8.1	Introduction	233
8.2	Contact and Residual Stresses	233
8.3	Material Defects and Lubricant Contamination	242
8.4	Bearing Fatigue Life and Contact Friction	251
8.5	Crack Development and Material Microstructure	254
	8.5.1 Crack Initiation and Crack Propagation	254
	8.5.2 Material Microstructure and Contact Fatigue Life	260
	8.5.3 Subsurface and Surface Material Cracking	261
8.6	Some Contemporary Contact Fatigue Models	265
	8.6.1 Mathematical Models for Fatigue Life of Bearings	266
	8.6.2 Mathematical Models for Fatigue Life of Gears	275
8.7	Closure	277
8.8	Exercises and Problems	279

9 Fracture Mechanics and Contact Fatigue   **295**

9.1	Introduction	295
9.2	Modeling the Vicinity of Crack Tips	296
	9.2.1 Problem Formulation and Solution for a Single Crack	297
	9.2.2 Multiple Cracks in an Elastic Plane	300
	9.2.3 Multiple Cracks in an Elastic Half-Plane	305
9.3	Perturbations for Multiple Cracks in a Half-Plane	308
	9.3.1 Perturbation Analysis for Multiple Cracks in an Elastic Plane	308
	9.3.2 Perturbation Analysis for Multiple Cracks in an Elastic Plane	318
	9.3.3 Perturbation Solution for Multiple Cracks in an Elastic Half-Plane	323
	9.3.4 Stress Intensity Factors for Multiple Cracks in an Elastic Half-Plane	330
9.4	Contact Problem for a Cracked Elastic Half-Plane	336
	9.4.1 Problem Formulation	337
	9.4.2 Problem Solution	342
	9.4.3 Contact Problem for Lubricated Solids	348
	9.4.4 Numerical Results	349
9.5	Directions of Fatigue Crack Propagation	356
9.6	Lubricant-Crack Interaction. Origin of Fatigue	360
	9.6.1 General Assumptions and Problem Formulation	362
	9.6.2 Problem in Dimensionless Variables	370
	9.6.3 Numerical Method	371

	9.6.4	Numerical Solutions for Subsurface and Surface Cracks	376
	9.6.5	Comparison of Analytical and Numerical Solutions for Small Cracks	394
	9.6.6	Origin of Contact Fatigue. Fatigue Life of Drivers versus Followers	398
9.7		Two-Dimensional Statistical Model of Contact Fatigue	417
	9.7.1	Initial Statistical Defect Distribution	418
	9.7.2	Crack Propagation versus Crack Initiation	420
	9.7.3	Applicability of Fracture Mechanics to Contact Fatigue	421
	9.7.4	Direction of Fatigue Crack Propagation	423
	9.7.5	Crack Propagation Calculations	424
	9.7.6	Crack Statistics	428
	9.7.7	Local Fatigue Damage Accumulation	430
	9.7.8	Survival Probability of Material as a Whole	431
	9.7.9	Variable Loading and Contact Fatigue	434
9.8		Analysis of the Pitting Model	436
	9.8.1	Analytical Analysis of the Pitting Model	436
	9.8.2	Numerical Analysis of the Pitting Model	440
9.9		Contact Fatigue of Rough Surfaces	445
	9.9.1	Contact Stresses in Rough Contacts	445
	9.9.2	Modeling Contact Fatigue in Rough Contacts	449
9.10		Three-Dimensional Model of Contact Fatigue	455
	9.10.1	Initial Statistical Defect Distribution	455
	9.10.2	Direction of Fatigue Crack Propagation	456
	9.10.3	Crack Propagation Calculations	457
	9.10.4	Crack Statistics	459
	9.10.5	Local and Global Fatigue Damage Accumulation	460
	9.10.6	Examples of Torsional and Bending Fatigue	463
9.11		Contact Fatigue of Radial Thrust Bearings	465
	9.11.1	Case of Axial Loading	466
	9.11.2	Case of Radial Loading	466
9.12		Closure	467
9.13		Exercises and Problems	469

10 Analysis of Fluid Lubricated Contacts   **479**

10.1		Introduction	479
10.2		Simplified Navier-Stokes and Energy Equations	480
10.3		Lightly Loaded Lubrication Regimes	484
	10.3.1	Problem Formulation	485
	10.3.2	Perturbation Solution of the EHL Problem	488
	10.3.3	Thermal EHL Problem	494
	10.3.4	Numerical Method for Lightly Loaded EHL Contacts	496
10.4		Pre-critical Lubrication Regimes	501

  10.4.1 Problem Formulation 504

  10.4.2 Asymptotic Analysis of the Problem 506

  10.4.3 Asymptotic Analysis of the System of Equations (10.84)-(10.88) 513

 10.5 Compressible Fluids in Heavily Loaded Contacts 515

 10.6 Over-critical Lubrication Regimes 517

  10.6.1 Problem Formulation 517

  10.6.2 Structure of the Solution 518

  10.6.3 Auxiliary Gap Function $h_H(x)$ 530

  10.6.4 Solutions in the ϵ_q-Zones 533

  10.6.5 Asymptotic Analysis of the Inlet and Exit ϵ_0-Zones . . 536

  10.6.6 Choosing Pre- or Over-critical Lubrication Regimes . . 542

  10.6.7 Analysis of the Ertel-Grubin Method 543

 10.7 Numerical Solution for EHL Contacts 547

  10.7.1 Numerical Procedure 548

  10.7.2 Some Numerical Results 551

 10.8 Numerical Solution of Asymptotic Equations 555

  10.8.1 Numerical Solution in the Inlet Zone 557

  10.8.2 Numerical Solution in the Exit Zone 561

  10.8.3 Some Numerical Results in the Inlet and Exit Zones . 565

  10.8.4 Numerical Precision and Stability Considerations . . . 571

 10.9 Analysis of EHL Contacts for Soft Solids 576

  10.9.1 Formulation of an EHL Problem for Soft Solids 577

  10.9.2 Qualitative Analysis of the EHL Problem 578

  10.9.3 Surface Velocities for Soft Solids 579

 10.10 Thermal EHL Problems 590

  10.10.1 Formulation of TEHL Problem 591

  10.10.2 Analytical Approximations for Newtonian Fluids in TEHL Contacts 595

  10.10.3 Numerical Solutions of Asymptotic TEHL Problems for Newtonian Lubricants 601

 10.11 Regularized Solution of Asymptotic Problems 604

 10.12 Regularization of the Isothermal EHL Problem 609

  10.12.1 Numerical Method for the Isothermal EHL Problem . 609

  10.12.2 Some Numerical Solutions of the Regularized Isothermal EHL Problem 617

 10.13 Numerical Validation of the Asymptotic Analysis 620

 10.14 Practical Use of the Asymptotic Solutions 625

 10.15 Approximations for Non-Newtonian Fluids 626

  10.15.1 Formulation of Isothermal EHL Problem for Non-Newtonian Fluids 627

  10.15.2 Isothermal EHL Problem for Pure Rolling. Pre- and Over-critical Lubrication Regimes. Some Numerical Examples 628

10.15.3 Isothermal EHL Problem for Relatively Large Sliding. Pre- and Over-critical Lubrication Regimes 651

10.15.4 Choosing Pre- and Over-critical Lubrication Regimes for Non-Newtonian Lubricants 661

10.15.5 Non-Newtonian Lubricants and Scale Effects 665

10.16 TEHL Problems for Non-Newtonian Lubricants 675

10.16.1 TEHL Problem Formulation for Non-Newtonian Fluids 675

10.16.2 Asymptotic Analysis of the Problem for Heavily Loaded Contacts 678

10.16.3 Heat Transfer in the Contact Solids 688

10.17 Regularization for Non-Newtonian Fluids 695

10.18 Friction in Heavily Loaded Lubricated Contacts 698

10.19 Closure 700

10.20 Exercises and Problems 702

11 Lubrication by Greases **709**

11.1 Introduction 709

11.2 Formulation of the EHL Problems for Greases 712

11.3 Properties of the Problem Solution for Greases 722

11.4 Greases in a Contact of Rigid Solids 724

11.4.1 Problem Formulation 724

11.4.2 Analysis of Possible Flow Configurations 726

11.4.3 Numerical Results 731

11.5 Regimes of Grease Lubrication without Cores 736

11.6 Closure 742

11.7 Exercises and Problems 743

12 Lubricant Degradation in EHL Contacts **745**

12.1 Introduction 745

12.2 EHL for Degrading Lubricants 746

12.3 Lubricant Flow Topology 754

12.4 Numerical Method for EHL Problems 756

12.4.1 Initial Approximation 757

12.4.2 Sliding Frictional Stress 757

12.4.3 Horizontal Component of the Lubricant Velocity and Flux 759

12.4.4 Lubricant Flow Streamlines 759

12.4.5 Separatrices of the Lubricant Flow 762

12.4.6 Solution of the Kinetic Equation and Lubricant Viscosity 764

12.4.7 Solution of the Reynolds Equation 766

12.5 Solutions for Lubricants without Degradation 770

12.6 EHL Solutions for Lubricants with Degradation 776

12.7 Lubricant Degradation and Contact Fatigue 790

          12.7.1 Model of Contact Fatigue 791
          12.7.2 Elastohydrodynamic Modeling for a Degrading Lubri-
                  cant 791
          12.7.3 Combined Model for Contact Fatigue and Degrading
                  Lubricant 793
          12.7.4 Numerical Results and Discussion 794
   12.8   A Qualitative Model of Lubricant Life 800
   12.9   Closure 802
   12.10 Exercises and Problems 804

13 Non-steady and Mixed Friction Problems   809
   13.1   Introduction 809
   13.2   Properly Formulated Non-steady EHL Problems 809
          13.2.1 A Non-steady Lubrication of a Non-conformal Contact 810
          13.2.2 Properly Formulated Non-steady Lubrication Prob-
                  lems for Journal Bearings 817
   13.3   Non-steady Lubrication of a Journal Bearing 827
          13.3.1 General Assumptions and Problem Formulation 829
          13.3.2 Case of Rigid Materials 830
          13.3.3 Contact Region Transformation 833
          13.3.4 Quadrature Formula and Discretization 835
          13.3.5 Iterative Numerical Scheme 837
          13.3.6 Analysis of Numerical Results 838
   13.4   Starved Lubrication and Lubricant Meniscus 851
          13.4.1 Problem Formulation 853
          13.4.2 Numerical Method 857
          13.4.3 Numerical Results 862
   13.5   Formulation and Analysis of a Mixed Lubrication Problem . . 871
          13.5.1 Problem Formulation 875
          13.5.2 Fluid Friction in Lightly and Heavily Loaded Lubri-
                  cated Contacts 881
          13.5.3 Boundary Friction 882
          13.5.4 Partial Lubrication of a Narrow Contact 887
   13.6   Dry Narrow Contact of Elastic Solids 892
          13.6.1 Examples of Dry Narrow Contacts 894
          13.6.2 Optimal Shape of Contacting Solids 896
   13.7   Closure 898
   13.8   Exercises and Problems 899

Index   905

Preface

The idea of creating this monograph was conceived when the authors collaborated on modeling of lubricant degradation. We considered that lubricant degradation is just one element in a series of inter-related research areas in tribology ranging from modeling of dry and lubricated contacts to fracture mechanics and contact fatigue. Therefore, the idea was to connect all these traditionally separate areas of research and produce a coherent path from modeling of various contact interactions to modeling of contact fatigue life. Then we reviewed the theoretical methodology used in these areas of tribology and realized that there are two basic theoretical techniques currently in use: (1) approximation methods originating from the Ertel-Grubin approach [1, 2] and their application to lubrication problems and (2) various numerical techniques applied to contact and lubrication problems as well as to problems of contact fatigue modeling. That prompted us to propose the asymptotic and perturbation methods and their numerical realization as an addition to the arsenal of the currently used methods of theoretical analysis. With respect to numerical methods applied to studying heavily loaded lubricated contacts for more than 60 years, it was widely recognized that practically all currently existing numerical methods suffer from instability. So far this problem of instability in heavily loaded lubricated contacts has not been resolved satisfactorily. Therefore, the other objective of this work was to propose a reasonably simple and naturally based regularization approach that would produce stable solutions.

Therefore, the entire monograph is based on the following ideas: (a) designing a well–founded passage from modeling of contact interactions through fracture mechanics to contact fatigue, (b) proposing a fundamentally sound statistical model of contact fatigue, (c) proposing an alternative to Ertel-Grubin and direct numerical methods for solution of problems for heavily loaded lubricated contacts (asymptotic methods followed by numerical analysis), (d) developing a general asymptotic approach to the solution of problems for heavily loaded contacts lubricated by fluids with various rheologies and greases, (e) designing a regularization approach to virtually eliminate instability of numerical solutions for isothermal heavily loaded lubricated contacts, (f) properly formulating and solving non-steady lubrication problems, (g) properly formulating and analyzing mixed friction/lubrication problems, and (h) modeling of lubricant degradation and application of these models to lubricated contacts and contact fatigue. It is noteworthy that friction plays a significant role in most of the models considered.

The authors are convinced that a better understanding of different phenomena in tribology can be achieved based upon a fundamental multidisciplinary approach that combines the existing concepts and theories of tribology with available analytical and numerical tools of mathematics. The fundamental approaches used in the monograph are the concepts and methods of contact problems of elasticity, linear fracture mechanics, structural and contact multi-axial fatigue modeling, elastohydrodynamic theory of lubrication, mixed and transient lubrication, and stress-induced polymer degradation. This monograph philosophy is based on the fact that such an approach makes the analysis straight forward and clear. The approaches such as asymptotic methods in many cases reveal analytical structures of the solutions as well as emphasize the dominant mechanisms involved in the phenomena, while numerical methods provide the details to such a structures and validate asymptotic approaches and solutions. Therefore, the monograph offers a number of fundamentally sound modeling approaches based on experimental data as well as some analytical (asymptotic in nature) and numerical approaches which are inter-related and inter-dependent. The objective of these approaches is a better understanding of the influence of various tribological parameters on friction, lubrication, and fatigue.

The book consists of thirteen chapters. In most cases the equations are derived from first principles. Each of the asymptotic and numerical methods is described in detail which makes it easier to apply them to different problems. In some cases the correctness of problem formulations is rigorously proven, and certain solution properties are derived. Some problems allow for relatively simple approximate (asymptotic) analytical solutions. The problem solutions are presented in the form of simple analytical formulas, graphs, and tables. Every chapter is provided with exercises of different levels of difficulty. The exercises highlight certain points that are important for understanding the material and mastering the appropriate skills.

Chapter 1 provides the basics of asymptotic and perturbation techniques. An introduction to classic contact problems of elasticity as well as a methodical asymptotic and numerical treatment of contact problems for rough/coated surfaces is provided in Chapter 2. In Chapter 3 an analysis of contact problems with nonlinear friction is provided. Similar analytical methods are used in other chapters. The next two chapters - Chapters 4 and 5 - discuss basic physical and rheological properties of Newtonian and non-Newtonian lubricating oils and introduce the reader to the topic of polymer additives and multi-grade lubricants. In Chapters 6 and 7, a detailed derivation and analysis of statistical kinetic models of stress-induced degradation of lubricants containing linear and star polymeric additives is proposed.

Chapter 8 provides an extensive review of experimental and theoretical data related to contact fatigue phenomena. This review aids the reader in understanding the major factors affecting contact fatigue, how some of the popular contact fatigue models incorporate these factors, and the advantages and drawbacks of these models. One of the major conclusions of Chapter 8 is a list of the most important parameters affecting contact fatigue. Moreover, it is concluded that contact fatigue life can be well approximated by the duration of the fatigue crack propagation period, and the crack nucleation period can be neglected. This prompted us to include an entire chapter (Chapter 9) on fracture mechanics and, in particular, on fracture mechanics of solids in dry and lubricated contacts. This chapter introduces expressions for approximating the stress intensity factors at crack tips in an analytical form. It explains why, in most cases, high cycle contact fatigue is of subsurface origin and why, in lubricated contacts, fatigue life of drivers is usually significantly different from fatigue life of followers. Some results of this chapter lead to the development of fundamentally sound two- and three-dimensional statistical models of structural and contact fatigue, which are analyzed in detail and compared to experimental data.

Chapter 10 is dedicated to asymptotic and numerical analysis of lightly and heavily loaded lubricated contacts. Some isothermal and thermal elastohydrodynamic problems of lubrication are considered. Validation of the asymptotic approaches is provided. The asymptotic analysis provides a framework for formulating problem solutions with simple analytical expressions for film thickness for both Newtonian and non-Newtonian lubricants in heavily loaded contacts. An important outcome of this methodology is the reduction of the number of problem input parameters. But most important, the analysis leads to a simple and effective regularization approach to numerical solution of the problem for isothermal heavily loaded lubricated contacts in both asymptotic and original formulations. Problems for grease lubricated contacts are formulated and discussed in Chapter 11. Stress-induced lubricant degradation in lubricated elastohydrodynamic contacts is considered in Chapter 12 where a number of interesting and important properties of lubricated contacts are revealed. Several non-steady problems for lubricated contacts are formulated and analyzed in Chapter 13 as well as problems for starved contacts and contacts under mixed lubrication conditions.

The book is offered as an enhancement of the tribology curriculum for senior undergraduate and graduate engineering and applied mathematics students as well as a reference/guide for researchers and practitioners in the field. For example, engineering students can be offered courses based upon chapters on fracture mechanics and contact fatigue, elastohydrodynamic lubrication as well as lubricant degradation and its effect on tribological parameters and contact fatigue. On the other hand, applied mathematics students can be offered a course on nontraditional applications of asymptotic and perturbation methods as well as on a regularization approach to solution of inherently numerically unstable problems for heavily loaded lubricated contacts. In each course, proofs can be omitted while the formulations of theorems and lemmas as well as remarks serve as an important source of information on problem solution properties. In addition, the content of the monograph can be used as a basis for designing an introductory/overview course on rheology, elastohydrodynamic lubrication, and contact fatigue.

Chapters 1-3 and 6-13 were prepared by I.I. Kudish while Chapters 4 and 5 were written by M.J. Covitch.

The authors acknowledge the contributions of Zachary Smith, Applied Mathematics student at Kettering University, for the preparation of many graphs and figures and Bruce Deitz, Kettering University librarian, for helping with literature searches. They also wish to acknowledge the Lubrizol Corporation for permission to publish chapters written by M.J. Covitch.

BIBLIOGRAPHY

[1] Ertel, A.M. 1945. Hydrodynamic Lubrication Analysis of a Contact of Curvilinear Surfaces. Dissertation. *Proc. of CNIITMASh*, Moscow, 1-64.

[2] Grubin, A.N. and Vinogradova, I.E. 1949. Investigation of the Contact of Machine Components. Ed. Kh. F. Ketova, *Central Sci. Research Inst. for Techn. and Mech. Eng.* (Moscow), Book No. 30 (DSIR translation No. 337).

1

Basics of Asymptotic Expansions and Methods

1.1 Introduction

In this chapter we will introduce some basic commonly used notations and operations related to asymptotic expansions. The focus will be on application of these notions and operations to practical problems. A more detailed description of the basics of asymptotic expansions can be found in a variety of books, for example, [1] - [4].

1.2 Ordering, Order Sequences, and Asymptotic Expansions

Asymptotic analysis is a study of the behavior of a certain mathematical object (function, or algebraic, differential, integral equation, integral, etc.) as a parameter or independent variable this object depends on is approaching a certain limit. The asymptotic behavior of a mathematical object as its parameter or variable approaches a certain limit is just the object limiting behavior. We will concentrate on asymptotic behavior as some parameter λ approaches zero or infinity. To discriminate between different asymptotic behaviors, we need to introduce ordering. Suppose we have two functions $f(x, \lambda)$ and $g(x, \lambda)$ determined on an interval $x \in I$ for $\lambda \geq 0$. In many cases it is useful to use two orderings: large O and small o. Let us define these orderings for $\lambda \to 0$. The definitions for $\lambda \to \infty$ are similar.

Large O

For fixed $x \in I$ and $\lambda \to 0$, we say that $f(x, \lambda) = O(g(x, \lambda))$ if there exists such a positive finite value M (generally, dependent on x) that $\mid f(x, \lambda) \mid \leq M \mid g(x, \lambda) \mid$ for $\lambda \to 0$.

Small o

For fixed $x \in I$ and $\lambda \to 0$, we say that $f(x, \lambda) = o(g(x, \lambda))$ if $\lim_{\lambda \to 0} \frac{f(x,\lambda)}{g(x,\lambda)} = 0$.

Often, $f \ll g$ as $\lambda \to 0$ (or $\lambda \ll 1$) is used as an equivalent notation.

If $\lim_{\lambda \to 0} \frac{f(x,\lambda)}{g(x,\lambda)} = 1$ for fixed $x \in I$ and $\lambda \to 0$, we say that $f(x,\lambda) \sim g(x,\lambda)$.

If the validity of the relationships $f(x,\lambda) = O(g(x,\lambda))$ or $f(x,\lambda) = o(g(x,\lambda))$ is independent of $x \in I$, then we say that the corresponding relationship is uniformly valid in I.

Here are some examples of the large O and small o definitions.

$$\lambda^2 = O(\lambda), \ \lambda \to 0,$$

$$1 - \cos(\lambda) = O(\lambda^2), \ \lambda \to 0,$$

$$\lambda = O(\lambda^2), \ \lambda \to \infty, \tag{1.1}$$

$$\ln(1 + \lambda) - \lambda = o(\lambda), \ \lambda \to 0.$$

The following asymptotic estimates are uniformly valid for $x \in I = (0, 1]$:

$$\frac{\lambda}{x^2+1} = O(\lambda), \ \lambda \to 0,$$

$$1 - \cos(\lambda x) = O(\lambda^2), \ \lambda \to 0, \tag{1.2}$$

while on the same interval the asymptotic estimates

$$\frac{\lambda}{x^2} = O(\lambda), \ \lambda \to 0,$$

$$\exp(-\frac{x}{\lambda}) = O(\lambda^2), \ \lambda \to 0 \tag{1.3}$$

are not uniform. The latter is due to the fact that depending on how fast x approaches 0 in relation to λ the above two expressions may be small, large, or of the order of unity as $\lambda \to 0$.

Generally, various operations such as addition, subtraction, multiplication, division, and integration can be performed on order relations. In many cases it is also possible to perform the operation of differentiation on order relations. However, in some cases a straightforward asymptotic estimate of a derivative may lead to an error. Below is an example when it is permissible to use the straightforward approach to the asymptotic estimation of a derivative while $\lambda \to 0$

$$\frac{\lambda}{x^2+1} = O(\lambda), \ x = O(1), \ \Rightarrow \ \frac{d}{dx}\frac{\lambda}{x^2+1} = \frac{O(\lambda)}{O(1)} = O(\lambda), \ x = O(1), \tag{1.4}$$

and an example when it clearly causes an error for $x \gg \lambda$, $x \in [0, 1]$

$$\cos(\frac{x}{\lambda}) = O(1), \ x = O(1), \ \Rightarrow \ \frac{d}{dx}\cos(\frac{x}{\lambda}) = \frac{O(1)}{O(1)} = O(1), \ x = O(1). \tag{1.5}$$

To produce a correct and accurate asymptotic estimate of a derivative of a function, there is usually need for more detailed knowledge of the function behavior. For more details on asymptotic differentiation of functions see [1].

1.3 Asymptotic Sequences and Expansions

A sequence of functions $\{\varphi_n(\lambda)\}$, $n = 1, 2, \ldots$ is called an asymptotic sequence if $\varphi_{n+1}(\lambda) = o(\varphi_n(\lambda))$ for $\lambda \to 0$ and $n = 1, 2, \ldots$. Asymptotic sequences may have finite or infinite number of members.

If the members of the asymptotic sequence $\{\varphi_n(x, \lambda)\}$ also depend on variable x the asymptotic sequence $\{\varphi_n(x, \lambda)\}$, $n = 1, 2, \ldots$ may or may not be uniformly asymptotic on an interval I. Below, are some examples of uniform on $[0, 1]$

$$\{(\lambda x)^n\}, \ n = 1, 2, \ldots, \ \lambda \to 0,$$
$$\{x^n \ln^{-n} \lambda\}, \ n = 1, 2, \ldots, \ \lambda \to 0, \tag{1.6}$$

and nonuniform on $(0, 1]$ asymptotic sequences

$$\{(\tfrac{\lambda}{x})^n\}, \ n = 1, 2, \ldots, \ \lambda \to 0,$$
$$\{\exp(-n\tfrac{x}{\lambda})\}, \ n = 1, 2, \ldots, \ \lambda \to 0. \tag{1.7}$$

Generally, various operations such as addition, subtraction, multiplication, division, and integration can be performed on asymptotic sequences. In many cases it is also possible to perform the operation of differentiation on the asymptotic sequence with respect to the parameter λ or variable x. However, in some cases differentiation of a uniformly valid asymptotic sequence/expansion may not produce a uniformly valid asymptotic sequence/expansion (see examples above and [1]).

Suppose $f(x, \lambda)$ is a function determined for $x \in [0, 1]$ and $\lambda \geq 0$ and $\{\varphi_n(\lambda)\}$, $n = 1, 2, \ldots$ is an asymptotic sequence as $\lambda \to 0$. Then the sum $\sum_{n=1}^{N} f_n(x)\varphi_n(\lambda)$ is called an asymptotic expansion of the function $f(x, \lambda)$ to N terms as $\lambda \to 0$ if

$$f(x, \lambda) - \sum_{n=1}^{N} f_n(x)\varphi_n(\lambda) = o(\varphi_N(\lambda)), \ \lambda \to 0, \tag{1.8}$$

for $N = 1, 2, \ldots$ If $N = \infty$, we get an infinite asymptotic series

$$f(x, \lambda) \sim \sum_{n=1}^{\infty} f_n(x)\varphi_n(\lambda), \ \lambda \to 0, \tag{1.9}$$

which may or may not be convergent. Nonetheless, in either case such a series or expansion may provide a very valuable information about the behavior of the function $f(x, \lambda)$ as $\lambda \to 0$. If an asymptotic series is divergent, then we should fix the number of terms N in the asymptotic expansion and consider the expansion behavior for $\lambda \to 0$. One has to keep in mind that the truncation error of the asymptotic expansion in (1.8) is of the order much smaller than function $\varphi_N(\lambda)$, which vanishes as $\lambda \to 0$. An example of a divergent asymptotic series for an integral is given below

$$\int\limits_0^\infty \frac{\lambda e^{-x}}{\lambda+x}dx \sim \sum_{n=0}^\infty \frac{(-1)^n n!}{\lambda^n}, \quad \lambda \to \infty. \tag{1.10}$$

It can be easily estimated that the remainder $R_N(\lambda)$ of the asymptotic series (1.10) satisfies the inequality

$$\mid R_N(\lambda) \mid = \mid \int\limits_0^\infty \frac{\lambda e^{-x}}{\lambda+x}dx - \sum_{n=0}^N \frac{(-1)^n n!}{\lambda^n} \mid < \frac{N!}{\lambda^N}. \tag{1.11}$$

It can be shown that $R_N(\lambda) \to \infty$ for a fixed λ and $N \to \infty$, i.e., the series diverges. On the other hand, $R_N(\lambda) \to 0$ for a fixed N and $\lambda \to \infty$. Therefore, the truncated asymptotic expansion from (1.10) can be successfully used for approximation of this integral for large λ.

A more detailed discussion of the behavior and utilization of convergent and divergent asymptotic expansions can be found in [1] - [4].

An asymptotic expansion is said to be uniformly valid in interval I if the relationship (1.8) holds uniformly in I. Obviously, for that to be true it is sufficient that $f_{n+1}(x) = O(f_n(x))$ as $\lambda \to 0$ and $x \in I$.

By repeated application of definition (1.8), we can uniquely determine the coefficients $\{f_n(x)\}$ of the asymptotic expansion as follows

$$f_k(x) = \lim_{\lambda \to 0} \frac{f(x,\lambda) - \sum\limits_{n=1}^{k-1} f_n(x)\varphi_n(\lambda)}{\varphi_k(\lambda)}, \quad k = 1, 2, \ldots. \tag{1.12}$$

1.4 Asymptotic Methods

Most asymptotic methods can be subjected to a simple classification on regular and singular asymptotic methods. Regular asymptotic methods are applicable to the problems in which the solution can be represented by a uniformly valid asymptotic expansion in the entire solution region. Physically, this situation corresponds to the case when the contribution to the problem solution of the physical mechanism which gave rise to the small parameter of the problem remains small in the entire solution region. Application of regular asymptotic methods is relatively simple.

The situation gets more complex when there is a necessity to apply singular asymptotic methods. In such cases (with few exceptions [2] - [4]) the solution is searched in the form of several non-uniformly valid asymptotic expansions. The reason for that is the change of the leading physical mechanisms contributing to problem solution in different solution regions. Each of these asymptotic expansions is valid in its own region. These regions overlap and allow for matching solutions in these regions. That is the basic concept of the most often used method of matched asymptotic expansions. The particular realization of this method depends on the specifics of the problem. Moreover, after the problem is solved by the method of matched asymptotic expansions, it is easy to construct a uniformly valid approximate solution (see [2]).

We assume that the reader is acquainted with the basics of asymptotic approaches. Therefore, we will not get into details of various asymptotic methods. Many examples of application of different regular and singular asymptotic expansion methods are given in [1] - [4]. The problems considered range from estimating functions and integrals to asymptotic solution of problems for ordinary and partial differential equations. In the following chapters we will use both the regular and singular asymptotic methods. In most cases when we need to use a singular asymptotic method we will use the matched asymptotic expansions method.

1.5   Exercises and Problems

1. For $\lambda \to 0$ determine the order of the following expressions: (i) $\ln(1 + \arctan \lambda)$, (ii) $\tan(\frac{\pi}{2} - \lambda^2)$, (iii) $\sqrt[3]{\frac{\ln(1+\lambda)}{1+\lambda} + \frac{e^\lambda}{1-\lambda}}$.

2. For $\lambda \to 0$ determine four-term asymptotic expansions of functions from Problem 1.

3. For $\lambda \to 0$ list the following functions λ^2, $\lambda^{-1/4}$, $\frac{\lambda^2}{\sin \lambda}$, $\lambda \ln \lambda$, $\lambda^{-3/2}$ in the decreasing order of magnitude.

4. For $\lambda \to 0$ determine asymptotic expansions o functions: (i) $\frac{\ln(1+\lambda x^2)}{\arcsin(\lambda x)}$, (ii) $\{1 + \tan(\sqrt{\lambda}x) + e^{\lambda x}\}^{-1}$, (iii) $\frac{\sin(\lambda x) - \lambda x}{\lambda^3 x^3}$. Are these asymptotic expansions uniformly valid for all x? In which regions are these asymptotic expansions non-uniformly valid, i.e., cease to remain asymptotic expansions?

5. For $\lambda \to 0$ find three-term asymptotic expansions of all three solutions of the following algebraic equations: (i) $(1+\lambda)x^3 - (6+\lambda^2)x^2 + (11-\lambda)x - 6 + 2\lambda = 0$, (ii) $\lambda x^3 - x + 1 + \lambda^2 = 0$.

BIBLIOGRAPHY

[1] Erdeiyi, A. 1956. *Asymptotic Expansions*. New York: Dover Publications.

[2] Van-Dyke, M. 1964. *Perturbation Methods in Fluid Mechanics*. New York-London: Academic Press.

[3] Kevorkian, J. and Cole, J.D. 1985. *Perturbation Methods in Applied Mathematics*. New York: Springer-Verlag.

[4] Nayfeh, A.H. 1984. *Introduction to Perturbation Techniques*. New York: John Wiley & Sons.

2

Contact Problems for Coated and Rough Surfaces

2.1 Introduction

A number of experimental studies [1, 2] revealed that the normal displacement in a contact of coated/rough surfaces due to coating/asperities presence is a nonlinear function of local pressure, and it can be approximated by a power function of pressure. Originally, a linear mathematical model accounting for the surface structure of two elastic solids in contact was introduced by I. Shtaerman [3]. He assumed that the effect of surface asperities present in a contact of two elastic solids can be essentially replaced by the presence of a thin coating simulated by an additional normal displacement of solids' surfaces proportional to a local pressure. Later, a similar nonlinear problem formulation that accounted for the above–mentioned experimental fact was proposed by L. Galin. In a series of papers [4] - [12] the problem was studied by numerical and asymptotic methods.

The purpose of this chapter is to analyze the problem analytically and numerically. The existence and uniqueness of a solution of a contact problem for elastic bodies with coated/rough surfaces is established based on the variational inequalities approach. Four different equivalent formulations of the problem including three variational ones were considered. A comparative analysis of solutions of contact problems for different values of input parameters (such as the indenter shape, parameters characterizing coating/roughness, elastic parameters of the substrate material) is done with the help of calculus of variations, the Zaremba-Giraud principle of maximum for harmonic functions, and numerical methods.

In particular, the results include the relationships between the pressure and displacement distributions for coated/rough and smooth solids as well as the relationships for solutions of the problems for coated/rough solids with fixed and free contact boundaries. The presentation of the material follows papers [8], [10] - [12].

2.2   Some Classic Results for Smooth Elastic Solids

In this section we present some classic contact problem formulations and results for the case of isotropic homogeneous elastic materials. We limit ourselves to only the most basic formulas for spatial and plane problems and to
the results for isotropic homogeneous elastic materials occupying half-space
and half-plane, which will be used later. The requirements necessary for solution of these problems are usually reduced to minimal smoothness of contact
surfaces. There are couple more complex cases that are considered in Sections
10.9 and 13.5 but not considered here. Besides the solutions presented below,
there is a wide variety of solutions for contact problems not related to the
analysis presented in this monograph.

2.2.1   Formulas and Results for Elastic Half-Space

Suppose an elastic isotropic homogeneous material with elastic modulus E and
Poisson's ratio ν occupies a half-space $z \leq 0$, and it is bounded by the plane
$z = 0$. Assuming that in a rectangular coordinate system (x, y, z) the plane
$z = 0$ is loaded by pressure $p(x, y)$ distributed over region Ω in the (x, y)-plane
($p(x, y) = 0$ outside of Ω) the displacements (u, v, w) of the material points
along (x, y, z) axes are [13, 14]

$$u(x, y, z) = \tfrac{1+\nu}{2\pi E} \{ z\tfrac{\partial V}{\partial x} - (1 - 2\nu) \int\limits_{-\infty}^{z} \tfrac{\partial V}{\partial x} dz \},$$

$$v(x, y, z) = \tfrac{1+\nu}{2\pi E} \{ z\tfrac{\partial V}{\partial y} - (1 - 2\nu) \int\limits_{-\infty}^{z} \tfrac{\partial V}{\partial y} dz \},$$

$$(2.1)$$

$$w(x, y, z) = \tfrac{1+\nu}{2\pi E} \{ z\tfrac{\partial V}{\partial z} - 2(1 - \nu)V \},$$

$$V = \int\int\limits_{\Omega} \tfrac{p(\xi,\eta)d\xi d\eta}{R}, \ \ R = \sqrt{(x - \xi)^2 + (y - \eta)^2 + z^2}.$$

Obviously, at the surface $z = 0$ we have

$$w(x, y, 0) = -\tfrac{1-\nu^2}{\pi E} \int\int\limits_{\Omega} \tfrac{p(\xi,\eta)d\xi d\eta}{R}, \ \ R = \sqrt{(x - \xi)^2 + (y - \eta)^2}. \quad (2.2)$$

If the plane $z = 0$ is loaded by an axially symmetric pressure $p(x, y)$ distributed over a circular region with radius b (outside of which it is zero), then
in a cylindrical coordinate system (r, φ, z) the normal displacement of the
plane $z = 0$ is equal to

$$w(x, y, 0) = -\tfrac{8(1-\nu^2)}{\pi E} \int\limits_{0}^{b} \tfrac{\rho}{r+\rho} K(\tfrac{2\sqrt{r\rho}}{r+\rho}) p(\rho)d\rho, \quad (2.3)$$

where K is the full elliptic integral of the second kind.

Let us assume that the tangential displacements of the half-space surface are negligibly small. Then the formulation of a typical contact problem for a rigid indenter with a smooth bottom of shape $z = f(x, y)$, which is normally pressed into the elastic half-space without friction, is given below [14]

$$-w(x, y, 0) = \delta + \beta_x x + \beta_y y - f(x, y), \quad \int\int_\Omega p(\xi, \eta)d\xi d\eta = P, \tag{2.4}$$

where δ is the normal displacement of the rigid indenter, β_x and β_y are the angles of the indenter rotation about the x- and y-axes caused by the moments M_x and M_y, respectively, P is the normal load applied to the indenter.

To complete the contact problem formulation, it is necessary to fix the boundary or part of the boundary of the contact region while on the rest of the boundary to require that pressure vanishes. For a circular contact with fixed boundary $r = b$ and $\beta_x = \beta_y = 0$, $f(x, y) = f(r)$, the problem solution has the form [15]

$$p(r) = \frac{E}{\pi(1-\nu^2)} \Big\{ \frac{1}{\sqrt{b^2-r^2}} [\delta - f(0) - b \int_0^b \frac{f'(\rho)d\rho}{\sqrt{b^2-\rho^2}}] $$

$$+ \int_r^b \frac{1}{\sqrt{s^2-r^2}} [s \int_0^s \frac{f'(\rho)d\rho}{\sqrt{s^2-\rho^2}}]'ds \Big\}, \tag{2.5}$$

$$\delta = \frac{1-\nu^2}{E} \frac{P}{2b} + f(0) + \frac{1}{b} \int_0^b \sqrt{b^2 - \rho^2} f'(\rho)d\rho,$$

while for for a circular contact with free boundary $r = b$ (i.e., $p(b) = 0$) and $\beta_x = \beta_y = 0$, $f(x, y) = f(r)$, the problem solution has the form [15]

$$p(r) = \frac{E}{\pi(1-\nu^2)} \frac{1}{b} \int_r^b \{t \int_0^t \frac{f'(\rho)d\rho}{\sqrt{t^2-\rho^2}} \}' \frac{dt}{\sqrt{t^2-r^2}}, \tag{2.6}$$

$$\int_0^b r\{r \int_0^r \frac{f'(\rho)d\rho}{\sqrt{r^2-\rho^2}} \}'dr = \frac{1-\nu^2}{E} \frac{P}{2b}.$$

Solutions of a number of contact problems with and without axial symmetry and with and without friction can be found in [14].

2.2.2 Formulas and Results for Elastic Half-Plane

Let us assume that an isotropic homogeneous elastic material with elastic modulus E and Poisson's ratio ν i is bounded by a horizontal plane $z = 0$ and occupies the half-space $z \leq 0$. The half-space boundary $z = 0$ is subjected to pressure $p(x)$ and tangential stress $\tau(x)$, which are uniform along the y-axis parallel to the plane $z = 0$. Then we can consider this as a case of plane deformation when none of the problem parameters depend on the y-coordinate and the whole problem can be considered in just (x, z) plane. The elastic displacements (u, w) of the half-plane boundary $z = 0$ along the (x, z)-axes are determined by formulas [13]

$$\frac{\pi E}{2(1-\nu^2)} u + C_u = -\frac{1-2\nu}{2(1-\nu)} \pi \int\limits_{-\infty}^{x} p(t)dt + \int\limits_{-\infty}^{\infty} \tau(t) \ln |t - x| \, dt,$$

$$(2.7)$$

$$\frac{\pi E}{2(1-\nu^2)} w + C_w = \int\limits_{-\infty}^{\infty} p(t) \ln |t - x| \, dt + \frac{1-2\nu}{2(1-\nu)} \pi \int\limits_{-\infty}^{x} \tau(t)dt,$$

where C_u and C_w are infinite constants. If the elastic material is represented by an infinite thick layer of thickness h, then we still can use formulas (2.7) where C_u and C_w are constants proportional to $\ln \frac{h}{a}$, $\frac{h}{a} \gg 1$ (a is a characteristic size of the contact region) [15].

Assuming that the surface tangential displacements can be neglected the formulation of a typical contact problem for a rigid indenter with a smooth bottom of shape $z = f(x)$, which is pressed into an elastic half-plane with the normal P and tangential T forces per unit length, is as follows [13]

$$-w(x, 0) = \delta + \beta_x x - f(x), \quad \int\limits_{\Omega} p(x)dx = P,$$

$$(2.8)$$

where δ is the normal displacement of the rigid indenter, β_x is the angle of the indenter rotation about the x-axis caused by the moment M_x, Ω is the contact region, and

$$\int\limits_{\Omega} \tau(x)dx = T.$$

$$(2.9)$$

To complete the contact problem formulation, it is necessary to fix the boundaries of the contact region Ω partially or completely while on the rest of the boundaries to require that pressure vanishes. If the contact region Ω is represented by an interval, then there are only three distinct possibilities: Contact Problem 1 - the end points of the contact interval $(-a, a)$ are fixed; Contact Problem 2 - the boundaries of the contact are free, i.e., $p(\pm a) = 0$; and Contact Problem 3 - one of the boundaries, for example, $x = -a$ is fixed while the other one is free, i.e., $p(a) = 0$. Let us consider contact problems without friction, i.e., $\tau(x) = 0$. Then assuming that angle β_x is known and is included in function $f(x)$ the solution of Contact Problem 1

$$\frac{2(1-\nu^2)}{\pi E} \int\limits_{-a}^{a} p(t) \ln \frac{1}{|t-x|} dt = \delta - f(x), \quad \int\limits_{-a}^{a} p(x)dx = P,$$

$$(2.10)$$

takes the form

$$p(x) = \frac{1}{\pi\sqrt{a^2 - x^2}} \left\{ P + \frac{E}{2(1-\nu^2)} \int\limits_{-a}^{a} \frac{f'(t)\sqrt{a^2 - t^2}\,dt}{t - x} \right\}.$$

$$(2.11)$$

Calculation of constant δ can be done by substituting $p(x)$ from formula (2.11) into the first equation from (2.10) and setting $x = 0$. The solution of Contact Problem 2, which is reduced to

$$\frac{2(1-\nu^2)}{\pi E} \int\limits_{-a}^{a} p(t) \ln \frac{1}{|t-x|} dt = \delta - f(x), \ \ p(\pm a) = 0, \ \ \int\limits_{-a}^{a} p(x) dx = P, \quad (2.12)$$

is determined by the formulas

$$p(x) = \frac{E}{2\pi(1-\nu^2)} \sqrt{a^2 - x^2} \int\limits_{-a}^{a} \frac{f'(t) dt}{\sqrt{a^2-t^2}\,(t-x)},$$

$$(2.13)$$

$$P = \frac{E}{2(1-\nu^2)} \int\limits_{-a}^{a} \frac{t f'(t) dt}{\sqrt{a^2-t^2}}, \ \ \int\limits_{-a}^{a} \frac{f'(t) dt}{\sqrt{a^2-t^2}} = 0.$$

Finally, Contact Problem 3, which is reduced to equations

$$\frac{2(1-\nu^2)}{\pi E} \int\limits_{-a}^{a} p(t) \ln \frac{1}{|t-x|} dt = \delta - f(x), \ \ p(a) = 0, \ \ \int\limits_{-a}^{a} p(x) dx = P, \quad (2.14)$$

has the solution

$$p(x) = \frac{E}{2\pi(1-\nu^2)} \sqrt{\frac{a-x}{a+x}} \int\limits_{-a}^{a} f'(t) \sqrt{\frac{a+t}{a-t}} \frac{dt}{t-x},$$

$$(2.15)$$

$$P = \frac{E}{2(1-\nu^2)} \int\limits_{-a}^{a} f'(t) \sqrt{\frac{a+t}{a-t}} dt.$$

Solutions of the contact problem with linear friction under conditions of full sliding when $\tau(x) = \lambda p(x)$ (λ is the friction coefficient) can be found in [16] while an approximate solution of the contact problem with stick and slip is presented in [13].

2.3 Spatial Rough Contacts Modeled by Nonlinear Coating

In this section some spatial contact problems for rough elastic solids are formulated, and the properties of their solutions are studied analytically.

2.3.1 Problem Formulation

Let us consider an isotropic elastic half-space $z \leq 0$ with Young's modulus E and Poisson's ratio ν. The surface of the half-space $z = 0$ is assumed to be

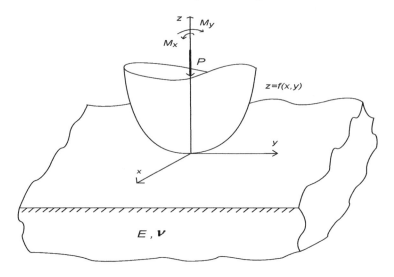

FIGURE 2.1
General view of the indenter pressed into an elastic half-space.

coated or to have some asperities. A rigid indenter bounded by the surface $z = f(x, y)$ is indented in the elastic half-space (see Fig. 2.1) by a normal force P.

Let us consider the equations and inequalities governing the problem. According to Shtaerman [3] we will assume that the total normal displacement of the half-space surface is a sum of the surface displacement of a smooth elastic half-space w_E [10]

$$w_E = -\frac{1}{\pi E'} \int\limits_{z=0} \int \frac{p(\xi, \eta) d\omega}{R},$$

(2.16)

$$R = \sqrt{(x - \xi)^2 + (y - \eta)^2}, \quad d\omega = d\xi d\eta,$$

and of an extra displacement due to the presence of surface asperities/coating w_R [1]

$$w_R = -k\varphi(p),$$

(2.17)

where $p = p(x, y)$ is the contact pressure (x and y are the orthogonal coordinates in the plane $z = 0$), k and $\varphi(p)$ are a constant and function of pressure p characterizing the normal displacement of the surface $z = 0$ solely due to the presence of coating/surface asperities, E' is the effective elastic modulus, $E' = E/(1 - \nu^2)$.

The contact region Ω can be determined as follows:

$$\Omega = \{(x, y) \mid p(x, y) > 0\}.$$

(2.18)

The boundary conditions in the plane $z = 0$ serving for determination of Ω and $p(x, y)$ can be presented in the form [10]

$$-Ap = -\delta - \beta_x x - \beta_y y + f(x, y), \; p(x, y) > 0, \tag{2.19}$$

$$-Ap \leq -\delta - \beta_x x - \beta_y y + f(x, y), \; p(x, y) = 0, \tag{2.20}$$

$$Ap = k\varphi(p) + \frac{1}{\pi E'} \underset{z=0}{\int \int} \frac{p(\xi, \eta) d\omega}{R}, \tag{2.21}$$

where A is a nonlinear integral operator, δ is the normal displacement of the rigid indenter, β_x and β_y are the angles of the indenter rotation about the x– and y–axes caused by the moments M_x and M_y, respectively.

Conditions (2.19) occur at points (x, y) of the surface $z = 0$ of the elastic half-space that are in contact with the indenter, and conditions (2.20) describe the absence of penetration of the surface of the indenter into the elastic half-space outside of the contact region Ω.

Therefore, for the given functions $\varphi(p)$, $f(x, y)$ and constants δ, β_x, β_y, k, and E', the problem is reduced to determination of the contact region Ω and the non-negative pressure $p(x, y)$ satisfying conditions (2.19)-(2.21).

2.3.2 Existence and Uniqueness of the Problem Solution

Let us formulate some necessary assumptions and definitions used in the further analysis. We will assume that

$$f(x, y) \geq 0, \; g(x, y) = \delta + \beta_x x + \beta_y y - f(x, y),$$
$$f(x, y), \; g(x, y) \in C(\Omega_\delta), \tag{2.22}$$

$$diam(\Omega_\delta) < \infty, \; \Omega_\delta = \{(x, y) \mid g(x, y) \geq 0\}$$
$$if \; \beta_x^2 + \beta_y^2 < \infty, \; 0 \leq \delta < \infty, \tag{2.23}$$

$$\varphi(-p) = -\varphi(p), \; \varphi(0) = 0, \; \varphi(p_2) > \varphi(p_1)$$
$$if \; p_2 > p_1, \; \varphi(p) \in C([0, \infty)), \tag{2.24}$$

$$\varphi(p) \leq a + bp^\alpha, \; p \geq 0 \; (a, b, \alpha \; - \; constants, \; \alpha > 0.3), \tag{2.25}$$

$$K = \{p \mid 0 \leq p \leq \varphi^{-1}[g(x, y)/k], \; (x, y) \in \Omega_\delta\} \subset L_{\alpha+1}(\Omega_\delta), \tag{2.26}$$

$K \; - \;$ *set of almost everywhere uniformly bounded functions,*

where $diam(\Omega_\delta)$ is the diameter of region Ω_δ, $C(\Omega)$ is the linear space of continuous on Ω functions, $L_q(\Omega)$ is the Hilbert space of integrable with power q functions determined on Ω [17]. If $\varphi(p)$ is determined only for $p \geq 0$, then its definition can be extended to negative p the following way $\varphi(p) = -\varphi(-p)$,

$\varphi^{-1}(\cdot)$ is the inverse function to $\varphi(\cdot)$. Obviously, K is a convex closed set. As an inner product (f, g) of two functions f and g, we will use the integral

$$(f, g) = \int\limits_{z=0} \int f(x, y)g(x, y)dxdy.$$

Some simple features of the solution of the problem are presented in Lemma 2.3.1 and Lemma 2.3.2.

Lemma 2.3.1 *Suppose conditions (2.22)-(2.26) are satisfied. If $p(x, y)$ is a measurable almost everywhere bounded solution of problem (2.19)-(2.21) for $k > 0$, $\beta_x^2 + \beta_y^2 < \infty$, and $0 \leq \delta < \infty$, then (i) $\Omega \subset \Omega_\delta$, (ii) $p \in K$, (iii) $p \in C(\Omega)$.*

Proof. It follows from (2.24) that $Ap \geq 0$ for $p \geq 0$. Therefore, if $(x, y) \in \Omega$, then (2.19), (2.22), and (2.23) lead to a conclusion that $(x, y) \in \Omega_\delta$. That proves statement (i). Based on (2.21) and (2.24) the first equation in (2.19) can be represented in the form

$$p(x, y) = \varphi^{-1}\left\{ \frac{1}{k}\left[g(x, y) - \frac{1}{\pi E'} \int\limits_{z=0} \int \frac{p(\xi, \eta)d\omega}{R} \right] \right\}. \tag{2.27}$$

This leads to boundness of $p(x, y)$, i.e.,

$$0 \leq p(x, y) \leq \varphi^{-1}[g(x, y)/k]. \tag{2.28}$$

Boundness of region Ω follows from (2.22), (2.23) and statement (i) of the Lemma 1. Continuity of $p(x, y)$ in Ω follows from (2.26)-(2.28) and from boundness of Ω [18]. That proves statements (ii) and (iii).

Lemma 2.3.2 *Suppose $p(x, y)$ is a solution of (2.19) -(2.21). Then under conditions of Lemma 2.3.1 the following integrals*

$$P = \int\limits_{z=0} \int p(\xi, \eta)d\omega, \quad \{M_x, M_y\} = \int\limits_{z=0} \int \{\xi, \eta\}p(\xi, \eta)d\omega \tag{2.29}$$

are finite.

Proof. Proof of Lemma 2.3.2 follows from statement (ii) of Lemma 2.3.1.

Theorem 2.3.1 *Suppose $g(x, y)$ satisfies (2.22) and $p_0(x, y) \in K$ is a solution of (2.19)-(2.21) in Ω for $k > 0$, $0 < E' < \infty$, $\beta_x^2 + \beta_y^2 < \infty$, and $0 \leq \delta < \infty$. Then problem (2.19)-(2.21) is equivalent to the variational inequality*

$$(Ap_0, p - p_0) \geq (g, p - p_0), \quad \forall p \in K. \tag{2.30}$$

Proof. *Necessity.* Let $p_0(x, y) \in K$ is a solution of (2.19)-(2.21). By multiplying (2.19) and (2.20) by $p(x, y) - p_0(x, y)$ (where $p \in K$) and integrating over the plane $z = 0$ we will arrive at (2.30).

Sufficiency. Let $p_0(x, y) \in K$ is a solution of (2.30). Let us consider an arbitrary point $(x_0, y_0) \in \Omega$ so that $p_0(x_0, y_0) > 0$. Because of to continuity of $p_0(x, y)$ there exists a small region O centered at (x_0, y_0) in which $p_0(x, y) > 0$. Let us pick an arbitrary continuous finite on O function $\eta(x, y)$. Then for a small enough number μ the function $p = p_0 + \mu\eta \in K$. Setting $\mu = \mu_0$ and $\mu = -\mu_0$ in inequality (2.30) leads to the equation $(Ap_0, \eta) = (g, \eta)$. Conditions (2.19) follow from the latter based on the fact that $\eta(x, y)$ is continuous and arbitrary. Now, let us consider an arbitrary inner point (x_0, y_0) of $F\Omega$ so that $p_0(x_0, y_0) = 0$, where $F\Omega$ is a region on the plane $z = 0$ comprising points of Ω and its boundary. Then there exists a region O_1 centered at (x_0, y_0) in which $p_0(x, y) = 0$. Let us pick an arbitrary non-negative continuous and finite on O_1 function $\eta(x, y)$. Then using $p = p_0 + \eta \in K$ in (2.30) we obtain conditions (2.20). For the boundary points of Ω conditions (2.19) and (2.20), follow from continuity of $g(x, y)$ and $p_0(x, y)$.

Theorem 2.3.2 *Suppose conditions (2.22)-(2.26) are satisfied. Then for $k > 0$, $0 < E' < \infty$, $\beta_x^2 + \beta_y^2 < \infty$ and $0 \leq \delta < \infty$ solution of the variational inequality (2.30) is equivalent to minimization on K of the functional*

$$\Phi(p) = \int\limits_{z=0} \int \left\{ k \int\limits_0^{p(x,y)} \varphi(t)dt \right. \tag{2.31}$$

$$\left. + \frac{p(x,y)}{2\pi E'} \int\limits_{z=0} \int \frac{p(\xi,\eta)d\omega}{R} - p(x,y)g(x,y) \right\} dxdy.$$

Proof. Necessity follows from estimating from below of $\varphi(p) - \varphi(p_0)$, where $p, p_0 \in K$ and $p_0(x, y)$ is a solution of inequality (15). Positiveness of operator A and trivial Lemma 2.3.3 are used.

Sufficiency follows from the condition $F'(0) \geq 0$ for minimum of $F(\mu) = \Phi(p_0 + \mu(p - p_0))$ as a function of $\mu \in [0, 1]$, where $p_0 \in K$ provides minimum to (2.31) and $p \in K$ is an arbitrary function.

Lemma 2.3.3 *If $\varphi(p)$ satisfies (2.24), then*

$$\int\limits_u^v \varphi(p)dp \geq \varphi(u)(v - u) \ if \ v \geq u. \tag{2.32}$$

Thus, we established the equivalence of problem (2.19)-(2.21) to problem (2.30) and to the problem of minimization of functional in (2.31) on K. We will prove the existence and uniqueness of the solution of these problems based on the analysis of variational inequality (2.30) [17]. Also, it can be done for the functional in (2.31) [19].

Lemma 2.3.4 *If $k > 0$, $0 < E' < \infty$, and the function $\varphi(p)$ satisfies (2.24), (2.25) then $A : K \to L_{(\alpha+1)/\alpha}(\Omega_\delta)$ is a bounded semi-continuous strictly monotonous operator.*

Proof. For $\alpha > 0.3$ boundedness of the operator A from (2.21) on K follows from (2.24), (2.25), boundness of Ω_δ, and the fact that $K \subset L_{\alpha+1}(\Omega_\delta)$ [15]. Semi-continuity of A [17], i.e., continuity of function $\mu \to (A(p_1 + \mu p_2), p_3)$ for $p_1, p_2, p_3 \in K$ is obvious. Strict monotonicity of the operator A follows from (2.24) and from the fact that it is positive.

Theorem 2.3.3 *Suppose conditions (2.22)-(2.26) are satisfied. Then the solution of variational inequality (2.30) in K exists and unique for $k > 0$, $0 < E' < \infty$, $\beta_x^2 + \beta_y^2 < \infty$ and $0 \le \delta < \infty$.*

Proof. The existence of the problem solution follows from Lemma 2.3.4 and Proposition 2.5 and Theorem 8.1 from [18]. The solution uniqueness follows from Theorem 8.3 from [18].

Theorem 2.3.4 *If $k > 0$, $0 < E' < \infty$, and condition (2.24) is satisfied, then the solution of problem (2.19)-(2.21) is unique for given: (a) force P and moments M_x and M_y (see (2.29)), (b) force P and rotation angles β_x, β_y, (c) displacement δ and moments M_x and M_y, and (d) displacement δ and rotation angles β_x, β_y.*

Proof. Let us consider case (a). Suppose there are two solutions $p_1(x, y)$ and $p_2(x, y)$ of problem (2.19)-(2.21) with distinct values of δ, β_x, and β_y, i.e., with $(\delta_1, \beta_{x1}, \beta_{y1})$ and $(\delta_2, \beta_{x2}, \beta_{y2})$, respectively, for the given values of P, M_x, and M_y. The two indicated sets of value of $(\delta, \beta_x, \beta_y)$ determine two functionals $\Phi_1(p)$ and $\Phi_2(p)$ from (2.31). Let us consider the difference $C = \Phi_1(p_2) - \Phi_1(p_1) + \Phi_2(p_1) - \Phi_2(p_2)$. Theorems 2.3.1 and 2.3.2 lead to $C > 0$. On the other hand, from (2.31) we get

$$C = (P_1 - P_2)(\delta_1 - \delta_2) + (M_{x1} - M_{x2})(\beta_{x1} - \beta_{x2})$$

$$+(M_y - M_{y2})(\beta_{y1} - \beta_{y2}), \tag{2.33}$$

where P_i, M_{xi} and M_{yi} are integrals of $p_i(x, y)$, $(i = 1, 2)$, according to (2.29). In case (a) $P_1 = P_2$, $M_{x1} = M_{x2}$ and $M_{y1} = M_{y2}$ that leads to $C = 0$. That contradicts the original assumption of non-uniqueness. In a similar fashion can be proven statements (b)-(d).

2.3.3   Solution Properties

Let us establish monotonicity of some of the parameters of the problem.

Theorem 2.3.5 *If $k > 0$, $0 < E' < \infty$, and conditions (2.24) are satisfied, then parameters P, M_x, and M_y are strictly monotonically increasing functions of δ, β_x, and β_y, respectively.*

Proof. Proof follows from inequality (2.33):

$$C = (P_1 - P_2)(\delta_1 - \delta_2) + (M_{x1} - M_{x2})(\beta_{x1} - \beta_{x2})$$

$$+ (M_{y1} - M_{y2})(\beta_{y1} - \beta_{y2}) \geq 0. \tag{2.34}$$

To perform some further qualitative analysis of problem (2.19)-(2.21), let us reformulate it. With the help of the Papkovich-Neiber representation [14] problem (2.19)-(2.21) can be reduced to determination of a harmonic in half-space $z < 0$ function $\Psi(x, y, z)$. The physical meaning of functions $\Psi(x, y, z)$ and $\partial \Psi(x, y, -0)/\partial z$ is clear from the fact that the vertical displacement of the half-space boundary $z = 0$ caused by elastic deformations of the body is $w_E = -(2/E')\Psi(x, y, 0)$ and the contact pressure $p = \partial \Psi(x, y, -0)/\partial z$. Therefore, problem (2.19)-(2.21) can be reduced to the following problem for Ψ:

$$\triangle \Psi = 0, \ z < 0, \tag{2.35}$$

$$k\varphi\left(\frac{\partial \Psi}{\partial z}\right) + \frac{2}{E'}\Psi = g(x, y), \ \frac{\partial \Psi}{\partial z} > 0, \ z = -0, \tag{2.36}$$

$$k\varphi\left(\frac{\partial \Psi}{\partial z}\right) + \frac{2}{E'}\Psi \geq g(x, y), \ \frac{\partial \Psi}{\partial z} = 0, \ z = -0, \tag{2.37}$$

where Δ is the Laplace operator.

The equivalence of problems (2.19)-(2.21) and (2.35)-(2.37) becomes obvious if we take into consideration that any harmonic in half-space $z < 0$ function Ψ can be represented in the form [14]

$$\Psi(x, y, z) = \frac{1}{2\pi} \int\limits_{z=0} \int \frac{p(\xi, \eta)d\omega}{\sqrt{(x-\xi)^2 + (y-\eta)^2 + z^2}}. \tag{2.38}$$

Clearly $\Psi(x, y, z) \geq 0$ if $p(x, y) \geq 0$ at any point (x, y).

Below we will use the principle of Zaremba-Giraud [20, 21] to establish certain solution properties. For harmonic functions this principle can be formulated as follows.

Theorem 2.3.6 *Suppose Ψ is a harmonic function for $z < 0$. If $\Psi(x, y, z) < \Psi(x, y, 0)$ for all $z < 0$, then the derivative $\partial \Psi(x, y, -0)/\partial z > 0$.*

Now, let us consider some properties of the problem at hand.

Theorem 2.3.7 *Suppose Ψ_1 and Ψ_2 are solutions of (2.35)-(2.37) for $0 < E' < \infty$ and $k\varphi = k\varphi_1, k\varphi_2 > 0$ and $g = g_1, g_2$. Then for each of the following conditions: (i) $k\varphi_1 = k\varphi_2$ and $g_1 > g_2$; (ii) $k\varphi_2 > k\varphi_1$ and $g_1 = g_2$ the inequality $\Psi_1(x, y, z) > \Psi_2(x, y, z)$ is valid for $z \leq 0$.*

Proof. Let us introduce a harmonic function $\Psi = \Psi_1 - \Psi_2$ and assume that $\Psi \leq 0$. According to the principle of maximum, $\min\limits_{z \leq 0} \Psi$ is reached at some point Q in plane $z = 0$. Therefore, $\Psi(Q) \leq 0$. According to boundary conditions

(2.36) and (2.37) the entire plane $z = 0$ can be subdivided into three regions Σ_1, Σ_2, and Σ_3:

$$\Sigma_1 : \frac{\partial \Psi_1}{\partial z} > 0, \frac{\partial \Psi_2}{\partial z} > 0; \; \Sigma_2 : \frac{\partial \Psi_1}{\partial z} > 0, \frac{\partial \Psi_2}{\partial z} = 0;$$

$$\Sigma_3 : \frac{\partial \Psi_1}{\partial z} = 0, \frac{\partial \Psi_2}{\partial z} > 0, \; \Sigma_4 : \frac{\partial \Psi_1}{\partial z} = 0, \frac{\partial \Psi_2}{\partial z} = 0. \tag{2.39}$$

Suppose conditions (i) are satisfied. Let us assume that $Q \in \Sigma_1$. Then according to (2.36) at point Q we get

$$k\{\varphi(\tfrac{\partial \Psi_1}{\partial z}) - \varphi(\tfrac{\partial \Psi_2}{\partial z})\} = g_1 - g_2 - \tfrac{2}{E'}\Psi > 0. \tag{2.40}$$

Here we use the assumption that $\Psi(Q) \leq 0$. Because of monotonicity of function φ and $k > 0$, we have $\frac{\partial \Psi_1}{\partial z} \geq \frac{\partial \Psi_2}{\partial z}$ that contradicts the principle of Zaremba-Giraud [20, 21] for harmonic functions. If $Q \in \Sigma_2 \bigcup \Sigma_4$, then again we have $\frac{\partial \Psi_1}{\partial z} \geq \frac{\partial \Psi_2}{\partial z}$. If $Q \in \Sigma_3$, then from (2.36) and (2.37) we have $\Psi_1 \geq \frac{E'}{2}g_1$ and

$$\Psi_2 = \tfrac{E'}{2}\{g_2 - k\varphi(\tfrac{\partial \Psi_2}{\partial z})\} < \tfrac{E'}{2}g_2 < \tfrac{E'}{2}g_1 \leq \Psi \tag{2.41}$$

that contradicts the original assumption $\Psi(Q) \leq 0$. Therefore, there are no points where $\Psi(Q) \leq 0$. Statement (ii) can be proven in a similar fashion.

Lemma 2.3.5 *Suppose $\delta_1 > \delta_2$. Then for $\delta = \delta_1$ the total force P applied to the indenter and the elastic displacement in the half-space are greater than those for $\delta = \delta_2$.*

Proof. Proof follows from Theorem 2.3.7.

Lemma 2.3.6 *Suppose there are two indenters one inside of the other, i.e., $f_1(x,y) < f_2(x,y)$. Then the total force P applied to the indenter bounded by function f_1 and the elastic displacement in the half-space are greater than those for the case of an indenter bounded by function f_2.*

Proof. Proof follows from Theorem 2.3.7 and the asymptotic representation of $\Psi(x,y,z) = P/(2\pi R) + o(R^{-1})$ for $R = (x^2 + y^2 + z^2)^{1/2} \to \infty$. Obviously, if $k\varphi_2 \geq k\varphi_1$ and/or $g_1 \geq g_2$, then $\Psi_1 \geq \Psi_2$ and $P_1 \geq P_2$.

Lemma 2.3.7 *Suppose there are two functions $\mid w_R \mid = k_1\varphi_1, k_2\varphi_2$ and $k_2\varphi_2 > k_1\varphi_1$. Then for $k\varphi = k\varphi_2$ the total force P applied to the indenter, its displacement δ, and the elastic displacement in the half-space are not smaller than those for $k\varphi = k\varphi_1$. Contact region Ω_2 includes contact region Ω_1.*

Proof. Proof of the first and third statements is based on Theorem 2.3.7. To prove the second statement, one should consider regions $\Sigma_1, \Sigma_2, \Sigma_3$ and to use Theorem 2.3.7 and the principle of Zaremba-Giraud [20, 21] for harmonic functions. To prove the fourth statement, let us assume that there exists a point Q on plane $z = 0$ at which $\frac{\partial \Psi_1(Q)}{\partial z} > 0$ and $\frac{\partial \Psi_2(Q)}{\partial z} = 0$. Then at Q from (2.36) and (2.37) we have

$$k_1 \varphi_1 \frac{\partial \Psi_1}{\partial z} + \frac{2}{E'} \Psi_1 = g \leq \frac{2}{E'} \Psi_2. \tag{2.42}$$

From these inequalities based on Theorem 2.3.7 and conditions (2.24), we get

$$0 < k_1 \varphi_1 \frac{\partial \Psi_1}{\partial z} + \frac{2}{E'} (\Psi_1 - \Psi_2) \leq 0, \tag{2.43}$$

which is impossible. Therefore, from inequality $\frac{\partial \Psi_1(Q)}{\partial z} > 0$ follows inequality $\frac{\partial \Psi_2(Q)}{\partial z} > 0$.

Theorem 2.3.8 *Suppose Ψ_1 and Ψ_2 are solutions of (2.35)-(2.37) for two values of Young's modulus $E' = E'_1, E'_2$; $0 < E'_2 < E'_1 < \infty$ and $k > 0$. Then the inequality $\Psi_1(x, y, z) > \Psi_2(x, y, z)$ is valid for $z \leq 0$.*

Proof. Let us introduce a harmonic function $\Psi = \Psi_1 - \Psi_2$ and assume that $\Psi \leq 0$, which is opposite to the Theorem statement. According to the principle of maximum, $\min_{z \leq 0} \Psi$ is reached at some point Q in the plane $z = 0$. Therefore, $\Psi(Q) \leq 0$. Similar to the proof of Theorem 6 the·entire plane $z = 0$ can be subdivided into three regions

$$\Sigma_1 : \frac{\partial \Psi_1}{\partial z} > 0, \ \frac{\partial \Psi_2}{\partial z} > 0; \ \Sigma_2 : \frac{\partial \Psi_1}{\partial z} > 0, \ \frac{\partial \Psi_2}{\partial z} = 0;$$
$$\Sigma_3 : \frac{\partial \Psi_1}{\partial z} = 0, \ \frac{\partial \Psi_2}{\partial z} > 0. \tag{2.44}$$

Let us assume that $Q \in \Sigma_1$. Then according to (2.36) at point Q we get

$$k \{\varphi(\frac{\partial \Psi_1}{\partial z}) - \varphi(\frac{\partial \Psi_2}{\partial z})\} = -\frac{2}{E'_1} \Psi + (\frac{2}{E'_2} - \frac{2}{E'_1}) \Psi_2 \geq 0 \tag{2.45}$$

that follows from the fact that $E'_1 > E'_2$ and the assumption that $\Psi(Q) \leq 0$. Using monotonicity of function φ for $k > 0$, we have $\frac{\partial \Psi_1}{\partial z} \geq \frac{\partial \Psi_2}{\partial z}$ that contradicts the principle of Zaremba-Giraud [20, 21]. If $Q \in \Sigma_2$, then again we have $\frac{\partial \Psi_1}{\partial z} \geq \frac{\partial \Psi_2}{\partial z}$. If $Q \in \Sigma_3$, then from (2.36) and (2.37) we have

$$k \varphi(\frac{\partial \Psi_2}{\partial z}) \leq \frac{2}{E'_1} \Psi_1 - \frac{2}{E'_2} \Psi_2 \leq 0 \tag{2.46}$$

because $E'_1 > E'_2$ and according to the assumption $\Psi(Q) \leq 0$. Therefore, from (2.24) we have $\frac{\partial \Psi_2}{\partial z} \leq 0$ that contradicts the assumption that $Q \in \Sigma_3$. Thus, there are no points where $\Psi(Q) \leq 0$.

Lemma 2.3.8 *Suppose $0 < E'_2 < E'_1 < \infty$ and $k > 0$. Then the total force P applied to the indenter for $E' = E'_1$ is greater than the one for $E' = E'_2$.*

Proof. Proof follows from Theorem 2.3.8 and the asymptotic representation of $\Psi(x, y, z) = P/(2\pi R) + o(R^{-1})$ for $R = (x^2 + y^2 + z^2)^{1/2} \to \infty$.

2.3.4 Problems with Fixed Contact Region and Their Relation to Contact Problems with Free Boundaries

Let us consider a family of contact problems with fixed contact region. These problems can be formulated in the form similar to (2.35)-(2.37), i.e.,

$$\Delta \Psi_S = 0, \ z < 0, \tag{2.47}$$

$$k\varphi(\tfrac{\partial \Psi_S}{\partial z}) + \tfrac{2}{E'} \Psi_S = g(x,y), \ z = -0, \ (x,y) \in S, \tag{2.48}$$

$$\tfrac{\partial \Psi_S}{\partial z} = 0, \ \tfrac{2}{E'} \Psi_S \geq g(x,y), \ z = -0, \ (x,y) \in CS, \tag{2.49}$$

where S is the contact region, $CS = \{(x,y) \mid (x,y) \ni S\}$ is the region complementary to S, and subscript S indicates a solution of the problem for a contact region S.

It can be shown that problem (2.47) has a unique solution for which analogues of Theorem 2.3.5 and 2.3.7 (statement (i)) as well as Lemmas 2.3.5 and 2.3.6 are valid. Proof of the analogue of statement (i) of Theorem 2.3.7 is conducted in a fashion similar to the one for Theorem 2.3.7 by replacing regions Σ_1, Σ_2, and Σ_3 by

$$\Sigma_1 : \ \tfrac{\partial \Psi_1}{\partial z} > 0, \ \tfrac{\partial \Psi_2}{\partial z} > 0; \ \Sigma_2 : \ \tfrac{\partial \Psi_1}{\partial z} \geq 0, \ \tfrac{\partial \Psi_2}{\partial z} \leq 0;$$

$$\Sigma_3 : \ \tfrac{\partial \Psi_1}{\partial z} = 0, \ \tfrac{\partial \Psi_2}{\partial z} > 0; \ \Sigma_4 : \ \tfrac{\partial \Psi_1}{\partial z} < 0, \ \tfrac{\partial \Psi_2}{\partial z} \geq 0; \tag{2.50}$$

$$\Sigma_5 : \ \tfrac{\partial \Psi_1}{\partial z} < 0, \ \tfrac{\partial \Psi_2}{\partial z} < 0.$$

Besides that, if $p_S(x,y) \geq 0$ at all points in S (see Theorem 2.3.11), then $\Psi_S(x,y,z) \geq 0$ for $z \leq 0$ (see (2.38)) and the analogues of Theorem 2.3.7 (statement (ii)) and Theorem 2.3.8 as well as Lemmas 2.3.7 and 2.3.8 are valid.

Let us establish some relations between problems (2.35)-(2.37) and (2.47).

Theorem 2.3.9 *Suppose $k > 0$, $0 < E' < \infty$, and Ω is the contact region for problem (2.35)-(2.37). Then for any region $S \neq \Omega$ the solution of problem (2.47) satisfies the inequalities $\Psi_S < \Psi$, $P_S < P$, i.e., $\Psi = \sup_S \Psi_S$, $P = \sup_S P_S$.*

Proof. Proof is similar to that for Theorem 2.3.7 with the replacement of regions Σ_1, Σ_2, and Σ_3 by

$$L_1 = \Sigma_1 \bigcap S, \ L_2 = \Sigma_1 \bigcap CS, \ L_3 = \Sigma_2 \bigcap S, \ L_4 = \Sigma_2 \bigcap CS,$$

$$\tag{2.51}$$

$$\Sigma_1 : \ \tfrac{\partial \Psi}{\partial z} > 0, \ \Sigma_2 : \ \tfrac{\partial \Psi}{\partial z} = 0.$$

Theorem 2.3.10 *Suppose $k > 0$, $0 < E' < \infty$, $p(x,y)$ is a solution of problem (2.35)-(2.37), and Ω is the contact region. Then for any region S the solution of problem (2.47) satisfies the inequality $p_S(x,y) > p(x,y)$ for all $(x,y) \in S \bigcap \Omega$.*

Proof. For all $(x, y) \in S \cap \Omega$ we have

$$k\varphi(\tfrac{\partial \Psi_S}{\partial z}) + \tfrac{2}{E'}\Psi_S = g = k\varphi(\tfrac{\partial \Psi}{\partial z}) + \tfrac{2}{E'}\Psi. \tag{2.52}$$

Taking into account that $\Psi_S < \Psi$ (see Theorem 2.3.9) and monotonicity of function $\varphi(p)$ we arrive at $p_S = \tfrac{\partial \Psi_S}{\partial z} > \tfrac{\partial \Psi}{\partial z} = p$ for $(x, y) \in S \cap \Omega$.

Theorem 2.3.11 *Suppose $k > 0$, $0 < E' < \infty$, Ω is the contact region for problem (2.35)-(2.37), and $p_S(x, y)$ and S are the pressure and contact region for problem (2.47). Then for $p_S(x, y)$ to be non-negative it is necessary and sufficient that $S \subseteq \Omega$.*

Proof. *Necessity.* Let us assume the opposite, i.e., suppose there exists such a region S that $S \cap C\Omega \neq \emptyset$ and $p_S(x, y) \geq 0$ at all points of S. According to Theorem 8, $\Psi_S(x, y, -0) < \Psi(x, y, -0)$, $p_S(x, y) = \partial \Psi_S(x, y, -0)/\partial z \geq 0$ and $p(x, y) = \partial \Psi_S(x, y, -0)/\partial z = 0$ at all points $(x, y) \in S \cap \Omega$. That contradicts the principle of Zaremba-Giraud [20, 21] for harmonic functions.

Sufficiency follows from Theorem 2.3.10.

Lemma 2.3.9 *Suppose $S \cap C\Omega \neq \emptyset$. Then under the conditions of Theorem 2.3.11 there exists at least one point $(x_*, y_*) \in S \cap C\Omega$ at which $p_S(x_*, y_*) < 0$.*

Proof. Proof follows from the considerations similar to the ones in the proof of the necessity in Theorem 2.3.11.

Let us introduce a definition of the region $B_\epsilon S$, the ϵ–belt of region S, as a set of points from S that are at a distance from the boundary of S not greater than ϵ.

Lemma 2.3.10 *Suppose the conditions of Theorem 2.3.10 are satisfied. Then for small enough $h > 0$ in the region $S \backslash B_h S$ the following inequalities*

$$0 < \varphi(p_S) - \varphi(p) \leq \tfrac{C}{h} mes(\Omega \backslash S), \ 0 < \Psi - \Psi_S < \tfrac{kE'C}{2h} mes(\Omega \backslash S) \tag{2.53}$$

hold, where constant $C = C(g, k, E', \Omega)$ is independent from S and $mes(\Omega \backslash S)$ is the measure (area) of the region $\Omega \backslash S$.

Proof. Let us subtract the first equation in (2.47) from (2.36). Based on formula (2.38) and the inequality $p_S > p$ in S, we get

$$0 < \varphi(p_S) - \varphi(p) \leq \tfrac{1}{\pi k E'} \int\limits_{z=0} \int \tfrac{p(\xi, \eta) d\omega}{R}, \ (x, y) \in S. \tag{2.54}$$

It follows from Lemma 2.3.1 that the integral in the latter inequality for $(x, y) \in S \backslash B_h S$ is not greater than $\max\limits_{\Omega} \varphi^{-1}(g/k) h^{-1} mes(\Omega \backslash S)$. Similarly, the second inequality can be obtained with the help of (2.36), (2.47), and the inequality for $\varphi(p_S) - \varphi(p)$.

Theorem 2.3.12 *Suppose condition (2.24) and conditions of Theorem 2.3.10 are satisfied. Then for any $h > 0$ the solution of problem (2.47) converges to the solution of problem (2.35)-(2.37) uniformly in $S \backslash B_h S$ in the C-norm of continuous functions as $mes(\Omega \backslash S) \to 0$.*

Proof. Proof follows from Lemma 2.3.3 and from continuity and strict monotonicity of function $\varphi^{-1}(\cdot)$.

Theorem 2.3.13 *Suppose conditions of Theorem 2.3.12 are satisfied, $p(x, y)$ and Ω are the solution of problem (2.35)-(2.37), and the bounded contact region, respectively, and the boundary of the contact region Ω is smooth enough. Then there exists a sequence of regions $\{S_n\}$ such that $S_n \in \Omega$, $S_n \to \Omega$ as $n \to \infty$ and $p(x, y) = \lim\limits_{n \to \infty} p_n(x, y)$ in the C-norm for $(x, y) \in \Omega$, where $p_n(x, y)$ are the solutions of problem (2.47) in regions S_n.*

Proof. For a bounded contact region Ω with a smooth enough boundary, we can determine a number sequence $\{H_n\}$ and regions $\{S_n\}$ as follows

$$S_n = \Omega \backslash B_{h_n}, \ h_n = H_n^2, \ \lim_{n \to \infty} H_n = 0, \ \lim_{n \to \infty} H_n^{-1} mes(\Omega \backslash S_n) = 0. \quad (2.55)$$

Regions $M_n = S_n \backslash B_{H_n}$ converge to Ω as $n \to \infty$. Based on Theorem 2.3.12 we can conclude that $p_{S_n}(x, y)$ converge to $p(x, y)$ for $(x, y) \in \Omega$.

2.3.5 Problems for Rough and Smooth Surfaces

Now, let us compare the solutions of contact problems for elastic half-spaces with rough and smooth (without asperities) surfaces. The formulations of contact problems for elastic half-space with smooth surface follows from (2.35)-(2.37) and (2.47) if $k = 0$.

Theorem 2.3.14 *Suppose $0 < E' < \infty$ and Ψ_0 and Ψ_k are the solutions of problem (2.35)-(2.37) for $k = 0$ and $k > 0$, respectively. Then $\Psi_0(x, y, z) > \Psi_k(x, y, z)$ for $z \le 0$ and $P_0 > P_k$.*

Proof. Proof is similar to the one for Theorem 2.3.7 and Lemma 2.3.3.

Lemma 2.3.11 *For fixed force P applied to indenter, its displacement δ is greater for the case of a rough $(k > 0)$ surface of the half-space than for the smooth $(k = 0)$ one.*

Proof. Proof follows from Theorems 2.3.5 and 2.3.14.

Lemma 2.3.12 *Suppose Ω_0 and Ω_k are the contact regions for solutions of problem (2.35)-(2.37) for $k = 0$ and $k > 0$, respectively. Then $\Omega_0 \subseteq \Omega_k$.*

Proof. Proof is similar to the one for Lemma 2.3.6.

Let us consider the relationship between the pressure distributions for rough surfaces with free and fixed contact boundaries obtained from solutions of problems (2.35)-(2.37) and (2.47). Suppose $k > 0$ and $p(x, y)$ and $p_S(x, y)$ are the pressure distributions and Ω and S are the contact regions for problems (2.35)-(2.37) and (2.47), respectively. Moreover, suppose that the limits $\Omega_0 = \lim_{k \to 0} \Omega$, $p_0(x, y) = \lim_{k \to 0} p(x, y)$, $p_{S0}(x, y) = \lim_{k \to 0} p_S(x, y)$ exist and represent the solutions of problems (2.35)-(2.37) and (2.47), respectively.

Lemma 2.3.13 *Suppose $0 < E' < \infty$, $k = 0$, and $p_0(x, y)$ and Ω_0 are the pressure and contact region for problem (2.35)-(2.37) while $p_{S0}(x, y)$ and S are the pressure and contact region for problem (2.47). Then $p_{S0}(x, y) \geq p_0(x, y)$ for all $(x, y) \in S \bigcap \Omega_0$.*

Proof. Proof follows from Theorem 2.3.10.

Lemma 2.3.14 *Suppose the conditions of Lemma 2.3.13 are satisfied and $\Omega_0 \subset S$. Then there exists at least one point $(x_*, y_*) \in S \backslash \Omega_0$ at which $p_{S0}(x_*, y_*) < 0$.*

Proof. Suppose for $k = 0$ functions Ψ_0 and Ψ_{S0} are the solutions of problems (2.35)-(2.37) and (2.47), respectively. According to Lemma 2.3.13 and Theorem 2.3.14, we have $\Psi_0(x, y, z) > \Psi_{S0}(x, y, z)$ for $z \leq 0$ and $p_{S0}(x, y) = \frac{\partial \Psi_{S0}(x, y, -0)}{\partial z} \geq p_0(x, y) = \frac{\partial \Psi_0(x, y, -0)}{\partial z}$ for all $(x, y) \in S \bigcap \Omega_0$. That contradicts the principle of Zaremba-Giraud [20, 21] because $\frac{\partial \Psi(x, y, -0)}{\partial z} \geq 0$ for all $(x, y) \in S \bigcap \Omega_0$. Therefore, the minimum point $(x_*, y_*) \in S \bigcap \Omega_0$ and $\frac{\partial \Psi(x_*, y_*, -0)}{\partial z} = \frac{\partial \Psi_{S0}(x_*, y_*, -0)}{\partial z} - \frac{\partial \Psi_0(x_*, y_*, -0)}{\partial z} < 0$. Taking into account the fact that $\frac{\partial \Psi_0(x, y, -0)}{\partial z} = 0$ for all points $(x, y) \in S \bigcap \Omega_0$ we conclude that there exists at least one point $(x_*, y_*) \in S \backslash \Omega_0$ at which $p_{S0}(x_*, y_*) < 0$.

In studies [22, 23, 24] it is shown that problem (2.47)-(2.49) can be solved exactly for $k = 0$ in cases when the contact region S has an elliptic or circular shape. The solution can be represented in the form

$$p_{S0}(x, y) = \int\limits_S \int g(\xi, \eta) N(x, y, \xi, \eta) d\omega, \qquad (2.56)$$

where $g(x, y)$ is determined in (2.22) and $N(x, y, \xi, \eta)$ is a certain singular function.

Let us consider the following iteration solution of problem (2.19)-(2.21) for $k = 0$. Suppose a circular region $S_0 = S$ is chosen in such a way that $\Omega_0 \subset S$, where Ω_0 is the contact region (see (2.18)) for problem (2.19)-(2.21). For the given function $g(x, y)$ let us take $S_1 = \Omega_\delta \subset S$ (see (2.23)) and set $g(x, y) = 0$ for $(x, y) \in S \backslash S_1$. After that we can perform regular iteration steps. Suppose the region S_n is determined. Then, calculate $p_{S0}^n(x, y)$ for $(x, y) \in S_n$ according to formula (2.56) and set $g(x, y) = 0$ if $p_{S0}^n(x, y) < 0$. The new region in which the corrected $g(x, y) > 0$ for all $(x, y) \in S_n$ we will call S_{n+1}. The iterations stop as soon as $S_n = S_{n+1}$.

Theorem 2.3.15 *The described iteration process for $k = 0$ converges to the solution of problem (2.19)-(2.21) for $k = 0$.*

Proof. If $p_0(x, y) \geq 0$ is the solution of problem (2.19)-(2.21), then from (2.19)- (2.21) we get $g(x, y) \geq 0$ for all $(x, y) \in \Omega_0$ and $\Omega_0 \subseteq S_1 = \Omega_\delta$ (see (2.23)), where Ω_0 is the contact region for problem (2.19)-(2.21) (see (2.18)). Besides, according to Lemma 2.3.13 we have $p_{S0}^n(x, y) \geq p_0(x, y) \geq 0$ for all points $(x, y) \in \Omega_0$. According to Lemma 2.3.14 it is established that $p_{S0}^n(x, y) < 0$ only at some points $(x, y) \in S_{n+1} \backslash \Omega_0$. Therefore, $\Omega_0 \sqsubseteq S_n$ and $S_{n+1} \sqsubseteq S_n$ for any $n \geq 0$. Obviously, the sequence of S_n converges as $n \to \infty$. If $S_* = \lim\limits_{n \to \infty} S_n$, then $\Omega_0 \sqsubseteq S_*$ and $p_{S*}(x, y) = \lim\limits_{n \to \infty} p_{S0}^n(x, y) \geq 0$ for all points $(x, y) \in S_*$. Let us show that $S_* = \Omega_0$. Let us assume the opposite, i.e., $S_* \backslash \Omega_0 \neq \emptyset$. Then according to Lemma 2.3.14 in $S_* \backslash \Omega_0$ there exists a point (x, y) at which $p_{S*}(x, y) < 0$. That contradicts the fact that $p_{S*}(x, y) \geq 0$. Therefore, $S_* = \Omega_0$. Taking into account the existence and uniqueness of the solution of problem (2.19)-(2.21), we conclude that $p_{S*}(x, y)$ is the solution of problem (2.19)-(2.21).

Lemma 2.3.15 *If $N(x, y, \xi, \eta) \geq 0$ for (ξ, η), $(x, y) \in S$, then the described iteration process is monotonic, i.e., $\Omega_0 \sqsubseteq S_{n+1} \sqsubseteq S_n \sqsubseteq S$ and $p_0(x, y) \leq p_{S0}^{n+1}(x, y) \leq p_{S0}^n(x, y) \leq \ldots \leq p_{S0}^1(x, y)$ for $n \geq 0$.*

Proof. Proof of the first statement follows from Theorem 2.3.15. The second statement follows from formula (2.48) with $N(x, y, \xi, \eta) \geq 0$ for $(x, y) \in S$ applied to two functions $g_n(x, y) \geq 0$ and $g_{n+1}(x, y) = g_n(x, y)$ if $p_{S0}^n(x, y) \geq 0$ for $(x, y) \in S_n$ and $g_{n+1}(x, y) = 0$ if $p_{S0}^n(x, y) < 0$ for $(x, y) \in S_n$.

If problem (2.19)-(2.21) is solved numerically, then the computational time for a single iteration is proportional to a number of nodes in S at which the pressure is determined. In reality, after each iteration the time required to perform the next one decreases because $S_{n+1} \sqsubseteq S_n$. This time can be reduced even further if each iteration of $p_{S0}^n(x, y)$ would be calculated according to (2.48) only in a narrow region adjacent to the boundary of S_n and, also, in the regions where $p_{S0}^n(x, y)$ was small in the preceding iterations. In this case, to ensure convergence function $p_{S0}^n(x, y)$ should be recalculated in the entire region S_n once in several iterations. That would enable us to determine the regions where $p_{S0}^n(x, y)$ is small. This iteration process represents a very efficient way of solving problem (2.19)-(2.21).

It can be shown that analogues of Theorem 2.3.13 and Lemma 2.3.9 are valid for solutions of problem (2.47) for $k = 0$ and $k > 0$ if the pressure functions are non-negative in a fixed contact region S for both cases.

2.3.6   A Different Approach to Problems (2.35)-(2.37)

Theorems similar to Theorems 2.3.3 and 2.3.5 for smooth surfaces ($k = 0$) were presented in [22]. The results stated in Theorems 2.3.7, 2.3.9-2.3.11, and

Lemmas 2.3.6, 2.3.7 were published in [25, 11]. In [11] Theorem 2.3.7 and Lemmas 2.3.5-2.3.7 were proved based on a variational inequality approach for function Ψ. Briefly this approach is represented below.

Let us introduce a set of functions $v \in V$ vanishing as R^{-1} while $R = (x^2 + y^2 + z^2)^{1/2} \to \infty$, which satisfy the following conditions

$$\frac{\partial v(x,y-0)}{\partial z} \geq 0, \; k\varphi(\frac{\partial v(x,y-0)}{\partial z}) + \frac{2}{E'}v(x,y,0) \geq g(x,y),$$

$$I(v,v) < C_1, \; J(v,v) < C_2, \tag{2.57}$$

where C_1 and C_2 are positive constants and

$$I(u,v) = \int\limits_{z<0} \int \int [\frac{\partial u}{\partial x}\frac{\partial v}{\partial x} + \frac{\partial u}{\partial y}\frac{\partial v}{\partial y} + \frac{\partial u}{\partial z}\frac{\partial v}{\partial z}]dxdydz,$$

$$J(u,v) = \int\limits_{z=0} \int \frac{\partial u}{\partial z}\varphi(\frac{\partial v}{\partial z})dxdy. \tag{2.58}$$

Theorem 2.3.16 *If $0 < E' < \infty$, $k > 0$, and the region where $\frac{\partial \Psi(x,y,-0)}{\partial z} \geq 0$ is bounded, then the problem for the variational inequality*

$$F(k,\Psi,v) = I(\Psi, v - \Psi) + \frac{kE'}{2}[J(\Psi,v) - J(\Psi,\Psi)] \geq 0,$$

$$\forall v \in V, \; \Psi \in V \tag{2.59}$$

is equivalent to problem (2.35)-(2.37).

Proof. Proof can be conducted in a fashion similar to that for Theorem 2.3.1.

Theorem 2.3.17 *Suppose V_1 and V_2 are closed convex sets of functions satisfying (2.49), function $\varphi(p)$ satisfies (2.24), $0 < E' < \infty$, $k_1 \geq k_2 > 0$, and $g_2 \geq g_1$. Besides, let us assume that $\Psi_1 \in V_1$ and $\Psi_2 \in V_2$ are solutions of variational inequalities $F(k_1, \Psi_1, w_1) \geq 0$, $\forall w_1 \in V_1$, and $F(k_2, \Psi_2, w_2) \geq 0$, $\forall w_2 \in V_2$, respectively. If there exist functions $w_1 \in V_1$ and $w_2 \in V_2$ such that*

$$w_1 + w_2 = \Psi_1 + \Psi_2,$$

$$\varphi(\frac{\partial w_1}{\partial z}) + \varphi(\frac{\partial w_2}{\partial z}) = \varphi(\frac{\partial \Psi_1}{\partial z}) + \varphi(\frac{\partial \Psi_2}{\partial z}),$$

$$I(w_1 - \Psi_2, w_1 - \Psi_1) = 0, \tag{2.60}$$

$$\int\limits_{z=0} \int [\frac{\partial \Psi_1}{\partial z} - \frac{\partial \Psi_2}{\partial z}][\varphi(\frac{\partial w_1}{\partial z}) - \varphi(\frac{\partial \Psi_1}{\partial z})]dxdy \leq 0,$$

then $w_1 = \Psi_1$ and $w_2 = \Psi_2$ for $z \leq 0$.

Proof. Proof is similar to Theorem 8.9 from Lions [17].

Theorem 2.3.18 *Suppose conditions (2.22)-(2.26) are satisfied. If Ψ_1 and Ψ_2 are the solutions of problem (2.35)-(2.37) for $0 < E' < \infty$ and $k = k_1 > 0$, $g(x,y) = g_1(x,y)$ and $k = k_2 > 0$, $g(x,y) = g_2(x,y)$, respectively, and $k_1 \geq k_2$ and $g_2(x,y) \geq g_1(x,y)$, then $\Psi_2(x,y,z) \geq \Psi_1(x,y,z)$.*

Proof. Proof of the theorem is based on Lemma 2.3.16.

Lemma 2.3.16 *Suppose V_1, V_2, $\varphi(p)$, k_1, k_2, g_1, g_2 satisfy conditions of Theorem 2.3.15 and $\Psi_1 \in V_1$ and $\Psi_2 \in V_2$ are solutions of problem (2.35)-(2.37) and*

$$v_1 = \min(\Psi_1, \Psi_2), \quad v_2 = \max(\Psi_1, \Psi_2). \tag{2.61}$$

Then $v_1 \in V_1$, $v_2 \in V_2$ and conditions (24) are valid for $w_1 = v_1$ and $w_2 = v_2$.

The existence and uniqueness of the solution of a plane contact problem with/without friction and some of its properties are discussed in the next chapter and in [30]. Some asymptotic and numerical solutions of these problems illustrating the obtained theoretical results are presented in the next chapter and in [8, 27].

2.4 Asymptotic Analysis of Plane Rough Contacts Modeled by Nonlinear Coating

Let us consider a plane problem for an infinite rigid cylindrical indenter contacting a thick layer of elastic isotropic homogeneous material bounded by two horizontal planes $z = 0$ and $z = -h$. The shape of the cylinder's base is described by a convex function $z = f(x)$. In this case the contact area bounded by lines $x = -a$ and $x = b$ (see the analysis for convex indenters in Section 3.4) is caring a load P per unit length. The material of the layer has Young's modulus E and Poisson's ratio ν. The surface of the material layer $z = 0$ is assumed to be coated or to have some asperities. In this case we can propose a plane formulation of the problem analogous to the one given by (2.19)-(2.21). However, for a plane formulation the equations are simpler and are as follows [8, 15]

$$k\varphi(p) + \frac{2}{\pi E'} \int_{-a}^{b} p(t)[\ln \mid \frac{h}{x-t} \mid + a_0]dt = \delta - f(x), \tag{2.62}$$

$$\int_{-a}^{b} p(t)dt = P, \tag{2.63}$$

where k and $\varphi(p)$ are a constant and function of pressure p characterizing the normal displacement of the surface $z = 0$ solely due to the presence of coating/surface asperities, E' is the effective elastic modulus, $E' = E/(1-\nu^2)$, δ is the normal displacement of the rigid indenter, h is the layer thickness, a_0 is a constant the value of which depends on the boundary conditions at the lower surface of the layer $z = -h$ (the values of a_0 for different conditions can be found in [15]).

In this problem formulation it is assumed that the rotation angle β_x is given, and it is included in the given function $f(x)$ (see (2.19)-(2.21)). Moreover, it is assumed that load P is also given. It is possible to consider three types of problems of this kind: (1) the problem for an indenter with sharp edges, i.e., a problem with fixed contact region $[-a, b]$; (2) the problem for an indenter without sharp edges, i.e., a problem with free boundaries of the contact that are determined by the equations $p(a) = p(b) = 0$; and (3) the problem with one fixed and one free boundaries, for example, with fixed boundary $x = -a$ and free boundary b that is determined by the equation $p(b) = 0$. There are no boundary conditions for Contact Problem 1, while the boundary conditions for Contact Problems 2 and 3 are as follows:

$$p(-a) = p(b) = 0 \ for \ Contact \ Problem \ 2, \tag{2.64}$$

$$p(b) = 0 \ for \ Contact \ Problem \ 3. \tag{2.65}$$

To conduct the further analysis we introduce the dimensionless variables using the characteristic length a_* and pressure p_* ($p_* a_* = \frac{2}{\pi} P$):

$$x' = \frac{x}{a_*}, \ a' = \frac{a}{a_*}, \ b' = \frac{b}{b_*}, \ p' = \frac{p}{p_*}, \ \delta' = \frac{E'}{p_* a_*}[\delta - (\ln \frac{h}{a_*} + a_0)P],$$

$$f' = \frac{E'}{p_* a_*} f, \ \lambda \varphi'(p', \lambda) = \frac{kE'}{p_* a_*} \varphi(p), \tag{2.66}$$

where λ is a nonnegative parameter characterizing the properties of roughness/coating. For Contact Problems 1 and 3 a_* can be any nonzero value, for example, $a_* = (a + b)/2$ for Contact Problem 1 and $a_* = a$ or $a_* = b$ for Contact Problem 3 if $x = a$ or $x = b$ is the fixed boundary, respectively. For Contact Problem 2 it is convenient to take a_* and p_* as solutions of the classic contact problem of elasticity for solids without coating and asperities.

For simplicity in the further analysis the primes at the dimensionless variables are omitted. By making a substitution

$$x = \frac{1}{2}[b - a + (b + a)y]$$

we convert the contact interval $[-a, b]$ to $[-1, 1]$. Therefore, without reducing the generality in dimensionless form Contact Problem 1 is represented by equations

$$\lambda \varphi(p, \lambda) + \frac{2}{\pi} \int_{-1}^{1} p(t) \ln \frac{1}{|y-t|} dt = \delta - f(y), \tag{2.67}$$

$$\int_{-1}^{1} p(t) dt = \frac{\pi}{2}, \tag{2.68}$$

Contact Problem 2 is represented by equations

$$\lambda\varphi(p,\lambda) + \tfrac{b+a}{\pi} \int\limits_{-1}^{1} p(t) \ln \tfrac{1}{|y-t|} dt = \delta - f(\tfrac{b-a}{2} + \tfrac{b+a}{2}y), \qquad (2.69)$$

$$p(-1) = p(1) = 0, \qquad (2.70)$$

$$\int\limits_{-1}^{1} p(t)dt = \tfrac{\pi}{b+a}, \qquad (2.71)$$

and Contact Problem 3 with the fixed left dimensionless boundary $x = -1$ of the contact is described by equations

$$\lambda\varphi(p,\lambda) + \tfrac{2}{\pi}\beta \int\limits_{-1}^{1} p(t) \ln \tfrac{1}{|y-t|} dt = \delta - f(\beta y + \beta - 1), \; \beta = \tfrac{b+1}{2}, \qquad (2.72)$$

$$p(1) = 0, \qquad (2.73)$$

$$\int\limits_{-1}^{1} p(t)dt = \tfrac{\pi}{2\beta}. \qquad (2.74)$$

To conduct the further analysis we will assume that $\varphi(p,\lambda)$ is a piece-wise differentiable function of p, which satisfies the following conditions:

$$\varphi(p,\lambda) > 0, \; \varphi'_p(p,\lambda) > 0 \; for \; p,\lambda > 0,$$

$$\varphi(p,\lambda) = A(\lambda)p^\alpha + \dots, \; \alpha > 0 \; for \; p \to \infty \; or \; p \to 0, \qquad (2.75)$$

$$A(\lambda) = A(0) + \dots, \; A(0) > 0 \; for \; \lambda \ll 1.$$

2.4.1   Plane Contact Problems with Fixed Boundaries

First, let us consider the application of the regular perturbation method to solution of Contact Problem 1 for $\lambda \gg 1$. In this case the major contribution to the displacement of the indenter comes from the deformation of asperities. In general, the solution of (2.67), (2.68) should be searched in the form of series

$$p(y) = \sum\limits_{n=0}^{\infty} \mu_n(\lambda)p_n(y), \; |y| \le 1, \; \delta = \sum\limits_{n=0}^{\infty} \gamma_n(\lambda)\delta_n \qquad (2.76)$$

over sets of unknown functions $\{\mu_n(\lambda)\}$ and $\{\gamma_n(\lambda)\}$. Based on representations (2.76) function $\varphi(p,\lambda)$ can be expanded as follows

$$\varphi(p,\lambda) = \sum\limits_{n=0}^{\infty} \mu_n^0(\lambda)\varphi_n(p_0,\dots,p_n). \qquad (2.77)$$

By substituting representations (2.76), (2.77) in equations (2.67), (2.68) and step by step balancing terms in these equations, we will find the specific expressions for functions $\{\mu_n(\lambda)\}$, $\{\gamma_n(\lambda)\}$, and $\{\mu_n^0(\lambda)\}$ as well as equations for functions $\{p_n(y)\}$ and constants $\{\delta_n\}$. Let us consider the above procedure in the case when function $\varphi(p, \lambda)$ is a differentiable function of both arguments and functions $\{\mu_n(\lambda)\}$, $\{\gamma_n(\lambda)\}$, and $\{\mu_n^0(\lambda)\}$ can be determined immediately as follows:

$$\mu_n(\lambda) = \mu_n^0(\lambda) = \lambda^{-n}, \ \gamma_n(\lambda) = \lambda^{1-n}, \ n = 1, 2, \dots. \tag{2.78}$$

Then from equations (2.76) and (2.77) we get

$$\varphi(p_0(y), \infty) = \delta_0, \ \int_{-1}^{1} p_0(y)dy = \tfrac{\pi}{4},$$

$$\varphi_p'(p_0(y), \infty)p_1(y) + \varphi_{\lambda^{-1}}'(p_0(y), \infty) + \tfrac{2}{\pi} \int_{-1}^{1} p_0(t) \ln \tfrac{1}{|y-t|} dt$$

$$= \delta_1 - f(y), \tag{2.79}$$

$$\int_{-1}^{1} p_1(y)dy = 0, \dots.$$

Based on (2.75) we conclude that the solutions of equations (2.79) are

$$p_0(y) = \tfrac{\pi}{4}, \ \delta_0 = \varphi(\tfrac{\pi}{4}, \infty),$$

$$p_1(y) = [\varphi_p'(\tfrac{\pi}{4}, \infty)]^{-1} \Big\{ \tfrac{1}{2} - \ln 2 + \tfrac{1}{2} \int_{-1}^{1} f(t)dt$$

$$+ \tfrac{1}{2}(1-y)\ln(1-y) + \tfrac{1}{2}(1+y)\ln(1+y) - f(y) \Big\}, \tag{2.80}$$

$$\delta_1 = \tfrac{3}{2} - \ln 2 + \tfrac{1}{2} \int_{-1}^{1} f(t)dt + \varphi_{\lambda^{-1}}'(\tfrac{\pi}{4}, \infty), \dots.$$

Formulas (2.76)-(2.78) and (2.80) determine a two-term uniformly valid on $[-1, 1]$ asymptotic solution of the problem.

From the expressions (2.80) for $p_0(y)$ and $p_1(y)$, it is clear that smoothness of $p(y)$ to a certain extent depends on smoothness of $f(y)$. Moreover, for a bounded $f(y)$ on $[-1, 1]$ the contact pressure $p(y)$ is also bounded on $[-1, 1]$ while $p'(y) \to \infty$ as $y \to \pm 1$.

Now, let us consider the case of $\lambda \ll 1$. This case can be analyzed by using the method of matched asymptotic expansions [28]. We will limit the analysis to consideration of just the leading terms of asymptotic expansions. To perform this analysis we will assume that conditions (2.75) are valid and on $[-1, 1]$ function $f'(y)$ satisfies the Hölder condition [16].

For small λ it is beneficial to represent equations in the equivalent form

$$p(y) = \frac{1}{2\sqrt{1-y^2}} \left\{ 1 + \frac{1}{\pi} \int\limits_{-1}^{1} \frac{f'(t)\sqrt{1-t^2}\,dt}{t-y} + \frac{\lambda}{\pi} \int\limits_{-1}^{1} \frac{\varphi'_p(p(t),\lambda)p'(t)\sqrt{1-t^2}\,dt}{t-y} \right\},$$

(2.81)

$$\delta = \ln 2 + \frac{1}{\pi} \int\limits_{-1}^{1} \frac{f(t)dt}{\sqrt{1-t^2}} + \frac{\lambda}{\pi} \int\limits_{-1}^{1} \frac{\varphi(p(t),\lambda)dt}{\sqrt{1-t^2}}$$

by inverting the integral operator in (2.67), (2.68) [16].

We will call the outer region the interval within which the solution of the problem is close to the one obtained by setting $\lambda = 0$ in equations (2.81). That leads to the outer solution of the problem [8]

$$p_0(y) = \frac{1}{2\sqrt{1-y^2}} \left\{ 1 + \frac{1}{\pi} \int\limits_{-1}^{1} \frac{f'(t)\sqrt{1-t^2}\,dt}{t-y} \right\}, \quad \delta_0 = \ln 2 + \frac{1}{\pi} \int\limits_{-1}^{1} \frac{f(t)dt}{\sqrt{1-t^2}}, \quad (2.82)$$

which approximates the problem solution at points y sufficiently far away from the end points of the contact $y = \pm 1$, i.e., for $y \pm 1 = O(1)$, $\lambda \ll 1$. Function $p_0(y)$ and constant δ_0 from (2.82) represent the solution of the classic contact problem without account for roughness/coating. However, the outer solution does not approximate the solution of the problem within small inner zones $y \pm 1 = o(1)$, $\lambda \ll 1$, where the contributions of the deformations of the bulk of material and asperities to the indenter displacement are of the same order of magnitude. Suppose $\varepsilon = \varepsilon(\lambda, \alpha) \ll 1$ is the characteristic size of the inner zones adjacent to points $y = \pm 1$.

Based on the fact that $f'(y)$ satisfies the Hölder condition and that the indenter possesses sharp edges at $y = \pm 1$ from (2.82), we obtain

$$p_0(y) = \varepsilon^{-1/2} \frac{N_-}{2\sqrt{2r}} + o(\varepsilon^{-1/2}), \quad r = \frac{x+1}{\varepsilon} = O(1), \quad \lambda \ll 1,$$

$$p_0(y) = \varepsilon^{-1/2} \frac{N_+}{2\sqrt{-2s}} + o(\varepsilon^{-1/2}), \quad s = \frac{x-1}{\varepsilon} = O(1), \quad \lambda \ll 1, \quad (2.83)$$

$$N_\pm = 1 + \frac{1}{\pi} \int\limits_{-1}^{1} f'(t)\sqrt{\frac{1\mp t}{1\pm t}}\,dt, \quad N_\pm = O(1), \quad \lambda \ll 1,$$

where r and s are the inner variables in the inner zones adjacent to points $y = -1$ and $y = 1$, respectively. Assuming that the solutions of the problem $p(y)$ in the inner zones match the inner asymptotic expansions of the outer solution from (2.83), we obtain

$$p(y) = O(\varepsilon^{-1/2}) \ for \ r, s = O(1), \ \lambda \ll 1.$$

Taking into account the latter estimate, we can search for the solutions of our equations (2.81) in the form

$$p(y) = p_0(y) + o(1) \ for \ y + 1 \gg \varepsilon \ and \ y - 1 \gg \varepsilon, \ \lambda \ll 1,$$

$$p(y) = \varepsilon^{-1/2} q(r) + o(\varepsilon^{-1/2}), \ r = O(1), \ \lambda \ll 1, \tag{2.84}$$

$$p(y) = \varepsilon^{-1/2} g(s) + o(\varepsilon^{-1/2}), \ s = O(1), \ \lambda \ll 1,$$

$$\delta = \delta_0 + \gamma_1(\lambda, \alpha)\delta_1 + o(\delta_1), \ \delta_1 \ll 1, \ \lambda \ll 1, \tag{2.85}$$

where $p_0(y)$ and δ_0 are determined by (2.82) while functions $q(r)$ and $g(s)$ and constants δ_1 and $\gamma_1(\lambda, \alpha)$ must be determined from the problem solution.

Let us consider the inner zone adjacent to the point $y = -1$. By rewriting equation (2.81) in variable r, using the assumptions (2.75) and representations from (2.84), and allowing λ to vanish while keeping $r = O(1)$, we obtain the equation for the leading term $q(r)$ of the asymptotic expansion of the solution in the inner zone $r = O(1)$, $\lambda \ll 1$, in the form

$$q(r) = \frac{1}{2\sqrt{2r}} \left\{ N_- + \frac{\alpha A(0)}{\pi} \int\limits_0^\infty \frac{q'(t)q^{\alpha-1}(t)\sqrt{2t}dt}{t-r} \right\}, \tag{2.86}$$

while for the characteristic size of the inner zone we get

$$\varepsilon(\lambda, \alpha) = \lambda^{\frac{2}{\alpha+1}}. \tag{2.87}$$

In a similar fashion using the estimate for ε from (2.87), we derive the equation for the leading term $g(s)$ of the asymptotic expansion of the solution in the inner zone $s = O(1)$, $\lambda \ll 1$, in the form

$$g(s) = \frac{1}{2\sqrt{-2s}} \left\{ N_+ + \frac{\alpha A(0)}{\pi} \int\limits_{-\infty}^0 \frac{g'(t)g^{\alpha-1}(t)\sqrt{-2t}dt}{t-s} \right\}. \tag{2.88}$$

Based on estimates (2.83) and representations (2.84) for the inner solutions to match the outer ones, we must require that

$$q(r) \rightarrow \frac{N_-}{2\sqrt{2r}}, \ r \rightarrow \infty, \tag{2.89}$$

and

$$g(s) \rightarrow \frac{N_+}{2\sqrt{-2s}}, \ s \rightarrow -\infty. \tag{2.90}$$

The latter matching conditions are self-consistent with the fact that integrals in equations (2.86) and (2.88) vanish as $r \rightarrow \infty$ and $s \rightarrow -\infty$, respectively.

Let us consider the indenter displacement δ. From the second equation in (2.81) and equations (2.82) and (2.85), it is obvious that

$$\delta - \delta_0 = \frac{\lambda}{\pi} \int\limits_{-1}^1 \frac{\varphi(p(t), \lambda)dt}{\sqrt{1-t^2}} = O(\varepsilon, \lambda), \tag{2.91}$$

where in (2.91) the term of the order of ε is due to the contributions of the inner zones while the term of the order λ represents the contribution of the outer region. Let us consider all possible cases in detail. If $0 < \alpha < 1$, then the contributions of the inner zones are much smaller then the one of the outer region (see (2.87)), and, therefore,

$$\delta_1 = \frac{1}{\pi} \int_{-1}^{1} \frac{\varphi(p_0(t),0)dt}{\sqrt{1-t^2}}, \quad \gamma_1(\lambda,\alpha) = \lambda. \tag{2.92}$$

If $\alpha = 1$, then the contributions to δ of the inner and outer regions are of the same order of magnitude. Therefore, using the the outer solution $p_0(y)$ from (2.82) and inner solutions $q(r)$ and $g(s)$ from (2.84), (2.86) and (2.84), (2.88), respectively, we can form a uniformly valid asymptotic approximation $p_u(y)$ of the problem solution

$$p_u(y) = \varepsilon^{-1} \frac{8\sqrt{1-x^2}}{N_+ N_-} p_0(y) q(\tfrac{y+1}{\varepsilon}) g(\tfrac{y-1}{\varepsilon}). \tag{2.93}$$

Then using representations (2.81), (2.82), and (2.85) we obtain

$$\delta_1 = \frac{1}{\pi} \int_{-1}^{1} \frac{\varphi(p_u(t),0)dt}{\sqrt{1-t^2}}, \quad \gamma_1(\lambda,\alpha) = \lambda. \tag{2.94}$$

For $\alpha > 1$ the main contributions to δ come from the inner zones (see (2.75), (2.82), (2.84), and (2.91)). Therefore, we find that

$$\delta_1 = \frac{A(0)}{\pi} \left\{ \int_{0}^{\infty} \frac{q^\alpha(r)dr}{\sqrt{2r}} + \int_{-\infty}^{0} \frac{g^\alpha(s)ds}{\sqrt{-2s}} \right\}, \quad \gamma_1(\lambda,\alpha) = \varepsilon. \tag{2.95}$$

It is important to notice that the indenter displacement δ in the presence of roughness is greater than without it. That becomes apparent from the fact that $\delta_1 > 0$ for small and large λ (see formulas (2.75), (2.80), (2.92)-(2.95)).

Some numerical results and their comparison with asymptotic solutions will be discussed in a separate section below.

2.4.2   Plane Contact Problems with Free Boundaries

For simplicity let us consider the case of even $f(y)$ so that $a = b$. The further analysis of the problem is done for a specific case of an indenter of parabolic shape $f(x) = x^2$. Then equations (2.69) and (2.70) of Contact Problem 2 can be rewritten in two equivalent forms as follows

$$\lambda \varphi(p,\lambda) + \frac{2b}{\pi} \int_{-1}^{1} p(t) \ln \left| \frac{1-t}{y-t} \right| dt = b^2(1-y^2), \tag{2.96}$$

$$p(y) = b\sqrt{1-y^2} + \frac{\lambda}{2\pi b} \sqrt{1-y^2} \int_{-1}^{1} \frac{\varphi'_p(p(t),\lambda)p'(t)dt}{\sqrt{1-t^2}(t-y)}, \tag{2.97}$$

$$\frac{\lambda}{\pi} \int_{-1}^{1} \frac{t\varphi'_p(p(t),\lambda)p'(t)dt}{\sqrt{1-t^2}} = 1 - b^2.$$

The expression for the indenter displacement δ easily follows from (2.69) after pressure $p(y)$ and contact boundary b are determined.

It can be easily verified that for $\varphi(p, \lambda) = p^2$ the exact solution of Contact Problem 2 is as follows

$$p(y) = \tfrac{1}{b}\sqrt{1 - y^2}, \; b = \sqrt{\tfrac{1}{2}[\sqrt{1 + 4\lambda} + 1]}, \tag{2.98}$$

the asymptotic expansions of which are

$$p(y) = \lambda^{-1/4}(1 - \tfrac{1}{4}\lambda^{-1/2} + \ldots)\sqrt{1 - y^2},$$
$$b = \lambda^{1/4}(1 + \tfrac{1}{4}\lambda^{-1/2} + \ldots), \; \lambda \gg 1, \tag{2.99}$$

$$p(y) = (1 - \tfrac{1}{2}\lambda + \tfrac{7}{8}\lambda^2 - \tfrac{33}{16}\lambda^3 + \ldots)\sqrt{1 - y^2},$$
$$b = 1 + \tfrac{1}{2}\lambda - \tfrac{5}{8}\lambda^2 + \tfrac{21}{16}\lambda^3 + \ldots, \; \lambda \ll 1. \tag{2.100}$$

First, let us consider application of the method of regular perturbations to solution of our problem for $\lambda \gg 1$. In this case the leading term of the solution is determined by deformation of asperities. For a special form of function

$$\varphi(p, \lambda) = p^\alpha, \tag{2.101}$$

it is immediately clear that the problem solution can be represented by the following asymptotic expansions in powers of λ:

$$p(y) = \sum_{n=0}^{\infty} \lambda^{-(2n+1)\gamma} p_n(y), \; | y | \le 1,$$
$$b = \sum_{n=0}^{\infty} \lambda^{-(2n-1)\gamma} b_n, \; \gamma = \tfrac{1}{\alpha+2}. \tag{2.102}$$

Substituting these expressions into equations (2.96) and (2.71) and equating the coefficients in the terms with the same powers of λ, we obtain equations for functions $p_n(y)$ and constants b_n in the form

$$p_0^\alpha(y) + b_0^2(y^2 - 1) = 0, \; \int_{-1}^{1} p_0(y)dy = \tfrac{\pi}{2b_0},$$

$$\alpha p_0^{\alpha-1}(y)p_1(y) + 2b_0b_1(y^2 - 1) + \tfrac{2b_0}{\pi} \int_{-1}^{1} p_0(t) \ln \tfrac{1-t}{|y-t|} dt = 0, \tag{2.103}$$

$$\int_{-1}^{1} p_1(t)dt = -\tfrac{\pi b_1}{2b_0^2}, \ldots.$$

The solutions of these equations are

$$p_0(y) = b_0^{2/\alpha}(1 - y^2)^{1/\alpha}, \ b_0 = \left\{ \frac{\sqrt{\pi}(\alpha+2)}{4} \frac{\Gamma[(\alpha+2)/(2\alpha)]}{\Gamma(1/\alpha)} \right\}^{\frac{\alpha}{\alpha+2}},$$

$$p_1(y) = \frac{2}{\alpha b_0} \left\{ b_1 - \frac{b_0^{2/\alpha}}{\pi}(1 - y^2)^{-1} \int\limits_{-1}^{1} (1 - t^2)^{1/\alpha} \ln \frac{1-t}{|y-t|} dt \right\} p_0(y), \qquad (2.104)$$

$$b_1 = \frac{4 b_0^{(\alpha+1)/\alpha}}{\pi^2(\alpha+2)} \int\limits_{-1}^{1} (1 - y^2)^{(1-\alpha)/\alpha} \int\limits_{-1}^{1} (1 - t^2)^{1/\alpha} \ln \frac{1-t}{|y-t|} dt dy, \dots$$

where $\Gamma(\cdot)$ is the gamma-function [29]. Formulas (2.102) and (2.104) represent the two-term uniformly valid on $[-1, 1]$ asymptotic solution of the problem for any $\alpha > 0$. It can be shown that for $\alpha = 2$ this expansion coincides with the expansion (2.99) of the exact solution.

Now, let us consider the case of $\lambda \ll 1$. It will be shown that for $\alpha > 3$ the solution can be obtained by application of the regular perturbations methods. For $\alpha > 3$ everywhere in the contact region $[-1, 1]$, the problem solution is mainly determined by the deformation of the bulk of the material. Therefore, for function φ from (2.101) the solution can be found in the form

$$p(y) = \sum_{n=0}^{\infty} \lambda^n p_n(y), \ | \ y \ | \leq 1, \ b = \sum_{n=0}^{\infty} \lambda^n b_n. \qquad (2.105)$$

Using equations (2.97) and (2.71) and acting in the fashion similar to the case of $\lambda \gg 1$, we obtain

$$p_0(y) = \sqrt{1 - y^2}, \ b_0 = 1,$$

$$p_1(y) = -\left\{ (\alpha - 1)b_1 + \frac{\alpha y}{2\pi} \int\limits_{-1}^{1} \frac{(1-t^2)^{(\alpha-3)/2} dt}{1-t} \right\} p_0(y), \qquad (2.106)$$

$$b_1 = \frac{1}{2\sqrt{\pi}} \frac{\Gamma[(\alpha-1)/2]}{\Gamma(\alpha/2)}, \dots$$

Formulas (2.105) and (2.106) determine the uniformly valid on $[-1, 1]$ two-term asymptotic expansion of the problem solution for $\alpha > 3$. That follows from the fact that the integral in (2.106) converges for $\alpha > 3$ and, therefore, the ratio $p_1(y)/p_0(y)$ is bounded on $[-1, 1]$. Moreover, for $\alpha = 2$ the latter expansion is reduced to the expansion (2.99) of the exact problem solution.

Let us turn now to the case of singular perturbations for $\lambda \ll 1$ and $0 < \alpha \leq 3$. We will analyze equations (2.97) by using the method of matched asymptotic expansions [8]. To analyze the inner zone, it is sufficient to use just a one-term asymptotic approximation in the outer region (see the expressions for $p_0(y)$ and b_0 in (2.106)):

$$p_0(y) = \sqrt{1 - y^2}, \ b_0 = 1. \qquad (2.107)$$

By introducing the inner variables $r = (y + 1)/\varepsilon$ and $s = (y - 1/\varepsilon)$ ($\varepsilon = \varepsilon(\lambda, \alpha)$ is the characteristic size of the inner zones), we immediately obtain the estimates $p_0(y) = \varepsilon^{1/2}\sqrt{2r} + o(\varepsilon^{1/2})$ for $r = O(1)$ and $p_0(y) = \varepsilon^{1/2}\sqrt{-2s} + o(\varepsilon^{1/2})$ for $s = O(1)$, $\lambda \ll 1$. To ensure matching of the problem solutions in the inner zones with the outer one represented by $p_0(y)$, when $y \to \mp 1$, it is necessary that $p(y) = O(\varepsilon^{1/2})$ as $r = O(1)$ and $s = O(1)$, $\lambda \ll 1$. Therefore, for $0 < \alpha < 3$ we will be searching the solution of (2.97) in the form

$$p(y) = p_0(y) + o(1), \ p_0(y) = O(1) \ for \ y \pm 1 = O(1) \ and \ \lambda \ll 1,$$

$$p(y) = \varepsilon^{1/2}q(r) + o(\varepsilon^{1/2}), \ q(r) = O(1) \ for \ r = O(1) \ and \ \lambda \ll 1,$$

$$p(y) = \varepsilon^{1/2}g(s) + o(\varepsilon^{1/2}), \ g(s) = O(1) \ for \ s = O(1) \ and \ \lambda \ll 1,$$

$$b = b_0 + \gamma_1(\lambda, \alpha)b_1 + o(\gamma_1), \ \gamma_1 \ll 1, \ b_1 = O(1) \ for \ \lambda \ll 1,$$

(2.108)

where $p_0(y)$ and b_0 satisfy (2.107), $q(r)$ and $g(s)$ are new unknown function representing the inner solutions, γ_1 is a function of the small parameter λ which has to be determined in the process of problem solution, and b_1 is the unknown constant.

By rewriting the first equation in (2.97) in variable r, substituting in it representations from (2.108), and letting λ approach 0 while $r = O(1)$ we obtain an equation for $q(r)$ and the characteristic size of the inner zones ε

$$q(r) = \sqrt{2r}\left\{1 + \frac{\alpha}{2\pi}\int\limits_0^\infty \frac{q'(t)q^{\alpha-1}(t)dt}{\sqrt{2t}(t-r)}\right\}, \ \varepsilon = \lambda^{2/(3-\alpha)}.$$

(2.109)

It can be shown that $\varepsilon^{1/2}q(r)$ as $r \to \infty$ matches $p_0(y)$ as $y \to -1$. Moreover, for $\alpha = 2$ the solution of equation (2.109) is $q(r) = \sqrt{2r}$, which is in agreement with (2.99).

Similarly, we get an equation for $g(s)$:

$$g(s) = \sqrt{-2s}\left\{1 - \frac{\alpha}{2\pi}\int\limits_{-\infty}^0 \frac{g'(t)g^{\alpha-1}(t)dt}{\sqrt{-2t}(t-s)}\right\}.$$

(2.110)

Obviously, we have matching of the inner solutions $q(r)$ and $g(s)$ with the inner asymptotic expansions of solution $p_0(y)$ in the outer region

$$q(r) \to \sqrt{2r}, \ r \to \infty; \ g(s) \to \sqrt{-2s}, \ s \to -\infty.$$

(2.111)

Because for even $f(y)$ the function of pressure $p(y)$ is even, it is clear that $g(s) = q(-r)$. Therefore, it is sufficient to consider the outer region and just one inner zone adjacent to the point $y = -1$.

For $\alpha = 3$ it is not possible to derive the equation for the leading term of the inner asymptotic $q(r)$. However, from the two-term outer solution of (2.97)

$$p(y) = p_0(y)\left\{1 + \gamma_1(\lambda, 3)b_1 - \lambda\frac{3}{2\pi}[2 + y\ln\frac{1-y}{1+y}]\right\},$$

(2.112)

by making the terms within the braces of the same order of magnitude, i.e., by comparing 1 with the term proportional to λ, we obtain $\varepsilon(\lambda, 3) = \exp(-1/\lambda)$. Therefore, for $\alpha < 3$ the size of the inner zone is a power function of λ while for $\alpha = 3$ it is exponentially small.

Now, we are ready to analyze the second equation in (2.97) for b. Substituting the expressions from (2.108)-(2.110) into this equation, we find that

$$\int_{-1}^{1} \frac{tp'(t)p^{\alpha-1}(t)dt}{\sqrt{1-t^2}} = O(\varepsilon^{\frac{\alpha-1}{2}}, 1), \quad \lambda \ll 1, \tag{2.113}$$

where term $\varepsilon^{\frac{\alpha-1}{2}}$ represents the contributions of the inner zones to the value of the integral while 1 is the contribution of the outer region.

Using the fact that $p(y)$ is even and using equations (2.97) and (2.108)-(2.110) for $0 < \alpha < 1$, we obtain

$$b_1 = \frac{\alpha}{\pi} \int_{0}^{\infty} \frac{q'(t)q^{\alpha-1}(t)dt}{\sqrt{2t}}, \quad \gamma_1 = \varepsilon. \tag{2.114}$$

Therefore, for $0 < \alpha < 1$ the variation of the size of the contact region is primarily determined by the inner solutions and it is of the order of magnitude of ε.

For $1 < \alpha \leq 3$ it is clear from (2.113) that the size of the contact region is primarily determined by the outer region. Therefore, $\gamma_1(\lambda, \alpha) = \lambda$ and b_1 is determined by formula (2.106).

In practice, in most cases the asperities are small and, therefore, it can be expected that $\lambda \ll 1$. In such a case for $\alpha > 1$, we have a very simple approximate two-term formula for the contact boundary

$$b = 1 + \frac{\lambda}{2\sqrt{\pi}} \frac{\Gamma[(\alpha-1)/2]}{\Gamma(\alpha/2)}. \tag{2.115}$$

It can be shown that in all above cases the size of the contact area with asperities is larger than the one for smooth surfaces and the maximum pressure is lower for a rough contact than for the smooth one.

2.4.3   Padé Approximations

Let us consider application of Padé approximations [30] to improvement of asymptotic results. Padé approximations of function $f(x)$ use rational functions

$$[g/q] = \sum_{j=0}^{g} a_j x^j / \sum_{j=0}^{q} b_j x^j, \tag{2.116}$$

where coefficients a_j and b_j are determined by the comparison of the centered at zero Taylor expansions of functions $f(x)$ and $[g/q]$.

We will illustrate the approach on the example of Contact Problem 2 for $\varphi(p, \lambda) = p^2$ when the exact solution of problem is known and it is given by

TABLE 2.1

The exact and approximated values of parameter b calculated based on asymptotic expansions for $\lambda \ll 1$ and $\lambda \gg 1$ as well as based on different Padé approximations.

λ	*Exact value of b*	*Asymptotic of b, $\lambda \ll 1$*	*Asymptotic of b, $\lambda \gg 1$*	$[1/1]$	$[1/2]$	$[2/1]$
0.5	1.169	1.258	1.138	1.077	1.004	1.174
1	1.272	2.188	1.250	1.111	1.120	1.298
2	1.414	10	1.399	1.143	1.867	1.519

(2.98). The advantage of the Padé approximation is due to the fact that in many cases using the Padé approximations with the same or less number of terms than in the original series approximation it is possible to get much more precise results. Let us use the four–term asymptotic expansion of parameter b given in (2.100) to obtain $[1/1]$, $[1/2]$, and $[2/1]$ [30]:

$$[1/1] = \frac{1+\frac{3}{2}\lambda}{1+\frac{5}{4}\lambda}, \quad [1/2] = \frac{1+\frac{1}{2}\lambda}{1+\frac{9}{14}\lambda - \frac{17}{56}\lambda^2}, \quad [2/1] = \frac{1+\frac{13}{5}\lambda+\frac{17}{40}\lambda^2}{1+\frac{21}{10}\lambda}. \tag{2.117}$$

For comparison the values of b calculated for different values of λ based on the exact solution from (2.98), asymptotic expansions (2.99), (2.100), and three different Padé approximations are presented in Table 2.1.

Clearly, the best precision is exhibited by the asymptotic approximation of b obtained for $\lambda \gg 1$ and by the Padé approximation $[2/1]$. It is understandable why $[2/1]$ is the best among the considered Padé approximations because in spite of the fact that it was constructed based on the asymptotic expansion (2.100) obtained for $\lambda \ll 1$ it resembles better the behavior of b for larger λ (see asymptotic expansion (2.99)). The same considerations work in other cases as well.

Therefore, based on asymptotic expansions and Padé approximations the problem solution can be obtained sufficiently accurately for all $\lambda \geq 0$.

2.4.4 Asymptotic and Numerical Analysis of Plane Rough Contacts with One Fixed and Another Free Boundaries

This section is devoted to the analysis of Contact Problem 3, i.e., of the contact problem for a rigid indenter with one sharp and another smooth edges, which is in contact with a rough half-plane. Contact Problem 3 is described by equations (2.72)-(2.74) in which constant β and pressure $p(y)$ are unknown. These equations are convenient for the problem analysis for $\lambda \gg 1$. For $\lambda \ll 1$ it is more appropriate to resolve equation (2.72) for $p(x)$ and to rewrite the equations in the form

$$p(y) = \frac{1}{2\pi\beta} \sqrt{\frac{1-y}{1+y}} \int_{-1}^{1} \sqrt{\frac{1+t}{1-t}} \frac{\psi(\beta,t)dt}{t-y}, \quad \int_{-1}^{1} \sqrt{\frac{1+t}{1-t}} \psi(\beta,t)dt = \pi,$$

$$\tag{2.118}$$

$$\psi(\beta,t) = \beta f'_{\beta t}(\beta t + \beta - 1) + \lambda \varphi'_p(p,\lambda)p'(t).$$

We will present the analysis for the case when the shape of the indenter bottom is given by

$$f(x) = Ax^2, \quad A > 0, \tag{2.119}$$

while the contribution of the surface roughness to the normal displacement $\varphi(p,\lambda)$ is represented by the function from equation (2.101). In (2.119) A is a constant.

First, let us consider application of the method of regular perturbations to solution of our problem for $\lambda \gg 1$. In this case the roughness term $\lambda\varphi(p,\lambda)$ dominates other terms in equation (2.72). Therefore, for $A = O(1)$, $\lambda \gg 1$, the problem solution can be searched in the form

$$p(y) = \sum_{k=0}^{\infty} \lambda^{-\frac{k+1}{\alpha+2}} p_k(y), \quad \beta = \sum_{k=0}^{\infty} \lambda^{-\frac{k-1}{\alpha+2}} \beta_k, \tag{2.120}$$

where functions $p_k(y)$ and constants β_k are of the order of magnitude one. Taking into account the expressions for $f(x)$ and $\varphi(p,\lambda)$ from (2.119) and (2.101), substituting (2.120) into equations (2.72)-(2.74), and equating terms of the same order of magnitude, we obtain

$$p_0^\alpha(y) = A\beta_0^2(3 - 2y - y^2), \quad p_0(1) = 0, \quad \int_{-1}^{1} p_0(y)dy = \frac{\pi}{2\beta_0}, \tag{2.121}$$

$$\alpha p_0^{\alpha-1}(y)p_1(y) = 2A\beta_0[\beta_1(3 - 2y - y^2) + y - 1], \quad p_1(1) = 0,$$

$$\tag{2.122}$$

$$\int_{-1}^{1} p_1(y)dy = -\frac{\pi\beta_1}{2\beta_0^2},$$

$$\alpha p_0^{\alpha-1}(y)p_2(y) + \frac{\alpha(\alpha-1)}{2}p_0^{\alpha-2}(y)p_1^2(y) + \frac{2}{\pi}\beta_0 \int_{-1}^{1} p_0(t) \ln \left| \frac{1-t}{y-t} \right| dt$$

$$= A[(2\beta_0\beta_2 + \beta_1^2)(3 - 2y - y^2) - 2\beta_1(1 - y)], \quad p_2(1) = 0, \tag{2.123}$$

$$\int_{-1}^{1} p_2(y)dy = \frac{\pi}{2\beta_0^2}[\frac{\beta_1^2}{\beta_0} - \beta_2].$$

By solving equations (2.121)-(2.123) we arrive at the formulas

$$p_0(y) = \{A\beta_0^2(1 - y)(3 + y)\}^{\frac{1}{\alpha}}, \quad \beta_0 = \left\{ \frac{3\alpha^2 \sqrt{\pi}}{A^{\frac{1}{\alpha}} 2^{\frac{2\alpha+1}{\alpha}}} \frac{\Gamma(\frac{3\alpha}{2})}{\Gamma(\frac{1}{\alpha})} \right\}^{\frac{\alpha}{\alpha+2}}, \tag{2.124}$$

$$p_1(y) = \tfrac{2}{\alpha\beta_0}(\beta_1 - \tfrac{1}{y+3})p_0(y),$$

$$\beta_1 = \tfrac{\alpha A^{\frac{1}{\alpha}}}{\pi(\alpha+2)}(2\beta_0)^{\frac{\alpha+2}{\alpha}}\left\{\tfrac{\sqrt{\pi}}{\alpha}\tfrac{\Gamma(\frac{1}{\alpha})}{\Gamma(\frac{\alpha+2}{2\alpha})} - 1\right\},$$

$$(2.125)$$

$$p_2(y) = \tfrac{1}{\alpha\beta_0}\left\{2\beta_2 + \tfrac{\beta_1^2}{\beta_0} - \tfrac{2\beta_1}{\beta_0}\tfrac{1}{y+3}\right.$$

$$-\tfrac{2}{\pi A}\tfrac{1}{3-2y-y^2}\int\limits_{-1}^{1} p_0(t)\ln\left|\tfrac{1-t}{y-t}\right|\,dt - \tfrac{2(\alpha-1)}{\alpha\beta_0}(\beta_1 - \tfrac{1}{y+3})^2\Big\}p_0(y),$$

$$\beta_2 = \tfrac{\beta_1^2}{\beta_0} + \tfrac{1}{\alpha+2}\left\{\tfrac{\alpha-1}{\alpha+1}2^{\frac{\alpha+2}{\alpha}}A^{\frac{1}{\alpha}}\beta_0^{\frac{2}{\alpha}}F(1, 2 - \tfrac{1}{\alpha}; 2 + \tfrac{1}{\alpha}; -1)\right. \qquad (2.126)$$

$$+\tfrac{4}{\pi^2}A^{\frac{2-\alpha}{\alpha}}\beta_0^{\frac{\alpha+4}{\alpha}}\int\limits_{-1}^{1}[(1-y)(3+y)]^{\frac{1-\alpha}{\alpha}}\,dy\int\limits_{-1}^{1}[(1-t)(3+t)]^{\frac{1}{\alpha}}$$

$$\times\ln\left|\tfrac{1-t}{y-t}\right|\,dt\Big\},$$

where $F(a, b; c; z)$ is the hypergeometric function [29].

Therefore, formulas (2.120), (2.124)-(2.126) determine the three-term uniformly valid on the entire interval $[-1, 1]$ asymptotic expansion of the solution of equations (2.72)-(2.74) for $\lambda \gg 1$. It is clear from the structure of the approximate solution (2.120), (2.124)-(2.126) that in the vicinity of the boundaries of the contact the pressure behaves as follows

$$p(y) \to const, \ y \to -1; \ p(y) \to const(1 - y)^{\frac{1}{\alpha}}, \ y \to 1. \qquad (2.127)$$

Now, let us consider the problem solution for $\lambda \ll 1$. In case of functions $f(x)$ and $\varphi(p, \lambda)$ determined by equations (2.119) and (2.101), respectively, equations (2.118) can be rewritten in the form

$$p(y) = A\beta\sqrt{1 - y^2} + A(\beta - 1)\sqrt{\tfrac{1-y}{1+y}}$$

$$+\tfrac{\lambda\alpha}{2\pi\beta}\sqrt{\tfrac{1-y}{1+y}}\int\limits_{-1}^{1}\sqrt{\tfrac{1+t}{1-t}}\tfrac{p^{\alpha-1}(t)p'(t)dt}{t-y}, \qquad (2.128)$$

$$\lambda\alpha\int\limits_{-1}^{1}\sqrt{\tfrac{1+t}{1-t}}p^{\alpha-1}(t)p'(t)dt = \pi[1 - A\beta(3\beta - 2)]. \qquad (2.129)$$

For $\lambda \ll 1$ solution of equations (2.128) and (2.129) we will find by applying the method of matched asymptotic expansions [28]. Then the solution of the problem in the external region is

$$p_0(y) = A\beta_0\sqrt{1 - y^2} + A(\beta_0 - 1)\sqrt{\tfrac{1-y}{1+y}}, \ \beta_0 = \tfrac{1}{3}\{1 + \sqrt{\tfrac{A+3}{A}}\}. \qquad (2.130)$$

Obviously, formulas (2.130) coincide with the solution of the corresponding contact problem for smooth solids [16].

It is important to recognize that the proximity of the value of constant A to 1 can vary, which should be reflected in the solution method. We will limit our analysis by just the case when $A-1 = O(1)$, $\lambda \ll 1$. Let us introduce the local variables $r = \frac{y+1}{\epsilon_r}$ and $s = \frac{y-1}{\epsilon_s}$, where $\epsilon_r(\lambda)$ and $\epsilon_s(\lambda)$ are the characteristic sizes of the inner zones adjacent to the points $y = -1$ and $y = 1$, respectively. Now, we can determine the asymptotic expansions of $p_0(y)$ in the inner zones

$$p_0(y) = \frac{2A(\beta_0-1)}{\sqrt{2r}}\epsilon_r^{-1/2} + o(\epsilon_r^{-1/2}), \quad r = O(1), \quad \lambda \ll 1,$$

$$(2.131)$$

$$p_0(y) = \frac{A(3\beta_0-1)\sqrt{-2s}}{2}\epsilon_s^{1/2} + o(\epsilon_s^{1/2}), \quad s = O(1), \quad \lambda \ll 1.$$

Following the principle of matching of asymptotic expansions [28] and using asymptotic expansions from (2.131) we will be looking for the solution for the function of pressure $p(y)$ in the inner zones in the form

$$p(y) = \epsilon_r^{-1/2}q(r) + o(\epsilon_r^{-1/2}), \quad q(r) = O(1), \quad r = O(1), \quad \lambda \ll 1,$$

$$(2.132)$$

$$p(y) = \epsilon_s^{1/2}g(s) + o(\epsilon_s^{1/2}), \quad g(s) = O(1), \quad s = O(1), \quad \lambda \ll 1,$$

where functions $q(r)$ and $g(s)$ should be determined from the solution of the problem in the inner zones.

The integral involved in equation (2.128) can be represented as a sum of three integrals over the external region encompassing almost the entire interval $[-1, 1]$ except for two small inner zones adjacent to points $y = -1$ and $y = 1$ and the latter small inner zones, i.e.,

$$\int_{-1}^{1} \Phi(t,y,p)dt = \left\{ \int_{-1}^{-1+\eta_r} + \int_{-1+\eta_r}^{1-\eta_s} + \int_{1-\eta_s}^{1} \right\}\Phi(t,y,p)dt,$$

$$(2.133)$$

$$\Phi(t,y,p) = \alpha\sqrt{\frac{1+t}{1-t}}\frac{p^{\alpha-1}(t)p'(t)}{t-y},$$

where $-1 + \eta_r$ and $1 - \eta_s$ are the boundaries of the inner zones with the external region. Moreover, constants $\eta_r = \eta_r(\lambda) \ll 1$ and $\eta_s = \eta_s(\lambda) \ll 1$ are such that $\frac{\eta_r(\lambda)}{\epsilon_r(\lambda)} \to \infty$ and $\frac{\eta_s(\lambda)}{\epsilon_s(\lambda)} \to \infty$ as $\lambda \to \infty$.

Estimating each of the integrals in (2.133) with the help of (2.130) and (2.132) we obtain the following asymptotic representations

$$\int_{-1}^{-1+\eta_r} \Phi(t,y,p)dt = \frac{\alpha}{2}\epsilon_r^{-\frac{\alpha+1}{2}}\int_{0}^{\infty} Q(t,r)dt + \dots,$$

$$Q(t,r) = \frac{q^{\alpha-1}(t)q'(t)\sqrt{2t}}{t-r}, \quad r = O(1),$$

$$\int_{-1+\eta_r}^{1-\eta_s} \Phi(t,y,p)dt = \int_{-1+\eta_r}^{1-\eta_s} \Phi(t,y,p_0)dt + \dots = O(1),$$

$$y \pm 1 = O(1),$$

$$\int_{1-\eta_s}^{1} \Phi(t, y, p)dt = 2\alpha \epsilon_s^{\frac{\alpha-3}{2}} \int_{-\infty}^{0} G(t, s)dt, \tag{2.134}$$

$$G(t, s) = \frac{g^{\alpha-1}(t)g'(t)}{\sqrt{-2t(t-s)}}, \quad s = O(1).$$

Similar estimates are valid for the integral in equation (2.129).

Let us assume that the contribution of the integral over the external/outer region in (2.134) is small and the integrals over the inner zones vanish as $r \to \infty$ and $s \to -\infty$, respectively. Now, we can derive the equations in the inner zones. Let us consider the vicinity of the point $y = -1$. Substituting relationships (2.131), (2.132), and (2.134) in equation (2.128) and using the above assumptions, we arrive at the equation

$$q(r) = \frac{1}{\sqrt{2r}} \left\{ 2A(\beta_0 - 1) + \frac{\alpha}{2\pi\beta_0} \int_{0}^{\infty} \frac{q^{\alpha-1}(t)q'(t)\sqrt{2t}dt}{t-r} \right\}. \tag{2.135}$$

Based on the fact that the local and integral terms in equation (2.128) are of the same order of magnitude in the process of the derivation of equation (2.135), we also obtain the size of the left inner zone

$$\epsilon_r = \lambda^{\frac{2}{\alpha+1}}. \tag{2.136}$$

Similarly, in the right inner zone we get

$$g(s) = \sqrt{-2s} \left\{ \frac{A(3\beta_0 - 1)}{2} + \frac{\alpha}{2\pi\beta_0} \int_{-\infty}^{0} \frac{g^{\alpha-1}(t)g'(t)dt}{\sqrt{-2t(t-s)}} \right\}, \tag{2.137}$$

$$\epsilon_s = \lambda^{\frac{2}{3-\alpha}}. \tag{2.138}$$

It is important to stress that all above assumptions are valid for $0 < \alpha < 3$. Moreover, it can be shown that if the following conditions are satisfied

$$q(r) \to \frac{2A(\beta_0-1)}{\sqrt{2r}}, \quad r \to \infty; \quad g(s) \to \frac{A(3\beta_0-1)\sqrt{-2s}}{2}, \quad s \to -\infty, \tag{2.139}$$

then the integrals involved in equations (2.135) and (2.137) vanish for $r \to \infty$ and $s \to -\infty$, respectively.

Notice that the inner zone sizes from (2.136) and (2.138) coincide with the ones obtained for the inner zones for Contact Problems 1 and 2, respectively.

For $\alpha = 3$ using the matched asymptotic expansions method, it is not possible to derive an asymptotically valid equation in the right inner zone. However, it is possible to estimate the characteristic size of this zone as $\epsilon_r = e^{-1/\lambda}$.

In case of $\alpha > 3$ in the vicinity of the right contact boundary $y = 1$, the inner zone does not exist. Therefore, everywhere in the contact except for the narrow inner zone in the vicinity of $y = -1$ the approximate solution can be obtained by regular asymptotic expansions. The main term of this approximate solution is determined by formulas (2.130). The next term of such an expansion can be obtained with the help of the uniformly valid on $[-1, 1]$ pressure $p_u(y)$, which takes into account the solution for $q(r)$ in the left inner zone (which is still determined by equation (2.135)).

Let us analyze the relationship (2.129) for β. For $0 < \alpha < 3$ with the help of (2.101), (2.132), (2.136), and (2.138), the evaluation of the integral in (2.129) provides us with the estimate

$$\int_{-1}^{1} \sqrt{\frac{1+t}{1-t}} p^{\alpha-1}(t)p'(t)dt = O(\lambda^{\frac{1-\alpha}{1+\alpha}}, 1, \lambda^{\frac{\alpha-1}{3-\alpha}}), \ \lambda \ll 1, \qquad (2.140)$$

where the value of $\lambda^{\frac{1-\alpha}{1+\alpha}}$ is the contribution of the left inner ϵ_r-zone, the value of 1 is the contribution of the external region while the value of $\lambda^{\frac{\alpha-1}{3-\alpha}}$ is the contribution of the right inner ϵ_s-zone. Therefore, constant β will be searched in the form

$$\beta = \beta_0 + \delta_1(\lambda)\beta_1 + o(\delta_1), \ \delta_1(\lambda) \ll 1, \ \beta_1 = O(1), \ \lambda \ll 1, \qquad (2.141)$$

where β_0, obviously, is the larger solution of the equation $A\beta_0(3\beta_0 - 2) = 1$, which follows from (2.130) for $y = 1$, and constants δ_1 and β_1 should be determined from the further analysis.

For $0 < \alpha < 1$ we have $\lambda^{\frac{1-\alpha}{1+\alpha}} \ll 1 \ll \lambda^{\frac{\alpha-1}{3-\alpha}}$ and the relationships (2.101), (2.132), (2.136), (2.138), and (2.141) lead to the formulas

$$\beta_1 = -\frac{\alpha}{\pi A(3\beta_0-1)} \int_{-\infty}^{0} \frac{g^{\alpha-1}(t)g'(t)dt}{\sqrt{-2t}}, \ \delta_1(\lambda) = \epsilon_s. \qquad (2.142)$$

Therefore, for $0 < \alpha < 1$ the value of β_1 is determined by the solution of the problem in the vicinity of $y = 1$.

For $\alpha = 1$ we have $\lambda^{\frac{1-\alpha}{1+\alpha}} = \lambda^{\frac{\alpha-1}{3-\alpha}} = 1$ and the contributions of the inner zones and the external region to the value of the integral in (2.140) are of the same order of magnitude. Therefore, by introducing the uniformly valid on $[-1, 1]$ solution,

$$p_u(y) = \frac{p_0(y)q(\frac{y+1}{\lambda})g(\frac{y-1}{\lambda})}{A^2(\beta_0-1)(3\beta_0-1)} \sqrt{\frac{1+y}{1-y}} \qquad (2.143)$$

and using (2.132), (2.136), (2.138), and (2.141), we obtain the expression

$$\beta_1 = -\frac{\alpha}{2\pi A(3\beta_0-1)} \int_{-1}^{1} \sqrt{\frac{1+t}{1-t}} p_u^{\alpha-1}(t)p_u'(t)dt, \ \delta_1(\lambda) = \lambda. \qquad (2.144)$$

For $1 < \alpha < 3$ we have $\lambda^{\frac{1-\alpha}{1+\alpha}} \gg 1 \gg \lambda^{\frac{\alpha-1}{3-\alpha}}$, and the relationships (2.101), (2.132), (2.136), (2.138), and (2.141) lead to the formulas

$$\beta_1 = -\frac{\alpha}{4\pi A(3\beta_0-1)} \int_{0}^{\infty} \sqrt{2t} q^{\alpha-1}(t)q'(t)dt, \ \delta_1(\lambda) = \epsilon_r. \qquad (2.145)$$

Therefore, for $1 < \alpha < 3$ the value of β_1 is determined by the solution of the problem in the vicinity of $y = -1$, which is on the opposite from $y = 1$ end of the contact.

It can be shown that for $\alpha \geq 3$ the values of β_1 and δ_1 are also determined by formulas (2.145).

It is important to realize that these results without significant changes can be extended on the cases of other indenter shapes and different relationships between the normal asperity displacement and contact pressure.

2.5 Some Numerical Methods and Results for Plane Rough Contacts

First, let us illustrate the quality of some of the above asymptotic approximations by comparing the values of the indenter displacement δ and the contact region radius b calculated based on these approximations and on solutions of the original problems obtained by some authors. In [7] for $f(x) = f(0)$, $\alpha = 0.5$, and $\lambda = 0.44445$, Contact Problem 4 was solved in its original formulation and it was obtained that $\delta - f(0) = 0.934$. For the same input data, the two-term asymptotic formulas (2.85) and (2.92) give 0.945. In [6] for $f(x) = x^2$, $\lambda = 12.2$, and $\alpha = 1$, the semi-width of the contact region was determined to be $b = 2.34$ while based on the two-term asymptotic formulas (2.102) and (2.104) we obtain $b = 2.33$.

Now, let us develop the numerical approaches to the above problems and consider some numerical solutions for problems in the original formulations and their asymptotic analogs as well and compare these solutions. Some of the proposed below numerical methods are based on the approaches used in the presented asymptotic methods. We will follow the procedures presented in [27, 31].

2.5.1 Numerical Methods for Plane Contact Problems in the Original Formulations

In developing numerical methods for contact problems for solids with rough surfaces, we will follow the approach proposed in [31] for solution of contact problems for rough solids with friction. On the interval $[-1, 1]$ let us introduce two systems of nodes (see [27, 31]): the integer nodes $\{y_i\}$ $(i = 0, \ldots, N)$, $y_0 = -1$, $y_N = 1$ and the semi-integer nodes $\{y_{j+1/2}\}$ $(j = 0, \ldots, N-1)$, $y_{j+1/2} = (y_j + y_{j+1})/2$. In addition, we will introduce the following notations $u_i = u(y_i)$ and $u_{i+1/2} = u(y_{i+1/2})$. For solution of Contact Problems 1-3 we will be using a quadrature

$$\int_{-1}^{1} \frac{u(t)dt}{t-y_{j+1/2}} = \sum_{i=1}^{N-1} \int_{y_{i-1/2}}^{y_{i+1/2}} \frac{u(t)dt}{t-y_{j+1/2}} \approx \sum_{i=1}^{N-1} \frac{y_{i+1/2}-y_{i-1/2}}{y_i-y_{j+1/2}} u_i \qquad (2.146)$$

related to the ones proposed in [31]. The benefits of using this kind of a quadrature are due to the fact that in [31] it is proven that solutions to the discrete analogs of a number of problems for a linear integral equation with singular kernel converge to the exact solutions of these equations in the space of continuous functions.

Equations of Contact Problems 1 and 3 follow from equation (2.69) for the specific values of constants a and b. Therefore, we will be using this equation as a proxy for all Contact Problems 1-3. Equation (2.69) is nonlinear. Therefore, to solve it we need to use an iteration process that involves linearization of all nonlinear functions for p in the vicinity on the known iterates $\{p^k\}$, a^k, and b^k, where k is the iteration number. Differentiating equation (2.69) with respect to y and satisfying it at points $y = y_{j+1/2}$, $j = 0,\dots,N-1$, using quadrature (2.146) leads to a system of nonlinear algebraic equations. With the help of linearization of these equations about p_j^k, a^k, and b^k, we obtain

$$\lambda\varphi_p'(p_{j+1/2}^k)\frac{p_{j+1}^{k+1}-p_j^{k+1}}{y_{j+1}-y_j} + \frac{b^k+a^k}{\pi}\sum_{i=1}^{N-1}\frac{y_{i+1/2}-y_{i-1/2}}{y_i-y_{j+1/2}}p_i^{k+1} + \triangle_b^{k+1}\{F_j^k$$

$$+f_{xb}''(\frac{b^k-a^k}{2} + \frac{b^k+a^k}{2}y_{j+1/2})\} + \triangle_a^{k+1}\{F_j^k + f_{xa}''(\frac{b^k-a^k}{2} \qquad (2.147)$$

$$+\frac{b^k+a^k}{2}y_{j+1/2})\} = -f_x'(\frac{b^k-a^k}{2} + \frac{b^k+a^k}{2}y_{j+1/2}), \; j = 0,\dots,N-1,$$

where we introduced the following notations

$$\triangle_b^{k+1} = b^{k+1} - b^k, \; \triangle_a^{k+1} = a^{k+1} - a^k,$$

$$(2.148)$$

$$F_j^k = \frac{1}{\pi}\sum_{i=1}^{N-1}\frac{y_{i+1/2}-y_{i-1/2}}{y_i-y_{j+1/2}}p_i^k.$$

The system of linear algebraic equations (2.147) and (2.148) is not complete. For Contact Problem 1 we need to add to this system the discrete analog of equation (2.71) and the equations following from $a = b = 1$:

$$\sum_{i=1}^{N}(y_i - y_{i-1})(p_{i-1}^{k+1} + p_i^{k+1}) = \frac{2\pi}{b^k+a^k}, \; \triangle_b^{k+1} = 0, \; \triangle_a^{k+1} = 0. \qquad (2.149)$$

For Contact Problem 2 we need to add to this system the discrete analogs of equations (2.70) and (2.71):

$$\sum_{i=1}^{N}(y_i - y_{i-1})(p_{i-1}^{k+1} + p_i^{k+1}) + \frac{2\pi}{(b^k+a^k)^2}(\triangle_b^{k+1} + \triangle_a^{k+1}) = \frac{2\pi}{b^k+a^k},$$

$$(2.150)$$

$$p_0^{k+1} = 0, \; p_N^{k+1} = 0.$$

For Contact Problem 3 to complete the system we need to add the discrete analogs of equations (2.73), (2.71) and the equation following from $a = 1$:

$$\sum_{i=1}^{N}(y_i - y_{i-1})(p_{i-1}^{k+1} + p_i^{k+1}) + \frac{2\pi}{(b^k + a^k)^2}\triangle_b^{k+1} = \frac{2\pi}{b^k + a^k},$$

(2.151)

$$\triangle_a^{k+1} = 0, \ p_N^{k+1} = 0.$$

Therefore, for each of Contact Problems 1-3 we obtained a system of $N + 3$ linear algebraic equations for values of $\{p_i^{k+1}\}$, $i = 0, \ldots, N$ and values of $\triangle_a^{k+1} = a^{k+1} - a^k$ and $\triangle_b^{k+1} = b^{k+1} - b^k$.

The convergence of these iteration processes for Contact Problems 1-3 is practically uniform for any not very small value of parameter λ. For Contact Problems 2 and 3 the role of the regularization terms similar to the ones proposed in [31] is played by the terms proportional to $\triangle_a^{k+1}$, $\triangle_b^{k+1}$, and $\triangle_b^{k+1}$, respectively. The latter terms vanish as the iteration processes converge.

The iteration processes are run as follows. Suppose the kth iterates $\{p_j^k\}$ $(j = 0, \ldots, N)$, a^k, and b^k are known. Then we solve one of the systems of equations (2.147), (2.148), and (2.149) for Contact Problem 1, (2.150) for Contact Problem 2, or (2.151) for Contact Problem 3, respectively, for values of $\{p_j^{k+1}\}$ $(j = 0, \ldots, N)$, $\triangle_a^{k+1}$, and $\triangle_b^{k+1}$. The latter two values allow to determine the new iterates $a^{k+1} = a^k + \triangle_a^{k+1}$ and $b^{k+1} = b^k + \triangle_b^{k+1}$. Further, the iterations continue until the desired precision is reached. The initial approximations can be taken, for example, as the first terms of the corresponding regular asymptotic expansions.

Now, let us consider the numerical schemes for the case of small roughness, i.e., $\lambda \ll 1$. For small λ it is beneficial to use the problem equations in the resolved form. Namely, for Contact Problems 1 and 3 we will be using equations (2.81) and (2.118) (see next section), respectively. Equation (2.97) is obtained for Contact Problem 2 in the case of a parabolic indenter with $f(x) = x^2$. In the case of a general function $f(x)$, equations (2.69)-(2.71) for Contact Problem 2 can be rewritten in the following equivalent form

$$p(y) = \frac{\sqrt{1-y^2}}{\pi(b+a)}\left\{\lambda \int_{-1}^{1} \frac{\varphi_p'(p(t))p'(t)dt}{\sqrt{1-t^2}(t-y)} + \int_{-1}^{1} \frac{f_t'(\frac{b-a}{2} + \frac{b+a}{2}t)dt}{\sqrt{1-t^2}(t-y)}\right\},$$

(2.152)

$$\int_{-1}^{1}\left\{\lambda\varphi_p'(p(t))p'(t) + f_t'(\frac{b-a}{2} + \frac{b+a}{2}t)\right\}\sqrt{\frac{1-t}{1+t}}dt = -\pi,$$

(2.153)

$$\int_{-1}^{1}\left\{\lambda\varphi_p'(p(t))p'(t) + f_t'(\frac{b-a}{2} + \frac{b+a}{2}t)\right\}\sqrt{\frac{1+t}{1-t}}dt = \pi.$$

For solution of Contact Problems 1-3 represented by equations (2.81), (2.152), (2.153), and (2.118), we will use the following quadrature:

$$\int_{-1}^{1}\frac{u(t)dt}{t-y_j} = \sum_{i=1}^{N}\int_{y_{i-1}}^{y_i}\frac{u(t)dt}{t-y_j} \approx \sum_{i=1}^{N}\frac{y_i - y_{i-1}}{y_{i-1/2} - y_j}u_{i-1/2}$$

(2.154)

related to the ones used in [31]. In addition to that, we will employ the approximation $u_{i-1/2} \approx (u_{i-1} + u_i)/2$.

Satisfying equations (2.81), (2.155), (2.153), and (2.118) at nodes $\{y_j\}$ based on quadrature (2.154) on each iteration, we obtain the following systems of linear equations:

$$2\sqrt{1-y_j^2}\,p_j^{k+1} - \frac{\lambda}{\pi}\sum_{i=1}^{N}\varphi_p'(p_{i-1/2}^k)\frac{p_i^{k+1}-p_{i-1}^{k+1}}{y_i-y_{i-1}}\frac{\sqrt{1-y_{i-1/2}^2}(y_i-y_{i-1})}{y_{i-1/2}-y_j}$$

$$= 1 + \frac{1}{\pi}\sum_{i=1}^{N}f_y'(y_{i-1/2})\frac{\sqrt{1-y_{i-1/2}^2}(y_i-y_{i-1})}{y_{i-1/2}-y_j},$$

(2.155)

$$p_j^{k+1} - \lambda\frac{\sqrt{1-y_j^2}}{\pi(b^k+a^k)}\sum_{i=1}^{N}\varphi_p'(p_{i-1/2}^k)\frac{p_i^{k+1}-p_{i-1}^{k+1}}{y_i-y_{i-1}}\frac{y_i-y_{i-1}}{\sqrt{1-y_{i-1/2}^2}(y_{i-1/2}-y_j)}$$

$$= \frac{\sqrt{1-y_j^2}}{\pi(b^k+a^k)}\sum_{i=1}^{N}f_y'(\frac{b^k-a^k}{2}+\frac{b^k+a^k}{2}y_{i-1/2})\frac{y_i-y_{i-1}}{\sqrt{1-y_{i-1/2}^2}(y_{i-1/2}-y_j)},$$

(2.156)

$$\sqrt{1+y_j}\,p_j^{k+1} - \lambda\frac{\sqrt{1-y_j}}{\pi(b^k+1)}\sum_{i=1}^{N}\varphi_p'(p_{i-1/2}^k)$$

$$\times\frac{p_i^{k+1}-p_{i-1}^{k+1}}{y_i-y_{i-1}}\sqrt{\frac{1+y_{i-1/2}}{1-y_{i-1/2}}}\frac{y_i-y_{i-1}}{y_{i-1/2}-y_j}$$

(2.157)

$$= \frac{\sqrt{1-y_j}}{\pi(b^k+1)}\sum_{i=1}^{N}f_y'(\frac{b^k-1}{2}+\frac{b^k+1}{2}y_{i-1/2})\sqrt{\frac{1+y_{i-1/2}}{1-y_{i-1/2}}}\frac{y_i-y_{i-1}}{y_{i-1/2}-y_j},$$

where $j = 0,\ldots,N$ and the superscripts k and $k+1$ indicate the iteration number.

To complete systems of equations (2.156) and (2.157), we need to add the discrete analogs of equations (2.153) and second equation in (2.118):

$$\sum_{i=1}^{N}\{f_y'(\frac{b^{k+1}-a^{k+1}}{2}+\frac{b^{k+1}+a^{k+1}}{2}y_{i-1/2})$$

$$+\lambda\varphi_p'(p_{i-1/2}^{k+1})\frac{p_i^{k+1}-p_{i-1}^{k+1}}{y_i-y_{i-1}}\}\sqrt{\frac{1-y_{i-1/2}}{1+y_{i-1/2}}}(y_i-y_{i-1}) = -\pi,$$

(2.158)

$$\sum_{i=1}^{N}\{f_y'(\frac{b^{k+1}-a^{k+1}}{2}+\frac{b^{k+1}+a^{k+1}}{2}y_{i-1/2})$$

$$+\lambda\varphi_p'(p_{i-1/2}^{k+1})\frac{p_i^{k+1}-p_{i-1}^{k+1}}{y_i-y_{i-1}}\}\sqrt{\frac{1+y_{i-1/2}}{1-y_{i-1/2}}}(y_i-y_{i-1}) = \pi,$$

$$\sum_{i=1}^{N}\{f_y'(\frac{b^{k+1}-1}{2}+\frac{b^{k+1}+1}{2}y_{i-1/2})$$

(2.159)

$$+\lambda\varphi_p'(p_{i-1/2}^{k+1})\frac{p_i^{k+1}-p_{i-1}^{k+1}}{y_i-y_{i-1}}\}\sqrt{\frac{1+y_{i-1/2}}{1-y_{i-1/2}}}(y_i-y_{i-1}) = \pi.$$

The iteration processes for Contact Problems 1-3 work as follows. First, we accept certain initial approximations for a^0, b^0, and $p^0(y)$, for example, $a^0 = b^0 = 1$, $p^0(y) = 1/\sqrt{1-y^2}$, $p^0(y) = \sqrt{1-y^2}$, $p^0(y) = \sqrt{\frac{1-y}{1+y}}$ for Contact Problems 1-3, respectively. Assuming that iterates $\{p_j^k\}$ ($j = 0, \ldots, N$) and a^k, b^k are known from the above linear systems of equations (2.155)-(2.157), we determine the new set of iterates $\{p_j^{k+1}\}$ ($j = 0, \ldots, N$). After that for Contact Problems 2 and 3, we solve nonlinear equations (2.158) and (2.159) for a^{k+1}, b^{k+1}, and b^{k+1}, respectively. Further, the iteration process continues until it converges.

After the two types of the iteration processes proposed for moderate and small λ, converged the value of constant δ can be determined from equation (2.69) for particular a and b and some value of y (for example, for $y = -1$) by using any quadrature formula involving the set of obtained values $\{p_j\}$ ($j = 0, \ldots, N$) (for example, see formula (2.167)).

In addition to the above two approaches for moderate and small λ, let us consider the numerical schemes, which work well for moderate and large λ. We will use the simple iteration method and the following quadrature:

$$\int_{-1}^{1} u(t) \ln | t - y_j | \, dt = \sum_{i=0}^{N-1} \int_{y_i}^{y_{i+1}} u(t) \ln | t - y_j | \, dt$$

$$\approx \sum_{i=0}^{N-1} u(y_{i+1/2})[(y_{i+1} - y_j) \ln | y_{i+1} - y_j |$$ (2.160)

$$-(y_i - y_j) \ln | y_i - y_j | - y_{i+1} + y_i].$$

By introducing the new function $g(y)$ and using quadrature (2.160) from equations (2.69) and (2.71), we obtain the new set of iterative numerical schemes

$$g_j^k = 1 - \frac{1}{\delta^k}\{f(\frac{b^k - a^k}{2} + \frac{b^k + a^k}{2} y_j)$$

$$-\frac{b^k + a^k}{2\pi} \sum_{i=0}^{N-1} (p_i^k + p_{i+1}^k)[(y_{i+1} - y_j) \ln | y_{i+1} - y_j |$$ (2.161)

$$-(y_i - y_j) \ln | y_i - y_j | - y_{i+1} + y_i]\}, \quad j = 0, \ldots, N,$$

$$p_j^{k+1} = (1 - \sigma)p_j^k + \sigma\varphi^{-1}(\frac{\delta^{k+1} g_j^k}{\lambda}), \quad j = 0, \ldots, N,$$ (2.162)

$$(b^{k+1} + a^{k+1})\{(1 - \sigma) \sum_{i=0}^{N-1} (p_i^k + p_{i+1}^k)(y_{i+1} - y_i)$$

$$+\sigma \sum_{i=0}^{N-1} [\varphi^{-1}(\frac{\delta^{k+1} g_i^k}{\lambda}) + \varphi^{-1}(\frac{\delta^{k+1} g_{i+1}^k}{\lambda})](y_{i+1} - y_i)\} = 2\pi,$$ (2.163)

where σ is the relaxation parameter, $0 < \sigma < 1$. For Contact Problem 1 to equations (2.161)-(2.163), we need to add the conditions

$$a^{k+1} = b^{k+1} = 1. \tag{2.164}$$

For Contact Problem 2 based on equations (2.69) and (2.70), we need to add the following equations

$$g_0^k = g_N^k = 0, \tag{2.165}$$

$$f(b^{k+1}) - f(-a^{k+1}) + \frac{b^k + a^k}{2\pi} \sum_{i=0}^{N-1} (p_i^{k+1} + p_{i+1}^{k+1})[(y_{i+1} + 1)$$

$$\times \ln | y_{i+1} + 1 | -(y_i + 1) \ln | y_i + 1 | -(y_{i+1} - 1) \ln | y_{i+1} - 1 | \tag{2.166}$$

$$+(y_i - 1) \ln | y_i - 1 |],$$

$$\delta^{k+1} = f(-a^{k+1}) + \frac{b^k + a^k}{2\pi} \sum_{i=0}^{N-1} (p_i^{k+1} + p_{i+1}^{k+1})[(y_{i+1} + 1)$$

$$\times \ln | y_{i+1} + 1 | -(y_i + 1) \ln | y_i + 1 | -y_{i+1} + y_i], \tag{2.167}$$

while for Contact Problem 3 based on equations (2.69) and (2.73), we need to add equation (2.167) and

$$a^{k+1} = 1, \ g_N^k = 0. \tag{2.168}$$

The just proposed iterative schemes work as follows. For Contact Problems 1-3 the initial approximations for moderate and large λ can be chosen based on the one-term asymptotic expansions of their solutions. In particular, for $\varphi(p) = p^\alpha$ and $f(y) = Ay^2$ for Contact Problems 1-3, we can take, respectively,

$$p^0(y) = \tfrac{\pi}{4}, \ \delta^0 = \lambda(\tfrac{\pi}{4})^\alpha \ for \ A = 0,$$

$$p^0(y) = \tfrac{1}{\xi}[Ad_0^2(1 - y^2)]^{1/\alpha}, \ a^0 = b^0 = \xi d_0,$$

$$d_0 = \left\{ \frac{\sqrt{\pi}(\alpha+2)}{4A^{1/\alpha}} \frac{\Gamma(\frac{\alpha+2}{2\alpha})}{\Gamma(\frac{1}{\alpha})} \right\}^{\alpha/(\alpha+2)}, \tag{2.169}$$

$$p^0(y) = \tfrac{1}{\xi}[Ad_1^2(1 - y)(3 + y)]^{1/\alpha}, \ b^0 = \xi d_1,$$

$$d_1 = \left\{ \frac{3\sqrt{\pi}\alpha^2}{2^{(2\alpha+1)/\alpha} A^{1/\alpha}} \frac{\Gamma(\frac{3\alpha}{2})}{\Gamma(\frac{1}{\alpha})} \right\}^{\alpha/(\alpha+2)}, \ \xi = \lambda^{1/(\alpha+2)}.$$

First, we determine the values of g_j^k from equations (2.161). Then for Contact Problem 1 we solve equations (2.163) and (2.164) for δ^{k+1}; for Contact Problem 2 we solve equations (2.162), (2.163), (2.165)-(2.167) for δ^{k+1}, a^{k+1}, and b^{k+1}; for Contact Problem 3 we solve equations (2.162), (2.163), (2.167), and

(2.168) for δ^{k+1}, and b^{k+1}. After that, using equation (2.162) we determine the new iterates p_j^{k+1}, $j = 0, \ldots, N$. These iterative processes continue until the desired precision is reached.

Obviously, the numerical schemes proposed for small and moderate values of λ on one hand and the schemes proposed for moderate and large λ on the other hand are the numerical analogs developed for small and large λ (see the previous two sections and the next section). Also, for small λ in [27] a number of numerical schemes are proposed that directly replicate the regular and matched asymptotic methods developed in this chapter.

The numerical simulations showed that for the the absolute precision of 0.001 and in case of the last set of schemes (2.161)-(2.168) for the relaxation parameter $\sigma = 0.15$ the solution converges after 3-8 iterations.

2.5.2 Numerical Methods for Plane Contact Problems in the Asymptotic Formulations

Let us introduce some fixed number R and split the segment $[0, \infty)$ into two: $[0, R]$ and (R, ∞). We will assume that R is sufficiently large so that for $r > R$ the solutions $q(r)$ of asymptotically valid equations (2.86) and (2.109) can be replaced by their asymptotic expansions (see equations (2.89) and (2.111)) $\frac{N_-}{2\sqrt{2r}}$ and $\sqrt{2r}$, respectively. As in the preceding subsection, on interval $[0, R]$ we introduce two systems of nodes r_i ($i = 0, \ldots, N$), $r_0 = 0$, $r_N = R$, and $r_{i+1/2}$ ($i = 0, \ldots, N-1$), $r_{i+1/2} = (r_i + r_{i+1})/2$, where N is a sufficiently large positive integer. Therefore, the result of solution of equations (2.86) and (2.109) is the set of values $q_i = q(r_i)$ ($i = 0, \ldots, N$).

While developing the numerical schemes for equations (2.86) and (2.109) we will follow the approach used for asymptotic analysis for $\lambda \ll 1$ (also see [27]).

2.5.2.1 Numerical Method for Asymptotic Problems with Fixed Boundary

Let us introduce such a number $R_0 = r_n$ that for $R_0 \leq r_i \leq R$ the solution of equation (2.86) can be obtained using the method of simple iteration while for $0 \leq r_i \leq R_0$ on each iteration the solution would satisfy a system of linear algebraic equations. Similar to the scheme obtained for the original problem for $\lambda \ll 1$ (see the preceding section) we get

$$\frac{2\pi\sqrt{2r_j}}{\alpha} q_j^{k+1} - \sum_{i=1}^{n-1} (q_{i-1/2}^k)^{\alpha-1} \frac{(q_i^{k+1} - q_{i-1}^{k+1})\sqrt{2r_{i-1/2}}}{r_{i-1/2} - r_j}$$

$$= \frac{\pi N_-}{\alpha} + \sum_{i=n}^{N} (q_{i-1/2}^k)^{\alpha-1} \frac{(q_i^k - q_{i-1}^k)\sqrt{2r_{i-1/2}}}{r_{i-1/2} - r_j} \qquad (2.170)$$

$$+ (\tfrac{N_-}{2})^\alpha G(\tfrac{\alpha+1}{2}, R, \tfrac{r_j}{R}), \ j = 0, \ldots, n-1,$$

$$q_j^{k+1} = (1-\sigma)q_j^k + \frac{\sigma}{2\sqrt{2r_j}}\{N_-$$

$$+\frac{\alpha}{\pi}[\sum_{i=1}^{n-1}(q_{i-1/2}^{k+1})^{\alpha-1}\frac{(q_i^{k+1}-q_{i-1}^{k+1})\sqrt{2r_{i-1/2}}}{r_{i-1/2}-r_j}$$

$$+\sum_{i=n}^{N}(q_{i-1/2}^k)^{\alpha-1}\frac{(q_i^k-q_{i-1}^k)\sqrt{2r_{i-1/2}}}{r_{i-1/2}-r_j}+(\tfrac{N_-}{2})^{\alpha}G(\tfrac{\alpha+1}{2},R,\tfrac{r_j}{R})]\},$$

(2.171)

$$j=n,\ldots,N-1,\quad q_N^{k+1}=\frac{N_-}{2\sqrt{2r_N}},$$

where superscript k is the iteration number, σ is the relaxation parameter $0<\sigma<1$, and function G is the integral

$$\int_R^\infty \frac{q^{\gamma-1}(t)q'(t)\sqrt{2t}dt}{t-r}$$

obtained for $q(r)=\frac{1}{\sqrt{2r}}$ and equal to [32]

$$G(\alpha,R,\tfrac{r}{R})=\frac{1}{(2R)^\alpha}\Big\{\frac{1}{\alpha}-\ln(1-\tfrac{r}{R})-\alpha\sum_{k=1}^\infty\frac{1}{k(k+\alpha)}(\tfrac{r}{R})^k\Big\}.$$

(2.172)

The iteration process runs as follows. As the initial approximation we take $q_j^0=\frac{N_-}{2\sqrt{2r_j}}$, $j=1,\ldots,N$, $q_0^0=\frac{N_-}{2\sqrt{2r_1}}$. Assuming that the values of q_j^k, $j=0,\ldots,N$, are known from equations (2.170) and (2.172) are determined the set of values q_j^{k+1}, $j=0,\ldots,n-1$. After that from equations (2.171) and (2.172) the rest of the values q_j^{k+1}, $j=n,\ldots,N$, are determined. The iteration process runs until it converges with the desired precision.

After the iterations converged for even function $f(x)$ and $\alpha>1$, we calculate the value of the first-order approximation for the indenter displacement (see (2.85) and (2.95))

$$\delta_1=\frac{2A(0)}{\pi}\Big\{\sum_{i=1}^{N-1}\frac{q_{i+1/2}^{\alpha-1}}{\sqrt{2r_{i+1/2}}}(r_{i+1}-r_i)+\frac{1}{\alpha-1}(\tfrac{N_-}{2})^\alpha(2R)^{(1-\alpha)/2}\Big\}.$$

(2.173)

2.5.2.2   Numerical Method for Asymptotic Problem with Free Boundary

Now, following exactly the same approach as for the derivation of the scheme (2.170)-(2.172) for solution of equation (2.109), we derive the following numerical scheme

$$q_j^{k+1}-\frac{\alpha\sqrt{2r_j}}{2\pi}\sum_{i=1}^{n-1}\frac{(q_{i-1/2}^k)^{\alpha-1}(q_i^{k+1}-q_{i-1}^{k+1})}{\sqrt{2r_{i-1/2}}(r_{i-1/2}-r_j)}$$

$$=N_-\sqrt{2r_j}+\frac{\alpha\sqrt{2r_j}}{2\pi}\Big\{\sum_{i=n}^{N}\frac{(q_{i-1/2}^k)^{\alpha-1}(q_i^{k+1}-q_{i-1}^{k+1})}{\sqrt{2r_{i-1/2}}(r_{i-1/2}-r_j)}$$

(2.174)

$$+N_-^\alpha G(\tfrac{3-\alpha}{2},R,\tfrac{r_j}{R})\Big\},\quad j=0,\ldots,n-1,$$

$$q_j^{k+1} = (1-\sigma)q_j^k + \sigma\sqrt{2r_j}\Big\{N_-$$

$$+\frac{\alpha}{2\pi}\Big[\sum_{i=1}^{n-1}(q_{i-1/2}^{k+1})^{\alpha-1}\frac{q_i^{k+1}-q_{i-1}^{k+1}}{\sqrt{2r_{i-1/2}}(r_{i-1/2}-r_j)}$$

$$+\sum_{i=n}^{N}(q_{i-1/2}^k)^{\alpha-1}\frac{q_i^k-q_{i-1}^k}{\sqrt{2r_{i-1/2}}(r_{i-1/2}-r_j)}+N_-^\alpha G(\tfrac{3-\alpha}{2},R,\tfrac{r_j}{R})\Big]\Big\},$$

(2.175)

$$j=n,\dots,N-1,\quad q_N^{k+1}=N_-\sqrt{2r_N},$$

where the superscript k is the iteration number, σ is the relaxation parameter $0 < \sigma < 1$, and function G is determined by equation (2.172). Equations (2.174) and (2.175) are derived for the general indenter shape described by function $f(x)$ for which in equation (2.109) the unity in braces should be replaced by constant N_-. For even function $f(x)$ we have

$$N_- = \frac{1}{2\pi b_0}\lim_{y\to -1+0}\int_{-1}^{1}\frac{f_t'(b_0 t)dt}{\sqrt{1-t^2}(t-y)}.$$

(2.176)

In case of $f(x) = x^2$ we have $N_- = 1$.

The order of calculations based on equations (2.174), (2.175), and (2.172) is similar to the one for the scheme based on (2.170)-(2.172). The initial approximation can be taken as $q_j^0 = N_-\sqrt{2r_j}$, $j = 0,\dots,N$.

After the iteration process converged for even $f(x)$ and $0 < \alpha < 1$, we can determine the first term in the asymptotic expansion (2.108) and (2.114) of b by using the formula

$$b_1 = 2\alpha b_0\Big\{\sum_{i=1}^{N}q_{0,i-1}^{\alpha-1}\frac{q_i-q_{i-1}}{\sqrt{2r_{i-1/2}}}+\frac{N_-^\alpha(2R)^{(\alpha-1)/2}}{1-\alpha}\Big\}$$

(2.177)

$$\Big/\Big\{\pi-\sum_{i=1}^{N}y_{i-1/2}f_{yy}''(b_0 y_{i-1/2})\sqrt{\frac{1-y_{i-12}}{1+y_{i-1/2}}}(y_i-y_{i-1})\Big\},$$

where nodes y_i and $y_{i-1/2}$ were introduced in the preceding section.

2.5.3 Numerical Results for Plane Contact Problems for Rough Surfaces

Computer simulations showed that it takes few iterations for the solutions based on all of the above numerical schemes for problems in the original and asymptotic formulations to converge. All iteration schemes produced stable numerical solutions with the prescribed precision. The solutions of the problems in the original formulation obtained based on the numerical schemes designed primarily for small/moderate, moderate, and moderate/large values of λ were practically identical.

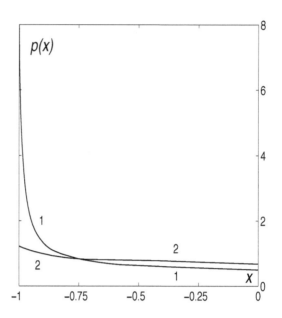

FIGURE 2.2
Left half of the pressure distribution $p(x)$ under a flat indenter for $\alpha = 0.8$
(curve 1: $\lambda = 0.05$, curve 2: $\lambda = 1.5$).

The results presented in this section are obtained for $\varphi(p) = p^\alpha$. Evenly
distributed nodes $y_i = -1 + 2i/N$ and $r_i = iR/N$, $i = 0, \ldots, N$, were used.
The relaxation parameter σ was taken equal to 0.15.

Solution of Contact Problem 1 exhibited a relatively strong dependence
of the values of $p(\pm 1)$ on the number of nodes N. These values practically
remained unchanged starting with $N = 150$. Therefore, solution of Contact
Problem 1 in the original and asymptotic formulations was done for $N = 200$
while solution of Contact Problem 2 in the original and asymptotic formulations was done for $N = 100$.

First, let us consider the solution of Contact Problem 1 for an indenter with
a flat bottom $f(x) = 0$. Obviously, in this case the function of pressure distribution $p(x)$ is even. Figure 2.2 shows the left half of the pressure distribution
in such a contact for $\alpha = 0.8$, $\lambda = 0.05$ and $\lambda = 1.5$. The curves in Fig.
2.2 illustrate the fact that for any positive λ pressure in a contact of rough
surfaces is finite as well as the tendency of the pressure at the contact boundaries (i.e., $p(\pm 1)$) to be a monotonically decreasing function of λ. Pressure
changes significantly only in the vicinity of the contact boundaries $x = \pm 1$.
Moreover, as λ approaches 0 pressure values $p(\pm 1)$ increase and approach

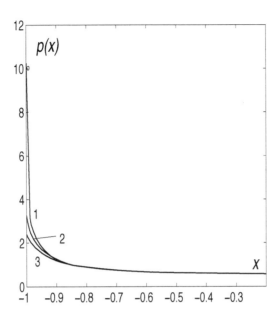

FIGURE 2.3
Left half of the pressure distribution $p(x)$ under a flat indenter for $\lambda = 0.1$
(curve 1: $\alpha = 0.5$, curve 2: $\alpha = 1$, curve 3: $\alpha = 1.5$).

infinity. Figure 2.3 demonstrates the asymptotic behavior of pressure under
a flat indenter $f(x) = 0$ near the left boundary $x = -1$ of the contact for
$\lambda = 0.1$ and $\alpha = 0.5$, $\alpha = 1.3$, $\alpha = 1.5$. The values of $p(\pm 1)$ are of the order
of $\epsilon^{-1/2} = \lambda^{-1/(\alpha+1)}$ (see (2.84) and (2.87)). Curves in Fig. 2.4 are graphs of
$q(r)$, which are the solutions of equation (2.86) obtained for $\alpha = 0.5$ (curve
1) and $\alpha = 3$ (curve 2). They represent a detailed behavior of pressure $p(x)$
in the vicinity of the contact boundary $x = -1$. The comparison of contact
pressure $p(x)$ obtained for $\alpha = 0.8$ and $\lambda = 0.1$ as the solutions of Contact
Problem 1 in the original (curve 1) and asymptotic (curve 2) formulations is
presented in Fig. 2.5. Curve 2 for the pressure distribution obtained from the
asymptotic solution is calculated based on the formula $p(x) = \epsilon^{-1/2}q(\frac{x+1}{\epsilon})$.
It is worth mentioning that a better agrement with curve 1 can be obtained
by using a uniformly valid approximation for pressure $p_u(x)$ calculated based
on the formula $p_u(x) = \frac{4}{\epsilon}q(\frac{x+1}{\epsilon})q(\frac{1-x}{\epsilon})$. Also, for comparison in Fig. 2.6 for
$\alpha = 0.8$ and $\lambda = 0.86835$ are given the distributions of pressure $p(x)$ obtained
from the numerical solution of Contact Problem 1 in the original formulation
(curve 1) and based on the two-term asymptotic solution (2.76) and (2.80)
obtained for $\lambda \gg 1$ (curve 2). In spite of the fact that $\lambda = 0.86835$ hardly

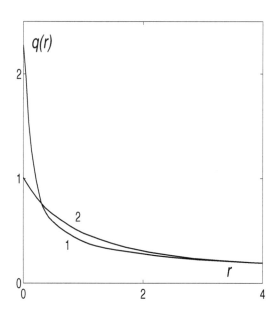

FIGURE 2.4
Graphs of the main term $q(r)$ of the asymptotic expansion of pressure $p(x)$
under a flat indenter in the vicinity of $x = -1$ obtained for $\alpha = 0.5$ (curve 1)
and $\alpha = 3$ (curve 2).

can be considered as a large number the agreement between the two curves
is relatively good. For large values of λ the solutions of Contact Problem 1 in
the original and asymptotic formulations are very close to each other. Graphs
in Fig. 2.7 obtained for $\alpha = 0.8$ provide a comparison of the behavior of the
indenter rigid displacement δ as a function of λ calculated based on the asymp-
totic formula $\delta = \ln 2 + \frac{\lambda}{\pi 2^{2\alpha}}\Gamma^2(\frac{1-\alpha}{2})/\Gamma(1-\alpha)$ obtained for $\lambda \ll 1$ (curve 1),
asymptotic formula $\delta = \lambda(\frac{\pi}{4})^{\alpha} + \frac{3}{2} - \ln 2$ obtained for $\lambda \gg 1$ (curve 2), and the
numerical solution of Contact Problem 1 in the original formulation (curve
3).

Now, let us consider some results for Contact Problem 2 in case of a
parabolic indenter $f(x) = x^2$. A series of graphs of pressure distribution $p(y)$
obtained from the original problem formulation for $\alpha = 0.8$ and $\lambda = 0$ (curve
1, in this case $p(y) = \sqrt{1 - y^2}$), $\lambda = 0.3$ (curve 2), $\lambda = 3$ (curve 3), $\lambda = 10$
(curve 4) are presented in Fig. 2.8. It is clear from Figs. 2.8 and 2.10 that as
λ increases the maximum pressure $p(0)$ decreases and for large λ it has the
value of the of magnitude of $\lambda^{-1/(\alpha+2)}$ (see (2.102)). In Figure 2.9 graphs of
the main term of the pressure distribution $q(r)$ are given for $\alpha = 0.5$ (curve

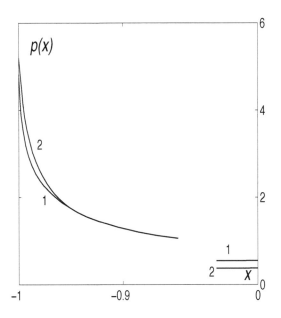

FIGURE 2.5

Graphs of pressure $p(x)$ obtained for $\alpha = 0.8$ and $\lambda = 0.1$ as the solutions of Contact Problem 1 in the original (curve 1) and asymptotic for $\lambda \ll 1$ (curve 2) formulations.

1), $\alpha = 1$ (curve 2), and $\alpha = 2$ (curve 3). It is important to realize that for $\alpha = 2$ function $q(r) = \sqrt{2r}$ (see (2.98) and (2.100)) is the one-term asymptotic representation of the Hertzian pressure $\sqrt{1 - y^2}$. These graphs show the rate with which functions $q(r)$ approach the asymptotic of the Hertzian pressure $q(r) = \sqrt{2r}$ as $r \to \infty$. Figure 2.10 shows the pressure distributions $p(y)$ obtained for $\alpha = 0.8$ and $\lambda = 4$ from the numerical solution of Contact Problem 2 in the original formulation (curve 1) and from the two-term asymptotic expansion (2.102) and (2.104) obtained for $\lambda \gg 1$ (curve 2). Clearly, the curves are close to one another. The comparison of the relationships $b(\lambda)$ obtained for $\alpha = 0.5$ from the solution of Contact Problem 2 in the asymptotic formulation and the two-term asymptotic expansion (2.108) and (2.114) valid for $\lambda \ll 1$ and based on numerical solution of equation (2.109) (curve 1), from the two-term asymptotic expansion (2.102) and (2.104) valid for $\lambda \gg 1$ (curve 2), from the numerical solution of equation (2.109) and second equation in (2.97), where for pressure $p(y)$ was used the uniformly valid approximation $p_u(y) = \frac{\epsilon}{2} q(\frac{y+1}{\epsilon}) q(\frac{1-y}{\epsilon})$, and from numerical solution of Contact Problem 2 in the original formulation (curve 4) is given in Fig. 2.11. Clearly, for small

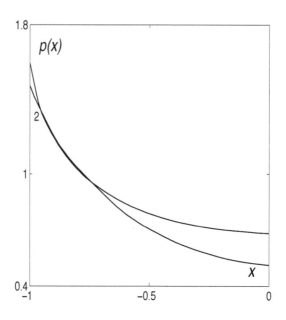

FIGURE 2.6
Graphs of the left half of the pressure distribution $p(x)$ obtained for $\alpha = 0.8$
and $\lambda = 0.86835$ as the solutions of Contact Problem 1 in the original (curve
1) and asymptotic for $\lambda \gg 1$ (curve 2) formulations.

λ curves 1 and 3 (asymptotic approximations for $\lambda \ll 1$) are close to curve
4 while for large λ curve 2 (asymptotic approximation for $\lambda \gg 1$) is close to
curve 4.

TABLE 2.2
The values of coefficients b_0,
b_1, and b_2 in the asymptotic
expansion (2.120) vs. parameter α (after Kudish [12]).
Reprinted with permission
from Springer.

α	b_0	b_1	b_2
0.5	0.777	0.346	0.020
0.8	0.898	0.442	0.711
1.0	0.994	0.500	1.032

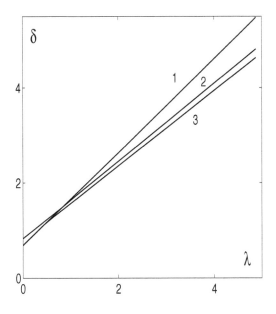

FIGURE 2.7
Graphs of the rigid displacement δ for a flat indenter as a function of λ calculated based on the asymptotic formula $\delta = \ln 2 + \frac{\lambda}{\pi 2^{2\alpha}}\Gamma^2(\frac{1-\alpha}{2})/\Gamma(1-\alpha)$ obtained for $\lambda \ll 1$ (curve 1), asymptotic formula $\delta = \lambda(\frac{\pi}{4})^\alpha + \frac{3}{2} - \ln 2$ obtained for $\alpha = 0.8$ and $\lambda \gg 1$ (curve 2), and the numerical solution of Contact Problem 1 in the original formulation (curve 3).

Let us consider a numerical example of a solution of Contact Problem 3 for the case when function $\varphi(p, \lambda)$ satisfies (2.101). We will assume that $\lambda \gg 1$, which makes the uniformly valid on $[-1, 1]$ asymptotic expansion based on (2.120) and (2.124) an approximate solution of Contact Problem 3. Let us assume that $f(x) = Ax^2$ and $A = 0.6$. The values of coefficients b_0, b_1, and b_2 as functions of α involved in the asymptotic expansion of $b = b_0\lambda^{1/(\alpha+2)} + b_1 + b_2\lambda^{-1/(\alpha+2)}$ are given in Table 2.2. It is obvious from Table 2.2 that all coefficients are of the order of magnitude 1. Moreover, due to the fact that all these coefficients are positive, we can conclude that the contact size increases with the increase of roughness, i.e., with the increase of parameter λ. Assuming that $b(\lambda)$ is a monotonically increasing function of λ for all $\lambda > 0$ (which is reasonable from the physical point of view), it is easy to determine the necessary (but not sufficient) condition of the asymptotic expansion (2.120) validity in the form $\lambda > \lambda_*(\alpha)$, where the value of $\lambda_*(\alpha)$ satisfies the equation $\frac{db}{d\lambda} = 0$. Based on the data from Table 2.2, we will find $\lambda_*(0.8) = 0.72$ and

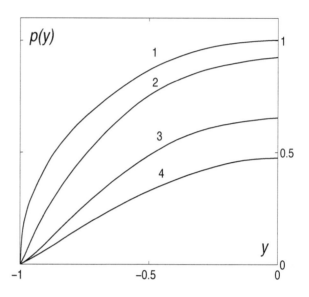

FIGURE 2.8
Graphs of the left half of the pressure distribution $p(y)$ obtained from the
original problem formulation for $\alpha = 0.8$ and $\lambda = 0$ (curve 1), $\lambda = 0.3$ (curve
2), $\lambda = 3$ (curve 3), $\lambda = 10$ (curve 4).

$\lambda_*(1) = 1.06$.
    The pressure distributions $p(y)$ for several values of λ and $\alpha = 0.5$ (solid
curves) and $\alpha = 1$ (dashed curves) are presented in Fig. 2.12. Each graph in
Fig. 2.12 is marked by a number that represents the value of λ for which it
was obtained. For comparison the graph of the main term of the asymptotic
solution for pressure $p_0(y)$ in the external region, which correspond to the
case of no roughness (curve marked with $\lambda = 0$) is also presented in Fig. 2.12.
It follows from these graphs that for $y = -1$ and $\lambda > 0$ contrary to the case of
$\lambda = 0$ the pressure $p(y)$ is finite and for $\lambda \gg 1$ it is of the order of magnitude of
$\lambda^{-\frac{1}{\alpha+1}}$ (see formulas (2.120) and (2.124)). Moreover depending on the values
of α and λ the pressure distribution $p(y)$ is either monotonic or may have a
number of extrema.

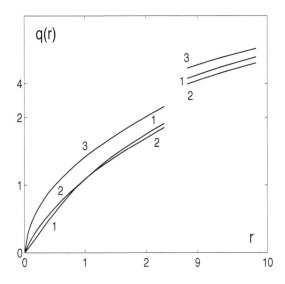

FIGURE 2.9

Graphs of the main term $q(r)$ of the asymptotic expansion of pressure $p(x)$ obtained for $\alpha = 0.5$ (curve 1), $\alpha = 1$ (curve 2), and $\alpha = 2$, which corresponds to the asymptotic of the Hertzian pressure distribution $p_0(x) = \sqrt{1 - x^2}$ (curve 3).

2.6 Asymptotic Analysis of Axially Symmetric Rough Contacts

In this section using asymptotic methods, we will analyze two types of axially symmetric contact problems for a rigid indenter pressed against an elastic half-space with rough surface: the problems with fixed and free boundaries. This analysis will involve application of methods of regular and matched asymptotic expansions.

Let us formulate the axially symmetric problems for which the contact boundaries are represented by a circle of radius b. Assuming that the angles $\beta_x = \beta_y = 0$ and the force P applied to the solids in contact is given and using the integral operator in (2.21) in polar coordinates we can formulate two problems. The case of fixed circular boundary is represented by Contact Problem 4 as follows

$$k\varphi(p) + \frac{8}{\pi E'} \int_0^b \frac{\rho}{r+\rho} K\left(\frac{2\sqrt{r\rho}}{r+\rho}\right) p(\rho)d\rho = \delta - f(r), \quad \int_0^b \rho p(\rho)d\rho = \frac{P}{2\pi}, \quad (2.178)$$

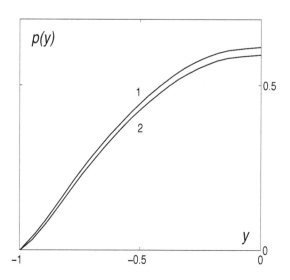

FIGURE 2.10
Graphs of the left half of the pressure distribution $p(y)$ obtained for $\alpha = 0.8$ and $\lambda = 4$ from the solution of Contact Problem 2 in the original formulation (curve 1) and from the the two-term asymptotic expansion (2.102) and (2.104) for $\lambda \gg 1$ (curve 2).

while the case of free circular boundary is represented by Contact Problem 5 as follows

$$k\varphi(p) + \frac{8}{\pi E'} \int\limits_0^b \frac{\rho}{r+\rho} K\left(\frac{2\sqrt{r\rho}}{r+\rho}\right) p(\rho) d\rho = \delta - f(r),$$

(2.179)

$$p(b) = 0, \ \int\limits_0^b \rho p(\rho) d\rho = \frac{P}{2\pi},$$

where r is the distance from the center of the contact, $z = f(r)$ is the shape of the indenter, K is the full elliptic integral of the second kind [29] while the meaning of parameters k, δ, and function φ is explained earlier.

Let us introduce the following dimensionless variables and parameters

$$(r', \rho') = \frac{1}{b_*}(r, \rho), \ p' = \frac{p}{p_*}, \ (\lambda \varphi', \delta', f') = \frac{E'}{p_* b_*}(k\varphi, \delta, f),$$

(2.180)

where b_* is the characteristic radius of the contact (for Contact Problem 4, which corresponds to a case of a rigid indenter with sharp edges $b_* = b$, while for Contact Problem 5, which corresponds to a case of an indenter with smooth edges $b_* = b_H$, b_H, is the Hertzian radius of the contact), p_* is the

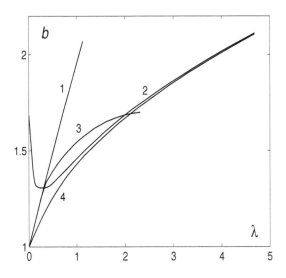

FIGURE 2.11
Graphs of $b(\lambda)$ obtained for $\alpha = 0.5$ from the solution of Contact Problem 2 in the original and asymptotic formulations for $\lambda \ll 1$ and $\lambda \gg 1$. The graph based on a numerical solution of (2.109) (curve 1), from the two-term asymptotic expansion (2.102) and (2.104) (curve 2), from the numerical solution of (2.109) and the second equation in (2.97), where for pressure $p(y)$ was used the uniformly valid approximation $p_u(y) = \frac{\epsilon}{2}q(\frac{y+1}{\epsilon})q(\frac{1-y}{\epsilon})$, and from a numerical solution of Contact Problem 2 in the original formulation (curve 4).

characteristic pressure, $p_* = \frac{3P}{2\pi b_*^2}$. Note that for $b_* = b_H = [\frac{3(1-\nu^2)PR}{4E}]^{1/3}$ we have $p_* = p_H$, where p_H is the maximum Hertzian pressure.

Using these dimensionless variables, Contact Problem 4 can be reduced to the equations (primes are omitted)

$$\lambda\varphi(p) + \frac{8}{\pi}\int_0^1 \frac{\rho}{r+\rho}K(\frac{2\sqrt{r\rho}}{r+\rho})p(\rho)d\rho = \delta - f(r), \quad \int_0^1 \rho p(\rho)d\rho = \frac{1}{3}. \qquad (2.181)$$

The solution of this problem is represented by the indenter displacement δ and the function of pressure $p(r)$.

At the same time, in dimensionless variables (2.180) Contact Problem 5 can be reduced to the equations (primes are omitted)

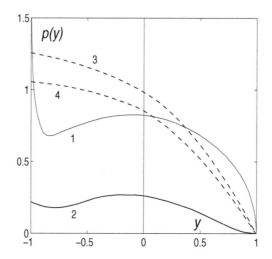

FIGURE 2.12
Graphs of the pressure distribution $p(y)$ in Contact Problem 3 obtained for
$\alpha = 0.5$ (solid curves: curve 1: $\lambda = 0$, curve 2: $\lambda = 8$) and $\alpha = 1$ (dashed
curves: curve 3: $\lambda = 10$, curve 4: $\lambda = 15$).

$$\lambda\varphi(p) + \frac{8}{\pi}b \int_0^1 \frac{\rho}{r+\rho}K(\frac{2\sqrt{r\rho}}{r+\rho})p(\rho)d\rho = \delta - f(br),$$

$$(2.182)$$

$$p(1) = 0, \ \int_0^1 \rho p(\rho)d\rho = \frac{1}{3b^2}.$$

The solution of this problem is represented by the contact radius b, the indenter displacement δ, and pressure $p(r)$. Using the fact that [29]

$$\frac{1}{1+\rho}K(\frac{2\sqrt{\rho}}{1+\rho}) = K(\rho)$$

and that $p(1) = 0$ the indenter displacement δ can be eliminated from equations (2.182), which leads to a new formulation of Contact Problem 5 as
follows

$$\lambda\varphi(p) + \frac{8}{\pi}b \int_0^1 \left\{ \frac{\rho}{r+\rho}K(\frac{2\sqrt{r\rho}}{r+\rho}) - \rho K(\rho) \right\}p(\rho)d\rho = f(b) - f(br),$$

$$(2.183)$$

$$p(1) = 0, \ \int_0^1 \rho p(\rho)d\rho = \frac{1}{3b^2},$$

$$\delta = f(b) + \lambda\varphi(p) + \tfrac{8}{\pi}b\int\limits_0^1 \rho K(\rho)p(\rho)d\rho.$$

Equations (2.181) and (2.183) of Contact Problems 2 and 3, respectively, are convenient for solving these problems for the cases of high roughness, i.e., for $\lambda \gg 1$.

For the cases of small roughness (i.e., for $\lambda \ll 1$), it is advantageous to transform these equations to another form by solving for $p(r)$ involved in the integrals. In particular, for Contact Problem 4 we get [8]

$$p(r) = \frac{1}{2\pi\sqrt{1-r^2}}\left\{\delta - f(0) - \int\limits_0^1 \frac{f'(\rho)d\rho}{\sqrt{1-\rho^2}} - \lambda\varphi(p(0))\right.$$

$$\left. -\lambda\int\limits_0^1 \frac{\varphi'_p(p(\rho))p'(\rho)d\rho}{\sqrt{1-\rho^2}}\right\} + \frac{1}{2\pi}\int\limits_r^1 \frac{1}{\sqrt{s^2-r^2}}\left[s\int\limits_0^s \frac{f'(\rho)d\rho}{\sqrt{s^2-\rho^2}}\right]' ds$$

$$+\frac{\lambda}{2\pi}\int\limits_r^1 \frac{1}{\sqrt{s^2-r^2}}\left[s\int\limits_0^s \frac{\varphi'_p(p(\rho))p'(\rho)d\rho}{\sqrt{s^2-\rho^2}}\right]' ds, \qquad (2.184)$$

$$\delta = \tfrac{2\pi}{3} + f(0) + \int\limits_0^1 \sqrt{1-\rho^2}f'(\rho)d\rho$$

$$+\lambda\varphi(p(0)) + \lambda\int\limits_0^1 \sqrt{1-\rho^2}\varphi'_p(p(\rho))p'(\rho)d\rho.$$

By this transformation equations of Contact Problem 5 can be converted to

$$p(r) = p_H(b,r) + \frac{\lambda}{2\pi b}\int\limits_r^1 \frac{1}{\sqrt{t^2-r^2}}\left\{t\int\limits_0^t \frac{\varphi'_p(p(\rho))p'(\rho)d\rho}{\sqrt{t^2-\rho^2}}\right\}' dt,$$

$$\lambda\int\limits_0^1 r\left\{r\int\limits_0^r \frac{\varphi'_p(p(\rho))p'(\rho)d\rho}{\sqrt{r^2-\rho^2}}\right\}' dr = \frac{2\pi}{3}\frac{1-b^3H(b)}{b},$$

$$\qquad (2.185)$$

$$p_H(b,r) = \frac{1}{2\pi b}\int\limits_r^1 \frac{1}{\sqrt{t^2-r^2}}\left\{t\int\limits_0^t \frac{f'_\rho(b\rho)d\rho}{\sqrt{t^2-\rho^2}}\right\}' dt,$$

$$H(b) = \frac{3}{2\pi b^2}\int\limits_0^1 r\left\{r\int\limits_0^r \frac{f_\rho(b\rho)d\rho}{\sqrt{r^2-\rho^2}}\right\}' dr.$$

It is interesting that for an indenter of a parabolic shape $f(br) = \tfrac{1}{2}\pi b^2 r^2$ and $\varphi(p) = p^2$ Contact Problem 5 has an exact analytical solution in the form

$$p(r) = \tfrac{1}{b^2}\sqrt{1-r^2}, \quad b = \sqrt[3]{\tfrac{1}{2}\left\{\sqrt{1+\tfrac{8\lambda}{\pi}}+1\right\}}. \qquad (2.186)$$

In the further analysis for $\varphi(p)$, we will use the function

$$\varphi(p) = p^\alpha, \ \alpha > 0, \tag{2.187}$$

where α is a constant.

2.6.1 Asymptotic Analysis of Axially Symmetric Rough Contacts with Fixed Boundaries

First, let us consider Contact Problem 4 for the case of large roughness, i.e., for $\lambda \gg 1$. That would require application of methods of regular asymptotic expansions [8]. Under these condition the main term of the solution for pressure $p(r)$ is determined by the interaction of the rigid indenter with the layer of roughness; the effects of elasticity of the half-space are secondary. Therefore, the problem solution will be searched in the form

$$p(r) = \sum_{n=0}^{\infty} \lambda^{-n} p_n(r), \ 0 \le r \le 1; \ \delta = \sum_{n=0}^{\infty} \lambda^{1-n} \delta_n, \tag{2.188}$$

where $p_n(r)$ and δ_n are new unknown functions and constants that are of the order of 1 uniformly on interval $[0,1]$ as $\lambda \to \infty$.

Let us substitute asymptotic expansions (2.188) in equations (2.181) and take into account the expression for $\varphi(p)$ from equation (2.187). By equating the terms with the same powers of λ, we obtain a series of solutions

$$p_0(r) = \tfrac{2}{3}, \ \delta_0 = (\tfrac{2}{3})^\alpha, \tag{2.189}$$

$$p_1(r) = \tfrac{1}{\alpha}(\tfrac{3}{2})^{\alpha-1}\left\{ \delta_1 - f(r) - \tfrac{16}{3\pi} \int\limits_0^1 \tfrac{\rho}{r+\rho} K(\tfrac{2\sqrt{r\rho}}{r+\rho}) d\rho \right\},$$
$$\tag{2.190}$$

$$\delta_1 = 2 \int\limits_0^1 rf(r)dr + \tfrac{32}{3\pi} \int\limits_0^1 rdr \int\limits_0^1 \tfrac{\rho}{r+\rho} K(\tfrac{2\sqrt{r\rho}}{r+\rho}) d\rho, \dots.$$

Equations (2.189) and (2.190) together with the expansions (2.188) determine a two-term uniformly valid on $[0,1]$ approximate solution of Contact Problem 4 for $\lambda \gg 1$, $\alpha > 0$ and a general sufficiently smooth function $f(r)$.

For $\lambda \gg 1$ the contact pressure is finite within the entire contact $[0,1]$, which is drastically different from the case of pressure behavior in the vicinity of the contact boundary $r = 1$ for a smooth (not rough) elastic half-space for which $p(r) \to \infty$ as $r \to 1$. Moreover, for $\lambda \gg 1$ the contact pressure $p(r)$ is practically constant in the entire contact. The indenter displacement $\delta = \lambda\delta_0 + \dots$, $\delta_0 > 0$, in a rough contact is much higher than in a corresponding smooth one.

For large roughness it follows from (2.188)-(2.190) that the main term of the pressure distribution is constant and is independent of the indenter shape and the half-space roughness characteristics while the indenter displacement

δ is greater than the one for the smooth contact and it is almost completely determined by function $\varphi(p)$, i.e., by the specific parameters characterizing roughness. Also, these equations show that the elastic displacements of the half-space affect the solution starting only with the next terms in approximations of $p(r)$ and δ. These observations allow to make a conclusion that for a rigid indenter of a general shape (i.e., not necessarily axially symmetric) for $\lambda \gg 1$ the pressure distribution is almost constant and can be approximated by an expansion similar to the one in 2.188). The pressure is almost completely determined by the area of the indenter footprint while its displacement is predominantly determined by the specifics of the roughness deformation.

Now, let us consider solution of equations (2.184) for the case of small roughness, i.e., for $\lambda \ll 1$. For simplicity we will consider the case of an indenter with a flat bottom $f(r) = f(0)$. Under these conditions it is impossible to solve the problem using regular asymptotic expansions. It is due to the fact that the presence of roughness is essential only within a small vicinity of the contact boundary (where pressure is high and surface displacement caused by roughness becomes relatively high) while away from the contact boundary roughness can be neglected. As before, we will call the external region the one where roughness can be neglected. In this region the main terms of the solution follow from equations (2.184)

$$p_0(r) = \frac{1}{3\sqrt{1-r^2}}, \quad \delta_0 = f(0) + \frac{2\pi}{3} \tag{2.191}$$

and coincide with the solution of the contact problem for this indenter and a smooth elastic half-space.

In the inner zone of a characteristic size ε, $\varepsilon = \varepsilon(\lambda, \alpha) \ll 1$, adjacent to $r = 1$, we can introduce a local variable $R = \frac{r-1}{\varepsilon}$. Keeping in mind that the asymptotic solution in the inner zone must match the asymptotic of the external solution from (2.191) at the boundary of the inner and external regions we will search for solution of the problem in the form

$$p(r) = p_0(r) + o(1), \quad p_0(r) = O(1) \ for \ r - 1 = O(1),$$

$$p(r) = \varepsilon^{-1/2}q(R) + o(\varepsilon^{-1/2}), \quad q(R) = O(1) \ for \ R = O(1), \tag{2.192}$$

$$\delta = \delta_0 + \gamma_1(\lambda, \alpha)\delta_1 + o(\gamma_1), \quad \delta_1 = O(1), \quad \gamma_1 = o(1), \quad \lambda \ll 1,$$

where $p_0(r)$ and δ_0 are determined by (2.191). Substituting asymptotic expansions (2.192) into equations (2.184) and estimating its terms in the inner zone we arrive at the following equation

$$q(R) = \frac{1}{3\sqrt{-2R}}\left\{1 - \frac{3\alpha}{2\pi} \int\limits_{-\infty}^{0} \frac{q'(\rho)q^{\alpha-1}(\rho)d\rho}{\sqrt{-2\rho}}\right\}$$

$$+ \frac{\alpha}{4\pi} \int\limits_{R}^{0} \frac{ds}{\sqrt{s-R}}\left\{\int\limits_{-\infty}^{s} \frac{q'(\rho)q^{\alpha-1}(\rho)d\rho}{\sqrt{s-\rho}}\right\}'. \tag{2.193}$$

Moreover, estimating the terms of equation (2.184) provides us with the estimate of the characteristic size of the inner zone

$$\varepsilon(\lambda, \alpha) = \lambda^{\frac{2}{\alpha+1}}. \tag{2.194}$$

From equations (2.184), expansions (2.192), and the estimate for ε from (2.194), it is easy to come up with the following results for δ_1 and γ_1:

$$\delta_1 = \frac{3-\alpha}{1-\alpha}, \; \gamma_1(\lambda, \alpha) = \lambda, \; 0 < \alpha < 1,$$

$$\delta_1 = \frac{1}{3} + \int\limits_0^1 \sqrt{1 - \rho^2} p'_u(\rho) d\rho, \; \gamma_1(\lambda, \alpha) = \lambda,$$

$$p_u(r) = \lambda^{-1/2} \sqrt{\frac{2}{1+r}} q(\frac{r-1}{\lambda}), \; \alpha = 1, \tag{2.195}$$

$$\delta_1 = \alpha \int\limits_{-\infty}^0 \sqrt{-2\rho} q'(\rho) q^{\alpha-1}(\rho) d\rho, \; \gamma_1(\lambda, \alpha) = \varepsilon, \; \alpha > 1.$$

As in the case of $\lambda \gg 1$ for small λ the contact pressure is also finite in the entire contact $[0, 1]$. However, in the vicinity of the contact boundary $r = 1$, the value of pressure $p(r) = O(\lambda^{-\frac{1}{\alpha+1}})$ is high for $\lambda \ll 1$. At the same time, due to the fact that $\delta = \delta_0 + \delta_1 \gamma_1 + \ldots, \; \delta_1 > 0$, the indenter displacement in a rough contact is greater than in a corresponding smooth one.

It is easy to extend the obtained results and the entire approach on the case of a general function $f(r)$.

2.6.2 Asymptotic Analysis of Axially Symmetric Rough Contacts with Free Boundaries

Let us consider Contact Problem 5 for the case of a rigid indenter of a paraboloidal shape described by

$$f(br) = \frac{\pi b^2}{2} r^2. \tag{2.196}$$

First, we will consider the case of large roughness, i.e., $\lambda \gg 1$. For $\lambda \gg 1$, the problem can be solved by using regular asymptotic methods. In particular, it is not difficult to see that the solution can be found in the form of the following expansions

$$p(r) = \lambda^{-2\gamma} \sum_{n=0}^{\infty} \lambda^{-3n\gamma} p_n(r), \; 0 \le r \le 1,$$

$$b = \lambda^{\gamma} \sum_{n=0}^{\infty} \lambda^{-3n\gamma} b_n, \; \gamma = \frac{1}{2(\alpha+1)}, \tag{2.197}$$

where $p_n(r)$ and δ_n are new unknown functions and constants, which are of the order of 1 uniformly on interval $[0, 1]$ as $\lambda \to \infty$. After $p(r)$ and b are determined, it is easy to find the indenter displacement from the last equation of (2.183).

Let us substitute representations (2.197) in equations (2.183) and take into account the expression for $\varphi(p)$ from (2.187). By equating terms with the same powers of λ after simple algebraic manipulations we get a series of solutions

$$p_0(r) = (\tfrac{\pi}{3}\tfrac{\alpha+1}{\alpha})^{2\gamma}(1-r^2)^{1/\alpha}, \ \ b_0 = (\tfrac{2}{\pi})^{\gamma}(\tfrac{2}{3}\tfrac{\alpha+1}{\alpha})^{\alpha\gamma},$$

$$p_1(r) = \tfrac{2}{\alpha b_0}\Big\{b_1 + 2(\tfrac{\pi}{2})^{(1-2\alpha)/\alpha}b_0^{2/\alpha}(1-r^2)^{-1}$$

$$\times \int_0^1 \rho(1-\rho^2)^{1/\alpha}\Big[K(\rho) - \tfrac{1}{r+\rho}K(\tfrac{2\sqrt{r\rho}}{r+\rho})\Big]d\rho\Big\}p_0(r),$$

(2.198)

$$b_1 = \tfrac{6}{\alpha+1}(\tfrac{\pi}{2})^{2(1-\alpha)/\alpha}b_0^{2(\alpha+2)/\alpha}\Big\{\int_0^1 r(1-r^2)^{(1-\alpha)/\alpha}dr$$

$$\times \int_0^1 \tfrac{\rho(1-\rho^2)^{1/\alpha}}{r+\rho}K(\tfrac{2\sqrt{r\rho}}{r+\rho})d\rho - \tfrac{\alpha}{2}\int_0^1 \rho(1-\rho^2)^{1/\alpha}K(\rho)d\rho\Big\}, \dots.$$

The obtained two-term solution approximation is uniformly valid on the entire interval $[0,1]$ occupied by the contact. For $\lambda \gg 1$ and $\alpha = 2$, the two-term asymptotic expansion of the exact solution given by equations (2.186) coincides with the two-term asymptotic represented by formulas (2.197) and (2.198).

For $\lambda \gg 1$ the contact radius $b = \lambda^{\frac{1}{2(\alpha+1)}}b_0 + \dots$, $b_0 > 0$ and, therefore, the rough contact is much larger than the corresponding smooth contact. At the same time, for $\lambda \gg 1$ contact pressure $p(r)$ in a rough contact is given by $p(r) = \lambda^{-\frac{1}{\alpha+1}}p_0(r) + \dots \ll 1$. That indicates that pressure in a rough contact is smaller than the one in a corresponding smooth contact.

Now, let us consider the case of small roughness, i.e., $\lambda \ll 1$. Substituting $f(r)$ from equation (2.196) in equations (2.185), we obtain

$$p_H(b,r) = b\sqrt{1-r^2}, \ H(b) = 1,$$

(2.199)

and equations for $p(r)$ and b are reduced to the form

$$p(r) = b\sqrt{1-r^2} + \tfrac{\lambda}{2\pi b}\int_r^1 \tfrac{1}{\sqrt{t^2-r^2}}\Big\{t\int_0^t \tfrac{\varphi_p'(p(\rho))p'(\rho)d\rho}{\sqrt{t^2-\rho^2}}\Big\}'dt,$$

(2.200)

$$\lambda\int_0^1 r\Big\{r\int_0^r \tfrac{\varphi_p'(p(\rho))p'(\rho)d\rho}{\sqrt{r^2-\rho^2}}\Big\}'dr = \tfrac{2\pi}{3}\tfrac{1-b^3}{b}.$$

We will start analysis of the above equations with regular asymptotic methods. It will be shown that this approach to solution of the problem works if

$\alpha > 3$ and $\lambda \ll 1$. Similar to the case of a plane problem let us try to find the solution in the form

$$p(r) = \sum_{n=0}^{\infty} \lambda^n p_n(r), \ 0 \le r \le 1; \ b = \sum_{n=0}^{\infty} \lambda^n b_n. \qquad (2.201)$$

Then substituting (2.201) into (2.200) and following the usual procedure we get the expressions for the first two terms of the asymptotic expansions

$$p_0(r) = \sqrt{1 - r^2}, \ b_0 = 1, \qquad (2.202)$$

$$p_1(r) = \frac{\alpha}{2\pi} \left\{ \frac{2\sqrt{1-r^2}}{\alpha^2-1} - \frac{1}{2} \int_r^1 \frac{1}{\sqrt{s^2-r^2}} \left[s \int_0^{s^2} \frac{(1-t)^{(\alpha-2)/2} dt}{\sqrt{s\sqrt{2-t}}} \right]' ds \right\}, \qquad (2.203)$$

$$b_1 = \frac{1}{\pi} \frac{\alpha}{\alpha^2-1}, \dots$$

Analyzing the behavior of the inner integral in (2.203) it can be shown that $p_1(r)/p_0(r)$ is finite/bounded on the entire interval $[0, 1]$ if $\alpha > 3$. That provides the condition for the validity of the presented approach.

It is obvious that the main term of the solution coincides with the Hertzian solution for a contact problem for a smooth elastic half-space. The value of b_1 is positive for $\alpha > 3$. Therefore, the Hertzian contact $[0, 1]$ belongs to the interior of the rough contact $[0, 1 + \lambda b_1 + \dots]$. Calculations show that $p_1(r) \le 0$, which suggests that in a rough contact pressure $p(r)$ is lower than in a corresponding smooth contact.

For $0 < \alpha \le 3$ and $\lambda \ll 1$ the method of regular asymptotic expansions is not applicable. We need to use the method of matched asymptotic expansions because pressure is close to the Hertzian one only outside of a narrow ring of the characteristic width ε adjacent to the contact boundary while within the ring the effects of asperity displacement and the displacement of the bulk of the elastic material are of the same order of magnitude. Let us introduce the inner variable $R = \frac{r-1}{\varepsilon}$, $\varepsilon = \varepsilon(\lambda, \alpha)$. For $0 < \alpha < 3$ the problem solution will be searched in the form

$$p(r) = p_0(r) + o(1), \ p_0(r) = O(1) \ for \ r - 1 = O(1), \ \lambda \ll 1,$$

$$p(r) = \varepsilon^{1/2} q(R) + o(\varepsilon^{1/2}), \ q(R) = O(1) \ for \ R = O(1), \ \lambda \ll 1, \qquad (2.204)$$

$$b = b_0 + \gamma_1(\lambda, \alpha) b_1 + o(\gamma_1), \ b_1 = O(1), \ \gamma_1 = o(1), \ \lambda \ll 1,$$

where $p_0(r)$ and b_0 are determined by the Hertzian solution given in (2.202). Substituting asymptotic expansions (2.204) into equations (2.200) and estimating its terms in the inner zone, we arrive at the following equation

$$q(R) = \sqrt{-2R} + \frac{\alpha}{4\pi} \int_R^0 \frac{1}{\sqrt{s-R}} \left\{ \int_{-\infty}^s \frac{q'(t)q^{\alpha-1}(t)dt}{\sqrt{s-t}} \right\}' ds, \qquad (2.205)$$

while the estimate of the characteristic size of the inner zone

$$\varepsilon(\lambda, \alpha) = \lambda^{\frac{2}{3-\alpha}}. \tag{2.206}$$

For $\alpha = 3$ this approach does not allow to obtain the equation in the inner zone. The two-term regular expansion (2.201)-(2.203) reveals that the inner zone in this case is exponentially small $\varepsilon = O(\exp(-\frac{1}{\lambda}))$ for $\lambda \ll 1$. This is the reason why the above approach does not work for $\alpha = 3$.

Now, by using (2.202) and (2.204) and estimating terms of the second equation in (2.200), we obtain

$$b_1 = -\frac{\alpha}{2\pi} \int\limits_{-\infty}^{0} \frac{q'(R)q^{\alpha-1}(R)dR}{\sqrt{-2R}}, \quad \gamma_1(\lambda, \alpha) = \varepsilon, \ 0 < \alpha < 1,$$

$$b_1 = -\frac{1}{4\pi}\left\{1 + 2\int\limits_{0}^{1} \frac{p'_u(\rho)d\rho}{\sqrt{1-\rho^2}}\right\}, \quad \gamma_1(\lambda, \alpha) = \lambda, \tag{2.207}$$

$$p_u(r) = \lambda^{1/2}\sqrt{\frac{1+r}{2}}q(\frac{r-1}{\lambda}), \quad \alpha = 1,$$

$$b_1 = \frac{1}{\pi}\frac{\alpha}{\alpha^2-1}, \quad \gamma_1(\lambda, \alpha) = \lambda, \ 1 < \alpha \leq 3.$$

For $\alpha = 2$ the solution of equation (2.205) is $q(R) = \sqrt{-2R}$ and $b_1 = \frac{2}{3\pi}$, which is in agreement with the asymptotic expansion of the exact solution given in (2.186).

Calculations show that $q(R)$ is a monotonically decreasing non–negative function of R. That implies that for $0 < \alpha < 1$ the value of $b_1 > 0$ (see (2.207)). Similarly, from (2.203) and the last expression for b_1 in (2.207) it follows that $b_1 > 0$. Therefore, for $\lambda \ll 1$ the size of a rough contact is also larger than a corresponding smooth one.

It is easy to extend the obtained results and the entire approach on the case of a general function $f(r)$.

2.6.3 Numerical Results for Axially Symmetric Rough Contacts

For numerical solution of Contact Problems 4 and 5 can be developed direct analogs of numerical schemes described in Section 2.5.

Let us illustrate the quality of some of the obtained asymptotic approximations by comparing the values of the radius of the contact region $b = 3.116$ obtain by solving the original problem (2.183), (2.187), and (2.196) for $\alpha = \frac{2}{3}$ and $\lambda = 2.889$ presented in [5] and $b = 3.099$ calculated based on the two-term asymptotic expansion (2.197) and (2.198). Obviously, for larger values of λ the agreement is even better.

The qualitative behavior of solutions of Contact Problems 4 and 5 near contact boundary is very similar to the one for Contact Problems 1 and 2.

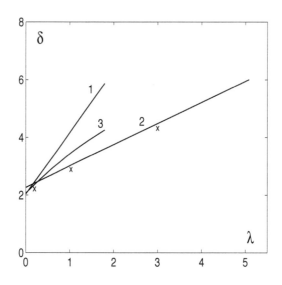

FIGURE 2.13
Graphs of the rigid displacement δ for a flat indenter as a function of λ
calculated based on the asymptotic formula for $\lambda \ll 1$ (curve 1), asymptotic
formula for $\lambda \gg 1$ (curve 2), and the uniformly valid solution $p_u(r)$ valid
for $\lambda \ll 1$ (curve 3). The crosses indicate the numerical solutions of Contact
Problem 4 in the original formulation.

Therefore, we will limit our analysis of Contact Problems 4 by just considering
Fig. 2.13. In Fig. 2.13 the comparison of the dependence of the indenter rigid
displacement δ as a function of λ obtained for $f(r) = 0$ and $\alpha = 0.8$ is presented. The graphs are based on: the two-term asymptotic expansion (2.191),
(2.192), and (2.195) developed for $\lambda \ll 1$ (curve 1), the two-term asymptotic
expansion (2.188)-(2.190) developed for $\lambda \gg 1$ (curve 2), and the expression
for δ from equation (2.184) in which pressure $p(r)$ is replaced the uniformly
valid approximation $p_u(r) = \sqrt{\frac{2}{\epsilon}q(\frac{r-1}{\epsilon})}/\sqrt{1+r}$, which involves the numerical
solution $q(R)$ of equation (2.193) (curve 3). In Fig. 2.13 small crosses indicate
the values of δ obtained from the direct solution of Contact Problem 4 in the
original formulation. It is interesting to notice that the values of δ obtained
from (2.188)-(2.190) approximate δ for all $\lambda > 0$ with the precision not lower
than 6%.

   The pressure distributions $p(r)$ obtained from solution of Contact Problem
5 in the original formulation for $f(br) = \frac{\pi b^2 r^2}{2}$, $\alpha = 0.8$ and $\lambda = 0$ (curve 1),
$\lambda = 0.1$ (curve 2), $\lambda = 0.9$ (curve 3), and $\lambda = 3$ (curve 4) are represented in
Fig. 2.14. Obviously, for $\lambda = 0$ the pressure distribution coincides with the

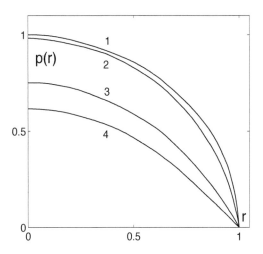

FIGURE 2.14

Graphs of the pressure distribution $p(r)$ obtained from the numerical solution of Contact Problem 5 in its original formulation for $f(br) = \frac{\pi b^2 r^2}{2}$, $\alpha = 0.8$ and $\lambda = 0$ (curve 1), $\lambda = 0.1$ (curve 2), $\lambda = 0.9$ (curve 3), and $\lambda = 3$ (curve 4).

Hertzian one $p(r) = \sqrt{1 - r^2}$. These graphs of $p(r)$ demonstrate the behavior of pressure similar to the one observed in Contact Problem 2.

2.7   An Example of an Application to Roller Bearings

To conclude this chapter let us briefly consider an example of an application of the obtained results to evaluation of the effect of roughness on the maximum pressure and contact size for a roller in contact with inner ring of a roller bearing. To do that we will estimate the values of constants k and α involved in the term $w_R = -k\varphi(p) = -kp^\alpha$ of the governing equation (2.17) assuming that the surfaces have been worked in and asperities deform elastically. Then based on the analysis and data in [1] we have

$$k = h_{max}\left\{\frac{5\sqrt{R}}{\eta_0\nu_0(1-\nu_0)k_1 h_{max}E'}\right\}^\alpha, \quad \alpha = \frac{2}{2\nu_0+1}, \tag{2.208}$$

where h_{max} and R are the maximum asperity height and mean radius of the top asperity curvature, respectively, η_0, ν_0, and k_1 are the constants that values depend on the type of surface finishing mechanical operations (usually, $\nu_0(1-\nu_0)k_1 \approx 0.4$ for all values of ν_0), and E' is the effective elastic modulus.

TABLE 2.3
Values of coefficient k for three classes of roughness.

Roughness class	h_{max} $[\mu m]$	R $[\mu m]$	$k \cdot 10^{10}$ $[m^{1.8}/N^{0.4}]$
9	1.31	0.65	0.57
10	0.62	0.34	0.28
11	0.36	0.18	0.16

TABLE 2.4
Values of the dimensionless maximum contact pressure p_{max}/p_H and semi-width of the contact w/a_H for three classes of roughness and two loads P.

P $[N/m]$	$9.375 \cdot 10^4$			$1.5 \cdot 10^5$		
Roughness class	9	10	11	9	10	11
k	0.430	0.210	0.120	0.300	0.150	0.080
p_{max}/p_H	0.934	0.967	0.984	0.954	0.977	0.987
w/a_H	1.266	1.157	1.102	1.205	1.121	1.073

For grinding finishing treatment we can take [1] $\eta_0 = 3$ and $\nu_0 = 2$. Therefore, from (2.208) we have $\alpha = 0.4$ and for a steel roller on a steel ring ($E = 2.12 \cdot 10^{11}$ $\frac{N}{m^2}$, $\nu = 0.3$) the values of k are given for three classes of roughness in Table 2.3 [1]. Based on the data from Table 2.3 for these classes of roughness and the effective roller and ring radius $R' = 8 \cdot 10^{-3}$ m for two levels of load P, we determine the values of λ as well as using the above analysis the dimensionless maximum contact pressure p_{max}/p_H and the semi-width of the contact w/a_H presented in Table 2.4 (p_H and a_H are the Hertzian maximum pressure and semi-width of the contact). It follows from Table 2.4 that as the load P increases the effect of roughness on dimensionless maximum contact pressure p_{max}/p_H and the semi-width of the contact w/a_H diminish.

2.8   Closure

Contact problems for coated/rough elastic solids are formulated and analyzed. It is shown that the solution of these problems exists and is unique for a number of different problem formulations. Several complementary to one another approaches to analyzing these problems are proposed. Some of the solution properties such as its monotonic dependence on input parameters are established. Solutions for coated/rough and smooth solids are compared and some relationships between such solutions are established. The approximate solutions of the plane and axially symmetric problems with fixed and free boundaries are obtained using the regular and matched asymptotic methods. In some cases these solutions are expressed in a simple convenient form for manual calculations. A number of numerical methods for the problems in the original and asymptotic formulations are developed and realized. The analysis of some of the numerical results is provided. An example of an application of the chapter results to a roller bearing is given.

2.9   Exercises and Problems

1. Obtain formulas (2.15) and (2.13) for solutions of plane contact problems with one free and another with fixed boundaries and with both free boundaries, respectively, from the expression (2.11) for the solution of the plane problem with fixed boundaries.

2. Provide a detailed proof of Theorem 2.3.2.

3. Consider applicability of Theorem 2.3.10 to the case of a plane problem for a parabolic indenter with $f(x) = x^2$ and a smooth elastic half-plane (without roughness). Consider solutions for contact problems with different fixed contact regions and compare them with the Hertzian solution $p(x) = \sqrt{1 - x^2}$.

4. Let the function characterizing the displacement due to roughness be $\varphi(p, \lambda) = p^\alpha$. Consider applicability of Theorem 2.3.14 to the cases of:

(a) plane problem with a parabolic indenter $f(x) = x^2$ and (i) $\alpha = 2$ by using the exact solution (2.98), (ii) $\alpha > 3$ and $\lambda \ll 1$ by using the asymptotic solution (2.105) and (2.106), (iii) $\lambda \gg 1$ by using the asymptotic solution (2.102) and (2.104);

(b) axially symmetric problem with a parabolic indenter $f(r) = \frac{\pi}{2}r^2$ and (i) $\alpha = 2$ by using the exact solution (2.186), (ii) $\alpha > 3$ and $\lambda \ll 1$ by using the asymptotic solution (2.201)-(2.203), (iii) $\lambda \gg 1$ by using the asymptotic solution (2.197) and (2.198).

5. For a plane rough contact of a flat indenter and for the displacement due to roughness described by $\varphi(p, \lambda) = p^\alpha$, $0 < \alpha < 1$, determine analytically the two-term asymptotic approximation of the indenter displacement δ.

6. Using Table 2.2 for three levels of surface roughness, $\lambda = 0.01$, $\lambda = 0.05$, and $\lambda = 0.1$, determine the approximate position of the free contact boundary in Contact Problem 3.

BIBLIOGRAPHY

[1] Dyomkin, N.B. 1969. Calculation and Experimental Study of Rough Contact Surfaces. *Proc. Sci. Conf. Contact Problems and Their Engineering Applications*, Moscow, Russia, 264-271.

[2] Tsukada, T. and Anno, Y. 1977. On the Approach Between a Sphere and a Rough Surface in Contact. *Bull. JSME* 11, No. 3:149-150.

[3] Shtaerman, I.Ya. 1949. *Contact Problems in Theory of Elasticity* Moscow-Leningrad: Gostekhizdat.

[4] Popov, G.Ya. and Savruk, V.V. 1971. Contact Problems in the Theory of Elasticity for Solids with Surface Structure and Circular Contact Region. *J. Mech. Solids* No. 3:80-87.

[5] Rabinovich, A.S. 1975. Axially Symmetric Contact Problem for Elastic Rough Solids. *J. Mech. Solids* No. 4:163-166.

[6] Rabinovich, A.S. 1974. Plane Contact Problem for Elastic Rough Solids. *J. Mech. Solids* No. 3:165-172.

[7] Martynenko, M.D. and Romanchik, V.S. 1977. On Solution of an Integral Equation for a Contact Problem of Elasticity for Rough Solids. *J. Appl. Math. Mech.* 41, No. 2:238-243.

[8] Alexandrov, V.M. and Kudish, I.I. 1979. Asymptotic Analysis of Plane and Axially Symmetric Contact Problems Taking into Account the Surface Structure of Interacting Solids. *J. Mech. Solids* No. 1:58-70.

[9] Klarbring, A., Mikelic, A., and Shillor, M. 1988. Frictional Comtact Problems with Normal Compliance. *Intern. J. Eng. Sci.* 26, No. 8:811-832.

[10] Kudish, I.I. 2000. Contact Interactions for an Elastic Half-Space with Rough Surface. *Dynamic Sys. and Appl.* 9, No. 2:229-246.

[11] Kudish, I.I. 1985. Contact Problems of Elasticity Taking into Account Surface Roughness. Proc. VNIPP *Studying, Calculating, and Designing of Rolling Bearings*, SPECINFORMCENTR VNIPP, No. 1:125-142.

[12] Kudish, I.I. 1983. Solving One Two-Dimensional Contact Problems for Rough Bodies. *Soviet Appl. Mech.* 19, No. 11:1005-1012.

[13] Galin, L.A. 1980. *Contact Problems in the Theory of Elasticity and Viscoelasticity*. Moscow: Nauka.

[14] Lurye, A.I. 1955. *Spatial Proplems in Elasticity*. Moscow: Gostekhizdat.

[15] Vorovich, I.I., Alexandrov, V.M., and Babesko, V.A. 1974. *Non --Classic Mixed Problems in the Theory of Elasticity*. Moscow: Nauka.

[16] Muskhelishvili, N.I. 1966. *Some Fundamental Problems of the Mathematical Theory of Elasticity*. Moscow: Nauka.

[17] Lions, I.L. 1969. *Quelques Methodes de Revolution des Problemes aux Limites non Lineaires*. Paris: Dunod Gauthier-Villars.

[18] Mikhaylov, V.P. 1979. *Differential Equations in Partial Derivatives*. Moscow: Nauka.

[19] Rabinovich, V.L. and Spector, A.A. 1985. Solution of Some Classes of Spatial Problems with Free Boundary. *J. Mech. Solids* No. 2:93-100.

[20] Giraud, G. 1932. Generalization des Problemes sur les Operations du Type Eliptique. *Bull. Sci. Math.* 56:248-272, 281-312, 316-352.

[21] Bitzadze, A.V. 1966. *Boundary Problems for Elliptic Equations of the Second Order*. Moscow: Nauka.

[22] Galin, L.A. 1946. Spatial Contact Problems of the Elasticity Theory for Circular Stamps. *J. Appl. Math. Mech.* 11, No. 2:281-284.

[23] Leonov, M.Ya. 1955. General Problem on Pressure Determination Under a Circular Stamp Indented in an Elastic Half-Space. *J. Appl. Math. Mech.* 17, No. 1:87-98.

[24] Mossakovsky, V.I., Kachalivskaya, N.E., and Golikova, S.S. 1985. *Contact Problems of Mathematical Theory of Elasticity*. Kiev: Naukova Dumka.

[25] Goldstein, R.V. and Spector, A.A. 1978. Variational Estimates of Solutions of Some Mixed Spatial Problems with Free Boundaries in the Theory of Elasticity. *J. Mech. Solids* No. 2:82-94.

[26] Kudish, I.I. 1987. Two-Dimensional Contact Problem on the Indentation of a Rough Bar by a Rigid Die with Friction. *J. Appl. Mech.* 23, No. 4:356-362.

[27] Kudish, I.I. 1986. Numerical Methods of Solving a Class of Nonlinear Integral and Integro-Differential Equations. *Zhurnal Vychislitel'noy Matematiki i Matematicheskoy Fiziki* 26, No. 10:1493-1511.

[28] Van-Dyke, M. 1964. *Perturbation Methods in Fluid Mechanics*. New York-London: Academic Press.

[29] *Handbook on Mathematical Functions with Formulas, Graphs, and Mathematical Tables*, Eds. M. Abramowitz and I. Stegun, National Bureau of Standards, Appl. Math. Series, 55, 1964.

[30] Baker, G.A.,Jr. and Graves-Morris, P. 1981. *Pade Approximants*. Reading: Addison-Wesley Publishing Co.

[31] Belotserkovsky, S.M. and Lifanov, I.K. 1993. *Method of Discrete Vortices*. Boca Raton: CRC Press.

[32] Pykhteev, G.N. 1980. *Exact Methods for Calculation of Integrals of Cauchy Type*. Novosibirsk: Nauka.

3

Contact Problems with Friction

3.1 Introduction

Studying of contact problems with friction has a long history. One of the first is for a plane contact problem in the case of full sliding [1] when frictional stress is linearly proportional to contact pressure. The case of a frictional contact with stick and slip is considered in [2]. In practice, in many cases frictional stress is a nonlinear function of pressure. Depending on material properties the apparent friction coefficient, which is equal to the ratio of the frictional stress and pressure may increase, decrease, or be practically constant as contact pressure increases. In this chapter we will take a look at some aspects of frictional contacts.

Let us consider a plane contact problem with friction under the conditions of full sliding. We will introduce a coordinate system with the x-axis along the contact and the z-axis directed perpendicular to the x-axis from the lower solid into the upper solid. Let us assume that a force P and moment M are acting on a rigid/elastic indenter of shape $z = f(x)$, which is interacting with an elastic layer (see Fig. 3.1). Following the classic problem formulation [1, 2], we obtain the equations [3]

$$\frac{1}{G'} \int\limits_a^x \tau(t)dt + \frac{2}{\pi E'} \int\limits_a^b p(t) \ln \mid \frac{1}{t-x} \mid dt = \delta - f(x), \ a < x < b, \qquad (3.1)$$

$$\int\limits_a^b p(t)dt = P, \qquad (3.2)$$

where G' and E' are the effective elasticity moduli, $\frac{1}{G'} = \frac{(1+\nu_1)(1-2\nu_1)}{E_1} - \frac{(1+\nu_2)(1-2\nu_2)}{E_2}$, $\frac{1}{E'} = \frac{1-\nu_1^2}{E_1} + \frac{1-\nu_2^2}{E_2}$, E_k and ν_k $(k = 1, 2)$ are the Young's moduli and Poison's ratios, respectively, $\tau(x)$ and $p(x)$ are contact frictional stress and pressure, δ is the rigid displacement of the contact solids (for an elastic indenter δ is the relative displacement of two points belonging to different solids located on the z-axis far away from each other), $[a, b]$ is the contact region, $a < b$. Function $f(x)$ may take into account the fact that the indenter is inclined at an angle of α to the z-axis.

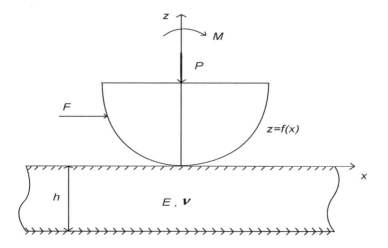

FIGURE 3.1
General view of an indenter pressed into an elastic layer.

Also, it is necessary to add to equations (3.1) and (3.2) the relationship between the frictional stress τ and contact pressure $p(x)$

$$\tau(x) = \psi(p(x)), \qquad (3.3)$$

where $\psi(p)$ is a certain known function of pressure p.

After the solution of the problem is obtained, it allows to calculate the frictional force F and moment M as follows

$$F = \int_a^b \tau(t)dt, \quad M = \int_a^b tp(t)d. \qquad (3.4)$$

Let us introduce the following dimensionless variables and parameters

$$(x', a', b') = \frac{(x,a,b)}{a_*}, \quad (p', \tau') = \frac{(p,\tau)}{p_*}, \quad (\eta\psi', f') = (\psi, f)\frac{E'}{ap_*},$$

$$\delta' = \delta\frac{E'}{ap_*} - \ln\frac{h}{a_*}, \qquad (3.5)$$

where a_* and p_* are the characteristic half-width of and pressure in the contact, $p_* = 2P/(\pi a_*)$, and η is a dimensionless parameter, which characterizes the ratio of the normal displacement caused by friction and normal pressure, i.e., if $\eta = 0$, then there is no friction, while if $\eta > 0$, then the friction is present. Then equations (3.1)-(3.3) can be reduced to (primes are omitted)

$$\eta \int_a^x \psi(p(t), \eta)dt + \frac{2}{\pi} \int_a^b p(t)\ln\left|\frac{1}{t-x}\right| dt = \delta - f(x), \qquad (3.6)$$

$$\int\limits_a^b p(t)dt = \tfrac{\pi}{2}. \tag{3.7}$$

We will assume that function $\psi(p,\eta)$ is a sufficiently smooth function, which satisfies the conditions

$$\psi(p,\eta) = O(1) \; for \; p = O(1),$$

$$\psi(p,\eta) = p^\gamma + o(p^\gamma) \; for \; p \to 0 \; or \; p \to \infty, \; \gamma > 0, \; \eta > 0. \tag{3.8}$$

In what follows we consider only the most important for practice case of small friction, i.e., the case when $\eta \ll 1$. It is possible to consider three types of problems for equations (3.6) and (3.7). The first of them is Contact Problem 1 for a rigid indenter with sharp edges, i.e., a problem with fixed contact region $[a,b]$ where without losing generality we can assume that $a = -1$ and $b = 1$. That leads to the following problem formulation:

$$\eta \int\limits_a^x \psi(p(t),\eta)dt + \tfrac{2}{\pi} \int\limits_a^b p(t) \ln \mid \tfrac{1}{t-x} \mid dt = \delta - f(x), \tag{3.9}$$

$$\int\limits_a^b p(t)dt = \tfrac{\pi}{2}. \tag{3.10}$$

For small η equations (3.9) and (3.10) can be rewritten in the equivalent form [3]

$$p(x) = \tfrac{1}{2\sqrt{1-x^2}} \left\{ 1 + \tfrac{1}{\pi} \int\limits_{-1}^1 \tfrac{f'(t)\sqrt{1-t^2}dt}{t-x} \right\}$$

$$+ \tfrac{\eta}{2\pi\sqrt{1-x^2}} \int\limits_{-1}^1 \tfrac{\psi(p(t),\eta)\sqrt{1-t^2}dt}{t-x}, \tag{3.11}$$

$$\delta = \ln 2 + \tfrac{1}{\pi} \int\limits_{-1}^1 \tfrac{f(t)dt}{\sqrt{1-t^2}} + \tfrac{\eta}{\pi} \int\limits_{-1}^1 \psi(p(t),\eta) \arccos t \, dt. \tag{3.12}$$

Contact Problem 2 is the problem for a rigid or elastic indenter without sharp edges, i.e., a problem with free boundaries of the contact that are determined by the equations $p(a) = p(b) = 0$. In this case in scaling formulas (3.5) $a_* = a_H = 2\sqrt{\tfrac{R'P}{\pi E'}}$ and $p_* = p_H = \sqrt{\tfrac{E'P}{\pi R'}}$ are the Hertzian contact half-width and maximum pressure in the contact of the same indenter with the same half-plane but without friction ($\tfrac{1}{R'} = \tfrac{1}{R_1} + \tfrac{1}{R_2}$, R_1 and R_2 are contact solids radii). To simplify this problem we can introduce a substitution $x = \tfrac{b+a}{2} + \tfrac{b-a}{2}y$, which converts the interval with unknown end points $x = a$ and $x = b$ to a fixed interval $[-1,1]$. Using these boundary conditions and the above substitution equations (3.9) and (3.10) can be rewritten in the equivalent form [3]

$$p(y) = \tfrac{\sqrt{1-y^2}}{2\pi} \left\{ \tfrac{2}{b-a} \int\limits_{-1}^1 \tfrac{f'_t(\tfrac{b+a}{2} + \tfrac{b-a}{2}t)dt}{\sqrt{1-t^2}(t-y)} + \eta \int\limits_{-1}^1 \tfrac{\psi(p(t),\eta)dt}{\sqrt{1-t^2}(t-y)} \right\}, \tag{3.13}$$

$$\frac{2}{b-a} \int_{-1}^{1} \frac{f'_t(\frac{b+a}{2} + \frac{b-a}{2}t)dt}{\sqrt{1-t^2}} = -\eta \int_{-1}^{1} \frac{\psi(p(t),\eta)dt}{\sqrt{1-t^2}}, \tag{3.14}$$

$$\frac{2}{b-a}\left\{ \int_{-1}^{1} \frac{tf'_t(\frac{b+a}{2} + \frac{b-a}{2}t)dt}{\sqrt{1-t^2}} - \pi \right\} = -\eta \int_{-1}^{1} \frac{t\psi(p(t),\eta)dt}{\sqrt{1-t^2}}. \tag{3.15}$$

After the problem is solved the displacement δ can be determined from the equation

$$\delta = f(a) + \frac{2}{\pi} \int_{a}^{b} p(t) \ln \frac{1}{|a-t|}dt. \tag{3.16}$$

Contact Problem 3 is the problem for a rigid indenter with one fixed and one free boundaries, for example, with fixed boundary $x = a = -1$ and free boundary $x = b$ that is determined by the equations $p(b) = 0$. It can be considered in the fashion similar to the one in the preceding chapter for coated/rough solids with the same boundary conditions. Here it will not be considered at all.

3.2 Plane Contacts with Nonlinear Friction and Fixed Boundaries

In this section we will consider application of regular and matched asymptotic expansions [3, 4] to solution of Contact Problem 1.

3.2.1 Application of Regular Perturbations to Solution of a Plane Problem for a Frictional Contact with Fixed Boundaries

Let us first consider the case when it is possible to apply regular asymptotic methods. For $\eta \ll 1$ the solution of Contact Problem 1 described by equations (3.11) and (3.12) will be searched in the following form

$$p(x) = \sum_{k=0}^{\infty} \eta^k p_k(x), \ |x| < 1, \ \delta = \sum_{k=0}^{\infty} \eta^k \delta_k, \tag{3.17}$$

where $p_k(x)$ and δ_k are new unknown functions and constants such that $\frac{p_{k+1}(x)}{p_k(x)} = O(1)$ uniformly for all $|x| < 1$. By substituting (3.17) into equations (3.11) and (3.12) and equating terms with the same powers of η, we obtain

$$p_0(x) = \frac{1}{2\sqrt{1-x^2}}\left\{ 1 + \frac{1}{\pi} \int_{-1}^{1} \frac{f'(t)\sqrt{1-t^2}dt}{t-x} \right\}, \ \delta_0 = \ln 2 + \frac{1}{\pi} \int_{-1}^{1} \frac{f(t)dt}{\sqrt{1-t^2}}, \tag{3.18}$$

$$p_1(x) = \frac{1}{2\pi\sqrt{1-x^2}} \int_{-1}^{1} \frac{\psi(p_0(t),0)\sqrt{1-t^2}dt}{t-x}, \tag{3.19}$$

$$\delta_1 = \frac{1}{\pi} \int\limits_{-1}^{1} \psi(p_0(t), 0) \arccos t \, dt, \ldots$$

Obviously, the expressions for $p_0(x)$ and δ_0 coincide with the solution of the corresponding contact problem without friction. At the same time, function $p_1(x)$ and constant δ_1 are determined by functions $p_0(x)$ and $\psi(p_0(x), 0)$. The latter function is controlled by the friction law.

The question about the validity of the regular asymptotic expansions in application to Contact Problem 1 is equivalent to establishing the conditions for which the ratios $p_{k+1}(x)/p_k(x)$ are uniformly bounded for all $\mid x \mid < 1$. With the help of estimates (3.8) and equations (3.18) and (3.19), we can show that $p_1(x)/p_0(x)$ is bounded for

$$0 < \gamma < 1. \tag{3.20}$$

Inequality (3.20) represents the condition of the validity of the regular asymptotic expansions method in application to Contact Problem 1.

Therefore, for γ from (3.20) formulas (3.17)-(3.19) determine a two-term asymptotic solution of Contact Problem 1 uniformly valid on the interval $(-1, 1)$. Using formulas (3.4), (3.5), (3.17)-(3.19), we obtain a two-term asymptotic representation for the dimensionless friction force

$$F = \eta \int\limits_{-1}^{1} \psi(p(t), \eta) dt = \eta \int\limits_{-1}^{1} \psi(p_0(t), 0) dt$$

$$+ \eta^2 \left\{ \int\limits_{-1}^{1} \psi'_p(p_0(t), 0) p_1(t) dt + \int\limits_{-1}^{1} \psi'_\eta(p_0(t), 0) dt \right\} + \ldots \tag{3.21}$$

3.2.2 Application of Matched Asymptotic Expansions to Solution of a Plane Problem for a Frictional Contact with Fixed Boundaries

In the preceding section, we learned that the regular asymptotic expansions method is no longer applicable to Contact Problem 1 in the vicinity of $x = \pm 1$ for $\gamma \geq 1$. Therefore, for $\gamma \geq 1$ solution of Contact Problem 1 requires application of the matched asymptotic expansions method.

For $\eta \ll 1$ in the external region represented by the interval $[-1, 1]$ with the exclusion of small zones located next to points $x = \pm 1$ the solution of Contact Problem 1 we will seek in the form

$$p(x) = p_0(x) + o(1), \ p_0(x) = O(1) \ for \ x \pm 1 = O(1), \ \eta \ll 1. \tag{3.22}$$

We will look for the displacement δ in the form

$$\delta = \delta_0 + \mu(\eta, \gamma)\delta_1 + o(\mu), \ \delta_1 = O(1), \ \mu(\eta, \gamma) \ll 1 \ for \ \eta \ll 1. \tag{3.23}$$

As in the preceding section, $p_0(x)$ and δ_0 coincide with the solution of the corresponding contact problem without friction and are determined by (3.18).

Taking into account that the indenter edges are sharp, we conclude that [1]

$$N_\pm = 1 \mp \frac{1}{\pi} \int\limits_{-1}^{1} f'(t) \sqrt{\frac{1\pm t}{1\mp t}} dt, \quad N_\pm = O(1) \ for \ \eta \ll 1. \tag{3.24}$$

Let us introduce the local variables $r = \frac{x+1}{\epsilon}$ and $s = \frac{x-1}{\epsilon}$ in the inner zones located next to the points $x = \pm 1$. The characteristic size of these inner zones $\epsilon = \epsilon(\eta, \gamma)$ is small for $\eta \ll 1$. In the inner zones the contributions of pressure $p(x)$ and frictional stress $\tau(x)$ to the normal displacement of the contact surface are of the same order of magnitude. The main terms of the asymptotic expansions of $p_0(x)$ in the inner zones are

$$p_0(x) = \frac{N_-}{2\sqrt{2r}}\epsilon^{-1/2} + \dots, \quad r = O(1) \ for \ \eta \ll 1,$$
$$p_0(x) = \frac{N_+}{2\sqrt{-2s}}\epsilon^{-1/2} + \dots, \quad s = O(1) \ for \ \eta \ll 1. \tag{3.25}$$

To provide matching of the solutions of the problem in the inner zones with the asymptotic expansions of $p_0(x)$ from (3.25), we will be looking for the solution in the inner zones in the form

$$p(x) = \epsilon^{-1/2}q(r) + o(\epsilon^{-1/2}), \quad q(r) = O(1), \quad r = O(1) \ for \ \eta \ll 1, \tag{3.26}$$

$$p(x) = \epsilon^{-1/2}g(s) + o(\epsilon^{-1/2}), \quad g(s) = O(1), \quad s = O(1) \ for \ \eta \ll 1. \tag{3.27}$$

Rewriting equation (3.11) in variable r, substituting in it the expansion (3.26), estimating the integrals involved, and taking into account the fact that in the inner zones the terms of the normal displacement depending on pressure p and frictional stress τ are of the same order of magnitude, we obtain

$$q(r) = \frac{1}{2\sqrt{2r}}\left\{N_- + \frac{1}{\pi}\int\limits_{0}^{\infty} \frac{q^\gamma(t)\sqrt{2t}dt}{t-r}\right\}. \tag{3.28}$$

In the derivation of equation (3.28), we also used the assumptions from (3.8) and estimates (3.25).

In addition to that we determine the characteristic size of the inner zones as

$$\epsilon = \eta^{\frac{2}{\gamma-1}}. \tag{3.29}$$

A similar analysis of the inner zone adjacent to $x = 1$ provides the equation for $g(s)$ in the form

$$g(s) = \frac{1}{2\sqrt{-2s}}\left\{N_+ + \frac{1}{\pi}\int\limits_{-\infty}^{0} \frac{g^\gamma(t)\sqrt{-2t}dt}{t-s}\right\}. \tag{3.30}$$

Obviously, the outlined method is valid for $\gamma > 1$.

For $\gamma = 1$ using the above method, it is not possible to obtain equations for $q(r)$ and $g(s)$ in the inner zones. However, using the exact solution of Contact Problem 1 for $\psi(p, \eta) = p$ [1] or the two-term solution of equation (3.11) in the external region it can be shown that

$$\epsilon = e^{-1/\eta}. \tag{3.31}$$

Now, let us consider calculation of the dimensionless friction force. Using the assumptions (3.8) we obtain the estimate

$$\int_{-1}^{1} \psi(p(t), \eta)dt = O(\epsilon^{\frac{2-\gamma}{2}}, 1), \tag{3.32}$$

where the term of the order of $\epsilon^{\frac{2-\gamma}{2}}$ is the contribution of the inner zones while the term of the order of 1 comes from integration over the external region. It follows from (3.32) that for $0 < \gamma < 2$ the dimensionless friction force is mainly determined by the pressure in the external region

$$F = \eta \int_{-1}^{1} \psi(p_0(t), 0)dt. \tag{3.33}$$

For $\gamma = 2$ from estimate (3.32) follows that the contributions of the external region and the inner zones to the friction force are of the same order of magnitude. Therefore, by introducing the uniformly valid on $[-1, 1]$ approximate solution $p_u(x)$ the dimensionless friction force is equal to

$$F = \eta \int_{-1}^{1} \psi(p_u(t), 0)dt,$$

$$p_u(x) = \frac{8\eta^{-2}}{N_- N_+} \sqrt{1 - x^2} p_0(x) q(\frac{x+1}{\eta^2}) g(\frac{x-1}{\eta^2}). \tag{3.34}$$

Finally, for $\gamma > 2$ the dimensionless friction force is predominantly determined by the solution of the problem in the inner zones. Using (3.8) we obtain

$$F = \eta^{\frac{1}{\gamma-1}} \int_{0}^{\infty} [q^\gamma(r) + g^\gamma(-r)]dr. \tag{3.35}$$

As an example let us consider Contact Problem 1 for an indenter with flat bottom $f(x) = const$ and $\psi(p, \eta) = p^\gamma$, $\gamma > 0$. Then for $0 < \gamma < 2$ from (3.18) and (3.33), we obtain $p_0(x) = \frac{1}{2\sqrt{1-x^2}}$ and

$$F = \eta 2^{-\gamma} \sqrt{\pi} \frac{\Gamma(1 - \frac{\gamma}{2})}{\Gamma(\frac{3-\gamma}{2})}, \tag{3.36}$$

where $\Gamma(\cdot)$ is the gamma-function [5]. The graph of the friction force F as a function of γ is given in Fig. 3.2. Obviously, the friction force F reaches its minimum value at approximately $\gamma = 1$ and increases significantly as γ increases.

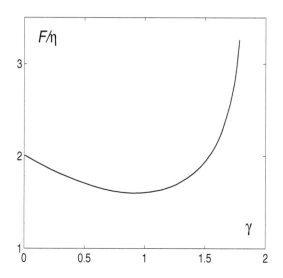

FIGURE 3.2
Friction force F under a flat indenter as a function of γ.

3.3 Plane Contacts with Nonlinear Friction and Free Boundaries

In this section we will consider application of regular and matched asymptotic expansions to solution of Contact Problem 2.

3.3.1 Application of Regular Perturbations to Solution of a Plane Problem for a Frictional Contact with Free Boundaries

For $\eta \ll 1$ let us consider application of the regular asymptotic expansions method to Contact Problem 2 described by equations (3.13)-(3.15). The solution will be searched in the form

$$p(y) = \sum_{k=0}^{\infty} \eta^k p_k(y), \ |\, y\,|\leq 1, \ a = \sum_{k=0}^{\infty} \eta^k \alpha_k, \ b = \sum_{k=0}^{\infty} \eta^k \beta_k, \qquad (3.37)$$

where $p_k(x)$, α_k, and β_k are new unknown functions and constants such that $\frac{p_{k+1}(x)}{p_k(x)} = O(1)$ uniformly for all $|\, x\,|\leq 1$.

Let us expand $f'_t(\frac{b+a}{2} + \frac{b-a}{2}t)$ in the Taylor series centered about $x =$

$\frac{\beta_0+\alpha_0}{2} + \frac{\beta_0-\alpha_0}{2}t$. By substituting this Taylor expansion and the representations (3.37) into equations (3.13)-(3.15) and equating terms with the same powers of η we obtain

$$p_0(y) = \frac{\sqrt{1-y^2}}{\pi(\beta_0-\alpha_0)} \int_{-1}^{1} \frac{f_t'(\frac{\beta_0+\alpha_0}{2} + \frac{\beta_0-\alpha_0}{2}t)dt}{\sqrt{1-t^2}(t-y)}, \tag{3.38}$$

$$p_1(y) = \frac{\alpha_1-\beta_1}{\beta_0-\alpha_0}p_0(y) + \frac{\sqrt{1-y^2}}{2\pi}\left\{ \frac{2}{(\beta_0-\alpha_0)^2}\left[\frac{(\beta_1-\alpha_1)(A_++A_-)}{2} \right.\right.$$

$$\tag{3.39}$$

$$\left.\left. +(\beta_1+\alpha_1+(\beta_1-\alpha_1)y)\int_{-1}^{1} \frac{f_{tt}''(\frac{\beta_0+\alpha_0}{2} + \frac{\beta_0-\alpha_0}{2}t)dt}{\sqrt{1-t^2}(t-y)} \right] + \int_{-1}^{1} \frac{\psi(p_0(t),0)dt}{\sqrt{1-t^2}(t-y)} \right\},$$

$$\alpha_1 = \frac{A_1B_+ - B_1A_+}{\Delta}, \quad \beta_1 = \frac{B_1A_+ - A_1B_-}{\Delta}, \quad \Delta = A_+B_- - A_-B_+,$$

$$A_\pm = \int_{-1}^{1} \sqrt{\frac{1\pm t}{1\mp t}} f_{tt}''(\frac{\beta_0+\alpha_0}{2} + \frac{\beta_0-\alpha_0}{2}t)dt,$$

$$A_1 = \frac{(\beta_0-\alpha_0)^2}{2} \int_{-1}^{1} \frac{\psi(p_0(t),0)dt}{\sqrt{1-t^2}}, \tag{3.40}$$

$$B_\pm = \int_{-1}^{1} t\sqrt{\frac{1\pm t}{1\mp t}} f_{tt}''(\frac{\beta_0+\alpha_0}{2} + \frac{\beta_0-\alpha_0}{2}t)dt,$$

$$B_1 = \frac{(\beta_0-\alpha_0)^2}{2} \int_{-1}^{1} \frac{t\psi(p_0(t),0)dt}{\sqrt{1-t^2}}.$$

It is important to realize that constants α_0 and β_0 are the solutions of equations (3.14) and (3.15) for $\eta = 0$ such that $\beta_0 > \alpha_0$.

Obviously, the expressions for $p_0(x)$, α_0, and β_0 coincide with the solution of the corresponding contact problem without friction. At the same time, function $p_1(x)$ and constants α_1 and β_1 are determined by functions $p_0(x)$ and $\psi(p_0(x),0)$ the latter of which is controlled by the friction law.

The validity of this approach is equivalent to the fact that the last integral in (3.39) is bounded for $|y| \le 1$. Based on (3.8) and the above integrals it follows that the obtained results are valid for $\gamma > 1$. Therefore, for $\gamma > 1$ formulas (3.37)-(3.40) determine the two-term uniformly valid asymptotic expansion of the solution of Contact Problem 2 for $\eta \ll 1$.

The two-term formula for the dimensionless friction force has the form

$$F = \frac{\eta(\beta_0-\alpha_0)}{2} \int_{-1}^{1} \psi(p_0(t),0)dt + \frac{\eta^2(\beta_0-\alpha_0)}{2}\left\{ \int_{-1}^{1} \psi_p'(p_0(t),0)p_1(t)dt \right.$$

$$\tag{3.41}$$

$$\left. + \int_{-1}^{1} \psi_\eta'(p_0(t),0)dt + \frac{\beta_1-\alpha_1}{\beta_0-\alpha_0}\int_{-1}^{1} \psi(p_0(t),0)dt \right\},$$

where $p_0(y)$ is determined by equation (3.38).

3.3.2   Application of Matched Asymptotic Expansions to Solution of a Plane Problem for a Frictional Contact with Free Boundaries

Let us use the method of matched asymptotic expansions to solve Contact Problem 2 for $\eta \ll 1$ and $0 < \gamma \leq 1$. Similar to the case of an indenter with sharp edges when we had to apply the matched asymptotic expansions we will introduce the external region, which occupies almost the entire interval $[-1, 1]$ except for small inner zones of characteristic size ϵ, which are located next to points $y = \pm 1$. For $\eta \ll 1$ in the external region, the solution will be searched in the form

$$p(y) = p_0(y) + o(1), \ p_0(y) = O(1) \ for \ y \pm 1 = O(1), \ \eta \ll 1. \qquad (3.42)$$

Then function $p_0(y)$ is determined by equation (3.38) while constants α_0 and β_0 are the solutions of equations (3.14) and (3.15) for $\eta = 0$ such that $\beta_0 > \alpha_0$. Let us introduce two constants

$$N_\pm = \frac{1}{\pi(\beta_0 - \alpha_0)} \lim_{y \to \mp 1} \int_{-1}^{1} \frac{f_t'(\frac{\beta_0 + \alpha_0}{2} + \frac{\beta_0 - \alpha_0}{2} t) dt}{\sqrt{1 - t^2}(t - y)}, \ N_\pm = O(1), \ \eta \ll 1. \qquad (3.43)$$

By introducing the inner variables $r = \frac{x+1}{\epsilon}$ and $s = \frac{x-1}{\epsilon}$ in the inner zones adjacent to $y = \mp 1$, respectively, we obtain the main terms of asymptotic expansions of $p_0(y)$ in the inner zones

$$p_0(y) = N_+ \sqrt{2r} \epsilon^{1/2} + \dots, \ r = O(1) \ for \ \eta \ll 1,$$
$$\qquad (3.44)$$
$$p_0(x) = N_- \sqrt{-2s} \epsilon^{1/2} + \dots, \ s = O(1) \ for \ \eta \ll 1.$$

To provide matching of the solutions of the problem in the inner zones with the asymptotic expansions of $p_0(x)$ from (3.44), we will be looking for the solution in the inner zones in the form

$$p(y) = \epsilon^{1/2} q(r) + o(\epsilon^{1/2}), \ q(r) = O(1), \ r = O(1) \ for \ \eta \ll 1, \qquad (3.45)$$

$$p(x) = \epsilon^{1/2} g(s) + o(\epsilon^{1/2}), \ g(s) = O(1), \ s = O(1) \ for \ \eta \ll 1, \qquad (3.46)$$

while constants a and b we will search as follows

$$a = \alpha_0 + \eta \alpha_1 + o(\eta), \ b = \beta_0 + \eta \beta_1 + o(\eta),$$
$$\qquad (3.47)$$
$$\alpha_1 = O(1), \ \beta_1 = O(1), \ \eta \ll 1.$$

Then using the assumption (3.8) and equations (3.43)-(3.47) in equation (3.13) and equating terms of the same order of magnitude for $\eta \ll 1$ in the inner zones we derive asymptotically valid equations for $q(r)$ and $g(s)$

$$q(r) = \sqrt{2r} \Big\{ N_+ + \frac{1}{2\pi} \int_0^{\infty} \frac{q^\gamma(t) dt}{\sqrt{2t}(t - r)} \Big\}, \qquad (3.48)$$

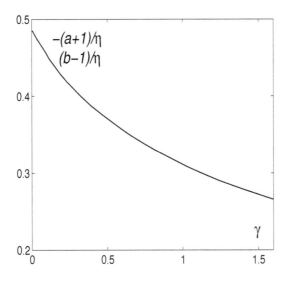

FIGURE 3.3
Boundaries of the contact region $-(a+1)/\eta$ and $(b-1)/\eta$ under a parabolic indenter as a function of γ.

$$g(s) = \sqrt{-2s}\left\{N_- + \tfrac{1}{2\pi}\int\limits_{-\infty}^{0}\frac{g^\gamma(t)dt}{\sqrt{-2t}(t-s)}\right\}. \tag{3.49}$$

In addition to that we determine the characteristic size of the inner zones as

$$\epsilon = \eta^{\frac{2}{1-\gamma}}. \tag{3.50}$$

Obviously, the outlined analysis is valid for $0 < \gamma < 1$. For $\gamma = 1$ it is not possible to conduct an analysis similar to the above one. However, it is possible to get an estimate for the characteristic size of the inner zones $\epsilon = e^{-1/\eta}$ for $\eta \ll 1$.

It is easy to verify that for $0 < \gamma < 1$ as well as for $\gamma > 1$ constants α_1 and β_1 are calculated according to formulas (3.40) while the main term of the dimensionless friction force is determined by

$$F = \tfrac{\eta(\beta_0 - \alpha_0)}{2}\int\limits_{-1}^{1}\psi(p_0(t),0)dt. \tag{3.51}$$

As an example let us consider Contact Problem 2 for a parabolic indenter $f(x) = x^2$ and $\psi(p,\eta) = p^\gamma$, $\gamma > 0$. Then from equations (3.14) and (3.15) for $\eta = 0$ we get $\alpha_0 = -1$ and $\beta_0 = 1$. Moreover, from formula (3.38) we get $p_0(y) = \sqrt{1-y^2}$, which allows for calculation of constants α_1 and β_1 from

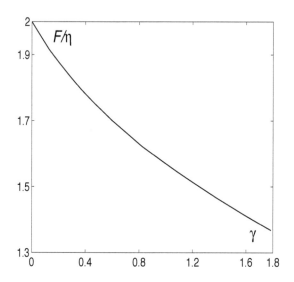

FIGURE 3.4
Friction force F under a parabolic indenter as a function of γ.

(3.40)

$$\alpha_1 = \beta_1 = -\frac{1}{\sqrt{\pi}\gamma} \frac{\Gamma(\frac{\gamma+1}{2})}{\Gamma(\frac{\gamma}{2})}, \tag{3.52}$$

where $\Gamma(\cdot)$ is a Gamma-function [5]. Therefore, the two-term expressions for contact boundaries a and b as well as the main term of the dimensionless friction force are

$$a = -1 - \frac{\eta}{\sqrt{\pi}\gamma} \frac{\Gamma(\frac{\gamma+1}{2})}{\Gamma(\frac{\gamma}{2})} + O(\eta^2), \ b = 1 - \frac{\eta}{\sqrt{\pi}\gamma} \frac{\Gamma(\frac{\gamma+1}{2})}{\Gamma(\frac{\gamma}{2})} + O(\eta^2),$$

$$F = \eta \frac{\sqrt{\pi}\gamma}{\gamma+1} \frac{\Gamma(\frac{\gamma}{2})}{\Gamma(\frac{\gamma+1}{2})} + O(\eta^2), \ \eta \ll 1. \tag{3.53}$$

The graph of $-(a+1)/\eta$ and $(b-1)/\eta$ as a function of γ is given in Fig. 3.3 while the dependence of the friction force F on γ is presented in Fig. 3.4.

3.4 Plane Frictional Rough Contacts Modeled by Nonlinear Coating

In this section we will consider plane analogs of problems considered above. In particular, we will study plane contact problems for an infinite cylindrical indenter contacting with friction a thick layer of elastic isotropic homogeneous material bounded by two horizontal planes $z = 0$ and $z = -h$. The friction stress τ is different from zero only where pressure p is different from zero and it is determined by a nonlinear relationship $\tau = \sigma\psi(p)$, where σ is a positive constant, ψ is a function of contact pressure p. The shape of the indenter base is described by function $z = f(x)$. The contact area bounded by lines $x = a$ and $x = b$ is caring a load P per unit length. The material of the layer has Young's modulus E and Poisson's ratio ν. The surface of the material layer $z = 0$ is assumed to be coated or to have some asperities. In this case we can propose a plane formulation of the problem analogous to the one given by (2.19)-(2.21). For the plane formulation the equations are as follows [6, 7]:

$$Ap = \delta + \beta x - f(x), \quad x \in \Omega, \tag{3.54}$$

$$Ap \geq \delta + \beta x - f(x), \quad x \ni \Omega, \tag{3.55}$$

$$Ap = \eta \int_{-\infty}^{x} \psi(p(t))dt + k\varphi(p) + \frac{2}{\pi E'} \int_{-\infty}^{\infty} p(\xi)[-\ln|\tfrac{x-\xi}{h}| + a_0]d\xi, \tag{3.56}$$

where A is a nonlinear operator, η is the effective friction coefficient, $\eta = \sigma\frac{(1-2\nu)(1+\nu)}{E}$, k and $\varphi(p)$ are a constant and function of pressure p characterizing the normal displacement of the surface $z = 0$ solely due to the presence of coating/surface asperities, E' is the effective elastic modulus, $E' = E/(1-\nu^2)$, δ is the normal displacement of the rigid indenter, h is the layer thickness, a_0 is a constant the value of which depends on the boundary conditions at the lower surface of the layer $z = -h$ (the values of a_0 for different conditions can be found in [9]). The contact region Ω is determined as follows

$$\Omega = \{x \mid p(x) > 0\}. \tag{3.57}$$

It will be shown that solution of equations (3.54)-(3.56) is equivalent to solution of a certain variational inequality. The existence and uniqueness of the problem solution is established. In addition to that, some qualitative properties of the problem solution and the topology of the contact region for convex indenters are determined.

The problem formulation for indenters with sharp edges (i.e., for problems with fixed contact edges) follows from (3.54)-(3.56) if the contact region Ω is replaced by a given region S. Because of an arbitrary shape of region S pressure $p(x)$ at some points of S may assume negative values.

Therefore, for the given functions $\psi(p)$, $\varphi(p)$, $f(x)$ and constants δ, β, η, k, and E' (and also region S for the problem with fixed region boundary), the problem is reduced to finding the pressure function $p(x)$ that satisfies (3.54)-(3.56).

The complexity of this problem is due to the fact that operator A is nonlinear and the boundaries of the contact region Ω are unknown.

Let us formulate some assumptions which will be used in the further analysis. We will assume that

$$f(x) \geq 0, \ g(x) = \delta + \beta x - f(x), \ f(x), g(x) \in C(\Omega_\delta), \qquad (3.58)$$

$$\sup(\Omega_\delta) - \inf(\Omega_\delta) < \infty, \ \Omega_\delta = \{x \mid g(x) \geq 0\}$$
$$ \hspace{8cm} (3.59)$$
$$if \ \mid \beta \mid < \infty, \ 0 \leq \delta < \infty,$$

$$\varphi(-p) = -\varphi(p), \ \varphi(0) = 0, \ \varphi(p_2) > \varphi(p_1)$$
$$ \hspace{8cm} (3.60)$$
$$if \ p_2 > p_1, \ \varphi(p) \in C([0, \infty)),$$

$$\varphi(p) \leq a + bp^\alpha, \ p \geq 0 \ (a, b, \alpha \ constants, \ \alpha > 0.3), \qquad (3.61)$$

$$\psi(p) \geq 0 \ \forall p \geq 0, \ \psi(0) = 0, \ \psi(p) \in C([0, \infty)),$$
$$ \hspace{8cm} (3.62)$$

$$\psi(p) \leq a_1 + b_1 p^\gamma, \ p \geq 0 \ (a_1, b_1, \gamma \ constants, \ \gamma > 0),$$

$$K = \{p \mid 0 \leq p \leq \varphi^{-1}[g(x)/k], \ x \in \Omega_\delta\}$$

$$\subset L_2(\Omega_\delta) \bigcap L_{\alpha+1}(\Omega_\delta) \bigcap L_\gamma(\Omega_\delta),$$
$$ \hspace{8cm} (3.63)$$

$$K \ - \ set \ of \ almost \ everywhere \ uniformly \ bounded$$
$$functions,$$

$$(\int_{-\infty}^{x} [\psi(p_2(t)) - \psi(p_1(t))]dt, p_1 - p_1) \geq 0 \ \forall p_1, p_2 \in K, \qquad (3.64)$$

where as before C is a space of continuous functions and the inner product in L_2 is defined as follows

$$(f, g) = \int_{-\infty}^{\infty} f(x)g(x)dx.$$

Obviously, K is a convex closed set. Let us notice that inequality (3.64) is valid for a family of functions ψ one of representatives of which is function $\psi(p) = p$.

3.4.1   Existence and Uniqueness of the Problem Solution

Below we will formulate a number of lemmas and theorems similar to the ones in the preceding chapter.

Lemma 3.4.1 *Suppose conditions (3.58)-(3.63) are satisfied. If $p(x)$ is a measurable almost everywhere bounded solution of problem (3.54)-(3.56) for $\eta \geq 0, \ k > 0, \mid \beta \mid < \infty, \ 0 \leq \delta < \infty$ and $h \geq e^{-a_0} \sup_{x,y \in \Omega} \mid x - y \mid$ then: (i) $\Omega \subset \Omega_\delta$, (ii) $p \in K$, (iii) $p \in C(\overline{\Omega})$.*

Proof. For $h \geq e^{-a_0} \sup\limits_{x,y \in \Omega} |x-y|$ from (3.60) and (3.62) follows that operator A from (3.56) is nonnegative, i.e., $Ap \geq 0$. Therefore, for $x \in \Omega$ (3.54) leads to $x \in \Omega_\delta$, i.e., the first statement of Lemma is valid. Based on (3.60) from (3.54) follows that

$$p(x) = \varphi^{-1}\left\{\frac{1}{k}\left[g(x) - \eta \int\limits_{-\infty}^{x} \psi(p(t))dt - \frac{2}{\pi E'} \int\limits_{-\infty}^{\infty} p(\xi)(-\ln|\tfrac{x-\xi}{h}| + a_0)d\xi\right]\right\}.$$

The statements of the lemma follow [9, 10] from the established solution properties and conditions (3.58)-(3.63).

Lemma 3.4.2 *Suppose $p(x)$ is a solution of (3.54) -(3.56). Then under conditions of Lemma 3.4.1 the following integrals*

$$P = \int\limits_{-\infty}^{\infty} p(x)dx, \quad M = \int\limits_{-\infty}^{\infty} xp(x)dx, \quad F_S = \eta \int\limits_{-\infty}^{\infty} \psi(p(x))dx \qquad (3.65)$$

are finite.

Theorem 3.4.1 *Suppose $g(x)$ satisfies (3.58) and $p_0(x) \in K$ is a solution of (3.54)- (3.56) in Ω for $\eta \geq 0$, $k > 0$, $0 < E' < \infty$, $|\beta| < \infty$, $0 \leq \delta < \infty$ and $h \geq e^{-a_0} \sup\limits_{x,y \in \Omega} |x - y|$. Then problem (3.54)-(3.56) is equivalent to the variational inequality*

$$(Ap_0, p - p_0) \geq (g, p - p_0), \quad \forall p \in K. \qquad (3.66)$$

Proof. *Necessity.* Let $p_0 \in K$ is a solution of problem (3.54)- (3.56) for $\eta \geq 0$, $k \geq 0$, $|\beta| < \infty$, $0 \leq \delta < \infty$. Then by multiplying the relationships (3.54), (3.56) by $p(x) - p_0(x)$ for $\forall p \in K$ and by integrating the products we obtain inequality (3.66).

Sufficiency. Let $p_0 \in K$ is a solution of the variational inequality (3.66). First, let us consider any interior point $x_0 \in \Omega$ at which $p_0(x_0) > 0$. Because of continuity of $p_0(x)$ there exists a vicinity O centered at point x_0 within which $p_0(x) > 0$. Let us chose an arbitrary bounded continuous function $\zeta(x)$ with $\sup \zeta(x) \in O$. Then for sufficiently small μ we have $p = p_0 + \mu\zeta \in K$. Setting μ consequently equal to μ_0 and $-\mu_0$ in (3.66) we get the equality $(Ap_0, \zeta) = (g, \zeta)$. Based on the fact that $\zeta(x)$ is arbitrary and continuous function, we conclude that equation (3.54) is valid.

Now, let us consider an arbitrary point x_0 which belongs to the interior of the complementary set $C\Omega = R^1 \backslash \Omega$. Obviously, $p_0(x_0) = 0$. Then there exists such a vicinity O_1 centered at x_0 that $p_0(x) = 0$ $\forall x \in O_1 \subset C\Omega$. Let us choose an arbitrary bounded continuous nonnegative function $\zeta(x)$ with $\sup \zeta(x) \subset O_1$. Then choosing $p = p_0 + \zeta \in K$ from (3.66) due to arbitrary nature of $\zeta(x)$ we obtain (3.55).

For $p = p_0$ and $x_0 \in \overline{\Omega}\backslash\Omega$ from continuity of $p_0(x)$ and $g(x)$ follows the validity of (3.54) and (3.55) at these x_0 as well.

Lemma 3.4.3 *If conditions (3.59)-(3.64) are satisfied and $\eta \geq 0$, $k > 0$, $0 < E' < \infty$, and $h \geq e^{-a_0} \sup_{x,y \in \Omega} | x - y |$ then operator $A : K \to L_{(\alpha+1)/\alpha}(\Omega_\delta)$ is a bounded semi-continuous strictly monotonous operator.*

Proof. From (3.61) and (3.62) we get [8] $A : K \to L_{(\alpha+1)/\alpha}(\Omega_\delta)$. The fact that A is bounded on K follows from (3.61) and (3.62), boundness of Ω_δ, and the inclusion $K \subset L_2(\Omega_\delta) \bigcap L_{\alpha+1}(\Omega_\delta) \bigcap L_\gamma(\Omega_\delta)$. Semi-continuity of operator A [9], i.e., continuity of the function $\mu \to (A(p_1 + \mu p_2), p_3) \; \forall p_1, p_2, p_3 \in K$ follows from conditions (3.60) and (3.62), boundness of Ω_δ and the fact that functions $p \in K$. Strict monotonicity of operator A [9] follows from (3.60), (3.64), and the inequality $h \geq e^{-a_0} \sup_{x,y \in \Omega} | x - y |$.

Theorem 3.4.2 *Suppose conditions (3.58)-(3.64) are satisfied. Then the solution of variational inequality (3.66) exists and unique in K for $\eta \geq 0$, $k > 0$, $0 < E' < \infty$, $| \beta | < \infty$, $0 \leq \delta < \infty$ and $h \geq e^{-a_0} \sup_{x,y \in \Omega} | x - y |$.*

Proof. The existence of the solution follows from Lemma 3.4.3, Proposition 2.5, and Theorem 8.1 from [9] while the uniqueness follows from Theorem 8.3 [9].

Theorem 3.4.3 *If $\eta \geq 0$, $k > 0$, $0 < E' < \infty$, $| \beta | < \infty$, $0 \leq \delta < \infty$ and $h \geq e^{-a_0} \sup_{x,y \in \Omega} | x - y |$ and condition (3.60) are satisfied, then the solution of problem (3.54)-(3.56) is unique for: (i) given force $P \geq 0$ and moment M, (ii) given force $P \geq 0$ and angle β, (iii) given displacement δ and moment M, (iv) given displacement δ and angle β.*

Proof. Let us prove statement (i). Suppose the given values of P and M correspond two different solutions $p_1(x)$ and $p_2(x)$ of problem (3.54)-(3.56), which possess different sets of parameters δ and β, i.e., (δ_1, β_1) and δ_2, β_2. Otherwise, there is a unique solution. Based on Theorem 3.4.1 we conclude that to these two sets of (δ, β) correspond two variational inequalities (3.66) for functions p_1 and p_2. Assuming that $p = p_2$ in inequality (3.66) written for the set (p_1, δ_1, β_1) and vise versa assuming that $p = p_1$ in inequality (3.66) written for the set (p_2, δ_2, β_2) and adding them leads to the inequality

$$(k[\varphi(p_1) - \varphi(p_2)] + \eta \int_{-\infty}^{x} [\psi(p_1(t)) - \psi(p_2(t))]dt$$

$$-\frac{2}{\pi E'} \int_{-\infty}^{\infty} [p_2(\xi) - p_1(\xi)](- \ln | \tfrac{x-\xi}{h} | + a_0)d\xi, p_2 - p_1) \geq (g_1 - g_2, p_2 - p_1).$$

From the uniqueness of the solution of problem (3.54)-(3.56) for the given set of (δ, β) follows that $p_1(x) \neq p_2(x)$. Therefore, using conditions (3.60), (3.64), and $h \geq e^{-a_0} \sup_{x,y \in \Omega} | x - y |$ from the latter inequality, we obtain

$$(g_1 - g_2, p_2 - p_1)$$

$$= (\delta_2 - \delta_1)(P_2 - P_1) + (\beta_2 - \beta_1)(M_2 - M_1) > 0,$$

(3.67)

where P_i and M_i are the first two integrals in (3.65) calculated for $p(x) = p_i(x)$, $i = 1, 2$. From the theorem conditions we have $P_1 = P_2 = P$ and $M_1 = M_2 = M$. Therefore, the right–hand side of inequality (3.71) is equal to zero. Obviously, we arrived at a contradiction. Statements (ii)-(iv) are proved in a similar fashion.

It is obvious that the formulated statements are valid for the problem with fixed boundaries of the contact region S as long as pressure $p(x)$ is nonnegative in S.

3.4.2 Qualitative Properties of the Solution

Let us establish some solution properties.

Theorem 3.4.4 *If $\eta \geq 0$, $k > 0$, $0 < E' < \infty$, $\mid \beta \mid < \infty$, $0 \leq \delta < \infty$ and $h \geq e^{-a_0} \sup_{x,y \in \Omega} \mid x - y \mid$ and conditions (3.61) and (3.64) are satisfied, then (i) for fixed angle β the applied force P is a strictly monotonically increasing function of the displacement δ and (ii) for fixed displacement δ moment M is a strictly monotonically increasing function of angle β.*

Proof. Let us consider two solutions of problem (3.54)-(3.56) obtained for the sets of (δ_1, β_1) and (δ_2, β_2), which correspond to the sets of the values of force and moment (P_1, M_1) and (P_2, M_2), respectively. Because of uniqueness of the problem solution for each set of values (δ, β), we arrive at inequality (3.71) (see proof of the preceding theorem). The Theorem statements directly follow from inequality (3.71).

Corollary 3.4.1 *Under the conditions of Theorem 3.4.4 for fixed angle β, the friction force F_S from (3.65) is a monotonically increasing function of the displacement δ.*

Let us consider an equivalent reformulation of problem (3.54)-(3.56) for unknown contact region Ω as follows

$$\triangle \chi = 0,$$

$$\left\{ \eta \int_{-\infty}^{x} \psi(\tfrac{\partial \chi}{\partial z}) dt + k\varphi(\tfrac{\partial \chi}{\partial z}) + \tfrac{2}{E'}\chi \right\} |_{z=0} = \delta + \beta x - f(x)$$

$$\text{if } \tfrac{\partial \chi}{\partial z} |_{z=0} > 0,$$

$$\left\{ \eta \int_{-\infty}^{x} \psi(\tfrac{\partial \chi}{\partial z}) dt + \tfrac{2}{E'}\chi \right\} |_{z=0} \geq \delta + \beta x - f(x) \text{ if } \tfrac{\partial \chi}{\partial z} |_{z=0} = 0,$$

(3.68)

where function χ is a logarithmic potential determined according to formulas
[6]

$$\chi = \chi_0 + a_0 + \ln h, \quad \chi_0(x,z) = \frac{1}{\pi} \int\limits_{-\infty}^{\infty} p(t) \ln \frac{1}{R} dt,$$

$$(3.69)$$

$$R = \sqrt{(x-t)^2 + z^2}, \quad \frac{\partial \chi(x,-0)}{\partial z} = p(x).$$

Theorem 3.4.5 *Let χ_1 and χ_2 be solutions of problem (3.68) and (3.69) for $\eta = 0$ and $\eta > 0$, respectively. Then inequality $\chi_1(x,z) > \chi_2(x,z)$ is valid for $k > 0$, all x and $z \leq 0$.*

Proof. The proof is done in the fashion similar to the proof of Theorem 2.3.7. By considering sets Σ_1, Σ_2, Σ_3, Σ_4 and by using the principle of Zaremba-Zhiro [11, 12, 13], we arrive at a contradiction.

Now, let us consider the topology of the contact region Ω in the case of a convex indenter without sharp edges.

Lemma 3.4.4 *Let $p(x)$ be a solution of problem (3.54)-(3.56) and $f(x)$ be a finite strictly concave function in any bounded region. Then within any interval ω outside of contact region Ω function $F(x) = Ap(x) - \delta - \beta x + f(x)$ is strictly concave.*

Proof. Let ω be an interval of nonzero length such that $\omega \bigcap \Omega = \emptyset$. Then for any $\theta \in [0,1]$ and $x \neq y$, $x,y \in \omega$, we have

$$\theta F(x) + (1-\theta)F(y) - F(\theta x + (1-\theta)y) = \theta f(x) + (1-\theta)f(y) - f(\theta x + (1-\theta)y)$$

$$+ \frac{2}{\pi E'} \int\limits_{-\infty}^{\infty} p(\xi) \Big\{ -\theta \ln |\frac{x-\xi}{h}| - (1-\theta)\ln |\frac{x-\xi}{h}| + \ln |\frac{\theta x + (1-\theta)y - \xi}{h}| \Big\} d\xi.$$

Based on the validity of the inequality

$$\theta \ln |x - \xi| + (1-\theta)\ln |y - \xi| \leq \ln |\theta x + (1-\theta)y - \xi|$$

and the fact that $f(x)$ is a strictly concave function and $(\xi - x)(\xi - y) > 0$ for $\xi \in \Omega$ and $x \neq y$, $x,y \in \omega$ we get the strict concavity of function $F(x)$, i.e.,

$$\theta F(x) + (1-\theta)F(y) \geq F(\theta x + (1-\theta)y) \ \forall x,y \in \omega, \ \forall \theta \in [0,1],$$

where the sign of equality occurs only for $x = y$.

A contact problem with fixed contact region S can be formulated as follows

$$Ap_S = \delta + \beta x - f(x), \ x \in S,$$

$$p_S = 0, \ Ap_S \geq \delta + \beta x - f(x), \ x \ni S,$$

$$Ap_S = \eta \int\limits_{-\infty}^{x} \psi(p_S(t))dt + k\varphi(p_S)$$

$$(3.70)$$

$$+ \frac{2}{\pi E'} \int\limits_{-\infty}^{\infty} p_S(\xi)[-\ln |\frac{x-\xi}{h}| + a_0]d\xi.$$

Depending on contact region S the solution of the problem, pressure $p_S(x)$, may assume not only nonnegative but also negative values. Note that the above statements are also valid for solutions of problem (3.70) as long as pressure $p_S(x)$ remains nonnegative in S.

Theorem 3.4.6 *Let $p(x)$ be a solution of problem (3.54)-(3.56) and $f(x)$ be a finite strictly concave function in any bounded region. Then the contact region Ω is an interval that may contain a finite or infinite set of isolated points not belonging to Ω.*

Proof. Let us assume the opposite, i.e., there exists a closed interval ω of nonzero length such that $\omega \cap \Omega = \emptyset$ and $\inf\limits_{x \in \Omega} < \inf\limits_{x \in \omega}, \sup\limits_{x \in \omega} < \sup\limits_{x \in \Omega}$. It is obvious that boundness of ω follows from boundness of Ω_δ. Then in terms of function $F(x)$ problem (3.54)-(3.56) is reduced to the following one

$$F(x) = 0, \ p(x) > 0; \ F(x) \geq 0, \ p(x) = 0. \tag{3.71}$$

Boundness and closeness of ω leads to the conclusion that function $F(x)$ attains its maximum at some point $x_m \in \omega$, i.e., $\max\limits_{x \in \omega} F(x) = F(x_m)$. Let us show that point x_m cannot belong to the interior of ω. Let us assume the opposite, i.e., let us assume that x_m belongs to the interior of ω. Then due to continuity of $F(x)$ (see Lemma 3.4.1) in a small vicinity of $x = x_m$ function $F(x)$ suppose to be a convex function which contradicts Lemma 3.4.4. Therefore, x_m belongs to the boundary of ω.

It is obvious that points $\inf\limits_{x \in \omega}$ and $\sup\limits_{x \in \omega}$ belong to the boundary of Ω. Therefore, from (3.71) and continuity of function $F(x)$, we have $F(\inf\limits_{x \in \omega}) = F(\sup\limits_{x \in \omega}) = 0$. That leads to the inequality $F(x) \leq 0$ for $x \in \omega$. Taking into account (3.71) we have $F(x) = 0 \ \forall x \in \omega$. The latter equality contradicts the strict concavity of function $F(x)$ within interval ω of nonzero length.

It is important to notice that the statements of Lemma 3.4.4 and Theorems 3.4.4 and 3.4.6 are valid. Also, these statements can be easily transferred on the case of an indenter with sharp edges (fixed contact regions).

3.5 Asymptotic and Numerical Analysis of Contacts with Friction for Large Roughness

Let us consider the case of an indenter with fixed boundaries impressed in an elastic layer/half-plane with large roughness $\lambda \gg 1$. The contact problems with free boundaries can be considered in a similar fashion (see Chapter 2). We will assume that $\varphi(p, \lambda)$ and $\psi(p, \eta)$ are sufficiently smooth functions and $\varphi(p, \lambda)$ is a monotonically increasing function of pressure p.

3.5.1 Asymptotic Analysis of Contacts with Friction for Large Roughness

Assuming that the contact region is singly connected, in dimensionless variables similar to (3.5) the analysis of the contact problem for an indenter with the bottom shape $z = f(x)$ and contact region $[-1, 1]$ can be reduced to solution of the equations

$$\eta \int_{-1}^{x} \psi(p(t))dt + \lambda\varphi(p) + \frac{2}{\pi} \int_{-1}^{1} p(t) \ln \mid \frac{1}{x-t} \mid dt = \delta - f(x), \qquad (3.72)$$

$$\int_{-1}^{1} p(t)dt = \frac{\pi}{2}. \qquad (3.73)$$

For $\lambda \gg 1$ the normal displacement caused by asperities dominates the displacement of the bulk elastic solids. In this case the major term in equation (3.72) is the term determined by asperity displacement. It means that the displacement of the indenter δ and contact pressure $p(x)$ are predominantly determined by the deformation of asperities (see Section 2.4.1). Therefore, using the regular asymptotic expansions, we will try to find the problem solution in the form

$$p(x) = \sum_{l=0}^{\infty} \lambda^{-l} p_l(x), \mid x \mid \leq 1; \ \delta = \sum_{l=0}^{\infty} \lambda^{1-l} \delta_l. \qquad (3.74)$$

Substituting expansions (3.74) into equations (3.72) and (3.73) and expanding functions φ and ψ in powers of λ for the first terms of the asymptotic expansions we get

$$\varphi(p_0(x)) = \delta_0, \ \int_{-1}^{1} p_0(t)dt = \frac{\pi}{2}, \qquad (3.75)$$

$$\varphi'_p(p_0(x))p_1(x) + \varphi'_{\lambda-1}p_0(x) + \eta \int_{-1}^{x} \psi(p_0(t))dt$$

$$+\frac{2}{\pi} \int_{-1}^{1} p_0(t) \ln \mid \frac{1}{x-t} \mid dt = \delta_1 - f(x), \ \int_{-1}^{1} p_1(t)dt = 0, \dots \qquad (3.76)$$

Using the monotonicity of function φ from equations (3.75) and (3.76), we obtain

$$p_0(x) = \frac{\pi}{4}, \ \delta_0 = \varphi(\frac{\pi}{4}), \qquad (3.77)$$

$$p_1(x) = [\varphi'_p(\frac{\pi}{4})]^{-1} \Big\{ \frac{1}{2} - \ln 2 + \frac{1}{2} \int_{-1}^{1} f(t)dt - \eta\psi(\frac{\pi}{4})x$$

$$+\frac{1}{2}(1-x)\ln(1-x) + \frac{1}{2}(1+x)\ln(1+x) - f(x) \Big\}, \qquad (3.78)$$

$$\delta_1 = \frac{3}{2} - \ln 2 + \frac{1}{2} \int_{-1}^{1} f(t)dt + \varphi'_{\lambda-1}(\frac{\pi}{4}) + \eta\psi(\frac{\pi}{4}), \dots$$

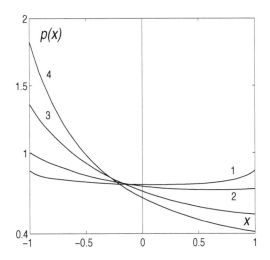

FIGURE 3.5
Pressure distribution $p(x)$ under a flat indenter for $\lambda = 10$, $\gamma_1 = 1$, $\gamma_2 = 0.5$, and different values of the friction coefficient η (curve 1: $\eta = 0$, curve 2: $\eta = 2$, curve 3: $\eta = 3$, curve 4: $\eta = 5$) (after Kudish [14]). Reprinted with permission from Springer.

Formulas (3.74), (3.77), and (3.78) determine the two-term asymptotic solution of the problem for $\lambda \gg 1$ uniformly valid on the interval $[-1, 1]$. It is obvious that in the case of a rough half-plane (contrary to the case of smooth half-plane, i.e., $\lambda = 0$) pressure $p(x)$ at the contact boundaries $x = \pm 1$ is bounded. On the other hand, for finite derivatives $f'(\pm 1)$ and $f''(\pm 1)$ we have unbounded $p'(x)$ and $p''(x)$ as $x \to \pm 1$. It means that for $\lambda \gg 1$ in the middle of the contact pressure varies relatively slowly while in the small vicinity of the contact boundaries $x = \pm 1$ pressure changes very quickly. In addition, the above formulas show that as the friction coefficient η increases the indenter displacement δ increases while pressure $p(x)$ increases in the direction of the friction force F and decreases in the opposite direction.

3.5.2 Some Numerical Solutions for Rough Contacts with Friction

Above we analyzed the qualitative behavior of contact problems with friction for rough solids. Now, let us consider quantitatively the solution of the contact problem with fixed boundaries.

For $\lambda \gg 1$ a solution of the system of equations (3.72) and (3.73) can be obtained by using the simple iteration approach. Let us introduce a set of nodes

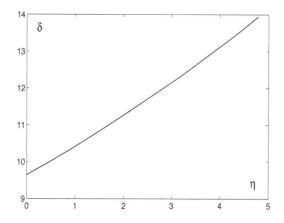

FIGURE 3.6
Dependence of the rigid indenter displacement δ on the friction coefficient η for $\lambda = 10$, $\gamma_1 = 1$, and $\gamma_2 = 0.5$ (after Kudish [14]). Reprinted with permission from Springer.

$\{x_i\}$, $i = 0, \ldots, N$, such that $x_0 = -1$ and $x_N = 1$ and new function $u(x)$ the values of which at nodes $\{x_i\}$ are u_i, $i = 0, \ldots, N$. Then the numerical scheme takes the following form

$$u_j^k = 1 - \tfrac{1}{\delta^k}\{f(x_j) - \tfrac{1}{\pi}\sum_{i=0}^{N-1}(p_i^k + p_{i+1}^k)[(x_{i+1} - x_j)$$

$$\times \ln|x_{i+1} - x_j| -(x_i - x_j)\ln|x_i - x_j| -x_{i+1} + x_i] \tag{3.79}$$

$$-\eta\sum_{i=0}^{j-1}\psi(p_{i+1/2}^k)(x_{i+1} - x_i)\}, \; j = 0, \ldots, N,$$

$$p_j^{k+1} = (1 - \zeta)p_j^k + \zeta\varphi^{-1}(\tfrac{\delta^{k+1}u_j^k}{\lambda}), \; j = 0, \ldots, N, \tag{3.80}$$

$$(1 - \zeta)\sum_{i=0}^{N-1}(p_i^k + p_{i+1}^k)(x_{i+1} - x_i) + \zeta\sum_{i=0}^{N-1}\left\{\varphi^{-1}(\tfrac{\delta^{k+1}u_i^k}{\lambda})\right.$$

$$\left. +\varphi^{-1}(\tfrac{\delta^{k+1}u_{i+1}^k}{\lambda})\right\}(x_{i+1} - x_i) = \pi. \tag{3.81}$$

In equations (3.79)-(3.81) superscripts k and $k + 1$ are the iteration numbers while the subscripts are the node numbers, $p_{i+1/2}^k = (p_i^k + p_{i+1}^k)/2$.

The iteration process (3.79)-(3.81) involves three steps. On the first step based on the known kth iterates δ^k and p_i^k, $i = 0, \ldots, N$, from equation

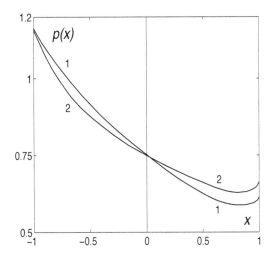

FIGURE 3.7
Pressure distribution $p(x)$ under a flat indenter obtained for $\eta = 2$, $\lambda = 10$, $\gamma_1 = 1$, $\gamma_2 = 0.5$ (curve 1 is based on the asymptotic solution, curve 2 is based on numerical solution of the problem in the original formulation).

(3.79) we determine the set of values u_i^k, $i = 0, \ldots, N$. Then solving equation (3.81) we determine the new iterate δ^{k+1}. On the third step the set of new iterates p_i^{k+1}, $i = 0, \ldots, N$, is determined from equation (3.80). The iteration process stops as the desired precision is reached.

As the initial approximations for the problem solution we can take

$$p_i^0 = \tfrac{\pi}{4}, \ i = 0, \ldots, N, \ \delta^0 = \lambda\varphi(\tfrac{\pi}{4}). \tag{3.82}$$

To illustrate the problem solution let us consider the case of an indenter with a flat bottom ($f(x) = const$) and angle $\beta = 0$. Moreover, we will assume that $\psi(p) = p^{\gamma_1}$, $\gamma_1 > 0$, and $\varphi(p) = p^{\gamma_2}$, $\gamma_2 > 0$. In Figure 3.5 graphs of pressure distribution $p(x)$ for rough contact with friction for $\lambda = 10$, $\gamma_1 = 1$, $\gamma_2 = 0.5$ and different values of the friction coefficient η are presented. Figure 3.5 shows that the pressure is finite in the contact region and its asymmetry increases for larger values of the friction coefficient η. Moreover, a series of simulations showed that variation of parameter γ_1 from 0.5 to 1 and same other parameters causes the problem solutions differ by no more than 2%. Figure 3.6 illustrates the dependence of the rigid indenter displacement δ as a function of the friction coefficient η. Obviously, this dependence of δ on η is close to a linear one and $\delta(\eta)$ grows monotonically with the increase of the friction coefficient η. In addition, simulations show that for larger values of the roughness coefficient the pressure distributions $p(x)$ get flatter while the indenter displacement δ

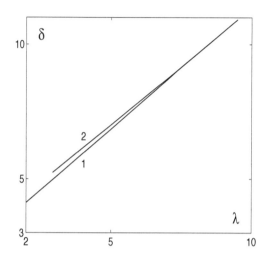

FIGURE 3.8
Dependence of the flat indenter displacement δ on the roughness parameter λ obtained for $\eta = 2$, $\gamma_1 = 1$, $\gamma_2 = 0.5$ (curve 1 is based on the asymptotic solution, curve 2 is based on numerical solution of the problem in the original formulation).

monotonically increases (see results of the preceding chapter for problems for rough contacts without friction).

For comparison of the asymptotic and numerical solutions graphs of pressure $p(x)$ under a flat indenter and indenter displacement as a function of λ are presented in Figures 3.7 and 3.8, respectively. The data for the graphs in both figures are obtained for $\gamma_1 = 1$, $\gamma_2 = 0.5$, and $\eta = 2$ from the two-term asymptotic formulas (3.74) (curves marked with 1), (3.77), and (3.78) and the direct solution of equations (3.72) and (3.73) (curves marked with 2). The pressure distributions in Fig. 3.7 are obtained for $\lambda = 10$. It is obvious from Fig. 3.7 and 3.8 that the asymptotic solution is close to the numerical one. The maximum relative difference between the pressure $p(x)$ and indenter displacement δ obtained asymptotically and numerically is about 7%.

3.6 Closure

Contact problems with nonlinear friction in the case of sliding are formulated and analyzed. The problem solution is analyzed for small friction coefficients analytically by using the regular and matched asymptotic expansions. Problems for fixed and free contact boundaries are considered. Some simple analytical formulas for the friction force and contact boundaries are obtained. The existence and uniqueness of a continuous solution of the contact problem for rough solids with nonlinear friction as well as some of its properties are established. A numerical method is proposed and realized for the case of contact problem for rough surfaces with friction. Some qualitative and quantitative observations of problem solutions are made.

3.7   Exercises and Problems

1. Obtain equations (3.11), (3.12), and (3.13)-(3.15) for problems with fixed (Contact Problem 1) and free (Contact Problem 2) boundaries, respectively.

2. Show that condition (3.20) provides for bounded ratio of $p_1(x)/p_0(x)$ and, therefore, for regular asymptotic expansion of the solution of Contact Problem 1 for $\eta \ll 1$.

3. Evaluate the two-term asymptotic representation for friction force F from formula (3.21) as a function of η for $\psi(p) = p^\gamma$, $\gamma = 0.5$ and $\gamma = 0.8$. Establish the value of η for which the term proportional to η^2 in (3.21) is smaller than the term proportional to η, which provides the range of applicability of this approximate expression. Compare the results to F from formula (3.36).

4. Using formulas (3.37)-(3.40) determine analytically the two-term asymptotic solution of Contact Problem 2 for $\psi(p) = p^\gamma$, $\gamma = 2$. Compare the results with formulas (3.52) and (3.53). Provide the same information as in Exercise 3 but for the friction force F determined by formula (3.41).

5. Provide the asymptotic and numerical solutions to the contact problem with free boundaries, friction, and high roughness (i.e., $\lambda \gg 1$) along the lines of Sections 3.5, 2.4, and 2.5.

BIBLIOGRAPHY

[1] Muskhelishvili, N.I. 1966. *Some Fundamental Problems of the Mathematical Theory of Elasticity*. Moscow: Nauka.

[2] Galin, L.A. 1980. *Contact Problems in the Theory of Elasticity and Viscoelasticity*. Moscow: Nauka.

[3] Alexandrov, V.M. and Kudish, I.I. 1981. Asymptotic Methods in Contact Problems with Nonlinear Friction. *J. Appl. Mech.* 17, No. 6:76-84.

[4] Van-Dyke, M. 1964. *Perturbation Methods in Fluid Mechanics*. New York-London: Academic Press.

[5] *Handbook on Mathematical Functions with Formulas, Graphs, and Mathematical Tables*, Eds. M. Abramowitz and I. Stegun, National Bureau of Standards, Applied Mathematics Series, 55, 1964.

[6] Shtaerman, I.Ya. 1949. *Contact Problems in Theory of Elasticity*. Moscow-Leningrad: Gastekhizdat.

[7] Kudish, I.I. 2000. Contact Interactions for an Elastic Half-Space with Rough Surface. *Dynamic Sys. and Appl.* 9, No. 2:229-246.

[8] Vorovich, I.I., Alexandrov, V.M., and Babesko, V.A. 1974. *Non − −Classic Mixed Problems in the Theory of Elasticity*. Moscow: Nauka.

[9] Lions, I.L. 1969. *Quelques Methodes de Revolution des Problemes aux Limites non Lineaires*. Paris: Dunod Gauthier-Villars.

[10] Mikhaylov, V.P. 1979. *Differential Equations in Partial Derivatives*. Moscow: Nauka.

[11] Giraud, G. 1932. Generalization des Problemes sur les Operations du Type Eliptique. *Bull. Sci. Math.* 56: 248-272, 281-312, 316-352.

[12] Courant, R. and Hilbert, D. 1989. *Methods of Mathematical Physics. Volume II. Partial Differential Equations*. New York: John Wiley and Sons.

[13] Bitzadze, A.V. 1966. *Boundary Problems for Elliptic Equations of the Second Order*. Moscow: Nauka.

[14] Kudish, I.I. 1987. Two-Dimensional Contact Problem on the Indentation of a Rough Bar by a Rigid Die with Friction. *Soviet Appl. Mech.* 23, No. 4:356-362.

4

Rheology of Lubricating Oils

4.1 Introduction

The transfer of power from its source of generation to its final destination where useful work occurs is the subject of much study and engineering development. In the case of internal combustion engines, liquid fuel is combusted by compression ignition (diesel) or spark ignition (gasoline). The gas expansion pressure within the combustion chamber drives a piston, which, in turn, translates reciprocal motion into rotary motion via a connecting rod that is attached to a crankshaft. Once the crankshaft is set in motion, pulleys, chains, clutches, and gears continue the chain of energy transfer to accomplish the ultimate purpose of the engine, be it transportation, pumping, electricity generation, or the like.

At every step of the energy transfer process, solid contacts are set in relative motion, and they often sustain considerable loads acting both normal and tangential to the direction of motion. Unless the contacts are separated by a lubricating film (solid, liquid or gas), a tremendous amount of frictional heat builds up in the contact zone that can lead to equipment failure in a relatively short period of time. Therefore, the lubricant is one of the most important components of an energy transfer device.

This monograph, in part, is devoted to selected performance attributes of liquid lubricants, particularly mineral oils and synthetic fluids. To fully describe the design features of modern lubricating oils is beyond the scope of this work. However, some elements need to be discussed to serve as the basis for further discussion.

Lubricating oils consist of base oils, which are selected on the basis of viscosity, thermal/oxidative stability, and cost. Chemical additives are also present and provide a number of useful functions. Viscosity modifiers (improvers) are oil-soluble polymers that impart useful rheological characteristics over a wide range of temperatures. Pour point depressants are added to prevent paraffin waxes (present in most refined mineral oils) from impeding oil flow at low temperatures. Dispersants, detergents, anti-wear agents, anti-oxidants, corrosion inhibitors, foam inhibitors, and friction modifiers do what their names imply.

Since viscosity is one of the most important design parameters of a loaded tribological contact, each application requires a lubricating oil with a particular set of rheological properties. To maintain proper lubrication under all operating conditions, the viscosity is required to meet stringent specifications at different temperatures, pressures, and shear rates. It should also be recognized that lubricants are subject to degradation during use. Exposure to high local temperatures, corrosive acids, unburned fuel, contaminants (liquids, solids, and gases), and severe mechanical forces can play havoc on the properties of a carefully designed lubricating oil. The useful life of the lubricant is often limited by its ability to maintain critical performance specifications during use.

4.2   Rheology Relationships for Lubricating Oils

4.2.1   Definition of Viscosity

The resistance to flow exerted by a fluid under the influence of external force is a phenomenological definition of viscosity. Mathematically, the dynamic viscosity μ is the proportionality constant between shear stress τ and shear rate γ otherwise known as Newton's law:

$$\tau = \mu\gamma. \tag{4.1}$$

Common units of measure of viscosity are summarized in Table 4.1.

TABLE 4.1
Viscosity units of measure.

	units	comments
Dynamic Viscosity	$mPa \cdot s$	SI unit
Dynamic Viscosity	cP	Centipoise, $1\ cP = 1\ mPa \cdot s$
Kinematic Viscosity	mm^2/s	SI unit
Kinematic Viscosity	cSt	Centistoke, $1\ cSt = 1\ mm^2/s$

Dynamic viscosity is calculated by independently measuring shear stress and shear rate. There are a large number of instruments, known collectively as viscometers, for measuring dynamic viscosity. In most cases, a fixed surface is opposed by a parallel surface moving at a certain velocity. Popular configurations include concentric cylinder and parallel plate devices. The surfaces are separated by a known gap, which is filled with the test lubricant. Shear stress is defined as the force required to maintain velocity divided by the surface area wetted by the lubricant. Shear rate, with units of s^{-1}, equals velocity divided by gap distance. In a constant rate viscometer, the velocity of the moving surface is set by the operator, and the torque on the fixed surface is measured. In a constant stress viscometer, the operator specifies the torque applied to the moving surface, and velocity is monitored.

Tribological contacts consist of two opposing surfaces in relative motion. One can be fixed or it can be moving at a different velocity or in a different direction than the other. Under a given set of conditions, all of the physical quantities relating to dynamic viscosity are correlated (force, contact area, velocity, and gap). Therefore, for all practical purposes, dynamic viscosity is the most appropriate parameter to use to characterize the load-carrying capacity of a lubricant.

Kinematic viscosity η is another common measure of viscosity, which is related to the dynamic viscosity μ as follows: $\mu = \rho\eta$, where ρ is the fluid density. The kinematic viscosity is determined by measuring the time for a fixed volume of fluid to flow through a vertical capillary under the influence of gravity. Consider two fluids of equal dynamic viscosity but of different densities. The force exerted by gravity on a unit volume of fluid is proportional to mass (or density, mass per unit volume). Therefore, the flow time through the capillary will be greater for the higher density fluid, and its kinematic viscosity will be lower than that of the lower density lubricant. To convert kinematic viscosity to dynamic viscosity, the former needs to be multiplied by lubricant density determined at the same temperature. For example, the kinematic viscosity and density of castor oil [20] at $40°C$ is 244 mm^2/s and 0.9464 g/cm^3 respectively. Dynamic viscosity equals 231 $mPa \cdot s$. Because of the low cost and simplicity of measuring kinematic viscosity, it is widely used to classify lubricating fluids into viscosity grades (such as SAE 5W-30 or ISO 46 for example). Its practical applicability in tribology, however, is limited to cases of freely draining contacts under low applied loads.

4.2.2 The Effect of Temperature on Viscosity

We all know from practical experience that a common way to lower fluid viscosity is to increase its temperature. Since viscosity is such a critical design parameter, it is important to be able to quantify its response to temperature. Although a number of theoretical and empirical mathematical models have been proposed for more than 100 years, the most common equation for lubricating oils was proposed by Walther [1, 2].

$$\log[\log(\eta + \theta)] = A - B \log T, \qquad (4.2)$$

where η is the kinematic viscosity, T is the absolute temperature in degrees Kelvin and A, B and θ are empirical constants (θ is usually between 0.6 mm^2/s and 0.8 mm^2/s and can be ignored for high viscosity fluids).

This equation is applicable over a limited temperature range. At low temperatures, paraffinic wax molecules (which are present in most mineral oil base stocks) nucleate and grow to form crystals. The crystals are often needle-like or plate-like having a high aspect ratio. The temperature at which wax begins to precipitate is known as the wax appearance temperature, also called the cloud point because the liquid takes on a turbid appearance [16]. The presence of this second phase causes the viscosity to increase (see Subsection 4.2.5) at a rate greater than predicted by extrapolation of (4.2) from higher temperatures. The upper temperature limit of the Walther equation is close to the initial boiling point. When the vapor pressure of the lowest molecular weight, most volatile fractions of the base oil approaches atmospheric pressure, microscopic gas bubbles begin forming. Just like wax crystals at low temperature, gas bubbles at high temperature form a second phase which affects viscosity.

Typical cloud point temperatures for API Group I and II base oils are $-20°C$ to $0°C$ as measured by ASTM Method D5771. The lowest boiling temperature of similar base stocks ranges from about $280°C$ to about $400°C$ as determined by ASTM D2887. Modern lubricating oils are rated for operation from about $-40°C$ to about $150°C$, depending upon viscosity grade. Thus, the Walther equation is more practically useful for describing the viscosity temperature relationship of mineral oil lubricants at high temperatures than at low temperatures.

It is common practice in the lubrication industry to measure kinematic viscosity at both $40°C$ and $100°C$. With two data points, constants A and B can be determined, and kinematic viscosity at any other temperature can be calculated. An example is shown in Fig. 4.1 for an SAE 15W-40 engine oil with kinematic viscosities of 14.76 mm^2/s and 108.3 mm^2/s at $100°C$ and $40°C$, respectively. By solving a system of two simultaneous equations (4.2) satisfied at the above two points, we get the values of the Walther constants A and B equal to 7.970 and 3.070, respectively, with θ assigned a value of 0.7 mm^2/s. Clearly, the measured viscosity in the cloud point region is substantially higher than its predicted value, whereas agreement is reasonably good at slightly higher sub-ambient temperatures. The percentage difference between measured and predicted kinematic viscosity data for four lubricating fluids at several temperatures is summarized in Table 4.2. At $0°C$, the measured viscosity agrees within 12% of the extrapolated value. In the cloud point region $(-20°C)$, the measured viscosity is more than double the predicted value. Similar trends are observed for the other fluids.

The lubrication industry adopted a term known as Viscosity Index (VI) to describe the response of viscosity to changes in temperature. It is an empirical quantity, which is calculated from kinematic viscosity values measured at $40°C$ and $100°C$ according to ASTM method D2270. The viscosity of high VI fluids, such as paraffinic oils, changes less with temperature than that of low VI fluids, such as naphthenic oils. Before the advent of instruments capable of measuring viscosity at low temperatures, VI was viewed as an indicator of flow behavior at sub-ambient temperatures. Today, VI is used primarily as a classification parameter for mineral and synthetic base oils [21].

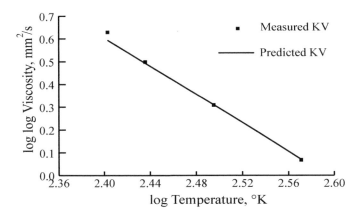

FIGURE 4.1

Measured vs. predicted kinematic viscosity for an SAE 15W-40 engine oil.

TABLE 4.2

Percent difference between measured and predicted kinematic viscosity η as a function of temperature for three engine oils and one hydraulic fluid.

Temperature, $^{\circ}C$	15W-40 (1)	15W-40 (2)	5W-30	ISO 32
100	0%	0%	0%	0%
40	0%	0%	0%	0%
0	12%	9%	8%	10%
−20	115%	83%	19%	−
−30	−	−	62%	77%

4.2.3   The Effect of Shear Rate on Viscosity

Many formulated fluids, like multi-grade engine oils, do not strictly follow Newton's law and are aptly called non-Newtonian lubricants. In most cases, viscosity decreases as shear rate increases (pseudoplastic behavior), but certain fluids exhibit dilatant flow in which viscosity increases with shear rate. For the purposes of this discussion, only pseudoplastic non-Newtonian fluids will be considered.

The viscosity-shear rate behavior of non-Newtonian fluids consists of three zones (see Fig. 4.2). At both very low and very high shear rates, viscosity appears to be independent of shear rate. These regions are often referred to as the lower and upper Newtonian plateaus, respectively. At intermediate ranges of shear rate, viscosity drops as shear rate increases, often following a

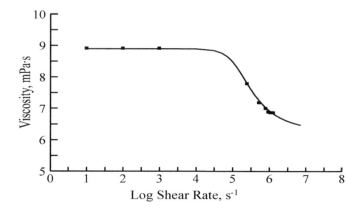

FIGURE 4.2
Shear rate dependence of viscosity (at $373°K$) for an SAE 5W-30 motor oil (data from Sorab et. al [4]).

simple power law relationship. This region is called the power law zone.

The phenomenological response of viscosity to shear rate can be understood in terms of simple molecular dynamics of a polymer dissolved in a solvent. At equilibrium, a polymer takes on a random coil configuration in solution. The volume occupied by the coil consists of both polymer and associated solvent. Einstein [2] first demonstrated that the relative viscosity of a fluid consisting of spheres suspended in a continuous fluid matrix increases in proportion to the volume fraction of the second phase (see equation 4.8). When the fluid is set in motion, the polymer coil begins to align in the direction of flow, thus reducing its apparent coil dimension in the plane normal to the flow direction. At very low flow rates, the polymer relaxation time (i.e., the characteristic time it takes to return to its equilibrium configuration when subjected to an external force) is faster than the rate at which the molecule is deformed in the flow field, and the coil size in solution is relatively unperturbed. As flow rate increases, the time scale of fluid deformation approaches that of the polymer relaxation time, defining the onset of the power-law zone. Eventually, at very high shear rates, the polymer chain is extended to its maximum length. This condition describes the onset of the upper Newtonian region.

Consider two fluids containing equal amounts by weight of ethylene-propylene copolymers A and B with weight-average molecular weights of 80,000 and 300,000, respectively (see Subsection 4.3.1 for a definitions of polymer molecular weight). The time required for a polymer to return to its equilibrium coil size after it is subjected to an external force is dependent upon the cooperative motions of the molecular bonds along the polymer chain. The greater number of bonds per chain, the longer it will take to relax. Thus, polymer B will have a longer relaxation time than polymer A. This means that the onset of power law region will occur at lower shear rates for polymer B than for polymer A. Another consequence of higher molecular weight is higher coil radius. Therefore, the lower Newtonian plateau viscosity of the polymer B fluid will be higher than for the polymer A fluid. The former will also undergo a greater degree of viscosity loss in the power law region because a higher molecular weight polymer in solution is more deformable than its lower molecular weight analogue (see Fig. 4.1).

4.2.4 The Effect of Pressure on Viscosity

Lubricating oils often experience very high pressures, especially in heavily loaded tribological contacts such as bearings and gears and in hydraulic power transmission systems. Pressures in ball and roller bearings have been reported to be up to $2 - 3\ GPa$ [5], $30 - 2000\ MPa$ in gasoline engine journal bearings [17], and as high as $69\ MPa$ in hydraulic systems [6, 7]. If lubricants were perfectly incompressible, viscosity would be unaffected by pressure. Because of the existence of free volume in the liquid state of most organic substances, lubricating oils are somewhat compressible, and the internal resistance to flow increases with pressure. Since elastohydrodynamic film thickness is governed, in large part, by viscosity, the opposing surfaces in a heavily loaded contact can be better supported by a lubricant under pressure than under ambient conditions.

TABLE 4.3
Pressure viscosity coefficients.

Lubricating fluid	Temperature ($^\circ C$)	$\alpha_p \times 10^5$ (KPa^{-1})	Reference
Diethyl-2 hexyl sebacate	0	1.40	[6]
	75	0.64	
Paraffinic mineral oil	0	2.00	[6]
VI = 99	75	1.30	
Paraffinic mineral oil	37.8	2.23	[6]
VI = 93	98.9	1.70	
Naphthenic mineral oil	37.8	3.60	[6]
VI = 30	98.9	2.13	
Polyisobutylene in paraffinic mineral oil	37.8	3.26	[7]
Polymethylmethacrylate in paraffinic mineral oil	37.8	2.45	[7]

To a first approximation, viscosity increases exponentially in proportion to pressure according to the Barus equation

$$\mu = \mu_a \exp(\alpha_p p), \tag{4.3}$$

where α_p is the pressure viscosity coefficient and μ_a is the viscosity at atmospheric pressure. The value of α_p depends upon the chemistry of the fluid as well as temperature. At very high pressures, viscosity increases at a slower rate than predicted, and it can be modeled by a power law relationship

$$\mu = \mu_a(1 + Cp)^n, \tag{4.4}$$

where C is a constant at a given temperature and n is equal to 16 [6] for most lubricating oils. Some examples of pressure viscosity coefficients reported for various lubricating oils may be found in Table 4.3

It is instructive to gain an order-of-magnitude appreciation for the extent to which viscosity increases under pressure. Using the pressure-viscosity coefficient data in Table 4.3 for the 93 VI mineral oil and the 30 VI naphthenic oils, the viscosity at $98.9°C$ increases by a factor of 67% and 90%, respectively, under the influence of 30 MPa pressure.

4.2.5   The Effect of Suspended Contaminants on Viscosity

Lubricating oils are formulated to deal with the reality that they will become contaminated with foreign substances during use. Particulate matter and aqueous acids generated by combustion of fossil fuels, unburned fuel, oxidative degradation of the lubricant, wear particles, and airborne particulates are common examples. Dispersants serve to keep these insoluble particles suspended in the oil and prevent them from forming deposits which can interfere with lubricant flow, heat transfer, and proper operation.

Under extreme conditions, heavy soot contamination of diesel engine oils can thicken the oil to such an extent that the oil cannot be drained out of the vehicle. To prevent this from occurring, a number of strategies have been found to be effective: change oil more frequently or improve the soot dispersing capability of the lubricant via additive design and formulation optimization. The additive design approach inhibits particles from forming large aggregates, which keeps the shape factor low (see equation (4.10)).

High operating temperatures can lead to thermal/oxidative degradation of the lubricating fluid. Introducing heteroatoms like oxygen and nitrogen into the base oil creates polar species, which are not soluble in most mineral oils or polyalphaolefin synthetics, often leading to a significant rise in viscosity. To enable lubricants to withstand high temperatures, certain chemical stabilizers such as anti-oxidants and detergents are added. Other approaches are to convert to more inherently-stable base fluids or increase oil change frequency.

Frequent filter inspection and replacement is an effective method for minimizing lubricant contamination with airborne particles such as dust.

Whether the contaminant is a solid or a liquid, it causes the viscosity of the fluid to increase if it is insoluble in the lubricating oil. The cumulative effect of insoluble contaminants on engine oil viscosity is illustrated by examining viscosity data from a Mack T-11 heavy duty diesel engine (see Fig. 4.3).

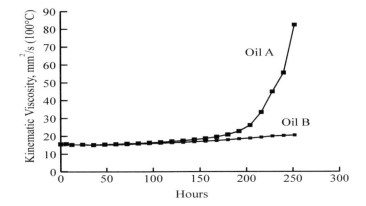

FIGURE 4.3
Kinematic viscosity $(100°C)$ of two SAE 15W-40 oils as a function of time in the Mack T-11 engine test.

Although the amount of insoluble soot present in Oils A and B at the end of the test was nearly equal (7.4% and 8.3%, respectively), the dispersant additives used in Oil B were far more effective at controlling viscosity rise than the additive package in Oil A.

Other factors, such as fuel dilution and polymer molecular weight degradation, are responsible for lowering the viscosity of engine oils in field service. Unburned fuel is soluble in the lubricant and, due to its low viscosity, depresses viscosity. High molecular weight viscosity index additives often undergo some level of molecular weight degradation, especially in the initial period of use. For example, an SAE 5W-30 engine oil formulated with a 35 SSI hydrogenated styrene-butadiene viscosity modifier (improvers) was run in a New York City taxi-cab fleet test [18], and kinematic viscosity was measured at $100°C$ on samples taken during the 8,000 mile oil drain period (see Fig. 4.4). The initial viscosity loss is attributed to polymer shear; but the steady accumulation of insoluble matter in the oil eventually counteracts the shearing effect, and viscosity begins to rise. An extensive study of polymer molecular weight degradation in heavy duty engine field service [19] confirmed that engine operation mechanically lowers polymer molecular weight as measured by gel permeation chromatography (GPC).

The problem of the effects of non-interacting suspended spherical particles on solution viscosity was first modeled by Einstein [8]

$$\mu_r = 1 + 2.5\varphi, \tag{4.5}$$

where μ_r is the relative viscosity, the ratio of solution dynamic viscosity to that of the solvent, φ is the volume fraction of second phase. This expression

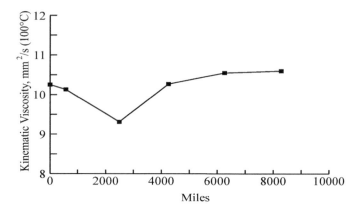

FIGURE 4.4
Kinematic viscosity $(100°C)$ of an SAE 5W-30 oil as a function of time in a
Chevrolet Caprice taxi equipped with a 4.3 liter V-6 engine.

was later modified by Brinkman [9] to

$$\mu_r = (1 - \varphi)^{-2.5}. \tag{4.6}$$

The main limitation to these models is the assumption that the dispersed
phase consists of non-interacting spheres. In reality, the morphology of diesel
engine soot is highly complex; and liquid contaminants, although generally
spherical, aggregate into clusters. Therefore, there have been a number of
successful attempts to model complex two-phase lubricating fluids by introducing a shape factor s into the Brinkman equation

$$\mu_r = (1 - s\varphi)^{-2.5}. \tag{4.7}$$

Four mathematical expressions that have been proposed for the shape factor
are summarized in Table 4.4.
Complex particle morphologies trap lubricating fluid within pores and between aggregates. If the particle is partially soluble in the oil phase, it might
be highly swollen. If the oil or additive molecules are attracted to the particle
surface, a tightly absorbed solvent layer is formed. In all cases, the effective
volume fraction of a particle is composed of two contributions: (1) the insoluble substance itself and (2) its associated oil. The latter factor can be ignored
in the case of suspended dense non-interacting hard spheres; but most contaminant particles are far from spherical, and they interact with the lubricating
oil. The shape factor in equation (4.10) is basically a correction factor, which
accounts for non-spherical particle morphology and associated oil. Numerically, it is dependent upon particle shape, state of aggregation, particle/oil
compatibility, temperature, and viscosity measurement conditions.

TABLE 4.4
Mathematical forms of the shape factor in equation (4.10), where ρ_s is the density of solid second phase, ρ_0 is the density of oil (continuous) phase, α is the mass fraction of solids in the swollen solid/oil particle, s_j and α are shape factor terms, p is the soot aggregate particle size, p_0 is the primary soot particle size. See references for complete definition of terms.

s	Application	Reference
$(\rho_s/(\alpha\rho_0)) - ((\rho_s/\rho_0) - 1)$	ASTM Sequence IIIC Gasoline Engine	[10]
α	diesel engine soot	[11]
s_j	clusters of uniform spheres	[12]
p^3/p_0^3	diesel engine soot	[13]

4.2.6 Density

To properly account for the volumetric and elastic response of lubricating oils to changes in temperature and pressure, it is necessary to introduce mathematical expressions for the influence of temperature and pressure on density ([22]):

$$\rho_i = \frac{\rho_0}{1+\beta(T_i-T_0)}, \tag{4.8}$$

where ρ_i is the final density, ρ_0 is the initial density, β is the volumetric temperature expansion coefficient ($63.3 \cdot 10^{-5}$ $°C^{-1}$ for oil), T_i is the final temperature ($°C$), and T_0 is the initial temperature ($°C$),

$$\rho_i = \frac{\rho_0}{1-(p_i-p_0)/E}, \tag{4.9}$$

where p_i is the final pressure (Pa), p_0 is the initial pressure (Pa), and E is the bulk modulus fluid elasticity coefficient (1.5 GPa for oil).

To illustrate that density, like viscosity, varies more with changes in temperature than pressure, equations (4.8) and (4.9) were used to calculate density at various temperatures and pressures for a mineral oil with $15°C$ density of 938 kg/m^3 (see Table 4.5).

TABLE 4.5
The influence of temperature and pressure on density (kg/m^3) of mineral oil.

$Pressure, MPa$	$15°C$	$75°C$	$125°C$	$200°C$
0.02	938.0	903.7	876.9	839.7
1.06	937.3	903.0	876.3	839.1
2.10	936.7	902.4	875.7	838.5
3.01	936.1	901.9	875.2	838.0

4.2.7 Thermal Conductivity

One of the most important functions of a lubricant is to remove heat from the tribological contact zone. High temperatures catalyze wear by lowering lubricant viscosity, which reduces oil film thickness that, in turn, shifts the wear regime from hydrodynamic to mixed to boundary. The ability of a lubricant, or any material for that matter, to remove heat is governed by Fourier's Law:

$$q = k\frac{dT}{ds}, \tag{4.10}$$

where q is the heat flux per unit area (measured in watts W/m^2), k is the coefficient of thermal conductivity ($W/(m°C)$), dT is the temperature difference across the contact ($°C$), and ds is the lubricant film thickness (m).

The value of k decreases linearly with temperature. A comparison of the cooling capacity of mineral oil and polyalphaolefin (PAO) synthetic fluids may be found in Table 4.6.

TABLE 4.6

Coefficient of thermal conductivity k as a function of temperature for an engine oil [24] and polyalphaolefin (PAO) synthetic fluids [23].

Temperature, $°C$	mineral oil	PAO 2	PAO 6	PAO 40
0	0.147	0.138	0.149	0.164
50	0.142	0.135	0.147	0.163
100	0.137	0.132	0.144	0.162
150	0.133	0.129	0.142	0.161

Another equation that can be used to estimate the coefficient of thermal conductivity k of petroleum fluids [23] relates k to density ρ at $15.6°C$:

$$k = \frac{A}{\rho}(1 - BT), \tag{4.11}$$

where k is in units of $W/(m \cdot° C)$, ρ is in units of kg/m^3, and T is the temperature in $°C$ while constants $A = 132\ W \cdot kg/(m^4 \cdot° C)$ and $B = 0.00125°C^{-1}$.

To estimate the effect of pressure on k, use equation (4.9) to compute ρ_i at the new pressure and substitute this value for ρ in equation (4.11).

4.3 Polymer Thickening and Shear Stability

4.3.1 Molecular Weight

As stated in Subsection 4.2.3, polymer molecular weight is a major factor influencing solution rheology. Before proceeding further, a few definitions are in order.

Polymers are chain-like molecules consisting of monomer units of molecular weight w_m linked together through a process known as polymerization consisting of the following steps: initiation, propagation, and termination. The number of monomer units per chain is known as the degree of polymerization (DP, l). A polymer can contain one monomer unit (such as polyethylene or polyisobutylene) or several (such as ethylene-propylene copolymers or polyalkylmethacrylates). One of the differences between a polymer and a pure organic compound is the definition of molecular weight. The molecular weight of the latter is calculated by adding up the atomic weights of the atoms in the molecule. For example, the molecular weight of isobutylene ($C4H8$) is 56.12 ($4 \times 12.01 + 8 \times 1.01$). The process of polymerization produces a statistical distribution of chain lengths. For convenience, consider a polymer consisting of a single monomer. The molecular weight w_i of each polymer chain i is $w_m l_i$, where l_i is the number of monomer units in chain i. Let n_i equal the number of chains of molecular weight w_i. Then the number-average and weight-average molecular weights of the polymer M_n and M_w, respectively, are defined as follows:

$$M_n = \sum_i w_i n_i / \sum_i n_i, \qquad (4.12)$$

$$M_w = \sum_i w_i^2 n_i / \sum_i w_i n_i, \qquad (4.13)$$

where the terms are summed over all values of l_i. The value of M_n correlates with the colligative properties of polymer solutions (vapor pressure lowering, freezing point elevation, osmotic pressure) and is useful for stoichiometric calculations (reactivity of chain ends, etc.). The rheological properties of polymer solutions are better correlated with M_w due in large part to the greater influence of high molecular weight species on bulk flow properties compared to lower molecular weight chains.

4.3.2 Dilute Solution Rheology

For many well-behaved polymer solutions, dilute solution viscosity is adequately described by the Huggins equation

$$\mu_{sp} = [\eta] + k'[\eta]^2 c, \qquad (4.14)$$

where μ_{sp} is the specific viscosity equal to $(\mu - \mu_0)/\mu_0$, c is the polymer concentration (g/dl), k' is Huggins constant, and $[\eta]$ is the intrinsic viscosity, related to molecular size at infinite dilution.

The intrinsic viscosity term $[\eta]$ is related to M_w by the Mark-Houwink relationship

$$[\eta] = k'M_w^a, \tag{4.15}$$

where k' and a are constants. Experimentally determined values of k' and a have been reported for ethylene-propylene copolymers [14] as well as for a number of other chemistries [15].

4.3.3   Shear Stability

It follows that large macromolecules are more efficient thickeners than low molecular weight polymers. If so, why not formulate multi-viscosity lubricants with the lowest concentration of the highest molecular weight polymers available for sale? The answer is simple. High molecular weight polymers are more susceptible to mechanical degradation than their lower molecular weight analogues. When a chain molecule passes through a tribological contact, it is elongated in the direction of flow. Shear and elongational forces are transmitted to every fluid element of the lubricant, and these forces are additive along the polymer backbone. The longer the backbone, the greater the force and the likelihood of polymer rupture. Polymer degradation can lead to lubricant viscosity dropping below its minimum design limit. In practice, the minimum design limit is understood to be the minimum viscosity of the lubricant's viscosity grade. For example, the kinematic viscosity of an SAE 5W-30 engine oil must be between 9.3 and 12.5 mm^2/s at $100°C$. If the viscosity of this oil stays within these limits over its useful life, it meets "stay-in-grade" performance. If the viscosity drops below 9.3 mm^2/s, several corrective actions can be taken. The oil can be formulated to a higher kinematic viscosity target, or a more shear-stable (lower molecular weight) viscosity modifier can be selected.

To maintain acceptable viscosity retention over the useful life of a lubricating oil, the polymeric viscosity modifier must be designed to strike a balance between thickening efficiency (for economic reasons) and shear stability. Chapters 5 - 7 are dedicated to further treatment of this problem.

To quantify the viscosity loss of lubricating oils during use, various industry working groups have developed laboratory methods for simulating shear degradation of polymer-containing fluids. Some lubrication applications impose greater levels of viscosity degradation than others and, therefore, demand the use of lower molecular weight polymer additives. A brief summary of some of the most common shear stability tests may be found in Table 4.7.

When a lubricant is tested in the field or in one of the laboratory tests in Table 4.7, the viscosity change can be expressed in terms of percentage of viscosity loss or SSI (shear stability index). The latter is a measure of the amount of viscosity contributed by the viscosity modified that is lost by mechanical degradation and is computed as follows:

$$SSI = \frac{NOV - SOV}{NOV - BBV} \times 100,$$ (4.16)

where NOV is the new oil viscosity, SOV is the sheared oil viscosity, and BBV is the base blend viscosity (base oil(s) + performance additive(s) without the viscosity modifier present), all measured at the same temperature.

Since the tendency of different field conditions and laboratory tests to degrade polymer molecular weight varies widely, it is important that the correct SSI value is used in a given application. Consider, for example, a single viscosity modifier that is used to formulate a passenger car motor oil, a heavy duty diesel engine oil, and an automotive gear oil. The lubricants are tested in field service; and NOV, SOV, and BBV are measured. SSI values of 15, 25, and 80 are calculated. This illustrates that there is not a unique shear stability index value that can be ascribed to a given viscosity modifier additive. The test conditions used to calculate SSI must be clearly understood to properly match a polymer to a given lubricant application.

TABLE 4.7
Laboratory tests that simulate viscosity loss due to mechanical shear.

Test	Test method	Application	Relative tendency to affect mechanical degradation
Kurt Orbahn diesel injector rig	ASTM D6278 ASTM D7109 CEC L-14	compression ignition engine oils	moderate
Sequence VIII spark ignition engine	ASTM D6709	spark ignition engine oils	mildest
KRL tapered bearing rig	CEC L-45 DIN 51 350 VW 1437	transmission, gear, and hydraulic system oils	most severe
Sonic shear	ASTM D2603 ATM D5621 JAASO M347	engine, hear, transmission, and hydraulic system oils	mild to severe
FZG gear rig	CEC L-37	gear oils	severe

Usually, the characteristic SSI associated with a viscosity modifier refers to ASTM method D6278 (30 cycles), but it is always wise to confirm. A subtle but important aspect of SSI calculations is to recognize that the only component that contributes to viscosity loss is the polymer itself. Many viscosity modifiers (improvers) are sold as viscous mixtures of polymers dissolved in mineral oil. The oil component needs to be factored into determination of the base blend viscosity (BBV) in equation (4.16) in order to calculate the true SSI.

4.4 Closure

The concept of viscosity is introduced and shown to be sensitive to temperature and pressure. Classical empirical formulas for modeling these relationships are provided. For fluids containing polymeric additives, viscosity becomes dependent upon shear rate as well. Polymer molecular weight averages are defined and offered as useful measures of thickening efficiency and shear stability. Rheological aspects of lubricant degradation are presented, and mechanisms for viscosity change in automotive and industrial lubricants are discussed.

4.5   Exercises and Problems

1. A Newtonian lubricating fluid having a viscosity of 46.8 $mPa \cdot s$ at $40°C$ fills the gap between two large parallel plates, the upper plate moving at a velocity of 2.16 m/s relative to the fixed lower plate. The shear stress required to maintain this velocity is 123.3 Pa. What is the distance between the plates?

Answer: Shear rate = velocity / gap. Gap = 0.82 mm.

2. A Newtonian transmission oil designed to operate at temperatures less than $85°C$ was used in a new aerodynamic transmission prototype that developed oil temperatures as high as $114°C$. The viscosity of this fluid at $25°C$ and $80°C$ is 66.6 and 11.0 mm^2/s respectively. What is the lowest viscosity this oil is expected to have in the prototype transmission? Assume $\theta = 0.7$.

Answer: Applying equation 2.2.2, and solving simultaneous equations results in $A = 8.1015$ and $B = 3.168$. The viscosity at $114°C$ is equal to 5.6 mm^2/s

3. A rookie tribologist measures the viscosity of an SAE 20W-50 racing oil at two different shear rates. The results were identical. He concludes that the fluid does not contain polymer additives. Why is his conclusion faulty?

Answer: He could have taken two measurements in the lower Newtonian regime (see Fig. 4.2) where a polymer-containing fluid appears to be Newtonian.

4. The polyisobutylene solution in Table 4.3 has a viscosity of 84.3 $mPa \cdot s$ at $100°F$. At the same temperature, what percentage increase in viscosity would be expected at a pressure of 50 MPa?

Answer: Applying equation (4.3) and $\alpha = 3.26 \cdot 10^{-5}$ from Table 4.3, the answer is 410.

5. An engineer examines an oil sample taken from a diesel locomotive. The engine was originally filled with 15.10 mm^2/s $(100°C)$ RightTrack II SAE 15W-40 oil. The used oil sample is black in color, and the lab reported that it contains 2.6% soot. There is virtually no dissolved fuel in the sample, yet the kinematic viscosity is unchanged. Provide at least one possible explanation why viscosity did not change.

Answer: (1) Viscosity loss due to polymer degradation exactly balanced viscosity rise due to soot contamination or (2) the oil contained a low molecular weight, highly shear-stable viscosity modifier, but the dispersant additives in the oil prevented significant viscosity increase (as in Oil B in Fig. 4.3).

6. What is the weight-average molecular weight of an A-B block copolymer molecule where the A-block consists of 320 styrene ($C8H8$) monomer units and the B-block consists of 1000 dodecylmethacrylate ($C16O2H30$) monomer units?

Answer: The molecular weight of styrene is 104.08 and dodecylmethacrylate is 254.16. The molecular weight of the block copolymer is 287,000. In this case, we are dealing with a single molecule where M_n and M_w are, by definition, equal.

7. You are asked to formulate a stay-in-grade SAE 10W-40 engine oil with a 35 SSI viscosity modifier. Given the $100°C$ kinematic viscosity of the oil (base oils and performance additive) without viscosity modifier is 4.64 mm^2/s and the viscosity range of this viscosity grade is 12.5 $mm^2/s - 16.3$ mm^2/s, what is the minimum new oil viscosity required to stay in grade? Is it feasible?

Answer: 16.7 mm^2/s. Since this target viscosity is greater than the maximum for this viscosity grade, it is not feasible to satisfy design requirements for this lubricant.

BIBLIOGRAPHY

[1] Walther, C. 1931. The Evaluation of Viscosity Data. *Erdol u. Teer* 7:382-384.

[2] Standard Test Method for Viscosity-Temperature Charts for Liquid Petroleum Products. 2004. *ASTM D*341 − −03, ASTM International, West Conshohocken, PA, USA, 1-5.

[3] A. Einstein, A. 1911. Correction to my paper "Eine neue Bestimmung der Molekldimensionen" or "A New Determination of Molecular Dimensions. *Ann. Phys. (Leipzig)* 34:591-592.

[4] Sorab, J., Holdeman, H.A., and Chui, G.K. 1993. Viscosity Prediction for Multigrade Oils. In *Tribological Insights & Performance Characteristics*, M.J. Covitch and S.C. Tung, eds. *Soc. Automotive Engr.* SP-966:241-252.

[5] E.V. Zaretsky, ed. 1997. *Tribology for Aerospace Applications*. STLE Publication SP, p. 37.

[6] Briant, J., Denis, J., and Parc, G., eds. 1989. *Rheological Properties of Lubricants*, Paris: Editions Technip.:115-120.

[7] So, B.Y.C. and Klaus, E.E. 1980. Viscosity-Pressure Correlation of Liquids, *ASLE Trans.* 23, No. 4:409-421.

[8] Einstein, A. 1906. "Eine neue Bestimmung der Molekldimensionen" or "A New Determination of Molecular Dimensions". *Ann. Phys.* 19:289-306.

[9] Brinkman, H.C. 1952. The Viscosity of Concentrated Suspensions and Solutions. *J. Chem. Phys.* 20:571.

[10] Spearot, J.A. 1974. Viscosity of Severely Oxidized Engine Oil. *Am. Chem. Soc. Div. Petrol. Chem.*:598-619.

[11] Yasutomi, S., Maeda, Y., and Maeda, T. 1981. Kinetic Approach to Engine Oil. 3. Increase in Viscosity of Diesel Engine Oil Caused by Soot Contamination. *Ind. Eng. Chem. Prod. Res. Div.* 20:540-544.

[12] Graham, A.L., Steele, R.D., and Bird, R.B. 1984. Particle Clusters in Concentrated Suspensions. 3. Prediction of Suspension Viscosity. *Ind. Eng. Chem. Fundam.* 23:420-425.

[13] Covitch, M.J., Humphrey, B.K., and Ripple, D.E. 1985. Oil Thickening in the Mack T-7 Engine Test - Fuel Effects and the Influence of Lubricant Additives on Soot Aggregation. *Soc. Automot. Eng. Tech. Paper Ser.* No. 852126.

[14] Crespi, G., Valvassori, A., and Flisi, U. 1977. Olefin Copolymers. *Stereo Rubbers*:365-431.

[15] Brandrup, J. and Immergut, E.H. 1975. *Polymer Handbook*, 2nd ed., Section IV, pp. 1-33. New York: John Wiley & Sons.

[16] Webber, R.M., George, H.F., and Covitch, M.J. 2000. Physical Processes Associated with Low Temperature Mineral Oil Rheology: Why the Gelation Index is Not Necessarily a Relative Measure of Gelation. *Soc. Automot. Eng. Tech. Paper Ser.* No. 2000-01-1806.

[17] Bates, T.W. 1990. Oil Rheology and Journal Bearing Performance: A Review. *Lub. Sci.* 2:157-176.

[18] Covitch, M.J., Weiss, J., and Kreutzer, I.M. 1999. Low-Temperature Rheology of Engine Lubricants Subjected to Mechanical Shear: Viscosity Modifier Effects. *Lub. Sci.* 11-4:337-364.

[19] Covitch, M.J., Wright, S.L., Schober, B.J., McGeehan, J.A., and Couch, M. 2003. Mechanical Degradation of Viscosity modifiers in Heavy Duty Diesel Engine Lubricants in Field Service. *Soc. Automot. Eng. Tech. Paper Ser.* No. 2003-01-3223.

[20] Forsythe, W.E., ed. 2003. *Smithsonian Physical Tables, Nineth Rev. Ed.*, p. 322, Knovel, Norwich, New York: Smithsonian Institution.

[21] April 2008. Appendix E - Base Oil Interchangeability Guidelines for Passenger Car Motor Oils and Diesel Engine Oils. *Amer. Petrol. Inst.*, Section E.1.3.

[22] Density of Fluids - Changing Pressure and Temperature. 2009. *The Engineering Toolbox*. www.EngineeringToolBox.com

[23] Booser, E.R., ed. 1997. *Tribology Data Handbook*. New York: CRC Press, p. 39.

[24] Hamrock, B.J. Lubricant Properties. 1994. *Fundamentals of Fluid Friction Lubrication*. New York: McGraw-Hill, 64-65.

5

Properties of Multi-grade Lubricating Oils

5.1 Introduction

In Chapter 4, polymeric viscosity modifiers (improvers) were shown to lower the dependence of lubricant viscosity on temperature, thus enabling the development of multi-grade lubricants. A practical consequence of formulating with high molecular weight polymers is the expectation that a certain amount of permanent viscosity loss will occur during use, depending upon the polymer's shear stability index. As long as the fluid maintains a minimum level of viscosity over its expected life, the tribological system should be adequately protected against premature wear, assuming the other performance additives remain active. Thus, degradation of polymer molecular weight is an important subject; and the development of mathematical models of polymer degradation is presented in Chapters 6 and 7.

This chapter serves as an introduction to the subsequent chapters by first defining what is meant by the term "viscosity grade." A discussion of several viscosity grading systems is presented, which defines the rheological parameters that distinguish one viscosity grade from another. The historical development of multi-grade lubricating oils is briefly reviewed in the context of industry trends to improve performance and increase energy efficiency.

The second half of this chapter is devoted to a discussion of the relationship between polymer architecture and molecular weight degradation. Although a thorough treatise on this subject is beyond the scope of this monograph, an overview of the most commercially significant classes of viscosity modifiers is presented. This understanding will serve as a segue to subsequent chapters of this book that discuss the degradation mechanisms of linear and star polymers in detail.

5.2 Multi-grade Lubricating Oils

5.2.1 Engine Startability at Low Temperature

Prior to 1955, engine oils did not contain viscosity modifiers. Known as mono-grade lubricants, they were categorized according to high-temperature viscosity; seven SAE* grades (from 10 to 70) were defined by viscosities measured at temperatures from $130°F$ to $210°F$, depending upon viscosity grade [1, 2]. Various attempts to assign "winter" grades such as SAE 10W and 20W were made, recognizing the need to ensure good starting performance in cold weather. The ability of the crankshaft of an engine to begin rotating is highly influenced by the thin lubricating oil film residing in the bearings. If the oil viscosity is too high, the starting motor is torque-limited, and the engine does not turn over. If the oil viscosity is low enough, the crankshaft turns and rotates at high speed, imposing a high shear rate on the thin lubricating film.

When polymeric additives (improvers) began to be introduced to improve low temperature engine start-ability, it was common practice to extrapolate kinematic viscosity data measured at $100°F$ and $210°F$ to $0°F$. However, it became clear that extrapolated viscosities did not match experimental cold cranking (starting) performance of engines at $0°F$ (see, for example, Figure 4.1). A new low temperature, high shear rate viscosity test was needed, and an ASTM[†] working group was formed to develop what came to be known as the Cold Cranking Simulator (CCS) test, ASTM D2602-67T. Improvements in the method over the years led to the current method, ASTM D5293. Various experimental studies confirmed that this relatively simple and inexpensive test correlated well with low temperature start-ability studies of passenger car engines [3, 4]. The term "Multi-Viscosity" was coined in the mid 1950s to refer to oils that can be identified by both a "W" grade and a high temperature viscosity grade, later changed to "multi-grade." These versatile engine oils, enabled by viscosity modifiers, were truly all-season lubricants. No longer did the vehicle owner have to schedule seasonal oil changes.

The first version of the current Engine Oil Viscosity Classification standard, SAE J300a, was published in 1967 and included both CCS minimum and maximum limits at $0°F$ and kinematic viscosity specifications at $210°F$. The SAE 10, 60, and 70 grades were dropped, and only 5W, 10W, and 20W low temperature grades were included. The 15W grade was added in 1975.

---

*Society of Automotive Engineers
[†]American Society for Testing and Materials

5.2.2 Engine Oil Pumpability at Low Temperature

The next significant change occurred in 1974 when it was formally recognized within SAE J300b that oil pumpability was another important performance feature of multi-grade oils. The engine lubricant must flow through a screen covering the oil inlet tube of the oil pump. If the oil becomes too viscous or gelatinous at low temperatures, its ability to be drawn into the pump and pumped throughout the engine becomes compromised. Whereas cranking imparts shear rates in excess of $10,000 \ s^{-1}$ to the oil, the shear rates involved in pumping are three orders of magnitude lower [2]. Therefore, a different low temperature rheology test was needed, and an ASTM subcommittee was formed in 1971 [5, 6] to explore this issue. Based upon oil pumpability failure reports and laboratory studies, it became clear that engine design, oil wax content, and lubricant additives influence pumpability. Two types of failures were identified: flow-limited and air-binding. When lubricant viscosity is excessively high, it doesn't flow at a sufficient rate to properly protect the piston ring belt and valve train regions; this represents the flow-limited condition. The air-binding mode occurs when the oil sets up as a gelatinous mass at low temperature, and the oil does not flow at all. The oil pump cavitates, and oil starvation can lead to premature engine failure.

The first pumping viscosity test adopted by the lubricant industry was called the Borderline Pumping Temperature test (BPT, ASTM D3829), which was adopted in 1979. It utilized a relatively simple device known as the Mini-Rotary Viscometer (MRV), which measured both yield stress and viscosity on sample oils cooled according to a specified temperature program. Yield stress (see Chapter 4) was found to correlate to air-binding failure, and viscosity was a good predictor of flow-limited conditions.

The 1980 revision of SAE J300 incorporated the CCS and MRV methods into the engine oil viscosity classification standard [2], specifying maximum viscosities and BPT values for each "W" grade. This document also added two more "W" grades (0W and 25W) and served as the model for subsequent versions of J300.

Shortly afterward, oil pumpability failures began to surface during the winters of 1980 and 1981 in the northern United States. The oils implicated in these failures passed the BPT test, and an industry study was commissioned to investigate. It was discovered that the rate at which the oil is cooled through the wax crystallization region highly influences both yield stress and viscosity. By changing the cooling program to the so-called TP1 cycle, a good correlation between laboratory results and oil pumpability in the field was finally established. This new version of the MRV test was called ASTM D4684 and inserted into J300 in 1986-1989. It includes maximum viscosity limits for each W grade and protection against air-binding failure by imposing a maximum 35 Pa yield stress.

5.2.3   Bearing Protection

The next evolution of the engine oil viscosity classification system began in 1977 when SAE recognized the need to protect automotive bearings and piston rings from excessive wear, particularly as the industry was developing lower viscosity engine oils to improve vehicle fuel economy [7]. An ASTM task force was assembled shortly thereafter to "develop a test method which measures high temperature, high shear (HTHS) viscosity." What emerged from this work were three instruments and test methods that were eventually incorporated into SAE J300 in 1991 (see Table 5.1). They all measure viscosity under the same conditions of temperature and average shear rate: $150°C$ and $10^6 \ s^{-1}$.

TABLE 5.1
ASTM methods to measure high temperature high shear rate viscosity of engine oils.

ASTM Method	Description
D4683	Tapered Bearing Simulator
D4741	Ravenfield
D5481	Cannon High-Pressure Capillary Rheometer

With the input of engine manufacturers, minimum high temperature high shear (HTHS) viscosities were assigned to each high temperature SAE viscosity grade to protect bearings and rings from excessive wear.

5.2.4   Engine Oil Viscosity Classification SAE J300

With an understanding of the historical development of the SAE Engine Oil Viscosity Classification System, it is instructive to present several examples of how to use this standard, summarized in Table 5.2. In these examples, all kinematic viscosity values are reported at $100°C$.

Example 5.1. SAE 30 is an example of a mono-grade oil that, by convention, does not contain a polymeric viscosity modifier. It has no low temperature viscosity specifications but must conform to both kinematic viscosity and HTHS limits. A lubricant with measured viscosities of 10.45 mm^2/s and 3.1 $mPa \cdot s$ HTHS would qualify as SAE 30.

Example 5.2. An SAE 5W-30 multi-grade oil must satisfy low temperature cranking and pumping limits in addition to the high temperature rheological requirements of an SAE 30 mono-grade. Assume that this oil has the same high temperature viscosity values as in *Example* 5.1. As long as the pumping viscosity at $-35°C$ is less than 60,000 $mPa \cdot s$ and the starting viscosity is less than 6,600 $mPa \cdot s$ at $-30°C$, this oil *might* qualify as a 5W-30 oil. If the CCS viscosity is in the lower half of the range, however, it might qualify as a 0W-30 oil. Therefore, it is often necessary to measure CCS and MRV at several temperatures to determine the correct W grade. SAE J300 labeling requirements state that "only the lowest W grade satisfied [by both CCS and MRV] may be referred to on the label." In other words, multiple viscosity grades cannot be assigned to a single engine oil.

Example 5.3. An oil analysis report on a synthetic SAE 15W-40 lubricant contains the following information: kinematic viscosity is equal to $15.23\ mm^2/s$; HTHS is equal to $3.8\ mPa \cdot s$; CCS is equal to $3,260\ mPa \cdot s$ at $-20°C$; MRV is equal to $12,400\ mPa \cdot s$ at $-25°C$ and yield stress $< 35\ Pa$. Since the CCS value was in the lower half of the range for a 15W oil, the CCS and MRV tests were re-run at $-25°C$ and $-30°C$, respectively, with the following results: CCS is equal to $5,705\ mPa \cdot s$ and MRV is equal to $28,600\ mPa \cdot s$ and yield stress $< 35\ Pa$. Conclusion: this oil should have been labeled SAE 10W-40.

TABLE 5.2

Engine Oil Viscosity Classification System, SAE J300 Rev. JAN2009. *
For 0W-40, 5W-40, and 10W-40 grades, ** for 15W-40, 20W-40, 25W-40, 40 grades.

SAE Viscosity Grade	Low-Temperature ($°C$) Cranking Viscosity $mPa \cdot s$ Max	Low-Temperature ($°C$) Pumping Viscosity $mPa \cdot s$ Max with No Yield Stress	Low-Shear-Rate Kinematic Viscosity (mm^2/s) at $100°C$ Min/Max	High-Shear-Rate Viscosity $mPa \cdot s$ at $150°C$ Min
0W	6,200 at −35	60,000 at −40	3.8/none	−
5W	6,600 at −30	60,000 at −35	3.8/none	−
10W	7,000 at −25	60,000 at −30	4.1/none	−
15W	7,000 at −20	60,000 at −25	5.6/none	−
20W	9,500 at −15	60,000 at −2	5.6/none	−
25W	13,000 at −10	60,000 at −15	9.3/none	−
20	−	−	5.6/< 9.3	2.6
30	−	−	9.3/< 12.5	2.9
40	−	−	12.5/< 16.3	3.5*
40	−	−	12.5/< 16.3	3.7**
50	−	−	16.3/< 21.9	3.7
60	−	−	21.9/< 26.1	3.7

5.2.5 Automotive Gear Oil Viscosity Classification SAE J306

Another major viscosity grade classification system, SAE J306, was developed for automotive gear oils (see Table 5.3). It is structured similarly to SAE J300 in that it comprises both high temperature kinematic viscosity limits and low temperature viscosity specifications. There are two main differences, however. First, Brookfield viscosity (ASTM method D2983) is used to measure low temperature viscosity. It is a low shear rate rheological technique based upon the Brookfield constant rate viscometer. Yield stress is not measured. Second, J306 requires the gear lubricant to be sufficiently shear stable to stay in grade based upon the 20-hour KRL tapered bearing shear test (see Chapter 4, Table 4.5). There is no corresponding requirement in SAE J300 for engine oils, although many original equipment manufacturers impose stay in grade criteria for crankcase lubricants as well.

TABLE 5.3
Automotive Gear Lubricant Viscosity Classification Standard, SAE J306 Rev. JUN2005.

SAE Viscosity Grade	Max. Temperature for D2983 Viscosity of 150,000 $mPa \cdot s$ ($\circ C$)	Min. Kinematic Viscosity at $100°C$	Max. Kinematic Viscosity at $100°C$
70W	−55	4.1	−
75W	−40	4.1	−
80W	−26	7.0	−
85W	−12	11.0	−
80	−	7.0	< 11.0
85	−	11.0	< 13.5
90	−	13.5	< 18.5
110	−	18.5	< 24.0
140	−	24.0	< 32.5
190	−	32.5	< 41.0
250	−	41.0	−

Example 5.4. An automotive gear oil has the following rheological characteristics: kinematic viscosity = 15.7 mm^2/s at $100°C$; sheared kinematic viscosity = 13.9 mm^2/s at $100°C$; Brookfield viscosity at $-40°C$ is equal to 123,000 $mPa \cdot s$. This gear oil qualifies as SAE 75W-90.

Example 5.5. If the gear oil of *Example* 5.4 were to shear out of grade in the 20-hour KRL test to 12.9 mm^2/s, it would have to be re-formulated to a higher new-oil kinematic viscosity, or a more shear stable viscosity modifier (lower SSI) would have to be substituted for the original polymer. Otherwise, it would be labeled as an SAE 75W oil.

5.3 Viscosity Modifiers

The role of viscosity improving polymers in multi-grade lubricants was introduced in Chapter 4. In this section, the various chemical types of polymers most often used in modern lubricating fluids are discussed with comments about the advantages and deficiencies of each. It is beyond the scope of this monograph to cover the chemistry in detail; it will be limited to fundamental repeat unit chemical structures.

Another way to characterize polymers is by architecture. A linear polymer is similar in structure to a single length of rope. If it contains more than one monomer, it is called a copolymer. Within the set of linear copolymers, there are several classes worth mentioning. Random linear copolymers consist of monomers A and B, which are distributed randomly along the polymer chain; the chance of encountering monomer A is the same at any chain position. Tapered linear block copolymers display a gradient in monomer concentration, with monomer A being more abundant at one end of the chain than at the other. A linear block copolymer has discrete sequences of pure monomer A and pure monomer B. Although both di-block and tri-block linear copolymers are commercially available, only di-block copolymers are used as viscosity improvers. The tri-block structure favors network formation that, at high enough concentrations, can adversely affect lubricant flow.

Another aspect of chain configuration that affects rheological properties is branching. Side-chain branching is caused by monomers that, when polymerized, contribute chemical branches that protrude away from the main chain at every monomer position. These branches can consist of one carbon atom as is the case with propylene to 20 or more carbon atoms in alkylmethacrylates. Another type of branching is long-chain branching, which resembles the branch of a tree. From the main linear chain, other chains branch out from random positions. Long–chain branching can be built into a polymer deliberately, or it can occur as a side effect of the polymerization process.

Star or radial copolymers are a subset of long-chain branching. Relatively long–chain "arms" containing one reactive end group are coupled to form a tight core from which the arms radiate. The arms can consist of one or more monomers arranged randomly or otherwise (tapered, block, etc.). Star polymers can also consist of a mixture of arm structures. Rheological properties are affected by the average number of arms per star, monomer chemistry and distribution within the arms, and arm molecular weight.

5.3.1 Olefin Copolymer (OCP) Viscosity Modifiers

Olefin copolymer viscosity modifiers are made from ethylene and propylene [8], among the highest volume and lowest cost petrochemical feedstocks in the polymer industry. The basic chemical structure of OCP viscosity modifiers

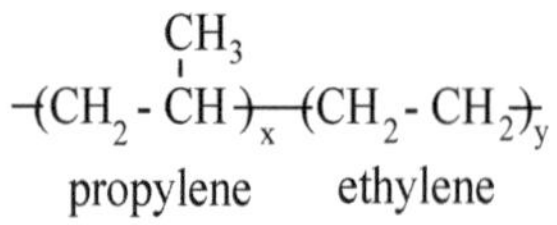

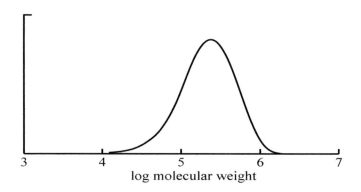

CH$_3$

$\text{-(CH}_2\text{-}\overset{|}{\text{CH}}\text{)}_{\overline{x}}\text{(CH}_2\text{-}\text{CH}_2\text{)}_{\overline{y}}$

propylene   ethylene

FIGURE 5.1
Chemical structure of OCP Viscosity modifiers.

log molecular weight

FIGURE 5.2
Molecular weight distribution for OCP viscosity modifier.

(also known as EP copolymers) is shown in Fig. 5.1.

Shear stability is controlled by adjusting molecular weight, molecular weight distribution and, to a lesser degree, ethylene content [8].

High molecular weight olefin copolymers command the highest share of the engine oil viscosity improver market due to their good rheological properties and favorable cost position. Ultra low molecular weight, shear-stable versions are used in automotive gear oils.

The relative ratio of ethylene and propylene can have a profound effect on lubricant rheology. When the ethylene content is less than about 55 weight %, the polymer is completely amorphous, meaning that it is 100% non-crystalline. Introducing a controlled amount of crystallinity by increasing ethylene content can improve low temperature rheology. The mechanism involves intra-molecular association of crystalline chain segments at low temperatures, which leads to shrinkage of the polymer coil in solution [10]. Since viscosity is proportional to average polymer coil diameter, low temperature flow is often improved, but not always. Inter-molecular crystallization can occur, particularly when polymer concentration is high, and it is not easily broken up when

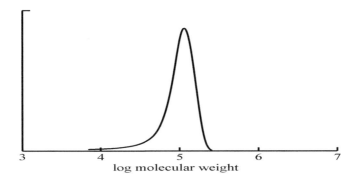

FIGURE 5.3

Molecular weight distribution of the OCP polymer of Fig. 5.2 after 250 cycles in the Orbahn degradation test (ASTM D6278).

diluted with oil. These temporary cross-links can cause gelation to occur at low temperatures, leading to oil pumping problems and unusual flow characteristics [12]. Ver Strate and coworkers [11] created a blocky OCP molecule with a high ethylene block in the middle and amorphous blocks at the ends of the chain that is claimed to reduce inter-molecular crystallization. Another type of crystallization can also take place at low temperatures. Paraffinic base oil wax can co-crystallize with long ethylene sequences of the OCP with negative rheological consequences, especially regarding oil pumpability. In many cases, the oil can be formulated with an effective pour point depressant (PPD) to correct this problem, but not always.

The molecular weight distribution of a typical OCP viscosity modifier measured by Gel Permeation Chromatography (GPC) is shown in Fig. 5.2. The ratio of M_w/M_n is typically 2.0 - 2.8. The shape of this GPC curve is similar to molecular weight distributions of other linear polymers, although the M_w/M_n ratios might be different. When a linear polymer is subjected to mechanical degradation (ASTM method D6278), the high molecular weight molecules are broken down preferentially to the lower molecular weight chains, and the distribution narrows and become less symmetric (see Fig. 5.3).

A mathematical treatment of the mechanical degradation process of OCP viscosity modifiers (improvers) is developed in Chapter 6.

5.3.2 Polyalkylmethacrylate (PMA) Viscosity Modifiers

Polyalkylmethacrylate viscosity modifiers (improvers) are synthesized from ester vinyl monomers as shown in Fig. 5.4, where the R groups represent linear or branched hydrocarbons containing 1 to 30 carbon atoms.

The rheological performance of this class of polymers can be finely tuned to a specific application by varying the alkyl side-chain group (R) length

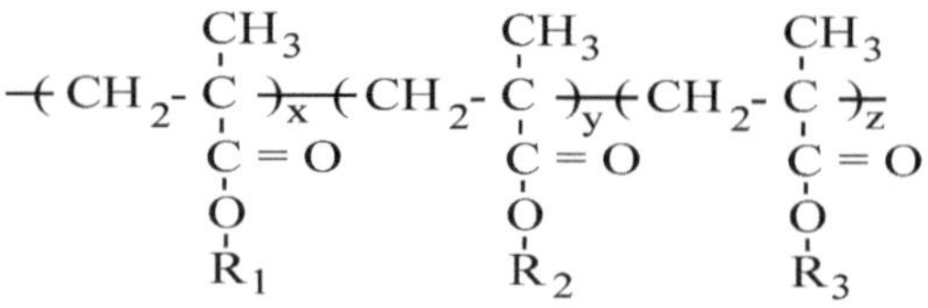

$$\overset{\displaystyle CH_3}{-\!\!\!\left(CH_2\text{-}\underset{\underset{\underset{R_1}{O}}{\overset{|}{C}=O}}{\overset{|}{C}}\right)_{\!\!x}}\overset{\displaystyle CH_3}{\left(CH_2\text{-}\underset{\underset{\underset{R_2}{O}}{\overset{|}{C}=O}}{\overset{|}{C}}\right)_{\!\!y}}\overset{\displaystyle CH_3}{\left(CH_2\text{-}\underset{\underset{\underset{R_3}{O}}{\overset{|}{C}=O}}{\overset{|}{C}}\right)_{\!\!z}}$$

FIGURE 5.4
Chemical structure of PMA Viscosity modifiers.

$$-\!\!\!\left(CH_2CH\right)_{\!x}\!\left(CH-CH\right)_{\!x}$$

styrene   maleate ester

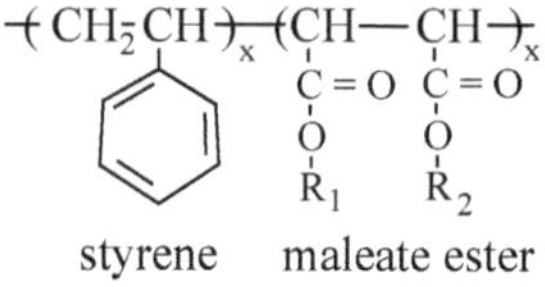

FIGURE 5.5
Chemical structure of SME Viscosity modifiers.

distribution and level of branching. Long side-chains favor oil solubility but lower thickening efficiency. The minimum average side-chain length is often controlled by solubility limitations in mineral oils, although short side-chain PMAs are more efficient thickeners than their long side-chain cousins.

An important design feature of polyalkylmethacrylates is the ability to control base oil wax crystallization at low temperatures [9]. A PMA can act as a pour point depressant (PPD) if the side-chain distribution is capable of co-crystallizing with paraffinic wax molecules present in the base oil(s). Very low polymer concentrations (often less than 0.1 mass %) can effectively reduce the lowest use temperature of a lubricant by 20 degrees Celsius or more.

PMAs are polymerized in mineral oil under conditions that permit a wide range of molecular weights to be easily obtained. Therefore, PMA viscosity modifiers have been developed for use in nearly all lubricant applications.

5.3.3   Styrene-Alkylmaleate Ester (SME) Copolymers Viscosity Modifiers

Styrene-alkylmaleate ester copolymers are similar to PMA viscosity modifiers in some respects, but some subtle differences exist. As shown in Fig. 5.5, alkyl ester side-chains (R) extend from the main chain, and one of the main design parameters is side-chain length and distribution, as in PMAs. The SME backbone consists of alternating styrene and maleate monomers, and there are two ester side-chains for every maleate monomer. Ester group placement on adjacent carbon atoms can, in certain lubricating oils, offer certain advantages with respect to low temperature rheology and pour point depressant

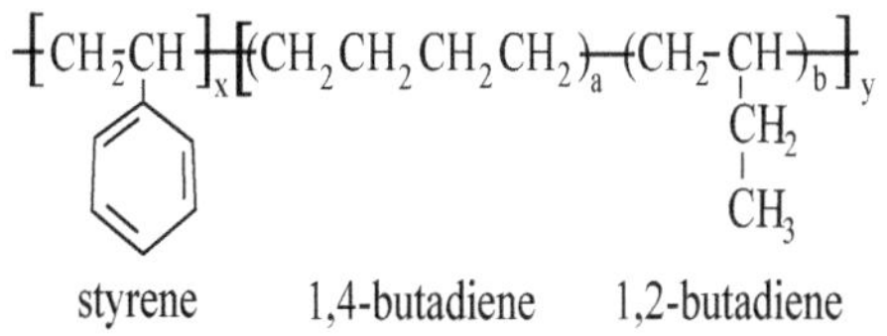

styrene   1,4-butadiene   1,2-butadiene

FIGURE 5.6

Chemical structure of Hydrogenated Styrene Butadiene A-B Block Copolymer Viscosity Modifiers.

styrene   1,4-isoprene   1,2-isoprene 3,4-isoprene

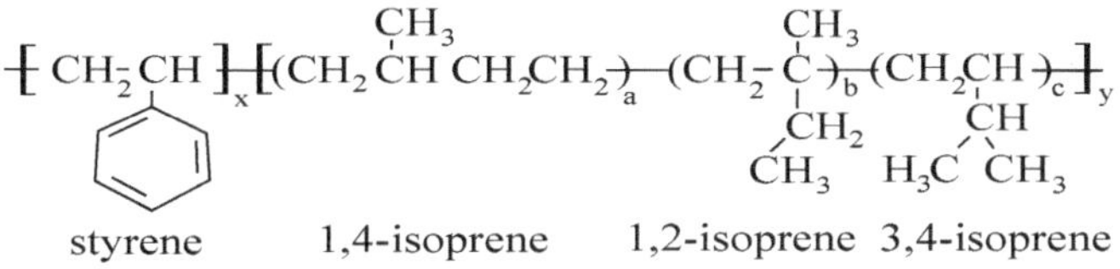

FIGURE 5.7

Chemical structure of Hydrogenated Styrene Isoprene A-B Block Copolymer Viscosity Modifiers.

performance.

5.3.4   Hydrogenated Styrene Diene (HSD) Copolymer Viscosity Modifiers

Hydrogenated styrene diene copolymers are A-B block copolymers with a block (A) of polystyrene at one end of the molecule and a block (B) of hydrogenated butadiene or isoprene at the other. Since polystyrene has poor solubility in mineral oil, these copolymers form micelles in oil solution. The micelle core is rich in polystyrene, and the olefin segments extend outward into the oil phase. Architecturally, the micelle resembles a star polymer, except that the core is not covalently bonded and can disintegrate under high temperature and high shear-rate conditions. This is a reversible phenomenon, whereby the micelle structure re-forms at lower temperatures and shear rates. Since thickening efficiency is often expressed in terms of kinematic viscosity boost at $100°C$, where the micelle is intact, styrene diene copolymers are more efficient thickeners at equal molecular weight than comparable OCP or PMA viscosity modifiers.

The chemical structures of styrene butadiene and styrene isoprene block copolymers are shown in Figs. 5.6 and 5.7, respectively.

As a brief introduction to the chemical nomenclature of conjugated dienes, refer to Fig. 5.8. Both butadiene and isoprene consist of four main carbon

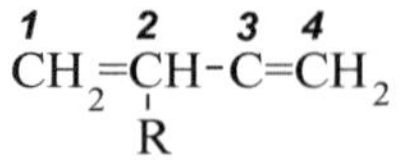

$$\overset{1}{C}\underset{2}{H}_2 = \overset{2}{C}H - \overset{3}{\underset{R}{C}} = \overset{4}{C}H_2$$

FIGURE 5.8

Carbon atom numbering convention for linear conjugated dienes. For butadiene, R = hydrogen (H); for isoprene, R = methyl (CH_3).

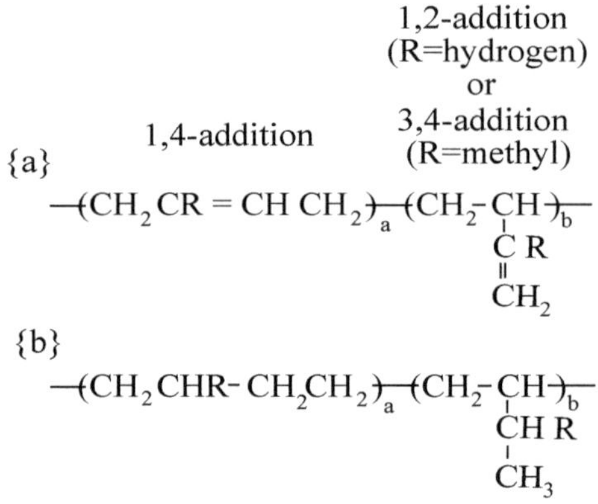

{a}

1,4-addition

1,2-addition
(R=hydrogen)
or
3,4-addition
(R=methyl)

$$-(CH_2\,CR = CH\,CH_2)_a\,(CH_2\text{-}CH)_b$$
$$\overset{}{\underset{CR}{|}}$$
$$\overset{\|}{CH_2}$$

{b}

$$-(CH_2\,CHR\text{-}CH_2CH_2)_a\,(CH_2\text{-}CH)_b$$
$$\overset{}{\underset{CH\,R}{|}}$$
$$\overset{}{\underset{CH_3}{|}}$$

FIGURE 5.9

Polymerization of conjugated dienes, {a} before hydrogenation, {b} after hydrogenation.

atoms numbered 1 through 4. Double bonds connect atoms 1 and 2 as well as 3 and 4. Double bonds are considered to be conjugated when they are separated by one single bond.

Conjugated dienes like butadiene and isoprene can polymerize in two ways as shown in Fig. 5.9. (1) The monomer attaches to the growing polymer chain at carbon 1 and propagates through carbon 4, repositioning the double bond to be between carbons 2 and 3. This is known as 1,4 addition and places all four main carbon atoms along the main chain. (2) The monomer attaches to the growing polymer chain at carbon 1 or 4 and propagates through carbon 2 or 3, respectively. This is called 1,2 or 3,4 addition polymerization, respectively, and creates linear or branched side-chains with only two carbon atoms situated along the main chain; the residual double bonds are located in the side chains. In butadiene polymerization, 1,2 addition is identical to 3,4 addition because of the symmetry of the molecule. By adjusting polymerization conditions (temperature, solvent chemistry, pressure, initiator chemistry), the rela-

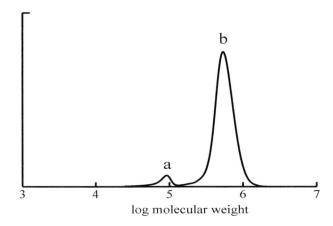

FIGURE 5.10

Molecular weight distribution of a Hydrogenated Diene Star Polymer. Peak a = non-coupled arms and peak b = coupled star molecules.

tive amounts of 1,4 2,3 and 3,4 addition products can be controlled. After the polymerization stage, the copolymer is reacted with hydrogen (hydrogenated) to convert greater than 95% of residual double bonds to single bonds, a step that is necessary to provide acceptable thermal/oxidative stability during use.

Tapered block copolymers have also been made as a variation on the theme of A-B block copolymers. Rather than having two distinct blocks joined at one point along the main chain, monomer concentration varies along the length of the copolymer chain in a continuous rather than step-wise manner. Tapered block copolymers are imperfect micelle formers and, therefore, are less sensitive to high temperature high shear rate rheological conditions than discrete A-B block copolymers.

5.3.5   Star Copolymer Viscosity Modifiers

Star copolymer viscosity modifiers offer unique rheological characteristics relative to linear copolymers of equal shear stability. The star architecture consists of linear polymer chains (called "arms") constrained at one end in a "core", which are covalently bonded via a multi-functional linking agent. A star polymer of equal shear stability to a linear copolymer of similar monomer composition has significantly higher thickening efficiency. We hypothesize that the close proximity of arms near the core acts like a cage that traps more oil per chain end than exists for un-constrained chains. As mentioned earlier in Chapter 4, the thickening effect of a polymer molecule is proportional to its effective volume in oil solution, which includes both polymer and associated solvent (oil).

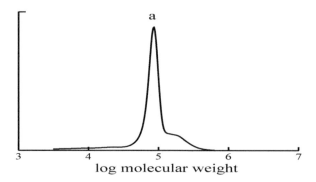

FIGURE 5.11

Molecular weight distribution of the star polymer of Figure 5.9 after 250 cycles in the Orbahn degradation test (ASTM D6278).

Because the linking step is not 100% efficient, there is a small population of orphaned arms in most star polymers, which is easily observed by GPC (see Fig. 5.10).

There are two chemical classes of star polymers reported in the lubricant literature. The first to be commercialized as a lubricant additive was based upon hydrogenated styrene diene copolymer arms [14] and the second consisted of methacrylate monomers polymerized with living free radical catalyst technology [15]. There are many variations of star polymers that are possible. The arms can be A-B block copolymers, tapered blocks, or random. Mixtures of arms varying in chemistry and chain length in each star molecule have also been reported in the patent literature. Thickening efficiency, shear stability, and low temperature rheological properties can be optimized for any application by balancing structural parameters such as the number of arms per star, arm molecular weight distribution, and monomer chemistry distribution and placement.

Star polymers respond to mechanical shearing forces differently than linear polymers as is discussed in depth in Chapter 7. As illustrated in Fig. 5.11, due to degradation the concentration of low molecular weight arm-like chains grows at the expense of the coupled arm stars.

5.3.6   Lubricant Applications

There are a number of criteria used to select specific viscosity modifiers for use in any given lubricant application. Table 5.4 compares the importance of these criteria for families of lubricants segmented by application. Shear stability is a primary consideration and is controlled by polymer molecular weight. In Chapter 4, Table 4.7, it is shown that certain applications are more prone to mechanically degrade polymers than others. Gear oils, for example, are formulated with low molecular weight polymers to handle severe shear stresses resulting from high sliding motion in intermeshing gears and heavily loaded bearings. Mechanical degradation forces in internal combustion engines are far less intense, permitting the use of higher molecular weight polymers. Figure 5.12 is a compilation of the molecular weight ranges of commercial viscosity modifiers in relationship to the shear stability requirements of common lubricant applications. High molecular weight OCPs are exclusively used in engine oils, whereas low molecular weight OCPs are well-suited for premium gear oils. PMA and SME viscosity modifiers span most lubricant types, whereas HSD and Star HSD are most appropriate for engine lubricants.

TABLE 5.4

Tribological factors of importance in various lubricant applications (1 = low in importance, 3 = important, 5 = very important).

	Excellent Shear Stability	Excellent Low Temperature Rheology	Excellent Load–Bearing Capacity at High Temperatures
Engine Oils	3	3	5
Transmission Fluids	3	5	3
Gear Oils	5	5	5
Hydraulic Fluids	5	5	3
Grease	1	3	4

Another important factor is low temperature rheology. To meet Brook-field or MRV specifications (see Subsections 5.2.1 and 5.2.2), certain polymer chemistries are more effective than others. Side-chain distributions of PMA and SME polymers can be tailored to provide outstanding low temperature viscosity, whereas fewer adjustable parameters are available with OCP chemistry (ethylene/propylene content). Polymers with side-chains that co-crystallize with waxy components of base oil improve the ability of a lubricant to flow at low temperatures. In addition, chemical/structural factors that promote intra-chain associations at low temperatures reduce the polymer coil radius and reduce its contribution to viscosity. The short side-chain distributions of HSD and Star polymers are adjustable during the polymerization process, which offers the opportunity to improve low temperature rheology as well. This is another reason why PMA viscosity modifiers are suited for use in so many lubricant applications.

The load–bearing capability of a lubricant operating at high temperatures is of significance in the hydrodynamic lubrication regime, especially when high normal loads pressurize the lubricant film in the contact zone. Lubricating oils

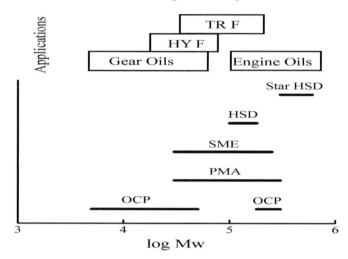

FIGURE 5.12

Molecular weight ranges of commercial viscosity modifiers in relation to the shear stability needs of various lubricant applications (TR F = Transmission Fluids, HY F = Hydraulic Fluids).

with high pressure viscosity coefficients (see Chapter 4, Subsection 4.2.4) and high viscosity provide the necessary level of protection against wear; but operating efficiency suffers if viscosity is too high. Referring again to Table 5.4, it is clear that load–bearing protection is a common feature of all lubricating oils, but especially in those applications where high local loads are present (gears, bearings, engine valve trains, etc.). Viscosity modifiers are good at elevating high temperature viscosity, but they are also prone to mechanical degradation. Therefore, low molecular weight polymers with good shear stability are often chosen for applications where viscosity stability is of paramount interest.

In certain applications, physical form of the polymer is extremely important. The best example is grease. The grease manufacturer prefers oil-free viscosity modifiers, because introducing excess oil harms stiffness and grease consistency. He rarely has equipment to handle and grind large blocks of OCP rubber. Therefore, polymers available in pellet or crumb form are preferred (HSD, Star HSD).

Finally, cost considerations can often out weigh the technical aspects of polymer selection. Lubricant manufacturers market both lower cost lubricants and higher-tier or premium products. The former are designed to meet performance specifications at lowest cost; the latter offer enhanced performance features provided by higher-cost additives and base fluids.

5.4   Closure

The rheological definitions of multi-grade engine oils and gear oils are defined within the context of the use of polymeric viscosity modifiers/improvers in the lubrication industry. Several examples are offered to aid the reader in applying SAE J300 and J306 viscosity classification systems to common lubrication problems. The second half of the chapter reviews the chemistry of the most significant classes of viscosity modifiers and discusses the relationship between polymer architecture and response to mechanical degradation forces acting upon the lubricant during operation.

5.5 Exercises and Problems

1. An automotive gear oil is obtained from a local auto parts store and is labeled as an SAE 80W-90 fluid. To make sure that the fluid meets the requirements of its viscosity grade, a sample is sent out for analysis. The report contains the following information: Kinematic viscosity at $100°C$ equals $13.6 \ mm^2/s$ and Brookfield viscosity at $-26°C$ equals $98,600 \ mPa \cdot s$. Why should you suspect that the oil might not qualify as SAE 80W-90? How can you be sure?

Answer: The kinematic viscosity is so close to the minimum of the 90-grade range that it will most likely fall below $13.5 \ mm^2/s$ after shearing in the 20-hour KRL bench test. Run the bench test to be sure.

2. Why is it important to measure the cold starting viscosity of an automotive engine lubricant when you can simple extrapolate high temperature kinematic viscosity late to low temperature?

Answer: Below the wax appearance temperature, the presence of suspended wax particles causes viscosity to increase above the baseline formed by extrapolation from high temperature viscosity measurements.

3. What are the similarities and differences between oil solutions of styrene-isoprene A-B block copolymers and styrene-isoprene star polymers?

Answer: Similarity: (1) Both polymers form star-like structures when dissolved in oil. Differences: (1) The A-B block copolymer micelle star "core" is not permanent; it is disrupted under high temperature, high shear conditions but reforms at lower temperatures and shear rates (HTHS). The star polymer core is covalently bonded and is not broken down under HTHS conditions. (2) The star copolymer core is highly oil-rich, which increases its thickening efficiency. The A-B block copolymer's styrene-rich core repels oil leading to much lower thickening power. (3) The A-B block copolymer is a linear polymer and less resistant to permanent viscosity loss due to mechanical degradation than the star polymer.

4. Referring to Figs. 5.7 and 5.8, what is the chemical structure of side-chains formed by 1,2 addition of 2,3-dimethylbutadiene $[CH_2 = C(CH_3) - C(CH_3) = CH_2]$?

Answer: the side chains would have the same structure as the side chain of 3,4-isoprene.

5. Star polymers S1 and S2 were added to the same synthetic base oil at equal concentration. Kinematic viscosity of the solutions are $11.67 \ mm^2/s$ and $16.71 \ mm^2/s$. The ASTM D6278 Orbahn shear bench test was run on both solutions with the following results (% viscosity loss): 1.6% and 10.4%, respectively. Provide an explanation to explain why S1 was more resistant to mechanical degradation than S2.

Answer: S1 is a lower molecular weight star polymer than S2 judging from its lower thickening efficiency. Lower molecular polymers are generally more resistant to mechanical degradation than higher molecular weight polymers of the same chemical class.

BIBLIOGRAPHY

[1] Rein. S.W. 1968. Low Temperature Viscosity of Automotive Engine Oils. *Lubrication* 54(7), 53-64.

[2] McMillan, M.L. 1977. Engine Oil Viscosity Classifications - Past, Present and Future. *Soc. Automot. Eng. Tech. Paper Ser.* No. 770373.

[3] May, C.J. et al. 1998. Cold Starting and pumpability Studies in Modern Engines. *ASTM Res. Report RR − −D02 − −1442*.

[4] Thompson, M., Von Eberan-Eberhorst, C.G.A., Rossi, A., and Holdack-Janssen, H. 1983. Low Temperature Starting and Pumpability Requirements of Eupopean Automotive Engines - Part 2. *Soc. Automot. Eng. Tech. Paper Ser.* No. 831717.

[5] Spearot, J.A. 1992. Engine Oil Viscosity Classification Low-Temperature Requirements - Current Status and Future Needs. In Rhodes, R.G., ed. *Low Temp. Lubricant Rheology Measurement and Relevance to Engine Operation* ASTM STP 1143. Am. Soc. for Testing and Materials. 161-181. Philadelphia.

[6] Shaub, H. A History of ASTM Accomplishments in Low Temperature Engine Oil Rheology: 1966-1991. In Rhodes, R.G., ed. *Low Temp. Lubricant Rheology Measurement and Relevance to Engine Operation* ASTM STP 1143. Am. Soc. for Testing and Materials. 1-19. Philadelphia.

[7] Spearot, J.A., ed. 1989. *High − −Temperature, High − −Shear Oil Viscosity, Measurement and Relationship to Engine Operation* ASTM STP 1068. Am. Soc. for Testing and Materials, Philadelphia.

[8] Covitch, M.J. 2009. Olefin Copolymer Viscosity Modifiers. In Rudnick, L.R., ed. *Lubricant Additives Chemistry and Applications Second Ed.* CRC Press. 283 - 314.

[9] Kinker, B.G. 2009. Polymethacrylate Viscosity modifiers and Pour Point Depressants. in Rudnick, L.R., ed. *Lubricant Additives Chemistry and Applications Second Ed.* CRC Press. 315 - 337.

[10] Kapuscinski, M.M., Sen, A. and Rubin, I.D. 1989. Solution Viscosity Studies on OCP VI Improvers in Oils. *Soc. Automot. Eng. Tech. Paper Ser.* No. 892152.

[11] Ver Strate, G. and Struglinski, M.J. 1991. Polymers as Lubricating Oil Viscosity Modifiers. In Schulz, D.N. and Glass, J.E. eds. *Polymers as Rheology Modifiers* ACS Symposium Series 462, Am. Chem. Soc. Philadelpia. 256-272.

[12] Rhodes, R.B. 1993. Low-Temperature Compatibility of Engine Lubricants and the Risk of Engine Pumpability Failure. *Soc. Automot. Eng. Tech. Paper Ser.* No. 932831.

[13] George, H.F. and Hedrick, D.P. 1993. Comparative Rheology of Commercial Viscosity Modifier Concentrates. *Soc. Automot. Eng. Tech. Paper Ser.* No. 932834.

[14] Briant, J., Denis, J., and Parc, G., eds. 1989. *Rheological Properties of Lubricants* 175-176, Paris: Editions Technip.

[15] Schober, B.J., Vickerman, R.J., Lee, O.-D., Dimitrakis, W.J., and Gajanayake, A. 2008. Controlled Architecture Viscosity modifiers for Driveline Fluids: Enhanced Fuel Efficiency and Wear Protection. 14th Annual Fuels and Lubes Asia Conference, Seoul, Korea, March 5-7, 2008. Fuel and Lubes Asia Inc., Muntinlupa, Philippines.

6

Modeling of Lubricant Polymer Molecular Weight, Viscosity, and Degradation in Kurt Orbahn Shear Stability Test for Viscosity Improvers (VI) with Linear Structure

6.1 Introduction

In lubricated contacts, oil film thickness is a critical design feature of a lubricant and it is controlled, in large part, by viscosity. Should viscosity drop below a critical value, opposite surfaces can begin to come into contact resulting in premature wear and fatigue damage. High molecular weight polymers, known as viscosity modifiers (VM) or viscosity improvers (VI), are added to lubricating oils to boost viscosity at high temperatures while minimizing thickening contribution at low temperatures. Viscosity increase is proportional to both VM concentration and molecular weight. To preserve lubricant film thickness between the moving surfaces over time, it is desirable to minimize molecular weight degradation of the polymer additive (see Chapter 4).

There exists a large body of experimental and theoretical studies of polymer degradation kinetics. Most of the theoretical studies of changes in the length distribution of polymer molecules over time with few exceptions (see [1, 2]) were concerned with polymer degradation caused by radiation (for example, [3]), thermal degradation of a polymer dissolved in a fluid (for example, [4, 5, 6, 7]), or non-isothermal crystallization of polymers [8]. Usually, in studies of thermal degradation, the mechanical stresses acting upon fluid were not considered. In studies of thermal degradation, the kinetic equation used was written for the chain–length distribution of the polymer molecules. Ziff and McGrady developed some analytical approaches to solution of this equation in certain special cases [5, 6]. In most analytical cases, the polymer scission rate was accepted to be a relatively simple function (for example, a power function) while the conditional probability for a polymer molecule to be broken into two fragments of specific chain lengths was somewhat simplistic. As a result, the probabilistic polymer molecular weight distribution was always single-modal. Moreover, there is a lack of analysis of the existence and uniqueness of the solution to the initial-value problem for the kinetic equation as well as of the general properties of the molecular weight distribution function and of its moments in the case of stress-induced degradation.

A simplistic model and numerical solution for stress-induced degradation were presented by Kudish and Ben-Amotz [9]. It was shown that in some cases an initially single-modal molecular weight distribution becomes a two-modal distribution due to mechanically induced polymer scission. These results are in a qualitative agreement with the results of a study by Odell et al. [10].

In this chapter a probabilistic kinetic equation describing the mechanism of stress-induced degradation of linear polymer molecules dissolved in a lubricating oil is derived [11, 12, 13]. The expressions for the probability of stress-induced scission and the density of the conditional probability of scission of polymer molecules involved in the kinetic equation are derived. The existence and uniqueness of the solution to the initial-value problem for the kinetic equation as well as some general properties of the moments of the polymer molecular weight distribution are established. These properties include conservation of the polymer total weight and increase of the number of polymer molecules along the fluid flow streamlines. Moreover, a numerical study of the kinetic equation is performed. In two cases the results of the numerical simulations are compared with independently obtained test data, showing very good agreement between the numerical and experimental data. Also, some results for two additional sets of input parameters that produced different cases of the probability of scission are presented. In the latter cases, in the early stages of the degradation process, the polymer molecular weight distribution exhibits a multi-modal type of behavior while at later stages of the degradation process the polymer molecular weight distribution becomes single modal.

6.2   Kinetic Equation for Stress-Induced Degradation of Lubricants with Linear Polymeric Additives (VI)

Polymer degradation may be caused by the combination of environmental parameters such as molecular weight, pressure, shear strain rate, and temperature as well as lubricant viscosity and polymer chemical characteristics (bond dissociation energy, bead radius, bond length, etc.). In this section a fundamental approach to the problem of modeling stress-induced polymer degradation is proposed. Polymer degradation is modeled on the basis of a kinetic equation for the density of the statistical distribution of polymer molecules as a function of molecular weight or chain length.

Let us assume that there is a relatively low concentration of long polymer molecules in solution. The polymer concentration is low enough so that the polymer molecules do not interact with each other. The fluid solution is set in

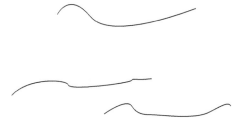

FIGURE 6.1

A view of linear polymer molecules stretched in a lubricant flow.

motion by external forces. In the process of this motion, the polymer molecules become stretched along the flow streamlines (see Fig. 6.1), and some of the polymer molecules may break into fragments due to both mechanical and thermal stresses. Let us derive a kinetic equation describing the process of polymer molecule scission in time t and space $\mathbf{x} = (x, y, z)$. The size of polymer molecule chains will be characterized by length l or by molecular weight w. The chain length l is defined as the number of monomer units in the polymer chain, also known as the degree of polymerization. We will consider homogeneous polymers with monomer molecular weight w_m, where $w = w_m l$. In what follows both these characteristics of polymer chains are used.

Let us introduce function $W(t, \mathbf{x}, l)$, which denotes the density of the distribution of polymer molecule weight at the time moment t in a unit volume centered at $\mathbf{x}$ as a function of the molecule chain length l. It is determined in such a way that $W(t, \mathbf{x}, l) \triangle l \triangle v$ is the weight of polymer molecules at the time moment t located in a small fluid volume $\triangle v$ centered at $\mathbf{x}$ with molecule chain lengths from l to $l + \triangle l$. In a similar manner, one can introduce the density of the distribution of the number of polymer molecules $n(t, \mathbf{x}, l)$. Since the weight of one polymer molecule of the length l is $w = w_m l$ one has:

$$W(t, \mathbf{x}, l) = w_m l n(t, \mathbf{x}, l). \tag{6.1}$$

Polymer molecules dissolved in a fluid are stretched along the fluid flow streamlines and move together with this fluid. Simultaneously, polymer molecules may undergo stress-induced scission. This process has a stochastic nature and can be characterized in probabilistic terms. Suppose $R(t, \mathbf{x}, l)$ is the probability of polymer scission varying between 0 and 1. We will treat $R(t, \mathbf{x}, l)$ as the probability of dissociation of a polymer molecule of a chain length l located at a point $\mathbf{x}$ at a time moment t. Suppose that the characteristic time of one act of polymer fragmentation is τ_f. Obviously, for any $l \geq 0$ one has

$$R(t, \mathbf{x}, l) \geq 0. \tag{6.2}$$

In practice, if polymer molecule scission occurs then the probability of scission into multiple fragments is always much smaller than that for scission into just two fragments. It is further assumed that scission of polymer molecules into multiple (more than two) fragments does not take place at all. It is natural to assume that for a polymer molecule of chain length L the probability to be broken into fragments of chain lengths l and $L - l$ depends on L and l. Assuming that polymer scission has occurred we can introduce the conditional probability density $p_c(t, \mathbf{x}, l, L)$, $l \leq L$, such that $p_c(t, \mathbf{x}, \lambda, L)\triangle l$ is the probability of a polymer molecule of chain length L at the time moment t located at the point $\mathbf{x}$ to break into two fragments of chain lengths λ and $L - \lambda$, $l \leq \lambda \leq l + \triangle l$. Then, for any $l \geq 0$ and $L \geq 0$:

$$p_c(t, \mathbf{x}, l, L) \geq 0, \ \int_0^L p_c(t, \mathbf{x}, l, L)dl = 1, \ L > 0,$$

$$\tag{6.3}$$

$$p_c(t, \mathbf{x}, l, L) = p_c(t, \mathbf{x}, L - l, L).$$

Taking into account the above considerations we can conclude that the total probability for a polymer molecule of chain length Λ, $L \leq \Lambda \leq L + \triangle L$ to be broken into two fragments of lengths λ and $\Lambda - \lambda$, $l \leq \lambda \leq l + \triangle l$, $L \leq \Lambda \leq L + \triangle L$, is $R(t, \mathbf{x}, \Lambda)p_c(t, \mathbf{x}, \lambda, \Lambda)\triangle l\triangle L$.

Below we consider the polymer molecule scission along the fluid flow streamlines given by the following equation:

$$\frac{d\mathbf{x}}{dt} = \mathbf{u}(t, \mathbf{x}),$$

$$\tag{6.4}$$

where $\mathbf{u} = (u_x, u_y, u_z)$ is the vector of the velocity of the fluid flow. For the sake of simplicity, we will assume that the fluid flow velocity $\mathbf{u}$ is known.

To derive the kinetic equation of stress-induced polymer scission, let us consider the weight of polymer molecules in a small fluid volume $\triangle v(t)$ at the time moment t located at the point $\mathbf{x}$. In the process of fluid motion, this fluid volume $\triangle v(t)$ at the time moment $t + \triangle t$ assumes the position at $\mathbf{x}(t + \triangle t)$ and volume $\triangle v(t + \triangle t)$. Therefore, we need to derive an equation to describe polymer molecular weight balance at these two time moments, which would take into account fluid motion and polymer scission.

Denote by $\triangle_-$ and $\triangle_+$ the weight of polymer molecules leaving and entering, respectively, the interval $[l, l + \triangle l]$ of polymer molecule chain lengths in a fluid volume $\triangle v(t)$ over a unit period of time. During the time period $\triangle t$, the number of polymer molecules leaving the indicated chain length interval due to scission is equal to $R(t, \mathbf{x}, l)n(t, \mathbf{x}, l)\triangle l\triangle v(t)\triangle t/\tau_f$. It follows from (6.1) that the weight of these molecules is

$$\triangle_- = R(t, \mathbf{x}, l)n(t, \mathbf{x}, L)w_m l\triangle l\triangle v(t)\tfrac{\triangle t}{\tau_f}$$

$$\tag{6.5}$$

$$= R(t, \mathbf{x}, l)W(t, \mathbf{x}, L)\triangle l\triangle v(t)\tfrac{\triangle t}{\tau_f}.$$

Now, let us determine the weight of polymer molecules entering the chain length interval $[l, l + \triangle l]$ due to scission of molecules of chain length greater than $l + \triangle l$. In the volume $\triangle v(t)$ the number of polymer molecules of chain length between L and $L + \triangle L$ that over the time period $\triangle t$ are broken into two fragments one of which has the chain length between l and $l + \triangle l$ is equal to (see (6.3))

$$R(t, \mathbf{x}, \Lambda)[p_c(t, \mathbf{x}, \lambda, \Lambda) + p_c(t, \mathbf{x}, \Lambda - \lambda, \Lambda)]n(t, \mathbf{x}, \Lambda)\triangle l\triangle v(t)\tfrac{\triangle t}{\tau_f}$$

$$= 2R(t, \mathbf{x}, \Lambda)p_c(t, \mathbf{x}, \lambda, \Lambda)n(t, \mathbf{x}, \Lambda)\triangle l\triangle v(t)\tfrac{\triangle t}{\tau_f}. \tag{6.6}$$

The weight of the fragments of these molecules entering the interval $[l, l+\triangle l]$ is $2R(t, \mathbf{x}, \Lambda)p_c(t, \mathbf{x}, \lambda, \Lambda)n(t, \mathbf{x}, \Lambda)w_m l\triangle t\triangle l\triangle L\triangle v(t)$. Taking into account the contribution of all such molecules of length greater than $l + \triangle l$ and using (6.1) one gets

$$\triangle_+ = \tfrac{\triangle t}{\tau_f}2l\triangle l\triangle v(t)\int\limits_{l+\triangle l}^{\infty} R(t, \mathbf{x}, L)p_c(t, \mathbf{x}, l, L)W(t, \mathbf{x}, L)\tfrac{dL}{L}. \tag{6.7}$$

Therefore, the balance equation for the polymer molecule scission along the flow streamlines is

$$W(t + \triangle t, \mathbf{x} + \triangle\mathbf{x}, \lambda)\triangle l\triangle v(t + \triangle t)$$

$$= W(t, \mathbf{x}, \lambda)\triangle l\triangle v(t) + \triangle_+ - \triangle_- + F(t, \mathbf{x}, \lambda)\triangle l\triangle v(t)\triangle t, \tag{6.8}$$

where $F(t, \mathbf{x}, \lambda)\triangle l\triangle t\triangle v(t)$ is the weight of the polymer molecules of chain lengths from l to $l + \triangle l$ added to or removed from the fluid volume $\triangle v(t)$ over the time period t.

The mass of a fluid particle $\triangle m = \rho(t, \mathbf{x}(t))\triangle v(t)$ along the flow streamlines is conserved ($\rho(t, \mathbf{x})$ is the fluid density at the point $\mathbf{x}$ and time moment t) if the total mass of the added/removed polymer molecules (characterized by the function F) is negligibly small in comparison with the mass of the fluid particle.

Let us assume that $R(t, \mathbf{x}, l)$, $p_c(t, \mathbf{x}, l, L)$, $W(t, \mathbf{x}, l)$, $F(t, \mathbf{x}, l)$, and $\rho(t, \mathbf{x})$ are continuous functions, and $W(t, \mathbf{x}, l)$ and $\rho(t, \mathbf{x})$ are continuously differentiable functions with respect to t and $\mathbf{x}$. Dividing (6.8) by $\triangle l\triangle t\triangle v(t)$, using (6.4)-(6.7) and allowing $\triangle t$ to approach zero one can derive the following kinetic equation:

$$\rho\tfrac{d}{dt}\tfrac{W}{\rho} = \tfrac{2l}{\tau_f}\int\limits_{l}^{\infty} R(t, \mathbf{x}, L)p_c(t, \mathbf{x}, l, L)W(t, \mathbf{x}, L)\tfrac{dL}{L}$$

$$- \tfrac{1}{\tau_f}R(t, \mathbf{x}, l)W(t, \mathbf{x}, l) + F(t, \mathbf{x}, l), \tag{6.9}$$

where

$$\tfrac{d}{dt} = \tfrac{\partial}{\partial t} + \mathbf{u}\tfrac{\partial}{\partial\mathbf{x}}, \tag{6.10}$$

which describes the process of polymer molecule scission. If the fluid is incompressible ($\rho(t, \mathbf{x}) = const$), then (6.9) can be reduced to

$$\frac{dW}{dt} = \frac{2l}{\tau_f} \int\limits_l^\infty R(t, \mathbf{x}, L) p_c(t, \mathbf{x}, l, L) W(t, \mathbf{x}, L) \frac{dL}{L}$$

$$- \frac{1}{\tau_f} R(t, \mathbf{x}, l) W(t, \mathbf{x}, l) + F(t, \mathbf{x}, l). \tag{6.11}$$

If $W(t, \mathbf{x}, \lambda) \triangle l$ represents the polymer weight per unit mass of the (compressible/incompressible) fluid centered at $\mathbf{x}$ at the time moment t with molecule chain lengths between l and $l + \triangle l$, then the kinetics of polymer scission is also described by equation (6.11). For the density distribution of the number of polymer molecules $n(t, \mathbf{x}, l) = W(t, \mathbf{x}, l)/(w_m l)$, equation (6.11) can be reduced to

$$\tau_f \frac{dn}{dt} = 2 \int\limits_l^\infty R(t, \mathbf{x}, L) p_c(t, \mathbf{x}, l, L) n(t, \mathbf{x}, L) dL - R(t, \mathbf{x}, l) n(t, \mathbf{x}, l)$$

$$+ \frac{\tau_f}{w_m l} F(t, \mathbf{x}, l). \tag{6.12}$$

Kinetic (6.9), (6.11), and (6.12) should be supplemented by the initial conditions on the density of polymer molecular weight W and the density of the number of polymer molecules n along each of the fluid streamlines $\mathbf{x}(t)$ in the form

$$W(0, \mathbf{x}(0), l) = W^0(l) \tag{6.13}$$

and

$$n(0, \mathbf{x}(0), l) = n^0(l). \tag{6.14}$$

The initial-value problems for the kinetic (6.9), (6.11), and (6.12) are represented by pairs of equations (6.9), (6.13); (6.11), (6.13); and (6.12), (6.14), respectively.

6.3   Probability of Scission of Linear Polymeric Molecules in Lubricants Subjected to Stress

Only few facts are known about the behavior of the probability of scission R. From thermodynamics it is obvious that R is a non-increasing function of U/kT, and, simultaneously, it is an increasing function of lubricant viscosity μ, shear strain rate S, and polymer chain length l. Here U is the dissociation energy of a polymer molecule bond, T is the lubricant temperature, and k is Boltzmann's constant ($k = 1.38 \times 10^{-23} J/K$). The analysis of this section follows the paper by Kudish et al. [13].

To derive the polymer probability of scission $R(t, \mathbf{x}, l)$, we can notice that it should depend on the intensity of polymer scission $I(l)$. The intensity of scission I, in turn, depends on the work E produced by the frictional force applied to a polymer molecule by the surrounding lubricant flow. Obviously, in a lubricant the shear stress τ is equal to μS and it is applied to the side surface of a polymer molecule stretched along the flow streamline. The side surface of such a molecule is proportional to $a_* l_* l$, where a_* and l_* are the polymer molecule bead radius and bond length, respectively, and $l_* l$ is the total polymer molecule length. Therefore, the shear force acting on the polymer molecule of chain length l is proportional to $a_* l_* \mu S l$. The extent to which the polymer molecule is stretched at the moment when scission occurs is proportional to a certain fraction of the polymer total length $l_* l$. Finally, we come up with the expression for the work E required for scission of a polymer molecule to occur

$$E = C a_* l_*^2 \mu(p, T) S l^2, \tag{6.15}$$

where p is the lubricant pressure, C is a dimensionless shield constant. We will assume that the process of scission takes place if $E > U_A$ and that it does not take place if $E \leq U_A$, where $U_A = U/N_A$ is a $C - C$ bond dissociation energy, U is a $C - C$ bond dissociation energy per polymer mole, and N_A is the Avogadro number, $N_A = 6.022 \times 10^{23} mole^{-1}$. Obviously, the total thermal energy of the polymer molecule of chain length l is proportional to lkT. Therefore, the dimensionless intensity of scission can be determined according to the formulas

$$I(l) = 0 \ if \ E \leq U_A, \ I(l) = \frac{2\alpha(E - U_A)}{lkT} \ if \ E > U_A, \ U_A = \frac{U}{N_A}, \tag{6.16}$$

where α is a dimensionless coefficient of proportionality. The values of the parameters C and α will be determined from the test data.

Now, let us derive an equation for the probability $Q(l)$ of the absence of scission of a polymer molecule of length l, i.e., for $Q(l) = 1 - R(l)$. By introducing the conditional probability $Q_0(l, \triangle l)$ of the absence of scission for a polymer molecule by $\triangle l$ longer than a molecule of chain length l as follows

$$Q_0(l, \triangle l) = 1 - I(l)\triangle l, \tag{6.17}$$

we can derive a simple equation for $Q(l)$ in the form

$$Q(l + \triangle l) = Q(l)[1 - I(l)\triangle l]. \tag{6.18}$$

Taking the limit in (6.18) as $\triangle l$ approaches 0 we obtain a differential equation for $Q(l)$ as follows

$$\frac{dQ}{dl} = -I(l)Q. \tag{6.19}$$

The solution of (6.19) is as follows

$$Q(l) = Q(0) \exp[-\int_0^l I(L)dL], \tag{6.20}$$

where $Q(0)$ is the probability of the absence of scission in case when $l = 0$ or $I(l) = 0$. The latter is obviously equal to $1 - \exp[-U_A/(kT)]$. Therefore, from (6.20) we obtain

$$R(l) = 1 - [1 - \exp(-\tfrac{U_A}{kT})]\exp[-\int_0^l I(L)dL]. \qquad (6.21)$$

In most practical cases $U_A/(kT) \gg 1$. That leads to a simplified expression following from (6.15), (6.16), and (6.21):

$$R(t, \mathbf{x}, l) = 0 \ if \ l \leq L_*,$$

$$R(t, \mathbf{x}, l) = 1 - (\tfrac{l}{L_*})^{\frac{2\alpha U_A}{kT}}\exp[-\tfrac{\alpha U_A}{kT}(\tfrac{l^2}{L_*^2} - 1)] \ if \ l \geq L_*. \qquad (6.22)$$

In (6.22) the quantity L_* is calculated according to formulas

$$L_* = \sqrt{\tfrac{\mu_a}{\mu}}L_0, \ L_0 = \sqrt{\tfrac{U_A}{Ca_*l_*^2\mu_a S}}, \ U_A = \tfrac{U}{N_A}, \qquad (6.23)$$

where L_0 is the characteristic polymer chain length determined using the initial viscosity μ_a of the lubricant before scission and L_* is the characteristic polymer chain length adjusted to the lubricant viscosity μ at a current level.

To better understand the behavior of the molecular weight distribution $W(t, l)$, we need to analyze the variations of the scission probability R as a result of variation of system parameters. The expression from (6.22) is a monotonically increasing continuous function with respect to chain length l. For different polymer additives and lubrication regimes by varying parameters U, T, C, α, a_*, l_*, μ, and S, the center of the distribution of the probability of scission (where $R(l) = 1/2$) can be positioned at any predetermined $l = l_c$ with any slope within a certain range of values. It can be shown from (6.22) that as L_* increases (i.e., μ_a/μ increases, and/or U increases, and/or $Ca_*l_*^2\mu_a S$ decreases) the distribution of the probability of scission $R(l)$ shifts away from $l = 0$ and vice versa. The position of the center of the distribution of $R(l)$ shifts toward $l = 0$ as $\alpha U/(kT)$ and L_*^{-1} increase and vice versa. The slope of $R(l)$ at the center of the distribution, where $R(l) = 1/2$ is controlled by the values of $\alpha U_A/(kT)$ and L_*: the slope increases with $\alpha U/(kT)$ and L_*^{-1}. Finally, the values of the shield constant C and parameter α have to be chosen to provide the best fit of the numerical results to the experimental data.

It is important to emphasize that parameters U, a_*, and l_* are uniquely defined by the nature of the polymer additive and parameters T, μ, and S are determined by the lubrication regime and the nature of the lubricant. Therefore, for a chosen lubricant, polymer additive, and lubrication conditions only two parameters C and α can be used for adjusting the model to match the test data.

6.4 Density of the Conditional Probability of Scission of Linear Polymeric Molecules in Lubricants

Now, let us derive the density of the conditional probability of scission $p_c(t, \mathbf{x}, l, L)$. To do that we need to introduce the conditional probability of scission $P_c(t, \mathbf{x}, l, L)$ according to the formula

$$P_c(t, \mathbf{x}, l, L) = \int_0^l p_c(t, \mathbf{x}, s, L) ds. \tag{6.24}$$

Here $P_c(t, \mathbf{x}, l, L)$ is the conditional probability for a polymer molecule of length L located at the point $\mathbf{x}$ at the time moment t to break into two fragments the left of which is of length not greater than l assuming that the molecule is stretched from left to right.

It is natural to expect that the conditional probability P_c depends on the distribution of the deformation energy along polymer molecules. Moreover, P_c should not depend on the level of the deformation energy but its relative distribution along polymer molecules.

First, we derive the equation for $P_c(t, \mathbf{x}, l, L)$. Suppose a polymer molecule undergoes scission. The following two events are mutually exclusive: (I) a molecule of length L breaks into two fragments one (the left) of which is of length not greater than l, and (II) a molecule of length L breaks into two fragments one (the left) of which has the length from the interval $[l, l + \triangle l]$. The probabilities of events (I) and (II) are $P_c(t, \mathbf{x}, l, L)$ and $[1 - P_c(t, \mathbf{x}, l, L)]Q_1(t, \mathbf{x}, l, L, \triangle l)$, respectively. Here $Q_1(t, \mathbf{x}, l, L, \triangle l)$ is the probability of the polymer molecule with length L to break into two fragments the left of which has the length from the interval between l and $l + \triangle l$. It is natural to assume that the probability $Q_1(t, \mathbf{x}, l, L, \triangle l)$ is proportional to the deformation energy stored in the segment $[l, l + \triangle l]$ of the molecule. Obviously, this deformation energy is equal to the absolute value of work of the external force $F_1(t, \mathbf{x}, l, L)$ applied to the fragment of length $\triangle l$. Therefore, we have

$$P_c(t, \mathbf{x}, l + \triangle l, L) = P_c(t, \mathbf{x}, l, L)$$

$$+[1 - P_c(t, \mathbf{x}, l + \triangle l, L)]C_1 \mid F_1(t, \mathbf{x}, l + \triangle l, L) \mid \triangle l, \tag{6.25}$$

where C_1 is a coefficient of proportionality. Taking the limit of (6.25) as $\triangle l \to 0$ we obtain a differential equation:

$$\frac{dP_c}{dl} = C_1(1 - P_c) \mid F_1 \mid . \tag{6.26}$$

The general solution to this equation is

$$P_c(t, \mathbf{x}, l, L) = 1 + C_2 \exp[-C_1 U_1(t, \mathbf{x}, l, L)], \quad l \leq L/2,$$

$$U_1(t, \mathbf{x}, l, L) = \int_0^l | F_1(t, \mathbf{x}, s, L) | \, ds,$$

(6.27)

where C_2 is an arbitrary constant. It is natural to assume that

$$P_c(t, \mathbf{x}, 0, L) = 0, \quad P_c(t, \mathbf{x}, L, L) = 0.$$

(6.28)

Thus, $C_2 = -1$. By our assumption a molecule breaks into two fragments; therefore, the probability of either the left or the right fragment to be of length not greater than $L/2$ is equal to $P_c(t, \mathbf{x}, L/2, L)$. Since at least one of the fragments has length not greater than $L/2$, one gets

$$2P_c(t, \mathbf{x}, L/2, L) = 1.$$

(6.29)

This implies

$$C_1 = \frac{\ln(2)}{U_1(t, \mathbf{x}, L/2, L)}.$$

(6.30)

Finally, the expression for P_c can be presented in the form

$$P_c(t, \mathbf{x}, l, L) = 1 - \exp\left[- \ln 2 \frac{U_1(t, \mathbf{x}, l, L)}{U_1(t, \mathbf{x}, L/2, L)} \right].$$

(6.31)

Using (6.24) and differentiating (6.31) we obtain the formula for $p_c(t, \mathbf{x}, l, L)$:

$$p_c(t, \mathbf{x}, l, L) = \ln 2 \frac{|F_1(t, \mathbf{x}, l, L)|}{U_1(t, \mathbf{x}, L/2, L)} \exp\left[- \ln 2 \frac{U_1(t, \mathbf{x}, l, L)}{U_1(t, \mathbf{x}, L/2, L)} \right].$$

(6.32)

The expressions from (6.31) and (6.32) possess a remarkable property: P_c and p_c depend only on $| F_1(t, \mathbf{x}, l, L) | /U_1(t, \mathbf{x}, L/2, L)$ and $U_1(t, \mathbf{x}, l, L)/U_1(t, \mathbf{x}, L/2, L)$, and, therefore, P_c and p_c are independent from the level of the external force F. Also, it means that the density of the conditional probability $p_c(t, \mathbf{x}, l, L)$ is independent from t and $\mathbf{x}$. In other words, if a polymer molecule undergoes scission, then the fragmentation process runs exactly the same way at any time moment and location. Using the symmetry of $p_c(l, L) = p_c(L - l, L)$, the fact that $\int_0^L p_c(l, L)dl = 1$ (see (6.3)), and making the substitution $l_1 = L - l$ in the following integral, we obtain that the average length of the fragments resulted from scission of polymer molecules of length L is equal to $L/2$, i.e.,

$$\int_0^L l p_c(l, L)dl = \frac{L}{2}.$$

(6.33)

Moreover, according to (6.29) the conditional probability $P_c(l, L)$ reaches its maximum value equal of $1/2$ when $l = L/2$, i.e., the most probable event of scission occurs when a molecule of length L gets broken in two fragments of equal length of $L/2$.

In most cases the acceleration of polymer molecules can be neglected in comparison with the viscous forces exerted upon them by the flow. Therefore, polymer molecules may be considered to be in a stationary deformed state. For a linear polymer made of a homogeneous material, the force F_1 applied to polymer molecules is

$$F_1 = C_3 a_* l_* \mu S(L - 2l),\tag{6.34}$$

where C_3 is a dimensionless coefficient of proportionality. Therefore, from (6.27) and (6.34) we have

$$U_1 = C_3 a_* l_* \mu Sl(L - l).\tag{6.35}$$

Finally, the expression for $p_c(t, \mathbf{x}, l, L)$ from (6.32) has the form:

$$p_c(t, \mathbf{x}, l, L) = \ln 2 \frac{4|L - 2l|}{L^2} \exp\left[-\ln 2 \frac{4l(L-l)}{L^2}\right].\tag{6.36}$$

The density of the conditional probability $p_c(t, \mathbf{x}, l, L)$ from (6.36) does not contain any adjustable parameters and is independent of t and $\mathbf{x}$. Moreover, $P_c(l, L)$ reaches its maximum equal to $1/2$ at $l = L/2$, where $p_c(l, L) = 0$ (see the expressions from (6.31), (6.35), and (6.36)). The latter conclusion is supported by experimental observations.

6.5   Lubricant Viscosity and Distribution of Polymeric Molecules

The lubricant viscosity μ depends on the distribution of the polymer additive. To take account of that, we will use the empirical Huggins and Mark-Houwink equations [14, 15] (also see Chapter (4))

$$\mu = \mu_a \frac{1 + c_p[\eta] + k_H(c_p[\eta])^2}{1 + c_p[\eta]_a + k_H(c_p[\eta]_a)^2}, \quad [\eta] = k' M_W^\beta,$$

$$M_W = \left\{\int_0^\infty w^\beta W(t, \mathbf{x}, w)dw \Big/ \int_0^\infty W(t, \mathbf{x}, w)dw\right\}^{1/\beta},\tag{6.37}$$

where μ_a is the initial lubricant viscosity, c_p and $[\eta]$ are the polymer concentration and the intrinsic viscosity, respectively, $[\eta]_a$ is the intrinsic viscosity at the initial time moment, k_H is the Huggins constant, k' and β are the Mark-Houwink constants, $w = w_m l$ is the polymer molecular weight, w_m is the monomer molecular weight.

6.6 Some Properties of the Kinetic Equation of Lubricant Degradation

In this section the existence and uniqueness theorem for the initial value problem for the kinetic equation is provided, and the main properties of the solution to this problem are established and analyzed. In addition to that we will consider some basic properties of kinetic equations (6.11) and (6.12). We have to remember that the polymer degradation process occurs while polymer molecules move along the flow streamlines. Therefore, along a fixed flow streamline the whole process depends on just time t. To analyze the degradation process let us fix an arbitrary fluid flow streamline $\mathbf{x} = \mathbf{x}(t)$. Below, we will use the notation $f(t, l) = f(t, \mathbf{x}(t), l)$ for functions n, n_0, W, W_0, R, and F. Similarly, we will use $p_c(t, l, L)$ for $p_c(t, \mathbf{x}(t), l, L)$. The analysis of this section follows the material of the paper by Kudish et al. [12].

Using the above notations one has the following initial-value problem for the density of the distribution of polymer molecular weight (see (6.12) and (6.14)):

$$\tau_f \frac{dn}{dt} = 2 \int\limits_l^\infty R(t, L) p_c(t, l, L) n(t, L) dL - R(t, l) n(t, l) + \frac{\tau_f}{w_{ml}} F(t, l),$$

$$(6.38)$$

$$n(0, l) = n^0(l).$$

A similar to (6.38) initial-value problem can be obtained for $W(t, l)$ from (6.11) and (6.13).

To simplify the further analysis, let us introduce the following notations $\tau = \frac{t}{\tau_f}$ and $F_*(\tau, l) = \frac{\tau_f}{w_{ml}} F(t, l)$. Then problem (6.38) can be rewritten in the form

$$\frac{dn}{d\tau} = 2 \int\limits_l^\infty R(\tau, L) p_c(\tau, l, L) n(\tau, L) dL - R(\tau, l) n(\tau, l) + F_*(\tau, l),$$

$$(6.39)$$

$$n(0, l) = n^0(l).$$

Now, we can formulate the basic properties of solutions to the kinetic equation, including the properties of some of the moments of the functions $n(\tau, l)$ and $W(\tau, l)$. To do that we will formulate and prove a theorem of existence and uniqueness of the solution of an initial-value problem similar to the one presented in [12].

From (6.22) and (6.23) follows that $R(\tau, l)$ is an increasing function on $[0, +\infty)$:

$$R(\tau, L) \geq R(\tau, l) \ for \ L \geq l \geq 0. \tag{6.40}$$

This reflects the natural property of polymer molecules according to which for longer molecules the probability to be broken is higher.

For some positive constant T_* let us introduce a function

$$A(l) = \exp[-\int\limits_0^l R_m(\eta)d\eta],$$

(6.41)

$$R_m(l) = \max_{\tau \in [0,T_*]} [R(\tau,l) \max_{\eta \in [0,l]} p_c(\tau,\eta,l)].$$

Lemma 6.6.1 *If $L_* > 0$ and $R(\tau,l)$, $p_c(\tau,l,L)$, and $A(l)$ are determined by (6.22), (6.36), and (6.41), respectively, then*

1.

$$0 < A(l) \le 1, \ l \ge 0,$$

(6.42)

2. There exists a constant $A_0 > 0$ such that

$$A(l) = \frac{A_0}{l^{\alpha_0}} + o(l^{-\alpha_0})) \ for \ l \to \infty, \ \alpha_0 = \ln 16.$$

(6.43)

Proof. Proof of item 1 in Lemma 6.6.1 follows from (6.22), (6.36), and (6.41). Item 2 follows from the relation (6.36) and the fact that according to (6.22) for large l function $R(\tau,l)$ is exponentially close to 1.

For some constant $m > 0$ let us consider the space $H_m([0,+\infty))$ of continuous functions $n(l)$ on $[0,+\infty)$ with a finite norm

$$\| n \|_m = \max_{l \ge 0} \frac{|n(l)|}{A^m(l)}$$

(6.44)

and the spaces $H^0_{T_*,K,m}$ and $H^1_{T_*,K,m}$ of continuous and one time continuously differentiable with respect to τ functions for $\tau \in [0,T_*]$ into $H_m([0,+\infty))$, respectively, with the finite norm

$$\| n \|_{T_*,K,m} = \max_{\tau \in [0,T_*]} e^{-K\tau} \| n \|_m,$$

(6.45)

where K is a positive constant the value of which will be specified later.

Theorem 6.6.1 *Assume that conditions (6.2) and (6.3) are satisfied, $L_* > 0$, and for some positive constant $m > 1/\ln 16$:*

$$n^0(l) \in H_m([0,+\infty)), \ F_*(\tau,l) \in H^0_{T_*,K,m}.$$

(6.46)

Then:

1. Problem (6.39) is linear and is uniquely solvable on $[0,T_*] \times [0,+\infty)$ and the solution $n(\tau,l)$ belongs to $H^1_{T_*,K,m}$.

2. The following estimate holds for $K > \frac{2(1-e^{-KT_*})}{m}$:

$$\| n \|_{T_*,K,m} \le \frac{1}{1-q}\{\| n^0 \|_m + \frac{mq}{2} \| F_*(\tau,l) \|_{T_*,K,m}\},$$

(6.47)

$$q = \frac{2(1-e^{-KT_*})}{Km} < 1.$$

3. In a finite time interval $[0, T_*]$, *the solution n of problem* (6.39) *depends continuously on the initial distribution* n_0 *and on the source term* $F_*(\tau, l)$, *that is if* n_1 *and* n_2 *are solutions obtained for* n_1^0, $F_{*1}(\tau, l)$ *and* n_2^0, $F_{*2}(\tau, l)$, *respectively, then*

$$\| n_1 - n_2 \|_{T_*, K, m} \leq \frac{1}{1-q} \{\| n_1^0 - n_2^0 \|_m$$

$$+ \frac{mq}{2} \| F_{*1}(t, l) - F_{*2}(t, l) \|_{T_*, K, m}\}.$$

(6.48)

Proof. Integrating Eqs. (6.39) one obtains

$$n(\tau, l) = Bn(\tau, l), \ Bn(\tau, l) = n_0(\tau, l) + 2 \int_0^\tau \int_0^\infty R(s, L) p_c(s, l, L)$$

$$\times \exp\left[-\int_s^\tau R(\theta, l) d\theta\right] n(s, l) dL ds, \ n_0(\tau, l) = n^0(l)$$

(6.49)

$$\times \exp\left[-\int_0^\tau R(s, l) ds\right] + \int_0^\tau F(s, l) \exp\left[-\int_s^\tau R(\theta, l) d\theta\right] ds,$$

where B is the integral operator. Then using (6.42), (6.44), (6.45), and (6.49), and estimating $| Bg - Bh |$, one obtains that operator B is a contraction mapping on functions g and h from $H_{T_*, K, m}^0$, i.e.,

$$\| Bg - Bh \|_{T_*, K, m} \leq q \| g - h \|_{T_*, K, m},$$

$$q = \frac{2(1 - e^{-KT_*})}{Km}, \ 0 < q < 1.$$

(6.50)

Similarly, one can obtain that

$$\| Bn_0 - n_0 \|_{T_*, K, m} \leq q \| n_0 \|_{T_*, K, m},$$

(6.51)

where function n_0 is determined from the last equation in (6.49). By taking

$$\| n_0 \|_{T_*, K, m} \leq \frac{1-q}{q} R_*$$

(6.52)

(R_* is an arbitrary positive constant) and taking into account that within the closed ball $\| n - n_0 \|_{T_*, K, m} \leq R_*$, the following estimate

$$\| Bn - n_0 \|_{T_*, K, m} \leq \| Bn - Bn_0 \|_{T_*, K, m} + \| Bn_0 - n_0 \|_{T_*, K, m}$$

$$\leq q \| n - n_0 \|_{T_*, K, m} + (1 - q) R_* \leq R_*,$$

holds, one can conclude that operator B is a constriction mapping of the closed ball $\| n - n_0 \|_{T_*, K, m} \leq R_*$ onto itself. Therefore, the continuous solution of (6.49) exists and is unique. Based on the fact that all functions in the right-hand side of (6.49) are continuous we conclude that the solution $n(\tau, l)$ is a differentiable function with respect to τ. Moreover, the iterative process $n_{k+1} = Bn_k$, $k = 0, 1, \ldots$ converges to this solution. That proves Item 1.

Item 2 follows from (6.49) and the estimate

$$\| n \|_{T_*, K, m} = \| Bn \|_{T_*, K, m} \leq \| n_0 \|_m + q \| n \|_{T_*, K, m}$$

$$+ \tfrac{qm}{2} \| F_* \|_{T_*, K, m} .$$

(6.53)

Item 3 follows from estimate (6.47) and the fact that equations (6.49) are linear.

Remark 6.6.1 *In conjunction with (6.22), (6.23), and (6.36) for the probability of scission R and the density of conditional probability of scission p_c, the results of Theorem 6.6.1 indicate that the density of the number of polymer molecules $n(\tau, l)$ as well as the density of the molecular weight distribution $W(\tau, l)$ are differentiable functions of τ, $\tau \in [0, \infty)$, and continuous functions of l, $l \geq 0$. Moreover, the densities $n(\tau, l)$ and $W(\tau, l)$ continuously depend on the initial densities $n^0(l)$ and $W^0(l)$, respectively, and on the source term $F_*(\tau, l)$.*

Corollary 6.6.1 *The assumptions of Theorem 6.6.1 for some positive constant R_* and $T_* < \infty$, and $\| n - n_0 \|_{T_*, K, m} \leq R_*$ lead to the following statements:*

1. The solution $n(\tau, l)$ of (6.49) exists and is unique in the norm

$$\| n \|_c = \max_{\tau \in [0, T_*]} \max_{l \geq 0} | n(\tau, l) |$$

(6.54)

of the space $C^1([0, T_*]) \times C([0, +\infty))$ of one time continuously differentiable with respect to τ and continuous with respect to l functions.

2.

$$n(\tau, l) \leq \tfrac{R_*}{q} e^{K\tau} A^m(l).$$

(6.55)

3. There exists a positive constant R_0 such that

$$n(\tau, l) \leq \tfrac{R_0 e^{K\tau}}{l^{m\alpha_0}} [1 + o(l^{-\alpha_0})] \; for \; l \to \infty, \; \alpha_0 = \ln 16.$$

(6.56)

Proof. Proof of the existence and uniqueness of the solution of (6.49) follows from the fact that norms defined by (6.44), (6.45), and (6.54) are equivalent on the interval $\tau \in [0, T_*]$, $T_* < \infty$. Inequality (6.55) follows from (6.44), (6.45), and (6.52). Item 3 follows from (6.55) and the asymptotic (6.43).

Remark 6.6.2 *The asymptotic estimate (6.56) indicates the fact that for the integral in (6.37) for the lubricant viscosity μ to converge the positive constant m should be sufficiently large, for example, $m\alpha_0 > \beta + 2$, $\beta > 0$.*

Corollary 6.6.2 *Under the assumptions of Theorem 6.6.1, if $W^0(l) = F_*(\tau, l) \equiv 0$ for $l \geq l_* \geq 0$, then $W(\tau, l) = 0$ for $l \geq l_*$ and $\tau > 0$.*

Theorem 6.6.2 *Assume that all conditions of Theorem 6.6.1 are satisfied and $m\alpha_0 > 2$.*

1. If $W^0(l)$ and $F_*(\tau, l)$ are nonnegative functions on $[0, +\infty)$, then

$$W(\tau, l) \geq 0, \ (\tau, l) \in [0, +\infty) \times [0, +\infty). \tag{6.57}$$

2. If $m\alpha_0 > 1$, and $\int\limits_0^\infty F_*(\tau, l)dl \geq 0$, then

$$\tfrac{d}{d\tau} \int\limits_0^\infty n(\tau, l)dl \geq 0. \tag{6.58}$$

3. If $m\alpha_0 > 2$, then

$$\tfrac{d}{d\tau} \int\limits_0^\infty W(\tau, l)dl < 0 \ if \ \int\limits_0^\infty lF_*(\tau, l)dl < 0,$$

$$\tfrac{d}{d\tau} \int\limits_0^\infty W(\tau, l)dl = 0 \ if \ \int\limits_0^\infty lF_*(\tau, l)dl = 0, \tag{6.59}$$

$$\tfrac{d}{d\tau} \int\limits_0^\infty W(\tau, l)dl > 0 \ if \ \int\limits_0^\infty lF_*(\tau, l)dl > 0.$$

4. Moreover, if $n^0(l) \in H_m([0, +\infty))$ and $F_*(\tau, l) \in H^0_{T_*, K, m}([0, +\infty))$, i.e.,

$$\int\limits_0^\infty n^0(l)l^j dl < \infty \ and \ \int\limits_0^\infty l^j F_*(\tau, l)dl < \infty \tag{6.60}$$

$$for \ all \ 0 \leq j < m\alpha_0 - 1$$

for some positive constant $m\alpha_0 > 1$, then

$$\int\limits_0^\infty n(\tau, l)l^j dl < \infty \ for \ all \ 0 \leq j < m\alpha_0 - 1 \tag{6.61}$$

and

$$\tfrac{d}{d\tau} \int\limits_0^\infty n(\tau, l)l^j dl \geq 0 \ if \ \int\limits_0^\infty l^j F_*(\tau, l)dl \geq 0, \ and \ 0 \leq j < 1$$

$$\tag{6.62}$$

$$\tfrac{d}{d\tau} \int\limits_0^\infty n(\tau, l)l^j dl \leq 0 \ if \ \int\limits_0^\infty l^j F_*(\tau, l)dl \leq 0 \ and \ 1 \leq j < m\alpha_0 - 1.$$

Proof. Since the solution to the kinetic equation is the function to which the aforementioned iterative process converges, using (6.3) one immediately gets (6.57).

From (6.39) one obtains

$$\tfrac{d}{d\tau} \int\limits_0^\infty n(\tau,l)dl = 2 \int\limits_0^\infty dl \int\limits_l^\infty R(\tau,L)p_c(\tau,l,L)n(\tau,L)dL$$

$$- \int\limits_0^\infty R(\tau,l)n(\tau,l)dl + \int\limits_0^\infty F_*(\tau,l)dl. \tag{6.63}$$

Changing the order of integration in (6.63) and using (6.3) one gets

$$\tfrac{d}{d\tau} \int\limits_0^\infty n(\tau,l)dl = \int\limits_0^\infty R(\tau,l)n(\tau,l)dl + \int\limits_0^\infty F_*(\tau,l)dl, \tag{6.64}$$

which leads to (6.58).

From (6.3) one has

$$\int\limits_0^L p_c(\tau,l,L)(2l - L)dl = 0. \tag{6.65}$$

Therefore, from (6.39) one gets

$$\tfrac{d}{dt} \int\limits_0^\infty W(\tau,l)dl = 2 \int\limits_0^\infty ldl \int\limits_l^\infty R(\tau,L)p_c(\tau,l,L)W(t,L)\tfrac{dL}{L}$$

$$- \int\limits_0^\infty R(\tau,l)W(\tau,l)dl + \int\limits_0^\infty w_m l F_*(\tau,l)dl. \tag{6.66}$$

Then changing the order of integration in (6.66) and using (6.65) one gets (6.59).

For $0 \le l \le L$, $L \ge 0$ the following inequalities are valid:

$$2l^j - L^j = 2l^j - (L - l + l)^j \ge l^j - (L - l)^j \ for \ 0 \le j < 1,$$

$$2l^j - L^j = 2l^j - (L - l + l)^j \le l^j - (L - l)^j \ for \ j \ge 1. \tag{6.67}$$

From (6.3) one gets

$$\int\limits_0^L p_c(\tau,l,L)l^j dl = \int\limits_0^L p_c(\tau,L-l,L)l^j dl = \int\limits_0^L p_c(\tau,l,L)(L-l)^j dl. \tag{6.68}$$

Therefore,

$$\int\limits_0^L p_c(\tau,l,L)[l^j - (L-l)^j]dl = 0. \tag{6.69}$$

From (6.39) one has

$$\tfrac{d}{d\tau} \int\limits_0^\infty n(\tau,l)l^j dl = 2 \int\limits_0^\infty l^j dl \int\limits_l^\infty R(\tau,L)p_c(\tau,l,L)n(\tau,L)dL$$

$$-\int\limits_0^\infty R(\tau,l)n(\tau,l)l^j\,dl + \int\limits_0^\infty l^j F_*(\tau,l)dl$$

$$(6.70)$$

$$= \int\limits_0^\infty R(\tau,L)n(\tau,L)dL \int\limits_0^L p_c(\tau,l,L)(2l^j - L^j)dl + \int\limits_0^\infty l^j F_*(\tau,l)dl.$$

Therefore, estimates (6.61) and (6.62) follow from (6.56), (6.57), (6.59), and (6.60).

Remark 6.6.3 *Formulas (6.58) and (6.59) reflect the properties of the polymer molecule degradation process in an isolated system (i.e., when $F_*(\tau,l) = 0$) during which the total number of molecules per lubricant unit mass increases while the total weight of polymer molecules per lubricant unit mass is preserved.*

Corollary 6.6.3 *Under the assumptions of Theorem 6.6.2:*

1. If W_1 and W_2 are the solutions obtained for a certain function $F_*(\tau,l)$ and the initial distributions $W_1^0(l)$ and $W_2^0(l)$, then $W_1^0(l) \le W_2^0(l)$ on $[0,+\infty)$ implies that $W_1(\tau,l) \le W_2(\tau,l)$ on $[0,T_*] \times [0,+\infty)$.

2. If W_1 and W_2 are the solutions obtained for a certain initial distribution $W^0(l)$ and source functions $F_{*1}(\tau,l)$ and $F_{*2}(\tau,l)$, then $F_{*1}(\tau,l) \le F_{*2}(\tau,l)$ on $[0,T_*] \times [0,+\infty)$ implies that $W_1(\tau,l) \le W_2(\tau,l)$ on $[0,T_*] \times [0,+\infty)$.

Proof. The proof of Items 1 and 2 is obvious.

Corollary 6.6.4 *Under the assumptions of Theorem 6.6.2 for $w_m > 0$, $\mu_a > 0$, $c_p > 0$, $k' > 0$, $k_H > 0$, $\beta > 0$, $m\alpha_0 > \beta + 2$, and*

$$\int\limits_0^\infty F_*(\tau,l)dl = 0 \text{ and } \int\limits_0^\infty l^\beta F_*(\tau,l)dl \le 0 \qquad (6.71)$$

the value of viscosity $\mu(\tau)$ determined by (6.37) is a monotonically decreasing function of τ, $\tau \ge 0$.

Proof. Proof follows from the monotonicity of the dependence of μ from (6.37) on $c_p[\eta]$ and the monotonicity of the dependence of $[\eta]$ on the integral

$$\int\limits_0^\infty w^\beta W(\tau,w)dw = w_m^{\beta+2}\int\limits_0^\infty l^{\beta+1}n(\tau,l)dl$$

that according to (6.63) is a monotonically decreasing finite function of τ (see Lemma 6.6.1) while according to (6.59) the integral

$$\int\limits_0^\infty W(\tau,w)dw$$

is conserved if the first of conditions in (6.71) is satisfied. The former is due to the fact that $m\alpha_0 > \beta + 2$.

Remark 6.6.4 *The statement of Corollary 6.6.4 indicates that in an isolated lubrication system as a result of lubricant degradation the lubricant viscosity $\mu(\tau)$ is a monotonically decreasing function of time τ.*

Theorem 6.6.3 *The equations of the problem (6.11), (6.13), and (6.37) are homogeneous; i.e., if a function $W(\tau, l)$ is a solution of (6.11), (6.13), and (6.37) for the initial condition $W^0(l)$, then for any positive constant c function $cW(\tau, l)$ is also a solution of (6.11), (6.13), and (6.37) for the initial condition $cW^0(l)$.*

Proof. The proof is obvious from (6.11), (6.13), and (6.37).

Remark 6.6.5 *Theorem 6.6.3 provides the foundation for interpretation and comparison of the numerical and experimental data obtained using the gel permeation chromatography (GPC) tests - the major technique for molecular weight measurements of soluble polymers.*

Finally, let us consider some properties of the probability of scission R and the density p_c of the conditional probability of scission.

Lemma 6.6.2 *Let $\gamma = \frac{2\alpha U}{N_A kT} > 0$, then for $l > \sqrt{\frac{U}{Ca_* l_*^2 N_A \mu S}}$:*
1. The probability of scission $R(\tau, l)$ determined by (6.22) is a monotonically increasing function of l and a monotonically decreasing function of L_*.
2. The probability of scission $R(\tau, l)$ determined by (6.22) is a monotonically increasing function of γ.

Proof. The proof follows from the expressions of the derivatives of R with respect to l, L_*, and γ, respectively.

Remark 6.6.6 *Based on Lemma 6.6.2 we can expect that the probability of scission $R(\tau, l)$ is a monotonically decreasing function of the lubricant temperature T and a monotonically increasing function of the shear stress μS. In other words, it can be expected that when the temperature T is lower and the shear stress μS is higher, the rate of lubricant degradation is higher.*

Lemma 6.6.3 *For $l \in [0, L]$ the density $p_c(\tau, l, L)$ of the conditional probability of scission in (6.36) is a decreasing function of l and $\max_{0 \leq l \leq L} p_c(\tau, l, L) = \frac{4 \ln(2)}{L}$.*

Proof. The proof follows from the expression of the derivative of $p_c(\tau, l, L)$ with respect to l.

6.7   A Limiting Case of the Kinetic Equation

Let us consider a special case of polymer molecules, which may break only into two fragments of equal length. This case is of some interest because under this assumption the kinetic equation allows for an explicit solution in quadratures. Also, this case represents an idealization of a common situation when polymer molecules predominantly break in halves. Therefore, let us assume that

$$p_c(\tau, l, L) = \delta(l - L/2), \tag{6.72}$$

where δ is the delta-function. Then equations (6.39) are reduced to

$$\frac{dW}{d\tau} = 2R(\tau, 2l)W(\tau, 2l) - R(\tau, l)W(\tau, l) + w_m l F_*(\tau, l),$$

$$W(0, l) = W^0(l). \tag{6.73}$$

Let us introduce new functions

$$f(\tau, l) = W(\tau, l) \exp \int_0^\tau R(s, l)ds,$$

$$g(\tau, l) = w_m l F_*(\tau, l) \exp \int_0^\tau R(s, l)ds. \tag{6.74}$$

For functions $f(\tau, l)$ and $g(\tau, l)$ defined in (6.74) one gets the following equations:

$$\frac{df}{d\tau} = A(\tau, l)f(\tau, 2l) + g(\tau, l), \quad f(0, l) = W^0(l),$$

$$A(\tau, l) = 2R(\tau, 2l) \exp\{\int_0^\tau [R(s, l) - R(s, 2l)]ds\}. \tag{6.75}$$

Constructing iterative solutions

$$f_{j+1}(\tau, l) = f_0(\tau, l) + \int_0^\tau A(s, l)f_j(s, l)ds, \quad j = 0, 1, \ldots,$$

$$f_0(\tau, l) = W^0(l) + \int_0^\tau g(s, l)ds, \tag{6.76}$$

one gets

$$f(\tau, l) = W^0(l) + \int_0^\tau w_m l F_*(s, l) \exp\left[\int_0^s R(s_1, l)ds_1\right]ds$$

$$+ \sum_{j=1}^\infty \int_0^\tau A(s_1, l)ds_1 \int_0^{s_1} A(s_2, 2l)ds_2 \ldots \int_0^{s_{j-1}} A(s_j, 2^j l)W_f(s_j, 2^j l)ds_j. \tag{6.77}$$

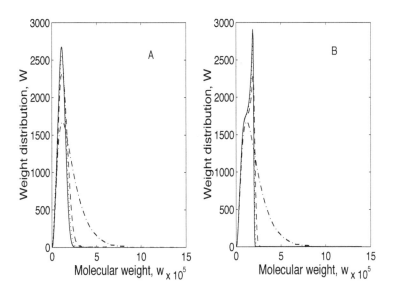

FIGURE 6.2
Polymer molecular weight distributions at different time moments during testing of the lubricant with OCP10 VM (A) and obtained from numerical modeling (B) for $U = 347kJ/mole$, $C = 0.044$, and $\alpha = 0.008$: dashed-dotted curve - initial molecular weight distribution, dashed curve - after 30 cycles, dotted curve - after 100 cycles, solid curve - after 250 cycles (after Kudish, Airapetyan, and Covitch [13]). Reprinted with permission from the STLE.

6.8 Numerical Method for Solution of the Kinetic Equation of Lubricant Degradation

The goal of this section is to describe a method for numerical solution of equations (6.39). Integrating the first equation in (6.39) along the flow streamline $\mathbf{x} = \mathbf{x}(t)$ from $\mathbf{x}(0)$ to $\mathbf{x}(t)$ one gets the following integral analog of (6.39):

$$n(\tau, l) = n^0(l)G(0, \tau, l)$$

$$+2 \int_0^\tau G(s, \tau, l)ds \int_l^\infty R(s, L)p_c(s, l, L)n(s, L)dL \tag{6.78}$$

$$+ \int_0^\tau G(s, \tau, l)F_*(s, l)ds, \ \ G(s, \tau, l) = \exp\left[-\int_s^\tau R(\theta, l)d\theta\right].$$

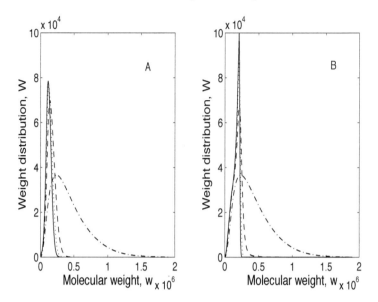

FIGURE 6.3

Polymer molecular weight distributions at different time moments during testing of the lubricant with OCP2 VM (A) and obtained from numerical modeling (B) for $U = 347kJ/mole$, $C = 0.044$, and $\alpha = 0.008$: dashed-dotted curve - initial molecular weight distribution, dashed curve - after 30 cycles, dotted curve - after 100 cycles, solid curve - after 250 cycles (after Kudish, Airapetyan, and Covitch [13]). Reprinted with permission from the STLE.

Let us assume that initially the polymer molecules dissolved in the lubricant are subdivided into N groups with the chain lengths l from the intervals $[l_N, l_{N-1}], \dots , [l_2, l_1]$, where $l_1 > l_2 > \dots > l_N > l_{N+1} = 0$. The number of molecules inside the groups is given by $N_1, N_2, \dots, N_N$. Thus, the initial density of the distribution of polymer molecules is given by the vector $(n_1^0, n_2^0, \dots, n_N^0)$, where

$$n_1^0 = \frac{N_1}{l_1 - l_2}, \; n_2^0 = \frac{2N_2}{l_1 - l_3}, \; \dots n_{N-1}^0 = \frac{2N_{N-1}}{l_{N-2} - l_N}, \; n_N^0 = \frac{2N_N}{l_{N-1} - l_N}. \qquad (6.79)$$

Denote by h a step with respect to time τ and consider the vector $(\tau_1, \tau_2, \dots, \tau_M)$, where $\tau_i = h(i - 1)$, $i = 1, 2, \dots, M$. Let us introduce the matrix $[n(i,j)]$, $i = 1, \dots, M$, $j = 1, \dots, N$, where $n(i,j)$ is the density of the distribution of the number of molecules belonging to the group with chain lengths from $[l_j, l_{j-1}]$ at the time moment τ_i. It follows from (6.78) that for the numerical evaluation of $n(i, j_0)$ for some j_0 one has to know $n(i,j)$ for $j < j_0$ only.

Therefore, one starts with the molecules of the first group with the chain

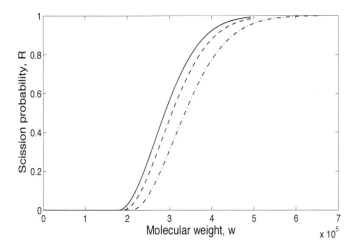

FIGURE 6.4

Distributions of the scission probability R versus polymer molecular weight $w = w_m l$ at the initial (solid curve) and final (dashed curve for the lubricant with OCP10 VM and dashed-dotted curve for the lubricant with OCP2 VM) time moments ($U = 347\ kJ/mole$, $C = 0.044$, and $\alpha = 0.008$) (after Kudish, Airapetyan, and Covitch [13]). Reprinted with permission from the STLE.

length l_1. Thus one gets:

$$n(i,1) = n^0(1)G_{0,i,1} + \int_0^{\tau_i} G(s,\tau_i,l_1)F(s,l_1)ds, \qquad (6.80)$$

where $G_{i,j,k}$ is obtained by numerical integration in the second equation in (6.78) with $s = \tau_i$, $\tau = \tau_j$, and $l = l_k$.

Then for the second group of polymer molecules with chain lengths from $[l_2, l_1]$ one numerically evaluates the integral that leads to the following formula:

$$n(i,2) = n^0(2)G_{0,i,2} + (l_1 - l_2) \int_0^{\tau_i} G(s,\tau_i,l_2)R(s,l_1)p_c(s,l_2,l_1)ds$$

$$+ \int_0^{\tau_i} G(s,\tau_i,l_2)F_*(s,l_2)ds. \qquad (6.81)$$

Continuing calculations the same way one consequently evaluates

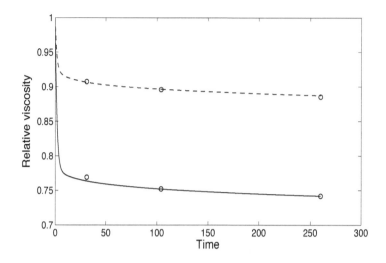

FIGURE 6.5

Loss of the lubricant viscosity μ caused by polymer molecule degradation versus number of cycles for the lubricant with OCP10 VM (dashed curve) and for the lubricant with OCP2 VM (solid curve), $U = 347\ kJ/mole$, $C = 0.044$, and $\alpha = 0.008$. Circles indicate the experimentally measured relative viscosity of the lubricants with OCP10 and OCP2 VM, respectively (after Kudish, Airapetyan, and Covitch [13]). Reprinted with permission from the STLE.

$$n(i,j) = n^0(j)G_{0,i,j} + \int_0^{\tau_i} G(s,\tau_i,l_j)[R(s,l_1)p_c(s,l_j,l_1)$$

$$n(s,l_1)(l_2 - l_1) \times + \sum_{k=2}^{j-1} R(s,l_k)p_c(s,l_j,l_k)n(s,l_k)(l_{k+1} - l_{k-1})]ds \qquad (6.82)$$

$$+ \int_0^{\tau_i} G(s,\tau_i,l_j)F(s,l_j)ds, \ \ j = 3,\ldots,N.$$

At every time step τ_i we recalculate the value of the lubricant viscosity μ following Huggins and Mark-Houwink equations (6.37) (see also [14, 15]). Integrals involved in (6.37) are calculated using the trapezoidal rule.

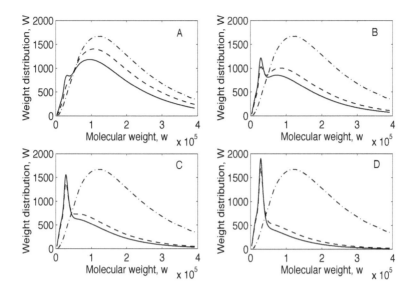

FIGURE 6.6
Polymer molecular weight distributions at different time moments obtained from numerical modeling for $U = 5 \ kJ/mole$, $C = 0.044$, and $\alpha = 1$: dashed-dotted curve - initial molecular weight distribution, (A): dashed curve - after $\tau = 0.72$, solid curve - after $\tau = 1.44$; (B): dashed curve - after $\tau = 2.16$, solid curve - after $\tau = 2.84$; (C): dashed curve - after $\tau = 3.6$, solid curve - after $\tau = 4.32$; (D): dashed curve - after $\tau = 5.04$, solid curve - after $\tau = 6$ (after Kudish, Airapetyan, and Covitch [13]). Reprinted with permission from the STLE.

6.9 Numerical Solutions of the Kinetic Equation. Comparison of the Numerical Solutions with Test Data

In this section the integro-differential kinetic equation for polymer degradation is solved numerically for $F_*(\tau, l) \equiv 0$ and analyzed for a number of different input data. The effects of pressure, shear strain rate, temperature, and lubricant viscosity on the process of lubricant degradation are considered. A comparison of numerically calculated molecular weight distributions with experimental ones obtained in bench tests showed that they are in excellent agreement with each other.

To validate the application of the kinetic equation to the process of stress-induced linear polymer scission, we compare the numerically calculated $W(\tau, l)$ with two series of experimental data obtained from a study of a me-

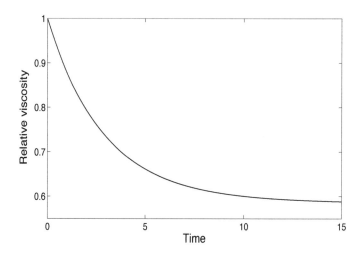

FIGURE 6.7

Loss of the lubricant viscosity μ caused by polymer molecule degradation versus time for $U = 5 \, kJ/mole$, $C = 0.044$, and $\alpha = 1$ (after Kudish, Airapetyan, and Covitch [13]). Reprinted with permission from the STLE.

chanically induced degradation of a dilute mineral oil solution of an ethylene/propylene copolymer using the Kurt Orbahn fuel injector bench test (see [11]). The results of these tests are presented in Figs. 6.2A and 6.3A as a series of polymer molecular weight distributions at four different time moments including the initial ones when $W(0, l) = W^0(l)$. Each cycle in the Orbahn injector test corresponds to a single pass of each fluid element through the shear flow field of the fuel injector.

The details of the numerical procedure are given in Section 6.8. For comparison with experimental data obtained for lubricants with OCP10 and OCP2 VM in particular calculations we assumed the following polymer parameter values (see Billmeyer [14], p. 16): $U = 347 \, kJ/mole$, $l_* = 0.154 \, nm$. The value of the bead radius a_* was derived from crystal structure of polyethylene (see Billmeyer [14], p. 126). The bead radius of an ethylene chain is $0.247 \, nm$. The OCP polymers are made up of ethylene and propylene units. For a propylene monomer unit, the bead radius is larger than the one for an ethylene unit since it includes methyl group protruding from the backbone carbon. To estimate the additional distance that a single methyl group would add to the bead radius of the polyethylene, we take into account two carbon atoms. That gives an additional distance of $0.5 \cdot 0.255 \, nm = 0.127 \, nm$. As a result, the approximate value of the bead radius of the polyethylene polypropylene copolymer a_* is $0.247 \, nm + 0.127 \, nm = 0.374 \, nm$, i.e., $a_* = 0.374 \, nm$. The other parameters characterizing the process are taken

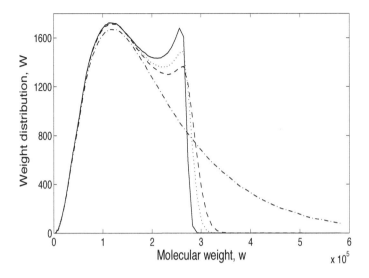

FIGURE 6.8

Polymer molecular weight distributions at different time moments obtained from numerical modeling for $U = 700\ kJ/mole$, $C = 0.044$, and $\alpha = 10^{-3}$: dashed-dotted curve - initial molecular weight distribution, dashed curve - after $\tau = 240$, dotted curve - after $\tau = 480$, solid curve - after $\tau = 1000$ (after Kudish, Airapetyan, and Covitch [13]). Reprinted with permission from the STLE.

as follows: $T = 310K$, $\mu_a = 0.00919\ Pa \cdot s$, $S = 5,000\ s^{-1}$.

In the two series of experimental data, the initial viscosities of the formulated lubricants were practically equal to the above given μ_a while the initial distributions of the polymer molecular weights $W^0(w)$ as well as their concentrations c_p were different. The initial distributions of the molecular weights are given in Figs. 6.2 and 6.3 and the concentrations of OCP10 and OCP2 VM were $c_p = 0.86\ g/dL$ and $c_p = 0.611\ g/dL$, respectively. When considering numerically the process of scission for these two different initial molecular weight distributions the same values for parameters τ_f, C, α, and k_H, k', β were used. The values for the latter parameters were taken as follows $k_H = 0.2$, $k' = 2.7 \cdot 10^{-4}\ dL/g(g/mole)^{-\beta}$, and $\beta = 0.7$ (see Crespi et al. [15]). The best match of numerical and test results is provided by $C = 0.044$ and $\alpha = 0.008$. The durations θ of one cycle in these cases are determined as follows: $\tau = 0.92\ (t = 0.92\tau_f)$ for the lubricant with OCP10 VM and $\tau = 0.32\ (t = 0.32\tau_f)$ for the lubricant with OCP2 VM. The numerical values of θ depend on a number of parameters, including the lubricant velocity $\mathbf{u}$. Therefore, if the velocity $\mathbf{u}$ is unknown or chosen different from the one in the actual test conditions then the value of θ is relative and serves only for the

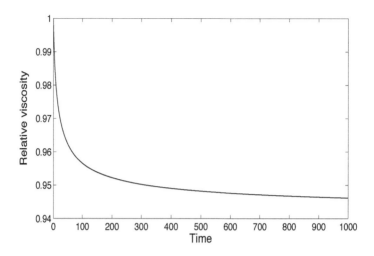

FIGURE 6.9
Loss of the lubricant viscosity μ caused by polymer molecule degradation versus time for $U = 700 \ kJ/mole$, $C = 0.044$, and $\alpha = 10^{-3}$ (after Kudish, Airapetyan, and Covitch [13]). Reprinted with permission from the STLE.

purpose of proper time scaling.

For l, according to the given experimental data, 232 nodes forming a non-uniform grid were chosen. For τ a uniform grid was used with the number of nodes depending on the length of the considered time interval (from 250 nodes in simulations on a small time scale to 10,000 nodes in simulations on a large time scale).

For the two series of input data the distributions of R versus molecular weight $w = w_m l$ at the initial and final (corresponding to 250 cycles) time moments are given in Fig. 6.4. The numerically obtained two series of polymer molecular weight distributions $W(\tau, w)$ versus polymer molecular weight w at the initial time moment and three different time moments (after 30, 100, and 250 cycles, respectively) are given in Figs. 6.2B and 6.3B. The initial molecular weight distributions in Figs. 6.2B and 6.3B were the same as for the experimental data in Figs. 6.2A and 6.3A, respectively. The comparison of the experimentally and numerically obtained molecular weight distributions presented in Figs. 6.2A, 6.2B and Figs. 6.3A, 6.3B shows excellent agreement if we take into account that the precision of the experimental measurements of the polymer molecular weight is about $10\% - 15\%$. In particular, the general shapes of the numerically obtained curves in Figs. 6.2B and 6.3B are very close to their experimental counterparts, the absolute values of the maxima of the molecular weight distributions and their locations are close to the experimental ones.

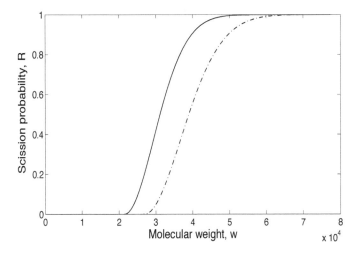

FIGURE 6.10
Distributions of the scission probability R versus polymer molecular weight $w = w_m l$ at the initial (solid curve) and final (dashed-dotted curve) time moments ($U = 5\ kJ/mole$, $C = 0.044$, and $\alpha = 1$) (after Kudish, Airapetyan, and Covitch [13]). Reprinted with permission from the STLE.

Let us consider the effect of polymer degradation on the lubricant viscosity loss. Studies of dilute polymer solutions [14] suggest that the lubricant viscosity μ depends not only on the lubricant pressure p and temperature T but also on the polymer molecular weight distribution W. For higher polymer concentrations, there exist many empirical relationships between lubricant viscosity μ and the polymer molecular weight distribution W. In this section we used the relations from [14, 15] (see Section 6.5). The calculated viscosity losses of the lubricants with OCP10 and OCP2 VM caused by polymer degradation as well as those measured experimentally at the temperature of $100°C$ are presented in Fig. 6.5. As it follows from Fig. 6.5 the calculated and experimental viscosity losses are in excellent agreement. During the first several cycles, the viscosity drops precipitously and then changes insignificantly after that. The viscosity loss of the lubricant with OCP2 VM is more than two times higher than for the lubricant with OCP10 VM because (a) for both lubricants most of the polymer molecules have molecular weight greater than $2 - 2.5 \times 10^5$ *daltons*, and, therefore, are in the region where the scission probability $R(w) \geq 1/2$ (see Fig. 6.4) and, therefore, the scission takes place and (b) for the lubricant with OCP2 VM the molecular weight of the most of the polymer molecules is about two times higher than that for the lubricant with OCP10 VM, which leads to faster (successive) scission.

It is worthwhile to understand the dependence of the polymer molecular

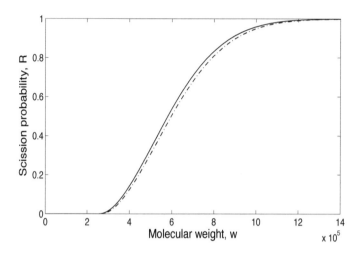

FIGURE 6.11
Distributions of the scission probability R versus polymer molecular weight
$w = w_m l$ at the initial (solid curve) and final (dashed-dotted curve) time
moments ($U = 700 \ kJ/mole$, $C = 0.044$, and $\alpha = 10^{-3}$) (after Kudish,
Airapetyan, and Covitch [13]). Reprinted with permission from the STLE.

weight distribution on function R for two reasons: (a) to explore possible
distinct degradation patterns and (b) to confirm the legitimacy of the comparison between the experimental and numerical results. Below, two additional cases of polymer degradation are considered for which only two parameters U and α are different from the input data described above. In Case 1:
$U = 5kJ/mole$, $\alpha = 1$ and in Case 2: $U = 700 \ kJ/mole$, $\alpha = 10^{-3}$. In both
cases $c_p = 0.86 \ g/dL$. The numerical results for these cases are presented in
Figs. 6.6, 6.7 and Figs. 6.8, 6.9, respectively. In Case 1 the bond dissociation
energy U is relatively small while the parameter α is relatively large. Therefore, the probability of scission R is steep and its center is positioned close
to $w = 0$ (see Fig. 6.10). The fact that the center of the distribution of R is
positioned relatively close to $w = 0$ leads to fast degradation of all polymer
molecular with molecular weight $w > 2 \cdot 10^4 \ daltons$ (see Fig. 6.10). That
results in a relatively disperse scission process. Moreover, because of such a
behavior of R at the early stages of the scission process a second maximum of
$W(\tau, w)$ is created to the left of the maximum of the initial molecular weight
distribution $W^0(w)$ (see Fig. 6.6), which grows with time while the original
maximum of the distribution of $W(\tau, w)$, related to the maximum of $W^0(w)$,
slowly disappears (see Fig. 6.6). For this case the calculated lubricant viscosity
loss is presented in Fig. 6.7.

In Case 2 the bond dissociation energy U is relatively large while the parameter α is small. Therefore, the center of the probability scission distribution R is shifted away from $w = 0$ and is positioned to the right of that for Case 1 (see Fig. 6.8). In Case 2 the slope of R is significantly smaller than in Case 1. Because $R > 0$ essentially only for $w > 2.5 \cdot 10^5$ *daltons*, the scission process runs slower than in Case 1. In particular, the noticeable departure of the polymer molecular weight distribution $W(\tau, w)$ from the initial distribution $W^0(w)$ occurs only after about $\tau = 240$ (see Fig. 6.11). At early stages of the scission process, a second maximum of $W(\tau, w)$ is created to the right of the maximum of the initial molecular weight distribution $W^0(w)$. After that the molecular distribution remains bi-modal because the threshold of the scission process L_* is located to the right of the maximum of the initial molecular weight distribution $W^0(w)$. In this case the viscosity loss (see Fig. 6.9) is smaller than in any of the above considered cases.

Now, let us characterize the process of polymer scission with respect to other parameters such as T, a_*, l_*, μ, S, α, and C. Most of the conclusions can be drawn based on the analysis of the characteristic polymer chain length L_* from (6.23) and the expression for the probability of scission $R(\tau, l)$ from (6.22) (see Lemmas 6.6.2 and 6.6.3). When the lubricant temperature T increases, the probability of polymer scission R is determined by the two competing processes: decrease of the slope of R and shift of R toward larger values of w because of usual decrease of the lubricant viscosity μ with temperature T. Therefore, the probability of polymer scission R decreases with temperature T (see Remark 6.6.6). The increase of the lubricant viscosity μ and μ_a (for example, due to increase of pressure p) and/or of the shear strain rate S cause L_* to decrease, which, in turn, accelerates scission (see Remark 6.6.6). The physical explanation of this mechanism is clear because an increase in μ and/or S leads to a corresponding increase in stretching forces acting upon polymer molecules due to increased friction between the polymer molecules and the surrounding lubricant. Finally, an increase in the value of the product $Ca_* l_*^2$ leads to the same effect as the increase in μ_a or S (see Lemma 6.6.2). An increase in the value of $\alpha U/(kT)$ leads to a steeper distribution of the probability of scission R. That results in faster scission of polymer chains (see Lemma 6.6.2).

Let us consider the effect of polymer degradation on the lubricant viscosity. Assuming that the lubricant viscosity at the initial time moment at $100°C$ is $\mu_a = 0.00919$ $Pa \cdot s$ for four cases considered above the graphs of μ/μ_a as a function of time are presented in Figs. 6.5, 6.7, and 6.9. The maximum drop in lubricant viscosity (about 35%) is observed in Case 1. In Case 2 the rate of the lubricant viscosity loss is the slowest among the four above considered cases. Moreover, the viscosity loss varies from about 5% to 35% (see Fig. 6.5, 6.7, and 6.9). In all cases the lubricant viscosity decreases due to the ongoing polymer additive degradation, which, in turn, leads to a decrease in film thickness in lubricated contacts.

As it is shown above the process of polymer molecule, scission depends strongly on the details of the behavior of the probability of polymer scission

R. Taking this into account and the fact that we were able to practically match the numerical and test data significantly elevates the confidence in the kinetic model itself and its practical applications to lubrication processes.

6.10   Closure

The process of molecular scission for a polymer additive dissolved in a lubricant is analyzed quantitatively based on a probabilistic kinetic equation for the density of the polymer molecular weight distribution. The polymer molecular scission process is affected by the temperature and the lubricant flow dynamics. The expressions for the probability of scission and the density of the conditional probability of scission are derived based on energy considerations. The existence and uniqueness of the initial-value problem for the kinetic equation is proven. Some basic properties of solutions of the kinetic equation are established. In particular, it is shown that in an isolated system the polymer weight per unit mass of the lubricant is conserved in time along the lubricant flow streamlines while the number of polymer molecules per unit of the lubricant mass is a non-decreasing function of time. It is shown that the qualitative and quantitative behavior of numerically calculated polymer molecular weight distributions and viscosity loss are in excellent agreement with the corresponding test data. It is further shown that the polymer scission process is strongly dependent on the behavior of the probability of scission R. This analysis reveals that depending on the particular properties of the probability of polymer scission R the polymer molecular weight distribution may exhibit a single- or multi-modal shape. Moreover, this analysis additionally validates the use of the kinetic equation for the analysis of lubricants with degrading polymer additives. Because of polymer additive degradation the lubricant viscosity decreases at a rate dependent on the nature of base stock, polymer additive, and the degradation regime. Based on the experimental data and the calculated results the viscosity loss varies from about 5% to 35% for two lubricants formulated with olefin copolymer viscosity modifiers of different average molecular weights.

6.11   Exercises and Problems

1. Show that for $l \geq 0$ the probability of scission $R(t, l)$ from equation (6.22) is a smooth monotonically increasing function of l.

2. What is the role of parameter L_* in the process of polymer molecule degradation?

3. Assuming that the polymer solution viscosity μ remains constant show that for $l > L_*$ the probability of scission $R(t, l)$ is a monotonically decreasing function of the polymer solution temperature T.

4. Show that the average length of the fragments resulted from scission of polymer molecules of length L is $L/2$.

5. Show that in an isolated system the molecular weight $W(\tau, \mathbf{x}, l)$ of a polymer solution is conserved along the flow streamlines while the number of polymer molecules $n(\tau, \mathbf{x}, l)$ tend to increase. Explain what takes place when over time some polymer molecules added or removed from the system.

6. Explain how the position of the midpoint of the probability of scission distribution $R(\tau, l)$ (i.e., the polymer molecule length l for which $R(\tau, l) = 1/2$) affects the rate of polymer molecule scission. What is the role of the slope of $R(\tau, l)$ in the process of polymer molecule scission?

7. Provide an equation based explanation as to why the gel permeation chromatography (GPC) test for measuring polymer molecular weight distribution $W(\tau, l)$ is supported by the presented model.

8. Explain and describe the tendency of the polymer solution viscosity μ to experience loss in time at different rates by relating the viscosity behavior to the behavior of the probability of scission $R(\tau, l)$.

9. Consider two polymer solutions with the same polymer additive but different average length of polymer chains. Which solution is more resistant to the viscosity loss: the solution with longer or shorter average polymer chain length? Explain your conclusion by relating the behavior of the viscosity to the behavior of the probability of scission $R(\tau, l)$.

BIBLIOGRAPHY

[1] Herbeaux, J.-L., Flamberg, A., Koller, R.D., and Van Arsdale, W.E. 1998. Assesment of Shear Degradation Simulators. *SAE Tech. Paper* No. 982637.

[2] Herbeaux, J.-L. 1996. *Mechanochemical Reactions in Polymer Solutions*. Ph.D. Dissertation, University of Houston, Houston (and references therein).

[3] Saito, O. 1958. On the Effect of High Energy Radiation to Polymers. I, Cross-Linking and Degradation. *J. Phys. Soc. Jpn.* 13:198-206.

[4] Montroll, E.W. and Simha, R. 1940. Theory of Depolymerization of Long Chain Molecules. *J. Chem. Phys.* 8:721-727.

[5] Ziff, R.M. and McGrady, E.D. 1985. The Kinetics of Cluster Fragmentation and Depolimerization. *J. Phys. A : Math. Gen.* 18:3027-3037.

[6] Ziff, R.M. and McGrady, E.D. 1986. Kinetics of Polymer Degradation. *AchS, Macromolecules* 19:2513-2519.

[7] McGrady, E.D. and Ziff, R.M. 1988. Analytical Solutions to Fragmentation Equations with Flow. *AchS J.* 34, No. 12:2073-2076.

[8] Burger, M. and Capasso, V. 2001. Mathematical Modeling and Simulation of Non-isothermal Crystallization of Polymers. *J. Math. Models and Meth. Appl. Sci.* 11, No. 6:1029-1053.

[9] Kudish, I.I. and Ben-Amotz, D. 1999. Modeling Polymer Molecule Scission in EHL Contacts. *The Advancing Frontier of Engineering Tribology, Proc. 1999 STLE/ASME H.S. Cheng Tribology Surveillance*, Eds.: Q. Wang, J. Netzel, and F. Sadeghi, pp. 176-182.

[10] Odell, J.A., Keller, A., and Rabin, Y. 1988. Flow-Induced Scission of Isolated Macromolecules. *J. Chem. Phys.* 88, No. 6:4022-4028 (and references therein).

[11] Covitch, M.J. 1998. How Polymer Architecture Affects Permanent Viscosity Loss of Multigrade Lubricants. *SAE Tech. Paper* No. 982638.

[12] Kudish, I.I., Airapetyan, R.G., and Covitch, M.J. 2002. Modeling of Kinetics of Strain Induced Degradation of Polymer Additives in Lubricants. *J. Math. Models and Meth. Appl. Sci.* 12, No. 6:835-856.

[13] Kudish, I.I., Airapetyan, R.G., and Covitch, M.J. 2003. Modeling of Kinetics of Stress Induced Degradation of Polymer Additives in Lubricants and Viscosity Loss. *STLE Tribology Trans.* 46, No. 1:1-11.

[14] Billmeyer, F.W., Jr. 1962. *Textbook of Polymer Science*, New York: John Wiley & Sons.

[15] Crespi, G., Vassori, A., Slisi, U. 1977. Olefin Copolymers. In *The Stereo Rubbers*. Ed. W.M. Saltman. New York: John Wiley & Sons. 365-431.

7

Modeling of Lubricant Degradation in Kurt Orbahn Shear Stability Tests for Viscosity Improvers Based on Star Polymers

7.1 Introduction

The kinetics of stress–induced degradation of a star polymer additive dissolved in a mineral oil lubricant is modeled. The polymer degradation is modeled based on a new system of kinetic integro-differential equations for the distribution densities of star polymer molecules with different number of arms and arm chain lengths. Some properties of the solution are established. Among these properties are the existence and uniqueness of the solution of the initial-value problem for the above–mentioned system of integro-differential kinetic equations for star polymer degradation. A numerical method for solution of the problem is proposed and realized. Some of the numerically simulated molecular weight distributions are compared with the independently obtained experimental ones. The lubricant viscosity losses due to polymer degradation are determined and compared with the experimentally measured ones. The theoretical and experimental data are in very good agreement.

7.2 System of Kinetic Equations for Degradation of Lubricants with Star Polymeric Additives

An important class of VMs used in lubricant industry is represented by star polymers, notably, hydrogenated poly(styrene-isoprene) copolymer stars. A star polymer molecule is composed of a small organic core to which a number of linear polymer arms are attached (see Fig. 7.1). In this section a system of kinetic equations for the densities of polymer molecular weight distributions describing the mechanism of stress–induced lubricant degradation is derived. Some general properties of the moments of the polymer molecular weight distributions are established. Among these properties are the existence and

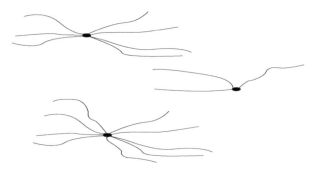

FIGURE 7.1
A view of star polymer molecules stretched in a lubricant flow.

uniqueness of the solution of the initial-value problem for the above–mentioned system of integro-differential kinetic equations for star polymer degradation. Also, these properties include conservation of the total polymer mass and increase of the total number of polymer molecules along the flow streamlines, decrease of the molecular weight and non-increase of the number of star polymer molecules with the number of polymer arms greater or equal to a chosen number in an isolated system. A numerical study of the system of kinetic equations is performed. The comparison of the theoretically predicted and experimentally measured molecular weight distributions of star polymers as well as of the predicted and measured viscosity losses shows that they are in very good agreement. A parametric analysis of the model is performed. The analysis that follows is based on [1].

Let us consider the process of stress–induced degradation of a lubricant formulated with a star polymeric viscosity modifier. In practice, star polymer additives are represented by a mixture of star polymer molecules with different number of arms. The number of such arms of a star polymer molecule may vary, for example, from 1 to 45. It is recognized that if a star polymer molecule has just one arm it can be considered as a linear polymer due to the fact that its core is small. It is assumed that all arms are made of the same polymer material with linear structure.

In practice, the composition of VM grade star polymer molecules is such that the molecular weight of one core is much smaller than the molecular weight of its arms. This should be understood statistically because the molecular weights of the cores as well as the molecular weights of the arms are random values and have specific probabilistic distributions. Therefore, we will analyze the kinetics of arm scission disregarding the effect of core presence. Moreover, we will assume that the concentration of the polymer additive in a lubricant is small and the arms of different star molecules as well as of the same ones do not interact with each other in any way. Finally, we will assume that the arms of star polymer molecules are stretched along the lubricant flow streamlines.

In the further analysis, it is also assumed that the polymer additive degrades due to stress induced scission, i.e., by rupture of polymer molecules that is caused by frictional stresses applied to them. Therefore, there are two different elementary mechanisms of polymer scission related to the following events: (a) an arm is ripped off the core completely and (b) an arm is shred in pieces. We will assume that at any given moment only one elementary process such as an arm of a polymer molecule is completely ripped off the core of the molecule or an arm is shred in pieces can take place. It is not a restrictive assumption because at any next moment another arm can be ripped off the core or shred in pieces. In particular, we assume that the latter two events never happen simultaneously. Moreover, taking into account the fact that shredding polymer arms in more than two pieces simultaneously is a very seldom event we will assume that if an arm is shredded it is shredded in just two pieces. These assumptions significantly simplify the development of the model and its analysis.

Let us assume that a lubricant is a compressible fluid with the mass density $\rho(t, \mathbf{x})$, where t is the time and $\mathbf{x} = (x, y, z)$ is the coordinate vector of the point of interest. The fluid is flowing with the velocity $\mathbf{u} = (u_x, u_y, u_z)$. The size of polymer arms we will characterize by their chain length l. We will assume than the polymer arm lengths are much smaller than the typical scale of fluid flow variations. Now, we can introduce the density $n_i(t, \mathbf{x}, l)$ of the statistical distribution of the number of single arms attached to cores of polymer molecules with i arms at the time moment t in a small lubricant volume $\triangle v(t)$ centered at the point $\mathbf{x}$ with the chain length between l and $l+\triangle l$, $i = 1, \ldots, I$, $(I > 1)$, i.e., $n_i(t, \mathbf{x}, l)\triangle l\triangle v(t)$ is the number of such arms attached to star polymer molecules with i arms. Here I is the maximum number of arms a star polymer molecule may have. Similarly, we can introduce the density $n_0(t, \mathbf{x}, l)$ of the statistical distribution of the number of single arms at the time moment t in a small lubricant volume $\triangle v(t)$ centered at the point $\mathbf{x}$ with the chain length between l and $l + \triangle l$, i.e., $n_0(t, \mathbf{x}, l)\triangle l\triangle v(t)$ is the number of single arms.

During the process of lubricant degradation, there is an exchange between groups of star molecules with different number of arms. Some of the arms get ripped off the cores as a whole and such molecules enter the group of star molecules with the number of arms smaller by one while the removed arms enter the group of single arms. Similarly, when arms of star molecules get shredded into pieces such star molecules continue to stay in their own groups while the shredded arm pieces provide an addition to the group of single arms.

Let us assume that $R_i(t, \mathbf{x}, l)$ is the probability of scission of a polymer arm for molecules with i arms, i.e., the probability of an arm of chain length l of a molecule with i arms located at the time moment t at the point $\mathbf{x}$ to break into two pieces over a unit time period. In addition to that we will assume that $R_{ai}(t, \mathbf{x}, l)$ is the probability of an arm of chain length l of a molecule with i arms located at the time moment t at the point $\mathbf{x}$ to be ripped off the core of a star polymer molecule as a whole. Obviously, longer arms have higher probabilities to be shredded in pieces or ripped off completely. Therefore, we will assume that

$$0 \leq R_i(t, \mathbf{x}, l) \leq 1, \ R_i(t, \mathbf{x}, l) \leq R_i(t, \mathbf{x}, L) \ for \ l \leq L,$$

$$0 \leq R_{ai}(t, \mathbf{x}, l) \leq 1, \ R_{ai}(t, \mathbf{x}, l) \leq R_{ai}(t, \mathbf{x}, L) \ for \ l \leq L. \tag{7.1}$$

We will assume that the characteristic time of one polymer chain rupture is τ_f.

Similarly, we will assume that $p_c(t, \mathbf{x}, l, L)$ is the density of the conditional probability of an arm of chain length L located at the time moment t at the point $\mathbf{x}$ to break into two pieces of chain lengths from the intervals $[l, l + \Delta l]$ and $[L - l, L - l - \Delta l]$, i.e., $p_c(t, \mathbf{x}, l, L)\Delta l$ is the probability of the above event and

$$p_c(t, \mathbf{x}, l, L) \geq 0, \ p_c(t, \mathbf{x}, L - l, L) = p_c(t, \mathbf{x}, l, L), \ 0 \leq l \leq L;$$

$$\int_0^L p_c(t, \mathbf{x}, l, L)dl = 1 \ for \ L > 0. \tag{7.2}$$

The particular expressions for $R_i(t, \mathbf{x}, l)$, $R_{ai}(t, \mathbf{x}, l)$, $i = 1, \ldots, I$, and $p_c(t, \mathbf{x}, l, L)$ are given in Section 7.3.

Then we can determine the rates $\triangle_+ \Delta l$ and $\triangle_- \Delta l$ at which the number of single polymer arms enter and leave the chain length interval $[l, l + \Delta l]$ due to fragmentation, respectively. Following the derivation of Section 6.2 (see [2, 3, 4]), we obtain

$$\triangle_+ = \frac{1}{\tau_f} 2 \int_l^\infty R(t, \mathbf{x}, L)p_c(t, \mathbf{x}, l, L)n_0(t, \mathbf{x}, L)dL,$$

$$\triangle_- = \frac{1}{\tau_f} R(t, \mathbf{x}, l)n_0(t, \mathbf{x}, l), \ i = 0, \ldots, I. \tag{7.3}$$

It is assumed here that the scission probability R_0 for single arms is identical to the scission probability R_1 of polymer molecules with just one arm and it is denoted by R.

Further, we will assume that at any given time moment only one event may take place: fragmentation of a polymer arm or ripping off a whole polymer arm of a core. Obviously, the conditional probabilities of the latter two events to happen are

$$\frac{R_i}{R_i + R_{ai}}, \ \frac{R_{ai}}{R_i + R_{ai}}, \tag{7.4}$$

respectively. Under these conditions we can determine the rate of the number of single polymer arms contribution $\triangle_{ci} \Delta l$ at which the single polymer arms ripped off the star polymer molecules with i arms as a whole enter the group of single arms with chain lengths from the interval $[l, l + \Delta l]$, i.e.,

$$\triangle_{ci} = \frac{1}{\tau_f} \frac{R_i(t, \mathbf{x}, l)R_{ai}(t, \mathbf{x}, l)}{R_i(t, \mathbf{x}, l) + R_{ai}(t, \mathbf{x}, l)} n_i(t, \mathbf{x}, l), \ i = 0, \ldots, I. \tag{7.5}$$

Similarly, we will determine the rate at which the number of fragments of single polymer arms $\triangle_{fi}\triangle l$ enter the group of single polymer arms with the chain length from the interval $[l, l + \triangle l]$,

$$\triangle_{fi} = \frac{1}{\tau_f} \int_l^\infty \frac{R_i^2(t,\mathbf{x},L)p_c(t,\mathbf{x},l,L)}{R_i(t,\mathbf{x},L)+R_{ai}(t,\mathbf{x},L)} n_i(t,\mathbf{x},L), \; i = 0, \ldots, I. \tag{7.6}$$

That occurs due to fragmentation of the polymer arms of the star polymer molecules with i arms.

Moreover, in the same fashion we determine the rates at which the number of polymer arms attached to the star molecules with i arms $\triangle_{+i}\triangle l$ and $\triangle_{-i}\triangle l$ enter and leave, respectively, the chain length interval $[l, l + \triangle l]$ due to fragmentation

$$\triangle_{+i} = \triangle_{fi} = \frac{1}{\tau_f} \int_l^\infty \frac{R_i^2(t,\mathbf{x},L)p_c(t,\mathbf{x},l,L)}{R_i(t,\mathbf{x},L)+R_{ai}(t,\mathbf{x},L)} n_i(t,\mathbf{x},L),$$
$$\tag{7.7}$$
$$\triangle_{-i} = \frac{1}{\tau_f} \frac{R_i^2(t,\mathbf{x},l)}{R_i(t,\mathbf{x},l)+R_{ai}(t,\mathbf{x},l)} n_i(t,\mathbf{x},l), \; i = 0, \ldots, I.$$

Now, let us introduce the rates $\triangle_{ci}^0 \triangle l$ and $\triangle_{ci+1}^* \triangle l$ at which the number of arms from the chain length interval $[l, l + \triangle l]$ attached to the star molecules with i and $i + 1$ arms leave and enter the group of the star molecules with i arms, respectively, due to the removal of one of the arms as a whole. The rate $\triangle_{ci}^0 \triangle l$ is a sum of two rates, i.e.,

$$\triangle_{ci}^0 = \triangle_{ci}^{0a} + \triangle_{ci}^{0b}, \; i = 0, \ldots, I. \tag{7.8}$$

The rate $\triangle_{ci}^{0a}\triangle l$ is due to the arms leaving the group of the star molecules with i arms because of the removal of a whole arm of chain length l, i.e.,

$$\triangle_{ci}^{0a} = \frac{1}{\tau_f} \frac{R_i(t,\mathbf{x},l)R_{ai}(t,\mathbf{x},l)}{R_i(t,\mathbf{x},l)+R_{ai}(t,\mathbf{x},l)} n_i(t,\mathbf{x},l), \; i = 0, \ldots, I. \tag{7.9}$$

and the rate $\triangle_{ci}^{0b}\triangle l$ is due to the arms leaving the group of the star molecules with i arms because of the removal of a whole arm of a chain length different from l. Let us determine the probability of the event to find a star molecule with i arms one of which has the chain length of l and with one of the other $i - 1$ arms removed as a whole. Then $N_l = n_i(t,\mathbf{x},l)\triangle l$ is the number of polymer arms of chain length l among all star molecules with i arms. Let N_i and N_L be the total number of arms of all star molecules with i arms and the number of the arms removed as a whole from the star molecules with i arms of chain lengths different from l, respectively, at the time moment t located at $\mathbf{x}$. It is easy to derive the expressions for the above quantities

$$N_i = \int_0^\infty n_i(t,\mathbf{x},L)dL, \; N_L = \int_0^\infty \frac{R_i(t,\mathbf{x},L)R_{ai}(t,\mathbf{x},L)n_i(t,\mathbf{x},L)dL}{R_i(t,\mathbf{x},L)+R_{ai}(t,\mathbf{x},L)}$$
$$\tag{7.10}$$
$$-\triangle l \frac{R_i(t,\mathbf{x},l)R_{ai}(t,\mathbf{x},l)n_i(t,\mathbf{x},l)}{R_i(t,\mathbf{x},l)+R_{ai}(t,\mathbf{x},l)}, \; i = 0, \ldots, I.$$

Now, we are ready to determine the probability p_{lLi} of the event to find among the star molecules with i arms a star molecule with one arm of chain length l and other $i - 1$ arms of some other chain lengths. Using combinatorics we find the probability p_{lLi} as follows

$$p_{lLi} = \frac{N_l N_L C_{N_i-2}^{i-2}}{N_l C_{N_i-1}^{i-1}} = (i-1)\frac{N_L}{N_i-1}. \tag{7.11}$$

We have to remember that the probabilistic approach developed in this section is valid only if the numbers of star molecules and their arms are large. Therefore, we can neglect 1 in comparison with N_i, $i = 1, \ldots, I$, and represent the later expression for p_{lLi} in the final form

$$p_{lLi} = (i-1)\frac{N_L}{N_i} \quad i = 1, \ldots, I. \tag{7.12}$$

After that we determine the rate $\triangle_{ci}^{0b}\triangle l$ from the equation

$$\triangle_{ci}^{0b} = \frac{1}{\tau_f}\{(i-1)n_i(t,\mathbf{x},l)\left[\int\limits_0^\infty \frac{R_i(t,\mathbf{x},L)R_{ai}(t,\mathbf{x},L)n_i(t,\mathbf{x},L)dL}{R_i(t,\mathbf{x},L)+R_{ai}(t,\mathbf{x},L)}\right.$$

$$\left. -\triangle l\frac{R_i(t,\mathbf{x},l)R_{ai}(t,\mathbf{x},l)n_i(t,\mathbf{x},l)}{R_i(t,\mathbf{x},l)+R_{ai}(t,\mathbf{x},l)}\right]/\int\limits_0^\infty n_i(t,\mathbf{x},L)dL\}, \quad i = 1, \ldots, I. \tag{7.13}$$

Therefore, from (7.8), (7.9), and (7.13) we obtain

$$\triangle_{ci}^0 = \frac{1}{\tau_f}\left\{(i-1)n_i(t,\mathbf{x},l)\left[\int\limits_0^\infty \frac{R_i(t,\mathbf{x},L)R_{ai}(t,\mathbf{x},L)n_i(t,\mathbf{x},L)dL}{R_i(t,\mathbf{x},L)+R_{ai}(t,\mathbf{x},L)}\right.\right.$$

$$\left. -\triangle l\frac{R_i(t,\mathbf{x},l)R_{ai}(t,\mathbf{x},l)n_i(t,\mathbf{x},l)}{R_i(t,\mathbf{x},l)+R_{ai}(t,\mathbf{x},l)}\right]/\int\limits_0^\infty n_i(t,\mathbf{x},L)dL \tag{7.14}$$

$$\left. +\frac{R_i(t,\mathbf{x},l)R_{ai}(t,\mathbf{x},l)}{R_i(t,\mathbf{x},l)+R_{ai}(t,\mathbf{x},l)}n_i(t,\mathbf{x},l)\right\}, \quad i = 1, \ldots, I.$$

Finally, we can conclude that

$$\triangle_{ci+1}^* = \triangle_{ci+1}^{0b}, \quad i = 1, \ldots, I,$$

$$\triangle_{ci+1}^* = \frac{1}{\tau_f}\left\{in_{i+1}(t,\mathbf{x},l)\left[\int\limits_0^\infty \frac{R_{i+1}(t,\mathbf{x},L)R_{ai+1}(t,\mathbf{x},L)n_{i+1}(t,\mathbf{x},L)dL}{R_{i+1}(t,\mathbf{x},L)+R_{ai+1}(t,\mathbf{x},L)}\right.\right.$$

$$\left. -\triangle l\frac{R_{i+1}(t,\mathbf{x},l)R_{ai+1}(t,\mathbf{x},l)n_{i+1}(t,\mathbf{x},l)}{R_{i+1}(t,\mathbf{x},l)+R_{ai+1}(t,\mathbf{x},l)}\right]/\int\limits_0^\infty n_{i+1}(t,\mathbf{x},L)dL\right\}, \tag{7.15}$$

Therefore, taking into account the above relationships and the lubricant flow we can write the following equations describing the exchanges between different groups of star polymer molecules and the group of single arms

$$n_0(t + \triangle t, \mathbf{x} + \triangle \mathbf{x}, l) \triangle l \triangle v(t + \triangle t) = n_0(t, \mathbf{x}, l) \triangle l \triangle v(t)$$

$$+ (\triangle_+ - \triangle_-) \triangle l \triangle v(t) \triangle t + \sum_{i=1}^{I} \triangle_{ci} \triangle l \triangle v(t) \triangle t$$

$$+ \sum_{i=1}^{I} \triangle_{fi} \triangle l \triangle v(t) \triangle t + F_0(t, \mathbf{x}, l) \triangle l \triangle v(t) \triangle t,$$

$$n_i(t + \triangle t, \mathbf{x} + \triangle \mathbf{x}, l) \triangle l \triangle v(t + \triangle t) = n_i(t, \mathbf{x}, l) \triangle l \triangle v(t)$$

$$+ (\triangle_{+i} - \triangle_{-i}) \triangle l \triangle v(t) \triangle t - \triangle_{ci}^0 \triangle l \triangle v(t) \triangle t + \triangle_{ci+1}^* \triangle l \triangle v(t) \triangle t$$

$$+ F_i(t, \mathbf{x}, l) \triangle l \triangle v(t) \triangle t, \; i = 1, \ldots, I-1,$$

$$n_I(t + \triangle t, \mathbf{x} + \triangle \mathbf{x}, l) \triangle l \triangle v(t + \triangle t) = n_I(t, \mathbf{x}, l) \triangle l \triangle v(t)$$

$$+ (\triangle_{+I} - \triangle_{-I}) \triangle l \triangle v(t) \triangle t - \triangle_{cI}^* \triangle l \triangle v(t) \triangle t$$

$$+ F_I(t, \mathbf{x}, l) \triangle l \triangle v(t) \triangle t,$$

(7.16)

where $F_i(t, \mathbf{x}, l) \triangle l \triangle v(t) \triangle t, \; i = 0, \ldots, I$, are the numbers of single arms and the arms attached to star polymer molecules with i arms, respectively, of length from l to $l + \triangle l$ added to or removed from the fluid volume $\triangle v(t)$ centered at $\mathbf{x}$ during the time period $\triangle t$, $\triangle \mathbf{x}$ is the displacement of the small fluid volume $\triangle v(t)$ along the fluid flow streamline $\mathbf{x} = \mathbf{x}(t)$ over a time period $\triangle t$, $\triangle \mathbf{x} = \mathbf{u} \triangle t$. The fluid flow streamlines $\mathbf{x}(t)$ are the solutions of the system of differential equations

$$\frac{d\mathbf{x}}{dt} = \mathbf{u}(t, \mathbf{x}), \; \mathbf{x}(0) = \mathbf{x}_0, \qquad (7.17)$$

where $\mathbf{x}_0$ is a starting point of a flow streamline.

Taking into account that along the flow streamlines the mass $\triangle m(t)$ of a fluid particle is conserved, i.e. $\triangle m(t) = \triangle m(t + \triangle t) = const$, and $\triangle m(t) = \rho(t, \mathbf{x}) \triangle v(t)$ after simple rearrangement of equations (7.16), dividing them by $\triangle t \triangle l \triangle m$, and taking the limits as $\triangle t \to 0$ and $\triangle l \to 0$ we obtain a system of integro-differential equations

$$\tau_f \rho \frac{d}{dt} \frac{n_0}{\rho} = 2 \int_l^{\infty} R(t, \mathbf{x}, L) p_c(t, \mathbf{x}, l, L) n_0(t, \mathbf{x}, L) dL$$

$$- R(t, \mathbf{x}, l) n_0(t, \mathbf{x}, l) + \sum_{i=1}^{I} M_{ai}(t, \mathbf{x}, l) n_i(t, \mathbf{x}, l)$$

$$+ \sum_{i=1}^{I} \int_l^{\infty} M_i(t, \mathbf{x}, L) p_c(t, \mathbf{x}, l, L) n_i(t, \mathbf{x}, L) dL + \tau_f F_0(t, \mathbf{x}, l),$$

(7.18)

$$M_i(t, \mathbf{x}, l) = \frac{R_i^2(t,\mathbf{x},l)}{R_i(t,\mathbf{x},l)+R_{ai}(t,\mathbf{x},l)}, \quad M_{ai}(t, \mathbf{x}, l) = \frac{R_i(t,\mathbf{x},l)R_{ai}(t,\mathbf{x},l)}{R_i(t,\mathbf{x},l)+R_{ai}(t,\mathbf{x},l)},$$

$$\tau_f \rho \frac{d}{dt} \frac{n_i}{\rho} = \int\limits_l^\infty M_i(t, \mathbf{x}, L) p_c(t, \mathbf{x}, l, L) n_i(t, \mathbf{x}, L) dL$$

$$-R_i(t, \mathbf{x}, l) n_i(t, \mathbf{x}, l)$$

$$-(i-1)n_i(t, \mathbf{x}, l) \int\limits_0^\infty M_{ai}(t, \mathbf{x}, L) n_i(t, \mathbf{x}, L) dL / \int\limits_0^\infty n_i(t, \mathbf{x}, L) dL \qquad (7.19)$$

$$+i n_{i+1}(t, \mathbf{x}, l) \int\limits_0^\infty M_{ai+1}(t, \mathbf{x}, L) n_{i+1}(t, \mathbf{x}, L) dL / \int\limits_0^\infty n_{i+1}(t, \mathbf{x}, L) dL$$

$$+\tau_f F_i(t, \mathbf{x}, l), \quad i = 1, \dots, I-1,$$

$$\tau_f \rho \frac{d}{dt} \frac{n_I}{\rho} = \int\limits_l^\infty M_I(t, \mathbf{x}, L) p_c(t, \mathbf{x}, l, L) n_I(t, \mathbf{x}, L) dL$$

$$-R_I(t, \mathbf{x}, l) n_I(t, \mathbf{x}, l)$$

$$(7.20)$$

$$-(I-1)n_I(t, \mathbf{x}, l) \int\limits_0^\infty M_{aI}(t, \mathbf{x}, L) n_I(t, \mathbf{x}, L) dL / \int\limits_0^\infty n_I(t, \mathbf{x}, L) dL +$$

$$\tau_f F_I(t, \mathbf{x}, l),$$

describing the kinetics of stress-induced scission of star polymer molecules. The fact that $M_i + M_{ai} = R_i$ is taken into account in (7.18)-(7.20). Furthermore, in (7.18)-(7.20) the derivative with respect to t is the total derivative along the flow streamline and it is introduced as follows

$$\frac{d}{dt} = \frac{\partial}{\partial t} + \mathbf{u} \frac{\partial}{\partial \mathbf{x}}. \qquad (7.21)$$

Obviously, the initial conditions for (7.18)-(7.20) are

$$n_i(0, \mathbf{x}_0, l) = n_i^0(l), \quad i = 0, \dots, I, \qquad (7.22)$$

where $n_{i0}(l)$, $i = 0, 1, \dots, I$, are known functions of the arm chain length l. The way to find the latter functions from the experimentally obtained distribution of molecular weight versus total lengths of arms of polymer molecules l is described in Section 7.5.

In addition to densities $n_i(t, \mathbf{x}, l)$ of the number of arms of star polymer molecules with i arms, we will also use the densities $W_i(t, \mathbf{x}, l)$ of the molecular weight of arms of star polymer molecules with i arms determined by formulas

$$W_i(0, \mathbf{x}_0, l) = w_m l n_i^0(0, \mathbf{x}_0, l), \quad i = 0, \dots, I, \qquad (7.23)$$

where w_m is the monomer molecular weight. Obviously, the system of kinetic equations for the densities $W_i(t, \mathbf{x}, l)$, $i = 0, \ldots, I$, can be obtained from (7.18)-(7.20) by multiplying them by $w_m l$ and using (7.23).

Let us consider some special cases. In case, when all polymer molecules are represented by just single arms we have $n_i(t, \mathbf{x}, l) = 0$, $i = 1, \ldots, I$ and the problem is reduced to the kinetic equation (7.18) for linear polymer molecules studied in Chapter 6. In case when $M_i = M$ and $M_{ai} = M_a$, $i = 1, \ldots, I$, equations (7.18)-(7.20) can be reduced to the following system

$$\tau_f \rho \frac{d}{dt} \frac{n_0}{\rho} = 2 \int_l^\infty R(t, \mathbf{x}, L) p_c(t, \mathbf{x}, l, L) n_0(t, \mathbf{x}, L) dL$$

$$- R(t, \mathbf{x}, l) n_0(t, \mathbf{x}, l)$$

$$+ M_a(t, \mathbf{x}, l) n_{iT}(t, \mathbf{x}, l) + \int_l^\infty M(t, \mathbf{x}, L) p_c(t, \mathbf{x}, l, L) n_{iT}(t, \mathbf{x}, L) dL \qquad (7.24)$$

$$+ \tau_f F_0(t, \mathbf{x}, l),$$

$$M(t, \mathbf{x}, l) = \frac{R^2(t, \mathbf{x}, l)}{R(t, \mathbf{x}, l) + R_a(t, \mathbf{x}, l)}, \quad M_a(t, \mathbf{x}, l) = \frac{R(t, \mathbf{x}, l) R_a(t, \mathbf{x}, l)}{R(t, \mathbf{x}, l) + R_a(t, \mathbf{x}, l)},$$

$$\tau_f \rho \frac{d}{dt} \frac{n_i}{\rho} = \int_l^\infty M(t, \mathbf{x}, L) p_c(t, \mathbf{x}, l, L) n_i(t, \mathbf{x}, L) dL$$

$$- R(t, \mathbf{x}, l) n_i(t, \mathbf{x}, l)$$

$$- (i-1) n_i(t, \mathbf{x}, l) \int_0^\infty M_a(t, \mathbf{x}, L) n_i(t, \mathbf{x}, L) dL / \int_0^\infty n_i(t, \mathbf{x}, L) dL \qquad (7.25)$$

$$+ i n_{i+1}(t, \mathbf{x}, l) \int_0^\infty M_a(t, \mathbf{x}, L) n_{i+1}(t, \mathbf{x}, L) dL / \int_0^\infty n_{i+1}(t, \mathbf{x}, L) dL$$

$$+ \tau_f F_i(t, \mathbf{x}, l), \quad i = 1, \ldots, I-1,$$

$$\tau_f \rho \frac{d}{dt} \frac{n_I}{\rho} = \int_l^\infty M(t, \mathbf{x}, L) p_c(t, \mathbf{x}, l, L) n_I(t, \mathbf{x}, L) dL$$

$$- R(t, \mathbf{x}, l) n_I(t, \mathbf{x}, l)$$

$$\qquad (7.26)$$

$$- (I-1) n_I(t, \mathbf{x}, l) \int_0^\infty M_a(t, \mathbf{x}, L) n_I(t, \mathbf{x}, L) dL / \int_0^\infty n_I(t, \mathbf{x}, L) dL$$

$$+ \tau_f F_I(t, \mathbf{x}, l),$$

where the definition of the function $n_{1T}(t, \mathbf{x}, l)$ is given later in equation (7.82). Furthermore, if $M_a = 0$ the the system (7.24)-(7.26) can be significantly simplified and reduced to the following linear system

$$\tau_f \rho \frac{d}{dt} \frac{n_0}{\rho} = 2 \int\limits_l^\infty R(t, \mathbf{x}, L) p_c(t, \mathbf{x}, l, L) n_0(t, \mathbf{x}, L) dL$$

$$-R(t, \mathbf{x}, l) n_0(t, \mathbf{x}, l) \qquad (7.27)$$

$$+ \int\limits_l^\infty R(t, \mathbf{x}, L) p_c(t, \mathbf{x}, l, L) n_{iT}(t, \mathbf{x}, L) dL + \tau_f F_0(t, \mathbf{x}, l),$$

$$\tau_f \rho \frac{d}{dt} \frac{n_i}{\rho} = \int\limits_l^\infty R(t, \mathbf{x}, L) p_c(t, \mathbf{x}, l, L) n_i(t, \mathbf{x}, L) dL \qquad (7.28)$$

$$-R(t, \mathbf{x}, l) n_i(t, \mathbf{x}, l) + \tau_f F_i(t, \mathbf{x}, l), \quad i = 1, \dots, I - 1,$$

$$\tau_f \rho \frac{d}{dt} \frac{n_I}{\rho} = \int\limits_l^\infty R(t, \mathbf{x}, L) p_c(t, \mathbf{x}, l, L) n_I(t, \mathbf{x}, L) dL \qquad (7.29)$$

$$-R(t, \mathbf{x}, l) n_I(t, \mathbf{x}, l) + \tau_f F_I(t, \mathbf{x}, l).$$

7.3 Probability of Scission and Density of Conditional Probability of Scission

The scission probabilities R_i and R_{ai} as well as the density of the conditional probability p_c are the functions that actually control the process of lubricant degradation. Therefore, it is extremely important to use the right expressions for their calculation. The expression for the density of the conditional probability p_c is energy based and is proposed in Section 6.4 (also see paper by Kudish et al. [3]) in the form

$$p_c(t, \mathbf{x}, l, L) = \ln(2) \frac{4|L - 2l|}{L^2} \exp\left[-\ln(2) \frac{4l|L - l|}{L^2} \right]. \qquad (7.30)$$

Taking into account the assumption that polymer arms do not interact with each other we can take the expressions for the probabilities of scission $R_i(l)$, $i = 0, \dots, I$, in the form identical to the one for polymer additives with linear structure (see Section 6.3 and also paper [3])

$$R_i(l) = R(l), \quad i = 0, \dots, I,$$

$$R(l) = 0 \text{ if } l \le L_*, \qquad (7.31)$$

$$R(l) = 1 - \left(\frac{l}{L_*}\right)^{\frac{2\alpha U_A}{kT}} \exp\left[\frac{\alpha C a_* l_*^2 \tau_S (L^2 - l^2)}{kT} \right] \text{ if } l > L_*,$$

where L_* is the characteristic length of a polymer arm that is determined from the equation [3]

$$L_* = \sqrt{\frac{U_A}{Ca_* l_*^2 \tau_S}}, \quad U_A = \frac{U}{N_A}, \tag{7.32}$$

where U is a $C - C$ (carbon-carbon) bond dissociation energy per mole, N_A is Avogadro's number, $N_A = 6.022 \cdot 10^{23} mole^{-1}$, C and α are the shield constants, k is Boltzmann's constant, $k = 1.381 \cdot 10^{-23} J/K$, T is the lubricant absolute temperature, a_* and l_* are the polymer molecule bead radius and bond length, respectively, τ_S is the shear stress, $\tau_S = \mu S$ (S is the shear rate).

To derive the expression for the probabilities $R_{ai}(l)$ of ripping off a whole arm of length l, we will use the above assumption that polymer arms do not interact with each other. We will assume that the dissociation energy of an arm from a core of a star polymer molecule is also equal to U_A and that the chain length of the core is $\triangle a$. In most cases the core size $\triangle a$ is very small in comparison with the arm chain lengths l. Then based on the chain length l of a star molecule we can represent the probability $R_{ai}(l)$ as the probability of a polymer arm of chain length l to break into two pieces of lengths l and 0 in the form

$$R_{ai}(l) = R_a(l), \quad i = 1, \ldots, I,$$

$$R_a(l) = \triangle a R(l) p_c(l, l) = \triangle a \frac{4 \ln(2)}{l} R(l). \tag{7.33}$$

In most practical cases we have $\triangle a / L_* \ll 1$. As a result of that $R_a(l) \ll R(l)$ for $l \geq 0$. Therefore, the event of ripping off a whole arm is very rare in comparison with arm fragmentation and we can assume that

$$R_{ai}(l) = 0, \quad i = 1, \ldots, I. \tag{7.34}$$

7.4 Forming Star Polymeric Molecules from Linear Polymeric Molecules and Cores

After the solution of the system of (7.18)-(7.20) has been obtained, we can determine the density $p_{(i)}(t, \mathbf{x}, l_1, l_2, \ldots, l_i)$ of the probability to find a star polymeric molecule with i, $i > 1$, arms of arm chain lengths $l_1, l_2, \ldots, l_i$. Assuming that the arms of such molecules are distributed chaotically with respect to their chain lengths we obtain

$$p_{(i)}(t, \mathbf{x}, l_1, l_2, \ldots, l_i) = \prod_{j=1}^{i} f_i(t, \mathbf{x}, l_j),$$

$$f_i(t, \mathbf{x}, l_j) = n_i(t, \mathbf{x}, l_j) / \int_0^\infty n_i(t, \mathbf{x}, l) dl. \tag{7.35}$$

Now, we can determine the density $f_{(i)}(t, \mathbf{x}, l)$ of the probability to find a star polymeric molecule with i, $i > 1$, arms of the cumulative (total) arm chain length $l = l_1 + l_2 + \ldots + l_i$ according to the convolution formula

$$f_{(i)}(t, \mathbf{x}, l) = \int\limits_{l_1+\ldots+l_i=l} \int p_{(i)}(t, \mathbf{x}, l_1, l_2, \ldots, l_i) dl_1 dl_2 \ldots dl_{i-1}$$

$$= \int\limits_0^l f_i(t, \mathbf{x}, l_{i-1}) dl_{i-1} \int\limits_0^{l-l_{i-1}} f_i(t, \mathbf{x}, l_{i-2}) dl_{i-2} \ldots$$

$$\times \int\limits_0^{l-l_{i-1}-\ldots-l_3} f_i(t, \mathbf{x}, l_2) dl_2 \int\limits_0^{l-l_{i-1}-\ldots-l_2} f_i(t, \mathbf{x}, l_1)$$

$$\times f_i(t, \mathbf{x}, l - l_{i-1} - \ldots - l_1) dl_1, \ i = 2, \ldots, I.$$

(7.36)

In addition to (7.36), we will introduce the following definitions

$$f_{(0)}(t, \mathbf{x}, l) = f_0(t, \mathbf{x}, l), \ f_{(1)}(t, \mathbf{x}, l) = f_1(t, \mathbf{x}, l).$$

(7.37)

By changing the order of integration, we obtain that

$$\int\limits_0^\infty f_{(i)}(t, \mathbf{x}, l) dl = 1, \ i = 0, \ldots, I.$$

(7.38)

Also, it is important to realize that $f_{(1)}(t, \mathbf{x}, l)$ is determined by just the second equation in (7.35) but not by (7.36).

The density $W_{(i)}(t, \mathbf{x}, l)$ of the distribution of molecular weight of the star polymer molecules with i arms of the total arm length l is proportional to the total arm length l of such molecules and to the density $f_{(i)}(t, \mathbf{x}, l)$ of the probability to find such star polymer molecules, i.e.,

$$W_{(i)}(t, \mathbf{x}, l) = \alpha_{wi} w_m l f_{(i)}(t, \mathbf{x}, l).$$

(7.39)

where α_{wi} is a coefficient of proportionality that has to be determined and w_m is the monomer molecular weight. Taking into account (7.39) and the fact that the total molecular weight of the star polymer molecules with i arms is equal to

$$w_m \int\limits_0^\infty l n_i(t, \mathbf{x}, l) dl,$$

we obtain the equation

$$\int\limits_0^\infty W_{(i)}(t, \mathbf{x}, l) dl = w_m \int\limits_0^\infty l n_i(t, \mathbf{x}, l) dl,$$

(7.40)

that provides the expression for the coefficient α_{wi} in the form

$$\alpha_{wi} = \frac{\int\limits_0^\infty l n_i(t, \mathbf{x}, l) dl}{\int\limits_0^\infty l f_{(i)}(t, \mathbf{x}, l) dl}, \ i = 0, \ldots, I.$$

(7.41)

Therefore, using (7.39) and (7.41) we obtain the expression for the density $W_{(i)}(t, \mathbf{x}, l)$ of the distribution of molecular weight of the star polymer molecules with i arms of total arm length l

$$W_{(i)}(t, \mathbf{x}, l) = w_m \int\limits_0^\infty Ln_i(t, \mathbf{x}, L)dL \frac{lf_{(i)}(t, \mathbf{x}, l)}{\int\limits_0^\infty Lf_{(i)}(t, \mathbf{x}, L)dL}, \quad i = 0, \dots, I. \qquad (7.42)$$

Now, let us determine the density $n_{(i)}(t, \mathbf{x}, l)$ of the number of the star polymer molecules with i arms of the total arm length l. Similar to the above analysis, we can conclude that the densities $n_{(i)}(t, \mathbf{x}, l)$ and $f_{(i)}(t, \mathbf{x}, l)$ are proportional to each other, i.e.,

$$n_{(i)}(t, \mathbf{x}, l) = \beta_{wi} f_{(i)}(t, \mathbf{x}, l), \qquad (7.43)$$

where β_{wi} is a coefficient of proportionality that has to be determined from the fact that the total number of the star polymer molecules with i arms is equal to the total number of their arms divided by i. Therefore, using (7.38) we have

$$n_{(i)}(t, \mathbf{x}, l) = \tfrac{1}{i} \int\limits_0^\infty n_i(t, \mathbf{x}, L)dL f_{(i)}(t, \mathbf{x}, l), \quad i = 1, \dots, I. \qquad (7.44)$$

It is important to emphasize that (7.36), (7.37), (7.42), and (7.44) are the result of the approximation based on the assumption that star polymer arms are distributed chaotically. Nonetheless, these approximations possess the main necessary properties of the problem solution. For example, as it will be shown in Section 7.7, in cases when $R_{ai}(t, \mathbf{x}, l) = 0$, $i = 1, \dots, I$, the total number of arms of the star polymer molecules $\int\limits_0^\infty n_i(t, \mathbf{x}, L)dL$ with i arms, $i = 1, \dots, I$, is conserved in time t. Therefore, the total numbers $\int\limits_0^\infty n_{(i)}(t, \mathbf{x}, L)dL$ of the star polymer molecules with i arms, $i = 1, \dots, I$, must be conserved in time t too and, obviously, they are. Similarly, because of polymer degradation the total molecular weight $\int\limits_0^\infty W_{(i)}(t, \mathbf{x}, L)dL$ of the star polymer molecules with i arms, $i = 1, \dots, I$, decreases with time t as it's supposed to.

The above formulas and their analysis provide confidence that the process of determination of the lubricant viscosity μ defined by (7.51) and (7.52) is sufficiently accurate because the density $W_{(i)}(t, \mathbf{x}, l)$ possesses the right properties.

Equations (7.36), (7.42), and (7.44) for $f_{(i)}(t, \mathbf{x}, l)$, $n_{(i)}(t, \mathbf{x}, l)$, and $W_{(i)}(t, \mathbf{x}, l)$ are very convenient for numerical calculations. In particular, the Fourier transform of (7.36) reduces the expression for Fourier transform of $f_{(i)}(t, \mathbf{x}, l)$ to a product of Fourier transforms of $f_i(t, \mathbf{x}, l)$ that can be easily and efficiently determined and, then, the expression can be inverted and the values of $f_{(i)}(t, \mathbf{x}, l)$, $n_{(i)}(t, \mathbf{x}, l)$ and $W_{(i)}(t, \mathbf{x}, l)$ can be obtained (see Section 7.8).

7.5 Approximation of the Initial Experimentally Obtained Distribution of Star Polymeric Additives

By considering the process of creating star polymer molecules, it is understood that currently, to a certain extent, this process does not control the actual proportions of star molecules with different number of arms. In other words, the result of this process is stochastic, i.e., the obtained star polymer molecules are represented by a mixture of star polymer molecules with different number of arms. However, it is clear that in this mixture should be a number of just linear polymer molecules (single arms), which did not get attached to cores of star molecules. Based on the fact that the actual number of such single polymer arms is large, one can conclude that the molecular weight of single polymer arms should be normally distributed. That can be clearly seen from the experimentally obtained initial distributions of polymer molecular weight $f_{m0}(l)$ versus total chain length l of arms of the star polymer molecules in lubricants with RI2 and RI3 polymer additives (see Figs. 7.2 and 7.3), namely, from the shape of the distributions in the vicinity of the small maxima located close to $l = 0$. In this analysis and later, it is assumed that the molecular weight of cores of the star polymer molecules is negligibly small in comparison with the molecular weight of single arms. However, one has to keep in mind that depending on some details of the process in the final star polymer distribution (the mixture) the portion represented by the single arm distribution may be given by a slightly distorted normal distribution. One of the possible reasons for that is the presence of a number of polymer molecules represented by cores with attached to them single arms of a relatively short chain lengths and the presence of just cores without any arms attached to them.

Therefore, the approximation process consists of three steps: (1) approximation of the mean μ_l and the standard deviation σ_l of single arms based on the initial molecular weight distribution of single arms together with polymer molecules represented by cores with attached to them single arms, (2) approximation of the rest of the initial molecular weight distribution based on the obtained mean μ_l and the standard deviation σ_l, and (3) approximation of the initial distributions of single arms attached to cores of star polymer molecules with i arms, $i = 1, \ldots, I$.

Step (1) is based on the experimental data in the vicinity of the small significant maximum of the initial experimental molecular weight distribution $f_{m0}(l)$ closest to $l = 0$. By a significant maximum, we call the maximum with filtered out "noise" located right at $l = 0$ and in a very small vicinity of $l = 0$ and produced by the presence of single cores in the mixture and other contaminants. To filter out this noise in the discretely given distribution $f_{m0}(l_j)$, $j = 1, \ldots, N + 1$, $(l_1 > l_2 > \ldots > l_{N+1} = 0)$, we can introduce a new function $g(l_j)$, $j = m + 1, \ldots, N + 1 - m$, as follows:

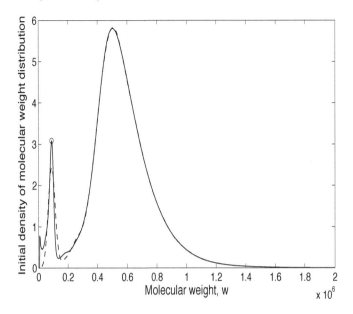

FIGURE 7.2

Comparison of the density of the experimental initial distribution of molecular weight $f_{m0}(w)$ (solid curve) with its least squares approximation $f_{m0a}(w)$ (dashed curve) for RI2 star polymer. A circle on the curve of $f_{m0}(w)$ indicates the position of the mean μ_l of the molecular weights of polymer single arms and star polymer molecules with just one arm (after Kudish, Airapetyan, Hayrapetyan, and Covitch [4]). Reprinted with permission from the STLE.

$$g(l_j) = \tfrac{1}{2m+1} \sum_{r=j-m}^{j+m} f_{m0}(l_r), \ j = m+1, \ldots, N+1-m,$$

where m is a relatively small in comparison with positive integer N. For $m \ll N$ the behavior of the functions $g(l_j)$ and $f_{m0}(l_j)$ is very similar. However, small oscillations of $f_{m0}(l_j)$ are dampened while the significant extrema and other features are preserved. By the proper choice of constant m function $g(l_j)$ can be made very smooth and non-oscillating. Therefore, the approximate position of the first significant maximum of $f_{m0}(l_j)$ is searched as the position of the maximum, l_{gmax}, of $g(l_j)$ closest to $l = 0$. After the position of the above maximum l_{gmax} of $g(l_j)$ is found a similar search for the first significant maximum of $f_{m0}(l_j)$ is done in a small vicinity of $l = l_{gmax}$.

It is assumed that the molecular weight distribution of single arms together with polymer molecules represented by cores with attached to them single arms follows the distribution $A_1 D(l, \mu_l, \sigma_l)$, where A_1 is proportional to the

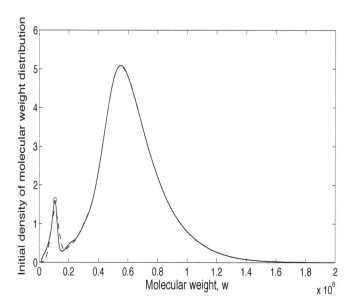

FIGURE 7.3
Comparison of the density of the experimental initial distribution of molecular
weight $f_{m0}(w)$ (solid curve) with its least squares approximation $f_{m0a}(w)$
(dashed curve) for RI3 star polymer. A circle on the curve of $f_{m0}(w)$ indicates
the position of the mean μ_l of the molecular weights of polymer single arms
and star polymer molecules with just one arm (after Kudish, Airapetyan,
Hayrapetyan, and Covitch [4]). Reprinted with permission from the STLE.

molecular weight of such molecules, i.e., $A_1 = \sqrt{2\pi}\sigma_l f_{m0}(\mu_l)$ and

$$D(l, \mu_l, \sigma_l) = \frac{1}{\sqrt{2\pi}\sigma_l} \exp\left[-\frac{(l-\mu_l)^2}{2\sigma_l^2}\right] \qquad (7.45)$$

is the normal distribution with the mean μ_l and the standard deviation σ_l.
In particular, the value of the mean μ_l is taken as the value of l at which the
small significant maximum of $f_{m0}(l)$ closest to $l = 0$ is reached. The standard
deviation σ_l is calculated from the equations

$$A_1 = \sqrt{2\pi}\sigma_l f_{m0}(\mu_l), \quad A_1 = 2\int_0^{\mu_l} f_{m0}(l)dl. \qquad (7.46)$$

On step (2) the approximation of the rest of the polymer molecular weight
distribution is based on forming star polymer molecules from single arms distributed according to the normal distribution from (7.45). This process is
controlled by formulas (7.35)-(7.37), where $n_i(t, \mathbf{x}, l) = D(l_j, \mu_l, \sigma_l)/(w_m l_j)$,
$j = 1, \ldots, i$. Because of the assumption that the weight of cores is negligibly

small in comparison with the characteristic weight of one arm the distributions of single arms and star polymer molecules with just one arm are indistinguishable. Therefore, the approximation of the star polymer molecular weight distribution is based on the expression

$$f_{m0}(l) \approx f_{m0a}(l) = w_m \sum_{i=1}^{I} \alpha_i l f_{(i)}(l),$$ (7.47)

where functions $f_{(i)}(l)$ are determined in (7.35)-(7.37) while $n_i(t, \mathbf{x}, l) = D(l_j, \mu_l, \sigma_l)/(w_m l_j)$, $j = 1, \ldots, i$, $i = 1, \ldots, I$, and $f_{m0a}(l)$ is the result of the approximation. Using the least squares approximation method, we can determine the set of nonnegative coefficients α_i, $i = 1, \ldots, I$, to minimize the discrepancy between $f_{m0a}(l)$ and $f_{ma}(l)$. To obtain a better approximation we can vary the maximum number of arms I attached to cores of star polymer molecules.

As it was mentioned earlier, in the framework of this model the initial distributions of the single arms and polymer molecules represented by cores with attached to them single arms are indistinguishable because the molecular weight of the core is assumed to be negligibly small. Therefore, we can assume that the initial distributions of the number of single arms $n_{00}(l)$ and the number of polymer molecules represented by cores with attached to them single arms $n_{10}(l)$ are as follows:

$$n_{00}(l) = 0, \quad n_{i0}(l) = \frac{\alpha_i}{l} D(l, \mu_l, \sigma_l),$$ (7.48)

where the set of constants α_i is obtained as a result of the above least squares approximation.

The approximation of the initial distributions of single arms attached to cores of star polymer molecules with i arms, $i = 2, \ldots, I$, is performed on Step (3). It is done based on the fact that polymer molecules with i arms are formed by attaching to cores i arms of lengths l each of which has the same normal distribution $D(l, \mu_l, \sigma_l)$ of molecular weight. Therefore, the initial distributions $n_{i0}(l)$, $i = 2, \ldots, I$, of the numbers of single arms attached to polymer molecules with i arms are

$$n_{i0}(l) = \frac{\alpha_i}{l} D(l, \mu_l, \sigma_l) \int_0^\infty L f_{(i)}(L) dL, \quad i = 2, \ldots, I,$$ (7.49)

where the mean μ_l and the standard deviation σ_l are determined on Step 1 while the values of parameters α_i, $i = 1, \ldots, I$, are determined on Step 2. The presence of the integral in (7.49) is because the total molecular weight of all molecules with i arms is equal to the total molecular weight of all arms that form such molecules.

The described approximation of the initial distribution of the polymer molecular weight is interesting in its own right. It gives the, so to speak, "spectral analysis" of the mixture of polymer molecules with different number of arms that can serve various purposes. For example, this analysis provides the opportunity to monitor and to evaluate the effectiveness of the technological process of star polymer making (also see Section 7.9).

7.6 Lubricant Viscosity and Distribution of Polymeric Molecules

The lubricant viscosity μ depends on the distribution of the polymer additive. In particular, for dilute solutions of linear polymeric molecules the empirical Huggins and Mark-Houwink equations [5, 6] can be applied (also see Chapter (4)). A recent study by Okumoto et al. [7] of four-arm star polystyrene in benzene revealed (see equation (5) and Fig. 7 from [7]) that one can apply the empirical Huggins and Mark-Houwink equations [5, 6, 7] to dilute star polymer solutions with practically the same power β (see below) as for the case of linear polymer molecules made of the same material.

Therefore, calculation of the lubricant viscosity μ can be done in two similar but different ways. The first of them is based on the usage of the function $W_T(t, \mathbf{x}, w)$ that corresponds to treating the arms of the star polymer molecules as separate polymer molecules with linear structure. This approach to determining the lubricant viscosity μ is described by the following equations

$$\mu = \mu_a \frac{1 + c_p[\eta] + k_H(c_p[\eta])^2}{1 + c_p[\eta]_a + k_H(c_p[\eta]_a)^2}, \quad [\eta] = k'M_V^\beta,$$

$$M_V = \left\{ \int_0^\infty w^\beta W_T(t, \mathbf{x}, w)dw \Big/ \int_0^\infty W_T(t, \mathbf{x}, w)dw \right\}^{1/\beta}, \tag{7.50}$$

where μ_a is the initial lubricant viscosity, c_p and $[\eta]$ are the polymer concentration and the intrinsic viscosity, respectively, $[\eta]_a$ is the intrinsic viscosity at the initial time moment, k_H is the Huggins constant, k' and β are the Mark-Houwink constants, $w = w_m l$ is the polymer molecular weight. The definition of the function $W_T(t, \mathbf{x}, l)$ that represents the function of the total molecular weight distribution is given in (7.81).

The second way of representing the dependence of the lubricant viscosity μ on the distribution of the molecular weight is based on considering the mixture of various star polymer molecules. It is adequate for star polymer molecules as it takes into account the distribution of the molecular weight of whole molecules not just their arms. This approach is based on the formulas

$$\mu = \mu_a \frac{1 + c_p[\eta] + k_H(c_p[\eta])^2}{1 + c_p[\eta]_a + k_H(c_p[\eta]_a)^2}, \quad [\eta] = k'M_V^\beta,$$

$$M_V = \left\{ \int_0^\infty w^\beta W_{cT}(t, \mathbf{x}, w)dw \Big/ \int_0^\infty W_{cT}(t, \mathbf{x}, w)dw \right\}^{1/\beta}, \tag{7.51}$$

$$W_{cT}(t, \mathbf{x}, l) = \sum_{i=0}^{I} W_{(i)}(t, \mathbf{x}, l),$$

(7.52)

$$W_{(0)}(t, \mathbf{x}, l) = W_0(t, \mathbf{x}, l), \quad W_{(1)}(t, \mathbf{x}, l) = W_1(t, \mathbf{x}, l),$$

where the densities of molecular weight distributions of star polymer molecules given by the functions $W_{(i)}(t, \mathbf{x}, l)$, $i = 2, \ldots, I$, are defined in (7.42).

The lubricant viscosities calculated for star polymer solution based on (7.51) and (7.52) and for polymer solution with linear structure based on (7.50) will be compared in Section 7.7.

7.7 Some Properties of the System of Kinetic Equations

In this section the main properties of the initial-value problem for kinetic equations (7.18)-(7.22) are established. We will assume that the vectors of the fluid velocity $\mathbf{u}(t, \mathbf{x})$ and the fluid flow streamlines $\mathbf{x}(t)$ are known. As we consider the initial-value problem (7.18)-(7.22), we will take advantage of the fact that additive degradation occurs along the flow streamlines described by (7.17). It means that along every streamline $\mathbf{x} = \mathbf{x}(t)$ the dependence of functions $n_i(t, \mathbf{x}, l)$ on $t, \mathbf{x}, l$ is reduced to the dependence of functions n_i on just t and l, i.e., along the streamlines the dependence of n_i on $\mathbf{x}$ can be omitted. Therefore, below we will use the notation $f(t, l) = f(t, \mathbf{x}(t), l)$ for functions $n_i(t, \mathbf{x}, l)$, $W_i(t, \mathbf{x}, l) = w_m l n_i(t, \mathbf{x}, l)$, $F_i(t, \mathbf{x}, l)$, $i = 0, \ldots, I$, etc. Similarly, we will use $p_c(t, l, L)$ for $p_c(t, \mathbf{x}(t), l, L)$ and $R(t, l)$ for $R(t, \mathbf{x}(t), l)$. Additionally, for convenience we will use the following notations $\tau = \frac{t}{\tau_f}$ and $F_{*i}(\tau, l) = \tau_f F_i(\tau, l)$, $i = 0, \ldots, I$.

First, let us consider some results on the existence and uniqueness of the solution of the systems of kinetic equations describing the stress-induced polymer degradation. It is sufficient to consider the problem for an incompressible lubricant fluid with unit density ($\rho(\tau, \mathbf{x}) = 1$) because for any density $\rho(\tau, \mathbf{x}) > 0$ by using the substitutions $n_{\rho i} = n_i/\rho$, $i = 0, \ldots, I$, the case of a compressible fluid is reduced to the case of an incompressible one with $\rho(\tau, \mathbf{x}) = 1$. We will be considering only the cases when $M_i = M = R^2/(R + R_a)$ and $M_{ai} = M_a = RR_a/(R + R_a)$, $i = 1, \ldots, I$, where functions $R(\tau, l)$ and $R_a(\tau, l)$ are determined by (7.31)-(7.33), respectively. It is obvious that

$$M + M_a = R, \ 0 \leq M \leq R \leq 1, \ 0 \leq M_a \leq R \leq 1, \quad (7.53)$$

and $R(\tau, l)$ is a monotonically increasing function of l, i.e.,

$$R(L) \geq R(l), \ L \geq l \geq 0. \quad (7.54)$$

For some constant $m > 0$ let us consider the space $H_m([0, +\infty))$ of continuous functions $n(l)$ on $[0, +\infty)$ with a finite norm

$$\| n \|_m = \max_{l \geq 0} \frac{|n(l)|}{A^m(l)}, \tag{7.55}$$

and for any positive constant T_* the spaces $H^0_{K,T_*,m}$ and $H^1_{K,T_*,m}$ of continuous and one time continuously differentiable functions $n(\tau)$ from $[0, T_*]$ into $H_m([0, +\infty))$, respectively, with the finite norm

$$\| n \|_{K,T_*,m} = \max_{\tau \in [0,T_*]} e^{-K\tau} \| n \|_m, \tag{7.56}$$

and, finally, the spaces $H^0_{T_*,m}$ and $H^1_{T_*,m}$ of continuous and one time continuously differentiable functions $n(\tau)$ from $[0, T_*]$ into $H_m([0, +\infty))$, respectively, with the finite norm

$$\| n \|_{T_*,m} = \max_{\tau \in [0,T_*]} \| n \|_m, \tag{7.57}$$

where K is a positive constant which will be specified later. The definition of function $A(l)$ is given in equation (6.41) and its properties are stated in Lemma 6.6.1.

Now, we can formulate the theorem of existence and uniqueness of the solution of the initial-value problem for the system of kinetic equations (7.27)-(7.29) and (7.22) in which the above replacements of t and F_i by τ and F_{*i}, respectively, were made.

Theorem 7.7.1 *Assume that* $M_a(\tau, l) = 0$ *for* $\tau \geq 0$, $l \geq 0$, *conditions (7.2) are satisfied,* $L_* > 0$, *and for some constants* $m > 1/\alpha_0$ ($\alpha_0 = \ln 16$) *and* $K > 2(1 - e^{-KT_*})/m$, *and an arbitrary positive constant* T_*

$$n^0_0(l), \ldots, n^0_I(l) \in H_m([0, +\infty)),$$

$$F_{*0}(\tau, l), \ldots, F_{*I}(\tau, l) \in H^0_{K,T_*,m}. \tag{7.58}$$

Then:
1. Problem (7.27)-(7.29) and (7.22) is linear and is uniquely solvable on $[0, T_*] \times [0, +\infty)$ *and the solution set* $n_0(\tau, l), \ldots, n_I(\tau, l)$ *belongs to* $H^1_{K,T_*,m}$.
2. The following estimates hold:

$$\| n_0 \|_{K,T_*,m} \leq \frac{1}{1-q} \{ \| n^0_0 \|_m + \frac{q}{2-q} \sum_{j=1}^{I} \| n^0_j \|_m +$$

$$\frac{mq}{2} [\| F_{*0}(\tau, l) \|_{K,T_*,m} + \frac{q}{2-q} \sum_{j=1}^{I} \| F_{*j}(\tau, l) \|_{K,T_*,m}] \}, \tag{7.59}$$

$$\| n_i \|_{K,T_*,m} \leq \frac{2}{2-q} \{ \| n^0_i \|_m + \frac{mq}{2} \| F_{*i}(\tau, l) \|_{K,T_*,m} \},$$

$$i = 1, \ldots, I, \quad q = \frac{2(1 - e^{-KT_*})}{Km} < 1.$$

3. *The solution of the problem $n_0(\tau, l), \ldots, n_I(\tau, l)$ continuously depends on an initial set of density distributions $n_0^0(l), \ldots, n_I^0(l)$ and on the set of source terms $F_{*0}(\tau, l), \ldots, F_{*I}(\tau, l)$, that is if $n_0^{(1)}(\tau, l), \ldots, n_I^{(1)}(\tau, l)$ and $n_0^{(2)}(\tau, l), \ldots, n_I^{(2)}(\tau, l)$ are the solutions obtained for $n_0^{0(1)}(l), \ldots, n_I^{0(1)}(l)$; $F_{*0}^{(1)}(\tau, l), \ldots, F_{*I}^{(1)}(\tau, l)$ and $n_0^{0(2)}(l), \ldots, n_I^{0(2)}(l)$; $F_{*0}^{(2)}(\tau, l), \ldots, F_{*I}^{(2)}(\tau, l)$, respectively, then*

$$\| n_0^{(1)} - n_0^{(2)} \|_{K, T_*, m} \leq \tfrac{1}{1-q} \{ \| n_0^{0(1)} - n_0^{0(2)} \|_m$$

$$+ \tfrac{q}{2-q} \sum_{j=1}^{I} \| n_j^{0(1)} - n_j^{0(2)} \|_m$$

$$+ \tfrac{mq}{2} [\| F_{*0}^{(1)}(\tau, l) - F_{*0}^{(2)}(1)(\tau, l) \|_{K, T_*, m}$$

$$+ \tfrac{q}{2-q} \sum_{j=1}^{I} \| F_{*j}^{(1)}(\tau, l) - F_{*j}^{(2)}(\tau, l) \|_{K, T_*, m}] \},$$

$$\| n_i^{(1)} - n_i^{(2)} \|_{K, T_*, m} \leq \tfrac{2}{2-q} \{ \| n_i^{0(1)} - n_i^{0(2)} \|_m$$

$$+ \tfrac{mq}{2} \| F_{*i}^{(1)}(\tau, l) - F_{*i}^{(2)}(\tau, l) \|_{K, T_*, m} \}, \quad i = 1, \ldots, I.$$

(7.60)

Proof. Because of linearity of (7.27)-(7.29) and (7.22) the proof of items 1-3 is done in a similar manner to the proof of Theorem 6.6.1.

Remark 7.7.1 *In Theorem 7.7.1 (as well as in Theorem 6.6.1) constant T_* may be taken as any positive constant or positive infinity (see Theorem 7.7.1) while positive constant K should satisfy the inequality $K > \frac{2(1 - e^{-KT_*})}{m}$. For any $T_* > 0$, this inequality is satisfied if $K > 2/m$.*

In case when $M_a(\tau, l) > 0$ for $\tau \geq 0$, $l \geq 0$, the system of equations (7.24)-(7.26) and (7.22) is nonlinear and to prove the existence and uniqueness of its solution we first prove the existence and uniqueness of the solution of the following initial-value problem

$$\tfrac{dn}{d\tau} = \int_l^{\infty} M(\tau, L) p_c(\tau, l, L) n(\tau, L) dL - R(\tau, l) n(\tau, l)$$

$$-(i - 1) n(\tau, l) \int_0^{\infty} M_a(\tau, L) n(\tau, L) dL / \int_0^{\infty} n(\tau, L) dL + G(\tau, l),$$

(7.61)

$$n(0, l) = n^0(l),$$

where i is a positive constant, $i \geq 1$, $G(\tau, l)$ is a given function, and functions M and M_a are the same functions as in (7.24)-(7.26). By integrating (7.61), we obtain

$$n(\tau, l) = Bn(\tau, l), \tag{7.62}$$

$$Bn(\tau, l) = n_{0*}(\tau, l)$$

$$+ \int_0^\tau \exp\left[-\int_s^\tau K(\theta, l)d\theta\right] \int_l^\infty M(s, L)p_c(s, l, L)n(s, L)dLds,$$

$$n_{0*}(\tau, l) = n^0(l) \exp\left[-\int_0^\tau K(s, l)ds\right]$$

$$+ \int_0^\tau G(s, l) \exp\left[-\int_s^\tau K(\theta, l)d\theta\right]ds, \tag{7.63}$$

$$K(\tau, l) = R(\tau, l) + (i - 1)f(n(\tau, l)),$$

$$f(n(\tau, l)) = \int_0^\infty M_a(\tau, L)n(\tau, L)dL / \int_0^\infty n(\tau, L)dL.$$

The existence and uniqueness of the solution of equation (7.61) will be derived from the analysis of the following iterative process

$$n_{(j+1)} = Bn_{(j)}, \quad j = 0, 1, \ldots, \quad n_{(0)} = n_{0*}, \tag{7.64}$$

where operator B and function n_0 are defined in (7.62).

Lemma 7.7.1 *If $i \geq 1$, functions M, M_a, and p_c are determined by (7.24), (7.30)-(7.33), and $n(\tau, l) \geq 0$, $G(\tau, l) \geq 0$ are continuous functions for $\tau \geq 0$, $l \geq 0$, then for $\tau \geq 0$ and $l \geq 0$*

$$Bn \geq e^{-i\tau}\left[n^0(l) + \int_0^\tau G(s, l)ds\right]. \tag{7.65}$$

Proof. Based on the conditions of Lemma 7.7.1, inequalities for M and M_a from (7.53) we obtain $f(n(\tau, l)) \leq 1$ and, therefore, $K(\tau, l) \leq 1 + i - 1 = i$ (see (7.63)). Moreover, from (7.62) and the fact that $p_c(\tau, l, L) \geq 0$ and $K(\tau, l) \leq i$ we obtain

$$Bn \geq n^0(l)e^{-i\tau} + \int_0^\tau e^{-i(\tau-s)}G(s, l)ds.$$

The latter estimate leads to inequality (7.65).

Lemma 7.7.2 *If conditions of Lemma 7.7.1 are satisfied, $m > 1/\alpha_0$ ($\alpha_0 = \ln 16$), and $\int_0^\infty n^0(l)dl > 0$, then for any two continuous nonnegative functions*

$p,\ q \in H_{T_*,m}$ such that $\int\limits_0^\infty p(\tau,l)dl \geq \int\limits_0^\infty n^0(l)dl,\ \int\limits_0^\infty q(\tau,l)dl \geq \int\limits_0^\infty n^0(l)dl,\ n^0 \in$ H_m, for $\tau \geq s \geq 0$ and $l \geq 0$

$$\int\limits_s^\tau \mid f(p(\theta,l)) - f(q(\theta,l)) \mid d\theta \leq \beta_i \mid e^{i\tau} - e^{is} \mid \parallel p - q \parallel_{T_*,m},$$

(7.66)

$$\beta_i(t) = \tfrac{1}{i} \int\limits_0^\infty A^m(l)dl/[\int\limits_0^\infty n^0(l)dl + \int\limits_0^\tau G(s,l)ds] < \infty.$$

Proof. The proof follows from the definition of the norm (7.55), (7.57), and the inequalities (7.65), $M_a(\tau,l) \leq 1,\ \int\limits_0^\infty p(\tau,l)dl \geq \int\limits_0^\infty n^0(l)dl$, and $\mid f(p) -$ $f(q) \mid \leq \{\int\limits_0^\infty M_a(\tau,l) \mid p(\tau,l) - q(\tau,l) \mid dl\ +\ f(q(\tau,l)) \int\limits_0^\infty \mid p(\tau,l) - q(\tau,l) \mid$ $dl\}/ \int\limits_0^\infty p(\tau,l)dl \leq 2 \int\limits_0^\infty \mid p(\tau,l) - q(\tau,l) \mid dl/ \int\limits_0^\infty p(\tau,l)dl.$

Lemma 7.7.3 *Under the conditions of Lemmas 7.7.1 and 7.7.2 for $m > 1/\alpha_0$ and for sufficiently small $T_* > 0$ operator B from (7.62) is a contraction mapping from $H_{T_*,m}$ into $H_{T_*,m}$, i.e. if $p,\ q \in H_{T_*,m}$ then*

$$\parallel Bp - Bq \parallel_{T_*,m} \leq q_0 \parallel p - q \parallel_{T_*,m},\ 0 < q_0 < 1.$$

(7.67)

Proof. Using the definition of operator B given in (7.62) and the fact that $\mid e^{-x} - e^{-y} \mid \leq \mid x - y \mid$ for nonnegative x and y, one obtains

$$\mid Bp - Bq \mid \leq (i-1)\{n^0(l)g(0,\tau,l) + \int\limits_0^\tau G(s,l)g(s,\tau,l)ds$$

$$+ \int\limits_0^\tau g(s,\tau,l) \int\limits_l^\infty M(s,L)p_c(s,l,L)q(s,L)dLds\}$$

$$+ \int\limits_0^\tau \int\limits_l^\infty M(s,L)p_c(s,l,L) \mid p(\theta,l) - q(\theta,l) \mid dLds,$$

$$g(s,\tau,l) = \int\limits_s^\tau \mid f(p(\theta,l)) - f(q(\theta,l)) \mid d\theta.$$

Taking maximum of the right-hand side and, then, of the left-hand side of the latter inequality with respect to l and τ in accordance with the norm definition from (7.55), (7.57), and using (7.66) and the inequality $A(l) \leq 1$, one arrives at the estimate

$$\parallel Bp - Bq \parallel_{T_*,m} \leq q_0 \parallel p - q \parallel_{T_*,m},$$

(7.68)

$$q_0 = (i-1)\beta_i(0)[(e^{iT_*} - 1) \parallel n^0 \parallel_m$$

(7.69)

$$+ (T_* e^{iT_*} - \tfrac{e^{iT_*}-1}{i})(\parallel G \parallel_{T_*,m} + \tfrac{\|q\|_{T_*,m}}{m})] + \tfrac{T_*}{m}.$$

Therefore, for sufficiently small T_* one has $0 < q_0 < 1$.

Lemma 7.7.4 *If constants $i \geq 1$, $m > 1/\alpha_0$, functions M, M_a, and p_c are determined by (7.24), (7.30)-(7.33), and $n^0(\tau, l) \geq 0$, $n^0 \in H_m$, $G(\tau, l) \geq 0$, $G \in H_{T_*, m}$, then*

$$\| Bn_{(0)} - n_{(0)} \|_{T_*, m} \leq (i - 1 + \tfrac{1}{m})(T_* \| n^0 \|_m + \tfrac{T_*^2}{2} \| G \|_{T_*, m}). \qquad (7.70)$$

Proof. The proof is conducted in the fashion similar to the proof of Lemma 7.7.3.

Theorem 7.7.2 *If $\int\limits_0^\infty n^0(l)dl > 0$, then under the conditions of Lemma 4.6.4 for an arbitrary positive constant R_*:*

1. The nonnegative solution $n(\tau, l)$ of problem (7.62), (7.63) exists and is unique for $\tau \in [0, T_*]$ and $l \geq 0$ in the closed set $V = \{n \in H^1_{T_*, m} : \| n - n_{(0)} \|_{T_*, m} \leq R_*\}$ if the positive constant T_* is sufficiently small that $q_0 < 1$ (see (7.69)) and

$$(i - 1 + \tfrac{1}{m})(T_* \| n^0 \|_m + \tfrac{T_*^2}{2} \| G \|_{T_*, m}) \leq (1 - q_0)R_*. \qquad (7.71)$$

2. For sufficiently small $T_* < m$

$$\| n \|_{T_*, m} \leq \tfrac{m}{m - T_*}(\| n^0 \|_m + T_* \| G \|_{T_*, m}). \qquad (7.72)$$

3. Solutions $n_1(\tau, l)$ and $n_2(\tau, l)$ of problem (7.62) and (7.63) for two initial conditions $n_1^0(l)$, $\int\limits_0^\infty n_1^0(l)dl > 0$ and $n_2^0(l)$, $\int\limits_0^\infty n_2^0(l)dl > 0$ and two source terms $G_1(\tau, l)$ and $G_2(\tau, l)$ satisfy the inequality

$$\| n_2 - n_1 \|_{T_*, m} \leq \tfrac{1}{1 - q_0}(\| n_2^0 - n_1^0 \|_m + T_* \| G_2 - G_1 \|_{T_*, m}). \qquad (7.73)$$

Proof. Based on Lemmas 7.7.3 and 7.7.4 for any functions p, $q \in V$ one obtains $\| Bp - Bq \|_{T_*, m} \leq q_0 \| p - q \|_{T_*, m}$, $0 < q_0 < 1$ and $\| Bn_{(0)} - n_{(0)} \|_{T_*, m} \leq (1 - q_0)R_*$ (see (7.70)). Therefore, based on inequality (7.68) and (7.69)

$$\| Bn - n_{(0)} \|_{T_*, m} \leq \| Bn - n_{(0)} \|_{T_*, m} + \| Bn_{(0)} - n_{(0)} \|_{T_*, m}$$

$$\leq q_0 \| n - n_{(0)} \|_{T_*, m} + (1 - q_0)R_* \leq R_*,$$

and operator B from (7.62), (7.63) we conclude that operator B is a contraction mapping of the closed set V onto V. Finally, one can determine that the solution of equations (7.62), (7.63) as well as of (7.61) exists and is unique in V. Moreover, every iterate $n_{(k)}(\tau, l)$, $k \geq 1$, of the converging iterative process (7.64) is nonnegative if $n_{(0)}(\tau, l) \geq 0$. The latter is true because $n^0(l) \geq 0$ and $G(\tau, l) \geq 0$ for $\tau \geq 0$ and $l \geq 0$ (see (7.62)). The solution of the problem $n(\tau, l)$ is a differentiable function of τ as all functions in (7.62) and (7.63) are continuous and all integrals converge. That proves item 1.

Item 2 follows from the estimate (see (7.62), (7.63))

$$\| n \|_{T_*,m} \leq \| Bn \|_{T_*,m} \leq \| n^0 \|_m + \tfrac{T_*}{m} \| n \|_{T_*,m} + T_* \| G \|_{T_*,m}).$$

Item 3 can be proven in a similar fashion.

Obviously, the system of equations (7.24)-(7.26), (7.22) can be reduced to sequential solution of the following equations

$$\tfrac{dn_i}{d\tau} = \int\limits_l^\infty M(\tau, L) p_c(\tau, l, L) n_i(\tau, L) dL - R(\tau, l) n_i(\tau, l)$$

$$-(i-1) n_i(\tau, l) \int\limits_0^\infty M_a(\tau, L) n_i(\tau, L) dL / \int\limits_0^\infty n_i(\tau, L) dL + G_i(\tau, l),$$

$$n_i(0, l) = n_i^0(l), \ G_I(\tau, l) = F_I(\tau, l) \ for \ i = I, \tag{7.74}$$

$$G_i(\tau, l) = F_i(\tau, l) + i n_{i+1}(\tau, l) \int\limits_0^\infty M_a(\tau, L) n_{i+1}(\tau, L) dL$$

$$/ \int\limits_0^\infty n_{i+1}(\tau, L) dL \ for \ i = I - 1, \ldots, 1,$$

$$\tfrac{dn_0}{d\tau} = \int\limits_l^\infty M(\tau, L) p_c(\tau, l, L) n_0(\tau, L) dL - R(\tau, l) n_0(\tau, l) + G_0(\tau, l),$$

$$n_0(0, l) = n_0^0(l),$$

$$G_0(\tau, l) = F_0(\tau, l) + M_a(\tau, l) n_{1T}(\tau, l) \tag{7.75}$$

$$+ \int\limits_l^\infty M(\tau, L) p_c(\tau, l, L) n_{1T}(\tau, L) dL,$$

starting with $i = I$ and ending with $i = 0$. Here n_{1T} is the sum of all n_i with respect to i from $i = 1$ to $i = I$ that is determined in (7.82). By integrating (7.74) and (7.75) we obtain

$$n_i(\tau, l) = B_i n_i(\tau, l), \ i = 0, \ldots, I, \ B_i n_i(\tau, l) = n_{i0*}(\tau, l)$$

$$+ \eta_i \int\limits_0^\tau \exp\left[- \int\limits_s^\tau K_i(\theta, l) d\theta \right] \int\limits_l^\infty M(s, L) p_c(s, l, L) n_i(s, L) dL ds,$$

$$n_{i0*}(\tau, l) = n_i^0(l) \exp\left[- \int\limits_0^\tau K_i(s, l) ds \right] \tag{7.76}$$

$$+ \int\limits_0^\tau G_i(s, l) \exp\left[- \int\limits_s^\tau K_i(\theta, l) d\theta \right] ds, \ \eta_0 = 2, \ \eta_i = 1, \ i = 1, \ldots, I,$$

$$K_i(\tau, l) = R(\tau, l) + (i-1)f(n_i(\tau, l)), \quad i = 1, \ldots, I,$$

$$K_0(\tau, l) = R(\tau, l), \tag{7.77}$$

$$f(n_i(\tau, l)) = \int_0^\infty M_a(\tau, L) n_i(\tau, L) dL / \int_0^\infty n_i(\tau, L) dL, \quad i = 1, \ldots, I.$$

Theorem 7.7.3 *If* $m > 1/\alpha_0$, $K > 2(1 - e^{-KT_*})/m$ *functions* M, M_a, *and* p_c *are determined by (7.24), (7.30)-(7.33),* $n_i^0(l) \geq 0$, $n_i^0 \in H_m$, $i = 0, \ldots, I$, $\int_0^\infty n_i^0(l) dl > 0$, $i = 1, \ldots, I$, $F_0 \in H_{T_*, m}$, $F_{*i}(\tau, l) \geq 0$, $i = 1, \ldots, I$, $F_{*i} \in H_{T_*, m}$, $i = 0, \ldots, I$, *then for an arbitrary positive constant* R_*:

1. The nonnegative solutions $n_0(\tau, l), \ldots, n_I(\tau, l)$ *of (7.74), (7.75) exist and are unique for* $\tau \in [0, T_*]$ *and* $l \geq 0$ *in the closed sets* $V_i = \{ n_i \in H_{T_*, m}^1 : \| n_i - n_{i((0)} \| \leq R_* \}$, $i = 1, \ldots, I$, $V_0 = \{ n_0 \in H_{K, T_*, m}^1 : \| n_0 - n_{0((0)} \| \leq R_* \}$, *if the positive constant* T_* *is sufficiently small that* $q_{i0} < 1$, $i = 1, \ldots, I$ *(where* q_{i0}, $i = 1, \ldots, I$, *are determined based on (7.69) when* $n = n_i$ *and* $G = G_i$) *and*

$$(i - 1 + \tfrac{1}{m})(T_* \| n_i^0 \|_m + \tfrac{T_*^2}{2} \| G_i \|_{T_*, m}) \leq (1 - q_{i0}) R_*, \tag{7.78}$$

$$T_* < m, \quad i = 1, \ldots, I.$$

2. For sufficiently small $T_* < m$ *function* $n_0(\tau, l)$ *satisfies the first inequality in (7.59) and*

$$\| n_i \|_{T_*, m} \leq \tfrac{m}{m - T_*} (\| n_i^0 \|_m + T_* \| G_i \|_{T_*, m}), \quad i = 1, \ldots, I. \tag{7.79}$$

3. The solution sets $n_i^{(1)}(\tau, l)$ *and* $n_i^{(2)}(\tau, l)$, $i = 0, \ldots, I$, *of equations (7.74), (7.75) for two sets of nonnegative initial conditions* $n_i^{0(1)}(l)$ *and* $n_i^{0(2)}(l)$, $i = 0, \ldots, I$, *from* H_m *such that* $\int_0^\infty n_i^{0(1)}(l) dl > 0$, $\int_0^\infty n_i^{0(2)}(l) dl > 0$, $i = 1, \ldots, I$, *and two sets of source terms* $F_{*i}^{(1)}(\tau, l)$ *and* $F_{*i}^{(2)}(\tau, l)$, $i = 0, \ldots, I$, *from* $H_{T_*, m}$ *such that* $F_{*i}^{(1)}(\tau, l) \geq 0$ *and* $F_{*i}^{(2)}(\tau, l) \geq 0$ *for* $\tau \geq 0$, $l \geq 0$, $i = 1, \ldots, I$, *satisfy the first inequality in (7.60) and*

$$\| n_i^{(2)} - n_i^{(1)} \|_{T_*, m} \leq \tfrac{1}{1 - q_{0i}} (\| n_i^{0(2)} - n_i^{0(1)} \|_m \tag{7.80}$$

$$+ T_* \| G_i^{(2)} - G_i^{(1)} \|_{T_*, m}), \quad i = 1, \ldots, I,$$

where $G_i^{(1)}(\tau, l)$ *and* $G_i^{(2)}(\tau, l)$, $i = 1, \ldots, I$, *are determined by (7.74).*

Proof. Based on Theorem 7.7.2, sequentially, for $i = I$ through $i = 1$, one gets the existence and uniqueness of $n_I(\tau, l) \geq 0$ in V_I. Therefore, $G_{I-1}(\tau, l) \geq$

$F_{*I-1}(\tau, l) \geq 0$ because $n_I(\tau, l) \geq 0$, $G_{I-1}(\tau, l)$ belongs to V_{I-1} because $F_{*I-1}(\tau, l)$ and $n_I(\tau, l)f(n_I(\tau, l))$ belong to V_{I-1}, and according to Theorem 4.6.2 the solution $n_{I-1}(\tau, l) \geq 0$ exists and is unique in V_{I-1}. The same way one can prove that the solutions $n_i(\tau, l) \geq 0$ exist and are unique in V_i for $i = 1, \ldots, I$. Based on these results and the fact that $F_{*0}(\tau, l)$ belongs to V_0, one concludes that $G_0(\tau, l)$ belongs to V_0 and, therefore, according to Theorem 6.6.1 solution $n_0(\tau, l) \geq 0$ exists and is unique in V_0. Inequalities (7.78)-(7.80) follow from the corresponding estimates of Theorems 7.7.2 and 6.6.1.

Remark 7.7.2 *Theorem 7.7.3 states that the problem described by (7.74) and (7.75) is formulated correctly, its solution set is bounded, and the problem solution depends on the initial values and the source terms continuously on $\tau \in [0, T_*]$, $T_* > 0$, and $l \geq 0$.*

Remark 7.7.3 *Obviously, norms determined in (7.56) and (7.57) are equivalent as long as $K > 0$ and $T_* < \infty$, i.e., the results obtained in Lemmas 7.7.1-7.7.4 and Theorems 7.7.1-7.7.3 in the first norm are valid in the second and vice versa.*

Lemma 7.7.5 *Assuming that conditions (7.1) and (7.2) are satisfied, $n_{i0}(l) = 0$, and $F_{*i}(\tau, l) = 0$ for $\tau \geq 0$, $l \geq 0$, and $i = 0, \ldots, I$, then $n_i(\tau, l) = 0$ for $\tau > 0$, $l \geq 0$, and $i = 0, \ldots, I$.*

Let us consider some of the properties of the system of kinetic equations (7.18)-(7.20), (7.50). The densities $n_T(\tau, l)$ and $W_T(\tau, l)$ of the distributions of all arms of all star polymer molecules and of the molecular weight of these arms as well as the density of the total number of added/removed per unit time polymer arms $F_T(\tau, l)$ are given by the formulas

$$n_T(\tau, l) = \sum_{i=0}^{I} n_i(\tau, l), \quad W_T(\tau, l) = w_m l n_T(\tau, l),$$

$$(7.81)$$

$$F_T(\tau, l) = \sum_{i=0}^{I} F_{*i}(\tau, l).$$

Similarly, we can introduce the densities $n_{iT}(\tau, l)$ and $W_{iT}(\tau, l)$ of the distributions of all arms of the star polymer molecules with the number of arms from i through I and of the molecular weight of these arms, respectively, as well as the density of the total number of added/removed per unit time polymer arms $F_{iT}(\tau, l)$ attached to the polymer molecules with the number of arms from i through I. The later densities are given by the formulas

$$n_{iT}(\tau, l) = \sum_{j=i}^{I} n_j(\tau, l), \quad W_{iT}(\tau, l) = w_m l n_{iT}(\tau, l),$$

$$(7.82)$$

$$F_{iT}(\tau, l) = \sum_{j=i}^{I} F_{*j}(\tau, l).$$

Assuming that $R_i(\tau, l) = R(\tau, l)$ and $R_{ai}(\tau, l) = R_a(\tau, l)$, $i = 0, \ldots, I$, from (7.18)-(7.20) (also see (7.24)-(7.26)) follows that

$$\rho \frac{d}{d\tau} \frac{n_T}{\rho} = 2 \int\limits_{l}^{\infty} R(\tau, L) p_c(\tau, l, L) n_0(\tau, L) dL - R(\tau, l) n_0(\tau, l)$$

$$+2 \int\limits_{l}^{\infty} M(\tau, L) p_c(\tau, l, L) n_{1T}(\tau, L) dL - M(\tau, l) n_{1T}(\tau, l) + F_T(\tau, l), \qquad (7.83)$$

$$M = \frac{R^2}{R + R_a},$$

$$\rho \frac{d}{d\tau} \frac{n_{iT}}{\rho} = \int\limits_{l}^{\infty} M(\tau, L) p_c(\tau, l, L) n_{iT}(\tau, L) dL - R(\tau, l) n_{iT}(\tau, l)$$

$$-(i-1) n_i(\tau, l) \int\limits_{0}^{\infty} M_a(\tau, L) n_i(\tau, L) dL / \int\limits_{0}^{\infty} n_i(\tau, L) dL + F_{iT}(\tau, l), \qquad (7.84)$$

$$M_a = \frac{R R_a}{R + R_a}, \quad i = 1, \ldots, I - 1.$$

Let us introduce the total number of polymer arms $N_{*T}(\tau)$ and the total molecular weight $W_{*T}(\tau)$ of star polymer molecules in a unit volume

$$N_{*T}(\tau) = \int\limits_{0}^{\infty} n_T(\tau, l) dl, \quad W_{*T}(\tau) = \int\limits_{0}^{\infty} W_T(\tau, l) dl, \qquad (7.85)$$

as well as the total number of arms $N_{*iT}(\tau)$ and the total molecular weight $W_{*iT}(\tau)$ of star polymer molecules in a unit volume with the number of arms varying from i through I

$$N_{*iT}(\tau) = \int\limits_{0}^{\infty} n_{iT}(\tau, l) dl, \quad W_{*iT}(\tau) = \int\limits_{0}^{\infty} W_{iT}(\tau, l) dl. \qquad (7.86)$$

To be able to characterize the process of star polymer scission in more detail we will also introduce the number $N_{*0}(\tau)$ and the molecular weight $W_{*0}(\tau)$ of single polymer arms in a unit volume

$$N_{*0}(\tau) = \int\limits_{0}^{\infty} n_0(\tau, l) dl, \quad W_{*0}(\tau) = w_m \int\limits_{0}^{\infty} l n_0(\tau, l) dl, \qquad (7.87)$$

and the number $N_{*I}(\tau)$ and the molecular weight $W_{*I}(\tau)$ of the arms of the star polymer molecules with I arms in a unit volume

$$N_{*I}(\tau) = \int\limits_{0}^{\infty} n_I(\tau, l) dl, \quad W_{*I}(\tau) = w_m \int\limits_{0}^{\infty} l n_I(\tau, l) dl. \qquad (7.88)$$

Theorem 7.7.4 *Assume that $R_i(\tau, l) = R(\tau, l)$ and $R_{ai}(\tau, l) = R_a(\tau, l)$, $i = 0, \ldots, I$, $m\alpha_0 > 1$, and conditions of Theorem 7.7.4, (7.1) and (7.2) are satisfied.*

1. If $n_{i0}(l) \geq 0$ and $F_{*i}(\tau, l) \geq 0$ for all $l \geq 0$ and $i = 0, \ldots, I$ then

$$n_i(\tau, l) \geq 0 \ for \ all \ \tau \in [0, T_*], \ l \geq 0, \ i = 0, \ldots, I, \tag{7.89}$$

2.

$$\rho \frac{d}{d\tau}\left[\frac{N_{*0}}{\rho}\right] = \int_0^\infty R(\tau, l)n_0(\tau, l)dl + \int_0^\infty M(\tau, l)n_{1T}(\tau, l)dl$$

$$+ \int_0^\infty F_T(\tau, l)dl, \tag{7.90}$$

3.

$$\rho \frac{d}{d\tau}\left[\frac{1}{\rho}\int_0^\infty n_0(\tau, l)l^j dl\right] \leq \int_0^\infty M_a(\tau, l)n_{1T}(\tau, l)l^j dl$$

$$+ \int_0^\infty M(\tau, L)n_{1T}(\tau, L) \int_0^L p_c(\tau, l, L)l^j dl dL \tag{7.91}$$

$$+ \int_0^\infty F_T(\tau, l)l^j dl \ for \ all \ 1 \leq j < m\alpha_0 - 1,$$

4.

$$\rho \frac{d}{d\tau}\left[\frac{N_{*I}}{\rho}\right] = -I \int_0^\infty M_a(\tau, l)n_I(\tau, l)dl + \int_0^\infty F_{*I}(\tau, l)dl, \tag{7.92}$$

5.

$$\rho \frac{d}{d\tau}\left[\frac{1}{\rho}\int_0^\infty n_I(\tau, l)l^j dl\right] \leq - \int_0^\infty M_a(\tau, l)n_I(\tau, l)l^j dl$$

$$- \int_0^\infty M(\tau, L)n_I(\tau, L) \int_0^L p_c(\tau, l, L)l^j dl dL$$

$$-(I-1) \int_0^\infty M_a(\tau, l)n_I(\tau, l)dl \int_0^\infty n_I(\tau, l)l^j dl / \int_0^\infty n_I(\tau, l)dl \tag{7.93}$$

$$+ \int_0^\infty F_{*I}(\tau, l)l^j dl \ for \ all \ 1 \leq j < m\alpha_0 - 1,$$

6. If

$$\int_0^\infty n_{i0}(l)l^j dl < \infty, \ \int_0^\infty F_{*i}(\tau, l)l^j dl < \infty \ for \ all \ 0 \leq j < m\alpha_0 - 1,$$

$$i = 0, \ldots, I, \tag{7.94}$$

for some positive constant $m > 1/\alpha_0$, then

$$\rho \frac{d}{d\tau} \left[\frac{1}{\rho} \int\limits_0^\infty n_T(\tau,l) l^j \, dl \right] \leq \int\limits_0^\infty F_T(\tau,l) l^j \, dl \tag{7.95}$$

$$for \ j = 0 \ and \ 1 \leq j < m\alpha_0 - 1.$$

Proof. The proof of item 1 follows from the convergence of the iterative method of solution of the system of (7.83), (7.84) and it is similar to the one used in Section 6.6 (also see [2]) in the sequence from $i = I$ to $i = 0$. Items 2-6 directly follow from the equations of the system by integrating the equations with 1, l, and l^j with respect to l from 0 to ∞, reversing the order of integration, and by using the integral relation from (7.2). In the relationships (7.91), (7.93), and (7.95) the equalities take place if and only if $R(\tau,l) = R_a(\tau,l) = 0$.

Corollary 7.7.1 *Under the assumptions of Theorem 7.7.4 for $m > 2/\alpha_0$, we have*
   1.

$$\frac{d}{d\tau} \frac{N_{*T}}{\rho} \geq 0 \ if \ \int\limits_0^\infty F_T(\tau,l) dl \geq 0, \tag{7.96}$$

   2.

$$\frac{d}{d\tau} \frac{W_{*T}}{\rho} < 0 \ if \ \int\limits_0^\infty l F_T(\tau,l) dl < 0,$$

$$\frac{d}{d\tau} \frac{W_{*T}}{\rho} = 0 \ if \ \int\limits_0^\infty l F_T(\tau,l) dl = 0, \tag{7.97}$$

$$\frac{d}{d\tau} \frac{W_{*T}}{\rho} > 0 \ if \ \int\limits_0^\infty l F_T(\tau,l) dl > 0,$$

   3.

$$\frac{d}{d\tau} \frac{N_{*0}}{\rho} \geq 0 \ if \ \int\limits_0^\infty F_{*0}(\tau,l) dl \geq 0, \ \ \frac{d}{d\tau} \frac{W_{*0}}{\rho} \geq 0 \ if \ \int\limits_0^\infty l F_{*0}(\tau,l) dl \geq 0, \tag{7.98}$$

   4.

$$\frac{d}{d\tau} \frac{N_{*I}}{\rho} \leq 0 \ if \ \int\limits_0^\infty F_{*I}(\tau,l) dl \leq 0,$$

$$\frac{d}{d\tau} \frac{W_{*I}}{\rho} \leq 0 \ if \ \int\limits_0^\infty l F_{*I}(\tau,l) dl \leq 0, \tag{7.99}$$

   5.

$$\frac{d}{d\tau} \frac{N_{*iT}}{\rho} \leq 0 \ if \ \int\limits_0^\infty F_{iT}(\tau,l) dl \leq 0,$$

$$\frac{d}{d\tau} \frac{W_{*iT}}{\rho} \leq 0 \ if \ \int\limits_0^\infty l F_{iT}(\tau,l) dl \leq 0, \ \ i = 1, \ldots, I. \tag{7.100}$$

Proof. The proof of items 1-5 follows from Theorem 7.7.4 when $j = 0$ and $j = 1$. In the relationships (7.97)-(7.100) involving derivatives of various W-values, the equalities take place if and only if $R(\tau, l) = R_a(\tau, l) \equiv 0$ and the corresponding integrals of $F_{*i}(\tau, l)$ are also equal to zero for all $\tau \geq 0$. In relationships (7.99) and (7.100) involving the N-values, the equalities take place if and only if $R_a(\tau, l) \equiv 0$ and the corresponding integrals of $F_{*i}(\tau, l)$ are also equal to zero for all $\tau \geq 0$.

Corollary 7.7.2 *Under the assumptions of Theorem 7.7.4 for $w_m > 0$, $\mu_a > 0$, $c_p > 0$, $k' > 0$, $k_H > 0$, $\beta > 0$, $m\alpha_0 > \beta + 2$, and*

$$\int_0^\infty F_T(\tau, l)dl = 0 \text{ and } \int_0^\infty l^{\beta+1} F_T(\tau, l)dl \leq 0 \qquad (7.101)$$

the value of $\mu(\tau)$ determined by (7.50) is a monotonically decreasing function of τ, $\tau \geq 0$.

Proof. The proof follows from the monotonicity of the dependence of μ from (7.50) on $c_p[\eta]$ and the monotonicity of the dependence of $[\eta]$ on the integral $\int_0^\infty w^\beta W_T(\tau, w)dw = w_m^{\beta+2} \int_0^\infty l^{\beta+1} n_T(\tau, l)dl$ that according to (7.95) is a monotonically decreasing function of τ while according to (7.97) the integral $\int_0^\infty W_T(\tau, w)dw$ is conserved if conditions (7.101) are satisfied.

Remark 7.7.4 *The statement of Corollary 7.7.3 indicates that in an isolated lubrication system as a result of lubricant degradation the lubricant viscosity $\mu(\tau)$ is a monotonically decreasing function of time τ.*

Remark 7.7.5 *The interpretation of relationships (7.96) and (7.97) is as follows: in an isolated system the total number of polymer arms per unit mass of lubricant increases with time and the total molecular weight of the polymer molecules per unit mass of lubricant is conserved. At the same time the relationships (7.98)-(7.100) can be interpreted as follows: in an isolated system the total number of single polymer arms per unit mass of lubricant and their molecular weight increase with time, the total number of arms of the star molecules with I arms per unit mass of lubricant and their molecular weight decrease with time, and the total number of arms of all star molecules with number of arms from i through I per unit mass of lubricant and their molecular weight decrease with time, respectively.*

Theorem 7.7.5 *Assume that $R_i(\tau, l) = R(\tau, l)$ and $R_{ai}(\tau, l) = R_a(\tau, l) = 0$, $i = 0, \ldots, I$, $m > /2\alpha_0$, and conditions (7.1) and (7.2) are satisfied for all $\tau \geq 0$ and $l \geq 0$. Then:*
  1. The relationships

$$\rho \frac{d}{d\tau} \left[\frac{1}{\rho} \int_0^\infty n_i(\tau, l) dl \right] = \int_0^\infty F_{*i}(\tau, l) dl,$$

$$\rho \frac{d}{d\tau} \left[\frac{1}{\rho} \int_0^\infty W_i(\tau, l) dl \right] = -\frac{1}{2} \int_0^\infty R(\tau, l) W_i(\tau, l) dl + w_m \int_0^\infty l F_{*i}(\tau, l) dl, \quad (7.102)$$

$$i = 1, \ldots, I,$$

are valid.

2. If $n_i^{(1)}(\tau, l)$ and $n_i^{(2)}(\tau, l)$, $i = 0, \ldots, I$, are solutions obtained for a certain set of source functions $F_{*i}(\tau, l)$, $i = 0, \ldots, I$, and the initial distributions $n_{i0}^{(1)}(l) \leq n_{i0}^{(2)}(l)$, for all $l \geq 0$ and $i = 0, \ldots, I$, then $n_i^{(1)}(\tau, l) \leq n_i^{(2)}(\tau, l)$, for all $\tau \geq 0$, $l \geq 0$ and $i = 0, \ldots, I$.

3. If $n_i^{(1)}(\tau, l)$ and $n_i^{(2)}(\tau, l)$, $i = 0, \ldots, I$, are the solutions obtained for a certain set of the initial distributions $n_{i0}(l)$, $i = 0, \ldots, I$, and two sets of source functions $F_{*i}^{(1)}(\tau, l)$ and $F_{*i}^{(2)}(\tau, l)$ such that $F_{*i}^{(1)}(\tau, l) \leq F_{*i}^{(2)}(\tau, l)$ for all $t \geq 0$, $l \geq 0$, and $i = 0, \ldots, I$, then $n_i^{(1)}(\tau, l) \leq n_i^{(2)}(\tau, l)$, for all $\tau \geq 0$, $l \geq 0$ and $i = 0, \ldots, I$.

4. If the initial conditions $n^0(l) = n_T^0(l) = \sum_{i=0}^{I} n_i^0(l)$ and the external source terms $F(\tau, l) = \sum_{i=0}^{I} F_{*i}(\tau, l)$ for equations (6.39) and (7.83) are equal, then the solutions $n(\tau, l)$ and $n_T(\tau, l) = \sum_{i=0}^{I} n_i(\tau, l)$ of these equations are identical.

Proof. The proof of relationships (7.102) can be obtained by integrating equations (7.19) and (7.20). Items 2 and 3 follow from the linearity of equations (7.18)-(7.20) and solutions nonnegativity (7.89) (also see Kudish, Airapetyan, and Covitch [2]). Item 4 follows from the uniqueness of the solutions of (6.39) and (7.83) as well as from (7.81) and (7.82).

Remark 7.7.6 *The interpretation of equations (7.102) is obvious: if no whole arms are ripped off the star molecules with i arms then the number of such arms changes only due to the external source functions $F_{*i}(\tau, l)$, $i = 1, \ldots, I$. Moreover, if $F_{*i}(\tau, l) = 0$ for all $\tau \geq 0$ and a particular i then the number of arms of polymer molecules with i arms is conserved. For for all $\tau \geq 0$ the molecular weight of star polymer molecules with i $(i > 1)$ arms monotonically decreases with time τ.*

Corollary 7.7.3 *Under the assumptions of Theorem 7.7.5 and Corollary 7.7.2 if the initial viscosities μ_a of the non-degraded solutions of linear and star polymers, described by (6.38) and (7.83), respectively, are equal and the viscosities are determined according to equations (6.37) and (7.50), (7.81), respectively, then they are equal at any time moment $\tau \geq 0$.*

Lemma 7.7.6 *Let $\beta > 0$ and $m\alpha_0 > 2 + \beta$ be certain constants and $n_i(\tau, l) \geq 0$, $i = 0, \ldots, I$, for $\tau \geq 0$, $l \geq 0$. Then function $W_{cT}(\tau, l)$ for $\tau \in [0, T_*]$, $l \geq 0$, determined by (7.42) and (7.52) satisfies the following relationships:*

1.

$$W_{cT} = w_m l[n_0(\tau, l) + n_1(\tau, l) + \sum_{i=2}^{I} \tfrac{1}{i} f_{(i)}(\tau, l) \int_0^\infty n_i(\tau, L) dL], \qquad (7.103)$$

where functions $f_{(i)}(\tau, l)$, $i = 2, \ldots, I$, are determined by (7.36).

2.

$$\int_0^\infty W_{cT}(\tau, l) dl = \int_0^\infty W_T(\tau, l) dl, \ \int_0^\infty l^\beta W_{cT}(\tau, l) dl \geq \int_0^\infty l^\beta W_T(\tau, l) dl, \qquad (7.104)$$

where $W_T(\tau, l)$ is determined by (7.81).

Proof. Using (7.36) for $i \geq 2$ and changing the order of integration we obtain

$$\int_0^\infty l^{1+\beta} f_{(i)}(\tau, l) dl = \int_0^\infty f_i(\tau, l_1) dl_1 \int_0^\infty f_i(\tau, l_2) dl_2 \ldots$$

$$\times \int_0^\infty f_i(\tau, l_{i-1}) dl_{i-1} \int_0^\infty f_i(\tau, l_i)(l_1 + l_2 + \ldots + l_i)^{1+\beta} dl_i. \qquad (7.105)$$

For $\beta = 0$ from (7.105) and (7.36) follows that

$$\int_0^\infty l f_{(i)}(\tau, l) dl = i \int_0^\infty l f_i(\tau, l) dl = i \int_0^\infty l n_i(\tau, l) dl / \int_0^\infty n_i(\tau, l) dl, \ i = 2, \ldots, I.$$

Taking into account the fact that for $\beta > 0$ and $l_j \geq 0$, $j = 1, \ldots, i$,

$$(l_1 + l_2 + \ldots + l_i)^{1+\beta} \geq l_1^{1+\beta} + l_2^{1+\beta} + \ldots + l_i^{1+\beta},$$

(which can be shown by using the mathematical induction) we obtain that

$$\int_0^\infty l^{1+\beta} f_{(i)}(\tau, l) dl \geq \int_0^\infty f_i(\tau, l_1) dl_1 \int_0^\infty f_i(\tau, l_2) dl_2 \ldots$$

$$\times \int_0^\infty f_i(\tau, l_{i-1}) dl_{i-1} \int_0^\infty f_i(\tau, l_i)[l_1^{1+\beta} + l_2^{1+\beta} + \ldots + l_i^{1+\beta}] dl_i$$

$$= i \int_0^\infty l^{1+\beta} n_i(\tau, l) dl / \int_0^\infty n_i(\tau, l) dl, \ i = 2, \ldots, I.$$

After that, items 1 and 2 follow from (7.35)-(7.37) and (7.81).

Remark 7.7.7 *Based on (7.50)-(7.52) and item 2 from Lemma 7.7.6 we can conclude that given the base stock with certain viscosity μ_a and any initial* *densities $n_i(0, l) = n_i^0(l)$, $i = 0, \ldots, I$, of the star polymer arm distributions the lubricant viscosity μ of the star polymer solution is always higher than the lubricant viscosity μ of the linear polymer solution with the cumulative density of star polymer arms*

$$n_T(0, l) = \sum_{i=0}^{I} n_i^0(l).$$

This relationship between the lubricant viscosities of star and linear polymer molecules holds for at least some initial period of time because $n_i(\tau, l)$, $i = 0, \ldots, I$, are continuous functions of time t.

Theorem 7.7.6 *The equations of problem (7.18)-(7.20) and (7.50) are homogeneous, i.e., if a set of functions $n_i(\tau, l)$, $i = 0, \ldots, I$, is a solution of (7.18)-(7.20) and (7.50) for the initial conditions $n_{i0}(l)$, $i = 0, \ldots, I$, then for any positive constant c the set of functions $cn_i(\tau, l)$, $i = 0, \ldots, I$, is also a solution of (7.18)-(7.20) and (7.50) for the initial conditions $cn_{i0}(l)$, $i = 0, \ldots, I$.*

Proof. The proof is obvious from (7.18)-(7.20) and (7.50).

Theorem 7.7.7 *The equations of problem (7.18)-(7.20), (7.51), and (7.52) are homogeneous, i.e., if a set of functions $n_i(\tau, l)$, $i = 0, \ldots, I$, is a solution of (7.18)-(7.20), (7.51), and (7.52) for the initial conditions $n_{i0}(l)$, $i = 0, \ldots, I$, then for any positive constant c the set of functions $cn_i(\tau, l)$, $i = 0, \ldots, I$, is also a solution of (7.18)-(7.20), (7.51), and (7.52) for the initial conditions $cn_{i0}(l)$, $i = 0, \ldots, I$.*

Proof. The proof is obvious from (7.18)-(7.20), (7.51), and (7.52).

Remark 7.7.8 *Theorems 7.7.6 and 7.7.7 provide the foundation for interpretation and comparison of the numerical and experimental data obtained using the gel permeation chromatography technique (GPC) tests.*

7.8   Numerical Method for Solution of the System of Kinetic Equations

In this section we will develop a method for numerical solution of the system of equations (7.18)-(7.20),(7.51) and (7.52). Let us introduce a time interval $\triangle\tau$ and consider a sequence of time moments $\tau_q = q\triangle\tau$, $q = 0, 1, \ldots, Q_t$. We will assume that polymer molecules dissolved in the lubricant are subdivided into N groups with the chain lengths l from the intervals $[l_{N+1}, l_N], \ldots, [l_2, l_1]$, where $l_1 > l_2 > \ldots > l_{N+1} = 0$. The number of polymer arms of chain length l, $l_{j+1} \leq l \leq l_j$, of molecules with i arms at the time moment τ_q is denoted by $N_i(q, j)$. Therefore, the distribution density of the star polymer arms assumes the values $n_i(q,j) = N_i(q,j)/(l_j - l_{j+1})$ while the distribution density of the molecular weight of star polymer arms assumes the values $w_i(q,j) = w_m n_i(q,j) l_j$, where w_m is the monomer molecular weight. Assuming that the density of the conditional probability $p_c(\tau, \mathbf{x}, l, L)$ does not depend on time τ and the spatial variable $\mathbf{x}$ (see (7.30) let us also introduce the notations:

$$R_i(q,j) = R_i(\tau_q, l_j), \quad R_{ai}(q,j) = R_{ai}(\tau_q, l_j),$$

$$p_c(i,j) = p_c(\tau_q, \mathbf{x}, l_i, L_j),$$

$$M(i,q,j) = \frac{R_i^2(\tau_q, l_j)}{R_i(\tau_q, l_j) + R_{ai}(\tau_q, l_j)}, \quad M_a(i,q,j) = \frac{R_i(\tau_q, l_j) R_{ai}(\tau_q, l_j)}{R_i(\tau_q, l_j) + R_{ai}(\tau_q, l_j)}, \quad (7.106)$$

$$\triangle l = [(l_1 - l_2)/2, l_2 - l_3, \ldots, l_{N-2} - l_{N-1}, (l_{N-1} - l_N)/2],$$

$$d(0) = 2, \quad d(i) = 1 \ for \ i = 1, \ldots, I.$$

Assuming that $\rho = const$, $F_{*i}(\tau, \mathbf{x}, l) = 0$ for $i = 0, 1, \ldots, I$, and using finite differences, from the system (7.18)-(7.20) one gets

$$\frac{W_i(q+1,j) - W_i(q,j)}{\triangle\tau} = \frac{d(i) l_j}{4} \sum_{m=1}^{j} M(i,q,m) p_c(m,j)(W_i(q+1,m)$$

$$+ W_i(q,m))\frac{\triangle l_m}{l_m} - \frac{1}{2}(M(i,q,j) + M_a(i,q,j))(W_i(q+1,j)$$

$$+ W_i(q,j)) - \frac{i-1}{2}(W_i(q+1,j) + W_i(q,j))J(i,q) \qquad (7.107)$$

$$+ i W_{i+1}(q,j) J(i+1,q),$$

$$i = 0, \ldots, I, \quad q = 1, \ldots, Q_t, \quad j = 1, \ldots, N,$$

$$J(i,q) = 0 \ if \ i > 1 \ or \ \sum_{j=1}^{N} W_i(q,j)\frac{\triangle l_j}{l_j} = 0,$$

$$\qquad (7.108)$$

$$J(i,q) = \frac{\sum_{j=1}^{N} R_{ai}(q,j) W_i(q,j)\frac{\triangle l_j}{l_j}}{\sum_{j=1}^{N} W_i(q,j)\frac{\triangle l_j}{l_j}} \ if \ \sum_{j=1}^{N} W_i(q,j)\frac{\triangle l_j}{l_j} \neq 0.$$

After solving these equations with respect to the unknowns $W_i(q+1,j)$, we obtain

$$W_i(q+1,j) = A_+^{-1}(i,q,j)\Big\{A_-(i,q,j) W_i(q,j) +$$

$$\qquad (7.109)$$

$$d(i) l_j \triangle\tau \sum_{m=1}^{j-1} M(i,q,m) p_c(j,m)[W_i(q,m) + W_i(q+1,m)]\frac{\triangle l_m}{l_m}$$

$$+2i\Delta\tau W_{i+1}(q,j)J(i+1,q)\Big\},$$

$$i = 0,\ldots,I,\ q = 1,\ldots,Q_t,\ j = 1,\ldots,N,$$

$$A_{\pm}(i,q,j) = 2 \pm \Delta\tau[M(i,q,j) + M_a(i,q,j) + (i-1)J(i,q)$$

$$-d(i)\Delta l_j M(i,q,j)p_c(j.j)],\tag{7.110}$$

$$i = 0,\ldots,I,\ q = 1,\ldots,Q_t,\ j = 1,\ldots,N.$$

The numerical algorithm is based on the above formulas and it requires three loops. In the external loop q runs from 1 to $Q_t - 1$. Inside this loop i runs from I to 0. In the inner most loop j runs from 1 to N. After calculation of the functions $W_i(q,j)$ for given q is done, we use (7.106) to evaluate the probabilities $M(i,q,m)$ and $M_a(i,q,m)$ for i from 0 to I and j from 1 to N. The lubricant viscosity $\mu(q)$ is calculated based on the trapezoidal rule and (7.51) and (7.52).

After calculation of functions $n_k(q,j)$ one determines the values $n_{(k)}(\tau,\mathbf{x},l)$ of the distribution density of the number of star polymer molecules with i arms. Taking the Fourier Transform of both sides of equation (7.36) we get

$$\mathbf{F}f_{(k)}(\tau,\mathbf{x},\xi) = \int\limits_0^\infty e^{-il\xi}f_{(k)}(\tau,\mathbf{x},l)dl$$

$$= \int\limits_0^\infty e^{-il\xi}dl \int\limits_0^l f_k(\tau,\mathbf{x},l_{k-1})dl_{k-1} \int\limits_0^{l-l_{k-1}} f_k(\tau,\mathbf{x},l_{k-2})dl_{k-2}\ldots$$

$$\times \int\limits_0^{l-l_{k-1}-\ldots-l_3} f_k(\tau,\mathbf{x},l_2)dl_2 \int\limits_0^{l-l_{k-1}-\ldots-l_2} f_k(\tau,\mathbf{x},l_1)$$

$$\times f_k(\tau,\mathbf{x},l-l_{k-1}-\ldots-l_1)dl_1\tag{7.111}$$

$$= \int\limits_0^\infty e^{-il_{k-1}\xi}f_k(\tau,\mathbf{x},l_{k-1})dl_{k-1} \int\limits_0^\infty e^{-il_{k-2}\xi}f_k(\tau,\mathbf{x},l_{k-2})dl_{k-2}\ldots$$

$$\int\limits_0^\infty e^{-il_1\xi}f_k(\tau,\mathbf{x},l_1)dl_1 \int\limits_0^\infty e^{-il\xi}f_k(\tau,\mathbf{x},l)dl$$

$$= \left\{\int\limits_0^\infty e^{-il\xi}f_k(\tau,\mathbf{x},l)dl\right\}^k,$$

where i is the imaginary unit, $i^2 = -1$.

Thus, to determine the values of functions $f_{(k)}(\tau,\mathbf{x},l)$, $k = 2,\ldots I$, we use the direct and inverse Fast Fourier Transforms. The procedure based on (7.111) is very fast, accurate, and efficient. After that, calculation of densities $W_{(i)}(\tau,\mathbf{x},l)$ and $n_{(i)}(\tau,\mathbf{x},l)$) of molecular weight and number of the distributions of star polymer molecules from (7.42) and (7.44) is done using the trapezoidal rule.

7.9 Numerical Results for Lubricants with Star Polymeric Additives and their Comparison with Kurt Orbahn Test Data

To validate the developed model of stress-induced degradation of star polymeric additive, we will compare the results of numerical simulation with the experimental results obtained for commercially available hydrogenated styrene isoprene star polymer additives RI2 and RI3 from the Kurt Orbahn fuel injector bench tests (ASTM method D6278) presented by Covitch in [8]. In all calculations, the lubricant viscosity μ is calculated based on equations (7.51) and (7.52).

The results of the Kurt Orbahn tests are represented in Figs. 7.4A and 7.5A as a series of histograms of star polymer molecular weight distributions that include the histogram of the initial distributions of the molecular weight $f_{m0}(l)$ and three additional histograms of the distributions of molecular weight after 30, 100, and 250 cycles. Each cycle in the Orbahn injector is represented by a single pass of the tested fluid volume through a fuel injector.

The densities (not histograms) of the experimental initial distributions of molecular weight $f_{m0}(l)$ of star polymeric additive and their least squares approximations $f_{m0a}(l)$ are presented in Figs. 7.2 and 7.3. A circle on the curves of $f_{m0}(l)$ in Figs. 7.2 and 7.3 indicates the position of the mean μ_l of the chain lengths of polymer single arms and star polymer molecules with just one arm. The numerical approximation of the experimentally obtained actual initial distributions of polymer molecular weight for lubricants with RI2 and RI3 star polymeric additives shows that the maximum number of star polymer molecule arms I can be taken equal to 25. For $I = 25$ the values of the approximation coefficients α_i for $i > I$ are either equal to zero or by at least two orders of magnitude are smaller than the values of α_i for $i \le I$. The approximation parameters α_i, $i = 1, \ldots, I$, for the lubricants with RI2 and RI3 additives are given in Table 7.1. The precision of the approximation of the experimental density distribution $f_{m0}(l)$ by $f_{m0a}(l)$ is very high except for the star polymer molecules with one arm. For molecules with two or more arms, the precision of the approximation is better than 0.95% while the precision of approximation for molecules with one arm is lower. The relatively poor precision of approximation for molecules with one arm is due to several reasons: (1) relatively poor precision of experimental measurements for polymer molecules with small chain lengths, (2) possible presence of a large number of single cores, and (3) presence of various low molecular weight contaminants.

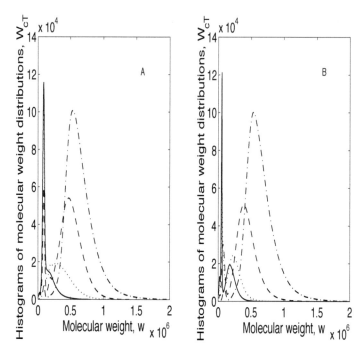

FIGURE 7.4

Histograms of star polymer molecular weight distributions at different time moments during the Kurt Orbahn testing of lubricants with RI2 star polymer (A) and obtained from numerical modeling (B). Here dashed-dotted curves - histogram of the initial molecular weight distribution, dashed curves - after 30 cycles, dotted curves - after 100 cycles, solid curves - after 250 cycles (after Kudish, Airapetyan, Hayrapetyan, and Covitch [4]). Reprinted with permission from the STLE.

The values of the sets of coefficients α_i, $i = 1, \ldots, I$, from Table 7.1 characterize the manufacturing process used for making RI2 and RI3 star polymers. The values of α_i in Table 7.1 indicate that for the most part in the initial mixture of RI2 and RI3 star polymer molecules the number of arms for most molecules ranges from 1 to 14.

Based on available experimental data the star polymer and operating parameters were taken as follows: $U = 347 \ kJ/mole$, $l^* = 0.154 \ nm$, $a^* = 0.356 \ nm$, $k_H = 0.945$, $k' = 5.05 \cdot 10^{-5}$, $\beta = 0.721$, $\triangle a = 0$, $T = 310K$, $S = 5 \cdot 10^6 s^{-1}$. The chain length l step sizes $\triangle l$ were chosen to be $\triangle l = 3.847 \cdot 10^2$ for single arms and $\triangle l = 3.0777 \cdot 10^3$ for star polymer molecules of RI2 additive and $\triangle l = 3.661 \cdot 10^2$ for single arms and $\triangle l = 2.929 \cdot 10^3$ for star polymer molecules of RI3 additive. The numbers of l-nodes used for approximation of single arms and star polymer molecules were taken equal to 1,000

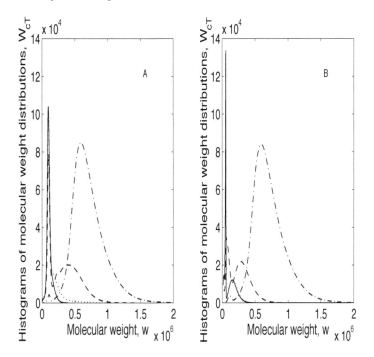

FIGURE 7.5

Histograms of star polymer molecular weight distributions at different time moments during the Kurt Orbahn testing of lubricants with RI3 star polymer (A) and obtained from numerical modeling (B). Here dashed-dotted curves - histogram of the initial molecular weight distribution, dashed curves - after 30 cycles, dotted curves - after 100 cycles, solid curves - after 250 cycles (after Kudish, Airapetyan, Hayrapetyan, and Covitch [4]). Reprinted with permission from the STLE.

and $2,000$, respectively. It is important to realize that the above values of $\triangle l$ are much smaller than the characteristic chain length L_*. For example, at $\tau = 75$ (for RI2 additive) and for $\tau = 22.5$ (for RI3 additive) $L_* = 4.8707 \cdot 10^4$ and $L_* = 4.9691 \cdot 10^4$, respectively. The time step size $\triangle \tau$ was chosen to be $\triangle \tau = 7.5 \cdot 10^{-2}$ for RI2 additive and $\triangle \tau = 2.25 \cdot 10^{-2}$ for RI3 additive $(1,000$ number of τ-nodes were used). As the numerical results were obtained for the case when $R_a(\tau, l) = F_{*i}(\tau, l) = 0$, $i = 0, \ldots, I$, the number of arms $n_i(\tau, l)$ of star polymer molecules with i arms, $i = 1, \ldots, I$, must be conserved (see Section 7.7). The latter conditions were maintained with the precision of about 2%.

A series of simulations allowed determining the values of the shield constants C and α and the time θ corresponding to one Orbahn test cycle that provide the best approximation of the experimental data. The values of these parameters were $C = 0.055$, $\alpha = 2.8525 \cdot 10^{-4}$, and $c_p = 0.566\ g/dL$, $\mu_a = 0.01046\ Pa \cdot s$, $\theta = 0.036$ $(t = 0.036\tau_f)$ for RI2 and $c_p = 0.485\ g/dL$, $\mu_a = 0.01005\ Pa \cdot s$, $\theta = 0.09$ $(t = 0.09\tau_f)$ for RI3. The numerical values of θ depend on a number of parameters, including the lubricant velocity **u**. Therefore, if the velocity **u** is unknown or chosen different from the actual test conditions then the value of θ is relative and serves only for the purpose of proper time scaling.

The numerically simulated histograms of the distributions of the molecular weight for star polymers with the initial distributions of molecular weight $W_{cT}(0, \mathbf{x}, l) = f_{m0a}(l)$ are presented in Figs. 7.4B and 7.5B for time moments $\tau = 1.08$, $\tau = 3.6$, $\tau = 9$ and $\tau = 2.7$, $\tau = 9$, $\tau = 22.5$ that correspond to 30, 100, and 250 of Orbahn test cycles for lubricants with RI2 and RI3 additives, respectively. Here $f_{m0a}(l)$ are the least squares approximations of the same as in Figs. 7.4A and 7.5A histograms of the experimental initial distributions of molecular weight $f_{m0}(l)$ of star polymeric additives. It is worth mentioning that in the simulations described below the spatial variable **x** is irrelevant. As one can see the numerically obtained data are in a very good agreement with the experimental ones. It is important to keep in mind that, generally, the precision of the molecular weight measurements is about $10\% -$ 20% and the measurement error is somewhat higher for polymer molecules with very short and very long chain lengths. This inaccuracy of measurements is practically irrelevant for polymer molecules with arms of very long chain length as there are only few of them. At the same time it may be significant for polymer molecules with arms of small chain length as they are encountered in large quantities. That explains why the discrepancy between the numerically and experimentally obtained molecular weight distributions increases with time as the polymer molecules get smaller (see Figs. 7.4 and 7.5).

In Figs. 7.6 and 7.7 the comparison of the numerically calculated and measured lubricant dynamic viscosity (further referred to as just viscosity) for RI2 and RI3 additives is presented: the continuous curves are the result of numerical simulation and the circles are the experimental data. As it follows from Figs. 7.6 and 7.7 the theoretically predicted and experimentally measured lubricant viscosity loss (of up to 27.5%) caused by lubricant degradation are practically identical. The results of Corollary 7.7.1 as well as Corollary 7.7.2 (see Figs. 7.6 and 7.7) were observed in numerical simulations.

Now, let us consider the dependence of the stress-induced degradation on the parameters of the model applicable to both ways of determining the lubricant viscosity, i.e., based on equations (7.50) and (7.51), (7.52). The examination of the parameters of the model shows that the stress-induced lubricant degradation depends on the initial distribution of star polymer molecular weight $f_{m0}(l)$. It follows from the model (see equations (7.18)-(7.20) and (7.50), (7.31), (7.32) and (7.51), (7.52), (7.31), (7.32) for the lubricant viscosity μ and the probability of scission $R(\tau, l)$) that the intensity of polymer scission increases for higher initial lubricant viscosity μ_a, higher polymer con-

TABLE 7.1

Sets of coefficients α_i involved in the least squares approximations $f_{m0a}(w)$ of the initial molecular weight distributions for RI2 and RI3 star polymers in equations (7.47)-(7.49) (after Kudish, Airapetyan, Hayrapetyan, and Covitch [4]). Reprinted with permission from the STLE.

k	α_k for RI2	α_k for RI3
1	1.7139	0.9390
2	0.1538	0.3221
3	0.6282	0.8260
4	0.9910	1.6646
5	4.1155	5.1356
6	6.9281	5.5196
7	4.8567	3.5926
8	3.6161	2.4593
9	1.8834	1.3173
10	1.4549	0.9509
11	0.5106	0.3974
12	0.5978	0.3859
13	0.0695	0.0719
14	0.2587	0.1475
15	0	0
16	0.0849	0.0437
17	0.0315	0.0117
18	0.0089	0.0048
19	0.0308	0.0072
20	0	0
21	0.0085	0.0016
22	0.0042	0
23	0	0
24	0.0018	0
25	0.0014	0.0003

centration c_p, and for star polymer molecules with longer arm chains. The nature of the base stock is reflected in the initial lubricant viscosity μ_a and, therefore, in the initial rate of lubricant degradation (see (7.50), (7.31), (7.32) and (7.51), (7.52), (7.31), (7.32)). The dependence of lubricant degradation with star polymeric additives like RI2 and RI3 on other parameters such as U, C, α, T, a_*, l_*, and S is similar to degradation patterns of a lubricant with linear polymeric additive such as OCP2 and OCP10 that was described

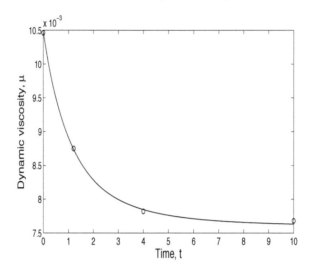

FIGURE 7.6
Numerically simulated loss of viscosity μ for the lubricant with RI2 star polymer additive caused by stress-induced polymer degradation versus time. Circles indicate the experimentally measured viscosity of the lubricant with RI2
star polymer additive (after Kudish, Airapetyan, Hayrapetyan, and Covitch
[4]). Reprinted with permission from the STLE.

in detail in Chapter 6. Namely, most of the conclusions can be drawn based
on the analysis of the characteristic polymer chain length L_* from (7.32) and
the expression for the probability of scission $R(\tau, l)$ from (7.31) (see Lemma
6.6.2). When the lubricant temperature T increases the probability of polymer scission R is determined by the two competing processes: decrease of the
slope of R and shift of R toward larger chain lengths l due to a usual decrease
of the lubricant viscosity μ with lubricant temperature T. Depending on the
particular lubricant nature, the probability of polymer scission R may increase
or decrease with lubricant temperature T. The increase of the lubricant viscosity μ and μ_a (for example, because of increase of pressure p) and/or of
the shear strain rate S cause L_* to decrease, which, in turn, accelerates scission. The physical explanation of this mechanism is clear because an increase
in μ and/or S leads to a corresponding increase in stretching forces acting
upon polymer molecules because of increased friction between the polymer
molecules and the surrounding lubricant. Finally, an increase in the value of
the product $Ca_* l_*^2$ leads to the same effect as the increase in μ and/or S. An
increase in the value of $aU/(kT)$ leads to a steeper distribution of the probability of scission R. That results in faster/slower scission of longer/shorter
polymer chains.

   In cases when $R_{ai}(\tau, l) = 0$ for all $i = 1, \ldots, I$ it follows from (6.12) and

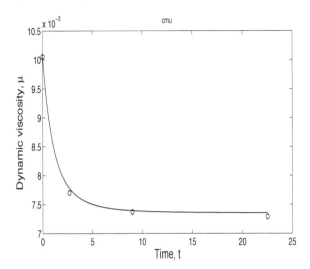

FIGURE 7.7
Numerically simulated loss of viscosity μ for the lubricant with RI3 star polymer additive caused by stress-induced polymer degradation versus time. Circles indicate the experimentally measured viscosity of the lubricant with RI3
star polymer additive (after Kudish, Airapetyan, Hayrapetyan, and Covitch
[4]). Reprinted with permission from the STLE.

(7.27)-(7.29) that the rates of viscosity loss calculated according to (6.37)
and (7.50) are identical for the same initial molecular weight distributions in
isolated systems, i.e. when $F(\tau, l) = F_{*i}(\tau, l) = 0$, $i = 0, \ldots, I$.

The test conditions and the initial molecular weight distributions of linear
polymeric additives OCP2, OCP10 and star polymeric additives RI2, RI3
were different from each. Therefore, it is not possible to compare directly
the patterns of star and linear polymer degradation on measured data that
is presented in Chapter 6 and this Chapter. It seems that star polymers of
the same molecular weight as linear polymers are better shielded from stress-
induced degradation than the linear ones, i.e., for given molecular weight star
polymer molecules undergo scission at a slower rate than polymer molecules
with linear structure (see [8]). It is due to the fact that in most cases the arms
of a star polymer molecule are significantly shorter than the chain length of
a linear polymer molecule of the same molecular weight and, therefore, the
probability for them to brake is lower (see equations (7.31), (7.32) for $R(\tau, l)$
and L_*, respectively). In particular, the latter means that the viscosity loss
of a star polymer solution occurs more gradually than the viscosity loss of a
linear polymer solution. A formal indication of the aforementioned difference
in the degradation rates is the lower value of the shield constant α and the
higher shear rate S for star polymer molecules than for the linear ones (see [2]

and [3]) that directly affect the linear and star polymer molecule probability of scission $R(l)$.

7.10   Closure

A mathematical model based on a system of coupled kinetic equations for stress-induced degradation of star polymers dissolved in a liquid lubricant is derived, analyzed, and tested against a set of experimental data. Some properties of the solutions of the system of kinetic equations are established. Among these properties are the existence and uniqueness of the solution of the initial-value problem for the above–mentioned system of integro-differential kinetic equations for star polymer degradation. A numerical method for solution of the system of kinetic equations is proposed. The numerically obtained results for star polymer molecular weight distributions and viscosity loss caused by degradation are in very good agreement with the experimental ones. It is shown that lubricant degradation patterns depend on the nature of the base stock, properties of the polymeric additive and its initial molecular distribution as well as on operating conditions. The developed method of approximation of the initial star polymer molecular weight distribution and the model of lubricant degradation enhance our understanding of the unique degradation characteristics of star polymers in lubricated contacts. Also, the above method provides the means of analyzing the initial composition of star polymer additive.

7.11   Exercises and Problems

1. List the major assumptions made for the derivation of the system (7.18)-(7.20), (7.51), and (7.52) that describe the degradation behavior of a star polymer solution.

2. What is the assumption used for forming star polymer molecules from linear polymer molecules?

3. Give a detailed description of the approach used for the approximation of the initial distribution of star polymer molecules. What is the major assumption used and how is it related to the shape of the experimentally obtained initial distributions of star polymermolecules in polymer solutions?

4. In an isolated system describe a possible exchange of molecules among the groups of star polymer molecules with ten, five, and one arms in a degrading polymeric solution. What happens in time with the number and molecular weight of the groups of star polymer molecules with: (a) the maximum number of arms in the solution and (b) just one arm?

5. Explain the mechanism of the solution viscosity loss due to star polymer degradation. Illustrate your answer with relevant formulas.

6. Compare two polymer solutions: one with linear (see preceding chapter) and another with star polymer additives. Assume that that the initial distribution of polymer molecular weights in both solutions is the same. Which of the two polymer solutions is more resistant to viscosity loss due to polymer degradation? Explain your conclusion. What is the role of the probability of scission $R(t, l)$ in your analysis?

BIBLIOGRAPHY

[1] Kudish, I.I. 2007. Modeling of Lubricant Performance in Kurt Orbahn Test for Viscosity Modifiers Based on Star Polymers. *Intern. J. Math. and Comp. Model.* 46, No. 5-6:632-656.

[2] Kudish, I.I., Airapetyan, R.G., and Covitch, M.J. 2002. Modeling of Kinetics of Strain Induced Degradation of Polymer Additives in Lubricants. *J. Math. Models and Meth. Appl. Sci.* 12, No. 6:835-856.

[3] Kudish, I.I., Airapetyan, R.G., and Covitch, M.J. 2003. Modeling of Kinetics of Stress Induced Degradation of Polymer Additives in Lubricants and Viscosity Loss. *STLE Tribology Trans.* 46, No. 1:1-11.

[4] Kudish, I.I., Airapetyan, R.G., Hayrapetyan, G.R., and Covitch, M.J. 2005. Kinetics Approach to Modeling of Stress Induced Degradation of Lubricants Formulated with Star Polymer Additives. *STLE Tribology Trans.* 48:176-189.

[5] Billmeyer, F.W., Jr. 1962. *Textbook of Polymer Science.* New York: John Wiley & Sons.

[6] Crespi, G., Vassori, A., and Slisi, U. 1977. Olefin Copolymers. In *The Stereo Rubbers.* Ed. W.M. Saltman. New York: John Wiley & Sons. 365-431.

[7] Mitsuhiro O., Yo N., Takashi N., and Akio T. 1998. Excluded-Volume Effects in Star Polymer Solutions: Four-Arm Star Polysteryne in Benzene. *Macromolecules* 31:1615-1620.

[8] Covitch, M.J. 1998. How Polymer Architecture Affects Permanent Viscosity Loss of Multigrade Lubricants. *SAE Tech. Paper* No. 982638.

8

Analysis of Some Experimental and Theoretical Data Related to Contact Fatigue. Review of Select Contact Fatigue Models

8.1 Introduction

The purpose of the review of experimental and theoretical data presented in Sections 8.2, 8.3, and 8.5 is to describe the modern understanding of experimental results and theoretical modeling of the contact fatigue phenomenon and other phenomena related to contact fatiguethat occur in different materials under various loading and environmental conditions. Section 8.4 illustrates the improvement in estimating fatigue life of radial ball bearings by introducing the dependence of fatigue life on friction. This review creates a realistic basis for the further analysis of the existing mathematical models of fatigue life for bearings and gears in Section 8.6.

8.2 Review of the Relationships between Contact Fatigue Life and Normal and Tangential Contact and Residual Stresses

Examination of various machine joints reveals several important fatigue and fracture mechanisms. These can be subsurface initiated fatigue, surface distress, corrosion and others, which can limit joint durability.

For any load carrying surface, it can be assumed that there are three or four layers. The surface contains the "outer layers" of oxides, absorbed films, reaction films, etc. The thickness of these surface layers is usually less than 1 micrometer. The near surface region contains various deformed layers, including the outer layer, where the microstructure may differ from that of the subsequent subsurface layers. Microstructure differences may arise from surface treatments, such as grinding and honing, heat treatment, etc. Consequently, in these regions hardness and residual stresses may be significantly different than in the subsurface region. Depth of the near surface region may be on the order of $50 - 100 \ \mu m$. The subsurface region located below the one just mentioned, which may extend $100 - 1000 \ \mu m$ (depending on the part's size) from the near surface region, is generally not affected by the mechanical processes or operationally induced physical and microstructural alterations. However, microstructure, hardness, and residual stress may differ from the core material and are generally the result of macro-processes, such as forging, heat treating, and surface hardening.

In the explanation and prediction of various material fatigue properties, it is no longer possible to ignore an initial statistical distribution of non-metallic inclusions, micro-cracks and other material defects, nature of the surface, lubricant type, lubricant films, contamination, filtration techniques, etc. The simplest and most studied among the aforementioned phenomena is the phenomenon of subsurface initiated fatigue. However, an agreement between researchers on the basic mechanisms of subsurface initiated fatigue has not been reached yet because of their complexity. It is much more difficult to understand the mechanisms of the surface initiated fatigue. This complexity is due to the direct influence of many additional factors, such as lubricant penetration into surface cracks, asperity interactions, surface chemical reactions, etc.

Therefore, our first goal is to identify and observe the most important features of the contact fatigue phenomena from experiment. The discussion will be limited to surface and subsurface initiated pitting with a focus on the mechanical aspects of the phenomena, putting aside chemical and any other environmental conditions. Thereafter, an analysis of existing contact fatigue models will follow with a discussion of their advantages and shortcomings.

This approach will identify the basic features, parameters, and mechanical properties necessary in the development of a successful physically sound and mechanistic-based contact fatigue model.

In a number of experimental studies researches have observed a strong dependence of contact fatigue life on normal contact stress. This has been observed in laboratory tests as well as in practice. Currently, it seems obvious that pitting life is a decreasing function of the normal contact stress.

There have been several experimental studies published concerning the influence of the contact tangential stress on fatigue life. In papers by Pinegin et al. [1] and Orlov et al. [2], a decrease of samples' pitting life caused by friction was observed. These researches discovered that a strong relationship exists between the number of cycles to pitting and a linear combination of the contact normal and frictional stresses.

Rolling contact fatigue tests of steel specimens have been carried out using a four-roller-type test machine by Soda and Yamamoto [3]. The experiments have indicated a relative decrease and increase in fatigue life when the traction has been applied in the same (follower) or opposite (driver) direction to the direction of relative motion, respectively. In these experiments both drivers and followers moved in the positive direction. Therefore, in the case of a

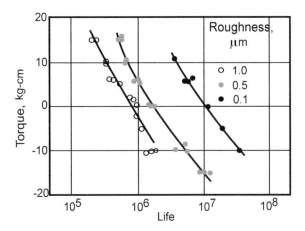

FIGURE 8.1
Rolling-contact fatigue life of a 0.45% C steel roller versus tangential traction (represented by torque) for various roughness of mating JIS SUJ2 (equivalent to AISI 52100) roller. Roughness is represented by peak-to-valley distance (after Soda and Yamamoto [3]). Reprinted with permission from the STLE.

driver the applied tangential stress (torque) was negative while in the case of a follower it was positive. In their study, all the cracks observed in the followers were surface cracks; however, the origin of these cracks was unclear. Life of the driver was $30 - 80$ times longer than that of the followers, and no fractures were observed on their surfaces. The fatigue life varied significantly depending on the direction and the magnitude of the tangential traction (see Figs. 8.1, 8.2, and Table 3 from Soda and Yamamoto [3]). Figure 8.1 shows that a 10–fold increase in roughness (which can be treated as an increase in tangential traction as well) results in a 10–fold decrease in fatigue life. It has been revealed that propagation of fatigue cracks on the later stages (after they were initiated) is mainly affected by frictional stresses.

The role of contact frictional stress on fatigue life of elastic solids has been studied by Kudish [4]. More specifically, on the basis of independently obtained bearing test results, some regression models have been proposed and implemented. A strong dependence of pitting life of ball bearings on friction in lubrication layers has been established. Bearing pitting life has been found to be a decreasing monotonic function of the contact friction forces. That fact has been related to a strong dependence of the normal stress intensity factors k_1 at fatigue crack tips on the friction coefficient. At the same time, a weak dependence of the tangential stress intensity factors k_2 at fatigue crack tips

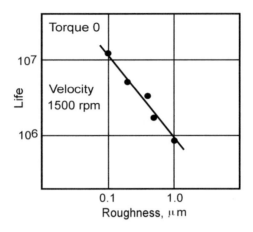

FIGURE 8.2
Rolling contact fatigue life versus roughness of harder mating roller (after
Soda and Yamamoto [3]). Reprinted with permission from the STLE.

on the friction coefficient has been established. The above–mentioned relationships were arrived at by Kudish [5] by using asymptotic methods under
the assumption that contact fatigue is of a subsurface origin.

The practical implication of this result is that fatigue failure is caused by
fatigue crack propagation. Based on the fact that $k_1^\pm$ are strongly dependent
on friction while $k_2^\pm$ exhibit a weak dependence on friction one can conclude
that the dominant mode of crack propagation, which leads to fatigue failure,
cannot be Mode II, but it is Mode I. Mode I is the crack propagation mode
under the action of normal stress while Mode II is the crack propagation mode
under the action of shear stress. This conclusion is extremely important when
selecting the crack propagation mode to be used in fatigue modeling.

There have been studies focused on the effect of the residual stress unavoid-
ably created by virtually all types of material treatments (thermal, thermo-
chemical, mechanical work-hardening, etc.) on contact fatigue of materials.
Experimental methods have been used in nearly all studies, and the available
literature sources do not offer an unambiguous treatment of this subject. For
instance, in Spektor et al. [6], the authors do not agree that residual stress
has a significant effect on fatigue of bearing steels. Some authors Mattson [7],
Kudryavtsev [8], Serensen [9], Scott et al. [10], and Almen and Black [11] believe that retardation of fatigue is favorably influenced by compressive residual
stresses and unfavorably influenced by tensile residual stresses. Other studies, for example, Brozgol [12], indicate that the compressive residual stress is
intolerable and that the small tensile residual stress is beneficial. But most
researchers support the opinion that compressive residual stresses are bene-

ficial and tensile residual stresses are detrimental to contact fatigue. It has been shown by Radhakrishnan and Baburamani [13] that with an increasing degree of pre-strain the crack nucleation period is delayed and the rate of crack propagation is decreased. The effect of residual stresses on fatigue life has been analyzed in detail by many researchers: Morrow and Millan [14], Mattson and Roberts [15], and Rowland [16], who have shown, that, in general, a significant residual stress is beneficial if compressive and detrimental if tensile.

In Averbach et al. [17], fatigue cracks have propagated through carburized cases in M-50NiL and CBS-1000M steels at constant stress intensity ranges, K, and at a constant cyclic peak load. A residual compressive stress of the order of 140 MPa has developed in the M-50NiL cases. It has been observed that the fatigue crack propagation rates, da/dN, slowed significantly. The residual stress in the cases in CBS-1000M steel was predominantly tensile, and the fatigue crack propagation rates da/dN were higher than in M-50NiL steel. From the data shown in Figs. 8.3 - 8.5 (Averbach et al., [17]), one can conclude that the fatigue crack growth rate is almost independent from the local hardness of the material and correlates well with the behavior of the residual stress.

The results of the study conducted by Kumar [18] show that a combination of plastic strains and low values of residual stress is conducive to subsurface crack initiation and growth.

In Lei et al. [19], the influence of residual stress in a plastically deformed rim on the cyclic crack growth produced by repeated rolling contacts was evaluated theoretically. One of the main conclusions of the study was that for cracks perpendicular to the solid surface residual stress significantly increased the fracture Mode II component, and it introduced a near threshold fracture Mode I component of crack growth. The underlying assumption was that the crack propagation direction is independent from the material stress state and is, as has been noted above, perpendicular to the solid surface.

Clark [20] has demonstrated that surface carburization accompanied by an increase of residual compressive stress increases fatigue life of bearing rings dramatically.

In Bhargava et al. [21] and Bhargava et al. [22], the researches presented a two-dimensional (plane strain) elastic-plastic finite element model of a rolling frictionless contact subjected to cyclic loading. As a result of these studies, the distributions of normal and shear residual stresses and strains were predicted at various depths of a tested rim. Furthermore, they have shown that a steady strain-stress state is usually attained after the first two loading cycles and that the residual stress does not change thereafter. In Ham et al. [23], a similar study for a frictional contact was conducted. The authors discovered that the maximum residual stress occurs below the surface at the depth of 0.4 of the contact half-width under pure rolling conditions and the residual stress is zero at the surface. In contrast, under mixed rolling and sliding conditions the maximum residual stress occurs at the material surface. Under mixed rolling

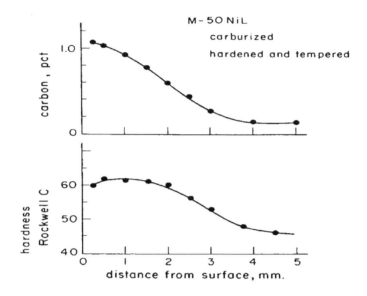

FIGURE 8.3

Hardness profile and carbon contents of carburized case in M-50NiL (after Averbach et al. [17]). Reprinted with permission from the AMMS.

and sliding conditions with a friction coefficient equal to 0.2 the maximum of the circumferential residual stress was about two times the value of that under pure rolling conditions. A similar problem for the cases of rail and bearing steels has been reported in studies by Ham et al. [24] and Kumar et al. [18]. In Kulkarni et al. [25], a three-dimensional elastic plastic finite element model has been used for studying of the residual stress in a repeated, frictionless rolling contact.

Bereznitskaya and Grishin [26] have shown that various methods of machining create different compressive and tensile surface and subsurface residual stresses. By using the appropriate machining to create compressive residual stresses, the specimen fatigue life was increased by 100%.

The influence of the residual stress, induced by plastic deformation, on rolling contact fatigue has been studied by Cretu and Popinceanu [27]. A general method for finding the optimum residual stress distribution, based on the consideration of equivalent stress, has been described. A positive influence of the compressive residual stress has been indicated. The B-10 fatigue life (that corresponds to survival probability equal to 0.9) of the pre-stressed test specimens was more than twice of that for the baseline obtained for not pre-

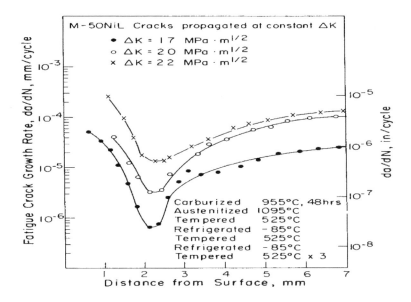

FIGURE 8.4

Fatigue crack propagation through carburized case in M-50NiL at constant values of the stress intensity factor K (after Averbach et al. [17]). Reprinted with permission from the AMMS.

stressed specimen group. The fatigue tests, carried out on four groups of ball bearings, have also confirmed both the existence of an optimum compressive residual stress distribution and the possibility of inducing this distribution by a pre-stress cycle.

In Voskamp [28], the changes of the residual stress gradients beneath the surface have been studied. Increases in the compressive residual stresses have been shown to be accompanied by changes in the metal microstructure. Possible residual stress distributions are shown in Fig. 8.6, Voskamp [28]. One important observation is that usually in thin near-surface layers the residual stresses are compressive, and below these layers, in the bulk of the material, they are much smaller and tensile. Voskamp has indicated that among various residual stress distributions there are two different types: (a) the distribution with the maximum residual stress on the surface of material and (b) the distribution with the maximum residual stress in a near surface layer. This fact may affect the location of pitting initiation. Sveshnikov [29] has observed that an increase in compressive residual stress leads to an increase in contact fatigue life.

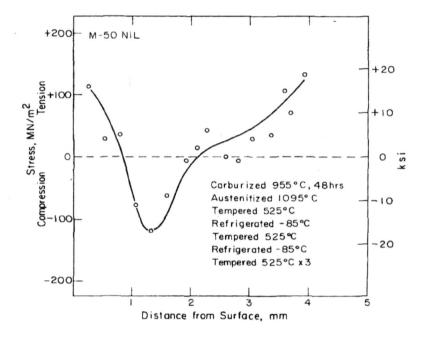

FIGURE 8.5

Residual stress in carburized M-50NiL (after Averbach et al. [17]). Reprinted with permission from the AMMS.

The material microstructure and residual stress distribution as functions of the depth below surface have been studied by Dudragne et al. [30] under pure rolling conditions. Also, the influence of the finishing operations on specimen fatigue life has been considered. The relations between the material microstructure, residual stress, and finishing operations (such as grinding, sanding, polishing, etc.) have been identified. The dependence of the number of cycles to fatigue failure and damage intensity on the distance from the specimen surface has been revealed. Most of the fatigue failures observed had subsurface origins.

Broszeit et al. [31] have shown that to determine the point in the material with the highest stress under contact loading one has to consider the stress state in almost the entire half-space. This is especially true in the case of compressive residual stress.

The influence of residual stress on the rolling element fatigue life has been studied by Kepple and Mattson [32]. They have shown experimentally that large tensile residual stresses decrease fatigue life of specimens. Furthermore, they have shown that fatigue life of specimens with high compressive residual stresses is practically independent from the magnitude of these residual stresses and is comparable with the fatigue life of specimens with zero residual

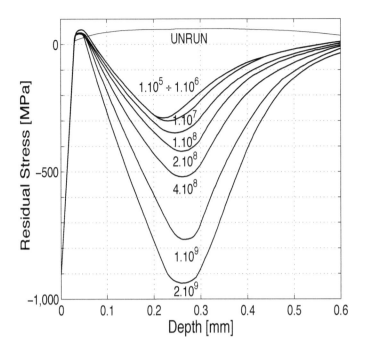

FIGURE 8.6

Residual stress as measured in the circumferential direction versus depth beneath the inner ring raceway of tested 6309 DGBB for various numbers of revolutions. Test conditions: Maximum contact stress: 3,300 MPa, inner ring rotational speed: 6,000 *rpm*, operating temperature: 53°*C* (after Voskamp [28]). Reprinted with permission from the ASME.

stresses.

In Janowski et al. [33], it has been stated that an increase in compressive residual stress leads to an increase in sample wear resistance.

The measurements of cyclic stress-strain hysteresis loop shapes of hardened SAE 52100 bearing steel have been presented by Hahn et al. [34]. The kinematics of the hardening behavior and properties of the 52100 grade have been employed in a two-dimensional elastic-plastic finite element model for pure rolling. The model has been used to evaluate distortions, cyclic plasticity, and residual stresses produced by repeated contacts. The results have been compared with extensive measurements of residual stresses and retained austenite transformation of bearing raceways reported by different authors. These comparisons revealed that the relatively modest circumferential residual stresses (from 50 *MPa* to 150 *MPa*) observed in the early portion of life ($N = 10^6$ loading cycles) are of mechanical origin. Higher residual stresses (from $100\ MPa$ to $780\ MPa$), observed in the later portion of life ($10^8 \leq N \leq 10^{10}$ loading cycles), have been associated with the material volume expansion. In addition, it has been determined that steel hardens noticeably during approximately the first 10 loading cycles.

The above review leads to the conclusion that fatigue life is affected not only by the normal contact stress but also by frictional effects as well. Both frictional and residual stresses have a strong influence on fatigue life. Also, most observations indicate that increase in frictional and tensile residual stresses, as well as decrease in compressive residual stress lead to decrease of contact fatigue life.

8.3   Review of the Relationships for Contact Fatigue Life versus Material Defects and Lubricant Contamination

The purpose of this review is to establish how various inclusions such as oxides and sulfides, small holes, and lubricant contamination affect contact fatigue. In order to create a solid foundation for fatigue modeling, it is necessary to understand where fatigue cracks initiate and what kind of stresses (tensile, compressive, shear stresses) cause their initiation and propagation. To achieve such an understanding, an analysis of some of the available experimental data is presented below.

The effect of non-metallic inclusions on fatigue strength of metals has been reviewed, and the major factors involved have been discussed by Murakami et al. [35], Nishioka [36], Yokobori [37], and Watanabe [38]. The important aspects are the inclusion shape, adhesion of inclusions to the material matrix, elastic constants of the materials of the matrix and the inclusions, and the inclusion sizes. These factors are related to the stress concentration factors and the stress distribution around inclusions. Many efforts have been made to quantitatively evaluate the stress concentration factors in the vicinity of inclusions by assuming that their shape is spherical or ellipsoidal. However, these assumptions are rough estimates because even a slight deviation from a spherical geometry can greatly affect the stress concentration factors. Adhesion of inclusions to the material matrix is usually not perfect and there are often some gaps between them; i.e., there are intrinsic cracks in the material. The stress concentration factors are practically useless in such cases. Therefore, the influence of non-metallic inclusions must be analyzed from the viewpoint of small cracks. Many researches report that the origin of fatigue fracture in high strength steels is not always located at the surface, but often at some distance away from the material surface. For Mode I loading equations for the normal stress intensity factor k_1 and the stress intensity factor threshold k_{th} for surface and subsurface cracks have been obtained as functions of the material hardness and of the square root of the area of the defect projected onto the plane perpendicular to the maximum tensile stress. From Tables 1 and 2, Murakami et al. [35], one can see that all failures were initiated at subsurface inclusions. Therefore, it has been concluded that non-metallic inclusions have a strong influence on fatigue life of high strength steels. In Murakami et al. [39], a predictive method for the fatigue failure estimation has been suggested. The approach is based on the material Vickers hardness and the maximum size of non-metallic inclusions. One of the main assumptions used by Murakami et al. [35, 39] is that one can replace an inclusion by a crack with the same area in the plane perpendicular to the maximum tensile stress. Similar results have been obtained by Murakami and Endo [40, 41] and by Murakami et al. [42]. Approximate expressions for the stress intensity factors k_1, k_2, and k_3 in terms of the square root of the projected crack area have been established by Murakami [43]. Murakami [44] has also found that the threshold stress intensity factor k_{th} and the critical diameter of a void in material are functions of the carbon content in steel.

In bearings, the strong dependence of contact fatigue on the number and size of non-metallic inclusions has been indicated by Spektor et al. [6]. They have shown experimentally that the main source of fatigue cracks is oxide non-metallic inclusions. Observations are supported by calculations (Andrews and Brooksbank [45]) of residual stresses resulting from different coefficients of thermal expansion of oxide non-metallic inclusions and martensite. Namely, oxides generate tensile stresses in the surrounding matrix. Incidentally, sulfide inclusions generate compressive stresses in the surrounding matrix, and they have rarely been observed to affect fatigue life of high strength steels. Coefficients of thermal expansion for different materials are presented in Fig. 8.7 (Brooksbank and Andrews [46]). This data explain why oxide non-metallic inclusions are more harmful than sulfide ones.

The Timken Company research has found that fatigue life is limited by large inclusions-stringers that still exist in nowadays highly publicized clean steels (Stover and Kolarik [47]). Stringers are represented by clusters of individual oxide particles. The total length of these stringers has been correlated with bearing fatigue life.

Moyer et al. [48] considered several different modes of contact fatigue. They were (1) inclusion–initiated fatigue, (2) microspalling or peeling damage, (3) point surface origin, (4) geometric stress concentration. Generally, subsurface fatigue initiated near non-metallic inclusions is associated with long bearing life. However, micro-stresses can be of a higher magnitude under high enough surface asperities than usual subsurface stresses. Microspalling and point surface origins are more specific mechanisms that are usually caused by gross extensions of several abnormalities such as large debris particles in the lubrication system, etc. Geometric stress concentration can be caused by improper design and bearing misalignment.

Laboratory testing and mathematical modeling of bearings have been con-

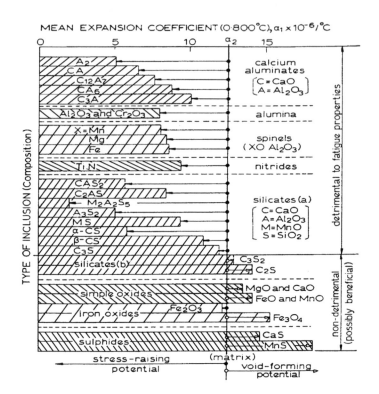

FIGURE 8.7

Stress-raising properties of inclusions in 1% C-Cr bearing steels (after Brooksbank and Andrews [46]). Reprinted with permission from the Elsevier Science Publishing.

ducted under conditions related to the field use over a long period of time by Tayeh and Woehrle [49]. They have confirmed that significant improvements in bearing fatigue performance are correlated with improvements in steel quality.

Fatigue failures of shot-peened carburized gears tend to be caused by subsurface non-metallic inclusions because fatigue strength of gear surfaces is improved by shot-peening operations (Toyoda et al. [50]). Toyoda et al. [50] have proposed that an empirical equation for fatigue strength should involve the Vickers hardness at a crack initiation site, the projected area of inclusions, and the residual stress. Also, they have found that decreasing the size of non-metallic inclusions for strengthening of carburized steels can be achieved by

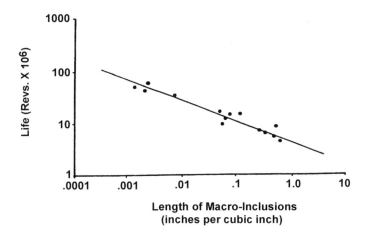

FIGURE 8.8

Bearing life-inclusion length correlation (after Stover and Kolarik II [51], COPYRIGHT The Timken Company 2009).

reducing the oxygen content.

In Stover and Kolarik [51, 52], a power relationship between the fatigue life of bearings and the ultrasonically measured oxide inclusion length in steel has been determined, and it is represented in Fig. 8.8. These findings suggest that bearing fatigue life is a power function of the steel oxygen content (see Fig. 8.8 and 8.9). The Timken Company's extensive use of an ultrasonic method for measuring the internal cleanliness of steels has led to significant life improvements of bearing steels. In Stover et al. [52], it has been stated that the size, frequency, and distribution of aluminum oxide inclusions (stringers) in steels can be assessed by using an ultrasonic detection method. As shown in Table I, [51], the minimum detected size of such inclusions is 0.533 mm.

In tests conducted by Clarke et al. [53], numerous micro-cracks oriented perpendicular to the direction of rotation have been observed. These cracks can be created only by propagation of cracks according to Mode I fracture mechanism. Authors have concluded that a strong correlation exists between a pit formation and the presence of near surface stringer–type aluminum oxide inclusions.

Studies by Bokman et al. [54] and Bokman and Pershtein [55] have shown that the non-metallic inclusion size distribution and crack distribution in aluminum alloys, both before and after plastic straining, are close to a logarithm-normal probabilistic distribution.

To elucidate the effects of small defects on the torsion fatigue strength, reversed torsion tests have been carried out by Endo and Murakami [56] on

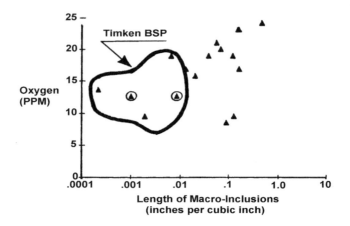

FIGURE 8.9

Oxygen versus total length of macro-inclusions, ingot cast steel (after Stover
and Kolarik II [51], COPYRIGHT The Timken Company 2009).

steel specimens containing a small hole of 40, 50, 80, 100, 200, and 500 μm
in diameter. When the hole diameter d was greater than 200 μm, the hole
became the origin of fracture. However, when the diameter d was smaller
than 100 μm, cracks initiated from other surface defects rather than the hole
edge; the latter cracks became the origin of fracture. Variation of the values of
the normal stress intensity factor for cracks emanating from a hole edge under
bending and torsion conditions are given in Table 8.1 (a is the hole radius, r is
the length of a crack from one side of a hole, σ is the acting stress). Table 8.1
indicates that cracks of size 1.1a behaved almost as solitary cracks of the same
size.

In Lankford and Kusenberger [57], electron microscopy of surface replicas
was employed to study the metallurgical micromechanics of fatigue crack initiation at surface and shallow subsurface inclusions in 4340 steel. They have
determined that cracks initiate through debonding of an inclusion from the
material matrix created by stresses at a tensile pole of the inclusion. The relationship between the location of the crack initiation and the direction of the
tensile stress near a surface inclusion is shown in Figs. (8.10a)-(8.10d), Lankford and Kusenberger [57]. All observed initiated cracks were oriented perpendicular to the direction of the tensile stress. A similar situation has been
observed regarding the relative directions of the initiated surface crack caused
by subsurface inclusions and active tensile stresses (Endo and Murakami [56]).
Based on the experimental observations and the available theoretical stress
analysis a model of crack initiation at inclusions in high strength steels was

TABLE 8.1

Stress intensity variation
of cracks emanating from
hole edge under bending
and torsion (after Endo
and Murakami [56]).
Reprinted with permission
from the ASME.

r/a	$F_1 = k_1/(\sigma\sqrt{\pi r})$	
	Bending	Torsion
0.1	2.78	3.57
0.2	2.41	2.99
0.4	1.97	2.31
0.6	1.73	1.94
0.8	1.57	1.72
1.0	1.47	1.58

proposed.

Some microstructural observations of rotating-bending fatigue specimens and a theoretical study of the stress concentrations at inclusions and voids have been performed by Eid and Thomason [58]. As a result, a basic understanding of the mechanism of the fatigue crack nucleation in high–strength steels under high cycle conditions has been developed. The authors have pointed out that under a given load fatigue cracks nucleate predominantly at the site of damaged alumina inclusions. Fatigue cracks did not nucleate while the alumina inclusions remained undamaged and firmly bonded to the material matrix. Even if the bond between the alumina inclusions and the material matrix remains intact, the local fracture of the inclusions themselves can produce surface holes that are responsible for the development of the fatigue crack nucleation. Manganese sulphide duplex inclusions and cementite particles were found to play no part in the nucleation of fatigue cracks in this type of high cycle fatigue specimens. All cracks shown in Fig. 8.11, Eid and Thomason [58], are perpendicular to the direction of the acting tensile stress.

In Lankford [59], the following stages of fatigue have been considered: (1) deterioration of a bond between a calcium aluminate inclusion and a material matrix and debonding of an inclusion from a material matrix, (2) fatigue crack initiation in a material matrix, (3) fatigue crack propagation. Debonding dramatically changed the distribution of stresses compared to the case of perfectly bonded inclusions. Fatigue crack initiation occurred at both bonded and debonded inclusions. Lankford indicates that only those cracks located near debonded inclusions appear to be able to establish a Mode II crack growth under further loading.

Several different rating methods of non-metallic inclusions in steels have been discussed by Koyanagi and Kinoshi [60]. Rousseau et al. [61] have stated

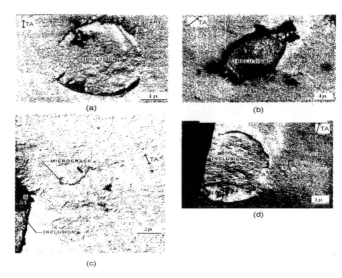

FIGURE 8.10
Electron micrographs showing early stages of fatigue micro-crack initiation near inclusions. (a) Partially debonded inclusion at which no apparent matrix deformation has yet occurred. (b) Earliest stage of fatigue crack initiation-nucleation of point surface defects (arrow) several microns from inclusion/matrix boundary, near equator of debonded inclusion. (c) Micro-crack formation via surface defect link-up, nearest end (arrow) of micro-crack segment lies at least 3 μm away from the debonded inclusion/matrix boundary (B). (d) Debonded inclusion at which point extrusions have linked-up to form tight closed fatigue crack (arrows), which joins debond seam at (A) (after Lankford and Kusenberger [57]). Reprinted with permission from the AMMS.

that all usual inclusion rating methods lead to approximately equivalent results, and the ASTM Recommended Practice for Determining Inclusion Content in Steel (E 45-63) is the simplest one. The steel fatigue resistance depends strongly on the steel–making practice, which determines steel cleanliness. However, fatigue properties do not depend upon the overall cleanliness of the steel. Under fatigue contact conditions hard and brittle oxide inclusions and titanium carbonitrides may induce the formation of butterflies and have often been found to be the origins of fatigue fracture in rotating beam specimens. Rousseau and his coauthors have never observed any butterflies around spheroidized cementite carbides nor around plastic (soft) inclusions such as manganese sulfides. Similar results have been observed for soft silicates.

Rolling contact fatigue tests have been conducted by Kinoshi and Koyanagi [62] to evaluate the effect of non-metallic inclusions in bearing steels similar to AISI 52100. Fatigue test results showed that alumina-type inclusions are

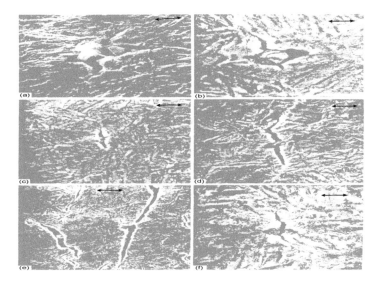

FIGURE 8.11

Scanning electron micrographs showing: (a) and (b) nucleation and growth of micro-cracks at points of maximum stress concentration on boundary of hole formed at damaged alumina inclusions, 4×10^6 cycles ($\times 3000$) and 4.3×10^6 cycles ($\times 3200$), respectively; (c) fatigue-crack nucleation away from inclusion sites at 4.5×10^6 cycles ($\times 1600$); (d) and (c) propagation of micro-cracks from alumina-nucleated holes and subsequent linking to form macro-cracks, 4.7×10^6 cycles ($\times 1650$) and 5×10^6 cycles ($\times 790$), respectively; (f) surface of specimen, having failed at 5.5×10^6 cycles, with no evidence of crack nucleation due to the highly-elongated manganese sulfide inclusions ($\times 3300$). Arrows indicate direction of cyclic bending stress (after Eid and Thomason [58]). Reprinted with permission from the Elsevier Science Publishing.

very harmful to fatigue life. However, the sulfide inclusions not only have no harmful influence, but increase fatigue life in the air-melted and, even, the vacuum-melted steels. The general relation between fatigue life and the alumina-type inclusion number in a volume unit is given in Fig. 8.12, Kinoshi and Koyanagi [62]. The general relation between fatigue life and sulfur contents in a volume unit is given in Figure 8.13, Kinoshi and Koyanagi [62].

In Moyer [63], some wear tests have been performed with different kinds of bearings in the presence of debris. Six factors have been identified that affect bearing performance when debris is present: debris size and its distribution, lubrication system, lubrication film thickness, levels of filtration, bearing materials, and contact size. All these parameters have been varied in the experiments. The results indicate that:

(1) The smallest debris particles cause mild wear. The intermediate particles

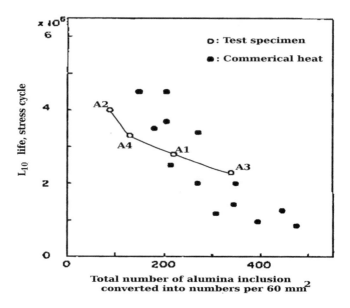

FIGURE 8.12

Relation between fatigue life and alumina type inclusion number (after Ki-noshi and Koyanagi [60]). Reprinted with permission from the ASTM.

$(5 - 40 \ \mu m)$ cause wear that is directly related to the amount of the debris present. These particles can cause severe surface fatigue for small bearings. The earliest fatigue failures seem to be directly related to the largest particles (with size greater than 40 μm) in lubrication system.

(2) The primary need of a filter in a lubrication system is to remove the largest debris particles that cause significant reduction in bearing life.

(3) There is a direct correlation between the debris hardness and wear. The higher the debris hardness the more severe is life reduction (especially when debris is harder than the bearing surfaces).

(4) The presence or absence of surface cracks depends on the particular material structure.

The above analyzed data show that no crack initiation has been observed at the locations of sulfide inclusions. In almost all studies, crack initiation has been observed at the locations of oxide inclusions. These cracks were initiated at the surface or beneath the material surface. Almost all experimental stud-ies showed that a fatigue crack propagates only under cyclic tensile stresses normal to its faces, indicating that the crack driving force is Mode I crack propagation mechanism. This is the reason why life prediction models need to consider not only the crack location but also its orientation. Other im-portant phenomena such as asperity interactions, contamination of lubricant,

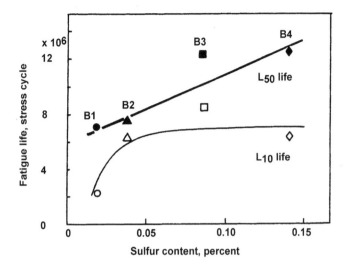

FIGURE 8.13

Relation between fatigue life and sulfur contents (after Kinoshi and Koyanagi [60]). Reprinted with permission from the ASTM.

and rigid particles imbedded into the contact surfaces (usually, as a result of machining), which may drastically reduce the material fatigue life, should also be incorporated into mechanistic based life prediction models.

8.4 Test–Based Least Squares Analysis of the Relationship between Bearing Fatigue Life and Contact Friction

In this section we will consider the improvement in estimating of radial ball bearing fatigue life by taking into account the traction force between bearing raceways and balls. According to Lundberg and Palmgren [64] fatigue life of radial ball bearings subjected to only radial load is calculated from the formula

$$L = L_{LP} = \left(\tfrac{C}{P}\right)^3, \tag{8.1}$$

where C is the bearing dynamic capacity, and P is the normal load applied to the most loaded ball of a bearing. Load P can be approximately calculated from the formula [81]

$$P = k\tfrac{F_r}{z}, \tag{8.2}$$

where F_r is the radial load applied to the bearing, z is the number of balls in the bearing, and k is a dimensionless coefficient. For $z \geq 4$ coefficient k varies between 4 and 4.5. For simplicity, in further calculations we will take $k = 5$.

Equations (8.1) and (8.2) for fatigue life L are in no way affected by the friction forces between balls and raceways of the bearing. However, in practice relatively small friction forces exist in contacts between balls and raceways. Let us try to approximately incorporate friction forces in calculation of the bearing fatigue life L. To do that we will come up with the regression formula for L that somehow incorporates friction.

It has been shown experimentally and will be shown theoretically that the friction coefficient λ has a significant influence on the subsurface stress field and, as a result of that, on the stress intensity factors and at the tips of subsurface fatigue cracks. At the same time, these stress intensity factors are insensitive to the changes in pressure caused by the friction coefficient λ. It is because for the friction coefficient λ of up to the values of $0.1 - 0.2$ the contact pressure is practically not affected by λ. On the other hand, it will be shown in Section 9.4 that the local details of the pressure $p(x)$ and frictional stress $\tau(x)$ distributions affect the stress intensity factors insignificantly (see the stress intensity factors and [4]).

Now, let us consider friction in the contacts of balls with raceways. With a relatively small error fluid friction in a lubricated contact can be represented by the formula

$$\tau = \frac{\mu(u_2 - u_1)}{h}, \tag{8.3}$$

where μ is the fluid viscosity ($\mu = \mu(p)$, p is contact pressure), u_1 and u_2 are contact surface linear velocities, and h is the gap between the contact surfaces, i.e., the film thickness. Therefore, the contact friction force F_{tr} is equal to

$$F_{tr} = \int\int_\Omega \tau dx dy, \tag{8.4}$$

where Ω is the contact region and x and y are the coordinates in the contact region Ω. It is well known (see [82]) that in heavily loaded lubricated contacts outside of the inlet and exit zones the pressure distribution is approximately equal to the Hertzian one in otherwise similar dry contacts. Therefore, for simplicity we will assume that the contact Ω is circular and the pressure distribution can be approximated by the Hertzian pressure $p(\rho) = p_H \sqrt{1 - (\rho/a_H)^2}$, where p_H and a_H are the maximum Hertzian pressure and Hertzian radius of the contact. Then, in dimensionless variables in a circular contact, the friction force F_{tr} is given by

$$F_{tr} = \frac{\pi V s_0}{6 H_0} \int_0^1 \mu(p)\rho d\rho, \quad p = \sqrt{1 - \rho^2},$$

$$V = \frac{24\mu_0 (u_1 + u_2) R_x^2}{p_H a_H^3}, \quad H_0 = \frac{2 R_x h_e}{a_H^2}, \tag{8.5}$$

where s_0 is the slide-to-roll ratio ($s_0 = 2(u_2 - u_1)/(u_2 + u_1)$), μ_0 is the ambient viscosity, R_x is the effective radius of curvature of a raceway and a ball ($1/R_x = 1/R_r \pm 1/R_b$, R_r and R_b are the radii of curvature of the raceway and the ball), h_e is the lubrication exit film thickness. In (8.5) for F_{tr} we used the fact that with high degree of precision in heavily loaded lubricated contacts $h = h_e$ within the Hertzian contact Ω. Knowing that the bearing fatigue life L suppose to depend on F_{tr}, we will try to approximate this relationship by the regression formula

$$L = L_{LP} \exp(a_0 + a_1 F_{tr}), \tag{8.6}$$

where constants a_0 and a_1 will be determined by using the least squares method.

TABLE 8.2
Comparison of the experimental,
Lundberg-Palmgren, and bearing
lives (in hours) obtained from the
regression analysis for a number of
different single row radial bearings
(after Kudish [4]). Reprinted with
permission from Allerton Press.

Bearing Type	L_{10exp}	L_{LP}	L
311SH	424	104	529
118	514	49	266
309	419	112	458
50309	346	112	465
70-60309	414	112	461
311	421	104	738
212	409	75	433
121	359	57	313
60120	536	46	242
118	342	49	247

However, the exact value of the slide-to-roll ratio s_0 and the particular dependence of the lubricant viscosity μ on pressure p used in tests were unknown. Also, for the lubricant temperature T we used the average temperature of the outer rings of the tested bearings. Therefore, without losing much of precision it is possible to reduce the expression for F_{tr} in (8.5) to just V/H_0. Based on this we will use the following regression formula

$$L = L_{LP} \exp(b_0 + b_1 F_{tr0}), \quad L_{LP} = (\tfrac{C}{P})^3, \quad F_{tr0} = \tfrac{V}{H_0}, \tag{8.7}$$

where constants b_0 and b_1 will be determined by using the least squares method. In all calculation of H_0 we used $H_0 = (VQ)^{0.75}$ that is related to Ertel-Grubin's formula [83] (see also Kudish [5]).

Equations (8.1), (8.2), (8.5), and (8.7) were used for a single row ball bearing analysis. The regression results for fatigue life L and the corresponding test results L_{10exp} for the 90% survival probability obtained for a tested group of 30 series with twenty bearing in each (made at different plants in the course of one year) as well as the Lundberg-Palmgren life L_{LP} are given in Table 8.2. The ball diameter D_b was in the range: $12.7 \ mm \leq D_b \leq 25.4 \ mm$.

The described regression technique made it possible to significantly reduce the scatter of the calculated Lundberg-Palmgren bearing fatigue life L_{LP}. It follows from Table 8.2 that the maximum scatter was reduced from 10 to 2 fold. In addition, the regression results showed that in all cases constants a_0 and b_0 were positive and constants a_1 and b_1 in all cases (except one) were negative. The above regression analysis, on the one hand, confirms the nature of the dependence of the bearing fatigue life on friction and, on the other hand, shows a monotonic decrease of the bearing fatigue life L with increase of friction F_{tr} (and $F_{tr0} = V/H_0$). A similar in the nature of the dependence of the bearing fatigue life on friction was observed in experimental studies of a model ball bearing [1] and of specimens [2].

8.5 Crack Initiation and Crack Propagation Phenomena. Material Microstructure and Contact Fatigue Life. Surface and Subsurface Cracks

This section is devoted to the analysis of crack initiation and propagation processes, and the influence of material microstructure on contact fatigue life. Also, some theoretical results are considered for surface and subsurface cracks.

Contradicting opinions exist on what process actually determines contact fatigue: crack initiation or crack propagation. The importance of surface and subsurface crack behavior and their influence on contact fatigue are not well understood. The extent to which the material microstructure affects contact fatigue needs to be better understood. All these topics are tackled in this part of the review.

8.5.1 Crack Initiation and Crack Propagation Stages

Roller bearing fatigue life has been considered by Bhargava et al. [84]. Bearing life calculations have been based on Mode II stress intensity factor as a driving force for propagation of small cracks. Comparisons with measured total lives provide insights into the validity of the analysis and the importance of growth relative to pre-initiation and crack initiation stages.

In O'Regan et al. [85], Mode II crack growth driving force has been evaluated for small subsurface cracks subjected to repeated pure rolling contact loads. The numerical model takes into account the following factors: (i) elastic stresses between contacting solids, (ii) friction resistance of sliding crack faces, and (iii) residual circumferential tensile and compressive stresses produced by plastic deformations. The residual stresses produced by plastic deformations of the rim have a significant effect on crack growth driving force.

In tests by Clarke et al. [53], numerous micro-cracks oriented perpendicular to the direction of rotation were observed. Also, material replicas have shown that micro-cracks, which eventually propagate to become macropits, are present as early as at 5% of the total fatigue life.

Otsuka et al. [86] and Shieh [87] have conducted Mode II fatigue experiments with steel. The extension of the pre-existing "crack" was observed along two distinct branches. The branch forming first, the tensile branch, was oriented perpendicular to the direction of the maximum tensile stress. The second branch, the shear branch, grew in the direction determined by the maximum shear stress. It is interesting that these authors noticed the tendency for the tensile branch to dominate, especially at high stress levels.

Kaneta et al. [88] studied numerically a material stress state near a semicircular surface crack under the action of normal Hertzian and tangential contact stresses. They noticed that the transition from a crack to a pit is not induced by the shear mode crack growth. The tensile mode crack growth is necessary for pit formation and it dominates as the crack propagates under action of the lubricant pressure.

Kaneta et al. [89] have shown that, in essence, propagation of a subsurface crack is associated with fracture of Mode II. Mode I (tensile mode) fracture can occur only under action of large normal and frictional contact stresses.

An experimental study of fatigue life and fatigue crack initiation, orientation, and propagation has been performed by Wu and Yang [90]. It was discovered that for grade 304 stainless steel the initial cracks appear induced by traction. Furthermore, crack propagation tends to be normal to traction. These statements are valid for axial and two-dimensional cyclic specimen loadings. Analysis of tested specimens has shown that the duration of the crack initiation is small in comparison with the duration of the crack propagation.

Kapelski et al. [91] studied alumina and silicon carbide ceramics under nonlubricated static and dynamic loading. It was observed that cracking is very sensitive to the coefficient of friction and usually occurs in the very beginning of cyclic loading. Most cracks were produced by normal tensile stresses, i.e., by Mode I fracture mechanism (Figs. 8.14 and 8.15; Kapelski et al. [91]).

Propagation of linear and branched cracks under cyclic loading has been studied by Kitagawa et al. [92] experimentally and with the aid of finite element methods. Two different cases of stable crack propagation have been observed: (a) under axial cyclic loading a Mode I crack asymptotically tends to propagate in the direction in which $k_2 = 0$ and $k_1 = \max(k_1)$ and (b) under a two-dimensional shear cyclic loading a crack branches into two cracks perpen-

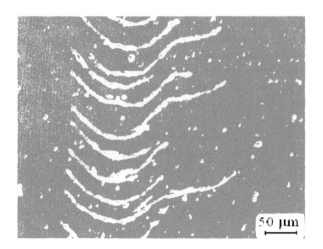

FIGURE 8.14
Crack array after 2 revolutions (Al_2O_3 ball on SiC disc) (after Kapelski et al. [91]). Reprinted with permission from the Elsevier Science Publishing.

dicular to each other. One of these cracks asymptotically tends to propagate in the direction in which $k_2 = 0$ and $k_1 = \max(k_1)$. The other propagates in the direction in which $k_2 = 0$ as well. The crack growth in the direction of the initial crack (in which the shear stresses are of maximum magnitude) is suppressed. It has been pointed out that propagation of different bent cracks (under various cyclic mixed mode loading) and even branched cracks is in a very good agreement with that determined according to Mode I fracture mechanism.

Nisitani and Goto [93], Tanaka et al. [94], and Ichikawa [95] have studied the ratio of durations of the crack initiation and crack propagation periods experimentally. It has been pointed out that the crack initiation period is short in comparison with the crack propagation period. This statement can be illustrated for steel by Fig. 8.16 (Tanaka et al. [94]) and for an Al-Ni alloy by Fig. 8.17 (Ichikawa [95]), where N is the number of loading cycles, p is the probability of crack initiation, crack propagation, or specimen failure, N_i is the number of loading cycles necessary for crack initiation, N_o is the number of loading cycles necessary for a crack to reach the size of $2l = 0.6$ mm, N_p is the number of loading cycles necessary for a crack to grow to failure, N_f is the total number of loading cycles to failure. The graphs of cumulative probability of failure for two stress levels are given in Fig. 8.16 (Tanaka et al. [94]). These graphs show that the crack propagation period is not less than $80\% - 90\%$ of the total life.

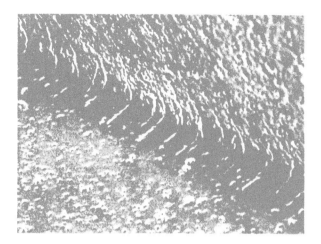

FIGURE 8.15
View of a section of a SiC disk (cut perpendicularly to the surface and tangentially to the wear track) (after Kapelski et al. [91]). Reprinted with permission from the Elsevier Science Publishing.

The contrary opinion has been presented in papers by Lundberg and Palmgren [64] and Ioannides and Harris [96], where contact fatigue models have been developed under the assumption that crack initiation overshadows crack propagation. However, this point of view is not supported by any experimental studies.

In Wert et al. [97], cracking, shown in Fig. 8.18, was observed to be perpendicular to the direction of sliding. The cracks shown in these figures are of the length not more than 20 μm.

In Kunio et al. [98], a study on initiation and early growth of a fatigue crack associated with a non-metallic inclusion in a high strength steel has been conducted. The following results have been pointed out:

(1) Poor adhesion between the material matrix and an aluminum oxide inclusion results in formation of an inclusion pit as a simple stress raiser.

(2) Metallurgical and micro-fractographic observations revealed that initiation and early growth of fatigue cracks initiated at non-metallic inclusions are of the shear rather than the tensile origin. Also, authors mentioned that the crack path appears to be perpendicular to the axial direction of the specimen as shown in Fig. 8.19 by Kunio et al. [98]. However, a higher magnification of the surface close to the inclusion pit revealed that a crack of the length of a few micrometers makes an angle of about 45 degrees with the axial direction at the inclusion.

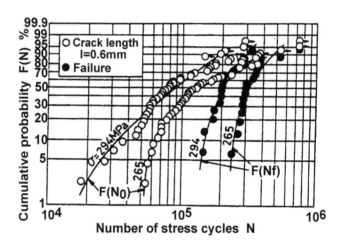

FIGURE 8.16

Distributions of lives for crack initiation and for final failure (after Tanaka et al. [94]). Reprinted with permission from the JSME.

The analysis of two- and three-dimensional finite element models of elastic plastic pure rolling contact was presented by Kumar [99]. The results (compared with those obtained from experiments on $7075 - T6$ Al alloy) show that a combination of plastic strain and low values of residual stress is conducive to subsurface crack initiation and growth.

Shao et al. [100] have reported the history of subsurface crack growth in a bearing steel cylinder subjected to a repeated rolling contact in the range from $N = 1.4 \cdot 10^6$ to $N = 4.9 \cdot 10^6$ loading cycles. This represents 89% of the total contact life.

Chen et al. [101] have evaluated the effects of the residual stress generated by a repeated rolling contact on Mode II stress intensity factor range $\triangle k_2$ that drives the subsurface crack growth in a rim. The value of $\triangle k_2$ has been evaluated according to the linear fracture mechanics for the case of a repeated two-dimensional, frictionless rolling contact and small planar subsurface cracks. Cracks with different inclinations have been studied. The comparison of theoretically and experimentally obtained data supports the opinion that location of the crack initiation is associated with location of the maximum of k_2 for cracks inclined at an angle $\alpha = 20°$ but not $\alpha = 0°$ as it was assumed in the Lundberg-Palmgren [64] bearing fatigue model.

Littmann and Moyer [102] have considered and visually classified differ-

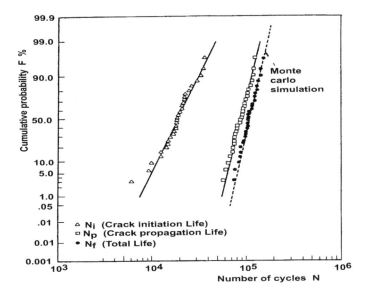

FIGURE 8.17

Cumulative probability of failure versus number of load cycles (after Ichikawa [95]). Reprinted with permission from the JSME.

ent contact fatigue failure modes such as: (1) inclusion origin, (2) geometric stress concentration, (3) surface point origin, (4) peeling or general superficial spalling, (5) subcase fatigue. The occurrence of one of the above mentioned modes is competitive, and depends on particular contact and cyclic loading conditions.

In Yamashita and Mura [103], a simple dislocation dipole model is proposed for contact fatigue initiation. Two different types of cracks have been observed at both edges of the contact region. Both crack types are induced by frictional stress causing tensile stresses at the crack initiation locations.

A model taking into account asperity presence in the contact is proposed by Yamashita et al. [104]. The overall influence of the combination of the contact pressure between smooth surfaces and between an asperity against a smooth surface on crack initiation and propagation has been studied. Some aspects of this phenomenon such as damage accumulation, micro-crack initiation, existence of non-propagating micro-cracks, inclination of micro-cracks due to forward movement of the material surface layer, and concentration of stress caused by asperity interaction have been considered. The averaging of all stresses over a small semicircular area has been used.

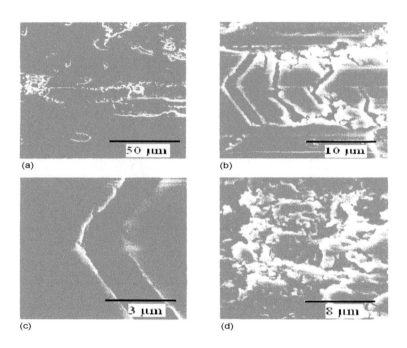

FIGURE 8.18
Wear track on disordered Cu_3Au after 250 μm of sliding with sapphire. (a) Overall view showing plastic deformation, cracking and delamination. (b) Cracking perpendicular to the direction of sliding. (c) Example of brittle nature of fracture. (d) Fine debris that may be compacted to provide a lamellae (after Wert et al. [97]). Reprinted with permission from the Elsevier Science Publishing.

8.5.2 Material Microstructure and Contact Fatigue Life

Drul' et al. [105] have mentioned that an increase in the austenite grain size leads to a very slow increase in contact fatigue life. Analyzing data presented by Ibale Rodriguez and Sevillano [106] and Schaper and Bosel [107], one can conclude that grains in the range from 1.2 μm to 48.5 μm have little influence on fatigue crack propagation.

Tsushima and Kashimura [108] have obtained some results indicating a positive influence of residual austenite on bearing fatigue life. This effect can be accounted for through the parameters of material fatigue resistance involved in the crack propagation laws (see Yarema [109]).

There has been almost no knowledge accumulated about influence of doping elements on fatigue life, and it is not clear (see Tsushima and Kashimura [108])

FIGURE 8.19
SEM photograph of the crack initiated at the typical non-metallic inclusion on the surface. Series B specimen, $N = 2.5 \cdot 10^4$ stress cycles at the stress level of 804 MPa (after Kunio et al. [98]). Reprinted with permission from Kluwer Academic Publishers.

how to incorporate them in fatigue modeling.

No appreciable influence of austenite grain boundaries on the crack paths was observed by Kunio et al. [98] in the case of inter-granular cracking experiments on high strength steels.

8.5.3   Subsurface and Surface Material Cracking

Keer et al. [110, 111] have presented some numerical results for a cracked dry elastic half-plane loaded by a Hertzian contact stress. A horizontal subsurface crack and a surface vertical crack were considered. The stress intensity factors at the crack tips have been calculated as functions of crack depth beneath the half-plane surface and their position relative to the loaded region. The cases of single subsurface and surface cracks, as well as the case of interaction between subsurface and surface cracks, have been considered. The authors have not taken into account the interaction of crack faces. In general, the behavior of the stress intensity factors is as follows. The normal stress intensity factor k_1 is different from zero only when a crack is behind the contact region. After a crack passes the contact region k_1 increases, reaches its maximum, and

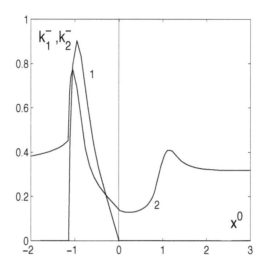

FIGURE 8.20
Distribution of the stress intensity factors $k_1^-(x^0)$ (curve 1) and $k_2^-(x^0)$ (curve 2) for a surface crack at its subsurface tip in the presence of a lubricant versus the position of the crack center x^0 relative to the contact center obtained for $\delta = 0.2$, $\alpha = \pi/6$, $q^0 = -0.5$, $\lambda = 0.1$, and $p(x) = \sqrt{1 - x^2}$ (after Kudish [115]). Reprinted with permission from the STLE.

then diminishes with distance. The absolute value of the shear stress intensity factor k_2 tends to be largest when the contact is approaching a crack. The value of k_2 reaches two extrema (one maximum and one minimum) when a crack is right beneath the boundaries of the contact region.

The same problem has been studied and similar results have been received by Kudish [4, 5]. In these papers only subsurface cracks have been considered. The main differences are that the interaction of crack faces has been taken into account and the approximate solutions were obtained in an analytical form (see Section 9.4). It has been pointed out that the friction coefficient has a very strong influence on the normal stress intensity factors $k_1^\pm$ and a relatively weak influence on the shear stress intensity factors $k_2^\pm$.

The same problem as in Keer et al. [110, 111] and Kudish [4, 5] has been studied for an inclined surface crack by Keer and Bryant [112]. Lubricant presence has been introduced implicitly by taking into account two parameters: an assumed percentage of an open part of the crack upon which fluid pressure acts upon and a friction coefficient between the crack faces. Crack growth and number of cycles for the crack to grow from its initial length to the final length have also been computed. The authors noticed that although calculated fa-

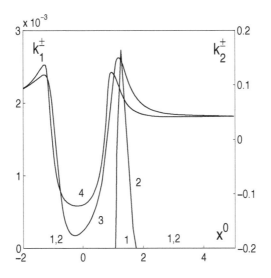

FIGURE 8.21

Distribution of the stress intensity factors $k_1^-(x^0)$ (curve 1), $k_1^+(x^0)$ (curve 2), $k_2^-(x^0)$ (curve 3), and $k_2^+(x^0)$ (curve 4) for a subsurface crack in the presence of a lubricant versus the position of the crack center x^0 relative to the contact center obtained for $\delta = 0.2$, $y^0 = -0.2$, $\alpha = \pi/6$, $q^0 = -0.1$, $\lambda = 0.1$, and $p(x) = \sqrt{1 - x^2}$ (after Kudish [115]). Reprinted with permission from the STLE.

tigue lives are crude the numbers produced agree in order of magnitude with experimentally observed lifetimes. The results of this study, however, do not explain the observed straight crack growth due to a high lubricant pressure penetrated inside the crack.

The growth behavior of surface cracks formed on lubricated rolling-sliding contact surfaces has been studied by Murakami et al. [113] by calculating the mixed-mode stress intensity factors k_1, k_2, and k_3. It has been assumed that the surface of an elastic half-space is loaded by normal Hertzian and tangential stresses. A surface vertical or inclined crack has been assumed to be the shape of a semicircle in the plane perpendicular to the direction of the pressure motion. Friction between the crack faces and the normal interactions between the crack faces have been ignored. It has been shown that the lubricant pressure may contribute to opening of a crack, and the stress intensity factors k_2 and k_3 are not influenced by the fluid pressure. It has also been noticed that the direction and the magnitude of the frictional force associated with the moving Hertzian contact pressure are the controlling factors which cause the crack mouth to open or to close and the lubricant to penetrate or not to penetrate the crack. It has been pointed out that surface cracks in a pure rolling case are free from the oil hydraulic pressure effect.

The same problem has been studied by Kaneta et al. [88, 114] and similar results have been obtained. In Kaneta et al. [89], the authors have considered the stress state around a circular planar subsurface crack parallel to the surface of a half-space. The half-space was subjected to Hertzian normal and tangential contact stresses. Friction and normal interaction between the crack faces have been taken into account. The stress intensity factors have been calculated for various sets of the problem parameters.

A two-dimensional problem for surface and subsurface straight cracks in an elastic half-plane covered by a layer of lubricant and subjected to normal and tangential (frictional) contact and residual stresses has been studied numerically by Kudish [115]. The conditions leading to the lubricant-crack interaction have been considered. It has been shown that for a surface crack filled with lubricant the value of the normal stress intensity factor k_1 may be of several orders of magnitude higher than the one for a similar subsurface crack while the values of the shear stress intensity factors k_2 do not differ drastically (compare Figs. 8.20 and 8.21 (see [116]), where δ is the crack semi-length, y^0 is the depth of the crack center beneath the material surface, α is the angle between the crack and the material surface, q^0 is the residual stress, λ is the friction coefficient, and $p(x)$ is the distribution of contact pressure). This analysis allowed for establishing the fact that under regular conditions fatigue pitting is of a subsurface origin. In addition, it was possible to explain the usual difference between fatigue lives of drivers and followers (see Section 9.6).

In Miller and Keer [117], a numerical solution to a two-dimensional problem for a rigid die sliding with friction along a half-plane containing a near surface imperfection in the form of a circular void or a rigid inclusion has been presented. The calculations have been performed for various combinations of the inclusion and void size, location, and the coefficient of friction between the die and the half-plane surface. It has been established that the presence of a subsurface void or inclusion can cause a significant alteration in the contact stress distribution. It has also been shown that a void gives a larger increase in shear stress than an inclusion.

The analysis presented in Sections 8.2-8.5 leads to the conclusion that fatigue life is affected by not only normal contact stresses. Frictional and residual stresses have an extremely strong influence on fatigue life. No fatigue crack initiation has been observed at sulfide inclusions. In almost all studies, fatigue crack initiation has been observed at the sites of oxide inclusions. These fatigue cracks may be initiated at the surface or beneath the material surface. Many authors have pointed out that crack initiation is much shorter than crack propagation. Therefore, for practical purposes, it is usually acceptable to neglect the crack initiation life and assume that the crack propagation life is approximately equal to the total life. Now, it is the common opinion that there is no difference between the general mechanisms of torsion (or tensile) fatigue and contact fatigue. Therefore, the same crack propagation laws can be applied to the investigation of contact fatigue. Almost all experimental studies show that a fatigue crack can propagate only under cyclic tensile stresses normal to its faces. Therefore, the crack driving force is of Mode I crack propagation mechanism. This is the reason why it is important to take into account not only the crack location but also its orientation.

It is also necessary to take into account such important phenomena as asperity interactions, contamination of lubricant, and rigid particles imbedded into the contact surfaces (usually, as a result of machining).

It must be understood that under certain conditions contact fatigue leads to a pitting phenomenon but under different conditions contact fatigue can lead to wear phenomenon. These two processes are competitive and, usually, run simultaneously. The occurrence of pitting and delay of wear (or vice versa) depends on many factors and, usually, cannot be predicted initially except for some special cases.

From the above review of the considered literature, it is obvious that the following parameters must be taken into account when developing an adequate contact fatigue model:

(1) Contact normal and frictional stresses

(2) Residual stress

(3) Initial statistical crack and material defect (oxide, etc., inclusions) distribution with respect to their location, orientation, and size

(4) Material fatigue crack resistance (parameters controlling fatigue crack propagation rate)

(5) Fracture toughness

(6) Surface roughness

(7) Lubricant contamination by statistically distributed solid particles

(8) Statistical contamination of material surfaces due to some mechanical treatments

8.6 Review of Some Contemporary Contact Fatigue Models

The research efforts in the field of contact fatigue in application to bearings and gears started with the groundbreaking work of Lundberg and Palmgren [64]. Their approach has its own limitations (discussed below), which caused over the years a continued effort in improving and developing new contact fatigue models [65] - [79] based on various assumptions. For example, some of the models are concerned only with the fatigue crack nucleation stage while other models consider fatigue crack propagation, some models consider Mode

I others Mode II crack propagation mechanisms, some are statistical while other models are deterministic, etc.

This section presents a critical analysis of some of the historically most popular statistical mathematical models of fatigue life applicable to bearings and gears [64, 80, 96], [120]-[133], [136]-[142], i.e. the analysis of the assumptions, advantages, shortcomings, and some contradicting aspects of these models. As a result of this analysis, some conclusions that adequately reflect the experimental and theoretical data discussed in preceding sections are drawn concerning the necessary features of a physically sound and successful mathematical model of contact fatigue.

8.6.1 Mathematical Models for Fatigue Life of Bearings

Most of the analysis presented in this section follows the material of paper by Kudish and Burris [118].

8.6.1.1 The Lundberg-Palmgren Model

Almost 60 years ago Lundberg and Palmgren [64] proposed the first mathematical model for prediction of bearing fatigue life. Their model states that for an inner bearing ring subjected to N cycles of a repeated stress, the probability of survival $S(N)$ with respect to subsurface initiated fatigue is given by

$$\ln \frac{1}{S(N)} \propto \frac{N^e \tau_0^c V}{z_0^h}, \tag{8.8}$$

where τ_0 is the maximum orthogonal shear stress; z_0 is the depth at which τ_0 occurs; V is the stressed volume; e, c, and h are empirical constants. The stressed volume V is proportional to the product of the contact area half-length, the length of ring raceway, and z_0. The latter equation is equivalent to

$$S(N) = \exp\left[-\left(\frac{N}{N^*}\right)^e\right], \quad N^* \propto \tau_0^{-c/e} z_0^{h/e} V^{-1/e}. \tag{8.9}$$

Therefore, fatigue life to failure follows the Weibull distribution with zero low bound, characteristic life N^*, and the dispersion exponent e.

Solving equation (8.9) for fatigue life N and using the fact that τ_0 is proportional to the maximum Hertzian pressure p_H we obtain

$$N = \frac{C}{p_H^{c/e}}, \tag{8.10}$$

where constant C depends on z_0, V, and S in an obvious way.

The main assumptions of this empirical model are

(i) The crack initiation period is much greater than the crack propagation period.

(ii) Fatigue failure is always initiated by the maximal orthogonal shear stress at the depth where it occurs.

(iii) The subsurface stress distribution is assumed to be a function of the normal contact (Hertzian) pressure only.

The Lundberg-Palmgren model has some inherent shortcomings, the most important of which are

(a) The model, based on a well-known weak dependence of shear stress on friction, does not adequately correspond to the experimentally observed strong dependence of bearing fatigue life on contact friction stress.

(b) Bearing fatigue life as a function of non-metallic inclusions, voids, preexisting and propagating surface and/or subsurface cracks cannot be described by the model.

(c) The model is unable to predict surface-initiated fatigue failure.

(d) The model does not take into account such factors as (1) contact tangential stress, (2) residual stress, (3) material cleanliness, (4) fatigue and fracture material properties, (5) surface roughness, (6) lubricant and surface contamination, etc.

(e) The model is based on a different fatigue mechanism than the models describing the common torsion/tensile (structural) fatigue tests.

(f) It is unreasonable to expect that equation (8.9) for the survival probability $S(N)$ reflects a correct dependence of $S(N)$ on the stressed volume V (see discussion of this item in Section 8.6). To show that, let us consider the scale effect by loading two roller bearings made of an identical material and of identical design but different size with the same maximum Hertzian pressure p_H ($p_H = \sqrt{\frac{E'P}{\pi R'}}$, where P is applied normal load and radius R' is proportional to the shaft diameter D) and different Hertzian contact half-width a_H ($a_H = 2\sqrt{\frac{R'P}{\pi E'}}$). Assuming that the shaft diameters for the smaller and larger bearing are D_1 and D_2, respectively, to maintain the same maximum Hertzian pressure p_H in the larger bearing as in the smaller one the load applied to a single contact (and, therefore, to the whole bearing) supposed to be $P_2 = qP_1$, where P_1 is the load applied to the smaller bearing and $q = D_2/D_1$. In such a case the Hertzian contact semi-width of the larger bearing is $a_{H2} = qa_{H1}$. In addition, the maximum orthogonal shear stresses in both bearings are equal $\tau_{02} = \tau_{01}$ and the depth where the maximum orthogonal shear stress occurs in the larger bearing is $z_{02} = qz_{01}$. Therefore, based on (8.8) for the survival probabilities of the two bearings $S_2(N)$ and $S_1(N)$ after the same number of stress cycles N, we obtain

$$S_2(N) = [S_1(N)]^{\alpha}, \ \alpha = q^{3-h}.$$

For roller bearings Lundberg and Palmgren proposed $h = 7/3$. For roller bearings for the survival probabilities $S_1(N) = 0.5$ and $S_1(N) = 0.9$ and several values of the ratio $q = D_2/D_1$, the survival probabilities $S_2(N)$ are presented in Table 8.3. In the view of the experimentally established relatively weak dependence of structural fatigue life of the sample size [119], it seems unreasonable to believe that the survival probability $S_2(N)$ of the larger bearing decreases so much in comparison with the survival probability $S_1(N)$ for the smaller bearing while the size of the larger bearing increases. Similar situation takes place in the case of two ball bearing of different size if the load P_2 applied to the larger bearing is equal to $q^2 P_1$.

TABLE 8.3
Survival probability $S_2(N)$ for the second roller bearings as a function of $S_1(N)$ and the ratio of the number of loading cycles N_2/N_1 for the two bearings with the same survival probability $S(N)$ versus the size ratio q.

q	$S_2(N)$ for $S_1(N) = 0.5$	$S_2(N)$ for $S_1(N) = 0.9$	N_2/N_1
2	0.333	0.846	0.663
4	0.174	0.767	0.440
8	0.0625	0.656	0.292
10	0.0401	0.613	0.256
20	0.0061	0.460	0.169
30	0.0012	0.362	0.133

The other way to look at this situation is by solving equation (8.8) for N we obtain

$$N \propto [\ln \tfrac{1}{S(N)}]^{1/e} [\tfrac{z_0^h}{\tau_0^c V}]^{1/e}.$$

Therefore, for the above two roller bearings under the described conditions assuming that $V_2 = q^3 V_1$ for the same survival probability $S(N)$, we get that the corresponding numbers of cycles N_2 and N_1 are related as follows

$$N_2 = N_1 q^{\frac{h-3}{e}}.$$

For roller bearings Lundberg and Palmgren proposed $e = 1.125$. The values of N_2/N_1 are given in Table 8.3 for some values of the size ratio q for roller bearings. Again, from the experimental point of view it seams unreasonable to believe that for the given survival probability $S(N)$ the number of loading cycles N_2 of the larger bearing decreases so much in comparison with the number of cycles N_1 for the smaller bearing while the size of the larger bearing increases (see [119]). Similar results can be obtained for ball bearings.

The Lundberg-Palmgren theory no longer reflects the latest achievements in steel making, design, and current changes in the environmental conditions of the bearing applications. Furthermore, this model provides an estimate of bearing fatigue life that is usually significantly below the actual life.

8.6.1.2   The ISO Standard for Bearing Fatigue Life

One of the first attempts to improve the Lundberg-Palmgren model has been presented in the ISO standard for rolling bearing life prediction. There, fatigue life L calculations have been altered by introducing variable fatigue life adjustment factors into the Lundberg-Palmgren model (see ISO Standard [80])

$$L = a_1 a_2 a_3 (\tfrac{C}{P})^p, \qquad (8.11)$$

where a_i $(i = 1, 2, 3)$ are the life adjustment factors, p is the given exponent ($p = 3$ for ball and $p = 10/3$ for roller bearings), C is the basic dynamic load capacity of a bearing, P is the equivalent load applied to a bearing. The values of constants a_i, p, and C are empirical. To some extent, these factors reflect improvements in bearing steel making and lubrication as well as in roughness of bearing working surfaces. However, the ISO model still does not address the dependence of bearing fatigue life on steel cleanliness, material fracture and fatigue parameters, residual stress, etc.

8.6.1.3 The Bearing Fatigue Models Developed by Tallian and Colleagues

Another approach to improving the Lundberg-Palmgren model has been described in papers by Chiu et al. [120]-[122], Tallian et al. [123], and Tallian [124]- [133]. The fatigue failure model proposed by Chiu et al. [120] looks at fatigue damage as a result of crack propagation under the action of a cyclic load. The model explicitly assumes that failure has occurred with the first spalling and implicitly assumes that crack propagation is caused by shear stresses. Since the mathematical framework of the model is presented in a very general form it is very difficult to draw any concrete conclusions in practical cases.

The later studies by Tallian [124] and by Chiu et al. [120]-[122], which have been conducted to explain and predict bearing fatigue initiated at surface and subsurface defects, are based on the same general assumptions and follow a similar mathematical procedure of representation as the earlier works of these authors. The proposed models are based on a crack propagation law and take into account the material matrix elastic and plastic properties, defect type and concentration, and geometry of macro-stress field. However, the absence of clearly stated assumptions and mechanical mechanisms, absence of methods for model developing, as well as a large number of empirical approximations and various details and coefficients make these models extremely difficult to use.

A theoretical study of bearing fatigue life with contaminated lubricant and its comparison with experimental results of different authors have been presented by Tallian [125, 126]. Wear caused by abrasive and debris particles has been estimated. An improved mathematical model for predicting bearing fatigue life has been developed by Tallian [127]. The main goal of this work is to evaluate the influence of surface traction, lubrication regime, and asperity interactions on bearing fatigue failure. The comparison of predicted fatigue life with bearing test data obtained by other authors has shown satisfactory agreement.

Tallian's [127] model has been further improved in [128]. The purpose of the latter paper is to introduce several modifying factors for film thickness, surface roughness, and friction. The approach is mostly based on empirical data. It is assumed that bearing fatigue life is determined by two independent and competing fatigue mechanisms: surface and subsurface initiated fatigue. This fact is expressed by means of the equation

$$N^{-\beta} = N_s^{-\beta} + N_{ss}^{-\beta}, \tag{8.12}$$

where N_s and N_{ss} are fatigue lives related to surface and subsurface initiated cracks, respectively, and β is the Weibull dispersion parameter. Several new factors responsible for surface and subsurface defect distributions, surface roughness, material fatigue resistance, surface friction stresses, etc., have been introduced in the contact fatigue formula developed by Tallian [127]. A comparison of the results based on the improved model with a few experimental data showed an agreement.

The main shortcomings of these models are as follows:

(a) The models lack clearly stated employed mechanical mechanisms of fatigue failure and are based on a number of empirical approximations (see above).

(b) One of the main assumptions of the models is that the crack initiation period is dominant (which contradicts some experimental data).

(c) The models do not take into account a relationship between fatigue life and the initial defect/crack distribution.

A unified model for bearing fatigue life prediction has been presented by Tallian in [128]. This model has been expanded to accommodate the threshold stress limit (in other words, the fatigue limit) in Tallian [129]. Further enhancements of the model led Tallian [130, 131] to a description of a summarized general model with the modifying factors which depend on material properties and operating conditions. A numerical procedure has been also presented. The shortcomings mentioned above are also pertinent to these models.

In Tallian [132], the author gave a comprehensive analysis of the existing models for bearing and gear fatigue life prediction as well as an analysis of some research fatigue models. He formulated a list of important parameters that must be included in a new fatigue model. In Tallian [133], the author presented a new fatigue model. The main assumptions of the model were formulated. Among these assumptions were

(1) The fatigue crack initiation period is negligibly small compared to the crack propagation period.

(2) The surface crack propagation equation employed in the model is similar to Paris law.

(3) Fatigue cracks propagate according to the shear stress mechanism (Mode II).

(4) Fatigue cracks initiate only from surface defects. Subsurface material defects are neglected.

(5) The distribution of defect severity is approximated by a power law.

(6) The defect density per contact surface element is Poisson distributed.

This model takes into consideration such parameters as contact geometry, material fatigue parameters, fatigue limit stress, defect severity measure, applied normal and frictional stresses, and stressed volume. The probability $p(N)$ for the volume $\triangle V$ to survive not less than N loading cycles is obtained in the form

$$p(N) = \exp[-m\triangle VF(N)], \tag{8.13}$$

where m is the defect volume density and $F(N)$ is the defect severity distribution. For calculating the global probability of survival $S(N)$, the model also employs the traditional Lundberg-Palmgren (Weibull) approach. That requires integration of the local probability of survival $p(N)$ over the stressed volume V and leads to the following formula for

$$S(N) = \exp(-CN^\beta V), \tag{8.14}$$

where C is a parameter depending solely on the applied load, material fatigue resistance, defect density, and severity, and β is the Weibull dispersion exponent. The model uses some approximations of numerically computed functions.

This approach has certain advantages. The fact that assumptions (8.8), (8.9), (8.13), and (8.14) are realistic and right on target make them clearly advantageous to the model. They allow to lay down the right foundation for fatigue modeling. The model takes into account a spatial material defect distribution versus defect size, normal and frictional stresses, and fatigue limit stress.

The main shortcomings of the model are as follows:

(a) The usage of Mode II crack propagation mechanism in the model is not supported by the experimental data on cracking discussed in Sections 8.3 and 8.5 (see also Kudish and Burris [134]).

(b) The variations in the crack severity distribution due to changes in crack sizes in the process of cyclic loading are not taken into account.

(c) It is unclear what orientation of surface cracks is used for calculations; surface crack orientation is one of the parameters strongly affecting stress intensity factors (see Chapter 9 and Kudish [135]).

(d) The surface crack/lubricant interactions such as the "wedge" and other effects are not taken into account.

(e) It is unreasonable to expect that the probability of survival S(N) would depend on the stressed volume V exponentially (see Romaniv et al. [119]). To clarify this point let us assume that for a certain bearing loaded by a certain load (with the stressed volume V) after N loading cycles the survival probability is $S(N) = S_0$. Now, let us chose a bearing with basically the same internal geometry but with a twice larger shaft diameter compared to the first one and loaded by exactly the same load as the first one. If the bearing shaft diameters are large enough, then the maximum Hertzian pressures for these bearings are almost equal and the stressed volume for the second one is approximately equal to $2V$. According to equation (8.14), for the larger bearing after N loading cycles the survival probability $S(N)$ will be approximately equal to S_0^2. Say, for the smaller bearing the survival probability is $S(N) = 0.1$, then for the larger one it is $S(N) = 0.01$. If the stressed volume V of the larger bearing is 10 times larger, then under the identical loading conditions, its survival probability would be $S(N) = 10^{-10}$. If the stressed volume V of the larger bearing is 100 times larger, then under the identical loading conditions its survival probability $S(N) = 10^{-100}$. This contradicts the experimental data on dependence of fatigue life on volume of a solid subjected to a cycling loading (see Romaniv et al. [119]). Also, see the discussion on the dependence of the survival probability $S(N)$ on the stressed volume V in item (e) of Section 8.6.

It follows from the analysis of this section that these models not adequately describe the actual fatigue processes that take place in bearings.

8.6.1.4   The SKF Bearing Fatigue Models

Several attempts to improve and to expand the Lundberg-Palmgren model has been made by SKF researches. The SKF bearing fatigue models have been presented in papers by Ioannides and Harris [96], Ioannides et al. [136], Hamer et al. [137], Ko et al. [138], Dwyer-Joyce et al. [139], Sayles and Ioannides [140]. The general features and the basic principles of these models have been described by Ioannides and Harris [96]. The main idea is that from a material point of view there is no logical justification to differentiate between fatigue caused by structural and rolling contact loading. It is assumed that material commences to fail when stress at a given location exceeds the material ability to withstand this stress. In comparison to the Lundberg-Palmgren model, this model is improved in several ways although both are purely empirical models.

The main assumptions of the model are as follows:

(1) Pitting is considered to occur when the first spall/crack is created.

(2) Bearing fatigue life can be approximated based solely on the crack initiation period.

(3) The Weibull weakest link theory is used in a similar way as in the Lundberg-Palmgren model.

(4) The following function is proposed for the survival probability S(N):

$$\ln \tfrac{1}{S} = A N^e I, \quad I = \int_V H(\sigma - \sigma_u) \frac{(\sigma - \sigma_u)^c}{z^h} dV, \tag{8.15}$$

where A, e, c, and h are empirical constants, σ is the equivalent stress that occurs at the depth z beneath the surface, σ_u is the fatigue limit stress (stress threshold), $H(x)$ is the step function ($H(x) = 1$, $x > 0$; $H(x) = 0$, $x \leq 0$), V is the stressed volume. The stressed volume V coincides with the half-space or the region, where $\sigma > \sigma_u$. Equation (8.15) is obtained by using a corresponding relationship for the survival probability of an arbitrarily small volume element $\triangle V$, however, large enough to contain many defects.

(5) The next step is a transition to the following expression for $S(N)$:

$$S = \int\limits_{I_{min}}^{I_{max}} \exp[AN^e I] f(I) dI, \tag{8.16}$$

where $f(I)$ is the probability density of the random integral I. The random character of the integral I is related to a statistically distributed stress threshold σ_u. Several manipulations have been made with the latter expression for S and, finally, the survival probability S has been approximated using the first two terms of its McLaurin series. The described theory has been applied to the analysis of fatigue tests in the following cases: (a) rotating beams, (b) beams in torsion, (c) flat beams in reversed bending, and (d) rolling bearings.

The discussed approach has some advantages and shortcomings. The main advantage of the model is the possibility to take into account: (a) the effective material stress state by the appropriate definition of the stress σ and (b) the threshold stress σ_u.

The main shortcomings of the model are reflected below:

(a) The model does not allow to properly take into account the strong influence of frictional and residual stresses on contact fatigue which is observed experimentally (see Section 8.2 and Kudish and Burris [134]). Instead it uses an effective/equivalent stress that has very little to do with the actual stress at a particular point.

(b) The model, as well as the Lundberg-Palmgren model, does not establish any relationship between fatigue life and the initial distribution of various material defects (such as inclusions, cracks, voids) (see Section 8.3 and Kudish and Burris [134]).

(c) Assumption (8.9) concerning the significance of the crack initiation period is not supported by experimental data (see Section 8.5 and Kudish and Burris [134]).

(d) Assumption (8.11) concerning the Weibull weakest link theory is inadequate to the nature of fatigue failure because it suggests that the failure occurs as soon as the first fatigue crack appears. That contradicts experimental data discussed in Section 8.3 (see Kudish and Burris [134]).

(e) The relationship for the survival probability S in (8.15) is expressed in the form of an integral over the material volume V. It has been discussed in Section 8.3 and Kudish and Burris [134] that fatigue cracks originate near inclusions, which practically always are small in comparison with the characteristic contact size and are discretely (not continuously) distributed over the material volume. Therefore, the relationship for survival probability expressed in terms of an integral over the material volume V is unable to adequately describe such a discrete in space inclusion distribution (which would lead to the integral equal to zero) and such a localized phenomenon as an initially small pitting spall. To explain that better let us assume that after a certain

number of loading cycles there is an initial small pitting spall at a single material point. According to the definition of a definite integral (Royden [143])
its value does not depend on any changes in the integrand at any single point.
Therefore, (8.15) as well as (8.16) for the survival probability S give $S = 1$,
i.e., the probability of fatigue failure is zero. That contradicts the assumption
(8.11) from which we obtain that for the case at hand $S < 1$ because according to the Weibull weakest link theory failure occurs at the time the first
spall is created. The same contradiction arises when pitting is initiated at a
single point at the working surface or at any set of points in the material of
measure zero, for example, a finite set of points (see Royden [143]). Also, see
the discussion of the dependence of the survival probability S on the stressed
volume V presented in this section earlier.

(f) The relationship between the probability density distributions $f(I)$ and
$g(\sigma_u)$ (which is the probability density distribution of the threshold σ_u, see
assumption 5) is never defined by the authors.

(g) The model does not add to the understanding of the fatigue phenomenon
because it of an empirical nature and does not uncover any particular mechanisms governing contact fatigue.

It can be shown that for $\sigma_u = 0$ and V proportional to the product of
a contact half-width, bearing ring raceway length, and z_0 (z_0 is the depth
at which the maximum orthogonal shear stress σ_0 occurs) the Lundberg-
Palmgren formula for $\ln(1/S)$ is one of the approximations for the Ioannides-
Harris formulas used in applications.

Ioannides et al. [136], using various fatigue stress criteria extended the above
model to studying effects of lubricant films, roughness, lubricant contamination, and internal plastic stress on bearing fatigue life. A comparison of predicted bearing lives with those obtained using the current at that time ISO
bearing life methods has been performed. It has been shown that the effect of
life adjustment factors of the ISO methods in the analyzed cases is clearly inadequate, both quantitatively and qualitatively. Experimental and theoretical
studies of raceway deformations caused by entrained particles have shown occurrence of severe denting. In particular, some conclusions of this paper have
been made using the maximum shear stress τ, $\sigma = \tau$, and $\sigma_u = \tau_0$. However,
other conclusions have been made using $\sigma = \tau + a p_h$ ($a = 0.3$), where p_h is the
hydrostatic stress. The authors' claim that (8.15) and (8.16) represent a local
characteristic of the shear stress field is baseless because of the integral form
of these equations. Moreover, the use of the two terms of approximation involving exclusively averaged values represents almost the Lundberg-Palmgren
model (see above).

Hamer et al. [137] have added the qualitative analysis of both dent size
and associated subsurface residual stress effects to the original life predicting
model from Ioannides and Harris [96]. The conclusions of this paper concerning
bearing fatigue life are made using the fatigue criterion $\sigma = \tau + a p_h$ (see above
and Ioannides et al. [136]). The slip field analysis of debris indentation into
bearing surface has been simulated by a rigid die impressed into a rigid-plastic

half-space by a normal force coupled with a traction force. The most striking outcome of this analysis is that when failure is initiated by surface dents and the associated with them residual stress the expected fatigue life may not increase when load decreases as rapidly as it is predicted by the conventional models. The reduction in the expected life is very sensitive to the ratio of dent size to roller radius. This may have important implications in terms of particle maximum size as well as the appropriate levels of lubricant filtration.

The analysis of the previous paper has been continued in Ko et al. [138]. Squashing ductile contaminant particles in bearings has been simulated using finite element methods. The calculated dent shapes based on this elastic-plastic analysis and the experimentally measured ones have been found in a good agreement with each other. The calculated residual stress fields associated with dents have been used to estimate rolling bearing life reduction. The significance of methods for determining filtration requirements has been emphasized. One of the important initial suggestions of the paper is that the existence of tensile residual stresses generated by dents may have a significant influence on crack initiation and, ultimately, on fatigue life. However, contrary to that, for bearing life estimation the authors have used the model from Ioannides and Harris [96], Ioannides et al. [136], Hamer et al. [137], Ko et al. [138], which is based on a different (not tensile) stress $\sigma = \tau + ap_h$, where τ and p_h are the maximum shear and hydrostatic stresses.

The subject of the Dwyer-Joyce et al. [139] and Sayles and Ioannides [140] papers is to extend the latter model in which the authors are explaining the influence of friable debris in concentrated surface contacts on bearing fatigue life. It has been noticed experimentally that the presence of friable debris can lead to significant wear and surface roughening of rolling elements. The deformation process in a contaminated contact can take one of the three forms depending on debris and rolling elements material properties:

(a) Debris particles get plastically deformed into platelets with or without rolling element plastic deformation.

(b) Debris particles fracture with or without rolling element plastic deformation. The fragments can get imbedded into rolling element surfaces.

(c) Debris particles undergo little or no deformation and can get imbedded into rolling element surfaces.

It is important that all shortcomings of the basic Ioannides and Harris [96] model are also pertinent to all other models presented by Ioannides et al. [136], Hamer et al. [137], Ko et al. [138], Dwyer-Joyce et al. [139], and Sayles and Ioannides [140].

8.6.2 Mathematical Models for Fatigue Life of Gears

The model presented by Coy et al. [141] has been developed for gear fatigue life prediction, and, in essence, it coincides with the Lundberg-Palmgren theory. The only difference is the adjustments of the Lundberg-Palmgren theory to gear geometry.

Another model for gear fatigue prediction has been developed by Blake [142]. The main assumptions of the Blake model are

(1) Only surface initiated cracks are taken into account.

(2) The probabilistic crack distribution related to crack size is used.

(3) The modeling is based on the assumption that fatigue life can be approximated by the crack propagation period and the crack initiation period is relatively short.

(4) It is assumed that cracks grow from their initial size to a critical size, which depends on the material stress state at cracks' boundaries, material fatigue resistance, and fracture toughness.

(5) A crack size satisfies a crack propagation law involving a number of cycles, the stress intensity factor range, and parameters characterizing material fatigue resistance.

(6) Propagation of a fatigue crack occurs only if its stress intensity factor is between the stress intensity threshold and fracture toughness. Three types of crack propagation laws (Paris' and composite Paris' laws as well as McEvily-Foreman relationship) were used.

(7) The crack propagation driving force is Mode II, i.e., it is driven by shear stress. The expression for the shear stress intensity factor, which is dependent on the Hertzian normal loading and friction coefficient, is obtained by using a regression of numerical solutions of a crack mechanics problem.

(8) The expression for the pitting probability is given in the form of an integral over the material volume.

(9) Asperity interactions are taken into account.

Undoubtedly, this model is the most advanced one among all of the discussed models of fatigue life. It has certain advantages in comparison with all previously described models. As advantages we can mention the following:

(a) The model takes into account an initial statistical defect distribution with respect to defect size.

(b) The model assumes that the crack propagation period is the main part of the total fatigue life that corresponds to experimental data (see Section 8.5 and Kudish and Burris [134]).

(c) The model considers fatigue crack growth caused by repeated loading and predicts fatigue life based on critical crack size.

The shortcomings of the model are as follows:

(1) The proposed model is not designed to describe or predict subsurface initiated pitting.

(2) The model does not take into account the crack propagation direction relative to the acting contact and residual stresses, i.e., crack orientation is fixed and is independent of the material stress state. The crack propagation direction is extremely important for the proper calculation of stress intensity factors (see Section 8.3, Chapter 9, and Kudish [135], Kudish and Burris [134]).

(3) The assumption that the driving force for contact fatigue is of Mode II is not supported by experimental data (see Section 8.5 and Kudish and Burris [134]).

(4) The crack distribution does not vary in time; i.e., the model does not take into account the changes in the crack distribution caused by fatigue crack growth.

(5) The survival probability is expressed in the form of an integral over the material volume, which cannot describe the nature of a local and discrete pitting phenomenon (see the relevant preceding discussion).

As a result of the undertaken analysis, one can conclude that some modifications of the existing fatigue life models or a new fatigue life model are needed. In the next chapter a new model that takes into account the most important parameters affecting contact fatigue life will be presented.

8.7   Closure

From the presented analysis, it is obvious that, at best, some of the existing contact fatigue models are only partially based on the fundamental physical and mechanical mechanisms governing the phenomenon. Most of them are based on assumptions some of which are not supported by experimental data. Some models involve a number of various approximations that usually do not reflect the actual processes occurring in material.

Despite that, a clear experimental understanding of the most essential physical quantities affecting contact fatigue life is now achieved. These quantities are

(a) normal stress and size of the contact,

(b) friction coefficient/stress,

(c) distribution of residual stress versus depth,

(d) initial statistical defect (crack) distribution versus defect size, and location,

(e) material fracture toughness,

(f) material hardness versus depth,

(g) material fatigue parameters as functions of materials hardness, etc.

The above analysis of the models leads to the conclusion that a comprehensive mathematical model of contact fatigue failure should be based on clearly stated mechanical principles following from the theory of elastic or elastohydrodynamic contact interactions and fracture mechanics. Such a model should take into consideration all the parameters just indicated in items (a)-(g).

In addition, a fatigue pitting model should be open and general enough to allow for accommodation of such effects as

- detailed lubrication conditions,

- machining and finishing conditions, surface roughness,

- abrasive contamination of lubricant,

- residual surface contamination,

- non-steady loading regimes, etc.

The advantage of such a comprehensive model would be that the effect of variables such as steel cleanliness, contact stresses, residual stresses, etc., on pitting fatigue life could be examined as single or composite entities.

8.8 Exercises and Problems

1. List the set of material and operational parameters which strongly affect contact fatigue and should be taken into account in an adequate contact fatigue model. Describe the effect of each of these parameters on contact fatigue.

2. What are the usual material surface and subsurface defects which may become the origins of contact fatigue? Which material defects are more/less dangerous? What are the roles of the defect nature, defect size, and position?

3. What is the most probable effect the level of friction has on fatigue life of rolling bearings as it follows from the presented regression analysis?

4. Suppose the subsurface stress field in the vicinity of a fatigue crack tip can be approximated based on just normal k_1 and shear k_2 stress intensity factors. Which stress intensity factor is very sensitive to and almost independent from variations in the frictional stress applied to the contact area of the solid?

5. What is the typical behavior of the residual stress versus the distance from the surface of the solid? What is the perceived role of tensile and compressive residual stress in developing contact fatigue? What effect does tensile and compressive residual stress have on the stress intensity factors k_1 and k_2 characterizing the stress state in the vicinity of a crack? Which – tensile or compressive – stress promotes fatigue crack growth?

6. What is the typical duration of the fatigue crack initiation (nucleation) phase in comparison with the crack propagation stage?

7. What is the role of lubricant contaminants in accelerating contact fatigue failure?

8. List all of the assumptions used by Lundberg and Palmgren in their theory of contact fatigue life of rolling bearings. Describe the significance of each of the assumptions.

9. What is the criterion of contact fatigue failure assumed by Lundberg and Palmgren in their model? Does it correspond well to the experimental evidence that multiple fatigue cracks exist in material long before it contact fatigue failure?

10. What is the dependence of the probability of survival $S(N)$ on the subsurface maximum orthogonal shear stress τ_0 and stressed volume V? Research the dependence of test specimen fatigue life on their size. Show that the experimentally established weak dependence of specimen fatigue life on their size is in direct contradiction with the Lundberg-Palmgren conclusion that the bearing survival probability $S(N)$ is an exponentially decreasing function of the stressed volume V.

11. Conduct a mental testing of two groups of roller bearings under the same stress conditions assuming that the bearing size of the second group is 50–fold greater than the bearing size of the first group (such as in case of regular car and tank gun bearings). For the same probability of survival

for both groups, determine the relationship between their fatigue lives based on the Lundberg-Palmgren model (8.8), where $h = 7/3$ and $e = 1.125$. Can these results be reconciled with practically observed weak dependence of test specimen fatigue life on their size?

12. Describe the ability of the Lundberg-Palmgren model to predict: (a) the experimentally observed strong dependence of bearing fatigue life on contact frictional and residual stresses; (b) bearing fatigue life as a function of non-metallic inclusions, voids, preexisting and propagating surface, and/or subsurface cracks; (c) surface-initiated fatigue failure; (d) the dependence of bearing fatigue life on material cleanliness, fatigue and fracture material properties, surface roughness, lubricant and surface contamination, etc.; (e) the dependence of bearing fatigue life and the survival probability on the stressed volume.

13. Characterize the basic fatigue mechanism accepted in the Lundberg-Palmgren model and compare it with the fatigue mechanism usually accepted in torsion/tensile (structural) fatigue tests.

14. Describe the ability of the Ioannides-Harris model to predict: (a) the experimentally observed strong influence of frictional and residual stresses on contact fatigue, (b) any relationship between fatigue life and the initial distribution of various material defects (such as inclusions, cracks, voids), (c) the relationship of the survival probability S on the material volume V.

15. In the Ioannides-Harris model the survival probability $S(N)$ is expressed in the form of an integral of a certain function over the material volume (see (8.15)). Is such an integral relationship for $S(N)$ able to represent any discrete fatigue damage localized in small volumes near inclusions/cracks that always are small in comparison with the characteristic contact size? Is the assumption that contact fatigue life is determined by the occurrence of first crack reconcilable with the integral form of $S(N)$ representation?

16. What are the details of the Weibull weakest link theory in application to contact fatigue failure? What is the experimental evidence with respect to the presence of fatigue cracks in material prior to its fatigue failure?

17. Is the assumption of the Ioannides-Harris model with respect to the duration of crack initiation versus crack propagation period different from the one assumed by the Lundberg-Palmgren model? What is the experimental evidence on the duration of crack initiation versus crack propagation period?

18. Describe the main assumptions of the Blake model with respect to type of cracks taken into account, crack initiation, and crack propagation, deterministic or statistical approach, the criterion of crack failure, the survival probability, and its form of representation.

19. Describe the main advantages of Blake's model in comparison with the Lundberg-Palmgren and Ioannides-Harris models.

20. Describe the shortcomings of Blake's model.

21. What is the fatigue crack growth mode considered in Blake's model? What is the experimental evidence on stability of Modes I and II of fatigue crack growth?

22. Because of fatigue crack growth the crack distribution versus crack size varies in time. Does any of the considered above models take into account the changes in the crack distribution caused by fatigue crack growth?

23. Suppose the initial crack distribution versus their size l_0 is described by the function $f(0, l_0) = \exp(-l_0^2)$. Let us assume that each crack half-length $l(N)$ grows with the number N of applied stress cycles according to the equations $\frac{dl}{dN} = \frac{l^n}{n-1}$, $l(0) = l_0$ while the crack distribution $f(N, l)$ satisfies the equation $f(N, l) = f(0, l_0)\frac{dl_0}{dl}$. Cracks do not coalesce. Determine the crack distributions $f(N, l)$ for $N = 10$ and $N = 100$. (Hint: In the equation for $f(N, l)$, express l_0 in terms of l.) Determine the crack distributions $f(N, l)$ for $N = 10$ and $N = 100$. Consider two cases of parameter n: $n = 4$ and $n = 6$. Graph three functions $f(N, l)$ (for $N = 0$, $N = 10$, and $N = 100$) in the same set of coordinate axes as functions of l, $l \geq 0$, for two cases of $n = 4$ and $n = 6$. Compare the graphs of $f(N, l)$ for each case of parameter n and different number of stress cycles N.

Using this example explain the deficiency of all described contact fatigue models with respect to the assumption that the crack distribution versus their size is independent on the number of applied stress cycles.

BIBLIOGRAPHY

[1] Pinegin, S.V., Shevelev, I.A., Gudchenko, V.M., Sedov, V.I., and Blokhin, Y.N. 1972. In *The Influence of External Factors on Rolling Contact Strength*. Moscow: Nauka.

[2] Orlov, A.V., Chermensky, O.N., and Nesterov, V.M. 1980. *Testing of Design Materials for Contact Fatigue*. Moscow: Mashinostroenie.

[3] Soda, N. and Yamamoto, T. 1982. Effect of Tangential Traction and Roughness on Crack Initiation/Propagation during Rolling Contact. *ASLE Trans.* 25, No. 2:198-206.

[4] Kudish, I.I. 1986. On the Influence of Friction Stresses on the Fatigue Failure of Machine Parts. *Soviet J. Fric. and Wear* 7, No. 5:812-822.

[5] Kudish, I.I. 1986. Study of the Effect of Subsurface Defects in a Elastic Material on Its Fracture. *Study, Calculation, and Design of Roller Bearings*, Spetsinformtsentr NPO VNIPP, No. 2:57-77.

[6] Spektor, A.G., Zelbet, B.M., and Kiseleva, S.A. 1980. *Structure and Properties of Bearing Steels*. Moscow: Metallurgya.

[7] Mattson, R.L. 1961. Fatigue, Residual Stresses, and Surface-Layer Strengthening by Work Hardening. In *Fatigue of Metals*. Moscow: Inostrannaya Literatura.

[8] Kudryavtsev, I.V. 1951. *Internal Stresses as a Strength Reserve in Machine Construction*. Moscow: Mashgiz.

[9] Serensen, S.V. 1952. Fatigue Resistance in Connection with Strain-Hardening and Design Factors. In *Increasing the Fatigue Strength of Machine Parts by Surface Treatment* Moscow: GNTI (State Scientific and Technical Publishing).

[10] Scott, R.L., Kepple, R.K., and Miller, M.H. 1962. The Effect of Processing-Induced Near-Surface Residual Stress on Ball Bearing Fatigue. In *Rolling Contact Phenomena Proc. Symp. held at the GM Research Laboratories, Warren, MI*, October 1960, Amsterdam: Elsevier Publishing.

[11] Almen, I. and Black, P. 1963. *Residual Stress and Fatigue of Metals*. New York: McGraw-Hill.

[12] Brozgol, I.M. 1973. *Effect of Fine Finishing of the Roller Path on the Quality of Bearings*. Survey. Moscow: NIINavtoprom.

[13] Radhakrishnan, V.M. and Baburamani, P.S. 1976. Initiation and Propagation of Fatigue Crack in Pre-strained Material. *Intern. J. Fract* 12, No. 3:369-380.

[14] Morrow, J.D. and Millan, J.F. 1961. Influence of Residual Stress on Fatigue of Steel. *SAE* TR-198.

[15] Mattson, R.L. and Roberts, J.G. 1960. Effect of Residual Stresses Induced by Strain Peening upon Fatigue Strength. In *Internal Stresses and Fatigue in Metals*. New York: Elsevier.

[16] Rowland, E.S. 1964. Effect of Residual Stress on Fatigue. In *Fatigue, an Interdisciplinary Approach*. Syracuse University Press, 229-240.

[17] Averbach, B.L., Bingzhe L., Pearson, P.K., Fairchild, R.E., and Bamberger, E.N. 1985. Fatigue Crack Propagation in Carburized High Alloy Bearing Steels. *Metallurg. Trans.*. Ser. A, 16A:1253-1265.

[18] Kumar, A.M., Hahn, G.T., Bhargava, V., and Rubin, C.A. 1989. Elastoplastic Finite Element Analyses of Two-Dimensional Rolling and Sliding Contact Deformation of Bearing Steel. *ASME J. Tribology* 111:309-314.

[19] Lei, T.S., Bhargava, V., Hahn, G.T., and Rubin, C.A. 1986. Stress Intensity Factors for Small Cracks in the Rim of Disks and Rings Subjected to Rolling Contact. *ASME J. Lubr. Techn.* 108, No. 4:540-544.

[20] Clark, J.C. 1985. Fracture Tough Bearings for High Stress Applications. *Proc. 21st Joint Propulsion Conf.* Paper AIAA-85-1138.

[21] Bhargava, V., Hahn, G.T., and Rubin, C.A. 1985. An Elastic-Plastic Finite Element Model of Rolling Contact. Part 1: Analysis of Single Contacts. *ASME J. Appl. Mech.* 52:67-74.

[22] Bhargava, V., Hahn, G.T., and Rubin, C.A. 1985. An Elastic-Plastic Finite Element Model of Rolling Contact. Part 2: Analysis of Repeated Contacts. *ASME J. Appl. Mech.* 52:75-82.

[23] Ham, G., Rubin, C.A., Hahn, G.T., and Bhargava, V. 1988. Elastoplastic Finite Element Analysis of Repeated, Two-Dimensional Rolling-Sliding Contacts. *ASME J. Tribology.* 110, No. 1:44-49.

[24] Ham, G.L., Hahn, G.T., Rubin, C.A., and Bhargava, V. 1989. Finite Element Analysis of the Influence of Kinematic Hardening in Two-Dimensional, Repeated, Rolling-Sliding Contact. *STLE Tribology Trans.* 32, No. 3:311-316.

[25] Kulkarni, S., Hahn, G.T., Rubin, C.A., and Bhargava, V. 1990. Elastoplastic Finite Element Analysis of Three-Dimensional, Pure Rolling Contact at the Shakedown Limit. *ASME J. Appl. Mech.* 57:57-65.

[26] Bereznitskaya, M.F. and Grishin, P.M. 1989. The Residual Stress Forming in the Surface Layers by the Complex Machining. *J. Physical − Chemical Mech. Mat.*, 104-105.

[27] Cretu, Sp.S. and Popinceanu, N.G. 1985. The Influence of Residual Stresses Induced by Plastic Deformation on Rolling Contact Fatigue. *Wear* 105:153-170.

[28] Voskamp, A.P. 1985. Material Response to Rolling Contact Loading. *ASME J. Tribology* 107:359-366.

[29] Sveshnikov, D.A. 1964. The Increase of Fatigue Strength of the Carburized and Cyanide Parts by Shot Peening. *Phys. Metallurgy and Thermal Treatment* 4:47-49.

[30] Dudragne, G., Fougeres, R., and Theolier, M. 1981. Analysis Method for Both Internal Stresses and Microstructural Effect under Pure Rolling Fatigue Conditions. *ASME J. Lubr. Techn.* 103, No. 4:521-525.

[31] Broszeit, E., Adelmann, J., and Zwirlein, O. 1984. Influence of Internal Stresses on the Stressing of Material in Components Subjected to Rolling-Contact Loads. *ASME J. Lubr. Techn.* 106, No. 4:499-504.

[32] Kepple, R.K. and Mattson, R.L. 1970. Rolling Element Fatigue and Macroresidual Stress. *ASME J. Lubr. Techn.* 92, No. 1:76-82.

[33] Janowski, S., Senatorski, J., and Szyrle, W. 1985. Research on Influence of Residual Stresses on the Wear Resistance. In *4th European Tribology Congress, EUROTRIB* 85, Ecully-France, Vol. 4, Paper 5.5:1-4.

[34] Hahn, G.T., Bhargava, V., Rubin, C.A., Chen, Q., and Kim, K. 1987. Analysis of the Rolling Contact Residual Stresses and Cyclic Plastic Deformation of SAE 52100 Steel Ball Bearing. *ASME J. Tribology* 109:618-626.

[35] Murakami, Y., Kodama, S., and Konuma, S. 1989. Quantitative Evaluation of Effects of Non-metallic Inclusions on Fatigue Strength of High Strength Steels. I: Basic Fatigue Mechanism and Evaluation of Correlation between the Fatigue Fracture Stress and the Size and Location of Non-metallic Inclusions. *Intern. J. Fatigue* 11, No. 5:291-298.

[36] Nishioka, K. 1957. On the Effect of Inclusion upon the Fatigue Strength. *Material Testing (Soc. Mater. Sci. Jpn.)* 6, No. 45:382-385.

[37] Yokobori, T. 1961. Fatigue of Metals and Fracture Mechanics. *J. Sci. in Machines* 13, No. 7:973-976.

[38] Watanabe, J. 1962. How Does the Cleanliness Influence the Mechanical Properties of Metals. *Bull. Jpn. Inst. Metals* 1, No. 2:129-133.

[39] Murakami, Y., Kodama, S., and Konuma, S. 1989. Quantitative Evaluation of Effects of Non-metallic Inclusions on Fatigue Strength of High Strength Steels. II: Fatigue Limit Evaluation Based on Statistics for Extreme Values of Inclusion Size. *Intern. J. Fatigue* 11, No. 5:299-307.

[40] Murakami, Y. and Endo, M. 1986. Effects of Hardness and Crack Geometries on Kth of Small Cracks Emanating from Small Defects. In *The Behavior of Short Fatigue Cracks*, EZGF Pub. 1 (Eds. K.J. Miller and E.R. de los Rios), 275-293.

[41] Murakami, Y. and Endo, T. 1986. Effects of Hardness and Crack Geometry on Kth of Small Cracks. *Zairyo (Soc. Mater. Sci. Jpn.)* 35, No. 395:911-917.

[42] Murakami, Y., Morinaga, H., and Endo, T. 1985. Effects of Geometrical Parameter and Mean Stress on Kth of Specimens Containing a Small Defect. *Zairyo (Soc. Mater. Sci. Jpn.)* 34, No. 385:1153-1159.

[43] Murakami, Y. 1985 Analysis of Stress Intensity Factors of Mode I, II and III for Inclined Surface Cracks of Arbitrary Shape. *Eng. Fract. Mech.* 22, No. 1:101-114.

[44] Murakami, Y. 1980. Effects of Small Defects on Fatigue Strength of Metals. *Intern. J. Fatigue* 2, No. 1:23-30.

[45] Andrews, K.W. and Brooksbank, D. 1972. Stresses Associated with Inclusions in Steel: A Photoelastic Analogue and the Effects of Inclusions in Proximity. *J. Iron and Steel Inst.* 210, Part 10:765-776.

[46] Brooksbank, D. and Andrews, K.W. 1972. Stress Fields around Inclusions and Their Relation to Mechanical Properties. *J. Iron and Steel Inst.* 210, Part 4:246-255.

[47] Stover, J.D. and Kolarik II, R.V. 1987. Air-Melted Steel with Ultra-Low Inclusion Stringer Content Further Improves Bearing Fatigue Life. *SAE Conf. Proc., The 4th Intern. Conf. on Automotive Eng.* Melbourne, Australia, 208.1-208.7.

[48] Moyer, C.A., Nixon, H.P., and Bhatia, R.R. 1988. Tapered Roller Bearing Performance for the 1990's. *SAE Tech. Paper* No. 881232 :1-10.

[49] Tayeh, G.S. and Woehrle, H.R. 1987. Improvements in Bearing Steel Cleanliness and the Effect on Integral Wheel Bearing Systems. *SAE Tech. Paper* No. 871983 :1-9.

[50] Toyoda, T., Kanazawa, T., and Matsumoto, K. 1990. A Study of Inclusions Causing Fatigue Cracks in Steels for Carburized and Shot-Peened Gears. *JSAE Review* 11, No. 1:50-54.

[51] Stover, J.D. and Kolarik II, R.V. 1987. The Evaluation of Improvements in Bearing Steel Quality Using an Ultrasonic Macro-Inclusion Detection Method. *The Timken Company Technical Note* January 1987 :1-12.

[52] Stover, J.D., Kolarik II, R.V., and Keener, D.M. 1989. The Detection of Aluminum Oxide Stringers in Steel Using an Ultrasonic Measuring Method. *Mechanical Working and Steel Processing XXVII :* Proc.

31*st Mechanical Working and Steel Processing Conf.*, Chicago, Illinois, October 22-25, 1989, Iron and Steel Soc., Inc., 431-440.

[53] Clarke, T.M., Miller, G.R., Keer, L.M., and Cheng, H.S. 1985. The Role of Near-Surface Inclusions in the Pitting of Gears. *ASLE Trans.* 28, No. 1:111-116.

[54] Bokman, M.A. and Pershtein, E.M. 1986. The Study of Micro-crack Growth under the Plastic Strain Conditions. *Plant Laboratory* Moscow, No. 2:68-70.

[55] Bokman, M.A., Pshenichnov, Yu.P., and Pershtein, E.M. 1984. The Microcrack and Non-metallic Inclusion Distribution in the Alloy D16 after a Plastic Strain. *Plant Laboratory*, Moscow, No. 11:71-74.

[56] Endo, M. and Murakami, Y. 1987. Effects of an Artificial Small Defect on Torsional Fatigue Strength of Steels. *ASME J. Eng. Materials and Techn.* 109, No. 2:124-129.

[57] Lankford, J. and Kusenberger, F.N. 1973. Initiation of Fatigue Cracks in 4340 Steel. *Metallurgical Trans.* 4:553-559.

[58] Eid, N.M.A. and Thomason, P.F. 1979. The Nucleation of Fatigue Cracks in a Low-Alloy Steel under High-Cycle Fatigue Conditions and Uniaxial Loading. *Acta Metallurgica* 27:1239-1249.

[59] Lankford, J. 1976. Inclusion-Matrix Debonding and Fatigue Crack Initiation in Low Alloy Steel. *Intern. J. Fract.* 12:155-156.

[60] Kinoshi, M. and Koyanagi, A. 1975. Effect of Nonmetallic Inclusions on Rolling-Contact Fatigue Life in Bearing Steels. *Bearing Steels : The Rating of Nonmetallic Inclusion*, ASTM STP 575, ASTM, 138-149.

[61] Rousseau, D., Seraphin, L., and Tricot, R. 1975. Nonmetallic Inclusion Rating and Fatigue Properties of Ball Bearing Steels. *Bearing Steels : The Rating of Nonmetallic Inclusion*, ASTM STP 575, ASTM, 49-65.

[62] Koyanagi, A. and Kinoshi, M. 1975. Several Rating Methods of Nonmetallic Inclusions in Bearing Steels in Japan. *Bearing Steels : The Rating of Nonmetallic Inclusion*, ASTM STP 575, ASTM, 22-37.

[63] Moyer, C.A. 1987. The Influence of Debris on Rolling Bearing Performance: Identifying the Relevant Factors. *SAE Tech. Paper* No. 871687, :1-10.

[64] Lundberg, G. and Palmgren, A. 1947. Dynamic Capacity of Rolling Bearings. *Acta Polytechnica (Mech. Eng. Ser. 1), Royal Swedish Acad. Eng. Sci.* 7:5-32.

[65] Schlicht, H., Schreiber, E., and Zwirlein, O., 1986, Fatigue and Failure Mechanism of Bearings, *Inst. Mech. Eng. Conf. Publ.* 1:85-90.

[66] Leng, X., Chen, Q., and Shao, E., 1988, Initiation and Propagation of Case Crushing Cracks in Rolling Contact Fatigue. *Wear* 122: 33-43.

[67] Zhou, R.S., Cheng, H.S., and Mura, T., 1989. Micropitting in Rolling and Sliding Contact under Mixed Lubrication. *ASME J. Tribology*, 111:605-613.

[68] Cheng, W., Cheng, H.S., Mura, T., and Keer, L.M., 1994. Micromechanics Modeling of Crack Initiation under Contact Fatigue. *ASME J. Tribology* 116:2-8.

[69] Cheng, W., and Cheng, H.S., 1995. Semi-analytical Modeling of Crack Initiation Dominant Contact Fatigue for Roller Bearings. *Proc. of the 1995 Joint ASME/STLE Tribology Conf., Orlando, FL.*

[70] Otsuka, A., Sugawara, H., and Shomura, M., 1996, A Test Method for Mode II Fatigue Crack Growth Relating for a Model for Rolling Contact Fatigue. *Fatigue Fract. Eng. Mater. Struct.* 19, No. 10:1265-1275.

[71] Melander, A., 1997. A Finite Element Study of Short Cracks with Different Inclusion Types under Rolling Contact Fatigue Load. *Int. J. Fatigue*, 19, No. 1:13-24.

[72] Vincent, A., Lormand, G., Lamagnere, P., Gosset, I., Girodin, D., Dudragne, G., and Fougeres, R., 1998. From White Etching Areas Formed around Inclusions to Crack Nucleation in Bearing Steels under Rolling Contact. *Bearing Steels : Into the 21st Century, ASTM STP No. 1327*, J. Hao and W. Green, eds., ASTM Special Technical Publication, West Conshohoken, PA, 109-123.

[73] Lormand, G., Meynaud, G., Vincent, A., Baudry, G., Girodin, D., and Dudragne, G., 1998. From Cleanliness to Rolling Fatigue Life of Bearings - A New Approach. *Bearing Steels : Into the 21st Century, ASTM STP No. 1327*, J. Hao and W. Green, eds., ASTM Special Technical Publication, West Conshohoken, PA, 55-69.

[74] Jiang, Y., and Sehitoglu, H., 1999, A Model for Rolling Contact Failure. *Wear* 224:38-49.

[75] Ringsberg, J.W., 2001. Life Prediction of Rolling Contact Fatigue Crack Initiation. *Int. J. Fatigue*, 23, No. 7:575-586.

[76] Shimizu, S., 2002. Fatigue Limit Concept and Life Prediction Model for Rolling Contact Machine Elements. *STLE Tribology Trans.*, 45, No. 1:39-46.

[77] Miyashita, Y., Yoshimura, Y., Xu, J.-Q., Horikoshi, M., and Mutoh, Y., 2003. Subsurface Crack Propaagation in Rolling Contact Fatigued Syntered Alloy. *JSME Int. J., Ser. A*, 46, No. 3:341-347.

[78] Liu, Y., Stratman, B., and Mahadevan, S., 2006. Fatigue Crack Initiation Life Prediction in Railroad Wheels. *Int. J. Fatigue* 28:747-756.

[79] Liu, Y., and Mahadevan, S., 2007. A Unified Multiaxial Fatigue Damage Model for Isotropic and Anisotropic Materials. *Int. J. Fatigue* 29:347-359.

[80] ISO, 1989. Rolling Bearings - Dynamic LZoad Ratings and Rating Life. *Draft International Standard ISO/DIS* 281, *ISO*, Geneva, Switzerland.

[81] Kovalev, M.P. and Narodetsky, M.Z. 1980. *Analysis of High Precision Ball Bearings*. Moscow: Mashinostroenie.

[82] Kudish, I.I. 1996. Asymptotic Analysis of a Problem for a Heavily Loaded Lubricated Contact of Elastic Bodies. Pre- and Over-critical Lubrication Regimes for Newtonian Fluids. *Dynamic Systems and Applications*, Dynamic Publishers, Atlanta, 5, No. 3:451-476.

[83] Grubin, A.N. and Vinogradova, I.E. 1949. Investigation of the Contact of Machine Components. Ed. Kh. F. Ketova, *Central Sci. Research Inst. for Techn. and Mech. Eng.* (Moscow), Book No. 30 (DSIR translation No. 337).

[84] Bhargava, V., Hahn, G.T., and Rubin, C.A. 1986. Analysis of Cyclic Crack Growth in High Strength Roller Bearings. *Theor. and Appl. Fract. Mech.* 5:31-38.

[85] O'Regan, D., Hahn, G.T., and Rubin, C.A. 1985. The Driving Force for Mode II Crack Growth under Rolling Contact. *Wear* 101:333-346.

[86] Otsuka, A., Mori, K., and Miyata, T. 1975. The Conditions of Fatigue Crack Growth in Mixed Mode Condition. *Eng. Fract. Mech.* 7:429-439.

[87] Shieh, W.T. 1977. Compressive Maximum Shear Crack Initiation and Propagation. *Eng. Fract. Mech.* 9:37-54.

[88] Kaneta, M., Yatsuzuka, H., and Murakami, Y. 1985. Mechanism of Crack Growth in Lubricated Rolling/Sliding Contact. *ASLE Trans.* 28, No. 3:407-414.

[89] Kaneta, M., Murakami, Y., and Okazaki, T. 1986. Growth Mechanism of Subsurface Crack Due to Hertzian Contact. *ASME J. Lubr. Techn.* 108, No. 1:134-139.

[90] Wu, Han C. and Yang, C.C.. 1987. On the Influence of Strain-Path in Multiaxial Fatigue Failure. *ASME J. Eng. Materials and Techn.* 109, No. 2:107-113.

[91] Kapelski, G., Platon, F., and Boch, P. 1988. Unlubricated Wear and Friction Behavior of Alumina and Silicon Carbide Ceramics. *Proc.*

15th Leeds – Lyon Symp. on Tribology, University of Leeds, UK, *Tribological Design of Machine Elements*, Eds. D. Dowson, C.M. Taylor, M. Godet, and D.Berthe, 349-354.

[92] Kitagawa Hideo, Yuuki Ryoji, and Tohgo Keiichiro. 1981. Fatigue Crack Growth Behavior from a K1 and K2 Mixed Mode Crack. *Trans. Jpn. Soc. Mech. Eng.* Ser. A, 47, No. 424:1283-1292.

[93] Nisitani, H. and Goto, M. 1984. Effect of Stress Ratio on the Propagation of Small Crack of Plain Specimens under High and Low Stress Amplitudes. *Trans. Jpn. Soc. Mech. Eng.* Ser. A, 50, No. 453:1090-1096.

[94] Tanaka, T., Sakai, T., and Okada, K. 1984. A Statistical Study of Fatigue Life Distribution Based on the Coalescence of Cracks from Surface Defects. *Trans. Jpn. Soc. Mech. Eng.* Ser. A, 50, No. 454:1166-1173.

[95] Ichikawa, M. 1984. Some Problems in Probabilistic Fracture Mechanics. *Trans. Jpn. Soc. Mech. Eng.* Ser. A, 50, No. 456:1435-1442.

[96] Ioannides, E. and Harris, T.A. 1985. A New Fatigue Life Model for Rolling Bearings. *ASME J. Tribology* 107:367-378.

[97] Wert, J.J., Srygley, F., Warren, C.D. and McReynolds, R.D. 1989. Influence of Long-Range Order on Deformation Induced by Sliding Wear. *Wear* 134:115-148.

[98] Kunio, T., Shimizu, S., Yamada, K., Sakura, K., and Yamamoto, T. 1981. The Early Stage of Fatigue Crack Growth in Martensitic Steel. *Intern. J. Fract.* 17, No. 2:111-119.

[99] Kumar, A. 1989. *Analysis of Subsurface Deformation under Rolling Contact*, Ph.D. Dissertation, Vanderbilt University.

[100] Shao, E., Huang, X., Wang, C., Zhu, Y., and Chen, Q. 1988. A Method of Detecting Rolling Contact Crack Initiation and the Establishment of Crack Propagation Curves. *ASLE Tribology Trans.* 31, No. 1:6-11.

[101] Chen, Q., Hahn, G.T., Rubin, C.A., and Bhargava, V. 1988. The Influence of Residual Stresses on Rolling Contact Mode II Driving Force in Bearing Raceways. *Wear* 126:17-30.

[102] Littmann, W.E. and Moyer, C.A. 1963. Competitive Modes of Failure in Rolling Contact Fatigue. *Automotive Eng. Congress, Detroit, MI*, *SAE* 620A, 1-6.

[103] Yamashita, N. and Mura, T. 1983. Contact Fatigue Crack Initiation under Repeated Oblique Force. *Wear* 91:235-250.

[104] Yamashita, N., Mura, T., and Cheng, H.S. 1985. Effect of Stress Induced by Spherical Asperity on Surface Pitting in Elasto-hydrodynamic Contacts. *ASLE Trans.* 28, No. 1:11-20.

[105] Drul', O.R., Levitsky, M.O., and Brodyak, D.I. 1987. The Influence of the Austenite Grain Size on the Cyclic Crack Resistance of the 40X Steel. *Physical – Chemical Mech. Materials.* Moscow, 116-118.

[106] Ibale Rodriguez, J.M. and Sevillano Gil, J. 1984. Fatigue Crack Path in Medium-High Carbon Ferrite-Perlite Structures. *Proc. 6th Intern. Conf. Fract. (ICF6)*, New Delphi, 4-10 Dec, 1984, Vol. 3, Oxford, e.a., 2073-2079.

[107] Schaper, M. and Bosel, D. 1985. Rasterelektronenmikro-skopische in-situ-Untersuchungen der Ermudungsribausbreitung in Metallishen Werkstoffen. *Prakt. Metallogr.* 22, No. 4:197-203.

[108] Tsushima, N. and Kashimura, H. 1984. Improvement of Rolling Contact Fatigue Life of Bearing Steels. *SAE Tech. Paper* No. 841123 :1-9.

[109] Yarema, S.Ya. 1981. Methodology of Determining the Characteristics of the Resistance to Crack Development (Crack Resistance) of Materials in Cyclic Loading. *J. Soviet Material Sci.* 17, No. 4:371-380.

[110] Keer, L.M., Bryant, M.D., and Haritos, G.K. 1980. Subsurface Cracking and Delamination. In *Solid Contact and Lubrication*, Eds. H.S. Cheng and L.M. Keer, ASME, 79-95.

[111] Keer, L.M., Bryant, M.D., and Haritos, G.K. 1982. Subsurface and Surface Cracking Due to Hertzian Contact. *ASME J. Lubr. Techn.* 104, No. 3:347-351.

[112] Keer, L.M. and Bryant, M.D. 1983. A Pitting Model for Rolling Contact Fatigue. *ASME J. Lubr. Techn.* 105, No. 2:198-205.

[113] Murakami, Y., Kaneta, M., and Yatsuzuka, H. 1985. Analysis of Surface Crack Propagation in Lubricated Rolling Contact. *ASLE Trans.* 28, No. 1:60-68.

[114] Kaneta, M., Suetsugu, M., and Murakami, Y. 1986. Mechanism of Surface Crack Growth in Lubricated Rolling/Sliding Spherical Contact. *ASME J. Appl. Mech.* 53, No. 2:354-360.

[115] Kudish, I.I. 2002. Lubricant-Crack Interaction, Origin of Pitting, and Fatigue of Drivers and Followers. *STLE Tribology Trans.* 45, No. 4:583-594.

[116] Kudish, I.I. and K.W. Burris, 2004. Modeling of Surface and Subsurface Crack Behavior under Contact Load in the Presence of Lubricant. *Intern. J. of Fracture*, 125:125-147.

[117] Miller, G.R. and Keer, L.M. 1983. Interaction between a Rigid Indentor and a Near-Surface Void or Inclusion. *ASME J. Appl. Mech.* 50:615-620.

[118] Kudish, I.I. and Burris, K.W. 2000. Modern State of Experimentation and Modeling in Contact Fatigue Phenomenon. Part II. Analysis of the Existing Statistical Mathematical Models of Bearing and Gear Fatigue Life. New Statistical Model of Contact Fatigue. *STLE Tribology Trans.* 43, No. 2:293-301.

[119] Romaniv, O.N., Yarema, S.Ya., Nikiforchin, G.N., Makhutov, N.A., and Stadnik, M.M. 1990. *Fracture Mechanics and Strength of Materials. Vol. 4. Fatigue and Cyclic Crack Resistance of Construction Materials.* Kiev: Naukova Dumka.

[120] Chiu, Y.P., Tallian, T.E., McCool, J.I., and Martin, J.A. 1969. A Mathematical Model of Spalling Fatigue Failure in Rolling Contact. *ASLE Trans.* 12:106-116.

[121] Chiu, Y.P., Tallian, T.E., and McCool, J.I. 1971. An Engineering Model of Spalling Fatigue Failure in Rolling Contact. I. The Subsurface Model. *Wear* 17:433-446.

[122] Chiu, Y.P., Tallian, T.E., and McCool, J.I. 1971. An Engineering Model of Spalling Fatigue Failure in Rolling Contact. II. The Surface Model. *Wear* 17:447-461.

[123] Tallian, T.E., Chiu, Y.P., and Van Amerongen, E. 1978. Prediction of Traction and Micro-geometry Effects on Rolling Contact Fatigue Life. *ASME J. Lubr. Techn.* 100, No. 2:156-166.

[124] Tallian, T.E. 1971. An Engineering Model of Spalling Fatigue Failure in Rolling Contact. III. Engineering Discussion and Illustrative Examples. *Wear* 17:463-480.

[125] Tallian, T.E. 1976. Prediction of Rolling Contact Fatigue Life in Contaminated Lubricant: Part I - Mathematical Model. *ASME J. Lubr. Techn.* 98, No.2:251-257.

[126] Tallian, T.E. 1976. Prediction of Rolling Contact Fatigue Life in Contaminated Lubricant: Part II - Experimental. *ASME J. Lubr. Techn.* 98, No. 2:384-392.

[127] Tallian, T.E. 1981. Rolling Bearing Life Modifying Factors for Film Thickness, Surface Roughness, and Friction. *ASME J. Lubr. Techn.* 103, No. 4:509-520.

[128] Tallian, T.E. 1982. A Unified Model for Rolling Contact Life Prediction. *ASME J. Lubr. Techn.* 104, No. 3:336-346.

[129] Tallian, T.E. 1986. Unified Rolling Contact Life Model with Fatigue Limit. *Wear* 107:13-36.

[130] Tallian, T.E. 1988. Rolling Bearing Life Prediction. Corrections for Material and Operating Conditions. Part I: General Model and Basic Life. *ASME J. Lubr. Techn.* 110, No. 1:1-6.

[131] Tallian, T.E. 1988. Rolling Bearing Life Prediction. Corrections for Material and Operating Conditions. Part II: The Correction Factors. *ASME J. Lubr. Techn.* 110, No. 1:7-13.

[132] Tallian, T.E. 1992. Simplified Contact Fatigue Life Prediction Model - Part I: Review of Published Models. *ASME J. Tribology* 114, No. 1:207-213.

[133] Tallian, T.E. 1992. Simplified Contact Fatigue Life Prediction Model - Part II: New Model. *ASME J. Tribology* 114, No. 1:214-222.

[134] Kudish, I.I. and Burris, K.W. 2000. Modern State of Experimentation and Modeling in Contact Fatigue Phenomenon. Part I. Contact Fatigue versus Normal and Tangential Contact and Residual Stresses. Nonmetallic Inclusions and Lubricant Contamination. Crack Initiation and Crack Propagation. Surface and Subsurface Cracks. *STLE Tribology Trans.* 43, No. 2:187-196.

[135] Kudish, I.I. 1987. Contact Problem of the Theory of Elasticity for Pre-Stressed Bodies with Cracks. *J. Appl. Mech. Techn. Phys.* 28, No. 2:295-303.

[136] Ioannides, E., Jacobson, B., and Tripp, J.H. 1989. Prediction of Rolling Bearing Life under Practical Operating Conditions. In *Tribological Design of Machine Elements*, Eds. D. Dowson, C.M. Taylor, M. Godet, and D. Berthe, Amsterdam: Elsevier Science Publishers B.V., 181-187.

[137] Hamer, J.C., Lubrecht, A.A., Ioannides, E., and Sayles, R.S. 1989. Surface Damage on Rolling Elements and Its Subsequent Effects on Performance and Life. In *Tribological Design of Machine Elements*, Eds. D. Dowson, C.M. Taylor, M. Godet, and D. Berthe, Elsevier Science Publishers B.V., Amsterdam, 189-197.

[138] Ko, C.N. and Ioannides, E. 1989. Debris Denting - The Associated Residual Stresses and Their Effect on the Fatigue Life of Rolling Bearing: An FEM Analysis. In *Tribological Design of Machine Elements*, Eds. D. Dowson, C.M. Taylor, M. Godet, and D. Berthe, Elsevier Science Publishers B.V., Amsterdam, 199-207.

[139] Dwyer-Joyce, R.S., Hamer, J.C., Sayles, R.S., and Ioannides, E. 1990. Surface Damage Effects Caused by Debris in Rolling Bearing Lubricants, with an Emphasis on Friable Materials. In *Rolling Element Bearings − Towards the 21st Century*, Seminar organized by the Tribology Group of the Inst. of Mech. Eng., November 16, 1-8.

[140] Sayles, R.S. and Ioannides, E. 1988. Debris Damage in Rolling Bearings and Its Effects on Fatigue Life. *ASME J. Tribology* 110:26-31.

[141] Coy, J.J., Townsend, D.P., and Zaretsky, E.V. 1976. Dynamic Capacity and Surface Fatigue Life for Spur and Helical Gears. *ASME J. Lubr. Techn.* 98:267-276.

[142] Blake, J.W. 1989. *A Surface Pitting Life Model for Spur Gears*, Ph.D. Dissertation, Northwestern University.

[143] Royden, H.L. 1988. *Real Analysis*. New York: Prentice Hall.

9

Some Fracture Mechanics Problems Related to Contact Fatigue. Contact Fatigue Modeling for Smooth Elastic Surfaces

9.1 Introduction

There have been developed some theoretical models of crack behavior under action of contact load Kaneta et al. [1, 2], Murakami et al. [3], Kudish [4, 5], Xu and Hsia [6], Panasyuk, Datsyshin, and Marchenko [7] and references therein). Most of these models take into account only subsurface cracks and assume that cracks are completely open. Ultimately, the way the crack configuration is accounted for affects the stress intensity factors and the stress fields near crack tips, crack propagation, and fatigue life. Moreover, an adequate solution of the problem cannot be obtained without some boundary conditions, which prevent crack faces from overlapping and prevent tensile stress from developing on crack faces. Therefore, there is still a need in an accurate predictive model for subsurface cracks.

In Sections 9.2-9.4 and 9.6 we will consider the behavior of surface and sub-surface cracks in a loaded elastic plane and half-plane. The contact interaction of crack faces is regarded as one of the most interesting aspects of the problem. The problem is reduced to a system of integro-differential equations with boundary conditions in the form of alternating equations and inequalities. Subsurface cracks are allowed to have multiple cavities. Because of the nature of some of the boundary conditions to these equations (see (9.197), (9.200), and (9.201)) it is impossible to use the numerical technique similar to the ones developed by Gupta and Erdogan [8], Erdogan and Arin [9], Panasyuk et al., [10], Savruk [11]. The numerical method that can incorporate the above mentioned boundary conditions is complex and it is described in Section 9.6. The main challenge in the application of the aforementioned numerical methods is related to the existence of multiple cavities and regions where crack faces contact each other directly. The situation is exacerbated by the necessity to know the location of various crack cavities. Therefore, approximate analytical solutions in such a situation are very valuable.

It has been reported in periodic literature that properly designed bearings and gears in most cases experience a subsurface originated fatigue failure.

At the same time it is understood that various surface defects can be very harmful by significantly reducing fatigue life. In this chapter we will try to get some insight into the mechanisms controlling contact fatigue. Therefore, the purpose of this chapter is to present solutions of some problems of fracture mechanics related to contact fatigue and to develop a new model of contact fatigue that is derived based on well defined physical mechanisms that follow from contact mechanics of dry and lubricated solids as well as from fracture mechanics and that would be based on the facts established in Chapter 8.

Section 9.2 is dedicated to derivation of the main equations that describe crack behavior in a loaded elastic plane and half-plane. Section 9.3 provides solutions of fracture mechanics problems for small straight cracks in an elastic plane and half-plane. In Section 9.4 the latter solution is adjusted to the case of an elastic half-plane loaded with contact pressure and frictional stress as well as by the residual stress and weakened by small cracks. This solution for a half-plane and its analysis are used for developing a new contact fatigue model. The direction of fatigue crack propagation is considered in Section 9.5. A comparative study of surface and subsurface cracks in lubricated contacts that provides the basis for the analysis of surface and subsurface originated fatigue is given in Section 9.6. The behavior of surface and subsurface cracks in solids subjected to moving contact stresses is considered. In particular, an analysis of a surface crack-lubricant interaction is presented in Section 9.6, which allows to explain some mechanisms of subsurface and surface originated fatigue and the usually observed differences between fatigue lives of drivers and followers.

Taking into consideration the results of the analysis in Chapter 8, in Section 9.7 a two-dimensional statistical model of contact fatigue is developed in detail including an elaborate discussion of the assumptions laid down in its foundation. In Section 9.8 an analytical and numerical analysis of the pitting model is presented including some analytical formulas for contact fatigue life. A case of non-steady (periodic) loading regimes is discussed in Section 9.7. The extension of this contact fatigue model on the case of three dimensions is given in Section 9.10. Finally, some applications of the developed model to fatigue life analysis of bearings is presented in Section 9.11.

9.2   Modeling of an Elastic Material Stress State in the Vicinity of Crack Tips

Mathematical models for elastic plane and half-plane weakened by cracks will be proposed. Contact interactions between crack faces will be regarded as the most interesting aspects of the problem. The problems will be reduced to systems of integro-differential equations with nonlinear boundary conditions in the form of alternating equations and inequalities. In Section 9.3 an asymptotic (perturbation) method for the case of small cracks will be applied to solution of the problem and some numerical examples for small cracks will be presented.

The main purpose of the section is to present formulations of the problems for cracks in an elastic plane and for subsurface cracks in an elastic half-plane.

9.2.1 Problem Formulation and Solution for a Single Crack in an Elastic Plane

Suppose an elastic plane with Young's modulus E and Poisson's ratio ν is weakened by a straight cut (crack) of half-length l. Below, the derivation of the classic problem equations follows Panasyuk, Savruk, Datsyshin [10] and Savruk [11]. Let us introduce a coordinate system in the plane in such a way that the x-axis is directed along the crack and the y-axis is perpendicular to the crack faces (see Fig. 9.1). We will assume that the stresses $\sigma_y^\pm$ and $\tau_{xy}^\pm$ at the crack faces are given (signs plus and minus indicate the boundary values of the corresponding stresses on opposite crack faces) and at infinity the stresses are equal to zero. Assuming that

$$\sigma_y^\pm(x,0) - i\tau_{xy}^\pm(x,0) = P(x) \pm q(x), \ \mid x \mid < l, \tag{9.1}$$

where $P(x)$ and $q(x)$ are known functions, we obtain

$$\begin{aligned} P(x) &= \tfrac{1}{2}(\sigma_y^+ + \sigma_y^-) - \tfrac{i}{2}(\tau_{xy}^+ + \tau_{xy}^-), \\ q(x) &= \tfrac{1}{2}(\sigma_y^+ - \sigma_y^-) - \tfrac{i}{2}(\tau_{xy}^+ - \tau_{xy}^-). \end{aligned} \tag{9.2}$$

First, we will consider an auxiliary problem. In addition to the assumption that $P(x)$ and $q(x)$ are known functions, let us assume that on the crack faces $\mid x \mid \leq l$ the jumps of the stresses and the derivatives of the jumps of the normal $v(x) = v^+(x) - v^-(x)$ and tangential $u(x) = u^+(x) - u^-(x)$ crack face displacements are given, i.e.,

$$\begin{aligned} \sigma_y^+ - \sigma_y^- - i(\tau_{xy}^+ - \tau_{xy}^-) &= 2q(x), \\ \tfrac{\partial}{\partial x}[u^+ - u^- + i(v^+ - v^-)] &= \tfrac{i(\kappa+1)}{2\mu}g'(x), \ \mid x \mid < l \end{aligned} \tag{9.3}$$

and at the crack tips $(x = \pm l)$ the jumps of the displacements of the crack faces are equal to zero, i.e.,

$$g(-l) = g(l) = 0, \tag{9.4}$$

where $\mu = E/[2(1+\nu)]$ is the material elastic modulus while $\kappa = 3 - 4\nu$ for plane deformation and $\kappa = (3-\nu)/(1-\nu)$ for generalized plane stress state.

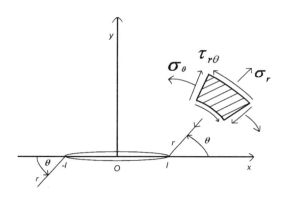

FIGURE 9.1
The general view of a straight crack in an elastic plane.

By introducing two analytic functions $\Phi(z)$ and $\Psi(z)$ of the complex variable $z = x + iy$ (see Muskhelishvili [12])

$$\sigma_x + \sigma_y = 2[\Phi(z) + \overline{\Phi}(z)], \quad \sigma_y - \sigma_x + 2i\tau_{xy} = 2[\bar{z}\Phi'(z) + \Psi(z)],$$

$$2\mu(u + iv) = \kappa\varphi(z) - z\overline{\Phi}(z) - \overline{\psi}(z), \tag{9.5}$$

$$\Phi(z) = \varphi'(z), \quad \Psi(z) = \psi'(z), \quad \Phi'(z) = \varphi''(z)$$

and the function
$$\Omega(z) = \overline{\Phi}(z) + z\overline{\Phi}'(z) + \overline{\Psi}(z), \tag{9.6}$$

we obtain

$$\sigma_x + \sigma_y = 2[\Phi(z) + \overline{\Phi}(z)], \quad \sigma_y - i\tau_{xy} = \Phi(z) + \Omega(\bar{z}) + (z - \bar{z})\overline{\Phi}'(z),$$

$$2\mu(u' + iv') = \kappa\Phi(z) - \Omega(\bar{z}) - (z - \bar{z})\overline{\Phi}'(z), \tag{9.7}$$

where u' and v' are the partial derivatives with respect to x of the displacements u and v in the directions along and orthogonal to the crack faces, respectively, and $\bar{z}$ means the complex conjugate of z.

We will assume that functions $\Phi(z)$ and $\Psi(z)$ are piece-wise analytical in the elastic plane including at infinity. In addition, we will assume that

$$\lim_{z \to x}(z - \bar{z})\Phi'(z) = 0 \; for \; |x| < l. \tag{9.8}$$

Taking into account (9.7) the boundary conditions from (9.3) can be rewritten in the form

$$\Phi^+(x) + \Omega^-(x) - \Phi^-(x) - \Omega^+(x) = 2q(x), \ |x| < l,$$

$$\kappa\Phi^+(x) - \Omega^-(x) - \kappa\Phi^-(x) + \Omega^+(x) = i(\kappa+1)g'(x), \ |x| < l. \tag{9.9}$$

By solving (9.9) for $\Phi^+(x) - \Phi^-(x)$ and $\Omega^+(x) - \Omega^-(x)$, we obtain two conjugate problems for $\Phi(z)$ and $\Omega(z)$ in the form

$$\Phi^+(x) - \Phi^-(x) = i[g'(x) - i\tfrac{2q(x)}{\kappa+1}] = iQ(x), \ |x| < l,$$

$$\Omega^+(x) - \Omega^-(x) = i[Q(x) + 2iq(x)], \ |x| < l. \tag{9.10}$$

It is well known (see Panasyuk, Savruk, Datsyshin [10] and Savruk [11]) that functions

$$\Phi(z) = \tfrac{1}{2\pi} \int\limits_{-l}^{l} \tfrac{Q(t)dt}{t-z}, \ \Omega(z) = \tfrac{1}{2\pi} \int\limits_{-l}^{l} \tfrac{Q(t)+2iq(t)}{t-z} dt \tag{9.11}$$

solve the conjugate problems (9.10) and, also, vanish at infinity. Therefore, using the expression for function $\Phi(z)$ we obtain

$$\Psi(z) = \tfrac{1}{2\pi} \int\limits_{-l}^{l} [\tfrac{\overline{Q}(t) - 2i\overline{q}(t)}{t-z} - \tfrac{tQ(t)}{(t-z)^2}] dt. \tag{9.12}$$

Finally, functions from (9.11) and (9.12) give the solution of the problem presented by (9.3) as well as by (9.10).

Now, let us consider a straight crack $|x| < l$ ($y = 0$) (similar to the analyzed above) in an elastic half-plane. The crack faces are loaded by stresses from (9.1) and at infinity the stresses vanish. We will try to find the complex potentials $\Phi(z)$ and $\Omega(z)$ in the form of (9.11) while assuming that the complex function $g(x)$, that characterizes the jumps of the crack faces displacement, is unknown. By satisfying the boundary condition

$$\sigma_y^+(x,0) + \sigma_y^-(x,0) - i[\tau_{xy}^+(x,0) + \tau_{xy}^-(x,0)] = 2P(x), \ |x| < l \tag{9.13}$$

that follows from (9.1), we obtain a singular integral equation

$$\int\limits_{-l}^{l} \tfrac{Q(t)+iq(t)}{t-x} dt = \pi P(x), \ |x| < l. \tag{9.14}$$

From (9.4) it follows that

$$\int\limits_{-l}^{l} g'(t)dt = 0. \tag{9.15}$$

The condition from (9.15) provides the uniqueness to the solution of (9.14) that can be expressed in the form

$$g'(x) = -i\tfrac{\kappa-1}{\kappa+1} q(x) + \tfrac{1}{\pi\sqrt{l^2-x^2}} \Big[-\int\limits_{-l}^{l} \tfrac{\sqrt{l^2-t^2}P(t)}{t-x} dt + i\tfrac{\kappa-1}{\kappa+1} \int\limits_{-l}^{l} q(t)dt \Big]. \tag{9.16}$$

The expression for the function from (9.16) allows to finalize the formulas for the complex potentials from (9.11) and (9.12).

To consider the stress field in the small vicinity of the crack tips $z = \pm l$, we need to consider the stresses at the points $z = \pm(l + z_r)$, $z_r = re^{i\theta}$, where r and θ are the polar coordinates in the polar systems with poles at $z = \pm l$ and the polar axis along the x-axis. In a small vicinity of the crack tips, i.e., for $\mid z_r \mid \ll l$, one can obtain the following asymptotic expressions for the stress field (see Panasyuk, Savruk, Datsyshin [10])

$$(\sigma_y, \sigma_x, \tau_{xy}) = \frac{k_1^{\pm}}{4\sqrt{2r}}(5\cos\tfrac{\theta}{2} - \cos\tfrac{5\theta}{2}, 3\cos\tfrac{\theta}{2} + \cos\tfrac{5\theta}{2},$$

$$- \sin\tfrac{\theta}{2} + \sin\tfrac{5\theta}{2})$$

$$+ \frac{k_2^{\pm}}{4\sqrt{2r}}(- \sin\tfrac{\theta}{2} + \sin\tfrac{5\theta}{2}, -7\sin\tfrac{\theta}{2} - \sin\tfrac{5\theta}{2},$$

$$3\cos\tfrac{\theta}{2} + \cos\tfrac{5\theta}{2}) + O(r^0), \quad r \to 0,$$

$$(9.17)$$

where $k_1^{\pm}$ and $k_2^{\pm}$ are the crack normal and shear stress intensity factors that can be determined from formulas (see Panasyuk, Savruk, Datsyshin [10])

$$k_1^{\pm} - ik_2^{\pm} = \mp \lim_{x \to \pm l} \sqrt{\tfrac{l^2 - x^2}{l}} g'(x). \tag{9.18}$$

In (9.17) and (9.18) the upper sign is related to the crack tip at $z = l$ and the low sign is related to the crack tip at $z = -l$.

Clearly, (9.17) demonstrate that the stress field in the small vicinity of a crack is controlled by the stress intensity factors $k_1^{\pm}$ and $k_2^{\pm}$. Therefore, to know the stress field in the vicinity of crack tips we need to know the values of the stress intensity factors $k_1^{\pm}$ and $k_2^{\pm}$.

9.2.2 Problem Formulation for Multiple Cracks in an Elastic Plane

The above analysis will serve as the basis for the consideration of the case of multiple cracks in an elastic plane with Young's modulus E and Poisson's ratio ν. To derive this system we will assume that an elastic plane is weakened by $N + 1$ straight cuts (cracks) of finite half-lengths l_k, $k = 0, 1, \ldots, N$ (see Fig. 9.2). Let us introduce a rectangular coordinate system (x^0, y^0) in the plane. The crack centers are located at the points with coordinates (x_k^0, y_k^0), $k = 0, 1, \ldots, N$. For each crack a local coordinate system is introduced in such a way that the x_k-axis is aligned with the crack and the y_k-axis is directed perpendicular to the crack faces. Cracks are oriented at angles α_k to the positive direction of the x^0-axis $(k = 0, 1, \ldots, N)$, $\alpha_0 = 0$. The crack faces are loaded with the normal $\sigma_k^{\pm}$ and tangential $\tau_k^{\pm}$ stresses, which can be represented by the equation

$$\sigma_k^{\pm} - i\tau_k^{\pm} = P_k(x_k) \pm q_k(x_k), \mid x_k \mid < l_k, \quad k = 0, 1, \ldots, N, \tag{9.19}$$

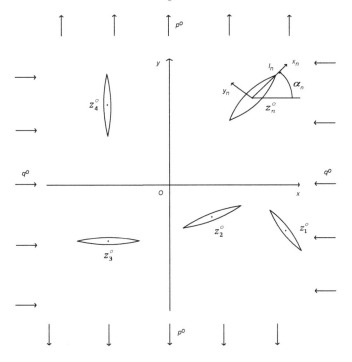

FIGURE 9.2
The general view of multiple straight cracks in an elastic plane. The local and global coordinate systems are shown.

where signs plus and minus correspond to the upper and lower faces of the kth crack, $P_k(x_k)$ and $q_k(x_k)$ are functions of normal and tangential stresses, respectively, which have to be determined, i is the imaginary unit. The stresses at infinity of the elastic plane are assumed to be equal to zero.

First, let us consider the problem for an infinite half-plane with just one crack $\mid x_k \mid \leq l_k$, $y_k = 0$ in the local coordinate system (x_k, y_k). Assuming that the jumps of the crack displacements $g_k(x_k)$ and stresses $q_k(x_k)$ are given and are related to the stresses and crack face displacements through the boundary conditions

$$\sigma_k^+ - \sigma_k^- - i(\tau_k^+ - \tau_k^-) = 2q_k(x_k),$$

$$\frac{\partial}{\partial x_k}[u_k^+ - u_k^- + i(v_k^+ - v_k^-)] = \frac{i(\kappa+1)}{2\mu}g_k'(x_k), \mid x_k \mid < l_k, \tag{9.20}$$

$$k = 0, 1, \ldots, N,$$

based on (9.11) and (9.12) we obtain the following formulas for complex potentials $\Phi(z_k)$ and $\Psi(z_k)$

$$\Phi(z_k) = \frac{1}{2\pi} \int\limits_{-l_k}^{l_k} \frac{Q_k(t_k)dt_k}{t_k - z_k}, \quad \Psi(z_k) = \frac{1}{2\pi} \int\limits_{-l_k}^{l_k} \left[\frac{\overline{Q}_k(t_k) - 2i\overline{q}_k(t_k)}{t_k - z_k} \right.$$

$$\left. - \frac{t_k Q_k(t_k)}{(t_k - z_k)^2} \right] dt_k, \quad z_k = x_k + i y_k, \quad Q_k(t_k) = g_k'(t_k) - i\frac{2q_k(t_k)}{\kappa + 1}. \tag{9.21}$$

Due to linearity of the problem functions

$$\Phi(z) = \frac{1}{2\pi} \sum_{k=0}^{N} \int\limits_{-l_k}^{l_k} \frac{Q_k(t_k)e^{i\alpha_k}dt_k}{T_k - z}, \quad T_k = te^{i\alpha_k} + z_k^0,$$

$$\Psi(z) = \frac{1}{2\pi} \sum_{k=0}^{N} e^{-2i\alpha_k} \int\limits_{-l_k}^{l_k} \left[\frac{\overline{Q}_k(t_k) - 2i\overline{q}_k(t_k)}{T_k - z} e^{-i\alpha_k} - \frac{\overline{T}_k Q_k(t_k)}{(\overline{T}_k - z)^2} e^{i\alpha_k} \right] dt_k, \tag{9.22}$$

that are obtained by superposition of complex potentials from (9.21), describe the stress state of the elastic plane caused by the jumps of the crack displacements $g_k(x_k)$ and stresses $q_k(x_k)$ applied to the straight crack faces with centers at $z_k^0 = x_k^0 + i y_k^0$ and oriented at angles α_k and having half-lengths l_k, $k = 0, 1, \ldots, N$.

Now, let us develop the solution of the problem for an elastic plane weakened by a number of straight cracks. Suppose that the cracks occupying the intervals $|x_k| \leq l_k$, $y_k = 0$ with the centers at (x_k^0, y_k^0), $k = 0, 1, \ldots, N$, are loaded by the stresses from (9.19) and at infinity the stresses are equal to zero. The complex potentials for such a problem we will try to find in the form of (9.22) assuming that the crack face displacement jumps $g_k(x_k)$, $k = 0, 1, \ldots, N$, are unknown. Taking into account that the jumps of the stresses on crack faces $q_k(x_k)$, $k = 0, 1, \ldots, N$, are already incorporated in (9.22), we have to satisfy the boundary conditions

$$\sigma_k^+(x_k, 0) + \sigma_k^-(x_k, 0) - i[\tau_k^+(x_k, 0) + \tau_k^-(x_k, 0)] = 2P_k(x_k),$$

$$|x_k| < l_k, \quad k = 0, 1, \ldots, N. \tag{9.23}$$

Using (9.6) and (9.7) from the previous section and (9.22) for complex potentials $\Phi(z)$ and $\Psi(z)$, we can calculate the normal $\sigma_k^\pm$ and tangential $\tau_k^\pm$ stresses on crack faces and satisfy the boundary conditions from (9.23). The latter results in a system of $N + 1$ singular integro-differential equations

$$\sum_{k=0}^{N} \int\limits_{-l_k}^{l_k} \left[Q_k(t_k)K_{nk}(t_k, x_n) + \overline{Q}_k(t_k)L_{nk}(t_k, x_n) \right.$$

$$\left. + \frac{iq_k(t_k)}{\overline{T}_k - \overline{X}_n} e^{i(\alpha_k - 2\alpha_n)} \right] dt_k = \pi P_n(x_n), \quad |x_k| < l_k, \quad k = 0, 1, \ldots, N, \tag{9.24}$$

where the kernels in (9.24) are determined by formulas

$$K_{nk}(t_k, x_n) = \frac{e^{i\alpha_k}}{2}\left[\frac{1}{T_k - X_n} + \frac{e^{-2i\alpha_n}}{\overline{T}_k - \overline{X}_n}\right],$$

$$L_{nk}(t_k, x_n) = \frac{e^{-i\alpha_k}}{2}\left[\frac{1}{\overline{T}_k - \overline{X}_n} - \frac{T_k - X_n}{(\overline{T}_k - \overline{X}_n)^2}e^{-2i\alpha_n}\right], \qquad (9.25)$$

$$T_k = t_k e^{i\alpha_k} + z_k^0, \ X_n = x_n e^{i\alpha_n} + z_n^0.$$

For convenience in further analysis the index in t_k is omitted.

The kernels in (9.25) are regular functions except for the case when $n = k$. For $n = k$ the kernels are as follows $K_{nk}(t, x_n) = 1/(t - x_n), \ L_{nk}(t, x_n) = 0$. Therefore, (9.24) can be rewritten in the form

$$\int_{-l_n}^{l_n} \frac{g_n'(t)dt}{t - x_n} + \sum_{k=0, k\neq n}^{N} \int_{-l_k}^{l_k}\left[g_k'(t)K_{nk}(t, x_n) + \overline{g}_k'(t)L_{nk}(t, x_n)\right]dt$$

$$= \pi P_n(x_n) + \frac{2i}{\kappa+1}\sum_{k=0}^{N}\int_{-l_k}^{l_k}\left\{q_k(t)\left[K_{nk}(t, x_n) - \frac{\kappa+1}{2(\overline{T}_k - \overline{X}_n)}e^{i(\alpha_k - 2\alpha_n)}\right]\right. \qquad (9.26)$$

$$\left. -\overline{q}_k(t)L_{nk}(t, x_n)\right\}dt, \ | x_n | < l_n, \ n = 0, 1, \ldots, N.$$

In case of self-balanced load, i.e., when $q_k(x_k) = 0, \ k = 0, 1, \ldots, N$, in (9.19), the system of (9.26) becomes simpler

$$\int_{-l_n}^{l_n} \frac{g_n'(t)dt}{t - x_n} + \sum_{k=0, k\neq n}^{N} \int_{-l_k}^{l_k}\left[g_k'(t)K_{nk}(t, x_n) + \overline{g}_k'(t)L_{nk}(t, x_n)\right]dt$$

$$= \pi P_n(x_n), \ | x_n | < l_n, \ n = 0, 1, \ldots, N. \qquad (9.27)$$

One of the most important cases of self-balanced load is the case of frictionless contact between the crack faces. Further, we will be considering only the case of plane deformation when $(\kappa+1)/(2\mu) = 4/E'$, where $E' = E/(1 - \nu^2)$ is the effective elasticity modulus. The case of generalized plane stress state can be considered similarly.

We need to obtain equations for the normal $v_k(x_k) = v_k^+(x_k) - v_k^-(x_k)$ and tangential $u_k(x_k) = u_k^+(x_k) - u_k^-(x_k)$ jumps of the respective displacements of the crack faces and the normal contact (internal) stresses $p_{nk}(x_k)$ acting between the crack faces. For cracks, the normal stresses σ_k are continuous functions. In addition, we will assume that the friction between the crack faces is zero. Therefore, we will require that $q_k(x_k) = 0, \ | x_k | \leq l_k, \ k = 0, 1, \ldots, N$. Generally, the normal $v_k^{\pm}$ and tangential $u_k^{\pm}$ displacements of opposite crack faces are different. However, at the crack tips of cracks they coincide, i.e.,

$$v_k(\pm l_k) = 0, \ u_k(\pm l_k) = 0, \ k = 0, 1, \ldots, N. \qquad (9.28)$$

The boundary conditions for surface cracks will be considered in detail in Section 9.6.

The stresses $P_k(x_k)$ can be represented as follows

$$P_k(x_k) = p_{nk}(x_k) + P_k^0(x_k), \ k = 0, 1, \ldots, N, \tag{9.29}$$

where $P_k^0(x_k)$ are the external normal stresses applied to the crack faces.

Using for crack faces the condition of non-overlapping we obtain

$$p_{nk}(x_k) = 0 \ for \ v_k(x_k) > 0,$$

$$p_{nk}(x_k) \leq 0 \ for \ v_k(x_k) = 0, \ k = 0, 1, \ldots, N. \tag{9.30}$$

The interpretation of the boundary conditions in (9.30) is simple: in the regions where the crack faces are in direct contact, the jump of the normal crack face displacement is zero and the normal contact stress is non-positive while in the regions where the crack faces are not in direct contact with one another the jump of the normal crack face displacement is positive and the normal contact stress is zero.

Based on the fact that the boundary conditions for the functions of crack faces tangential $u_k(x_k)$ and normal $v_k(x_k)$ displacement jumps are of different form (see (9.28) and (9.30)), it is convenient to split (9.27) for functions $g_k(x_k)$ into separate equations for functions $v_k(x_k)$ and $u_k(x_k)$ of the form

$$\int_{-l_n}^{l_n} \frac{v_n'(t)dt}{t - x_n} + \sum_{k=0, k \neq n}^{N} \int_{-l_k}^{l_k} [v_k'(t)A_{nk}^r(t, x_n) - u_k'(t)B_{nk}^r(t, x_n)]dt$$

$$= \pi Re(P_n(x_n)), \tag{9.31}$$

$$\int_{-l_n}^{l_n} \frac{u_n'(t)dt}{t - x_n} + \sum_{k=0, k \neq n}^{N} \int_{-l_k}^{l_k} [v_k'(t)A_{nk}^i(t, x_n) - u_k'(t)B_{nk}^i(t, x_n)]dt$$

$$= -\pi Im(P_n(x_n)), \ | \ x_n \ | < l_n, \ n = 0, 1, \ldots, N, \tag{9.32}$$

$$A_{nk} = \overline{K_{nk} + L_{nk}}, \ B_{nk} = -i(\overline{K_{nk} - L_{nk}}), \ A_{nn} = B_{nn} = 0,$$

$$(A_{nk}^r, B_{nk}^r) = Re(A_{nk}, B_{nk}), \ (A_{nk}^i, B_{nk}^i) = Im(A_{nk}, B_{nk}).$$

Equations (9.20), (9.25), (9.28)-(9.32) represent a closed system for determining the functions of crack faces displacement jumps $u_k(x_k)$, $v_k(x_k)$, and the normal contact stress $p_{nk}(x_k)$ applied to the crack faces ($k = 0, 1, \ldots, N$) for the given values of constants ν, E', l_k, α_k, and functions $P_k^0(x_k)$, $k = 0, 1, \ldots, N$. After the solution of the problem has been obtained, the stress intensity factors $k_{1n}^{\pm}$ and $k_{2n}^{\pm}$ are determined according to equations (see (9.18) from the previous section)

$$k_{1n}^{\pm} + ik_{2n}^{\pm} = \mp \frac{E'}{4} \lim_{x_n \to \pm l_n} \sqrt{\frac{l_n^2 - x_n^2}{l_n}} [v_n'(x_n) + iu_n'(x_n)], \ 0 \leq n \leq N. \tag{9.33}$$

9.2.3 Problem Formulation for Multiple Cracks in an Elastic Half-Plane

To obtain the system of equations for the case of an elastic half-plane weakened by N cracks, we need to assume that in case of $N + 1$ cracks in an elastic plane one of the cracks is of an infinite length, i.e., $l_0 = \infty$, $\alpha_0 = 0$, and all other N cracks are located below the x^0-axis. Therefore, for the case of plane deformation, the system of (9.26) can be rewritten in the form

$$\int_{-\infty}^{\infty} \frac{g_0'(t)dt}{t-x^0} + \sum_{k=1}^{N} \int_{-l_k}^{l_k} [g_k'(t)K_{0k}(t,x^0) + \overline{g}_k'(t)L_{0k}(t,x^0)]dt$$

$$= \pi P_0(x^0) + \frac{i}{2(1-\nu)} \sum_{k=0}^{N} \int_{-l_k}^{l_k} \left\{ q_k(t)\left[K_{0k}(t,x^0) - \frac{2(1-\nu)}{\overline{T}_k-x^0}e^{\alpha_k} \right] \right.$$

$$\left. - \overline{q}_k(t)L_{0k}(t,x^0) \right\}dt,$$

$$\int_{-l_n}^{l_n} \frac{g_n'(t)dt}{t-x_n} + \int_{-\infty}^{\infty} [g_0'(t)K_{n0}(t,x_n) + \overline{g}_0'(t)L_{n0}(t,x_n)]dt \qquad (9.34)$$

$$+ \sum_{k=1,k\neq n}^{N} \int_{-l_k}^{l_k} [g_k'(t)K_{nk}(t,x_n) + \overline{g}_k'(t)L_{nk}(t,x_n)]dt = \pi P_n(x_n)$$

$$+ \frac{i}{2(1-\nu)} \sum_{k=0}^{N} \int_{-l_k}^{l_k} \left\{ q_k(t)\left[K_{nk}(t,x_n) - \frac{2(1-\nu)}{\overline{T}_k-\overline{X}_n}e^{i(\alpha_k - 2\alpha_n)} \right] \right.$$

$$\left. - \overline{q}_k(t)L_{nk}(t,x_n) \right\}dt, \ |x_n| < l_n, \ n = 1,\ldots,N.$$

By solving [10, 11] the first of equations (9.34) for $g_0(x^0)$, we obtain

$$g_0(x^0) = \frac{1}{\pi} \int_{-\infty}^{\infty} P_0(t) \ln|t - x^0| dt - i\frac{1-2\nu}{2(1-\nu)} \int_{-\infty}^{x^0} q_0(t)dt$$

$$- \frac{1}{2\pi} \sum_{k=1}^{N} \int_{-l_k}^{l_k} \overline{g}_k'(t)W_k(t,x^0)dt - \frac{1}{\pi} \sum_{k=1}^{N} \int_{-l_k}^{l_k} [q_k(t)e^{i\alpha_k} \ln|x^0 - T_k| \qquad (9.35)$$

$$+ \overline{q}_k(t)\frac{i}{4(1-\nu)}W_k(t,x^0)]dt + g_{00}, \ W_k(t,x^0) = ie^{-i\alpha_k}\frac{\overline{T}_k-T_k}{\overline{T}_k-x^0},$$

where g_{00} is an arbitrary complex constant. For subsurface cracks, i.e., for $|y_k^0| > l_k |\sin\alpha_k|$, kernels $W_k(t,x^0)$ from (9.35) are regular functions while for surface cracks, i.e., for $|y_k^0| = l_k |\sin\alpha_k|$, these kernels are singular.

The expression for $g_0(x^0)$ from (9.35) being substituted in (9.34), after changing the order of integration and calculation of some integrals [10] leads to a system of integro-differential equations for an elastic half-plane weakened by N cracks

$$\int_{-l_n}^{l_n} \frac{g_n'(t)dt}{t-x_n} + \sum_{k=1}^{N} \int_{-l_k}^{l_k} [g_k'(t)R_{nk}(t,x_n) + \overline{g}_k'(t)S_{nk}(t,x_n)]dt$$

$$+ \sum_{k=1}^{N} \int_{-l_k}^{l_k} [q_k(t)Q_{nk}(t,x_n) - \overline{q}_k(t)G_{nk}(t,x_n)]dt$$

$$= \pi P_n(x_n) + \int_{-\infty}^{\infty} [P_0(t)M_n(t,x_n) + \overline{P}_0(t)N_n(t,x_n)]dt$$

$$+ i \int_{-\infty}^{\infty} [q_0(t)H_{n0}(t,x_n) - \overline{q}_0(t)L_{n0}(t,x_n)]dt, \quad | x_n |< l_n,$$

$$n = 1,\ldots,N,$$

$$R_{nk}(t,x_n) = (1 - \delta_{nk})K_{nk}(t,x_n) + \frac{e^{i\alpha_k}}{2} \left\{ \frac{1}{X_n - \overline{T}_k} + \frac{e^{-2i\alpha_n}}{\overline{X}_n - T_k} \right.$$

$$\left. + (\overline{T}_k - T_k)\left[\frac{1 + e^{-2i\alpha_n}}{(\overline{X}_n - T_k)^2} + \frac{2e^{-2i\alpha_n}(T_k - X_n)}{(\overline{X}_n - T_k)^3} \right] \right\}, \tag{9.36}$$

$$S_{nk}(t,x_n) = (1 - \delta_{nk})L_{nk}(t,x_n) + \frac{e^{-i\alpha_k}}{2} \left[\frac{T_k - \overline{T}_k}{(X_n - \overline{T}_k)^2} \right.$$

$$\left. + \frac{1}{\overline{X}_n - T_k} + \frac{e^{-2i\alpha_n}(T_k - X_n)}{(\overline{X}_n - T_k)^2} \right],$$

$$Q_{nk}(t,x_n) = \frac{ie^{i\alpha_k}}{4(1-\nu)} \left[-\frac{1}{T_k - X_n} + (3 - 4\nu)\left(\frac{e^{-2i\alpha_n}}{\overline{T}_k - \overline{X}_n} - \frac{1}{T_k - X_n} \right) \right.$$

$$\left. + \frac{e^{-2i\alpha_n}}{T_k - \overline{X}_n} + \frac{T_k - \overline{T}_k}{(\overline{X}_n - T_k)^2} - e^{-2i\alpha_n}\frac{(T_k - \overline{T}_k)(2X_n - \overline{X}_n - T_k)}{(\overline{X}_n - T_k)^3} \right],$$

$$G_{nk}(t,x_n) = \frac{ie^{-i\alpha_k}}{4(1-\nu)} \left\{ -\frac{1}{\overline{T}_k - \overline{X}_n} + \frac{e^{-2i\alpha_n}(T_k - X_n)}{(\overline{T}_k - \overline{X}_n)^2} - \frac{T_k - \overline{T}_k}{(X_n - \overline{T}_k)^2} \right.$$

$$\left. + (3 - 4\nu)\left[\frac{1}{\overline{X}_n - T_k} + \frac{e^{-2i\alpha_n}(T_k - X_n)}{(\overline{X}_n - T_k)^2} \right] \right\},$$

$$M_n(t,x_n) = \frac{i}{2} \left[-\frac{1}{t - X_n} + \frac{e^{-2i\alpha_n}}{t - \overline{X}_n} \right],$$

$$N_n(t,x_n) = \frac{i}{2} \left[\frac{1 - e^{-2i\alpha_n}}{t - \overline{X}_n} - \frac{e^{-2i\alpha_n}(\overline{X}_n - X_n)}{(t - \overline{X}_n)^2} \right], \tag{9.37}$$

$$H_{n0}(t,x_n) = \frac{1}{2} \left[\frac{1}{t - X_n} - \frac{e^{-2i\alpha_n}}{t - \overline{X}_n} \right],$$

$$T_k = te^{i\alpha_k} + z_k^0, \quad X_n = x_n e^{i\alpha_n} + z_n^0,$$

where δ_{nk} is the Kronecker tensor ($\delta_{nk} = 0$ for $n \neq k$, $\delta_{nk} = 1$ for $n = k$).

Let us consider a specific case of an elastic half-plane weakened by N straight cracks. We will consider the crack faces frictionless, i.e., $q_k(x_k) = 0$, $k = 1, \ldots, N$. On the interval $[x_i, x_e]$ the half-plane boundary $y^0 = 0$ is loaded by the pressure distribution $p(x^0)$ and by the frictional stress $\tau(x^0)$. Moreover, at infinity the half-plane is loaded by a tensile or compressive (residual) stress $\sigma_{x^0}^\infty = q^0$, which is directed along the x^0-axis. The stresses $P_k(x_k)$ can be represented as follows (see (9.29)):

$$P_k(x_k) = p_{nk}(x_k) + P_k^0(x_k), \ \ k = 1, \ldots, N, \tag{9.38}$$

$$P_0(x^0) = -p(x^0), \ \ q_0(x^0) = i\tau(x^0), \ \ Im(p) = Im(\tau) = 0, \tag{9.39}$$

and from (9.36) and (9.22) we get

$$P_k^0 = p_k^0 - i\tau_k^0 = -\frac{1}{\pi} \int_{x_i}^{x_e} [p(t)D_k(t, x_k) + \tau(t)G_k(t, x_k)]dt$$

$$-\frac{q^0}{2}(1 - e^{-2i\alpha_k}), \tag{9.40}$$

$$D_k(t, x_k) = \frac{i}{2}\left[-\frac{1}{t-X_k} + \frac{1}{t-\overline{X}_k} - \frac{e^{-2i\alpha_k}(\overline{X}_k - X_k)}{(t-\overline{X}_k)^2}\right],$$

$$G_k(t, x_k) = \frac{1}{2}\left[\frac{1}{t-X_k} + \frac{1-e^{-2i\alpha_k}}{t-\overline{X}_k} - \frac{e^{-2i\alpha_k}(t-X_k)}{(t-\overline{X}_k)^2}\right], \ \ k = 1, \ldots, N, \tag{9.41}$$

where $P_k^0(x_k)$ are the external normal stresses applied to the crack faces and $Im(p_{nk}(x_k)) = 0$, $k = 1, \ldots, N$.

It is easy to see that (9.36)-(9.41) can be reduced to (9.38)-(9.40) (also see (9.31)):

$$\int_{-l_k}^{l_k} \frac{v_k'(t)dt}{t-x_k} + \sum_{m=1}^{N} \int_{-l_m}^{l_m} [v_m'(t)A_{km}^r(t, x_k) - u_m'(t)B_{km}^r(t, x_k)]dt$$

$$= \pi p_{nk}(x_k) - \int_{x_i}^{x_e} [p(t)D_k^r(t, x_k) + \tau(t)G_k^r(t, x_k)]dt - \pi q^0 \sin^2 \alpha_k,$$

$$\tag{9.42}$$

$$\int_{-l_k}^{l_k} \frac{u_k'(t)dt}{t-x_k} + \sum_{m=1}^{N} \int_{-l_m}^{l_m} [v_m'(t)A_{km}^i(t, x_k) - u_m'(t)B_{km}^i(t, x_k)]dt$$

$$= \int_{x_i}^{x_e} [p(t)D_k^i(t, x_k) + \tau(t)G_k^i(t, x_k)]dt + \pi\frac{q^0}{2}\sin 2\alpha_k,$$

where

$$A_{nk} = \overline{R_{nk} + S_{nk}}, \ \ B_{nk} = -i(\overline{R_{nk} - S_{nk}}),$$

$$(A_{nk}^r, B_{nk}^r, D_n^r, G_n^r) = Re(A_{nk}, B_{nk}, \overline{D}_n, \overline{G}_n), \tag{9.43}$$

$$(A_{nk}^i, B_{nk}^i, D_n^i, G_n^i) = Im(A_{nk}, B_{nk}, \overline{D}_n, \overline{G}_n),$$

where kernels R_{nk}, S_{nk}, K_{nk}, L_{nk} and variables T_k and X_n are determined by (9.37). Obviously, the crack faces displacement jumps $u_k(x_k)$, $v_k(x_k)$, and the normal contact stress $p_{nk}(x_k)$ applied to the crack faces $(k = 1, \ldots, N)$ for the given values of constants ν, E', q^0, l_k, α_k, $k = 1, \ldots, N$, and functions $p(x^0)$ and $\tau(x^0)$ are determined by (9.42), where kernels A_{nk}, B_{nk}, D_n, and G_n are determined by (9.37), (9.41), and (9.43) and stresses $P_k(x_k)$, $k = 1, \ldots, N$, are determined by (9.38)-(9.40). After the solution of the problem has been obtained, the stress intensity factors $k_{1n}^{\pm}$ and $k_{2n}^{\pm}$ are determined according to (9.33). The boundary conditions for subsurface cracks are as follows (see (9.28) and (9.30))

$$v_k(\pm l_k) = 0, \ u_k(\pm l_k) = 0, \ k = 1, \ldots, N, \tag{9.44}$$

$$p_{nk}(x_k) = 0 \ for \ v_k(x_k) > 0, \ p_{nk}(x_k) \leq 0 \ for \ v_k(x_k) = 0,$$
$$k = 1, \ldots, N. \tag{9.45}$$

The boundary conditions for surface cracks will be considered in detail in Section 9.6.

9.3   Perturbation Analysis of the Problems for Small Multiple Straight Cracks in an Elastic Plane or Half-Plane

This section is devoted to perturbation analysis of the problems for multiple cracks in elastic plane and half-plane formulated in Section 9.2. The main difference of these solutions from the well–known ones is the fact that cracks may form such configurations that their faces may be in partial of full contact with each other. Ultimately, the crack configurations affect their stress intensity factors, and, therefore, the stress state of the material near the crack tips.

9.3.1   Perturbation Solution for Multiple Cracks in an Elastic Plane

The main purpose of this section is to develop an efficient asymptotic (analytical) method for solution of the problem for small cracks. These solutions will be used in Section 9.3.1 to obtain the asymptotic expressions for the stress intensity factors $k_{1n}^{\pm}$ and $k_{2n}^{\pm}$ for "small" cracks and to analyze their behavior.

Suppose at infinity an elastic plane weakened by N straight cracks is subjected a biaxial tension and/or compression by stresses $\sigma_x^\infty = q^0$ and $\sigma_y^\infty = p^0$, where q^0 and p^0 are certain constants. Then stresses P_n^0 are determined as follows

$$P_n^0 = p_n^0 - i\tau_n^0 = -\frac{p^0}{2}[1 + \eta + (1 - \eta_0)e^{-2i\alpha_n}], \ \eta_0 = \frac{q^0}{p^0},$$

$$n = 1, \ldots, N.$$

(9.46)

Let us introduce the dimensionless variables

$$(x_n^{0'}, y_n^{0'}) = (x_n^0, y_n^0)/\tilde{b}, \ (p_n^{0'}, \tau_n^{0'}, p_n') = (p_n^0, q_n^0, p_n)/\tilde{q},$$

$$(x_n', t') = (x_n, t)/l_n, \ (v_n', u_n') = (v_n, u_n)/\tilde{v}_n,$$

(9.47)

$$(k_{1n}^{\pm '}, k_{2n}^{\pm '}) = (k_{1n}^{\pm}, k_{2n}^{\pm})/(\tilde{q}\sqrt{l_n})$$

and scale the problem at hand. In (9.47) it is assumed that $\tilde{v}_n = 4\tilde{q}l_n/E'$, where $\tilde{q}$ and $\tilde{b}$ are the characteristic stress and linear scale of the problem. In this particular case parameters $\tilde{q}$ and $\tilde{b}$ are taken as follows

$$\tilde{q} = p^0 \ if \ p^0 \neq 0 \ or \ \tilde{q} = q^0 \ if \ p^0 = 0, \ \tilde{b} = \min_{n \neq k} \mid z_k^0 - z_n^0 \mid.$$

(9.48)

By introducing the dimensionless parameters $\delta_n = l_n/\tilde{b}$ and for simplicity omitting primes at the dimensionless variables, we arrive at the following system

$$\int_{-1}^{1} \frac{v_k'(t)dt}{t - x_k} + \sum_{m=1}^{N} \delta_m \int_{-1}^{1} [v_m'(t)A_{km}^r(t, x_k) - u_m'(t)B_{km}^r(t, x_k)]dt$$

$$= \pi p_{nk}(x_k) + \pi p_k^0,$$

(9.49)

$$\int_{-1}^{1} \frac{u_k'(t)dt}{t - x_k} + \sum_{m=1}^{N} \delta_m \int_{-1}^{1} [v_m'(t)A_{km}^i(t, x_k) - u_m'(t)B_{km}^i(t, x_k)]dt$$

$$= \pi\tau_k^0, \ \mid x_k \mid < 1, \ k = 1, \ldots, N,$$

$$A_{nk} = \overline{K_{nk} + L_{nk}}, \ B_{nk} = -i(\overline{K_{nk} - L_{nk}}), \ A_{nn} = B_{nn} = 0,$$

(9.50)

$$(A_{nk}^r, B_{nk}^r) = Re(A_{nk}, B_{nk}), \ (A_{nk}^i, B_{nk}^i) = Im(A_{nk}, B_{nk}),$$

$$T_k = \delta_k te^{i\alpha_k} + z_k^0, \ X_n = \delta_n x_n e^{i\alpha_n} + z_n^0,$$

(9.51)

$$v_k(\pm 1) = 0, \ u_k(\pm 1) = 0,$$

(9.52)

$$p_{nk}(x_k) = 0 \ for \ v_k(x_k) > 0, \ p_{nk}(x_k) \leq 0 \ for \ v_k(x_k) = 0,$$

(9.53)

$$p_k^0 - i\tau_k^0 = -\frac{1}{2}[1 + \eta_0 + (1 - \eta_0)e^{-2i\alpha_k}],$$

(9.54)

$$k_{1k}^{\pm} + ik_{2k}^{\pm} = \mp \lim_{x_k \to \pm 1} \sqrt{1 - x_k^2}[v_k'(x_k) + iu_k'(x_k)],$$

(9.55)

where kernels K_{km} and L_{km} in dimensionless variables are also described by equations (9.25). Therefore, the dimensionless problem for multiple straight cracks in an elastic plane is described by equations (9.25) and (9.49)-(9.55).

1. Now, let us consider the case of "small" cracks, i.e., the case when the crack lengths are small in comparison with the distances between them. Therefore, we will assume that

$$\delta_0 = \max_{1 \le k \le N} \delta_k \ll 1.$$

In this case we have

$$z_n^0 - z_k^0 \gg \delta_0 \; for \; n \neq k, \; n, k = 1, \ldots, N. \tag{9.56}$$

In (9.56) the estimate $g \gg h$ for complex values g and h should be understood in the following sense $g\bar{g} \gg h\bar{h}$. Obviously, if $g\bar{g} \gg h\bar{h}$ then $g \gg h$ and vice versa. Similarly, we can determined the relationship $g \sim h$. Therefore, from 9.56 follows that

$$T_k - X_n \gg \delta_0 \; for \; n \neq k, \; n, k = 1, \ldots, N. \tag{9.57}$$

Analyzing the structure of kernels A_{nk} and B_{nk} from equations (9.25), (9.50), and (9.51), we come to the conclusion that these kernels can be represented in the form of regular for all x_n and t power series in $\delta_k \ll 1$ and $\delta_m \ll 1$:

$$\{A_{km}(t, x_k), B_{km}(t, x_k)\}$$

$$= \sum_{j+n=0; \; j,n \ge 0}^{\infty} (\delta_k x_k)^j (\delta_m t)^n \{A_{kmjn}, B_{kmjn}\}. \tag{9.58}$$

In equations (9.58) the values of A_{kmjn} and B_{kmjn} are independent of δ_k, δ_m, x_k, and t. The values of A_{kmjn} and B_{kmjn} are certain functions of constants α_k, α_m, x_k^0, y_k^0, x_m^0, and y_m^0.

The solution of the system of equations (9.49) we will try to find in the form of asymptotic series in powers of $\delta_0 \ll 1$:

$$\{v_k, u_k, p_{nk}\} = \sum_{j=0}^{\infty} \delta_0^j \{v_{kj}, u_{kj}, p_{nkj}\} \tag{9.59}$$

using the regular perturbation method [13, 14]. Using (9.58) and (9.59) and equating the terms with the same powers of $\delta_0 \ll 1$ in (9.49), we obtain the following series of equations

$$\int_{-1}^{1} \frac{v_{kh}'(t)dt}{t - x_k} = \pi p_{nkh}(x_k) + \delta_{h0}\pi p_k^0$$

$$+ \varepsilon_h \pi \sum_{j,l \ge 0, m \ge 1}^{j+l+m=h} (m-1)(\tfrac{\delta_k}{\delta_0}x_k)^j C_{kjlm-1}^r, \tag{9.60}$$

$$\int_{-1}^{1} \frac{u'_{kh}(t)dt}{t-x_k} = \delta_{h0}\pi\tau_k^0 + \varepsilon_h\pi \sum_{j,l\geq0,m\geq1}^{j+l+m=h} (m-1)(\tfrac{\delta_k}{\delta_0}x_k)^j C_{kjlm-1}^i,$$

(9.61)

$$h = 0, 1, \ldots,$$

$$C_{kjlm} = \frac{1}{\pi} \sum_{n=1}^{N} (\tfrac{\delta_n}{\delta_0})^{m+1} (A_{knjm}V_{nlm-1} - B_{knjm}U_{nlm-1}),$$

(9.62)

$$\{V, U\}_{nlm} = \int_{-1}^{1} t^m \{v_{nl}(t), u_{nl}(t)\}dt, \quad \varepsilon_h = 1 - \delta_{h0}.$$

(9.63)

The summation in (9.60) and (9.61) is done with respect to all j, l, and m that satisfy the conditions: $j \geq 0$, $l \geq 0$, $m \geq 1$, $j+l+m = h$, and δ_{nk} is the Kronecker tensor ($\delta_{nk} = 0$ for $n \neq k$, $\delta_{nk} = 1$ for $n = k$). In the process of the derivation of (9.60)-(9.63), we used integration by parts and the boundary conditions

$$v_{kh}(\pm 1) = 0, \ u_{kh}(\pm 1) = 0, \ k = 1, \ldots, N, \ h = 0, 1, \ldots,$$

(9.64)

that follow from (9.52) and (9.59).

Let us analyze the system of alternating equalities and inequalities from (9.53) for $\delta_0 \ll 1$. By substituting the expressions for v_k and u_k into (9.53), we obtain

$$\sum_{j=0}^{\infty} \delta_0^j p_{nkj}(x_k) = 0 \ if \ \sum_{j=0}^{\infty} \delta_0^j v_{kj}(x_k) > 0,$$

(9.65)

$$\sum_{j=0}^{\infty} \delta_0^j p_{nkj}(x_k) \leq 0 \ if \ \sum_{j=0}^{\infty} \delta_0^j v_{kj}(x_k) = 0.$$

Let us assume that $v_{k0}(x_k) > 0$. Then for $\delta_0 \ll 1$ from the first inequality in (9.65) follows that $p_{nkj}(x_k) = 0$ for all $j \geq 0$. Now, let us consider the opposite case when $v_{k0}(x_k) = 0$. Then only one of the two cases may occur: (a) $p_{nk0}(x_k) < 0$ or (b) $p_{nk0}(x_k) = 0$. In case (a) independent of the values of $p_{nkj}(x_k)$ for $j \geq 1$, we have $p_{nk}(x_k) < 0$ for $\delta_0 \ll 1$. Under these conditions from the last condition in (9.65), we find $v_{kj}(x_k) = 0$ for all $j \geq 0$. In case (b) we get $p_{nk0}(x_k) = 0$ and $v_{k0}(x_k) = 0$, and the analysis of the relationships for $p_{nkj}(x_k)$ and $v_{kj}(x_k)$ for $j \geq 1$ should be done in a similar way but in the higher order of approximation with respect to $\delta_0 \ll 1$.

2. Let us consider the stress state of the elastic plane with cracks in the zero approximation. In this case the problem is described by (9.60), (9.61), (9.64) for $h = 0$ and by the boundary conditions

$$p_{nk0}(x_k) = 0 \ for \ v_{k0}(x_k) > 0, \ p_{nk0}(x_k) \leq 0 \ for \ v_{k0}(x_k) = 0,$$

(9.66)

that follows from the above analysis of (9.65).

Let us assume that $v_{k0}(x_k) > 0$ for $|x_k| < 1$. Using this assumption from (9.66) we will get $p_{nk0}(x_k) = 0$ for $|x_k| < 1$ while (9.60) and (9.64) lead to

[15, 16] $v_{k0}(x_k) = -p_k^0 \sqrt{1-x_k^2} \geq 0$. From the latter we conclude that $p_k^0 < 0$. It is easy to see that for $p_k^0 \geq 0$ and $h = 0$ (9.60), (9.64), and (9.66) are satisfied by functions $v_{k0}(x_k) = 0$ and $p_{nk0}(x_k) = -p_k^0 \leq 0$ for $|x_k| < 1$. Solving (9.61) and (9.64) we obtain [15, 16] $u_{k0}(x_k) = -\tau_k^0 \sqrt{1-x_k^2}$. Therefore, in the zero approximation the problem solution is

$$v_{k0}(x_k) = -p_k^0 \theta(-p_k^0)\sqrt{1-x_k^2}, \quad u_{k0}(x_k) = -\tau_k^0 \sqrt{1-x_k^2},$$

$$p_{nk0}(x_k) = -p_k^0 \theta(p_k^0), \tag{9.67}$$

where $\theta(x)$ is the Heavyside function ($\theta(x) = 0$ for $x \leq 0$ and $\theta(x) = 1$ for $x > 0$).

From the above analysis of (9.65) it follows that for $p_k^0 < 0^*$ and $|x_k| < 1$ we have $v_{k0}(x_k) > 0$ and $p_{nkj}(x_k) = 0$ for all $j \geq 0$ and $|x_k| < 1$ while any signs of the values of $v_{kj}(x_k)$ for $j \geq 1$ do not violate the first inequality in (9.65). Therefore, using (9.64), (9.65), and (9.67) from (9.60) and (9.61) follows [15, 16] that

$$v_{k1}(x_k) = u_{k1}(x_k) = 0, \tag{9.68}$$

$$v_{k2}(x_k) = -C_{k001}^r \sqrt{1-x_k^2}, \quad u_{k2}(x_k) = -C_{k001}^i \sqrt{1-x_k^2}, \tag{9.69}$$

$$v_{k3}(x_k) = -\frac{\delta_k}{2\delta_0} C_{k101}^r x_k \sqrt{1-x_k^2},$$

$$u_{k3}(x_k) = -\frac{\delta_k}{2\delta_0} C_{k101}^i x_k \sqrt{1-x_k^2}, \tag{9.70}$$

$$v_{k4}(x_k) = [F_k^r + \tfrac{H_k^r}{6}(1+x_k^2)]\sqrt{1-x_k^2},$$

$$u_{k4}(x_k) = [F_k^i + \tfrac{H_k^i}{6}(1+x_k^2)]\sqrt{1-x_k^2}, \tag{9.71}$$

$$F_k = -3C_{k003} - C_{k021}, \quad H_k = -(\tfrac{\delta_k}{\delta_0})^2 C_{k201}.$$

In (9.68)-(9.71) it is taken into account that $C_{k011} = C_{k002} = C_{k012} = C_{k102} = C_{k111} = 0$.

For $p_k^0 > 0$ from the analysis of (9.65), it follows (see (9.67)) that $v_{kj}(x_k) = 0$ and for all $j \geq 0$ and $|x_k| < 1$ while any signs of the values of $p_{nkj}(x_k)$ for $j \geq 1$ do not violate the second inequality in (9.65). Therefore, for $p_k^0 > 0$ from (9.60), (9.62), and (9.63), we obtain

$$p_{nk1}(x_k) = 0, \tag{9.72}$$

$$p_{nk2}(x_k) = -C_{k001}^r, \tag{9.73}$$

$$p_{nk3}(x_k) = -\frac{\delta_k}{\delta_0} C_{k101}^r x_k, \tag{9.74}$$

---

*Here and further the intermediate cases such as $p_k^0 \cong \delta_0^2 C_{k001}^r + \ldots$ are not analyzed. The analysis of such cases can be done in a similar manner.

$$p_{nk4}(x_k) = F_k^r + H_k^r x_k^2. \tag{9.75}$$

Functions $u_{kh}(x_k)$ for $h = 0, \ldots, 4$ are still determined by (9.67)-(9.71).

3. Now, let us consider the case of $p_k^0 = 0$ when in the zero approximation $v_{k0}(x_k) = p_{nk0}(x_k) = 0$ for $|x_k| < 1$. In this case, in (9.65) the summation starts with $j = 1$. Therefore, by conducting a similar to the performed above analysis of (9.65) for $p_k^0 = 0$, we come up with the following alternating equalities and inequalities:

$$p_{nk1}(x_k) = 0 \ for \ v_{k1}(x_k) > 0, \ p_{nk1}(x_k) \le 0 \ for \ v_{k1}(x_k) = 0. \tag{9.76}$$

It is obvious that for $h = 1$ the functions from (9.68) and (9.72) satisfy (9.60), (9.61), (9.64), and (9.76). Therefore, we need to consider the next order of approximation.

4. In case of $p_k^0 = 0$, using (9.67), (9.68), and (9.72) and acting in the fashion similar to the one used above from (9.65), we obtain

$$p_{nk2}(x_k) = 0 \ for \ v_{k2}(x_k) > 0, \ p_{nk2}(x_k) \le 0 \ for \ v_{k2}(x_k) = 0, \tag{9.77}$$

It follows from (9.69) and (9.77) that for $C_{k001}^r < 0$ we have $v_{k2}(x_k) > 0$ and $p_{nkj}(x_k) = 0$ for all $j \ge 0$ and $|x_k| < 1$. It is easy to see that for $h = 2$ and $C_{k001}^r \ge 0$ functions $v_{k2}(x_k) = 0$ and $p_{nkj}(x_k) = -C_{k001}^r$ satisfy (9.60) and (9.77). Therefore,

$$v_{k2}(x_k) = -C_{k001}^r \theta(-C_{k001}^r)\sqrt{1 - x_k^2},$$
$$p_{nk2}(x_k) = -C_{k001}^r \theta(C_{k001}^r). \tag{9.78}$$

Moreover, for $p_k^0 = 0$ and $C_{k001}^r < 0$ we have $p_{nkj}(x_k) = 0$ for all $j \ge 0$ and $|x_k| < 1$ while functions $v_{kj}(x_k)$ satisfy (9.67), (9.68), (9.78), (9.70), and (9.71). At the same time, functions $u_{kj}(x_k)$ satisfy (9.67)-(9.71). In case of $p_k^0 = 0$ and $C_{k001}^r > 0$, we have $v_{kj}(x_k) = 0$ for all $j \ge 0$ and $|x_k| < 1$ while functions $p_{nkj}(x_k)$ satisfy (9.67), (9.71)-(9.75).

5. Now, let us consider the case when $p_k^0 = C_{k001}^r = 0$ and $v_{kj}(x_k) = p_{nkj}(x_k) = 0$ for $|x_k| < 1$ and $j = 0, 1, 2$. In this case from (9.65) we obtain

$$p_{nk3}(x_k) = 0 \ for \ v_{k3}(x_k) > 0, \ p_{nk3}(x_k) \le 0 \ for \ v_{k3}(x_k) = 0, \tag{9.79}$$

Let us assume that $C_{k101}^r < 0$. In this case it can be shown that the interval $(-1, 1)$ occupied by the kth crack is subdivided into two subintervals $(-1, b_{k3})$ and $(b_{k3}, 1)$ on which $v_{k3}(x_k) = 0$ and $v_{k3}(x_k) > 0$, respectively. The unknown constant b_{k3} can also be found in the form of a regular asymptotic expansion in powers of δ_0 in the form

$$b_{k3} = \sum_{j=0}^{\infty} \delta_0^j \beta_{kj}, \ \beta_{kj} = O(1) \ for \ \delta_0 \ll 1 \ and \ j \ge 0, \tag{9.80}$$

where the coefficients β_{kj} are unknown and should be determined in the process of solution. Obviously, $|\beta_{k0}| \le 1$.

By making the substitutions

$$x_k = \tfrac{1}{2}[1 + b_{k3} + (1 - b_{k3})y], \quad t = \tfrac{1}{2}[1 + b_{k3} + (1 - b_{k3})\tau], \tag{9.81}$$

in (9.49), (9.52), and (9.53) and by using (9.58), (9.59), and (9.81), we obtain

$$\int_{-1}^{1} \frac{v'_{k3}(\tau)d\tau}{\tau - y} = \pi \frac{1 - \beta_{k0}}{2} \{p_{nk3}(y) + \frac{\delta_k}{2\delta_0} C^r_{k101}[1 + b_{k3} + (1 - b_{k3})y]\}, \tag{9.82}$$

$$v_{k3}(\pm 1) = 0,$$

$$\int_{-1}^{1} \frac{u'_{k4}(\tau)d\tau}{\tau - y} = \pi \frac{1 - \beta_{k0}}{2} \{p_{nk4}(y) - \frac{\beta_{k1}}{1 - \beta_{k0}} p_{nk3} - F^r_k - \frac{H^r_k}{4}[1 + b_{k3} \tag{9.83}$$

$$+(1 - b_{k3})y]^2 - \frac{\delta_k}{\delta_0} C^r_{k101} \beta_{k1} [\frac{\beta_{k0}}{1 - \beta_{k0}} + y]\}, \quad v_{k4}(\pm 1) = 0,$$

The unknown coefficients β_{kj} are determined by the equation $p_{nk}(b_{k3}) = 0$. Therefore, using (9.59) and substitutions from (9.81) we find that coefficients β_{kj} are determined by equations

$$p_{nkj}(-1) = 0 \; for \; j \geq 3. \tag{9.84}$$

Obviously, $p_{nk3}(y) = 0$ for $|y| < 1$. Therefore, from (9.82) we obtain [15, 16]

$$v_{k3}(y) = -\frac{1 - \beta_{k0}}{4} \frac{\delta_k}{\delta_0} C^r_{k101}[1 + \beta_{k0} + \frac{1 - \beta_{k0}}{2} y]\sqrt{1 - y^2}\theta(1 - y^2). \tag{9.85}$$

Further, to determine the expression for $p_{nk3}(y)$ we will substitute (9.85) in (9.82) and make use of the integral

$$\int_{-1}^{1} \frac{d\tau}{\sqrt{1 - \tau^2}(\tau - y)} = 0, \; |y| \leq 1, \tag{9.86}$$

$$\int_{-1}^{1} \frac{d\tau}{\sqrt{1 - \tau^2}(\tau - y)} = -\frac{\pi}{\sqrt{y^2 - 1}} sign(y), \;\; |y| \geq 1.$$

As a result of the above calculations, we will get

$$p_{nk3}(y) = \frac{\delta_k}{\delta_0} C^r_{k101} \frac{sign(y)}{\sqrt{y^2 - 1}}[-(1 + \beta_{k0})y \tag{9.87}$$

$$+(1 - \beta_{k0})(\tfrac{1}{2} - y^2)]\theta(y^2 - 1).$$

By substituting the expression for $p_{nk3}(y)$ from (9.87) into the first of (9.84), we will determine the value of the constant β_{k0}

$$\beta_{k0} = -\tfrac{1}{3}. \tag{9.88}$$

Now, by using the value of β_{k0} from (9.88) in (9.85) and (9.87) and by converting the obtained expression back to the variable x_k (see (9.81) for b_{k3} replaced by β_{k0}), we obtain

$$v_{k3}(x_k) = -\frac{\sqrt{3}\delta_k}{18\delta_0} C^r_{k101}(3x_k + 1)\sqrt{1 + 2x_k - 3x_k^2}\theta(1 + 2x_k - 3x_k^2),$$

$$p_{nk3}(x_k) = -\frac{\sqrt{3}\delta_k}{9\delta_0} C^r_{k101}(3x_k - 2)\sqrt{\frac{3x_k+1}{x_k-1}}\theta(3x_k^2 - 2x_k - 1).$$

(9.89)

Similarly, for $p_k^0 = C^r_{k001} = 0$ and $C^r_{k101} > 0$, we obtain

$$\beta_{k0} = \frac{1}{3}.$$

(9.90)

$$v_{k3}(x_k) = \frac{\sqrt{3}\delta_k}{18\delta_0} C^r_{k101}(1 - 3x_k)\sqrt{1 - 2x_k - 3x_k^2}\theta(1 - 2x_k - 3x_k^2),$$

$$p_{nk3}(x_k) = -\frac{\sqrt{3}\delta_k}{9\delta_0} C^r_{k101}(3x_k + 2)\sqrt{\frac{3x_k-1}{x_k+1}}\theta(3x_k^2 + 2x_k - 1).$$

(9.91)

For $p_k^0 = C^r_{k001} = 0$ and $C^r_{k101} \neq 0$ the signs of the values of $v_{k4}(y)$ for $|y| < 1$ do not violate the second inequality in (9.65). It can be shown that for $C^r_{k101} < 0$ the solution of (9.83) has the form [15, 16]

$$v_{k4}(y) = \frac{2}{3}[F_k^r + \frac{H_k^r}{27}(5 + 6y + 4y^2)$$

$$- \frac{\delta_k}{4\delta_0} C^r_{k101}\beta_{k1}(1 - 2y)]\sqrt{1 - y^2}.$$

(9.92)

Furthermore, by using the integral from (9.86) as well as equations (9.83) and (9.84) for $j = 4$, we obtain

$$\beta_{k1} = \frac{4}{3}[F_k^r + \frac{H_k^r}{9}]/(\frac{\delta_k}{\delta_0} C^r_{k101}).$$

(9.93)

Finally, using (9.81), (9.83), (9.92) and (9.93), we find that

$$v_{k4}(x_k) = \frac{\sqrt{3}}{9}[F_k^r + \frac{H_k^r}{9}(3x_k + 1)](3x_k + 1)$$

$$\times \sqrt{1 + 2x_k - 3x_k^2}\theta(1 + 2x_k - 3x_k^2),$$

(9.94)

$$p_{nk4}(x_k) = \frac{\sqrt{3}}{27} H_k^r(-1 - 6x_k + 9x_k^2)\sqrt{\frac{3x_k+1}{x_k-1}}\theta(3x_k^2 - 2x_k - 1).$$

Similarly, for $p_k^0 = C^r_{k001} = 0$ and $C^r_{k101} > 0$ for β_{k1}, $v_{k4}(x_k)$, and $p_{nk4}(x_k)$, we obtain (9.93) and (9.94) in which the signs of β_{k1} and x_k are changed to opposite ones.

The expressions for $u_{kj}(x_k)$ are still determined by (9.67)-(9.71).

1.6. Finally, let us consider the case when $p_k^0 = C^r_{k001} = C^r_{k101} = 0$. In this case $v_{kj}(x_k) = p_{nkj}(x_k) = 0$ for $|x_k| < 1$ and $j = 0, 1, 2, 3$. Therefore, for $v_{k4}(x_k)$ and $p_{nk4}(x_k)$ from (9.65) we get an analog of (9.66) in the form

$$p_{nk4}(x_k) = 0 \; for \; v_{k4}(x_k) > 0, \; p_{nk4}(x_k) \leq 0 \; for \; v_{k4}(x_k) = 0.$$

(9.95)

From (9.60), (9.62)-(9.65) for $h = 4$ and from (9.95) we can conclude that $v_{k4}(x_k)$ and $p_{nk4}(x_k)$ are even functions of x_k. First, let us derive the conditions when $v_{k4}(x_k) = 0$ for $\mid x_k \mid < 1$. From (9.60) we obtain (9.75) that with the help of (9.71) leads to

$$F_k^r \leq 0, \ H_k^r \leq 0 \ or \ F_k^r + H_k^r \leq 0, \ H_k^r > 0, \tag{9.96}$$

that satisfy (9.95).

Now, let us consider the case when on the interval $(-1, 1)$, occupied by the kth crack, there exists an open crack segment. Suppose this segment is $(-b_{k4}, b_{k4})$. In this case constant b_{k4} satisfies one of the two conditions: (a) $p_{nk4}(b_{k4}) = 0$, $b_{k4} < 1$ or (b) $b_{k4} = 1$. From (9.95) follows that $v_{k4}(x_k) > 0$ for $\mid x_k \mid < b_{k4}$ and in case (a) $p_{nk4}(x_k) \leq 0$ for $b_{k4} \leq \mid x_k \mid \leq 1$.

Constant b_{k4} also can be represented in the form of a regular asymptotic series in powers of δ_0. For further analysis we will assume that

$$b_{k4} = \gamma_{k0} + O(\delta_0), \ \delta_0 \ll 1. \tag{9.97}$$

Then in case (a) the solution of (9.60) for $h = 4$ and conditions $v_{k4}(\pm\gamma_{k0}) = 0$ has the form [15, 16]

$$v_{k4}(x_k) = \tfrac{1}{2}[2F_k^r + \tfrac{H_k^r}{3}(\gamma_{k0}^2 + 2x_k^2)]\sqrt{\gamma_{k0}^2 - x_k^2}\theta(\gamma_{k0}^2 - x_k^2),$$

$$p_{nk4}(x_k) = \mid x_k \mid \{[F_k^r + H_k^r(x_k^2 - \tfrac{\gamma_{k0}^2}{2})]/\sqrt{x_k^2 - \gamma_{k0}^2}\}\theta(x_k^2 - \gamma_{k0}^2). \tag{9.98}$$

To obtain (9.98) we used the fact that $p_{nk4}(x_k) = 0$ for $\mid x_k \mid \leq \gamma_{k0}$. Using the condition $p_{nk4}(\gamma_{k0}) = 0$ we find

$$\gamma_{k0}^2 = -2\tfrac{F_k^r}{H_k^r}. \tag{9.99}$$

By substituting the value of γ_{k0} from (9.99) in (9.98), we find

$$v_{k4}(x_k) = -\tfrac{H_k^r}{3}[-\tfrac{2F_k^r}{H_k^r} - x_k^2]^{3/2}\theta(-\tfrac{2F_k^r}{H_k^r} - x_k^2),$$

$$p_{nk4}(x_k) = H_k^r \mid x_k \mid [\tfrac{2F_k^r}{H_k^r} + x_k^2]^{3/2}\theta(\tfrac{2F_k^r}{H_k^r} + x_k^2). \tag{9.100}$$

From (9.99) and (9.100) as well as from the inequality $\gamma_{k0}^2 \geq 0$, we determine the conditions

$$F_k^r > 0, \ H_k^r < 0, \ 2F_k^r + H_k^r < 0 \tag{9.101}$$

that make (9.95) correct.

Now, let us consider case (b). It is easy to see that in this case the solution from (9.98) is still valid if in (9.98) $\gamma_{k0} = 1$. Therefore, from (9.98) we calculate

$$v_{k4}(x_k) = [F_k^r + \tfrac{H_k^r}{6}(1 + x_k^2)]\sqrt{1 - x_k^2}. \tag{9.102}$$

From (9.95) follows that $v_{k4}(x_k) > 0$ for $|x_k| < 1$. The latter inequality is satisfied in the following cases

$$F_k^r > 0, \ H_k^r < 0, \ 2F_k^r + H_k^r \geq 0 \ or \ H_k^r > 0, \ 6F_k^r + H_k^r \geq 0 \ or$$

$$H_k^r = 0, \ F_k^r > 0. \tag{9.103}$$

The analysis of (9.96), (9.101), and (9.103) shows that there is one more parametric region

$$F_k^r < 0, \ H_k^r > 0, \ F_k^r + H_k^r > 0, \ 6F_k^r + H_k^r \leq 0 \tag{9.104}$$

that is not covered by the above analysis. It can be shown that within the parametric region described by (9.104) only one crack configuration can be realized. Namely, on the interval $(-1, 1)$ there exist two symmetric subintervals $(-1, -a_{k4})$ and $(a_{k4}, 1)$ within which the crack is open, i.e., $v_{k4}(x_k) > 0$. By representing constant a_{k4} for $\delta_0 \ll 1$ by an asymptotic expansion and taking into account only the main term of the expansion

$$a_{k4} = \sigma_{k0} + O(\delta_0), \ \delta_0 \ll 1 \tag{9.105}$$

from (9.60) and conditions $v_{k4}(\pm 1) = v_{k4}(\pm \sigma_{k0}) = 0$ (see (9.64) and (9.95)) we obtain [15, 16]

$$v_{k4}(x_k) = \{(F_k^r + H_k^r \tfrac{1+\sigma_{k0}^2}{6})[E(\xi, \zeta) - \tfrac{E(\zeta)}{K(\zeta)} F(\xi, \zeta)]$$

$$+ \tfrac{\sqrt{\omega(x_k)}}{|x_k|}[F_k^r + \tfrac{H_k^r}{3}(\tfrac{1+\sigma_{k0}^2}{2} + x_k^2)]\}\theta(\omega(x_k)),$$

$$p_{nk4}(x_k) = \tfrac{sign(x_k^2 - \sigma_{k0}^2)}{\sqrt{-\omega(x_k)}} \{(H_k^r \tfrac{1+\sigma_{k0}^2}{2} - F_k^r)[\tfrac{E(\zeta)}{K(\zeta)} - x_k^2] \tag{9.106}$$

$$- \tfrac{H_k^r}{3}[2(1 + \sigma_{k0}^2)\tfrac{E(\zeta)}{K(\zeta)} - \sigma_{k0}^2 - 3x_k^4]\}\theta(-\omega(x_k)),$$

$$\xi = \arcsin\left[\tfrac{1}{x_k}\sqrt{\tfrac{x_k^2 - \sigma_{k0}^2}{1 - \sigma_{k0}^2}}\right], \ \zeta = \sqrt{1 - \sigma_{k0}^2},$$

$$\omega(x_k) = (1 - x_k^2)(x_k^2 - \sigma_{k0}^2),$$

where $F(\xi, \zeta)$ and $E(\xi, \zeta)$ are elliptic and $K(\zeta)$ and $E(\zeta)$ are full elliptic integrals of the first and second kind, respectively [17]. From (9.106) and the condition $p_{nk4}(\sigma_{k0}) = 0$ we obtain an equation for σ_{k0} in the form

$$\frac{E(\sqrt{1-\sigma_{k0}^2})}{K(\sqrt{1-\sigma_{k0}^2})} = \sigma_{k0}^2 \frac{3\sigma_{k0}^2 - 1 + 6F_k^r/H_k^r}{\sigma_{k0}^2 + 1 + 6F_k^r/H_k^r}, \tag{9.107}$$

the numerical solution of which is given in Fig. 9.3.

As before, the expressions for $u_{kj}(x_k)$ are determined by (9.67)-(9.71).

Finally, we can conclude that we obtained the five-term analytic asymptotic solution of the problem for multiple straight cracks in an elastic plane, which takes into account possible contact interaction of crack faces.

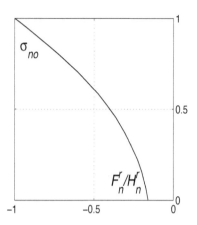

FIGURE 9.3
Graph of a numerical solution for σ_{k0} as a function of F_k^r/H_k^r.

9.3.2 Analysis of the Stress Intensity Factors for Multiple Cracks in an Elastic Plane

In this subsection we will derive very efficient asymptotic (analytical) representations for the stress intensity factors $k_{1k}^{\pm}$ and $k_{2k}^{\pm}$ for "small" cracks and analyze their behavior. The information obtained from this analysis will be used in Section 9.4 for explanation and prediction of the direction of fatigue crack propagation. To derive the expressions for the stress intensity factors, we will use (9.55) and the five-term asymptotic approximations obtained above.

1. Using (9.59), (9.67)-(9.71) from (9.55) we obtain

$$k_{1k}^{\pm} = -p_k^0 - \delta_0^2 C_{k001}^r \mp \tfrac{1}{2}\delta_k\delta_0^2 C_{k101}^r + \delta_0^4(F_k^r + \tfrac{H_k^r}{2}) + \ldots,$$

$$(9.108)$$

$$p_k^0 < 0; \ k_{1k}^{\pm} = 0, \ p_k^0 > 0.$$

For $p_k^0 = 0$ using (9.59), (9.67), (9.68), (9.70), (9.71), and (9.78) from (9.55), we determine

$$k_{1k}^{\pm} = -\delta_0^2 C_{k001}^r \mp \tfrac{1}{2}\delta_k\delta_0^2 C_{k101}^r + \delta_0^4(F_k^r + \tfrac{H_k^r}{2}) + \ldots,$$

$$p_k^0 = 0, \ C_{k001}^r < 0;$$

$$(9.109)$$

$$k_{1k}^{\pm} = 0, \ p_k^0 = 0, \ C_{k001}^r > 0.$$

For $p_k^0 = C_{k001}^r = 0$ using (9.59), (9.67)-(9.69), (9.89), (9.90), and (9.94) from (9.55) we find

$$k_{1k}^+ = k_{1k}, \ k_{1k}^- = 0, \ p_k^0 = C_{k001}^r = 0, \ C_{k101}^r < 0,$$

$$k_{1k}^+ = 0, \ k_{1k}^- = k_{1k}, \ p_k^0 = C_{k001}^r = 0, \ C_{k101}^r > 0, \tag{9.110}$$

$$k_{1k} = \frac{2\sqrt{6}}{9}\delta_k\delta_0^2 \mid C_{k101}^r \mid +\frac{4\sqrt{6}}{9}\delta_0^4(F_k^r + \frac{4H_k^r}{9}) + \ldots.$$

Similarly, for $p_k^0 = C_{k001}^r = C_{k101}^r = 0$ using (9.59), (9.67)-(9.70), (9.100), (9.102), and (9.106) from (9.55), we obtain

$$k_{1k}^{\pm} = O(\delta_0^5) \ under \ conditions \ of \ (9.96) \ or \ (9.101),$$

$$k_{1k}^{\pm} = \delta_0^4(F_k^r + \frac{H_k^r}{2}) + \ldots \ under \ conditions \ of \ (9.103),$$

$$k_{1k}^{\pm} = \frac{\delta_0^4}{\sqrt{1-\sigma_{k0}^2}}\left\{F_k^r\left[1 - \frac{E(\sqrt{1-\sigma_{k0}^2})}{K(\sqrt{1-\sigma_{k0}^2})}\right] + \frac{H_k^r}{6}\left[3 - \sigma_{k0}^2\right.\right. \tag{9.111}$$

$$\left.\left.-(1+\sigma_{k0}^2)\frac{E(\sqrt{1-\sigma_{k0}^2})}{K(\sqrt{1-\sigma_{k0}^2})}\right]\right\} + \ldots \ under \ conditions \ of \ (9.104).$$

Finally, using (9.67)-(9.71) from (9.55) we find

$$k_{2k}^{\pm} = -\tau_k^0 - \delta_0^2 C_{k001}^i \mp \frac{1}{2}\delta_k\delta_0^2 C_{k101}^i + \delta_0^4(F_k^i + \frac{H_k^i}{2}) + \ldots \tag{9.112}$$

To conclude the derivation of the above expressions for $k_{1k}^{\pm}$ and $k_{2k}^{\pm}$, let us specify the expressions for the constants C_{kjlm} that are calculated based on (9.63) and (9.64) as well as on the expressions for $v_{kh}(x_k)$ and $u_{kh}(x_k)$ derived above. These expressions have the following form

$$C_{k001} = -\frac{1}{2}\sum_{m=1}^{N}(\frac{\delta_m}{\delta_0})^2[p_m^0\theta(-p_m^0)A_{km01} - \tau_m^0 B_{km01}]. \tag{9.113}$$

Moreover, constant C_{k101} can be determined from (9.113) by replacing A_{km01} and B_{km01} by A_{km11} and B_{km11}, respectively.

2. Now, let us analyze the expressions for the stress intensity factors. Equations (9.108)-(9.112) show that with the precision of $O(\delta_0^2)$, $\delta_0 \ll 1$, each crack behaves like a single one in the corresponding stress field. Starting with the order of magnitude $O(\delta_0^2)$ and higher cracks feel the presence of other cracks, i.e., they begin to interact. Moreover, for $p_k^0 = 0$ and various locations of other cracks the configuration of the kth crack can be very complex. It means that the crack may have multiple segments where the crack faces are in contact with each other.

For $\delta_0 \leq 0.2$, the discrepancy between the values of the stress intensity factors $k_{1k}^{\pm}$ obtained by a direct numerical solution of the problem and by using

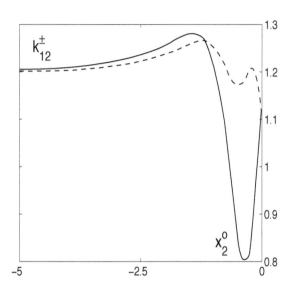

FIGURE 9.4
The dependence of the normal stress intensity factors $k_{12}^{\pm}$ on the coordinate x_2^0 for two parallel cracks with $p^0 = 1$, $q^0 = 0$, $z_1^0 = (0,0)$, $\alpha_k = 0$, $\delta_k = 0.4$, $y_2^0 = -1$ $(k = 1, 2)$: k_{12}^+ - solid curve, k_{12}^- - dashed curve.

the asymptotic formulas from (9.108)-(9.112) is no greater than 2% while the discrepancy between similarly obtained values of the stress intensity factors $k_{2k}^{\pm}$ is no greater than 5% (see Subsection 9.3.4). The asymptotic solutions of the problem can be improved by using the Páde approximations [18]. The behavior of the stress intensity factors $k_{1k}^{\pm}$ and $k_{2k}^{\pm}$ as functions of x_2^0 (the solid curve corresponds to sign + while the dashed curve corresponds to sign $-$) for two parallel cracks with $p^0 = 1$, $q^0 = 0$, $z_1^0 = (0,0)$, $\alpha_k = 0$, $\delta_k = 0.4$, $y_2^0 = -1$ $(k = 1, 2)$ is shown in Fig. 9.4 and 9.5.

The analysis of formulas (9.110) and (9.113) leads to the conclusion that the necessary condition for crack faces to be in contact with each other is $\eta_0 \leq 0$. Let us consider the border case when $\eta_0 = 0$, $\alpha_1 = 0$, $\alpha_2 = \pi/2$ while the other parameters correspond to the ones used in Fig. 9.4 and 9.5. In this case the behavior of the stress intensity factors $k_{1k}^{\pm}$ and $k_{2k}^{\pm}$ is different from the one described above. For fixed z_1^0, y_2^0 and all x_2^0 the normal stress intensity factors $k_{11}^{\pm}$ at the tips of the upper crack is greater than 1 and its maximum equal to 1.014 is reached at $x_2^0 \approx -0.4$ (see Fig. 9.6). Under these conditions the values of k_{11}^+ and k_{11}^- as well as the values of k_{21}^+ and k_{21}^- are practically equal, i.e., they have from four to six equal significant digits after the decimal point, respectively. For the given above parameters, the graph of

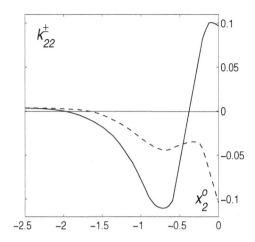

FIGURE 9.5
The dependence of the normal stress intensity factors $k_{22}^{\pm}$ on the coordinate x_2^0 for two parallel cracks with $p^0 = 1$, $q^0 = 0$, $z_1^0 = (0,0)$, $\alpha_k = 0$, $\delta_k = 0.4$, $y_2^0 = -1$ ($k = 1, 2$): k_{22}^+ – solid curve, k_{22}^- – dashed curve.

$k_{21}^{\pm}(x_2^0)$ is presented in Fig. 9.7 by the dashed curve.

The analysis of $k_{12}^{\pm}$ for the crack with $\alpha_2 = \pi/2$ (perpendicular to the tensile stress $p^0 = 1$) can be made based on Figs. 9.8 and 9.9. Far away and in the vicinity of the center of the upper crack, i.e., for $x_2^0 \leq -2.25$ and $-0.45 \leq x_2^0 \leq 0$ tensile stresses are created, i.e., $k_{12}^{\pm} > 0$. Between these two zones of tensile stresses, a zone with compressive stresses is located. Moreover, as the lower crack ($y_2^0 = -1$) approaches along the x_2^0-axis from $-\infty$ to 0 (see Fig. 9.8), the stress intensity factors $k_{12}^{\pm}$ increase from zero, reach their maximum, and after that decrease to zero.[†] These zones are followed by the zone with compressive stresses in which $k_{12}^{\pm} = 0$. Within the central zone with tensile stresses, the stress intensity factors $k_{12}^{\pm}(x_2^0)$ monotonically increase from zero to their maximum values that are reached at $x_2^0 = 0$ (see Fig. 9.9). In the vicinity of the points where $k_{12}^{\pm}(x_2^0)$ vanishes the faces of the lower crack are partially in contact with each other. The graphs of $k_{22}^{\pm}(x_2^0)$ for the considered set of the problem parameters are given in Fig. 9.7.

The curves that correspond to the case of two cracks in Figs. 9.7-9.9 are labeled with number 1.

[†] In Figures 9.7, 9.8, and 9.9 the solid curves correspond to $k_{22}^+(x_2^0)$ and $k_{12}^+(x_2^0)$ while the dashed curves correspond to $k_{22}^-(x_2^0)$ and $k_{12}^-(x_2^0)$.

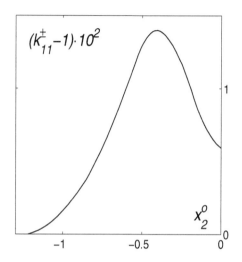

FIGURE 9.6
The dependence of the normal stress intensity factors $k_{11}^{\pm}$ on the coordinate x_2^0 for $\eta_0 = 0$, $\alpha_1 = 0$, $\alpha_2 = \pi/2$ while other parameters are the same as in Fig. 9.4.

Some results calculated for three cracks with $y_2^0 = -2$, $y_2^0 = 0$, and $p^0 = 1$, $q^0 = 0$, $z_1^0 = (0,1)$, $z_3^0 = (0,-1)$, $\alpha_1 = \alpha_3 = 0$, $\alpha_2 = \pi/2$, $\delta_1 = \delta_3 = 0.4$, $\delta_2 = 0.2$ are given in Figs. 9.7-9.9 (the curves that correspond to $y_2^0 = -2$ are labeled with number 2 while the curves that correspond to $y_2^0 = 0$ are labeled with number 3). The behavior of the stress intensity factors $k_{1k}^{\pm}(x_2^0)$, $k = 1,3$ is similar to the one presented in Fig. 9.6 but differs by a more narrow range of variation from 0.9526 to 0.9562. The value 0.9526 correspond to the case of two cracks when the crack with $\alpha_2 = \pi/2$ is absent. In the case of symmetrically located cracks ($y_2^0 = 0$), we have $k_{12}^+ \equiv k_{12}^-$.

The behavior of the stress intensity factors $k_{22}^{\pm}(x_2^0)$ for $y_2^0 = -2$ is shown in Fig. 9.7 and it is similar to the behavior of $k_{22}^{\pm}(x_2^0)$ for the case of two cracks. The behavior of $k_{22}^-(x_2^0) = -k_{22}^+(x_2^0)$ for $y_2^0 = 0$ that is presented in Fig. 9.7, is different from the ones considered above not only quantitatively but also qualitatively.

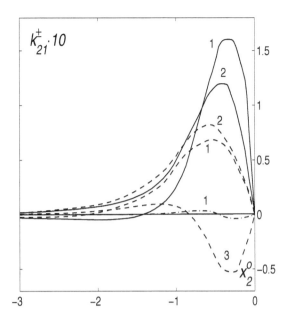

FIGURE 9.7
The dependence of the shear stress intensity factors $k_{2n}^{\pm}$ on the coordinate x_2^0 for two cracks (curves marked with 1) for the same conditions as in Fig. 9.6 and for three cracks (curves marked with 2) for $y_2^0 = -2$, and $p^0 = 1$, $q^0 = 0$, $z_1^0 = (0, 1)$, $z_3^0 = (0, -1)$, $\alpha_1 = \alpha_3 = 0$, $\alpha_2 = \pi/2$, $\delta_1 = \delta_3 = 0.4$, $\delta_2 = 0.2$: $k_{21}^{\pm}$ for the case of two cracks - point-dashed curve. Subscripts correspond to curve numbers.

9.3.3 Perturbation Solution for Multiple Cracks in an Elastic Half-Plane

The main purpose of this subsection is to develop an efficient asymptotic (analytical) method for solution of the problem for "small" cracks in an elastic half-plane. These solutions will be used in Subsection 9.3.4 to obtain the asymptotic expressions for the stress intensity factors $k_{1k}^{\pm}$ and $k_{2k}^{\pm}$ for "small" cracks and to analyze their behavior.

In practically important cases the elastic solid boundaries are loaded with distributed stresses. Such cases are not considered in [9, 10, 11] where the analysis is focused on consideration of point forces. Moreover, it is important to analyze the general situation when crack faces may be in partial or full contact. The solution of such a problem will be used for fatigue failure modeling.

Let us consider subsurface cracks. Surface cracks are considered in Section

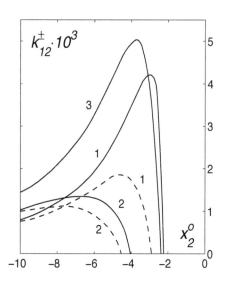

FIGURE 9.8
The dependence of the shear stress intensity factors $k_{12}^{\pm}$ on the coordinate x_2^0 within the range $x_2^0 \leq -2.25$ for the crack with $\alpha_2 = \pi/2$. Curves marked with 1 (for two cracks) and with 2 and 3 (for three cracks) correspond to $y_2^0 = -1$, $y_2^0 = -2$, and $y_2^0 = 0$, respectively.

9.6. Using the dimensionless variables from (9.47) and

$$\{p', \tau', q^{0'}\} = \{p, \tau, q^0\}/\tilde{q}, \quad \{a, b\} = \{x_i, x_e\}/\tilde{b}, \tag{9.114}$$

equations (9.37), (9.40)- (9.43) can be reduced to (see (9.49))

$$\int_{-1}^{1} \frac{v_k'(t)dt}{t-x_k} + \sum_{m=1}^{N} \delta_m \int_{-1}^{1} [v_m'(t)A_{km}^r(t, x_k) - u_m'(t)B_{km}^r(t, x_k)]dt$$

$$= \pi p_{nk}(x_k) + \pi p_k^0(x_k),$$

$$\int_{-1}^{1} \frac{u_k'(t)dt}{t-x_k} + \sum_{m=1}^{N} \delta_m \int_{-1}^{1} [v_m'(t)A_{km}^i(t, x_k) - u_m'(t)B_{km}^i(t, x_k)]dt \tag{9.115}$$

$$= \pi \tau_k^0(x_k),$$

$$p_k^0(x_k) - i\tau_k^0(x_k) = -\frac{1}{\pi} \int_{a}^{b} [p(t)\overline{D}_k(t, x_k) + \tau(t)\overline{G}_k(t, x_k)]dt + P_k^\infty,$$

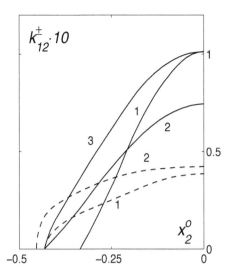

FIGURE 9.9
The dependence of the shear stress intensity factors $k_{12}^{\pm}$ on the coordinate x_2^0 within the range $-0.5 < x_2^0 \leq 0$ for the crack with $\alpha_2 = \pi/2$. Curves marked with 1 (for two cracks) and with 2 and 3 (for three cracks) correspond to $y_2^0 = -1$, $y_2^0 = -2$, and $y_2^0 = 0$, respectively.

where

$$P_k^\infty = p_k^\infty - i\tau_k^\infty = -\tfrac{1}{2}q^0(1 - e^{-2i\alpha_k}),$$

$$v_k(\pm 1) = u_k(\pm 1) = 0 \ for \ |y_k^0| > \delta_k \ |\sin\alpha_k|, \ k = 1, \ldots, N,$$

$$A_{nk} = \overline{R_{nk} + S_{nk}}, \ B_{nk} = -i(\overline{R_{nk} - S_{nk}}), \tag{9.116}$$

$$(A_{nk}^r, B_{nk}^r, D_k^r, G_k^r) = Re(A_{nk}, B_{nk}, \overline{D}_k, \overline{G}_k),$$

$$(A_{nk}^i, B_{nk}^i, D_k^i, G_k^i) = Im(A_{nk}, B_{nk}, \overline{D}_k, \overline{G}_k),$$

$$D_k(t, x_k) = \tfrac{i}{2}\left[-\frac{1}{t-X_k} + \frac{1}{t-\overline{X}_k} - \frac{e^{-2i\alpha_k}(\overline{X}_k - X_k)}{(t-\overline{X}_k)^2} \right],$$

$$G_k(t, x_k) = \tfrac{1}{2}\left[\frac{1}{t-X_k} + \frac{1-e^{-2i\alpha_k}}{t-\overline{X}_k} - \frac{e^{-2i\alpha_k}(t-X_k)}{(t-\overline{X}_k)^2} \right], \tag{9.117}$$

$$T_k = \delta_k t e^{i\alpha_k} + z_k^0, \ X_n = \delta_n x_n e^{i\alpha_n} + z_n^0, \ k, n = 1, \ldots, N, \tag{9.118}$$

and kernels R_{km} and S_{km}, which depend on the kernels K_{km} and L_{km} in dimensionless variables are also described by (9.37) and (9.25), respectively.

For simplicity primes at dimensionless variables are omitted. In equations (9.47) and (9.114) the characteristic values $\tilde{q}$ and $\tilde{b}$ are related to the specific distributions of the pressure $p(x^0)$ and the frictional stress $\tau(x^0)$.

1. Let us assume that all cracks are small in comparison with the size $\tilde{b}$ and, simultaneously, cracks are small in comparison to the distances between them, i.e., estimates (see (9.56))

$$z_n^0 - z_k^0 \gg \delta_0 \ for \ n \neq k, \ n, k = 1, \ldots, N \tag{9.119}$$

are satisfied for $\delta_0 = \max_{1 \leq k \leq N} \delta_k \ll 1$. The assumption that all cracks are subsurface and much smaller in size than their distances to the half-plane surface

$$z_k^0 - \overline{z}_k^0 \gg \delta_0 \ for \ k = 1, \ldots, N \tag{9.120}$$

is crucial for simplifying the problem solution. Moreover, the latter assumption with the precision of $O(\delta_0^2)$, $\delta_0 \ll 1$ provides the conditions for considering each crack as a single crack in an elastic plane (see Subsection 9.3.1) the faces of which are loaded by certain stresses related to contact pressure $p(x^0)$, contact frictional stress $\tau(x^0)$, and the residual stress q^0.

Obviously, the above assumptions lead to the following inequalities:

$$z_n^0 - \overline{z}_k^0 \gg \delta_0 \ for \ n, k = 1, \ldots, N,$$

$$T_k - X_n \gg \delta_0, \ \overline{T}_k - X_n \gg \delta_0,$$

$$T_k - \overline{T}_k \gg \delta_0, x^0 - X_k \gg \delta_0, \ x^0 - \overline{X}_k \gg \delta_0 \tag{9.121}$$

$$for \ |t| \leq 1, \ |x_k| \leq 1, \ |x_n| \leq 1, \ |x^0| < \infty,$$

$$n \neq k, \ n, k = 1, \ldots, N.$$

Based on (9.37), (9.25), (9.116)-(9.118) and using (9.121) for $\delta_0 \ll 1$, we obtain that all kernels are regular functions of t, x_n, and x_k, and they can be represented by power series in $\delta_k \ll 1$ and $\delta_n \ll 1$ in the form

$$\{A_{km}(t, x_k), B_{km}(t, x_k)\}$$

$$= \sum_{j+n=0; j, n \geq 0}^{\infty} (\delta_k x_k)^j (\delta_m t)^n \{A_{kmjn}, B_{kmjn}\}, \tag{9.122}$$

$$\{D_k(t, x_k), G_k(t, x_k)\} = \sum_{j=0}^{\infty} (\delta_k x_k)^j \{D_{kj}, G_{kj}\}. \tag{9.123}$$

In (9.122) and (9.123) the values of A_{kmjn} and B_{kmjn} are independent of δ_k, δ_m, x_k, and t while the values of $D_{kj}(t)$ and $G_{kj}(t)$ are independent of δ_k and x_k. The values of A_{kmjn} and B_{kmjn} are certain functions of constants

α_k, α_m, x_k^0, y_k^0, x_m^0, and y_m^0 while the values of $D_{kj}(t)$ and $G_{kj}(t)$ are certain functions of α_k, x_k^0, and y_k^0.

Let us assume that functions $p(x^0)$ and $\tau(x^0)$ can also be represented by power series in $\delta_0 \ll 1$, i.e.,

$$\{p(x^0), \tau(x^0)\} = \sum_{j=0}^{\infty} \delta_0^j \{q_j(x^0), \tau(x^0)\}. \tag{9.124}$$

Then from (9.115), (9.123), and (9.124) we obtain

$$p_k^0(x_k) - i\tau_k^0(x_k) = -\sum_{h=0}^{\infty} \delta_0^h \sum_{j,l \geq 0}^{j+l=h} (\tfrac{\delta_k}{\delta_0} x_k)^j Y_{kjl},$$

$$P_k^{\infty} = p_k^{\infty} - i\tau_k^{\infty} = -\tfrac{1}{2} q^0 (1 - e^{-2i\alpha_k}), \tag{9.125}$$

$$Y_{kjl} = \tfrac{1}{\pi} \int_a^b [q_l(t)\overline{D}_{kj}(t) + \tau_l(t)\overline{G}_{kj}(t)] dt.$$

The inner summation in (9.125) is done with respect to all integer j and l that satisfy the inequalities $j \geq 0$, $l \geq 0$, $j + l \leq h$.

The solution of the problem we will try to find using regular perturbation methods in the form of asymptotic series in powers of $\delta_0 \ll 1$

$$\{v_k, u_k, p_{nk}\} = \sum_{j=0}^{\infty} \delta_0^j \{v_{kj}, u_{kj}, p_{nkj}\}. \tag{9.126}$$

Using (9.122)-(9.126) and equating the terms with the same powers of $\delta_0 \ll 1$ in (9.115), we obtain the following series of equations

$$\int_{-1}^{1} \frac{v_{kh}'(t)dt}{t-x_k} = \pi p_{nkh}(x_k) + \delta_{h0}\pi p_k^0$$

$$+ \varepsilon_h \pi \sum_{j,l \geq 0; m \geq 1}^{j+l+m=h} (m-1)(\tfrac{\delta_k}{\delta_0} x_k)^j C_{kjlm-1}^r - \pi \sum_{j,l \geq 0}^{j+l=h} (\tfrac{\delta_k}{\delta_0} x_k)^j Y_{kjl}^r, \tag{9.127}$$

$$\int_{-1}^{1} \frac{u_{kh}'(t)dt}{t-x_k} = \delta_{h0}\pi \tau_k^0 + \varepsilon_h \pi \sum_{j,l \geq 0; m \geq 1}^{j+l+m=h} (m-1)(\tfrac{\delta_k}{\delta_0} x_k)^j C_{kjlm-1}^i \tag{9.128}$$

$$- \pi \sum_{j,l \geq 0}^{j+l=h} (\tfrac{\delta_k}{\delta_0} x_k)^j Y_{kjl}^i,$$

$$v_{kh}(\pm 1) = 0, \quad u_{kh}(\pm 1) = 0, \quad h = 0, 1, \ldots, \quad k = 1, \ldots, N, \tag{9.129}$$

where constants C_{kjlm} and Y_{kjl} are determined by (9.58), (9.62), (9.63), and (9.125), respectively. The superscripts r and i correspond to the real and imaginary parts of the quantity at hand.

It is obvious that the structure of equations (9.127)-(9.129) is similar to the one of equations (9.60), (9.61), and (9.64) for the case of multiple cracks in an elastic plane. Let us consider a special case of the problem when

$$Y_{k01} = Y_{k11} = 0, \tag{9.130}$$

and the problem analysis is somewhat simpler. In this case the analysis coincides with the one from Subsection 9.3.1 and the solution of the problem can be obtained from the solution of the problem from Subsection 9.3.1 by simple substitutions. Therefore, we will present only the final solutions without considering the details of the derivation. Let us introduce the following notations

$$D_{k001} = p_k^\infty + i\tau_k^\infty - Y_{k00}, \ D_{k101} = -Y_{k10},$$

$$F_k^0 = -C_{k001} + Y_{k02}, \ H_k^0 = (\tfrac{\delta_k}{\delta_0})Y_{k20}. \tag{9.131}$$

Based on the results of Subsection 9.3.1 and with the use of (9.130) and (9.131), it can be shown that the tree-term asymptotic solution of the problem under the consideration has the form: for $D_{k001}^r \neq 0$

$$v_{k0}(x_k) = -D_{k001}^r \mu_v \sqrt{1 - x_k^2}, \ p_{nk0}(x_k) = -D_{k001}^r \mu_p,$$

$$v_{k1}(x_k) = -\tfrac{\delta_k}{2\delta_0} D_{k101}^r \mu_v x_k \sqrt{1 - x_k^2}, \ p_{nk1}(x_k) = -\tfrac{\delta_k}{\delta_0} D_{k101}^r \mu_p x_k,$$

$$v_{k2}(x_k) = [F_k^{0r} + \tfrac{H_k^{0r}}{6}(1 + 2x_k^2)]\mu_v \sqrt{1 - x_k^2}, \tag{9.132}$$

$$p_{nk2}(x_k) = [F_k^{0r} + H_k^{0r} x_k^2]\mu_p,$$

$$\mu_v = \theta(-D_{k001}^r), \ \mu_p = \theta(D_{k001}^r);$$

for $D_{k001}^r = 0$, $D_{k101}^r \neq 0$

$$v_{k0}(x_k) = p_{nk0}(x_k) = 0,$$

$$v_{k1}(x_k) = \tfrac{\sqrt{3}\delta_k}{18\delta_0} \mid D_{k101}^r \mid (1 - 3z)\sqrt{1 - 2z - 3z^2}\theta(1 - 2z - 3z^2),$$

$$p_{nk1}(x_k) = -\tfrac{\sqrt{3}\delta_k}{9\delta_0} \mid D_{k101}^r \mid (3z + 2)\sqrt{\tfrac{3z-1}{z+1}}\theta(3z^2 + 2z - 1),$$

$$v_{k2}(x_k) = \tfrac{\sqrt{3}}{9}[F_k^{0r} + \tfrac{H_k^{0r}}{9}(1 - 3z)](1 - 3z) \tag{9.133}$$

$$\times \sqrt{1 - 2z - 3z^2}\theta(1 - 2z - 3z^2),$$

$$p_{nk2}(x_k) = \tfrac{\sqrt{3}}{27} H_k^{0r}(-1 + 6z + 9z^2)\sqrt{\tfrac{3z-1}{z+1}}\theta(3z^2 + 2z - 1),$$

$$z = x_k sign(D_{k101}^r);$$

for $D_{k001}^r = 0$, $D_{k101}^r = 0$

$$v_{k0}(x_k) = v_{k1}(x_k) = p_{nk0}(x_k) = p_{nk1}(x_k) = 0,$$

$$v_{k2}(x_k) = 0, \ p_{nk2}(x_k) = F_k^{0r} + H_k^{0r} x_k^2$$

for $F_k^{0r} \leq 0$, $H_k^{0r} \leq 0$ *or* $F_k^{0r} + H_k^{0r} \leq 0$, $H_k^{0r} > 0$;

$$v_{k2}(x_k) = -\frac{H_K^{0r}}{3}(-z)^{3/2}\theta(-z),$$

$$p_{nk2}(x_k) = H_k^{0r} \mid x_k \mid z^{1/2}\theta(-z),$$

$$z = x_k^2 + \frac{2F_k^{0r}}{H_k^{0r}} \ for \ F_k^{0r} > 0, \ H_k^{0r} < 0, \ 2F_k^{0r} + H_k^{0r} < 0; \qquad (9.134)$$

$$v_{k2}(x_k) = [F_k^{0r} + \frac{H_k^{0r}}{6}(1 + 2x_k^2)]\sqrt{1 - x_k^2}, \ p_{nk2}(x_k) = 0$$

for $F_k^{0r} > 0$, $H_k^{0r} < 0$, $2F_k^{0r} + H_k^{0r} \geq 0$ *or*

$$H_k^{0r} > 0, \ 6F_k^{0r} + H_k^{0r} > 0 \ or \ F_k^{0r} > 0, \ H_k^{0r} = 0;$$

$$v_{k2}(x_k) = \{-(F_k^{0r} + H_k^{0r}\frac{1+\sigma_{k0}^2}{6})[E(\xi, \zeta) - \frac{E(\zeta)}{K(\zeta)}F(\xi, \zeta)]$$

$$+\frac{\sqrt{\omega(x_k)}}{|x_k|}[F_k^{0r} + \frac{H_k^{0r}}{3}(\frac{1+\sigma_{k0}^2}{2} + x_k^2)]\}\theta(\omega(x_k)),$$

$$p_{nk2}(x_k) = \frac{sign(x_k^2 - \sigma_{k0}^2)}{\sqrt{-\omega(x_k)}}\{(H_k^{0r}\frac{1+\sigma_{k0}^2}{2} - F_k^{0r})[\frac{E(\zeta)}{K(\zeta)} - x_k^2]$$

$$-\frac{H_k^{0r}}{3}[2(1 + \sigma_{k0}^2)\frac{E(\zeta)}{K(\zeta)} - \sigma_{k0}^2 - 3x_k^4]\}\theta(-\omega(x_k)),$$

$$\xi = \arcsin\left[\frac{1}{x_k}\sqrt{\frac{x_k^2 - \sigma_{k0}^2}{1 - \sigma_{k0}^2}}\right], \ \zeta = \sqrt{1 - \sigma_{k0}^2}, \qquad (9.135)$$

$$\omega(x_k) = (1 - x_k^2)(x_k^2 - \sigma_{k0}^2)$$

for $F_k^{0r} < 0$, $H_k^{0r} > 0$, $F_k^{0r} + H_k^{0r} > 0$, $6F_k^{0r} + H_k^{0r} \leq 0$;

$$u_{k0}(x_k) = -D_{k001}^i\sqrt{1 - x_k^2},$$

$$u_{k1}(x_k) = -\frac{\delta_k}{2\delta_0}D_{k101}^i x_k\sqrt{1 - x_k^2},$$

$$u_{k2}(x_k) = [F_k^{0r} + \frac{H_k^{0r}}{6}(1 + 2x_k^2)]\sqrt{1 - x_k^2},$$

where the constant σ_{k0} satisfies (9.107).

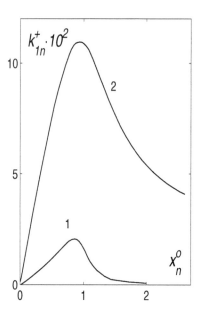

FIGURE 9.10
The dependence of the normal stress intensity factor k_{1n}^+ on the coordinate x_n^0 for the case of the boundary of half-plane loaded with pure tangential stress $\tau(x^0) = -\lambda\sqrt{1 - (x^0)^2}$, $\lambda = 0.1$, $p(x^0) = q^0 = 0$: $\alpha_n = 0$ - curve marked with 1, $\alpha = \pi/2$ - curve marked with 2.

9.3.4 Analysis of the Stress Intensity Factors for Multiple Cracks in an Elastic Half-Plane

In this section we will analyze numerically the asymptotic representations for the stress intensity factors $k_{1k}^\pm$ and $k_{2k}^\pm$ for "small" cracks that follow from (9.55), (9.125), (9.131)-(9.135). The comparison of the asymptotic and numerical results will be presented in Section 9.6. The results of this analytical analysis will serve as a part of the foundation for the development of 2D and 3D statistical models of contact fatigue (see Sections 9.7 and 9.10) and as a part of the basis for understanding that under normal conditions the contact fatigue initiation site is located beneath the surface (see Kudish [19, 20]). Moreover, based on these results the usual difference in fatigue lives of drivers and followers involved in sliding motion will be explained (see Subsection 9.6.6 and Kudish [19, 20]).

It follows from the obtained tree-term solution that pressure $p(x^0)$ and contact frictional stress $\tau(x^0)$ applied to the half-plane boundary as well as the residual stress q^0 applied at infinity determine the problem solution starting

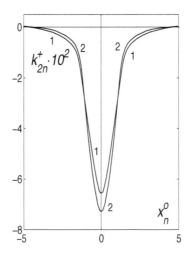

FIGURE 9.11
The dependence of the shear stress intensity factor k_{2n}^+ on the coordinate x_n^0
for the case of the boundary of half-plane loaded with pure tangential stress
$\tau(x^0) = -\lambda\sqrt{1 - (x^0)^2}$, $\lambda = 0.1$, $p(x^0) = q^0 = 0$: $\alpha_n = 0$ - curve marked with
1, $\alpha = \pi/2$ - curve marked with 2.

with the zero term (i.e., the term proportional to δ^0). Interaction between
cracks takes place starting with terms proportional to δ_0^2. The crack faces
may be fully or partially open. The specific crack behavior depends on its
relative location with respect to the loaded "contact" region as well as on the
stress level.

Let us consider several examples for some specific stresses applied to the
half-plane boundary and crack orientations and locations. The following parameters $a = -1$, $b = 1$, $\delta_0 = 0.1$ will be kept constant.

Example 1. Let us assume that the half-plane boundary is subjected to only
frictional stress $\tau(x^0) = -\lambda\sqrt{1 - (x^0)^2}$, $\lambda = 0.1$, $p(x^0) = q^0 = 0$. Here λ is
the friction coefficient. In this case the stress intensity factors $k_{1n}^\pm$ and $k_{2n}^\pm$ are
proportional to λ.

The graphs of the stress intensity factors k_{1n}^+ and k_{2n}^+ for $y^0 = -0.2$ and
two orientation angles $\alpha_n = 0$ (curves marked with 1) and $\alpha_n = \pi/2$ (curves
marked with 2) are presented in Fig. 9.10 and 9.11, respectively. From Fig.
9.10 follows that for $x_n^0 < 0$ the crack faces are in contact with each other.
For $\alpha_n = \pi/2$ at the crack tip $x_n^0 = 1$ the stress intensity factor k_{1n}^+ (curve 2)
is significantly higher than the one for $\alpha_n = 0$ (curve 1). This is mostly due to
the fact that for $\alpha_n = \pi/2$ the crack tip is closer to the half-plane boundary
than the one for $\alpha_n = 0$. Samples of the behavior of the stress intensity

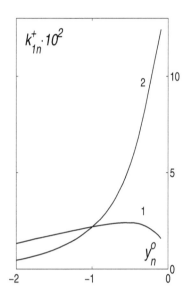

FIGURE 9.12
The dependence of the normal stress intensity factor k_{1n}^+ on the coordinate y_n^0 for the case of the boundary of half-plane loaded with pure tangential stress $\tau(x^0) = -\lambda\sqrt{1-(x^0)^2}$, $\lambda = 0.1$, $p(x^0) = q^0 = 0$: $\alpha_n = 0$ - curve marked with 1, $\alpha = \pi/2$ - curve marked with 2.

factors k_{1n}^+ and k_{2n}^+ versus the crack center depth y_n^0 beneath the boundary of the half-plane obtained for $x_n^0 = 0.9$ and $\alpha_n = 0$ (curves 1) and $\alpha_n = \pi/2$ (curves 2) are presented in Figs. 9.12 and 9.13, respectively. It is interesting to mention some features of these solutions. The value of $k_{1n}^+(y_n^0)$ for $\alpha_n = \pi/2$ is a monotonically decreasing function of depth y_n^0, and it vanishes as depth y_n^0 increases. The stress intensity factor $k_{1n}^+(y_n^0)$ for $\alpha_n = 0$ also vanishes at infinity after reaching its maximum at $y_n^0 \approx -0.5$. The numerical calculations show that $k_{1n}^+(y_n^0)$ does not vanish at any finite depth y_n^0. This situation is quite different from the case of the presence of pressure $p(x^0)$ (see below).

Example 2. Let us assume that the boundary of the elastic half-plane is loaded with pressure $p(x^0) = 0.5/\sqrt{1-(x^0)^2}$ and frictional stress $\tau(x^0) = -\lambda p(x^0)$, where λ is the friction coefficient. The dependence of the stress intensity factors $k_{1n}^\pm$ and $k_{2n}^\pm$ on x_n^0 and y_n^0 obtained for $\lambda = 0.1$ and $\lambda = 0.2$ is presented in Figs. 9.14-9.18. The curves that correspond to $q^0 = 0$ and $\lambda = 0.1$ are labeled with number 1 while the ones that correspond to $q^0 = 0$ and $\lambda = 0.2$ are labeled with number 2. In Figs. 9.14 and 9.15, the graphs of functions k_{1n}^+ and k_{2n}^+ are given for $y_n^0 = -0.2$ and $\alpha_n = \pi/2$. In Fig. 9.15

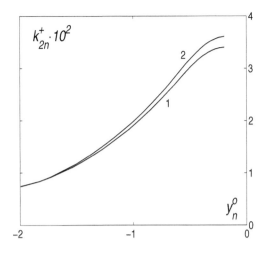

FIGURE 9.13
The dependence of the shear stress intensity factor k_{2n}^+ on the coordinate y_n^0 for the case of the boundary of half-plane loaded with pure tangential stress $\tau(x^0) = -\lambda\sqrt{1-(x^0)^2}$, $\lambda = 0.1$, $p(x^0) = q^0 = 0$: $\alpha_n = 0$ - curve marked with 1, $\alpha = \pi/2$ - curve marked with 2.

the curves that correspond to the residual stresses $q^0 = -0.015708$ and $q^0 = 0.0314$ are labeled with numbers 3 and 4, respectively. Obviously, the shear stress intensity factor $k_{2n}^+(x^0)$ reaches its extremal values in the immediate vicinity of $x_n^0 = a$ and $x_n^0 = c$. As it follows from Fig. 9.15 the normal stress intensity factor $k_{1n}^+(x_n^0)$ strongly depends on the values of λ (i.e., on the level of frictional stress) and the residual stress q^0. Namely, the normal stress intensity factor $k_{1n}^+(x_n^0)$ monotonically increases when the coefficient of friction λ and tensile residual stress q^0 increase and vice versa. Also, it is important to note that to the left of a certain point x_n^0 cracks in both cases are closed ($k_{1n}^+(x_n^0) = 0$). That is caused by the resultant compressive stress field in this zone which is created not only by pressure $p(x^0)$ but also by the frictional stress $\tau(x^0)$ directed to the left. It can be shown that for $\lambda = 0$ the stress field is compressive everywhere and, therefore, $k_{1n}^\pm(x^0) = 0$ for all x_n^0.

For $\alpha_n = \pi/2$, $q^0 = 0$, $\lambda = 0.1$, $x_n^0 = 2.1$, and $\alpha_n = \pi/2$, $q^0 = 0$, $\lambda = 0.2$, $x_n^0 = 3.5$, the graphs of $k_{1n}^+(y_n^0)$ are labeled by numbers 1 and 2, respectively, and are presented in Fig. 9.16. These graphs show that the normal stress intensity factor $k_{1n}^+(y_n^0)$ monotonically decreases as the depth of the crack center y_n^0 increases. In this case, starting from a certain depth the crack faces close completely and, therefore, $k_{1n}^+(y_n^0)$ becomes equal to zero.

The graphs of $k_{1n}^\pm(x_n^0)$ and $k_{2n}^+(x_n^0)$ obtained for $y_n^0 = -0.2$ and $\alpha_n = 0$

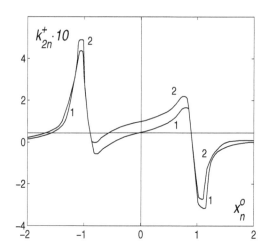

FIGURE 9.14
The dependence of the shear stress intensity factor k_{2n}^+ on the coordinate x_n^0 for the case of the boundary of half-plane loaded with normal $p(x^0) = 0.5/\sqrt{1 - (x^0)^2}$ and frictional $\tau(x^0) = -\lambda p(x^0)$ stresses, $y_n^0 = -0.2$, $\alpha_n = \pi/2$, $q^0 = 0$: $\lambda = 0.1$ - curve marked with 1, $\lambda = 0.2$ - curve marked with 2.

are presented in Fig. 9.17 and 9.18, respectively. In Fig. 9.18 the solid curve corresponds to $k_{1n}^+(x_n^0)$ while the dashed curve to $k_{1n}^-(x_n^0)$. The comparison of Figs. 9.14 and 9.17 reveals a strong dependence of $k_{2n}^+(x_n^0)$ on crack orientation, i.e., on the angle α_n, and a weak dependence of $k_{2n}^+(x_n^0)$ on the value of the coefficient of friction λ ($\lambda = 0.1$ - curve 1 and $\lambda = 0.2$ - curve 2). In Fig. 9.18 the graphs of $k_{1n}^+(x_n^0)$ and $k_{1n}^-(x_n^0)$ are close to each other. Moreover, the comparison of the graphs from Figs. 9.15 and 9.18 shows that the values of the stress intensity factor $k_{1n}^+(x_n^0)$ for $\alpha_n = \pi/2$ are by two orders of magnitude higher than the ones obtained for $\alpha_n = 0$. However, qualitatively the behavior of the stress intensity factors $k_{1n}^+(x_n^0)$ for cases of $\alpha_n = \pi/2$ and $\alpha_n = 0$ is similar.

Example 3. Suppose the half-plane boundary is loaded by the constant pressure $p(x^0) = \pi/4$ and frictional stress $\tau(x^0) = -\lambda p(x^0)$ while the residual stress $q^0 = 0$. In Figs. 9.19-9.24 the graphs of k_{1n}^+ and k_{2n}^+ are given as functions of x_n^0 and y_n^0 for $\lambda = 0.1$ (curves labeled with 1) and $\lambda = 0.2$ (curves labeled with 2). Moreover, in Figs. 9.23 and 9.24 the graphs are obtained for $x_n^0 = 3$, $x_n^0 = 2$, and $x_n^0 = 3.25$, $x_n^0 = 2$, respectively. In Figs. 9.19, 9.21, and 9.23, the graphs correspond to the angle $\alpha_n = 0$ while the graphs in Figs. 9.20, 9.22, and 9.24 correspond to the angle $\alpha_n = \pi/2$. In Figs. 9.19-9.22, the graphs are obtained for $y_n^0 = -0.2$.

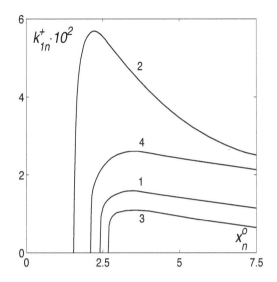

FIGURE 9.15
The dependence of the normal stress intensity factor k_{1n}^{+} on the coordinate x_n^0 for the case of the boundary of half-plane loaded with normal $p(x^0) = 0.5/\sqrt{1-(x^0)^2}$ and frictional $\tau(x^0) = -\lambda p(x^0)$ stresses, $y_n^0 = -0.2$, $\alpha_n = \pi/2$: $\lambda = 0.1$, $q^0 = 0$ - curve marked with 1, $\lambda = 0.2$, $q^0 = 0$ - curve marked with 2, $\lambda = 0.1$, $q^0 = -0.015708$ - curve marked with 3, $\lambda = 0.2$, $q^0 = 0.0314$ - curve marked with 4.

The behavior of the stress intensity factors $k_{1n}^{+}(x_n^0)$ and $k_{2n}^{+}(x_n^0)$ in Example 3 and the results of the solution of a contact problem for an elastic half-plane weakened by straight cracks (see Section 9.4) are similar. It is important to mention that the above stress intensity factors are similar qualitatively and quantitatively. Therefore, the stress intensity factors $k_{1n}^{\pm}$ and $k_{2n}^{\pm}$ exhibit a weak dependence on the details of the distributions of $p(x^0)$ and $\tau(x^0)$. The latter means that for many practical applications the distributions of $p(x^0)$ and $\tau(x^0)$ can be replaced by their averaged values

$$\int_a^b p(x^0)dx^0/(b-a) \ and \int_a^b \tau(x^0)dx^0/(b-a),$$

respectively. The dimensional analytical expressions for the main terms of the asymptotic representations of $k_{1n}^{\pm}$ and $k_{2n}^{\pm}$ calculated based on averaged $p(x^0)$ and $\tau(x^0)$ are given in (9.235) and (9.236).

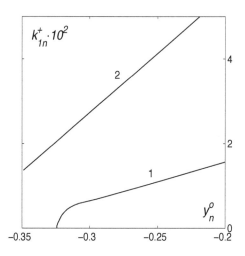

FIGURE 9.16
The dependence of the normal stress intensity factor k_{1n}^+ on the coordinate y_n^0 for the case of the boundary of half-plane loaded with normal $p(x^0) = 0.5/\sqrt{1 - (x^0)^2}$ and frictional $\tau(x^0) = -\lambda p(x^0)$ stresses, $\alpha_n = \pi/2$, $q^0 = 0$: $\lambda = 0.1$, $x_n^0 = 1.1$ - curve marked with 1, $\lambda = 0.2$, $x_n^0 = 3.5$ - curve marked with 2.

9.4 Contact Problem for an Elastic Half-Plane Weakened by Straight Cracks

It can be shown that in lightly loaded and in certain cases of heavily loaded lubricated contacts the elastohydrodynamic problem for elastic solids weakened by cracks can be split into two: the elastohydrodynamic problem of lubrication and the problem of finding the stress field in an elastic solids weakened by cracks. The latter problem is much simpler than the whole lubrication problem for solids with cracks. The goal of this section is to qualitatively and quantitatively analyze the contact problem with dry/fluid friction for elastic solids weakened by cracks. In addition to that, we will analyze what effect the details of the frictional stress distribution have on stress intensity factors. Dry and fluid friction will be considered.

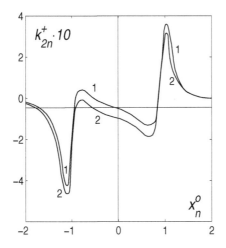

FIGURE 9.17
The dependence of the shear stress intensity factor k_{2n}^+ on the coordinate x_n^0 for the case of the boundary of half-plane loaded with normal $p(x^0) = 0.5/\sqrt{1-(x^0)^2}$ and frictional $\tau(x^0) = -\lambda p(x^0)$ stresses, $y_n^0 = -0.2$, $\alpha_n = 0$, $q^0 = 0$: $\lambda = 0.1$, - curve marked with 1, $\lambda = 0.2$ - curve marked with 2.

9.4.1 Problem Formulation

Let us consider a contact problem for a rigid indenter with the bottom of shape $y^0 = f(x^0)$ pressed into an elastic half-plane (see Fig. 9.25). The elastic half-plane with effective elastic modulus E' ($E' = E/(1-\nu^2)$, E and ν are the half-plane Young's modulus and Poisson's ratio) that is weakened by N straight cracks. The crack faces are frictionless. Let us introduce a global coordinate system with the x^0-axis directed along the half-plane boundary and the y^0-axis perpendicular to the half-plane boundary and pointed in the direction outside the material. The half-plane occupies the area of $y^0 \leq 0$. At infinity the half-plane is loaded by a tensile or compressive (residual) stress $\sigma_{x^0}^\infty = q^0$ which is directed along the x^0-axis. Besides the global coordinate system we will introduce local orthogonal coordinate systems for each straight crack of half-length l_k in such a way that their origins are located at the crack centers with complex coordinates $z_k^0 = x_k^0 + iy_k^0$, $k = 1, \ldots, N$, the x_k-axes are directed along the crack faces and the y_k-axes are directed perpendicular to them. The cracks are inclined to the positive direction of the x^0-axis at angles α_k, $k = 1, \ldots, N$. All cracks are considered to be subsurface and their faces are frictionless. The faces of every crack may be in partial or full contact with each other.

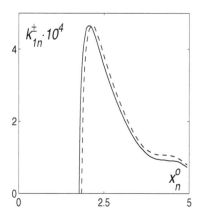

FIGURE 9.18
The dependence of the normal stress intensity factors $k_{1n}^{\pm}$ on the coordinate x_n^0 for the case of the boundary of half-plane loaded with normal $p(x^0) = 0.5/\sqrt{1-(x^0)^2}$ and frictional $\tau(x^0) = -\lambda p(x^0)$ stresses, $y_n^0 = -0.2$, $\alpha_n = 0$, $q^0 = 0$, $\lambda = 0.1$: k_{1n}^{+} - solid curve, k_{1n}^{-} - dashed curve.

   The indenter is loaded by a normal force P and may be separated from the half-plane by a layer of lubricant. The indenter creates a pressure $p(x^0)$ and frictional stress $\tau(x^0)$ distributions. The frictional stress $\tau(x^0)$ between the indenter and the boundary of the half-plane is determined by the contact pressure $p(x^0)$ through a certain relationship. We will consider the cases when the frictional stress $\tau(x^0)$ is created by dry and fluid friction.
   Based on (9.20), (9.35), and (9.39) and using the anti-symmetry of the normal displacements $v_0^{+}(x^0) = -v_0^{-}(x^0)$ for cracks with frictionless faces we obtain

$$v_0^{+} = -\frac{2}{\pi E'} \int_{x_i^0}^{x_e^0} p(t) \ln \frac{1}{|x^0-t|} dt - \frac{1-2\nu}{1-\nu} \frac{1}{E'} \int_{x_i^0}^{x_e^0} \tau(t) dt$$

$$(9.136)$$

$$\frac{1}{4\pi} \sum_{k=1}^{N} \int_{-l_k}^{l_k} [v_k'(t) W_k^r(t,x^0) - u_k'(t) W_k^i(t,x^0)] dt + v_{0*},$$

where v_{0*} is an arbitrary constant, and W_k^r and W_k^i are the real and imaginary parts of the kernel from (9.35) (see also (9.140)). Therefore, in dimensionless variables from (9.47) the equations of the problem for an indenter without sharp edges follow from the equations of Subsection 9.2.3 and have the fol-

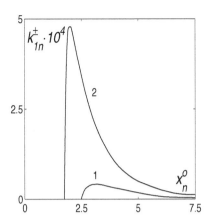

FIGURE 9.19
The dependence of the normal stress intensity factor k_{1n}^{+} on the coordinate x_n^0 for the case of the boundary of half-plane loaded with normal $p(x^0) = \pi/4$ and frictional $\tau(x^0) = -\lambda p(x^0)$ stresses, $y_n^0 = -0.2$, $\alpha_n = 0$, $q^0 = 0$: $\lambda = 0.1$ - curve marked with 1, $\lambda = 0.2$ - curve marked with 2.

lowing form:

$$\int\limits_{a}^{x^0} \tau(t)dt + \frac{2}{\pi}\int\limits_{a}^{b} p(t)\ln\frac{1}{|x^0-t|}dt - \frac{2}{\pi}\sum_{k=1}^{N}\int\limits_{-1}^{1}[v_k'(t)W_k^r(t,x^0)$$

$$(9.137)$$

$$-u_k'(t)W_k^i(t,x^0)]dt = c - f(x^0), \quad \int\limits_{a}^{b} p(t)dt = \frac{\pi}{2},$$

$$p(a) = p(b) = 0, \tag{9.138}$$

$$\int\limits_{-1}^{1}\frac{v_k'(t)dt}{t-x_k} + \sum_{m=1}^{N}\delta_m\int\limits_{-1}^{1}[v_m'(t)A_{km}^r(t,x_k) - u_m'(t)B_{km}^r(t,x_k)]dt$$

$$= \pi p_{nk}(x_k) + \pi p_k^0(x_k),$$

$$\int\limits_{-1}^{1}\frac{u_k'(t)dt}{t-x_k} + \sum_{m=1}^{N}\delta_m\int\limits_{-1}^{1}[v_m'(t)A_{km}^i(t,x_k) - u_m'(t)B_{km}^i(t,x_k)]dt$$

$$= \pi\tau_k^0(x_k),$$

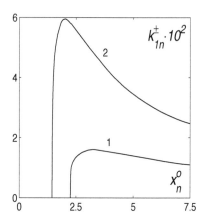

FIGURE 9.20
The dependence of the normal stress intensity factor k_{1n}^+ on the coordinate
x_n^0 for the case of the boundary of half-plane loaded with normal $p(x^0) = \pi/4$
and frictional $\tau(x^0) = -\lambda p(x^0)$ stresses, $y_n^0 = -0.2$, $\alpha_n = \pi/2$, $q^0 = 0$: $\lambda = 0.1$
- curve marked with 1, $\lambda = 0.2$ - curve marked with 2.

$$p_k^0 - i\tau_k^0 = -\frac{1}{\pi}\int_a^b [p(t)\overline{D}_k(t, x_k) + \tau(t)\overline{G}_k(t, x_k)]dt$$

$$-\tfrac{1}{2}q^0(1 - e^{-2i\alpha_k}),$$

$$v_k(\pm 1) = 0, \ u_k(\pm 1) = 0,$$

$$p_{nk}(x_k) = 0 \ for \ v_k(x_k) > 0, \ p_{nk}(x_k) \leq 0 \ for \ v_k(x_k) = 0,$$

$$k = 1, \dots, N,$$

where

$$W_k(t, x^0) = ie^{-i\alpha_k}\frac{\overline{T}_k - T_k}{\overline{T}_k - x^0}, \ W_k^r = Re(W_k), \ W_k^i = Im(W_k), \qquad (9.140)$$

$$A_{km} = \overline{R_{km} + S_{km}}, \ B_{km} = -i(\overline{R_{km} - S_{km}}),$$

$$(A_{km}^r, B_{km}^r, D_k^r, G_k^r) = Re(A_{km}, B_{km}, \overline{D}_k, \overline{G}_k), \qquad (9.141)$$

$$(A_{km}^i, B_{km}^i, D_k^i, G_k^i) = Im(A_{km}, B_{km}, \overline{D}_k, \overline{G}_k),$$

(9.139)

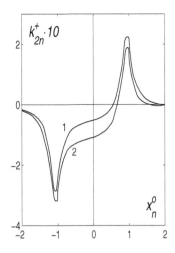

FIGURE 9.21
The dependence of the shear stress intensity factor k_{2n}^+ on the coordinate x_n^0 for the case of the boundary of half-plane loaded with normal $p(x^0) = \pi/4$ and frictional $\tau(x^0) = -\lambda p(x^0)$ stresses, $y_n^0 = -0.2$, $\alpha_n = 0$, $q^0 = 0$: $\lambda = 0.1$ - curve marked with 1, $\lambda = 0.2$ - curve marked with 2.

$$D_k(t, x_k) = \frac{i}{2}\left[-\frac{1}{t-X_k} + \frac{1}{t-\overline{X}_k} - \frac{e^{-2i\alpha_k}(\overline{X}_k - X_k)}{(t-\overline{X}_k)^2}\right],$$

$$G_k(t, x_k) = \frac{1}{2}\left[\frac{1}{t-X_k} + \frac{1-e^{-2i\alpha_k}}{t-\overline{X}_k} - \frac{e^{-2i\alpha_k}(t-X_k)}{(t-\overline{X}_k)^2}\right],$$

(9.142)

$$T_k = te^{i\alpha_k} + z_k^0, \quad X_n = x_n e^{i\alpha_n} + z_n^0, \quad k, n = 1, \ldots, N \qquad (9.143)$$

and kernels R_{km}, S_{km}, K_{km}, and L_{km} in dimensionless variables are also described by (9.37) and (9.25), respectively. For simplicity primes at the dimensionless variables are omitted. The characteristic values $\tilde{q}$ and $\tilde{b}$ that are used for scaling are the maximum Hertzian pressure p_H and the Hertzian contact half-width a_H

$$p_H = \sqrt{\frac{E'P}{\pi R}}, \quad a_H = 2\sqrt{\frac{RP}{\pi E'}}, \qquad (9.144)$$

where R can be taken as the indenter curvature radius at the center of its bottom. In (9.136)-(9.141) a and b are the dimensionless contact boundaries x_i^0 and x_e^0, c is the indenter vertical dimensionless displacement, δ_k is the dimensionless crack half-length, $\delta_k = l_k/\tilde{b}$.

In a similar fashion we can formulate a contact problem for an indenter with sharp edges at $x = a$ and $x = b$. In the latter case the contact region

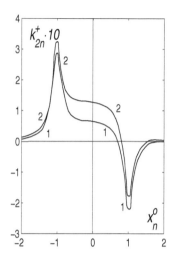

FIGURE 9.22
The dependence of the shear stress intensity factor k_{2n}^+ on the coordinate x_n^0 for the case of the boundary of half-plane loaded with normal $p(x^0) = \pi/4$ and frictional $\tau(x^0) = -\lambda p(x^0)$ stresses, $y_n^0 = -0.2$, $\alpha_n = \pi/2$, $q^0 = 0$: $\lambda = 0.1$ - curve marked with 1, $\lambda = 0.2$ - curve marked with 2.

boundaries $x = a$ and $x = b$ are fixed (known) and the problem is reduced to (9.137), (9.139)-(9.143).

Therefore, for the given shape of the indenter $f(x^0)$, the frictional stress functions $\tau(x^0)$, the orientation angles α_k, and positions of the cracks z_k^0, $k = 1,\ldots,N$, the solution of the problem is represented by the pressure distribution $p(x^0)$, the boundaries of the contact region a and b (for an indenter without sharp edges) and the indenter vertical displacement c, as well as by the crack faces displacement jumps $u_k(x_k)$, $v_k(x_k)$, and the normal contact stress $p_{nk}(x_k)$ applied to the crack faces ($k = 1,\ldots,N$). After the solution of the problem has been obtained, the stress intensity factors $k_{1k}^\pm$ and $k_{2k}^\pm$ are determined according to (9.33).

9.4.2 Problem Solution

Solution of this problem is associated with formidable difficulties represented by the nonlinearities caused by the presence of the free contact and crack boundaries and the interaction between different cracks and between cracks and contact pressure $p(x^0)$ and frictional stress $\tau(x^0)$ as well as the residual stress q^0. Under general conditions solution of this problem can be done only numerically. However, the problem can be effectively solved with the use of

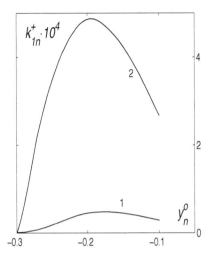

FIGURE 9.23
The dependence of the normal stress intensity factor k_{1n}^+ on the coordinate y_n^0 for the case of the boundary of half-plane loaded with normal $p(x^0) = \pi/4$ and frictional $\tau(x^0) = -\lambda p(x^0)$ stresses, $y_n^0 = -0.2$, $\alpha_n = 0$, $q^0 = 0$: $\lambda = 0.1$ - curve marked with 1, $\lambda = 0.2$ - curve marked with 2.

just analytical methods in the case when all cracks are small in comparison with the characteristic size $\tilde{b}$ of the contact region, i.e., when $\delta_0 = \max\limits_{1 \le k \le N} \delta_k$.
In this case the influence of the presence of cracks on the contact pressure is of the order of $O(\delta_0)$ and with the precision of $O(\delta_0)$ the crack system in the half-plane is subjected to the action of the contact pressure $p_0(x^0)$ and frictional stress $\tau_0(x^0)$ that are obtained in the absence of cracks. The further simplification of the problem is achieved under the assumption that cracks are small in comparison to the distances between them, i.e., (see (9.56))

$$z_n^0 - z_k^0 \gg \delta_0, \; n \ne k, \; n, k = 1, \dots, N. \tag{9.145}$$

The latter assumption with the precision of $O(\delta_0^2)$, $\delta_0 \gg 1$, provides the conditions for considering each crack as a single crack in an elastic half-plane (see Subsection 9.3.1) while the crack faces are loaded by certain stresses related to the contact pressure $p_0(x^0)$, contact frictional stress $\tau_0(x^0)$, and the residual stress q^0. The crucial assumption for simple and effective analytical solution of the considered problem is the assumption that all cracks are subsurface and much smaller in size than their distances to the half-plane surface

$$z_k^0 - \bar{z}_k^0 \gg \delta_0, \; k = 1, \dots, N. \tag{9.146}$$

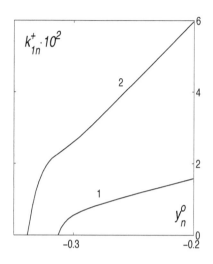

FIGURE 9.24
The dependence of the normal stress intensity factor k_{1n}^+ on the coordinate y_n^0 for the case of the boundary of half-plane loaded with normal $p(x^0) = \pi/4$ and frictional $\tau(x^0) = -\lambda p(x^0)$ stresses, $y_n^0 = -0.2$, $\alpha_n = \pi/2$, $q^0 = 0$: $\lambda = 0.1$ - curve marked with 1, $\lambda = 0.2$ - curve marked with 2.

Essentially, that assumption permits to consider each crack as a single crack in a plane (not a half-plane) with faces loaded by certain stresses related to $p_0(x^0)$, $\tau_0(x^0)$, and q^0. In fact it can be shown that $p(x^0) = p_0(x^0) + O(\delta_0^2)$ and $\tau(x^0) = \tau_0(x^0) + O(\delta_0^2)$ for $\delta_0 \ll 1$.

First, we will consider the pressure solutions of the problems for indenters with and without sharp edges. After that we will consider the stressed state of the material in the vicinity of the crack tips for both cases.

Let us consider the solution of the contact problem for an indenter with sharp edges for $\delta_0 \ll 1$. In this case the kernels from (9.139)-(9.143), (9.25), and (9.37) can be represented by power series in $\delta_0 \ll 1$ (see Section 9.3), and the problem solution can be found in the form of the series

$$\{p, \tau, c, p_{nk}\} = \sum_{j=0}^{\infty} \delta_0^j \{p_j, \tau_j, c_j, p_{nkj}\},$$

$$\{v_k, u_k\} = \sum_{j=0}^{\infty} \delta_k^j \{v_{kj}, u_{kj}\}.$$

(9.147)

By taking into account the fact that for $\delta_0 \ll 1$ kernels $W_k(t, x^0)$ can be

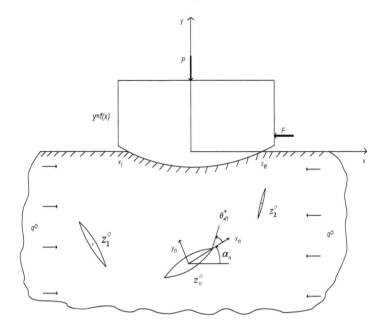

FIGURE 9.25
The general view of a rigid indenter in contact with a cracked elastic half-plane.

represented by the power series

$$W_k(t, x^0) = \sum_{j=0}^{\infty} (\delta_k t)^j W_{kj}(x^0), \tag{9.148}$$

the analysis of (9.137) for $\delta_0 \ll 1$ leads to the following equations for p_j and τ_j:

$$\int_a^{x^0} \tau_j(t)dt + \frac{2}{\pi} \int_a^b p_j(t) \ln \frac{1}{|x^0-t|} dt = d_j(x^0), \quad \int_a^b p_j(t)dt = \delta_{j0} \frac{\pi}{2},$$

$$j = 0, 2, \ldots,$$

$$p_1(x^0) = \tau_1(x^0) = 0, \tag{9.149}$$

$$d_0(x^0) = c_0 - f(x^0), \quad d_j(x^0) = c_j - c_{j-1}^W(x^0), \quad j = 2, \ldots,$$

$$c_j^W = -\frac{1}{\pi} \sum_{k=1}^{N} \left(\frac{\delta_k}{\delta_0}\right)^{j-1} \sum_{h=0}^{j} [V_{khj-h} W_{kj-h}^r - U_{khj-h} W_{kj-h}^i],$$

$$\{V,U\}_{klm} = \int_{-1}^{1} t^m \{v'_{kl}, u'_{kl}\}dt = -m \int_{-1}^{1} t^{m-1}\{v_{kl}, u_{kl}\}dt,$$

where $\delta_{j0} = 1$ if $j = 0$ and $\delta_{j0} = 0$ otherwise, W^r_{km} and W^i_{km} are the real and imaginary parts of kernel W_{km}, respectively. The boundary conditions (9.129) that are valid for small subsurface cracks were used in the last expression in (9.149).

Now, let us consider the example of Coulomb dry friction which in dimensional variables can be represented by

$$\tau(x^0) = -\lambda p(x^0), \tag{9.150}$$

where λ is the coefficient of friction. In (9.150) we assume that $\lambda \geq 0$, and, therefore, the frictional stress is directed to the left. Equation (9.150), being rewritten in dimensionless variables used in (9.137) and (9.138), takes the form

$$\tau(x^0) = -\eta p(x^0), \quad \eta = \lambda \frac{1-2\nu}{1-\nu}. \tag{9.151}$$

Then, the solution of (9.149) for $j = 0$ can be expressed in the form [16]

$$p_0(x^0) = \frac{\cos \pi \gamma}{2} \Big\{ d'_0(x^0) \sin \pi \gamma$$

$$+ \frac{1}{(b-x^0)^{1/2-\gamma}(x^0-a)^{1/2+\gamma}} \Big[1 - \frac{\cos \pi \gamma}{\pi} \int_a^b \frac{(b-t)^{1/2-\gamma}(t-a)^{1/2+\gamma}d'_0(t)dt}{t-x^0} \Big] \Big\}, \tag{9.152}$$

$$\gamma = -\frac{1}{\pi} \arctan \frac{(1-2\nu)\lambda}{2(1-\nu)},$$

where constant c_0 can be obtained from the first of equations (9.149) by setting $x^0 = a$ and $d'_0(x^0)$ is the derivative of $d_0(x^0)$ with respect to x^0. Similarly can be obtained the solutions for $j = 2, \ldots$. The simplest example of a solution of such a kind can be determined in the case of an indenter with a flat horizontal bottom when $f(x^0) = const$. Then

$$p_0(x^0) = \frac{\cos \pi \gamma}{2(b-x^0)^{1/2-\gamma}(x^0-a)^{1/2+\gamma}}. \tag{9.153}$$

Let us consider the process of solution of the contact problem for an indenter without sharp edges, i.e., the problem with unknown boundaries a and b. Therefore, we need to solve equations (9.137)-(9.143), (9.25), and (9.37) for $\delta_0 \ll 1$. As before, in this case the kernels from (9.139)-(9.143), (9.25), and (9.37) can be represented by power series in $\delta_0 \ll 1$ (see Section 9.3). For convenience, in equations (9.137)-(9.143), (9.25), and (9.37) it is desirable to make a substitution

$$x^0 = \frac{b+a}{2} + \frac{b-a}{2}y. \tag{9.154}$$

After that the problem solution can be sought in the form

$$\{p(y), \tau(y), a, b, c, p_{nk}(y)\}$$

$$= \sum_{j=0}^{\infty} \delta_0^j \{p_j(y), \tau_j(y), a_j, b_j, c_j, p_{nkj}(y)\}, \tag{9.155}$$

$$\{v_k, u_k\} = \sum_{j=0}^{\infty} \delta_k^j \{v_{kj}, u_{kj}\},$$

where constants a_j, b_j, c_j and functions p_j, τ_j, v_{kj}, u_{kj}, p_{nkj} have to be determined in the process of solution. We will limit ourselves to determining only the first two terms of the expansions in (9.155) in the case of Coulomb's friction law given by (9.150) and (9.151). In a fashion similar to the case of an indenter with sharp edges, we get [16]

$$p_0(x^0) = \frac{\cos \pi \gamma}{2} \left\{ \frac{\cos \pi \gamma}{\pi} (b_0 - x^0)^{1/2+\gamma} (x^0 - a_0)^{1/2-\gamma} \right.$$

$$\times \int_{a_0}^{b_0} \left(\frac{t-a_0}{b_0-t}\right)^{\gamma} \frac{f'(t)dt}{\sqrt{(b_0-t)(t-a_0)(t-x^0)}} - f'(x^0) \sin \pi \gamma \Big\}, \tag{9.156}$$

$$\gamma = -\frac{1}{\pi} \arctan \frac{(1-2\nu)\lambda}{2(1-\nu)},$$

$$\int_{a_0}^{b_0} \left(\frac{t-a_0}{b_0-t}\right)^{\gamma} \frac{tf'(t)dt}{\sqrt{(b_0-t)(t-a_0)}} = \frac{\pi}{\cos \pi \gamma}, \quad \int_{a_0}^{b_0} \left(\frac{t-a_0}{b_0-t}\right)^{\gamma} \frac{f'(t)dt}{\sqrt{(b_0-t)(t-a_0)}} = 0, \tag{9.157}$$

$$p_1(x^0) = 0, \quad a_1 = b_1 = c_1 = 0, \tag{9.158}$$

where equations (9.157) serve for determination of the constants a_0 and b_0 while constant c_0 can be determined from the first of equations (9.137) if $x^0 = a_0$ and constants a and b are replaced by constants a_0 and b_0. The simplest example of a solution of such a kind can be determined in the case of an indenter with a parabolic bottom when $f(x^0) = (x^0 + d)^2$, while $a_0 = -b_0$. Here constant d should be determined from the problem solution. Then from (9.156) and (9.157) we obtain

$$p_0(x^0) = \cos \pi \gamma \left(\frac{b_0-x^0}{b_0+x^0}\right)^{\gamma} \sqrt{b_0^2 - (x^0)^2}, \tag{9.159}$$

$$b_0 = (1 - 4\gamma^2)^{-1/2}, \quad d = -2\gamma b_0.$$

The second equation in (9.156) indicates that constant γ is a monotonically decreasing function of the friction coefficient λ and a monotonically increasing function of Poison's ration ν. We have to keep in mind that the usual ranges of the friction coefficient λ and Poison's ratio ν are $0 \leq \lambda \leq 0.25$ and $0 \leq \nu \leq 0.5$, respectively. It can be shown that γ is always small. For example, for steels $\nu = 0.3$ and constant γ varies between 0 and -0.0227 while λ varies between

0 and 0.25. Therefore, the pressure distributions from (9.153) and (9.159) can be closely approximated by

$$p_0(x^0) = \frac{1}{2\sqrt{(b-x^0)(x^0-a)}}, \ a < x^0 < b, \tag{9.160}$$

for an indenter with sharp edges and

$$p_0(x^0) = \sqrt{1 - (x^0)^2}, \ | \ x^0 \ |\leq 1, \ b_0 = 1, \tag{9.161}$$

for an indenter without sharp edges, respectively.

It is worth mentioning that cracks affect the contact boundaries a and b and the pressure distribution $p(x^0)$ as well as each other starting with the terms of the order of δ_0^2.

In both cases of an indenter with and without sharp edges the further problem solution concerning the material stressed state in the vicinity of crack tips is reduced to the application of the methodology used for solution of the problem for an elastic half-plane weakened by small straight cracks that is described in Subsections 9.3.2 and 9.3.3. Some examples of the behavior of the stress intensity factors $k_{1k}^{\pm}$ and $k_{2k}^{\pm}$ are presented below.

9.4.3   Contact Problem for Lubricated Solids

Let us consider a heavily loaded lubricated contact. In this case it is possible to neglect the influence of lubricant on the distribution of contact pressure and the indenter displacement (see Chapter 10 and Kudish [13]). Namely, the pressure distribution can be considered close to the Hertzian pressure distribution in a dry frictionless contact. This assumption considerably simplifies solution of the problem, nonetheless, preserving its characteristic and significant features.

We will consider the lubricant to be a Newtonian fluid. Then under isothermal conditions with high degree of accuracy in dimensionless variables the frictional stress τ can be obtained from the relation

$$\tau = -\frac{\mu s}{h}, \tag{9.162}$$

where μ is the lubricant viscosity, s is the constant slide-to-roll ratio proportional to the sliding velocity, and h is the gap between the indenter and the half-plane. The lubricant viscosity μ and gap h depend on pressure p, i.e., $\mu = \mu(p)$ and $h = h(p)$. In dimensionless variables the lubricant viscosity $\mu(p)$ can be approximated by a modified exponential relationship

$$\mu(p) = \exp(Qp) \ for \ p < p_*,$$

$$\mu(p) = \exp(Qp_*) \ for \ p \geq p_*, \ Q = \alpha_p \tilde{q}, \tag{9.163}$$

where α_p is the pressure coefficient of viscosity and p_* is the pressure at which the lubricant viscosity reaches its maximum. In heavily loaded contacts, by using the fact that pressure p is close to the Hertzian one, the dimensionless gap h can be approximated by the formula (see Chapter 10):

$$h(x^0) = 1 + \frac{1}{H_0}\{| x^0 | \sqrt{(x^0)^2 - 1}$$
$$- \ln [| x^0 | + \sqrt{(x^0)^2 - 1}]\}\theta[(x^0)^2 - 1], \quad (9.164)$$

where H_0 is the dimensionless lubrication film thickness at the exit from the contact $(H_0 = 2Rh_e/a_H^2$, h_e is the dimensional film thickness at the exit from the contact, a_H is determined by (9.144)), and $\theta(x)$ is the unit step function $(\theta(x) = 0$, $x \leq 0$ and $\theta(x) = 1$, $x > 0)$. For the purpose of this study for a fully flooded lubrication regime, the value of H_0 can be approximated by the Ertel-Grubin formula (see Chapter 10, Ertel [22], Grubin [23])

$$H_0 = 0.272(VQ)^{3/4}, \quad V = \frac{24\mu_0(u_1+u_2)R^2}{a_H^3 p_H}, \quad (9.165)$$

where μ_0 is the lubricant ambient viscosity, u_1 and u_2 are the linear velocities of the points of the half-plane boundary and the points of the bottom of the indenter, respectively. In (9.162) constant s depends on a number of factors such as the initial viscosity of the lubricant, elastic constants of the half-plane material, regime of lubrication (load applied to the indenter, amount of available lubricant, etc.).

To be able to compare the stressed fields in the vicinity of the crack tips caused by contact stresses in dry and lubricated contacts, we will choose the value of s as follows

$$s = \lambda \frac{\pi}{2} \frac{1-2\nu}{1-\nu} / \int\limits_a^b \frac{\mu(p(x^0))dx^0}{h(x^0)}. \quad (9.166)$$

The value of s from (9.166) was obtained from the requirement that in dry and lubricated contacts the friction forces are equal. Because in a heavily loaded lubricated contact the pressure distribution is close to the Hertzian one using expression (9.164), we can reduce (9.166) to

$$s = \lambda \frac{\pi}{2} \frac{1-2\nu}{1-\nu} / \int\limits_a^b \frac{\mu(\sqrt{1-(x^0)^2})\theta(1-(x^0)^2)dx^0}{h(x^0)}. \quad (9.167)$$

Further, by utilizing the approach developed in Subsections 9.3.3 and 9.3.4 we obtain a two-term asymptotic representation of the solution of the problem for the material stressed field in the vicinity of small cracks. This solution is given by (9.125), (9.131)–(9.134). After that the stress intensity factors $k_{1k}^\pm$ and $k_{2k}^\pm$ can be determined from equation (9.55).

9.4.4 Numerical Results

Let us analyze the behavior of the solution of the problem under consideration for different values of the friction coefficient. First, let us consider the case of an

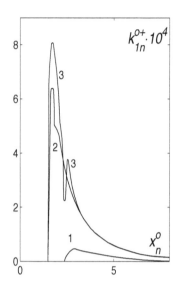

FIGURE 9.26
The dependence of the normal stress intensity factor k_{1n}^{0+} on x_n^0 for $\alpha_n = 0$: $\lambda = 0.1$ - curve marked with 1, $\lambda = 0.2$ - curve marked with 2, fluid friction that corresponds to dry friction with $\lambda = 0.2$ - curve marked with 3.

indenter with sharp edges. We will assume that $a = -1$ and $b = 1$. Keeping in mind that for λ ranging from 0 to 0.25 the value of the constant γ from (9.152) is small we can approximately accept as the solution the pressure distribution from (9.160). Some analysis of the behavior of the stress intensity factors $k_{1n}^{\pm}$ and $k_{2n}^{\pm}$ for this case is presented in Example 2 of Subsection 9.3.4.

Let us focus on the case of an indenter without sharp edges the shape of the bottom of which is described by $f(x^0) = (x^0 + d)^2$. Then $a_0 = -b_0$ and the approximate expressions for the distribution of pressure $p_0(x^0)$ and the right boundary of the contact b_0 are given by (9.159). Let us analyze in detail the behavior of the stress intensity factors $k_{1n}^{\pm}$ and $k_{2n}^{\pm}$. It is worth mentioning that in dimensional variables both and are proportional to $\sqrt{l_n}$. The further analysis is done in dimensionless variables. Let $k_{1n}^{0\pm}$ and $k_{2n}^{0\pm}$ be the stress intensity factors at the tips of the nth crack when the residual stress is zero, i.e., $q^0 = 0$. A single-term approximations for stress intensity factors $k_{1n}^{0\pm}$ and $k_{2n}^{0\pm}$ follow from (9.55), (9.125), (9.131)-(9.134) and have the form

$$k_{1n}^{0\pm} = \sqrt{l}Y_{n00}^r \theta(Y_{n00}^r), \quad k_{2n}^{0\pm} = \sqrt{l}Y_{n00}^i,$$

$$Y_{n00} = \frac{1}{\pi} \int\limits_{-b_0}^{b_0} p_0(t)[\overline{D}_{n0}(t) - \lambda \overline{G}_{n0}(t)]dt, \tag{9.168}$$

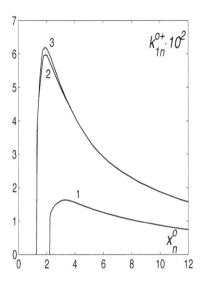

FIGURE 9.27
The dependence of the normal stress intensity factor k_{1n}^{0+} on x_n^0 for $\alpha_n = \pi/2$: $\lambda = 0.1$ - curve marked with 1, $\lambda = 0.2$ - curve marked with 2, fluid friction that corresponds to dry friction with $\lambda = 0.2$ - curve marked with 3.

where functions $D_{n0}(t)$ and $G_{n0}(t)$ are the main terms of the asymptotic series for $D_n(t, x_n)$ and $G_n(t, x_n)$ for $\delta_0 \ll 1$ (see (9.123)). Then in the presence of the residual stress q^0, we have

$$k_{1n}^{\pm} = \sqrt{l}[Y_{n00}^r + q^0 \sin^2 \alpha_n]\theta[Y_{n00}^r + q^0 \sin^2 \alpha_n],$$

$$k_{2n}^{\pm} = \sqrt{l}[Y_{n00}^i - \tfrac{q^0}{2} \sin 2\alpha_n]. \tag{9.169}$$

It follows from (9.169) that the residual stress does not affect $k_{1n}^{\pm}$ and $k_{2n}^{\pm}$ for horizontal cracks ($\alpha_n = 0$), and it produces the maximum effect on cracks oriented at angles $\alpha_n = \pi/2$ and $\alpha_n = \pi/4$.

Below we consider the numerical results for $k_{1n}^{0\pm}$ and $k_{2n}^{0\pm}$ (case of absent residual stress, $q^0 = 0$) for $y_n^0 = -0.2$ and $\delta_n = 0.1$ obtained by using formulas (9.168). In Figs. 9.26-9.29 curves marked with 1 correspond to $\lambda = 0.1$ while curves marked with 2 correspond to $\lambda = 0.2$. Based on (9.159) for $\lambda = 0.1$ we obtain $\gamma = -0.0159$, $b_0 = 1.0005$, $d = 0.0318$ and for $\lambda = 0.2$ we obtain $\gamma = -0.0317$, $b_0 = 1.002$, $d = 0.0635$.

The behavior of the normal stress intensity factor k_{1n}^{0+} for the cases of a horizontal $\alpha_n = 0$ (Fig. 9.26) and a vertical $\alpha_n = \pi/2$ (Fig. 9.27) cracks is similar. At the same time, in case of $\alpha_n = \pi/2$ the value of k_{1n}^{0+} is by more

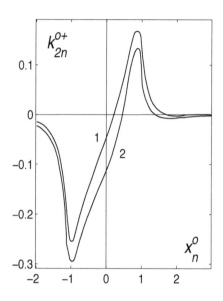

FIGURE 9.28
The dependence of the shear stress intensity factor k_{2n}^{0+} on x_n^0 for $\alpha_n = 0$:
$\lambda = 0.1$ - curve marked with 1, $\lambda = 0.2$ - curve marked with 2.

than an order of magnitude greater than in case of $\alpha_n = 0$. Moreover, the
value of k_{1n}^{0+} increases significantly with the friction coefficient λ and reaches
its maximum outside the contact region $[-b_0, b_0]$ in the vicinity of the contact
boundary that is opposite to the direction of the frictional stress $\tau = -\lambda p$.
The latter is due to the fact that in the region of $x^0 > b_0$ the frictional stress τ
creates tensile stresses. In the absence of friction ($\lambda = 0$), the whole half-plane
is subjected to compressive stresses. As a result of that all cracks are closed
and the normal stress intensity factors $k_{1n}^{0+} = 0$.

The behavior of the shear stress intensity factor k_{2n}^{0+} is different. The values
of k_{2n}^{0+} are significantly affected by the angle of crack orientation α_n. On
the other hand, the changes in the values of k_{2n}^{0+} caused by variations of the
friction coefficient λ are insignificant. The graphs of $k_{2n}^{0+}(x_n^0)$ for a horizontal
$\alpha_n = 0$ (Fig. 9.28) and a vertical $\alpha_n = \pi/2$ (Fig. 9.29) cracks indicate that
that the shear stress intensity factor k_{2n}^{0+} reaches its extremal values in the
close vicinity of the boundaries of the contact region, i.e., when crack centers
x_n^0 are close to points $x^0 = -b_0$ and $x^0 = b_0$.

Now, let us consider the behavior of k_{1n}^{0+} and k_{2n}^{0+} as functions of depth y_n^0
of the crack center below the half-plane surface. The graphs of $k_{1n}^{0+}(y_n^0)$ for
$\delta_n = 0.1$, $\lambda = 0.2$ and a vertical ($\alpha_n = \pi/2$) and horizontal ($\alpha_n = 0$) cracks

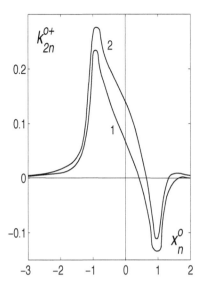

FIGURE 9.29
The dependence of the shear stress intensity factor k_{2n}^{0+} on x_n^0 for $\alpha_n = \pi/2$:
$\lambda = 0.1$ - curve marked with 1, $\lambda = 0.2$ - curve marked with 2 (after Kudish
[14]). Reprinted with permission of Springer.

are given in Figs. 9.30 and 9.31, respectively. Namely, curves marked with 1
and 2 correspond to $x_n^0 = 2.5$ and $x_n^0 = 5$, respectively. For a vertical crack
$k_{1n}^{0+}(y_n^0)$ is a monotonically increasing function of y_n^0, i.e., as the crack gets
farther away from the half-plane surface the value of $k_{1n}^{0+}(y_n^0)$ gets smaller (see
Fig. 9.30). For a horizontal crack $k_{1n}^{0+}(y_n^0)$ reaches its extremal value at some
depth beneath the surface (see Fig. 9.31). The graphs of $k_{1n}^{0+}(y_n^0)$ in Figs. 9.30
and 9.31 show that the effect of the contact frictional stress τ is limited to
only a thin near surface wedge-shaped material region of size comparable to
contact half-width b_0. Everywhere outside of this region the material is under
action of compressive stresses, i.e., $k_{1n}^{0+} = 0$.

The behavior of k_{2n}^{0+} as a function of y_n^0 is different for different x_n^0. In
Figs. 9.32 and 9.33 the graphs of $k_{2n}^{0+}(y_n^0)$ obtained for $\delta_n = 0.1$, $\lambda = 0.2$
are presented. Namely, in Fig. 9.32 the curve marked with 1 corresponds to
$\alpha_n = \pi/2$, $x_n^0 = 2.5$, the curve marked with 2 corresponds to $\alpha_n = \pi/2$, $x_n^0 =$
5, and the curve marked with 3 corresponds to $\alpha_n = 0$, $x_n^0 = 2.5$, while in
Fig. 9.33 all curves correspond to $\alpha_n = \pi/2$ and the curve marked with 1
is obtained for $x_n^0 = -2$, the curve marked with 2 is obtained for $x_n^0 = -1$,
and the curve marked with 3 is obtained for $x_n^0 = 1$. It is obvious from these

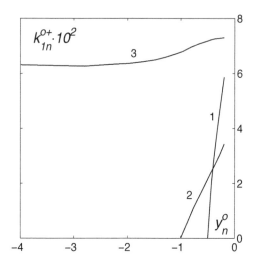

FIGURE 9.30
The dependence of the normal stress intensity factor k_{1n}^{0+} on y_n^0 for $\lambda = 0.2$, $\alpha_n = \pi/2$: $x_n^0 = 2.5$ - curve marked with 1, $x_n^0 = 5$ - curve marked with 2, fluid friction that corresponds to dry friction with $\lambda = 0.2$ and $x_n^0 = 2.5$ - curve marked with 3.

graphs that $k_{2n}^{0+}(y_n^0)$ is essentially different from zero only in the near surface layer of the material where it may posses several extrema, change its sign, etc.

Let us consider the case when the residual stress q^0 is different from zero. The residual stress influence on k_{1n}^{+} results in its increase for a tensile residual stress $q^0 > 0$ or its decrease for a compressive residual stress $q^0 < 0$ of the material region with tensile stresses and the level of these stresses. From formulas (9.169) and Fig. 9.34 (obtained for $y_n^0 = -0.2$, $\alpha_n = \pi/2$, and $\delta_n = 0.1$) follows that for all x_n^0 and for increasing residual stress q^0 (see the curves marked with 3 and 5 that correspond to $\lambda = 0.1$, $q^0 = 0.04$, and $\lambda = 0.2$, $q^0 = 0.02$, respectively) the stress intensity factor k_{1n}^{+} is a non-decreasing function of q^0. Moreover, if at some material point $k_{1n}^{+}(q_1^0) > 0$ for some residual stress q_1^0, then $k_{1n}^{+}(q_2^0) > k_{1n}^{+}(q_1^0)$ for $q_2^0 > q_1^0$ (compare curves marked with 1 and 2 with curves marked with 3 and 4 as well as with curves marked with 5 and 6). Similarly, for all x_n^0 when the magnitude of the compressive residual stress ($q^0 < 0$) increases (see curves marked with 4 and 6 that correspond to $\lambda = 0.1$, $q^0 = -0.01$ and $\lambda = 0.2$, $q^0 = -0.03$, respectively) and at some material point $k_{1n}^{+}(q_1^0) > 0$ for some residual stress q_1^0 then $k_{1n}^{+}(q_2^0) < k_{1n}^{+}(q_1^0)$ for $q_2^0 < q_1^0$. Based on the fact that the normal stress intensity factor k_{1n}^{0+} is positive only in the near surface material layer, we can make a conclusion that this layer increases in size and, starting with a certain

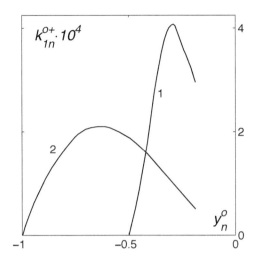

FIGURE 9.31
The dependence of the normal stress intensity factor k_{1n}^{0+} on y_n^0 for $\lambda = 0.2$, $\alpha_n = 0$: $x_n^0 = 2.5$ - curve marked with 1, $x_n^0 = 5$ - curve marked with 2.

value of tensile residual stress $q^0 > 0$, crack propagation becomes possible at any depth beneath the half-plane surface. For increasing compressive residual stresses, the thickness of the material layer where $k_{1n}^+ > 0$ decreases.

Let us consider the case of fluid (lubrication) friction based on relationships (9.162), (9.163), and (9.167). The numerical results for dry (see (9.151)) and fluid (see (9.162)) friction revealed that the behavior of the stress intensity factors k_{1n}^{0+} for these cases is close qualitatively as well as quantitatively (see curves marked with 3 in Fig. 9.26 and 9.27). These numerical results were obtained for $\lambda = 0.2$, $s = -0.0129$, $a_0 = -10$, $b_0 = 1$, $y_n^0 = -0.2$, $\delta_n = 0.1$, $Q = 10$, $p_* = 0.25$ (see (9.163)), $H_0 = 0.1$ and two angles: $\alpha_n = 0$ and $\alpha_n = \pi/2$. Therefore, the behavior of k_{1n}^{0+} is insensitive to even relatively large variations in the behavior of the contact frictional stress τ as long as the friction force (integral of τ over the contact region) applied to the surface of the half-plane is the same.

In the single-term approximation the behavior of k_{1n}^{0-} and k_{2n}^{0-} is identical to the one of k_{1n}^{0+} and k_{2n}^{0+}, respectively. Generally, the difference between k_{1n}^{0-} and k_{1n}^{0+} as well as between k_{2n}^{0-} and k_{2n}^{0+} is of the order of magnitude of $\delta_0 \ll 1$.

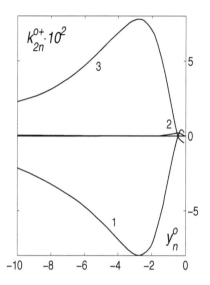

FIGURE 9.32
The dependence of the shear stress intensity factor k_{2n}^{0+} on y_n^0 for $\lambda = 0.2$, $\delta_n = 0.1$: $\alpha = \pi/2$, $x_n^0 = 2.5$ - curve marked with 1, $\alpha = \pi/2$, $x_n^0 = 5$ - curve marked with 2, $\alpha = 0$, $x_n^0 = 2.5$ - curve marked with 3.

9.5   Directions of Fatigue Crack Propagation

At the present time a number of different criteria are used for prediction of the resistance to fatigue crack propagation. Among them are the deformation criteria [24, 25], the force criteria [26] - [29], the energy–based criteria [30], and others [26, 31, 32]. The process of fatigue failure is usually subdivided into three major stages: the nucleation period, the period of slow pre-critical fatigue crack growth, and the short period of fast unstable crack growth ending in material loosing its integrity. The durations of the first two stages of fatigue failure depend on a number of parameters such as material properties, specific environment, stress state, temperature, etc. Usually, the nucleation period is short (see Chapter 8). In [11, 31] is given a literature review and some of the methods for calculation of crack trajectories in brittle solids. We are interested in the cases when the main part of the process of fatigue failure is due to slow pre-critical crack growth. In these cases $\max (k_{1n}^+, k_{1n}^-) < K_f$, where K_f is the material fracture toughness.

Numerous experimental studies [27] - [29] established the fact that at rela-

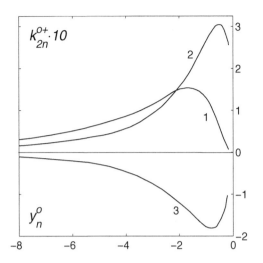

FIGURE 9.33
The dependence of the shear stress intensity factor k_{2n}^{0+} on y_n^0 for $\lambda = 0.2$, $\delta_n = 0.1$, $\alpha = \pi/2$: $x_n^0 = -2$ - curve marked with 1, $x_n^0 = -1$ - curve marked with 2, $x_n^0 = 1$ - curve marked with 3.

tively low loads materials undergo the process of pre-critical failure while the rate of crack growth dl_k/dN is dependent on $k_{1n}^{\pm}$, k_{th}, and K_f [26, 28, 30][‡]

$$\frac{dl_k}{dN} = f(k_{1n}^+, k_{1n}^-, k_{th}, K_f), \qquad (9.170)$$

where N is the number of loading cycles, k_{th} is the stress intensity factor threshold, and f is a certain function. A number of such equations of pre-critical crack growth and their analysis are presented in [33]. To determine the trajectory of the pre-critical crack growth, it is necessary to integrate (9.170) together with the initial condition

$$l_k \mid_{N=0} = l_{k0}. \qquad (9.171)$$

However, equations (9.170) and (9.171) are not sufficient for determining the trajectory of the crack growth. We also need to know the direction of crack propagation.

[‡]It is assumed that the sizes of the zones near crack tips where plastic deformations occur are small in comparison with crack sizes.

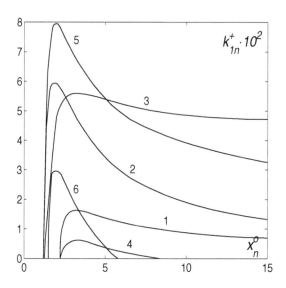

FIGURE 9.34
The dependence of the normal stress intensity factor k_{1n}^+ on x_n^0 for $\delta_n = 0.1$, $\alpha = \pi/2$, $y_n^0 = -0.2$ and different levels of the residual stress q^0. Curves 1 and 2 are obtained for $q^0 = 0$ and $\lambda = 0.1$ and $\lambda = 0.2$, respectively. Curves 3 and 4 are obtained for $\lambda = 0.1$, $q^0 = 0.04$ and $\lambda = 0.2$, $q^0 = 0.02$, respectively, while curves 5 and 6 are obtained for $\lambda = 0.1$, $q^0 = -0.01$ and $\lambda = 0.2$, $q^0 = -0.03$, respectively (after Kudish [14]). Reprinted with permission of Springer.

The angles $\theta_{k*}^{\pm}$ of deflection of the initial direction of crack propagation can be determined from the equation [10, 11]

$$\theta_{k*}^{\pm} = 2\arctan\frac{k_{1n}^{\pm} - \sqrt{(k_{1n}^{\pm})^2 + 8(k_{2n}^{\pm})^2}}{4k_{2n}^{\pm}}, \tag{9.172}$$

which follows from the assumption that the crack propagation occurs in the direction perpendicular to the direction of the maximum rupture stress.

Calculations based on (9.172) lead to a zigzagged–shaped crack trajectory. The actual iteration process $\alpha_k^{\pm(j+1)}(N + \triangle N) = \alpha_k^{\pm(j)}(N + \triangle N) + \theta_{k*}^{\pm j}$, $\alpha_k^{\pm(0)}(N + \triangle N) = \alpha_k^{\pm}$, $j = 0, 1, \ldots$ (where $\theta_{k*}^{\pm j}$ are determined according to (9.172)) converges to the angles that satisfies the equations

$$k_{2n}^{\pm} = 0. \tag{9.173}$$

Equations (9.173) follows from the fact that $\lim\limits_{j\to\infty} \theta_{k*}^{\pm j} = 0$. The latter means that the actual direction of crack propagation is symmetric with respect to applied normal tensile load. In [34] this hypothesis is used to determine crack trajectories.

A number of direct experimental observations (see Chapter 8 and [35]) show that in the process of their growth fatigue cracks orient in such a way that asymptotically the directions of their growth approach the directions prescribed by (9.173). Now, let us find the limiting in this sense directions of straight cracks, i.e., cracks for which $\alpha_k^{\pm} = \alpha$. Assuming that the angle α_k can also be represented by an asymptotic series in powers of $\delta_0 \ll 1$, i.e.,

$$\alpha_k = \sum_{j=0}^{\infty} \delta_0^j \alpha_{kj} \qquad (9.174)$$

and using (9.54), (9.112) from (9.173) we obtain a series of equations for α_{kj} in the form

$$\tau_k^0(\alpha_{k0}) = 0, \ \alpha_{k1} = 0, \ \alpha_{k2} = 2C_{k001}^i / \frac{\partial^2 \tau_k^0(\alpha_{k0})}{\partial \alpha_k^2}, \dots, \ \eta_0 \neq 1, \qquad (9.175)$$

$$C_{k001}^i(\alpha_{10}, \dots, \alpha_{k0} \dots, \alpha_{N0}) = 0, \dots, \ \eta_0 = 1; \ k = 1, \dots, N, \qquad (9.176)$$

where k is the number of cracks involved. The meaning of parameter η_0 is given in (9.46).

The analysis of (9.54), (9.175), and (9.176) leads to the following conclusions. For $\eta_0 \neq 1$ from (9.54) and (9.175) we obtain two solutions: the angles $\alpha_{k0} = 0$ and $\alpha_{k0} = \pi/2$. As a result of that for α_{k2} we also obtain two values $\alpha_{k2} = \alpha_{k2}(\alpha_{k0})$. Therefore, for a asymmetric stress field ($\eta_0 \neq 1$) there are two possible orientations of a crack the angle between which is equal to $\pi/2$ with the precision of $O(\delta_0^2)$, $\delta_0 \ll 1$.§ For $\eta_0 = 1$, i.e., for a symmetric stress field, we get a system of N nonlinear equations (9.176) the solution of which depends on the mutual location and lengths of cracks involved. In general, there may be several such solutions of equations (9.176). From (9.58), (9.62), and (9.63) follows that the system of equations (9.176) can be reduced to

$$\sum_{m=1}^{N} (\frac{\delta_m}{\delta_0})^2 A_{km01}^i = 0,$$

$$A_{km01}^i = Im(e^{i\alpha_m} \frac{\partial A_{km}}{\partial T_m} + e^{-i\alpha_m} \frac{\partial A_{km}}{\partial \overline{T}_m}) |_{\delta_k = \delta_m = 0}, \ k = 1, \dots, N. \qquad (9.177)$$

This mechanism of crack propagation may serve as an explanation of an explosive radial fracture of non-tempered glass [36].

Taking into account the further terms in the expansion from (9.174) requires consideration of slightly curved cracks. However, this complication does not

---

§The indicated mechanism of crack propagation is a plausible explanation of the experimentally established patterns of crack propagation in different materials under asymmetric loading [35].

add any significant knowledge to our understanding of the mechanism of fracture.

The possibility of several limiting directions of crack propagation can be interpreted as a model of crack branching. Obviously, the possibility of branching occurs only when along the potential directions, characterized by angles α_k, of crack propagation $k_{1n}(\alpha_k) > k_{th}$. By accepting this model of branching based on the above analysis, we can conclude that different loadings (stress fields) may lead to different crack branching. For example, for a nonuniform stress field ($\eta_0 \neq 1$) there are two possible almost perpendicular to each other crack propagation directions α_{k1} and α_{k2}. If originally a crack propagates along the direction given by angle α_{k1} then it may branch along angle α_{k2} if at some moment $k_{1n}(\alpha_{k2})$ becomes greater than $k_{1n}(\alpha_{k1})$. The experimental observations of such branching situations are described in [35, 36].

Based on the fact that usually the stress field is asymmetric cracks may change the direction of their growth only by an angle close to $\pi/2$.

9.6   Lubricant-Crack Interaction. Origin of Fatigue. Contact Fatigue Lives of Drivers versus Followers

It has been reported in periodic literature that occasionally bearings and gears experience a surface or near surface originated fatigue failure. One of the most important questions which may allow to understand better surface and subsurface originated fatigue is: How important is the possible interaction of lubricant with surface crack-like defects? There were developed some theoretical models of crack behavior under action of contact load Kaneta et al. [1, 2], Murakami et al. [3], Kudish [4, 5], Xu and Hsia [6], Panasyuk, Datsyshin, and Marchenko [7] and references therein). Recently, Murakami et al. [37] conducted a series of experiments and a theoretical study of surface crack behavior. It was shown experimentally that crack propagation is at least as important as crack initiation as fatigue cracks were observed long before the actual failure occurred. Moreover, the authors showed a dominant role of contact frictional stress and its direction in the experimentally observed crack propagation patterns.

Most of these models take into account only subsurface cracks. There are few models that do take into account surface cracks and their interaction with lubricant in a relatively simplistic way, for example, Bower [38] and Murakami et al. [37]. Both Bower [38] and Murakami et al. [37] assumed that surface cracks cannot develop multiple cavities completely or partially filled with lubricant, which can join into a single cavity or get separated one from another. The way the crack configuration is accounted for ultimately affects the stress intensity factors near crack tips, crack propagation, and fatigue life.

Moreover, an adequate solution of the problem cannot be obtained without some boundary conditions that do not allow crack faces to overlap and tensile stress to develop on crack faces. Thus, there is still a need in an accurate predictive model for surface crack-lubricant interaction.

This section is devoted to a new formulation and numerical method of solution of a problem for subsurface and surface crack-lubricant interaction. The interaction of lubricant with elastic solids within cavities of surface cracks and contact interaction of crack faces are regarded as the most interesting aspects of the problem. The problem is reduced to a system of integro-differential equations with boundary conditions in the form of alternating equations and inequalities. For surface cracks a number of boundary conditions describing the lubricant behavior within crack cavities is introduced. Among these conditions are the ones concerning development of several separate cavities some of which are completely and others partially filled with lubricant, pressure rise in the cavities completely filled with lubricant, etc. Subsurface cracks are also allowed to have multiple cavities. Because of the nature of some of the boundary conditions (see (9.184), (9.188)-(9.190)), it is impossible to use the numerical techniques similar to the ones developed by Gupta and Erdogan [8], Erdogan and Arin [9], Panasyuk et al., [10], Savruk [11]. The main challenge in the application of the aforementioned methods is the necessity to determine the contact stress p_n applied to the crack faces of the crack segments in contact with one another and of the closed cavities completely filled with lubricant simultaneously with the normal $v(x)$ and tangential $u(x)$ displacement jumps of crack faces as well as the existence of multiple cavities. The situation is exacerbated by the necessity to know the location of various crack cavities. The main purpose of this section is five–fold:

(a) to present a new formulation of the problem for subsurface/surface cracks interacting with lubricant,

(b) to develop an efficient numerical method for solution of the problem,

(c) to compare the earlier obtained simple asymptotic analytical solutions for small subsurface cracks with the numerical ones,

(d) to determine the origin of fatigue failure, i.e., whether it is of surface or subsurface origin, and

(e) to explain the usual difference in fatigue life of follows and drivers involved in sliding motion.

The numerical method developed in this section allows for determination of all three functions characterizing the crack stress-strain state simultaneously. The numerical results indicate that depending on crack orientation the normal stress intensity factors for surface cracks may be two or more orders of magnitude higher than the ones for similar subsurface cracks. It is shown that the numerical and approximate analytical (asymptotic) solutions obtained by Kudish [14] for small subsurface cracks are in good agreement with each other. The results of the analysis based on the problem formulation and numerical method employed in this section lead to the understanding that under normal conditions the contact fatigue initiation site is located beneath the surface (see Kudish [19], Kudish and Burris [20]). Moreover, based on these results the usual difference in fatigue lives of drivers and followers in sliding motion was explained (see Kudish [19], Kudish and Burris [20]). Some examples of numerical results for surface and subsurface cracks will be presented.

9.6.1   General Assumptions and Problem Formulation

Now, let us consider the boundary of an elastic half-plane with Young's modulus E and Poisson's ratio ν loaded by a moving normal $p(x^0)$ and tangential $\tau(x^0)$ contact stresses in the global (x^0, y^0) coordinate system (see Fig. 9.35). The stresses are transmitted through an incompressible lubrication film. The half-plane is pre-stressed by the compressive residual stress q^0 and is weakened by a straight surface or subsurface crack of half-length l, which makes angle α with the positive direction of the x^0-axis. Friction is neglected everywhere on crack faces except for the negligibly small region next to the surface crack mouth, an opening of the crack toward the lubrication layer. The equations for the normal $v(x) = v^+(x) - v^-(x)$ and tangential $u(x) = u^+(x) - u^-(x)$ jumps of the respective displacements of the crack faces and the normal contact stress $p_n(x)$ acting between the crack faces were derived in a local coordinate system (x, y) related to a crack in Section 9.4 and are as follows (Panasyuk et al. [10], Savruk [11], Kudish [5]):

$$\int_{-l}^{l} \frac{v'(t)dt}{t-x} + \int_{-l}^{l} [v'(t)A^r(t,x) - u'(t)B^r(t,x)]dt$$

$$\tag{9.178}$$

$$= \tfrac{4\pi}{E'} p_n(x) - \tfrac{4}{E'} \int_{x_i}^{x_e} [p(t)D^r(t,x) + \tau(t)G^r(t,x)]dt - \tfrac{4\pi}{E'} q^0 \sin^2 \alpha,$$

$$\int_{-l}^{l} \frac{u'(t)dt}{t-x} + \int_{-l}^{l} [v'(t)A^i(t,x) - u'(t)B^i(t,x)]dt$$

$$\tag{9.179}$$

$$= \tfrac{4}{E'} \int_{x_i}^{x_e} [p(t)D^i(t,x) + \tau(t)G^i(t,x)]dt - \tfrac{4\pi}{E'} q^0 \sin 2\alpha,$$

where the kernels of (9.178) and (9.179) are determined by the formulas

$$A = \overline{R+S}, \ B = i(\overline{S-R}), \ (A^r, B^r, D^r, G^r)$$

$$= Re(A, B, \overline{D}, \overline{G}), \ (A^i, B^i, D^i, G^i) = Im(A, B, \overline{D}, \overline{G}),$$

$$R(t,x) = \tfrac{e^{i\alpha}}{2} \Big\{ \tfrac{1}{X-T} + \tfrac{e^{-2i\alpha}}{\overline{X}-T} + (\overline{T}-T)\Big[\tfrac{1+e^{-2i\alpha}}{(\overline{X}-T)^2} \tag{9.180}$$

$$+ \tfrac{2e^{-2i\alpha}(T-X)}{(\overline{X}-T)^3} \Big] \Big\}, \ S(t,x) = \tfrac{e^{-i\alpha}}{2}\Big[\tfrac{T-\overline{T}}{(X-\overline{T})^2} + \tfrac{1}{\overline{X}-T} + \tfrac{e^{-2i\alpha}(T-X)}{(\overline{X}-T)^2} \Big],$$

$$D(t,x) = \tfrac{i}{2}\Big[-\tfrac{1}{t-X} + \tfrac{1}{t-\overline{X}} - \tfrac{e^{-2i\alpha}(\overline{X}-X)}{(t-\overline{X})^2} \Big],$$

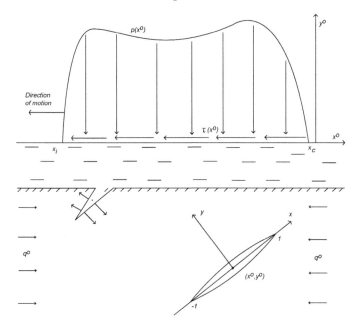

FIGURE 9.35
The general view of the loaded elastic half-plane with surface and subsurface cracks. The global coordinate system connected with the half-plane boundary and the local coordinate system connected with a crack are given.

$$G(t,x) = \tfrac{1}{2}\left[\frac{1}{t-X} + \frac{1-e^{-2i\alpha}}{t-\overline{X}} - \frac{e^{-2i\alpha}(t-X)}{(t-\overline{X})^2}\right],$$

$$T = te^{i\alpha} + z^0, \; X = xe^{i\alpha} + z^0, \; z^0 = x^0 + iy^0,$$

where x is the x-coordinate of the crack faces in the local coordinate system with the origin at the crack center located at the point (x^0, y^0) in the global coordinate system (see Fig. 9.35), α and l are the angle between the x-axis of the local and x^0-axis of the global coordinate systems and crack half-length, respectively, v, u, and p_n are jumps of the normal and tangential displacements of the crack faces and the normal stress applied to the crack faces, respectively, E' is the effective elasticity modulus of the half-plane material, $E' = E/(1-v^2)$, x_i and x_e are the coordinates of the beginning and end of the loaded region in the global coordinate system, and $\overline{X}$ is the complex conjugate of X.

Because of the relative motion of the crack with respect to the contact region $[x_i, x_e]$ loaded by $p(x_0)$ and $\tau(x^0)$, the stress state of the material around the crack changes significantly. That causes changes in the crack configuration as well as in the boundary conditions describing these crack configurations at different locations. Our goal is to establish all such boundary conditions that

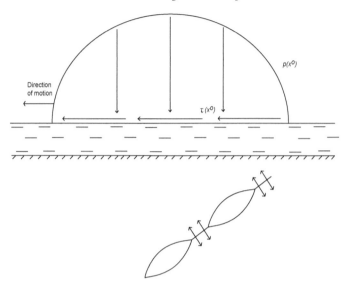

FIGURE 9.36
The general view of a subsurface crack with several cavities and closed segments.

must be added to equations (9.178)-(9.180) and to describe all possible crack configurations. These conditions depend on whether the crack is of a surface or subsurface nature, and whether its faces are in contact with each other or not.

First, let us consider a subsurface crack. Obviously, at the subsurface crack tips the jumps of the normal and tangential displacements of its faces are equal to zero

$$v(\pm l) = u(\pm l) = 0. \tag{9.181}$$

At the open segments of the subsurface crack the normal stress is zero, and it is non-positive at the segments of the crack that are in contact with each other (see Fig. 9.36)

$$p_n(x) = 0, \ v(x) > 0; \ p_n(x) \leq 0, \ v(x) = 0 \ for \ y^0 < -l\sin\alpha. \tag{9.182}$$

Now, let us consider the case of a surface crack the mouth ($x = l$) of which emerges at the half-plane boundary at the point with the global x^0-coordinate equal to $x^0 + l\cos\alpha sign(\alpha)$. At the subsurface tip of the surface crack, the jump of the tangential displacements of its faces is equal to zero. At the mouth of such a surface crack, the derivative of this jump along the crack is finite. Thus, we have

$$u(-l) = 0, \ \frac{du(l)}{dx} \ is \ finite. \tag{9.183}$$

A more accurate statement is: the shear stress intensity factor k_2^+ is zero (see (9.192)).

We will say that the surface crack mouth is open if $v(l) > 0$ and closed if $v(l) = 0$. Let us assume that the crack in question has several segments within which its faces are in contact with each other as well as a number of cavities containing some lubricant. All of these cavities except maybe one are not connected to the lubricant layer covering the half-plane boundary nor to each other. Moreover, we will assume that the lubricant within the cavities is in the state of quasi-hydrostatic equilibrium.

Let us establish some additional conditions at every surface crack segment mentioned above. Physical considerations imply that at every point of the crack its faces do not overlap, i.e., $v(x) \geq 0$. Let us define the crack cavities - the singly connected, non-intersecting sets of points x such that $C_i(x^0) = supp[v(x)]$, $i = 1, \ldots$, for which $v(x) > 0$. The numeration of these cavities C_i starts from the half-plane boundary. In each of the cavities C_i the lubricant is under action of an inherent constant lubricant pressure $p_n(x) = -p_n^i$, $i = 1, \ldots$, which generally are unknown in advance. Let at some position of the surface crack its mouth be open, i.e., cavity C_1 is connected to the layer of the lubricant covering the half-plane boundary. Hence, taking into account smallness of cavity C_1, it is natural to assume that the pressure within it is equal to that at its mouth, i.e.,

$$p_n^i = -p(x^0 + l\cos\alpha\,sign(\alpha)), \ v(l) > 0; \ y^0 = -l\sin\alpha. \tag{9.184}$$

Next, let us consider the cavities C_i ($i > 1$), which are not connected to the boundary of the lubricated half-plane. These cavities are defined by the set of indices

$$I(x^0) = \{i, \ C_i \cap \{(x, y), \ y = 0\} = \emptyset, \ i = 1, \ldots\}. \tag{9.185}$$

Assuming that the lubricant tensile strength limit is zero we obtain that the stresses in the crack cavities $p_n^i \leq 0$, $i \in I(x^0)$. Let us consider cavity C_i, $i \in I(x^0)$. While the distributions of contact pressure $p(x^0)$ and frictional stress $\tau(x^0)$ move along the half-plane boundary, the half-plane stress-strain state changes, so does the configuration of cavity C_i. Volume V^i of cavity C_i cannot be smaller than volume V_0^i of the lubricant trapped within it. This follows from the lubricant incompressibility. Thus, $V^i \geq V_0^i$, where the cavity volume is

$$V^i = V^i(x^0) = \int_{C_i} v(x)dx.$$

If cavity C_i is lubricant free, then the normal stress applied to the crack faces within this cavity is $p_n^i = 0$ (the lubricant vapor pressure in the cavity and the lubricant surface tension are neglected). Therefore, we arrive at the system of alternating equations and inequalities (see Fig. 9.37)

$$p_n^i = 0, \ V^i > V_0^i; \ p_n^i \leq 0, \ V^i = V_0^i; \ i \in I(x^0) \ for \ y^0 = -l\sin\alpha. \tag{9.186}$$

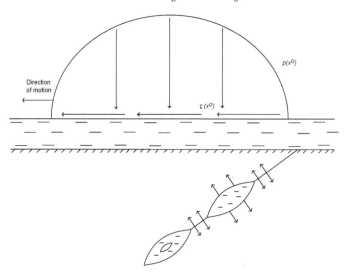

FIGURE 9.37
Two cavities of a surface crack: fully filled with lubricant and partially filled
with lubricant.

The method for determining the lubricant volumes V_0^i is given below. It is
necessary to mention that conditions (9.186) hold as long as the neighboring
cavities do not merge with C_i, i.e.,

$$C_i \cap C_j = \emptyset, \ i \neq j; \ i, j \in I(x^0). \tag{9.187}$$

Relations (9.183)-(9.187) represent the necessary boundary and additional
conditions from which one can obtain the unknown stresses p_n^i acting on the
crack faces of cavities C_i, $i \in I(x^0)$, if the lubricant volumes V_0^i are known.

We will assume that initially (far away from the incoming load where
$\max x^0 = -\infty$) the surface crack is completely filled with lubricant or completely closed. Let us show that it always can be assumed. Suppose the crack
is partially open and partially filled with lubricant. If the surface crack is open
toward the incoming lubricant subjected to high pressure, then the void in
the crack gets immediately filled up with lubricant as soon as the load comes
close to the crack mouth. If the surface crack is oriented in the direction away
from the incoming load, then as the load approaches close enough to the crack
it gradually closes the crack starting from its subsurface tip until it closes up
completely. In the latter case the fact that the surface crack was initially fully
or partially filled with lubricant does not change its behavior. Therefore, for
certainty we will assume that initially the surface crack is always completely
filled with lubricant or completely closed.

Now, let us consider the formulas for determining the lubricant volumes V_0^i
within cavities C_i. As the external load moves along the half-plane boundary

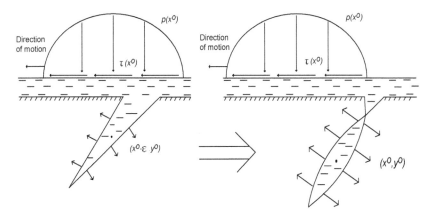

FIGURE 9.38
Development of several new cavities from an old one caused by a small transition of the contact load (i.e., by a change in the stress-strain state of the material near a surface crack).

the crack approaches the loaded region $[x_i, x_e]$ and simultaneously changes its configuration. Let us consider the behavior of the surface crack during its relative motion. Suppose the surface crack is completely open and filled with lubricant when its center is located at points $x = x^0 - \epsilon$, $(\epsilon \to +0, \epsilon > 0)$, and its faces get in contact with each other when the crack center reaches point $x = x^0$, so that k cavities C_i, $i = 1, \ldots, k \in I(x^0)$, are formed simultaneously (see Fig. 9.38). Volumes $V^i(x^0)$ of these cavities coincide with lubricant volumes V_0^i enclosed within them

$$V_0^i = \int_{C_i} v(x)dx, \quad p_n^i = -\lim_{\epsilon \to +0} p(x^0 - \epsilon + l \cos \alpha \, sign(\alpha)),$$

$$(9.188)$$

$$i = 1, \ldots, k \in I(x^0), \quad I(x^0 - \epsilon) = \emptyset, \quad \epsilon \to +0 \; for \; y^0 = -l \sin \alpha.$$

The additional condition in (9.188) imposed on p_n^i follows from the fact that the pressure in the lubricant varies continuously.

Further, if at some location of the surface crack its k cavities C_i, $i = i_0 + 1, \ldots, i_0 + k \in I(x^0 - \epsilon)$, merge simultaneously into a single cavity (see Fig. 9.39) C_j, $j \in I(x^0)$, as $\epsilon \to +0$, $\epsilon > 0$, we obtain the following relationships

$$V_0^j = \sum_{i=i_0+1}^{i_0+k} V_0^i, \quad j \in I(x^0),$$

$$C_l(x^0 - \epsilon) \cap C_m(x^0 - \epsilon) = \emptyset, \quad l \neq m, \; l, m \in I(x^0 - \epsilon), \; \epsilon > 0, \qquad (9.189)$$

$$C_j(x^0) = \lim_{\epsilon \to +0} \bigcap_{i=i_0+1}^{i_0+k} C_i(x^0 - \epsilon) \; for \; y^0 = -l \sin \alpha.$$

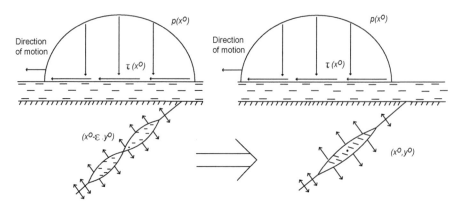

FIGURE 9.39
Merger of several old cavities into a new one caused by a small transition of
the contact load (i.e., by a change in the stress-strain state of the material
near a surface crack).

Also, it is possible that at some point crack cavity C_j gives rise to k cavities
C_i, $i = i_0 + 1, \ldots, i_0 + k \in I(x^0)$, simultaneously. This is a more complex
situation, which allows for two different scenarios. If there were no void in
the original cavity C_j (i.e., volumes $V^j = V_0^j$), then because of continuity the
cavities formed also have no voids. On the other hand, if the original cavity
C_j contains a void (i.e., $V^j > V_0^j$), then each of the newly formed cavities C_i
also has a void and the lubricant volume from cavity C_j is distributed among
the cavities C_i directly proportional to their volumes.

Now, let us consider the case of two adjacent crack cavities C_j and C_{j+1}
separated by a "neck" - just one point x_j. That means that for small enough
$\epsilon > 0$ the points on the crack faces with coordinates $x = x_j + \epsilon \in C_j$ and $x =
x_j - \epsilon \in C_{j+1}$ belong to different cavities. Suppose that cavity C_j is completely
filled with lubricant and cavity C_{j+1} is partially filled with lubricant (see
Fig. 9.40). Moreover, suppose that a small displacement of the external load
moving with a constant speed U_l causes the cavity neck to open up. Then,
the lubricant starts to slowly flow from cavity C_j with the higher pressure
equal to $-p_n^j$ into cavity C_{j+1} with the lower pressure $-p_n^{j+1}$. The lubricant
viscosity in such a slow flow can be neglected and the process can be described
by the equation

$$U_l \frac{dV^j(x^0)}{dx^0} = -v(x_j)\sqrt{\tfrac{2}{\rho}(p_n^j - p_n^{j+1})}, \tag{9.190}$$

which follows from the fact that the lubricant flux through the open neck is
equal to the rate of decrease of the lubricant volume in cavity C_j completely
filled with the lubricant

$$U_l \frac{dV^j(x^0)}{dx^0} = -v(x_j)U(x^0),$$

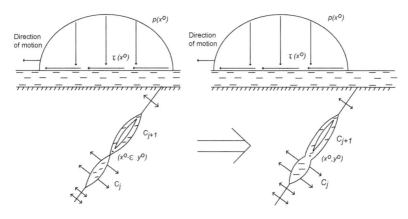

FIGURE 9.40
Two adjacent cavities of a surface crack separated by a neck which are fully and partially filled with lubricant, respectively.

and from Bernoulli's equation

$$p_n^j(x^0) = p_n^{j+1}(x^0) + \frac{\rho U^2(x^0)}{2},$$

where ρ is the lubricant volume density and $U(x^0)$ is the lubricant speed in the neck cross-section of the crack. This process continues until one of the following occurs: (a) pressures in the sub-cavities equalize (i.e., $p_n^j = p_n^{j+1}$), (b) the crack faces get squeezed and a new cavity is created from cavities C_j and C_{j+1}, which is completely filled with lubricant, (c) the crack faces get squeezed and a new neck separating the sub-cavities is created (i.e., $v(x_j) = 0$), or (d) cavity C_j stops contracting, i.e., $U(x^0) = 0$. This finalizes all the necessary conditions within the cavities of a surface crack containing lubricant.

When considering the surface crack segments with faces in contact with each other it is necessary that

$$p_n(x) \leq 0, \; v(x) = 0 \; for \; y^0 = -l \sin \alpha. \tag{9.191}$$

These relations follow from the fact that no external forces act on the segments of the crack faces in question.

Thus, for the given values of constants E, ν, x_i, x_e, α, $z^0 = (x^0, y^0)$, l, q^0, U_l, and functions $p(x^0)$, $\tau(x^0)$, the problem solution consists of functions $v(x)$, $u(x)$, and $p_n(x)$. Having solved the problem the stress intensity factors $k_1^{\pm}$ and $k_2^{\pm}$ at the crack tips can be obtained from the formulas (9.33)

$$k_1^{\pm} + ik_2^{\pm} = \mp \frac{E'}{4} \lim_{x \to \pm l} \sqrt{\frac{l^2 - x^2}{l}} [v'(x) + iu'(x)], \tag{9.192}$$

where the upper (lower) signs correspond to the crack tip with $x = l$ ($x = -l$).

9.6.2 Problem in Dimensionless Variables

Let us introduce the dimensionless variables

$$\{\bar{x}^0, \bar{y}^0, a, c\} = \{x^0, y^0, x_i, x_e\}/b_0, \quad \{\bar{p}, \bar{q}^0, \bar{p}_n, \bar{p}_n^i\} = \{p, q^0, p_n, p_n^i\}/q_0,$$

$$(\bar{x}, \bar{y}, \bar{t}) = (x, y, t)/l, \quad (\bar{v}, \bar{u}) = (v, u)/v_0, \quad (\bar{k}_1^\pm, \bar{k}_2^\pm) = (k_1^\pm, k_2^\pm)/(q_0\sqrt{l}),$$

$$v_0 = 4q_0l/E', \quad \delta = \frac{l}{b_0}, \quad \kappa = \frac{q_0}{E'}, \quad Eu = \frac{q_0}{\rho U_l^2}, \quad \beta = \frac{8\kappa^2}{Eu}, \quad \gamma = \delta^2\beta.$$

Here q^0 and b^0 are characteristic pressure and half-width of the loaded region, respectively. Then the problem equations can be presented in the form

$$\int_{-1}^{1} \frac{v'(t)dt}{t-x} + \delta \int_{-1}^{1} [v'(t)A^r(t,x) - u'(t)B^r(t,x)]dt \qquad (9.193)$$

$$= \pi p_n(x) - \int_{a}^{c} [p(t)D^r(t,x) + \tau(t)G^r(t,x)]dt - \pi q^0 \sin^2\alpha,$$

$$\int_{-1}^{1} \frac{u'(t)dt}{t-x} + \delta \int_{-1}^{1} [v'(t)A^i(t,x) - u'(t)B^i(t,x)]dt$$

$$(9.194)$$

$$= \int_{a}^{c} [p(t)D^i(t,x) + \tau(t)G^i(t,x)]dt - \frac{\pi}{2}q^0 \sin 2\alpha,$$

$$X = \delta x e^{i\alpha} + z^0, \quad T = \delta t e^{i\alpha} + z^0, \quad z^0 = x^0 + iy^0. \qquad (9.195)$$

The boundary and additional conditions are described by

$$v(\pm 1) = u(\pm 1) = 0 \ for \ y^0 < -\delta \sin\alpha, \qquad (9.196)$$

$$p_n(x) = 0 \ for \ v(x) > 0, \ p_n(x) \le 0 \ for \ v(x) = 0,$$

$$(9.197)$$

$$for \ y^0 < -\delta \sin\alpha, \ v(-1) = u(-1) = 0,$$

$$\lim_{x \to 1} \sqrt{1-x^2}v'(x) = \lim_{x \to 1} \sqrt{1-x^2}u'(x) = 0 \ for \ y^0 = -\delta \mid \sin\alpha \mid, \qquad (9.198)$$

$$p_n^1 = -p(x^0 + \delta \cos\alpha \, sign(\alpha)) \ if \ v(1) > 0 \ for \ y^0 = -\delta \mid \sin\alpha \mid, \qquad (9.199)$$

$$p_n^i = 0, \ V^i > V_0^i; \ p_n^i \le 0, \ V^i = V_0^i; \ i \in I(x^0)$$

$$(9.200)$$

$$for \ y^0 = -\delta \mid \sin\alpha \mid,$$

$$\frac{dV^j(x^0)}{dx^0} = -v(x_j)\sqrt{\frac{p_n^j - p_n^{j+1}}{\gamma}} \ if$$

$$v(x_j) > 0, \ V_0^j = V^j(x^0); \ V_0^{j+1} < V^{j+1}(x^0) \qquad (9.201)$$

$$for \ y^0 = -\delta \mid \sin\alpha \mid,$$

For simplicity, here and further bars over dimensionless variables are omitted. The dimensionless kernels in equations (9.193), (9.194) and in the rest of the conditions are identical to the ones presented earlier in this section if parameter l is replaced by δ.

Thus, for the given values of constants δ, α, $z^0 = (x^0, y^0)$, a, c, q^0, and functions $p(x^0)$ and $\tau(x^0)$ the problem solution consists of functions $v(x)$, $u(x)$, and $p_n(x)$. The stress intensity factors $k_1^{\pm}$ and $k_2^{\pm}$ can be obtained from the equations

$$k_1^{\pm} + ik_2^{\pm} = \mp \lim_{x \to \pm 1} \sqrt{1 - x^2}[v'(x) + iu'(x)], \tag{9.202}$$

where the upper signs correspond to the crack tip with $x = 1$ and the lower signs correspond to the crack tip with $x = -1$.

9.6.3 Numerical Method

To develop a numerical scheme let us define two sets of nodes

$$\xi_k = \cos\frac{2k-1}{2n}\pi, \ k = 1,\ldots,n; \ \xi_0 = 1, \ \xi_{n+1} = -1,$$
$$\eta_m = \cos\frac{m}{n}\pi, \ m = 1,\ldots,n-1; \ \eta_0 = 1, \ \eta_n = -1, \tag{9.203}$$

where nodes $-1 < \xi_k < 1$ and $-1 < \eta_m < 1$ are the roots of the Chebyshev orthogonal polynomials of the first $T_n(\xi_k) = 0$ and second $U_{n-1}(\eta_m) = 0$ kind.

The solution of the problem can be searched in the form

$$\frac{du}{dx} = \frac{U(x)}{\sqrt{1-x^2}}, \ \frac{dv}{dx} = \frac{V(x)}{\sqrt{1-x^2}}, \tag{9.204}$$

where $U(x)$ and $V(x)$ are new unknown functions. For functions $U(x)$ and $V(x)$ one can use the approximations by Chebyshev orthogonal polynomials

$$\{U(x), V(x)\} = \frac{1}{n}\sum_{k=1}^{n}(-1)^{k+1}\{U(\xi_k), V(\xi_k)\}\frac{T_n(x)\sqrt{1-\xi_k^2}}{x-\xi_k}. \tag{9.205}$$

According to (9.202), (9.204), and (9.205) after the solution is completed the stress intensity factors $k_1^{\pm}$ and $k_2^{\pm}$ can be found from the equations

$$k_1^{\pm} = \mp V(\pm 1), \ k_2^{\pm} = \mp U(\pm 1),$$

$$\{U(-1), V(-1)\} = \frac{1}{n}\sum_{k=1}^{n}(-1)^{k+n}\{U(\xi_k), V(\xi_k)\}\tan\frac{2k-1}{4n}\pi, \tag{9.206}$$

$$\{U(1), V(1)\} = \frac{1}{n}\sum_{k=1}^{n}(-1)^{k}\{U(\xi_k), V(\xi_k)\}\cot\frac{2k-1}{4n}\pi.$$

Using (9.204) and (9.205) normal $v(\eta_m)$ and tangential $u(\eta_m)$ displacement jumps of the crack faces are determined by

$$u(-1) = v(-1) = 0,$$

$$\{u(\eta_m), v(\eta_m)\} = \sum_{k=1}^{n} a_{km}\{U(\xi_k), V(\xi_k)\}, \ m = 0, \ldots, n-1, \tag{9.207}$$

and volumes V^i of cavities C_i are determined by the formula

$$V^i = \sum_{k=1}^{n} \sum_{m=m_b+1}^{m_e-1} a_{km}(\xi_m - \xi_{m+1})V(\xi_k), \tag{9.208}$$

where m_b and m_e are the node η_m numbers at the beginning and the end of cavity C_i. In case of cavity C_1 with an open mouth (i.e., when $v(1) > 0$) $m_b = 0$ and the summation in (9.208) starts with $m = 0$ instead of $m = 1$. The values of coefficients a_{km} are independent of $U(\xi_k)$, $V(\xi_k)$, and $p_n(\eta_m)$ and can be calculated just once from the formulas

$$a_{km} = a_{km+1} + \frac{(-1)^{k+1}\sqrt{1-\xi_k^2}}{n\prod\limits_{i=1}^{n}(1-\xi_i)} \int\limits_{\frac{m}{n}\pi}^{\frac{m+1}{n}\pi} \prod_{i=1,i\neq k}^{n}(\cos x - \xi_i)dx,$$

$$a_{kn} = 0, \ k = 1, \ldots, n, \ m = 0, \ldots, n-1. \tag{9.209}$$

Using approximations from (9.204) and (9.205) in (9.193) and (9.194) and using the Gauss-Chebyshev quadrature, we arrive at the discrete analogues of the latter equations (Gupta and Erdogan [8], Erdogan and Arin [9], Panasyuk et al. [10], Savruk [11])

$$\frac{1}{n}\sum_{k=1}^{n}\left[\frac{1}{\xi_k-\eta_m} + \delta A^r(\xi_k,\eta_m)\right]V(\xi_k) - \frac{\delta}{n}\sum_{k=1}^{n}B^r(\xi_k,\eta_m)U(\xi_k)$$

$$= p_n(\eta_m) - \frac{1}{\pi}\int\limits_{a}^{c}[p(t)D^r(t,\eta_m) + \tau(t)G^r(t,\eta_m)]dt - q^0\sin^2\alpha,$$

$$\frac{\delta}{n}\sum_{k=1}^{n}A^i(\xi_k,\eta_m)V(\xi_k) + \frac{1}{n}\sum_{k=1}^{n}\left[\frac{1}{\xi_k-\eta_m} - \delta B^i(\xi_k,\eta_m)\right]U(\xi_k) \tag{9.210}$$

$$= \frac{1}{\pi}\int\limits_{a}^{c}[p(t)D^i(t,\eta_m) + \tau(t)G^i(t,\eta_m)]dt + \frac{1}{2}q^0\sin 2\alpha,$$

where variables $U(\xi_k)$, $V(\xi_k)$, $k = 1, \ldots, n$, and $p_n(\eta_m), m = 1, \ldots, n-1$, are the unknowns that should be determined from the system (9.210) together with the appropriate additional conditions following from (9.195)-(9.201) and the rest of the conditions. Given the known functions $p(x^0)$ and $\tau(x^0)$ the integrals in the right-hand sides of (9.210) can be calculated with high precision using any adaptive integration method. The additional conditions from

(9.196)-(9.201) describing the normal stress and the displacement jumps of the crack faces within the cavities partially or fully filled with lubricant and within cavity C_1 with an open mouth, as well as the aforementioned parameters within the segments of the crack faces being in contact with each other can be discretized as follows:

$$p_n(\eta_m) = 0 \; if \; \sum_{k=1}^{n} a_{km} V(\xi_k) > 0;$$

$$p_n(\eta_m) \leq 0 \; if \; \sum_{k=1}^{n} a_{km} V(\xi_k) = 0, \qquad (9.211)$$

$$m = m_b + 1, \ldots, m_e - 1 \; for \; y^0 \leq -\delta \mid \sin\alpha \mid,$$

$$p_n^1(\eta_m) = -p(x^0 + \delta \cos\alpha sign(\alpha)) \; if \; \sum_{k=1}^{n} a_{km} V(\xi_k) > 0, \qquad (9.212)$$

$$m = 0, \ldots, m_e - 1 \; for \; y^0 = -\delta \mid \sin\alpha \mid,$$

$$p_n^i(\eta_m) = 0 \; if \; \sum_{k=1}^{n} \sum_{m=m_b+1}^{m_e-1} a_{km}(\xi_m - \xi_{m+1}) V(\xi_k) > V_0^i,$$

$$p_n^i(\eta_m) \leq 0 \; if \; \sum_{k=1}^{n} \sum_{m=m_b+1}^{m_e-1} a_{km}(\xi_m - \xi_{m+1}) V(\xi_k) = V_0^i, \qquad (9.213)$$

$$m = m_b + 1, \ldots, m_e - 1, \; i \in I(x^0) \; for \; y^0 = -\delta \mid \sin\alpha \mid.$$

The conditions for determining the inherited lubricant volumes and, in some cases, the normal stresses (see (9.186), etc.) in closed cavities C_i can be discretized and presented in the form of linear equations using linearization and Newton's method.

Finally, the problem can be reduced to a system of $3n - 1$ linear algebraic equations for $3n - 1$ unknowns $U(\xi_k)$, $V(\xi_k)$, $k = 1, \ldots, n$, and $p_n(\eta_m)$, $m = 1, \ldots, n - 1$. In particular, the first $2(n - 1)$ equations come from (9.210). The next equation for $V(\xi_k)$ depending on the nature of the crack (whether it is surface or subsurface) is represented by one of the equations (9.214) or (9.215). For a surface crack with an open mouth, we have

$$\sum_{k=1}^{n} (-1)^k V(\xi_k) \cot \frac{2k-1}{4n} \pi = 0 \; for \; y^0 = -\delta \mid \sin\alpha \mid, \qquad (9.214)$$

which represents the fact that for a surface crack with an open mouth the singularity of $v(x)$ at $x = 1$ is weaker than it is indicated by (9.204), i.e., the normal stress intensity factor $k_1^+ = 0$ (see Savruk [11] and (9.206)). For a subsurface crack the integral of dv/dx (see (9.204)) from -1 to 1 is zero, and the same conclusion is correct for a surface crack with closed mouth, i.e., with $v(1) = 0$. Therefore,

$$\sum_{k=1}^{n} V(\xi_k) = 0 \; for \; y^0 < -\delta \, | \sin \alpha \, | \;\; or$$

(9.215)

$$for \; y^0 = -\delta \, | \sin \alpha \, | \;\; and \;\; \sum_{k=1}^{n} a_{km} V(\xi_k) = 0.$$

Similarly, the additional equations for $U(\xi_k)$

$$\sum_{k=1}^{n} U(\xi_k) = 0 \; for \; y^0 < -\delta \, | \sin \alpha \, | \;\; or$$

(9.216)

$$for \; y^0 = -\delta \, | \sin \alpha \, | \;\; and \;\; \sum_{k=1}^{n} a_{km} V(\xi_k) = 0,$$

describe the case of a subsurface crack or a surface one with the closed mouth and the equation

$$\sum_{k=1}^{n} (-1)^k U(\xi_k) \cot \frac{2k-1}{4n} \pi = 0 \; for \; y^0 = -\delta \, | \sin \alpha \, |, \qquad (9.217)$$

describes the case of a surface crack with an open mouth (i.e., the case with the shear stress intensity factor $k_2^+ = 0$). The system also includes $n-1$ equations for $p_n(\eta_m)$ at each node $\eta_m, \; m = 1, \ldots, n-1$, in one of the forms given by equations (9.197), (9.199)-(9.201). The fact that some of the equations for $p_n^i(\eta_m)$ are presented in the form $V^i = V_0^i$ (see (9.200), (9.201)) is the main reason why the earlier developed numerical methods (see Gupta and Erdogan [8], Erdogan and Arin [9], Panasyuk et al. [10], Savruk [11]) do not work in the case of lubricant-surface crack interaction.

The problem is nonlinear because the locations and sizes of crack segments with faces in contact with each other and/or of the cavities are unknown in advance. The system reflects the current crack configuration (such as the number and location of cavities) and the current status of the surface crack cavities (such as partially or fully filled with lubricant). For a new load position it is necessary to perform some iterations with changed configuration and/or status of the crack cavities to obtain the problem solution.

To determine the crack configuration one must recognize each of the nodes $\eta_m, \; m = 1, \ldots, n-1$, as a "cavity" or "contact" node and to perform the necessary changes in the last $n+1$ or $n-1$ equations of the system which correspond to the values of $p_n(\eta_m)$. In order to do that the validity of the following conditions must be checked after each iteration:

$$v(\eta_m) = \sum_{k=1}^{n} a_{km} V(\xi_k) > 0 \; if \; p_n(\eta_m) = 0 \; or$$

(9.218)

$$v(\eta_m) = \sum_{k=1}^{n} a_{km} V(\xi_k) = 0 \; if \; p_n(\eta_m) > 0 \; or$$

$$v(1) = \sum_{k=1}^{n} a_{k0} V(\xi_k) > 0 \ and \ p_n(1) > 0 \ for \ y^0 = -\delta \mid \sin \alpha \mid,$$

for "cavity" nodes η_m and

$$v(\eta_m) = \sum_{k=1}^{n} a_{km} V(\xi_k) < 0 \ if \ p_n(\eta_m) = 0 \ or$$

$$v(\eta_m) = \sum_{k=1}^{n} a_{km} V(\xi_k) = 0 \ if \ p_n(\eta_m) < 0 \ or \qquad (9.219)$$

$$v(1) = \sum_{k=1}^{n} a_{k0} V(\xi_k) \leq 0 \ and \ p_n(1) > 0 \ for \ y^0 = -\delta \mid \sin \alpha \mid,$$

for "contact" nodes η_m. Here $p_n(1)$ can be obtained by using

$$p_n(1) = p_0 \ if \ p_0 \leq 0, \ p_n(1) = 0 \ if \ p_0 > 0,$$

$$p_0 = \sum_{m=1}^{n-1} (-1)^{m+1} (1 + \eta_m) p_n(\eta_m), \qquad (9.220)$$

$$p_n(x) = \sum_{m=1}^{n-1} p_n(\eta_m) \frac{U_{n-1}(x)}{(x - \eta_m) U'_{n-1}(\eta_m)}.$$

Moreover, for a surface crack the status of "cavity" nodes η_m must be identified by checking the nodes located in cavities partially/fully filled with lubricant. Depending on that some changes in the surface crack cavities status and in the last $n+1$ equations may be necessary. To make these changes after each iteration, the validity of the following conditions must be checked

$$\sum_{k=1}^{n} \sum_{m=m_b+1}^{m_e-1} a_{km}(\xi_m - \xi_{m+1}) V(\xi_k) > V_0^i \ if \ p_n^i(\eta_m) = 0,$$

$$\sum_{k=1}^{n} \sum_{m=m_b+1}^{m_e-1} a_{km}(\xi_m - \xi_{m+1}) V(\xi_k) = V_0^i \ if \ p_n^i(\eta_m) \leq 0, \qquad (9.221)$$

$$m = m_b + 1, \ldots, m_e - 1, \ i \in I(x^0) \ for \ y^0 = -\delta \mid \sin \alpha \mid,$$

$$\sum_{k=1}^{n} a_{km} V(\xi_k) > 0 \ if \ p_n^i(\eta_m) = -p(x^0 + \delta \cos \alpha \, sign(\alpha)),$$

$$\qquad (9.222)$$

$$m = 0, \ldots, m_e - 1, \ for \ y^0 = -\delta \mid \sin \alpha \mid.$$

Furthermore, after each iteration when the checking of the crack configuration and cavities status is complete functions $v(\eta_m)$, $m = 0, \ldots, n$, and $p_n(\eta_m)$, $m = 1, \ldots, n-1$, must be adjusted as follows: negative $v(\eta_m)$ and positive $p_n(\eta_m)$ must be set equal to zero. The new iteration can be performed based on the corrected crack configuration and cavities status. Therefore, if the crack configuration or cavities status were changed, then some of the last $n-1$ equations must be changed as described earlier. The iterations should be stopped when the solution reaches the desired precision.

One of the significant difficulties of the problem for surface cracks is that cavity C_1 with an open mouth (and, therefore, fully filled with lubricant) at some point may close its mouth and trap a certain amount of lubricant. The lubricant volume trapped in the cavity has significant effect on the material stress-strain state near the tip of the surface crack and on the stress intensity factor k_1^-, in particular. Therefore, it is extremely important to be able to determine the trapped volume as accurate as possible. That requires usage of an adaptive procedure for the increment of load displacement along the half-plane boundary. If necessary, the load displacement increment is halved until the crack parameters change by no more than the specified small value. In general, any changes in crack configuration and/or cavities status are allowed if the solution changes by no more than the specified small value. That is controlled by the increment of the load displacement. Taking into account that significant changes in crack configuration and status may occur as the crack mouth reaches the vicinity of the points $x = a$ and $x = c$ (the beginning and end of the loaded region), in the vicinity of these points the increment of the load displacement is reduced also.

9.6.4   Some Examples of Numerical Solutions for Subsurface and Surface Cracks

It is assumed that the pressure $p(x^0) = \sqrt{1 - (x^0)^2}$ and tangential stress $\tau(x^0) = -\lambda p(x^0)$ (λ is the friction coefficient) applied to the elastic half-plane boundary (see Fig. 9.35) move along it from right to left. The boundaries of the loaded lubricated region are defined as $a = -1$ and $b = 1$. The above assumptions correspond to $q_0 = p_H$ and $b_0 = a_H$ used in scaling the dimensional variables, i.e., maximum Hertzian pressure and half-width of the Hertzian contact, respectively.

All calculations were performed with double precision. The numerical solutions obtained for $n = 24$ and $n = 48$ number of nodes were found to be different by no more than 1.5%. Therefore, the numerical results presented below were obtained for $n = 24$. The initial increment of load movement was taken equal to $\triangle x^0 = 0.1$. The precision of the iterative solution in general was maintained at the level of 10^{-4} and of the lubricant volumes trapped within the surface crack cavities at the level of 10^{-8}. At any particular crack location the iteration process converged to the solution in $3 - 5$ iterations. The solution process is mainly slowed down by the necessity to conserve volumes of lubricant trapped within closed cavities of a surface crack, which, in turn, requires adaptation (substantial decrease down to 10^{-7}) of the initial increment of load movement $\triangle x^0$.

A number of researches believe that fatigue cracks propagate due to Mode

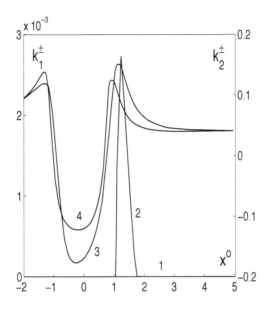

FIGURE 9.41
Distributions of the stress intensity factors $k_1^{\pm}$ and $k_2^{\pm}$ as functions of x^0 in case of a subsurface crack with $\delta = 0.2$, $\alpha = \pi/6$, $y^0 = -0.2$, $\lambda = 0.1$, and $q^0 = -0.1$ (k_1^{-} - curve 1, k_1^{+} - curve 2, k_2^{-} - curve 3, k_2^{+} - curve 4) (after Kudish [19]). Reprinted with permission of the STLE.

II failure mechanism and, usually, specifically this mechanism of fatigue crack propagation is theoretically modeled. There exists a widely spread belief that it is almost impossible to find zones of material with tensile stresses under rolling conditions. However, from the experimental point of view, this is ambiguous. For example, Bower [38] indicates that "...it has proved difficult to propagate stable Mode II fatigue cracks in laboratory experiments: the cracks almost invariably branch to propagate as Mode I branch cracks." We strongly believe that the predominant mechanism of crack propagation is Mode I. Therefore, we will pay more attention to considering the behavior of the stress intensity factors $k_1^{\pm}$.

First, let us consider an example for a subsurface crack. To illustrate the subsurface crack behavior we will assume that the y^0-coordinate of the crack centers is $y^0 = -0.2$, the crack dimensionless semi-length is $\delta = 0.2$. We assume that the friction coefficient $\lambda = 0.1$. For subsurface cracks oriented at angles $\alpha = \pi/2$ and $\alpha = \pi/6$ and the compressive residual stress $-0.5 \leq q^0 \leq -0.2$, the crack faces remain in full contact with each other (that is, $v(x) = 0$, $\mid x \mid \leq 1$) as the load moves along the half-plane boundary from

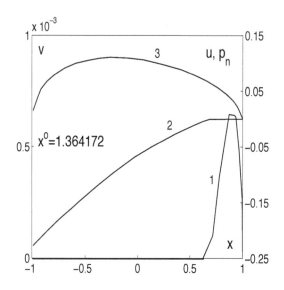

FIGURE 9.42
Distributions of the jumps of the normal $v(x)$ (curve 1) and tangential $u(x)$ (curve 3) displacements of the subsurface crack faces and of the normal stress $p_n(x)$ (curve 2) in case of a crack with data the same as in from Fig. 9.41 at the location of the subsurface crack center $x^0 = 1.364172$ (after Kudish [19]). Reprinted with permission of the STLE.

(i.e., the crack center moves from $x^0 = -\infty$ to $x^0 = +\infty$) and $k_1^{\pm}(x^0) = 0$. The reduction of the compressive residual stress to $q^0 = -0.1$ allows the crack with $\alpha = \pi/6$ to develop a small cavity in the vicinity of its forward tip $x = 1$ (see Figs. 9.41 and 9.42). In this case as the load moves toward the crack, ahead of the loaded region, under it, and far behind it, the crack is closed completely. However, when the center of the crack is located in the vicinity of $x^0 = 1.364172$ (that is, slightly behind the loaded region), the crack is partially open (see Fig. 9.42). This is also represented by a spike of k_1^+ in Fig. 9.41. At any crack position the stress intensity factor $k_1^- = 0$.

The further reduction of the compressive residual stress to $q^0 = -0.01$ allows the subsurface crack with angle $\alpha = \pi/6$ to open completely at some point. When the crack center is located at $x^0 = -2$ the crack center is far from the loaded region and the crack is completely closed, i.e., $k_1^{\pm} = 0$ (see Fig. 9.43). The crack faces remain in full contact with each other ($k_1^{\pm} = 0$) while the crack is under the loaded region $[-1, 1]$. As the load passes over the crack at some point x^0, the crack opens up partially in the vicinity of

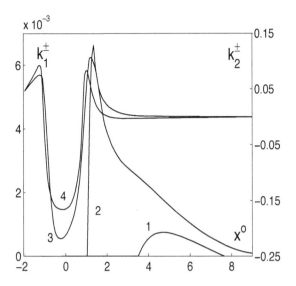

FIGURE 9.43
Distributions of the stress intensity factors $k_1^{\pm}$ and $k_2^{\pm}$ as functions of x^0 in case of a subsurface crack with $\delta = 0.2$, $\alpha = \pi/6$, $y^0 = -0.2$, $\lambda = 0.1$, and $q^0 = -0.01$ (k_1^- - curve 1, k_1^+ - curve 2, k_2^- - curve 3, k_2^+ - curve 4) (after Kudish [19]). Reprinted with permission of the STLE.

its forward tip $x = 1$ (i.e., $k_1^+ > 0$, $k_1^- = 0$), which is located away from the loaded region (see Fig. 9.44).¶ As the load moves farther away, the crack opens up completely, i.e., $k_1^{\pm} > 0$ (see Figs. 9.43, 9.45). The normal stress intensity factor k_1^+ reaches its maximum $k_1^+ = 0.66 \cdot 10^{-2}$ at $x^0 = 1.32605$. While the load moves farther, the crack closes up in the vicinity of its trailing tip $x = -1$ ($k_1^- = 0$) but remains open in the vicinity of its forward tip $x = 1$, i.e., $k_1^+ > 0$ (Fig. 9.46). The development of crack cavities is due to a tensile stress field surrounding the crack. The stress intensity factors $k_2^{\pm}(x^0)$ reach their extrema when the crack tips $x = \pm 1$ are in the vicinity of the boundaries of the loaded region $[-1, 1]$ (see Fig. 9.43).

For smaller friction coefficient $\lambda = 0.02$ (and the same $\alpha = \pi/6$ and $q^0 = -0.01$) behind the loaded region, the initially (at $x^0 = -2$) completely closed subsurface crack opens up only partially so that $k_1^+ > 0$ and $k_1^- = 0$. As the load moves further to the left, the crack closes up completely and remains

---

¶We will say that a point is in front of a contact if the contact frictional stress in the contact is directed toward the point, and we will say that a point is behind the contact otherwise.

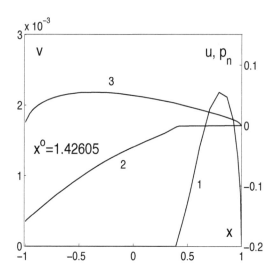

FIGURE 9.44
Distributions of the jumps of the normal $v(x)$ (curve 1) and tangential $u(x)$
(curve 3) displacements of the subsurface crack faces and of the normal stress
$p_n(x)$ (curve 2) in case of a crack with data the same as in from Fig. 9.43 at
the location of the subsurface crack center $x^0 = 1.42605$ (after Kudish and
Burris [20]). Reprinted with permission from Kluwer Academic Publishers.

closed.

It is worth mentioning that under realistic conditions of mixed lubrication
of rough surfaces when the pressure distribution has a highly irregular shape,
shallow subsurface cracks may have a number of intervals at which crack
faces are in contact with each other. That profoundly affects the values of the
normal and tangential stress intensity factors.

Now, let us consider a subsurface crack with $\delta = 0.2$, $\alpha = -\pi/6$, $y^0 =$
-0.2, $\lambda = 0.1$, and $q^0 = -0.01$. Initially, the crack is located at $x^0 = -2$
and it is completely closed (see Fig. 9.47). It remains closed until the loaded
region passes over the crack. Soon after that, the crack opens up partially and
developed a small cavity in the vicinity of its trailing tip $x = 1$ ($k_1^+ > 0$ and
$k_1^- = 0$) that is located closer to the loaded region. Then, the cavity growth
to a certain extent, reaches its maximum size (remaining partially open),
and gradually shrinks. The distribution of $k_1^+(x^0)$ possesses two maxima (see
Fig. 9.47). In the case of the same friction coefficient $\lambda = 0.1$ but higher
compressive residual stress $q^0 = -0.05$ as the loaded region passes over the
same crack ($\alpha = -\pi/6$), the initially completely closed crack (see Fig. 9.48)

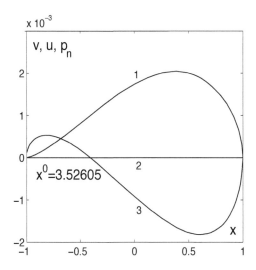

FIGURE 9.45
Distributions of the jumps of the normal $v(x)$ (curve 1) and tangential $u(x)$
(curve 3) displacements of the subsurface crack faces and of the normal stress
$p_n(x)$ (curve 2) in case of a crack with data the same as in from Fig. 9.43 at
the location of the subsurface crack center $x^0 = 3.52605$ (after Kudish and
Burris [20]). Reprinted with permission from Kluwer Academic Publishers.

opens up partially for a short period of time, and then it closes up completely.
However, soon after that, the crack opens up partially again for a brief period
of time and develops a tiny cavity (not shown in Fig. 9.48). Finally, as the load
continues moving to the left, the crack closes up and remains closed. When
the crack is partially open, $k_1^+ > 0$ and $k_1^- = 0$.

Let us consider a subsurface crack of semi-length $\delta = 0.2$ oriented at an
angle of $\alpha = 0$ and located at $y^0 = -0.2$ beneath the half-plane surface.
Ahead of the loaded region, the crack is partially open in the vicinity of its
forward tip $x = 1$ with $k_1^+ > 0$ and $k_1^- = 0$ (see Figs. 9.49 and 9.50). As the
load moves along the half-plane surface, at some point, the crack closes up.
Its faces remain in full contact with each other (i.e., $k_1^\pm = 0$) while the crack
is under the loaded region $[-1, 1]$ (see Figs. 9.51 and 9.52). As the load passes
at some point x^0, the crack opens up partially in the vicinity of its forward tip
$x = 1$ ($k_1^+ > 0$ and $k_1^- = 0$), which is located away from the loaded region (see
Fig. 9.53). As the loaded region moves away and the crack gets far enough, it
opens up completely, i.e., $k_1^\pm > 0$ (see Fig. 9.54). As the load moves farther,
the crack closes up in the vicinity of its forward tip $x = 1$ ($k_1^+ = 0$) but

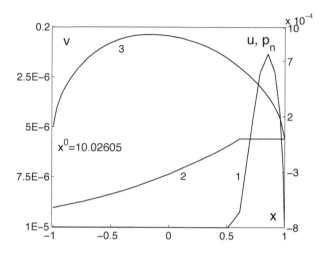

FIGURE 9.46
Distributions of the jumps of the normal $v(x)$ (curve 1) and tangential $u(x)$ (curve 3) displacements of the subsurface crack faces and of the normal stress $p_n(x)$ (curve 2) in case of a crack with data the same as in from Fig. 9.43 at the location of the subsurface crack center $x^0 = 10.02605$ (after Kudish and Burris [20]). Reprinted with permission from Kluwer Academic Publishers.

remains open in the vicinity of its trailing tip $x = -1$, i.e., $k_1^- > 0$ (see Fig. 9.55).

In all these cases with partially or completely open cracks except for the case of of $\alpha = -\pi/6$, the cracks begin to open up at the points adjacent to the crack tips. In the latter case the crack opens up at a point which is away from the crack tips. Then, the newly developed cavity grows in size and reaches the crack tip at $x = 1$. In all cases, the development of cavities is caused by a tensile stress field surrounding the cracks.

The normal stresses $p_n = 0$ inside the cavities of subsurface cracks and $p_n \leq 0$ outside of them where the crack faces are in contact with each other. The stress intensity factors $k_2^{\pm}$ reach their extrema at the crack tips, which are close to the boundaries $x^0 = \pm 1$ of the loaded region (see all the graphs of $k_2^{\pm}$ shown above). The distributions of $k_2^+(x^0)$ and $k_2^-(x^0)$ exhibit similar behavior and in most cases are close to each other.

For the cases presented in Figs. 9.41-9.49, the maximum values $k_1 = \max\{k_1^+(x^0), k_1^-(x^0)\}$ of the stress intensity factors $k_1^{\pm}$ are given in Table 9.1. For all considered subsurface cracks (except the one with $\alpha = 0$) and $\lambda > 0$, the numerical data shows that $k_1 = \max\{k_1^+(x^0)\} > \max\{k_1^-(x^0)\}$. This data indicate that if $\lambda > 0$ and $\alpha \neq 0$ then the maximum of $k_1^{\pm}$ is reached behind the loaded region and the smaller the magnitude of the compressive

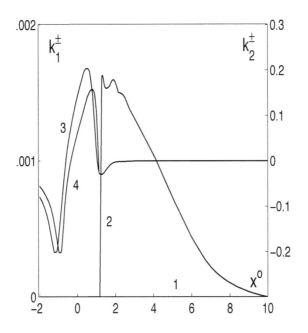

FIGURE 9.47
Distributions of the stress intensity factors $k_1^{\pm}$ and $k_2^{\pm}$ as functions of x^0 in case of a subsurface crack with $\delta = 0.2$, $\alpha = -\pi/6$, $y^0 = -0.2$, $\lambda = 0.1$, and $q^0 = -0.01$ (k_1^-: curve 1, k_1^+: curve 2, k_2^-: curve 3, k_2^+: curve 4) (after Kudish [19]). Reprinted with permission of the STLE.

residual stress q^0 the greater is the maximum of k_1.

TABLE 9.1
Maxima of the normal stress intensity factors k_1
for different subsurface cracks (after Kudish [19]).
Reprinted with permission of the STLE.

Data from	x^0-coordinate of the crack center	k_1
Fig. 9.49	1.219103	$0.246755 \cdot 10^{-2}$
Fig. 9.41	1.264172	$0.271905 \cdot 10^{-2}$
Fig. 9.43	1.326050	$0.659977 \cdot 10^{-2}$
Fig. 9.47	1.336572	$0.163820 \cdot 10^{-2}$
Fig. 9.48	1.349097	$0.270490 \cdot 10^{-4}$

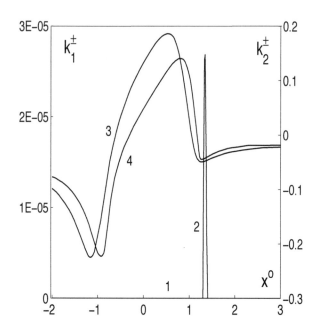

FIGURE 9.48

Distributions of the stress intensity factors $k_1^\pm$ and $k_2^\pm$ as functions of x^0 in case of a subsurface crack with $\delta = 0.2$, $\alpha = -\pi/6$, $y^0 = -0.2$, $\lambda = 0.1$, and $q^0 = -0.05$ (k_1^-: curve 1, k_1^+: curve 2, k_2^-: curve 3, k_2^+: curve 4) (after Kudish [19]). Reprinted with permission of the STLE.

A better understanding of to what extent variations in the friction coefficient λ affect the values of the stress intensity factors $k_1^\pm$ and $k_2^\pm$ for subsurface cracks can be gained from Figs. 9.56 and 9.57. These results are obtained based on the asymptotic solutions (9.168) and (9.169), which for $\delta \ll 1$ give $k_1^\pm = k_{1*} + O(\delta)$ and $k_2^\pm = k_{2*} + O(\delta)$. The data represented in Figs. 9.56 and 9.57 are obtained for a subsurface crack ($\delta = 0.05$, $\alpha = 1.047198$, $y^0 = -0.3$) for $q^0 = -0.001$ and two realistic values of the friction coefficient $\lambda = 0.04$ and $\lambda = 0.08$. Figure 9.56 shows a strong dependence of the normal stress intensity factors $k_1^\pm$ on the friction coefficient λ while Fig. 9.57 demonstrates a weak dependence of the shear stress intensity factors $k_2^\pm$ on the friction coefficient λ. This fact supports the experimentally observed strong dependence of contact fatigue life on frictionalstress and suggests that contact fatigue life is controlled not by shear but tensile stresses, i.e., contact fatigue is caused by Mode I fatigue crack growth.

The comparison of the asymptotically and numerically obtained values for

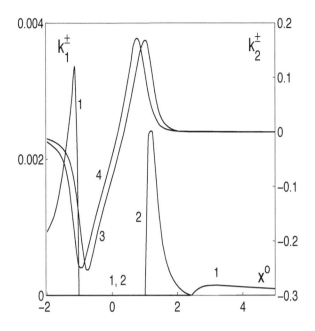

FIGURE 9.49
Distributions of the stress intensity factors $k_1^{\pm}$ and $k_2^{\pm}$ as functions of x^0 in case of a subsurface crack with $\delta = 0.2$, $\alpha = 0$, $y^0 = -0.2$, $\lambda = 0.1$, and $q^0 = -0.5$ (k_1^-: curve 1, k_1^+: curve 2, k_2^-: curve 3, k_2^+: curve 4).

the stress intensity factors $k_1^{\pm}$ and $k_2^{\pm}$ for small subsurface cracks is given in Subsection 9.6.5.

Let us consider some surface cracks of different orientation. When considering these types of cracks, we will assume that the friction coefficient $\lambda = 0.1$, the compressive residual stress $q^0 = -0.5$ (which is rather high), the crack center is at $y^0 = -0.1$, and, initially, the crack is at $x^0 = -2$.

First, let us consider a surface crack with the semi-length $\delta = 0.12$ that is oriented at an angle of $\alpha = 0.985111$ with the positive direction of the global x^0-axis. The crack remains closed near its subsurface tip $x = -1$, i.e., the normal stress intensity factor $k_1^-(x^0) = 0$ while the shear stress intensity factor $k_2^-(x^0)$ reaches its maximum (of approximately 0.63) and minimum (of approximately 0.13) values in the vicinity of the end points of the loaded region $x^0 = -1$ and $x^0 = 1$, respectively (see Fig. 9.58). Far from the loaded region $[-1, 1]$, at $x^0 = -2$, the crack mouth is open, and the lubricant (with low viscosity at ambient pressure) fills completely the small cavity, which is adjacent to the crack mouth. As the load moves toward the crack, the crack

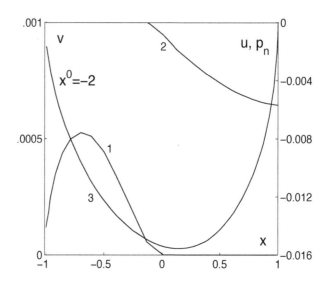

FIGURE 9.50
Distributions of the jumps of the normal $v(x)$ (curve 1) and tangential $u(x)$
(curve 3) displacements of the subsurface crack faces and of the normal stress
$p_n(x)$ (curve 2) in case of a crack with data the same as in from Fig. 9.49 at
the location of the subsurface crack center $x^0 = -2$.

mouth opens up wider, and the cavity slowly grows in size (see Fig. 9.59). Here
and further, in figures p stands for the lubricant pressure at the crack mouth.
The rate of this growth increases as the loaded region approaches the crack
mouth. However, the crack remains only partially open. As the loaded region
continues to move, the cavity slowly shrinks squeezing the lubricant out. By
the time the crack mouth approaches the right boundary of the loaded region,
the lubricant is completely squeezed out of the crack and the crack is closed.
As the load moves farther to the left, the crack remains completely closed for
a short period of time. Soon after the crack mouth leaves the loaded region,
again the crack opens up partially creating a small cavity adjacent to its
mouth (see Fig. 9.60). The cavity slightly grows in size, reaches its maximum
size (the crack remains open just partially), and then it slowly shrinks with
the crack mouth still being open. Within the crack closed segments the normal
stress p_n is negative.

Second, let us consider the situation when a surface crack with the semi-
length $\delta = 0.15$ that is oriented at an angle of $\alpha = 0.729728$ with the positive
direction of the global x^0-axis. Figure 9.61 depicts the behavior of the stress
intensity factors $k_1^-(x^0)$ and $k_2^-(x^0)$. Again, far from the loaded region (at

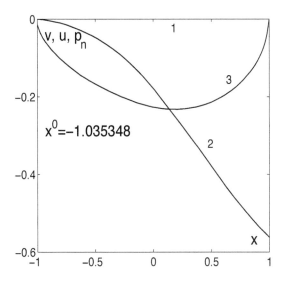

FIGURE 9.51
Distributions of the jumps of the normal $v(x)$ (curve 1) and tangential $u(x)$ (curve 3) displacements of the subsurface crack faces and of the normal stress $p_n(x)$ (curve 2) in case of a crack with data the same as in from Fig. 9.49 at the location of the subsurface crack center $x^0 = -1.035348$.

$x^0 = -2$), the crack mouth is open and the lubricant fills completely small cavity C_1, which is adjacent to the crack mouth. As the load moves toward the crack, the crack mouth opens up wider and wider, and cavity C_1 slowly grows in size. The rate of this growth increases drastically as the crack mouth moves into the inlet zone of the loaded (lubricated) region. At some point in the inlet zone, the crack opens up completely (see Fig. 9.62). It not only remains open for some time but also its volume V^1 increases. As the load moves further to the left, somewhere in the middle of the crack its faces come into contact forming a "neck" between two newly created cavities completely filled with lubricant. From this moment on, volume V^2 of cavity C_2 and the volume of lubricant V_0^2 trapped within it, remain constant and equal to $0.76143 \cdot 10^{-2}$. After that, the "neck" lengthens separating the two cavities completely (see Fig. 9.63). At this time, in the compressed crack cavity C_2, adjacent to the crack trailing tip $x = -1$, the pressure is constant and is higher than that in cavity C_1, where the pressure is equal to the one at the crack mouth. As the load continues to move to the left, the crack cavity C_1 decreases in size while volume V_2 of cavity C_2 and lubricant volume within it remain constant and equal. At some point, just before the crack mouth leaves the loaded

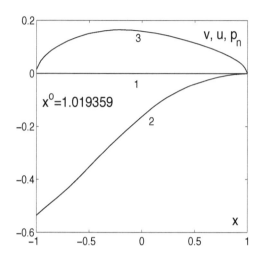

FIGURE 9.52
Distributions of the jumps of the normal $v(x)$ (curve 1) and tangential $u(x)$ (curve 3) displacements of the subsurface crack faces and of the normal stress $p_n(x)$ (curve 2) in case of a crack with data the same as in from Fig. 9.49 at the location of the subsurface crack center $x^0 = 1.019359$.

region, crack cavity C_1 rapidly closes up trapping some lubricant inside of a tiny cavity adjacent to the crack mouth (see Fig. 9.64). Soon after the crack mouth leaves the loaded region, it opens up rapidly (see Fig. 9.65). Cavity C_1, adjacent to the crack mouth, reaches its maximum size, and then shrinks slowly. At the same time, the volume of cavity C_2 remains constant, and the lubricant pressure within it slowly increases. During all the time of its existence, cavity C_2 remains attached to the subsurface crack tip $x = -1$, which causes $k_1^-(x^0) > 0$. Within the crack closed segments the normal stress p_n is negative.

Now, let us consider a surface crack with the dimensionless crack half-length $\delta = 0.2$ and the angle of crack orientation $\alpha = \pi/6$. Some results for this crack are shown in Figs. 9.66-9.68. The initial behavior of this crack is pretty much similar to the behavior of the previous one. When the crack is completely open and the cavity size reaches its maximum (at $x^0 = -1.064648$), the cavity volume is $V^1 = 4.564$. The difference occurs when the crack mouth approaches the right boundary $x^0 = 1$ of the loaded region and the crack already has two separate cavities. Volume V^2 of cavity C_2 is completely filled with lubricant and volume V_0^2 of this lubricant is equal to $0.26707 \cdot 10^{-3}$. At a certain time moment, cavity C_2 detaches itself from the trailing subsurface crack tip $x = -1$ (see Fig. 9.67) and remains detached (see Fig. 9.68). This

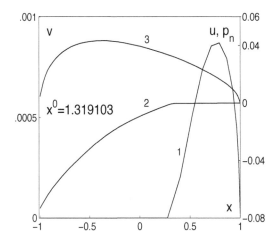

FIGURE 9.53
Distributions of the jumps of the normal $v(x)$ (curve 1) and tangential $u(x)$ (curve 3) displacements of the subsurface crack faces and of the normal stress $p_n(x)$ (curve 2) in case of a crack with data the same as in from Fig. 9.49 at the location of the subsurface crack center $x^0 = 1.319103$.

causes $k_1^-(x^0)$ to become equal to zero. The lubricant pressure within this cavity decreases as time goes by and the crack gets farther from the loaded region.

Finally, we will consider a surface crack with the dimensionless crack half-length $\delta = 0.3$ and the angle of crack orientation $\alpha = 0.339837$. Some results for this crack are depicted in Figs. 9.69 - 9.72. The initial behavior of this crack is similar to the one of the latter two. When the crack is completely open and the cavity size reaches its maximum (at $x^0 = -1.182031$) the cavity volume is $V^1 = 16.632$. The difference in the crack behavior is due to a peculiar shape of the crack cavity C_1 adjacent to the crack mouth (see Fig. 9.70). As the load continues to move to the left, at $x^0 = 0.188814$ cavity C_1 splits into two new cavities (see Fig. 9.71): a small new cavity C_1 with an open crack mouth and the newly created cavity C_2 separated from the mouth. At this time moment and later, the crack has three separate cavities: C_1 - the cavity with open mouth and C_2 and C_3 - two subsurface cavities containing some lubricant. Cavity C_3 is closer to the trailing subsurface crack tip $x = -1$ than cavity C_2. At this time moment, the volumes of the cavities and the lubricant within them are $V^1 = V_0^1 = 0.10682 \cdot 10^{-4}$, $V^2 = V_0^2 = 0.44138 \cdot 10^{-2}$, and $V^3 = V_0^3 = 0.36481 \cdot 10^{-2}$. As the the crack mouth approaches the right boundary of the loaded region, first, the lubricant pressure in cavity C_2 increases, reaches its maximum, and, then, monotonically decreases. The latter is due to the fact

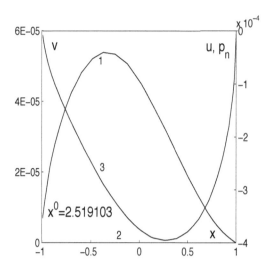

FIGURE 9.54
Distributions of the jumps of the normal $v(x)$ (curve 1) and tangential $u(x)$
(curve 3) displacements of the subsurface crack faces and of the normal stress
$p_n(x)$ (curve 2) in case of a crack with data the same as in from Fig. 9.49 at
the location of the subsurface crack center $x^0 = 2.519103$.

that this crack segment enters a region of the half-plane with tensile stresses.
At the same time, cavity C_3 remains completely filled with lubricant. At some
point, cavity C_3 detaches itself from the trailing subsurface crack tip $x = -1$
and remains detached. Therefore, from this moment on $k_1^- = 0$. Finally, at
$x^0 = 0.827918$ the pressure in cavity C_2 drops to zero. After that cavity C_2
becomes only partially filled with lubricant. For example, $V^2 = 0.74068 \cdot 10^{-2}$,
$V_0^2 = 0.44138 \cdot 10^{-2}$ at $x^0 = 0.852787$. Then, cavity C_2 continues to grow,
reaches its maximum size (see Fig. 9.72), and slowly shrinks in size, all the time
remaining just partially filled with lubricant at zero pressure. It is important
to remember that as soon as cavities C_2 and C_3 got separated from the crack
mouth the lubricant volumes V_0^2 and V_0^3 trapped within them remain the
same and equal to the values given above.

For smaller values of the angle $\alpha > 0$ (larger surface crack half-length δ),
the maximum of the stress intensity factor $k_1^-(x^0)$ is larger. The absolute
maximum of $k_1^-(x^0)$ is reached when the crack mouth just enters the loaded
region. It is mostly due to the lubricant penetrated the crack mouth and its
pressure applied to the crack faces. The second local maximum of $k_1^-(x^0)$ (if
exists) occurs behind the loaded region in the zone of tensile stress. The second
local maximum of $k_1^-(x^0)$ is much smaller than the first one. The numerical
data show that the absolute maximum of $k_1^-(x^0)$ is reached at a point to

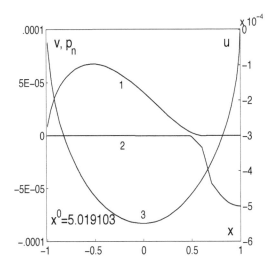

FIGURE 9.55
Distributions of the jumps of the normal $v(x)$ (curve 1) and tangential $u(x)$ (curve 3) displacements of the subsurface crack faces and of the normal stress $p_n(x)$ (curve 2) in case of a crack with data the same as in from Fig. 9.49 at the location of the subsurface crack center $x^0 = 5.019103$.

the right of the left endpoint of the loaded region, $x^0 = -1$. That is due to the competing influence of the lubricant pressure applied to the half-plane boundary and to the crack faces. The lubricant pressure applied to the half-plane boundary tries to squeeze the lubricant out of the crack and to close it down while the lubricant pressure applied to the crack faces tends to open the crack up. The described big spike of $k_1^-(x^0)$ at the entrance into the loaded region represents the mechanism of the "lubrication wedge effect" discussed in literature. The maximum values $k_1 = \max\{k_1^-(x^0)\}$ of the normal stress intensity factors for above considered cases of surface cracks are gathered in Table 9.2.

For surface cracks with $\alpha > 0$, which may open up and be completely filled with lubricant, the influence of the friction coefficient λ and the residual stress q^0 on the normal stress intensity factor k_1^- is negligibly small due to the dominant role of the high lubricant pressure penetrating such cracks.

Surface cracks with with angles $\alpha < 0$ are closed completely while the external (contact) load approaches them and passes over them. It is due to the fact that in front of the loaded region the material of the half-plane is compressed. Depending on the value of angle $\alpha < 0$, these cracks may open completely, but most often they open just partially behind the loaded region (i.e., for $x^0 > 1$) where the material of the half-plane is stretched by a tensile

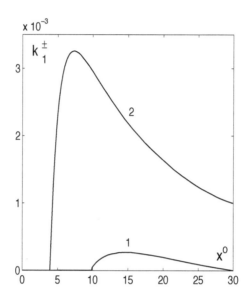

FIGURE 9.56
Distributions of the normal stress intensity factors $k_1^{\pm}$ as a function of x^0 in case of a subsurface crack with $\delta = 0.05$, $\alpha = 1.047198$, $y^0 = -0.3$, $q^0 = -0.001$ for friction coefficient $\lambda = 0.04$ (curve 1) and $\lambda = 0.08$ (curve 2) (after Kudish [19]). Reprinted with permission of the STLE.

TABLE 9.2
Maxima of the normal stress intensity factors k_1 for different surface cracks (after Kudish [19]). Reprinted with permission of the STLE.

Angle α	x^0-coordinate of surface crack center	x^0-coordinate of surface crack mouth	k_1
0.985111	any	any	0
0.728728	-0.865192	-0.746993	0.32761
0.523599	-0.964648	-0.814658	0.90725
0.339837	-1.082031	-0.882038	2.04985

stress. Behind the loaded region, for a surface crack partially open just near its mouth the stress intensity factor $k_1^- = 0$; otherwise, if a surface crack is completely open the stress intensity factor attains relatively small values typical for subsurface cracks (see Table 9.1).

   The distributions of the stress intensity factor $k_2^-(x^0)$ are similar for all considered cases of surface cracks. The distribution of the shear stress intensity factor $k_2^-(x^0)$ reaches its extrema when the trailing subsurface crack tip $x =$

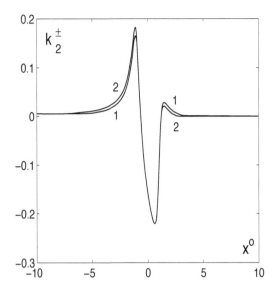

FIGURE 9.57
Distributions of the shear stress intensity factors $k_2^\pm$ as a function of x^0 in case of a subsurface crack with $\delta = 0.05$, $\alpha = 1.047198$, $y^0 = -0.3$, $q^0 = -0.001$ for friction coefficient $\lambda = 0.04$ (curve 1) and $\lambda = 0.08$ (curve 2) (after Kudish [19]). Reprinted with permission of the STLE.

-1 is in the vicinity of the endpoints of the loaded region $[-1, 1]$.

In all series of simulations during certain periods of time, both surface and subsurface cracks were just partially open. In most simulated cases, surface cracks had multiple cavities. These observations lead to the conclusion that the correct solutions in such cases cannot be obtained without using the inherent conditions for crack faces and lubricant-surface crack interaction introduced above. Moreover, the numerical results showed that usually the maximum of the stress intensity factor k_1^- for surface cracks is about two orders of magnitude higher than $k_1^\pm$ for similar subsurface cracks (see Figs. 9.43 and 9.66). In practice, it means that the presence of crack-like surface defects may reduce fatigue life of a machine part by several $(4 - 6)$ orders of magnitude. This is the result of the lubricant wedge effect. Based on this comparative analysis of stress intensity factors for subsurface and surface cracks and the mathematical model of contact fatigue (see Sections 9.7, 9.8, and 9.10, also see Kudish [39]), some important conclusions regarding fatigue behavior of contact solids can be made (see Sections 9.7, 9.8, and 9.10). In particular, in most cases pitting is initiated at some subsurface (not surface) point in

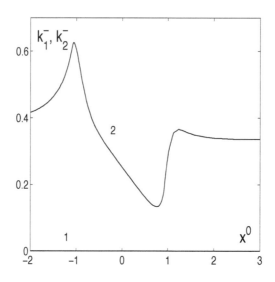

FIGURE 9.58
Distributions of the stress intensity factors k_1^- (curve 1) and k_2^- (curve 2) in case of a surface crack: $\alpha = 0.985111$, $\lambda = 0.1$, and $q^0 = -0.5$.

the material. In cases when pitting is originated at the surface, it happens not long before failure takes place and it is caused by contamination or wear particles. Moreover, this analysis of the behavior of surface and subsurface cracks provides an explanation of the difference between fatigue lives of drivers and followers involved in sliding motion.

9.6.5   Comparison of Analytical (Asymptotic) and Numerical Solutions for Small Subsurface Cracks

Let us consider a subsurface crack, which is small in comparison with the size of the loaded region ($\delta \ll 1$) and is located far from the half-plane boundary relative to its size ($|\, y^0\,| \gg \delta$). The two-term asymptotic expansions of the stress intensity factors, $k_1^{\pm}$ and $k_2^{\pm}$, can be calculated as follows (see Section 9.4 and also Kudish [13, 14]):

$$k_1^{\pm} = c_0^r \pm \tfrac{1}{2}\delta c_1^r + \ldots \; if \; c_0^r > 0, \; k_1^{\pm} = 0 \; if \; c_0^r < 0,$$

$$k_1^{\pm} = \tfrac{\sqrt{3}\delta}{9} c_1^r [\pm 7 - 3\theta(c_1^r)] \sqrt{\tfrac{1 \pm \theta(c_1^r)}{1 \pm 3\theta(c_1^r)}} + \ldots \; if \; c_0^r = 0 \; and \; c_1^r \neq 0,$$

$$\tag{9.223}$$

$$k_2^{\pm} = c_0^i \pm \tfrac{1}{2}\delta c_1^r + \ldots,$$

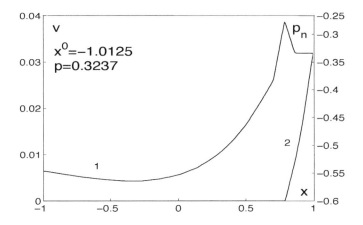

FIGURE 9.59
Distributions of the jump of the normal displacements of the surface crack faces $v(x)$ (curve 1) and of the normal stress $p_n(x)$ (curve 2) in case of $\alpha = 0.985111$, $\lambda = 0.1$, $q^0 = -0.5$ at the crack center location $x^0 = -1.0125$.

$$c_j = \frac{1}{\pi} \int\limits_a^c [p(x)\overline{D}_j(x) + \tau(x)\overline{G}_j(x)]dx + \frac{\delta_{j0}}{2} q^0(1 - e^{-2i\alpha}), \quad j = 1, 2,$$

$$c_j^r = Re(c_j), \quad c_j^i = Im(c_j),$$

where the kernels are determined according to the formulas

$$D_0(x) = \frac{i}{2}\left[-\frac{1}{x-z^0} + \frac{1}{x-\overline{z}^0} - \frac{e^{-2i\alpha}(\overline{z}^0-z^0)}{(x-\overline{z}^0)^2} \right], \quad G_0(x) = \frac{1}{2}\left[\frac{1}{x-z^0} \right.$$

$$+\frac{1-e^{-2i\alpha}}{x-\overline{z}^0} - \frac{e^{-2i\alpha}(x-z^0)}{(x-\overline{z}^0)^2} \right], \quad D_1(x) = \frac{ie^{-i\alpha}}{2(x-\overline{z}^0)^2}\left[1 - e^{-2i\alpha} \right.$$

$$\left. -\frac{2e^{-2i\alpha}(\overline{z}^0-z^0)}{x-\overline{z}^0} \right] + \frac{ie^{i\alpha}}{2}\left[-\frac{1}{(x-z^0)^2} + \frac{e^{-2i\alpha}}{(x-\overline{z}^0)^2} \right], \quad G_1(x) = \frac{e^{-i\alpha}}{2(x-\overline{z}^0)^2}\left[1 \right.$$

$$\left. -e^{-2i\alpha} - \frac{2e^{-2i\alpha}(x-z^0)}{x-\overline{z}^0} \right] + \frac{e^{i\alpha}}{2}\left[\frac{1}{(x-z^0)^2} + \frac{e^{-2i\alpha}}{(x-\overline{z}^0)^2} \right],$$

(9.224)

and $\theta(x)$ is the step function ($\theta(x) = -1$ for $x < 0$ and $\theta(x) = 1$ for $x \geq 0$) and δ_{0j} is the Kronecker symbol ($\delta_{0j} = 1$ for $j = 0$ and $\delta_{0j} = 0$ for $j \neq 0$). Let us compare the asymptotically ($k_{1a}^{\pm}$ and $k_{2a}^{\pm}$) and numerically ($k_{1n}^{\pm}$ and $k_{2n}^{\pm}$) obtained solutions of the problem for the case when $y^0 = -0.4$, $\delta = 0.1$, $\alpha = \pi/2$, $\lambda = 0.1$, and $q^0 = -0.005$. These solutions are represented in Fig. 9.73. It follows from Fig. 9.73 that the asymptotic and numerical solutions are almost identical except for the region where the numerically obtained $k_1^+(x^0)$ is close to zero. The difference is mostly caused by the fact

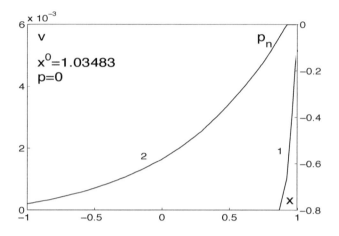

FIGURE 9.60
Distributions of the jump of the normal displacements of the surface crack
faces $v(x)$ (curve 1) and of the normal stress $p_n(x)$ (curve 2) in case of $\alpha =$
0.985111, $\lambda = 0.1$, $q^0 = -0.5$ at the crack center location $x^0 = 1.034830$.

that the used asymptotic solutions involve only two terms, i.e., the accuracy
of these asymptotic solutions is $O(\delta^2)$ for small δ. More accurate asymptotic
solutions are given in Section 9.4 and papers by Kudish [13, 14]. However,
according to the two-term asymptotic solutions the maximum values of $k_1^{\pm}$
differ from the numerical ones by no more than 1.4%. One can expect to get
much higher precision if $\delta < 0.1$ and $y^0 \gg \delta$.

The results of this section are obtained for the cases when the size of the
externally loaded lubricated contact region is greater or comparable with the
crack size. In the opposite cases when the size of the lubricated contact region is much smaller than the crack size, the situation can be modeled by the
presence of normal and tangential point forces applied to the contact through
the lubricant layer. That simplifies equations (9.178) and (9.179). In fact, the
integrals in the right-hand sides of (9.178) and (9.179) can be replaced by
local terms related to the position of the point forces. In these cases qualitatively and quantitatively surface crack-lubricant interaction may change
significantly. The anticipated change is due to the fact that for point forces
the sizes of the regions with compressive and tensile stresses may be comparable with the crack size or even smaller. Fortunately, in most of important
for practice cases during most of their fatigue lives cracks are much smaller
than the contact region and the above results hold.

Let us summarize the above analysis and its results. The new formulation
of the problem given above allows to link the material fatigue failure to two
causes: (1) propagation of subsurface cracks and (2) lubricant–surface crack

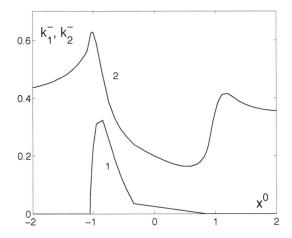

FIGURE 9.61
Distributions of the stress intensity factors k_1^- (curve 1) and k_2^- (curve 2) in case of a surface crack: $\alpha = 0.729728$, $\lambda = 0.1$, and $q^0 = -0.5$.

interaction that may lead to the surface–originated pitting. The lubricant-crack interaction and partial contact of crack faces are considered. The problem is reduced to a nonlinear problem with one-sided constraints. A new fast converging iterative numerical method for a surface or subsurface crack interacting with a lubrication layer is proposed. The method is based on the Gauss-Chebyshev quadratures and differs from the previously used methods as it allows to determine simultaneously the displacement jumps of and the normal stress applied to the crack faces. The method allows to determine the location of cavities fully or partially filled with lubricant and the segments within crack which faces are in contact with each other. Moreover, it provides for conservation of lubricant volumes trapped within closed surface crack cavities.

Using the new numerical method examples for one surface and one subsurface cracks are considered. It is demonstrated that under certain conditions the asymptotic analytical solutions for small subsurface cracks provide a very good approximation to the problem solution. It is shown that surface and subsurface cracks may develop multiple cavities as load moves along the half-plane boundary. For surface cracks the effect of lubricant-crack interaction dominates over the effects of frictional and residual stresses. It was shown that the solutions for partially open surface and subsurface cracks cannot be obtained without the proposed inherent conditions for these cracks. The numerical results indicate that the normal stress intensity factor k_1^- for surface cracks may be several orders of magnitude greater than the one for similar

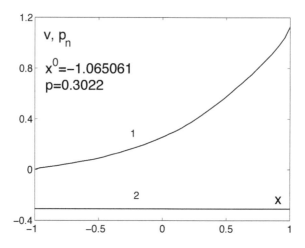

FIGURE 9.62
Distributions of the jump of the normal displacements of the surface crack
faces $v(x)$ (curve 1) and of the normal stress $p_n(x)$ (curve 2) in case of $\alpha =$
0.729728, $\lambda = 0.1$, $q^0 = -0.5$ at the crack center location $x^0 = -1.065061$.

subsurface cracks. That represents the mechanism of the "lubrication wedge
effect." The elevated pressure applied to crack faces leads to fast crack propagation and subsequent material failure.

9.6.6 Origin of Contact Fatigue. Contact Fatigue of Drivers versus Followers

In this section the mechanical model for lubricated elastic solids weakened by
cracks, which was studied in Subsections 9.6.4 and 9.6.6, is used to explain
mechanisms of surface and subsurface originated contact fatigue. Based on the
crack analysis, it is shown that pitting has predominantly subsurface origin.
Moreover, an explanation of the difference in fatigue behavior of followers and
drivers involved in sliding motion is presented.

In the early stages of the bearing and gear industries development when
bearing and gear parts were made from steel of lesser quality than nowadays, most of the fatigue failures were of subsurface origin. That was reflected
by Lundberg and Palmgren [40] in their model of bearing fatigue failure.
Since that time the quality of steel has been improved significantly. Lately,
it has been noticed that in many cases bearings and gears experience a surface or near surface originated fatigue failure. That cannot be explained by
the Lundberg-Palmgren model. This situation caused the development of a
number of new models by Ioannides and Harris [41], Coy et al. [42], Kaneta

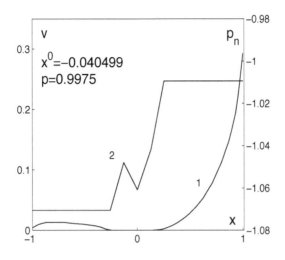

FIGURE 9.63
Distributions of the jump of the normal displacements of the surface crack faces $v(x)$ (curve 1) and of the normal stress $p_n(x)$ (curve 2) in case of $\alpha = 0.729728$, $\lambda = 0.1$, $q^0 = -0.5$ at the crack center location $x^0 = -0.040499$.

et al. [1, 2], Kudish [4, 5, 13, 39, 43, 44], Tallian [45], Blake [46], Kudish and Burris [47, 48]. Most of these models can predict only subsurface–originated fatigue. The models that take into consideration surface originated fatigue do not take into account lubricant–surface interaction at all or do it in a very simplistic and approximate manner.

It is well known that lubrication benefits contact surfaces by reducing friction and preventing direct surface contact. However, lubricants may be very detrimental to fatigue life of lubricated surfaces with certain crack–like defects. This happens when high pressure lubricant penetrates into surface cracks and literally causes rupture of material after very few loading cycles.

Some attempts of theoretical studying the phenomenon of subsurface and surface crack-lubricant interaction were made by Kaneta et al. [1, 2], Murakami et al. [3], Kudish [4, 5, 49], Xu and Hsia [6], and others. A review of the theoretical studies on crack propagation under action of contact loading is presented by Panasyuk, Datsyshin, and Marchenko [7] (see also references therein).

An experimental study of crack behavior in lubricated contacts is extremely hard to do. However, Murakami et al. [37] conducted a very well planned series of experiments and a related theoretical study of a surface/subsurface crack behavior. It was shown experimentally that crack propagation is at least as important as crack initiation, and the direction of the surface frictional stress

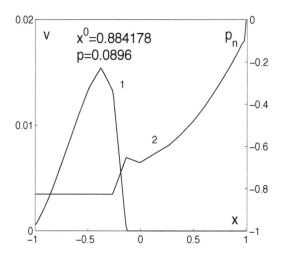

FIGURE 9.64
Distributions of the jump of the normal displacements of the surface crack faces $v(x)$ (curve 1) and of the normal stress $p_n(x)$ (curve 2) in case of $\alpha = 0.729728$, $\lambda = 0.1$, $q^0 = -0.5$ at the crack center location $x^0 = 0.884178$.

plays the defining role in determining crack propagation. We will analyze some of the observations of [37] in more detail later based on our theoretical results. These observations are as follows:

Observation (1). After an arrowhead crack was observed at $N = 1.15 \cdot 10^6$ cycles, the rotation direction was reversed while the follower remained unchanged. The original arrowhead immediately stopped propagating and after a further $N = 2.7 \cdot 10^5$ cycles the crack had still not grown, but new pits had appeared at other locations of the follower. The direction of the arrowheads of the new pits was opposite to that of the original crack. These results show that the usually used in some models assumptions that contact fatigue is determined by just the magnitude (regardless of the sign) of the shear, maximum orthogonal or Mises stress contradict the experimental data. In fact, the experiments show the direct dependence of fatigue crack growth and, therefore, contact fatigue on the directions of contact load motion and frictional force. In addition, these experiments prove that contact fatigue life is not determined by the appearance of the first fatigue crack as it is assumed by Lundberg and Palmgren [40] and Ioannides and Harris [41]. It is shown experimentally that fatigue cracks existed in the follower for at least about half of its fatigue life.

Observation (2). The arrowhead crack of the follower formed a pit a few cycles after it was observed.

Observation (3). The growth of cracks in the driver required many cycles

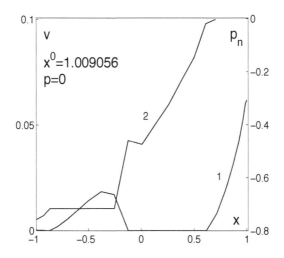

FIGURE 9.65
Distributions of the jump of the normal displacements of the surface crack faces $v(x)$ (curve 1) and of the normal stress $p_n(x)$ (curve 2) in case of $\alpha = 0.729728$, $\lambda = 0.1$, $q^0 = -0.5$ at the crack center location $x^0 = 1.009056$.

even though the cracks have a shape similar to the one observed in micro-pits.

Observation (4). The difference in crack growth behavior between the follower and the driver was caused by the difference in the mechanical conditions of crack growth, not in crack initiation.

The theoretical studies presented in these papers treat the surface crack-lubricant interaction in a very simplistic manner. For example, in Murakami et al. [37] surface cracks were not allowed to develop multiple cavities fully or partially filled with lubricant. Ultimately, the way the crack configuration is accounted for affects the predictions of crack stress intensity factors, propagation, and fatigue life. The latter experimental results were confirmed by Nelias et al. [50]. The authors of [50] also indicated that the orientation of butterflies and micro-cracks is not parallel to the main crack; that is, the main crack is not oriented along the shear stress. The defining role of the friction stress in the direction of fatigue cracks is also demonstrated in tests conducted by Hoeprich [51]. His experiments showed that in case of direct asperity contacts the crack initiation time is very small, shallow micropitting occurs on early stages of fatigue test, and cracks grow in the direction opposite to the direction of the friction stress applied to asperities.

The goals of this Section are

(A) To show that in most cases cracks may develop multiple cavities. These cavities affect the material stress state in the vicinity of a crack and cannot be

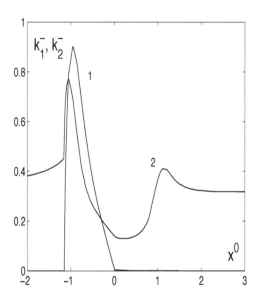

FIGURE 9.66
Distributions of the stress intensity factors k_1^- (curve 1) and k_2^- (curve 2) in case of a surface crack: $\alpha = \pi/6$, $\lambda = 0.1$, and $q^0 = -0.5$ (after Kudish and Burris [20]). Reprinted with permission from Kluwer Academic Publishers.

accounted for without the proper boundary conditions (see Kudish [4, 49]).

   (B) To show that normal stress intensity factors for surface cracks may be orders of magnitude higher than for similar subsurface ones.

   (C,D) Using the analytical solutions for small subsurface cracks and numerical solutions for surface cracks.

   (C) To determine the predominant origin of contact fatigue in the absence of contaminants and initial surface crack-like defects and to explain the cases of surface originated fatigue.

   (D) To explain the difference between fatigue behavior of followers and drivers involved in sliding motion.

9.6.6.1   General Assumptions

Let us assume that an elastic half-plane boundary is loaded by a moving normal $p(x^0)$ and tangential $\tau(x^0)$ contact stresses on an interval $[x_i, x_e]$ in the global coordinate system (x^0, y^0) with the x^0-axis directed along the half-plane boundary. The general view of the contact is given in Fig. 9.35. The stresses are transmitted through an incompressible lubrication film. The half-plane is pre-stressed by a compressive residual stress q^0 and is weakened by a

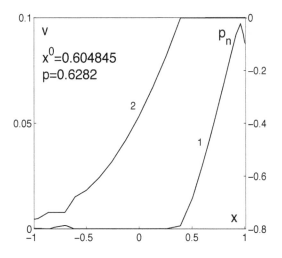

FIGURE 9.67
Distributions of the jump of the normal displacements of the surface crack faces $v(x)$ (curve 1) and of the normal stress $p_n(x)$ (curve 2) in case of $\alpha = \pi/6$, $\lambda = 0.1$, $q^0 = -0.5$ at the crack center location $x^0 = 0.604845$.

straight surface or subsurface crack. Friction is neglected everywhere on crack faces except for the negligibly small region next to the mouth of a surface crack. The analysis of crack behavior is done in a local coordinate system (x, y) connected with crack faces (see Fig. 9.35).

Under the above assumptions the stress-strain state of a crack in a quasi-static approximation is described by a system of integro-differential equations from Subsection 9.6.1 (also see Kudish [4, 49]) with respect to functions v, u, and p_n that are the normal and tangential displacement jumps of the crack faces and the normal stress applied to them, respectively. Surface and subsur-face cracks are allowed to have multiple cavities. For surface cracks a number of boundary conditions characterizing lubricant behavior within crack cavities including development of cavities some of which are fully and others partially filled with lubricant, pressure rise in the cavities fully filled with lubricant, etc., is taken into account. Because of relative motion of a crack with respect to the position of the loaded by $p(x^0)$ and $\tau(x^0)$ interval $[x_i, x_e]$ the stress state of the material surrounding cracks changes significantly. That causes changes in crack configuration and the boundary conditions describing these crack configuration at different locations. These boundary conditions depend on whether a crack is of a surface or subsurface nature, and whether its faces are in contact with each other. We will assume that lubricant trapped within cavities of a surface crack is in hydrostatic equilibrium and the pressure of

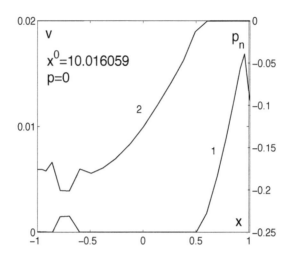

FIGURE 9.68
Distributions of the jump of the normal displacements of the surface crack faces $v(x)$ (curve 1) and of the normal stress $p_n(x)$ (curve 2) in case of $\alpha = \pi/6$, $\lambda = 0.1$, $q^0 = -0.5$ at the crack center location $x^0 = 10.016059$.

the lubricant vapor can be neglected.

In linear fracture mechanics the normal and tangential stresses in material in the vicinity of crack tips are given by formulas (Panasyuk et al. [10])

$$\sigma_1 = r^{-1/2}[k_1^{\pm} f_1(\theta) + k_2^{\pm} f_2(\theta)] + O(r^0),$$

$$\sigma_2 = r^{-1/2}[k_1^{\pm} g_1(\theta) + k_2^{\pm} g_2(\theta)] + O(r^0),$$

$$\tau_{12} = r^{-1/2}[k_1^{\pm} f_2(\theta) + k_2^{\pm} g_1(\theta)] + O(r^0) \ for \ r \to 0,$$

where σ_1, σ_2, and τ_{12} are the normal and tangential stresses in the local coordinate system, $k_1^{\pm}$ and $k_2^{\pm}$ are the normal and shear stress intensity factors that characterize the stress-strain state of material, r and θ are the distance from a crack tip to and the angle between the x-axis and the direction toward the point of interest, respectively, $f_1(\theta)$, $f_2(\theta)$, $g_1(\theta)$, and $g_2(\theta)$ are certain functions of angle θ (Panasyuk et al. [10]). The stress intensity factors control propagation of fatigue cracks. Assuming that the load (i.e., stresses $p(x^0)$ and frictional stress $\tau(x^0)$) are moving along the x^0 axis we can write down some of the Paris-like equations describing propagation of fatigue cracks that have the form (Yarema [33])

$$\frac{dl}{dN} = F(\max_{-\infty < x^0 < \infty} \triangle k_1^{\pm}), \tag{9.225}$$

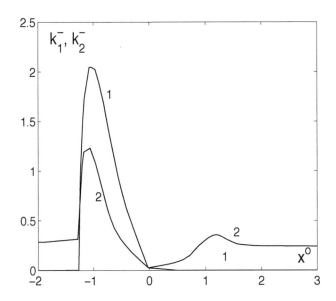

FIGURE 9.69
Distributions of the stress intensity factors k_1^- (curve 1) and k_2^- (curve 2) in case of a surface crack: $\alpha = 0.339837$, $\delta = 0.3$, and other parameters as in the previous case of a surface crack (after Kudish [19]). Reprinted with permission of the STLE.

where l is the fatigue crack half-length, N is the number of applied loading cycles, F is a certain monotonically increasing function such as, for example, $F(x) = Ax^n$ (A is a positive constant, n is a constant that usually varies between 6 and 12), and maximum of the variation of the normal stress intensity factors $\triangle k_1^{\pm}$ is taken with respect to x^0. Therefore, because of high values of n even a small increase in $\triangle k_1^{\pm}$ leads to a significant speed up of fatigue crack growth. In cases of smooth contact surfaces subjected to non-tensile residual stress $q^0 \leq 0$, the maximum variation of the normal stress intensity factors $\max\limits_{-\infty < x^0 < \infty} \triangle k_1^{\pm}$ is identical to just the maximum of the stress intensity factors $k_1^{\pm}$ over one loading cycle, i.e., $\max\limits_{-\infty < x^0 < \infty} \triangle k_1^{\pm} = \max\limits_{-\infty < x^0 < \infty} k_1^{\pm}$. In cases of tensile residual stress $q^0 > 0$, we have $\max\limits_{-\infty < x^0 < \infty} \triangle k_1^{\pm} \leq \max\limits_{-\infty < x^0 < \infty} k_1^{\pm}$ and, therefore, in (9.225) we have to use the the maximum variation of the normal stress intensity factors $\max\limits_{-\infty < x^0 < \infty} \triangle k_1^{\pm}$. However, for reasonably low tensile residual stress $q^0 > 0$ due to highly compressive stress field in front of the contact, we still have $\max\limits_{-\infty < x^0 < \infty} \triangle k_1^{\pm} = \max\limits_{-\infty < x^0 < \infty} k_1^{\pm}$.

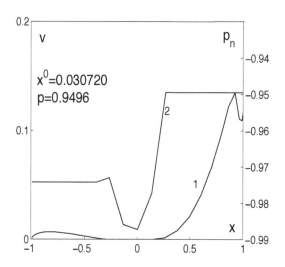

FIGURE 9.70
Distributions of the jump of the normal displacements of the surface crack
faces $v(x)$ (curve 1) and of the normal stress $p_n(x)$ (curve 2) in case of a
surface crack: $\alpha = 0.339837$, $\delta = 0.3$, and other parameters as in the previous
case of a surface crack at the location of the surface crack center $x^0 = 0.030720$
(after Kudish [19]). Reprinted with permission of the STLE.

This consideration leads to a simple conclusion. The tensile residual stress
$q^0 > 0$ increases the value of $\max\limits_{-\infty < x^0 < \infty} k_1^{\pm}$ and, therefore, when the above de-
scribed equality between $\max\limits_{-\infty < x^0 < \infty} \triangle k_1^{\pm}$ and $\max\limits_{-\infty < x^0 < \infty} k_1^{\pm}$ takes place pitting
occurs earlier than for non-tensile residual stress $q^0 \leq 0$.

For the given values of Young's modulus E and Poisson's ratio ν of the solid
material, boundaries of the loaded region x_i and x_e, angle of crack orientation
α, position of the crack center $z^0 = (x^0, y^0)$, crack half-length l, lubricant
density ρ, material residual stress q^0, contact pressure $p(x^0)$, and contact frictional stress $\tau(x^0)$, the problem solution consists of functions $v(x)$, $u(x)$, and
$p_n(x)$ and is determined by the equations presented in Section 9.4. In the
further analysis, we will be using the dimensionless variables and equations
introduced in Section 9.6. Having solved the problem the stress intensity factors and at the crack tips can be obtained from (9.202)

$$k_1^{\pm} + i k_2^{\pm} = \mp \lim_{x \to \pm 1} \sqrt{1 - x^2}[v'(x) + iu'(x)], \qquad (9.226)$$

where the upper signs correspond to the crack tip with $x = 1$ and the lower
signs correspond to the crack tip with $x = -1$.

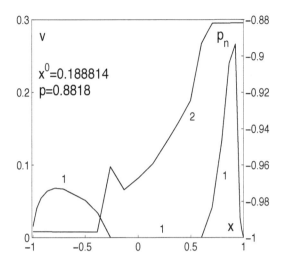

FIGURE 9.71
Distributions of the jump of the normal displacements of the surface crack faces $v(x)$ (curve 1) and of the normal stress $p_n(x)$ (curve 2) in case of a surface crack: $\alpha = 0.339837$, $\delta = 0.3$, and other parameters as in the previous case of a surface crack at the location of the surface crack center $x^0 = 0.188814$ (after Kudish [19]). Reprinted with permission of the STLE.

9.6.6.2 Some Data and Problem Properties Used in Analysis

It is assumed that the pressure $p(x^0) = \sqrt{1 - (x^0)^2}$ and tangential stress $\tau(x^0) = -\lambda p(x^0)$ applied to the boundary of the elastic half-plane (see Fig. 9.35) move along it from right to left. The contact tangential stress is caused by sliding one contact surface over another. The values of the friction coefficient λ used throughout the section cover the most important for practice regimes of elastohydrodynamic ($0.002 \leq \lambda \leq 0.08$) and mixed ($0.08 \leq \lambda \leq 0.2$) lubrication (see Hamrock [53], p. 8). The boundaries of the loaded lubricated region are $a = -1$ and $c = 1$.

It follows from equations (9.178)-(9.192), and (9.202) that if $c = -a$ and $p(-x^0) = p(x^0)$, $\tau(-x^0) = \tau(x^0)$ (for example, $\tau(x^0) = -\lambda p(x^0)$) for $a \leq x^0 \leq c$, then for surface and subsurface cracks

$$k_1^{\pm}(x^0, y^0; -\lambda, -\alpha) = k_1^{\pm}(-x^0, y^0; \lambda, \alpha),$$
$$k_2^{\pm}(x^0, y^0; -\lambda, -\alpha) = -k_2^{\pm}(-x^0, y^0; \lambda, \alpha).$$
(9.227)

Therefore, the behavior of cracks with $\alpha \leq 0$ can be understood and determined based on the latter formulas and the behavior of cracks with $\alpha \geq 0$.

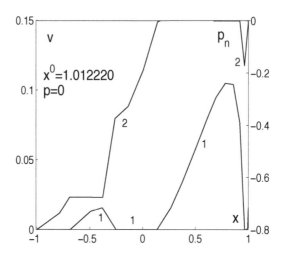

FIGURE 9.72

Distributions of the jump of the normal displacements of the surface crack faces $v(x)$ (curve 1) and of the normal stress $p_n(x)$ (curve 2) in case of a surface crack: $\alpha = 0.339837$, $\delta = 0.3$, and other parameters as in the previous case of a surface crack at the location of the surface crack center $x^0 = 1.012220$ (after Kudish [19]). Reprinted with permission of the STLE.

The results of the preceding subsection show that in most cases surface and subsurface cracks develop cavities and zones of direct contact of crack faces. These cavities may significantly affect the material stress state in the vicinity of a crack and cannot be accounted for without the proper boundary conditions. That substantiates item (A).

For subsurface cracks, the data from Table 9.1 indicates that if $\lambda > 0$ and $\alpha \neq 0$, then the maximum of $k_1^{\pm}$ is reached behind the loaded region and the smaller the magnitude of the compressive residual stress q^0 and/or greater the friction coefficient λ the greater is the maximum value k_1 of the normal stress intensity factor (see preceding subsection). Moreover, k_1 is significantly greater for subsurface cracks oriented in the direction opposite to the direction of the frictional stress than for cracks with similar parameters but oriented along the frictional stress (see cases for $\alpha = \pi/6$ and $\alpha = -\pi/6$ in the preceding subsection). For cracks located deeper than the discussed ones and subjected to the action of smaller friction, the maximum stress intensity factor k_1 is even smaller than in the considered cases. For example, for $\alpha = \pi/6$, $\delta = 0.2$, $y^0 = -0.275$, $\lambda = 0.1$, and $q^0 = -0.01$, the maximum stress intensity factor $k_1 = 1.4 \cdot 10^{-4}$ and for $\alpha = \pi/6$, $\delta = 0.2$, $y^0 \leq -0.3$, $\lambda \leq 0.1$, and

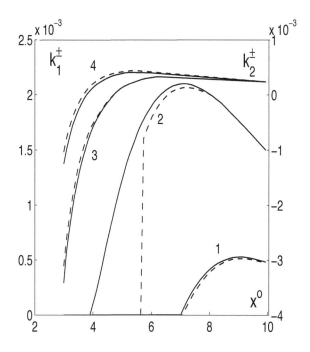

FIGURE 9.73
Comparison of the two-term asymptotic expansions $k_{1a}^{\pm}$ and $k_{2a}^{\pm}$ with the numerically calculated stress intensity factors $k_{1n}^{\pm}$ and $k_{2n}^{\pm}$ obtained for $y^0 = -0.4$, $\delta = 0.1$, $\alpha = \pi/2$, $\lambda = 0.1$, and $q^0 = -0.005$. Solid curves are numerical results while dashed curves are asymptotical results. (k_{1n}^{-} group 1, k_{1n}^{+} group 2, k_{2n}^{-} group 3, k_{2n}^{+} group 4) (after Kudish [19]). Reprinted with permission of the STLE.

$q^0 \leq -0.01$ we have $k_1 = 0$. That is why Nelias et al. [50] observed that on both driver and follower disks the depth of fatigue micro-cracks $| y^0 |$ was much smaller than the Hertzian contact half-width a_H ($| y^0 | < 1.5 \cdot 10^{-3} a_H$ for the driver and $3 \cdot 10^{-3} a_H \leq | y^0 | \ll a_H$ for the follower).

Let us reiterate some observations made in the preceding subsection. Figure 9.56 shows a strong dependence of the normal stress intensity factor $k_1^{\pm}$ on the value of the friction coefficient λ while Fig. 9.57 demonstrates a relatively weak dependence of the shear stress intensity factor $k_2^{\pm}$ on the friction coefficient λ. This fact supports the experimentally observed strong dependence of contact fatigue life on frictional stress and suggests that contact fatigue life is controlled not by a shear stress but a tensile one, that is, contact fatigue is caused by Mode I crack propagation.

Numerical results show that for surface cracks with $\alpha > 0$, which may open and be completely filled with lubricant, the influence of the values of the friction coefficient λ and the residual stress q^0 within reasonable bounds $(0.002 \leq \lambda \leq 0.2, \; -0.5 \leq q^0 \leq 0.05)$ on the normal stress intensity factor k_1^- at the subsurface tip of surface cracks is negligibly small because of the dominant role of the high lubricant pressure penetrating such cracks.

Surface cracks with $\alpha < 0$ are closed completely while the external (contact) load approaches them and passes over them. Depending on the value of the angle $\alpha < 0$ they may open up completely, but most often just partially, behind the loaded region $(x^0 > 1)$ where the material of the half-plane is under a tensile stress. For such a surface crack partially open just near its mouth, the stress intensity factor $k_1^- = 0$. Otherwise, if a surface crack is completely open the stress intensity factor k_1^- attains relatively small values typical for subsurface cracks (see Table 9.1).

The maximum values k_1 of the normal stress intensity factors for surface cracks are given in Table 9.2. For smaller values of angle α (larger crack half-length δ), the maximum of the stress intensity factor k_1^- is larger. The absolute maximum of k_1^- is reached when the crack mouth just enters the loaded region. It is mostly due to the lubricant pressure applied to the crack faces. The second local maximum of k_1^- (if it exists) occurs behind the loaded region and it is much smaller than the first one. The data from Table 9.2 show that for smaller angles α the absolute maximum of k_1^- is reached at a point to the right from the left boundary of the loaded region, $x^0 = -1$. That is due to the competing influence of the lubricant pressure applied to the half-plane boundary and to crack faces.

9.6.6.3 Discussion of Numerical Results

It follows from Table 9.1 that the maximum of the maxima of stress intensity factors $k_1^{\pm}$ for the above subsurface cracks is $0.66 \cdot 10^{-2}$. From Table 9.2, the positive minimum of the maxima of the stress intensity factors k_1^- for the considered surface cracks is 0.327612 while the maximum of the maxima of k_1^- is 2.04985. Therefore, the maximum of stress the intensity factors k_1^- for surface cracks may be at least 50 times greater than the maximum of $k_1^{\pm}$ for similar subsurface cracks. For similar surface and subsurface cracks with $\alpha = \pi/6$ subjected to the residual stresses $q^0 = -0.5$ and $q^0 = -0.01$ the maxima of the stress intensity factors $k_1^{\pm}$ are equal to 0.907249 and $0.66 \cdot 10^{-2}$ (see Tables 9.1 and 9.2), respectively. Hence, even under favorable conditions (reduced compressive residual stress $q^0 = -0.01$) the maximum of the stress intensity factors $k_1^{\pm}$ for subsurface cracks may be several orders of magnitude smaller than the maximum of k_1^- for similar surface cracks (also see comparison of the stress intensity factors for surface and subsurface cracks given in Tallian, Hoeprich, and Kudish [67]). In practice, it means that the presence of crack-like surface defects may reduce fatigue life of a machine part by several (4-8) orders of magnitude. That is the actual result of the lubricant wedge effect, and it substantiates the claim of item (B).

9.6.6.4 Origin of Contact Fatigue (Pitting). Contact Fatigue of Drivers and Followers Involved in Sliding Motion

Let us try to understand the origination of pitting (item (C)) and to explain why in fatigue tests of solids involved in sliding and made of identical elastic materials followers usually fail earlier than drivers (item (D)). The experimental studies of these phenomena are presented by Murakami et al. [37], Nelias et al. [50], and Soda and Yamomoto [52]. In particular, in [52] it was observed that fatigue life of drivers is 30-50 times greater than fatigue life of followers made of the same material and subjected to the same loading conditions (also see Fig. 8.1).

We will assume that contact surfaces are sufficiently smooth and at no time contaminants or debris particles are present in the loaded region unless it is stated otherwise. A driver we will define as one of the contacting solids for which the directions of sliding motion and frictional stress are opposite while a follower will be the solid for which the directions of sliding motion and frictional stress coincide. According to the earlier discussed experimental data, it is reasonable to assume that the directions of crack propagation are controlled by tensile stress (see discussion of Figs. 9.56 and 9.57 in Subsection 9.6.4 as well as Panasyuk et al. [10] and Kudish [39]). Therefore, the instantaneous angles of crack propagation with respect to the direction along which the normal stress intensity factors reach their maximum values are (Panasyuk et al. [10])

$$\theta^{\pm} = 2 \arctan \frac{k_1^{\pm} - \sqrt{(k_1^{\pm})^2 + 8(k_2^{\pm})^2}}{4k_2^{\pm}}. \tag{9.228}$$

Equation (9.228) was used by Datsyshyn and Panasyuk [34] for calculations of surface crack propagation paths.

It follows from (9.227) and (9.228) that

$$\theta^{\pm}(x^0, y^0; -\lambda, -\alpha) = -\theta^{\pm}(-x^0, y^0; \lambda, \alpha).$$

By assuming that locally, in the vicinity of crack tips, the stress field is symmetric and the stress is tensile we set $\theta^{\pm} = 0$. It means that we assume that the tips of fatigue cracks propagate in the directions $\alpha^{\pm}$ determined by the solutions of the equations

$$k_2^{\pm}(x^0, y^0; \alpha^{\pm}) = 0. \tag{9.229}$$

Moreover, such a direction is orthogonal to the direction along which the tensile stress and coincides with the direction along which the stress intensity factors $k_1^{\pm}$ reach their maximum values. For small subsurface cracks (i.e., for $\delta \ll 1$) located far from the half-plane boundary the single term asymptotic solutions give $k_1^{\pm} = k_1 + O(\delta)$, $k_2^{\pm} = k_2 + O(\delta)$, and $\alpha^{\pm} = \alpha + O(\delta)$ satisfy equations (see Subsection 9.6.1 and Kudish [14, 44, 39])

$$\tan(2\alpha) = -2y^0 \int\limits_a^b (x - x^0) T(x, x^0, y^0) dx / \Big\{ \tfrac{\pi}{2} q^0 + \int\limits_a^b [(x - x^0)^2$$

$$-(y^0)^2] T(x, x^0, y^0) dx \Big\}, \quad T(x, x^0, y^0) = \tfrac{y^0 p(x) + (x - x^0) \tau(x)}{[(x - x^0)^2 + (y^0)^2]^2}. \tag{9.230}$$

Taking into account just the one-term asymptotic solutions is sufficient because most of their lives fatigue cracks remain small (see Subsection 9.7.5). Therefore, the contributions provided by higher–order terms are negligibly small. On the other hand, the great benefit of the one-term asymptotic approximation is the fact that in this approximation straight cracks remain straight during their entire lives.

Equation (9.230) has two solutions: α and $\alpha + \pi/2$. Only one of these two propagation directions, say, α, provides maximum to k_1, which enables the subsurface fatigue crack to propagate. It follows from (9.227) and (9.230) that for even functions $p(x^0)$ and $\tau(x^0)$ the fatigue crack propagation direction changes with the reversal of the frictional stress $\tau(x^0)$, i.e.,

$$\alpha(-x^0, y^0; -\lambda) = -\alpha(x^0, y^0; \lambda), \tag{9.231}$$

and it is independent from the crack half-length δ. This result is supported by the experimental data from Murakami et al. [37] (see *Observation* (1)). For three cases of input data with $p(x^0) = \sqrt{1 - (x^0)^2}$ and $\tau(x^0) = -\lambda p(x^0)$ (Case 1: $\lambda = 0.04$, $q^0 = -0.0021$, Case 2: $\lambda = 0.03$, $q^0 = -0.0021$, Case 3: $\lambda = 0.04$, $q^0 = -0.0025$) with other parameters $y^0 = -0.2$, $a = -1$, $b = 1$ the calculation results for such angle α are presented in Fig. 9.74. These graphs of α are shown only for those x^0 for which the stress intensity factors k_1^+ and/or k_1^- are positive. Therefore, for $\lambda > 0$ in front, under, and close behind the loaded region in the material of the half-plane the resultant stress is compressive. Farther behind the loaded region the resultant stress is tensile, that is, $k_1^\pm > 0$, and in Cases 1-3 fatigue cracks may propagate in the directions prescribed by angles $\alpha = 0.307820$, $\alpha = 0.017825$, and $\alpha = 0.077251$, respectively, at which the stress intensity factors $k_1^\pm = k_1 + O(\delta)$, $\delta \ll 1$, reach their maximum values.

Therefore, if the stress field near each of fatigue cracks changes relatively slowly, then equation (9.229) have two solutions: $\alpha_1^\pm = \alpha_1^-$ and $\alpha_2^\pm = \alpha_2^-$. The actual direction of fatigue crack propagation $\alpha^\pm = \alpha + O(\delta)$, $\delta \ll 1$ is determined as one of the two angles from α_1^+ and α_2^+ for which the stress intensity factors $k_1^\pm = k_1 + O(\delta)$, $\delta \ll 1$, reach their maximum values Kudish [39]. Based on equation (9.231) and Fig. 9.74 the angles at which fatigue cracks propagate in the driver $\alpha_d^\pm = \alpha_d$ and follower $\alpha_f^\pm = \alpha_f$ (see also sketches of a driver and follower in Fig. 9.75) are related as follows

$$\alpha_d = -\alpha_f, \quad \alpha_d < 0, \quad \alpha_f > 0. \tag{9.232}$$

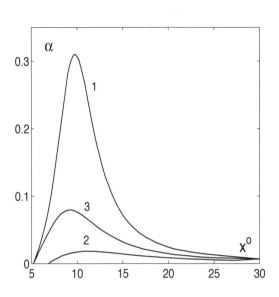

FIGURE 9.74
The distribution of the crack orientation angle α obtained from equation (9.229) for three cases: case 1 - $\lambda = 0.04$, $q^0 = -0.0021$, case 2 - $\lambda = 0.03$, $q^0 = -0.0021$, case 3 - $\lambda = 0.04$, $q^0 = -0.0025$; $y^0 = -0.2$, $a = -1$, $b = 1$ (after Kudish [19]). Reprinted with permission of the STLE.

In practically important applications, the speed U_l of the load motion along a driver and follower surfaces is much smaller than the speed of sound c_s in elastic materials. Therefore, one can ignore the negligibly small difference (of the order of $O(U_l^2/c_s^2)$) between the elastic stress state in the cases of moving and motionless load (see Galin [15]). Thus, the stress states in the driver ($\lambda > 0$) and follower ($\lambda < 0$) with subsurface cracks subjected to a repeated moving load are practically the mirror images of one another (see (9.227), (9.231) and Fig. 9.75).

The fatigue crack initiation period in steels is short in comparison with the total fatigue life (see Section 8.5 and [47, 39]). Suppose the materials of the driver and follower are identical, including the size and distribution of subsurface material defects. Taking into account the symmetry properties of the stress intensity factors $k_1^{\pm}(x^0, y^0; \lambda, \alpha)$ and $k_2^{\pm}(x^0, y^0; \lambda, \alpha)$ the anti-symmetry of the propagation angles $\alpha^{\pm}(x^0, y^0; \lambda)$ with respect to x^0 for positive and negative coefficients of friction λ (see (9.227)) in the driver ($\lambda > 0$) and follower ($\lambda < 0$) subsurface fatigue cracks nucleate simultaneously and propagate at the same rate as long as cracks remain subsurface. In other words, from the fatigue point of view the situations with the driver and follower are identical

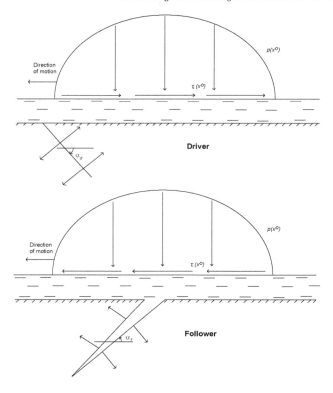

FIGURE 9.75
Sketches of a driver and follower (after Kudish [19]). Reprinted with permission of the STLE.

as long as the damage remains subsurface, that is, their fatigue lives should be equal. Thus, the experimentally observed by Murakami et al. [37] and Nelias et al. [50] difference in fatigue lives of drivers and followers cannot be explain by just initiation or propagation of subsurface cracks (see *Observation* (4)).

Fortunately, this situation can be resolved by considering surface cracks. Surface cracks may be (1) nucleated at the surface or (2) emerge at the surface as a result of subsurface crack propagation. The numerical results for surface cracks show that if $\lambda > 0$ and a surface crack is oriented in such a way that it opens completely while approaching the half-plane loaded region (i.e., $\alpha_f > 0$) the stress intensity factor k_1^- is much higher than the one for a similar subsurface crack. As a result, the surface fatigue crack propagates fast and fatigue life of the material of the half-plane may be several orders of magnitude shorter than for subsurface initiated fatigue. That can be easily seen from Paris-like equation (9.225) for fatigue crack propagation. In practice, for smooth surfaces without initial surface defects and contaminants this is not the case for both the follower and driver with normally long fatigue lives.

Under these conditions, they withstand a substantial number of loading cycles, at least 10^6. Hence, even if such surface originated fatigue cracks propagate they cause a shallow wear instead of relatively deep pitting. The difference between the phenomena of wear and micropitting on one hand and pitting on the other hand is in density and scale of the damage. Thus, surface cracks of type 1) cannot be responsible for the premature pitting in followers compared with pitting in drivers and the phenomenon of pitting is predominantly of a subsurface origin. This conclusion is in agreement with *Observations* $(1) - (3)$ made by Murakami et al. [37] as materials withstand a significant number of cycles (at least over 10^5) before failure.

At the same time, in case 2) it takes a relatively long time for a subsurface nucleated crack to propagate to the surface (see *Observation* (1)). As a subsurface nucleated crack reaches the surface and becomes a surface one, it may (depending on its orientation angle α, the magnitude and direction of the frictional stress $\tau(x^0)$, and the direction of the load motion) lead to a pit formation after just a relatively small number of loading cycles (due to the "lubricant wedge effect," see *Observation* (2)) or it may withstand continuing loading for a long time. Based on the preceding analysis for the driver $\alpha_d < 0$ (see a driver sketch in Fig. 9.75) the subsurface nucleated fatigue cracks, which at some point after a significant number of loading cycles become the surface ones, do not allow lubricant to penetrate them as their mouths do not open toward the approaching load. Therefore, such cracks may stop growing or continue growing slowly, almost like subsurface ones (see *Observations* (1) and (3)). On the other hand, for the follower $\alpha_f > 0$ (see a follower sketch in Fig. 9.75) the surface cracks which were nucleated as subsurface ones and after a significant number of loading cycles reached the contact surface open up completely while the external load approaches their mouths and allow for high pressure lubricant to penetrate them. Such cracks propagate extremely fast due to large magnitude of the stress intensity factor k_1^- (see above). Therefore, the premature pitting experimentally observed in followers is caused by fast propagation of subsurface nucleated cracks that emerge at the contact surface and experience the "lubricant wedge effect." This conclusion (see item (D)) is fully supported by the test results of Murakami et al. [37] (see *Observation* (2)). Moreover, this explains the well–known experimental fact that a short time before failure test rigs with failing specimens exhibit elevated levels of vibrations.

The difference in the morphology of the fatigue damage in Tests [1]-[4] and Test [5] of Murakami et al. [37] is due to the fact that in Tests [1]-[4] the final stage of crack growth was due to a subsurface initiated crack reaching surface and interacting with lubricant while in Test [5] (case of driver damage) such a subsurface originated crack that became a surface one had no opportunity to interact with lubricant. In the latter case after becoming a surface crack it was able to open up only behind the loaded region. Therefore, the "lubricant wedge effect" does not occur for such a crack, and it behaves in a fashion similar to a subsurface one and experiences the same (low) level of stress intensity factor k_1^-. This different crack development in followers and drivers causes the differences in the damage.

There are two other important conclusions that can be drawn from this analysis. First, for smooth surfaces in the absence of initial crack-like surface defects and/or contaminants, the pitting phenomenon is predominantly initiated at some subsurface material point (see item (C)). After nucleation fatigue cracks propagate toward the material surface. In drivers even after these fatigue cracks reach the material surface they continue to grow at the pace of subsurface cracks or stop growing completely. Contrary to that, in followers after subsurface originated cracks emerge at the material surface their interaction with lubricant causes a rapid development of a pit or pits due to the "lubricant wedge effect."

Second, the latter conclusion leads to a simple explanation of often experimentally observed surface–initiated fatigue. As it was shown for smooth surfaces in the absence of contaminants and initial surface crack-like defects long fatigue lives are associated with subsurface originated fatigue. Contrary to that, short fatigue lives are associated with rough surfaces and the presence of crack-like surface defects.

Based on the presented analysis we can conclude that the cases of experimentally observed high–cycle surface–originated fatigue are caused by contaminants that damage the specimen surface late in the course of subsurface–initiated fatigue and cause fast pitting. That means that surface–originated pitting actually takes place soon after contamination occurs but long after the loading cycling started. This follows from the fact that crack-like defects directed toward the incoming lubricant and contact load create very high normal stress intensity factors k_1^-, which, in turn, lead to fast crack propagation and onset of pitting. Therefore, the key to prolonging fatigue life is the proper filtration and absence of any stress concentrators such as debris, dents, and asperities.

Most of the discussed results are typical for case hardened steels for which the maximum compressive residual stress q^0 may reach about $-0.1p_H$. For through hardened steels it is common that in a surface material layer the residual stress is tensile and it may reach about $0.05p_H$. In the latter case for subsurface cracks, the stress intensity factors $k_1^\pm$ may be several times higher (according to (9.223) and (9.224) by about $q^0 \sin 2\alpha$) than for case hardened steels but still much smaller than for similar surface cracks (see Kudish [39]). Therefore, the above analysis pertains to both case and through hardened steels.

Crack face friction and crack roughness are expected to change the results of the above analysis only slightly. Especially it is true for calculation of the maximum stress intensity factor k_1 for surface cracks experiencing the "lubrication wedge effect" for which the impact of the crack face friction and roughness is negligibly small as the crack faces do not contact each other directly. In the latter case, it is due to the fact that k_1 is the result of interaction of high fluid pressure applied to crack faces and elasticity of the solid material.

Under the conditions of direct asperity contact Hoeprich [51] observed fast onset of shallow surface fatigue (micropitting) and the fatigue crack growth in the direction opposite to the direction of the friction stress. That may be caused by highly localized stress fields near asperities. The latter depend on asperities geometry and the irregularities in the distribution of the residual stress near asperities. These localized stress fields result in plastic deformations of asperities and their cracking. Based on the above analysis we can conclude that this cracking is of a surface or shallow subsurface origin. The onset of surface cracking may be caused by the presence of contaminants and wear debris. In both cases surface and shallow subsurface cracking occur in the environment of irregular localized high stress fields. Moreover, the cracks observed near the tooth dedendum may experience the "lubrication wedge effect" due to their orientation and the direction of load motion. It is due to the fact that surface cracks directed toward the coming load always open up for lubricant to penetrate them in spite of the frictional stress direction.

The numerical results indicate that even under favorable conditions the normal stress intensity factors for surface cracks may be several orders of magnitude higher than those for similar subsurface cracks. That represents the mechanism of the "lubrication wedge effect." The elevated pressure applied to the crack faces leads to a fast crack propagation and subsequent material failure. In practice, the initial presence of crack-like surface defects may translate into a reduction of a machine part fatigue life by several orders of magnitude. It is shown that the pitting phenomenon is predominantly of a subsurface origin. The fact that in most comparable cases fatigue lives of followers is shorter than the one of drivers is explained.

9.7 Two-Dimensional Statistical Model of Contact Fatigue

Taking into consideration the experimental and theoretical results discussed in preceding sections, the basic principles of a sound mathematical model of contact fatigue should be as follows:

(1) The existence of the initial statistical defect distribution in materials should be assumed.

(2) The existing material defects such as voids and inclusions may be replaced by cracks of equivalent size and orientation.

(3) The period of crack initiation may be assumed to be much shorter than the crack propagation period.

(4) The changes of the statistical crack distribution due to fatigue crack growth in time should be reflected in the model and should be determined for any time moment.

(5) The model should involve the material stress analysis under the action of contact normal and tangential stresses as well as the residual stress.

(6) The normal and tangential stress intensity factors at crack tips caused by the stress field should be determined.

(7) Subsurface and surface cracks should be taken into account.

(8) The fatigue crack growth should be considered according to Paris equation for fatigue crack growth or a similar equation.

(9) The probability of the fatigue damage should be calculated at every material point at any time moment.

(10) The probability of pitting (or probability of survival of the solid as a whole) should be formulated based on the above–mentioned local probabilities of fatigue damage.

The detailed derivation of the new statistical model of contact fatigue life and its qualitative analysis presented below is based on [39, 55]. The validation of the model and its applicability to calculation of bearing fatigue life and some particular data as well as the quantitative analysis of the model are given in the next section. The model assumptions and their validation as well as the model properties are discussed. A parametric study of the model is performed. Some analytical formulas for relationships between fatigue life and parameters characterizing the distribution of material defects, material fatigue resistance, and stress state are obtained and analyzed.

The model is different from the previously published models of contact fatigue (see Blake [46], Ioannides and Harris [41], Tallian [56, 57], etc.) in several important aspects such as statistical treatment of material defect distribution, which changes in time, stress analysis, crack propagation versus crack initiation, etc. The new statistical model of contact fatigue is based on contact and fracture mechanics and a statistical treatment of the initial distribution of material defects. The model takes into account normal, frictional and residual stresses, initial statistical distribution of defects versus their size, location, and orientation, material fatigue resistance parameters, etc.

To develop the model we will consider each of the assumptions used and the equations following from these assumptions in detail.

9.7.1   Initial Statistical Defect Distribution

Experiments have demonstrated that contact fatigue in steels is due to the presence of defects such as nonmetallic inclusions, carbides, etc. In particular, Dudragne et al. [58] and Murakami et al. [3] (see Tables 1, 2) have stated that in their tests all fatigue failures are initiated at subsurface nonmetallic inclusions. Spektor et al. [59] have experimentally demonstrated that the main source of fatigue cracks in bearings is oxide inclusions. Earlier, it is stated that the most likely source of fatigue crack initiation is represented by inclusions with low coefficient of thermal expansion (oxides, etc.). Moreover, it is reported that under certain operating conditions the normal stress intensity factor k_1 at the subsurface tips of surface cracks can be several orders of magnitude higher than the one at the tips of similar subsurface cracks (see Subsection 9.6.4). These high values of the normal stress intensity factor k_1 may lead to extremely fast propagation of surface cracks in comparison with the subsurface ones, i.e., to very short fatigue life. It is widely accepted that under normal operating conditions fatigue life of bearings and gears can be classified as a high cycle fatigue phenomenon (i.e., they can withstand at least $10^6 - 10^7$ loading cycles). Therefore, long fatigue lives observed in tests under normal operating conditions and short fatigue lives resulting from surface–initiated fatigue cracks are irreconcilable. Thus, one can conclude that in bearings and gears with normally long fatigue life fatigue cracks are initiated at subsurface defects. In accordance with these conclusions, it is assumed that material defects are far from each other and from the material surface (subsurface defects) and they do not interact.

Let us consider an elastic material with Young's modulus E and Poisson's ratio ν bounded by a plane and subjected to a moving cyclic load. The contact load applied to the material moves along the x-axis from $+\infty$ to $-\infty$ (see Fig. 9.35). We will introduce a rectangular coordinate system in material in such a way that the x-axis is directed along the direction of the motion of the loaded contact, and the z- and y-axes are perpendicular to the x-axis and are directed along and perpendicular to the surface, respectively. It is assumed that none of the stress field characteristics depend on the z coordinate. We will assume that there is an initial statistical defect distribution in material. In cases when some defects are clustered together, they can be replaced by one defect of an equivalent length. Therefore, we will also assume that every defect can be represented by a straight crack of half-length l_0. The crack distribution is described by the probabilistic density function $f(0, x, y, z, l_0)$ such that $f(0, x, y, z, l_0)dl_0 dx dy dz$ is the number of defects with the half-length between l_0 and $l_0 + dl_0$ in the material volume $dxdydz$ centered at (x, y, z) with dx, dy, and dz dimensions along the respective axes. The characteristic linear size of the volume $dxdydz$ is considered to be much greater than the typical size of the material inhomogenuity and, at the same time, much smaller than the characteristic size of the loaded contact region. Therefore, it is assumed that the defect population in any of the parallelepipeds $dxdydz$ is large enough to ensure an adequate and accurate representation of the phenomenon based on the probabilistic function f. Therefore, the defect distribution $f(0, x, y, z, l_0)$ is assumed to be discrete in material and represents a local characteristic of the material defectiveness.

The model can be developed for any particular initial distribution $f(0, x, y, z, l_0)$. The experiments done by Bokman, Pshenichnov, and Pershtein [60] demonstrated that the distribution of nonmetallic inclusions in aluminum with respect to their size resembles a log-normal one. If the initial defect distribution $f(0, x, y, z, l_0)$ versus defect initial half-length l_0 is taken as a log-normal

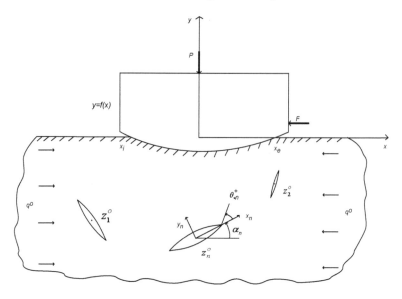

FIGURE 9.76
Schematic view of the contact.

distribution with the mean value μ_{ln} and standard deviation σ_{ln}

$$f(0, x, y, z, l_0) = 0 \; if \; l_0 \leq 0,$$

$$f(0, x, y, z, l_0) = \frac{\rho(0, x, y, z)}{\sqrt{2\pi}\sigma_{ln}l_0} \exp\left[-\frac{1}{2}\left(\frac{\ln(l_0) - \mu_{ln}}{\sigma_{ln}}\right)^2\right] \; if \; l_0 > 0,$$

(9.233)

the model development and analysis can be significantly simplified (see the following sections). In (9.233) $\rho(0, x, y, z)$ is the crack volume density at the initial time moment $N = 0$.

9.7.2 Crack Propagation versus Crack Initiation

In experimental studies of different steels, including bearing ones, by Han and Yang [61], Nisitani and Goto [62], Shao et al. [63], Clarke et al. [64], and Nélias et al. [65], it was established that the crack initiation stage is much shorter than the crack propagation stage. Therefore, soon after cycling loading starts small fatigue cracks are initiated near oxide inclusions (see Kudish and Burris [47]). In case of very small oxides, fatigue cracks may also initiate near carbides. Based on theoretical studies (see Murakami [66]) it is reasonable to assume that defects can be replaced by cracks of the same size. In two-dimensional crack mechanics usually cracks are modeled by finite cuts in continuous material with specified half-lengths.

9.7.3 Applicability of the Quasi-Brittle Fracture Mechanics to the Analysis of Contact Fatigue

The fatigue processes in material start as accumulation of dislocations near defects that rapidly grow into micro-cracks (see Section 8.4 and Kudish and Burris [47]). As the linear fracture mechanics (LEFM) is a natural extension of the dislocation theory, LEFM can be effectively used to approximately describe the mechanisms controlling contact fatigue. Theoretical data on stresses near voids and inclusions (Murakami [66]) indicate that even when a crack length is small in comparison with an adjacent inclusion/void the stress intensity factors at crack tips can be reasonably well approximated by the stress intensity factors for cracks of combined length of the inclusion/void and the small cracks adjacent to the inclusion/void. Therefore, for the purpose of stress intensity factor analysis, such combined structures of inclusions and cracks can be replaced by equivalent cracks of slightly larger size. It leads to the conclusion that in $f(0, x, y, z, l_0)$ one can assume that the variable l_0 represents crack half-lengths.

The next major assumption employed in the model by Kudish and Burris [39, 48, 55] is that LEFM for quasi-brittle materials is applicable to the analysis of fatigue cracks in steels. Let us verify that in most cases the radius r_p of the plasticity zone at a crack tip is much smaller than the crack half-length l. For a plane stress state when the plasticity zone at a crack tip is small in comparison with the crack length, we have

$$\frac{2r_p}{l} \ll 1, \ r_p = \frac{1}{2\pi}\left(\frac{k_1}{\sigma_p}\right)^2, \tag{9.234}$$

where k_1 is the normal stress intensity factor at a crack tip, σ_p is the material yield stress.

The inequality from (9.234) must be verified for typical cracks under typical loading conditions. Let us consider relatively small cracks, for example, with $l = 0.1a_H$ and a reasonable maximum Hertzian pressures p_H occurring in bearings and gears, $p_H < \sigma_p$. Here a_H and p_H are the semi-width (in the direction of the motion) and the maximum Hertzian pressure in a dry contact that are typical for most dry and heavily loaded lubricated contacts. Keeping in mind that k_1 is an increasing function of the friction coefficient λ (Kudish [48]) one can take $\lambda = 0.01$, which is about the right order of magnitude or higher than in real applications. For subsurface cracks Kudish [48] showed that k_1 is also a function of the crack orientation angle α and the maximum value of k_1 is reached for cracks oriented almost perpendicular to the material surface, i.e., at $\alpha = \pi/2$. In addition to that, Kudish [48] showed that for higher compressive residual stresses q^0, acting along the x-axis, the subsurface cracks with angles $\alpha \approx \pi/2$ have smaller normal stress intensity factors k_1 than the same cracks for lower compressive or tensile residual stresses. Taking this into account and using the fact that in practice the residual stresses are compressive or very small tensile, for the sake of making a simple estimate let us assume that $q^0 = 0$. Then for a crack with the half-length $l = 0.1a_H$

oriented at angle $\alpha = \pi/2$ to the x-axis and located at the depth of $y = 0.5a_H$ beneath the material surface, which is subjected to the Hertzian pressure q with the maximum p_H and the frictional stress τ with the friction coefficient $\lambda = 0.01$, one gets $k_1 = 5.5 \cdot 10^{-5} \, p_H l^{1/2}$ (Kudish [48]). Substituting this into (9.234) and using the fact that $p_H < \sigma_p$ one gets that $2r_p/l = 9 \cdot 10^{-9} \ll 1$. Therefore, for small cracks the assumption that $2r_p/l \ll 1$ holds and the methods of quasi-brittle fracture mechanics can be applied.

For longer cracks with the normal stress intensity factor $k_1 \approx K_f$ (where K_f is the material fracture toughness,) the latter inequality can be violated. For example, for $K_f = 76 \, MPa \cdot m^{1/2}$ and $\sigma_p = 1.8 \, GPa$ from (9.234) follows that $2r_p = 189 \, \mu m$, which is comparable with the size of the Hertzian contact. However, as it is shown below, the crack propagation rate is much higher for larger cracks than for smaller ones and, therefore, during almost the entire fatigue process cracks remain small and their behavior can be considered based on LEFM for quasi-brittle materials.

To make the further analysis simpler from now on, we will concentrate only on subsurface initiated fatigue assuming that there are no defects in a thin surface layer of material. Then at the tips of small subsurface cracks the normal k_1 and shear k_2 stress intensity factors can be approximated by asymptotic formulas (see Sections 9.3 and 9.4 and Kudish [48])

$$k_1 = \sqrt{l}[Y^r + q^0 \sin^2 \alpha]\theta[Y^r + q^0 \sin^2 \alpha],$$

$$k_2 = \sqrt{l}[Y^i - \tfrac{q^0}{2} \sin 2\alpha],$$

$$Y = \frac{1}{\pi} \int\limits_{-a_H}^{a_H} [q(t)\overline{D}_0(t) + \tau(t)\overline{G}_0(t)]dt, \quad \tau = -\lambda q,$$

$$\{Y^r, Y^i\} = \{Re(Y), Im(Y)\},$$

$$D_0(t) = \frac{i}{2}\left[-\frac{1}{t-X} + \frac{1}{t-\overline{X}} - \frac{e^{-2i\alpha}(\overline{X}-X)}{(t-\overline{X})^2}\right],$$

$$G_0(t) = \frac{1}{2}\left[\frac{1}{t-X} + \frac{1-e^{-2i\alpha}}{t-\overline{X}} - \frac{e^{-2i\alpha}(t-X)}{(t-\overline{X})^2}\right], \quad X = x + iy,$$

(9.235)

where i is the imaginary unit ($i^2 = -1$), $\theta(x)$ is a step function: $\theta(x) = 0$, $x \leq 0$ and $\theta(x) = 1$, $x > 0$. It is important to mention that according to (9.235) for subsurface cracks the quantities of $k_{10} = k_1 l^{-1/2}$ and $k_{20} = k_2 l^{-1/2}$ are functions of x and y and are independent from l.

If the pressure distribution $q(x)$ and the frictional stress $\tau(x)$ change relatively slowly, then by replacing $q(x)$ and $\tau(x)$ by $p_a = P/L$ and $-\eta p_a$ (where P is the normal force applied to the contact during one loading cycle and L is the contact width along the x-axis) the value of Y from (9.235) can be simplified and approximated by the formula

$$Y = \frac{p_a}{\pi} \left\{ -i \left[\frac{1}{2} \ln \frac{(a_H - X)(a_H + \overline{X})}{(a_H - \overline{X})(a_H + X)} + e^{2i\alpha} \frac{a_H(X - \overline{X})}{a_H^2 - X^2} \right] \right.$$

$$\left. -\lambda \left[\ln \left| \frac{a_H - X}{a_H + X} \right| - e^{2i\alpha} \ln \frac{X - a_H}{a_H + X} + e^{2i\alpha} \frac{a_H(X - \overline{X})}{a_H^2 - X^2} \right] \right\}.$$

$$(9.236)$$

If in (9.235) $q(x)$ is the Hertzian pressure distribution, then in (9.236) the following value for $p_a = p_H/4$ should be used.

9.7.4 Direction of Fatigue Crack Propagation

The resultant stress field in material is formed by the interaction of three types of stresses: stress produced by the contact pressure, frictional, and residual stresses. The residual stress is compressive in some regions of material and tensile in others. Usually, the friction force produced by the contact frictional stress applied to the material surface is small but not zero. In most cases for smooth surfaces the distribution of the contact pressure is close to the Hertzian one. Under the action of almost Hertzian contact pressure, residual, and small frictional stresses, the subsurface stress field is dominated by compressive stresses. However, because of the presence of the frictional and tensile residual stresses the resultant stress may be tensile in the regions behind the contact with respect to the direction of the friction force and in the zones where the residual stress is tensile. There are regions of a tensile resultant stress even in the case of zero residual stress. Such regions are subsurface and have a wedge-type shape. Moreover, these regions are located outside of the contact and extend in the direction opposite to the direction of the friction force. These conclusions are based on direct stress calculations (see Sections 9.3, 9.4, and Kudish and Burris [47] and Kudish [48]).

It is widely accepted that fatigue cracks get initiated by shear stresses. Experimental and theoretical studies suggest (see Section 9.5 and Kudish and Burris [47]) that soon after initiation fatigue cracks propagate in the direction determined by the local stress field, namely, perpendicular to the local maximum tensile stress. Along this direction the shear stress intensity factor $k_2 = 0$. Therefore, to find the direction of fatigue crack propagation (i.e., the angle of crack propagation α) it is necessary to solve the equation

$$k_2(N, l, \alpha, x, y, z) = 0, \qquad (9.237)$$

where N is the number of loading cycles. The fact that for small subsurface cracks $k_{20} = k_2 l^{-1/2}$ is independent from l together with (9.237) lead to the conclusion that for loading with constant amplitude angle α characterizing the direction of fatigue crack growth is independent from N. Moreover, angle α is a function of only crack location, i.e., $\alpha = \alpha(x, y, z)$. In particular, according to (9.237) (see formula (9.230)) at any point (x, y, z), there are two angles α_1

and α_2 along which a crack may propagate that satisfy the equation

$$\tan 2\alpha = -\frac{2y \int\limits_{-a_H}^{a_H} (t-x)T(t,x,y)dt}{\frac{\pi}{2}q^0 + \int\limits_{-a_H}^{a_H} [(t-x)^2 - y^2]T(t,x,y)dt},$$ (9.238)

$$T(t,x,y) = \frac{yq(t) + (t-x)\tau(t)}{[(t-x)^2 + y^2]^2}.$$

Along these directions k_1 reaches its extremum values. That becomes obvious by calculating

$$\frac{\partial k_{10}}{\partial \alpha} = \frac{2}{\pi} \cos 2\alpha \left\{ 2y \int\limits_{-a_H}^{a_H} (t-x)T(t,x,y)dt \right.$$

(9.239)

$$\left. + \tan 2\alpha \int\limits_{-a_H}^{a_H} [(t-x)^2 - y^2]T(t,x,y)dt + \frac{\pi}{2}q^0 \sin 2\alpha \right\}$$

and substituting α from (9.238) in formula (9.239) which leads to $\frac{\partial k_{10}}{\partial \alpha} = 0$.

The actual direction of crack propagation α is determined by one of these two angles α_1 and α_2 for which the value of the normal stress intensity factor $k_1(N,l,\alpha,x,y,z)$ is greater. If angle α is determined this way, it guarantees that cracks propagate in the direction perpendicular to the maximum tensile stress. That fact is supported by experimental data (see Section 8.3 and Kudish and Burris [47]). It is important to understand that if all three stresses, i.e., pressure $q(x)$, frictional stress $\tau(x)$, and residual stress q^0 increase/decrease proportionally angles α_1 and α_2 that satisfy (9.237) remain the same.

Under the simplified assumptions when the contact pressure and tangential stress are slowly changing functions the two solutions $\alpha = \alpha_1$ and $\alpha = \alpha_2$ of (9.237) can be found from equations

$$\tan 2\alpha_1 = \frac{p_a(A_1 + \lambda A_3)}{\frac{\pi}{2}q^0 - p_a(A_2 + \lambda A_4)}, \quad \alpha_2 = \alpha_1 \pm \frac{\pi}{2},$$

$$A_1 = y^2(\mid X - a_H \mid^{-2} - \mid X + a_H \mid^{-2}),$$ (9.240)

$$A_2 = y[(a_H - x) \mid X - a_H \mid^{-2} + (a_H + x) \mid X + a_H \mid^{-2}],$$

$$A_3 = \arctan \frac{a_H - x}{y} + \arctan \frac{a_H + x}{y} - A_2, \quad A_4 = \ln \frac{\mid X - a_H \mid}{\mid X + a_H \mid} + A_1.$$

9.7.5 Crack Propagation Calculations

The crack growth is produced by a superposition of contact normal and frictional as well as residual stresses. As it was described earlier, fatigue life is strongly dependent on friction. The propagation of fatigue cracks cannot be caused by shear stresses in the immediate neighborhood of crack tips as these stresses are insensitive to friction (see Sections 9.3, 9.4 and Kudish [48, 43]). On the other hand, the normal stress intensity factor k_1 is strongly dependent on friction (see Sections 9.3, 9.4 and Kudish [48, 43]). The illustration of this behavior of the normal k_1 and tangential k_2 stress intensity factors is presented in Figs. 9.14 and 9.15 in Section 9.7. Therefore, the fact that fatigue life is determined not by shear but tensile stresses is not surprising. In fact, propagation of fatigue cracks is caused by small tensile stresses resulting from a superposition of the stresses, caused by the pressure and frictional stress as well as by the residual stress.

Therefore, in general terms fatigue crack propagation can be described by the following initial-value problem

$$\frac{dl}{dN} = F(l, \max_{-\infty < x < \infty} \triangle k_1, k_{th}, K_f), \ l \mid_{N=0} = l_0, \tag{9.241}$$

where F is a given function, $\triangle k_1$ is the range of variations of k_1 during one loading cycle, k_{th} and K_f are the material stress intensity threshold and fracture toughness, respectively, N is the number of loading cycles. Equations (9.241) should be solved along every line parallel to the x-axis in the material stressed volume V along which $\max_{-\infty < x < \infty} \triangle k_1 > 0$. In the crack propagation equation (9.241) the material fatigue parameters may depend on austenite grain size and other material microstructural parameters. Typical graphs of a crack propagation rate dl/dN versus l is schematically presented in Fig. 9.77. It is clear from this graph that there are three distinct stages of crack development: (a) growth of small cracks, (b) propagation of well developed cracks, and (c) explosive and, usually, unstable growth of large cracks.

The phase of small crack growth is the slowest one and it represents the main portion of the entire crack propagation period. This situation usually causes confusion about the duration of the phases of crack initiation and propagation of small cracks as for a long period of time these small cracks are hardly detectable. The next phase, propagation of well developed cracks, usually takes less time than the phase of small crack growth. And, finally, the explosive crack growth takes almost no time.

About 17 crack propagation equations of type (9.241) are collected and analyzed by Yarema [33]. Any one of these equations can be used in the model to describe propagation of fatigue cracks. However, the simplest way to consider propagation of fatigue cracks is to assume that (9.241) can be used in the Paris form

$$\frac{dl}{dN} = g_0(\max_{-\infty < x < \infty} \triangle k_1)^n, \ l \mid_{N=0} = l_0, \tag{9.242}$$

where g_0 and n are the parameters of material fatigue resistance and l_0 is the crack initial half-length. Paris law represents a simple and natural assumption that the rate of fatigue crack propagation is greater for greater stress levels, i.e., dl/dN is proportional to a certain positive power of the normal stress intensity factor k_1. The fact that the threshold stress intensity factor k_{th} is not

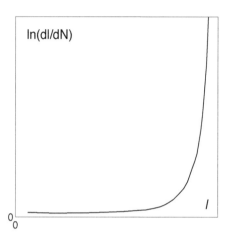

FIGURE 9.77
Schematic graph of a crack propagation rate dl/dN versus crack half-length l.

included in equation (9.242) allows to avoid the effect of "double dipping" of taking into account both the threshold stress intensity factor k_{th} and residual stress q^0 which conceptually plays the same role.

Let us illustrate the above phases of subsurface crack growth by analyzing Paris' law (9.242). Assuming that the amplitude of cyclic loading is constant and taking into account that for subsurface cracks $k_{10} = k_1 l^{-1/2}$ is independent from l the solution of the initial-value problem (9.242) can be represented in the form

$$l = l_0 \left\{ 1 - N(\tfrac{n}{2} - 1)g_0 \left[\max_{-\infty < x < \infty} \triangle k_{10} \right]^n / l_0^{\frac{2-n}{2}} \right\}^{\frac{2}{2-n}}, \quad n > 2. \qquad (9.243)$$

Under contact conditions when the residual stress is compressive $q^0 \le 0$ or even for small tensile residual stresses $q^0 > 0$ due to domination of applied contact pressure the resultant stress in the material for the most part is compressive. Therefore, we have $\triangle k_{10} = k_{10}$. It is necessary to notice that in material

$$N < l_0^{\frac{2-n}{2}} \left\{ (\tfrac{n}{2} - 1)g_0 \left[\max_{-\infty < x < \infty} k_{10} \right]^n \right\}^{-1}. \qquad (9.244)$$

Based on (9.243) one can formally determine the aforementioned phases of crack development. Namely, the phase of small crack growth is described by the relation

$$N \ll l_0^{\frac{2-n}{2}} \left\{ (\tfrac{n}{2} - 1)g_0 \left[\max_{-\infty < x < \infty} k_{10} \right]^n \right\}^{-1}, \qquad (9.245)$$

the phase of well–developed crack propagation is represented by

$$N \propto l_0^{\frac{2-n}{2}} \left\{ (\tfrac{n}{2} - 1)g_0 \left[\max_{-\infty<x<\infty} k_{10} \right]^n \right\}^{-1}, \tag{9.246}$$

and, finally, the phase of crack explosive growth is determined by

$$N \approx l_0^{\frac{2-n}{2}} \left\{ (\tfrac{n}{2} - 1)g_0 \left[\max_{-\infty<x<\infty} k_{10} \right]^n \right\}^{-1}. \tag{9.247}$$

In (9.246) N is of the same order of magnitude as the expression in the right hand side but not necessarily close to it, in (9.247) N is approximately equal to the expression in the right hand side. It is obvious from (9.244)-(9.246) that it takes many loading cycles for a small crack to grow into a well developed crack. Similarly, (9.244), (9.246), and (9.247) lead to the conclusion that for a well–developed crack it takes much less loading cycles to grow to a critical crack length l_c for which $k_1 = K_f$. According to (9.247), the phase of explosive crack growth takes just few loading cycles.

Based on this analysis it is clear that the number of loading cycles needed for a crack to reach its critical length is almost independent from the material fracture toughness K_f. This conclusion is supported by direct numerical simulations. Moreover, the fact that the small crack propagation phase represents the main period of crack growth allows to use (9.235) for stress intensity factors obtained for small subsurface cracks in Sections 9.3, 9.4 (see also Kudish [48]) for contact fatigue calculations. Formulas (9.235) were derived for constant residual stresses q^0. Further on, we will use an additional assumption that formulas (9.235) can also be used in the case of the residual stress q^0 varying slowly with the depth y beneath the surface, i.e., $q^0 = q^0(y)$.

For the further analysis it is necessary to determine the crack initial half-length l_{0c} which after N loading cycles reaches the critical size of l_c. Equation (9.243) gives

$$l_{0c} = \left\{ l_c^{\frac{2-n}{2}} + N(\tfrac{n}{2} - 1)g_0 \left[\max_{-\infty<x<\infty} k_{10} \right]^n \right\}^{\frac{2}{2-n}}, \tag{9.248}$$

where l_{0c} depends on N, y, and z. Obviously, for $n > 2$ and fixed x, y, and z the value of l_{0c} is a decreasing function of N.

It is important to keep in mind that $K = \max_{-\infty<x<\infty} \triangle k_{10}$ is a function of y and z. Obviously, if $K(y_1, z_1) < K(y_2, z_2)$, then $l_c(N, y_1, z_1) = [K_f/K(y_1, z_1)]^2 > [K_f/K(y_2, z_2)]^2 = l_c(N, y_2, z_2)$. For $n > 2$ and $N > 0$ from (9.248) follows that

$$l_{0c}(N, y_1, z_1) > l_{0c}(N, y_2, z_2) \tag{9.249}$$

$$if \quad \max_{-\infty<x<\infty} k_{10}(x, y_1, z_1) < \max_{-\infty<x<\infty} k_{10}(x, y_2, z_2).$$

Therefore, $l_{0c}(N, y, z)$ is minimal where $k_{10}(x, y, z)$ is maximal which, in turn, happens where the tensile stress reaches its maximum.

9.7.6   Crack Kinetics and Statistics

To describe crack statistics after the crack initiation phase is over, it is necessary to make certain assumptions. The simplest assumptions of this kind are the existing cracks do not heal and new cracks are not created. In other words, the number of cracks in any material volume is constant in time. Under normal operating conditions, this assumption holds well. The two reasons for this are: (a) the initial material defects (inclusions), etc., are scarce or observed in clusters, which can be replaced by equivalent cracks, and (b) fatigue cracks are initiated near defects almost immediately after the cyclic loading is applied (Kudish and Burris [48], Cherepanov [26]). The first stage of fatigue crack propagation is the period that begins immediately after cracks are initiated. This period corresponds to the slow growth of small fatigue cracks. As it is shown above, fatigue life is almost solely determined by the stage of small crack growth. Therefore, during almost the whole fatigue life, fatigue cracks remain small and do not interact with each other.

The preexisting defects which are located close to each other (grouped in a cluster) are replaced by one equivalent crack determined by the same value of the stress intensity factors as produced by the defect cluster. Based on a reasonable assumption that the defect distribution is initially scarce, the coalescence of cracks and changes in the general stress field are possible only when cracks have already reached relatively large sizes. However, this situation may happen only during the last stage of crack growth, which is insignificant for calculation of fatigue life. Therefore, we can also assume that over almost all life span of fatigue cracks their orientation does not change.

Let us introduce the density of crack distribution as function $f(N, x, y, z, l)$ of crack half-length l after N loading cycles in a small parallelepiped $dxdydz$ with the center at the point with coordinates (x, y, z). Then the assumption that the number of cracks in any material volume is constant in time leads to the equation ‖

$$f(N, x, y, z, l)dl = f(0, x, y, z, l_0)dl_0, \qquad (9.250)$$

which being solved for $f(N, x, y, z, l)$ gives

$$f(N, x, y, z, l)dl = f(0, x, y, z, l_0)\tfrac{dl_0}{dl}, \qquad (9.251)$$

where l_0 and dl_0/dl as functions of N and l can be obtain from the solution of (9.241). To give a simple illustration of the crack distribution $f(N, x, y, z, l)$ behavior let us use Paris' law (9.242) for crack propagation. Then from (9.243)

‖Other existing contact fatigue models that account for material defect distribution assume that the defect distribution versus their size is fixed and does not change under cyclic loading.

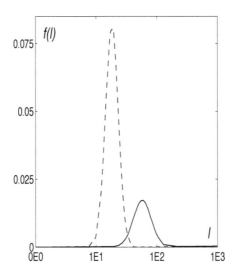

FIGURE 9.78
Schematic view of the evolution of the crack distribution $f(N_1, x, y, z, l)$ versus l with time N for $N = N_1$ (dashed curve) and $N = N_2$ (solid curve), $N_2 > N_1$ (after Kudish [39]). Reprinted with permission from the STLE.

one gets

$$l_0 = \{l^{\frac{2-n}{2}} + N(\tfrac{n}{2} - 1)g_0[\max_{-\infty<x<\infty} k_{10}]^n\}^{\frac{2}{2-n}},$$

$$\tfrac{dl_0}{dl} = \{1 + N(\tfrac{n}{2} - 1)g_0[\max_{-\infty<x<\infty} k_{10}]^n l^{\frac{n-2}{2}}\}^{\frac{n}{n-2}}.$$

(9.252)

Equations (9.251) and (9.252) lead to the expression for the crack distribution function f after N loading cycles

$$f(N, x, y, z, l) = f(0, x, y, z, l_0(N, l, y, z))\{1$$

$$+N(\tfrac{n}{2} - 1)g_0[\max_{-\infty<x<\infty} k_{10}]^n l^{\frac{n-2}{2}}\}^{\frac{n}{n-2}},$$

(9.253)

where $l_0(N, l, y, z)$ is determined by the first of the equations in (9.251).

A number of conclusions can be drawn from (9.253). Namely, the crack distribution function $f(N, x, y, z, l)$ depends on the initial crack distribution $f(0, x, y, z, l_0)$ and its changes with the number of applied loading cycles N in such a way that the crack volume density $\rho(N, x, y, z)$ remains constant.

Suppose the initial crack distribution $f(0, x, y, z, l_0)$ is a finite function of l_0, i.e., $f(0, x, y, z, l_0) > 0$ for $l_{10} < l_0 < l_{20}$ (l_{10} and l_{20} are certain constants, $0 \leq l_{10} < l_{20} < \infty$) and $f(0, x, y, z, l_0) = 0$ otherwise. Then for $N > 0$ the distribution $f(N, x, y, z, l)$ remains finite and positive for $l_1 < l < l_2$, where l_1 and l_2 are determined by (9.243) with $l_0 = l_{10}$ and $l_0 = l_{20}$, respectively. It can be easily shown that $l_2 - l_1 > l_{20} - l_{10}$, $l_1 > l_{10}$, and $l_2 > l_{20}$ for $N > 0$. Therefore, if $N_2 > N_1 \geq 0$, then $f(N_2, x, y, z, l)$ can be obtained from $f(N_1, x, y, z, l)$ by a corresponding stretching this distribution along the l-axis. Obviously, for $N = N_2 > N_1$, the crack distribution is wider than the one for $N = N_1$. A schematic view of such a distribution evolution with N is given in Fig. 9.78.

9.7.7 Local Fatigue Damage Accumulation

It is clear that if on a certain line through (x, y, z) parallel to the x-axis after N loading cycles half-lengths of all cracks $l < l_c = (k_f / k_{10})^2$ then there is no damage on this line and the material local survival probability $p(N, x, y, z) = 1$. On the other hand, if on this line after N loading cycles half-lengths of all cracks $l \geq l_c$, then all cracks reached their critical size and the material on this line (at the crack locations) is completely damaged - disintegrated. In this case $p(N, x, y, z) = 0$. The material local survival probability $p(N, x, y, z)$ is a certain monotonic measure of the portion of cracks with half-length l below the critical half-length l_c. Therefore, the local survival probability $p(N, x, y, z)$ is a monotonic function of the integral $\int_0^{l_c} f(N, x, y, z)dl$. The simplest function of such sort that provides for some basic probability properties can be represented by the following expressions

$$p(N, x, y, z) = \frac{1}{\rho} \int_0^{l_c} f(N, x, y, z, l)dl \; if \; f(0, x, y, z, l_0) \neq 0,$$

$$p(N, x, y, z) = 1 \; otherwise,$$

$$\rho = \rho(N, x, y, z) = \int_0^{\infty} f(N, x, y, z, l)dl,$$

$$\rho(N, x, y, z) = \rho(0, x, y, z).$$

(9.254)

Equations (9.254) determine the material local survival probability after N loading cycles as a ratio of the number of fatigue cracks with lengths below the critical $2l_c$ to the total number of cracks on the line through (x, y, z) parallel to the x-axis. Obviously, the local survival probability $p(N, x, y, z)$ is a monotonically decreasing function of N because fatigue crack lengths $2l$ tend to grow with the number of loading cycles N.

To calculate $p(N, x, y, z)$ from (9.254) one can use the expression for f determined by (9.253). However, it is more convenient to modify (9.254) by using (9.250). Namely, by changing the integration variable from l to l_0 and changing the upper limit of integration, one arrives at

$$p(N, x, y, z) = \frac{1}{\rho} \int_0^{l_{0c}} f(0, x, y, z, l_0) dl_0 \ if \ f(0, x, y, z, l_0) \neq 0,$$

$$(9.255)$$

$$p(N, x, y, z) = 1 \ otherwise,$$

where l_{0c} is determined by (9.248) and ρ is the initial volume density of cracks. Thus, to every material point (x, y, z) is assigned a certain survival probability, $0 \leq p(N, x, y, z) \leq 1$.

Equations (9.255) demonstrate that the material local survival probability $p(N, x, \ y, z)$ is mainly controlled by the initial crack distribution $f(0, x, y, z, l_0)$, material fatigue resistance parameters g_0 and n, and contact and residual stresses. Moreover, it easily follows from (9.255) that the material local survival probability $p(N, x, y, z)$ is a decreasing function of N because l_{0c} from (9.248) is a decreasing function of N for $n > 2$.

9.7.8 Survival Probability of Material as a Whole

To come up with the survival probability $P(N)$ of the material as a whole, we will assume that the material fails as soon as it fails at just one point. It is important to keep in mind that we assumed that the initial crack distribution in the material is discrete. Let $p_i(N) = p(N, x_i, y_i, z_i)$, $i = 1, \ldots, N_c$, where N_c is the total number of cracks in the material stressed volume V. Then based on the above assumption the material survival probability $P(N)$ is equal to

$$P(N) = \prod_{i=1}^{N_c} p_i(N). \tag{9.256}$$

Obviously, $P(N)$ from (9.256) satisfies the inequalities

$$[p_m(N)]^{N_c} \leq P(N) \leq p_m(N), \ p_m(N) = \min_V p(N, x, y, z). \tag{9.257}$$

In (9.257) the right inequality shows that the survival probability $P(N)$ is never greater than the minimum value $p_m(N)$ of the local survival probability $p(N, x, y, z)$ over the material stressed volume V. Let us show that in most cases at the relatively early stages of the fatigue process $P(N) = p_m(N)$. We need to make a reasonable assumption that it is a very rare occurrence when more than one initial pits appear simultaneously. If it is so, then the first pit is created by the cracks from a small volume element of the material with the least favorable conditions, i.e., with the smallest survival probability $p_m(N)$ (see (9.257)).

Let us provide an analytical substantiation for the assumption that the first pit is created by the cracks from a small material volume with the smallest survival probability $p_m(N)$. For simplicity, let us assume that the initial crack distribution across volume V is uniform and the residual stress is compressive or zero. We will consider two small material volumes: volume M with the center at which $\max\limits_{V} k_1$ is reached and any other small material volume S within which the normal stress intensity factor $k_1 = (1-D)\max\limits_{V} k_1$, $D > 0$. As the initially longest crack from M with the initial half-length l_{0max} approaches its critical size, it propagates at a progressively higher rate than other cracks do. Moreover, to reach its critical size this crack must undergo N_M number of loading cycles (see (9.247))

$$N_M \approx l_{0max}^{\frac{2-n}{2}}\{(\tfrac{n}{2} - 1)g_0[\max_{M,-\infty<x<\infty} k_{10}]^n\}^{-1}.$$

The initially longest crack from S with the same initial half-length l_{0max} to reach its critical size must undergo N_S number of loading cycles (see (9.247))

$$N_S \approx l_{0max}^{\frac{2-n}{2}}\{(\tfrac{n}{2} - 1)g_0[\max_{S,-\infty<x<\infty} k_{10}]^n\}^{-1}.$$

Dividing the latter equation by the former one, we obtain

$$N_S = N_M(1 - D)^{-n}.$$

Now, assuming that $D = 0.1$ and n varies from 6.67 to 9 the latter equation gives $N_S/N_M \approx 1 + nD \approx 1.67 - 1.90$. Therefore, it takes from 67% to 90% more loading cycles for the initially largest crack from S to reach its critical size than for the crack of the same initial size located in M.

Now, let us consider some cracks from M with the initial half–length $l_0 = (1 - K)l_{0max}$, $K > 0$. Using a similar analysis based on (9.247), we arrive at the formula

$$N_{MK} = N_M(1 - K)^{\frac{2-n}{n}},$$

where N_{MK} is the number of loading cycles it takes for a crack from M with the initial half-length $l_0 = (1 - K)l_{0max}$ to reach its critical size. Assuming that $K = 0.3$ and n varies between 6.67 and 9, the latter equation gives $N_{MK}/N_M \approx 1 + (n/2 - 1)K \approx 1.70 - 2.05$. Therefore, for a crack from M, which is initially by 30% shorter than the longest crack in M, it takes from 70% to 105% more loading cycles to reach its critical size than for the initially longest crack in M.

This simple analysis leads to the following conclusion. By the time all cracks in M with the initial half-length $0.7l_{0max} \leq l_0 \leq l_{0max}$ reach their critical size, none of the cracks in the rest of the material (where $\max\limits_{S,\,S\cap M=\emptyset} k_{10}$ is lower than 90% of $\max\limits_{M} k_{10} = \max\limits_{V} k_{10}$) reach their critical size. Therefore, at these relatively early stages of the fatigue process (which are of most interest for practical applications), the survival probability of any volume of the material except for the small parallelepiped M is equal to 1. Thus, at the early stages of the fatigue process the material survival probability $P(N)$ as a whole is

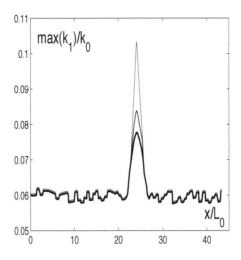

FIGURE 9.79
Illustration of the growth of the initially randomly distributed normal stress intensity factor k_1 with time N (after Kudish [39]). Reprinted with permission from the STLE.

determined by the survival probability of the parallelepiped M (see (9.256)), i.e., $P(N) = p_m(N)$.

Moreover, the above analysis also validates one of the main assumptions of the model that new cracks are not being created. Namely, if new cracks do get created in the process of loading, they are very small and have no chance to catch up with already existing and propagating cracks.

Therefore, under the normal circumstances the material survival probability determined by the product of the local survival probabilities for all defects (see (9.256)) is close to the survival probability of the "most dangerous" defect, i.e., to $p_m(N)$. The other way to illustrate this fact is to take into account that the value of the power n ($n = 6.67 - 9$) in Paris's equation for fatigue crack growth is high. This situation is depicted in Fig. 9.79 (see Tallian, Hoeprich, and Kudish [67]). The data for Fig. 9.79 are obtained for fatigue cracks randomly distributed over the material volume and subjected to a cyclic load. In Fig. 9.79 the values of the normal stress intensity factor k_1 are shown at different time moments (k_0 and L_0 are the characteristic normal stress intensity factor and geometric size of the solid). These graphs clearly show that the crack with the initially larger value of the normal stress intensity factor k_1 propagates much faster than all other cracks, i.e., the value of its k_1 increases much faster than the values of k_1 for all other cracks, which are almost dormant. As a result of that, the crack with the initially larger value of k_1 reaches its critical size way ahead of other cracks. This event determines the time and the place where fatigue occurs initially. Therefore, in spite of (9.256), for high values of n the material survival probability $P(N)$ is a local fatigue characteristic, and it is determined by the material defect with the initially highest value of the stress intensity factor k_1. The higher the power n is the more accurate the above–mentioned assumption is. Let us mention that this reasoning is not limited to the case of an initial crack distribution uniform across the material volume. This reasoning works in the general case as well.

Finally, we can conclude that at the early stages of the fatigue process the material survival probability $P(N)$ as a whole is determined by the minimum survival probability $p_m(N)$, i.e.,

$$P(N) = p_m(N). \tag{9.258}$$

If the initial crack distribution is taken in the log-normal form of (9.233) then according to (9.255) the expression for $p_m(N)$ is given by the formula

$$p_m(N) = \tfrac{1}{2} \min_V \{1 + erf[\tfrac{\ln l_{0c}(N,y,z)-\mu_{ln}(x,y,z)}{\sqrt{2}\sigma_{ln}(x,y,z)}]\},$$

where $erf(x)$ is the error integral. Based on monotonicity of function $erf(x)$, we have

$$p_m(N) = \tfrac{1}{2}\{1 + erf[\min_V \tfrac{\ln l_{0c}(N,y,z)-\mu_{ln}(x,y,z)}{\sqrt{2}\sigma_{ln}(x,y,z)}]\}. \tag{9.259}$$

If the initial distribution of defects is the same throughout the material (except for a thin surface layer where it is assumed that there are no defects), i.e., μ_{ln} and σ_{ln} are constants, then the expression (9.259) for $p_m(N)$ can be simplified as follows

$$p_m(N) = \tfrac{1}{2}\{1 + erf[\tfrac{\ln \min_V l_{0c}(N,y,z)-\mu_{ln}}{\sqrt{2}\sigma_{ln}}]\}. \tag{9.260}$$

According to (9.248), for $n > 2$ the value of $\min_V l_{0c}(N, y, z)$ is reached at the point(s) where $\max_V k_{10}$ occurs. Obviously, the survival probability $p_m(N)$ is a complex combined measure of applied stresses, initial crack distribution, material fatigue parameters, and the number of loading cycles.

To determine the fatigue life N of a contact for the given survival probability $P(N) = P_*$, it is necessary to solve the equation

$$p_m(N) = P_*. \tag{9.261}$$

Finally, the contact fatigue model is reduced to (9.235), (9.237), (9.248), (9.258)-(9.261) for the fatigue life N of the material as a whole. A detailed analysis of solutions of (9.261) is given in Subsections 9.8.1 and 9.8.2.

9.7.9 Stochastic Loading, Variable Loading Amplitude, Periodic Loading Regimes and Contact Fatigue

In this section a generalization of the model for the cases of variable and periodic loading and stochastic contact pressure and residual stress are proposed.

Let us consider a periodic cyclic regime of loading such that the amplitude of the maximum contact pressure assumes the following values: $q_{max}(N) = q_i$ for $kN_0 + n_i < N \leq kN_0 + n_{i+1}$, $i = 0, \ldots, j$, $k = 0, 1, \ldots$. Here N_0 is the period of cycling loading, n_i are nonnegative numbers such that $n_i < n_{i+1}$, $n_0 = 0$ and $n_{j+1} = N_0$.

The model of contact fatigue can be extended on this case of periodic cyclic loading. However, one of the difficulties in such a model generalization is the necessity of keeping the right order in the sequence of the solutions of the initial value problems for the crack growth equation (9.242) in agreement with the sequence of changing pressure $q_{max}(N)$. In practical cases the period of loading, N_0 is negligibly small in comparison with the number of cycles over which the survival probability $P(N)$ changes noticeably. Therefore, with sufficient for practice precision we can introduce averaging over a block of loading $(N, q_{max}(N))$. In this case the angles of crack propagation α_i corresponding to the loads q_i are determined by the equations (see (9.237))

$$k_2(N, l, \alpha_i, x, y, z) = 0, \quad i = 0, \ldots, j, \tag{9.262}$$

while the resultant/average angle of crack propagation α_m is determined by the equation

$$\tan \alpha_m = \frac{\sum\limits_{i=0}^{j} \triangle n_i k_{10i}^n \sin \alpha_i}{\sum\limits_{i=0}^{j} \triangle n_i k_{10i}^n \cos \alpha_i}, \quad \triangle n_i = n_{i+1} - n_i. \tag{9.263}$$

The average value Mk_{10}, which should be substituted for k_{10} in all formulas for the survival probability $P(N)$, can be determined as follows

$$Mk_{10} = \left\{ \frac{1}{N} [(\sum\limits_{i=0}^{j} \triangle n_i k_{10i}^n \cos \alpha_i)^2 + (\sum\limits_{i=0}^{j} \triangle n_i k_{10i}^n \sin \alpha_i)^2]^{1/2} \right\}^{1/n}. \tag{9.264}$$

In equations (9.263) and (9.264) the values of k_{10i} are calculated for angles α_i based on formulas (9.235). In case of a continuous variation of the maximum of contact pressure $q_{max}(N)$, sums in equations (9.263) and (9.264) should be replaced by the corresponding integrals. Obviously, for the case of cyclic loading with constant amplitude formulas for angle α_m and coefficient Mk_{10} based on equations (9.263) and (9.264) are reduced to the formulas derived for such a case earlier.

Let us consider the case of a stochastic contact pressure distribution $q_\xi(x)$ with the probabilistic density of distribution $g(\xi)$, which changes from one loading cycle to another. Then the analogs of equations (9.263) and (9.264) are as follows:

$$\int\limits_{-\infty}^{\infty} k_2(\xi, \alpha) g(\xi) d\xi = 0, \quad Mk_{10} = \left\{ \int\limits_{-\infty}^{\infty} k_{10}^n(\xi) g(\xi) d\xi \right\}^{1/n}, \tag{9.265}$$

where $k_{10}(\xi)$ and $k_2(\xi, \alpha)$ are the values of coefficients k_{10} and k_2 for $q(x) = q_\xi(x)$. After that the survival probability $P(N)$ is determined based on the earlier derived formulas and angle α obtained from the solution of the first equation in (9.265) and k_{10} replaced by Mk_{10} from (9.265).

The model of contact fatigue considered above can be generalized for the cases of stochastic residual stress. Let us suppose that the residual stress q^0 is a stochastic function of the depth y, i.e., it is fixed in any specific sample but varies in a series of samples from one sample to another. Suppose a particular realization of the residual stress $q^0_\xi(x, y, z)$ is characterized by the parameter ξ, and $g(\xi)$ is the density of the stochastic distribution of $q^0_\xi(x, y, z)$. Then, the survival probability $P_\xi(N)$ for this realization of residual stress can be calculated according to the aforementioned formulas. The average survival probability $P(N)$ is determined by

$$P(N) = \int\limits_{-\infty}^{\infty} g(\xi) P_\xi(N) d\xi. \tag{9.266}$$

If a stochastic contact pressure distribution $q_\xi(x)$ with the probabilistic density of distribution $g(\xi)$ is constant for each sample but varies from one sample to another one in the series of samples, then for a specific contact pressure distribution $q_\xi(x)$ the survival probability $P_\xi(N)$ is determined according to formulas from the preceding subsections. The average survival probability $P(N)$ is determined according to equation (9.266).

Furthermore, these models can be generalized on a three–dimensional case of contact or structural fatigue (see Section 9.10) as well as on the cases of non-zero stress intensity threshold k_{th}.

9.8 Analytical and Numerical Analysis of the Pitting Model. Some Examples

In this section we will analyze analytically and numerically the contact fatigue model derived in Section 9.7.

9.8.1 Analytical Analysis of the Pitting Model

First, let us consider the model behavior in some simple cases. If $f(0, x, y, z, l_0)$ is a uniform crack distribution over the material volume V (except for a thin surface layer where $f = 0$). Then based on (9.255) and inequality (9.249) it can be shown that $p(N, x, y, z)$ reaches its minimum at the points where k_{10} and the principal tensile stress reach their maximum values. This leads to the conclusion that the material local failure probability $(1 - p)$ reaches its maximum at the points with maximal tensile stress. Therefore, for a uniform initial crack distribution $f(0, x, y, z, l_0) = f(0, 0, 0, l_0)$ the survival probability $P(N)$ from (9.258) is determined by the material local survival probability at the points at which the maximal tensile stress is attained. Intuitively, this seems obvious.

However, the latter conclusion is not necessarily correct if the initial crack distribution $f(0, x, y, z, l_0)$ is not uniform over the material volume. Suppose, $k_{10}(x, y, z)$ is maximal at the point (x_m, y_m, z_m) and at the initial time moment $N = 0$ at some point (x_*, y_*, z_*) there exist cracks larger than the ones at the point (x_m, y_m, z_m), namely,

$$\int_0^{l_c} f \, dl_0 \mid_{(x_*, y_*, z_*)} < \int_0^{l_c} f \, dl_0 \mid_{(x_m, y_m, z_m)}.$$

Then after a certain number of loading cycles $N > 0$ the material damage at point (x_*, y_*, z_*) may be greater than at point (x_m, y_m, z_m), where l_{0c} reaches its maximum value. Therefore, fatigue failure may occur at point (x_*, y_*, z_*) instead of point (x_m, y_m, z_m), and the material weakest point is not necessarily is the material most stressed point.

If μ_{ln} and σ_{ln} depend on the coordinate of the material point (x, y, z), then there may be a series of points where in formula (9.260) for the given number of loading cycles N the minimum over the material volume V is reached. The coordinates of such point may change with N. This situation represents different potentially competing fatigue mechanisms such as pitting, flaking, etc. The occurrence of fatigue damage at different points in the material depends on the initial defect distribution, applied stresses, residual stress, etc. For example, if the initial defect distribution has a number of subsurface shallow defects of large size in the region of high compressive residual stresses, these defects will give rise to shallow subsurface cracks which will slowly propagate practically parallel to the material surface. There is a chance that after a certain number of loading cycles N some of them reach the critical size and create an unstable crack which may generate damage through flaking. At the same time, deeper in the material where the resultant stress is tensile and small another group of cracks may propagate toward the surface at a certain angle that depends on the stress state in the vicinity of the initial defects. If these cracks reach their critical size, they will generate a pit. It is just a matter of which process flaking or pitting runs faster.

In Section 8.6 and in Kudish and Burris [48], it was shown that in models based on assumptions similar to those of the Lundberg-Palmgren model the dependence of the survival probability on the stressed volume is exponential. This relationship is not realistic because it contradicts the experimental studies (see Romaniv et al. [68]), which show that, usually, there is a relatively weak dependence of fatigue life on the material stressed volume. In the model presented in Section 9.7, the stressed volume plays no explicit role. However, implicitly it does. In fact, the initial crack distribution $f(0, x, y, z, l_0)$ depends on the material volume. In general, in a larger volume of the material, there is a greater chance to find inclusions of greater size than in a smaller one. These larger inclusions represent a potential source of pitting and may cause a decrease in the material fatigue life of a larger material volume.

Now, let us establish the relationship between the mean μ_{ln} and standard deviation σ_{ln} of the initial log-normal crack distribution and the regular initial mean μ and standard deviation σ:

$$\mu_{ln} = \frac{M \ln(l)}{\rho}, \quad \sigma_{ln} = [\frac{M \ln^2(l)}{\rho} - \mu_{ln}^2]^{1/2},$$

$$\mu = \frac{Ml}{\rho}, \quad \sigma = [\frac{Ml^2}{\rho} - \mu^2]^{1/2}, \tag{9.267}$$

where $Mh(l)$ is the moment of function $h(l)$ defined by

$$Mh(l) = \int\limits_0^\infty h(l) f(0, x, y, z, l) dl. \tag{9.268}$$

Simple calculations show that

$$\mu = \exp[\mu_{ln} + 0.5\sigma_{ln}^2], \quad \sigma = \mu[\exp(\sigma_{ln}^2) - 1]^{1/2},$$

$$\mu_{ln} = \ln \frac{\mu^2}{\sqrt{\mu^2 + \sigma^2}}, \quad \sigma_{ln} = \sqrt{\ln[1 + (\frac{\sigma}{\mu})^2]}. \tag{9.269}$$

Let us assume that μ_{ln} and σ_{ln} are constants. Suppose the material failure occurs at point (x, y, z) with the failure probability $1 - P(N)$. By that we actually determine the point where in (9.260) the minimum over the material volume V is reached. Therefore, at this point in (9.260) the operation of minimum over the material volume V can be dropped. By solving (9.260) and (9.261) one gets

$$N = \{(\frac{n}{2} - 1)g_0[\max_{-\infty < x < \infty} k_{10}]^n\}^{-1}\{\exp[(1 - \frac{n}{2})(\mu_{ln}$$

$$+\sqrt{2}\sigma_{ln} erf^{-1}(2P_* - 1))] - l_c^{\frac{2-n}{2}}\}, \tag{9.270}$$

where $erf^{-1}(x)$ is the inverse function to the error integral $erf(x)$. Let us simplify this equation for the case of a material initially free of damage, i.e., when $P(0) = 1$. Discounting the very tail of the initial crack distribution we get $\max\limits_V l_0 \leq l_c$. Thus, for well–developed cracks (see (9.246)) and, in many cases, even for small cracks (see (9.245)) the second term in (9.248) for l_{0c} dominates the first one. It means that the dependence of l_{0c} on l_c and, consequently, on the material fracture toughness K_f can be neglected. Therefore, (9.270) can be approximated by

$$N = \{(\frac{n}{2} - 1)g_0[\max_{-\infty < x < \infty} k_{10}]^n\}^{-1}\{\exp[(1 - \frac{n}{2})(\mu_{ln}$$

$$+\sqrt{2}\sigma_{ln} erf^{-1}(2P_* - 1))]\}. \tag{9.271}$$

It follows from (9.235) that k_{10} is proportional to the maximum contact pressure q_{max} and, also, depends on the friction coefficient λ and the ratio of the residual stress q^0 and maximum pressure q_{max}. Making use of (9.269) and (9.271) and assuming that $q_{max} = p_H$ one arrives at a simple analytical formula (see also Kudish and Burris [48])

$$N = \frac{C_0}{(n-2)g_0 p_H^n} \left(\frac{\sqrt{\mu^2+\sigma^2}}{\mu^2}\right)^{\frac{n}{2}-1} \exp[(1 - \frac{n}{2})(\ln \frac{\mu^2}{\sqrt{\mu^2+\sigma^2}}$$

$$+\sqrt{2\ln[1 + (\frac{\sigma}{\mu})^2]}erf^{-1}(2P_* - 1))], \tag{9.272}$$

where C_0 depends only on the friction coefficient λ and the ratio of the residual stress q^0 and maximum Hertzian pressure p_H. Finally, assuming that $\sigma \ll \mu$ from (9.272), one can obtain the formula

$$N = \frac{C_0}{(n-2)g_0 p_H^n \mu^{\frac{n}{2}-1}} \exp[(1 - \frac{n}{2})\frac{\sqrt{2}\sigma}{\mu}erf^{-1}(2P_* - 1))]. \tag{9.273}$$

Actually, fatigue life formulas (9.272) and (9.273) can be represented in the form of the Lundberg-Palmgren formula (8.10), i.e.,

$$N = \frac{C_*}{p_H^n}, \tag{9.274}$$

where parameter n can be compared with constant c/e in the Lundberg-Palmgren formula (8.10). Formula (9.274) shows the usual dependence of fatigue life N on the maximum Hertzian stress p_H. The major difference between the Lundberg-Palmgren formula (8.10) and formula (9.274) based on this model of contact fatigue is the fact that in (9.274) constant C_* depends on material defect parameters μ, σ, coefficient of friction λ, residual stresses q^0, and probability of survival P_* in a certain way while in the Lundberg-Palmgren formula (8.10) constant C depends only on the depth z_0 of the maximum orthogonal stress, stressed volume V, and probability of survival $P(N) = P_*$.

Let us analyze formula (9.273). It demonstrates the intuitively obvious fact that the fatigue life N is inverse proportional to the value of the parameter g_0 that characterizes the material crack propagation resistance. So, for materials with lower crack propagation rate, the fatigue life is higher and vice versa. Equation (9.273) exhibits a usual for roller and ball bearings dependence of fatigue life N on the maximum Hertzian pressure p_H (compare formulas (8.10) and (9.274)). Thus, from the well–known experimental data for bearings, the range of n values is $20/3 \leq n \leq 9$. Keeping in mind that usually $\sigma \ll \mu$, for these values of n contact fatigue life N is practically inverse proportional to a positive power of the mean crack size, i.e., to $\mu^{n/2-1}$. Therefore, fatigue life N is a decreasing function of the initial mean crack (inclusion) size μ. This conclusion is valid for any value of the material survival probability P_* and is supported by the experimental data discussed in Kudish and Burris [48]. If $P_* > 0.5$ (see (9.261) and (9.273)), then $erf^{-1}(2P_* - 1) > 0$ and (keeping in mind that $n > 2$) fatigue life N is a decreasing function of the initial standard deviation of crack sizes σ. Similarly, if $P_* < 0.5$, then fatigue life N is an increasing function of the initial standard deviation of crack sizes σ. According to (9.273), for $P_* = 0.5$ fatigue life N is independent from σ, and, according to (9.272), for $P_* = 0.5$ fatigue life N is a slowly increasing function of σ. By differentiating $p_m(N)$ obtained from (9.260) with respect to σ, we can conclude that the dispersion of $P(N)$ increases with σ.

From (9.235) (also see Kudish [48]) follows that the stress intensity factor k_1 decreases as the magnitude of the compressive residual stress q^0 increases and/or the magnitude of the friction coefficient λ decreases. Therefore, it follows from formulas (9.271) and (9.272) that C_0 is a monotonically decreasing function of the residual stress q^0 and friction coefficient λ. Numerical simulations of fatigue life show that the value of C_0 is very sensitive to the details of the residual stress distribution q^0 versus depth. As the residual stress distribution q^0 and friction coefficient λ are practically solely dependent on the manufacturing operations (such as heating and surface finishing operations), the value of constant C_0 represents the measure of the manufacturing process quality and stability.

Being applied to bearings and/or gears the described statistical contact fatigue model can be used as a research and engineering tool in modeling pitting. In the latter case, some of the model parameters may be assigned certain fixed values based on the scrupulous analysis of steel quality and quality and stability of manufacturing processes.

9.8.2 Numerical Analysis of the Pitting Model. Some Examples

Let us choose a basic set of model parameters typical for bearing testing: maximum Hertzian pressure $p_H = 2\ GPa$, contact region half-width in the direction of motion $a_H = 0.249\ mm$, friction coefficient $\lambda = 0.002$ residual stress varying from $q^0 = -237.9\ MPa$ on the surface to $q^0 = 0.035\ MPa$ at the depth of 400 μm below it, fracture toughness K_f varying between 15 and 95 $MPa \cdot m^{1/2}$, $g_0 = 8.863\ MPa^{-n} \cdot m^{1-n/2} \cdot cycle^{-1}$, $n = 6.67$, mean of crack initial half-lengths $\mu = 49.41\ \mu m$ ($\mu_{ln} = 3.888 + ln(\mu m)$), crack initial standard deviation $\sigma = 7.61\ \mu m$ ($\sigma_{ln} = 0.1531$). Numerical results show that the fatigue life is practically independent from the material fracture toughness K_f, which supports the assumption used for the derivation of formulas (9.271)-(9.273). To illustrate the dependence of contact fatigue life on some of the model parameters, just one parameter from the basic set of parameters will be varied at a time and graphs of the pitting probability $1 - P(N)$ for these sets of parameters (basic and modified) will be compared. Figure 9.81 shows that as the initial values of the mean μ of crack half-lengths and crack standard deviation σ increase contact fatigue life N decreases. Similarly, contact fatigue life decreases as the magnitude of the tensile residual stress and/or friction coefficient increase (see Fig. 9.15). The results show that the fatigue life does not change when the magnitude of the compressive residual

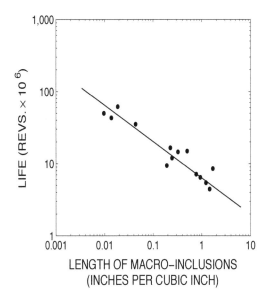

FIGURE 9.80
Bearing life-inclusion length correlation (after Stover and Kolarik II [52],
COPYRIGHT The Timken Company 2009).

stress is increased/decreased by 20% of its base value while the tensile portion of the residual stress distribution remains the same. Obviously, that is in agreement with the fact that tensile stresses control fatigue. Moreover, the fatigue damage occurs in the region with the resultant tensile stresses close to the boundary between tensile and compressive residual stresses. However, when the compressive residual stress becomes small enough the acting frictional stress may supersede it and create new regions with tensile stresses that potentially may cause acceleration of fatigue failure.

The numerical results support the conclusion that $ln(N)$ is practically a linear function of μ and σ. This behavior of fatigue life N versus μ is similar to the fatigue life-material defect relationship obtained experimentally by Stover and Kolarik II [52]. This fact supports the validity of the approach used for developing the new contact fatigue model.

Let us consider an example of the further validation of the new contact fatigue model for tapered roller bearings based on a series of approximate calculations of fatigue life. The main simplifying assumption made is that bearing fatigue life can be closely approximated by taking into account only the most loaded contact. The following parameters have been used for calculations: $p_H = 2.12\ GPa$, $a_H = 0.265\ mm$, $\lambda = 0.002$, $g_0 = 6.009\ MPa^{-n} \cdot m^{1-n/2} \cdot cycle^{-1}$, $n = 6.67$, the residual stress varied from $q^0 = -237.9\ MPa$ on the surface to $q^0 = 0.035\ MPa$ at the depth of $400\ \mu m$

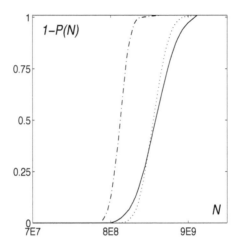

FIGURE 9.81
Pitting probability $1 - P(N)$ calculated for the basic set of parameters (solid curve) with $\mu = 49.41\ \mu m$, $\sigma = 7.61\ \mu m$ ($\mu_{ln} = 3.888 + \ln(\mu m)$, $\sigma_{ln} = 0.1531$), for the same set of parameters and the increased initial value of crack mean half-lengths (dash-dotted curve) $\mu = 74.12\ \mu m$ ($\mu_{ln} = 4.300 + \ln(\mu m)$, $\sigma_{ln} = 0.1024$), and for the same set of parameters and the increased initial value of crack standard deviation (dotted curve) $\sigma = 11.423\ \mu m$ ($\mu_{ln} = 3.874 + \ln(\mu m)$, $\sigma_{ln} = 0.2282$) (after Kudish [39]). Reprinted with permission from the STLE.

TABLE 9.3
Relationship between the tapered bearing fatigue life $N_{15.9}$ and the initial inclusion size mean and standard deviation (after Kudish [39]). Reprinted with permission from the STLE.

$\mu\ [\mu m]$	$\sigma\ [\mu m]$	$N_{15.9}\ [cycles]$
49.41	7.61	$2.5 \cdot 10^8$
73.13	11.26	$1.0 \cdot 10^8$
98.42	15.16	$5.0 \cdot 10^7$
147.11	22.66	$2.0 \cdot 10^7$
244.25	37.62	$6.0 \cdot 10^6$

below the surface, fracture toughness K_f varied between 15 and $95 MPa{\cdot}m^{1/2}$. The crack (inclusion) initial mean half-length μ varied between 49.41 and

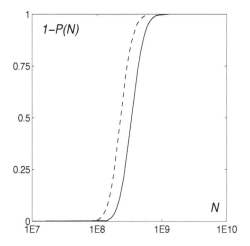

FIGURE 9.82
Pitting probability $1 - P(N)$ calculated for the basic set of parameters including $\lambda = 0.002$ (solid curve) and for the same set of parameters and the increased friction coefficient (dashed curve) $\eta = 0.004$ (after Kudish [39]). Reprinted with permission from the STLE.

244.25 μm ($\mu_{ln} = 3.888 - 5.498 + ln(\mu m)$), the crack initial standard deviation varied between $\sigma = 7.61$ and 37.61 μm ($\sigma_{ln} = 0.1531$). The results for fatigue life $N_{15.9}$ (for $P(N_{15.9}) = P_* = 0.159$) calculations are given in the Table 9.3 and practically coincide with the experimental data obtained by The Timken Company and presented in Fig. 19 by Stover, Kolarik II, and Keener [70] (in the present text this graph is given as Fig. 9.80). One must keep in mind certain differences in the numerically obtained data and the data presented in the above mentioned Fig. 9.80. In Fig. 9.80 fatigue life is given as a function of the cumulative inclusion length (sum of all inclusion lengths over a cubic inch of steel), and here fatigue life is calculated as a function of the mean inclusion length. However, the numerical data for fatigue life can be brought in perfect compliance with the experimental data by choosing the right measure of steel cleanliness (say, mean inclusion half-length μ) and the proper values for the parameters of the material fatigue resistance g_0 and n.

It is also interesting to point out that based on the results following from the new model, bearing fatigue life can be significantly improved for steels with the same cumulative inclusion length but smaller mean half-length μ (see Fig. 9.81). In other words, fatigue life of a bearing made from steel with large number of small inclusions is higher than of the one made of steel with small number of larger inclusions given that the cumulative inclusion length

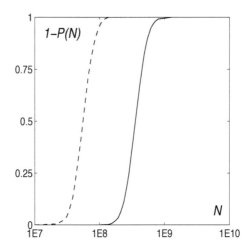

FIGURE 9.83
Pitting probability $1 - P(N)$ calculated for the basic set of parameters (solid curve) and for the same set of parameters and changed profile of residual stress q^0 (dashed curve) in such a way that at points where q^0 is compressive its magnitude is unchanged and at points where q^0 is tensile its magnitude is doubled (after Kudish [39]). Reprinted with permission from the STLE.

is the same in both cases. Moreover, bearing and gear fatigue lives with small percentage of failures (survival probability $P > 0.5$) for steels with the same cumulative inclusion length can also be improved several times if the width of the initial inclusion distribution is reduced, i.e., when the standard deviation σ of the initial inclusion distribution is made smaller (see Figure 9.81). Figures 9.82 and 9.83 show that the elevated values of the tensile residual stress are much more detrimental to fatigue life than greater values of the friction coefficient.

Finally, the described model is flexible enough to allow for replacement of the density of the initial crack distribution (see (9.233)) by a different function and of Paris's equation for fatigue crack propagation (see (9.242)) by another equation. Such modifications would lead to results on fatigue life varying from the presented above. However, the methodology, i.e., the way the formulas for fatigue life are obtain and the most important conclusions will remain the same.

This methodology has been applied to the engineering analysis of wear and contact fatigue in cases of lubricant contaminated by rigid abrasive particles and contact surfaces charged with abrasive particles [71] as well as to calculation of bearing wear and contact fatigue life [44].

In conclusion we can state that the new statistical contact fatigue model that takes into account the most important parameters of the contact fatigue phenomenon (such as normal and frictional contact and residual stresses, initial statistical defect distribution, orientation of fatigue crack propagation, material fatigue resistance, etc.) is derived and analyzed. The model allows for examination of the effect of variables such as steel cleanliness, applied stresses, residual stress, etc. on contact fatigue life as single or composite entities. Some analytical results illustrating the new model and its validation by the experimentally obtained fatigue life data for tapered bearings are presented. In addition, the new contact fatigue model gives a relatively simple tool for analyzing bearing and gear manufacturing process quality and stability as well as the steel cleanliness and their effect on bearing and gear performance.

9.9 Contact Fatigue of Rough Surfaces

Let us consider a problem about contact fatigue of an elastic solid with rough surface without taking into account possible adhesion in the asperity contacts as well as the competing phenomena such as wear and near surface fatigue damage; i.e., let us focus on the influence of surface roughness on subsurface originated contact fatigue - pitting. More specifically, let us assume that a rigid parabolic punch $y = \frac{x^2}{2R}$ with radius R is cyclically normally indented by force P in a dry elastic half-plane (with modulus E and Poisson's ration ν) bounded by a rough horizontal line. The roughness geometry of the half-plane boundary is described by function $\varphi_\star(x)$. We will assume that the cyclic contact of the punch occurs with the same area of the elastic solid so that roughness does not change from one loading cycle to another.

Obviously, the problem consists of two subproblems: determining contact stresses and determining contact fatigue. We will start with the former one.

9.9.1 Contact Stresses in Rough Contacts

In most cases frictional stress in a contact causes very small changes in contact pressure. Therefore, we will assume that frictional stress can be neglected in determining contact pressure. Also, it is assumed that the applied load P is high enough and the surface roughness is sufficiently small to provide for a continuous contact, i.e., $q(x) > 0$ for all $a < x < b$. That leads to the following formulation of the contact problem (see Section 2.2):

$$\varphi_\star(x) - \varphi_\star(b) + \frac{x^2 - b^2}{2R} + \frac{2}{\pi E'} \int\limits_a^b q(t) \ln \left| \frac{b-t}{x-t} \right| dt = 0, \qquad (9.275)$$

$$q(a) = q(b) = 0, \ \int\limits_a^b q(t)dt = P,$$

where $q(x)$ is the contact pressure, a and b are contact boundaries, $E' = \frac{E}{1-\nu^2}$ is the effective elastic modulus. By introducing new function $\varphi(x) = \varphi_*(x) - \varphi_*(b)$ in dimensionless variables

$$\{x', a', b'\} = \frac{1}{a_H}\{x, a, b\}, \ q' = \frac{q}{p_H}, \ w\varphi'(x') = \frac{2R}{a_H^2}\varphi(x),$$

problem (9.275) can be rewritten in the form (for simplicity primes are omitted)

$$w\varphi(x) + x^2 - b^2 + \frac{2}{\pi}\int\limits_a^b q(t)\ln\left|\frac{b-t}{x-t}\right| dt = 0, \ q(a) = q(b) = 0,$$

(9.276)

$$\int\limits_a^b q(t)dt = \frac{\pi}{2},$$

where a_H and p_H are the half-width and maximum pressure in a Hertzian contact of the same punch and a half-plane with smooth boundary, w is a dimensionless constant characterizing the level of roughness. To simplify problem analysis, we will introduce the substitution $x = \frac{a+b}{2} + \frac{b-a}{2}v, \ -1 \le v \le 1$. Then, according to Section 2.2 the exact solution of problem (9.276) has the form

$$q(v) = \frac{b-a}{2}\sqrt{1-v^2}\left\{1 + \frac{w}{2\pi}\left(\frac{2}{b-a}\right)^2 \int\limits_{-1}^1 \frac{\varphi'(\frac{a+b}{2}+\frac{b-a}{2}t)dt}{\sqrt{1-t^2}(t-v)}\right\},$$

(9.277)

$$w\int\limits_{-1}^1 \frac{\varphi'(\frac{a+b}{2}+\frac{b-a}{2}t)dt}{\sqrt{1-t^2}} = \pi\frac{b^2-a^2}{2},$$

(9.278)

$$w\int\limits_{-1}^1 \frac{\varphi'(\frac{a+b}{2}+\frac{b-a}{2}t)tdt}{\sqrt{1-t^2}} = \pi[\frac{(b-a)^2}{2} - 1].$$

Obviously, in this solution parameters a and b depend on function $\varphi'(y)$ in some nonlinear fashion. Therefore, it is difficult to extract any useful information from this solution in case of function $\varphi'(v)$ dependent on some random variables because operators of averaging (calculating moments) and integral operators in (9.277) and (9.278) are not commutative.

However, it is possible to analyze solution (9.277) and (9.278) in the case of relatively small roughness, i.e. in the case of $w \ll 1$. Expressing the solution in the form of asymptotic power series in w

$$a = -1 + wa_1 + w^2a_2 + \dots, \ b = 1 + wb_1 + w^2b_2 + \dots,$$

(9.279)

$$q(v) = q_0(v) + wq_1(v) + w^2q_2(v) + \dots, \ q_0(v) = \sqrt{1-v^2},$$

and substituting it in equations (9.277) and (9.278) after simple manipulations we obtain

$$a_1 = \frac{1}{2\pi} \int\limits_{-1}^{1} \sqrt{\frac{1-t}{1+t}} \varphi'(t) dt, \quad b_1 = \frac{1}{2\pi} \int\limits_{-1}^{1} \sqrt{\frac{1+t}{1-t}} \varphi'(t) dt, \tag{9.280}$$

$$a_2 = \frac{3a_1^2 - 2a_1 b_1 - b_1^2}{8} + \frac{b_1 + a_1}{4\pi} \int\limits_{-1}^{1} \sqrt{\frac{1-t}{1+t}} \varphi''(t) dt$$

$$+ \frac{b_1 - a_1}{4\pi} \int\limits_{-1}^{1} t \sqrt{\frac{1-t}{1+t}} \varphi''(t) dt, \quad b_2 = \frac{a_1^2 + 2a_1 b_1 - 3b_1^2}{8} \tag{9.281}$$

$$+ \frac{b_1 + a_1}{4\pi} \int\limits_{-1}^{1} \sqrt{\frac{1+t}{1-t}} \varphi''(t) dt + \frac{b_1 - a_1}{4\pi} \int\limits_{-1}^{1} t \sqrt{\frac{1+t}{1-t}} \varphi''(t) dt,$$

$$q_1(v) = \frac{b_1 - a_1}{2} \sqrt{1 - v^2} + \frac{\sqrt{1-v^2}}{2\pi} \int\limits_{-1}^{1} \frac{\varphi'(t) dt}{\sqrt{1-t^2}(t-v)},$$

$$q_2(y) = \frac{b_2 - a_2}{2} \sqrt{1 - v^2} \tag{9.282}$$

$$+ \frac{\sqrt{1-v^2}}{2\pi} \left\{ - \frac{b_1 - a_1}{2} \int\limits_{-1}^{1} \frac{\varphi'(t) dt}{\sqrt{1-t^2}(t-v)} + \int\limits_{-1}^{1} \frac{\varphi''(t)[\frac{b_1 + a_1}{2} + \frac{b_1 - a_1}{2} t] dt}{\sqrt{1-t^2}(t-v)} \right\}.$$

Now, let us consider the case of random roughness geometry $\varphi'(v)$. We will be concerned with just the mean characteristics of the process of contact interaction and, later, contact fatigue. Suppose we have an infinite set of independent random variables $\xi = \{\xi_0, \xi_1, \ldots, \xi_k, \ldots\}$ with densities of distribution $\{h_0(\xi_0), h_1(\xi_1), \ldots, h_k(\xi_k), \ldots\}$, respectively. The densities are such that

$$\int\limits_{-\infty}^{\infty} h_k(\xi_k) d\xi_k = 1, \quad \int\limits_{-\infty}^{\infty} \xi_k h_k(\xi_k) d\xi_k = 0, \quad k = 0, 1, \ldots. \tag{9.283}$$

Because of the independence of $\{\xi_0, \xi_1, \ldots, \xi_k, \ldots\}$, the density of the distribution of the whole set ξ is

$$g(\xi) = \prod\limits_{k=1}^{\infty} h_k(\xi_k). \tag{9.284}$$

Then the random roughness geometry $\varphi'(v)$ can be represented as follows:

$$\varphi'(v) = \sum\limits_{k=0}^{\infty} \xi_k T_k(v), \quad |v| \leq 1, \tag{9.285}$$

where $T_k(v)$ are Chebyshev orthogonal polynomials of the first kind [17] **
and

$$\xi_0 = \frac{1}{\pi} \int\limits_{-1}^{1} \frac{\varphi'(t)dt}{\sqrt{1-t^2}}, \quad \xi_k = \frac{2}{\pi} \int\limits_{-1}^{1} \frac{\varphi'(t)T_k(t)dt}{\sqrt{1-t^2}}, \quad k > 0. \tag{9.286}$$

Series (9.285) is convergent if its coefficients ξ_k approach zero sufficiently fast
as $k \to \infty$.

The second condition in (9.283) guarantees that statistically $\varphi'(v)$ varies
around $y = 0$, i.e., around the surface of a smooth half-plane. Substituting
representation (9.284) in formulas (9.280)-(9.282) gives

$$a_1 = \frac{\xi_0}{2} - \frac{\xi_1}{4}, \; b_1 = \frac{\xi_0}{2} + \frac{\xi_1}{4}, \; a_2 = \frac{1}{8}(-\xi_0\xi_1 + \frac{\xi_1^2}{4})$$

$$+\frac{\xi_0}{4\pi} \sum_{k=1}^{\infty} k\xi_k \int\limits_{-1}^{1} \frac{(1-t)U_{k-1}(t)dt}{\sqrt{1-t^2}} + \frac{\xi_1}{8\pi} \sum_{k=1}^{\infty} k\xi_k \int\limits_{-1}^{1} \frac{t(1-t)U_{k-1}(t)dt}{\sqrt{1-t^2}},$$

$$b_2 = -\frac{1}{8}(\xi_0\xi_1 + \frac{\xi_1^2}{4}) + \frac{\xi_0}{4\pi} \sum_{k=1}^{\infty} k\xi_k \int\limits_{-1}^{1} \frac{(1+t)U_{k-1}(t)dt}{\sqrt{1-t^2}}$$

$$\tag{9.287}$$

$$+\frac{\xi_1}{8\pi} \sum_{k=1}^{\infty} k\xi_k \int\limits_{-1}^{1} \frac{t(1+t)U_{k-1}(t)dt}{\sqrt{1-t^2}},$$

$$q_1(v) = \sqrt{1-v^2}\{\xi_1 + \frac{1}{2} \sum_{k=2}^{\infty} \xi_k U_{k-1}(v)\},$$

$$q_2(v) = \frac{\sqrt{1-v^2}}{2}\left\{b_2 - a_2 + \frac{\xi_1}{4}\left[-\sum_{k=2}^{\infty} \xi_k U_{k-1}(v)\right.\right. \tag{9.288}$$

$$\left.\left. +\frac{1}{\pi} \sum_{k=2}^{\infty} k\xi_k \int\limits_{-1}^{1} \frac{U_{k-1}(t)dt}{\sqrt{1-t^2}}\right] + \frac{1}{2\pi}(\xi_0 + \frac{\xi_1}{2}y) \sum_{k=2}^{\infty} k\xi_k \int\limits_{-1}^{1} \frac{U_{k-1}(t)dt}{\sqrt{1-t^2}(t-v)}\right\}.$$

In the second formula of (9.288), the expression within braces is given by
series of polynomials that can be determined using the formulas

---

**Chebyshev orthogonal polynomials of the first kind $T_k(v)$ and of the second kind $U_k(v)$
[17] are defined as follows $T_k(\cos\theta) = \cos k\theta$ and $U_k(\cos\theta) = \frac{\sin(k+1)\theta}{\sin\theta}$, respectively. These
polynomials satisfy the following properties $\int\limits_{-1}^{1} \frac{T_k(v)T_m(v)dv}{\sqrt{1-v^2}} = 0$ if $k \neq m$, $\int\limits_{-1}^{1} \frac{T_k^2(v)dv}{\sqrt{1-v^2}} = \frac{\pi}{2}$
if $k \neq 0$ and the integral is equal to π if $k = 0$; $\int\limits_{-1}^{1} \sqrt{1-v^2}U_k(v)U_m(v)dv = 0$ if $k \neq m$,

$\int\limits_{-1}^{1} \sqrt{1-v^2}U_k^2(v)dv = \frac{\pi}{2}$ if $k \geq 0$, $\int\limits_{-1}^{1} \frac{dt}{\sqrt{1-t^2}(t-v)} = 0$, and $\int\limits_{-1}^{1} \frac{T_k(t)dt}{\sqrt{1-t^2}(t-v)} = \pi U_{k-1}(v)$, $k \geq$
1. It is well known that almost any continuous function on interval $[-1, 1]$ can be expanded
in Chebyshev orthogonal polynomials $T_k(v)$ [54].

$$\int\limits_{-1}^{1} \frac{t^{2m+1}dt}{\sqrt{1-t^2}} = 0, \quad \int\limits_{-1}^{1} \frac{t^{2m}dt}{\sqrt{1-t^2}} = \frac{\pi(2m-1)!!}{(2m)!!},$$

$$\int\limits_{-1}^{1} \frac{t^m dt}{\sqrt{1-t^2}(t-v)} = \pi Q_k(v) \quad (m = 2k+1),$$

$$= \pi v Q_k(v) \quad (m = 2k+2),$$

(9.289)

$$Q_k(v) = \sum_{j=0}^{k} \frac{(2j-1)!!}{(2j)!!} v^{2k-2j}.$$

Let us introduce a weighted average $\langle f(\xi) \rangle = \int\limits_{-\infty}^{\infty} g(\xi)f(\xi)d\xi$, where $d\xi = d\xi_0 d\xi_1 \ldots d\xi_k \ldots$ Then using formulas (9.283), (9.284), (9.287), and (9.288) it is easy to obtain

$$\langle a_1 \rangle = \langle b_1 \rangle = \langle q_1(v) \rangle = 0,$$

(9.290)

$$\langle a_2 \rangle = -\frac{\langle \xi_1^2 \rangle}{32}, \quad \langle b_2 \rangle = \frac{\langle \xi_1^2 \rangle}{32}, \quad \langle q_2(v) \rangle = \frac{\langle \xi_1^2 \rangle}{32}\sqrt{1-v^2}.$$

Therefore, the statistically averaged pressure and contact boundaries are

$$\langle a \rangle = -1 - \omega^2 \frac{\langle \xi_1^2 \rangle}{32} + \ldots, \quad \langle b \rangle = 1 + \omega^2 \frac{\langle \xi_1^2 \rangle}{32} + \ldots,$$

(9.291)

$$\langle q(v) \rangle = \langle b \rangle \sqrt{1-v^2}.$$

Formulas (9.291) show that statistically the size of the contact region and pressure in a rough contact are larger than in a smooth one (for comparison see Chapter (2)).

9.9.2 Modeling Contact Fatigue in Rough Contacts

To consider subsurface–originated fatigue in the case of rough surfaces, we will follow the approach presented in Subsection (9.7.9) and, more specifically, formulas (9.265) designed for dealing with statistically distributed parameters of a loaded contact. We will assume that wear and other surface and near–surface fatigue phenomena (such as shallow flaking, etc.) of rough surfaces do not occur or are small enough and can be neglected. In other words, we will be concerned only with subsurface–originated pitting. To solve our fatigue problem for a rough contact, we need the angle $\langle \alpha \rangle$ of fatigue crack propagation and the value proportional to the stress intensity factor Mk_{10} (see formulas (9.265)).

For the further analysis, it is beneficial to have the explicit representations of the stress intensity factors k_1 and k_2 from (9.235), namely,

$$k_1 = \frac{l^{1/2}}{\pi} \int\limits_{a}^{b} \left\{ 1 - \frac{[(t-x)^2 - y^2]\cos 2\alpha - 2y(t-x)\sin 2\alpha}{(t-x)^2 + y^2} \right\} T_1(t, x, y)dt$$

$$+q^0 \sin^2 \alpha, \ k_2 = -\frac{l^{1/2}}{\pi} \int\limits_a^b \{[(t-x)^2 - y^2] \sin 2\alpha$$

$$+2y(t-x)\cos 2\alpha\} T_2(t,x,y)dt - \frac{q^0}{2} \sin 2\alpha, \qquad (9.292)$$

$$T_1(t,x,y) = \frac{yp(t)+(t-x)\tau(t)}{(t-x)^2+y^2}, \ T_2(t,x,y) = \frac{T_1(t,x,y)}{(t-x)^2+y^2}.$$

To conduct our analysis easier, we need to make the limits of integration in the integrals involved in (9.292) equal to constants independent from a and b. That is achieved by using the substitutions $x = \frac{b+a}{2} + \frac{b-a}{2}v$ and $t = \frac{b+a}{2} + \frac{b-a}{2}u$. Therefore, we have

$$k_1 = \frac{l^{1/2}}{\pi} \frac{b-a}{2} \int\limits_{-1}^{1} \left\{ 1 - \frac{[(\frac{b-a}{2})^2(u-v)^2 - y^2]\cos 2\alpha - 2\frac{b-a}{2}y(u-v)\sin 2\alpha}{(\frac{b-a}{2})^2(u-v)^2+y^2} \right\}$$

$$\times T_1(u,v,y)du + q^0 \sin^2 \alpha, \ T_1(u,v,y) = \frac{yp(u)+\frac{b-a}{2}(u-v)\tau(u)}{(\frac{b-a}{2})^2(u-v)^2+y^2}, \qquad (9.293)$$

$$k_2 = -\frac{l^{1/2}}{\pi} \frac{b-a}{2} \int\limits_{-1}^{1} \{[(\tfrac{b-a}{2})^2(u-v)^2 - y^2]\sin 2\alpha$$

$$+2\tfrac{b-a}{2}y(u-v)\cos 2\alpha\} T_2(u,v,y)du - \frac{q^0}{2} \sin 2\alpha, \qquad (9.294)$$

$$T_2(u,v,y) = \frac{T_1(u,v,y)}{(\frac{b-a}{2})^2(u-v)^2+y^2}.$$

First, let us determine the angle $\langle \alpha \rangle$ of fatigue crack propagation. To determine angle $\langle \alpha \rangle$ from solution of equation (see the first equation in (9.265))

$$\langle k_2 \rangle = \int\limits_{-\infty}^{\infty} g(\xi)k_2(x,y,\xi)d\xi = 0, \qquad (9.295)$$

we will search it in the form

$$\langle \alpha \rangle = \alpha^0 + \omega\langle \alpha \rangle_1 + \omega^2\langle \alpha \rangle_2 + \dots, \ \omega \ll 1, \qquad (9.296)$$

where α^0 is the crack propagation angle obtained for smooth surfaces, which is calculated according to the formula (see (9.230))

$$\tan 2\alpha^0 = -\frac{2y \int\limits_{-1}^{1}(t-x)T_2(t,x,y))dt}{\frac{\pi}{2}q^0 + \int\limits_{-1}^{1}[(t-x)^2-y^2]T_2(t,x,y)dt}, \qquad (9.297)$$

$$T_2(t,x,y) = \frac{yq(t)+(t-x)\tau(t)}{[(t-x)^2+y^2]^2},$$

while $\langle a \rangle_1$ and $\langle a \rangle_2$ are correction terms that have to be determined (q^0 is the residual stress). To calculate $\langle k_2 \rangle$ we first expand k_2 in the Taylor formula

$$k_2 = k_2 \mid_{\omega=0} + \frac{\partial k_2}{\partial \frac{b-a}{2}} \mid_{\omega=0} (\frac{b-a}{2} - 1) + \frac{\partial k_2}{\partial \alpha} \mid_{\omega=0} (\alpha - \alpha^0)$$

$$+ \frac{\partial k_2}{\partial q} \mid_{\omega=0} (q - q_0) + \frac{1}{2} \frac{\partial^2 k_2}{\partial \frac{b-a}{2}^2} \mid_{\omega=0} (\frac{b-a}{2} - 1)^2$$

$$+ \frac{1}{2} \frac{\partial^2 k_2}{\partial \alpha^2} \mid_{\omega=0} (\alpha - \alpha^0)^2 + \frac{\partial^2 k_2}{\partial \frac{b-a}{2} \partial \alpha} \mid_{\omega=0} (\frac{b-a}{2} - 1)(\alpha - \alpha^0) \qquad (9.298)$$

$$+ \frac{\partial^2 k_2}{\partial q \partial \frac{b-a}{2}} \mid_{\omega=0} (\frac{b-a}{2} - 1)(q - q_0)$$

$$+ \frac{\partial^2 k_2}{\partial q \partial \alpha} \mid_{\omega=0} (\alpha - \alpha^0)(q - q_0) + \ldots,$$

where $\frac{\partial k_2}{\partial q} \mid_{\omega=0}$ is a linear integral operator following from (9.292). Then, substituting the expressions for a, b, q, and $\alpha = \langle a \rangle$ from (9.279)-(9.282), (9.296) as well as using the fact that $\langle a_1 \rangle = \langle b_1 \rangle = \langle q_1 \rangle = 0$ and (9.283) equation (9.295) leads to the asymptotic expansion

$$\langle k_2 \rangle = k_2 \mid_{\omega=0} + \omega^2 \frac{\partial k_2}{\partial \frac{b-a}{2}} \mid_{\omega=0} \langle \frac{b_2 - a_2}{2} \rangle + \frac{\partial k_2}{\partial \alpha} \mid_{\omega=0} (\omega \langle a \rangle_1$$

$$+ \omega^2 \langle a \rangle_2) + \omega^2 \frac{\partial k_2}{\partial q} \mid_{\omega=0} \langle q_2 \rangle + \frac{\omega^2}{2} \frac{\partial^2 k_2}{\partial \frac{b-a}{2}^2} \mid_{\omega=0} \langle (\frac{b_1 - a_1}{2})^2 \rangle \qquad (9.299)$$

$$+ \frac{\omega^2}{2} \frac{\partial^2 k_2}{\partial \alpha^2} \mid_{\omega=0} \langle a \rangle_1^2 + \omega^2 \frac{\partial^2 k_2}{\partial \frac{b-a}{2} \partial q} \mid_{\omega=0} \langle \frac{b_1 - a_1}{2} q_1 \rangle + \ldots = 0.$$

Collecting all terms proportional to ω and ω^2 and equating them to zero results in equations

$$\langle a \rangle_1 = 0,$$

$$\frac{\partial k_2}{\partial \frac{b-a}{2}} \mid_{\omega=0} \langle \frac{b_2 - a_2}{2} \rangle + \frac{\partial k_2}{\partial \alpha} \mid_{\omega=0} \langle a \rangle_2 + \frac{\partial k_2}{\partial q} \mid_{\omega=0} \langle q_2 \rangle \qquad (9.300)$$

$$+ \frac{1}{2} \frac{\partial^2 k_2}{\partial \frac{b-a}{2}^2} \mid_{\omega=0} \langle (\frac{b_1 - a_1}{2})^2 \rangle + \frac{\partial^2 k_2}{\partial \frac{b-a}{2} \partial q} \mid_{\omega=0} \langle \frac{b_1 - a_1}{2} q_1 \rangle = 0.$$

Taking into account the expressions $\langle (\frac{b_1 - a_1}{2})^2 \rangle = \frac{\langle \xi_1^2 \rangle}{16}$, $\langle \frac{b_2 - a_2}{2} \rangle = \frac{\langle \xi_1^2 \rangle}{32}$, $\langle \frac{b_1 - a_1}{2} q_1 \rangle = \frac{3 \langle \xi_1^2 \rangle}{16} \sqrt{1 - v^2}$, we obtain

$$\langle a \rangle_1 = 0,$$

$$\langle a \rangle_2 = -\frac{1}{32} \frac{\langle \xi_1^2 \rangle}{\frac{\partial k_2}{\partial \alpha} \mid_{\omega=0}} \{ \frac{\partial k_2}{\partial \frac{b-a}{2}} \mid_{\omega=0} + \frac{\partial k_2}{\partial q} \mid_{\omega=0} \sqrt{1 - v^2} + \frac{\partial^2 k_2}{\partial \frac{b-a}{2}^2} \mid_{\omega=0} \qquad (9.301)$$

$$+ 6 \frac{\partial^2 k_2}{\partial \frac{b-a}{2} \partial q} \mid_{\omega=0} \sqrt{1 - v^2} \}.$$

Using the expansion of k_{10} similar to the one in (9.298) as well as the binomial expansion for $\alpha = \langle\alpha\rangle$, we obtain

$$\langle k_{10}^n\rangle = (k_{10}\mid_{\omega=0})^n\{1 + \omega^2 n[\tfrac{\partial k_{10}}{\partial\frac{b-a}{2}}\mid_{\omega=0}\langle\tfrac{b_2-a_2}{2}\rangle$$

$$+\tfrac{\partial k_{10}}{\partial\alpha}\mid_{\omega=0}\langle\alpha\rangle_2 + \tfrac{\partial k_{10}}{\partial q}\mid_{\omega=0}\langle q_2\rangle + \tfrac{1}{2}\tfrac{\partial^2 k_{10}}{\partial\frac{b-a}{2}^2}\mid_{\omega=0}\langle(\tfrac{b_1-a_1}{2})^2\rangle$$

$$+\tfrac{\partial^2 k_{10}}{\partial q\partial\frac{b-a}{2}}\mid_{\omega=0}\langle\tfrac{b_1-a_1}{2}q_1\rangle]/k_{10}\mid_{\omega=0}$$

$$+\omega^2\tfrac{n(n-1)}{2}[(\tfrac{\partial k_{10}}{\partial\frac{b-a}{2}}\mid_{\omega=0})^2\langle(\tfrac{b_1-a_1}{2})^2\rangle$$

$$+2\tfrac{\partial k_{10}}{\partial\frac{b-a}{2}}\mid_{\omega=0}\tfrac{\partial k_{10}}{\partial q}\mid_{\omega=0}\langle\tfrac{b_1-a_1}{2}q_1\rangle$$

$$+\langle(\tfrac{\partial k_{10}}{\partial q}\mid_{\omega=0}q_1)^2\rangle]/(k_{10}\mid_{\omega=0})^2 + \dots\},$$

(9.302)

where $\frac{\partial k_{10}}{\partial q}\mid_{\omega=0}$ is a linear integral operator following from (9.292). Using (9.283), (9.287), and (9.288), it is easy to obtain that

$$\langle(\tfrac{\partial k_{10}}{\partial q}\mid_{\omega=0}q_1)^2\rangle = \tfrac{\langle\xi_1^2\rangle}{16}(\tfrac{\partial k_{10}}{\partial q}\mid_{\omega=0}\sqrt{1-v^2})^2$$

$$+\tfrac{1}{4}\sum_{k=1}^{\infty}\langle\xi_k^2\rangle(\tfrac{\partial k_{10}}{\partial q}\mid_{\omega=0}\sqrt{1-v^2}U_{k-1}(v))^2\} + \dots.$$

(9.303)

Therefore, using formulas (9.301)-(9.303) and the second formula in (9.265), we find the value of Mk_{10}, which represents a replacement of k_{10} in a stochastic case

$$Mk_{10} = (\langle k_{10}^n\rangle)^{1/n}.$$

(9.304)

The rest of fatigue modeling is done in accordance with Section (9.7) and Subsection (9.7.9).

Finally, let us analyze the shift in the material point where fatigue damage occurs due to the surface roughness. As we showed in (9.302)-(9.304)

$$Mk_{10} = k_{10}(x, y)\mid_{\omega=0}[1 + \omega^2\psi(x, y)],$$

(9.305)

where the expression for function $\psi(x, y)$ is obvious from formulas (9.302)-(9.304). Moreover, function $\psi(x, y)$ as well as its derivatives are of the order of unity for $\omega \ll 1$. Let us assume that for smooth surfaces the point of fatigue damage (x^0, y^0) is located beneath the material surface. Obviously, the coordinates of this point satisfy the following equations

$$\tfrac{\partial k_{10}(x^0, y^0)\mid_{\omega=0}}{\partial x} = \tfrac{\partial k_{10}(x^0, y^0)\mid_{\omega=0}}{\partial y} = 0.$$

(9.306)

At (x^0, y^0) function $k_{10}(x, y)\mid_{\omega=0}$ reaches its maximum. Keeping in mind the sufficient conditions for the maximum (second derivative test) we will consider the most general case when

$$D^0 = \frac{\partial^2 k_{10}(x^0,y^0)|_{\omega=0}}{\partial x^2} \frac{\partial^2 k_{10}(x^0,y^0)|_{\omega=0}}{\partial y^2} - \left(\frac{\partial^2 k_{10}(x^0,y^0)|_{\omega=0}}{\partial x \partial y}\right) > 0. \tag{9.307}$$

The special case of $D^0 = 0$ will not be considered. To determine the point at which the damage occurs for rough surfaces, we need to solve equations

$$\frac{\partial M k_{10}(x,y)}{\partial x} = \frac{\partial M k_{10}(x,y)}{\partial y} = 0. \tag{9.308}$$

The solution of these equations will be searched in the form

$$x = x^0 + \omega x_1 + \omega^2 x_2 + \ldots, \quad y = y^0 + \omega y_1 + \omega^2 y_2 + \ldots, \quad \omega \ll 1, \tag{9.309}$$

where x_1, x_2, y_1, and y_2 have to be determined.

By expressing $k_{10}(x,y) |_{\omega=0}$ in the form of Taylor formula with retained third derivatives, using the representations (9.305) and (9.309), and assumptions (9.306), the solution of equations (9.308) is reduced to solution of two systems of linear algebraic equations. The first of them is

$$\frac{\partial^2 k_{10}(x^0,y^0)|_{\omega=0}}{\partial x^2} x_1 + \frac{\partial^2 k_{10}(x^0,y^0)|_{\omega=0}}{\partial x \partial y} y_1 = 0,$$

$$\frac{\partial^2 k_{10}(x^0,y^0)|_{\omega=0}}{\partial x \partial y} x_1 + \frac{\partial^2 k_{10}(x^0,y^0)|_{\omega=0}}{\partial y^2} y_1 = 0, \tag{9.310}$$

the unique solution of which (due to the assumption $D^0 > 0$, see (9.307)) is

$$x_1 = y_1 = 0. \tag{9.311}$$

Using (9.311) the second system can be represented in the form

$$\frac{\partial^2 k_{10}(x^0,y^0)|_{\omega=0}}{\partial x^2} x_2 + \frac{\partial^2 k_{10}(x^0,y^0)|_{\omega=0}}{\partial x \partial y} y_2$$

$$= -k_{10}(x^0,y^0) |_{\omega=0} \frac{\partial \psi(x^0,y^0)}{\partial x},$$

$$\frac{\partial^2 k_{10}(x^0,y^0)|_{\omega=0}}{\partial x \partial y} x_2 + \frac{\partial^2 k_{10}(x^0,y^0)|_{\omega=0}}{\partial y^2} y_2 \tag{9.312}$$

$$= -k_{10}(x^0,y^0) |_{\omega=0} \frac{\partial \psi(x^0,y^0)}{\partial y},$$

the solution of which is

$$x_2 = \frac{k_{10}(x^0,y^0)|_{\omega=0}}{D^0} \left\{ \frac{\partial \psi(x^0,y^0)}{\partial y} \frac{\partial^2 k_{10}(x^0,y^0)|_{\omega=0}}{\partial x \partial y} \right.$$

$$\left. - \frac{\partial \psi(x^0,y^0)}{\partial x} \frac{\partial^2 k_{10}(x^0,y^0)|_{\omega=0}}{\partial y^2} \right\},$$

$$y_2 = \frac{k_{10}(x^0,y^0)|_{\omega=0}}{D^0} \left\{ \frac{\partial \psi(x^0,y^0)}{\partial x} \frac{\partial^2 k_{10}(x^0,y^0)|_{\omega=0}}{\partial x \partial y} \right. \tag{9.313}$$

$$\left. - \frac{\partial \psi(x^0,y^0)}{\partial y} \frac{\partial^2 k_{10}(x^0,y^0)|_{\omega=0}}{\partial x^2} \right\}.$$

Therefore, the damage point for rough surfaces shifts by a distance proportional to ω^2 compared to the case of smooth surfaces.

Let us consider an example when

$$\varphi'(v) = \sum_{k=2}^{\infty} \xi_k T_k(v), \ |v| \leq 1. \tag{9.314}$$

Then $\xi_1 = \langle \xi_1^2 \rangle = 0$ and

$$\langle \alpha \rangle = \alpha^0 + o(\omega^2), \ Mk_{10} = k_{10}|_{\omega=0} \{1$$

$$+\omega^2 \frac{n-1}{8} \sum_{k=2}^{\infty} \langle \xi_k^2 \rangle (\frac{\partial k_{10}}{\partial q}|_{\omega=0} \sqrt{1-v^2} U_{k-1}(v))^2 \} + o(\omega^2), \ \omega \ll 1. \tag{9.315}$$

Finally, replacing $\max\limits_{-\infty < x < \infty} k_{10}^n$ in (9.270) by $\max\limits_{-\infty < x < \infty} Mk_{10}^n$, we obtain the formula

$$N_R = \frac{N_0}{1+\omega^2 \frac{n(n-1)}{8} \sum_{k=2}^{\infty} \langle \xi_k^2 \rangle (\frac{\partial k_{10}}{\partial q}|_{\omega=0}\sqrt{1-v^2} U_{k-1}(v))^2 + \ldots} \tag{9.316}$$

for the number of loading cycles N_R of a solid with rough surface until fatigue failure, which corresponds to the given probability of survival P_*. Here N_0 is the number of loading cycles N_0 of a smooth solid until fatigue failure with the same probability of survival, which can be determined by one of the formulas (9.270)-(9.274), the value of the denominator is calculated at the point at which $\max\limits_{-\infty < x < \infty} Mk_{10}^n$ is attained.

Assuming that random coefficients ξ_k are distributed normally $h_k(\xi_k) = \frac{1}{\sqrt{2\pi}\sigma_k} \exp(-\frac{\xi_k^2}{2\sigma_k^2})$ (σ_k is the standard deviation) we obtain $\langle \xi^2 \rangle = \sigma_k^2$. Based on (9.316) we have

$$N_R = \frac{N_0}{1+\omega^2 \frac{n(n-1)}{8} \sum_{k=2}^{\infty} \sigma_k^2 (\frac{\partial k_{10}}{\partial q}|_{\omega=0}\sqrt{1-v^2} U_{k-1}(v))^2 + \ldots}. \tag{9.317}$$

Therefore, we can conclude that fatigue life of rough surfaces behaves similarly to fatigue life of smooth surfaces, however, it is slightly smaller than the latter one. Because of positiveness of all terms in the denominator of formula (9.316), it is easy to get a series of useful inequalities for N_R. For example, if all σ_k^2 for $k \geq 2$ are smaller than $\sigma_m^2 \neq 0$, $m \geq 2$, then from (9.316), we have

$$N_R \leq \frac{N_0}{1+\omega^2 \frac{n(n-1)}{8} \sigma_m^2 (\frac{\partial k_{10}}{\partial q}|_{\omega=0}\sqrt{1-v^2} U_{m-1}(v))^2}. \tag{9.318}$$

The obtained results indicate that small surface roughness has little effect on subsurface originated contact fatigue - pitting. It is well known that local stresses caused by the direct interaction of asperities may be high in the vicinity of the solid surface but vanish fast as a point moves away from the solid surface. Therefore, we can conclude that surface roughness may have a significant effect only on surface and near surface originated fatigue such as wear and shallow flaking.

For relatively small roughness fatigue life of heavily loaded lubricated contacts (see Chapter (10)) is close to the one for dry contacts considered above given that other parameters are the same. More specifically, this statement is true as long as the height of asperities in a lubricated contact is smaller than the film thickness separating contact surfaces, i.e., direct contact of asperities does not occur.

9.10 Three-Dimensional Model of Contact Fatigue

The material of this section is the extension of the two-dimensional model of contact fatigue described in Section 9.7 on the the case of three dimensions. This model can also be used in analyzing structural fatigue of materials (see Kudish [55]). The development of the model is very similar to the one described in Section 9.7 and follows the same steps. Therefore, just a short description of the model is given. Two examples of the application of this model to structural fatigue are provided.

9.10.1 Initial Statistical Defect Distribution

We will assume that the material defects are far from each other and do not interact. Suppose there is a characteristic size L_σ in material that is determined by the typical variations of the material stresses and surface geometry. Let us also assume that there is a size L_f in material such that $L_d \ll L_f \ll L_\sigma$, where L_d is the typical distance between the material defects. In other words, we will assume that the defect population in any of such volumes L_f^3 is large enough to ensure an adequate statistical representation of the phenomenon. By doing so, we assume that any parameter variations on the scale of L_f are indistinguishable for the fatigue analysis purposes of and that any volume L_f^3 can be represented by its center, i.e., a point (x, y, z). Therefore, we can assume that there is an initial statistical defect distribution in the material. For the purposes of the further analysis, we will replace each defect by a penny-shaped crack with a diameter approximately equal to the diameter of the defect. The usage of penny-shaped subsurface and semi-circular surface cracks is advantageous to our analysis because in the accepted approximation cracks maintain their shape and their size is characterized by just one parameter. The orientation of these cracks will be considered later. The initial statistical distribution is described by the probabilistic density function $f(0, x, y, z, l_0)$, such that $f(0, x, y, z, l_0)dl_0 dx dy dz$ is the number of defects with the radii between l_0 and $l_0 + dl_0$ in the material volume $dx dy dz$ centered about point (x, y, z).

The defect distribution is a local characteristic of material defectiveness. The model can be developed for any specific initial distribution $f(0, x, y, z, l_0)$. Following the experimental findings [60] a log-normal initial defect distribution $f(0, x, y, z, l_0)$ versus the defect initial radius l_0 can be accepted (see Section 9.7 and equation (9.233))

$$f(0, x, y, z, l_0) = 0 \ if \ l_0 \le 0,$$

$$f(0, x, y, z, l_0) = \frac{\rho(0, x, y, z)}{\sqrt{2\pi}\sigma_{ln}l_0} \exp\left[-\frac{1}{2}\left(\frac{\ln(l_0) - \mu_{ln}}{\sigma_{ln}}\right)^2\right] \ if \ l_0 > 0,$$

(9.319)

where μ_{ln} and σ_{ln} are the mean value and standard deviation of the crack radii, respectively.

9.10.2   Direction of Fatigue Crack Propagation

As it was done in Section 9.7 it is assumed that the duration of the phase of crack initiation is negligibly small in comparison with the duration of the phase of crack propagation. It is also assumed that linear elastic fracture mechanics is applicable to small fatigue cracks. The details of the substantiation of these assumptions see in Section 9.7. Based on that in the vicinity of a crack the stress intensity factors completely characterize the material stress state. The normal k_1 and shear k_2 and k_3 stress intensity factors at the edge of a single crack of radius l can be represented in the form

$$k_1 = F_1(x, y, z, \alpha, \beta)\sigma_1\sqrt{\pi l}, \ k_2 = F_2(x, y, z, \alpha, \beta)\sigma_1\sqrt{\pi l},$$

$$k_3 = F_3(x, y, z, \alpha, \beta)\sigma_1\sqrt{\pi l},$$

(9.320)

where σ_1 is the maximum of the local principal stress, F_1, F_2, and F_3 are certain functions of the point coordinates (x, y, z) and the crack orientation angles α and β with respect to the coordinate planes. The coordinate system is introduced in such a way that the x- and y-axes are directed along the material surface, the z-axis is directed perpendicular to the material surface. It is well known [26] that for single cracks or even a population of small cracks distant from each other and from the solid boundary functions F_1, F_2, and F_3 are practically independent from l.

The resultant stress field in an elastic material is formed by stresses $\sigma_x(x, y, z)$, $\sigma_y(x, y, z)$, $\sigma_z(x, y, z)$, $\tau_{xz}(x, y, z)$, $\tau_{xy}(x, y, z)$, and $\tau_{zy}(x, y, z)$. There are regions in material subjected to tensile stress and other regions subjected to compressive stress. Conceptually, there is no difference between the phenomena of structural and contact fatigue as the local material response to the same stress in both cases is the same. What is different between these two cases is the stress distributions. As long as the stress levels do not exceed the limits of applicability of the quasi-brittle linear fracture mechanics when plastic zones at crack edges are small the rest of the material behaves like an elastic solid. The actual stress distributions in cases of structural and contact fatigue are taken in the proper account. In both cases there are zones with tensile stresses in material. In contact interactions where compressive stress is usually dominant, there are still zones in material subjected to tensile stress caused by contact frictional or tensile residual stress (see [14, 67]).

It is widely accepted that fatigue cracks get initiated by shear stresses. Experimental and theoretical studies suggest (see Kudish and Burris [47]) that soon after initiation fatigue cracks propagate in the direction determined by the local stress field, namely, perpendicular to the local maximum tensile (principle) stress. Therefore, we will assume that fatigue is caused by propagation of penny-shaped subsurface cracks under the action of principal maximum tensile stresses. As it has been shown in Subsection 9.6.6 failure due to surface cracks is an unusual event and a very rapid process, and it will not be considered in this model. Therefore, surface cracks are not included in the model. On a plane perpendicular to a principal stress the shear stresses are equal to zero, i.e., the shear stress intensity factors $k_2 = k_3 = 0$. To find the plane of fatigue crack propagation (i.e., the orientation angles α and β), which is perpendicular to the maximum principal tensile stress, it is necessary to find the directions of these principal stresses. The latter is equivalent to solution of the equations

$$k_2(N, \alpha, \beta, l, x, y, z) = 0, \quad k_3(N, \alpha, \beta, l, x, y, z) = 0. \tag{9.321}$$

It is important to remember that for the most part of their lives fatigue cracks created or existed near material defects remain small. Therefore, penny-shaped subsurface cracks conserve their shape but increase in size.

The fact that for steady cyclic loading for "small cracks" $k_{20} = k_2/\sqrt{l}$ and $k_{30} = k_3/\sqrt{l}$ are independent from the number of cycles N and crack radius l together with equation (9.321) lead to the conclusion that for cyclic loading with constant amplitude angles α and β characterizing the plane of fatigue crack growth are independent from N and l. Thus, angles α and β are functions of only crack location, i.e., $\alpha = \alpha(x, y, z)$ and $\beta = \beta(x, y, z)$. For most stress fields (excluding stress fields with special symmetry) at any point (x, y, z), there are possible to find several sets of solutions (α_m, β_m) to equations (9.321). The crack propagation angles α and β are determined by one of these sets of values (α_n, β_n) for which the value of the normal stress intensity factor $k_1(N, l, x, y, z)$ is positive and maximal. It is guaranteed that fatigue cracks propagate in the direction perpendicular to the maximum tensile stress if α and β are determined this way.

9.10.3   Crack Propagation Calculations

In general, propagation of a fatigue crack subjected to only normal tensile stress can be described by the initial-value problem (9.241). It should be solved at every point of the material stressed volume V at which $\max_{T}(k_1) > k_{th}$, where the maximum is taken over the duration of the loading cycle T and k_{th} is the material stress intensity threshold. Equation (9.241) may depend on the parameters of the material microstructure. Typical graph of crack propagation rate dl/dN versus l is schematically presented in Fig. 9.77. It is clear from this graph that there are three distinct stages of crack development: (a) growth of small cracks, (b) propagation of well–developed cracks, and (c) explosive and, usually, unstable growth of large cracks.

The phase of small crack growth is the slowest one and, it represents the main portion of the entire crack propagation period. This situation usually causes confusion about the duration of the phases of crack initiation and propagation of small cracks. The next phase, propagation of well–developed cracks, usually takes significantly less time than the phase of small crack growth. And, finally, the explosive crack growth takes almost no time.

A number of crack propagation equations of type (9.241) are collected and analyzed by Yarema [33]. Any one of these equations can be used in the model to describe propagation of fatigue cracks. We will use the simplest of them, which allows to take into account the residual stress and, at the same time, to avoid the above–mentioned effect of "double dipping" while taking into account the stress intensity threshold k_{th} (for the details of this discussion, see Subsection 9.7.5). Therefore, for fatigue crack propagation we will be using Paris's equation

$$\frac{dl}{dN} = g_0 (\max_{-\infty < x < \infty} \triangle k_1)^n, \ l \mid_{N=0} = l_0, \tag{9.322}$$

where g_0 and n are the parameters of material fatigue resistance and l_0 is the crack initial half-length. Notice, that in case of contact fatigue $\triangle k_1 = k_1$.

Fatigue cracks propagate until they reach their critical size with radius l_c for which $k_1 = K_f$ (K_f is the material fracture toughness), i.e., to $l_c = (K_f/k_{10})^2$. After that their growth becomes unstable and very fast. Usually, the phase of explosive crack growth takes just few loading cycles.

It can be shown that the number of loading cycles needed for a crack to reach its critical radius is almost independent from the material fracture toughness K_f. This conclusion is supported by direct numerical simulations. For the further analysis, it is necessary to determine the crack initial radius l_{0c}, which after N loading cycles reaches the critical size of l_c. Using the analysis of Subsection 9.7.5, we get (see formula (9.248))

$$l_{0c} = \{l_c^{\frac{2-n}{2}} + N(\tfrac{n}{2} - 1)g_0[\max_{-\infty < x < \infty} \triangle k_{10}]^n\}^{\frac{2}{2-n}}, \tag{9.323}$$

where l_{0c} depends on N, x, y, and z. Obviously, for $n > 2$ and fixed x, y, and z the value of l_{0c} is a decreasing function of N. It is important to keep in mind that $l_{0c}(N, x, y, z)$ is minimal where $k_{10}(x, y, z)$ is maximal, which, in turn, happens where the tensile stress reaches its maximum.

9.10.4   Crack Statistics

To describe crack statistics after the crack initiation phase is over, it is necessary to make certain assumptions. The simplest assumptions of this kind are: the existing cracks do not heal and new cracks are not created. In other words, the number of cracks in any material volume is constant in time. The pre-existing defects, which are located close to each other (grouped in a cluster), are replaced by one equivalent crack determined by the same value of the stress intensity factors as produced by the defect cluster. Based on a reasonable assumption that the defect distribution is initially scarce, the coalescence of cracks and changes in the general stress field are possible only when cracks have already reached relatively large sizes. However, this may happen only during the last stage of crack growth, which is insignificant for calculation of fatigue life. Therefore, we can also assume that over almost all life span of fatigue cracks their orientation does not change. This leads to the equation for the density of crack distribution $f(N, x, y, z, l)$ as a function of crack half-length l after N loading cycles in a small parallelepiped $dxdydz$ with the center at the point with coordinates (x, y, z) (see Subsection 9.7.6):

$$f(N, x, y, z, l)dl = f(0, x, y, z, l_0)dl_0, \tag{9.324}$$

which being solved for $f(N, x, y, z, l)$ gives

$$f(N, x, y, z, l)dl = f(0, x, y, z, l_0)\tfrac{dl_0}{dl}, \tag{9.325}$$

where l_0 and dl_0/dl as functions of N and l can be obtain from the solution of (9.322) in the form (see equation (9.252))

$$l_0 = \{l^{\frac{2-n}{2}} + N(\tfrac{n}{2} - 1)g_0[\max_{-\infty < x < \infty} \triangle k_{10}]^n\}^{\frac{2}{2-n}},$$

$$\tfrac{dl_0}{dl} = \{1 + N(\tfrac{n}{2} - 1)g_0[\max_{-\infty < x < \infty} \triangle k_{10}]^n l^{\frac{n-2}{2}}\}^{\frac{n}{n-2}}. \tag{9.326}$$

Equations (9.325) and (9.326) lead to the expression for the crack distribution function f after N loading cycles

$$f(N, x, y, z, l) = f(0, x, y, z, l_0(N, l, y, z))\{1$$

$$+ N(\tfrac{n}{2} - 1)g_0[\max_{-\infty < x < \infty} \triangle k_{10}]^n l^{\frac{n-2}{2}}\}^{\frac{n}{n-2}}, \tag{9.327}$$

where $l_0(N, l, y, z)$ is determined by the first of the equations in (9.326).

A number of important conclusions can be drawn from (9.327). Namely, the crack distribution function $f(N, x, y, z, l)$ depends on the initial crack distribution $f(0, x, y, z, l_0)$ and its changes with the number of applied loading cycles N in such a way that the crack volume density $\rho(N, x, y, z)$ remains constant. Because of crack growth the crack distribution $f(N, x, y, z, l)$ is wider with respect to l as the number of loading cycles N increases. Therefore, even if the number of cracks may change in the process of cyclic loading, it is safe to assume that the crack distribution $f(N, x, y, z, l)$ changes with time (i.e., with the number of loading cycles N). In spite of that all fatigue models that take into account material statistical defect distribution implicitly assume that the defect distribution does not change with loading. The detailed analysis of this crack distribution is given in Subsection 9.7.6.

9.10.5   Local and Global Fatigue Damage Accumulation

It is clear that if at a certain point (x, y, z) after N loading cycles radii of all cracks $l < l_c$ then there is no damage at this point and the material local survival probability $p(N, x, y, z) = 1$. On the other hand, if at this point after N loading cycles radii of all cracks $l \geq l_c$, then all cracks reached the critical size and the material at this point (at the crack locations) is completely damaged and the local survival probability $p(N, x, y, z) = 0$. It is reasonable to assume that the material local survival probability $p(N, x, y, z)$ is a certain monotonic measure of the portion of cracks with radius l below the critical radius l_c. Therefore, $p(N, x, y, z)$ can be represented by the following expressions (see equation (9.254)):

$$p(N, x, y, z) = \frac{1}{\rho} \int\limits_0^{l_c} f(N, x, y, z, l)dl \ \ if \ \ f(0, x, y, z, l_0) \neq 0,$$

$$p(N, x, y, z) = 1 \ otherwise,$$

$$\rho = \rho(N, x, y, z) = \int\limits_0^\infty f(N, x, y, z, l)dl, \tag{9.328}$$

$$\rho(N, x, y, z) = \rho(0, x, y, z).$$

Obviously, the local survival probability $p(N, x, y, z)$ is a monotonically decreasing function of N because fatigue crack radii l tend to grow with the number of loading cycles N.

To calculate $p(N, x, y, z)$ from (9.328) one can use the expression for f determined by (9.327). However, it is more convenient to modify it and use in the form

$$p(N, x, y, z) = \frac{1}{\rho} \int\limits_0^{l_{0c}} f(0, x, y, z, l_0)dl_0 \ \ if \ \ f(0, x, y, z, l_0) \neq 0,$$

$$p(N, x, y, z) = 1 \ otherwise, \tag{9.329}$$

where l_{0c} is determined by (9.323) and ρ is the initial volume density of cracks. Thus, to every material point (x, y, z) is assigned a certain survival probability, $0 \leq p(N, x, y, z) \leq 1$.

Equations (9.329) demonstrate that the material local survival probability $p(N, x, y, z)$ is mainly controlled by the initial crack distribution $f(0, x, y, z, l_0)$, material fatigue resistance parameters g_0 and n, and contact and residual stresses. Moreover, it easily follows from (9.329) that the material local survival probability $p(N, x, y, z)$ is a decreasing function of N because l_{0c} from (9.323) is a decreasing function of N for $n > 2$.

To come up with the survival probability $P(N)$ of the material as a whole at early stages of the fatigue process, we will follow the analysis of Subsection 9.7.8, which leads to the formula

$$P(N) = p_m(N), \ p_m(N) = \min_V p(N, x, y, z), \tag{9.330}$$

where the maximum is taken over the (stressed) volume of the solid.

If the initial crack distribution is taken in the log-normal form of equation (9.319), then

$$P(N) = p_m(N) = \tfrac{1}{2}\left\{1 + erf\left[\min_V \frac{\ln l_{0c}(N,y,z) - \mu_{ln}}{\sqrt{2}\sigma_{ln}}\right]\right\}, \tag{9.331}$$

where $erf(x)$ is the error integral. Obviously, the survival probability $p_m(N)$ is a complex combined measure of applied stresses, initial crack distribution, material fatigue parameters, and the number of loading cycles.

If the mean μ_{ln} and the standard deviation σ_{ln} depend on location in material, i.e., on (x, y, z), then depending on functions $\mu_{ln}(x, y, z)$ and $\sigma_{ln}(x, y, z)$ the minimum in formula (9.331) can occur at different points the location of which may or may not primarily be dictated by these functions.

In cases when the mean μ_{ln} and the standard deviation σ_{ln} are constants throughout, the material formula (9.331) can be significantly simplified

$$P(N) = p_m(N) = \tfrac{1}{2}\left\{1 + erf\left[\frac{\ln \min_V l_{0c}(N,y,z) - \mu_{ln}}{\sqrt{2}\sigma_{ln}}\right]\right\}. \tag{9.332}$$

In the further analysis we will assume that μ_{ln} and σ_{ln} are constants.

To determine fatigue life N of a contact for the given survival probability $P(N) = P_*$, it is necessary to solve the equation

$$p_m(N) = P_*. \tag{9.333}$$

The results of the analysis of this model are identical to the ones obtained in Section 9.8. We will reiterate just the final most important results. Suppose the material failure occurs at point (x, y, z) with probability $1 - P(N)$. By that we actually determine the point where in (9.332) the minimum over the material volume V is reached. Therefore, at this point in (9.332), the operation of minimum over the material volume V can be dropped. By solving (9.332) and (9.333), one gets

$$N = \{(\tfrac{n}{2} - 1)g_0[\max_{-\infty < x < \infty} \triangle k_{10}]^n\}^{-1}\{\exp[(1 - \tfrac{n}{2})(\mu_{ln}$$

$$+ \sqrt{2}\sigma_{ln}erf^{-1}(2P_* - 1))] - l_c^{\frac{2-n}{2}}\}, \tag{9.334}$$

where $erf^{-1}(x)$ is the inverse function to the error integral $erf(x)$. Let us simplify this equation for the case of a material initially free of damage, i.e., when $P(0) = 1$. Discounting the very tail of the initial crack distribution, we get $\max\limits_{V} l_0 \leq l_c$. Thus, for well–developed cracks and, in many cases, even for small cracks, the second term in (9.323) for l_{0c} dominates the first one. It means that the dependence of l_{0c} on l_c and, therefore, on the material fracture toughness K_f can be neglected. Therefore, equation (9.334) can be approximated by

$$N = \{(\tfrac{n}{2} - 1)g_0[\max\limits_{-\infty < x < \infty} \triangle k_{10}]^n\}^{-1}\{\exp[(1 - \tfrac{n}{2})(\mu_{ln}$$

$$+ \sqrt{2}\sigma_{ln}erf^{-1}(2P_* - 1))]\}.$$

(9.335)

Taking into account that k_{10} is proportional to the maximum contact pressure q_{max} and, also, depends on the friction coefficient λ and the ratio of residual stress q^0 and $q_{max} = p_H$ one arrives at a simple analytical formula

$$N = \frac{C_0}{(n-2)g_0 p_H^n}\left(\frac{\sqrt{\mu^2+\sigma^2}}{\mu^2}\right)^{\tfrac{n}{2}-1}\exp[(1 - \tfrac{n}{2})(\ln \frac{\mu^2}{\sqrt{\mu^2+\sigma^2}}$$

$$+ \sqrt{2\ln[1 + (\tfrac{\sigma}{\mu})^2]}erf^{-1}(2P_* - 1))],$$

(9.336)

where C_0 depends only on the friction coefficient λ and the ratio of the residual stress q^0 and the maximum Hertzian pressure p_H. In (9.336) we used the expressions for μ_{ln} and σ_{ln} as functions of the regular initial mean μ and standard deviation σ from (9.267). Finally, assuming that $\sigma \ll \mu$ from (9.336) one can obtain the formula

$$N = \frac{C_0}{(n-2)g_0 p_H^n \mu^{\tfrac{n}{2}-1}}\exp[(1 - \tfrac{n}{2})\frac{\sqrt{2}\sigma}{\mu}erf^{-1}(2P_* - 1))].$$

(9.337)

Also, formulas (9.336) and (9.337) can be represented in the form of the Lundberg-Palmgren formula (see formula (9.274) and the discussion in Subsection 9.8.1).

Formula (9.337) demonstrates the intuitively obvious fact that the fatigue life N is inverse proportional to the value of the parameter g_0 that characterizes the material crack propagation resistance. Equation (9.337) exhibits a usual for roller and ball bearings as well as for gears dependence of the fatigue life N on the maximum Hertzian pressure p_H. Thus, from the well–known experimental data for bearings the range of n values is $20/3 \leq n \leq 9$. Keeping in mind that usually $\sigma \ll \mu$, for these values of n contact fatigue life N is practically inverse proportional to a positive power of the mean crack size, i.e., to $\mu^{n/2-1}$. Therefore, fatigue life N is a decreasing function of the initial mean crack (inclusion) size μ. This conclusion is valid for any material survival probability P_* and is supported by the experimental data discussed in Kudish and Burris [48]. If $P_* > 0.5$, then $erf^{-1}(2P_* - 1) > 0$ and (keeping in mind that $n > 2$) fatigue life N is a decreasing function of the initial standard deviation of crack sizes σ. Similarly, if $P_* < 0.5$, then fatigue life N is an increasing function of the initial standard deviation of crack sizes σ. According to (9.337), for $P_* = 0.5$ fatigue life N is independent from σ, and, according to (9.336), for $P_* = 0.5$ fatigue life N is a slowly increasing function of σ. By differentiating $p_m(N)$ obtained from (9.332) with respect to σ, we can conclude that the dispersion of $P(N)$ increases with σ.

The stress intensity factor k_1 decreases as the magnitude of the compressive residual stress q^0 increases and/or the magnitude of the friction coefficient λ decreases. Therefore, in (9.336) and (9.337) the value of C_0 is a monotonically decreasing function of residual stress q^0 and friction coefficient λ.

Being applied to bearings and/or gears the described statistical contact fatigue model can be used as a research and engineering tool in modeling pitting. In the latter case, some of the model parameters may be assigned certain fixed values based on the scrupulous analysis of steel quality and quality and stability of manufacturing processes.

9.10.6 Examples of Torsional and Bending Fatigue

For simplicity we will assume that in a beam material the defect distribution is space-wise uniform and follows equation (9.319). To use the available formulas for torsional and bending loadings, we will also assume that the residual stress is zero.

First, let us consider torsional fatigue. Suppose a beam is made of an elastic material with elliptical cross section (a and b are the ellipse semi-axes, $b < a$) and directed along the y-axis. The beam is under action of torque M_y about the y-axis applied to its ends. The side surfaces of the beam are free of stresses. Then it can be shown (see Lurye [72], p. 398) that

$$\tau_{xy} = -\tfrac{2G\gamma a^2}{a^2+b^2} z, \ \tau_{zy} = \tfrac{2G\gamma b^2}{a^2+b^2} x, \ \sigma_x = \sigma_y = \sigma_z = \tau_{xz} = 0, \tag{9.338}$$

where G is the material shear elastic modulus, $G = E/[2(1 + \nu)]$ (E and ν are Young's modulus and Poisson's ratio of beam material), and γ is a dimensionless constant. By introducing the principal stresses σ_1, σ_2, and σ_3 that satisfy the equation $\sigma^3 - (\tau_{xy}^2 + \tau_{zy}^2)\sigma = 0$, we obtain that

$$\sigma_1 = -\sqrt{\tau_{xy}^2 + \tau_{zy}^2}, \ \sigma_2 = 0, \ \sigma_3 = \sqrt{\tau_{xy}^2 + \tau_{zy}^2}. \tag{9.339}$$

For the case of $a > b$ the maximum principal tensile stress $\sigma_1 = -\tfrac{2G\gamma a^2 b}{a^2+b^2}$ is reached at the surface of the beam at points $(0, y, \pm b)$ and depending on the sign of M_y it acts in one of the direction described by the directional cosines

$$\cos(\alpha, x) = \mp\tfrac{\sqrt{2}}{2}, \ \cos(\alpha, y) = \pm\tfrac{\sqrt{2}}{2}, \ \cos(\alpha, z) = 0, \tag{9.340}$$

where α is the direction along one of the principal stress axes. For the considered case of elliptic beam, the moments of inertia of the beam elliptic cross section about the x- and y-axes, I_x and I_z as well as the moment of torsion M_y applied to the beam are as follows (see Lurye [72], pp. 395, 399) $I_x = \pi a b^3/4$, $I_z = \pi a^3 b/4$, $M_y = G\gamma C$, $C = 4 I_x I_z/(I_x + I_z)$. Keeping in mind that according to Hasebe and Inohara [73] and Isida [74], the stress intensity factor k_1 for the edge crack of radius l and inclined to the surface of a half-plane at the angle of $\pi/4$ (see (9.340)) is $k_1 = 0.705 \mid \sigma_1 \mid \sqrt{\pi l}$, we obtain $k_{10} = \frac{1.41}{\sqrt{\pi}} \frac{|M_y|}{ab^2}$. By substituting the expression for k_{10} into equation (9.335), we obtain fatigue life of a beam under torsion

$$N = \frac{2}{(n-2)g_0} \{ \frac{1.257 a b^2}{|M_y|} \}^n g(\mu, \sigma), \tag{9.341}$$

$$g(\mu, \sigma) = (\frac{\sqrt{\mu^2+\sigma^2}}{\mu^2}) \sigma^{\frac{n-2}{2}} \exp[(1 - \frac{n}{2})\sqrt{2\ln[1 + (\frac{\sigma}{\mu})^2]} \tag{9.342}$$

$$\times erf^{-1}(2P_* - 1)\}.$$

Now, let us consider bending fatigue of a beam/console made of an elastic material with elliptical cross section (a and b are the ellipse semi-axes) and length L. The beam is directed along the y-axis and it is under the action of a bending force P_x directed along the x-axis and applied to its free end. The side surfaces of the beam are free of stresses. The other end $y = 0$ of the beam is fixed. Then it can be shown (see Lurye [72]) that

$$\sigma_x = \sigma_z = 0, \quad \sigma_y = -\frac{P_x}{I_z} x(L - y),$$

$$\tau_{xz} = 0, \quad \tau_{xy} = \frac{P_x}{2(1+\nu)I_z} \frac{2(1+\nu)a^2+b^2}{3a^2+b^2} \{a^2 - x^2 - \frac{(1-2\nu)a^2 z^2}{2(1+\nu)a^2+b^2}\}, \tag{9.343}$$

$$\tau_{zy} = -\frac{P_x}{(1+\nu)I_z} \frac{(1+\nu)a^2+\nu b^2}{3a^2+b^2} xz,$$

where I_z is the moment of inertia of the beam cross section about the z-axis. Again, by introducing the principal stresses that satisfy the equation $\sigma^3 - \sigma_y \sigma^2 - (\tau_{xy}^2 + \tau_{zy}^2)\sigma = 0$, we find that

$$\sigma_1 = \frac{1}{2}[\sigma_y - \sqrt{\sigma_y^2 + 4(\tau_{xy}^2 + \tau_{zy}^2)}], \quad \sigma_2 = 0,$$
$$\tag{9.344}$$
$$\sigma_3 = \frac{1}{2}[\sigma_y + \sqrt{\sigma_y^2 + 4(\tau_{xy}^2 + \tau_{zy}^2)}].$$

The tensile principal stress σ_1 reaches its maximum $\frac{4|P_x|L}{\pi a^2 b}$ at the surface of the beam at one of points $(\pm a, 0, 0)$ (depending on the sign of load P_x) and is acting along the y-axis - the axis of the beam. Based on equations (9.344) and the solution for the surface crack inclined to the surface of the half-space at angle of $\pi/2$ (see Hasebe and Inohara [73] and Isida [74]), we obtain $k_{10} = \frac{4.484}{\sqrt{\pi}} \frac{|P_x|L}{a^2 b}$. By substituting the expression for k_{10} into equation (9.335), we obtain bending fatigue life of a beam

$$N = \frac{2}{(n-2)g_0} \{ \frac{0.395 a^2 b}{|P_x|L} \}^n g(\mu, \sigma), \tag{9.345}$$

where function $g(\mu, \sigma)$ is determined by equation (9.342).

In both cases of torsion and bending, fatigue life is independent of the elastic characteristic of the beam material (see formulas (9.341), (9.345), and (9.342)), and it is dependent on fatigue parameters of the beam material (n and g_0), the initial defect distribution (i.e. on μ and σ), the geometry of the beam cross section (a and b), and its length L and applied loading (P_x or M_y).

It is important to realize that in the presence of a subsurface compressive residual stress the actual mechanism of fatigue may change. Instead of fatigue cracks being nucleated at the beam surface the cracks leading to fatigue failure may be nucleated somewhere beneath the beam surface and than propagate at a rate slower then the one for similar surface cracks until they reach the surface. After that a faster phase of fatigue related to propagation of surface cracks will be in effect. A similar scenario may realize when the initial defect distribution is not space-wise uniform, i.e., fast fatigue crack growth may start at a different point at the surface or beneath the surface of a beam.

In a similar fashion the model can be applied to contact fatigue if the stress field is known (see Subsection 9.8.2).

Let us summarize the above analysis. A new three-dimensional statistical fatigue model equally applicable to contact and structural fatigue is derived and analyzed. The model takes into account the most important parameters of the fatigue phenomenon such as acting stresses, initial statistical defect distribution, orientation of fatigue crack propagation, material fatigue resistance, etc. The model allows studying the effect of material cleanliness, applied stresses, etc., on fatigue life as single or composite entities. Some analytical results illustrating application of the new model to torsional and bending fatigue of beams are presented.

9.11 Application of the New Contact Fatigue Model to Radial Thrust Bearings

Let us consider a couple of examples of the application of the contact fatigue model introduced in Section 9.10 to radial thrust bearings subjected to axial and radial forces. First, we will briefly discuss the stress analysis of single row radial thrust bearings and, then, apply the above–mentioned contact fatigue analysis.

Let us consider a single row radial thrust bearing mounted on a shaft rotating with frequency n_s. The bearing has z_b number of balls of diameter D_b. The radii of the cross sections of the inner and outer rings are r_i and r_o, respectively, and the bearing radial gap is e. In calculation of normal loads between balls and raceways, we will neglect friction.

9.11.1   Case of Axial Loading

We will conduct a simple force analysis for a single row radial thrust bearing with z_b number of balls of diameter D_b. By introducing $r_m = r_i + r_o - D_b$, we can determine the nominal angle of contact α_0 as follows [75]:

$$\cos(\alpha_0) = 1 - \frac{e}{2r_m}, \qquad (9.346)$$

which depends only on the bearing geometry.

In case when the bearing is loaded by an axial force F_a the problem of finding the forces between the balls and the raceways can be reduced to solution of the equation [75]

$$C^* \left(\frac{\cos(\alpha_0)}{\cos(\alpha)} - 1 \right)^{3/2} \sin(\alpha) = F_a, \qquad (9.347)$$

where C^* is a known constant characterizing the bearing geometry and strength/elastic properties, and α is the actual contact angle that corresponds to the applied force F_a. After (9.347) is solved the force acting between each ball and the inner raceway is determined from the formula

$$P = \frac{F_a}{z_b \sin(\alpha)}. \qquad (9.348)$$

As soon as we found the normal force P acting between balls and the raceway, we can assume a certain friction coefficient λ, material fatigue properties, and the initial crack distribution and conduct a fatigue calculation based on the model described in Section 9.10. The specific calculations are left as an exercise for a reader.

9.11.2   Case of Radial Loading

Suppose a single row radial bearing with z_b balls and radial clearance e is loaded by a radial force F_r. Then the problem of analyzing the stress state of the bearing can be reduced to solution of the equation [75]:

$$\sum_{i=0}^{z_b-1} [(\delta_0 + \tfrac{e}{2})\cos(i\gamma) - \tfrac{e}{2}]^{3/2}\theta[(\delta_0 + \tfrac{e}{2})\cos(i\gamma) - \tfrac{e}{2}]\cos(i\gamma) = \frac{F_r}{C_\delta},$$
$$(9.349)$$

$$\gamma = \frac{2\pi}{z_b},$$

where δ_0 is the radial displacement of the most heavily loaded ball caused by the radial force F_r, C_δ is the known elastic characteristic of the bearing, γ is the angle between neighboring balls, $\gamma = 2\pi/z_b$, and θ is the unit step function ($\theta(x) = 0$, $x \leq 0$; $\theta(x) = 1$, $x > 0$). After solving (9.349) for δ_0 the force acting between the bearing balls and the raceways can be determined by

$$P_i = C_\delta[(\delta_0 + \tfrac{e}{2})\cos(i\gamma) - \tfrac{e}{2}]^{3/2}\theta[(\delta_0 + \tfrac{e}{2})\cos(i\gamma) - \tfrac{e}{2}],$$
$$(9.350)$$

$$i = 0, \ldots, z_b - 1.$$

The maximum force $P_{max} = \max_i P_i$ acting between the most loaded ball and the inner raceway can be either determined using the solution of equations (9.349) and (9.350) or it can be approximately calculated from the formula [75]

$$P_{max} = k_0 \frac{F_r}{z_b}, \qquad (9.351)$$

where k_0 is a dimensionless coefficient that depends on the radial clearance e and elastic characteristic C_δ of the bearing. For $z_b \geq 4$ coefficient k_0 varies between 4 and 4.5. As our goal is to just illustrate how to apply the described above contact fatigue model to determining bearing fatigue life, for simplicity, for further calculations we can take $k_0 = 4.25$.

As soon as we determined the maximum normal force P_{max} acting between balls and the raceway, we can assume a certain friction coefficient λ, material fatigue properties, and the initial crack distribution and conduct a fatigue calculation based on the model described in Section 9.10. The specific calculations are left as an exercise for a reader.

9.12   Closure

The chapter presents a detailed analysis of various plane crack mechanics problems such as problems for an elastic plane and half-plane weakened by a system of cracks. A number of problems for an elastic half-plane with cracks is considered for the cases when the external normal and tangential loads are applied to the boundary of the half-plane. Surface and subsurface cracks are considered. All cracks are considered to be straight cuts. The problems are analyzed by the regular asymptotic method and numerical methods. Solutions of the problems include the stress intensity factors. The regular asymptotic method is applied under the assumption that cracks are far from one another and from the boundary of the elastic solid. It is shown that the results obtained for subsurface cracks based on asymptotic expansions and numerical solutions are in very good agreement. The influence of the normal and tangential contact stresses applied to the boundary of a half-plane as well as the residual stress on the stress intensity factors for subsurface cracks are analyzed. It is determined that the frictional and residual stresses provide a significant if not the predominant contribution to the problem solution. Routinely, cracks possess a number of cavities and regions where their faces are in contact with each other. Based on the numerical solution of the problem for surface cracks in the presence of lubricant the physical nature of the "wedge effect" when lubricant under a sufficiently large pressure penetrates a surface crack and ruptures it is considered. Solution of this problem also provides the basis for the understanding of fatigue crack origination site (surface versus subsurface) and the difference of fatigue lives of drivers and followers. Using the analysis of the preceding chapter and the results obtained in this chapter, new two- and three-dimensional models of contact fatigue are developed. These models take into account the initial crack distribution, fatigue properties of the solids, and growth of fatigue cracks under the properly determined combination of normal and tangential contact and residual stresses. The formula for fatigue life based on these models can be reduced to a simple formula, which takes into account most of the significant parameters affecting contact fatigue. The properties of these contact fatigue models are analyzed and the results based on them are compared to the experimentally obtained results on contact fatigue for tapered bearings.

9.13 Exercises and Problems

1. What is the meaning of the normal $k_1^\pm$ and shear $k_2^\pm$ stress intensity factors? What are the definitions of $k_1^\pm$ and $k_2^\pm$ and how they are related to the crack configuration? Express $k_1^\pm$ and $k_2^\pm$ in terms of the normal $v(x)$ and tangential $u(x)$ crack faces displacement jumps.

2. What are the values of the stress intensity factors $k_1^\pm$ and $k_2^\pm$ when the complex crack faces displacement jump $g'(x) = o((1 - x^2)^{-1/2})$ as $x \to \pm 1$?

3. Derive equations (9.36) for an elastic half-plane weakened by N straight cracks from equations (9.31) and (9.32) for an elastic plane weakened by $N+1$ straight cracks.

4. Substantiate why the solution of the system of equations (9.49)-(9.55) for "small" cracks can be obtained in the form of power series (9.59) in δ_0.

5. Develop a two-term asymptotic solution of equations (9.49)-(9.55) for "small" cracks when $p_k^0 = \delta_0^2 C_{k001}^r + \ldots$, $C_{k001}^r = O(1)$ for $\delta_0 \ll 1$.

6. Assume that an elastic plane is weakened by a single straight crack with dimensionless semi-length $\delta_0 = 0.2$ with its center located at $(x^0, y^0) = (0, 0)$. The crack faces are loaded by the stresses from (9.54). Determine the crack faces normal $v(x)$ and tangential $u(x)$ displacement jumps as well as the normal $k_1^\pm$ and shear $k_2^\pm$ stress intensity factors for various combinations of acting stresses $\eta = -1$, -0.5, 0, 0.5, 1 and the angles of crack orientation $\alpha = 0$, $\pi/6$, $\pi/4$, $\pi/3$, $\pi/2$ based on (a) one-term expansions, (b) two-term expansions, (c) tree-term expansions, (d) four-term expansions, (e) five-term expansions. Plot and compare graphs of $v(x)$ and $u(x)$ obtained in (a)-(e).

7. Assuming that cracks propagate in the direction along which the shear stress intensity factors $k_2^\pm$ are equal to zero and the normal stress intensity factors $k_1^\pm$ attain their maximum value for $\eta = -1$, -0.5, 0, 0.5, 1 and $\delta_0 = 0.2$ determine: (a) the direction of a single crack propagation in an elastic plane from the conditions $k_2^-(\alpha_-) = 0$, $k_1^- = \max_\alpha k_1^-(\alpha)$ based on the one-term and three-term approximations for the stress intensity factors k_1^- and k_2^-; (b) the direction of a single crack propagation in an elastic plane from the conditions $k_2^+(\alpha_+) = 0$, $k_1^+ = \max_\alpha k_1^+(\alpha)$ based on the one-term and three-term approximations for the stress intensity factors k_1^+ and k_2^+; (c) compare the directions of the single crack propagation, i.e., compare angles α_- and α_+.

8. Assume that an elastic half-plane with a single straight crack of dimensionless semi-length $\delta_0 = 0.2$ located at the depth of $y^0 = -0.8$ below the half-plane surface is loaded by the Hertzian pressure $p(x^0) = \sqrt{1 - (x^0)^2}\theta[1 - (x^0)^2]$ and the tangential stress $\tau(x^0) = -\lambda p(x^0)$, where λ is the friction coefficient. Determine the direction of crack propagation for $-\infty < x^0 < \infty$ under the assumptions of Problem 7 based on the one-term approximations for the stress intensity factors $k_1^\pm$ and $k_2^\pm$. For specific calculations assume

that the friction coefficient $\lambda = 0.1$, Poisson's ratio $\nu = 0.3$, and the residual stress $q^0 = -0.2$, 0, 0.2. Analyze the range of the obtained angle $\alpha = \alpha^{\pm}$ as a function of the relative crack position x^0 with respect to the Hertzian pressure.

9. List and discuss all assumptions made to obtain the asymptotic solution of the contact problem for an elastic half-plane weakened by straight cracks.

10. Describe the three distinct stages of fatigue crack propagation. Substantiate the approach of approximating fatigue cracks by straight cuts over practically their entire life span.

11. What are the assumptions with respect to fatigue cracks and the initial fatigue crack distribution $f(0, x, y, z, l)$ used in the derivation of the contact fatigue models?

12. Does growth of fatigue cracks lead to changes in the initial fatigue crack distribution $f(0, x, y, z, l)$? What happens with the fatigue crack distribution $f(N, x, y, z, l)$ over time (i.e., over number of loading cycles N) if (a) fatigue cracks get arrested and (b) their lengths monotonically increase with N. Consider the properties of equation (9.251).

13. List the main assumptions used for the derivation of the local $p(N, x, y, z)$ and global $P(N)$ probabilities of survival. Explain why at least at early stages of fatigue $P(N) = p_m(N)$.

14. Assume that for survival probabilities P_1 and P_2 fatigue lives are N_1 and N_2, respectively. Based on equation (9.272) determine the ratio of the standard deviation σ and mean μ of the initial crack distribution $f(0, x, y, z, l)$.

15. Analytically and numerically analyze the behavior of $\ln(N)$ versus $\ln(\mu)$ and show that this relationship is very close to a linear one.

16. Describe the distinct differences in the formulations of the problems for subsurface and surface cracks. Explain the source of additional complexity of the problem for surface cracks fully or partially filled with lubricant.

17. Describe all possible scenarios of a surface crack cavity dynamics and the related to these scenarios boundary conditions imposed on the crack.

18. Describe the main assumptions and the idea behind the Gupta and Erdogan [8] and Savruk [11] numerical methods. Explain why in case of cracks with their faces partially in contact with each other the above numerical methods cannot be applied directly. What are the distinct differences between the above methods and the approach of this chapter.

19. For similar surface and subsurface cracks compare the order of magnitude of (a) the normal stress intensity factors k_1 and (b) the shear stress intensity factors k_2. How does the crack orientation angle α affect the orders of magnitude of the stress intensity factors k_1 and k_2. Give the assessment of the risk of fatigue failure of solids with various surface and subsurface cracks.

20. Describe the influence of the friction coefficient λ on the stress intensity factors k_1 and k_2 for surface and subsurface cracks.

21. Explain why in lubricated contacts usually followers have shorter fatigue lives than drivers. What is the role of surface and subsurface fatigue cracks in fatigue of followers and drivers?

22. Discuss the similarities and differences in fatigue of dry and lubricated contacts. Explain why fatigue lives of dry contacts are usually significantly shorter than of the lubricated ones subjected to the same operating conditions.

23. Pick a specific thrust bearing and determine its fatigue life distribution for applied (a) axial load F_a and (b) radial load F_r. You will have to assume a specific value for the friction coefficient λ, parameters of material fatigue properties, and the initial crack distribution to conduct a fatigue calculation based on the model described in Section 9.10.

BIBLIOGRAPHY

[1] Kaneta M., Murakami Y., and Okazaki T. 1986. Growth Mechanism of a Subsurface Crack Due to Hertzian Contact. *ASME J. Tribology* 108:134-139.

[2] Kaneta M., Yatsuzuka H., and Murakami Y. 1985. Mechanism of Crack Growth in Lubricated Rolling/Sliding Contact. *ASLE Trans.* 28:407-414.

[3] Murakami Y., Kaneta M., and Yatsuzuka H. 1985. Analysis of Surface Crack Propagation in Lubricated Rolling Contact. *ASLE Trans.* 28:60-68.

[4] Kudish, I.I. 1984. The Contact-Hydrodynamic Problem of Lubrication Theory for Elastic Bodies with Cracks. *J. Appl. Math. and Mech.* 48:799-808.

[5] Kudish, I.I. 1985. Failure Mechanisms of Elastic Bodies with Cracks in the Presence of a Lubricant. *J. Strength Materials* 17:64-70.

[6] Xu, Z.-Q. and Hsia, K.J. 1997. A Numerical Solution of a Surface Crack under Cyclic Hydraulic Pressure Loading. *ASME J. Tribology* 119:637-645.

[7] Panasyuk V.V., Datsyshin O.P., and Marchenko H.P. 1995. The Crack Propagation Theory Under Rolling Contact. *Eng. Fract. Mech.* 52, No. 1:179-191.

[8] Gupta G.D. and Erdogan F. 1974. The Problem of Edge Cracks in an Infinite Stripe. *ASME Trans.* Ser. E, 41:1001-1006.

[9] Erdogan F. and Arin K. 1975. A Half-Plane and Stripe with an Arbitrarily Located Crack. *Intern. J. Fract.* 11:191-204.

[10] Panasyuk V.V., Savruk M.P., and Datsyshin A.P. 1976. *Stress Distribution Near Cracks in Plates and Shells*. Kiev: Naukova Dumka.

[11] Savruk M.P. 1981. *Two − −Dimensional Problems of Elasticity for Bodies with Cracks*. Kiev: Naukova Dumka.

[12] Muskhelishvili, N.I. 1953. *Some Problems of Mathematical Theory of Elasticity*. Gronigen: Noordhoff.

[13] Kudish, I.I. 1986. Contact Problem of the Elasticity Theory for Bodies with Cracks. *J. Appl. Math. and Mech.* 50:1020-1033.

[14] Kudish, I.I. 1987. Contact Problem of the Theory of Elasticity for Pre-stressed Bodies with Cracks. *J. Appl. Mech. and Techn. Phys.* 28, 2:295-303.

[15] Galin, L.A. 1980. *Contact Problems in Elasticity and Visco − −Elasticity* Moscow: Nauka.

[16] Shtaerman, I.Ya. 1949. *Contact Problem of Elasticity* Moscow-Leningrad: Gostekhizdat.

[17] *Handbook of Mathematical Functions with Formulas, Graphs and Mathematical Tables*, Eds. M. Abramowitz and I.A. Stegun, National Bureau of Standards, 55, 1964.

[18] Baker, G.A., Jr. and Graves-Morris, P. 1981. *Pade Approximants*. Reading: Addison-Wesley.

[19] Kudish I.I. 2002. Lubricant-Crack Interaction, Origin of Pitting, and Fatigue of Drivers and Followers. *STLE Tribology Trans.* 45:583-594.

[20] Kudish I.I. and Burris, K.W. 2004. Modeling of Surface and Subsurface Crack Behavior under Contact Load in the Presence of Lubricant. *Intern. J. Fract.* 125:125-147.

[21] Kudish, I.I. 1996. Asymptotic Analysis of a Problem for a Heavily Loaded Lubricated Contact of Elastic Bodies. Pre- and Over-critical Lubrication Regimes for Newtonian Fluids. *Dynamic Systems and Applications*, Dynamic Publishers, Atlanta, 5, No. 3:451-476.

[22] Ertel, A.M. 1945. Hydrodynamic Lubrication Analysis of a Contact of Curvilinear Surfaces. *Proc. of CNIITMASh*, Moscow, 1-64.

[23] Grubin, A.N. 1949. The Fundamentals of the Hydrodynamic The-ory of Lubrication for Heavily Loaded Cylindrical Surfaces. *Proc. of CNIITMASh* 30, Moscow, 126-184.

[24] Panasyuk V.V. 1968. *Limit Equilibrium of Brittle Solids with Cracks*. Kiev: Naukova Dumka.

[25] Coffin, L.F. 1954. A Study of the Effects of Cyclic Thermal Stresses on a Ductile Metal. *ASME Trans.* 76, No. 6:931-950.

[26] Cherepanov, G.P. 1979. *Mechanics of Brittle Fracture*. New York: McGraw–Hill.

[27] Paris, P. and Erdogan, F. 1963. A Critical Analysis of Crack Propagation Laws. *ASME J. Basic Eng.* 85:528-534.

[28] Yarema, S.Ya. 1977. Analysis of Fatigue Crack Growth and Kinetic Dia-grams of Fatigue Failure. *Physico − −Chem. Mech. Materials* 13, No. 4:3-22.

[29] Yarema, S.Ya. and Mikitishin, S.I. 1975. Analytic Description of Diagrams of Material Fatigue Failure. *Physico——Chem. Mech. Materials* 11, No. 4:47-54.

[30] Cherepanov, G.P. 1987. *Mechanics of Rock Fracture in Drilling.* Moscow: Nedra.

[31] Parton, V.Z. and Morozov, E.V. 1985. *Mechanics of Elastoplastic Fracture.* Moscow: Nauka.

[32] Morozov, N.F. 1984. *Mathematical Problems in Crack Theory.* Moscow: Nauka.

[33] Yarema, S.Ya. 1981. Methodology of Determining the Characteristics of the Resistance to Crack Development (Crack Resistance) of Materials in Cyclic Loading. *J. Soviet Material Sci.* 17, No. 4:371-380.

[34] Datsyshyn, O.P. and Panasyuk, V.V. 2001. Pitting of the Rolling Bodies Contact Surface. *Wear* 251:1347-1355.

[35] Finkel, V.M. 1970. *Physics of Fracture.* Moscow: Metalugia.

[36] Solntzev, S.S. and Morozov, E.M. 1978. *Fracture of Glass.* Moscow: Mashinostroenie.

[37] Murakami Y., Sakae, C., Ichimaru, K., and Morita, T. 1997. Experimental and Fracture Mechanics Study of the Pit Formation Mechanism under Repeated Lubricated Rolling-Sliding Contact: Effects of Reversal of Rotation and Change of the Driving Roller. *ASME J. Tribology* 119, No. 4:788-796.

[38] Bower, A.F. 1988. The Influence of Crack Face Friction and Trapped Fluid on Surface Initiated Rolling Contact Fatigue Cracks. *ASME J. Tribology* 110, No. 4:704-711 (and references therein).

[39] Kudish, I.I. 2000. A New Statistical Model of Contact Fatigue. *STLE Tribology Trans.* 43:711-721.

[40] Lundberg, G. and Palmgren, A. 1947. Dynamic Capacity of Rolling Bearings. *Acta Polytechnica (Mech. Eng. Ser. 1), Royal Swedish Acad. Eng. Sci.* 7:5-32.

[41] Ioannides, E. and Harris, T.A. 1985. A New Fatigue Life Model for Rolling Bearings. *ASME J. Tribology* 107:367-378.

[42] Coy, J.J., Townsend, D.P., and Zaretsky, E.V. 1976. Dynamic Capacity and Surface Fatigue Life for Spur and Helical Gears. *ASME J. Lubr. Techn.* 98:267-276.

[43] Kudish, I.I. 1986. On the Influence of Friction Stresses on the Fatigue Failure of Machine Parts. *Soviet J. Frict. and Wear* 7:32-40.

[44] Kudish, I.I. 1990. Statistical Calculation of Rolling Bearing Wear and Pitting. *Soviet J. Frict. and Wear* 11:71-86.

[45] Tallian, T.E. 1988. Rolling Bearing Life Prediction. Corrections for Material and Operating Conditions. Part I: General Model and Basic Life. *ASME J. Tribology* 110:1-6.

[46] Blake, J.W. 1989. *A Surface Pitting Life Model for Spur Gears*, Ph.D. Dissertation, Northwestern University.

[47] Kudish, I.I. and Burris, K.W. 2000. Modern State of Experimentation and Modeling in Contact Fatigue Phenomenon. Part I. Contact Fatigue versus Normal and Tangential Contact and Residual Stresses. Nonmetallic Inclusions and Lubricant Contamination. Crack Initiation and Crack Propagation. Surface and Subsurface Cracks. *STLE Tribology Trans.* 43, No. 2:187-196.

[48] Kudish, I.I. and Burris, K.W. 2000. Modern State of Experimentation and Modeling in Contact Fatigue Phenomenon. Part II. Analysis of the Existing Statistical Mathematical Models of Bearing and Gear Fatigue Life. New Statistical Model of Contact Fatigue. *STLE Tribology Trans.* 43, No. 2:293-301.

[49] Kudish, I.I. 1995. Surface and Subsurface Cracks and Pores with Liquid under Action of a Contact Load in the Presence of a Lubricant. In *Fracture Mechanics* 25, ASTM STP 1220, Ed. F. Erdogan, ASTM, Philadelphia, 606-621.

[50] Nelias, D., Dumont, M.-L., Couhier, F., Dudragne, G., and Flamand, L. 1998. Experimental and Theoretical Investigation on Rolling Contact Fatigue of 52100 and M50 Steels under EHL and Micro-EHL Conditions. *ASME J. Tribology* 120, No. 2:184-190.

[51] Hoeprich, M. 2001. Analysis of Micropitting of Prototype Surface Fatigue Test Gear. *Tribotest J.* 7, No.4:333-347.

[52] Soda, N. and Yamamoto, T. 1982. Effect of Tangential Traction and Roughness on Crack Initiation/Propagation During Rolling Contact. *ASLE Trans.* 25, No. 2:198-206.

[53] Hamrock, B.J. 1991. *Fundamentals of Fluid Film Lubrication*. Cleveland: NASA Reference Publication 1225.

[54] Szegö, G. 1959. *OrthogonalPolynomials*. American Mathematical Society, Colloquim Publications, Vol. XXIII, New York.

[55] Kudish, I.I. 2007. Fatigue Modeling for Elastic Materials with Statistically Distributed Defects. *ASME J. Appl. Mech.* 74:1125-1133.

[56] Tallian, T.E. 1992. Simplified Contact Fatigue Life Prediction Model - Part I: Review of Published Models. *ASME J. Tribology* 114, No. 1:207-213.

[57] Tallian, T.E. 1992. Simplified Contact Fatigue Life Prediction Model - Part II: New Model. *ASME J. Tribology* 114, No. 1:214-222.

[58] Dudragne, G., Fougeres, R., and Theolier, M. 1981. Analysis Method for Both Internal Stresses and Microstructural Effect under Pure Rolling Fatigue Conditions. *ASME J. Lubr. Techn.* 103, No. 4:521-525.

[59] Spektor, A.G., Zelbet, B.M., and Kiseleva, S.A. 1980. *Structure and Properties of Bearing Steels*. Moscow: Metallurgia.

[60] Bokman, M.A., Pshenichnov, Yu.P., and Pershtein, E.M. 1984. The Microcrack and Non-metallic Inclusion Distribution in Alloy D16 after a Plastic Strain. *Plant Laboratory*, Moscow, 11:71-74.

[61] Wu, Han C. and Yang, C.C. 1987. On the Influence of Strain-Path in Multiaxial Fatigue Failure. *ASME J. Eng. Materials and Techn.* 109, No. 2:107-113.

[62] Nisitani, H. and Goto, M. 1984. Effect of Stress Ratio on the Propagation of Small Crack of Plain Specimens under High and Low Stress Amplitudes. *Trans. Jpn. Soc. Mech. Eng.*, Ser. A, 50, No. 453:1090-1096.

[63] Shao, E., Huang, X., Wang, C., Zhu, Y., and Chen, Q. 1988. A Method of Detecting Rolling Contact Crack Initiation and the Establishment of Crack Propagation Curves. *STLE Tribology Trans.* 31, No. 1:6-11.

[64] Clarke, T.M., Miller, G.R., Keer, L.M., and Cheng, H.S. 1985. The Role of Near-Surface Inclusions in the Pitting of Gears. *STLE Trans.* 28, No. 1:111-116.

[65] Nelias, D., Dumont, M.L., Champiot, F., Vincent, A., Girodin, D., Fougres, R., and Flamand, L. 1999. Role of Inclusions, Surface Roughness and Operating Conditions on Rolling Contact Fatigue. *ASME J. Tribology* 121, No. 1:240-251.

[66] Murakami, Y., ed. 1987. *Stress Intensity Factors Handbook* 1. Oxford: Pergamon Press.

[67] Tallian, T., Hoeprich, M., and Kudish, I. 2001. Author's Closure. *STLE Tribology Trans.* 44, No. 2:153-155.

[68] Romaniv, O.N., Yarema, S.Ya., Nikiforchin, G.N., Makhutov, N.A., and Stadnik, M.M. 1990. *Fracture Mechanics and Strength of Materials, Vol. 4. Fatigue and Cyclic Crack Resistance of Construction Materials*. Kiev: Naukova Dumka.

[69] Stover, J.D. and Kolarik II, R.V. 1987. The Evaluation of Improvements in Bearing Steel Quality Using an Ultrasonic Macro-Inclusion Detection Method. *The Timken Company Technical Note*, (January 1987), 1-12.

[70] Stover, J.D., Kolarik II, R.V., and Keener, D.M. 1989. The Detection of Aluminum Oxide Stringers in Steel Using an Ultrasonic Measuring Method. *Mechanical Working and Steel Processing XXVII* : *Proc.* 31*st Mech. Working and Steel Processing Conf.*, Chicago, Illinois, October 22-25, 1989, Iron and Steel Soc., Inc., 431-440.

[71] Kudish, I.I. 1991. Wear and Fatigue Pitting Taking into Account Contaminated Lubricants and Abrasive Particles Indented into Working Surfaces, *Journal of Friction and Wear, Vol.* 12, No. 3:713-725.

[72] Lurye, A.I. 1970. *Theory of Elasticity.* Moscow: Nauka.

[73] Hasebe, N. and Inohara, S. 1980. Stress Analysis of a Semi-infinite Plate with an Oblique Edge Crack. *Ing. Arch.* 49:51-62.

[74] Isida, M. 1979. Tension of a Half Plane Containing Array Cracks, Branched Cracks and Cracks Emanating from Sharp Notches. *Trans. Jpn. Soc. Mech. Engrs.* 45, No. 392:306-317.

[75] Kovalev, M.P. and Narodetsky, M.Z. 1980. *Analysis of High Precision Ball Bearings.* Moscow: Mashinostroenie.

10

Asymptotic and Numerical Analysis of
Lightly and Heavily Loaded Fluid Lubricated
Contacts of Elastic Solids

10.1 Introduction

The science of tribology is dedicated to understanding the fundamental physical and chemical factors that govern friction and wear. As energy conservation and equipment longevity continue to increase in importance, efforts to maximize unit fuel consumption while reducing wear-related failure are becoming more relevant.

Lubricated contacts may be characterized by the ratio of the average lubricant film thickness h and the composite root mean square surface roughness $\sigma = (\sigma_1^2 + \sigma_2^2)^{1/2}$ of both contacting surfaces in the concentrated contact region.

The classical modes of lubrication are depicted schematically in Fig. 10.1. A typical Stribeck curve is shown in Fig. 10.2, which describes wear regimes in terms of the effects of lubricant viscosity, sliding speed and normal load on friction coefficient. When the lubricant film is sufficiently thick ($h \gg \sigma$), the opposing surfaces never contact one another. The condition when such contact surfaces are almost non-deformed is known as hydrodynamic lubrication and is the predominant lubrication mode in lightly loaded journal bearings. Since normal loads are relatively small, there is negligible deformation of opposing contact surfaces. Lubrication dynamics are largely dictated by viscous properties of the lubricating film, and the coefficient of friction is directly proportional to viscosity when speed and load are fixed.

As the film thickness h approaches 3-5 times σ the asperities of the opposing surfaces are still not in contact (only rarely asperities brush against each other) [1, 2] normal forces become sufficiently large that elastic deformation of the contacting surfaces occur. This condition is known as elastohydrodynamic lubrication (EHL) because the lubrication regime is controlled by both viscous effects and elastic deformation of the contacting surfaces. The EHL regime is usually encountered in situations in which high loads act over relatively small contact areas such as in ball bearings, roller bearings, and gear teeth. For hard materials, higher loads are required to reach EHL than for softer, deformable

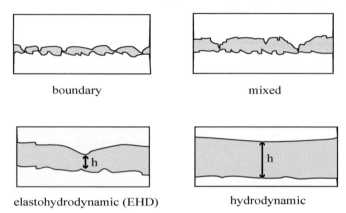

boundary

mixed

elastohydrodynamic (EHD)

hydrodynamic

FIGURE 10.1
Pictorial representation of lubrication regimes.

materials.

Asperity contact becomes more frequent as h approaches σ. In the mixed lubrication regime, the coefficient of friction begins to rise, and a significant amount of frictional heat is generated in the lubrication zone. Cyclic loading can lead to pitting failure, a subject that was discussed at length in Chapter 9.

As load further increases or speed decreases, the friction coefficient reaches a plateau. Unless soft, compliant surface films are formed by reaction of extreme pressure and/or anti-wear additives with the contact surfaces, asperities can become welded to one another leading to catastrophic failure. This regime is known as boundary Lubrication.

This chapter describes the analysis of various fluid lubrication problems associated with EHL for both hard and soft contacts, compressible fluids, and non-Newtonian lubricants. Analytical and numerical treatments of selected classic and newer problems of elastohydrodynamic lubrication are covered as well as some numerical methods, regularization approaches, and results for non-traditional and advanced EHL problems.

10.2 Simplified Navier-Stokes and Energy Equations

Let us consider in a Cartezian coordinate system (x, y, z) and time t a non-steady flow of incompressible viscous fluid with density ρ, viscosity μ, velocity components u, v, w, pressure p and the components of stress tensor p_{xx}, p_{yy}, p_{zz}, $p_{xy} = p_{yx}$, $p_{yz} = p_{zy}$, $p_{zx} = p_{xz}$. Such a flow satisfies the

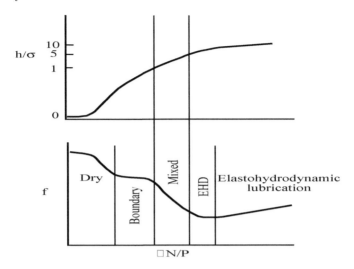

FIGURE 10.2
Stribeck curve relating coefficient of friction (ordinate) to lubricant viscosity
μ, speed u and normal load per unit area P.

equations [3]

$$\frac{\partial u}{\partial t} + u\frac{\partial u}{\partial x} + v\frac{\partial u}{\partial y} + w\frac{\partial u}{\partial z} = \frac{1}{\rho}\left(\frac{\partial p_{xx}}{\partial x} + \frac{\partial p_{yx}}{\partial y} + \frac{\partial p_{zx}}{\partial z}\right),$$

$$\frac{\partial v}{\partial t} + u\frac{\partial v}{\partial x} + v\frac{\partial v}{\partial y} + w\frac{\partial v}{\partial z} = \frac{1}{\rho}\left(\frac{\partial p_{xy}}{\partial x} + \frac{\partial p_{yy}}{\partial y} + \frac{\partial p_{zy}}{\partial z}\right), \qquad (10.1)$$

$$\frac{\partial v}{\partial t} + u\frac{\partial w}{\partial x} + v\frac{\partial w}{\partial y} + w\frac{\partial w}{\partial z} = \frac{1}{\rho}\left(\frac{\partial p_{zx}}{\partial x} + \frac{\partial p_{yz}}{\partial y} + \frac{\partial p_{zz}}{\partial z}\right).$$

The mass forces are omitted in equations (10.1) due to the fact that in lubricated contacts they are negligibly smaller than acting stresses. For a fluid with
Newtonian rheology for which the components of the stress tensor, pressure,
fluid viscosity, and the components of the tensor of shear rates are related as
follows [3]:

$$p_{xx} = -p + 2\mu\frac{\partial u}{\partial x}, \ p_{yy} = -p + 2\mu\frac{\partial v}{\partial y}, \ p_{zz} = -p + 2\mu\frac{\partial w}{\partial z},$$

$$p_{xy} = p_{yx} = \mu\left(\frac{\partial u}{\partial y} + \frac{\partial v}{\partial x}\right), \ p_{yz} = p_{zy} = \mu\left(\frac{\partial v}{\partial z} + \frac{\partial w}{\partial y}\right), \qquad (10.2)$$

$$p_{zx} = p_{xz} = \mu\left(\frac{\partial w}{\partial x} + \frac{\partial u}{\partial z}\right),$$

equations (10.1) are reduced to classic Navier-Stokes equations if the fluid
viscosity μ is constant. If the fluid viscosity varies from one point of the flow
to another, the flow equations assume the form (see equations (10.1) and

(10.384))

$$\rho\left(\frac{\partial u}{\partial t} + u\frac{\partial u}{\partial x} + v\frac{\partial u}{\partial y} + w\frac{\partial u}{\partial z}\right)$$

$$= -\frac{\partial p}{\partial x} + 2\frac{\partial}{\partial x}\left[\mu\frac{\partial u}{\partial x}\right] + \frac{\partial}{\partial y}\left[\mu\left(\frac{\partial u}{\partial y} + \frac{\partial v}{\partial x}\right)\right] + \frac{\partial}{\partial z}\left[\mu\left(\frac{\partial w}{\partial x} + \frac{\partial u}{\partial z}\right)\right],$$

$$\rho\left(\frac{\partial v}{\partial t} + u\frac{\partial v}{\partial x} + v\frac{\partial v}{\partial y} + w\frac{\partial v}{\partial z}\right) =$$

$$-\frac{\partial p}{\partial y} + \frac{\partial}{\partial x}\left[\mu\left(\frac{\partial u}{\partial y} + \frac{\partial v}{\partial x}\right)\right] + 2\frac{\partial}{\partial y}\left[\mu\frac{\partial v}{\partial y}\right] + \frac{\partial}{\partial z}\left[\mu\left(\frac{\partial v}{\partial z} + \frac{\partial w}{\partial y}\right)\right],$$

$$\rho\left(\frac{\partial w}{\partial t} + u\frac{\partial w}{\partial x} + v\frac{\partial w}{\partial y} + w\frac{\partial w}{\partial z}\right)$$

$$= -\frac{\partial p}{\partial z} + \frac{\partial}{\partial x}\left[\mu\left(\frac{\partial w}{\partial x} + \frac{\partial u}{\partial z}\right)\right] + \frac{\partial}{\partial y}\left[\mu\left(\frac{\partial v}{\partial z} + \frac{\partial w}{\partial y}\right)\right] + 2\frac{\partial}{\partial z}\left[\mu\frac{\partial w}{\partial z}\right].$$

(10.3)

For an incompressible fluid we have to add to equations (10.3) the continuity equation in the form [3]

$$\frac{\partial u}{\partial x} + \frac{\partial v}{\partial y} + \frac{\partial w}{\partial z} = 0. \tag{10.4}$$

To simplify these equations for the case of a lubricated contact of two solids, we need to specify the coordinate system. We will assume that the xy-plane is the plane equidistant from the solid surfaces and the z-axis is directed through the solids upward. For typical lubricated contacts, the characteristic size L_{xy} of the contact in the xy-plane is much larger than the lubrication film thickness L_z, i.e., $L_z/L_{xy} \ll 1$. Moreover, we will assume that the characteristic velocities in the xy-plane and along the z-axis are U_{xy} and U_z, respectively. Let us introduce the following dimensionless variables

$$\bar{t} = t\frac{U_{xy}}{L_{xy}}, \ (\bar{x}, \bar{y}) = \frac{1}{L_{xy}}(x, y), \ \bar{z} = \frac{z}{L_z}, \ \bar{p} = p\frac{L_z^2}{\mu_\star U_{xy}L_{xy}},$$

$$(\bar{u}, \bar{v}) = \frac{1}{U_{xy}}(u, v), \ \bar{w} = \frac{w}{U_z}, \ \bar{\mu} = \frac{\mu}{\mu_\star},$$

(10.5)

where $\mu_\star$ is the characteristic value of the fluid viscosity. Using the dimensionless variables in the continuity equation, we obtain

$$\frac{\partial \bar{u}}{\partial \bar{x}} + \frac{\partial \bar{v}}{\partial \bar{y}} + \frac{U_z}{U_{xy}}\frac{L_{xy}}{L_z}\frac{\partial \bar{w}}{\partial \bar{z}} = 0. \tag{10.6}$$

To balance the terms of this equation, we assume that

$$\frac{U_z}{U_{xy}}\frac{L_{xy}}{L_z} = 1. \tag{10.7}$$

Therefore, in dimensionless variables equations (10.3) and (10.6) can be reduced to

$$Re(\frac{\partial \bar{u}}{\partial t} + \bar{u}\frac{\partial \bar{u}}{\partial \bar{x}} + \bar{v}\frac{\partial \bar{u}}{\partial \bar{y}} + \bar{w}\frac{\partial \bar{u}}{\partial \bar{z}}) = -\frac{\partial \bar{p}}{\partial \bar{x}} + 2(\frac{L_z}{L_{xy}})^2 \{\frac{\partial}{\partial \bar{x}}[\bar{\mu}\frac{\partial \bar{u}}{\partial \bar{x}}]$$

$$+\frac{\partial}{\partial \bar{y}}[\bar{\mu}(\frac{\partial \bar{u}}{\partial \bar{y}} + \frac{\partial \bar{v}}{\partial \bar{x}})] + \frac{\partial}{\partial \bar{z}}[\bar{\mu}(\frac{\partial \bar{w}}{\partial \bar{x}})]\} + \frac{\partial}{\partial \bar{z}}[\bar{\mu}\frac{\partial \bar{u}}{\partial \bar{z}}],$$

$$Re(\frac{\partial \bar{v}}{\partial t} + \bar{u}\frac{\partial \bar{v}}{\partial \bar{x}} + \bar{v}\frac{\partial \bar{v}}{\partial \bar{y}} + \bar{w}\frac{\partial \bar{v}}{\partial \bar{z}}) = -\frac{\partial \bar{p}}{\partial \bar{y}} + (\frac{L_z}{L_{xy}})^2 \{\frac{\partial}{\partial \bar{x}}[\bar{\mu}(\frac{\partial \bar{u}}{\partial \bar{y}}$$

$$+\frac{\partial \bar{v}}{\partial \bar{x}})] + 2\frac{\partial}{\partial \bar{y}}[\bar{\mu}\frac{\partial \bar{v}}{\partial \bar{y}}]\} + \frac{\partial}{\partial \bar{z}}[\bar{\mu}\frac{\partial \bar{v}}{\partial \bar{z}}] + (\frac{L_z}{L_{xy}})^2 \frac{\partial}{\partial \bar{z}}[\bar{\mu}\frac{\partial \bar{w}}{\partial \bar{y}}],$$

$$(10.8)$$

$$Re(\frac{\partial \bar{w}}{\partial t} + \bar{u}\frac{\partial \bar{w}}{\partial \bar{x}} + v\frac{\partial \bar{w}}{\partial \bar{y}} + \bar{w}\frac{\partial \bar{w}}{\partial \bar{z}}) = -(\frac{L_{xy}}{L_z})^2 \frac{\partial \bar{p}}{\partial \bar{z}} + (\frac{L_z}{L_{xy}})^2 \{\frac{\partial}{\partial \bar{x}}[\bar{\mu}\frac{\partial \bar{w}}{\partial \bar{x}}]$$

$$+\frac{\partial}{\partial \bar{y}}[\bar{\mu}\frac{\partial \bar{w}}{\partial \bar{y}}]\} + \frac{\partial}{\partial \bar{x}}[\bar{\mu}\frac{\partial \bar{u}}{\partial \bar{z}}] + \frac{\partial}{\partial \bar{y}}[\bar{\mu}\frac{\partial \bar{v}}{\partial \bar{z}}] + 2\frac{\partial}{\partial \bar{z}}[\bar{\mu}\frac{\partial \bar{w}}{\partial \bar{z}}],$$

$$\frac{\partial \bar{u}}{\partial \bar{x}} + \frac{\partial \bar{v}}{\partial \bar{y}} + \frac{\partial \bar{w}}{\partial \bar{z}} = 0, \qquad (10.9)$$

where the effective Reynolds number $Re = \frac{\rho U_{xy} L_{xy}}{\mu_*}(\frac{L_z}{L_{xy}})^2$ and the ratio $\frac{L_z}{L_{xy}}$ are small. In fact, in a typical journal bearing $L_{xy} = 0.12\ m$, $L_z = 5 \times 10^{-5}\ m$, $U_{xy} = 5\ m/s$, $\rho = 850 N \cdot s^2/m^4$, $\mu_* = 0.5\ N \cdot s/m^2$ and, therefore, the Reynolds number $Re = 1.77 \cdot 10^{-4}$ and $\frac{L_z}{L_{xy}} = 4.17 \cdot 10^{-4}$. Besides that, we assume that all the dimensionless terms such as $\frac{\partial \bar{u}}{\partial \bar{x}}$, $\frac{\partial \bar{u}}{\partial \bar{y}}$, $\frac{\partial \bar{u}}{\partial \bar{z}}$, etc., are of the order of 1. From the physical point of view, it is clear that at least two mechanisms in the viscous flow are important, i.e., viscous shearing and pressure gradient. Therefore, we need to retain these terms in the simplified equations. Smallness of the Reynolds number Re and ratio L_z/L_{xy} does exactly that. Using these assumptions we obtain that all terms in the right-hand side of the first two equations from (10.3) except for two terms are small. In a similar fashion, evaluating terms of the third equation from (10.8), we obtain that all terms of the equation are negligibly small in comparison with term $\frac{\partial p}{\partial z}$. As a result of this analysis we retain in dimensional form only the largest terms of Navier-Stokes equations as follows

$$-\frac{\partial p}{\partial x} + \frac{\partial}{\partial z}[\mu\frac{\partial u}{\partial z}] = 0, \quad -\frac{\partial p}{\partial y} + \frac{\partial}{\partial z}[\mu(\frac{\partial v}{\partial z}] = 0, \quad \frac{\partial p}{\partial z} = 0. \qquad (10.10)$$

It is important to notice that the terms related to u and v retained in equations (10.10) come from the main terms of the stress tensor components p_{yz}, and p_{zx}. Therefore, the simplified equations of a flow of viscous non-Newtonian fluid in a lubricated contact can be obtained by replacing the second terms in the first two equations from (10.10) by the main terms of $\frac{\partial p_{xz}}{\partial z}$ and $\frac{\partial p_{yz}}{\partial z}$ for that rheology, respectively.

In case of non-isothermal conditions two additional assumptions need to be made. They are (a) the major source of heat generation is due to shear viscous motion in lubricants and (b) the main direction of heat flow is across the lubrication film thickness. These assumptions together with the previously stated ones after estimating terms of the equation allow for significant simplification of the general energy equation. For fluids with Newtonian and non-Newtonian rheologies such a simplified equation can be represented in the form

$$\frac{\partial}{\partial z}\left(\lambda \frac{\partial T}{\partial z}\right) = -p_{xz}\frac{\partial u}{\partial z} - p_{yz}\frac{\partial v}{\partial z}, \tag{10.11}$$

where T is the fluid temperature and λ is the heat conductivity. In equation (10.11), in the stress tensor only the major term components are retained. For example, for Newtonian fluids in equation (10.11) for p_{xz} and p_{yz} we use $p_{xz} = \mu\frac{\partial u}{\partial z}$ and $p_{yz} = \mu\frac{\partial v}{\partial z}$, respectively.

Some analysis of physical and rheological lubricant properties as well as some specific data on parameters characterizing lubricant behavior are given in Chapter (4).

10.3 Asymptotic Analysis of a Lightly Loaded Lubricated Contact of Elastic Solids

This section deals with steady plane isothermal and non-isothermal elastohydrodynamic lubrication (EHL) problems for lightly loaded smooth contacts. A lubricant is considered to be an incompressible non-Newtonian fluid of rather general rheology. In lightly loaded lubricated contacts, the influence of the elastic properties of the solids in contact on the pressure in the contact is relatively small. In other words, the problem solution in the entire lubricated region/contact is close to the solution of the problem for a lubricant flow through a narrow gap of a known shape.

The aim of the section is to consider lightly loaded lubricated contact by employing the method of regular perturbations [4, 5], which allows to study the solution behavior in contacts with fluid of rather complex rheology. As a result of application of the above–mentioned method the differential equations for the first two terms of the asymptotic expansion of the problem solution for pressure, gap, and lubrication film thickness are obtained. Moreover, some new two-term analytic formulas for the lubrication film thickness and frictional forces are derived. The obtained relationships are illustrated by examples of fluids with different rheologies. In the special case of Newtonian fluid, the leading terms of the aforementioned asymptotic expansions coincide with the classic Kapitza [6] results. A thermal EHL problem for a lightly loaded contact is briefly considered.

The formulas obtained for the lubrication film thickness and frictional forces can be used in treating experimental results for lightly loaded lubricated contacts as well as for making a judgment on application of a fluid of a particular rheology.

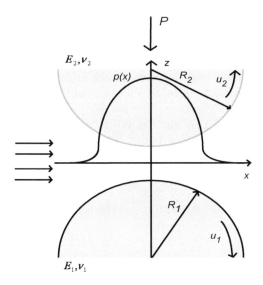

FIGURE 10.3
The general view of a lubricated contact.

10.3.1   Problem Formulation

Let us consider two infinite parallel cylinders made of elastic materials with Young's modulus E_1 and Poisson's ratio ν_1 for the lower cylinder of radius R_1 and with Young's modulus E_2 and Poisson's ratio ν_2 for the upper cylinder of radius R_2. In the coordinate system with the x-axis along the contact and the z-axis through the cylinders' centers the cylinders move with surface velocities u_1 and u_2, respectively. The cylinders are separated from one another by a thin layer of incompressible lubricant and pressed one into another by a force P acting along the z-axis. The general view of the lubricated contact is given in Fig. 10.3.

Let us formulate the main problem assumptions. We will assume that the lubrication film thickness is much smaller than the the size of the contact which, in turn, is much smaller than cylinders' radii. The variations of the film thickness $h(x)$ in the contact are small, i.e., $dh/dx \ll 1$. Moreover, we will assume that frictional effects caused by lubricant's viscosity dominate the inertia effects, i.e., the lubricant moves relatively slowly. We will assume that the lubrication process is isothermal and the lubricant viscosity μ depends only on pressure $p(x)$. These are the typical assumptions used in the theory of elastohydrodynamic lubrication (see Hamrock [7]).

Under the above assumptions for a fluid with a rather general rheology, the equations that relate the tangential stress τ to the strain rate $\partial u/\partial z$ in the fluid can be represented in the form [8]

$$\tau = \Phi(\mu \tfrac{\partial u}{\partial z}) \ or \ \mu \tfrac{\partial u}{\partial z} = F(\tau), \tag{10.12}$$

where Φ and F are functions inverses of each other, $u(x, z)$ is the fluid velocity, $\mu(p)$ is the lubricant viscosity being a smooth monotonically increasing
function of pressure p. Moreover, function $\Phi(x)$ is an arbitrary smooth monotonically increasing odd function and $\Phi(0) = F(0) = 0$. Therefore, taking into
account (10.12) after simplification of Navier-Stocks equations for the fluid
the EHL problem can be reduced to the following equations [7]

$$\tfrac{\partial}{\partial z} \Phi(\mu \tfrac{\partial u}{\partial z}) = \tfrac{\partial p}{\partial x}, \ \tfrac{\partial p}{\partial z} = 0, \ \tfrac{\partial u}{\partial x} + \tfrac{\partial w}{\partial z} = 0, \tag{10.13}$$

with no-slip and no penetration through the the cylinders' surfaces boundary
conditions

$$u(x, -\tfrac{h}{2}) = u_1, \ u(x, \tfrac{h}{2}) = u_2,$$

$$w(x, (-1)^j \tfrac{h}{2}) = (-1)^j \tfrac{u_i}{2} \tfrac{dh}{dx}, \ j = 1, 2, \tag{10.14}$$

where $w(x, z)$ is the z-component of the fluid velocity and $z = \pm h(x)/2$ are
the upper and lower cylinder surfaces, respectively.

The second equation in (10.13) leads to the conclusion that pressure $p = p(x)$ is independent from z. Integrating the second equation in (10.12) and
the first in (10.13) gives

$$u(x, z) = u_1 + \tfrac{1}{\mu} \int\limits_{-h/2}^{z} F(f + t \tfrac{dp}{dx}) dt, \tag{10.15}$$

where $f = f(x)$ is the unknown sliding frictional stress which simultaneously
represents the tangential stress τ in the midplane $z = 0$.

Using equation (10.15) and the second boundary condition in (10.14), we
obtain the equation for $f(x)$ in the form

$$\int\limits_{-h/2}^{h/2} F(f + t \tfrac{dp}{dx}) dt = \mu(u_2 - u_1). \tag{10.16}$$

By integrating the third (mass conservation) equation in (10.13) with respect to z from $-h/2$ to $h/2$ and using equation (10.15) and the last pair of
boundary conditions in (10.14), we derive the generalized Reynolds equation

$$\tfrac{d}{dx} \left[\tfrac{1}{\mu} \int\limits_{-h/2}^{h/2} dz \int\limits_{-h/2}^{z} F(f + t \tfrac{dp}{dx}) dt + u_1 h \right] = 0. \tag{10.17}$$

By changing the order of integration in equation (10.17) and using equation
(10.16), the Reynolds equation can be reduced to the final simplified form

$$\tfrac{d}{dx} \left[\tfrac{1}{\mu} \int\limits_{-h/2}^{h/2} z F(f + z \tfrac{dp}{dx}) dz - \tfrac{u_1 + u_2}{2} h \right] = 0. \tag{10.18}$$

We need to add to equations 10.16 and 10.18 the relationship

$$h(x) = h_e + \frac{x^2 - x_e^2}{2R'} + \frac{2}{\pi E'} \int_{x_i}^{x_e} p(t) \ln \frac{x_e - t}{|x - t|} dt, \qquad (10.19)$$

for gap $h(x)$ [7], which is obtained based on the above assumptions and by replacing the cylinders in the contact by elastic half-planes. In equation (10.19) h_e is the lubrication film thickness at the exit point of the contact x_e, i.e., $h_e = h(x_e)$, x_i is the inlet point of the contact, R' and E' are the effective radius and elasticity modulus, $1/R' = 1/R_1 \pm 1/R_2$ (signs $+$ and $-$ are chosen in accordance with cylinders' curvatures) and $1/E' = 1/E_1' + 1/E_2'$, $E_j' = E_j/(1 - \nu_j^2)$, $j = 1, 2$.

To make the problem complete we need to add to the above equations some conditions. It is reasonable to assume that within the lubricated contact pressure is much higher than the ambient pressure. Moreover, to prevent the lubricant from cavitation at the exit of the contact, we may require the pressure gradient to be zero [7]. Thus, we obtain the following boundary conditions on pressure

$$p(x_i) = p(x_e) = \frac{dp(x_e)}{dx} = 0. \qquad (10.20)$$

The last condition that must be added to the equations is the condition that the integral of pressure over the contact is equal to the applied normal force P, i.e.,

$$\int_{x_i}^{x_e} p(t) dt = P. \qquad (10.21)$$

Therefore, for given functions μ, Φ, F and constants x_i, u_1, u_2, R', E', and P the EHL problem is reduced to solution of equations (10.16), (10.18)-(10.21) for functions of sliding frictional stress $f(x)$, pressure $p(x)$, gap $h(x)$, and two constants - exit coordinate x_e and exit film thickness h_e.

After the problem is solved friction force $F_T(z)$ in the lubrication layer can be calculated from the formula

$$F_T(z) = \int_{x_i}^{x_e} \tau(x, z) dx = \int_{x_i}^{x_e} \Phi(\mu \frac{\partial u}{\partial z}) dx. \qquad (10.22)$$

Using equation (10.22) we can determine the friction forces $F_T^\pm = F_T(\pm h/2)$ on the cylinders' surfaces

$$F_T^\pm = F_S \pm F_R, \quad F_S = \int_{x_i}^{x_e} f(x) dx, \quad F_R = \frac{1}{2} \int_{x_i}^{x_e} h(x) \frac{dp}{dx} dx, \qquad (10.23)$$

where F_S and F_R are the sliding and rolling friction forces, respectively.

Let us introduce dimensionless variables related to the flow of a viscous fluid

$$\{x', a, c\} = \{x, x_i, x_e\} \frac{\vartheta}{2R'}, \quad \{z', h'\} = \{z, h\} \frac{1}{h_e}, \quad p' = p \frac{\pi R'}{\vartheta P},$$

$$\mu' = \frac{\mu}{\mu_a}, \quad F' = F \frac{2h_e}{\mu_a(u_1 + u_2)}, \quad \{f', \tau'\} = \{f, \tau\} \frac{2\pi}{h_e P} (\frac{R'}{\vartheta})^2, \qquad (10.24)$$

where $\vartheta = \vartheta(\mu_a, u_1, u_2, R', P, \ldots)$ is a certain dimensionless parameter that depends on the fluid specific rheology (i.e., function F) and is independent from the effective elastic modulus E' and μ_a is the lubricant viscosity at ambient pressure. Some specific forms of parameter ϑ will be considered later.

By integrating equation (10.18) with respect to x from x to x_e and taking into account equation (10.16) and the last of the boundary conditions (10.20), we obtain the problem in dimensionless variables (for simplicity here and further primes are omitted)

$$\frac{1}{\mu} \int_{-h/2}^{h/2} zF(f + z\frac{dp}{dx})dz = h - 1, \tag{10.25}$$

$$\int_{-h/2}^{h/2} F(f + z\frac{dp}{dx})dz = \mu s_0, \tag{10.26}$$

$$\gamma(h - 1) = x^2 - c^2 + \frac{1}{V}\frac{2}{\pi} \int_a^c p(t) \ln\frac{c-t}{|x-t|}dt, \tag{10.27}$$

$$p(a) = p(c) = 0, \tag{10.28}$$

$$\int_a^c p(t)dt = \frac{\pi}{2}, \tag{10.29}$$

$$\gamma = \vartheta^2 \frac{h_e}{2R'}, \quad V = \frac{R'E'}{\vartheta^2 P}, \quad s_0 = 2\frac{u_2-u_1}{u_2+u_1}. \tag{10.30}$$

In the above dimensionless variables for the given values of parameters a and V and functions μ and F, the EHL problem is reduced to solution of equations (10.25)-(10.30) for the functions of pressure $p(x)$ and gap $h(x)$ as well as for the values of the dimensionless lubrication film thickness γ and exit coordinate c.

10.3.2   Perturbation Solution of the EHL Problem

Below we will consider application of the method of regular perturbations [4, 5, 9] to solution of the EHL problem (10.25)-(10.30) for a lightly loaded contact. The problem is formulated in the region with unknown boundaries $z = \pm h/2$ and $x = c$ which makes the solution of the problem more complicated. Fortunately, this difficulty can be easily cured by introducing the following substitutions

$$x = \tfrac{1}{2}(c + a) + \tfrac{1}{2}(c - a)\xi, \quad z = \tfrac{1}{2}h\zeta, \quad f(x) = g(\xi), \quad h(x) = H(\xi), \tag{10.31}$$

which allow to reduce the problem equations to the following

$$\frac{H^2}{4\mu} \int_{-1}^{1} tF(g + \frac{Ht}{c-a}\frac{dp}{d\xi})dt = H - 1, \tag{10.32}$$

$$\int\limits_{-1}^{1} F(g + \tfrac{Ht}{c-a}\tfrac{dp}{d\xi})dt = 2\tfrac{\mu s_0}{H},$$ (10.33)

$$\gamma(H-1) = (\tfrac{c-a}{2})^2\xi^2 + \tfrac{c^2-a^2}{2}\xi + \tfrac{a^2+2ac-3c^2}{4}$$

(10.34)

$$+\tfrac{c-a}{\pi V}\int\limits_{-1}^{1} p(t)\ln\tfrac{1-t}{|\xi-t|}dt,$$

$$p(-1) = p(1) = 0,$$ (10.35)

$$\int\limits_{-1}^{1} p(t)dt = \tfrac{\pi}{c-a}.$$ (10.36)

We will consider the contact being lightly loaded if $V \gg 1$. The proposed definition of a lightly loaded lubricated contact coincides with the notion that the influence of the cylinders elasticity has a small effect on the EHL problem solution.

Equations (10.32) - (10.36) suggest that the solution can be searched in the form of regular power series in $V^{-1} \ll 1$

$$p(\xi) = \sum_{k=0}^{\infty} V^{-k}p_k(\xi), \; H(\xi) = \sum_{k=0}^{\infty} V^{-k}H_k(\xi),$$

(10.37)

$$g(\xi) = \sum_{k=0}^{\infty} V^{-k}g_k(\xi), \; c = \sum_{k=0}^{\infty} V^{-k}c_k, \; \gamma = \sum_{k=0}^{\infty} V^{-k}\gamma_k,$$

where functions $p_k(\xi)$, $H_k(\xi)$, $g_k(\xi)$ and constants c_k, γ_k has to be determined.

By using the Taylor series expansion of $F(g + \tfrac{Hv}{c-a}\tfrac{dp}{d\xi})$ centered at $(p_0, H_0, g_0, c_0, \gamma_0)$ and representations (10.37) in equations (10.32) - (10.36), we will be able to obtain a boundary-value problem for nonlinear differential equations for the set of functions and constants $(p_0, H_0, g_0, c_0, \gamma_0)$ and a series of boundary-value problems for linear differential equations for the set of functions and constants $(p_k, H_k, g_k, c_k, \gamma_k)$, $k = 1, \ldots$.

The main advantage of this approach in comparison with the numerical solution is the ability to obtain simple structural asymptotic formulas for the lubrication film thickness and friction forces. Using equations (10.30) and (10.37) we obtain the following two-term formulas for the film thickness

$$h_e = \tfrac{2R'}{\vartheta^2}(\gamma_0 + \tfrac{1}{V}\gamma_1 + \ldots) = \tfrac{2R'}{\vartheta^2}\gamma_0 + \tfrac{2P}{\pi E'}\gamma_1 + \ldots$$ (10.38)

and for the components of the friction force

$$F_S = \tfrac{P}{\vartheta}(a_0 + \tfrac{1}{V}a_1 + \ldots) = \tfrac{P}{\vartheta}a_0 + \tfrac{P^2\vartheta}{\pi R'E'}a_1 + \ldots,$$

(10.39)

$$F_R = \tfrac{P}{\vartheta}(b_0 + \tfrac{1}{V}b_1 + \ldots) = \tfrac{P}{\vartheta}b_0 + \tfrac{P^2\vartheta}{\pi R'E'}b_1 + \ldots,$$

$$a_0 = \frac{\gamma_0(c_0-a)}{\pi} \int\limits_{-1}^{1} g_0(\xi)d\xi, \quad b_0 = \frac{\gamma_0}{\pi} \int\limits_{-1}^{1} H_0(\xi)\frac{dp_0}{d\xi}d\xi,$$

$$(10.40)$$

$$a_1 = \frac{1}{\pi}[\gamma_0(c_0-a)\int\limits_{-1}^{1} g_1(\xi)d\xi + [\gamma_1(c_1-a)+\gamma_0 c_1]\int\limits_{-1}^{1} g_0(\xi)d\xi],$$

$$b_1 = \frac{1}{\pi}[\gamma_0\int\limits_{-1}^{1}[H_1(\xi)\frac{dp_0(\xi)}{d\xi}+H_0(\xi)\frac{dp_1}{d\xi}]d\xi + \gamma_1\int\limits_{-1}^{1} H_0(\xi)\frac{dp_0}{d\xi}d\xi].$$

In the latter formulas the numerical values of coefficients γ_0, γ_1, a_0, b_0, a_1, and b_1 depend on the specifics of functions F, Φ, μ, and constants a and s_0. It follows from equations (10.38) - (10.40) that lubrication film thickness h_e and components F_S and F_R of the friction force depend on E' only starting with the second terms proportional to V^{-1} for $V \gg 1$.

Let us consider some lubricants with different rheologies. First, we will consider a lubricant with power rheology [8, 9], which, under the problem assumptions in dimensional variables, is represented by

$$\tau = \mu^\alpha \mid \frac{\partial u}{\partial z} \mid^{\alpha-1} \frac{\partial u}{\partial z}, \quad \alpha > 0. \tag{10.41}$$

In this case in dimensionless variables $F(x) = 12\gamma^{(1+\alpha)/\alpha} \mid x \mid^{(1-\alpha)/\alpha} x$. By introducing the dimensionless variables (10.24), we obtain

$$\vartheta^2 = \frac{R'}{3\mu_a(u_1+u_2)}(\frac{P}{\pi R'})^{1/\alpha}, \tag{10.42}$$

while constant V is determined by the expression in (10.30). Therefore, for the power rheology formulas for the lubrication film thickness h_e and components F_S and F_R of the friction force (10.38) - (10.40) can be rewritten in the form

$$h_e = 6\mu_a(u_1+u_2)[\frac{\pi R'}{P}]^{1/\alpha}\gamma_0 + \frac{2P}{\pi E'}\gamma_1 + \dots,$$

$$F_S = P\sqrt{\frac{3\mu_a(u_1+u_2)}{R'}}[\frac{\pi R'}{P}]^{1/(2\alpha)}a_0$$

$$+\frac{P^2}{\pi R'E'}\sqrt{\frac{R'}{3\mu_a(u_1+u_2)}}[\frac{P}{\pi R'}]^{1/(2\alpha)}a_1 + \dots, \tag{10.43}$$

$$F_R = P\sqrt{\frac{3\mu_a(u_1+u_2)}{R'}}[\frac{\pi R'}{P}]^{1/(2\alpha)}b_0$$

$$+\frac{P^2}{\pi R'E'}\sqrt{\frac{R'}{3\mu_a(u_1+u_2)}}[\frac{P}{\pi R'}]^{1/(2\alpha)}b_1 + \dots$$

Obviously, in (10.43) coefficients γ_0, γ_1, a_0, b_0, a_1, and b_1 depend on the specifics of function μ and constants a, s_0, and α.

In case of a Newtonian fluid $\alpha = 1$ and

$$\vartheta^2 = P/[3\pi\mu_a(u_1+u_2)], \tag{10.44}$$

while function $F(x)$ in dimensionless variables has the form

$$F(x) = 12\gamma^2 x. \tag{10.45}$$

By solving equation (10.26), we find

$$f = \frac{1}{12\gamma^2} \frac{\mu s_0}{h}. \tag{10.46}$$

By using (10.46) in (10.23) with proper dimensions, we obtain the analogs of formulas (10.43) for Newtonian lubricants

$$h_e = \frac{\mu_a (u_1 + u_2) R'}{P} d_{00} + \frac{P}{E'} d_{01} + \dots,$$

$$F_S = (u_2 - u_1)\left\{ \sqrt{\frac{\mu_a P}{(u_1 + u_2)}} d_{10} + \frac{1}{R'E'} \sqrt{\frac{P^5}{\mu_a (u_1 + u_2)^3}} d_{11} + \dots \right\}, \tag{10.47}$$

$$F_R = \sqrt{\mu_a (u_1 + u_2) P} d_{20} + \frac{1}{R'E'} \sqrt{\frac{P^5}{\mu_a (u_1 + u_2)}} d_{21} + \dots,$$

where coefficients d_{00}, d_{01}, d_{10}, d_{11}, d_{20}, and d_{21} depend only on the fluid viscosity μ and inlet coordinate a and are independent of slide-to-roll ratio s_0.

Finally, let us consider a lubricant with Erying rheology [8] for which in dimensional variables

$$\tau = G_1 \mu_1 arsh(\frac{\mu}{G} \frac{\partial u}{\partial z}), \tag{10.48}$$

where $\mu_1 = \mu_1(p)$ is a certain dimensionless function of pressure, G and G_1 are constant shear stresses characterizing the fluid rheology. For this rheology constant ϑ can be taken from (10.44) or from

$$\vartheta^2 = \frac{4R'G}{\mu_a (u_1 + u_2)}. \tag{10.49}$$

After that it is clear that in formulas for the film thickness (10.38) and for the friction force components (10.40) coefficients γ_0, γ_1, a_0, b_0, a_1, and b_1 depend on constants a, s_0 and the specifics of functions μ and $P/(R'G_1\mu_1)$.

Now, let us consider the realization of the above method and some numerical results for the case of Newtonian rheology. In this case in the dimensionless variables we have

$$F(g + \frac{H\zeta}{c-a} \frac{dp}{d\xi}) = 12\gamma^2 (g + \frac{H\zeta}{c-a} \frac{dp}{d\xi}). \tag{10.50}$$

Following the described procedure, we obtain

$$\frac{dp_0}{d\xi} = \frac{\mu_0(c_0-a)(H_0-1)}{2\gamma_0^2 H_0^3}, \quad g_0 = \frac{\mu_0 s_0}{12\gamma_0^2 H_0},$$

$$\gamma_0(H_0-1) = \left(\frac{c_0-a}{2}\right)^2\xi^2 + \frac{c_0^2-a^2}{2}\xi + \frac{a^2+2ac_0-3c_0^2}{4},$$

$$p_0(-1) = p_0(1) = 0, \quad \int_{-1}^{1} p_0(t)dt = \frac{\pi}{c_0-a}, \qquad (10.51)$$

$$\frac{dp_1}{d\xi} = \frac{\mu_0(c_0-a)}{2\gamma_0^2 H_0^3}\left\{(H_0-1)\frac{d\ln\mu_0}{dp_0}p_1 + [3(3-H_0-\frac{2}{H_0})\right.$$

$$\left.+\frac{a(c_0-a)}{\gamma_0}(\xi-1)(2-\frac{3}{H_0})]\frac{c_1}{c_0-a} - 3(1-\frac{1}{H_0})\frac{\gamma_1}{\gamma_0} + (\frac{3}{H_0}-2)Q\right\},$$

$$g_1 = \frac{\mu_0 s_0}{12\gamma_0^2 H_0}\left\{\frac{d\ln\mu_0}{dp_0}p_1 - \frac{1}{H_0}\frac{\partial H_0}{\partial c_0}c_1 - (1+\frac{1}{H_0})\frac{\gamma_1}{\gamma_0} - \frac{Q}{H_0},\right.$$

$$H_1 = \frac{\partial H_0}{\partial c_0}c_1 - (H_0-1)\frac{\gamma_1}{\gamma_0} + Q, \quad Q(u) = \frac{c_0-a}{\pi}\int_{-1}^{1} p_0(t)\ln\frac{1-t}{|\xi-t|}dt, \qquad (10.52)$$

$$p_1(-1) = p_1(1) = 0, \quad \int_{-1}^{1} p_1(t)dt = -\frac{\pi c_1}{(c_0-a)^2}, \ldots.$$

Let us consider solution of equations (10.51) and (10.52) for the simplest case of constant viscosity $\mu(p) = \mu_0(p_0) = 1$. It is not difficult to integrate equations (10.51) (see also [6]) and get the solution in the form

$$p_0(x) = \frac{b^3}{8}\left\{\frac{b^2-3c_0^2}{b^2}[\arctan\frac{x}{b} + \frac{bx}{b^2+x^2}] - \frac{2b\gamma_0 x}{(b^2+x^2)^2} - c^0\right\},$$

$$c^0 = \frac{b^2-3c_0^2}{b^2}[\arctan\frac{a}{b} + \frac{ab}{b^2+a^2}] - \frac{2ab\gamma_0}{(b^2+a^2)^2}, \qquad (10.53)$$

$$b = \sqrt{\gamma_0 - c_0^2}, \quad x = [a + c_0 + (c_0-a)\xi]/2.$$

Constants γ_0 and c_0 involved in (10.53) are determined by a system of non-linear algebraic equations

$$\frac{b^2-3c_0^2}{b^2}[\arctan\frac{c_0}{b} - \arctan\frac{a}{b} + \frac{bc_0}{\gamma_0} - \frac{ab}{b^2+x^2}] - \frac{2ab\gamma_0}{(b^2+a^2)^2} - \frac{2bc_0}{\gamma_0} = 0,$$

$$\frac{b^2-3c_0^2}{b^2}[\frac{c_0}{b}(\arctan\frac{c_0}{b} - \arctan\frac{a}{b}) - \frac{a(c_0-a)}{b^2+a^2}] + \frac{a-c_0^2}{b^2+a^2} + \frac{2a\gamma_0(c_0-a)}{(b^2+a^2)^2} = 4\pi b^2.$$

$$(10.54)$$

The solution of system (10.54) can be represented in the form

$$p_1(x) = \frac{1}{b^3}\{\frac{3\gamma_1}{b^2}\varepsilon_\gamma(x) + \frac{c_1}{c_0-a}\varepsilon_c(x) + \frac{\gamma_0}{b^2}\varepsilon(x)\}, \quad \varepsilon_\gamma(x) = \frac{\gamma_0}{b^2}A_4(\rho)$$

$$-A_3(\rho), \quad \varepsilon_c(x) = \frac{6\gamma_0(ac_0-\gamma_0)}{b^4}A_4(\rho) + \frac{9\gamma_0-4ac_0}{b^2}A_3(\rho) - 3A_2(\rho)$$

$$+\frac{2a}{b}[2B_2(\rho) - \frac{3\gamma_0}{b^2}B_3(\rho)], \quad \varepsilon(x) = \frac{2b}{\pi}\int\limits_{a/b}^{\rho}\frac{1}{(1+t^2)^3}[\frac{3\gamma_0}{b\gamma^2(1+t^2)} - 2]$$

$$\times \int\limits_{a/b}^{c_0/b} q_0(z)\ln\frac{c_0/b-z}{|t-z|}dzdt, \quad A_{n+1}(\rho) = \frac{1}{2n}[\frac{\rho}{(1+\rho^2)^n} - \frac{ab^{2n-1}}{(b^2+a^2)^n}]$$

$$+\frac{2n-1}{2n}A_n(\rho), \quad A_1(\rho) = \arctan\rho - \arctan\frac{a}{b}, \quad B_n(\rho) = \frac{1}{2n}[(\frac{b^2}{b^2+a^2})^n$$

$$-\frac{1}{(1+\rho^2)^n}], \quad \rho = x/b, \quad x = [a + c_0 + (c_0-a)\xi]/2.$$

$$(10.55)$$

From equations (10.55) and the last three equations in (10.52), we find a system of linear algebraic equations for c_1 and γ_1

$$3\varepsilon_\gamma(c_0)\gamma_1 + \frac{b^2}{c_0-a}\varepsilon_c(c_0)c_1 = -\gamma_0\varepsilon(c_0),$$

$$3\int\limits_a^{c_0}\varepsilon_\gamma(x)dx\gamma_1 + \frac{b^2}{c_0-a}\left[\int\limits_a^{c_0}\varepsilon_c(x)dx + \frac{\pi b^3}{2}\right]c_1 = -\gamma_0\int\limits_a^{c_0}\varepsilon(x)dx.$$

$$(10.56)$$

By consequently solving systems (10.54) and (10.56), we find constants c_0, γ_0, c_1, and γ_1. After that we can find constants involved in equations (10.47) for the lubrication film thickness and the components of the friction force in the form

$$d_{00} = 6\pi\gamma_0, \quad d_{01} = \frac{2\gamma_1}{\pi}, \quad d_{10} = \frac{\alpha_1}{\sqrt{3\pi}b}, \quad d_{20} = \frac{3}{2\sqrt{3\pi}b\sqrt{2}}[\frac{b^2-c_0^2}{b^2}\alpha_1$$

$$+\frac{a\gamma_0}{b^2+a^2} - c_0], \quad d_{11} = -\frac{1}{3\pi^2\sqrt{3\pi}b^3}\{\gamma_1\alpha_2 + \frac{2c_1}{c_0-a}[\frac{b^2\alpha_1}{2}$$

$$+(ac_0 - \gamma_0)\alpha_2 - ab\beta_1] + \gamma_0\int\limits_{a/b}^{c_0/b}\frac{Q(t)dt}{(1+t^2)^2}\},$$

$$(10.57)$$

$$d_{21} = -\frac{1}{\pi^2\sqrt{3\pi}b^3}\{2\gamma_1(\frac{\gamma_0\alpha_3}{b^2} - \alpha_2) + \frac{c_1}{c_0-a}[\frac{4\gamma_0(ac_0-\gamma_0)\alpha_3}{b^2}$$

$$+(5\gamma_0 - 2ac_0)\alpha_2 - b^2\alpha_1 + 2ab(\beta_1 - \frac{2\gamma_0\beta_2}{b^2})] + \gamma_0\int\limits_{a/b}^{c_0/b}\frac{Q(t)}{(1+t^2)^2}$$

$$\times[\frac{2\gamma_0}{b^2(1+t^2)} - 1]dt\}, \quad \alpha_i = A_i(c_0/b), \quad \beta_i = B_i(c_0/b), \quad i = 1, 2, \ldots.$$

TABLE 10.1

Data for c_0, d_{00}, d_{10}, d_{20}, $p_{0max} = \max\limits_{a \leq x \leq c_0} p_0(x)$,

$H_{0min} = \min\limits_{a \leq x \leq c_0} H_0(x)$ versus the inlet coordinate a (after

Kudish [9]). Published with permission from Allerton
Press.

a	c_0	d_{00}	d_{10}	d_{20}	p_{0max}	H_{0min}
-0.031	0.014	0.076	3.601	0.037	63.850	0.951
-0.164	0.059	0.572	2.245	0.187	13.181	0.886
-0.554	0.123	1.734	1.792	0.529	4.543	0.835
-0.954	0.148	2.316	1.725	0.757	3.686	0.823
-5	0.170	2.962	1.767	1.285	2.762	0.816
-10	0.171	2.990	1.791	1.375	2.713	0.816

The numerical results for c_0, d_{00}, d_{10}, d_{20}, $p_{0max} = \max\limits_{a \leq x \leq c_0} p_0(x)$, $H_{0min} = \min\limits_{a \leq x \leq c_0} H_0(x)$ and c_1, d_{01}, d_{11}, d_{21} versus the inlet coordinate a are given in Tables 10.1 and 10.2, respectively. It follows from Tables 10.1 and 10.2 that as the amount of the entrained lubricant in a lightly loaded lubricated contact increases (which is equivalent to increasing $\mid a \mid$, $a < 0$) the film thickness increases while the maximum pressure p_{0max} decreases. The behavior of the rolling and sliding components of the friction force is non-monotonic.

TABLE 10.2

Data for c_1, d_{01}, d_{11}, and d_{21} versus the
inlet coordinate a (after Kudish [9]).
Published with permission from Allerton
Press.

a	c_1	d_{01}	d_{11}	d_{21}
-0.031	0.144	0.006	0.1032	-0.0080
-0.164	0.254	0.022	0.0118	-0.0131
-0.554	0.346	0.052	-0.0019	-0.0155
-0.954	0.366	0.065	-0.0034	-0.0147
-5	0.368	0.073	-0.0036	-0.0121
-10	0.369	0.074	-0.0036	-0.0119

10.3.3 Thermal EHL Problem

Now, let us take a look at a thermal EHL problem for a lightly loaded contact [9]. In addition to the assumptions made earlier, we need to assume that the heat generation in a lubrication layer is primarily due viscous shear in the lubricant. Moreover, based on the fact that the lubrication film thickness is much smaller than the contact size, we can assume that the generated heat is primarily transported across the film in the direction perpendicular to solid surfaces [7]. Using experimental studies we can assume that the lubricant viscosity is a certain function of lubricant pressure and temperature, i.e., $\mu = \mu(p, T)$, where $T = T(x, z)$ is the lubricant temperature.

Based on the above assumption the simplification of the fluid energy equation leads to the problem [7]

$$\frac{\partial}{\partial z}(\lambda \frac{\partial T}{\partial z}) = -\tau \frac{\partial u}{\partial z}, \; T(x, -\frac{h}{2}) = T_{w1}(x), \; T(x, \frac{h}{2}) = T_{w2}(x), \qquad (10.58)$$

where $T_{w1}(x)$ and $T_{w2}(x)$ are the temperatures of the lower and upper solids, respectively, and $\lambda = \lambda(p, T)$ is the lubricant coefficient of heat conductivity.

By introducing the following additional dimensionless variables

$$\{T', T'_{w1}, T'_{w2}\} = \{T, T_{w1}, T_{w2}\}/T_0 - 1, \; \lambda' = \lambda/\lambda_0 \qquad (10.59)$$

(where T_0 and λ_0 are the characteristic lubricant temperature and heat conductivity) and a new dimensionless parameter

$$\kappa = \frac{(u_1+u_2)P}{\pi \lambda_0 T_0 \vartheta^2} \qquad (10.60)$$

and by proceeding with derivation of equations in a manner similar to the one used in Subsection 10.3.1 we arrive at the system of equations (for simplicity here and further primes are omitted)

$$\int_{-h/2}^{h/2} \frac{z}{\mu} F(f + z \frac{dp}{dx}) dz = h - 1, \; p(a) = p(c) = 0, \qquad (10.61)$$

$$\int_{-h/2}^{h/2} \frac{1}{\mu} F(f + z \frac{dp}{dx}) dz = s_0, \qquad (10.62)$$

$$\frac{\partial}{\partial z}(\lambda \frac{\partial T}{\partial z}) = -\kappa \frac{\gamma^2}{\mu}(f + z \frac{dp}{dx}) F(f + z \frac{dp}{dx}),$$

$$T(x, -\frac{h}{2}) = T_{w1}(x), \; T(x, \frac{h}{2}) = T_{w2}(x), \qquad (10.63)$$

$$\gamma(h - 1) = x^2 - c^2 + \frac{1}{V} \frac{2}{\pi} \int_a^c p(t) \ln \frac{c-t}{|x-t|} dt, \qquad (10.64)$$

$$\int_a^c p(t) dt = \frac{\pi}{2}, \qquad (10.65)$$

For lightly loaded lubricated contact ($V \gg 1$) the process of solution of system (10.61)-(10.65) is similar to the one described for the isothermal contact. Namely, as in Subsection 10.3.2 we introduce new variables according to (10.31) and for $V \gg 1$ look for the solution in the form given by (10.37) and

$$T(\xi, \zeta) = \sum_{k=0}^{\infty} V^{-k} T_k(\xi, \zeta), \qquad (10.66)$$

where functions $p_k(\xi)$, $H_k(\xi)$, $g_k(\xi)$, $T_k(\xi, \zeta)$ and constants c_k, γ_k has to be determined. As a result we obtain systems of nonlinear and linear differential equations similar to the ones obtained in Subsection 10.3.2. Therefore, for the considered above rheologies all formulas (10.38), (10.40), (10.38), and (10.47) for film thickness h_e and components F_S and F_R of the friction force are still valid. However, in this case coefficients γ_0, γ_1, a_0, b_0, a_1, b_1, d_{00}, d_{10}, d_{20}, d_{01}, d_{11}, and d_{21} besides earlier mentioned parameters (see the preceding Subsection) also depend on λ, T_{W1}, and T_{W2}.

10.3.4   Numerical Method for Lightly Loaded EHL Contacts with Non-Newtonian Lubricant

In this subsection we will consider numerical approach to solution of the EHL problem for lightly loaded contacts with non-Newtonian lubricant. In addition we will compare the numerical and asymptotic solutions and determine the range of parameter V for which the asymptotic solution is sufficiently accurate.

The main advantage of the direct numerical solution is in the fact that the numerical procedure is insensitive to the details of non-Newtonian rheology of the fluid which may limit application of analytical methods. A numerical solution allows to determine the qualitative and quantitative behavior of the solution. However, it does not provide an opportunity to reveal the analytical structure of the solution.

The numerical method for solution of an EHL problem for lightly loaded contacts [10] is designed along the lines of the asymptotic method developed in the preceding subsection for $V \gg 1$. Because of the problem nonlinearity the solution process is iterative. In particular, it means that the term $\frac{1}{V} \int\limits_{-1}^{1} p(t) \ln \frac{1-t}{|\xi-t|} dt$ describing the surface displacement caused by elastic deformations of the contact solids has a small effect on the problem solution and, therefore, it is always taken from the preceding iteration. The equations of the problem are taken in the form (see equations (10.32)-(10.36))

$$\frac{d}{d\xi}\left[M_0(p, H)\frac{dp}{d\xi} - H\right] = 0, \ p(-1) = p(1) = \frac{dp(1)}{d\xi} = 0,$$

$$M_0(p, H) = \frac{H^2}{4\mu} \int\limits_{-1}^{1} tF(g + \frac{Ht}{c-a}\frac{dp}{d\xi})dt / \frac{dp}{d\xi}, \tag{10.67}$$

$$\int\limits_{-1}^{1} F(g + \frac{Ht}{c-a}\frac{dp}{d\xi})dt = 2\frac{\mu s_0}{H}, \tag{10.68}$$

$$\gamma(H - 1) = (\frac{c-a}{2})^2\xi^2 + \frac{c^2-a^2}{2}\xi + \frac{a^2+2ac-3c^2}{4}$$

$$+ \frac{c-a}{\pi V} \int\limits_{-1}^{1} p(t) \ln \frac{1-t}{|\xi-t|} dt, \tag{10.69}$$

$$\int\limits_{-1}^{1} p(t)dt = \tfrac{\pi}{c-a}. \tag{10.70}$$

The main idea of the method is to reduce solution of the problem to solution of the nonlinear system of equations

$$L_1(\gamma, c) = \tfrac{dp(1)}{d\xi} = 0,$$

$$L_2(\gamma, c) = \int\limits_{-1}^{1} p(t)dt - \tfrac{\pi}{c-a} = 0, \tag{10.71}$$

by the modified Newton's method [11] for two unknowns γ and c assuming that functions $p(\xi)$, $H(\xi)$, and $g(\xi)$ can be easily calculated. Calculation of functions $L_1(\gamma, c)$ and $L_2(\gamma, c)$ and their approximate derivatives with respect to γ and c for Newton's method is done by calculation expressions $[L_1(\gamma, c) - L_1(\gamma + \triangle\gamma, c)]/\triangle\gamma$, etc.

The general scheme of the iterative process is as follows. For given constants γ, c and function $p(\xi)$ from equation (10.69), we calculate the new approximation for gap $H(\xi)$ and after that from equation (10.68) using Newton's method we determine the new approximation for sliding frictional stress $g(\xi)$. That prepares us for determination of the new approximation of pressure $p(\xi)$ from equations (10.67).

Now, let us consider some details of the method. We will introduce partitions on the intervals $[-1, 1]$ along the ξ-axes $\{\xi_k\}$, $k = 1, \ldots, I$, $\xi_1 = -1$, $\xi_I = 1$ and along the ζ-axes $\{\zeta_n\}$, $n = 1, \ldots, J$, $\zeta_1 = -1$, $\zeta_J = 1$. Nodes $\{\xi_k\}$ are more dense in the zones where $dp/d\xi$ is large while the partition along the ζ-axis is uniform with the step size $\triangle\zeta = 2/J$. Using notations $p_k = q(\xi_k)$, $H_k = H(\xi_k)$, $g_k = g(\xi_k)$, and $\mu_k = \mu(p_k)$ and the approximation for $p(\xi)$ on the interval $[\xi_k, \xi_{k+1}]$

$$p(\xi) = p_k + \frac{p_{k+1} - p_k}{\xi_{k+1} - \xi_k}(\xi - \xi_k)$$

we obtain

$$\int\limits_{-1}^{1} p(t) \ln \tfrac{1-t}{|\xi-t|} dt \approx \sum\limits_{i=1}^{I} N_i(\xi_k), \quad k = 1, \ldots, I,$$

$$N_i(\xi) = [p_i - \xi_i \tfrac{p_{i+1}-p_i}{\xi_{i+1}-\xi_i}][K_1(\xi_{i+1}) - K_1(\xi_i) - K_2(\xi - \xi_{i+1})$$

$$+ K_2(\xi - \xi_i)] + \tfrac{p_{i+1}-p_i}{\xi_{i+1}-\xi_i}[K_3(\xi_{i+1}) - K_3(\xi_i) - K_4(\xi - \xi_{i+1}, \xi)$$

$$+ K_4(\xi - \xi_i, \xi)], \quad K_1(\xi) = (\xi - 1)[\ln(1 - \xi) - 1], \tag{10.72}$$

$$K_2(\xi) = \xi[1 - \ln |\xi|], \quad K_3(\xi) = \tfrac{(1-\xi)^2}{2}[\ln(1 - \xi) - \tfrac{1}{2}] + K_1(\xi),$$

$$K_4(\xi, \zeta) = \tfrac{\xi^2}{2}[\ln |\xi| - \tfrac{1}{2}] + \zeta K_2(\xi).$$

Similarly, using the trapezoidal rule for integration with respect to ζ from (10.68) we find

$$\{F[g_k - \tfrac{H_k}{c-a}\tfrac{dp(\xi_k)}{d\xi}] + F[g_k + \tfrac{H_k}{c-a}\tfrac{dp(\xi_k)}{d\xi}]\}\tfrac{\triangle\zeta}{2}$$

$$+ \sum_{i=1}^{J-1} F[g_k + \tfrac{(-1+i\triangle\zeta)H_k}{c-a}\tfrac{dp(\xi_k)}{d\xi}]\triangle\zeta = \tfrac{2\mu_k s_0}{H_k}. \tag{10.73}$$

Based on the last boundary condition in (10.67), the exact solution of equation (10.68) at $\xi = \xi_I$ is

$$g_I = \Phi(\tfrac{\mu_I s_0}{H_I}). \tag{10.74}$$

It is convenient to conduct the solution of equations (10.73) in the sequence from $k = I - 1$ to $k = 1$ by taking the initial approximation for g_k being equal to g_{k+1}.

The discretization of the integral in (10.71) with the help of the trapezoidal rule is obvious.

To find a new approximation of the pressure distribution $\{p_k\}$, we integrate the first equation in (10.67) with respect to ξ from $\xi_{k-1/2} = (\xi_{k-1} + \xi_k)/2$ to $\xi_{k+1/2} = (\xi_k + \xi_{k+1})/2$. By approximating the derivatives, we will obtain

$$\tfrac{M_0(\xi_{k+1/2})}{\xi_{k+1}-\xi_k} p_{k+1} - \left[\tfrac{M_0(\xi_{k+1/2})}{\xi_{k+1}-\xi_k} + \tfrac{M_0(\xi_{k-1/2})}{\xi_k-\xi_{k-1}}\right] p_k$$

$$+ \tfrac{M_0(\xi_{k-1/2})}{\xi_k-\xi_{k-1}} p_{k-1} = H_{k+1/2} - H_{k-1/2}, \quad k = 2, \ldots, I - 1, \tag{10.75}$$

$$p_1 = p_I = 0.$$

Here values of $H_{k+1/2}$ and $H_{k-1/2}$ are determined using the quadratic interpolation while $M_0(\xi_{k+1/2})$ and $M_0(\xi_{k-1/2})$ are calculated based on the information from the preceding iteration at $\xi_{k+1/2}$ and $\xi_{k-1/2}$, respectively. The latter involves calculation of $p_{k-1/2} = (p_{k-1} + p_k)/2$ and $p_{k+1/2} = (p_k + p_{k+1})/2$ as well as usage of $H_{k+1/2}$ and $H_{k-1/2}$. To avoid the difficulty in calculation of M_0 at a point where $dp/d\xi = 0$ at such a point we take a ratio of M determined at the values of p and H at the above point and at a small value of $dp/d\xi$ and divide it by the latter.

As a result equations (10.75) form a three-diagonal system of linear algebraic equations that can be easily solved for the set of $\{p_k\}$, $k = 1, \ldots, I$, by standard methods. After that the new sets of $dp(\xi_k)/d\xi$ and μ_k are easily found. That completes all necessary calculations for a single iteration. Such iterations are continued until the process converges with the desired precision.

Let us consider some numerical data for lubricants with power and Newtonian rheologies and constant viscosity $\mu = 1$. In dimensionless variables for a fluid with power rheology, we have

$$F(x) = (12\gamma^{\alpha+1})^{1/\alpha} \mid x \mid^{(1-\alpha)/\alpha}, \ \alpha > 0. \tag{10.76}$$

The partition along the ξ-axes is taken as follows: $\xi_k = \xi_{k-1} + d_0\lambda^{k-1}, k = 1, \ldots, I, \xi_1 = -1, \lambda = d_1^{1/(I-1)}, d_0 = 2(1-\lambda)/(1-\lambda^{I-1})$. In particular simulations we used $I = 81$ and $d_1 = 0.05$.

TABLE 10.3
The dimensionless sliding component $F_S(V, \alpha)$ of
the friction force for a fluid with power rheology
and constant viscosity (after Kudish and Yatzko
[10]). Published with permission from Allerton
Press.

V/α	0.75	1	1.25	1.5
128	0.003605	0.005502	0.004649	0.004053
64	0.003520	0.005474	0.004200	0.003021
32	0.003444	0.005436	0.003951	0.002748
16	0.003402	0.005407	0.003817	0.002537

The numerical method demonstrated good convergence for $V > 8$. For decreasing V in the region of $V \leq 8$, the convergence of the method gets worse and depends on the value of α. It is not surprising as the method was designed in a manner similar to the regular asymptotic method applicable only to lightly loaded lubricated contacts represented by values of $V \gg 1$. On the other hand, the numerical method has a wider range of applicability than the asymptotic one.

For $a = -10$ and $s_0 = 0.02$ the numerical results for the exit coordinate c, lubrication film thickness γ, and maximum pressure p_{max} versus parameter V are presented in Figs. 10.4 - 10.6, respectively. Curves 1, 2, 3, and 4 correspond to $\alpha = 0.75$, $\alpha = 1$, $\alpha = 1.25$, and $\alpha = 1.5$. It follows from Figs. 10.4 and 10.5 that $c(V)$ and $\gamma(V)$ monotonically decrease as parameter V increases and as $V \to \infty$ they approach their limiting values, which correspond to the case of rigid solids. The same figures show that c and γ monotonically increase with increase of α. The pressure distribution $p(x)$ has one extremum - maximum located to the right of $x = 0$. The maximum pressure p_{max} monotonically increases as a function of V and decreases with increase of α (see Fig. 10.6). As $V \to \infty$ the value of p_{max} approaches its limiting value equal to the one for rigid solids. Table 10.3 demonstrates monotonic increase of of the sliding component of the friction force F_S with parameter V and non-monotonic behavior as a function of α. To compare the numerical and two-term asymptotic solutions (see Subsection 10.3.2) let us consider the data obtained for a Newtonian fluid ($\alpha = 1$) with $a = -10$ and $s_0 = 0.02$, which is presented in Table 10.4. The data indicate the fact that for $V \geq 16$ the asymptotic solution is in excellent agreement with the numerical one. The above results show that the asymptotic formulas are very accurate and can be used for practical

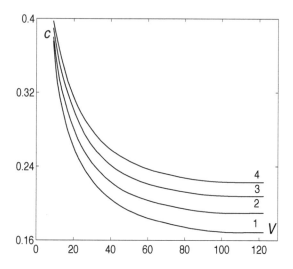

FIGURE 10.4
The dependence of the exit coordinate c on V for $\alpha = 0.75$ (curve 1), $\alpha = 1$
(curve 2), $\alpha = 1.25$ (curve 3), and $\alpha = 1.5$ (curve 4) (after Kudish and Yatzko
[10]). Published with permission from Allerton Press.

TABLE 10.4
The asymptotic and numerical values of c, γ, F_S,
F_R, and q_{max} for Newtonian fluid with constant
viscosity as functions of parameter V (after Kudish
and Yatzko [10]). Published with permission from
Allerton Press.

V	c	γ	F_S	F_R	q_{max}
		Asymptotic method			
128	0.1890	0.1644	0.005830	0.4466	2.6343
64	0.2071	0.1701	0.005826	0.4451	2.5523
32	0.2432	0.1816	0.005818	0.4423	2.4362
16	0.3154	0.2045	0.005801	0.4366	2.3597
		Numerical method			
128	0.1873	0.1642	0.005502	0.4381	2.6088
64	0.2029	0.1691	0.005474	0.4328	2.5559
32	0.2348	0.1798	0.005436	0.4235	2.4195
16	0.3014	0.2027	0.005407	0.4079	2.2103
8	0.3913	0.2318	0.005371	0.3862	1.9420

calculations for $V \geq 10$.

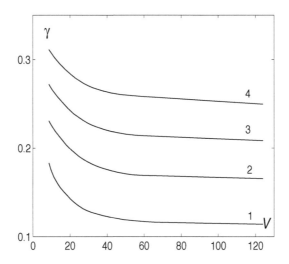

FIGURE 10.5
The dependence of the lubrication film thickness γ on V for $\alpha = 0.75$ (curve 1), $\alpha = 1$ (curve 2), $\alpha = 1.25$ (curve 3), and $\alpha = 1.5$ (curve 4) (after Kudish and Yatzko [10]). Published with permission from Allerton Press.

Cases of heavily loaded lubricated contacts are considered in the following sections.

10.4 Asymptotic Analysis of a Heavily Loaded Lubricated Contact of Elastic Solids. Pre-critical Lubrication Regimes for Newtonian Liquids

In the classic formulation the EHL problem has been studied in numerous papers and monographs by numerical and approximate analytical methods since the end of the 1940. Some references to these papers and monographs can be found in Dowson et al. [12], Ertel [13], Grubin [14], and Kudish et al. [15], the latest achievements in numerical methods of solving the considered problem are described by Houpert et al. [16], Bissett [17], Hamrock [18] (also see the bibliography in these papers), Lubrecht and Venner [19], Evans and Hughes [20], and others.

The lubrication regimes studied in this section correspond to the case of a relatively weak ("non-prevailing" in any zone of a contact region) dependence

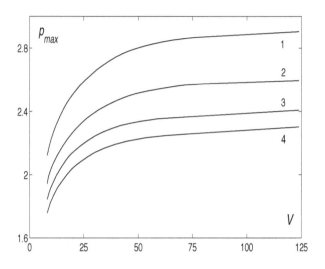

FIGURE 10.6
The dependence of the maximum pressure p_{max} on V for $\alpha = 0.75$ (curve 1), $\alpha = 1$ (curve 2), $\alpha = 1.25$ (curve 3), and $\alpha = 1.5$ (curve 4) (after Kudish and Yatzko [10]). Published with permission from Allerton Press.

of lubricant viscosity μ on pressure p. These regimes were studied earlier by Dowson et al. [12], Houpert et al. [16], Bissett [17], Hamrock [18], Lubrecht and Venner [19], and Evans and Hughes [20] using numerical methods. Ertel [13], Grubin [14], Archard et al. [21], and Crook [22] considered cases of rapidly growing with pressure lubricant viscosity using approximate analytical methods. The method used by Crook [22] differs from the methods employed by Ertel [13] and Grubin [14] only by more precise techniques under the same prior assumptions. The method proposed by Archard et al. [21] is the extension of the methods published by Ertel [13], Grubin [14], and Crook [22] for the case of a weak relationship between the viscosity μ and pressure p. The main difference between the methods published by Archard et al. [21] and Ertel [13], Grubin [14], and Crook [22] is in approximations of the gap function $h(x)$ and in a parabolic approximation for pressure $p(x)$ in the zone of large pressure. The purpose of the first modification is to take into account the pressure gradient along the lubricant flow and of the second one is to simplify the approximate calculations. All in all, these approximate analytical studies of the considered problem are based on certain prior assumptions which should be checked and are vital and necessary for application of the aforementioned analytical methods.

The main mathematical difficulties in the analytical analysis of the classic

EHL problem are

1. The essential non-linearity of the problem causing the possibility of existence solutions with qualitatively different structure depending on the values of the problem input parameters.

2. The integro-differential form of the problem equations causing the existence of a small parameter ω (possible definitions of the small parameter ω will be considered in Subsection 10.15.4) and boundary layers for heavily loaded lubrication regimes.

3. The proximity of the considered problem to the classic contact problem of elasticity described by an integral equation of the first kind numerical solutions of which are generally unstable with respect to small perturbations. (The latter problem is the limiting case for the EHL problem for the small parameter $\omega = 0$.)

4. The presence of the unknown dimensionless free exit boundary $x = c$ (exit point) and the dimensionless exit lubrication film thickness H_0.

5. The possibility to discriminate and to differentiate between the influence of different zones (inlet and exit zones) and regions (Hertzian region) of the lubricated contact on the EHL problem solution.

The purpose of this analysis is not to receive one more approximate formula for lubrication film thickness (which already exist in abundance) but to give a complete analysis of the aforementioned classic problem without any contradictions and prior assumptions about its solution. In addition to that, we will understand better the structure of the EHL problem solution and the reason of why there exists so many similar formulas for the lubrication film thickness.

The approach employed in the present and the following sections differs from the previously published ones in that it does not use any prior assumptions about the solution of the problem. A method of matched asymptotic expansions (see Van Dyke [4] and Cole [5]) is applied to studying the equations of the EHL problem. The early history of applying the methods of regular and matched asymptotic expansions to various EHL problems can be traced through papers of one of the authors (Kudish [23]-[27]). Kudish [25] investigated the lubrication problem for non-Newtonian fluids under iso- and non-isothermal conditions using the method of regular asymptotic expansions (Cole [5]). For the first time the method of matched asymptotic expansions was applied to the simplest lubrication problem with Newtonian fluid in an isothermal heavily loaded contact by Kudish [23, 24]. In the next papers (Kudish [25, 26]) this method has been further developed for the case of Newtonian fluids. Also, Kudish [25, 26] has established the conditions for the existence of the second pressure peak and for the dependence of its magnitude, width, and location on the inlet coordinate, temperature, etc. (compare with similar results by Kudish [25]). In papers [24, 27] Kudish investigated a lubrication problem for rough heavily loaded contact lubricated by a non-Newtonian fluid with general rheology. A combination of methods of regular and matched asymptotic expansions was used. In all these cases, the structure of the contact region and the solution behavior were studied and asymptotically valid formulas for the film thickness were obtained.

The method allows to determine the structure of the solution, to study the boundary layers (inlet and exit zones), to obtain asymptotically valid equations, and, most importantly, to get asymptotically accurate estimates for the lubrication film thickness and friction force. Furthermore, several difficulties of the numerical solution of the problem are avoided, such as the presence of a small parameter at the higher derivative in the problem equations and potential instabilities of the problem solution.

As it is our first encounter with matched asymptotic expansions in application to EHL problems we will start with the simplest possible problem. We will consider a plane problem of isothermal lubrication for heavily loaded elastic solids. The lubricant will be considered to be a Newtonian fluid. We will start by studying the so called pre-critical lubrication regimes, the so-called over-critical lubrication regimes are considered in Section 10.6. The structure of the contact area and of the solution will be determined. In the boundary layers (inlet and exit zones), two types of equivalent systems of equations for the major terms of asymptotic expansions for unknown quantities will be obtained. Furthermore, some asymptotic formulas for lubrication film thickness will be derived. The regimes of starved and fully flooded lubrication will be analyzed.

10.4.1 Problem Formulation

Let us consider a steady plane isothermal EHL problem for a heavily loaded contact of two moving smooth elastic cylinders (see Fig. 10.3) the equations for which were derived in the preceding section. The lubricant is an incompressible fluid with Newtonian rheology. To study the problem in the case of a heavily loaded contact, it is reasonable and convenient to choose the dimensionless variables

$$x' = \frac{x}{a_H}, \ a = \frac{x_i}{a_H}, \ c = \frac{x_e}{a_H}, \ p' = \frac{p}{p_H}, \ h' = \frac{h}{h_e}, \ \mu' = \frac{\mu}{\mu_a} \qquad (10.77)$$

and parameters

$$V = \frac{24\mu_a (u_1 + u_2) R'^2}{a_H^3 p_H}, \ H_0 = \frac{2R' h_e}{a_H^2}, \qquad (10.78)$$

which are scaled based on the parameters of the limiting case of a dry Hertzian contact of two elastic cylinders, namely, based on the maximum Hertzian pressure $p_H = \sqrt{\frac{E'P}{\pi R'}}$ and the Hertzian contact semi-width $a_H = 2\sqrt{\frac{R'P}{\pi E'}}$. The explanation of other parameters in (10.78) is given in Subsection 10.3.1. In dimensionless variables (10.77) and (10.78) equations (10.16), (10.18)-(10.21) of the classic EHL problem are reduced to the following system of integro-differential equations (further primes are omitted)

$$\frac{d}{dx} \left[\frac{H_0^2}{V} \frac{h^3}{\mu} \frac{dp}{dx} - h \right] = 0, \qquad (10.79)$$

$$p(a) = p(c) = \frac{dp(c)}{dx} = 0, \tag{10.80}$$

$$H_0(h-1) = x^2 - c^2 + \frac{2}{\pi} \int_a^c p(t) \ln \left| \frac{c-t}{x-t} \right| dt, \tag{10.81}$$

$$\int_a^c p(t)dt = \frac{\pi}{2}. \tag{10.82}$$

In the above equations the dimensionless lubricant viscosity is taken according to an exponential Barus law in the form

$$\mu = \mu(p, Q) = \exp(Qp), \tag{10.83}$$

where Q is the pressure viscosity coefficient.

In equations (10.79)-(10.82) the inlet coordinate a, the speed-load parameter V, and the pressure coefficient of viscosity Q at atmospherical pressure and contact temperature are considered to be known while the exit coordinate c and the lubrication film thickness H_0 together with the functions of pressure $p(x)$ and gap $h(x)$ in the contact region $[a, c]$ have to be determined from the solution of the problem.

Also, the problem equations (10.79)-(10.82) can be presented in another equivalent form. Solving equation (10.81) for pressure $p(x)$ and taking into account boundary conditions (10.80) we obtain the equivalent form (Vorovich et al. [28]) of our system expressed in terms of pressure $p(x)$

$$p(x) = R(x) \left[1 - \frac{1}{2\pi} \int_a^c \frac{dM(p,h)}{dt} \frac{dt}{R(t)(t-x)} \right],$$
$$\tag{10.84}$$

$$R(x) = \sqrt{(x-a)(c-x)},$$

$$\int_a^c \frac{dM(p,h)}{dt} \frac{dt}{R(t)} = \pi(a+c),$$
$$\tag{10.85}$$

$$\int_a^c \frac{dM(p,h)}{dt} \frac{tdt}{R(t)} = \pi [(\frac{c-a}{2})^2 + \frac{(a+c)^2}{2} - 1],$$

$$M(p, h) = \frac{H_0^3}{V} \frac{h^3}{\mu} \frac{dp}{dx}, \tag{10.86}$$

$$H_0(h-1) = x^2 - c^2 + \frac{2}{\pi} \int_a^c p(t) \ln \left| \frac{c-t}{x-t} \right| dt. \tag{10.87}$$

It is obtained by inverting the singular Cauchy integral in equations (10.79)-(10.81). The equivalence of the systems of equations (10.79)-(10.82) and (10.84)-(10.87) takes place under the following condition (Vorovich et al [28])

$$\frac{1}{\pi} \int_a^c M(p, h) \frac{dt}{R(t)} + \frac{2}{\pi} \int_a^c p(t) \ln \frac{1}{|c-t|} dt + c^2 - \frac{1}{2}(\frac{c-a}{2})^2 - (\frac{a+c}{2})^2$$
$$\tag{10.88}$$

$$= \ln \left| \frac{4}{c-a} \right|.$$

Equation (10.88) describes the condition necessary for the existence of a bounded solution for $p(x)$ (Vorovich et al. [28]), i.e., the solution that is equal to zero at $x = a$ and $x = c$.

Moreover, because of the equivalence of systems (10.79)-(10.81) and (10.84)-(10.87) in the latter system equation (10.87) for gap $h(x)$ can be replaced by (the Reynolds) equation

$$M(p,h) = H_0(h-1), \tag{10.89}$$

which follows from integration of equation (10.79) with the last boundary condition from (10.80) (also see (10.86)). Therefore, system of equations (10.84)-(10.87) is equivalent to the system of equations (10.84)-(10.86) and (10.90).

Another equivalent form of our original system (10.79)-(10.82) can be obtained by noticing that equation (10.79) is equivalent to (see (10.86))

$$\frac{dM(p,h)}{dx} = H_0 \frac{dh}{dx}, \tag{10.90}$$

and if the function of gap $h(x)$ is known then pressure $p(x)$ can be expressed in the form (see (10.79), (10.80), and (10.86))

$$p(x) = H_0 \int\limits_a^x \frac{h(t)-1}{W(p(t),h(t))} dt, \quad W(p,h) = M(p,h)/\frac{dp}{dx}. \tag{10.91}$$

Substituting equations (10.90) and (10.91) into equations (10.84) and (10.85) we obtain the representation of the problem in terms of gap $h(x)$ in the form

$$H_0 \int\limits_a^x \frac{h(t)-1}{W(p(t),h(t))} dt = R(x)\left[1 - \frac{H_0}{2\pi} \int\limits_a^c \frac{dh}{dt} \frac{dt}{R(t)(t-x)}\right], \tag{10.92}$$

$$R(x) = \sqrt{(x-a)(c-x)},$$

$$H_0 \int\limits_a^c \frac{dh}{dt} \frac{dt}{R(t)} = \pi(a+c), \quad H_0 \int\limits_a^c \frac{dh}{dt} \frac{tdt}{R(t)} = \pi[(\frac{c-a}{2})^2 + \frac{(a+c)^2}{2} - 1], \tag{10.93}$$

$$p(x) = H_0 \int\limits_a^x \frac{h(t)-1}{W(p(t),h(t))} dt, \quad W(p,h) = M(p,h)/\frac{dp}{dx}, \tag{10.94}$$

where $M(p,h)$ is determined by equation (10.86).

Two equivalent systems of equations (10.84)-(10.86), (10.90) and (10.92)-(10.94), (10.86) expressed in terms of pressure $p(x)$ and gap $h(x)$, respectively, in some cases are useful for numerical solution of EHL problems for heavily loaded contacts. However, the direct numerical solution of the system of equations (10.92)-(10.94), (10.86) exhibits some patterns of instability.

10.4.2 Asymptotic Analysis of the Problem for Heavily Loaded Lubricated Contact

We will consider equations (10.79)-(10.82) together with the equivalence condition (10.88).

Let us analyze a heavily loaded contact for which the presence of a small parameter ω in the problem equations is typical, for example, $\omega = V \ll 1$. In this case the deformation effects in the elastic solids almost everywhere in the contact prevail over the lubrication effects. That results in the smallness of the first term of equation (10.79) compared to the second term everywhere in the contact except for narrow boundary layers that are next to the inlet $x = a$ and exit $x = c$ points, respectively.

Thus, the contact will be called heavily loaded if

$$H_0(h-1) \ll 1 \ and \ \frac{d}{dx}\left[\frac{H_0^3}{V}\frac{h^3}{\mu}\frac{dp}{dx}\right] \ll 1$$

$$for \ x - a \gg \epsilon_q \ and \ c - x \gg \epsilon_g,$$

(10.95)

where $\epsilon_q = \epsilon_q(\omega) \ll 1$ and $\epsilon_g = \epsilon_g(\omega) \ll 1$ are characteristic sizes of the inlet and exit zones, which are boundary layers located next to points $x = a$ and $x = c$ (Van Dyke [4] and Cole [5]). The characteristic size of the inlet zone $\epsilon_q(\omega)$ is determined by the given inlet coordinate a (see (10.99)) and depends on ω. The characteristic size of the exit zone $\epsilon_g(\omega)$ is unknown and is determined by the exit coordinate c (see (10.100)). The region determined by inequalities $x - a \gg \epsilon_q$ and $c - x \gg \epsilon_g$ will be called external (Dyke [4] and Cole [5]) or Hertzian region (Vorovich et al. [28]).

We will begin the analysis of the problem with investigation of the external region temporarily assuming that the boundaries of the contact region $x = a$ and $x = c$ are known and fixed. Then with the help of estimate (10.95) equations (10.79)-(10.82) yield

$$x^2 - c^2 + \frac{2}{\pi}\int\limits_a^c p(t)\ln\left|\frac{c-t}{x-t}\right|dt = o(1),$$

$$x - a \gg \epsilon_q \ and \ c - x \gg \epsilon_g.$$

(10.96)

Equations (10.96) and (10.82) with the accuracy of up to $o(1)$ coincide with the equations of a classic contact problem of elasticity (Vorovich et al. [28]) and have the well–known solution

$$p_0(x) = \sqrt{(x-a)(c-x)} + \frac{1+2ac+(c-a)^2/4-(a+c)x}{2\sqrt{(x-a)(c-x)}},$$

$$x - a \gg \epsilon_q \ and \ c - x \gg \epsilon_g.$$

(10.97)

Thus, in the external region the problem solution has the form

$$p(x) = p_0(x) + o(1), \quad x - a \gg \epsilon_q \ and \ c - x \gg \epsilon_g.$$

(10.98)

Let us assume that for $\omega \ll 1$ the coordinate of the inlet point a is given by

$$a = -1 + \alpha_1\epsilon_q, \ \alpha_1 = O(1), \ \omega \ll 1,$$

(10.99)

where α_1 is a given non-positive constant. Then the exit coordinate c may be found in the form

$$c = 1 + \beta_1 \epsilon_g + o(\epsilon_g), \ \beta_1 = O(1), \ w \ll 1, \tag{10.100}$$

where β_1 is an unknown constant and it is subject to calculation during the solution process.

It follows from equations (10.97)-(10.100) that in the external region $p_0(x) = O(1)$ for $w \ll 1$. Besides, under the assumption

$$\epsilon_g = O(\epsilon_q) \ for \ w \ll 1 \tag{10.101}$$

the estimate $p_0(x) = O(\epsilon_q^{1/2})$, $w \ll 1$, is valid for in the inlet and exit zones, i.e., for $r = (x - a)/\epsilon_q = O(1)$ and $s = (x - c)/\epsilon_g = O(1)$, where r and s are local coordinates in the inlet and exit zones. According to the principle of matched asymptotic expansions (Van Dyke [4]), one can obtain the following estimate for $p(x)$ in the inlet and exit zones, i.e.,

$$p(x) = O(\epsilon_q^{1/2}) \ for \ r = O(1) \ and \ s = O(1), \ w \ll 1. \tag{10.102}$$

Based on estimate (10.102) the problem solution in the inlet and exit zones will be searched in the form

$$p(x) = \epsilon_q^{1/2} q(r) + o(\epsilon_q^{1/2}), \ q(r) = O(1) \ r = O(1), \ w \ll 1, \tag{10.103}$$

$$p(x) = \epsilon_g^{1/2} g(s) + o(\epsilon_g^{1/2}), \ g(s) = O(1) \ s = O(1), \ w \ll 1, \tag{10.104}$$

where $q(r)$ and $g(s)$ are the major terms of the pressure $p(x)$ asymptotic expansions in the inlet and exit zones.

Using (10.97) equations (10.81), (10.82), and (10.88) can be transformed into

$$H_0(h - 1) = \frac{2}{\pi} \int_a^c [p(t) - p_0(t)] \ln \left| \frac{c-t}{x-t} \right| \, dt, \tag{10.105}$$

$$\int_a^c [p(t) - p_0(t)] dt = 0, \tag{10.106}$$

$$\int_a^c M(p, h) \frac{dt}{R(t)} + 2 \int_a^c [p(t) - p_0(t)] \ln \frac{1}{|c-t|} dt = 0. \tag{10.107}$$

Integrating equation (10.79) with respect to x from x to c and using the last boundary condition in (10.80) equations (10.86) and (10.105) yield

$$M(p, h) = \frac{2}{\pi} \int_a^c [p(t) - p_0(t)] \ln \left| \frac{c-t}{x-t} \right| \, dt. \tag{10.108}$$

Resolving (10.108) for $p(x) - p_0(x)$ leads to the estimate (Vorovich et al. [28])

$$p(x) - p_0(x) = O(M(p, h)) \tag{10.109}$$

$$for \ x - a \gg \epsilon_q \ and \ c - x \gg \epsilon_g, \ w \ll 1,$$

For further analysis it is necessary to make the following temporary assumptions

$$M(p,h) \ll \epsilon_q^{3/2} \ for \ x - a \gg \epsilon_q \ and \ c - x \gg \epsilon_g, \ \omega \ll 1,$$

$$\int_a^c M(p,h)\frac{dt}{R(t)} \ll \epsilon_q^{3/2}, \ \omega \ll 1.$$

(10.110)

Obviously, estimate (10.95) and the second estimate in (10.110) are independent. The validity of the above assumptions will be examined after the analysis of the problem is completed.

Estimates (10.109) and (10.110) in the external region lead to

$$p(x) - p_0(x) \ll \epsilon_q^{3/2} \ for \ x - a \gg \epsilon_q \ and \ c - x \gg \epsilon_g, \ \omega \ll 1. \quad (10.111)$$

The integrals of function $p(x) - p_0(x)$ in (10.106) and (10.107) can be expressed by a sum of three integrals: over the inlet and exit zones and over the external region. Therefore, using estimates (10.110), (10.111), and expressions (10.99), (10.100), (10.103), and (10.104) equations (10.106) and (10.107) can be reduced to the form

$$\epsilon_q^{3/2} \int_0^\infty [q(t) - q_a(t)]dt + \epsilon_g^{3/2} \int_{-\infty}^0 [g(t) - g_a(t)]dt + \ldots = 0, \quad (10.112)$$

$$\epsilon_q^{3/2} \ln \frac{1}{2} \int_0^\infty [q(t) - q_a(t)]dt + \epsilon_g^{3/2} \int_{-\infty}^0 [g(t) - g_a(t)] \ln \frac{1}{|t|} dt$$

$$+ \epsilon_g^{3/2} \ln \frac{1}{\epsilon_g} \int_{-\infty}^0 [g(t) - g_a(t)]dt \ldots = 0,$$

(10.113)

where functions $q_a(r)$ and $g_a(s)$ are the major terms of the inner asymptotic expansions of the external asymptotic (10.97), (10.98) in the inlet and exit zones, which are determined by the equalities

$$q_a(r) = \sqrt{2r} + \frac{\alpha_1}{\sqrt{2r}}, \ g_a(s) = \sqrt{-2s} - \frac{\beta_1}{\sqrt{-2s}}. \quad (10.114)$$

Using the fact $\epsilon_g^{3/2} \ln \frac{1}{\epsilon_g} \gg \epsilon_q^{3/2}$ for $\omega \ll 1$ that follows from the assumption (10.101), from (10.112), and from (10.113), we obtained

$$\int_{-\infty}^0 [g(t) - g_a(t)]dt = 0, \quad (10.115)$$

$$\int_0^\infty [q(t) - q_a(t)]dt = 0, \quad (10.116)$$

$$\int_{-\infty}^0 [g(t) - g_a(t)] \ln \frac{1}{|t|} dt = 0. \quad (10.117)$$

To get the asymptotic expansions of the gap function h in the inlet and exit zones, the integral in (10.105) can be expressed by the sum of three integrals over the inlet and exit zones and over the external region. Thus, following the described estimating procedure and using equations (10.103), (10.104), (10.111), (10.114)-(10.117) the equations for the major terms of the asymptotic expansions $h_q(r)$ and $h_g(s)$ of the gap function h in the inlet and exit zones are obtained in the form

$$H_0[h_q(r) - 1] = \epsilon_q^{3/2} \frac{2}{\pi} \int\limits_0^\infty [q(t) - q_a(t)] \ln \frac{1}{|r-t|} dt + \dots,$$

(10.118)

$$H_0[h_g(s) - 1] = \epsilon_g^{3/2} \frac{2}{\pi} \int\limits_{-\infty}^0 [g(t) - g_a(t)] \ln \frac{1}{|s-t|} dt + \dots.$$

It can be shown that the contributions of the effects of elasticity and lubrication to the solution of the problem in the inlet and exit zones are of the same order of magnitude. The commensurability of the aforementioned effects leads to the commensurability of the major terms of the asymptotic expansions of the terms of equation (10.79) in the inlet zone. Therefore, using expressions (10.103), (10.104), (10.118) and estimates

$$\mu(p, Q) = O(1) \ for \ r = O(1) \ and \ s = O(1), \ \omega \ll 1 \qquad (10.119)$$

following from the accepted class of relations for $\mu(p, Q)$, the major term of the asymptotic for the lubrication film thickness H_0 becomes (Kudish [23])

$$H_0 = A(V\epsilon_q^2)^{1/3} + \dots, \ \omega \ll 1, \qquad (10.120)$$

where $A(\alpha_1) = O(1)$ for $\omega \ll 1$ and $A(\alpha_1)$ is an unknown nonnegative constant independent from ω and ϵ_q. Taking into account estimate (10.120) a similar analysis of the exit zone confirms the validity of the assumption $\epsilon_g^{3/2} \ln \frac{1}{\epsilon_g} \gg \epsilon_q^{3/2}$ for $\omega \ll 1$ and allows to set (see (10.101))

$$\epsilon_g = \epsilon_q. \qquad (10.121)$$

Now, it is useful to introduce the definitions of the regimes of starved and fully flooded lubrication, which are determined by the lubricant flux entering the contact or the position of the inlet coordinate a. These lubrication regimes can be easily expressed by means of the order of proximity of the inlet coordinate a to the left boundary $x = -1$ of the Hertzian dry contact. The conditions when

$$h(x) - 1 \ll 1 \ for \ all \ x \in [a, c], \ \omega \ll 1 \qquad (10.122)$$

are called regimes of starved and scanty lubrication, and the conditions when

$$h(x) - 1 = O(1) \ for \ x - a = O(\epsilon_q) \ and \ c - x = O(\epsilon_g), \ \omega \ll 1 \qquad (10.123)$$

are called regimes of fully flooded lubrication. A separate definition of starved and scanty lubrication regimes will be given below. From (10.118) and (10.121) follows that starved lubrication regimes (10.122) take place under the condition

$$\epsilon_q^{3/2} \ll H_0, \ \omega \ll 1, \tag{10.124}$$

and fully flooded lubrication regimes (10.123) – under the condition

$$\epsilon_q^{3/2} = O(H_0), \ \omega \ll 1. \tag{10.125}$$

Thus, using formula (10.120) for the exit film thickness H_0 we obtain the conditions for starved

$$\epsilon_q \ll V^{2/5}, \ \omega \ll 1, \tag{10.126}$$

and fully flooded

$$\epsilon_q = V^{2/5}, \ \omega \ll 1, \tag{10.127}$$

lubrication regimes. For fully flooded lubrication regimes equations (10.120) and (10.127) lead to the formula for the lubrication film thickness

$$H_0 = AV^{3/5}, \ \omega \ll 1. \tag{10.128}$$

Finally, using equations (10.118), (10.120), (10.126), and (10.127), the final equations for and for starved and fully flooded lubrication regimes can be written in the form, respectively,

$$h_q(r) = 1, \tag{10.129}$$

$$h_g(s) = 1; \tag{10.130}$$

$$A[h_q(r) - 1] = \frac{2}{\pi} \int\limits_0^\infty [q(t) - q_a(t)] \ln \frac{1}{|r-t|} dt, \tag{10.131}$$

$$A[h_g(s) - 1] = \frac{2}{\pi} \int\limits_{-\infty}^0 [g(t) - g_a(t)] \ln \frac{1}{|s-t|} dt. \tag{10.132}$$

The asymptotic analysis of equation (10.79) in the inlet zone and of the first boundary condition in (10.80) based on expressions (10.103), (10.118), and (10.120) yields the asymptotically valid equations in the inlet zone

$$\frac{dM_0(q,h_q,\mu_q,r)}{dr} = \frac{2}{\pi} \int\limits_0^\infty \frac{q(t) - q_a(t)}{t-r} dt, \ q(0) = 0, \tag{10.133}$$

$$M_0(q, h, \mu, r) = A^3 \frac{h^3}{\mu} \frac{dq}{dr}. \tag{10.134}$$

Using (10.104), (10.118), (10.120), and (10.121), a similar analysis of equation (10.79) in the exit zone and of the two last boundary conditions in (10.80) leads to the equations

$$\frac{dM_0(g,h_g,\mu_g,s)}{ds} = \frac{2}{\pi} \int\limits_{-\infty}^0 \frac{g(t) - g_a(t)}{t-s} dt, \ g(0) = \frac{dg(0)}{ds} = 0. \tag{10.135}$$

In equations (10.133) and (10.135) functions $\mu_q(r)$ and $\mu_g(s)$ are the major terms of the asymptotic expansions of $\mu(p, Q)$ in the inlet and exit zones.

To close the systems of obtained equations, it is necessary to add to them the conditions

$$q(r) \to q_a(r), \ r \to \infty, \tag{10.136}$$

$$g(s) \to g_a(s), \ s \to -\infty, \tag{10.137}$$

following from the principle of asymptotic matching (Van Dyke [4]).

Thus, using the proposed asymptotic analysis of the EHL problem formula (10.120) for the film thickness is obtained, as well as two closed systems of integro-differential equations: in the inlet zone - equations (10.133), (10.134), (10.136), (10.114), (10.115), and (10.129) for the starved lubrication regimes (or (10.131) for fully flooded lubrication regimes) for functions $q(r)$, $h_q(r)$, and constant A, and in the exit zone - equations (10.135), (10.134), (10.137), (10.114), (10.114), and (10.130) for the starved lubrication regimes (or (10.132) for fully flooded lubrication regimes) for functions $g(s)$, $h_g(s)$, and constant β_1 (see equation (10.100)).

Finally, it should be noted that estimate (10.95) involved in the definition of heavily loaded lubrication regimes follows from (10.97)-(10.100), and (10.120). Besides, estimates (10.110) follow from (10.95), (10.103), (10.104), and (10.120).

In some cases estimate (10.119) playing the validation role of the entire asymptotic analysis in the inlet and exit zones imposes some restrictions. For example, if the lubricant viscosity $\mu(p, Q)$ is determined by the generalized exponential function

$$\mu = \exp(Q p^m), \ Q \gg 1 \ for \ \omega \ll 1 \ (Q, \ m \geq 0), \tag{10.138}$$

then the restrictions on ϵ_q follow from (10.119) and take the form

$$\epsilon_q \ll Q^{-2/m} \ or \ \epsilon_q = O(Q^{-2/m}), \ \omega \ll 1. \tag{10.139}$$

These estimates should be taken into account together with estimates (10.126) and (10.127).

Estimates (10.126), (10.127), and (10.139) define, the so–called, pre-critical lubrication regimes. In general, for a given $\epsilon_q = \epsilon_q(V, Q)$ the pre-critical regimes are characterized by the following condition

$$\mu(\epsilon_q^{1/2}, Q) = O(1), \ \epsilon_q \ll 1, \ \omega \ll 1, \tag{10.140}$$

while the over-critical lubrication regimes are determined by the condition

$$\mu(\epsilon_q^{1/2}, Q) \gg 1, \ \epsilon_q \ll 1, \ \omega \ll 1. \tag{10.141}$$

Now, the detailed definitions for pre-critical starved and fully flooded lubrication regimes can be given. The pre-critical starved lubrication regimes are those for which the relationships (10.122) and (10.140) are valid. The fully flooded pre-critical lubrication regimes are those for which relations (10.123) and (10.140) are valid.

It should be noted that the described asymptotic analysis is valid if the small parameter $\omega = Q^{-1} \ll 1$ or $\omega = V \ll 1$ and conditions (10.140) (which in the case of viscosity μ from (10.138) coincides with conditions (10.139)) are valid.

After the asymptotic solutions of the problem in the inlet and exit zones are obtained, we can determine the uniformly valid approximate solution of the problem for pressure $p_u(x)$ and gap $h_u(x)$ in the form [4]

$$p_u(x) = \tfrac{\epsilon_q}{2} q(\tfrac{x-a}{\epsilon_q}) g(\tfrac{x-c}{\epsilon_q}), \quad h_u(x) = h_q(\tfrac{x-a}{\epsilon_q}) h_g(\tfrac{x-c}{\epsilon_q}). \tag{10.142}$$

where a and c are determined by formulas (10.99) and (10.100). It can be clearly seen that in the Hertzian region $p_u(x)$ and $h_u(x)$ are practically equal to the Hertzian pressure $\sqrt{1-x^2}$ and 1, respectively, while in the inlet and exit zones they practically coincide with $\epsilon_q^{1/2} q(r)$, $h_q(r)$ and $\epsilon_q^{1/2} g(s)$, $h_g(s)$, respectively. For starved lubrication with high precision $h_u(x) = 1$ in the entire contact region. For pre-critical regimes, these uniformly valid in the entire contact region functions can be used anywhere the numerical (i.e., also approximate) solutions for pressure $p(x)$ and gap $h(x)$ are used.

10.4.3 Asymptotic Analysis of the System of Equations (10.84)-(10.88)

For $\omega \ll 1$ in the inlet and exit zones, let us obtain the systems of asymptotically valid equations equivalent to those derived in Subsection 10.4.2. Suppose we have a heavily loaded lubricated contact and, therefore, estimate (10.95) is true and the inlet coordinate a satisfies equation (10.99).

Using (10.86) and the definition of a heavily loaded contact (10.95), from (10.84) in the external region, we can find

$$p(x) = p_0(x) + o(1), \quad p_0(x) = \sqrt{(x-a)(c-x)},$$
$$\tag{10.143}$$
$$x - a \gg \epsilon_q \text{ and } c - x \gg \epsilon_g.$$

Taking the coordinate of the exit point c according to (10.100) and using assumption (10.101), we obtain that $p_0(x) = O(\epsilon_q^{1/2})$ for $r = O(1)$ and $s = O(1)$, $\omega \ll 1$. As a result, in the inlet and exit zones, we will search the solution in the form (10.103) and (10.104).

Let us consider equation (10.84) in the inlet zone. The integral in (10.84) can be expressed as a sum of tree integrals: over the inlet and exit zones and over the external region. Formula (10.120) for H_0 and the following asymptotically valid equation for $q(r)$ in the inlet zone

$$q(r) = \sqrt{2r}\left[1 - \tfrac{1}{2\pi} \int\limits_0^\infty \tfrac{d}{dt} M_0(q, h_q, \mu_q, t) \tfrac{dt}{\sqrt{2t(t-r)}}\right] \tag{10.144}$$

can be obtained by estimating each of these integrals using expressions (10.99), (10.100), (10.103), and the first estimate in (10.110). A similar analysis in the exit zone leads to equality (10.121) and the following equation for $g(s)$

$$g(s) = \sqrt{-2s}\left[1 - \frac{1}{2\pi}\int_{-\infty}^{0}\frac{d}{dt}M_0(g, h_g, \mu_g, t)\frac{dt}{\sqrt{-2t}(t-s)}\right],\qquad(10.145)$$

where the functions $h_q(r)$, $h_g(s)$, $\mu_q(q)$, and $\mu_g(g)$ have the same meaning as earlier.

Using (10.95), (10.99), (10.100), (10.103), (10.104), (10.120), and (10.121) the asymptotic analysis of equations (10.85) and (10.86) leads to a system of linear algebraic equations for A^3 and β_1. The solution of this system has the form

$$1 = \pi\alpha_1/\int_{0}^{\infty}\frac{d}{dt}M_0(q, h_q, \mu_q, t)\frac{dt}{\sqrt{2t}},\qquad(10.146)$$

$$\beta_1 = \frac{1}{\pi}\int_{-\infty}^{0}\frac{d}{dt}M_0(g, h_g, \mu_g, t)\frac{dt}{\sqrt{-2t}}.\qquad(10.147)$$

Also, equation (10.146) can be represented in the form

$$A^3 = \pi\alpha_1/\int_{0}^{\infty}\frac{d}{dt}\left(\frac{h_q^3}{\mu_q}\frac{dq}{dt}\right)\frac{dt}{\sqrt{2t}}.\qquad(10.148)$$

Therefore, in the inlet zone the problem is reduced to the system of equations (10.134), (10.144), (10.146), and equations (10.129) or (10.131) (depending on whether the lubrication regime is starved or fully flooded) for functions $q(r)$, $h_q(r)$, and constant A. In the exit zone the problem is reduced to the system of equations (10.134), (10.145), (10.147) and equations (10.130) or (10.132) (depending on whether the lubrication regime is starved or fully flooded) for functions $g(s)$, $h_g(s)$, and constant β_1. Note that equality (10.117) is the condition of equivalence of the latter systems and the corresponding systems from Subsection 10.4.2.

It should be pointed out that asymptotic relationships (10.136) and (10.137) can be obtained from equations (10.134), (10.144)-(10.146). Obviously, the accepted assumption (10.95) is valid.

It can be shown that for $\mu = 1$ and regimes of starved lubrication the solutions of the asymptotically valid systems of equations of Subsections 10.4.2 and 10.4.3 in the inlet and exit zones have the following properties

$$q(r, \alpha_1) = |\alpha_1|^{1/2}q(r/|\alpha_1|, -1),$$

$$g(s, \alpha_1) = |\alpha_1|^{1/2}g(s/|\alpha_1|, -1),\qquad(10.149)$$

$$A(\alpha_1) = |\alpha_1|^{2/3}\theta(-\alpha_1)A(-1),\quad \beta_1(\alpha_1) = |\alpha_1|\beta_1(-1),$$

where $\theta(x)$ is a step function, $\theta(x) = 0$, $x \le 0$ and $\theta(x) = 1$, $x > 0$. It follows from (10.147) that the formulation of the problem makes sense only for $a \le -1$

($\alpha_1 \leq 0$), which means that $H_0 \geq 0$ ($A \geq 0$). Obviously, $H_0 = A = 0$ for $a = -1$ ($\alpha_1 = 0$).

Let us finalize the results of the section. A detailed analysis of the structure of the solution of the problem for heavily loaded lubricated contacts in pre-critical regimes is proposed. It is shown that the contact region consists of three characteristic zones: the zone with predominating "elastic solution" - the Hertzian region and the narrow inlet and exit zones (boundary layers) compared to the size of the Hertzian region. In the inlet and exit zones, the influences of the solids' elasticity and viscous flow of lubricant are of the same order of magnitude. In the inlet and exit zones two equivalent asymptotically valid systems of equations describing the solution of the problem are derived.

The solutions of these systems depend on a smaller number of initial problem parameters than the solution of the original problem (10.79)-(10.83). It was shown that regimes of starved and fully flooded lubrication can be realized in the framework of pre-critical regimes. In all cases the characteristic size of the inlet and exit zones in a heavily loaded lubricated contact can be considered small compared to the size of the Hertzian region. The asymptotic formula for the lubrication film thickness is obtained using the fact that the terms of the equation describing lubrication flow and elastic effects are of the same order of magnitude. The lubrication film thickness depends on the particular regime of lubrication and the solution behavior in the inlet zone. The definition of pre-critical regimes is given. Based on that definition the limitations of the applied method were derived. The explicit expressions for coefficient A in the film thickness formula and the exit coordinate β_1 in terms of pressure and gap are derived. This analysis of the problem provides a better understanding of the structure of the solution of the problem in pre-critical lubrication regimes.

The asymptotic method presented in this section will be extended to non-Newtonian lubricants in Section 10.15.

10.5 Asymptotic Analysis of Heavily Loaded Contacts of Elastic Solids Lubricated by Compressible Fluids

Let us consider a steady isothermal EHL problem for a heavily loaded contact of two infinite parallel cylinders lubricated by a compressible Newtonian fluid. Under the assumptions of this chapter in dimensionless variables, the problem can be reduced to the Reynolds equation

$$\frac{d}{dx}\{\rho[M(p,h) - H_0 h]\} = 0, \ \ M(p,h) = \frac{H_0^3}{V}\frac{h^3}{\mu}\frac{dp}{dx}, \tag{10.150}$$

which expresses the fact of fluid flux preservation through the gap between two solids. The rest of the problem equations coincide with equations (10.80)-

(10.82). In equation (10.152) $\rho = \rho(p)$ is the fluid density expressed by a monotonically increasing function of pressure such that $\rho(0) = 1$ (here the density is scaled by the density ρ_0 at the atmospheric pressure).

By integrating equation (10.152) with boundary conditions (10.80), it can be rewritten in the form

$$\rho[M(p, h) - H_0 h] = -H_0.$$

Solving the latter equation for $H_0(h - 1)$ and differentiating it the Reynolds equation can be rewritten in the form

$$\tfrac{d}{dx}[M_\rho(p, h) - H_0 h] = 0, \quad M_\rho(p, h) = M(p, h) + H_0[\tfrac{1}{\rho(p)} - 1]. \qquad (10.151)$$

Therefore, the EHL problem for compressible lubricants can be reduced to equations (10.151) and (10.80)-(10.82). The latter system of equations is equivalent to the system of equations (10.84)-(10.88) in which function $M(p, h)$ is replaced by function $M_\rho(p, h)$ from (10.151).

After that the asymptotic analysis conducted in Section 10.4 can be applied in its entirety to the EHL problem in case of a compressible Newtonian lubricant by replacing $M(p, h)$ by $M_\rho(p, h)$ from (10.151). As a result of this analysis, we obtain two systems: (10.133), (10.134), (10.136), (10.114), (10.115), and (10.129) for the starved lubrication regimes (or (10.131) for fully flooded lubrication regimes) for functions $q(r)$, $h_q(r)$, and constant A in the inlet zone and (10.135), (10.134), (10.137), (10.114), (10.114), and (10.130) for the starved lubrication regimes (or (10.132) for fully flooded lubrication regimes) for functions $g(s)$, $h_g(s)$, and constant β_1 as well as the asymptotic formula (10.120) for the film thickness H_0. In addition to that we obtain two system of equations (10.134), (10.144), (10.146), and equations (10.129) or (10.131) (depending on whether the lubrication regime is starved or fully flooded) for functions $q(r)$, $h_q(r)$, and constant A in the inlet zone and (10.134), (10.145), (10.147) and equations (10.130) or (10.132) (depending on whether the lubrication regime is starved or fully flooded) for functions $g(s)$, $h_g(s)$, and constant β_1 in the exit zone. The latter pairs of systems of equations are equivalent. In these systems function M_0 from (10.134) must be replaced by function $M_{\rho 0}$ as follows

$$M_{\rho 0}(p, h, \mu, x) = M_0(p, h, \mu, x) + (\tfrac{V^{2/5}}{\epsilon_q})^{5/6} A[\tfrac{1}{\rho(p)} - 1]$$

$$= A^3 \tfrac{h^3}{\mu} \tfrac{dp}{dx} + (\tfrac{V^{2/5}}{\epsilon_q})^{5/6} A[\tfrac{1}{\rho(p)} - 1]. \qquad (10.152)$$

In formula (10.120) for the the film thickness H_0 constant A besides other parameters depends on the behavior of lubricant density $\rho(p)$ in the inlet zone.

This asymptotic analysis is valid as long as in the Hertzian region

$$(\tfrac{V^{2/5}}{\epsilon_q})^{5/6}[\tfrac{1}{\rho(p)} - 1] \ll 1 \ \ for \ x - a \gg \epsilon_q \ and \ c - x \gg \epsilon_g, \qquad (10.153)$$

and in the inlet and exit zones

$$\left(\frac{V^{2/5}}{\epsilon_q}\right)^{5/6}\left[\frac{1}{\rho(\epsilon_q^{1/2})}-1\right]=O(1),\ \omega\ll 1. \tag{10.154}$$

Numerical solutions of the asymptotic equations for incompressible fluids show (see Section 10.8) that for $\alpha_1 < 0$ we have $A > 0$. Therefore, based on formula (10.146), the fact that $\rho(p)$ is a monotonically increasing function of p and that $dq(r)/dr > 0$, we can expect that the value of constant A for the case of a compressible fluid is smaller than the one in the corresponding case of an incompressible fluid.

10.6   Asymptotic Analysis of a Heavily Loaded Lubricated Contact of Elastic Solids. Over-critical Lubrication Regimes for Newtonian Liquids

In this section we will consider the same problem as in the previous one, i.e., the plane problem of isothermal lubrication for heavily loaded elastic bodies with Newtonian fluid. However, this time we will consider the over-critical lubrication regimes. The problem will be analyzed using the methods of matched asymptotic expansions. The structure of the contact area will be investigated. It will be ascertained that there exist two boundary layers (two inlet zones) with different characteristic sizes at one side of the external (Hertzian) region and two boundary layers (two exit zones) at its opposite side. In each of the larger inlet and exit zones as well as in the Hertzian region, the asymptotically valid equations for the major terms of asymptotic expansions of the unknown quantities will be considered and some solved for the case of a Barus lubricant viscosity. Some structural asymptotic formulas for lubrication film thickness will be obtained. The approximate Ertel-Grubin's and Greenwood's methods and their prior assumptions will be analyzed in detail.

10.6.1   Problem Formulation

Let us consider a classic plane isothermal problem for an incompressible Newtonian lubricant in a heavily loaded contact of smooth elastic cylinders. Under these conditions in terms of the dimensionless variables (10.77) and parameters (10.78), the problem is reduced to the classic system of integro-differential equations (10.79)-(10.82). In Section 10.4 the problem formulation is described in more detail as well as the main mathematical difficulties in analysis of this problem are given. For over-critical lubrication regimes the possibility to separate the influence of different regions of a lubricated contact on the solution of the problem is even more crucial then for pre-critical regimes considered in Section 10.4. The reason for that is the integral nature of the gap dependence

on pressure as well as the fact that in over-critical lubrication regimes the
solution structure is far from obvious.

The purpose of this study is to progress in the understanding of the considered classic problem for over-critical regimes without using any prior assumptions. We will apply the methods of matched asymptotic expansions (Cole [4])
to analyzing problem (10.79)-(10.82), which will allow to take into account the
mutual influence of elastic and hydrodynamic effects properly. The method
allows analyzing the structure of the solution, obtaining asymptotically valid
analytical solutions in the larger inlet and exit zones, getting analytical approximate solutions in the Hertzian region, and for different regimes obtaining
an asymptotically valid estimate for the lubrication film thickness H_0. However, this analysis stops short of establishing the asymptotically valid equations in smaller inlet and exit zones. Nonetheless, the analysis will provide a
solid basis for a number of important conclusions. Moreover, in Section 10.13
we will see that the asymptotically valid equations derived for pre-critical
lubrication regimes under certain conditions also describe over-critical lubrication regimes.

It is necessary to note that in papers by Kudish [23] - [27], [29] on heavily loaded lubrication contacts mostly pre-critical regimes were discussed for
which the estimate (10.140) (Kudish [24, 27, 29]) holds. The maximum characteristic size of the inlet zone for which the lubrication regime is still a
pre-critical one we will call the critical characteristic size of the inlet zone
$\epsilon_q = \epsilon_0$. To determine ϵ_0 we need to consider the definitions of the pre- and
over-critical regimes (10.140) and (10.141), respectively. The value of ϵ_0 is
such that for any $\epsilon_q \gg \epsilon_0$ estimate (10.141) is valid and for $\epsilon_q = \epsilon_0$ estimate
(10.140) is valid, respectively. If it appears that $\epsilon_0 \ll 1$ for the given problem
parameters, then it means that the viscosity μ depends on a large parameter
Q. Thus, if $\epsilon_0 \ll 1$, then for $\epsilon_q = O(\epsilon_0)$ we have pre-critical regimes and for
$\epsilon_q \gg \epsilon_0$ - over-critical regimes. For example, let $\mu = \exp(Qp^m)$, where m
is a positive constant. Then $\epsilon_0 = Q^{-2/m} \ll 1$ for $\omega = Q^{-1} \ll 1$ (see also
Kudish [23] - [27], [29]). Obviously, if $\epsilon_0 \gg 1$ or $\epsilon_0 = O(1)$ for $\omega \ll 1$ then the
over-critical regimes can not be realized and pre-critical lubrication regimes
occur.

It can be shown that the conditions that indicate the occurrence of over-
critical regimes are identical to those that are considered by Ertel [13], Grubin [14], Crook [22], and Greenwood [30]. Some classifications of lubrication
regimes are given by Greenwood [30] and Johnson [31].

10.6.2   Structure of the Solution. Asymptotic Solutions in the External Region

Let us assume that $\mu = \mu(p, Q)$ and for $p = O(1)$ the viscosity is very high,
i.e., $\mu(p, Q) \gg 1$ for some small parameter $\omega \ll 1$. The specific ways the small
parameter can be introduced will be considered later. One of these ways is to
define it as $\omega = Q^{-1} \ll 1$.

Let us consider a heavily loaded contact for which almost everywhere in the contact region the elastic deformation effects prevail over the effects caused by the lubrication viscous flow. That results in the smallness of the first term of equation (10.79) compared to 1 everywhere in the contact except for the narrow inlet and exit zones (which are located next to the inlet $x = a$ and exit $x = c$ points).

As in the preceding section a contact will be called heavily loaded if far away from the inlet and exit points estimates (10.95) are valid. We will take the inlet coordinate in the form similar to (10.99)

$$a = a_0 + \alpha_1 \epsilon_q + \ldots, \quad \alpha_1, \, a_0 = O(1), \, \omega \ll 1, \quad (10.155)$$

where a_0 is unknown and is defined by equation (10.159), α_1 is a given non-positive constant, and ϵ_q is a given characteristic size of the inlet zone, $\epsilon_q \ll 1$ for $\omega \ll 1$ (ϵ_q may depend on the problem parameters Q and V differently). If $\epsilon_q = O(min(V^{2/5}, \epsilon_0))$ then this is a pre-critical regime, which can be studied by the method described in the preceding section. Further, we will assume that

$$\epsilon_q(\omega) \gg \epsilon_0(\omega), \, \omega \ll 1. \quad (10.156)$$

That means that the lubrication regime is over-critical and the solution structure is different from the one analyzed in the preceding section. Thus, in order to analyze the problem, we should develop another asymptotic approach different from the one presented in Section 10.6.

The exit coordinate c will be searched in the form

$$c = c_0 + \beta_1 \epsilon_q + \ldots, \quad \beta_1, \, c_0 = O(1), \, \omega \ll 1, \quad (10.157)$$

where constant c_0 is unknown and is defined by equation (10.160) while constant β_1 is unknown and should be determined from the solution of the problem. The points with $x = a_0$ and $x = c_0$ are close to -1 and 1, respectively, and they represent the left and right end-points of the Hertzian region where the external bounded solution $p_{ext}(x)$ of the problem is asymptotically valid.

According to the principle of matched asymptotic expansions (Cole [4]) for $\epsilon_q = O(\epsilon_0)$, $\omega \ll 1$, in the inlet and exit zones the estimate $p(x) = O(\epsilon_q^{1/2})$ can be obtained. Therefore, it is natural to assume that for $\epsilon_q \gg \epsilon_0$ in the inlet and exit zones the estimate

$$p(x) = O(\epsilon_q^{1/2}), \, x - a_0 = O(\epsilon_q) \text{ and } x - c_0 = O(\epsilon_q) \text{ for } \omega \ll 1 \quad (10.158)$$

also holds.

Also, let us determine the left and right boundaries of the external (Hertzian) region $x = a_0$ and $x = c_0$

$$a_0 = -1 + \alpha_0 \epsilon_0^{1/2} \epsilon_q^{1/2} + \ldots, \quad \alpha_0 = O(1), \, \omega \ll 1, \quad (10.159)$$

$$c_0 = 1 + \beta_0 \epsilon_0^{1/2} \epsilon_q^{1/2} + \ldots, \quad \beta_0 = O(1), \, \omega \ll 1, \quad (10.160)$$

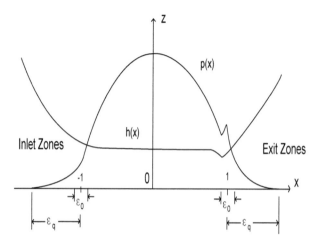

FIGURE 10.7
The schematic view of a heavily loaded contact in over-critical lubrication regime.

by satisfying the conditions of no singularity in the solution $p_{ext}(x)$ of the external problem for all $x \in [a_0, c_0]$, i.e., in particular, for $x = a_0$ and $x = c_0$. The values of constants α_0 and β_0 are unknown and must be determined in accordance with the mentioned conditions of no singularity at $x = a_0$ and $x = c_0$.

Using the pressure estimates (10.158) in all inlet and exit zones the solution of the problem in the inlet and exit ϵ_q-zones will be searched in the form

$$p(x) = \epsilon_0^{1/2} q(r) + o(\epsilon_0^{1/2}), \ q(r) = O(1),$$

$$r = \tfrac{x-a_0}{\epsilon_q} = O(1), \ \omega \ll 1,$$

(10.161)

$$p(x) = \epsilon_0^{1/2} g(s) + o(\epsilon_0^{1/2}), \ g(s) = O(1),$$

$$s = \tfrac{x-c_0}{\epsilon_q} = O(1), \ \omega \ll 1,$$

(10.162)

where r and s are the local independent variables in the inlet and exit ϵ_q-zones, which are adjacent to the inlet $x = a$ and exit $x = c$ points and have the characteristic size of ϵ_q, $q(r)$ and $g(s)$, are the unknown functions representing the leading terms of the pressure asymptotic expansions in the inlet and exit ϵ_q-zones.

Further, it can be shown that asymptotic expansions (10.161) and (10.162) cannot be matched with the asymptotic expansions of the external solution $p_{ext}(x)$. It is caused by ignoring the presence of small inlet and exit zones (ϵ_0-zones) with characteristic size of ϵ_0 located around points $x = a_0$ and $x = c_0$.

In the inlet and exit ϵ_0-zones, the effects of elasticity and lubrication flow are of the same order of magnitude. The solutions in the inlet ϵ_0- and ϵ_q-zones must match as well as the inner asymptotic of the external solution $p_{ext}(x)$ and the solution in the inlet ϵ_0-zone (see Fig. 10.7). Similarly, the solutions in the exit ϵ_0- and ϵ_q-zones must match as well as the inner asymptotic of the external solution $p_{ext}(x)$ and the solution in the exit ϵ_0-zone. Note that the absence of singularities in $p_{ext}(x)$ at $x = a_0$ and $x = c_0$ leads to the estimate $p_{ext}(x) = O(\epsilon_0^{1/2})$ for $x - a_0 = O(\epsilon_0)$ and $x - c_0 = O(\epsilon_0)$, $w \ll 1$. That allows for matching $p_{ext}(x)$ with the solutions in the inlet and exit ϵ_0-zones.

Based on the described behavior of the solution $p_{ext}(x)$ in the inlet and exit ϵ_0-zones, the solution of the problem can be searched in the form

$$p(x) = \epsilon_0^{1/2} q_0(r_0) + o(\epsilon_0^{1/2}), \quad q_0(r_0) = O(1),$$

$$r_0 = \tfrac{x-a_0}{\epsilon_0} = O(1), \quad w \ll 1,$$

$$(10.163)$$

$$p(x) = \epsilon_0^{1/2} g_0(s_0) + o(\epsilon_0^{1/2}), \quad g_0(s_0) = O(1),$$

$$s_0 = \tfrac{x-c_0}{\epsilon_0} = O(1), \quad w \ll 1,$$

$$(10.164)$$

where $q_0(r_0)$ and $g_0(s_0)$ are unknown functions, which should be determined, and r_0 and s_0 are the local point coordinates in the inlet and exit ϵ_0-zones, respectively.

For further analysis it is more convenient to re-scale the problem equations by using as the characteristic dimensional film thickness the central film thickness $h_0 = h(x = 0.5(a_0 + c_0))$ instead of the exit film thickness h_e. That will cause only one of the dimensionless problem equations to change as follows

$$H_0(h - 1) = x^2 - (\tfrac{a_0+c_0}{2})^2 + \tfrac{2}{\pi} \int\limits_a^c p(t) \ln \left| \tfrac{\frac{a_0+c_0}{2}-t}{x-t} \right| \, dt, \tag{10.165}$$

where the dimensionless central film thickness H_0 is given by the formula

$$H_0 = \tfrac{2R' h_0}{a_H^2}. \tag{10.166}$$

Based on asymptotic representations (10.157)-(10.164) for some $\eta(w)$ such that $\eta/\epsilon_0 \gg 1$ and $\eta/\epsilon_q \ll 1$ for $w \ll 1$, we have

$$\int\limits_a^c p(t) \ln \left| \tfrac{\frac{a_0+c_0}{2}-t}{x-t} \right| \, dt$$

$$= \left\{ \int\limits_a^{a_0-\eta} + \int\limits_{a_0-\eta}^{a_0+\eta} + \int\limits_{c_0-\eta}^{c_0+\eta} + \int\limits_{c_0+\eta}^c \right\} p(t) \ln \left| \tfrac{\frac{a_0+c_0}{2}-t}{x-t} \right| \, dt$$

$$= \epsilon_0^{1/2} \epsilon_q \int\limits_{\alpha_1-\alpha_0(\epsilon_0/\epsilon_q)^{1/2}-\alpha_2\epsilon_0/\epsilon_q+...}^{-\eta/\epsilon_q} q(r) \ln \left| \tfrac{\frac{c_0-a_0}{2}-\epsilon_q r}{x-a_0-\epsilon_q r} \right| \, dr$$

$$+\epsilon_0^{3/2} \int\limits_{-\eta/\epsilon_0}^{\eta/\epsilon_0} q_0(r_0) \ln \left| \frac{\frac{c_0-a_0}{2}-\epsilon_0 r_0}{x-a_0-\epsilon_0 r_0} \right| \, dr_0$$

$$+ \int\limits_{a_0+\eta}^{c_0-\eta} p_{ext}(t) \ln \left| \frac{\frac{a_0+c_0}{2}-t}{x-t} \right| \, dt$$

$$+\epsilon_0^{3/2} \int\limits_{-\eta/\epsilon_0}^{\eta/\epsilon_0} g_0(s_0) \ln \left| \frac{\frac{a_0-c_0}{2}-\epsilon_0 s_0}{x-c_0-\epsilon_0 s_0} \right| \, ds_0$$

$$+\epsilon_0^{1/2}\epsilon_q \int\limits_{\eta/\epsilon_q}^{\beta_1+\dots} g(s) \ln \left| \frac{\frac{a_0-c_0}{2}-\epsilon_q s}{x-c_0-\epsilon_q s} \right| \, ds + \dots$$

$$= \epsilon_0^{1/2}\epsilon_q \int\limits_{\alpha_1-\alpha_0(\epsilon_0/\epsilon_q)^{1/2}-\alpha_2\epsilon_0/\epsilon_q+\dots}^{0} q(r) \ln \left| \frac{\frac{c_0-a_0}{2}-\epsilon_q r}{x-a_0-\epsilon_q r} \right| \, dr$$

$$+\epsilon_0^{3/2} \int\limits_{-\eta/\epsilon_0}^{\eta/\epsilon_0} [q_0(r_0) - q(0)\theta(-r_0) - \epsilon_0^{-1/2} p_{ext}(a_0 + \epsilon_0 r_0)\theta(r_0)]$$

$$\times \ln \left| \frac{\frac{c_0-a_0}{2}-\epsilon_0 r_0}{x-a_0-\epsilon_0 r_0} \right| \, dr_0 + \int\limits_{a_0}^{c_0} p_{ext}(t) \ln \left| \frac{\frac{a_0+c_0}{2}-t}{x-t} \right| \, dt \qquad (10.167)$$

$$+\epsilon_0^{3/2} \int\limits_{-\eta/\epsilon_0}^{\eta/\epsilon_0} [g_0(s_0) - g(0)\theta(s_0) - \epsilon_0^{-1/2} p_{ext}(c_0 + \epsilon_0 s_0)\theta(-s_0)]$$

$$\times \ln \left| \frac{\frac{a_0-c_0}{2}-\epsilon_0 s_0}{x-c_0-\epsilon_0 s_0} \right| \, ds_0$$

$$+\epsilon_0^{1/2}\epsilon_q \int\limits_{0}^{\beta_1+\dots} g(s) \ln \left| \frac{\frac{a_0-c_0}{2}-\epsilon_q s}{x-c_0-\epsilon_q s} \right| \, ds + \dots,$$

where $\theta(x)$ is a step-function such that $\theta(x) = 0$ for $x \leq 0$ and $\theta(x) = 1$ for $x > 0$. Here we made a natural assumption (which is substantiated in Subsection 10.6.4) that in the inlet and exit ϵ_0-zones $q(r) = q(0) + O(\epsilon_0)$ and $g(s) = g(0) + O(\epsilon_0)$, respectively.

Now, let us obtain the asymptotic of the bounded solution of the problem in the external region $[a_0, c_0]$. However, first we need to find a singular solution of the problem. Suppose that in the latter case the boundaries of the external contact region $x = a_0$ and $x = c_0$ are fixed. Let us assume that in the external region the estimate

$$M(p, h) \ll \epsilon_0^{3/2} \; for \; x - a_0 \gg \epsilon_0, \; c_0 - x \gg \epsilon_0, \; \omega \ll 1, \qquad (10.168)$$

holds (see a similar estimate in (10.95), Kudish [29]), where $M(p, h) = \frac{H_0^3}{V} \frac{h^3}{\mu} \frac{dp}{dx}$. Estimating the integrals in (10.81) and (10.82), and using (10.99),

(10.156), (10.157), (10.159)-(10.168) for $\omega \ll 1$ in the external region we obtain the equations

$$x^2 - (\tfrac{a_0+c_0}{2})^2 + \tfrac{2}{\pi} \int\limits_{a_0}^{c_0} p_{ext}(t) \ln \left| \frac{\frac{a_0+c_0}{2}-t}{x-t} \right| dt$$

$$+\epsilon_0^{1/2}\epsilon_q \tfrac{2}{\pi} \int\limits_{\alpha_1 - \alpha_0(\epsilon_0/\epsilon_q)^{1/2}+\ldots}^{0} q(r) \ln \left| \frac{\frac{c_0-a_0}{2}-\epsilon_q r}{x-a_0-\epsilon_q r} \right| dr$$

$$+\epsilon_0^{1/2}\epsilon_q \tfrac{2}{\pi} \int\limits_{0}^{\beta_1+\ldots} g(s) \ln \left| \frac{\frac{a_0-c_0}{2}-\epsilon_q s}{x-c_0-\epsilon_q s} \right| ds + \ldots = 0,$$

(10.169)

$$\int\limits_{a_0}^{c_0} p_{ext}(t)dt$$

$$+\epsilon_0^{1/2}\epsilon_q \left\{ \int\limits_{\alpha_1-\alpha_0(\epsilon_0/\epsilon_q)^{1/2}+\ldots}^{0} q(r)dr + \int\limits_{0}^{\beta_1+\ldots} g(s)ds \right\} + \ldots = \tfrac{\pi}{2}.$$

It can be shown that equations (10.169) approximate the original problem (10.79)-(10.82) in the external region with the accuracy of $O(\epsilon_0^{3/2})$.

The solution of equations (10.169) will be searched by regular perturbation methods in the form

$$p_{ext}(x) = p_0(x) + \epsilon_0^{1/2}\epsilon_q p_1(x) + o(\epsilon_0^{1/2}\epsilon_q),$$

(10.170)

$$x - a_0 \gg \epsilon_0, \quad c_0 - x \gg \epsilon_0.$$

Substituting (10.170) in (10.169) and equating coefficients of the same powers of ϵ_0 and ϵ_q (which are considered to be independent, $\epsilon_q \gg \epsilon_0$) for $p_0(x)$ and $p_1(x)$, we obtain the following sets of equations:

$$x^2 - (\tfrac{a_0+c_0}{2})^2 + \tfrac{2}{\pi} \int\limits_{a_0}^{c_0} p_0(t) \ln \left| \frac{\frac{a_0+c_0}{2}-t}{x-t} \right| dt = 0, \quad \int\limits_{a_0}^{c_0} p_0(t)dt = \tfrac{\pi}{2}, \quad (10.171)$$

$$\int\limits_{a_0}^{c_0} p_1(t) \ln \left| \frac{\frac{a_0+c_0}{2}-t}{x-t} \right| dt$$

$$= - \int\limits_{\alpha_1-\alpha_0(\epsilon_0/\epsilon_q)^{1/2}+\ldots}^{0} q(r) \ln \left| \frac{\frac{c_0-a_0}{2}-\epsilon_q r}{x-a_0-\epsilon_q r} \right| dr$$

(10.172)

$$- \int\limits_{0}^{\beta_1+\ldots} g(s) \ln \left| \frac{\frac{a_0-c_0}{2}-\epsilon_q s}{x-c_0-\epsilon_q s} \right| ds,$$

$$\int\limits_{a_0}^{c_0} p_1(t)dt = - \int\limits_{\alpha_1-\alpha_0(\epsilon_0/\epsilon_q)^{1/2}+\ldots}^{0} q(r)dr - \int\limits_{0}^{\beta_1+\ldots} g(s)ds.$$

The singular solutions of systems (10.171) and (10.172) have the form [28] (also see Subsection 2.2.2)

$$p_0(x) = \sqrt{(x-a_0)(c_0-x)} + \frac{1+2a_0c_0+(c_0-a_0)^2/4-(c_0+a_0)x}{2\sqrt{(x-a_0)(c_0-x)}}, \qquad (10.173)$$

$$p_1(x) = \frac{1}{\pi\sqrt{1-y^2}} \left\{ -\frac{2}{c_0-a_0} \left[\int\limits_{\alpha_1-\alpha_0(\epsilon_0/\epsilon_q)^{1/2}+...}^{0} q(r)dr \right. \right.$$

$$+ \int\limits_{0}^{\beta_1+...} g(s)ds \Bigg] - \frac{1}{\pi} \int\limits_{-1}^{1} \frac{\sqrt{1-t^2}\,dt}{t-y} \int\limits_{\alpha_1-\alpha_0(\epsilon_0/\epsilon_q)^{1/2}+...}^{0} \frac{q(r)dr}{\frac{c_0-a_0}{2}(1+t)-\epsilon_q r} \qquad (10.174)$$

$$- \frac{1}{\pi} \int\limits_{-1}^{1} \frac{\sqrt{1-t^2}\,dt}{t-y} \int\limits_{0}^{\beta_1+...} \frac{g(s)ds}{\frac{c_0-a_0}{2}(t-1)-\epsilon_q s} \Bigg\}, \quad y = \frac{2x-a_0-c_0}{c_0-a_0}.$$

In the last two integrals in (10.174), the integrands have singularities at points $t = y$ and at two discrete points $r = 0$, $t = -1$ ($t \geq -1$, $r \leq 0$) and $s = 0$, $t = 1$ ($t \leq 1$, $s \geq 0$), respectively. We will show that it is possible to change the order of integration in the repeated integrals in (10.174) as in integrals of regular functions

$$\int\limits_{-1}^{1} \frac{\sqrt{1-t^2}\,dt}{t-y} \int\limits_{\alpha_1-\alpha_0(\epsilon_0/\epsilon_q)^{1/2}+...}^{0} \frac{q(r)dr}{\frac{c_0-a_0}{2}(1+t)-\epsilon_q r}$$

$$= \int\limits_{\alpha_1-\alpha_0(\epsilon_0/\epsilon_q)^{1/2}+...}^{0} q(r)dr \int\limits_{-1}^{1} \frac{\sqrt{1-t^2}\,dt}{(t-y)[\frac{c_0-a_0}{2}(1+t)-\epsilon_q r]},$$

$$\qquad (10.175)$$

$$\int\limits_{-1}^{1} \frac{\sqrt{1-t^2}\,dt}{t-y} \int\limits_{0}^{\beta_1+...} \frac{g(s)ds}{\frac{c_0-a_0}{2}(t-1)-\epsilon_q s}$$

$$= \int\limits_{0}^{\beta_1+...} g(s)ds \int\limits_{-1}^{1} \frac{\sqrt{1-t^2}\,dt}{(t-y)[\frac{c_0-a_0}{2}(t-1)-\epsilon_q s]}.$$

Let us first consider the integral involving $q(r)$. By introducing a small enough value $\epsilon < 0$ this integral can be transformed into a sum of three integrals

$$\int_{-1}^{1} \frac{\sqrt{1-t^2}\,dt}{t-y} \int_{\alpha_1-\alpha_0(\epsilon_0/\epsilon_q)^{1/2}+\cdots}^{0} \frac{q(r)\,dr}{\frac{c_0-a_0}{2}(1+t)-\epsilon_q r}$$

$$= \int_{\alpha_1-\alpha_0(\epsilon_0/\epsilon_q)^{1/2}+\cdots}^{\epsilon} q(r)\,dr \int_{-1}^{1} \frac{\sqrt{1-t^2}\,dt}{(t-y)[\frac{c_0-a_0}{2}(1+t)-\epsilon_q r]}$$

$$-\frac{q(0)}{\epsilon_q} \int_{-1}^{1} \frac{\sqrt{1-t^2}}{t-y} \ln\left| \frac{t+1}{t+1-\frac{2\epsilon\epsilon_q}{c_0-a_0}} \right| dt$$

$$+\int_{\epsilon}^{0} [q(r)-q(0)]\,dr \int_{-1}^{1} \frac{\sqrt{1-t^2}\,dt}{(t-y)[\frac{c_0-a_0}{2}(1+t)-\epsilon_q r]},$$

(10.176)

where in the first and the third integrals the order of integration is changed
(Gakhov [32]). Similarly, we obtain

$$\int_{-1}^{1} \frac{\sqrt{1-t^2}\,dt}{t-y} \int_{0}^{\beta_1+\cdots} \frac{g(s)\,ds}{\frac{c_0-a_0}{2}(t-1)-\epsilon_q s}$$

$$= \int_{-\epsilon}^{\beta_1+\cdots} g(s)\,ds \int_{-1}^{1} \frac{\sqrt{1-t^2}\,dt}{(t-y)[\frac{c_0-a_0}{2}(t-1)-\epsilon_q s]}$$

$$-\frac{g(0)}{\epsilon_q} \int_{-1}^{1} \frac{\sqrt{1-t^2}}{t-y} \ln\left| \frac{t-1+\frac{2\epsilon\epsilon_q}{c_0-a_0}}{t-1} \right| dt$$

(10.177)

$$+\int_{0}^{-\epsilon} [g(s)-g(0)]\,ds \int_{-1}^{1} \frac{\sqrt{1-t^2}\,dt}{(t-y)[\frac{c_0-a_0}{2}(t-1)-\epsilon_q s]}.$$

Using the expressions for the integrals

$$\int_{-1}^{1} \frac{\sqrt{1-t^2}\,dt}{(t-y)[\frac{c_0-a_0}{2}(t+1)-\epsilon_q r]} = -\frac{2\pi}{c_0-a_0}\left\{1-\frac{\sqrt{t_0^2-1}}{y-t_0}\right\},$$

$$t_0 = -1+\frac{2\epsilon_q r}{c_0-a_0} \le -1,$$

(10.178)

$$\int_{-1}^{1} \frac{\sqrt{1-t^2}\,dt}{(t-y)[\frac{c_0-a_0}{2}(t-1)-\epsilon_q s]} = -\frac{2\pi}{c_0-a_0}\left\{1+\frac{\sqrt{t_1^2-1}}{y-t_1}\right\},$$

$$t_1 = 1+\frac{2\epsilon_q s}{c_0-a_0} \ge 1,$$

and calculating the inner integrals in (10.176) and (10.177) by using substitution $t = (1-\tau^2)/(1+\tau^2)$ and the integral Bierens De Haan [33] (p. 194, No. 15)

$$\int\limits_0^\infty \frac{\ln(p^2+\tau^2)d\tau}{q^2-\tau^2} = -\frac{\pi}{q}\arctan\frac{q}{p}$$

we obtain [32, 34])

$$\int\limits_{-1}^1 \frac{\sqrt{1-t^2}}{t-y}\ln\left|\frac{t+1}{t-\tau_0}\right|dt = 2\pi\sqrt{1-y^2}\{\tfrac{1}{2}\arccos y$$

$$+ \arctan\sqrt{\frac{1-y}{1+y}\frac{\tau_0+1}{\tau_0-1}} - \arctan\sqrt{\frac{1-y}{1+y}}\} - \pi(1+\tau_0+\sqrt{\tau_0^2-1}) \qquad (10.179)$$

$$+\pi y\ln|\tau_0-\sqrt{\tau_0^2-1}|, \quad \tau_0=-1+\frac{2\epsilon\epsilon_q}{c_0-a_0}, \quad |y|<1, \ \epsilon<0,$$

$$\int\limits_{-1}^1 \frac{\sqrt{1-t^2}}{t-y}\ln\left|\frac{t-\tau_1}{t-1}\right|dt = 2\pi\sqrt{1-y^2}\{\arctan\sqrt{\frac{1-y}{1+y}}$$

$$- \arctan\sqrt{\frac{1-y}{1+y}\frac{\tau_1+1}{\tau_1-1}} + \frac{\pi}{2} - \tfrac{1}{2}\arccos y\} - \pi(1-\tau_1+\sqrt{\tau_1^2-1}) \qquad (10.180)$$

$$-\pi y\ln|\tau_1+\sqrt{\tau_1^2-1}|, \quad \tau_1=1-\frac{2\epsilon\epsilon_q}{c_0-a_0}, \quad |y|<1, \ \epsilon<0.$$

Using formulas (10.178), (10.179), and (10.180) in expression (10.174) for function $p_1(x)$ and taking the limit as $\varepsilon \to -0$ for any fixed value of y ($|y|<1$), we arrive at

$$p_1(x) = \frac{2}{\pi(c_0-a_0)\sqrt{1-y^2}}\left\{-\int\limits_{\alpha_1-\alpha_0(\epsilon_0/\epsilon_q)^{1/2}+\dots}^{0} q(r)\frac{\sqrt{t_0^2(r)-1}}{y-t_0(r)}dr\right.$$

$$+ \int\limits_0^{\beta_1+\dots} g(s)\frac{\sqrt{t_1^2(s)-1}}{y-t_1(s)}ds\right\} + \frac{2}{\pi\epsilon_q}[q(0)-g(0)]\{\tfrac{1}{2}\arccos y \qquad (10.181)$$

$$- \arctan\sqrt{\frac{1-y}{1+y}}\right\}.$$

However, taking into account the identity

$$\tfrac{1}{2}\arccos y - \arctan\sqrt{\frac{1-y}{1+y}} = 0$$

for any y ($|y|<1$), we find the final expression for $p_1(x)$ in the form

$$p_1(x) = \frac{2}{\pi(c_0-a_0)\sqrt{1-y^2}}\left\{-\int\limits_{\alpha_1-\alpha_0(\epsilon_0/\epsilon_q)^{1/2}+\dots}^{0} q(r)\frac{\sqrt{t_0^2(r)-1}}{y-t_0(r)}dr\right.$$

$$+ \int\limits_0^{\beta_1+\dots} g(s)\frac{\sqrt{t_1^2(s)-1}}{y-t_1(s)}ds\right\}, \qquad (10.182)$$

$$t_0=-1+\frac{2\epsilon_q r}{c_0-a_0}\le-1, \quad t_1=1+\frac{2\epsilon_q s}{c_0-a_0}\ge 1.$$

It can be shown by the direct substitution that expression (10.182) for $p_1(x)$ satisfies equations (10.172). To do that and some other calculations, we will use the following integrals:

$$\int_{-1}^{1} \frac{\ln|y-t|dt}{\sqrt{1-t^2}(t-x)} = -\frac{\pi \, sign(x)}{\sqrt{x^2-1}} \ln \frac{x-y}{x+sign(x)\sqrt{x^2-1}}, \quad |x|>1, \ |y|<1,$$

$$\int_{-1}^{1} \frac{\ln|y-t|dt}{\sqrt{1-t^2}(t-x)} = -\frac{\pi \, sign(x)}{\sqrt{x^2-1}} \ln \frac{xy-1+\sqrt{(y^2-1)(x^2-1)}}{|x|+\sqrt{x^2-1}}, \qquad (10.183)$$

$$|x|>1, \ |y|>1,$$

the expressions for which are obtained by using substitution $t=(1-\tau^2)/(1+\tau^2)$ and the integral Bierens De Haan [33] (p. 194, No. 15) $\int_0^\infty \frac{\ln(p^2+\tau^2)d\tau}{q^2-\tau^2}$ presented earlier as well as the other two integrals Bierens De Haan [33] (p. 193, No. 13 and p. 194, No. 16)

$$\int_0^\infty \frac{\ln(p^2+\tau^2)d\tau}{q^2+\tau^2} = \frac{\pi}{q} \ln(p+q), \quad \int_0^\infty \frac{\ln(p^2-\tau^2)^2 d\tau}{q^2+\tau^2} = \frac{\pi}{q} \ln(p^2+q^2).$$

Furthermore, it is interesting to note that the outlined procedure of finding $p_1(x)$ is equivalent but different from the one used by Kuznetsov in [35] for solution of a dry (without lubrication) contact problem with a known additional load applied outside of a contact.

Let us obtain a bounded solution $p_{ext}(x)$ of system (10.169) with the precision of $o(\epsilon_0^{1/2}\epsilon_q)$ using formula (10.170) and solutions (10.173) and (10.182). It can be done by choosing the proper values for constants α_0 and β_0 involved in expressions (10.159) and (10.160) for a_0 and c_0, respectively. It is obvious that for $p_{ext}(x)$ to be bounded for any y ($|y|<1$) it is necessary to require that the sum of the second terms in $p_0(x)$ and $\epsilon_0^{1/2}\epsilon_q p_1(x)$ be a regular function at $y=\pm1$ with the precision of $o(\epsilon_0^{1/2}\epsilon_q)$. That leads to the requirement that in $p_{ext}(x)=p_0(x)+\epsilon_0^{1/2}\epsilon_q p_1(x)+\ldots$ the expression representing the coefficient of $1/\sqrt{1-y^2}$ is equal to zero at $y=\pm1$ with the precision of $o(\epsilon_0^{1/2}\epsilon_q)$. The latter leads to

$$1-\left(\frac{c_0-a_0}{2}\right)^2 + \frac{c_0^2-a_0^2}{2} - \epsilon_0^{1/2}\epsilon_q \frac{2}{\pi}\left\{-\int_{\alpha_1+\ldots}^{0} q(r)\frac{\sqrt{t_0^2(r)-1}}{1+t_0(r)}dr\right.$$

$$\left.+\int_0^{\beta_1+\ldots} g(s)\frac{\sqrt{t_1^2(s)-1}}{1+t_1(s)}ds\right\} = 0, \ t_0 = -1+\frac{2\epsilon_q r}{c_0-a_0} \leq -1,$$

$$1-\left(\frac{c_0-a_0}{2}\right)^2 - \frac{c_0^2-a_0^2}{2} - \epsilon_0^{1/2}\epsilon_q \frac{2}{\pi}\left\{\int_{\alpha_1+\ldots}^{0} q(r)\frac{\sqrt{t_0^2(r)-1}}{1-t_0(r)}dr\right.$$

$$-\int\limits_{0}^{\beta_1+\dots} g(s)\frac{\sqrt{t_1^2(s)-1}}{1-t_1(s)}ds\Bigg\} = 0, \quad t_1 = 1 + \frac{2\epsilon_q s}{c_0-a_0} \geq 1.$$

Using expressions (10.159) and (10.160) for a_0 and c_0, respectively, and estimating the integrals involved, we obtain the asymptotic formulas for α_0 and β_0 in the form

$$\alpha_0 = \frac{2}{\pi}\int\limits_{\alpha_1}^{0}\frac{q(r)dr}{\sqrt{-2r}} + \dots, \quad \beta_0 = -\frac{2}{\pi}\int\limits_{0}^{\beta_1}\frac{g(s)ds}{\sqrt{2s}} + \dots. \tag{10.184}$$

Because $q(r) \geq 0$, $\alpha_1 < 0$, and $g(s) \geq 0$, $\beta_1 > 0$, we get $\alpha_0 \geq 0$ and $\beta_0 \leq 0$. Therefore, formulas (10.159), (10.160), and (10.184) for the boundary points a_0 and c_0 indicate that the external region $[a_0, c_0]$ is slightly narrower then the classic Hertzian region $[-1, 1]$ and belongs to the interior of the latter one.

Formula (10.182) for $p_1(x)$ can be rewritten further as follows

$$p_1(x) = \frac{2}{\pi(c_0-a_0)\sqrt{1-y^2}}\Bigg\{-\int\limits_{\alpha_1-\alpha_0(\epsilon_0/\epsilon_q)^{1/2}+\dots}^{0}\frac{q(r)t_0(r)dr}{\sqrt{t_0^2(r)-1}}$$

$$+\int\limits_{0}^{\beta_1+\dots}\frac{g(s)t_1(s)ds}{\sqrt{t_1^2(s)-1}} + y\Bigg[-\int\limits_{\alpha_1-\alpha_0(\epsilon_0/\epsilon_q)^{1/2}+\dots}^{0}\frac{q(r)dr}{\sqrt{t_0^2(r)-1}}$$

$$+\int\limits_{0}^{\beta_1+\dots}\frac{g(s)ds}{\sqrt{t_1^2(s)-1}}\Bigg] \tag{10.185}$$

$$+(y^2-1)\Bigg[\int\limits_{\alpha_1-\alpha_0(\epsilon_0/\epsilon_q)^{1/2}-\alpha_2\epsilon_0/\epsilon_q+\dots}^{0}\frac{q(r)dr}{\sqrt{t_0^2(r)-1}(y-t_0(r))}$$

$$-\int\limits_{0}^{\beta_1+\dots}\frac{g(s)ds}{\sqrt{t_1^2(s)-1}(y-t_1(s))}\Bigg]\Bigg\}.$$

Making use of (10.185) provides us with the asymptotic of $p_1(x)$ in the form

$$p_1(x) = \epsilon_q^{-1/2}\frac{(\alpha_0+\beta_0)y-\alpha_0+\beta_0}{2\sqrt{1-y^2}} + \frac{\sqrt{1-y^2}}{\pi}\Bigg\{\int\limits_{\alpha_1+\dots}^{0}\frac{q(r)dr}{\sqrt{t_0^2(r)-1}(y-t_0(r))}$$

$$-\int\limits_{0}^{\beta_1+\dots}\frac{g(s)ds}{\sqrt{t_1^2(s)-1}(y-t_1(s))}\Bigg\} + \dots, \tag{10.186}$$

$$t_0 = -1 + \frac{2\epsilon_q r}{c_0-a_0}, \quad t_1 = 1 + \frac{2\epsilon_q s}{c_0-a_0}.$$

We need to determine a more detailed asymptotic behavior of $p_1(x)$ in the vicinity of points $x = a_0$, $x = c_0$, i.e., in the vicinity of $y = \pm 1$. To achieve that we need to analyze the behavior of the integrals in (10.186). Considering the first integral in (10.186), we have

$$\int\limits_{\alpha_1+\ldots}^{0} \frac{q(r)dr}{\sqrt{t_0^2(r)-1}(y-t_0(r))} = \int\limits_{\alpha_1+\ldots}^{0} \frac{[q(r)-q(0)]dr}{\sqrt{t_0^2(r)-1}(y-t_0(r))}$$

$$+q(0)\left\{\int\limits_{-\infty}^{0} \frac{dr}{\sqrt{t_0^2(r)-1}(y-t_0(r))} - \int\limits_{-\infty}^{\alpha_1+\ldots} \frac{dr}{\sqrt{t_0^2(r)-1}(y-t_0(r))}\right\}.$$

(10.187)

In (10.187), the first integral in the right-hand side of the expression is a bounded regular function for any $\mid y \mid < 1$ because $q(r)$ is a continuously differentiable function, and, therefore, the integrand is a regular function for $r \in [\alpha_1 - \alpha_0(\epsilon_0/\epsilon_q)^{1/2}+\ldots, 0]$. The third term in the right-hand side of (10.187) is also a bounded regular function for any $\mid y \mid < 1$ because $q(0)$ is finite and the integrand is a regular function for any $\mid y \mid < 1$ and $r \leq \alpha_1 - \alpha_0(\epsilon_0/\epsilon_q)^{1/2}+\ldots$ while as $r \to -\infty$ the integrand vanishes as $(-r)^{-3/2}$. In addition we have

$$\int\limits_{-\infty}^{0} \frac{dr}{\sqrt{t_0^2(r)-1}(y-t_0(r))} = \frac{c_0-a_0}{2}\frac{\epsilon_q^{-1}}{\sqrt{1-y^2}}\left\{\frac{\pi}{2} - \arctan\sqrt{\frac{1+y}{1-y}}\right\}.$$

Using the expression for $t_0(r)$, as a result of the above considerations we obtain that

$$\int\limits_{\alpha_1+\ldots}^{0} \frac{q(r)dr}{\sqrt{t_0^2(r)-1}(y-t_0(r))} = F_r(y)$$

(10.188)

$$+\frac{2\epsilon_q^{-1}q(0)}{\sqrt{1-y^2}}\left\{\frac{\pi}{2} - \arctan\sqrt{\frac{1+y}{1-y}}\right\} + \ldots,$$

where $F_r(y)$ is a bounded regular function for all $\mid y \mid < 1$. A similar analysis can be done for the second integral in (10.186). Finally, the asymptotic of $p_1(x)$ in the vicinity of $y = \pm 1$ can be represented in the form

$$p_1(x) = \epsilon_q^{-1/2}\frac{(\alpha_0+\beta_0)y-\alpha_0+\beta_0}{2\sqrt{1-y^2}} + \frac{q(0)}{\epsilon_q}\left\{1 - \frac{2}{\pi}\arctan\sqrt{\frac{1+y}{1-y}}\right\}$$

(10.189)

$$+\frac{g(0)}{\epsilon_q}\left\{1 - \frac{2}{\pi}\arctan\sqrt{\frac{1-y}{1+y}}\right\} + \frac{\sqrt{1-y^2}}{\pi}G(y) + o(\epsilon_q^{-1}),$$

where $G(y)$ is a bounded regular function for all $\mid y \mid < 1$.

It is important to emphasize that the contributions of all other terms in the latter expression for $p_1(x)$ are of the order of $o(\epsilon_q^{-1})$, which provide a contribution to $p_{ext}(x)$ in the inlet and exit ϵ_0-zones of the order of $o(\epsilon_0^{1/2})$ (see formula (10.170) for $p_{ext}(x)$).

Using the expressions for a_0 and c_0, we find the asymptotic of $p_0(x)$ as follows

$$p_0(x) = \sqrt{1-y^2} - \epsilon_0^{1/2}\epsilon_q^{1/2}\frac{(\alpha_0+\beta_0)y-\alpha_0+\beta_0}{2\sqrt{1-y^2}} + o(\epsilon_0^{1/2}\epsilon_q^{1/2}).$$

(10.190)

Now, based on functions $p_0(x)$ and $p_1(x)$ we can determine the asymptotic representations of $p_{ext}(x)$ in the inlet and exit ϵ_0-zones as follows

$$p_{ext}(x) = \epsilon_0^{1/2}\{q(0) + \sqrt{2r_0}\theta(r_0)\} + o(\epsilon_0^{1/2}),$$

$$p_{ext}(x) = \epsilon_0^{1/2}\{g(0) + \sqrt{-2s_0}\theta(-s_0)\} + o(\epsilon_0^{1/2}).$$

(10.191)

By setting values of α_0 and β_0 to be equal to the expressions in (10.184), we eliminate singularities at $x = a_0$ and $x = c_0$ in the terms of $p_{ext}(x)$ of the orders of magnitude proportional to $\epsilon_q^{1/2}$ and $\epsilon_0^{1/2}$.

Therefore, for $r_0 = O(1)$ and $s_0 = O(1)$, $w \ll 1$, the first two leading terms of the asymptotic expansions (10.170) for $p_{ext}(x)$ in the inlet and exit ϵ_0-zones are of order of $\epsilon_0^{1/2}$, i.e., of the same order of magnitude as the problem solution in the inlet and exit ϵ_0-zones. As the result of that the latter asymptotic expansions of the external solution $p_{ext}(x)$ potentially allow for matching $p_{ext}(x)$ with the solutions in the inlet and exit ϵ_0-zones (see estimate (10.158)).

10.6.3   Auxiliary Gap Function $h_H(x)$. Behavior of $h(x)$ in the Inlet and Exit ϵ_q-Zones

Now, let us define an auxiliary gap function $h_H(x)$, which will be used to evaluate the asymptotic behavior of gap $h(x)$ in the inlet and exit ϵ_q-zones by the following equation

$$H_0[h_H(x) - 1] = x^2 - (\tfrac{c_0+a_0}{2})^2 + \tfrac{2}{\pi}\int\limits_{a_0}^{c_0} p_{ext}(t) \ln \mid \tfrac{\frac{c_0+a_0}{2}-t}{x-t} \mid dt$$

$$+\epsilon_0^{1/2}\epsilon_q \tfrac{2}{\pi} \int\limits_{\alpha_1 - \alpha_0(\epsilon_0/\epsilon_q)^{1/2}+\dots}^{0} q(r) \ln \mid \tfrac{\frac{c_0-a_0}{2}-\epsilon_q r}{x-a_0-\epsilon_q r} \mid dr \qquad (10.192)$$

$$+\epsilon_0^{1/2}\epsilon_q \tfrac{2}{\pi} \int\limits_{0}^{\beta_1+\dots} g(s) \ln \mid \tfrac{\frac{c_0-a_0}{2}+\epsilon_q s}{x-c_0-\epsilon_q s} \mid ds.$$

It follows from (10.170)-(10.172) that $h_H(x) = 1$, $x \in [a_0, c_0]$. Also, for further analysis we will need the asymptotic behavior of $h_H(x)$ in the inlet and exit ϵ_q- and ϵ_0-zones. To obtain this behavior we need to consider the corresponding integrals (see (10.192)) of $p_0(x)$ and $p_2(x)$. We will make use of the following integrals

$$\tfrac{2}{\pi}\int\limits_{-1}^{1} \sqrt{1-t^2} \ln \mid \tfrac{t}{y-t} \mid dt = -y^2 + \mid y \mid \sqrt{y^2-1} - \ln \mid\mid y \mid + \sqrt{y^2-1} \mid,$$

$$\tfrac{2}{\pi}\int\limits_{-1}^{1} \tfrac{1}{\sqrt{1-t^2}} \ln \mid \tfrac{t}{y-t} \mid dt = -\ln \mid\mid y \mid + \sqrt{y^2-1} \mid,$$

$$\frac{1}{\pi} \int\limits_{-1}^{1} \frac{t}{\sqrt{1-t^2}} \ln \mid \frac{t}{y-t} \mid dt = y - sign(y)\sqrt{y^2-1} \ for \ \mid y \mid \geq 1.$$

Based on the above integrals and (10.173), we get

$$x^2 - (\frac{c_0+a_0}{2})^2 + \frac{2}{\pi} \int\limits_{a_0}^{c_0} p_0(t) \ln \mid \frac{\frac{c_0+a_0}{2}-t}{x-t} \mid dt$$

$$= (\frac{c_0-a_0}{2})^2 \mid y \mid \sqrt{y^2-1} - \ln \mid\mid y \mid +\sqrt{y^2-1} \mid \qquad (10.193)$$

$$+ \frac{c_0^2-a_0^2}{2} sign(y)\sqrt{y^2-1} \ for \ \mid y \mid \geq 1.$$

Using the fact that the integrands in the integrals in expression (10.182) for $p_1(x)$ are regular functions and using the integrals in (10.183), we obtain

$$\int\limits_{a_0}^{c_0} p_1(t) \ln \mid \frac{\frac{c_0+a_0}{2}-t}{x-t} \mid dt =$$

$$\int\limits_{\alpha_1-\alpha_0(\epsilon_0/\epsilon_q)^{1/2}+...}^{0} q(r) \ln \mid \frac{[yt_0(r)-1]sign(y)-\sqrt{(y^2-1)(t_0^2(r)-1)}}{t_0(r)} \mid dr \qquad (10.194)$$

$$+ \int\limits_{0}^{\beta_1+...} g(s) \ln \mid \frac{[yt_1(s)-1]sign(y)+\sqrt{(y^2-1)(t_1^2(s)-1)}}{t_1(s)} \mid ds$$

$$for \ \mid y \mid \geq 1, \ t_0(r) = -1 + \frac{2\epsilon_q r}{c_0-a_0}, \ t_1(s) = 1 + \frac{2\epsilon_q s}{c_0-a_0}.$$

From (10.192) with the help of (10.157), (10.158), (10.161)-(10.168), (10.170), (10.173), and (10.182), we readily obtain the asymptotic expansions of $h_H(x)$ for $r = O(1)$ and $s = O(1)$, $\omega \ll 1$,

$$H_0[h_H(x) - 1] = -\epsilon_q^{3/2} \frac{4}{3} r\sqrt{-2r}\theta(-r) + o(\epsilon_q^{3/2}),$$

$$r = O(1), \ \omega \ll 1, \qquad (10.195)$$

$$H_0[h_H(x) - 1] = \epsilon_q^{3/2} \frac{4}{3} s\sqrt{2s}\theta(s) + o(\epsilon_q^{3/2}), \ s = O(1), \ \omega \ll 1. \qquad (10.196)$$

Now, let us get the asymptotic behavior of $h_H(x)$ in the inlet and exit ϵ_0-zones, i.e., for $r_0 = O(1)$ and $s_0 = O(1)$, $\omega \ll 1$. In the inlet ϵ_0-zone $y = -1 + 2\epsilon_0 r_0/(c_0 - a_0)$ and the two-term asymptotic expansion of the expression in the left-hand side of (10.193) has the form

$$x^2 - (\frac{c_0+a_0}{2})^2 + \frac{2}{\pi} \int\limits_{a_0}^{c_0} p_0(t) \ln \mid \frac{\frac{c_0+a_0}{2}-t}{x-t} \mid dt$$

$$\qquad (10.197)$$

$$= -2\alpha_0\epsilon_0\epsilon_q^{1/2}\sqrt{-2r_0} + \frac{2}{3}(-2r_0)^{3/2}\epsilon_0^{3/2} + o(\epsilon_0^{3/2})$$

while the asymptotic of the integral of $p_1(x)$ from (10.194) is given by

$$\int_{a_0}^{c_0} p_1(t) \ln \left| \frac{\frac{c_0+a_0}{2}-t}{x-t} \right| \, dt = 2a_0\epsilon_0\epsilon_q^{1/2}\sqrt{-2r_0} + o(\epsilon_0^{3/2}). \qquad (10.198)$$

Therefore, from expression (10.192) for $h_H(x)$ and the obtained asymptotic expressions (10.197) and (10.198) in the inlet ϵ_0-zone, we find that

$$H_0[h_H(x) - 1] = -\epsilon_0^{3/2}\frac{4}{3}r_0\sqrt{-2r_0}\theta(-r_0) + O(\epsilon_0^{3/2}),$$

$$r_0 = O(1), \quad \omega \ll 1. \qquad (10.199)$$

Similarly, in the exit ϵ_0-zone we have the asymptotic estimate

$$H_0[h_H(x) - 1] = \epsilon_0^{3/2}\frac{4}{3}s_0\sqrt{2s_0}\theta(s_0) + O(\epsilon_0^{3/2}),$$

$$s_0 = O(1), \quad \omega \ll 1, \qquad (10.200)$$

where the terms of the order of $O(\epsilon_0^{3/2})$ are produced by the last two integrals in expression (10.192) for $h_H(x)$. The fact that it is very difficult to obtain the exact analytical expressions for these terms prevents us from deriving the asymptotically valid equations in the inlet and exit ϵ_0-zones.

Let us consider the behavior of gap $h(x)$ in the inlet and exit ϵ_q-zone. By transforming equation (10.81) with the help of equation (10.192), we find

$$H_0[h(x) - 1] = H_0[h_H(x) - 1]$$

$$+ \frac{2}{\pi}\int_{a}^{c} p(t) \ln \left| \frac{\frac{c_0+a_0}{2}-t}{x-t} \right| \, dt - \frac{2}{\pi}\int_{a_0}^{c_0} p_{ext}(t) \ln \left| \frac{\frac{c_0+a_0}{2}-t}{x-t} \right| \, dt$$

$$(10.201)$$

$$- \epsilon_0^{1/2}\epsilon_q\frac{2}{\pi}\int_{\alpha_1-\alpha_0(\epsilon_0/\epsilon_q)^{1/2}+\dots}^{0} q(r) \ln \left| \frac{\frac{c_0-a_0}{2}-\epsilon_q r}{x-a_0-\epsilon_q r} \right| \, dr$$

$$- \epsilon_0^{1/2}\epsilon_q\frac{2}{\pi}\int_{0}^{\beta_1+\dots} g(s) \ln \left| \frac{\frac{c_0-a_0}{2}+\epsilon_q s}{x-c_0-\epsilon_q s} \right| \, ds.$$

Using expressions (10.99), (10.157), (10.161)-(10.168), (10.170), estimating the integrals over all zones of the contact, and taking into account estimates from (10.195) and (10.201), we obtain the asymptotic estimates of gap $h(x)$ in the inlet and exit ϵ_q-zones in the form

$$H_0[h(x) - 1] = -\epsilon_q^{3/2}\frac{4}{3}r\sqrt{-2r}\theta(-r) + o(\epsilon_q^{3/2}),$$

$$r = O(1), \quad \omega \ll 1, \qquad (10.202)$$

$$H_0[h(x) - 1] = \epsilon_q^{3/2}\frac{4}{3}s\sqrt{2s}\theta(s) + o(\epsilon_q^{3/2}), \quad s = O(1), \quad \omega \ll 1. \qquad (10.203)$$

As a matter of fact, asymptotic estimates (10.202) and (10.203) mean that gap $h(x)$ is a given function in the inlet and exit ϵ_q-zones if the film thickness H_0 is known. Essentially, it is determined by the external pressure $p_{ext}(x)$ in the Hertzian region. This pressure is close to the Hertzian one $\sqrt{1-x^2}\theta(1-x^2)$ in a dry contact of elastic cylinders.

In the inlet and exit ϵ_0-zones, we can obtain the estimate

$$H_0[h(x)-1] = O(\epsilon_0^{3/2}) \ for \ r_0 = O(1) \ and \ s_0 = O(1), \ \omega \ll 1. \quad (10.204)$$

10.6.4 Asymptotic Solutions for $q(r)$ and $g(s)$ in the Inlet and Exit ϵ_q-Zones

Based on estimates (10.202) and (10.203) one can conclude that in the inlet and exit ϵ_q-zones with a small error the lubricant flow is realized in the gap with almost given configuration. Using estimates (10.161) and (10.202), requiring that the two terms of equation (10.79) are of the same order of magnitude, and estimating these terms for $r = O(1)$, $\omega \ll 1$, we find the estimate for the central film thickness H_0 in the form

$$H_0 = A(V\epsilon_0^{-1/2}\epsilon_q^{5/2})^{1/3} + \ldots, \quad A = O(1), \quad \omega \ll 1, \quad (10.205)$$

where $A = A(\alpha_1) \geq 0$ is a constant independent from ω, ϵ_0, and ϵ_q.

Further, based on (10.161), (10.162), (10.202), (10.203), and (10.205) from equation (10.79) and the first and third conditions in (10.80) in the inlet and exit ϵ_q-zones, we obtain the asymptotically valid systems of equations for function $q(r)$

$$\frac{A^3 h_q^3}{\mu_q(q)}\frac{dq}{dr} = -\tfrac{4}{3}r\sqrt{-2r}\theta(-r), \ q(\alpha_1) = 0, \quad (10.206)$$

$$h_q(r) = 1 \ for \ \epsilon_q \ll (V\epsilon_0^{-1/2})^{1/2}, \ \omega \ll 1, \quad (10.207)$$

or

$$A[h_q(r)-1] = -\tfrac{4}{3}r\sqrt{-2r}\theta(-r) \ for \ \epsilon_q = (V\epsilon_0^{-1/2})^{1/2}, \ \omega \ll 1, \quad (10.208)$$

and for $g(s)$

$$\frac{A^3 h_q^3}{\mu_g(g)}\frac{dg}{ds} = \tfrac{4}{3}s\sqrt{2s}\theta(s) - \tfrac{4}{3}\beta_1\sqrt{2\beta_1}, \ g(\beta_1) = 0, \quad (10.209)$$

$$h_g(s) = 1 \ for \ \epsilon_q \ll (V\epsilon_0^{-1/2})^{1/2}, \ \omega \ll 1, \quad (10.210)$$

or

$$A[h_g(s)-1] = \tfrac{4}{3}s\sqrt{2s}\theta(s) \ for \ \epsilon_q = (V\epsilon_0^{-1/2})^{1/2}, \ \omega \ll 1, \quad (10.211)$$

where $h_q(x)$, $\mu_q(q)$ and $h_q(x)$, $\mu_q(q)$ are the major terms of the asymptotic expansions of gap $h(x)$ and viscosity $\mu(p)$ in the inlet and exit ϵ_q-zones, respectively. Relations (10.207), (10.210) and (10.208), (10.211) are valid for starved and fully flooded lubrication regimes, respectively.

It is important to note that the value of constant A is not determined and cannot be determined by solving the system of equations (10.206)-(10.208) in the inlet ϵ_q-zone as well as the value of β_1 is not determined by equations (10.209)-(10.211) in the exit ϵ_q-zone. These constants are determined by the inlet and exit ϵ_0-zones, respectively.

Let us consider a two-term asymptotic approximations of pressure $p(x)$ in the inlet and exit ϵ_q-zones. The next to the Hertzian contribution of the order of magnitude $O(\epsilon_q^{3/2})$ to the function of gap $h(x)$ in the inlet and exit ϵ_q-zones is the contribution of the displacement produced by pressure $\epsilon_0^{1/2}\epsilon_q p_1(x)$. The order of magnitude of this contribution is $\epsilon_0^{1/2}\epsilon_q$. Therefore, by estimating the terms of the Reynolds equation and by taking into account that the latter contribution is of the order of magnitude of $\epsilon_0^{1/2}\epsilon_q$, we obtain that such a two-term approximation of pressure should be searched as follows $p(x) = \epsilon_0^{1/2}q(r) + \epsilon_0\epsilon_q^{-1/2}q_1(r) + \ldots$ and $p(x) = \epsilon_0^{1/2}g(s) + \epsilon_0\epsilon_q^{-1/2}g_1(s) + \ldots$ (where $q_1(r)$ and $g_1(s)$ are new unknown functions of order 1) in the inlet and exit ϵ_q-zones, respectively. We will not engage in this process because $\epsilon_0 \ll \epsilon_q$ and the contribution of the second terms in the pressure representation in the inlet and exit ϵ_q-zones to the solution of the problem in the inlet and exit ϵ_0-zones is negligibly small.

Therefore, in the inlet and exit ϵ_q-zones the problem is reduced to integration of nonlinear first-order differential equations for the leading terms of the asymptotic representations. Let us consider an example of solutions of systems (10.206)-(10.208) and (10.209)-(10.211) for the case of $\mu = exp(Qp)$. We will have $\epsilon_0 = Q^{-2}$, $\omega = Q^{-1} \ll 1$, $\mu_q(q) = \exp(q)$, $\mu_g(g) = \exp(g)$, and

$$H_0 = A(VQ\epsilon_q^{5/2})^{1/3} + \ldots, \quad A = O(1), \quad \epsilon_q \ll (VQ)^{1/2}, \quad \omega \ll 1, \qquad (10.212)$$

$$H_0 = A(VQ)^{3/4} + \ldots, \quad A = O(1), \quad \epsilon_q = (VQ)^{1/2}, \quad \omega \ll 1. \qquad (10.213)$$

Then for starved and fully flooded lubrication regimes we, respectively, obtain

$$q(r) = -\ln\left\{1 + \frac{8\sqrt{2}}{15A^3}[(-r)^{5/2}\theta(-r) - (-\alpha_1)^{5/2}\theta(-\alpha_1)]\right\}, \qquad (10.214)$$

$$q(r) = -\ln\left\{1 + \frac{1}{2A}\left(\frac{9}{4A}\right)^{1/3}J_4[\tau(\alpha_1), \tau(r)]\right\},$$

$$\tau(r) = \left(\frac{4\sqrt{2}}{3A}\right)^{1/3}\sqrt{-r}\,\theta(-r),$$

$$J_4(\alpha, y) = \frac{1}{3}\left[\frac{2t^2}{3(1+t^3)} - \frac{t^2}{(1+t^3)^2} - \frac{1}{9}\ln\frac{(t+1)^2}{t^2-t+1}\right.$$

$$\left. + \frac{2\sqrt{3}}{9}\arctan\frac{2t-1}{\sqrt{3}}\right]\Big|_\alpha^y,$$

(10.215)

where $\theta(r)$ is a step function, $\theta(r) = 0$, $r \leq 0$ and $\theta(r) = 1$, $r > 0$. Obvious mechanical considerations lead to the fact that the solution $q(r)$ for $r \geq$

α_1 of equations (10.206)-(10.208) is bounded. Moreover, function $q(r)$ is a nondecreasing function of r and $q(r) = q(0)$ for $r > 0$. Therefore, the direct matching of function $q(r) = q(0)$ as $r \to \infty$ with the major term of the inner expansion of the external solution (10.209) is impossible without considering the asymptotic of the solution in the inlet ϵ_0-zone. From the explanation presented it follows that the major term of the two asymptotic expansions that match solution $\epsilon_0^{1/2} q_0(r_0)$ in the inlet ϵ_0-zone as $r_0 \to \pm\infty$ takes the form

$$q_a(r_0) = q(0) + \sqrt{2r_0}\theta(r_0). \tag{10.216}$$

In the particular case of $\mu = \exp(Qp)$ from the fact that $q(0)$ is bounded (i.e., the logarithms are determined) and from expressions (10.214) and (10.215) for different lubrication regimes, we obtain the low bounds for constant A as follows

$$A > \left(\tfrac{8\sqrt{2}}{15}\right)^{1/3} \mid \alpha_1 \mid^{5/6} \theta(-\alpha_1)$$

$$= 0.91 \mid \alpha_1 \mid^{5/6} \theta(-\alpha_1), \ \epsilon_q \ll (VQ)^{1/2}, \tag{10.217}$$

$$A > \left(\tfrac{3}{4\sqrt{2}}\right)^{1/2} \{-J_4[\tau(\alpha_1), 0]\}^{3/4}, \ \epsilon_q = (VQ)^{1/2}.$$

For $\alpha_1 = -\infty$ the last inequality in (10.217) is reduced to

$$A > \tfrac{(54\sqrt{3}\pi^3)^{1/4}}{27} = 0.272. \tag{10.218}$$

Furthermore, note that for vanishing r the estimate $dq/dr = O((-r)^{3/2})$ follows from (10.206). In view of $r = r_0\epsilon_0/\epsilon_q$ and from the last estimate for dq/dr as r vanishes, we obtain

$$\tfrac{dp}{dx} = O(\epsilon_0^2 \epsilon_q^{-5/2}), \ r_0 = O(1), \ w \ll 1. \tag{10.219}$$

For the same case of $\mu = \exp(Qp)$ in the exit ϵ_q-zone from equations (10.209)-(10.211) for $g(s)$, we find that for scanty lubrication regimes

$$g(s) = -\ln\left\{1 + \tfrac{8\sqrt{2}}{15A^3}[\beta_1^{5/2}\theta(\beta_1) - s^{5/2}\theta(s)\right.$$

$$\left. -\tfrac{5}{2}\beta_1^{3/2}\theta(\beta_1)[\beta_1 - s\theta(s)]]\right\}, \tag{10.220}$$

while for fully flooded lubrication regimes

$$g(s) = -\ln\left\{1 + \tfrac{1}{2A}\left(\tfrac{9}{4A}\right)^{1/3}[J_4[\tau(s), \tau(\beta_1)]\right.$$

$$\left. -\tfrac{4\sqrt{2}}{3A}\beta_1^{3/2}\theta(\beta_1)J_1[\tau(s), \tau(\beta_1)]]\right\}, \ \tau(s) = \left(\tfrac{4\sqrt{2}}{3A}\right)^{1/3}\sqrt{s}\theta(s),$$

$$\tag{10.221}$$

$$J_1(\alpha, y) = \left[\tfrac{4t^2}{9(1+t^3)} + \tfrac{t^2}{3(1+t^3)^2} - \tfrac{2}{27}\ln\tfrac{(t+1)^2}{t^2-t+1}\right.$$

$$\left. +\tfrac{4\sqrt{3}}{27}\arctan\tfrac{2t-1}{\sqrt{3}}\right]\mid_\alpha^y,$$

where $J_4(\alpha, y)$ is defined in (10.215).

Estimating $\frac{d^2 g(s)}{ds^2}$ for vanishing s such that $s = \frac{\epsilon_0}{\epsilon_q} s_0$, $s_0 = O(1)$, $w \ll 1$, based on (10.209)-(10.211) we find that

$$\frac{d^2 g(s)}{ds^2} = O(\epsilon_0^{1/2} \epsilon_q^{-1/2}), \quad s_0 = O(1), \quad w \ll 1. \tag{10.222}$$

Therefore, we can conclude that in the exit ϵ_0-zone

$$\frac{d^2 p(x)}{dx^2} = O(\epsilon_0 \epsilon_q^{-5/2}), \quad s_0 = O(1), \quad w \ll 1. \tag{10.223}$$

This estimate coincides with the estimate for $d^2 p/dx^2$ in the ϵ_0-inlet zone, which follows from (10.219).

The above analysis of the behavior of $p_{ext}(x)$ and $g(s)$ in the ϵ_0-exit zone indicates that the major term of the two asymptotic expansions that match the solution $\epsilon_0^{1/2} g_0(s_0)$ in the exit ϵ_0-zone as $s_0 \to \pm\infty$ has the form

$$g_a(s_0) = g(0) + \sqrt{-2s_0} \theta(-s_0). \tag{10.224}$$

10.6.5 Asymptotic Analysis of the Inlet and Exit ϵ_0-Zones

Let us consider the asymptotic relationships in the inlet and exit ϵ_0-zones in more detail. For this purpose it is necessary to estimate the integrals in equations (10.82) and (10.201). Let us introduce a function $\eta(w)$ so that $\epsilon_0 \ll \eta(w) \ll \epsilon_q$ for $w \ll 1$. We have

$$\int_a^c p(t) \ln \left| \frac{\frac{c_0 + a_0}{2} - t}{x - t} \right| dt = \left[\int_a^{a_0 - \eta} + \int_{a_0 - \eta}^{a_0 + \eta} + \int_{c_0 - \eta}^{c_0 + \eta} + \int_{c_0 + \eta}^{c} \right] p(t)$$

$$\times \ln \left| \frac{\frac{c_0 + a_0}{2} - t}{x - t} \right| dt.$$

From the last equality using expressions (10.161)-(10.168), (10.170), (10.216), and (10.224), we find

$$\int_a^c p(t) \ln \left| \frac{\frac{c_0 + a_0}{2} - t}{x - t} \right| dt$$

$$= \epsilon_0^{1/2} \epsilon_q \int_{\alpha_1 - \alpha_0 (\epsilon_0/\epsilon_q)^{1/2} - \alpha_2 \epsilon_0/\epsilon_q + \ldots}^{0} q(r) \ln \left| \frac{\frac{c_0 - a_0}{2} - \epsilon_q r}{x - a_0 - \epsilon_q r} \right| dr$$

$$+ \epsilon_0^{3/2} \int_{-\infty}^{\infty} \tilde{q}_0(r_1) \ln \left| \frac{\frac{c_0 - a_0}{2} - \epsilon_0 r_1}{x - a_0 - \epsilon_0 r_1} \right| dr_1 + \int_{a_0}^{c_0} p_{ext}(t) \ln \left| \frac{\frac{c_0 + a_0}{2} - t}{x - t} \right| dt \tag{10.225}$$

$$+ \epsilon_0^{3/2} \int_{-\infty}^{\infty} \tilde{g}_0(s_1) \ln \left| \frac{\frac{c_0 - a_0}{2} + \epsilon_0 s_1}{c_0 - x + \epsilon_0 s_1} \right| ds_1$$

$$+ \epsilon_0^{1/2} \epsilon_q \int_0^{\beta_1 + \ldots} g(s) \ln \left| \frac{\frac{c_0 - a_0}{2} + \epsilon_q s}{x - c_0 - \epsilon_q s} \right| ds + o(\epsilon_0^{3/2}), \quad w \ll 1,$$

where $\widetilde{q}_0(r_0) = q_0(r_0) - q_a(r_0)$ and $\widetilde{g}_0(s_0) = g_0(s_0) - g_a(s_0)$. The estimates $\epsilon_0 \ll \eta \ll \epsilon_q$ are used for deriving equation (10.225). In a similar way we obtain the asymptotic expression for the integral in equation (10.82).

Using this procedure of estimating the integral in equation (10.82) and using formulas (10.225), we can transform equations (10.201) and (10.82) as follows

$$H_0[h(x) - 1] = \epsilon_0^{3/2} \frac{2}{\pi} \int\limits_{-\infty}^{\infty} \widetilde{q}_0(r_1) \ln \left| \frac{\frac{c_0 - a_0}{2} - \epsilon_0 r_1}{x - a_0 - \epsilon_0 r_1} \right| dr_1$$

$$+\epsilon_0^{3/2} \frac{2}{\pi} \int\limits_{-\infty}^{\infty} \widetilde{g}_0(s_1) \ln \left| \frac{\frac{c_0 - a_0}{2} + \epsilon_0 s_1}{c_0 - x + \epsilon_0 s_1} \right| ds_1 + o(\epsilon_0^{3/2}), \tag{10.226}$$

$$\epsilon_0^{3/2} \int\limits_{-\infty}^{\infty} \widetilde{q}_0(r_1) dr_1 + \epsilon_0^{3/2} \int\limits_{-\infty}^{\infty} \widetilde{g}_0(s_1) ds_1 + o(\epsilon_0^{3/2}) = 0.$$

The matching conditions on the boundaries of the external region and the inlet and exit ϵ_q-zones, obviously, have the form (see (10.169))

$$\widetilde{q}_0(r_0) \to 0, \ r_0 \to \pm\infty, \tag{10.227}$$

$$\widetilde{g}_0(s_0) \to 0, \ s_0 \to \pm\infty. \tag{10.228}$$

Now, let us consider the inlet ϵ_0-zone. Using the asymptotic of $h_H(x)$ for $r_0 = O(1)$, $\omega \ll 1$, from (10.199) and (10.226) with the precision of $o(\epsilon_0^{3/2})$, $\omega \ll 1$, we find the expression for $h(x)$ in the form

$$H_0[h(x) - 1] = \epsilon_0^{3/2} \ln \frac{c_0 - a_0}{2\epsilon_0} \frac{2}{\pi} \int\limits_{-\infty}^{\infty} \widetilde{q}_0(r_1) dr_1 + O(\epsilon_0^{3/2}),$$

$$r_0 = O(1), \ \omega \ll 1. \tag{10.229}$$

Obviously, this expression for $h(x)$ does not allow for matching $q_0(r_0)$ with $q_a(r_0)$ as $r_0 \to \infty$. Thus, in order to match these solutions it is necessary for the right side in (10.229) to be equal to zero. In turn, from the second equation in (10.226) and the mentioned conclusion we obtain

$$\int\limits_{-\infty}^{\infty} [q_0(r_0) - q_a(r_0)] dr_0 = 0, \tag{10.230}$$

$$\int\limits_{-\infty}^{\infty} [g_0(s_0) - g_a(s_0)] ds_0 = 0. \tag{10.231}$$

With the help of (10.230) and (10.231) for the regimes of scanty and fully flooded lubrication in the inlet and exit ϵ_0-zones from the first equation in (10.226), we find that in the inlet and exit ϵ_0-zones $h(x) - 1 \ll 1$ and

$$H_0[h(x) - 1] = \epsilon_0^{3/2} \left\{ -\frac{4}{3} r_0 \sqrt{-2r_0} \theta(-r_0) \right.$$

$$\left. +\frac{2}{\pi} \int\limits_{-\infty}^{\infty} [q_0(t) - q_a(t)] \ln \frac{1}{|r_0 - t|} dt \right\} + O(\epsilon_0^{3/2}), \ r_0 = O(1), \ \omega \ll 1, \tag{10.232}$$

$$H_0[h(x) - 1] = \epsilon_0^{3/2} \left\{ \tfrac{4}{3} s_0 \sqrt{2 s_0} \theta(s_0) \right.$$

$$+ \tfrac{2}{\pi} \int\limits_{-\infty}^{\infty} [g_0(t) - g_a(t)] \ln \tfrac{1}{|s_0 - t|} dt \left. \right\} + O(\epsilon_0^{3/2}), \quad s_0 = O(1), \quad \omega \ll 1, \tag{10.233}$$

respectively. Unfortunately, it is very difficult if not impossible to specify analytically in more detail the terms $O(\epsilon_0^{3/2})$ in equations (10.232) and (10.233). This represents the main obstacle in obtaining the asymptotically valid equations for $q_0(r_0)$ and $g_0(s_0)$ in the inlet and exit ϵ_0-zones, respectively.

From equations (10.232), (10.233) and formula (10.205) for H_0, we obtain that in the inlet and exit ϵ_0-zones

$$h(x) - 1 \ll 1 \ for \ r_0 = O(1) \ and \ s_0 = O(1), \quad \omega \ll 1. \tag{10.234}$$

This is true as long as $\epsilon_0^{3/2} \ll H_0$ (see the formula for H_0 (10.205) and the size of $\epsilon_q = (V \epsilon_0^{-1/2})^{1/2}$ for fully flooded over-critical lubrication regimes from (10.208)), i.e.,

$$\epsilon_0 \ll V^{2/5}, \quad \omega \ll 1. \tag{10.235}$$

The latter inequality represents the condition that discriminates between the pre- and over-critical lubrication regimes. Namely, if

$$\epsilon_0 = O(V^{2/5}) \ or \ \epsilon_0 \gg V^{2/5}, \quad \omega \ll 1, \tag{10.236}$$

then the lubrication regime is pre-critical (see estimate (10.127) for $\epsilon_q = V^{2/5}$ for fully flooded pre-critical lubrication regimes and the definitions of pre-critical (10.140) and over-critical (10.141) regimes) while if estimate (10.235) is satisfied then the lubrication regime is over-critical.

Moreover, it is necessary to demand matching the asymptotic expansions of the first and second derivatives of pressure p on the boundaries of the inlet and exit ϵ_q- and ϵ_0-zones, i.e., the validity of estimates (10.219) and (10.222) on the boundaries of the inlet and exit ϵ_q- and ϵ_0-zones, respectively. With this in mind we can say that in the inlet and exit zones the distribution of pressure p depends on the independent variables $r = \frac{x - a_0}{\epsilon_q}$, $r_0 = \frac{x - a_0}{\epsilon_0}$ and $s = \frac{x - c_0}{\epsilon_q}$, $s_0 = \frac{x - c_0}{\epsilon_0}$, respectively. Therefore, in the inlet and exit zones the operator of differentiation d/dx can be represented as follows

$$\frac{d}{dx} = \frac{1}{\epsilon_q} \frac{d}{dr} + \frac{1}{\epsilon_0} \frac{d}{dr_0}, \quad \frac{d}{dx} = \frac{1}{\epsilon_q} \frac{d}{ds} + \frac{1}{\epsilon_0} \frac{d}{ds_0}.$$

Taking into account the estimates for dp/dx in the ϵ_0-inlet zone and for $d^2 p/dx^2$ in the ϵ_0-exit zone as well as estimates (10.202), (10.203) for gap h in the ϵ_q-inlet and exit zones and estimates (10.234) in the ϵ_0-inlet and exit zones zone, we obtain in the inlet zones

$$\frac{d}{dx}\left(\frac{h^3}{\mu} \frac{dp}{dx} \right) = \frac{\epsilon_0^{1/2}}{\epsilon_q^2} \frac{d}{dr}\left(\frac{h_q^3}{\mu_q} \frac{dq}{dr} \right) + \frac{\epsilon_0}{\epsilon_q^{5/2}} \frac{d}{dr_0}\left(\frac{1}{\mu_q(q_0)} \frac{dq_0}{dr_0} \right) + \ldots, \tag{10.237}$$

and in the exit zones

$$\frac{d}{dx}\left(\frac{h^3}{\mu}\frac{dp}{dx}\right) = \frac{\epsilon_0^{1/2}}{\epsilon_q^2}\frac{d}{ds}\left(\frac{h_g^3}{\mu_g}\frac{dg}{ds}\right) + \frac{\epsilon_0}{\epsilon_q^{5/2}}\frac{d}{ds_0}\left(\frac{1}{\mu_g(g_0)}\frac{dg_0}{ds_0}\right) + \dots . \tag{10.238}$$

In addition, in the ϵ_0-inlet and exit zones in expressions (10.237) and (10.238) we need to take into account that $r = \epsilon_0\epsilon_q^{-1}r_0$ and $s = \epsilon_0\epsilon_q^{-1}s_0$, respectively. Therefore, based on equations (10.206)-(10.208) and (10.209)-(10.211) for $q(r)$ and $g(s)$, we conclude that

$$\frac{H_0^3\epsilon_0^{1/2}}{V\epsilon_q^2}\frac{d}{dr}\left(\frac{h_q^3}{\mu_q(q)}\frac{dq}{dr}\right) = -\epsilon_0^{1/2}2\sqrt{-2r_0}\theta(-r_0),$$

$$\frac{H_0^3\epsilon_0^{1/2}}{V\epsilon_q^2}\frac{d}{ds}\left(\frac{h_g^3}{\mu_g(g)}\frac{dg}{ds}\right) = \epsilon_0^{1/2}2\sqrt{2s_0}\theta(s_0). \tag{10.239}$$

At last, we substitute (10.232), (10.233), (10.243), and (10.238) in equation (10.79) and in boundary conditions (10.80), and take into account equations (10.239) and in the inlet and exit ϵ_0-zones. That allows us to derive the asymptotically valid equations

$$\frac{d}{dr_0}M_0(q_0, 1, \mu_q(q_0), r_0) = \frac{2}{\pi}\int\limits_{-\infty}^{\infty}\frac{q_0(t)-q_a(t)}{t-r_0}dt + O(1),$$
$$\tag{10.240}$$

$$r_0 = O(1), \ \omega \ll 1,$$

$$\frac{d}{ds_0}M_0(g_0, 1, \mu_g(g_0), s_0) = \frac{2}{\pi}\int\limits_{-\infty}^{\infty}\frac{g_0(t)-g_a(t)}{t-s_0}dt + O(1),$$
$$\tag{10.241}$$

$$s_0 = O(1), \ \omega \ll 1,$$

where function M_0 is defined by the equality

$$M_0(p, h, \mu, x) = A^3\frac{h^3}{\mu}\frac{dp}{dx}, \tag{10.242}$$

functions μ_q and μ_g are the leading terms of the asymptotic expansions of viscosity $\mu(p, Q)$ in the inlet and exit ϵ_0-zones. In these equations terms $O(1)$ are certain functions of the order of 1 that vanish as $r_0 \to \pm\infty$ and $s_0 \to \pm\infty$.

Thus, the considered problem for $\omega \ll 1$ is reduced to a set of problems in different zones of the contact area. In the inlet ϵ_q-zone the asymptotic analogue of the original problem equations (10.79)-(10.82) is the system of equations (10.206)-(10.208) for function $q(r)$, in the inlet ϵ_0-zone the asymptotic analogue of equations (10.79)-(10.82) is the system of equations (10.216), (10.227), (10.230), (10.240), and (10.242) for function $q_0(r_0)$ and constant A

$$\frac{d}{dr_0}M_0(q_0, 1, \mu_q(q_0), r_0) = \frac{2}{\pi}\int\limits_{-\infty}^{\infty}\frac{q_0(t)-q_a(t)}{t-r_0}dt + O(1),$$

$$q_0(r_0) \to q_a(r_0) = q(0) + \sqrt{2r_0}\theta(r_0), \ r_0 \to \pm\infty, \tag{10.243}$$

$$\int\limits_{-\infty}^{\infty}[q_0(t) - q_a(t)]dt = 0,$$

in the Hertzian region the asymptotic analogue is represented by equations
(10.169) whose solution is represented by formulas (10.170), (10.173), and
(10.182), in the exit ϵ_0-zone the analogue of the original system is given
by equations (10.224), (10.228), (10.231), (10.241), and (10.242) for function
$g_0(s_0)$ and constant β_1

$$\tfrac{d}{ds_0} M_0(g_0, 1, \mu_g(g_0), s_0) = \tfrac{2}{\pi} \int\limits_{-\infty}^{\infty} \tfrac{g_0(t) - g_a(t)}{t - s_0} dt + O(1),$$

$$g_0(s_0) \to g_a(s_0) = g(0) + \sqrt{-2s_0}\,\theta(-s_0), \quad s_0 \to \pm\infty, \qquad (10.244)$$

$$\int\limits_{-\infty}^{\infty} [g_0(t) - g_a(t)] dt = 0,$$

and in the exit ϵ_q-zone the asymptotic analogue of the original problem equations (10.79)-(10.82) is the system of equations (10.209)-(10.211) for function
$g(s)$.

It is important to mention that the presence of terms $O(1)$ in equations
(10.243) and (10.244) is essential. For example, if $\mu_q(q) = \mu_g(q) = \exp(q)$
and these terms are removed from the latter equations then their only solutions are $q_0(r_0) = q_a(r_0)$, $A^3 \exp(-q(0)) = 0$ and $g_0(s_0) = g_a(s_0)$, A^3
$\times \exp(-g(0)) = 0$, respectively. As our analysis showed the values of constant A in the lubrication film thickness formulas (10.212) and (10.213) are
not determined by the inlet ϵ_q-zone, Hertzian region, or the exit zones but
determined by the problem solution in the inlet ϵ_0-zone. On the other hand,
lack of knowledge about the specifics of terms $O(1)$ in equations (10.243) and
(10.244) does not allow for actual numerical calculation of constant A involved
in the lubrication film thickness formulas (10.212) and (10.213) based on these
equations.

In Section 10.13 we will see that the asymptotic equations derived for pre-
critical regimes are actually valid for over-critical regimes assuming that in
this case $\alpha_1 \gg 1$.

Note, that in the external region the estimate $\mu \gg 1$ leads to the validity of
estimates (10.168) and $H_0(h-1) \ll 1$ in the external region. That validates
the entire analysis of the problem.

Obviously, the described analysis is valid if the small parameter of the
problem $\omega = Q^{-1} \ll 1$ or $\omega = V \ll 1$. The method is also applicable to
the problems under non-isothermal conditions and the problems with non-
Newtonian fluids. In these cases only solutions in the inlet and exit ϵ_q zones
and function M_0 from (10.242) must be adjusted for thermal conditions or
fluid non-Newtonian rheology. The rest of the analysis and formulas remain
intact.

The described analysis allows for several important conclusions.

1. Under various conditions when $\mu(p) = \exp(Qp)$ the EHL problem solution exhibits a sharp narrow (pin–like) spike located close to the exit point

$x = c$. The above analysis shows that this spike cannot be realized in the exit ϵ_q-zone where the pressure distribution described by function $g(s)$ is a monotonically decreasing function of s as well as it is not realized in the Hertzian region. Therefore, it can be realized only in the exit ϵ_0-zone. Theoretically, the pressure spike width and its height is supposed to decrease as the value of Q increases. It is clear from the fact that for $Q \gg 1$ the characteristic size of the exit ϵ_0-zone is $Q^{-2} \ll 1$ and the order of magnitude of pressure $p(x)$ in the exit zones is $\epsilon_0^{1/2} = Q^{-1} \ll 1$. However, in many numerical solutions the pressure spike height increases significantly as Q grows. Primarily, it is caused by pressure instability in the exit ϵ_0-zone. To describe this pressure spike numerically in a sufficiently accurate manner, the step size of the numerical calculations in the exit ϵ_0-zone has to be significantly smaller than its characteristic size ϵ_0. A similar conclusion can be made about the precision of the value of constant A and the step size in the inlet ϵ_0-zone. For example, if $\mu(p) = \exp(Qp)$ and $Q = 25$ (which is a pretty typical practical situation) we have $\epsilon_0 = Q^{-2} = 0.0016$. If we consider that the inlet and exit ϵ_0-zones should be covered by at least 20 step sizes, then the step size should about 0.00008. This requirement places a significant demand on computer memory and clock speed as the solution should be determined at more than 2.1/0.00008=26,250 nodes. Obviously, calculations with the step size (in the inlet and exit ϵ_0-zones) greater than 0.00008 will produce results with an error greater than 5%. The usually used step sizes for calculations in such cases are at least an order of magnitude larger. In more details the subject of precision and stability of numerical solutions of EHL problems is considered in the following sections.

2. The situation with solution precision is better in cases of pre-critical lubrication regimes but, still, for desired solution precision a proper care should be taken of the step size in numerical calculations. For example, for fully flooded pre-critical regimes with $V = 0.001$, the characteristic size of the inlet zone is $\epsilon_q = V^{2/5} = 0.063$. Therefore, to have at least 20 nodes in the inlet zone the step size supposed to be below 0.00315. Therefore, if the original EHL problem needs to be solved with such a precision and constant step size throughout the contact region then it would require to consider problem solution at more than 650 nodes.

3. Depending on the values of the parameters of a lubricated contact, the lubrication film thickness may be described by different formulas. Examples of that are the film thickness formulas (10.128) and (10.213) for fully flooded pre- and over-critical lubrication regimes, respectively. We learned that constant A in the film thickness formula (10.213) for over-critical lubrication regimes is determined by the problem solution in the inlet ϵ_0-zone. However, practically it cannot be determined from the problem solution in the inlet ϵ_0-zone as the actual equations are not completely specified. In spite of that formula (10.213) as well as formula (10.128) are well suited for use in curve fitting of experimentally obtained data. Also, they can be used in curve fitting of

sufficiently accurate numerically obtained data (see comment 1).

4. The other conclusion derived from the this analysis concerns a large number of existing formulas for film thickness. These formulas are obtained by curve fitting using either numerical or experimental data. The existence of numerous formulas for the film thickness of Newtonian lubricants can be mainly explained by several reasons: (a) by some inaccuracies in the input parameters of experimentally obtained data on film thickness, (b) by using numerically obtained film data obtained for different pre- and over-critical lubrication regimes and treating this data as a homogeneous pool of data for curve fitting while from the above analysis it is obvious that it is not (as the governing relationships are not the same), (c) the sets of data obtained by different numerical methods with various step sizes have very different precision.

10.6.6 Choosing Pre- or Over-Critical Lubrication Regimes and Small Parameter ω

In this section we consider two very different regimes of lubrication: pre- and over-critical lubrication regimes. Because the solutions for these two regimes are different and, in particular, the formulas for the lubrication film thickness H_0 are different, it is important to learn how to recognize which regime of lubrication is realized in each particular case.

Let us assume that the function of lubricant viscosity $\mu(p)$ is known. There are only two choices for the problem small parameter: $\omega = V$ or ω equal to some small parameter involved in the relationship for viscosity $\mu(p)$. Whatever choice of the small parameter ω is made the characteristic size of the inlet zone in fully flooded pre-critical lubrication regimes is equal to $\epsilon_q = V^{2/5}$ (see the expression for ϵ_q in (10.127)).

Now, we need to determine the critical size ϵ_0 of the inlet zone. It is determined as the maximum solution of equation (10.140). For practical purposes we will replace equation (10.140) by

$$\mu(\epsilon_0^{1/2}) = C, \qquad (10.245)$$

where C is a constant of the order of magnitude of 1, for example, $C = e$. Therefore, the critical size ϵ_0 of the inlet zone is the maximum among all solutions of equation (10.245). If $\mu(p)$ is a monotonically increasing function of p then ϵ_0 is a unique solution of the above equation.

After that, the value of ϵ_0 must be compared to the characteristic size $\epsilon_q = V^{2/5}$ of the inlet region in a pre-critical lubrication regime. If $\epsilon_0 \gg \epsilon_q$, then the lubrication regime is pre-critical, the problem small parameter ω can be taken equal to $V \ll 1$, and in the inlet and exit zones the lubricant viscosity can be assumed to be $\mu(p) = 1$. Moreover, the lubrication film thickness H_0 should be calculated according to the formula $H_0 = A(V\epsilon_q^2)^{1/3}$ for $\epsilon_q \ll V^{2/5}$ or $\epsilon_q = V^{2/5}$ (see (10.120)).

If $\epsilon_0 = O(\epsilon_q) = O(V^{2/5})$, then the problem small parameter ω again can be taken equal to $V \ll 1$ or to the parameter involved in the relationship of $\mu(p)$, which causes $\mu(p)$ to increase significantly as pressure p increases from 0 to 1. In this case the lubrication regime is still pre-critical and the above formula should be used for calculating the values of the film thickness H_0. However, the lubricant viscosity $\mu(p)$ is no longer can be assumed to be equal to 1 in the inlet and exit zones.

The above situation can be easily illustrated for the case of an exponential viscosity $\mu(p) = exp(Qp)$. In this case $\epsilon_0 = Q^{-2}$ and the estimate $\epsilon_0 \gg \epsilon_q$ (we need to remember that in a pre-critical regime $max(\epsilon_q) = V^{2/5}$) or $\epsilon_0 = O(\epsilon_q) = O(V^{2/5})$ means that a pre-critical regime is realized if $Q \ll V^{-1/5}$ or $Q = O(V^{-1/5})$ while $V \ll 1$. In other words, a pre-critical regime can occur only when the pressure viscosity coefficient Q is relatively small or moderate in value.

If $\epsilon_0 \ll \epsilon_q = V^{2/5}$, then the lubrication regime is over-critical, the small parameter ω can be chosen as a specific small parameter involved in the expression for the lubricant viscosity $\mu(p)$ that causes its significant growth from $\mu(p) = 1$ for $p = 0$ to $\mu(p) \gg 1$ for $p = 1$. In this case the value of the lubrication film thickness H_0 should be calculated for the specific values of ϵ_0 and ϵ_q based on formula $H_0 = A(V\epsilon_0^{-1/2}\epsilon_q^{5/2})^{1/3}$ (see formula (10.205)) for over-critical regimes (also, see criteria (10.235) and (10.236)).

The latter conditions also can be easily illustrated for the case of an exponential viscosity $\mu(p) = exp(Qp)$ for which $\epsilon_0 = Q^{-2}$ and estimate $\epsilon_0 \ll \epsilon_q = V^{2/5}$ means that an over-critical regime is realized if $Q \gg V^{-1/5}$ while $V \ll 1$. In other words, an over-critical regime can occur only when the pressure viscosity coefficient Q is sufficiently large.

The approaches for replacing the actual viscosity dependence $\mu(p)$ by a standardized exponential like dependence are discussed in Section 10.15.4.

10.6.7 Analysis of the Ertel-Grubin Method

The Ertel-Grubin method is an approximate engineering method developed in the forties by Ertel [13] and Grubin [14]. It was used for analysis of heavily loaded lubricated contacts. Initially, this method was developed for the case of an isothermal problem for Newtonian fluids. A more careful analysis of this problem under the same prior assumptions was done by Crook [22]. Later, the Ertel-Grubin method was generalized and applied to studying of a wide class of lubrication problems, including non-isothermal problems, the problems for non-Newtonian lubricants, etc. Some related studies are cited in [30].

Let us analyze the essence of the Ertel-Grubin method in the simplest case of an isothermal problem for smooth elastic cylinders and an incompressible Newtonian lubricant under heavily loaded conditions. Then the problem can be reduced to equations (10.79)-(10.82).

In the preceding sections it is shown that the conditions of heavily loaded contact occur for $V \ll 1$ or $Q \gg 1$. Further, we assume that $Q \gg 1$.

Let us formulate the basic assumptions used in the Ertel-Grubin method.

1. It is assumed that the fluid viscosity is determined by the relation $\mu = \exp(Qp)$.

2. At the left boundary of the Hertzian zone (represented by the interval $[-1, 1]$), i.e., in the inlet zone, it is assumed that the following estimate is valid

$$Qp(-1) \gg 1 \ for \ Q \gg 1. \tag{10.246}$$

Let us reproduce the formal scheme of the Ertel-Grubin method. From (10.79)-(10.81) after one integration with respect to x, we obtain

$$M(p, h) = H_0(h - 1). \tag{10.247}$$

Taking into account the exponential smallness of $M(p, h)$ in the Hertzian (external) region we find that in the Hertzian region $H_0(h - 1) \ll 1$. Let us take in (10.81) and (10.82) $a + 1 \ll 1$ and $c - 1 \ll 1$. From the equation $H_0(h - 1) = 0$ in the Hertzian region for $Q \gg 1$, we obtain the function of pressure $p(x)$, which with high accuracy is close to the Hertzian pressure $p(x) = \sqrt{1 - x^2}\theta(1 - x^2)$. Hence, with a corresponding accuracy from (10.81), we get $h(x) = h_0(x)$, where

$$H_0(h_0 - 1) = [|\,x\,|\,\sqrt{x^2 - 1} - \ln(|\,x\,| + \sqrt{x^2 - 1})]\theta(x^2 - 1). \tag{10.248}$$

Integrating equation (10.247) with respect to x from a to x and taking into account the accepted relationship for $\mu = \exp(Qp)$ and equation (10.248), we obtain

$$1 - \exp[-Qp(x)] = \frac{VQ}{H_0^2} \int\limits_a^x \frac{h_0(t)-1}{h_0^3(t)} dt. \tag{10.249}$$

Let us set $x = -1$ in equation(10.250) and use assumption (10.246). It results in an approximate equation

$$1 = \frac{VQ}{H_{0EG}^2} \int\limits_a^{-1} \frac{h_0(t)-1}{h_0^3(t)} dt \tag{10.250}$$

for H_{0EG}, which is the Ertel-Grubin lubrication film thickness. Ertel [13] and Grubin [14], by numerically integrating equation (10.250) for the inlet boundary $a = -\infty$ obtained the formula for the film thickness H_{0EG}

$$H_{0EG} = 0.254(VQ)^{8/11}. \tag{10.251}$$

A more careful asymptotic analysis of the integral in (10.250) with the help of estimate (10.202) produces the formula

$$H_{0EG} = A_0(VQ)^{3/4}, \quad A_0 = (\tfrac{2\sqrt{3}\pi^3}{19683})^{1/4} = 0.272. \tag{10.252}$$

The same result is received by Crook [22].

In a similar way for scanty lubrication regimes, we obtain

$$H_{0EG} = A_1(-\alpha_1)^{5/6}\theta(-\alpha_1)(VQ\epsilon_q^{5/2})^{1/3}, \quad A_1 = (\tfrac{8\sqrt{2}}{15})^{1/3} = 0.91. \quad (10.253)$$

A little more careful analysis leads to the inequality

$$1 > \tfrac{VQ}{H_0^2} \int\limits_a^{-1} \tfrac{h(t)-1}{h^3(t)} dt. \quad (10.254)$$

Using the assumptions of this section and previously obtained asymptotic estimates for $h(x)$ in the inlet ϵ_q- and ϵ_0-zones for the film thickness H_0 in a fully flooded over-critical lubrication regime, we obtain

$$H_0 > H_{0EG} = A_0(VQ)^{3/4}, \quad (10.255)$$

where coefficient A_0 is determined from solution of the equation obtained from inequality (10.217) in which the sign of inequality is replaced by the sign of equality.

Although Greenwood's [30] method represents a certain improvement of Ertel-Grubin's method, it is still based on Ertel-Grubin's assumptions. The results arising from the replacement of assumption (10.246) by $Qp(-1) = 0.2$ are discussed by Greenwood [30] where he mentioned several defects of the Ertel-Grubin method.

Obviously, the intended area of applicability of the Ertel-Grubin method is limited to only over-critical lubrication regimes. That can be seen from the agreement of the orders of magnitude of H_0 in (10.213), (10.212) and (10.252), (10.253) for $\omega = Q^{-1} \ll 1$, respectively. For pre-critical lubrication regimes the Ertel-Grubin method is not applicable at all because everywhere in the inlet zone we have $\mu = O(1)$, $\omega \ll 1$, and, hence, assumption (10.246) is not valid. Besides that, in the inlet zone the function of gap $h(x)$ is significantly different from $h_0(x)$ in (10.248).

The great merit of the Ertel-Grubin method was in its simplicity and ability to predict the value of lubrication film thickness (see estimate (10.205) for $\epsilon_0 = Q^{-2} \ll 1$ and (10.253)) of a correct order.

Nevertheless, the Ertel-Grubin method has significant defects. Among them, we can classify the assumption that the approximate expression (10.248) is valid for gap $h(x)$ everywhere in the inlet zone up to point $x = -1$, which is the beginning of the Hertzian region. That leads to ignoring the very existence of the inlet ϵ_0-zone. In other words, this assumption leads to a non-controllable error in calculation of constant A in formula (10.205) for the lubrication film thickness H_0.

Strictly speaking, assumption (10.246) makes the Ertel-Grubin method contradictory and flawed. Actually, according to (10.161), (10.214), and (10.215), we have $Qp(-1) = q(-\alpha_0(\epsilon_0/\epsilon_q)^{1/2}) + \ldots = q(0) + \ldots = O(1)$ for $Q \gg 1$. The latter contradicts assumption (10.246) that $Qp(-1) \gg 1$ for $Q \gg 1$, which can be reformulated as $q(0) \gg 1$ for $Q \gg 1$. In particular, the assumption represented by (10.246) results in an artificially lowered value of the lubrication film thickness calculated by the Ertel-Grubin method compared to the actual values. Therefore, it is not surprising that coefficient A_0 in the formula for film thickness in fully flooded lubrication regime in the Ertel-Grubin equation (10.252) coincides with the low bound for A presented in (10.218).

The fact that the lubrication film thickness values calculated using (10.252) and measured experimentally are close (in spite of the aforementioned artificially lowered value) can be explained by reduction of the actual film thickness due to heat generation, lubricant compressibility, and surface roughness. However, in cases of high temperatures, high rolling and sliding speeds, etc., the deviation between the these theoretical and experimental values of film thickness is significant.

Let us summarize the results of the section. The over-critical lubrication regimes are defined and studied. The detailed consideration is given to the structure of the solution of the problem for heavily loaded lubricated contact in over-critical regimes. It is shown that the contact region consists of five characteristic zones: the zone with prevailing "elastic solution" - the Hertzian region, two different small inlet and small exit zones compared to the size of the Hertzian region.

A two-term asymptotic approximation of the solution is found in the Hertzian region. The expression for the major term of the latter solution depends only on the total applied load and the coordinates of the beginning and end of this region. The expression for the second term is determined by the viscous lubricant flow in the inlet and exit gap between "non-deformable" solids the shapes of which are determined by the major asymptotic term of pressure in the Hertzian region. The unknown coordinates of the beginning and end of the Hertzian region are found as certain functions of pressure in the inlet and exit ϵ_q-zones that are located next to the inlet and exit coordinates of the lubricated contact.

The size of the ϵ_0-inlet zone is defined by the parameter controlling the growth rate of the lubricant viscosity with pressure. In the inlet and exit ϵ_0-zones, the influences of solids' elasticity and the viscous lubrication flow are of the same order of magnitude. The characteristic sizes of the inlet and exit ϵ_0-zones are equal. The second inlet zone (ϵ_q-inlet zone), which is located next to the ϵ_0-inlet zone (ϵ_0-inlet zone is located between the Hertzian region and the ϵ_q-inlet zone) has a characteristic size ϵ_q determined by the position of the inlet coordinate a. The size of the inlet ϵ_q-zone is much larger than the size of the inlet ϵ_0-zone. In the inlet ϵ_q-zone the solution of the problem is determined by the viscous lubricant flow in the inlet gap between almost non-deformable solids whose shape is determined by the first two terms of the pressure approximation in the Hertzian region.

In the inlet and exit ϵ_q-zones the asymptotically valid systems of equations describing the solution of the problem are derived and solved. The solutions of these systems depend on a smaller number of problem parameters than the solution of the original problem. It is shown that regimes of starved/scanty and fully flooded lubrication can be realized in the frame work of over-critical regimes. In all cases the characteristic size of the inlet ϵ_q-zone can be considered small compared to the size of the Hertzian region. The lubrication film thickness formula obtained in the analysis of the system depends on the particular regime of lubrication. The limitations of the applied method are derived.

A simple set of rules for determining whether the regime is pre- or over-critical is devised (see estimates (10.235) and (10.236)).

The position and nature of often observed pressure spike in the vicinity of the exit point $x = c$ are discussed. It is stated that the usually abnormal behavior of the numerically obtained pressure spike for large viscosity pressure coefficients Q is caused by solution instability. Certain requirements on the step size for numerical solution of the original EHL problem for pre- and over-critical lubrication regimes are established.

Formulas (10.128) and (10.213) for the lubrication film thickness in pre- and over-critical regimes are well suited for use in curve fitting of experimentally and numerically obtained data.

The existence of numerous film thickness formulas for Newtonian lubricants obtained theoretically and/or experimentally is explained, in particular, based on comparison of the governing relations for the film thickness for different pre- and over-critical regimes.

The described method can be easily extended on the case of non-Newtonian fluids. The proposed analysis reveals the structure of the solution of the problem in over-critical regimes. This analysis provides some aid in numerical solution of the problem.

The approximate Ertel-Grubin and Greenwood methods are analyzed, and the contradictions of their original assumptions and, hence, of the final results are demonstrated.

10.7 Numerical Solution for Moderately and Heavily Loaded EHL Contacts

There are a number of serious difficulties in studying EHL problems in general and for heavily loaded regimes, in particular. The difficulties common for all lubrication regimes are related to the problem nonlinearity and a priori unknown exit boundary. In addition to that for heavily loaded lubricated regimes the problem possesses narrow boundary layers near the inlet and exit points of the contact (inlet and exit zones) and in the rest of the contact region, the problem is close to an integral equation of the first kind. It is well known that generally a solution of an integral equation of the first kind is very sensitive to small perturbations and, therefore, unstable. The above indicates that on top of common difficulties typical for EHL problems for heavily loaded regimes we are faced with the necessity to maintain high precision in the inlet and exit zones and to overcome potential solution instability.

Over years a number of different numerical methods were developed [7]-[19], [36]. All these methods work well for lightly and moderately loaded lubricated contacts. However, most of them still have trouble to overcome the potential instability of the problem solution for heavily loaded contacts. The method proposed in this section (see [37]) makes an attempt to overcome this difficulty for moderately to heavily loaded contacts by introducing a naturally occurring regularization along the lines of the methods described in [38]. However, for sufficiently heavily loaded regimes, the method still tends to be unstable and tends to diverge.

10.7.1 Numerical Procedure for Moderately to Heavily Loaded EHL Contacts

We will consider the simplest case of a Newtonian fluid. On this case will be able to illustrate all significant features of an EHL problem solution for a moderately to heavily loaded contact. In dimensionless variables (10.77) and (10.78), the problem is reduced to the system of equations (10.79)-(10.82). We will assume that the lubricant viscosity is controlled by Barus exponential relationship

$$\mu = \exp(Qp), \tag{10.256}$$

where Q is the dimensionless pressure viscosity coefficient.

Therefore, for the given inlet coordinate a and parameters V and Q we need to determine the functions of pressure $p(x)$ and gap $h(x)$ as well as the exit coordinate c and lubrication film thickness H_0. It is easier to numerically solve the problem if the contact region $[a, c]$ with the unknown boundary c can be replaced by the interval $[-1, 1]$. It is achieved by introducing the substitution

$$x = \frac{c+a}{2} + \frac{c-a}{2}\xi, \tag{10.257}$$

where ξ is the new independent variable. Substitution (10.257) reduces the system to the form

$$\frac{2H_0^2}{V(c-a)} \frac{d}{d\xi}[h^3 \exp(-Qp)\frac{dp}{d\xi} - h] = 0, \tag{10.258}$$

$$p(-1) = p(1) = \frac{dp(1)}{d\xi} = 0, \tag{10.259}$$

$$H_0(h-1) = (\frac{c+a}{2} + \frac{c-a}{2}\xi)^2 - c^2 + \frac{c-a}{\pi}\int_{-1}^{1} p(t)\ln|\frac{1-t}{\xi-t}| dt, \tag{10.260}$$

$$\int_{-1}^{1} p(t)dt = \frac{\pi}{c-a}. \tag{10.261}$$

It is noteworthy that in the case of heavily loaded contacts $V \ll 1$ and/or $Q \gg 1$ (see the preceding two sections). In these sections it was shown that the pressure is close to the Hertzian one for purely dry contacts away from the inlet $\xi = -1$ and exit $\xi = 1$ points and the gap is approximately constant and equal to 1. At the same time within the inlet and exit zones the pressure and gap are equally governed by both the lubricant viscous flow and elastic displacements of the solid surfaces. Moreover, as it was shown by asymptotic methods in heavily loaded contacts the lubrication film thickness H_0 and exit coordinate c are predominantly determined by the problem solution within the inlet and exit zones, respectively. Therefore, the iterative method for solution of the problem has to take into account the fact that in the inlet and exit zones the viscous and elastic effects are of the same order.

To simplify the problem and to eventually reduce it to a system of linear algebraic equations, we will use the method of quasi-linearization [39] which basically represents linearization of the problem about its k-th iteration p_k, h_k, H_{0k}, c_k (k is the iteration number). Application of this method leads to the system

$$\frac{2H_{0k}^3}{V(c_k-a)} \frac{d}{d\xi} \left(h_k^3 e^{-Qp_k} \frac{d\triangle p_{k+1}}{d\xi} \right) - \frac{c_k-a}{\pi} \int\limits_{-1}^{1} \frac{\triangle p_{k+1}(t)dt}{t-\xi}$$

$$+ \frac{6H_{0k}^3}{V(c_k-a)} \frac{d}{d\xi} \left(h_k^2 e^{-Qp_k} \frac{\partial h}{\partial p} |_k \triangle p_{k+1} \frac{dp_k}{d\xi} \right)$$

$$- \frac{2H_{0k}^3 Q}{V(c_k-a)} \frac{d}{d\xi} \left(h_k^3 e^{-Qp_k} \triangle p_{k+1} \frac{dp_k}{d\xi} \right)$$

$$+ \triangle H_{0k+1} \left\{ \frac{6H_{0k}^2}{V(c_k-a)} \frac{d}{d\xi} \left(h_k^3 e^{-Qp_k} \frac{dp_k}{d\xi} \right) \right.$$

$$\left. + \frac{6H_{0k}^3}{V(c_k-a)} \frac{d}{d\xi} \left(h_k^2 e^{-Qp_k} \frac{\partial h}{\partial H_0} |_k \frac{dp_k}{d\xi} \right) \right\}$$

$$+ \triangle c_{k+1} \left\{ - \frac{2H_{0k}^3}{V(c_k-a)} \frac{d}{d\xi} \left(h_k^3 e^{-Qp_k} \frac{dp_k}{d\xi} \right) \right.$$

$$+ \frac{6H_{0k}^3}{V(c_k-a)} \frac{d}{d\xi} \left(h_k^2 e^{-Qp_k} \frac{\partial h}{\partial c} |_k \frac{dp_k}{d\xi} \right) - c_k - (c_k - a)\xi$$

$$- \frac{1}{\pi} \int\limits_{-1}^{1} \frac{p_k(t)dt}{t-\xi} \right\} = H_{0k} \left\{ \frac{dh_k}{d\xi} - \frac{2H_{0k}^3}{V(c_k-a)} \frac{d}{d\xi} \left(h_k^3 e^{-Qp_k} \frac{dp_k}{d\xi} \right) \right\},$$

$$\qquad (10.262)$$

$$\triangle p_{k+1}(-1) = \triangle p_{k+1}(1) = 0, \qquad (10.263)$$

$$\int\limits_{-1}^{1} \triangle p_{k+1}(t)dt + \frac{\pi}{(c_k-a)^2} \triangle c_{k+1} = \frac{\pi}{c_k-a} - \int\limits_{-1}^{1} p_k(t)dt. \qquad (10.264)$$

The equation for constant c follows from the requirement that the lubricant flux must be conserved, i.e., from equation (10.258) integrated with respect to ξ from ξ to 1 with the help of the third boundary condition in (10.259) and equation (10.260)

$$\frac{2H_0^2}{V(c-a)}h^3 e^{-Qp}\frac{dp}{d\xi} = h - 1.$$

Application of quasi-linearization to the latter equation gives

$$\frac{2H_{0k}^3}{V(c_k-a)}h_k^3 e^{-Qp_k}\frac{d\triangle p_{k+1}}{d\xi} - \frac{c_k-a}{\pi}\int\limits_{-1}^{1}\triangle p_{k+1}\ln\mid\frac{1-t}{t-\xi}\mid dt$$

$$+\frac{6H_{0k}^3}{V(c_k-a)}h_k^2 e^{-Qp_k}\frac{\partial h}{\partial p}\mid_k \triangle p_{k+1}\frac{dp_k}{d\xi}$$

$$-\frac{2H_{0k}^3 Q}{V(c_k-a)}h_k^3 e^{-Qp_k}\triangle p_{k+1}\frac{dp_k}{d\xi}$$

$$+\triangle H_{0k+1}\{\frac{6H_{0k}^2}{V(c_k-a)}h_k^3 e^{-Qp_k}\frac{dp_k}{d\xi}$$

$$+\frac{6H_{0k}^3}{V(c_k-a)}h_k^2 e^{-Qp_k}\frac{\partial h}{\partial H_0}\mid_k \frac{dp_k}{d\xi})\} \tag{10.265}$$

$$+\triangle c_{k+1}\{-\frac{2H_{0k}^3}{V(c_k-a)}h_k^3 e^{-Qp_k}\frac{dp_k}{d\xi})$$

$$+\frac{6H_{0k}^3}{V(c_k-a)}\frac{d}{d\xi}(h_k^2 e^{-Qp_k}\frac{\partial h}{\partial c}\mid_k \frac{dp_k}{d\xi})$$

$$-(1+\xi)(\frac{c_k+a}{2} + \frac{c_k-a}{2}\xi) + 2c_k - \frac{1}{\pi}\int\limits_{-1}^{1}p_k(t)\ln\mid\frac{1-t}{t-\xi}\mid dt\}$$

$$= H_{0k}\{dh_k - 1 - \frac{2H_{0k}^2}{V(c_k-a)}h_k^3 e^{-Qp_k}\frac{dp_k}{d\xi}\}.$$

In equations (10.262)-(10.265) function $\triangle p_{k+1} = p_{k+1} - p_k$ and constants $\triangle H_{0k+1} = H_{0k+1} - H_{0k}$ and $\triangle c_{k+1} = c_{k+1} - c_k$ are incremental changes in the corresponding quantities from iteration to iteration, $\frac{\partial h}{\partial p}\mid_k \triangle p_{k+1}$ is the result of application of the linear operator (the derivative of h_k with respect to p_k) to function $\triangle p_{k+1}$, $\frac{\partial h}{\partial H_0}\mid_k$ and $\frac{\partial h}{\partial c}\mid_k$ are the partial derivatives of h_k with respect to H_0 and c, respectively, calculated on the k-th iteration.

Further on, we introduce two sets of nodes on $[-1,1]$: integral nodes $\{\xi_i\}$, $i = 1,\ldots,I$, $\xi_1 = -1$, $\xi_I = 1$ and semi-integral nodes $\{\xi_{i+1/2} = (\xi_i + \xi_{i+1})/2\}$, $i = 1,\ldots,I-1$. After that we integrate equations (10.262) and (10.265) with respect to ξ over the intervals $[\xi_{j-1/2},\xi_{j+1/2}]$ and $[\xi_{I-1},\xi_I]$, respectively, and use the approximation

$$\frac{dp}{d\xi}(\xi_{i+1/2}) \approx \frac{p(\xi_{i+1})-p(\xi_i)}{\xi_{i+1}-\xi_i}$$

and the rectangle rule [38]

$$\int\limits_{-1}^{1}\frac{p(t)dt}{t-\xi_i} \approx \sum_{j=1}^{I-1}\frac{p(\xi_j)(\xi_{j+1}-\xi_j)}{\xi_{j+1/2}-\xi_i}, \quad i = 1,\ldots,I,$$

to obtain a system of linear algebraic equations for $\{\triangle p_{k+1}(\xi_i)\}$, $i = 1, \ldots, I$, and constants $\triangle H_{0k+1}$ and $\triangle c_{k+1}$.

As soon as values $\{\triangle p_{k+1}(\xi_i)\}$, $i = 1, \ldots, I$, $\triangle H_{0k+1}$, and $\triangle c_{k+1}$ are found from the solution of the above–mentioned system of linear algebraic equations we can determine the new iterates of our solution

$$p_{k+1} = p_k + \triangle p_{k+1}, \ H_{0k+1} = H_{0k} + \triangle H_{0k+1}, \ c_{k+1} = c_k + \triangle c_{k+1}.$$

After that the new function of gap h_{k+1} can be easily calculated using usual quadrature formulas, for example, the trapezoidal rule. To initiate this iterative process, we have to provide the initial approximations for p, H_0, and c. For moderately to heavily loaded regimes that can be done as follows

$$c_0 = 1, \ H_{00} = 0.272(VQ)^{3/4}, \ p_0(\xi) = \sqrt{1 - (\tfrac{1+a}{2} + \tfrac{1-a}{2}\xi)^2}$$

or by taking the values of p, H_0, and c from the known solution with the closest set of input parameters a, V, and Q. The iterative process runs until the solution converges with the desired precision.

It is important to mention that in equations (10.262) and (10.265) the terms proportional to $\triangle H_{0k+1}$ and $\triangle c_{k+1}$ represent the naturally occurring regularization terms for solution of this ill-posed lubrication problem. In [38] it is rigorously proven that a similarly designed numerical method for solution of one-dimensional linear first kind singular integral equations converges to the exact solution of the problem. In this sense the mathematical features of the EHL problem for heavily loaded regimes is very close to the first kind singular integral equations considered in [38].

The proposed numerical method possesses some advantages in comparison with other methods. Namely, it allows simultaneous determination of constants H_0 and c along with pressure $p(\xi)$, automatic regularization procedure by using the terms proportional to $\triangle H_{0k+1}$ and $\triangle c_{k+1}$, and the guarantee of the constant lubricant flux through the gap between the lubricated solids. The method converges well for lightly and moderately to heavily loaded regimes. However, it still does not converge for severely heavily loaded regimes. Some remedies related to stability of heavily to severely heavily loaded EHL contacts are proposed in the following sections.

10.7.2 Some Numerical Results for Heavily Loaded EHL Contacts

The numerical results described below are obtained for the fixed inlet coordinate $a = -2$, pressure coefficient $Q = 7$, and slide-to-roll ratio $s_0 = 2(u_2 - u_1)/(u_2 + u_1) = 0.01$. The results were obtained on a nonuniform set of nodes $\{\xi_i\}$ more dense near the exit point 1. The total number of nodes used was $I = 100$. For moderate values of parameters V and Q, the increase of the total number of nodes I from 100 to 200 does not noticeably change the solution of the problem. The relative error was maintained at the level of

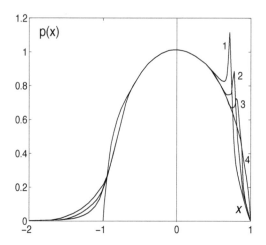

FIGURE 10.8
The profiles of the pressure distribution $p(x)$ obtained for $Q = 7$ and $V = 0.2$ (curve 1), $V = 0.1$ (curve 2), $V = 0.05$ (curve 3), Hertzian pressure distribution (curve 4) (after Airapetov, Kudish, and Panovko [37]). Reprinted with permission from Allerton Press.

10^{-4}. The results are presented in Fig. 10.8-10.12, where curves marked with 1 correspond to $V = 0.2$, curves marked with 2 correspond to $V = 0.1$, and curves marked with 3 correspond to $V = 0.05$. Figure 10.8 shows pressure distributions for different values of parameter V. Also, for comparison in Fig. 10.8 is shown the Hertzian distribution of pressure $\sqrt{1 - x^2}$ marked by 4. It is clear that the pressure distribution has a narrow spike close to the exit point $x = c$. In [25] by solving asymptotically valid equations in the exit zone it was determined that this narrow spike exists only when the lubricant viscosity μ is strongly dependent on pressure p. Specifically, in the latter study it was found that if the lubricant viscosity $\mu(p) = \exp(Qp^m), Q \gg 1$, then the narrow spike does not exist when $m \leq 0.25$ and it does exist when $m \geq 0.75$. The pressure distributions in Fig. 10.8 indicate that as V increases the pressure spike increases in hight and moves toward the center of the contact. The pressure distribution exhibits a similar behavior when $\mid a \mid$ ($a < 0$) increases but to a much lesser extent. Moreover, when varying the inlet coordinate a within the region $a < -2$ it practically does not affect the numerical results. For a significantly closer to the beginning of the Hertzian region $x = -1$ (for example, $a = -1.25$) there is a small difference in the pressure distribution in the inlet and exit zones, especially near the spike. However, this difference increases as a approaches to -1. These changes in pressure in the inlet zone affect the value of the film thickness H_0 and the exit coordinate c.

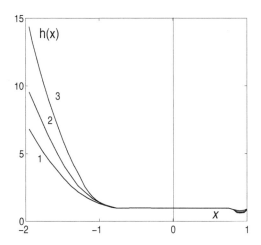

FIGURE 10.9
The profiles of the gap distribution $h(x)$ obtained for $Q = 7$ and $V = 0.2$ (curve 1), $V = 0.1$ (curve 2), $V = 0.05$ (curve 3) (after Airapetov, Kudish, and Panovko [37]). Reprinted with permission from Allerton Press.

The graphs of the gap distributions $h(x)$ are presented in Fig. 10.9. It is clear that in the cental part of the contact (Hertzian region) we have a practically constant gap $h(x) = 1$. Moreover, the minimum of the film thickness is by no more than 15% smaller than the gap value in the center of the contact. In more detail it can be seen from Fig. 10.10 where a blow up of the gap distributions are shown in the exit zone.

Variations of the dimensionless sliding $f_s(x) = \frac{V}{12H_0}\frac{\mu s_0}{h}$ and rolling $f_r(x)$ $= \frac{H_0 h}{2}\frac{dp}{dx}$ frictional stresses are presented in Fig. 10.11 and 10.12, respectively. Obviously, the shape of the distributions of the sliding frictional stress $f_s(x)$ resembles the shape of pressure $p(x)$ distributions. The rolling frictional stress $f_r(x)$ is mainly concentrated (different from zero) in the inlet and exit zones, and it is proportional to and, to a certain extent, resembles $\frac{1}{\mu}\frac{dp}{dx}$. For fixed values of the slide-to-roll ratio s_0, sufficiently small parameter V and/or sufficiently large parameter Q the ratio of the rolling and sliding frictional stresses $f_r(x)/f_s(x)$ is small in the entire contact.

For the values of parameter $V = 0.2$, 0.1, 0.005, the following results for the film thickness $H_0 = 0.3637$, $0.2342, 0.1479$ and the exit coordinate $c = 1.078$, 1.059, 1.044 were obtained. Based on the corrected Ertel-Grubin formula (see Section 10.6) $H_0 = 0.272(VQ)^{3/4}$ for the used values of V and Q we get $H_0 = 0.3501$, 0.2082, 0.1238, respectively. The comparison of the two sets of H_0 values reveals no surprise, i.e., $H_0^{Ertel-Grubin} <$

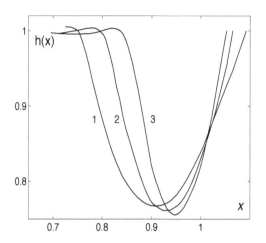

FIGURE 10.10
Magnification of the profiles of the gap distribution $h(x)$ in the vicinity of
the exit point obtained for $Q = 7$ and $V = 0.2$ (curve 1), $V = 0.1$ (curve 2),
$V = 0.05$ (curve 3) (after Airapetov, Kudish, and Panovko [37]). Reprinted
with permission from Allerton Press.

$H_0^{EHLnumerical}$. The explanation of this relationship between the Ertel-Grubin
$H_0^{Ertel-Grubin}$ and the numerically obtained $H_0^{EHLnumerical}$ is based on the
results of the asymptotic analysis of the over-critical regimes in Section
10.6. In these cases the dimensionless minimal film thickness is equal to
$h_{min} = 0.7653, 0.7587, 0.7545$, respectively. Therefore, based on this we get
$H_{min} = h_{min}H_0 = 0.2806, 0.1777, 0.1116$.

The results for a lightly loaded contact differ significantly from the ones for
heavily loaded conditions. It can clearly be seen from Fig. 10.13 the results of
which (pressure and gap) are obtained for $V = 20$ and $Q = 4$. In particular,
the pressure distribution $p(x)$ (curve 1) possesses a clear maximum at $x \approx 0.2$
and it differs significantly from the Hertzian pressure distribution (curve 3).
This pattern of pressure behavior is the continuation of the trend for the
pressure spike behavior indicated above in Fig. 10.8. At the same time, the
gap distribution $h(x)$ (curve 2) is no longer constant in the central part of the
contact region but rather resembles the gap determined by function $1 + \frac{x^2}{H_0}$.

Similar results were obtained for $Q = 5$ and $Q = 6$ and same other parameters. The pressure spike, exit coordinate c, and film thickness H_0 decrease
with decreasing value of Q.

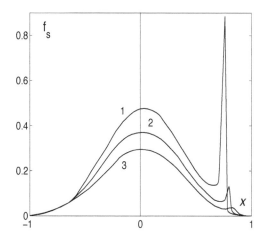

FIGURE 10.11
The profiles of the sliding frictional stress distribution $f_s(x)$ obtained for $Q = 7$ and $V = 0.2$ (curve 1), $V = 0.1$ (curve 2), $V = 0.05$ (curve 3) (after Airapetov, Kudish, and Panovko [37]). Reprinted with permission from Allerton Press.

10.8 Numerical Solution of Asymptotically Valid Equations for Pre-critical Regimes in the Inlet and Exit Zones for Newtonian Liquids

In the preceding section it was mentioned that the major difficulty encountered in numerical solution of EHL problems for heavily loaded contacts is instability of pressure $p(x)$ in the inlet and mostly exit zones. There are two interconnected sources of this instability in the original equations (10.79)-(10.83). These are due to fast growth of viscosity μ with pressure p (i.e., the presence of a large parameter in μ) and for high μ due to the proximity of these equations to a problem for an integral equation of the first kind

$$x^2 - c^2 + \frac{2}{\pi} \int\limits_a^c p(t) \ln |\frac{c-t}{x-t}| dt = 0, \quad \int\limits_a^c p(t)dt = \frac{\pi}{2},$$

the numerical solutions of which are often unstable. In case of an exponential viscosity $\mu(p) = \exp(Qp)$ the solution instability usually reveals itself for sufficiently large values of Q. For the same reason as the original equations of the EHL problem, the asymptotically valid in the inlet zone equations (10.133), (10.134), (10.136), (10.114), (10.115), and (10.129) for the starved

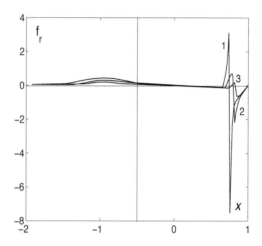

FIGURE 10.12
The profiles of the rolling frictional stress distribution $f_r(x)$ obtained for $Q = 7$
and $V = 0.2$ (curve 1), $V = 0.1$ (curve 2), $V = 0.05$ (curve 3) (after Airapetov,
Kudish, and Panovko [37]). Reprinted with permission from Allerton Press.

lubrication regimes (or (10.131) for fully flooded lubrication regimes) and
in the exit zone equations (10.135), (10.134), (10.137), (10.114), (10.114),
and (10.130) for the starved lubrication regimes (or (10.132) for fully flooded
lubrication regimes) are also susceptible to numerical instability. In fact, in
the above equations the large parameter Q is removed but now functions $q(r)$
and $g(s)$ approach infinity as $r \to \infty$ and $s \to -\infty$, respectively. Therefore,
the numerical solutions of these equations may still be unstable due to the
necessity to solve them for large r and s for which the equations are still close
to integral equations of the first kind.

The numerical instability can be avoided if instead of solving the above
equations we solve the pair of equivalent systems (10.134), (10.144), (10.146),
and (10.129) for the starved lubrication regimes (or (10.131) for fully flooded
lubrication regimes) and (10.134), (10.145), (10.147), and (10.130) for the
starved lubrication regimes (or (10.132) for fully flooded lubrication regimes)
in the inlet and exit zones, respectively. Theoretically, the above two types
of systems of asymptotically valid equations are equivalent. However, numerically the latter asymptotically valid equations in the inlet and exit zones
provide the opportunity to get stable converging solutions. It is due to the
fact that equations (10.144) and (10.145) are analytically (not numerically)
resolved with respect to the major term of the solution and provide stable
asymptotic pressure behavior as $r \to \infty$ and $s \to -\infty$, respectively. In this
form the equations guaranty the correct solution behavior at infinity (see rela-

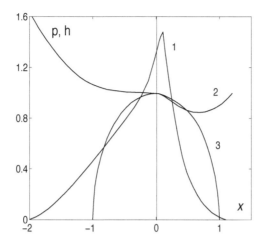

FIGURE 10.13
The profiles of the pressure distribution $p(x)$ (curve 1) and gap $h(x)$ (curve 2) obtained for $V = 20$ and $Q = 4$. Curve 3 is the Hertzian pressure distribution (after Airapetov, Kudish, and Panovko [37]). Reprinted with permission from Allerton Press.

tionships in (10.114), (10.136), and (10.137)) while for small and moderate r and s the equations are no longer close to integral equations of the first kind. Moreover, numerical solutions (iterations) of the latter systems converge faster for faster–growing lubricant viscosity.

10.8.1 Numerical Solution in the Inlet Zone

Let us consider the inlet region. Function M_0 we will represent in the form

$$M_0(q, h_q, \mu_q, r) = W(q, h_q, \mu_q, r)\frac{dq}{dr}, \ \ W(q, h_q, \mu_q, r) = \frac{A^3 h_q^3}{\mu_q(q)}. \quad (10.266)$$

By integrating equations (10.131) and (10.133) with respect to r and using the fact that $h_q(r) \to 1$ and $M_0(q, h_q, \mu_q, r) \to 0$ as $r \to \infty$, we obtain

$$h_q(r) = 1 + \frac{W(q, h_q, \mu_q, r)}{A}\frac{dq}{dr}. \quad (10.267)$$

Now, we introduce two sets of nodes $r_k = k\triangle r$ and $r_{k+1/2} = 1/2(r_k + r_{k+1})$, $k = 0, \ldots$, where $\triangle r$ is a positive step size. The equations will be solved by iterations. It is clear that constant A is a monotonically increasing function of $|\alpha_1|$, $\alpha_1 \leq 0$. That enables us to consider two slightly different approaches to numerical solution of the problem. The first is to solve the problem and determine A and $q(r)$ for the given value of $\alpha_1 < 0$. The other one is based on determining α_1 and $q(r)$ for the given value of A. The second approach allows us to solve the problems in the inlet and exit zones independently of each other.

Let us consider the second approach first. Assuming that $r_{N+1/2}$ is sufficiently large, we will approximate the integral in equation (10.144) at $r = r_k$ as follows

$$\int_0^\infty \frac{d}{dt} M_0(q, h_q, \mu_q, t) \frac{dt}{\sqrt{2t(t-r_k)}}$$

$$\approx \int_0^{r_{1/2}} \frac{d}{dt} [W \frac{dq}{dt}] \frac{dt}{\sqrt{2t(t-r_k)}} + \sum_{j=1}^N \int_{r_{i-1/2}}^{r_{i+1/2}} \frac{d}{dt} [W \frac{dq}{dt}] \frac{dt}{\sqrt{2t(t-r_k)}}$$

$$\approx \frac{d}{dr} [W \frac{dq}{dr}] |_{r_1} \int_0^{r_{1/2}} \frac{dt}{\sqrt{2t(t-r_k)}} + \sum_{j=1}^N \frac{d}{dr} [W \frac{dq}{dr}] |_{r_i} \int_{r_{i-1/2}}^{r_{i+1/2}} \frac{dt}{\sqrt{2t(t-r_k)}} \tag{10.268}$$

$$\approx \frac{1}{\Delta r^2} [W_{3/2}(q_2 - q_1) - W_{1/2}(q_1 - q_0)] \frac{1}{\sqrt{2r_k}} \ln | \frac{\sqrt{r_{1/2}}-\sqrt{r_k}}{\sqrt{r_{1/2}}+\sqrt{r_k}} | +$$

$$\frac{1}{\Delta r^2} \sum_{j=1}^N [W_{j+1/2}(q_{j+1} - q_j) - W_{j-1/2}(q_j - q_{j-1})] \frac{1}{\sqrt{2r_k}}$$

$$\times \ln | \frac{\sqrt{r_{j+1/2}}-\sqrt{r_k}}{\sqrt{r_{j-1/2}}-\sqrt{r_k}} \frac{\sqrt{r_{j-1/2}}+\sqrt{r_k}}{\sqrt{r_{j+1/2}}+\sqrt{r_k}} | .$$

Therefore, by satisfying equation (10.144) at nodes r_k and approximating the integral according to (10.268), we obtain the system of nonlinear equations for $i+1$-st iterates q_k^{i+1} in the form

$$q_k^{i+1} = \sqrt{2r_k} - \frac{1}{2\pi\Delta r^2} \sum_{j=1}^N [W_{j+1/2}^{i+1}(q_{j+1}^{i+1} - q_j^{i+1})$$

$$-W_{j-1/2}^{i+1}(q_j^{i+1} - q_{j-1}^{i+1})]\gamma_{jk}, \quad k = 0, \ldots, N+1, \tag{10.269}$$

$$\gamma_{jk} = \ln | \frac{\sqrt{r_{3/2}}-\sqrt{r_k}}{\sqrt{r_{3/2}}+\sqrt{r_k}} |, \quad j = 1,$$

$$\gamma_{jk} = \ln | \frac{\sqrt{r_{j+1/2}}-\sqrt{r_k}}{\sqrt{r_{j-1/2}}-\sqrt{r_k}} \frac{\sqrt{r_{j-1/2}}+\sqrt{r_k}}{\sqrt{r_{j+1/2}}+\sqrt{r_k}} |, \quad j > 1, \tag{10.270}$$

where i is the iteration number. By introducing a new set of unknowns $v_k = q_k^{i+1} - q_k^i$, we can propose the following approximation of equations (10.269)

$$q_k^i + v_k + \frac{1}{2\pi\Delta r^2} \sum_{j=1}^N [W_{j+1/2}^i(q_{j+1}^i - q_j^i) - W_{j-1/2}^i(q_j^i - q_{j-1}^i)$$

$$+W_{j+1/2}^i(v_{j+1} - v_j) - W_{j-1/2}^i(v_j - v_{j-1})]\gamma_{jk} = \sqrt{2r_k}, \tag{10.271}$$

$$k = 0, \ldots, N+1.$$

Therefore, we obtained a finite system of $N + 2$ linear algebraic equations in $N + 2$ unknowns v_k, $k = 0, \ldots, N + 1$. Here the influence factors γ_{jk} are determined from (10.270), N is a sufficiently large integer. Obviously, the approximation error gets smaller as number N increases and $\triangle r$ decreases.

After solution of system (10.271) for values v_k, $k = 0, \ldots, N + 1$, is done the new iterates of pressure q_k^{i+1}, $k = 0, \ldots, N + 1$, are calculated from the formula $q_k^{i+1} = q_k^i + v_k$.

The next step of the process after the new iterates of q_k^{i+1} are obtained is calculation of the new iterates of gap $h_q(r_{k+1/2})$. For regimes of starved lubrication $h_q(r_{k+1/2}) = 1$, $k = 0, \ldots, N$. For fully flooded lubrication regimes the values of gap $h_q(r_{k+1/2})$, $k = 0, \ldots, N$, can be found by solving the nonlinear equation (10.267) at nodes $r_{k+1/2}$, $k = 0, \ldots, N$, assuming that the values of pressure q_k, $k = 0, \ldots, N + 1$, are known. Solution of equation (10.267) is done iteratively based on a modified Newton's method

$$h_q^{i+1}(r_{k+1/2}) = h_q^i(r_{k+1/2}) - D/D_{h_q}, \ D = h_q^i(r_{k+1/2}) - 1$$

$$-\frac{W(q_{k+1/2}^{i+1}, h_q^i(r_{k+1/2}), \mu_q(q_{k+1/2}^{i+1}), r_{k+1/2})}{A} \frac{q_{k+1}^{i+1} - q_k^{i+1}}{\triangle r},$$

$$D_{h_q} = 1 - \alpha_* \frac{W_{h_q}(q_{k+1/2}^{i+1}, h_q^i(r_{k+1/2}), \mu_q(q_{k+1/2}^{i+1}), r_{k+1/2})}{A} \frac{q_{k+1}^{i+1} - q_k^{i+1}}{\triangle r},$$

$$q_{k+1/2}^{i+1} = \frac{q_{k+1}^{i+1} + q_k^{i+1}}{2}, \ k = 0, \ldots, N,$$

(10.272)

where α_* is a sufficiently small positive number (for example, $\alpha_* = 0.05$). The introduction of small parameter α_* is due to the fact that the value of derivative D_{h_q} changes its sign (may become equal or close to zero) at some point $r_{k+1/2} > 0$ while for large $r_{k+1/2}$ derivative D_{h_q} is close to 1. The iteration of $h_q(r_{k+1/2})$ should be done in the order from $k = N - 1$ to $k = 0$. The initial approximation for gap $h_q^0(r_{k+1/2})$ can be obtained from the equation

$$h_q^0(r_{k+1/2}) = 1 + \frac{W(q_{k+1/2}^1, 1, \mu_q(q_{k+1/2}^1), r_{k+1/2})}{A} \frac{q_{k+1}^1 - q_k^1}{\triangle r},$$

$$k = 0, \ldots, N.$$

(10.273)

Notice, that the iteration process based on

$$h_q^{i+1}(r_{k+1/2}) = 1$$

(10.274)

$$+\frac{W(q_{k+1/2}^{i+1}, h_q^i(r_{k+1/2}), \mu_q(q_{k+1/2}^{i+1}), r_{k+1/2})}{A} \frac{q_{k+1}^{i+1} - q_k^{i+1}}{\triangle r}, \ k = 0, \ldots, N$$

instead of (10.272), which also converges but at a slightly slower pace.

After the new iterates of q_k^{i+1} and $h_q^{i+1}(r_{k+1/2})$ are obtained the value of α_1 is determined from equation (10.146). Using the approximation of the integral in (10.146) similar to the one used in (10.268) we obtain the formula

$$\alpha_1 = \frac{1}{\pi \Delta r} \sum_{j=1}^{N} \delta_j \frac{W_{j+1/2}^{i+1}(q_{j+1}^{i+1}-q_j^{i+1}) - W_{j-1/2}^{i+1}(q_j^{i+1}-q_{j-1}^{i+1})}{\sqrt{2r_j}},$$

(10.275)

$$\delta_j = 2, \ j = 1; \ \delta_j = 1, \ j > 1.$$

The initial approximation of q_k can be taken based on a modified asymptotic of the Hertzian pressure such as

$$q_k^0 = \gamma r_k, \ 0 \le r_k \le r_*; \ q_k^0 = \sqrt{2r_k} + \frac{\alpha_1}{\sqrt{2r_k}}, \ r_k > r_*,$$

(10.276)

$$r_* = 0.79A, \ \gamma = 1.06A^{-1/2}, \ \alpha_1 = -0.53A.$$

For larger values of A it may be beneficial to take the solution of the problem obtained for A closest to the one at hand as the initial approximation for q_k. That would accelerate the iteration process convergence. The iteration process stops when the desired precision ε is reached.

Now, let us consider the first approach. The only change in the second approach is that coefficient A is no longer is constant but changes from iteration to iteration as follows

$$\frac{A^{i+1}}{A^i} = \left\{ \frac{\pi \Delta r \alpha_1}{\sum_{j=1}^{N} \delta_j [W_{j+1/2}^{i+1}(q_{j+1}^{i+1}-q_j^{i+1}) - W_{j-1/2}^{i+1}(q_j^{i+1}-q_{j-1}^{i+1})](2r_j)^{-1/2}} \right\}^{1/3},$$

(10.277)

$$\delta_j = 2, \ j = 1; \ \delta_j = 1, \ j > 1,$$

where A^i is the i-th iterate of coefficient A. All other formulas derived for the first approach remain in force except for the fact that in equations (10.271)-(10.276) A is replaced by A^i, A^{i+1}, and A^0, respectively. As we know, in this approach the value of α_1 can be used to get a better initial approximation for gap $h_q^0(r_{k+1/2})$ than the one determined by equation (10.273). It can be obtained from the equation (see formula (10.208))

$$h_q^0(r_{k+1/2}) = 1 - \frac{4}{3A}\eta_{k+1/2}\sqrt{-2\eta_{k+1/2}}\theta(-\eta_{k+1/2}),$$

(10.278)

$$\eta_{k+1/2} = r_{k+1/2} + \alpha_1, \ k = 0, \dots, N.$$

The convergence of the schemes based on the first and second approaches is about the same for small and moderate values of a. Moreover, for starved lubrication regimes they converge very fast. Usually, it takes about 10 iterations to get a solution with the absolute precision of 10^{-4}. For fully flooded lubrication regimes and large $| \, a \, |$, $(a < 0)$ the convergence is much slower due to slow convergence of the gap function $h_q(r_{k+1/2})$. For the same solution precision it may take more than 200 iterations, to reach the solution. However, the scheme based on the first approach converges faster than the one based on the second approach. This is due to the fact that usually the initial approximation $h_q^0(r)$ is better and for large $\mid a \mid$, $(a < 0)$ coefficient A is very insensitive to even large variations in α_1.

10.8.2 Numerical Solution in the Exit Zone

In a similar fashion a numerical scheme can be proposed in the exit zone. For convenience, in the exit zone we will introduce the new variable $r = -s$. Then, at $r = r_k$ using boundary condition $\frac{dg(0)}{dr} = 0$, the integral in equation (10.145) can be approximated as follows (for comparison see (10.268)):

$$\int_0^\infty \frac{d}{dt} M_0(g, h_g, \mu_g, t) \frac{dt}{\sqrt{2t(t-r_k)}}$$

$$\approx \int_0^{r_{1/2}} \frac{d}{dt} [W \frac{dg}{dt}] \frac{dt}{\sqrt{2t(t-r_k)}} + \sum_{j=1}^N \int_{r_{i-1/2}}^{r_{i+1/2}} \frac{d}{dt} [W \frac{dg}{dt}] \frac{dt}{\sqrt{2t(t-r_k)}}$$

$$\approx \frac{d}{dr} [W \frac{dq}{dr}] \mid_{0.5r_{1/2}} \int_0^{r_{1/2}} \frac{dt}{\sqrt{2t(t-r_k)}}$$

$$+ \sum_{j=1}^N \frac{d}{dr} [W \frac{dg}{dr}] \mid_{r_i} \int_{r_{i-1/2}}^{r_{i+1/2}} \frac{dt}{\sqrt{2t(t-r_k)}} \tag{10.279}$$

$$\approx \frac{2}{\triangle r^2} W_{1/2}(g_1 - g_0) \frac{1}{\sqrt{2r_k}} \ln \mid \frac{\sqrt{r_{1/2}}-\sqrt{r_k}}{\sqrt{r_{1/2}}+\sqrt{r_k}} \mid +$$

$$\frac{1}{\triangle r^2} \sum_{j=1}^N [W_{j+1/2}(g_{j+1} - g_j) - W_{j-1/2}(g_j - g_{j-1})] \frac{1}{\sqrt{2r_k}}$$

$$\times \ln \mid \frac{\sqrt{r_{j+1/2}}-\sqrt{r_k}}{\sqrt{r_{j-1/2}}-\sqrt{r_k}} \frac{\sqrt{r_{j-1/2}}+\sqrt{r_k}}{\sqrt{r_{j+1/2}}+\sqrt{r_k}} \mid .$$

By satisfying equation (10.145) at nodes r_k and approximating the integral according to (10.279), we obtain the system of nonlinear equations for $i+1$-st iterates g_k^{i+1} in the form

$$g_k^{i+1} + \frac{1}{\pi \triangle r^2} W_{1/2}^{i+1} (g_1^{i+1} - g_0^{i+1}) \ln \mid \frac{\sqrt{r_{1/2}}-\sqrt{r_k}}{\sqrt{r_{1/2}}+\sqrt{r_k}} \mid$$

$$+ \frac{1}{2\pi \triangle r^2} \sum_{j=1}^N [W_{j+1/2}^{i+1}(g_{j+1}^{i+1} - g_j^{i+1}) - W_{j-1/2}^{i+1}(g_j^{i+1} - g_{j-1}^{i+1})] \tag{10.280}$$

$$\times \ln \mid \frac{\sqrt{r_{j+1/2}}-\sqrt{r_k}}{\sqrt{r_{j-1/2}}-\sqrt{r_k}} \frac{\sqrt{r_{j-1/2}}+\sqrt{r_k}}{\sqrt{r_{j+1/2}}+\sqrt{r_k}} \mid = \sqrt{2r_k}, \ k = 0, \ldots, N+1,$$

where i is the iteration number. By introducing a new set of unknowns $v_k = g_k^{i+1} - g_k^i$ and linearizing the equations in the vicinity of g_k^i, we obtain a finite system of $N+2$ linear algebraic equations in $N+2$ unknowns v_k, $k = 0, \ldots, N+1$:

$$g_k^i + v_k + \frac{1}{\pi \triangle r^2}[W_{1/2}^i(g_1^i - g_0^i) + W_{1/2}^i(v_1 - v_0)]\ln\left|\frac{\sqrt{r_{1/2}} - \sqrt{r_k}}{\sqrt{r_{1/2}} + \sqrt{r_k}}\right|$$

$$+\frac{1}{2\pi \triangle r^2}\sum_{j=1}^{N}[W_{j+1/2}^i(g_{j+1}^i - g_j^i) - W_{j-1/2}^i(g_j^i - g_{j-1}^i)$$

$$+W_{j+1/2}^i(v_{j+1} - v_j) - W_{j-1/2}^i(v_j - v_{j-1})]\gamma_{jk} = \sqrt{2r_k},$$

$$k = 0, \ldots, N+1,$$

(10.281)

where N is a sufficiently large integer and constants γ_{jk} are determined in (10.269). Obviously, the approximation error gets smaller as number N increases and $\triangle r$ decreases.

After solution of system (10.281) for values v_k, $k = 0, \ldots, N+1$, is done the new iterates of pressure g_k^{i+1}, $k = 0, \ldots, N+1$, are calculated from the formula $g_k^{i+1} = g_k^i + v_k$.

Having the new iterates of g_k^{i+1} we can calculate the new iterates of gap $h_g(r_{k+1/2})$ in the fashion similar to the one used in the inlet zone. For regimes of starved lubrication $h_g(r_{k+1/2}) = 1$, $k = 0, \ldots, N$. For fully flooded lubrication regimes the values of the gap $h_g(r_k)$, $k = 0, \ldots, N$, can be found by solving the nonlinear (Reynolds) equation

$$h_g(r) = 1 - \frac{W(g, h_g, \mu_g, r)}{A}\frac{dg}{dr}$$

(10.282)

at nodes $r_{k+1/2}$, $k = 0, \ldots, N$. As in the inlet zone, solution of equation (10.282) can be done iteratively based on a modified Newton's method

$$h_g^{i+1}(r_{k+1/2}) = h_g^i(r_{k+1/2}) - D/D_{h_g}, \quad D = h_g^i(r_{k+1/2}) - 1$$

$$+\frac{W(g_{k+1/2}^{i+1}, h_g^i(r_{k+1/2}), \mu_g(g_{k+1/2}^{i+1}), r_{k+1/2})}{A}\frac{g_{k+1}^{i+1} - g_k^{i+1}}{\triangle r},$$

$$D_{h_g} = 1 + \alpha_*\frac{W_{h_g}(g_{k+1/2}^{i+1}, h_g^i(r_{k+1/2}), \mu_g(g_{k+1/2}^{i+1}), r_{k+1/2})}{A}\frac{g_{k+1}^{i+1} - g_k^{i+1}}{\triangle r},$$

(10.283)

$$g_{k+1/2}^{i+1} = \frac{g_{k+1}^{i+1} + g_k^{i+1}}{2}, \quad k = 0, \ldots, N,$$

(α_* is a sufficiently small positive number) or based on the iteration process

$$h_g^{i+1}(r_{k+1/2}) = 1$$

$$-\frac{W(g_{k+1/2}^{i+1}, h_g^i(r_{k+1/2}), \mu_g(g_{k+1/2}^{i+1}), r_{k+1/2})}{A}\frac{g_{k+1}^{i+1} - g_k^{i+1}}{\triangle r}, \quad k = 0, \ldots, N.$$

(10.284)

Opposite to the case of the inlet zone the iteration process (10.284) converges as fast as the one presented in (10.283).

The initial approximation for gap $h_g^0(r_{k+1/2})$ can be taken as $h_g^0(r_{k+1/2}) = 1$, $k = 0, \ldots, N$, or it can be obtained from the equation

$$h_g^0(r_{k+1/2}) = 1$$

$$-\frac{W(g_{k+1/2}^1, 1, \mu_g(g_{k+1/2}^1), r_{k+1/2})}{A} \frac{g_{k+1}^1 - g_k^1}{\triangle r}, \quad k = 0, \ldots, N.$$

(10.285)

After the new iterates of g_k^{i+1} and $h_g^{i+1}(r_{k+1/2})$ are obtained, the value of β_1 is determined from equation (10.147) in which s is replaced by $-r$. Using the approximation of the integral in (10.147) similar to the one used in the inlet zone, we obtain the formula

$$\beta_1 = 2 \frac{W_{1/2}(g_1^{i+1} - g_0^{i+1})}{\pi \triangle r \sqrt{2\triangle r}}$$

$$+ \frac{1}{\pi \triangle r} \sum_{j=1}^{N} \frac{W_{j+1/2}^{i+1}(g_{j+1}^{i+1} - g_j^{i+1}) - W_{j-1/2}^{i+1}(g_j^{i+1} - g_{j-1}^{i+1})}{\sqrt{2r_j}}.$$

(10.286)

The initial approximation of g_k can also be taken as a modified asymptotic of the Hertzian pressure, however, different from the one used in the inlet zone, i.e.,

$$g_k^0 = \gamma r_k^2, \ 0 \le r_k \le r_*; \ g_k^0 = \sqrt{2r_k} - \frac{\beta_1}{\sqrt{2r_k}}, \ r_k > r_*,$$

(10.287)

$$r_* = 0.3A, \ \gamma = 3.37A^{-3/2}, \ \beta_1 = 0.36A.$$

In the exit zone the patterns of convergence for starved and fully flooded lubrication regimes are similar to the ones in the inlet region. However, for fully flooded lubrication regimes the convergence of this scheme with the absolute precision of 10^{-4} is much faster than in the inlet region. This is mainly due to the fact that the function of gap $h_g(r_{k+1/2})$ is everywhere relatively close to 1. Therefore, the convergence process usually takes less than 20 iteration.

The other major difference between the solution convergence in the inlet and exit zones is the fact that for elevated values of Q_0 (for $\mu_q(q) = e^{Q_0 q}$ and $\mu_g(g) = e^{Q_0 g}$) and A in the inlet zone the solutions remain stable while in the exit zone they become unstable. This instability manifests itself in two different ways: the elevated sensitivity of $\max g(s)$ as a function of N and $\triangle r$ as well as in small oscillations of the derivative of pressure $dg(s)/ds$ and gap $h_g(s)$ between the position of the pressure spike and the exit point $s = 0$. The latter type of instability in the region adjacent to the exit point, when it manifests itself, is easily corrected by using simple averaging of the gap values, i.e., $h_g(r_{k-1/2}) = 0.5[h_g(r_{k-1/2}) + h_g(r_{k+1/2})]$. That slightly increases the discrepancy in the Reynolds equation. However, it still remains within acceptable limits. Because of this regularization treatment of gap $h_g(s)$, we

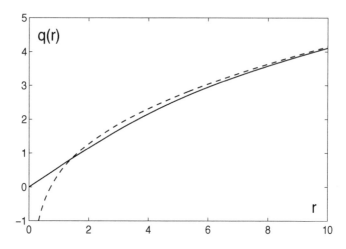

FIGURE 10.14
Main term of the asymptotic of the pressure distribution $q(r)$ (solid curve)
and the asymptote of the Hertzian pressure $q_a(r)$ (dashed curve) in the inlet
zone of a starved lubricated contact for $Q_0 = 1$ and $A = 2$.

need to take a special care of the values of $h_g(\triangle r/2)$ and $h_g(3\triangle r/2)$. That is
done by using the fact that $h_g(0) = 1$ and by employing a cubic interpolation
as follows

$$h_g(r) = \tfrac{1}{\triangle r^3}\{-\tfrac{8}{315}(r - r_{5/2})(r - r_{7/2})(r - r_{9/2})$$

$$+\tfrac{1}{5}h_g(r_{5/2})r(r - r_{7/2})(r - r_{9/2})$$

$$-\tfrac{2}{7}h_g(r_{7/2})r(r - r_{5/2})(r - r_{9/2})$$ (10.288)

$$+\tfrac{1}{9}h_g(r_{9/2})r(r - r_{5/2})(r - r_{7/2})\}, \; r = r_{1/2}, \; r_{3/2}.$$

The sensitivity of $\max g(s)$ as a function of N and $\triangle r$ is much harder to treat.
Unfortunately, the usual regularization techniques do not work. A simple and
very effective method of regularization for numerical solution of asymptotically valid problems in the inlet and exit zones will be proposed in Section
10.11. A practically identical regularization approach will be employed for
solution of the original EHL problem.

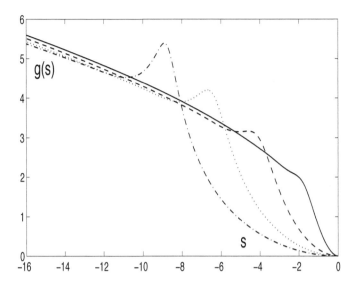

FIGURE 10.15
Main term of the asymptotic of the pressure distribution $g(s)$ in the exit zone of a starved lubricated contact for $Q_0 = 1$, $A = 1$ (solid curve), $A = 2$ (dashed curve), $A = 3$ (dotted curve), and $A = 4$ (dash-dotted curve).

10.8.3 Some Numerical Results for Pre-critical Regimes in the Inlet and Exit Zones

The above–described numerical methods converge well to stable solutions in the inlet zone for starved and fully flooded lubrication regimes. In the exit zone they also converge well for sufficiently small values of A and Q_0 (see the discussion below). Obviously, for starved lubrication regimes in the inlet and exit zones the equations are nonlinear if μ_q and μ_g depend on $q(r)$ and $g(s)$, respectively. Otherwise, the equations are linear and their solution is achieved in just one iteration.

Let us consider the case of an exponential viscosity for which $\mu(p) = \exp(Qp)$. First, we will consider regimes of starved lubrication for which $\epsilon_q^{1/2} = Q^{-1} \ll V^{1/5}$, $Q \gg 1$ and, therefore, we have $\mu_q(q) = e^q$ and $\mu_g(g) = e^g$ (i.e. $Q_0 = 1$). The absolute error of calculations was chosen to be not higher than $\varepsilon = 10^{-4}$. To check the convergence of the numerical scheme for $A = 2$, three series of calculations were done for $N = 400$, $\triangle r = 0.0625$, $N = 800$, $\triangle r = 0.03125$, and $N = 1,600$, $\triangle r = 0.015625$ ($N \triangle r = 25$). The solution precision was reached after 8-13 iterations. The maximum relative errors of the solutions obtained for $N = 400$, $N = 800$ as well as for

$N = 1,600$ in the values of α_1, β_1, $q(r)$, and $g(s)$ were found to be not greater than 0.58% and 0.33%. Therefore, the rest of calculations was done for $N = 800$, $\triangle r = 0.03125$. The relative error in the integral conditions $\int_0^\infty [q(r) - q_a(r)]dr = 0$ and $\int_{-\infty}^0 [g(s) - g_a(s)]ds = 0$ did not exceed 0.7%. The proximity of these integrals to zero serves as a gage whether the product $N \triangle r$ is large enough to provide sufficient solution precision. The graphs of $q(r)$ for $A = 2$ and $g(s)$ for four values of the parameter $A = 1$, 2, 3, and 4 are given in Fig. 10.14 and 10.15. For these values of A the corresponding values of α_1 are equal to -0.5016, -1.4045, -2.5244, and -3.9783 while the values of β_1 are equal to 0.3646, 0.8966, 1.5017, and 2.1552. Obviously, the values of $|\alpha_1|$ and β_1 increase as the value of A does. For the above values of A the curves of $q(r)$ resemble each other. Therefore, just one curve of $q(r)$ (solid curve) and for comparison the curve of the Hertzian pressure asymptote $q_a(r)$ from (10.114) (dashed curve) for $A = 2$ are given in Fig. 10.14. It can be clearly seen from Fig. 10.14 that $q(r)$ is a monotonically increasing function of r which approaches its asymptote $q_a(r)$ and does not exhibit any signs of instability or oscillations, in particular, for large r. Figure 10.15 demonstrates the behavior of all four curves of $g(s)$ (solid, dashed, dotted, and dash-dotted curves correspond to $A = 1$, 2, 3, and 4, respectively). Each of the curves of $g(s)$ possesses a local maximum (pressure spike) that shifts closer to the center of the contact (to $s = -\infty$) and increases in value as the values of $|\alpha_1|$ and A increase. As in the case of $q(r)$ for large s the behavior of $g(s)$ practically coincides with the behavior of its asymptote $g_a(s)$ and does not exhibit any oscillations or signs of instability.

Let us consider some starved lubrication regimes for which the viscosity in the inlet and exit zones can be considered constant. In these cases the problem equations are linear and their solution does not require iterations. For an exponential viscosity with $\mu(p) = \exp(Qp)$ and for regimes of starved lubrication with $\epsilon_q^{1/2} \ll Q^{-1} \ll V^{1/5}$, $Q \gg 1$, we have $\mu_q(q) = \mu_g(g) = 1$ and $Q_0 = 0$. Because of solutions in the inlet and exit zones approach their asymptotes slower than for the case of exponential viscosity, we will do calculations for $N = 1,600$ and $\triangle r = 0.03125$. We will consider solutions of the problem in the inlet and exit zones for $A = 1$, 2, 3, and 4 for which the values of α_1 are equal to -0.4759, -1.3905, -2.5721, and -3.9444 while the values of β_1 are equal to 0.5080, 1.4363, 2.6561, and 4.1274, respectively. The numerical solutions for $q(r)$ in the inlet zone qualitatively resemble the one in Fig. 10.14. They are monotonically increasing functions of r which approach $\sqrt{2r}$ as $r \to \infty$, however, at a much slower pace for larger values of A than for the corresponding cases of $\mu_q(q) = e^q$ described above. From the data presented above it follows that for the same values of constant A the values of α_1 for the cases of $\mu_q(q) = 1$ and $\mu_q(q) = e^q$ do not vary much. Effectively, the above behavior indicates that for the same values of ϵ_q and A the inlet zone is wider in the cases of constant viscosity $(\mu_q(q) = 1)$ compared to the

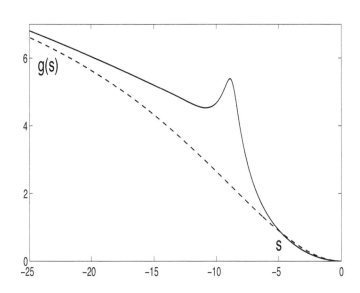

FIGURE 10.16
Main term of the asymptotic of the pressure distribution $g(s)$ in the exit zone
of a starved lubricated contact for $A = 4$ and viscosity $\mu_g(g) = 1$ (dashed
curve) and $\mu_g(g) = e^g$ (solid curve).

cases when $\mu_q(q) = e^q$. That requires an increased number of nodes N for
numerical solution in the former cases.

For starved lubrication regimes and constant viscosity $\mu_g(g) = 1$ in the exit
zone the solutions for $g(s)$ are monotonically decreasing functions of s, which
approach $\sqrt{-2s}$ as $s \to -\infty$. Also, it happens slower than in the corresponding
cases of $\mu_g(g) = e^g$. Therefore, for $\epsilon_q^{1/2} \ll Q^{-1} \ll V^{1/5}$, $Q \gg 1$ and $\mu_g(g) =
1$ the solution functions $g(s)$ differ significantly from the solutions for $g(s)$
obtained for $\epsilon_q^{1/2} = Q^{-1} \ll V^{1/5}$, $Q \gg 1$, $\mu_g(g) = e^g$, and depicted in Fig.
10.15. That can be clearly seen from Fig. 10.16 in which for $A = 4$ graphs of
two functions $g(s)$ for $\mu_g(g) = 1$ (dashed curve) and $\mu_g(g) = e^g$ (solid curve)
are presented. Moreover, from the presented data it is clear that for regimes
of starved lubrication for the same values of ϵ_q and constant A the exit zone
for constant viscosity is also larger than for exponential viscosity $(\mu_g(g) = e^g)$
and the difference in size increases as A increases. Therefore, the numerical
solution in the former case requires an increased number of nodes N.

It is important to realize that for starved lubrication regimes two different
solutions in both the inlet and exit zones can be converted into each other by
a simple transformation as long as for both solutions the value of $Q_0 A^{3/4}$ is
the same. Moreover, for small values of Q_0 for starved lubrication regimes it

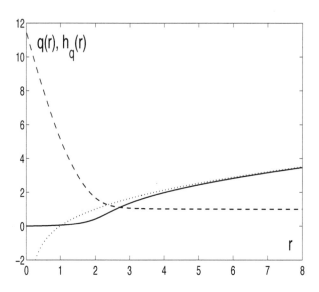

FIGURE 10.17
Main terms of the asymptotic distributions of pressure $q(r)$ (solid curve), gap $h_q(r)$ (dashed curve), and Hertzian pressure asymptote $q_a(r)$ (dotted curve) in the inlet zone of a fully flooded lubricated contact for $A = 0.525$ and $Q_0 = 1$.

is easy co come up with simple analytical formulas for A and β_1 as functions of α_1. By employing the following transformation

$$(\widetilde{r}, \widetilde{s}, \widetilde{\alpha_1}, \widetilde{\beta_1}) = \gamma^2(r, s, \alpha_1, \beta_1), \quad (\widetilde{q}, \widetilde{g}) = \gamma(q, g), \quad \widetilde{A} = \gamma^{4/3}A \qquad (10.289)$$

the asymptotically valid equations (10.134), (10.144), (10.146), and (10.129) in the inlet zone and equations (10.134), (10.145), (10.147), and (10.130) in the exit zone remain the same if variables $(r, \alpha_1, A, q, h_q, \mu_q, s, \beta_1, g, h_g, \mu_g)$ are replaced by $(\widetilde{r}, \widetilde{\alpha_1}, \widetilde{A}, \widetilde{q}, h_q, \widetilde{\mu_q}, \widetilde{s}, \widetilde{\beta_1}, \widetilde{g}, h_g, \widetilde{\mu_g})$. Here $\mu_q = \exp(Q_0 q)$, $\mu_g = \exp(Q_0 g)$ and $\widetilde{\mu_q} = \exp(Q_* \widetilde{q})$, $\widetilde{\mu_g} = \exp(Q_* \widetilde{g})$, $Q_* = Q_0 \gamma^{-1}$. Therefore, in the inlet zone for the input parameters Q_0 and A the original problem for $q(r)$ and α_1 is equivalent to the modified problem for $\widetilde{q}$ and $\widetilde{\alpha_1}$ for the input parameters $Q_* = Q_0 A^{3/4}$ and $\widetilde{A} = 1$ if $\gamma = A^{-3/4}$, respectively, i.e., these two problems are described by the same equations with the above mentioned replacements. The same situation takes place in the exit zone. Based on the latter and the physics of the lubrication process, we can conclude that parameters $\widetilde{\alpha_1}$ and $\widetilde{\beta_1}$ are certain monotonically increasing functions f_i and f_e of just one parameter Q_*, i.e., $\alpha_1 = A^{3/2}f_i(Q_0 A^{3/4})$ and $\beta_1 = A^{3/2}f_e(Q_0 A^{3/4})$. Therefore, for $Q_0 = 0$ by inverting the latter formulas for α_1 and β_1, we obtain

$$A = A_1 \mid \alpha_1 \mid^{2/3}, \quad \beta_1 = \beta_{10} \mid \alpha_1 \mid, \quad A_1 = 1.64, \quad \beta_{10} = 1.054, \qquad (10.290)$$

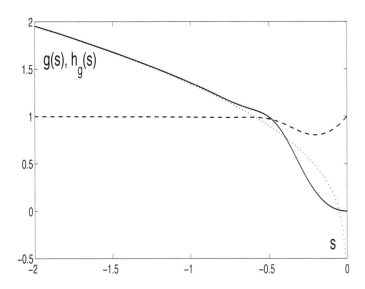

FIGURE 10.18
Main terms of the asymptotic distributions of pressure $g(s)$ (solid curve),
gap $h_g(s)$ (dashed curve), and the Hertzian pressure asymptote $g_a(s)$ (dotted
curve) in the exit zone of a fully flooded lubricated contact for $A = 0.525$ and
$Q_0 = 2.5$.

where constants A_1 and β_{10} are the values of A and β_1 obtained from the
numerical solution of the original systems of asymptotically valid in the inlet
and exit zones equations for $Q_0 = 0$ and $\alpha_1 = -1$. Formulas (10.290) can
be used for pre-critical starved lubrication regimes for $Q_0 = 0$ in conjunction
with the formula for the film thickness H_0 for Newtonian lubricant (10.120),
i.e.,

$$H_0 = 1.64(a+1)^{2/3}V^{1/3}, \tag{10.291}$$

where a is the inlet coordinate.

Moreover, for $\mu_q = \exp(Q_0 q)$ and $\mu_g = \exp(Q_0 g)$ as the value of the parameter $Q_* = Q_0 A^{3/4}$ increases in the inlet zone the solution approaches its
asymptotic a bit slower and in the exit zone the maximum of the pressure
spike increases and it moves away from the exit point toward the center of
the contact region.

Now, let us consider some results for pre-critical fully flooded lubrication
regimes. Let us consider some examples for exponential viscosity $\mu(p) = \exp(Qp)$. Then for $\epsilon_q = V^{2/5} = O(Q^{-2})$, $Q \gg 1$, we have $\mu_q(q) = e^{Q_0 q}$, $\mu_g(g) = e^{Q_0 g}$, and $Q_0 = QV^{1/5} = O(1)$. For fully flooded lubrication
regimes in the inlet zone the iterations starting with the initial approxima-

TABLE 10.5

The dependence of
coefficient A and the inlet
gap $h_q(\triangle r/2)$ on the inlet
coordinate α_1.

α_1	A	$h_q(\triangle r/2)$
0	0	1
-0.016	0.100	1.079
-0.059	0.200	1.238
-0.142	0.300	1.527
-0.322	0.400	2.168
-0.992	0.500	5.262
-1.930	0.525	11.342
-4.325	0.535	33.804

tion from (10.276) also converge to stable solutions, however, it takes more iterations. For example, for $Q_0 = 1$, $A = 0.525$, $N = 480$, $\triangle r = 0.03125$, and the absolute precisions $\varepsilon = 0.01$, $\varepsilon = 0.001$, and $\varepsilon = 0.0001$, the solutions in the inlet zone converged after 156, 333, and 517 number of iterations, respectively. The solutions obtained for these three levels of precision are as follows: $\alpha_1 = -1.8236$, $h_q(\triangle r/2) = 10.6001$, $\alpha_1 = -1.9072$, $h_q(\triangle r/2) = 11.2607$, and $\alpha_1 = -1.9160$, $h_q(\triangle r/2) = 11.3310$, respectively. For $Q_0 = 1$, $A = 0.525$, $N = 800$, and $\triangle r = 0.015625$, the graphs of $q(r)$, $h_q(r)$, and the asymptote $q_a(r)$ of the Hertzian pressure are given in Fig. 10.17.

For $Q_0 = 1$ and several values of coefficient A the values of the inlet coordinate α_1 and gap $h_q(\triangle r/2)$ are presented in Table 10.5. This information gives an idea of how quickly lubricant starvation develops (i.e., the proximity of the inlet coordinate α_1 to zero) and how it affects the lubrication film thickness H_0, which is directly proportional to coefficient A (see formula (10.128)). It is not unexpected that when coefficient A approaches its limiting value the gap at the inlet point increases (without bound). Therefore, for values of A close to its limiting value the number of iterations required to get a converged solution increases. Note, that $A = 0.525$ is relatively close to its limiting value which requires so many iterations. The evidence of that is in the fact that for $Q_0 = 1$, $A = 0.535$, $N = 800$, $\triangle r = 0.015625$, and $\varepsilon = 0.001$ we have $\alpha_1 = -4.325$ and $h_q(\triangle r/2) = 33.804$, i.e., the value of $h_q(\triangle r/2)$ almost tripled while the value of A is increased from 0.525 to 0.535 by just 0.01 (or 1.9%). Also, the data from Table 10.5 shows that the size of the entire inlet zone is usually very small, i.e., for $V \ll 1$ it is about $6V^{2/5} \ll 1$. Based on the numerical analysis of the fully flooded inlet zone and formula (10.128) for $\mid a \mid \gg 1$, $a < 0$, we can obtain some formulas for the lubrication film thickness H_0 as follows

$$H_0 = AV^{3/5}, \ A = 0.535, \ Q_0 = 1; \ A = 0.676, \ Q_0 = 2. \tag{10.292}$$

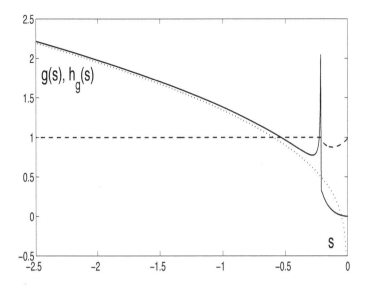

FIGURE 10.19
Main terms of the asymptotic distributions of pressure $g(s)$ (solid curve), gap $h_g(s)$ (dashed curve), and the Hertzian pressure asymptote $g_a(s)$ (dotted curve) in the exit zone of a fully flooded lubricated contact for $A = 0.525$ and $Q_0 = 10$.

In the exit zone, for fully flooded pre-critical regimes for $Q_0 = 1$, $A = 0.525$, $N = 800$, $\triangle r = 0.00390625$, and $\varepsilon = 0.0001$ the solution converges after 17 iterations, and we get $\beta_1 = 0.120$ and $\min h_g(s) = 0.773$. It is interesting that for pre-critical fully flooded lubrication regimes in the exit zone for $Q_0 \leq 2.5$ the pressure distribution $g(s)$ instead of having a pressure spike behaves monotonically with a mild "hump." However, for larger values of Q_0 we get a pressure spike which increases in height and becomes very thin as Q_0 increases. This pressure and gap behavior are illustrated in Fig. 10.18 and 10.19 the data for which is obtained for $A = 0.525$, $\varepsilon = 0.001$, $Q_0 = 2.5$, and $Q_0 = 10$, respectively. The general pressure hump/spike behavior is similar to the one of the pressure spike under starved lubrication conditions.

10.8.4 Numerical Precision and Stability Considerations

The availability of the asymptotically valid in the inlet and exit zones equations provides a unique opportunity to analyze in detail the sensitivity of the numerical solutions to the value of the step size $\triangle r$. Let us consider an incompressible fluid with Newtonian rheology with $\epsilon_q = Q_0 Q^{-1} \ll V^{2/5}$ or

$\epsilon_q = Q_0 Q^{-1} = O(V^{2/5})$, $\mu_q(q) = e^{Q_0 q}$, and $\mu_g(g) = e^{Q_0 g}$. For $A = 1, 2, 3$, $Q_0 = 1$, and absolute error $\varepsilon = 10^{-4}$ for both starved and fully flooded lubrication regimes in the inlet and exit zones the solution stabilizes for $\triangle r \leq 0.00335$ while $N \triangle r$ is kept approximately equal to 6.7. Therefore, the sufficient accuracy of the solution of the original EHL problem in the inlet and exit zones can be achieved if the step size $\triangle y$ would be smaller or equal to $\triangle y = 0.00335 \frac{2}{c-a} \epsilon_q \leq \frac{0.0067}{c-a} V^{2/5}$. For example, for a realistic values of $V = 0.01$, $a = -2$, and $c \approx 1$ this requirement translates into $\triangle y \leq 0.00071$. Therefore, if the span of the lubricated contact region is about 3, then the number of evenly spaced nodes covering the region should be about $4,238$ or higher.

TABLE 10.6

Data on the sensitivity of
the inlet coordinate α_1 as a
function of the step size $\triangle r$
for $A = 2$ and $Q_0 = 2.5$.

$\triangle r$ (N)	α_1
0.0700000 (100)	-1.217
0.0350000 (200)	-1.236
0.0175000 (400)	-1.250
0.0087500 (800)	-1.260
0.0043750 (1,600)	-1.268
0.0021875 (3,200)	-1.273

TABLE 10.7

Data on the sensitivity of the pressure spike
maximum $\max g(r)$ in the exit zone and the exit
coordinate β_1 as functions of the step size $\triangle r$ for
$A = 2$ and $Q_0 = 2.5$.

$\triangle r$ (N)	$\max g(s)$	β_1	C.N.
0.00600 (1,200)	14.799	1.015	$1.37 \cdot 10^9$
0.00415 (1,600)	17.746	1.050	$5.06 \cdot 10^9$
0.00335 (2,000)	19.970	1.758	$1.14 \cdot 10^{10}$
0.00280 (2,400)	21.780	1.880	$2.16 \cdot 10^{10}$
0.00240 (2,800)	23.469	2.021	$3.74 \cdot 10^{10}$
0.00210 (3,200)	24.814	1.109	$5.84 \cdot 10^{10}$
0.00187 (3,600)	26.278	1.118	$8.84 \cdot 10^{10}$
0.00168 (4,000)	27.708	1.140	$1.30 \cdot 10^{11}$

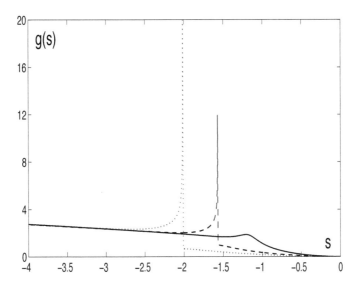

FIGURE 10.20
Main term of the asymptotic of the pressure distribution $g(s)$ in the exit zone
of a starved lubricated contact for $A = 1$ (solid curve), $A = 1.5$ (dashed
curve), $A = 2$ (dotted curve), and $\mu_g(g) = e^{Q_0 g}$, $Q_0 = 2.5$, $\triangle r = 0.00335$,
$N = 2,000$.

Let us consider the sensitivity of the asymptotic solutions for starved lubrication regimes (i.e., for $h_q(r) = h_g(s) = 1$) for the case of elevated Q_0. The
values of the inlet coordinate α_1 as a function of the step size $\triangle r$ and number
of nodes N are given in Table 10.6) for $A = 2$, $Q_0 = 2.5$, and the precision
$\varepsilon = 10^{-3}$. The values of the pressure spike maximum $\max g(s)$ in the exit zone,
the exit coordinate β_1, and the system's Jacobian condition number $C.N.$ as
functions of the step size $\triangle r$ and number of nodes N are given in Table 10.7
for $A = 2$, $Q_0 = 2.5$, and $\varepsilon = 10^{-3}$. The relationship between the step size
$\triangle r$ and the number of nodes N is kept in such a way that $N\triangle r \approx 6.7$. It
can be clearly seen that in the inlet zone α_1 as a function of $\triangle r$ reaches its
limiting/constant value for relatively large step sizes $\triangle r \leq 0.0021875$. In the
exit zone for large s pressure $g(s)$ is well behaved for all values of N, i.e., it
approaches its asymptote and it is stable. Typical examples of the behavior of
$g(s)$ for $A = 1$, 1.5, 2, $\mu_g(g) = e^{Q_0 g}$, $Q_0 = 2.5$, $\triangle r = 0.00335$, and $N = 2,000$
are presented in Fig. 10.20. It is clear from Fig. 10.20 that for sufficiently
large values of A and/or Q_0 the pressure spike is extremely thin and high (pin
like) while for relatively small values of A and Q_0 (for example, for $A = 1$
and $Q_0 = 2.5$) the pressure spike is reasonably wide and not excessively high.

For the values of A and Q_0 for which the pressure spike is thin and high its maximum $\max g(s)$ in the exit zone remains sensitive to the value of the step size $\triangle r$ even for $\triangle r < 0.00168$ and $N > 4{,}000$ (see Table 10.7). The system's Jacobian condition number $C.N.$ is high for all values of N and it increases from $1.37 \cdot 10^9$ to $1.3 \cdot 10^{11}$ as N increases from $1{,}200$ to $4{,}000$ and $\triangle r$ decreases from 0.006 to 0.00168. Compare these condition numbers with the condition number $C.N. = 2.81 \cdot 10^8$ for the case of a stable solution for $A = 1$, $\mu_g(g) = e^{Q_0 g}$, $Q_0 = 2.5$, $\triangle r = 0.00335$, and $N = 2{,}000$ (see Fig. 10.20) which is much lower (about two orders of magnitude) than the ones for the cases presented in Table 10.7 for which the solutions are unstable. This elevated solution sensitivity as well as small oscillations of the derivative of pressure $dg(s)/ds$ (not the distribution of pressure $g(s)$ itself) between the position of the pressure spike and the exit point $s = 0$ are the indications of instability. This instability is caused by a strong dependence of the viscosity $\mu_g(g) = e^{Q_0 g}$ on g or equally by high value of constant A. This is in perfect agreement with the fact that the values of the system's Jacobian condition number $C.N.$ are very high (see Table 10.7). The unstable conditions are described by inequality $Q_0 A^{3/4} > 3$ (see transformation (10.289) and its discussion). For values of A and Q_0 such that $Q_0 A^{3/4} \leq 3$, the solution in the exit zone is perfectly stable.

For fully flooded lubrication regimes the typical pressure behavior in the inlet and exit zones is similar to the ones for starved lubrication regimes. The solutions for pressure $q(r)$ and gap $h_q(r)$ in the inlet zone are stable for all combinations of values of Q_0 and A while the solutions for pressure $g(s)$ and gap $h_g(s)$ in the exit zone for elevated values of Q_0 demonstrate instability. The instability presents itself by small oscillations of gap $h_g(s)$ in the area between the pressure spike and the exit point $s = 0$ and by producing an excessively high pressure spike. The former is caused by small oscillations of the derivative $dg(s)/ds$ in the aforementioned area. However, the former and latter manifestations of instability are caused by strong dependence of viscosity $\mu_g(g)$ on pressure g. The only remedy to this instability situation in the exit zone is a proper regularization. The gap oscillations are remedied by a simple averaging regularization procedure described earlier in this section. That causes some loss of precision, i.e., the discrepancy in the Reynolds equation increases but still remains within acceptable range of ± 0.015. A more sophisticated approach to regularization will be described later.

At the same time, in the inlet zone the problem solution is stable for all combinations of values of Q_0 and A. Therefore, film thickness H_0 can be determined in a stable manner in the inlet zone as it depends on the solution for $A = A(\alpha_1)$ only in the inlet zone. It is sufficient to know film thickness H_0 to be able to determine the sliding frictional stress $f(x)$ and the sliding friction force F_S (see Section 10.20) as they mainly depend on pressure $p(x)$ in the Hertzian region which for heavily loaded lubricated contacts is close to the Hertzian pressure $\sqrt{1 - x^2}$. The latter can be considered as one of the advantages of the asymptotic approach in comparison with the solution of the original (non-asymptotic) problem, which involves dealing with a possibly numerically unstable solution in the exit zone.

After analyzing the numerical results obtained from solution of asymptotically valid equations in the inlet and exit zones, we can make a conclusion about the value of the step size $\triangle y$, which would provide the adequate precision for the solution of the original (non-asymptotic) EHL problem. As the previous consideration shows the major limitation to achieving the desired precision of the solution of the original EHL problem in the most of the contact region (excluding the exit zone) is the ability to resolve the inlet zone sufficiently accurately. Suppose the step size $\triangle r_*$ provides the sufficient precision to the solution of the asymptotically valid equations in the inlet zone. Then it is clear that to get the same solution details and precision from a numerical solution of the original EHL problem for pre-critical lubrication regimes it is necessary to use the step size $\triangle y = \frac{2}{c-a}\epsilon_q \triangle r_*$. Similarly, for over-critical lubrication regimes the step size for solution of the original EHL problem should be $\triangle y = \frac{2}{c-a}\epsilon_0 \triangle r_*$ (recall that for over-critical regimes $\epsilon_q \gg \epsilon_0$). Therefore, if $\epsilon_q \ll 1$ and $\epsilon_0 \ll 1$ to get an accurate numerical solution of the original EHL problem, we should require that step size $\triangle y \ll \triangle r_*$. For example, for $V = 10^{-3}$, $Q = 1$, $a = -2$, and $c \approx 1$ for pre-critical lubrication regimes the step size should not be greater than $\triangle y = 0.000092015$ assuming that $\triangle r_* = 0.0021875$. For the case of evenly spaced nodes, the latter step size translates into the necessity to have more than 10,868 nodes in the lubricated contact region $[a, c]$. We have to keep in mind that in the Hertzian region the requirements on the step size can be significantly relaxed. In spite of that it is clear that this requirement on the step size still puts severe demands on computer resources required for solution of EHL problems for line contacts with high precision. Problems for lubricated rough surfaces in line contacts require even more detailed description and that imposes even higher demands on computer memory and speed.

Keeping in mind that the properties of the EHL problems for smooth point (two-dimensional) contacts are similar to the problem for line contacts the absolute precision of $\epsilon = 10^{-4}$ can be guaranteed for $V = 10^{-3}$, $Q = 1$ if the step size is not greater than 0.000092015 and the number of equidistant nodes is not less than $10,868 \times 10,868 = 118,113,424$. Of course we need to realize that in the Hertzian region and in the direction perpendicular to lubricant flow the grid can be more coarse. Nonetheless, that makes it practically impossible to solve such problems adequately accurate. The situation with rough point EHL contacts is even worse.

However, in most numerical studies of the original EHL problem the step size is chosen independently of the above condition and it is much larger than the one which would provide sufficiently high accuracy of the solution. To determine numerical convergence it is customary to use the comparison of numerical solutions obtained for two step sizes $\triangle y_1$ and $\triangle y_2$ such that $\triangle y_2 = 0.5 \triangle y_1$. The above analysis shows that even if they differ by just a decimal of a percent it does not provide the necessary assurance of the solution sufficient precision.

To conclude this section it is important to emphasize the advantages of the asymptotic approach. They are due to the fact that the asymptotic equations contain smaller number of the original problem input parameters and the numerical solutions of these equations are always stable in the inlet zone and easy to obtain. In the exit zone the problem solution is stable in a certain range of the input parameters. Moreover, the presented asymptotic analysis provides simple structural formulas for film thickness. The subsequent numerical analysis of the asymptotic equations in the inlet zone supplies the only coefficient in the film thickness formulas that is not determined analytically but is obtained numerically. The values of this coefficient finalize the film thickness formulas.

10.9 Analysis of EHL Contacts for Soft Solids

The purpose of this section is to provide an opportunity of a new prospective on classic EHL problems. We propose and analyze a new formulation of EHL problems that explains and predicts the abnormal phenomena observed in experiments with hard and soft lubricated solids by M. Kaneta and associates [40, 41]. For the first time, they have described the formation of a dimple in a heavily loaded lubricated contact of a relatively soft elastic solid with a hard one. The occurrence of a dimple cannot be predicted by the classic EHL theory. Kaneta et al. [40, 41] believe that the abnormal behavior of lubricated elastic surfaces can be explained by invoking solidification of oil in contact. Another interpretation of the abnormal phenomena based on a non-steady approach to the EHL problem is offered by Cermk [42]. However, the results of this approach contradict the steady nature of the experimental data presented by Kaneta et al. [40, 41]. Recently a number of numerical studies on this subject were published by Kaneta and Young [43, 44]. The main idea of the latter studies is that the dimple phenomenon is caused by the heat generation in the lubrication layer and its dissipation in the solids. However, the formulation of the problem in which it is studied in [43, 44] has a serious defect. Namely, for sufficiently high slide-to-roll ratios, the lubricant flows in opposite directions near two contact surfaces. Therefore, at the separatrix between the two oppositely directed flows, a condition for continuity of heat flux must be imposed. That condition was not used by Kaneta and Young in their studies. In spite of the fact that heat generation is important under high sliding conditions the lack of the above–mentioned boundary condition invalidates their analysis. Moreover, in spite of the importance of lubricant heat generation and its dissipation in solids the Kaneta and Young [43, 44] solution does not and cannot predict all the features (a)-(d) of such a lubricated contact observed in experiments (see Subsection 10.9.3.2).

The new formulation of the problem [45] is based on the traditional for the EHL theory approach to steady lubricated contacts as contacts of two elastic solids separated by a fluid film without oil solidification phenomenon. A new formulation of a steady problem for lubricated rollers made of elastic materials with low Young's modulus is considered. The distinct feature of this formulation that differs it from other ones is that the linear velocities of the surfaces take into account the tangential displacements of elastic surfaces. Therefore, in Reynolds equation the surface linear velocities are presented by functions of the location in the contact region instead of constants. This formulation explains a deep depression (dimple) produced in contact surfaces of soft materials. These dimples occur in place of flat surfaces which suppose to occur according to the classic EHL theory. A qualitative and numerical analysis of the new EHL problem formulation is presented. The dependence of dimple sizes on problem parameters is considered.

10.9.1 Formulation of a Steady EHL Problem for Soft Solids

Let us consider a pretty much standard formulation of a plane EHL problem. Suppose two elastic circular cylinders, slowly rolling one over another, with parallel axes and radii, R_1 and R_2. Cylinder 1 is acted upon by force P along the z-axis through the cylinders centers in a Cartesian coordinate system with the x-axis along the lubrication layer. The linear velocities of the surface points, located far away from the contact zone, are u_1 for cylinder 1 and u_2 for cylinder 2.

The cylinders are fully separated by the lubrication layer $h(x)$ (see Fig. 10.21), which is formed by a Newtonian incompressible fluid with viscosity μ, which depends on lubricant pressure p. The lubrication process is isothermal. Because of slow motion of the contact surfaces in lubrication layer, the inertial forces are much smaller than viscous forces. Let us assume that the lubrication layer is relatively small with regard to the characteristic size of the contact, which, in turn, is much smaller than the cylinders' radii.

Under the above assumptions in the dimensional variables, the contact with Newtonian lubricant (which according to (10.12) corresponds to the rheological function $F(x) = x$) is described by equations equations very similar to equations (10.17)- (10.21)

$$\frac{d}{dx}\left[\frac{h^3}{12\mu}\frac{dp}{dx} - \frac{v_1+v_2}{2}h\right] = 0, \tag{10.293}$$

$$p(x_i) = p(x_e) = \frac{dp(x_e)}{dx} = 0, \tag{10.294}$$

$$h = h_e + \frac{x^2-x_e^2}{2R'} + \frac{2}{\pi E'}\int\limits_{x_i}^{x_e} p(t)\ln\mid\frac{x_e-t}{x-t}\mid dt, \tag{10.295}$$

$$\int\limits_{x_i}^{x_e} p(t)dt = P, \tag{10.296}$$

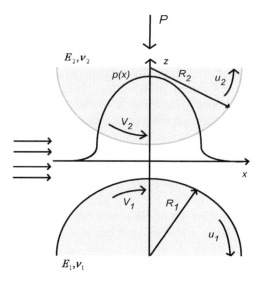

FIGURE 10.21
The general view of a lubricated contact.

where v_1 and v_2 are linear velocities of contact surfaces within the contact region, which may depend on x, x_i and x_e are the inlet and exit coordinates of the contact, h_e is the exit film thickness, R' and E' are the effective radius and elastic modulus of the solids (see Subsection 10.3.1).

10.9.2 Qualitative Analysis of the EHL Problem for Soft Solids

The Reynolds equation (10.293) can be rewritten in the form

$$\frac{dQ_x(x)}{dx} = 0, \quad Q_x(x) = U(x)h(x), \quad U(x) = \frac{v_1+v_2}{2} - \frac{h^2}{12\mu}\frac{dp}{dx}, \qquad (10.297)$$

where Q_x is the x-component of the lubricant flux and $U(x)$ is the average of the lubricant flow velocity. Integration of the above equation gives

$$U(x) = \frac{U_0}{h(x)}, \quad U_0 = U(x_i) = constant. \qquad (10.298)$$

For a heavily loaded contact in the central part of the contact region, Hertzian region (far from the inlet x_i and exit x_e points), the second term of the equation for $U(x)$ is negligibly small in comparison to the first one (see the second estimate in (10.95)). Therefore,

$$U(x) \approx \frac{v_1+v_2}{2}, \qquad (10.299)$$

and in the central part of the contact region equation (10.298) gives

$$\frac{v_1+v_2}{2} \approx \frac{U_0}{h(x)}. \tag{10.300}$$

Thus, the conclusion that the linear velocities of contact surfaces, v_1 and v_2, are dependent upon x follows from Kaneta's experimental fact that gap h is a function of x. This result is true only for the Hertzian zone in a heavily loaded lubricated contact.

If the shear moduli of elasticity G_k ($k = 1, 2$) of the solid materials in contact are large enough then the tangential displacements of contact surfaces are negligibly small. In this case the surface linear velocities, v_1 and v_2, are practically independent from x and can be accurately approximated by the "rigid" linear velocities, u_1 and u_2. Therefore, according to the classic formulation of the EHL problem the gap is flat in the Hertzian zone of the contact region, i.e., it is independent from x.

The experimental data presented by Kaneta et al. [41] shows that in certain cases of contacts between soft and hard elastic materials gap $h(x)$ significantly varies within the central part of the contact region. The above variations of $h(x)$ represent an abnormal shape of the contact surfaces in a heavily loaded EHL contact. As it follows from the latter equation the average surface velocity, $(v_1 + v_2)/2$, varies significantly with x. This phenomenon is caused by relatively large tangential displacements of soft elastic surfaces.

The conclusions made for incompressible lubricants are also valid for compressible lubricants and lubricants with non-Newtonian rheology. This is based on the fact that the lubricant compressibility and rheology affect the gap between the contact surfaces in the inlet and exit zones of a heavily loaded contact and practically do not affect the gap in the Hertzian region of the contact. The same conclusions are valid for non-isothermal EHL problems as well.

The described variations in surface velocities, v_1 and v_2, reflect tangential displacements of contact surfaces. Therefore, in order to describe significant variations in surface velocities, the tangential displacements of relatively soft elastic surfaces have to be taken into consideration.

10.9.3 Surface Velocities for Soft Solids

Let us take into account the elastic displacements of contact surfaces, U_k ($k = 1, 2$), in the tangential to the surfaces direction. Then under steady conditions the equations for surface linear velocities can be written as follows (see Galin [46]):

$$v_k = u_k(1 + \frac{dU_k}{dx}), \quad \frac{\pi E_k}{2(1-\nu_k^2)}\frac{dU_k}{dx} = -\frac{1-2\nu_k}{2-2\nu_k}\pi p + \int\limits_{x_i}^{x_e} \frac{\tau_k dt}{t-x}, \tag{10.301}$$

where τ_k $(k = 1, 2)$ are the tangential stresses created in a lubricant and acting upon the adjacent material surfaces

$$\tau_1 = \frac{\mu(v_2-v_1)}{h} - \frac{h}{2}\frac{dp}{dx}, \quad \tau_2 = -\frac{\mu(v_2-v_1)}{h} - \frac{h}{2}\frac{dp}{dx}. \qquad (10.302)$$

Using equations (10.301) and (10.302) the following linear singular integral equation can be derived for the slip velocity $s = v_2 - v_1$:

$$s = \left[\frac{(1-2\nu_1)u_1}{G_1} - \frac{(1-2\nu_2)u_2}{G_2}\right]p - \frac{2}{\pi}\left[\frac{(1-\nu_1)u_1}{G_1} + \frac{(1-\nu_2)u_2}{G_2}\right]\int_{x_i}^{x_e} \frac{\mu s}{h}\frac{dt}{t-x}$$

$$(10.303)$$

$$+\frac{1}{\pi}\left[\frac{(1-\nu_1)u_1}{G_1} - \frac{(1-\nu_2)u_2}{G_2}\right]\int_{x_i}^{x_e} h\frac{dp}{dt}\frac{dt}{t-x} + u_2 - u_1.$$

After equation (10.303) is solved for s from equations (10.301) and (10.302) one can determine

$$\frac{v_1+v_2}{2} = \frac{u_1+u_2}{2} - \frac{1}{2}\left[\frac{(1-2\nu_1)u_1}{G_1} + \frac{(1-2\nu_2)u_2}{G_2}\right]p$$

$$+\frac{1}{\pi}\left[\frac{(1-\nu_1)u_1}{G_1} - \frac{(1-\nu_2)u_2}{G_2}\right]\int_{x_i}^{x_e} \frac{\mu s}{h}\frac{dt}{t-x} \qquad (10.304)$$

$$-\frac{1}{2\pi}\left[\frac{(1-\nu_1)u_1}{G_1} + \frac{(1-\nu_2)u_2}{G_2}\right]\int_{x_i}^{x_e} h\frac{dp}{dt}\frac{dt}{t-x}.$$

Obviously, if $u_1 = u_2$, $G_1 = G_2$, and $v_1 = v_2$ the solution of equation (10.303) is $s(x) = 0$ and the expression (10.304) for the average surface velocity can be simplified. Furthermore, if the shear moduli G_1 and G_2 of contact materials are much larger than the maximum Hertzian pressure p_H then $v_k = u_k + O(p_H/G_1, p_H/G_2)$ and the classic EHL problem formulation is valid. However, if p_H/G_1 and/or p_H/G_2 are not very small (the shear moduli G_1 and G_2 are comparable with p_H), then the actual average of surface velocities $(v_1 + v_2)/2$ may significantly differ from the "rigid" one $(u_1 + u_2)/2$. Therefore, the new problem formulation described by equations (10.293) -(10.296), (10.303), and (10.304) must be used.

Similarly, a spacial EHL problem, which takes into account tangential deformations of elastic contact surfaces, can be formulated.

10.9.3.1 Dimensionless Variables and Numerical Method

By using the dimensionless variables and parameters

$$\{s', v'_{1,2}\} = \{s, v_{1,2}\}\frac{2}{u_1+u_2}, \quad s_0 = 2\frac{u_2-u_1}{u_2+u_1}, \quad m_{1,2} = \frac{p_H}{G_{1,2}},$$

$$\alpha_{1,2} = (1 - \nu_{1,2})m_{1,2}, \quad \beta_{1,2} = (1 - 2\nu_{1,2})m_{1,2}, \quad A_0 = \alpha_1 + \alpha_2,$$

$$(10.305)$$

$$\{A_1, B_1\} = \{\alpha_1, \beta_1\}(1 - \tfrac{s_0}{2}), \quad \{A_2, B_2\} = \{\alpha_2, \beta_2\}(1 + \tfrac{s_0}{2}),$$

$$A = A_0(A_1 + A_2), \quad B = B_1 - B_2,$$

$$C = A_0(A_1 - A_2), \ D = B_1 + B_2,$$

in addition to dimensionless variables (10.77) and (10.78) the problem can be reduced to the system of integro-differential equations (further the primes at dimensionless variables are omitted)

$$\frac{H_0^2}{V} \frac{h^3}{\mu} \frac{dp}{dx} = \tfrac{1}{2}[(v_1 + v_2)h - v_1(c) - v_2(c)], \ \ p(a) = p(c) = 0, \tag{10.306}$$

$$\frac{v_1 + v_2}{2} = 1 - \frac{D}{2}p + \frac{CV}{12\pi H_0} \int\limits_a^c \frac{\mu s}{h} \frac{dt}{t - x} - \frac{AH_0}{2\pi} \int\limits_a^c \frac{h\,dp}{t - x}, \tag{10.307}$$

$$s + \frac{AV}{2\pi H_0} \int\limits_a^c \frac{\mu s}{h} \frac{dt}{t - x} = s_0 + Bp + \frac{CH_0}{\pi} \int\limits_a^c \frac{h\,dp}{t - x}, \tag{10.308}$$

$$H_0(h - 1) = x^2 - c^2 + \tfrac{2}{\pi} \int\limits_a^c p(t) \ln \mid \tfrac{c - t}{x - t} \mid \, dt, \tag{10.309}$$

$$\int\limits_a^c p(t)dt = \tfrac{\pi}{2}. \tag{10.310}$$

In equation (10.309) for gap $h(x)$ we still neglected the influence of the surface tangential displacements.

The problem is completely described by equations (10.306)-(10.310). In addition to these equations the expression for the lubricant viscosity $\mu(p)$ as a function of pressure p and the values of constants a, V, A, B, C, and D must be given. The solution of the problem comprises pressure $p(x)$, gap $h(x)$, and slip velocity $s(x)$ distributions as well as coordinate of the exit boundary c and and film thickness H_0.

The described problem can be solved by using iterations. Let us give a general description of the calculation process for one iteration. Suppose the initial approximations for $p(x)$, $h(x)$, H_0, and c are known. First, we introduce a substitution

$$x = \frac{c + a}{2} + \frac{c - a}{2}y,$$

which maps interval $[a, c]$ onto $[-1, 1]$ to replace the problem with the unknown boundary $x = c$ by a problem with known boundaries. That allows us to introduce two partitions: $\{y_i\}$, $i = 1, \ldots, I$, $y_1 = -1$, $y_I = 1$ and $y_{i+1/2} = (y_i + y_{i+1})/2$, $i = 1, \ldots, I-1$. The solution of equation (10.308), $s_k = s(y_k)$, is determined at nodes $\{y_i\}$, $i = 1, \ldots, I$, from the system of I linear algebraic equations

$$s_k + \frac{AV}{24\pi H_0} \sum_{i=1}^{I-1} \frac{\mu_{i+1/2}(s_i + s_{i+1})}{h_{i+1/2}} \frac{y_{i+1} - y_i}{y_{i+1/2} - y_k} = Bp_k$$

$$+ \frac{CH_0}{\pi} \sum_{i=1}^{I-1} \frac{h_{i+1/2}(p_{i+1} - p_i)}{y_{i+1/2} - y_k}, \ \ k = 1, \ldots, I. \tag{10.311}$$

In equation (10.311) we used a quadrature formula similar to the ones used in [38]. Having the values of s_k and using the proper quadrature formulas for calculation of singular integrals [38] the average velocity $(v_1(y) + v_2(y))/2$ is determined from equation (10.307) at nodes $\{y_{i+1/2}\}$, $i = 1, \ldots, I-1$, and for $y_I = 1$ using a similar quadrature formula for calculation of singular integrals. After that equations (10.306), (10.309), and (10.310) are solved for $p(y)$, $h(y)$, H_0, and c by one of the well-known methods, for instance, the method presented in Section 10.7. The iterations stop when consecutive iterates satisfy the desired accuracy.

10.9.3.2   Numerical Results and Discussion

According to Kaneta et al. [41], certain specific features (abnormal phenomena) of a lubricated contact were observed experimentally and were linked to the existence of a dimple in EHL contacts:

(a) in some cases when a soft material is involved in a contact a dimple was produced while in other cases when both surfaces are made from a hard material no dimple was observed,

(b) no noticeable dimple was observed when a surface made from a hard material was moving and its counterpart made from a soft material was stationary,

(c) a dimple was observed in the contact region between hard and soft materials for slide-to-roll ratio $s_0 = 2$ and it was not observed for $s_0 = -2$;

(d) a dimple got more pronounced with an increase in slide-to-roll ratio s_0 and its position moved toward the inlet point while the minimum film thickness decreased only slightly.

For further analysis the lubricant viscosity, μ, will be used according to the exponential law, $\mu = \exp(Qp)$, where Q is the dimensionless pressure coefficient of viscosity. Moreover, it will be assumed that the inlet coordinate $a = -2$, the total number of computational nodes $I = 100$, and the relative precision of numerical solutions is 10^{-4}.

Let us consider each of the aforementioned features of the phenomenon and then find out whether the solution of an EHL problem in the new formulation (10.306)-(10.310) possesses these features. It is necessary to keep in mind that in this section the notions of soft and hard materials involve not only their elastic properties but also the applied loading (see formulas for parameters $m_{1,2}$, A, B, C, and D in (10.305)). The materials are called hard if $m_{1,2}$, A, B, C, and D are very small and soft otherwise.

Feature (a). Let us consider a lubricated contact of two hard materials. Then, according to (10.305) $m_{1,2} \ll 1$ and constants A, B, C, and D in equations (10.307) and (10.308) are small. Therefore, s and $(v_1 + v_2)/2$ are approximately equal to the "rigid" slip s_0 and 1, respectively. As a result, the EHL problem is identical to the classic one. In Sections 10.4 and 10.6 based on the classic formulation it has been shown that the film thickness is flat in the Hertzian zone of a heavily loaded contact. Two examples of such

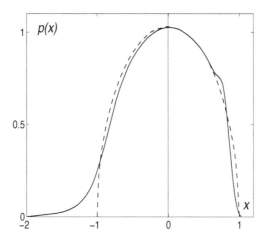

FIGURE 10.22
Distributions of pressure $p(x)$ (solid curve) in a lubricated contact ($V =$ 0.1, $Q = 4$) and the Hertzian pressure (dotted curve) in a dry contact (after Kudish [45]). Reprinted with permission from the ASME.

classic solutions for hard surfaces (with constants $A = B = C = D = 0$), and parameters $V = 0.1$, $Q = 4$, and $Q = 6.25$, respectively, are represented in Figs. 10.22, 10.23 and 10.24, 10.25. These solutions are independent from the slide-to-roll ratio s_0. According to the classic problem formulation gap distribution $h(x)$ is virtually flat in the central part of the contact as shown in Fig. 10.23 and 10.25.

A similar behavior of a solution for equations (10.306)-(10.310) is observed for hard materials ($\nu_{1,2} = 0.3$, $m_{1,2} = 0.003$, $A = 0.00003$, $B = -0.0037$, $C = -0.00004$, $D = 0.00312$, $s_0 = 2.375$, $V = 0.1$, and $Q = 4$). In particular, the average velocity $(v_1 + v_2)/2$ differs from 1 by less than 0.002, and in the Hertzian region the distribution of gap $h(x)$ differs from 1 by less than 0.008. A dimple is not observed in these solutions of classic and modified problems. On the contrary, for surfaces made from soft and hard materials (see Figs. 10.26 and 10.27)* with $\nu_1 = 0.3$, $\nu_2 = 0.1$, $m_1 = 0.003$, $m_2 = 0.25$, $A = 0.15832$, $B = -0.56463$, $C = -0.15908$, $D = 0.56287$, $s_0 = 3.125$, $V = 0.1$, and $Q = 4$ there is a pronounced dimple. The surface average velocity, $(v_1 + v_2)/2$, differs from 1 significantly in the central part of the contact. In Figure 10.27 and in the other graphs that follow the frictional stress which is

---

*In Figs. 10.27, 10.29, 10.31-10.33 the left scale is for values of gap $h(x)$ and average velocity $\frac{v_1(x)+v_2(x)}{2}$ while the right scale is for sliding frictional stress $f(x) = \frac{V}{12H_0}\frac{\mu s}{h}$.

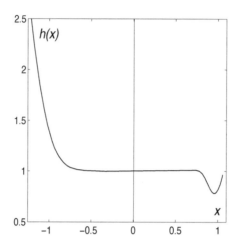

FIGURE 10.23

Gap distributions $h(x)$ in a lubricated contact ($V = 0.1$, $Q = 4$) (after Kudish [45]). Reprinted with permission from the ASME.

usually much greater than the rolling frictional stress in the Hertzian region of the contact is represented by just its sliding part.

Therefore, the EHL model described by equations (10.306)-(10.310) represents the experimentally observed by Kaneta et al. [41] **Feature (a)** adequately.

Feature (b). Let us consider a stationary surface made from a soft material, i.e., $u_1 = 0$. Then from equation (10.301) follows that $v_1 = 0$. In equations (10.301) for the other surface made from a hard material the actual surface velocity v_2 is practically equal to the "rigid" one u_2. In such cases the values of dimensionless parameters A, B, C, and D in (10.307) are small and $(v_1 + v_2)/2 \approx 1$. That brings us back to the classic EHL problem. Therefore, similarly to the classic formulation of the EHL problem the gap is flat in the Hertzian region of a heavily loaded contact and a dimple is not observed. This finding is in agreement with the experimental data by Kaneta et al. [41].

Feature (c). In the case of a contact of soft and hard materials represented in Figs. 10.26 and 10.27 which are obtained for $\nu_{1,2} = 0.3$, $m_1 = 0.003$, $m_2 = 0.15$, $A = -C = 0.3801$, $B = -D = -0.156$, $V = 0.1$, and $Q = 6.25$ for the slide-to-roll ratio $s_0 = 2$ a pronounced dimple is produced in the contact region. However, in Figs. 10.28 and 10.29 for the same materials and loading parameters as in Figs. 10.28 and 10.29 and the slide-to-roll ratio $s_0 = -2$ ($\nu_{1,2} = 0.3$, $m_1 = 0.003$, $m_2 = 0.15$, $A = C = 0.00076$, $B = D = 0.00312$, $V = 0.1$, and $Q = 6.25$) no noticeable dimple is observed. This finding is in agreement with the experimental data by Kaneta et al. [41].

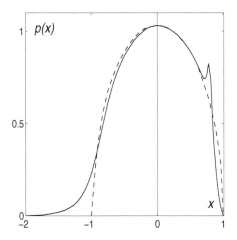

FIGURE 10.24
Distributions of pressure $p(x)$ (solid curve) in a lubricated contact ($V = 0.1$, $Q = 6.25$) and the Hertzian pressure (dotted curve) in a dry contact (after Kudish [45]). Reprinted with permission from the ASME.

Feature (d). As it follows from Figs. 10.32 and 10.33 a dimple gets more pronounced with increase of the slide-to-roll ratio s_0 and its center moves toward the inlet point. The graphs represented in Figs. 10.32 and 10.33 are obtained for $\nu_1 = 0.3$, $\nu_2 = 0.1$, $m_1 = 0.003$, $m_2 = 0.25$, $V = 0.1$, $Q = 4$, $s_0 = 0$ (i.e., $A = 0.06262$, $B = -0.21844$, $C = -0.06125$, $D = 0.22156$) and $s_0 = 2.375$ (i.e., $A = 0.13535$, $B = -0.48154$, $C = -0.1356$, $D = 0.48096$), respectively. In these graphs the minimal film thickness decreases slightly as s_0 increases. Therefore, the EHL model described by equations (10.306)-(10.310) represents **Feature (d)** experimentally observed by Kaneta et al. [41] adequately.

To give a better idea of how the depth of a dimple changes with the value of the slide-to-roll ratio s_0 some numerical data for two series of problem parameters (Series 1: $\nu_{1,2} = 0.3$, $m_1 = 0.003$, $m_2 = 0.25$, $V = 0.1$, $Q = 5$ and Series 2: $\nu_1 = 0.3$, $\nu_2 = 0.1$, $m_1 = 0.003$, $m_2 = 0.25$, $V = 0.1$, $Q = 4$) is presented in Tables 10.8 and 10.9, respectively. The depth of a dimple, d, is determined as the difference between the maximum gap in the Hertzian zone of the contact and the minimum gap in the inlet zone of the contact. The data are presented in absolute values and as a ratio to the minimum value of the gap in the inlet zone of the contact.

The new formulation of the EHL problem presented in this section gives an adequate generalization of the classic EHL problem. The new problem formulation is validated by an analytical and numerical analysis as well as by the

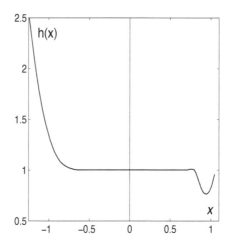

FIGURE 10.25
Distribution of gap $h(x)$ in a lubricated contact ($V = 0.1$, $Q = 6.25$) (after Kudish [45]). Reprinted with permission from the ASME.

experimental data obtained by Kaneta et al. [41]. The quantitative differences between the data described in this section and the data obtained by Kaneta et al. [41] are caused by a variety of reasons such as different contact geometry, different contact operating parameters, different elasticity parameters, oil compressibility, thermal deformations, and onset of plastic deformations. The onset of material plastic deformations can be realized under certain conditions, in particular, when the material is soft and frictional stresses are sufficiently high. The latter may happen when the pressure coefficient of viscosity is large. Different from Kaneta et al. [41] operating parameters are chosen due to numerical stability considerations.

TABLE 10.8
The dependence of dimple depth d on slide-to-roll ratio s_0 for Series 1 (after Kudish [45]). Reprinted with permission from the ASME.

s_0	-2	0	2	2.375
d	no dimple	no dimple	0.1753	0.2088
%	0	0	18.1%	21.9%

Moreover, it is important to mention that the occurrence of the abnormal

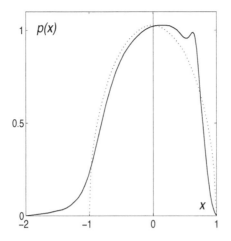

FIGURE 10.26
Distributions of pressure $p(x)$ (solid curve) in a lubricated contact ($\nu_1 = 0.3$, $\nu_2 = 0.1$, $m_1 = 0.003$, $m_2 = 0.25$, $A = 0.1583$, $B = -0.5646$, $C = -0.1591$, $D = 0.5629$, $s_0 = 3.125$, $V = 0.1$, $Q = 4$) and the Hertzian pressure (dotted curve) in a dry contact (after Kudish [45]). Reprinted with permission from the ASME.

phenomena (dimples) does not seem to be linked to oil solidification. In order to illustrate this fact let us perform a simple analysis. It is widely accepted that oil solidification is observed under high pressures regardless of contact kinematics. Such high pressures can be created in a lubricated contact region between a moving steel ball (with high elasticity modulus) and a stationary glass (with low elasticity modulus). Therefore, under such conditions oil solidification is possible. However, in the aforementioned case of high pressures a noticeable dimple is not produced. This has been described by the analysis of **Feature (b)** without taking into account the phenomenon of oil solidification.

TABLE 10.9
The dependence of dimple depth d on slide-to-roll ratio s_0 for Series 2 (after Kudish [45]). Reprinted with permission from the ASME.

s_0	-2	0	$.3752$	3.125
d	no dimple	0.0404	0.2364	0.3431
%	0	3.7%	22.5%	33.3%

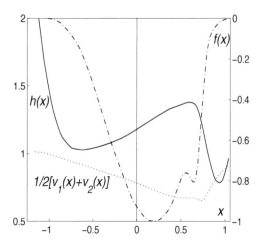

FIGURE 10.27
Distributions of gap $h(x)$ (solid curve), average velocity $\frac{v_1(x)+v_2(x)}{2}$ (dotted curve), and sliding frictional stress $f(x)$ (dash-dotted curve) in a lubricated contact ($\nu_1 = 0.3$, $\nu_2 = 0.1$, $m_1 = 0.003$, $m_2 = 0.25$, $A = 0.1583$, $B = -0.5646$, $C = -0.1591$, $D = 0.5629$, $s_0 = 3.125$, $V = 0.1$, $Q = 4$) (after Kudish [45]). Reprinted with permission from the ASME.

In general, in heavily loaded contacts an EHL problem solution for the new formulation (10.306)-(10.310) has properties similar to those of the classic EHL problem solution except for the dimple phenomenon. In particular, the similarities are noted for certain problem parameters when a spike of pressure is observed in the exit zone of the contact. Using asymptotic approach of the preceding sections, it can be shown that in fully flooded heavily loaded lubrication regimes (and in most other cases too), which are described by equations (10.306)-(10.310), the pressure distribution in the Hertzian zone of the contact region is close to the Hertzian one. Moreover, it can be shown that for these regimes lubrication film thickness H_0 is determined by elastic and lubrication processes in the inlet zone of the contact region. This fact explains why H_0 is practically independent from the slide-to-roll ratio s_0. For example, $H_0 = 0.18004$ for $s_0 = 0$ and $H_0 = 0.19123$ for $s_0 = 2.375$, while parameters $\nu_{1,2}$, $m_{1,2}$, V, and Q are the same as in Figs. 10.32 and 10.33, $H_0 = 0.22013$ for $s_0 = -2$ and $H_0 = 0.23224$ for $s_0 = 2$ while parameters $\nu_{1,2}$, $m_{1,2}$, V, and Q are the same as in Figs. 10.31 and 10.29. Therefore, for the aforementioned cases the solutions of the modified and classic EHL problems are practically identical in the inlet zone. As a result, the film thickness H_0 increases with Q and V according to the formulas obtained in the preceding sections. The difference between the solutions of the classic and new EHL problems mainly

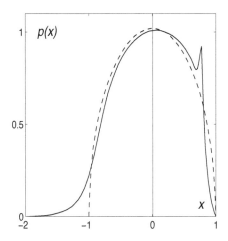

FIGURE 10.28
Distributions of pressure $p(x)$ (solid curve) in a lubricated contact ($\nu_{1,2} = 0.3$, $m_1 = 0.003$, $m_2 = 0.15$, $A = -C = 0.3801$, $B = -D = -0.156$, $V = 0.1$, $Q = 6.25$, $s_0 = 2$) and the Hertzian pressure (dotted curve) in a dry contact (after Kudish [45]). Reprinted with permission from the ASME.

occurs (if at all) in the Hertzian region and it depends solely on the values of parameters A, B, C, D, and s_0.

The material of the section presents a new adequate formulation of EHL problems for soft elastic materials. The properties of solutions of the new EHL problem (in particular, the existence of a dimple) are consistent with those observed in experiments. A qualitative analysis of a lubricated contact for soft elastic materials has shown that if the gap between lubricated contact surfaces varies substantially in the Hertzian region of a heavily loaded contact then the surface linear velocities must vary significantly as well. Therefore, the tangential displacements of elastic surfaces have been taken into proper account in a new formulation of the EHL problem, which has been analyzed both numerically and analytically. Furthermore, the section shows that oil solidification does not seem to relate to dimple existence.

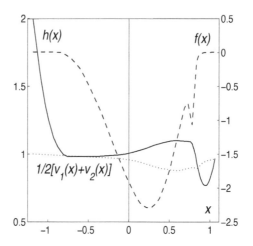

FIGURE 10.29
Distributions of gap $h(x)$ (solid curve), average velocity $\frac{v_1(x)+v_2(x)}{2}$ (dotted curve), and sliding frictional stress $f(x)$ (dash-dotted curve) in a lubricated contact ($\nu_{1,2} = 0.3$, $m_1 = 0.003$, $m_2 = 0.15$, $A = -C = 0.3801$, $B = -D = -0.156$, $V = 0.1$, $Q = 6.25$, $s_0 = 2$) (after Kudish [45]). Reprinted with permission from the ASME.

10.10 Thermal EHL Problems for Heavily Loaded Contacts

In this section we will consider a special case of a thermal EHL (TEHL) problem for Newtonian lubricant when the contact surface temperatures are known and equal. We will develop a powerful perturbation technique for simplification and analysis of the problem. This approach will allow us to reduce the whole thermal EHL problem (which includes the energy equation for lubricant) to an EHL problem for a modified Reynolds equation in a certain sense similar to the classic isothermal one. The latter problem can be solved numerically or studied by the appropriate asymptotic methods. Historically, this approach was first developed by Kudish in [26, 27, 48] - [50] for analysis of isothermal and thermal EHL problem for Newtonian and non-Newtonian fluid lubricants and greases. Later, it was employed for fluid lubricants with non-Newtonian rheologies in [51].

The general case of a thermal problem for a non-Newtonian lubricant with heat dissipation in contact solids in considered in the next section.

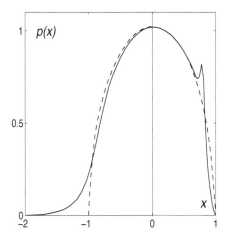

FIGURE 10.30

Distributions of pressure $p(x)$ (solid curve) in a lubricated contact ($\nu_{1,2} =$ 0.3, $m_1 = 0.003$, $m_2 = 0.15$, $A = C = 0.00076$, $B = D = 0.00312$, $V = 0.1$, $Q = 6.25$, $s_0 = -2$) and the Hertzian pressure (dotted curve) in a dry contact (after Kudish [45]). Reprinted with permission from the ASME.

10.10.1 Formulation and Analysis of a TEHL Problem for Newtonian Fluid in Heavily Loaded Contacts

Let us consider the behavior of a Newtonian fluid in a TEHL contact of two elastic cylinders separated by a lubrication film and loaded with a given normal force acting along their center line. The details of the problem set up are given in Section 10.3. We will assume that contact surface temperatures $T_{w1}(x)$ and $T_{w2}(x)$ are known and equal, i.e., $T_{w1}(x) = T_{w2}(x) = T_{w0}(x)$. It is a well–known experimental fact that the lubricant viscosity μ increases with lubricant pressure p and decreases with temperature T. To model this dependence of lubricant viscosity μ on temperature T, we will take[†]

$$\mu(p, T) = \mu^0(p) \exp[-\alpha_T(T - T_a)], \qquad (10.312)$$

where $\mu^0(p)$ is independent from lubricant temperature T and is a monotonically increasing with pressure p lubricant viscosity at the ambient temperature T_a and α_T is the temperature viscosity coefficient which may depend only on pressure p. With respect to lubricant heat conductivity λ for simplicity we

---

[†]Similarly, we can assume that $\mu(p, T) = \frac{\mu^0(p)}{1+\delta(T-T_a)}$. For this lubricant viscosity, the further analysis of the problem is similar but simpler.

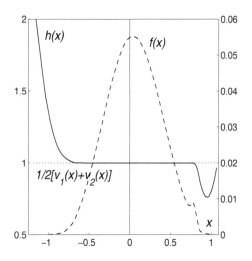

FIGURE 10.31
Distributions of gap $h(x)$ (solid curve), average velocity $\frac{v_1(x)+v_2(x)}{2}$ (dotted curve), and sliding frictional stress $f(x)$ (dash-dotted curve) in a lubricated contact ($\nu_{1,2} = 0.3$, $m_1 = 0.003$, $m_2 = 0.15$, $A = C = 0.00076$, $B = D = 0.00312$, $V = 0.1$, $Q = 6.25$, $s_0 = -2$) (after Kudish [45]). Reprinted with permission from the ASME.

will assume that it may depend on pressure p but is independent from temperature T, i.e., $\lambda = \lambda(p)$. Therefore, both μ^0 and λ are independent from the coordinate z across the lubrication layer.

To model this problem we will accept all assumptions concerning the relative sizes of the film thickness, contact, and solid radii as well as the direction and nature of the heat flow in lubricated contacts made in Section 10.3. In the case of Newtonian fluid in dimensional variables $F(x) = x$ and, therefore, from equations (10.13) and boundary conditions (10.14), we derive an equation for the sliding frictional stress $f(x)$ in the form

$$f(x) = \left\{ u_2 - u_1 - \frac{dp}{dx} \int\limits_{-h/2}^{h/2} \frac{zdz}{\mu} \right\} \Big/ \int\limits_{-h/2}^{h/2} \frac{dz}{\mu}. \tag{10.313}$$

Integrating the third equation in (10.13) with respect to z from $z = -h/2$ to $z = h/2$ with the boundary conditions for the velocity component w from (10.14) and using the expression for $f(x)$ from (10.313), we obtain the modified Reynolds equation and boundary conditions (see (10.20))

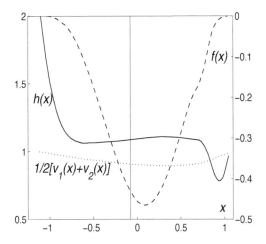

FIGURE 10.32

Distributions of gap $h(x)$ (solid curve), average velocity $\frac{v_1(x)+v_2(x)}{2}$ (dotted curve), and sliding frictional stress $f(x)$ (dash-dotted curve) in a lubricated contact ($v_1 = 0.3, v_2 = 0.1$, $m_1 = 0.003$, $m_2 = 0.25$, $A = 0.06262$, $B = -0.21844$, $C = -0.06125$, $D = 0.22156$, $V = 0.1$, $Q = 4$, $s_0 = 0$) (after Kudish [45]). Reprinted with permission from the ASME.

$$\frac{d}{dx}\left\{\left[\int\limits_{-h/2}^{h/2} \frac{z^2 dz}{\mu} - \left(\int\limits_{-h/2}^{h/2} \frac{zdz}{\mu}\right)^2 / \int\limits_{-h/2}^{h/2} \frac{dz}{\mu}\right]\frac{dp}{dx}\right.$$

$$\left. +(u_2 - u_1)\int\limits_{-h/2}^{h/2} \frac{zdz}{\mu} / \int\limits_{-h/2}^{h/2} \frac{dz}{\mu} - \frac{u_1+u_2}{2}h\right\} = 0, \tag{10.314}$$

$$p(x_i) = p(x_e) = \frac{dp(x_e)}{dx} = 0.$$

Using the reduced energy equation (10.58), the expressions for $F(x)$, and $f(x)$, we obtain the equations for the lubricant temperature T (see equations (10.58))

$$\frac{\partial^2 T}{\partial z^2} = -\frac{1}{\lambda\mu}\left\{u_2 - u_1 + \frac{dp}{dx}\left[z\int\limits_{-h/2}^{h/2} \frac{ds}{\mu} - \int\limits_{-h/2}^{h/2} \frac{sds}{\mu}\right]\right\}^2 / \left\{\int\limits_{-h/2}^{h/2} \frac{ds}{\mu}\right\}^2, \tag{10.315}$$

$$T(x, -\tfrac{h}{2}) = T_{w1}(x), \; T(x, \tfrac{h}{2}) = T_{w2}(x).$$

For heavily loaded TEHL contact we introduce dimensionless variables (10.77), (10.59), $\{T', T'_{wi}\} = \{T, T_{wi}\}/T_a - 1$ and parameters (10.78) as well

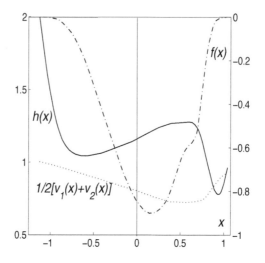

FIGURE 10.33

Distributions of gap $h(x)$ (solid curve), average velocity $\frac{v_1(x)+v_2(x)}{2}$ (dotted curve), and sliding frictional stress $f(x)$ (dash-dotted curve) in a lubricated contact ($v_1 = 0.3$, $v_2 = 0.1$, $m_1 = 0.003$, $m_2 = 0.25$, $A = 0.13542$, $B = -0.48166$, $C = -0.13569$, $D = 0.48084$, $V = 0.1$, $Q = 4$, $s_0 = 2.375$) (after Kudish [45]). Reprinted with permission from the ASME.

as two additional dimensionless parameters

$$\delta = \alpha_T T_a, \quad \kappa_N = \frac{\mu_0(u_1+u_2)^2}{4\lambda_a T_a}. \tag{10.316}$$

In these dimensionless variables the problem is reduced to the system (primes at the dimensionless variables are omitted)

$$\frac{d}{dx}\left\{\frac{12H_0^2}{V}\left[\int_{-h/2}^{h/2}\frac{z^2 dz}{\mu} - \left(\int_{-h/2}^{h/2}\frac{z dz}{\mu}\right)^2 \Big/ \int_{-h/2}^{h/2}\frac{dz}{\mu}\right]\frac{dp}{dx}\right. \tag{10.317}$$

$$\left. +s_0 \int_{-h/2}^{h/2}\frac{z dz}{\mu} \Big/ \int_{-h/2}^{h/2}\frac{dz}{\mu} - h\right\} = 0, \quad p(a) = p(c) = \frac{dp(c)}{dx} = 0,$$

$$\frac{\partial^2 T}{\partial z^2} = -\frac{\kappa_N}{\lambda\mu}\left\{s_0 + \frac{12H_0^2}{V}\frac{dp}{dx}\left[z\int_{-h/2}^{h/2}\frac{ds}{\mu} - \int_{-h/2}^{h/2}\frac{s ds}{\mu}\right]\right\}^2 \Big/ \left\{\int_{-h/2}^{h/2}\frac{ds}{\mu}\right\}^2, \tag{10.318}$$

$$T(x, -\tfrac{h}{2}) = T_{w1}(x), \quad T(x, \tfrac{h}{2}) = T_{w2}(x),$$

$$H_0(h-1) = x^2 - c^2 + \frac{2}{\pi}\int_a^c p(t)\ln\left|\frac{c-t}{x-t}\right| dt, \tag{10.319}$$

$$\int\limits_a^c p(t)dt = \tfrac{\pi}{2}. \tag{10.320}$$

In equations (10.317)-(10.320) the dimensionless parameter V is determined by the formula

$$V = \tfrac{24\mu_0(u_1+u_2)R'^2}{a_H^3 p_H} \tag{10.321}$$

coincides with parameter V determined by (10.78).

In the introduced dimensionless variables, we have

$$\mu(p,T) = \mu^0(p)\exp(-\delta T). \tag{10.322}$$

Therefore, the system can be rewritten in the final form

$$\tfrac{d}{dx}\{M(p,h) - H_0 h\} = 0, \ \ p(a) = p(c) = \tfrac{dp(c)}{dx} = 0, \tag{10.323}$$

$$M(p,h) = \tfrac{12H_0^3}{V}\left[\int\limits_{-h/2}^{h/2} z^2 e^{\delta T}dz\right.$$

$$-\left(\int\limits_{-h/2}^{h/2} z e^{\delta T}dz\right)^2 / \int\limits_{-h/2}^{h/2} e^{\delta T}dz\right]\tfrac{1}{\mu^0}\tfrac{dp}{dx} \tag{10.324}$$

$$+H_0 s_0 \int\limits_{-h/2}^{h/2} z e^{\delta T}dz / \int\limits_{-h/2}^{h/2} e^{\delta T}dz,$$

$$\tfrac{\partial^2 T}{\partial z^2} = -\kappa_N \tfrac{\mu^0}{\lambda}e^{\delta T}\left\{s_0 + \tfrac{12H^2}{V}\tfrac{1}{\mu^0}\tfrac{dp}{dx}\left[z\int\limits_{-h/2}^{h/2} e^{\delta T}ds\right.\right.$$

$$\left.\left. -\int\limits_{-h/2}^{h/2} e^{\delta T}sds\right]\right\}^2 /\left(\int\limits_{-h/2}^{h/2} e^{\delta T}ds\right)^2, \tag{10.325}$$

$$T(x,-\tfrac{h}{2}) = T_{w1}(x), \ T(x,\tfrac{h}{2}) = T_{w2}(x),$$

$$H_0(h-1) = x^2 - c^2 + \tfrac{2}{\pi}\int\limits_a^c p(t)\ln|\tfrac{c-t}{x-t}|\,dt, \tag{10.326}$$

$$\int\limits_a^c p(t)dt = \tfrac{\pi}{2}. \tag{10.327}$$

10.10.2 Some Analytical Approximations of TEHL Problems for Newtonian Fluids

Obviously, this is a highly complex nonlinear system of integro-differential equations. The goal of our analytical approach based on the regular perturbation method is for a sufficiently high slide-to-roll ratio s_0 to simplify the problem. To achieve this goal we will try to find a perturbation series solution for lubricant temperature T and use it in the generalized Reynolds equation to simplify it. Instead of using a small parameter for this purpose we will be using a small function of x which represents the ratio of the rolling and sliding frictional stresses

$$\nu(x) = \frac{H_0 h}{2f} \frac{dp}{dx}. \tag{10.328}$$

More specifically, we will assume that in the inlet and exit zones of the lubricated contact

$$\nu(x) = \frac{H_0 h}{2f} \frac{dp}{dx} \ll 1, \ \omega \ll 1, \tag{10.329}$$

where ω is a small parameter characterizing heavily loaded lubrication regimes. It can be either $\omega = V \ll 1$ or $\omega = Q^{-1} \ll 1$. It will be shown that if condition (10.329) holds in the inlet and exit zones then it holds in the central (Hertzian) region of the contact as well.

Let us expand $T(x, z)$, $T_{w1}(x)$, and $T_{w2}(x)$ in asymptotic power series in $\nu(x) \ll 1$ as follows

$$T(x, z) = T_0(x, z) + \nu(x)T_1(x, z) + O(\nu^2(x)),$$

$$T_0(x, z), \ T_1(x, z) = O(1), \ \omega \ll 1,$$

$$T_{w1}(x) = T_{w10}(x) + \nu(x)T_{w11}(x) + O(\nu^2(x)),$$

$$T_{w10}(x), \ T_{w11}(x) = O(1), \ \omega \ll 1, \tag{10.330}$$

$$T_{w2}(x) = T_{w20}(x) + \nu(x)T_{w21}(x) + O(\nu^2(x)),$$

$$T_{w20}(x), \ T_{w21}(x) = O(1), \ \omega \ll 1,$$

where $T_0(x, z)$ and $T_1(x, z)$ are the unknown functions that need to be determined while $T_{w10}(x)$, $T_{w20}(x)$, $T_{w11}(x)$, and $T_{w21}(x)$ are given functions.

Substituting expansions (10.330) in equations (10.325), we obtain a sequence of boundary-value problems

$$\frac{\partial^2 T_0}{\partial z^2} = -\kappa_N s_0^2 \frac{\mu^0}{\lambda} e^{\delta T_0} / \left(\int_{-h/2}^{h/2} e^{\delta T_0} ds \right)^2, \tag{10.331}$$

$$T_0(x, -\tfrac{h}{2}) = T_{w10}(x), \ T_0(x, \tfrac{h}{2}) = T_{w20}(x),$$

$$\frac{\partial^2 T_1}{\partial z^2} = -\kappa_N s_0^2 \frac{\mu^0}{\lambda} e^{\delta T_0} \left[\frac{4z}{h} + \delta T_1 - 2\delta \int_{-h/2}^{h/2} T_1 e^{\delta T_0} ds / \int_{-h/2}^{h/2} e^{\delta T_0} ds \right.$$

$$\left. - \frac{4}{h} \int_{-h/2}^{h/2} s e^{\delta T_0} ds / \int_{-h/2}^{h/2} e^{\delta T_0} ds \right] / \left(\int_{-h/2}^{h/2} e^{\delta T_0} ds \right)^2, \tag{10.332}$$

$$T_1(x, -\tfrac{h}{2}) = T_{w11}(x), \ T_1(x, \tfrac{h}{2}) = T_{w21}(x).$$

Here we used the fact that

$$\nu(x) = \frac{6H_0^2}{V s_0} \frac{h}{\mu^0} \frac{dp}{dx} \int\limits_{-h/2}^{h/2} e^{\delta T_0} ds \{1 + o(1)\}, \quad \omega \ll 1. \tag{10.333}$$

Using the conditions for our special case we obtain $T_{w10} = T_{w20} = T_{w0}$ and $T_{w11} = T_{w21} = 0$, where $T_{w0}(x)$ is a given function. It is easy to see that the solution of problem (10.331) has the form

$$T_0 = T_{w0} + \tfrac{2}{\delta} \ln\{\cosh(\tfrac{Bh}{2})/\cosh(Bz)\},$$

$$B = \tfrac{2}{h} \ln\{\eta + \sqrt{\eta^2 + 1}\}, \quad \eta = \tfrac{|s_0|}{2} \sqrt{\frac{\delta \kappa_N \mu^0}{2\lambda}} e^{-\delta T_{w0}}. \tag{10.334}$$

Function T_0 is even with respect to variable z. Therefore,

$$\int\limits_{-h/2}^{h/2} e^{\delta T_0} ds = e^{\delta T_{w0}} \frac{\sinh(Bh)}{B}, \quad \int\limits_{-h/2}^{h/2} s e^{\delta T_0} ds = 0. \tag{10.335}$$

Formulas (10.335) allow to simplify problem (10.332) for T_1 and reduce it to the form

$$\frac{\partial^2 T_1}{\partial z^2} = -\kappa_N s_0^2 \frac{\mu^0}{\lambda} e^{\delta T_0} \left[\frac{4z}{h} + \delta T_1\right]$$

$$-2\delta \int\limits_{-h/2}^{h/2} T_1 e^{\delta T_0} ds / \int\limits_{-h/2}^{h/2} e^{\delta T_0} ds \Big] / \Big(\int\limits_{-h/2}^{h/2} e^{\delta T_0} ds \Big)^2, \tag{10.336}$$

$$T_1(x, -\tfrac{h}{2}) = 0, \quad T_1(x, \tfrac{h}{2}) = 0.$$

It is easy to see that the solution of problem (10.336) is an odd function of z, which has the form

$$T_1 = \tfrac{2}{\delta}\{\frac{\tanh(Bz)}{\tanh(Bh/2)} - \frac{2z}{h}\}. \tag{10.337}$$

It is worth mentioning that in a similar fashion can be obtained the expressions for the further terms $T_k(x, z)$, $k \geq 2$. Below it is shown that to analyze the problem for $p(x)$, $h(x)$, H_0, and c it is not sufficient to approximate $T(x, z)$ by just $T_0(x, z)$ because by doing so some terms in the generalized Reynolds equation of the same order of magnitude as the retained ones will be lost. On the other hand, retaining both terms $T_0(x, z)$ and $T_1(x, z)$ in the approximation of $T(x, z)$ supplies a correct approximation for the generalized Reynolds equation with function $M(p, h)$ as follows (see (10.324))

$$M(p, h) = \frac{H_0^3}{V} \frac{h^3}{\mu^0} \frac{dp}{dx} R_T(x),$$

$$R_T(x) = e^{\delta T_{w0}} \frac{3(1+\eta^2)}{\ln^2(\eta+\sqrt{\eta^2+1})}\{1 + \ln(1 + \eta^2)$$

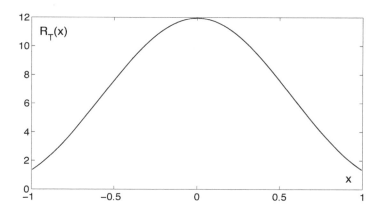

FIGURE 10.34
A typical graph of $R_T(x)$ as a function of x.

$$-\frac{\sqrt{\eta^2+1}}{\eta}\ln(\eta + \sqrt{\eta^2 + 1}) - 2\frac{\int_0^{\ln(\eta+\sqrt{\eta^2+1})}\ln(\cosh(t))dt}{\ln(\eta+\sqrt{\eta^2+1})}\Bigg\}$$

(10.338)

$$+O(\nu^2), \ \omega \ll 1, \ \beta = \frac{|s_0|}{2}\sqrt{\frac{\delta\kappa_N}{2\lambda}}, \ \eta = \beta\sqrt{\mu^0 e^{-\delta T_{w0}}},$$

where

$$\int_0^z \ln(\cosh(t))dt = \frac{z^2}{2} - z\ln 2 + \frac{\pi^2}{24} - \frac{1}{2}\sum_{k=1}^{\infty}\frac{(-1)^{k+1}}{k^2}e^{-2kz},$$

(10.339)

$$\int_0^z \ln(\cosh(t))dt = \frac{z^3}{6} + O(z^4), \ z \ll 1.$$

Here it is taken into account that $\int_{-h/2}^{h/2} ze^{\delta T}dz = O(\nu) \ll 1$ and the expression for $M(p, h)$ from formula (10.338) is based on just the first and the last terms in equation (10.324). In formula (10.338) function $R_T(x)$ represents the ratio of two functions $M(p, h)$ determined for non-isothermal and isothermal conditions (see formula (10.86)). A typical graph of function $R_T(x)$ obtained based on the Hertzian pressure $\sqrt{1 - x^2}$ and $T_{w0}(x) = 0$ is given in Fig. 10.34. Obviously, the proximity of the solutions of the isothermal and non-isothermal lubrication problems is determined by the behavior of function $R_T(x)$. The closer $R_T(x)$ is to unity the closer are these solutions to each other and vice versa the greater the value of $R_T(x)$ the farther these solutions are apart.

It can be shown that if $T_{w0} = 0$ and $\eta \to 0$ then $M(p,h) \to \frac{H_0^3}{V} \frac{h^3}{\mu^0} \frac{dp}{dx}$, which corresponds to the case of isothermal regime for Newtonian lubricant. On the other hand, for $\eta \to \infty$ (which represents the conditions of high heat generation in the lubrication layer), we have

$$M(p,h) \to \frac{H_0^3}{V} \frac{h^3}{\mu^0} \frac{dp}{dx} R_\infty(x),$$

$$(10.340)$$

$$R_\infty(x) = \frac{3(1+\eta^2)}{\ln^2(\eta+\sqrt{\eta^2+1})}, \quad \beta = \frac{|s_0|}{2}\sqrt{\frac{\delta \kappa_N}{2\lambda}}, \quad \eta = \beta\sqrt{\mu^0 e^{-\delta T_{w0}}}.$$

The comparison of formula (10.340) with the one for the isothermal case (10.86) shows that for $\eta \gg 1$ the effective viscosity is the function $\frac{\mu^0}{R_\infty(x)}$. Usually, surface temperature $T_{w0}(x)$ of the contacting solids experiences an increase as x varies from the inlet to the exit point. In most cases this increase in $T_{w0}(x)$ is modest and depends on the material properties of contacting solids and the amount of heat generated in the lubrication layer. Therefore, if for $\eta \gg 1$ the above–mentioned effective viscosity grows with η, then it grows slowly not in a usual exponential manner as the lubricant viscosity μ^0 at the ambient temperature does. That makes the case of $\eta \gg 1$ more similar to the case of constant viscosity.

The above properties of function $M(p,h)$ hold a promise of solution regularization in cases of fast growing with pressure isothermal viscosity (see Section 10.12).

However, one has to remember that for sufficiently large values of $R_T(x)$ the approximation of pressure by the Hertzian one (as it is done in the asymptotic analysis) in the region away from the contact inlet and exit points might no longer be valid due to the thermal elastic deformations of the solids in contact.

We succeeded in reducing the non-isothermal problem to the approximate one which resembles the isothermal EHL problem. The point made about the necessity of taking into account not only T_0 but also T_1 for correct determination of the film thickness H_0 can be clearly seen from the expression for $M(p,h)$ in (10.338) if $\nu(x) \ll 1$ in the inlet zone.

Let us analyze the equations of the approximate problem (10.323), (10.326), (10.327), and (10.338) in more detail in accordance with the asymptotic methods developed for heavily loaded EHL contacts in Sections 10.4 and 10.6. To do that we define a new function

$$G(p,h) = \frac{H_0^3}{V} \frac{h^3}{\mu^0} e^{\delta T_{w0}} \frac{dp}{dx} / M(p,h), \qquad (10.341)$$

which allows to consider and compare various lubrication regimes. Now, we will consider the behavior of function $G(p,h)$ in three cases: $\eta \ll 1$, $\eta \sim 1$, and $\eta \gg 1$. Using the above expression for $M(p,h)$, we have

$$G(p,h) = 1 + O(\eta), \quad \eta \ll 1, \qquad (10.342)$$

$$G(p,h) = O(1) + O(\eta), \quad \eta \sim 1, \qquad (10.343)$$

$$G(p,h) = \frac{\ln^2(\eta+\sqrt{\eta^2+1})}{3\eta^2}[1 + O(\tfrac{1}{\ln\eta})], \ \eta \gg 1. \tag{10.344}$$

Based on the analysis of Section 10.4 and estimates (10.342)-(10.344) we conclude that the pre-critical lubrications regimes are realized when in the inlet zone $x + 1 = O(\epsilon_q)$ we have

$$\mu^0(\epsilon_q^{1/2}) = O(1), \ \eta(\epsilon_q^{1/2}) \ll 1, \ \omega \ll 1, \tag{10.345}$$

$$\mu^0(\epsilon_q^{1/2}) = O(1), \ \eta(\epsilon_q^{1/2}) = O(1), \ \omega \ll 1, \tag{10.346}$$

$$\frac{\ln^2[\eta(\epsilon_q^{1/2})+\sqrt{\eta^2(\epsilon_q^{1/2})+1}]}{\eta^2(\epsilon_q^{1/2})}\mu^0(\epsilon_q^{1/2}) = O(1), \ \eta(\epsilon_q^{1/2}) \gg 1, \ \omega \ll 1, \tag{10.347}$$

while the critical size of the inlet zone ϵ_0 is determined as before as $\epsilon_0 = \max(\epsilon_q)$, where values of ϵ_q satisfy one of the estimates (10.345)-(10.347).

On the other hand, for pre-critical regimes of lubrication ($\epsilon_q \ll 1$, $\omega \ll 1$) it can be shown that if $\nu(x) \ll 1$, $\omega \ll 1$, in the inlet zone, then the same estimate holds in the exit zone and the Hertzian region. For pre-critical regimes the condition $\nu(x) \ll 1$, $\omega \ll 1$, in the inlet zone is equivalent to

$$\epsilon_q \ll \mid s_0 \mid^{6/5} V^{2/5}\{\mu^0(\epsilon_q^{1/2})e^{-\delta T_{w0}(\epsilon_q^{1/2})}\frac{B}{\sinh(Bh)}\}^{6/5}, \ \omega \ll 1, \tag{10.348}$$

where function B is determined by equation (10.334) and in this case it is calculated based on $\mu^0(\epsilon_q^{1/2})$ and $T_{w0}(\epsilon_q^{1/2})$. Using the asymptotic representations (10.118) for gap $h(x)$ in the inlet and exit zones, it is easy to analyze the pre-critical regimes of starved lubrication. In particular, for the film thickness H_0 we get the formula in (10.120) where constant A in addition to dependence on α_1 (see the relationship for a in (10.99)) also is influenced by μ^0, δ, λ, κ_N, and T_{w0}.

Let us introduce the following functions

$$M_0(q,h,\mu^0,\delta,\eta,x) = A^3\frac{h^3}{\mu^0}\frac{dp}{dx}R_T(x), \tag{10.349}$$

where $R_T(x)$ is determined by formula (10.338). Then the proposed asymptotic formula (10.120) for the film thickness allows to obtain two closed systems of integro-differential equations: in the inlet zone - equations (10.133), (10.349), (10.136), (10.114), (10.115), (10.129) for the starved lubrication regimes for functions $q(r)$, $h_q(r)$, and constant A, and in the exit zone - equations (10.135), (10.349), (10.137), (10.114), (10.114), (10.130) for the starved lubrication regimes for functions $g(s)$, $h_g(s)$, and constant β_1 (see equation (10.100)). It is important to remember that in the asymptotically valid equations in the inlet and exit zones function M_0 needs to be replaced by $M_0(q,h_q,\mu_q^0 e^{-\delta_q T_{w0q}},\delta_q,\eta_q,r)$ and $M_0(g,h_q,\mu_g^0 e^{-\delta_g T_{w0g}},\delta_g,\eta_g,s)$, respectively. Indexes q and g indicate the main terms of asymptotic expansions of the corresponding functions. Similarly, in the inlet and exit zones we can obtain the asymptotically valid equations (10.134), (10.144), and (10.146) for functions $q(r)$, $h_q(r)$, and constant A and equations (10.134), (10.145), and (10.147) for functions $g(s)$, $h_g(s)$, and constant β_1 with the above–mentioned replacements for functions M_0 and W_0.

Now, let us consider the over-critical lubrication regimes for which $\epsilon_q \gg \epsilon_0$. As before, the validity of the estimate $\nu(x) \ll 1$ in the inlet ϵ_q-zone guaranties its validity in the entire contact (see estimate (10.219) in the ϵ_0-inlet zone). It can be shown that the estimate $\nu(x) \ll 1$ in the inlet ϵ_q-zone is equivalent to

$$\epsilon_q \ll |s_0|^{3/2} \left(\tfrac{V}{\epsilon_0^{1/2}}\right)^{1/2} \{\mu^0(\epsilon_0^{1/2})e^{-\delta T_{w0}(\epsilon_q^{1/2})}\tfrac{B}{\sinh(Bh)}\}^{3/2}, \; \omega \ll 1, \quad (10.350)$$

where function B is determined by equation (10.334) and in this case it is calculated based on $\mu^0(\epsilon_0^{1/2})$ and $T_{w0}(\epsilon_q^{1/2})$. The further analysis of the over-critical lubrication regimes is done in accordance with the asymptotic method described in Section 10.6 In particular, the lubrication film thickness can still be determined from formula (10.205), where constant A depends on α_1, κ_N, μ^0, λ, and T_{w0}.

For fully flooded lubrication regimes in the inlet zone the estimate $\nu(x) \ll 1$, $\omega \ll 1$, is no longer valid. For the cases of pre-critical regimes when in the inlet zone $\nu(x) \sim 1$, $\omega \ll 1$, or in the cases of over-critical regimes when in the inlet ϵ_q-zone $\nu(x) \sim 1$, $\omega \ll 1$, a qualitative analysis can still be conducted and structurally the same formulas for the film thickness H_0 as in the cases when $\nu(x) \ll 1$, $\omega \ll 1$, can be obtained.

10.10.3 Numerical Solutions of Asymptotic TEHL Problems for Newtonian Lubricants

The general approaches to solution of asymptotically valid equations in the inlet and exit zones of non-isothermal lubricated contacts under pre-critical lubrications conditions coincide with the ones employed for solution of similar isothermal problems, which are described in Subsections 10.8.1 and 10.8.2. The main difference in these approaches is due to different calculation of functions M and W, which involves temperature effects. Besides that in the exit zone for $\mu_g = e^{Q_0 g}$, large values of Q_0, and small values of β the consequent iterates may experience oscillations about the exact problem solutions (these are not instability oscillations but oscillations from one iteration to another). To dampen these oscillations one may use the following approach $g_n^{k+1} = 0.5(g_n^k + g_n^{k+1})$, $n = 1, \ldots, N + 1$, where k is the iteration number. However, it is not necessary. Therefore, we will just illustrate on several examples of starved lubrication regimes the influence of non-isothermal conditions on the problem solution in the inlet and exit zones.

Let us consider starved lubrication conditions, i.e., $h_q(r) = h_g(s) = 1$ for lubricant viscosity, which in the inlet and exit zones is described by equations $\mu_q^0 = e^{Q_0 q}$ and $\mu_g^0 = e^{Q_0 g}$, respectively. In practice, the surface temperature $T_{w0}(x) \geq 0$ tends to grow with x. For simplicity, we will consider only the case

TABLE 10.10

Comparison of the inlet coordinates for isothermal α_1^0 and non-isothermal α_1^t solutions for different values of coefficient A and $Q_0 = 1$, $\beta = 0.1$, and $T_{w0} = 0$.

A	1	2	3
α_1^0	-0.5016	-1.4045	-2.5244
α_1^t	-0.5178	-1.4423	-2.5894

of $T_{w0} = 0$. Therefore, in our further analysis δ will not play an independent role besides its influence on the value of parameter β (see formula (10.338)).

Qualitatively, in the inlet zone of a non-isothermal lubricated contact pressure $q(r)$ behaves exactly the same way as in an isothermal contact. Quantitatively, these solutions are slightly different, which accounts for the presence of temperature effects. The numerical solutions are stable. We will not present graphs of such solutions as the similar ones were considered in detail in Subsection 10.8.3. Instead, for comparison the values of inlet coordinate α_1 obtained under isothermal conditions (labeled by α_1^0) and non-isothermal conditions (labeled by α_1^t) are presented in Table 10.10 for $Q_0 = 1$, $\beta = 0.1$ and several values of the coefficient A. The variations of α_1 as a function of parameter β are presented in Table 10.11 for $Q_0 = 1$ and $A = 2$.

TABLE 10.11

The dependence of α_1 on parameter β for $Q_0 = 1$, $A = 2$, and $T_{w0} = 0$.

β	0.01	0.03162	0.0707	0.1	0.3162
α_1	-1.4078	-1.4155	-1.4305	-1.4423	-1.5331

Data in Tables 10.10 and 10.11 show that as parameters β and/or A increase the inlet coordinate α_1 moves farther away from the contact center. Also, it can be viewed as follows: as the amount of heat generated in the lubrication layer (represented by parameter β) increases lubricant viscosity decreases and it takes more lubricant in the inlet zone to maintain the same film thickness H_0 (proportional to coefficient A) as in the case of a corresponding isothermal contact.

In the exit zone, qualitatively and quantitatively solutions of the non-isothermal and isothermal problems are similar. However, the pressure spike in non-isothermal solutions (if exists) is lower and smother than in the corresponding isothermal cases. That can be seen in Fig. 10.35 where for

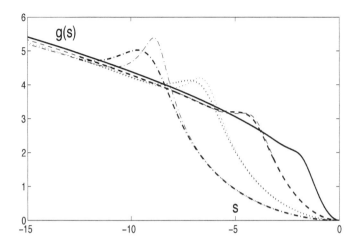

FIGURE 10.35
Graphs of $g(s)$ for isothermal (thin curves) and non-isothermal (thick curves)
problems obtained for $A = 1$ (solid curves), $A = 2$ (dashed curves), $A = 3$
(dotted curves), $A = 4$ (dash-dotted curves) and $Q_0 = 1$, $\beta = 0.1581$, $N = 1000$, and $\triangle r = 0.015625$.

$Q_0 = 1$, $\beta = 0.1581$, $N = 1000$, $\triangle r = 0.015625$, and $A = 1$, 2, 3, and 4
the comparison of the isothermal and non-isothermal solutions is presented.
It is clear from Fig. 10.35 that for this data pressure distributions $g(s)$ in
non-isothermal contacts are very close to the corresponding ones in isothermal contacts. Moreover, as parameter β increases the pressure spike gets
relatively smaller and it tends to move away from the exit point $s = 0$.
An example of such a behavior of $g(s)$ is shown in Fig. 10.36 obtained for
$A = 2$, $Q_0 = 2.5$, $N = 1600$, $\triangle r = 0.00415$ and several values of parameter
β from $\beta = 0.004472$ through $\beta = 0.3162$. For significantly larger values of β
the solutions for the exit zone pressure $g(s)$ become monotonic.

In cases of fully flooded lubrication regimes, we cannot use the above obtained approximation for function $M(p, h)$. However, from the numerical point
of view it is still useful as we can rearrange function $M(p, h)$ as the one obtained asymptotically multiplied by the ratio of the original function $M(p, h)$
and its asymptotic representation. After that the same numerical iterative
procedure as for the cases of starved lubrication regimes can be used while
the values of the above ratio should be taken from the previous iteration
and temperature T should be determined from the numerical solution of the
original problem (10.325) in the inlet and exit zones. We will not pursue
this any further as the behavior of the solution in the exit zone under fully
flooded lubrication conditions is clear from the features the solution exhibits

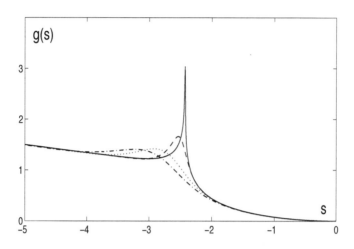

FIGURE 10.36
Graphs of $g(s)$ for non-isothermal problem obtained for $A = 2$, $Q_0 = 2.5$, $N = 1600$, $\triangle r = 0.0078125$, and $\beta = 0.0122$ (solid curve), $\beta = 0.0707$ (dashed curve), $\beta = 0.2236$ (dotted curve), $\beta = 0.3162$ (dash-dotted curve).

under starved lubrication conditions in non-isothermal contacts and under fully flooded lubrication conditions in isothermal contacts (see Subsection 10.8.3).

10.11   Numerical Solution of Asymptotic Isothermal EHL Problems Revisited. Regularization Approach

The history of numerical methods in application to solution of EHL problems for heavily loaded lubricated contacts started with the revolutionary work of A.I. Petrusevich [47] in the end of 1940th in which, for the first time, he demonstrated the existence of a pressure spike in the exit zone. Since that time a series of different numerical methods has been developed. Among them are various modifications of Newton's and gradient methods, multi-level multi-grid methods, methods involving usage of fast Fourier transform, etc. Each of these methods has its own advantages and shortcomings.

All of these methods have one common drawback: they produce inaccurate and unstable solutions for heavily loaded EHL contacts lubricated by fluids with high pressure viscosity coefficients, which was demonstrated in Subsection 10.8.4. This numerical instability is primarily caused by the proximity of the EHL problems for heavily loaded contacts to an integral equation of the first kind and by extremely strong nonlinearity of the problem resulting from high values of the pressure viscosity coefficient. The numerical instability manifests itself in oscillating either pressure or its derivative as well as gap in the vicinity of a pressure spike located in the exit zone of a lubricated contact, in strong dependence/sensitivity of the pressure spike maximum on the numerical step size/number of nodes used, high condition number of the approximating algebraic system of equations and its proximity to a system with singular matrix, etc. This is a typical signature of an ill-posed problem.

The extensive numerical experiments showed that just decreasing the numerical step size does not alleviate the solution instability as it occurs in most ill-posed problems. A detailed numerical analysis of asymptotically valid equations approximating the original EHL problem in the inlet and exit zones of heavily loaded lubricated contact allows to point out several facts. It has been observed that in the inlet zone of such a contact (primarily responsible for lubrication film formation) an accurate stable solution can be obtained for a sufficiently small step size. That has been achieved by solving numerically the problem equations resolved for pressure involved in the expression for the gap (see Subsection 10.8.1). Moreover, that means that the equally accurate solution of the original (non-asymptotic) EHL problem can be obtained only for sufficiently small step sizes in the inlet zone. In particular, to get an equally accurate solution of the original EHL problem its step size suppose to be equal to the characteristic size of the inlet zone (which depends on the problem input parameters) times the step size of the asymptotically valid problem sufficient for obtaining solution with desired precision (see Subsection 10.8.4).

In the exit zone of a heavily loaded EHL contact for sufficiently high pressure viscosity coefficients Q (proportional to constant Q_0), even for very small step sizes, the problem solution does not possess adequate precision and it remains unstable (see Subsection 10.8.4). Unfortunately, the usually used regularization methods do not improve solution stability. Luckily, the solution can be regularized by employing the procedure based on the naturally occurring process of heat generation in a lubrication film. A properly chosen sufficiently small heat generation in combination with a properly designed numerical scheme may lead to accurate and stable solutions in the exit zone (see Subsection 10.8.4).

Mathematically, for $T_{w0} = 0$ the above regularization is described by function $M(p, h)$ from equations (10.338) and (10.339)

$$M(p, h) = \frac{H_0^3}{V} \frac{h^3}{\mu} \frac{dp}{dx} R_T(x),$$

$$R_T(x) = \frac{3(1+\eta^2)}{\ln^2(\eta+\sqrt{\eta^2+1})} \left\{ 1 + \ln(1 + \eta^2) \right.$$

(10.351)

$$-\frac{\sqrt{\eta^2+1}}{\eta}\ln(\eta+\sqrt{\eta^2+1}) - 2\frac{\int\limits_{0}^{\ln(\eta+\sqrt{\eta^2+1})}\ln(\cosh(t))dt}{\ln(\eta+\sqrt{\eta^2+1})}\bigg\}, \quad \eta = \beta\sqrt{\mu},$$

where integral in (10.351) is calculated based on formulas (10.339), μ is the lubricant viscosity, and β is a certain nonnegative parameter. In all other respects the numerical schemes in the inlet and exit zones are identical to the ones presented in Subsections 10.8.1 and 10.8.2.

Such characteristics of a heavily loaded lubricated contact as lubrication film thickness, exit coordinate of a contact, and maximum of pressure spike are determined predominantly by the inlet and exit zones. Moreover, the main contributions to these parameters come from the portions of the inlet and exit zones where pressure is relatively low. The regions where pressure is high are practically irrelevant for determining film thickness and exit coordinate of a contact. That explains the success of employing the regularization procedure based on introduction of heat generation. In fact, for nominally small heat generation in the regions where pressure is small the effect of heat generation on film thickness is small. At the same time, in high pressure zones heat generation is significantly higher, which dampens pressure growth and, therefore, the unstable growth of pressure in the pressure spike area. This is the conceptual foundation of this regularization approach.

TABLE 10.12

The illustration of convergence/stability of non-isothermal solution in the exit zone for $\beta = 0.012247$.

Δr (N)	C.N.	β_1	$\max g(s)$ @ s_{max}
0.015625 (800)	$4.63 \cdot 10^7$	0.6317	4.5559 @ -2.4375
0.006000 (1,200)	$1.96 \cdot 10^8$	0.6562	4.7372 @ -2.4376
0.004150 (1,600)	$5.48 \cdot 10^8$	0.6394	4.7661 @ -2.4375

The level of heat generation (the value of constant β) should be chosen in such a way that the solution obtained based on it would be stable and, at the same time, close to the solution of the corresponding non-regularized (original) EHL problem. That can be achieved by comparing relatively stable solution characteristics such as the exit coordinates and positions of the pressure spike (see Table 10.7 in Subsection 10.8.4 and Table 10.12) obtained based on regularized and non-regularized solutions. The comparison of the maxima of pressure spikes has no part in this process because the maximum of pressure spike is an unstable characteristic of the original EHL problem.

The above regularization approach leads to good results in both the inlet and exit zones. Let us consider some examples in the exit zone for lubricant viscosity described by equation $\mu_g^0 = e^{Q_0 g}$. Small values of β provide a natural regularization of the problem. For $A = 2$ and $Q_0 = 2.5$ for which the

TABLE 10.13

The illustration of convergence/stability of non-isothermal solution in the exit zone for $\beta = 0.014142$.

$\triangle r$ (N)	$C.N.$	β_1	max $g(s)$ @ s_{max}
0.015625 (800)	$4.62 \cdot 10^7$	0.6308	4.5269 @ -2.4375
0.006000 (1,200)	$1.25 \cdot 10^9$	0.6421	4.6294 @ -2.4420
0.004150 (1,600)	$4.56 \cdot 10^9$	0.6455	4.6381 @ -2.4443

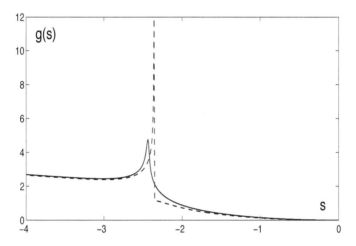

FIGURE 10.37

Graphs of $g(s)$ for non-isothermal problem (solid curve) and for isothermal problem (dashed curve) obtained for $A = 2$, $Q_0 = 2.5$, $\beta = 0.012247$, $N = 1600$, and $\triangle r = 0.0078125$.

isothermal non-regularized solution exhibits some instability signs (see Fig. 10.20 and Table 10.7) while for $A = 2$, $Q_0 = 2.5$ and $\beta = 0.012247$ and $\beta = 0.014142$ the solutions in general and the pressure spike, in particular, are stable. That can be seen from Tables 10.12 and 10.13, where the values of the condition number $C.N.$, the exit coordinate β_1, the maximum of pressure spike max $g(s)$ and its location s_{max} are presented for several values of parameters N (number of nodes) and $\triangle r$ (numerical step size). Also, it can be seen from the comparison of the graphs of $g(s)$ in Fig. 10.37 obtained for isothermal non-regularized and regularized/non-isothermal cases for $A = 2$, $Q_0 = 2.5$, $\beta = 0.012247$, $N = 1600$, and $\triangle r = 0.0078125$. Obviously, for small β the position of the pressure spike practically does not change and almost coincides with the position of the pressure spike following from the non-regularized problem.

For small values of β all parameters of the regularized/non-isothermal problem solution are only slightly different from the ones for the isothermal problem except for the fact that the pressure spike behaves in a stable manner and there are no signs of solution instability in the region between the pressure spike and the contact exit point (see discussions in Subsections 10.8.3 and 10.8.4). We can get an idea of what changes in the isothermal solution can be caused by the introduction of small heat generation. In particular, for $A = 2$, $Q_0 = 1$, $\beta = 0.012247$, $N = 1600$, and $\triangle r = 0.0078125$ the stable isothermal and regularized/non-isothermal solutions have the following parameters $\beta_1 = 0.8966$, $\max g(s) = 3.2809$ reached at $s_{max} = -4.5802$ and $\beta_1 = 0.8635$, $\max g(s) = 3.1714$ reached at $s_{max} = -4.6059$, respectively. Also, it can be seen from the comparison of the values of the condition number $C.N.$ for isothermal (see Table 10.7 in Subsection 10.8.4) and regularized/non-isothermal problems (see Table 10.12). In the latter case these values are by at least 10% lower which promotes stability of numerical solutions. Therefore, the introduction of thermal effects provides a natural way of numerical solution regularization. It is important to realize that to regularize the exit solution for larger values of Q_0 it is necessary to use a greater value of parameter β.

Also, it is important to point out that the pressure viscosity coefficient Q is equal to $Q_0 \epsilon_q^{-1/2}$. It means that, for example, for fully flooded pre-critical lubrication regimes $\epsilon_q = V^{2/5}$ and $Q = Q_0 V^{-1/5}$. Therefore, if $V = 0.001$ and for $Q_0 = 10$ we have a stable solution it means that a stable solution can be obtained for $Q = 39.8$. These kind of values of the pressure viscosity coefficient Q represent the upper bound of the range of Q in practice.

To conclude this section it is worth to make the following remark. It is important to keep in mind that from the physical point of view for large values of Q_0 the pressure spike itself does not have much of an influence on the rest of the solution as it practically does not affect any of the vital EHL parameters possibly with just one exception - the rolling frictional stress, which is small in most cases. It is caused by the fact that the pressure spike is narrow. Therefore, the choice of the value of parameter β for regularization of isothermal solutions can be made without much influence of the resulting from it size of the pressure spike. We can also stress that due to solution stability in the inlet zone to determine the lubrication film thickness H_0 and sliding frictional stress $f(x)$ and force F_S (see Section 10.20) it is sufficient to solve the EHL problem in just the inlet zone and to use the Hertzian pressure outside of it. It is a distinct advantage of the asymptotic approach.

10.12 Numerical Solution of the Original Isothermal EHL Problem Revisited. Regularization Approach and Stable Numerical Method

In Subsection 10.4.1 it has been shown that systems (10.79)-(10.82) and (10.84)-(10.87) are equivalent. In Section 10.7 it has been mentioned that for truly heavily loaded contacts the numerical procedure described in that section based on the former system of equations still tends to be unstable. In this section we will use the ideas developed for obtaining stable and accurate solutions of asymptotically valid equations to propose an improved (compared to Section 10.7) numerical scheme for heavily loaded EHL contacts in the original (non-asymptotic) formulation.

10.12.1 Numerical Method for the Original Isothermal EHL Problem Revisited. Regularization Approach

To improve numerical stability of problem solutions for Newtonian lubricants in heavily loaded contacts, we will employ equations (10.84)-(10.87)

$$p(x) = R(x)\Big[1 - \tfrac{1}{2\pi}\int\limits_a^c \tfrac{dM(p,h)}{dt}\tfrac{dt}{R(t)(t-x)}\Big],$$

$$R(x) = \sqrt{(x-a)(c-x)},$$
(10.352)

$$\int\limits_a^c \tfrac{dM(p,h)}{dt}\tfrac{dt}{R(t)} = \pi(a+c),$$

$$\int\limits_a^c \tfrac{dM(p,h)}{dt}\tfrac{tdt}{R(t)} = \pi[(\tfrac{c-a}{2})^2 + \tfrac{(a+c)^2}{2} - 1],$$
(10.353)

$$M(p,h) = \tfrac{H_0^3}{V}\tfrac{h^3}{\mu}\tfrac{dp}{dx},$$
(10.354)

$$H_0(h-1) = x^2 - c^2 + \tfrac{2}{\pi}\int\limits_a^c p(t)\ln\mid\tfrac{c-t}{x-t}\mid dt,$$
(10.355)

resolved for pressure $p(x)$ involved in the expression for gap $h(x)$ and the regularization approach based on the introduction of heat generation in the contact (i.e., on modification of function $M(p,h)$). The latter means that for large values of pressure viscosity coefficient Q (where $\mu = \exp(Qp)$) instead of using function $M(p,h)$ determined by (10.354) we will be using function $M(p,h)$ calculated as follows (see equations (10.338) and (10.339) for $T_{w0} = 0$)

$$M(p,h) = \tfrac{H_0^3}{V}\tfrac{h^3}{\mu}\tfrac{dp}{dx}R_T, \quad R_T = \tfrac{3(1+\eta^2)}{\ln^2(\eta+\sqrt{\eta^2+1})}\Big\{1 + \ln(1+\eta^2)\Big\}$$
(10.356)

$$-\frac{\sqrt{\eta^2+1}}{\eta}\ln(\eta+\sqrt{\eta^2+1})-2\frac{\int_0^{\ln(\eta+\sqrt{\eta^2+1})}\ln(\cosh(t))dt}{\ln(\eta+\sqrt{\eta^2+1})}\Bigg\},\ \eta=\beta\sqrt{\mu},$$

where the integral in (10.356) is calculated based on formulas (10.339) and the details of the dependence of β on s_0 and other parameters can be found in (10.338). The suitability of the above modification of function $M(p,h)$ is based on two facts: (a) for $\beta\to0$ function $M(p,h)$ from (10.356) approaches function $M(p,h)$ from (10.354), which corresponds to the isothermal case and (b) $M(p,h)$ from (10.356) provides sufficient regularization of solutions of the original isothermal EHL problem. The latter has been demonstrated in Section 10.11 for equations asymptotically valid in the inlet and exit zones.

By introducing the substitution $x=\frac{a+c}{2}+\frac{c-a}{2}y$ equations (10.352)-(10.356) can be reduced to equations within fixed boundaries $[-1,1]$ as follows

$$p(y)=\frac{c-a}{2}\sqrt{1-y^2}\left[1-\frac{1}{2\pi}\left(\frac{2}{c-a}\right)^3\int_{-1}^1\frac{d}{d\tau}(W\frac{dp}{d\tau})\frac{d\tau}{\sqrt{1-\tau^2}(\tau-y)}\right],\qquad(10.357)$$

$$\int_{-1}^1\frac{d}{d\tau}(W\frac{dp}{d\tau})\frac{d\tau}{\sqrt{1-\tau^2}}=\pi(a+c)(\frac{c-a}{2})^2,$$

$$(10.358)$$

$$\int_{-1}^1\frac{d}{d\tau}(W\frac{dp}{d\tau})\frac{\tau d\tau}{\sqrt{1-\tau^2}}=\pi\frac{c-a}{2}[(\frac{c-a}{2})^2-1],$$

$$H_0(h-1)=\frac{a^2+2ac-3c^2}{4}+\frac{c^2-a^2}{2}y+(\frac{c-a}{2})^2y^2$$

$$(10.359)$$

$$+\frac{c-a}{\pi}\int_{-1}^1 p(\tau)\ln\mid\frac{1-\tau}{y-\tau}\mid d\tau,$$

where $M(p,h)$ is replaced by function $W(p,h)$ as follows

$$W(p,h)=\frac{H_0^3}{V}\frac{h^3}{\mu}\qquad(10.360)$$

for small to moderate values of Q, and it is replaced by

$$W(p,h)=\frac{H_0^3}{V}\frac{h^3}{\mu}R_T,\ \ R_T=\frac{3(1+\eta^2)}{\ln^2(\eta+\sqrt{\eta^2+1})}\Big\{1+\ln(1+\eta^2)$$

$$(10.361)$$

$$-\frac{\sqrt{\eta^2+1}}{\eta}\ln(\eta+\sqrt{\eta^2+1})-2\frac{\int_0^{\ln(\eta+\sqrt{\eta^2+1})}\ln(\cosh(t))dt}{\ln(\eta+\sqrt{\eta^2+1})}\Big\},\ \eta=\beta\sqrt{\mu},$$

for large values of Q. The integral in (10.361) is defined in formulas (10.339).

Let us introduce two sets of nodes: $y_k=-1+(k-1)\triangle y,\ k=1,\ldots,N$, and $y_{k+1/2}=0.5(y_k+y_{k+1})$, where $\triangle y$ is the step size. Now, we can approximate the integral

$$\int_{-1}^{1} \frac{d}{d\tau}(W\frac{dp}{d\tau}) \frac{d\tau}{\sqrt{1-\tau^2}(\tau-y_k)} \approx \int_{-1}^{y_{3/2}} \frac{d}{d\tau}(W\frac{dp}{d\tau}) \frac{d\tau}{\sqrt{1-\tau^2}(\tau-y_k)}$$

$$+ \sum_{i=2}^{N-1} \int_{y_{i-1/2}}^{y_{i+1/2}} \frac{d}{d\tau}(W\frac{dp}{d\tau}) \frac{d\tau}{\sqrt{1-\tau^2}(\tau-y_k)}$$

$$+ \int_{y_{N-1/2}}^{1} \frac{d}{d\tau}(W\frac{dp}{d\tau}) \frac{d\tau}{\sqrt{1-\tau^2}(\tau-y_k)}$$

$$\approx \frac{d}{dy}(W\frac{dp}{dy}) \mid_{y_{5/4}} \int_{-1}^{y_{3/2}} \frac{d\tau}{\sqrt{1-\tau^2}(\tau-y_k)} \qquad (10.362)$$

$$+ \sum_{i=2}^{N-1} \frac{d}{dy}(W\frac{dp}{dy}) \mid_{y_i} \int_{y_{i-1/2}}^{y_{i+1/2}} \frac{d\tau}{\sqrt{1-\tau^2}(\tau-y_k)}$$

$$+ \frac{d}{dy}(W\frac{dp}{dy}) \mid_{y_{N-1/4}} \int_{y_{N-1/2}}^{1} \frac{d\tau}{\sqrt{1-\tau^2}(\tau-y_k)},$$

$$y_{5/4} = -1 + \frac{\Delta y}{4}, \quad y_{N-1/4} = 1 - \frac{\Delta y}{4}.$$

Taking into account that $\frac{dp(1)}{dy} = 0$ the second derivatives in (10.362) can be approximated as follows

$$\frac{d}{dy}(W\frac{dp}{dy}) \mid_{y_{5/4}} \approx \frac{1}{\Delta y^2} [W_{5/2}(p_3 - p_2) - W_{3/2}(p_2 - p_1)],$$

$$\frac{d}{dy}(W\frac{dp}{dy}) \mid_{y_i} \approx \frac{1}{\Delta y^2} [W_{i+1/2}(p_{i+1} - p_i) - W_{i-1/2}(p_i - p_{i-1})], \qquad (10.363)$$

$$\frac{d}{dy}(W\frac{dp}{dy}) \mid_{y_{N-1/4}} \approx -\frac{2}{\Delta y^2} W_{N-1/2}(p_N - p_{N-1}).$$

Integrals involved in (10.362) are calculated by employing the substitution $\tau = \frac{2t}{1+t^2}$ and have the form

$$I(a,b,y_k) = \int_a^b \frac{d\tau}{\sqrt{1-\tau^2}(\tau-y_k)} = \frac{1}{\sqrt{1-y_k^2}} J(a,b,y_k),$$

$$J(a,b,y_k) = \ln \mid \frac{t_b-t_{1k}}{t_a-t_{1k}} \frac{t_a-t_{2k}}{t_b-t_{2k}} \mid, \qquad (10.364)$$

$$t_a = \frac{1-\sqrt{1-a^2}}{a}, \quad t_b = \frac{1-\sqrt{1-b^2}}{b}, \quad t_{1k} = \frac{1-\sqrt{1-y_k^2}}{y_k}, \quad t_{2k} = \frac{1+\sqrt{1-y_k^2}}{y_k}.$$

In (10.364) t_a and t_b are solutions of the equations $\frac{2t}{1+t^2} = a$ and $\frac{2t}{1+t^2} = b$ chosen in such a way that both $\mid t_a \mid$ and $\mid t_b \mid$ are not greater than 1.

By satisfying equation (10.357) at nodes y_k, $k = 1, \ldots, N$, we obtain a system of N nonlinear algebraic equations

$$p_k + \frac{1}{2\pi\triangle y^2}(\frac{2}{c-a})^2[W_{5/2}(p_3 - p_2) - W_{3/2}(p_2 - p_1)]J(-1, y_{3/2}, y_k)$$

$$+ \frac{1}{2\pi\triangle y^2}(\frac{2}{c-a})^2 \sum_{i=2}^{N-1}[W_{i+1/2}(p_{i+1} - p_i) - W_{i-1/2}(p_i - p_{i-1})]$$

$$\times J(y_{i-1/2}, y_{i+1/2}, y_k) - \frac{1}{\pi\triangle y^2}(\frac{2}{c-a})^2 W_{N-1/2}(p_N - p_{N-1})$$

$$\times J(y_{N-1/2}, 1, y_k) = \frac{c-a}{2}\sqrt{1 - y_k^2}, \ k = 1, \ldots, N.$$

$$(10.365)$$

This kind of integral approximation was tested on dry contact problems (see Subsection 2.2.2) with the punch bottom described by function Cy^{2m} (C and m are positive constants). These contact problems possess well–known simple exact analytical solutions. The numerical solutions of these problems obtained using the integral approximation similar to (10.362)-(10.364) provide excellent approximations for the exact ones. The absolute error of the numerical solution depends on the values of constants C and m as well as the step size $\triangle y$.

To iteratively solve this system we introduce a set of new unknowns $z_k = p_k^{n+1} - p_k^n$, $k = 1, \cdots, N$ (n is the iteration number) and reduce the system of nonlinear equations (10.365) to a system of N linear algebraic equations

$$\frac{1}{2\pi\triangle y^2}(\frac{2}{c^n-a})^2 \sum_{i=2}^{N-1}[W_{i+1/2}^n(z_{i+1} - z_i) - W_{i-1/2}^n(z_i - z_{i-1})]\tilde{J}_{ik}$$

$$- \frac{1}{\pi\triangle y^2}(\frac{2}{c^n-a})^2 W_{N-1/2}^n(z_N - z_{N-1})\tilde{J}_{Nk} + z_k$$

$$+ \frac{1}{2\pi\triangle y^2}(\frac{2}{c^n-a})^2 \sum_{i=2}^{N-1}[W_{i+1/2}^n(p_{i+1}^n - p_i^n) - W_{i-1/2}^n(p_i^n - p_{i-1}^n)]\tilde{J}_{ik}$$

$$(10.366)$$

$$- \frac{1}{\pi\triangle y^2}(\frac{2}{c^n-a})^2 W_{N-1/2}^n(p_N^n - p_{N-1}^n)\tilde{J}_{Nk} + p_k^n$$

$$= \frac{c^n-a}{2}\sqrt{1 - y_k^2}, \ k = 1, \ldots, N,$$

$$\tilde{J}_{2k} = J(-1, y_{3/2}, y_k) + J(y_{3/2}, y_{5/2}, y_k),$$

$$\tilde{J}_{ik} = J(y_{i-1/2}, y_{i+1/2}, y_k), \ 2 < i < N, \ \tilde{J}_{Nk} = J(y_{N-1/2}, 1, y_k).$$

Now, let us derive formulas for calculation of film thickness H_0 and exit coordinate c. To do that we need to approximate the integrals involved in equations (10.358). Following the same approach as was employed for evaluation of the integral in (10.362), we obtain

$$\int\limits_{-1}^{1} \frac{d}{d\tau}(W\frac{dp}{d\tau})\frac{d\tau}{\sqrt{1-\tau^2}} \approx \frac{1}{\triangle y}\sum_{i=2}^{N-1}[W_{i+1/2}(p_{i+1} - p_i)$$

$$-W_{i-1/2}(p_i - p_{i-1})]L_i - \frac{2}{\Delta y}W_{N-1/2}(p_N - p_{N-1})L_N,$$

$$\int_{-1}^{1} \frac{d}{d\tau}\left(W\frac{dp}{d\tau}\right)\frac{\tau d\tau}{\sqrt{1-\tau^2}} \approx \frac{1}{\Delta y}\sum_{i=2}^{N-1}[W_{i+1/2}(p_{i+1} - p_i)] \tag{10.367}$$

$$-W_{i-1/2}(p_i - p_{i-1})]M_i - \frac{2}{\Delta y}W_{N-1/2}(p_N - p_{N-1})M_N,$$

$$L_2 = \frac{1}{\sqrt{1-y_{5/4}^2}} + \frac{1}{\sqrt{1-y_2^2}}, \quad L_N = \frac{1}{\sqrt{1-y_{N-1/4}^2}},$$

$$M_2 = \frac{y_{5/4}}{\sqrt{1-y_{5/4}^2}} + \frac{y_2}{\sqrt{1-y_2^2}}, \quad M_N = \frac{1}{\sqrt{1-y_{N-1/4}^2}}, \tag{10.368}$$

$$L_i = \frac{1}{\sqrt{1-y_i^2}}, \quad M_i = \frac{y_i}{\sqrt{1-y_i^2}}, \quad i = 3,\ldots,N-1.$$

These integral approximations are consistent with the ones used for approximation of integrals in formulas (10.275) and (10.286) and employed for solution of asymptotically valid equations.

Knowing the values of these integrals we divide the second equation in (10.358) by the first equation, which eliminates H_0 from the ratio and produces the equation

$$\gamma = \int_{-1}^{1} \frac{d}{d\tau}\left(W\frac{dp}{d\tau}\right)\frac{\tau d\tau}{\sqrt{1-\tau^2}} / \int_{-1}^{1} \frac{d}{d\tau}\left(W\frac{dp}{d\tau}\right)\frac{d\tau}{\sqrt{1-\tau^2}}, \tag{10.369}$$

$$\gamma = \frac{\frac{c-a}{2} - \frac{2}{c-a}}{c+a}. \tag{10.370}$$

The integrals in (10.369) are calculated based on formulas (10.367) and (10.368). Solving equation (10.370) for c as a quadratic equation provides two solutions

$$c = a\frac{1\pm2\sqrt{\gamma^2+\frac{1-2\gamma}{a^2}}}{1-2\gamma}. \tag{10.371}$$

To chose the correct sign in (10.371), we notice that when the problem parameters approach the ones for a dry (not lubricated) contact then $\gamma \to 0$ and we suppose to have $a \to -1$ and $c \to -a$. Expanding formula (10.371) for small γ we easily understand that to satisfy the above conditions we must chose sign minus. Therefore, for the exit coordinate we get the following formula

$$c = a\frac{1-2\sqrt{\gamma^2+\frac{1-2\gamma}{a^2}}}{1-2\gamma}. \tag{10.372}$$

After the value of c is determined we calculate film thickness H_0 from the first equation in (10.358) assuming that the integral in the left–hand side is approximated based on formulas (10.367) and (10.368). As a result of that we get a formula

$$H_0^3 = \pi(c+a)(\tfrac{c-a}{2})^2 \triangle y / \Big\{ \sum_{i=2}^{N-1} \Big[\tfrac{W_{i+1/2}}{H_0^3}(p_{i+1} - p_i)$$

$$-\tfrac{W_{i-1/2}}{H_0^3}(p_i - p_{i-1}) \Big] L_i - 2\tfrac{W_{N-1/2}}{H_0^3}(p_N - p_{N-1}) L_N \Big\}. \tag{10.373}$$

Finally, to calculate the gap between the lubricated solids $h(y)$, we can employ a discrete analog of equation (10.359) in the form

$$H_0^{n+1}(h_{k+1/2}^{n+1} - 1) = \tfrac{a^2 + 2ac^{n+1} - 3(c^{n+1})^2}{4} + \tfrac{(c^{n+1})^2 - a^2}{2} y_{k+1/2}$$

$$+(\tfrac{c^{n+1}-a}{2})^2 y_{k+1/2}^2 + \tfrac{c^{n+1}-a}{\pi} \sum_{i=1}^{n-1} N_i(y_{k+1/2}), \; k = 1, \ldots, N-1,$$

$$N_i(y) = [p_i^{n+1} - y_i \tfrac{p_{i+1}^{n+1} - p_i^{n+1}}{y_{i+1} - y_i}][K_1(y_{i+1}) - K_1(y_i) - K_2(y - y_{i+1})$$

$$+K_2(y - y_i)] + \tfrac{p_{i+1}^{n+1} - p_i^{n+1}}{y_{i+1} - y_i}[K_3(y_{i+1}) - K_3(y_i) - K_4(y - y_{i+1}, y) \tag{10.374}$$

$$+K_4(y - y_i, y)], \; K_1(y) = (y-1)[\ln(1-y) - 1],$$

$$K_2(y) = y[1 - \ln |\, y \,|], \; K_3(y) = \tfrac{(1-y)^2}{2}[\ln(1-y) - \tfrac{1}{2}] + K_1(y),$$

$$K_4(y, \varsigma) = \tfrac{y^2}{2}[\ln |\, y \,| - \tfrac{1}{2}] + \varsigma K_2(y).$$

Also, we can use iteration based approach. Instead of using the discrete analogue of equation (10.359), we can use the equation

$$h = 1 + \tfrac{1}{H_0} \tfrac{2}{c-a} W \tfrac{dp}{dy}, \tag{10.375}$$

which is obtained from the original Reynolds equation (10.79) rewritten in variable y and integrated once with respect to y with the use of the boundary condition $\tfrac{dp(1)}{dy} = 0$ (see boundary conditions (10.80)). Equation (10.375) can be satisfied at nodes $y_{k+1/2}$, $k = 1, \ldots, N-1$, and solved using a modified Newton's method similar to the one proposed in Section 10.8.1 (see iterative procedure described in (10.272)). However, it is easier to use a simple iteration process (for comparison see (10.274)), which leads to the following

$$h_{k+1/2}^{n+1} = 1 + \tfrac{1}{H_0^{n+1}} \tfrac{2}{c^{n+1}-a} W(p_{k+1/2}^{n+1}, h_{k+1/2}^n) \tfrac{p_{k+1}^{n+1} - p_k^{n+1}}{\triangle y},$$

$$p_{k+1/2}^{n+1} = 0.5(p_k^{n+1} + p_{k+1}^{n+1}), \; k = 0, \ldots, N-1, \tag{10.376}$$

where n is the iteration number.

Now, let us describe the steps of the iteration process as a whole. For certainty, we will assume that $\mu = \exp(Qp)$. The iteration process is designed in such a way that for the given values of parameters a, V, and Q we are searching for the sets of p_k, $k = 1, \ldots, N$, $h_{k+1/2}$, $k = 1, \ldots, N - 1$, and constants H_0 and c. First we take some initial approximations for values of p_k^0, $h_{k+1/2}^0$, H_0^0, and c^0. For example, these can be taken as follows

$$H_0^0 = 0.272(VQ)^{3/4}, \ Q > 0 \ and \ H_0 = 0.2, \ Q = 0, \ c^0 = 1,$$

$$p_k^0 = 0 \ if \ | \tfrac{a+c^0}{2} + \tfrac{c^0-a}{2} y_k | > 1,$$

$$p_k^0 = \sqrt{1 - y_k^2} \ otherwise, \ k = 1, \ldots, N,$$

$$h_{k+1/2}^0 = [| \, x_{k+1/2} \, | \sqrt{x_{k+1/2}^2 - 1} - \ln(| \, x_{k+1/2} \, |$$

$$+ \sqrt{x_{k+1/2}^2 - 1})] \theta(x_{k+1/2}^2 - 1), \ x_{k+1/2} = \tfrac{a+c^0}{2} + \tfrac{c^0-a}{2} y_{k+1/2},$$

$$k = 1, \ldots, N - 1,$$

(10.377)

or from the earlier obtained solution with the closest set of parameters a, V, and Q. Suppose we have the n-th iterates of p_k^n, $k = 1, \ldots, N$, $h_{k+1/2}^n$, $k = 1, \ldots, N - 1$, H_0^n and c^n. Then, the new iterates of p_k^{n+1}, $k = 1, \ldots, N$, are determined from the formula $p_k^{n+1} = z_k + p_k^{n+1}$, where values of z_k, $k = 1, \ldots, N$, are obtained from solution of a system of linear algebraic equations (10.366). After that the new iterate c^{n+1} is determined from the equation (see equations (10.367)-(10.369) and (10.372))

$$c^{n+1} = a \frac{1 - 2\sqrt{(\gamma^{n+1})^2 + \frac{1-2\gamma^{n+1}}{a^2}}}{1 - 2\gamma^{n+1}}, \qquad (10.378)$$

where the value of γ^{n+1}, based on formula (10.369), is calculated using approximations (10.367), (10.368), and the new iterates p_k^{n+1}, $k = 1, \ldots, N$. That follows by calculation of the new iterate H_0^{n+1} based on formula (10.373)

$$(H_0^{n+1})^3 = (H_0^n)^3 \pi (c^{n+1} + a)(\tfrac{c^{n+1}-a}{2})^2 \Delta y$$

$$/\Big\{ \sum_{i=2}^{N-1} [W_{i+1/2}^{n+1}(p_{i+1}^{n+1} - p_i^{n+1}) - W_{i-1/2}^{n+1}(p_i^{n+1} - p_{i-1}^{n+1})] L_i \qquad (10.379)$$

$$- 2W_{N-1/2}^{n+1}(p_N^{n+1} - p_{N-1}^{n+1}) L_N \Big\},$$

in which c, p_i, and $W_{i+1/2}$ must be replaced by c^{n+1}, p_i^{n+1}, and $W_{i+1/2}^{n+1}$, respectively. After that the new set of iterates $h_{k+1/2}^{n+1}$, $k = 1, \ldots, N - 1$, is obtained from equation (10.376). The iteration process continues until the desired precision is reached.

For large number of nodes N a multigrid technique in combination with interpolation can be employed to accelerate the solution process by reducing the average order of the systems of linear equations solved. That would require solving the problem for a cascade of denser grids. The interpolation is used to obtain the initial approximation for the solution on a denser grid using the data from the solution on a less dense grid.

The necessity of regularization is determined based on the properties of the obtained solution and, in particular, its sensitivity to the number of nodes N, presence of pressure p or gap h oscillations close to the exit point, etc. In case the regularization is needed, one has to replace function W from formula (10.360) by the one described in formula (10.361). The rest of the iteration procedure remains the same. The value of parameter β in (10.361) is adjusted empirically to provide stability and, at the same time, proximity to the undisturbed solution.

For very large values of Q (such as over 25-30) at the early stages of the iteration process, it is prudent to slow down the iterations due to the fact that the initial approximation used is usually very far from the problem solution and the pressure distribution depends strongly on even small variations in the gap distribution (because the problem is stiff). That can be accomplished by introducing interpolation/relaxation as follows

$$h_{k+1/2}^{n+1} = (1 - \alpha_*)h_{k+1/2}^n + \alpha_* h_{k+1/2}^{n+1}, \quad k = 0, \ldots, N - 1,$$

$$H_0^{n+1} = (1 - \alpha_*)H_0^n + \alpha_* H_0^{n+1}, \quad c^{n+1} = (1 - \alpha_*)c^n + \alpha_* c^{n+1}, \tag{10.380}$$

where α_* is a sufficiently small positive constant. For example, α_* can be taken equal to 0.1 or 0.15 at the initial stages of the iteration process. At the later stages of the iteration process, the value of this constant α_* can be increased and even taken equal to 1. It is customary that for fully flooded lubrication regimes at the later stages the rate of process convergence decreases in the inlet zone where $h(y) \gg 1$ due to significant nonlinearity of Reynolds equation in h.

Obviously, if formula (10.361) for function W is used the described method provides the solution to the thermal EHL problem in the regimes of starved lubrication.

The described regularized iteration approach works equally well for isothermal and non-isothermal EHL problems for lubricants with non-Newtonian rheology. The only adjustment that is required is the replacement of function W by the one corresponding to the problem at hand. The above–described regularization without any changes can be extended on spatial EHL problems by replacing terms $\partial p/\partial x$ and $\partial p/\partial y$ by terms (see (10.361)) $R_T \partial p/\partial x$ and $R_T \partial p/\partial y$, respectively.

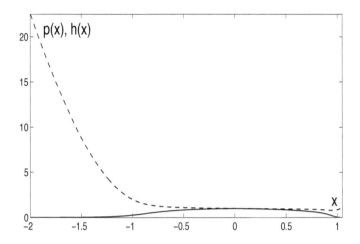

FIGURE 10.38
Graphs of pressure $p(x)$ (solid curve) and gap $h(x)$ (dashed curve) obtained
for $V = 0.1$, $Q = 0$, and $\beta = 0$.

10.12.2 Some Numerical Solutions of the Original Regularized Isothermal EHL Problem

The described regularization procedure allows to solve the EHL problems for
a wide range of pressure viscosity coefficient Q (from 0 to 40), which covers
the range of its variations in practice. In the further numerical analysis, we
will be using $a = -2$, $V = 0.1$, $N = 1601$, $\triangle y = 2/(N-1)$, and absolute
precision $\varepsilon = 0.001$. We will demonstrate the behavior of the solution of the
EHL problem on five examples of heavily loaded contacts: for (a) $Q = 0$,
(b) $Q = 5$, (c) $Q = 10$, (d) $Q = 20$, and (e) $Q = 35$ obtained for $\beta = $
0, 0, 0.0325, 0.045, and 0.85, respectively. For these cases the values of the
solution parameters H_0, c, $h(\triangle y/2)$, $\min h(y)$, and maximum of the pressure
spike $\max p(y)$ for the original EHL problem are gathered in Table 10.14 The
graphs of pressure $p(x)$ and gap $h(x)$ for cases (a)-(e) are presented in Figs.
10.38-10.42, respectively.

It is obvious from these graphs that as Q increases inlet gap $h(\triangle y/2)$ monotonically decreases while film thickness H_0, exit coordinate c, and the maximum of pressure spike $\max p(y)$ monotonically increase. That becomes clear
if you take into account two facts that as Q increases (i) a lubricated contact transitions from a fully flooded pre-critical to an over-critical lubrication
regime in which the characteristic size of the inlet and exit zones transitions
from the value proportional to $\epsilon_q = V^{2/5}$ to $\epsilon_q = (VQ)^{1/2}$ and (ii) the inlet
coordinate $\alpha_1 = (a+1)/\epsilon_q$ decreases. Moreover, as Q increases the pressure

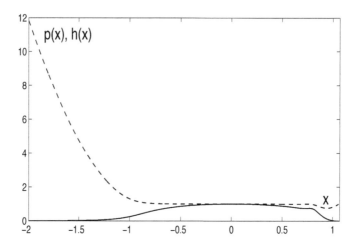

FIGURE 10.39
Graphs of pressure $p(x)$ (solid curve) and gap $h(x)$ (dashed curve) obtained for $V = 0.1$, $Q = 5$, and $\beta = 0$.

spike shifts toward the center of the contact. There are no signs of solution instability, the graphs of pressure $p(x)$ and gap $h(x)$ are smooth. All these results are repeatable with high precision for larger number of nodes N (smaller step sizes $\triangle y$).

TABLE 10.14
Solution parameters H_0, c, $h(\triangle y/2)$, $\min h(y)$, and $\max p(y)$ for $V = 0.1$ and different values of Q and β.

Q	β	H_0	c	$h(\triangle y/2)$	$\min h(y)$	$\max p(y)$
0	0	0.102	1.036	22.187	0.792	-
5	0	0.197	1.060	11.827	0.760	0.752
10	0.0325	0.278	1.075	8.624	0.762	1.462
20	0.4500	0.408	1.094	6.153	0.769	1.983
35	0.8500	0.558	1.113	4.761	0.777	2.096

It is important to remember that for fixed values of V the minimum value of parameter β which provides a stable problem solution is a monotonically increasing function of Q. Usually, even significant increase of the value of β (for example, doubling of β) changes the solution parameters by less then 0.75% while the pressure spike maximum $\max p(x)$ may change by couple percents. At the same time, the film thickness H_0 behaves very conservatively

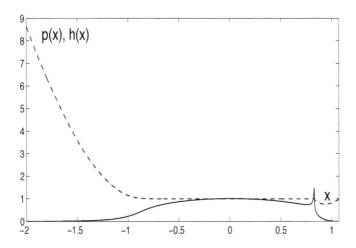

FIGURE 10.40
Graphs of pressure $p(x)$ (solid curve) and gap $h(x)$ (dashed curve) for regularized problem obtained for $V = 0.1$, $Q = 10$, and $\beta = 0.0325$.

by changing by less than 0.2%. In cases when the film thickness H_0 changes more significantly to preserve the value of H_0 it is advisable to use parameter β as a monotonically increasing function from zero in the inlet zone to a certain value in the exit zone where solution instability may occur. This is one of the reasons why it is preferable to solve separately the non-regularized asymptotic problem in the inlet zone and a regularized asymptotic problem in the exit zone.

Let us stress the fact that in the original non-regularized formulation numerical solutions of isothermal EHL problem for heavily loaded contacts tend to be unstable while being regularized in the described manner the solutions become stable. The conclusion that can be drawn from this fact is that for heavily loaded lubricated contacts the isothermal formulation of the EHL problem is inadequate and should be replaced by the thermal formulation of the EHL problem with minimum heat generation. The level of heat generations should be chosen based on numerical results.

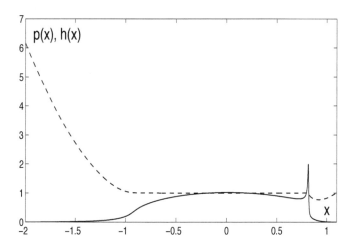

FIGURE 10.41
Graphs of pressure $p(x)$ (solid curve) and gap $h(x)$ (dashed curve) for regularized problem obtained for $V = 0.1$, $Q = 20$, and $\beta = 0.45$.

10.13 Numerical Validation of the Asymptotic Analysis. Some Additional Properties of the Original EHL Problem Solutions

To validate the asymptotic approach developed earlier for pre-critical lubrication regimes, let us compare the solutions of the EHL problem in the asymptotic and original formulations. Generally, it can be done in two different ways: on the conceptual and detailed numerical levels.

On the conceptual level it can be done as follows. If the asymptotic approach is valid, then for Newtonian fluids for fully flooded pre-critical lubrication regimes the formula for the film thickness $H_0 = AV^{3/5}$ (see formula (10.128)) should be correct. More specifically, for large enough $| a |$, $a < 0$, (which can be judged by the value of $h(a)$ in comparison with 1) the values of coefficient A and $\beta_1 = (c-1)/V^{2/5}$ (see formulas (10.100), (10.101), (10.127)) suppose to be functions of only $Q_0 = QV^{1/5}$. For the asymptotic method to be valid, the characteristic size of the inlet zone $\epsilon_q = V^{2/5}$ (see formula (10.127)) should be small and the lubrication regime suppose to be pre-critical. Therefore, for sufficiently large $| a |$, $a < 0$, practically fixed value of the parameter $Q_0 = QV^{1/5}$ (because Q_0 is fixed while V is small) and for different values of parameter V the value of coefficient $A = H_0 V^{-3/5}$ should be constant. Let us consider two examples. For $N = 1,601$, $\triangle y = 0.00125$, $V = 0.05$

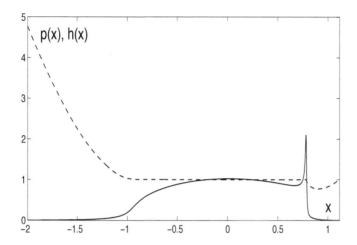

FIGURE 10.42
Graphs of pressure $p(x)$ (solid curve) and gap $h(x)$ (dashed curve) for regularized problem obtained for $V = 0.1$, $Q = 35$, and $\beta = 0.85$.

and $V = 0.1$ the values of $\epsilon_q = V^{2/5} = 0.302$ and $\epsilon_q = V^{2/5} = 0.398$ are relatively small. For $Q = 0$ we have $\mu(\epsilon_q^{1/2}, Q) = \exp(\epsilon_q^{1/2} Q) = 1$, which means that the lubrication regime is pre-critical (see definition of precritical regimes (10.140)). Notice that the solution of the original EHL problem for $a = -2$, $V = 0.05$, and $Q = 0$ gives $H_0 = 0.066$ and $c = 1.030$ while for $a = -2$, $V = 0.1$ the solution data is presented in Table 10.14. Therefore, for $a = -2$, $Q = 0$, $V = 0.05$, and $V = 0.1$ the values of coefficient $A = H_0 V^{-3/5}$ are equal to 0.401 and 0.404 while the values of $\beta_1 = (c - 1)/V^{2/5}$, are equal to 0.098 and 0.091, respectively. The differences between these pairs of values of A and β_1 are of about 0.75% and 7%. For smaller values of V the agreement between the solutions of the asymptotic and original EHL problems is even better. That validates the asymptotic approach on the conceptual level.

Validation of the asymptotic approach on the detailed numerical level involves several steps that include comparison of numerical solutions of asymptotic and original EHL problems for pre-critical lubrication regimes and matching some properties of the asymptotic and original EHL problems for both pre- and over-critical lubrication regimes. Let us start with the comparison of numerical solutions of asymptotic and original EHL problems for Newtonian fluids in pre-critical lubrication regimes. To make this comparison we will have to go through several steps. First, for the given values of V, Q, and a we determine a solution of the original EHL problem. Then we solve the asymptotic equations for $Q_0 = QV^{1/5}$ and $\alpha_1 = (a + 1)V^{-2/5}$ for the same values of parameters V, Q, and a. After that we compare the values

of the film thickness H_0 from formula (10.128) obtained from numerical solutions of the asymptotic and original EHL problems. Similarly, we compare
the values of exit coordinate c and $\min h$ obtained from numerical solution of
the asymptotic and original EHL problems. Let us do the comparison for a
series of three solutions obtained for $a = -2$, $V = 0.05$ ($\epsilon_q = V^{2/5} = 0.302$),
$Q = 0$ ($Q_0 = 0$), $Q = 1.821$ ($Q_0 = 1$), and $Q = 3.641$ ($Q_0 = 2$) (based on
the relationship $Q = Q_0 V^{1/5}$) with the corresponding asymptotic solutions.
For all these solutions $\epsilon_q = V^{2/5} = 0.302$ and $\alpha_1 = -3.314$ (see formula
$a = -1 + \alpha_1 \epsilon_q$). The solutions of the original EHL problem are represented in
Table 10.15. In Table 10.16 we present the values of $H_{0(asym)}$, $c_{(asym)}$, and
$\min h_g(s)$ obtained from numerical solutions of the asymptotically valid equations in the inlet and exit zones and formulas $H_0 = AV^{3/5}$ and $c = 1 + \beta_1 \epsilon_q$
for the same values of parameters Q, V, and $\alpha_1 = -3.314$. The original EHL
problem is solved for $N = 1,325$, $\triangle y = 0.001509434$ while the asymptotic
problems are solved for $N = 1000$, $\triangle r = 0.0075$. Notice that the step sizes
$\triangle y = 0.001509434$ and $\triangle r = 0.0075$ approximately satisfy the relationship
$\triangle y = \frac{2\epsilon_q}{c-a} \triangle r$ obtained in Subsection 10.8.4.

TABLE 10.15
Parameters H_0, c, and $\min h(y)$
obtained from solution of the original
EHL problem for $a = -2$, $V = 0.05$,
and different values of Q.

Q (Q_0)	H_0	c	$\min h(y)$
0 (0)	0.066	1.030	0.793
1.821 (1)	0.089	1.038	0.771
3.641 (2)	0.110	1.044	0.761

TABLE 10.16
Parameters $H_{0(asym)}$, $c_{(asym)}$, and
$\min h_g(s)$ obtained from solution of the
asymptotic EHL problem for $\alpha_1 = -3.314$
and different values of $Q(Q_0)$.

Q (Q_0)	$H_{0(asym)}$	$c_{(asym)}$	$\min h_g(s)$
0 (0)	0.0642	1.0291	0.791
1.821 (1)	0.0883	1.0347	0.770
3.641 (2)	0.1106	1.0391	0.760

From the data presented in Tables 10.15 and 10.16, it is clear that in spite of the fact that $\epsilon_q = 0.302$ is not very small the agreement between the solutions of the original and asymptotically valid equations of the EHL problem is very good. Moreover, the comparison of the values of the film thickness H_0, minimum gap min h, and exit coordinate c obtained from the numerical solution of the original EHL problem an the asymptotic ones shows that the difference is smaller or equal than 2.7%. The precision of the asymptotic solution becomes even better for smaller values of V.

Now, let us validate the asymptotic approach differently. The asymptotic equations for Newtonian fluids in pre-critical lubrication regimes (10.134), (10.144), (10.146), and (10.129) for the starved lubrication regimes (or (10.131) for fully flooded lubrication regimes) and (10.134), (10.145), (10.147), and (10.130) for the starved lubrication regimes (or (10.132) for fully flooded lubrication regimes) show that their solutions depend only on two parameters α_1 and $Q_0 = QV^{1/5}$. Therefore, if the asymptotic analysis for pre-critical lubrication regimes is valid, then we can expect that for different values of parameter V and the same values of parameters α_1 and Q_0 solutions of the original EHL problem suppose to exhibit a property that the value of coefficient $A = H_0 V^{-3/5}$ is constant. Let us examine these statement using two series of solutions of the original EHL problem obtained for $V = 0.05$, $a = -2$ ($\alpha_1 = -3.314$) and $V = 0.01$, $a = -1.525234$ ($\alpha_1 = -3.314$). In both series of calculations, we will assume that $N = 1,325$, $\triangle y = 0.001509434$ and $Q_0 = 1$ and $Q_0 = 2$. These values of Q_0 clearly indicate (see the definition of pre-critical lubrication regimes (10.140)) that the lubrication regime is pre-critical. The results of these calculations are presented in Table 10.17.

TABLE 10.17
Parameters H_0, c, $h(\triangle y/2)$, and min $h(y)$ obtained from solution of the original EHL problem for cases of $V = 0.05$, 0.01 and $Q_0 = 1$, 2.

a	V	Q (Q_0)	H_0	c	$h(\triangle y/2)$	min $h(y)$
-2.000	0.05	1.821 (1)	0.0887	1.0383	24.880	0.7708
-2.000	0.05	3.641 (2)	0.1097	1.0442	20.268	0.7607
-1.525	0.01	2.512 (1)	0.0337	1.0211	23.473	0.7703
-1.525	0.01	5.024 (2)	0.0419	1.0242	19.023	0.7595

From the data of Table 10.17 for $V = 0.05$, 0.01 and $Q_0 = 1$, we have $A = H_0 V^{-3/5} = 0.5349$ and 0.5336, respectively, and for $V = 0.05$, 0.01 and $Q_0 = 2$ we have $A = H_0 V^{-3/5} = 0.6617$ and 0.6640, respectively. Therefore, the difference between the corresponding values of constant A is less than 0.34%, which again validates the asymptotic analysis for pre-critical lubrication regimes.

Finally, let us find out if the asymptotic equations derived for Newtonian fluids in pre-critical regimes can be used for calculations for over-critical lubrication regimes. To do that we will consider the case of fully flooded over-critical lubrication regime (see the definition of over-critical lubrication regimes (10.141)) with viscosity $\mu = \exp(Qp)$ and inlet coordinate $a = -1 + \alpha_{10}(VQ)^{1/2}$. Then the same inlet coordinate in pre-critical lubrication regime would be $a = -1 + \alpha_1 V^{2/5}$, which means that $\alpha_1 = \alpha_{10} Q_0^{1/2}$, $Q_0 = QV^{1/5}$. As it was mentioned earlier, the solution of asymptotic equations for pre-critical lubrication regime depend only on the values of parameters α_{10} and Q_0. It means that we can expect to get film thickness $H_0 = AV^{3/5}$, where $A = A(\alpha_{10}, Q_0)$. Therefore, if over-critical lubrications regimes can be described by the asymptotic equations for pre-critical lubrication regimes, then we can expect that we get a good match of the values of the lubrication film thickness H_0 obtained from solution of the asymptotic equations for pre-critical lubrication regimes with the solution of the original EHL problem. Let us consider the case of $a = -2$, $V = 0.05$, and $Q = 10$ ($Q_0 = 5.4928$). This is clearly an over-critical lubrication regime because $e^{Q_0} = 242.94 \gg 1$ (see (10.141)). In this case $\alpha_1 = -3.314$ and from the solution of the asymptotic equations for pre-critical lubrication regimes obtained for $N = 1,000$ and $\triangle r = 0.0075$ we have $A = 1.0824$ and, therefore, $H_0 = AV^{3/5} = 0.1794$. The solution of the original EHL problem for the same values of parameters a, V, Q, and $N = 1,325$, $\triangle y = 0.001509434$ gives $H_0 = 0.1743$. The difference between these values of H_0 is 2.9%. Therefore, the asymptotic equations derived for pre-critical regimes can be used for calculations for over-critical lubrication regimes as well. Also, this conclusion allows us not to do any further analysis of heavily loaded over-critical lubrication regimes beyond the one conducted in Section 10.6.

Moreover, taking into account the fact that the solution of the asymptotic equations for pre-critical lubrication regimes depends only on parameter α_{10} and Q_0 if we compare formulas for the lubrication film thickness for pre-critical $H_0 = BV^{3/5}$ and over-critical $H_0 = A(VQ)^{3/4}$ lubrication regimes, we obtain that $B = AQ_0^{3/4}$ for $Q_0 \gg 1$. Therefore, for heavily loaded pre- and over-critical lubrication regimes (which are determined by the fact that the characteristic size of the inlet zone $\epsilon_q = V^{2/5} \ll 1$ and $\epsilon_q = (VQ)^{1/2} \ll 1$, respectively), one can create a map of level curves for $H_0 V^{-3/5} = f_*(Q_0)$, i.e., for any values of parameters V and Q for which $Q_0 = QV^{1/5} = const$ we have $H_0 V^{-3/5} = const$.

The analysis of the numerical results in this section validates the asymptotic approaches used for the cases of heavily loaded pre- and over-critical lubrication regimes. Moreover, it reveals that the asymptotic equations obtained for pre-critical lubrication regimes can be used for calculations of over-critical lubrication regimes.

10.14 How to Use the Numerical Solutions of the Asymptotic Equations in the Inlet and Exit Zones?

Now, let us consider how to practically use the numerical solutions of the asymptotic equations for pre-critical regimes in the inlet and exit zones and the analytical formulas for film thickness H_0. Suppose the lubricant viscosity is described by exponential Barus relationship $\mu(p) = \exp(Qp)$. The original EHL problem involves three parameters a, V, and Q. As we already know for heavily loaded EHL contacts $V \ll 1$ and/or $Q \gg 1$.

First, we need to understand whether we are dealing with a pre-critical or over-critical heavily loaded lubrication regime. Let us keep in mind that in a fully flooded pre-critical lubrication regime the characteristic size of the inlet zone is either $\epsilon_q = V^{2/5}$ if $Q_0 = QV^{1/5} = O(1)$ or $\epsilon_q = Q^{-2}$ otherwise while in a fully flooded over-critical lubrication regime the characteristic sizes of the inlet ϵ_q-zone and ϵ_0-zone are $\epsilon_q = (VQ)^{1/2}$ and $\epsilon_0 = Q^{-2}$ ($\epsilon_q \gg \epsilon_0$), respectively. By calculating the value of $Q_0 = QV^{1/5}$ we can determine the type of a regime we have at hand, i.e. (A) if $Q_0 \ll 1$ or $Q_0 = O(1)$ then the regime is pre-critical or (B) if $Q_0 \gg 1$ then the regime is over-critical.

In case (A) (pre-critical regimes) when $V^{2/5} \ll Q^{-2}$ we can accept $\epsilon_q = Q^{-2}$, $\alpha_1 = (a+1)Q^2$, $Q_0 = V^{1/5}Q \ll 1$ while when $V^{2/5} = O(Q^{-2})$ we can accept either $\epsilon_q = Q^{-2}$, $\alpha_1 = (a+1)Q^2$, $Q_0 = 1$ or $\epsilon_q = V^{2/5}$, $\alpha_1 = (a+1)V^{-2/5}$, $Q_0 = V^{1/5}Q = O(1)$. Besides that $r = \frac{x-a}{\epsilon_q}$ and $s = \frac{x-c}{\epsilon_q}$. After that we can solve numerically the asymptotic equations for a pre-critical lubrication regime and obtain two sets of solutions: α_1, A, $q(r)$, $h_q(r)$ in the inlet zone and β_1, $g(s)$, $h_g(s)$ in the exit zone. Based on these solutions we determine $H_0 = AV^{3/5}$ if $\epsilon_q = V^{2/5}$ and $H_0 = A(VQ^{-4})^{1/3}$ if $\epsilon_q = Q^{-2}$. The uniformly valid asymptotic approximations of pressure and gap distributions are determined by formulas (10.142), i.e.,

$$p_u(x) = \tfrac{\epsilon_q}{2} q(\tfrac{x-a}{\epsilon_q}) g(\tfrac{x-c}{\epsilon_q}), \quad h_u(x) = h_q(\tfrac{x-a}{\epsilon_q}) h_g(\tfrac{x-c}{\epsilon_q}), \tag{10.381}$$

where a and c are determined by formulas (10.99) and (10.100), i.e., $a = -1 + \alpha_1 \epsilon_q$ and $c = 1 + \beta_1 \epsilon_q$. In cases when $\alpha_1 \ll 1$ we have starved lubrication regimes ($h_q(r) = h_g(s) = 1$) and we can take $h_u(x) = 1$.

In case (B) (over-critical regimes) when $V^{2/5} \gg Q^{-2}$ we can expect to get $H_0 = A(VQ)^{3/4}$. For practical purposes we still can use solutions of the asymptotic equations obtained for the case of pre-critical lubrication regimes. At the same time we can expect that the lubrication film is formed in the inlet ϵ_q- and ϵ_0-zones of sizes $\epsilon_q = (VQ)^{1/2}$ and $\epsilon_0 = Q^{-2}$. The exit zones have the same respective sizes. The numerical step size should be chosen accordingly.

To summarize, it is important that as a byproduct of the asymptotic analysis right-away we get a structural formula for the film thickness H_0. Only the coefficient of proportionality A is unknown in this formula. To determine this coefficient one needs to solve the asymptotic equations in the inlet zone or to fit the structural formula for H_0 to the available experimental data.

10.15 Some Analytical Approximations and Numerical Solutions of EHL Problems for Non-Newtonian Fluids

The section is dedicated to the asymptotic analysis of steady heavily loaded isothermal and thermal lubricated line contacts under the no-slip condition and two opposite limiting cases of pure rolling and relatively high slide-to-roll ratio. The problem is considered for two general classes of non-Newtonian lubricant rheologies when the shear strain and stress can be expressed as certain explicit functions of shear stress and strain, respectively. Some approximations of the generalized Reynolds equation for non-Newtonian fluids that resemble the Reynolds equation for a Newtonian fluid are obtained. The main idea of the method in the isothermal case is analytical solution of the problem for the sliding shear stress and the consequent reduction the problem to asymptotic and/or numerical solution just for the pressure and gap functions. Similarly, for the non-isothermal problem the method is based on analytical solution of the problem for the sliding shear stress, temperature, and heat flux and the reduction of the problem to asymptotic and/or numerical solution for the pressure, gap, and varying from point to point slide-to-roll ratio and average surface velocity. The procedure for deriving formulas for the isothermal lubrication film thickness is based on the methods outlined in Sections 10.4, 10.6, and 10.10 and it is presented for the cases when the influence of the lubrication shear stresses on surface normal and tangential displacements can be neglected. A number of examples illustrating application of the described technique is given. Some general issues as to what is the range of parameters when the proposed approximation provides asymptotically correct solutions and how to define appropriately the pressure viscosity coefficient are sorted out in the case of an isothermal EHL problem for a non-Newtonian lubricant. It is shown that in certain cases the asymptotic procedure described below provides asymptotically correct solutions only in regimes of starved lubrication while in other cases it is valid for both starved and fully flooded lubrication regimes. The knowledge gained from the isothermal EHL problem is used in the analysis of TEHL problem that also takes into account the tangential displacement of the surfaces points in contact and heat transfer into the contact surfaces. A two-term asymptotic approximation of the latter problem is obtained analytically and finalized numerically. One of the applications of the developed technique is the analysis of the dimple phenomenon.

In general, the problem formulation is very similar to the ones considered before. The main differences are as follows:

(a) the lubricant is a non-Newtonian fluid,

(b) the contact solids may be made of sufficiently soft elastic materials which require proper accounting for surface tangential displacements,

(c) the heat generated in the lubrication layer is transferred to the contact surfaces and is dissipated in the contact solids.

These differences (a)-(c) make the EHL problem even more nonlinear and complex than the EHL problem for Newtonian fluid. Therefore, it is worth to first simplify the problem as much as possible and only after that to do the asymptotic and/or numerical analysis.

Let us consider a line lubricated contact for cylindrical solids of radii R_1 and R_2 made of elastic materials with Young's moduli E_1 and E_2 and Poisson's ratios ν_1 and ν_2, respectively. Far from the contact the cylinders' surfaces are steadily moving with linear velocities u_1 and u_2 and pressed against each other with a normal force P. The lubricant separating the solids is an incompressible non-Newtonian fluid. Under classical assumptions [7] of slow motion and narrow gap between the surfaces the rheology of the lubricant can be described by the equations

$$\mu \frac{\partial u}{\partial z} = F(\tau) \text{ or } \tau = \Phi(\mu \frac{\partial u}{\partial z}), \qquad (10.382)$$

where u is the lubricant velocity along the x-axis, $u = u(x, z)$, τ is the shear stress, $\tau = \tau(x, z)$, and μ is the lubricant viscosity that depends on lubricant pressure p and temperature T, $\mu = \mu(p, T)$, F and Φ are functions describing the lubricant rheology. The coordinate system is introduced in such a manner that the x-axis is directed along the contact in the direction of motion, the y-axis is directed along cylinders' axes, and the z-axis is directed along the line connecting the cylinders' centers (see Fig. 10.3). In equations (10.382) functions F and Φ are given odd smooth functions, which are inverses of each other and $F(0) = \Phi(0) = 0$.

10.15.1 Formulation of Isothermal EHL Problem for Non-Newtonian Fluids in Heavily Loaded Contacts

First we will consider an isothermal EHL problem for non-Newtonian lubricants in the case when the tangential surface displacements can be neglected. Under isothermal conditions the lubricant viscosity μ is independent of the lubricant temperature T, i.e., $\mu = \mu(p)$ and, therefore, μ is independent of z. Let us introduce the dimensionless variables for heavily loaded contact

$$\{x', a, c\} = \frac{1}{a_H}\{x, x_i, x_e\}, \ p' = \frac{p}{p_H}, \ \{h', z'\} = \frac{1}{h_e}\{h, z\},$$

$$\mu' = \frac{\mu}{\mu_a}, \ \{\tau', f', \Phi'\} = \frac{\pi R'}{P}\{\tau, f, \Phi\}, \ F' = \frac{2h_e}{\mu_a(u_1+u_2)}F, \qquad (10.383)$$

and parameters

$$s_0 = 2\frac{u_2 - u_1}{u_2 + u_1}, \quad V = \frac{24\mu_a(u_1 + u_2)R'^2}{a_H^3 p_H}, \quad H_0 = \frac{2R'h_e}{a_H^2}, \tag{10.384}$$

Based on the above assumptions and using the no-slip boundary conditions for the lubricant velocity u at the contact surfaces $z = \pm h/2$ in the dimensionless variables (10.383) and (10.384), the EHL problem can be reduced to the following system of equations [52] (further primes at the dimensionless variables are omitted)

$$\frac{d}{dx}\left[\frac{1}{\mu}\int\limits_{-h/2}^{h/2} zF(f + H_0 z\frac{dp}{dx})dz - h\right] = 0, \tag{10.385}$$

$$\frac{1}{\mu}\int\limits_{-h/2}^{h/2} F(f + H_0 z\frac{dp}{dx})dz = s_0, \tag{10.386}$$

$$p(a) = p(c) = \frac{dp(c)}{dx} = 0, \tag{10.387}$$

$$H_0(h - 1) = x^2 - c^2 + \frac{2}{\pi}\int\limits_a^c p(t)\ln|\frac{c-t}{x-t}|\,dt, \tag{10.388}$$

$$\int\limits_a^c p(t)dt = \frac{\pi}{2}. \tag{10.389}$$

In equations (10.383)-(10.389) function f is the sliding frictional stress, $f = f(x)$, h is the gap between the contact surfaces, $h = h(x)$, s_0 is the slide-to-roll ratio based on surface rigid velocities u_1 and u_2, a and c are the x-coordinates of the inlet and exit points of the contact, H_0 is the dimensionless exit film thickness. In equations (10.383) x_i and x_e are the dimensional x-coordinates of the inlet and exit points of the contact, p_H and a_H are the Hertzian maximum pressure and the Hertzian half-width of the contact, $p_H = \sqrt{\frac{E'P}{\pi R'}}$, $a_H = 2\sqrt{\frac{R'P}{\pi E'}}$, $\frac{1}{R'} = \frac{1}{R_1} \pm \frac{1}{R_2}$, $\frac{1}{E'} = \frac{1-\nu_1^2}{E_1} + \frac{1-\nu_2^2}{E_2}$, h_e is the dimensional lubrication film thickness at the exit from the contact, $h_e = h(x_e)$, μ_a is the lubricant viscosity at ambient temperature T_a. In most cases the rheology functions F and Φ and the lubricant viscosity μ also involve parameter V as well as some other parameters.

Therefore, for the given values of parameters a, V, s_0, and other parameters involved as well as for the given functions $\mu(p)$, $F(x)$, and $\Phi(x)$ the solution of the EHL problem is represented by parameters c, H_0 and functions $f(x)$, $p(x)$, and $h(x)$.

10.15.2 Isothermal EHL Problem for Pure Rolling Conditions. Pre- and Over-critical Lubrication Regimes. Some Numerical Examples

One of the most often encountered in practice condition is lubrication under almost purely rolling conditions. Because of that we will first analyze the problem in the cases of pure rolling conditions in heavily loaded EHL contacts. For the slide-to-roll ratio $s_0 = 0$ due to the fact that $F(x)$ is an odd function of x and $F(0) = 0$ the solution of equation (10.386) is obviously $f(x) = 0$. In this case the generalized Reynolds equation (10.386) is reduced to

$$\frac{d}{dx}\{M(\mu, p, h, \frac{dp}{dx}, V, H_0) - H_0 h\} = 0, \tag{10.390}$$

$$M(\mu, p, h, \frac{dp}{dx}, V, H_0) = \frac{H_0}{\mu} \int\limits_{-h/2}^{h/2} zF(H_0 z \frac{dp}{dx}) dz. \tag{10.391}$$

Therefore, the lubrication problem is reduced to solution of the system of equations (10.390), (10.391), (10.387)-(10.389).

Heavily loaded EHL contact conditions are usually caused by low surface velocities or high load applied to the contact and/or the kind of lubricant viscosity that experiences steep increase with pressure. In any of these cases the EHL problem contains a small parameter $\omega \ll 1$. Depending on what causes the lubricated contact to be heavily loaded (including properties of the lubricant viscosity μ) this small parameter can be defined differently. For example, when the lubricant viscosity μ increases with pressure p just moderately, then the heavy loaded conditions of the contact are caused by the smallness of parameter V. Therefore, in this case the small parameter ω can be taken equal to V, i.e., $\omega = V \ll 1$. In the cases when heavy loading is caused by fast increase of the lubricant viscosity μ with pressure p from 1 to $\mu(1) \gg 1$ the small parameter ω can be taken, for example, equal to the reciprocal of $ln(\mu(1))$, i.e., $\omega = 1/ln(\mu(1)) \ll 1$. In particular, for the exponential law $\mu = \exp(Qp)$ (where $Q = \alpha_p p_H$, α_p is the pressure viscosity coefficient), the latter definition of ω leads to $\omega = Q^{-1} \ll 1$.

Let us consider some examples of lubricants with Newtonian and non-Newtonian rheologies.

Example 1. For a lubricant of Newtonian rheology we have (see equation (10.391))

$$F(x) = \frac{12H_0}{V} x, \quad M = \frac{H_0^3}{V} \frac{h^3}{\mu} \frac{dp}{dx}. \tag{10.392}$$

Example 2. Let us consider a lubricant with Ostwald-de Waele (power law) rheology [8]. Then we have

$$F(x) = \frac{12H_0}{V_n} \mid x \mid^{(1-n)/n} x, \quad V_n = V(\frac{P}{\pi R'G})^{(n-1)/n}, \tag{10.393}$$

where G is the characteristic shear modulus of the fluid. It is easy to see that in this case equation (10.391) leads to

$$M = \frac{24n}{2n+1} \frac{H_0^{\frac{2n+1}{n}}}{V_n} \frac{1}{\mu} (\frac{h}{2})^{\frac{2n+1}{n}} \mid \frac{dp}{dx} \mid^{\frac{1-n}{n}} \frac{dp}{dx}. \tag{10.394}$$

Interestingly enough, for $n = 1$ the Ostwald-de Waele rheology is reduced to the Newtonian one and equation (10.394) for M is reduced to equation (10.392) for a lubricant with Newtonian rheology.

Example 3. Let us consider a lubricant with Reiner-Philippoff-Carreau rheology [8] described by the function

$$\Phi(x) = \frac{V}{12H_0} x \{ \eta + (1 - \eta)[1 + (\frac{V}{12H_0 G_0} \mid x \mid)^m]^{\frac{n-1}{m}} \},$$

$$\eta = \frac{\mu_\infty}{\mu_0}, \; G_0 = \frac{\pi R' G}{P},$$

(10.395)

where μ_0 and μ_∞ are the lubricant viscosities at the shear rates $\partial u / \partial z = 0$ and $\partial u / \partial z = \infty$, respectively, which are scaled according to formulas (10.383), and G is the characteristic shear stress of the fluid, $G > 0$, n and m are constants, $0 \leq n \leq 1$, $m > 0$.

Let us consider two limiting cases for equation (10.395). If the shear stress τ is relatively small, then the lubricant behaves like a Newtonian fluid, i.e.,

$$\Phi(x) = \frac{V}{12H_0} x \; if \; \frac{V}{12H_0 G_0} \mu \frac{\partial u}{\partial z} \ll 1.$$

(10.396)

In this case $F(x)$ is determined by equation (10.392) and, therefore, the Reynolds equation coincides with the one for Newtonian fluid. In the opposite case of relatively large shear stress τ, the lubricant behaves according to power law

$$\Phi(x) = (1 - \eta)(\frac{V}{12H_0})^n (\frac{|x|}{G_0})^{n-1} x \; if \; \frac{V}{12H_0 G_0} \mu \frac{\partial u}{\partial z} \gg 1.$$

(10.397)

Solving equation (10.397) for x we get the expression for $F(x)$ in the form similar to (10.393)

$$F(x) = \frac{12H_0}{V(1-\eta)^{1/n}} [\frac{|x|}{G_0}]^{(1-n)/n} x, \; if \; \frac{V}{12H_0 G_0} \mu \frac{\partial u}{\partial z} \gg 1.$$

(10.398)

It this case equation (10.391) leads to

$$M = \frac{24n}{2n+1} \frac{H_0^{\frac{2n+1}{n}}}{V(1-\eta)^{1/n} G_0^{(1-n)/n}} \frac{1}{\mu} (\frac{h}{2})^{\frac{2n+1}{n}} \mid \frac{dp}{dx} \mid^{\frac{1-n}{n}} \frac{dp}{dx}.$$

(10.399)

Now, let us consider the structure of a heavily loaded lubricated contact. In the central part of a heavily loaded lubricated contact, the Hertzian region, the pressure distribution $p(x)$ is well approximated by the Hertzian one equal to $\sqrt{1 - x^2}\theta(1 - x^2)$ while the gap $h(x)$ is practically equal to 1 (see the preceding sections). The behavior of the film thickness H_0 is predominantly determined by the EHL problem solution in the inlet zone that is a small vicinity of the inlet point $x = a$ while the location of the exit point $x = c$ is predominantly determined by the behavior of this solution in the exit zone - small vicinity of $x = c$. In preceding sections it was established that depending on the behavior of the lubricant viscosity and contact operating parameters there are possible two very different mechanisms of lubrication: pre- and over-critical lubrications regimes. These regimes reflect different mechanical processes that cause the contact to be heavily loaded. At the same time these regimes correspond to the different definitions of the small parameter ω discussed earlier (also see [29]). It was established earlier that the sizes of the inlet and exit zones are small in comparison with the size of the Hertzian region which is approximately equal to 2. Let us assume that the characteristic size of the inlet zone is $\epsilon_q = \epsilon_q(\omega) \ll 1$ for $\omega \ll 1$. Based on the principle of asymptotic matching of the solutions in the Hertzian region and inlet zone, in the inlet zone where $r = O(1)$, $r = (x - a)/\epsilon_q$, $\epsilon_q(\omega) \ll 1$ for $\omega \ll 1$, both the lubricant viscosity μ and gap h are of the order of one while $p(x) = O(\epsilon_q^{1/2})$, $\omega \ll 1$. We will assume that the lubricant viscosity μ is a non-decreasing function of pressure p. To characterize the above–mentioned lubrication regimes, let us introduce a critical size $\epsilon_0(\omega)$ of the inlet zone as the maximum value of $\epsilon_q(\omega)$ for which the following estimate holds (see Section 10.6)

$$\mu(\epsilon_q^{1/2}(\omega)) = O(1), \ \omega \ll 1. \tag{10.400}$$

Therefore, $\mu(\epsilon_q^{1/2}(\omega)) = O(1)$ for $\epsilon_q = O(\epsilon_0(\omega))$ and $\mu(\epsilon_q^{1/2}(\omega)) \gg 1$ for $\epsilon_q \gg \epsilon_0(\omega)$, $\omega \ll 1$. Now, it is possible to define the necessary conditions for pre- and over-critical lubrication regimes. A pre-critical lubrication regime is such a regime that the size of the inlet zone ϵ_q satisfies the estimate

$$\epsilon_q(\omega) = O(\min(1, \epsilon_0(\omega))), \ \omega \ll 1, \tag{10.401}$$

while an over-critical lubrication regime is a regime for which the size of the inlet zone ϵ_q satisfies the estimate

$$\epsilon_0(\omega) \ll \epsilon_q(\omega) \ll 1, \ \omega \ll 1. \tag{10.402}$$

Keeping in mind that $\epsilon_q(\omega) \ll 1$ for $\omega \ll 1$ from estimates (10.400)-(10.402), we obtain that over-critical regimes cannot be realized if $\epsilon_0(\omega) \gg 1$ for $\omega \ll 1$.

Clearly, the existence of the critical size $\epsilon_0(\omega)$ of the inlet zone depends on the particular behavior of the lubricant viscosity μ and the chosen small parameter ω. For example, if $\mu = \exp[Qp/(1+Q_1 p)]$, $Q, Q_1 \geq 0$, $Q \geq Q_1$, $Q \gg 1$ and $\omega = Q^{-1} \ll 1$, then $\epsilon_0(\omega) = Q^{-2}$. On the other hand, for the same as in the above example function μ, $Q \gg 1$, and $\omega = V \ll 1$ pre-critical regimes can be realized only for $\epsilon_q(V) = O(Q^{-2})$ while over-critical regimes are possible for $\epsilon_q(V) \gg Q^{-2}$, $Q \gg 1$.

10.15.2.1 Pre-critical Lubrication Regimes

In Section 10.4 it was shown that for pre-critical lubrication regimes it is beneficial to represent our system of equations (10.390), (10.391), (10.387)-(10.389) in the equivalent form (Vorovich et al. [28], see also (10.84)-(10.88))

$$p(x) = R(x)\left[1 - \frac{1}{2\pi}\int_a^c \frac{dM(p,h)}{dt}\frac{dt}{R(t)(t-x)}\right],$$
$$\tag{10.403}$$
$$R(x) = \sqrt{(x-a)(c-x)},$$

$$\int_a^c \frac{dM(p,h)}{dt} \frac{dt}{R(t)} = \pi(a+c),$$

$$\int_a^c \frac{dM(p,h)}{dt} \frac{tdt}{R(t)} = \pi[(\frac{c-a}{2})^2 + \frac{(a+c)^2}{2} - 1],$$

(10.404)

$$M(p,h) = \frac{H_0}{\mu} \int_{-h/2}^{h/2} zF(H_0 z \frac{dp}{dx})dz,$$

(10.405)

$$H_0(h-1) = x^2 - c^2 + \frac{2}{\pi} \int_a^c p(t) \ln |\frac{c-t}{x-t}| \, dt,$$

(10.406)

$$\frac{1}{\pi} \int_a^c M(p,h) \frac{dt}{R(t)} + \frac{2}{\pi} \int_a^c p(t) \ln \frac{1}{|c-t|} dt$$

(10.407)

$$+c^2 - \frac{1}{2}(\frac{c-a}{2})^2 - (\frac{a+c}{2})^2 = \ln |\frac{4}{c-a}|.$$

We will make the same assumption as in Section 10.4 that in a heavily loaded lubricated contact away from the contact boundaries $x = a$ and $x = c$ the first term W in equation (10.390) is negligibly small in comparison with the second one - $H_0 h$ (see similar assumption in (10.95)). Moreover, it was shown that in a pre-critical lubrication regime the inlet zone is represented by just one boundary layer [4] that has a homogeneous structure. It means that in the entire inlet zone the effects of the viscous fluid flow and elastic displacements are of the same order of magnitude and none of them prevails over another. In such cases in the inlet and exit zones, we have (see Section 10.4)

$$p = O(\epsilon_q^{1/2}), \quad \frac{dp}{dx} = O(\epsilon_q^{-1/2}), \quad h - 1 = O(\frac{\epsilon_q^{3/2}}{H_0}), \quad r = O(1),$$

$$p = O(\epsilon_g^{1/2}), \quad \frac{dp}{dx} = O(\epsilon_g^{-1/2}), \quad h - 1 = O(\frac{\epsilon_g^{3/2}}{H_0}), \quad s = O(1),$$

(10.408)

where $\epsilon_q = \epsilon_q(\omega) \ll 1$ and $\epsilon_g = \epsilon_g(\omega) \ll 1$ are characteristic sizes of the inlet and exit zones, which are boundary layers located next to points $x = a$ and $x = c$ (Van Dyke [4]). The characteristic size of the inlet zone $\epsilon_q(\omega)$ is determined by the given inlet coordinate (see (10.99))

$$a = -1 + \alpha_1 \epsilon_q, \quad \alpha_1 = O(1) \; for \; \omega \ll 1,$$

(10.409)

and depends on ω. The characteristic size of the exit zone $\epsilon_g(\omega)$ is unknown and is determined by the exit coordinate (see (10.100))

$$c = 1 + \beta_1 \epsilon_g, \quad \beta_1 = O(1) \; for \; \omega \ll 1.$$

(10.410)

In equations (10.409) and (10.410) α_1 is a given non-positive constant while β_1 is an unknown constant and is subject to calculation during the solution process. The region determined by the inequalities $x - a \gg \epsilon_q$ and $c - x \gg \epsilon_g$ will be called external (Van Dyke [4]) or Hertzian region (Vorovich et al. [28]).

Using the assumption that in a heavily loaded lubricated contact away from the contact boundaries $x = a$ and $x = c$ the first term W in equation (10.390) is negligibly small in comparison with the second one and temporarily assuming that the boundaries of the contact region $x = a$ and $x = c$ are known and fixed in the external region, we obtain (see equation 10.96))

$$x^2 - c^2 + \frac{2}{\pi} \int\limits_a^c p(t) \ln \left| \frac{c-t}{x-t} \right| dt = o(1)$$
(10.411)

$$for \; x - a \gg \epsilon_q \; and \; c - x \gg \epsilon_g.$$

Equations (10.411) and (10.389) with the accuracy of up to $o(1)$ provide us with the solution of a classic contact problem of elasticity (Vorovich et al. [28]), which has the form

$$p_0(x) = \sqrt{(x-a)(c-x)} + \frac{1+2ac+(c-a)^2/4-(a+c)x}{2\sqrt{(x-a)(c-x)}}$$
(10.412)

$$for \; x - a \gg \epsilon_q \; and \; c - x \gg \epsilon_g.$$

Thus, in the external region, the problem solution has the form

$$p(x) = p_0(x) + o(1) \; for \; x - a \gg \epsilon_q \; and \; c - x \gg \epsilon_g.$$
(10.413)

Based on estimates (10.408) we will try to find the solution of the problem in the inlet and exit zones in the form (see (10.103) and (10.104))

$$p(x) = \epsilon_q^{1/2} q(r) + o(\epsilon_q^{1/2}), \; q(r) = O(1) \; r = O(1), \; \omega \ll 1,$$
(10.414)

$$p(x) = \epsilon_g^{1/2} g(s) + o(\epsilon_g^{1/2}), \; g(s) = O(1) \; s = O(1), \; \omega \ll 1,$$
(10.415)

where $q(r)$ and $g(s)$ are the major terms of the pressure $p(x)$ asymptotic expansions in the inlet and exit zones. Obviously, the solutions from (10.414) and (10.415) must match the asymptotic representations of $p_0(x)$ at the boundaries of the inlet and exit zones with the Hertzian region (for details see Section 10.4).

To proceed further along the lines of the analysis outlined in the preceding sections and to obtain formulas for the lubrication film thickness H_0, we have to make an assumption about the behavior of the rheology function F or Φ. Therefore, we can assume that the rheology of the lubricant is such that

$$F(H_0 \epsilon_q^{-1/2} y(t)) = V^{-k}(\omega^{-l} \epsilon_q^{-1/2} H_0^{n+1})^{1/m} F_0(y(t)) + \ldots,$$
(10.416)

$$F_0(y(t)) = O(1) \; for \; y(t) = O(1),$$

where k, l, m, and n are constants, $m > 0$. Now, by making the necessary estimates and following the analysis of Section 10.4, we will show that we can take $\epsilon_g = \epsilon_q$, which will allow us to arrive at the formula for the film thickness

$$H_0 = A(V^{km}\omega^l\epsilon_q^{\frac{3m+1}{2}})^{\frac{1}{m+n+1}} + \ldots, \quad A(\alpha_1) = O(1), \quad \omega \ll 1, \qquad (10.417)$$

where $A(\alpha_1)$ is an unknown nonnegative constant independent from ω and ϵ_q, which is determined by the solution of the problem in the inlet zone. A further analysis along the lines of Section 10.4 in the inlet and exit zones lead to the following asymptotically valid equations:

$$\frac{dM_0(q,h_q,\mu_q,r)}{dr} = \frac{2}{\pi}\int_0^\infty \frac{q(t)-q_a(t)}{t-r}dt,$$
$$\qquad (10.418)$$

$$q(0) = 0, \quad q(r) \to q_a(r) \text{ as } r \to \infty,$$

$$\int_0^\infty [q(t) - q_a(t)]dt = 0, \qquad (10.419)$$

$$\frac{dM_0(g,h_g,\mu_g,s)}{ds} = \frac{2}{\pi}\int_{-\infty}^0 \frac{g(t)-g_a(t)}{t-s}dt,$$
$$\qquad (10.420)$$

$$g(0) = \frac{dg(0)}{ds} = 0, \quad g(s) \to g_a(s) \text{ as } s \to -\infty,$$

$$\int_{-\infty}^0 [g(t) - g_a(t)]dt = 0, \qquad (10.421)$$

where

$$M_0(p,h,\mu,x) = \frac{A^{\frac{m+n+1}{m}}}{\mu}\int_{-h/2}^{h/2} zF_0(z\frac{dp}{dx}), \qquad (10.422)$$

functions $\mu_q(r)$ and $\mu_g(s)$ are the major terms of the asymptotic expansions of $\mu(p,Q)$ in the inlet and exit zones and functions $q_a(r)$ and $g_a(s)$ are the major terms of the inner asymptotic expansions of the external asymptotic (10.412), (10.413) in the inlet and exit zones which are determined by the equalities (see (10.114))

$$q_a(r) = \sqrt{2r} + \frac{\alpha_1}{\sqrt{2r}}, \quad g_a(s) = \sqrt{-2s} - \frac{\beta_1}{\sqrt{-2s}}. \qquad (10.423)$$

The expressions for functions $h_q(r)$ and $h_g(s)$ depend on the lubrication conditions. In case of starved lubrication $\epsilon_q^{3/2} \ll H_0$, we have (see Section 10.4)

$$\epsilon_q \ll \omega^{\frac{2l}{3n+2}}, \quad \omega \ll 1, \qquad (10.424)$$

and the expressions for the gap in the inlet and exit zones are

$$h_q(r) = 1, \qquad (10.425)$$

$$h_g(s) = 1. \tag{10.426}$$

In case of fully flooded lubrication regimes, we can take

$$\epsilon_q = \omega^{\frac{2l}{3n+2}}, \ \omega \ll 1. \tag{10.427}$$

That leads to the following formula for the film thickness in a fully flooded lubricated contact

$$H_0 = A(V^{km}\omega^l)^{\frac{3}{3n+2}} + \dots, \ A(\alpha_1) = O(1), \ \omega \ll 1, \tag{10.428}$$

where $A(\alpha_1)$ is an unknown nonnegative constant independent from ω which is determined by the solution of the problem in the inlet zone. In this case the expressions for $h_q(r)$ and $h_g(s)$ have the form

$$A[h_q(r) - 1] = \frac{2}{\pi} \int_0^\infty [q(t) - q_a(t)] \ln \frac{1}{|r-t|} dt, \tag{10.429}$$

$$A[h_g(s) - 1] = \frac{2}{\pi} \int_{-\infty}^0 [g(t) - g_a(t)] \ln \frac{1}{|s-t|} dt. \tag{10.430}$$

The above analysis is valid for the small parameter of the problem $\omega = Q^{-1} \ll 1$ or $\omega = V \ll 1$ as long as $\mu(p, Q) = O(1)$ in the inlet and exit zones. More details on such restrictions can be found in Section 10.4.

Now, let us consider equation (10.403) and (10.404). In the inlet and exit zones the integrals in these equations can be expressed as a sum of tree integrals: over the inlet and exit zones, and over the external region. Making the necessary estimates we obtain the formula (10.417) for the film thickness H_0 and the following asymptotically valid equations for $q(r)$ and A in the inlet zone

$$q(r) = \sqrt{2r}\left[1 - \frac{1}{2\pi} \int_0^\infty \frac{d}{dt} M_0(q, h_q, \mu_q, t) \frac{dt}{\sqrt{2t}(t-r)}\right], \tag{10.431}$$

$$\pi\alpha_1 = \int_0^\infty \frac{d}{dt} M_0(q, h_q, \mu_q, t) \frac{dt}{\sqrt{2t}},$$

as well as asymptotically valid equations for $g(s)$ and β_1 in the exit zone

$$g(s) = \sqrt{-2s}\left[1 - \frac{1}{2\pi} \int_{-\infty}^0 \frac{d}{dt} M_0(g, h_g, \mu_g, t) \frac{dt}{\sqrt{-2t}(t-s)}\right], \tag{10.432}$$

$$\beta_1 = \frac{1}{\pi} \int_{-\infty}^0 \frac{d}{dt} M_0(g, h_g, \mu_g, t) \frac{dt}{\sqrt{-2t}},$$

where the functions $h_q(r)$, $h_g(s)$, $\mu_q(q)$, and $\mu_g(g)$ have the same meaning as earlier.

It is important to remember the general conditions of applicability of the proposed method. Therefore, we need to remember that (see Section 10.4) for a given $\epsilon_q = \epsilon_q(V, Q)$ the pre-critical regimes are characterized by the following condition

$$\mu(\epsilon_q^{1/2}, Q) = O(1), \quad \epsilon_q \ll 1, \quad \omega \ll 1, \quad (10.433)$$

while the over-critical regimes are determined by the condition

$$\mu(\epsilon_q^{1/2}, Q) \gg 1, \quad \epsilon_q \ll 1, \quad \omega \ll 1. \quad (10.434)$$

Let us consider pre-critical lubrication regimes for Examples 1-3 presented earlier when the small parameter $\omega = V \ll 1$. The other choice of the small parameter can be $\omega = Q^{-1} \ll 1$.

Example 1. For a Newtonian fluid in equation (10.416), we have $k = 0$, $l = m = n = 1$ or $k = 1$, $l = 0$, $n = 1$ for $\omega = V \ll 1$ and $k = 1$, $l = 0$, $m = n = 1$ for $\omega = Q^{-1} \ll 1$. Therefore, equations (10.417) and (10.428) are reduced to

$$H_0 = A_s(V\epsilon_q^2)^{1/3}, \quad A_s = O(1), \quad \epsilon_q \ll \epsilon_f = V^{2/5}, \quad \omega \ll 1, \quad (10.435)$$

for starved lubrication and

$$H_0 = A_f V^{3/5}, \quad A_f = O(1), \quad \epsilon_q = O(V^{2/5}), \quad \omega \ll 1, \quad (10.436)$$

for fully flooded lubrication. These formulas coincide with the ones derived in Section 10.4 for Newtonian fluids.

Example 2. For a fluid with the Ostwald-de Waele (power law) rheology described by equations (10.393) (see also (10.489)), we have $k = 0$, $l = m = n$ for $\omega = V_n \ll 1$ and $k = 1$, $l = 0$, $m = n$ for $\omega = Q^{-1} \ll 1$. Therefore, we obtain [48]

$$H_0 = A_s(V_n^n \epsilon_q^{\frac{3n+1}{2}})^{\frac{1}{2n+1}}, \quad A_s = O(1), \quad \epsilon_q \ll \epsilon_f = V_n^{\frac{2n}{3n+2}}, \quad \omega \ll 1, \quad (10.437)$$

for starved lubrication and

$$H_0 = A_f V_n^{\frac{3n}{3n+2}}, \quad A_f = O(1), \quad \epsilon_q = O(\epsilon_f), \quad \omega \ll 1, \quad (10.438)$$

for fully flooded lubrication.

For starved lubrication and $Q = 0$ we can obtain a simple formula for coefficient A_s, (see the procedure outlined in Subsection 10.8.4)

$$A_s = A_1 \mid \alpha_1 \mid^{\frac{3n+1}{2(2n+1)}}, \quad (10.439)$$

where $A_1 = A_1(n)$ is the value of A_s which should be obtained numerically for $Q = 0$ and $\alpha_1 = -1$.

Example 3. The case of a lubricant with Reiner-Philippoff-Carreau rheology described by equation (10.492) contains three distinct cases. In both cases when $\frac{V}{12H_0 G_0} \mid \mu \frac{\partial u}{\partial z} \mid \ll 1$ or $\frac{V}{12H_0 G_0} \mid \mu \frac{\partial u}{\partial z} \mid = O(1)$ for $\omega \ll 1$ the formulas for the film thickness H_0 coincides with the ones for a Newtonian fluid (see

Example 1). However, the coefficients of proportionality A_s and A_f in the latter case differ from the Newtonian case and depends on the values of parameters n, m and G_0. In case when $V/(12H_0G_0) \mid \mu \partial u / \partial z \mid \gg 1$ for $\omega \ll 1$ (here the small parameter ω and the corresponding powers k, l, m, and n are determined as in Example 2) the rheological function Φ is well approximated by the power law represented by formulas (10.489), where parameter V_n must be replaced by $V_n = V[G_0(1 - \eta)]^{\frac{1}{n}}/G_0 \ll 1$. In this case by solving equation (10.489) for x it is easy to determine function

$$F(x) = \frac{1}{(1 - \eta)^{1/n}} \frac{12H_0}{V} \left(\frac{\mid x \mid}{G_0} \right)^{\frac{1-n}{n}} x.$$

Therefore, the formulas for the film thickness H_0 coincide with the ones for a fluid with the power law rheology (see Example 2). The above conditions on $\frac{V}{12H_0G_0} \mid \mu \frac{\partial u}{\partial z} \mid$ impose some limitations on the problem parameters which are not restrictive in the limiting case of Newtonian behavior and pretty restrictive in the limiting case of "power law" behavior. The latter restrictions have the form of $\epsilon_q \gg (1 - \eta)^4 G_0^{4(1-n)} V^{-2}$ for starved lubrication conditions and $V \gg [(1 - \eta)G_0^{1-n}]^{3/2}$. Practically these restriction mean that the pure rolling operating conditions for which a fluid with Reiner-Philippoff-Carreau rheology demonstrates non-Newtonian behavior hardly exist. It is clear from the mechanical point of view: under conditions of pure rolling the shear rates and shear stresses remain relatively small, which prevents the fluid from exhibiting its non-Newtonian behavior.

Let us summarize the results of the section. A detailed analysis of the structure of the problem solution for heavily loaded contacts lubricated by non-Newtonian fluid in pre-critical regimes is proposed. It is shown that the contact region consists of three characteristic zones: the zone with predominately "elastic solution" - the Hertzian region and the narrow inlet and exit zones (boundary layers) compared to the size of the Hertzian zone. In the inlet and exit zones the influences of the solids' elasticity and the viscous lubricant flow are of the same order of magnitude. In the inlet and exit zones two equivalent asymptotically valid systems of equations describing the solution of the problem are derived.

The solutions of these systems depend on a smaller number of initial problem parameters than the solution of the original problem. In all cases the characteristic size of the inlet and exit zones can be considered small compared to the size of the Hertzian region. The asymptotic formulas for the lubrication film thickness are derived. The lubrication film thickness depends on a particular regime of lubrication and the solution behavior in the inlet zone. The definition of pre-critical regimes is given. Based on that definition the limitations of the applied method were derived.

10.15.2.2 Numerical Solution of Asymptotically Valid Equations for Non-Newtonian Fluids in the Inlet and Exit Zones in Pre-critical Regimes

It was mentioned earlier for the case of Newtonian fluids that one of the major difficulties encountered in numerical solution of EHL problems for heavily loaded contacts is instability of pressure $p(x)$ in the inlet and exit zones. As for Newtonian fluids for non-Newtonian fluids, there are two interconnected sources of this instability in the original equations (10.390), (10.391), (10.387)-(10.389) that are due to fast growth of viscosity μ with pressure p (i.e., the presence of a large parameter in μ) and for high μ due to the proximity of these equations to an integral equation of the first kind numerical solutions of which are inherently unstable. In case of an exponential viscosity $\mu(p) = \exp(Qp)$, the solution instability usually reveals itself for sufficiently large values of Q. For the same reason as the original equations of the EHL problem the asymptotically valid in the inlet zone equations (10.418), (10.419), (10.422), (10.423), and (10.425) for starved lubrication regimes or (10.429) for fully flooded lubrication regimes and in the exit zone equations (10.420)-(10.423), and (10.426) for starved lubrication regimes or (10.430) for fully flooded lubrication regimes are also at risk of being numerically unstable. The detailed explanation of why that is so you can can find in Section 10.8 on numerical solution of asymptotically valid equations for Newtonian fluids. To make the numerical solution of asymptotically valid equations stable, we will instead solve equations (10.431) and (10.432) for $q(r)$ and $g(s)$, respectively.

Let us introduce two sets of nodes $r_k = k\triangle r$, $k = 0,\dots,N{+}2$, and $r_{k+1/2} = 1/2(r_k + r_{k+1})$, $k = 0,\dots,N+1$, where $\triangle r$ is a positive step size and N is a sufficiently large positive integer. The equations will be solved by iteratively. It is clear that coefficient A is a monotonically increasing function of $|\alpha_1|$, $\alpha_1 \leq 0$. That enables us instead of determining the value of A for the given value of α_1 to determine the value of α_1 for the given value of A. In addition, such an approach allows us to solve the problem in the inlet and exit zones independently of each other. The numerical scheme for equation (10.431) and other related to it equations has the same form as the scheme in equations (10.271)-(10.275), where function $W(q, h_q, \mu_q, r)$ is determined by equations (10.422) and

$$M_0(q, h_q, \mu_q, r) = W(q, h_q, \mu_q, r)\tfrac{dq}{dr}. \qquad (10.440)$$

Therefore, function $W(q, h_q, \mu_q, r)$ is defined by the equation

$$W(p, h, \mu, x) = \frac{A^{\frac{m+n+1}{m}}}{\mu\frac{dp}{dx}} \int\limits_{-h/2}^{h/2} z F_0(z\tfrac{dp}{dx}). \qquad (10.441)$$

The initial approximation of q_k can be taken as the asymptotic of the Hertzian pressure, i.e., $q_k^0 = \sqrt{2r_k}$, $k = 0,\dots,N+2$. The iteration process stops when the desired precision ε is reached.

In the exit zone the numerical scheme is developed in a similar fashion.

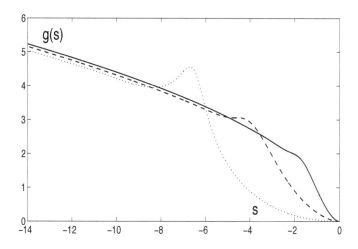

FIGURE 10.43
Main term of the asymptotic of the pressure distribution $g(s)$ in the exit zone of a starved lubricated contact for $Q_0 = 1$, $A = 3$, $n = 0.75$ (solid curve), $n = 1$ (dashed curve), and $n = 1.25$ (dotted curve).

Let us consider the example of power law fluid under pure rolling conditions (see Example 2 above). In this case the expression for function W has the form

$$W(p, h, \mu, x) = \frac{24n}{2n+1} \frac{A^{\frac{2n+1}{n}}}{\mu} \left(\frac{h}{2}\right)^{\frac{2n+1}{n}} \mid \frac{dp}{dx} \mid^{\frac{1-n}{n}}. \qquad (10.442)$$

We will discuss some results for regimes of starved lubrication for $\epsilon_q^{1/2} = Q^{-1} \ll V^{1/5}$, $Q \gg 1$ and exponential viscosity $\mu = \exp(Qp)$. In such a case the viscosity in the inlet and exit zones has the form $\mu_q(q) = e^{Q_0 q}$ and $\mu_g(g) = e^{Q_0 g}$ for $Q_0 = 1$. The absolute error of calculations was chosen to be $\varepsilon = 0.0001$. The convergence of this numerical schemes was established in the case of Newtonian fluid in Section 10.8. All of the following calculations were done for $N = 800$, $\triangle r = 0.03125$ and $N = 2,000$, $\triangle r = 0.0078125$ in the inlet and exit zones, respectively. For example, for $A = 3$ and $n = 0.75$, $n = 1$, and $n = 1.25$ the values of α_1 are equal to -2.2491, -2.5244, -2.7490, and the values of β_1 are equal to 1.2908, 1.5017, 1.5950, respectively. For the above values of n functions $q(r)$ are monotonically increasing functions of r, which approach their asymptotes $q_a(r)$, the latter, in turn, approach $\sqrt{2r}$ as $r \to \infty$. These curves do not exhibit any signs of instability or oscillations. Qualitatively and quantitatively the behavior of pressure distribution $q(r)$ for non-Newtonian fluids ($n \neq 1$) is very similar to the one for Newtonian lubricants ($n = 1$), which is illustrated in Fig. 10.14 and, therefore, the graphs of $q(r)$ for power law fluid are not presented here. For comparison three graphs

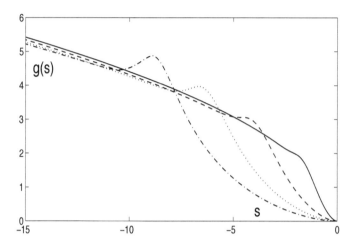

FIGURE 10.44
Main term of the asymptotic of the pressure distribution $g(s)$ in the exit zone
of a starved lubricated contact for $Q_0 = 1$, $n = 0.75$, $A = 1$ (solid curve),
$A = 2$ (dashed curve), $A = 3$ (dotted curve), and $A = 4$ (dash-dotted curve).

of functions $g(s)$ are given for $A = 3$ in Fig. 10.43, two of which are determined
for non-Newtonian fluid with $n = 0.75$ (solid curve) and $n = 1.25$ (dotted
curve) while the third one is determined for the case of Newtonian fluid with
$n = 1$ (dashed curve). It is clear from this figure that the pressure spike
increases with n. Note that β_1 increases monotonically with n and the pressure
spike increases in value and moves toward the contact center as n increases.
As in the case of $q(r)$ for large r the behavior of $g(s)$ for large s resembles the
behavior of its asymptote $g_a(s)$ and does not exhibit any oscillations or signs
of instability.

Figures 10.44 and 10.45 demonstrate the behavior of three curves of $g(s)$
for $Q_0 = 1$, $n = 0.75$ and $n = 1.25$ (solid, dashed, and dotted curves correspond to $A = 1$, 2, 3, respectively). In addition to that, in Fig. 10.44 for
$n = 0.75$ and $A = 4$ the pressure distribution $g(s)$ in the exit zone is given
by a dash-dotted curve. For $n = 1.25$ and $A = 4$ the pressure distribution
$g(s)$ exhibits the same signs of instability as in the cases of Newtonian fluid
described in detail in Subsection 10.8.3, which can be cured by the regularization described in Section 10.11. In particular, for $Q_0 = 1$, $n = 0.75$,
$A = 1$, 2, 3, and 4, we have $\alpha_1 = -0.4366$, -1.2408, -2.2491, -3.4009 and
$\beta_1 = 0.2879$, 0.7509, 1.2908, 1.8803, respectively, while for $Q_0 = 1$, $n = 1.25$,
$A = 1$, 2, 3, and 4, we have $\alpha_1 = -0.5526$, -1.5357, -2.7490, -4.1261 and
$\beta_1 = 0.3850$, 0.9510, 1.5950, 2.6671, respectively. For $n = 0.75$, $A = 1$, 2,
and $n = 1.25$, $A = 1$, the pressure distributions $g(s)$ in the exit zone

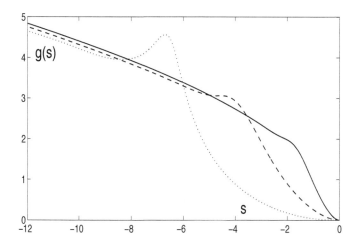

FIGURE 10.45
Main term of the asymptotic of the pressure distribution $g(s)$ in the exit zone
of a starved lubricated contact for $Q_0 = 1$, $n = 1.25$, $A = 1$ (solid curve),
$A = 2$ (dashed curve), and $A = 3$ (dotted curve).

are monotonic, i.e., they possess no local maxima. Contrary to that, for
$n = 0.75$, $A = 3$, 4 and $n = 1.25$, $A = 2$, 3, 4, the pressure distributions
$g(s)$ do possess local maxima, which shift closer to the center of the lubricated contact (to $s = -\infty$) and increase in value as $\mid \alpha_1 \mid$ and A increase. As
parameter n increases the height of the pressure spike increases and it shifts
closer to the contact center. As in the case of $q(r)$ for large s, the behavior of
$g(s)$ resembles the behavior of its asymptote $g_a(s)$ and does not exhibit any
oscillations or signs of instability. Also, for starved lubrication regimes, we can
get a transformation similar to (10.289) presented for the case of Newtonian
fluid.

Let us consider some results for pre-critical fully flooded lubrication regimes.
Qualitatively, the behavior of pressure $q(r)$ and $g(s)$ and gap $h_q(r)$ and $h_g(s)$
is similar to the one for Newtonian fluids described in Subsection 10.8.3
and the behavior of $q(r)$ and $g(s)$ considered above for starved lubrication
regimes. For $A = 0.4$, $Q_0 = 1$, and $n = 1.25$ ($\alpha_1 = -0.8045$, $h_q(\triangle r/2) =$
4.9248, $\beta_1 = 0.0858$, $\min h_g(s) = 0.7860$) and $A = 0.4$, $Q_0 = 5$, and $n = 1.25$
($\alpha_1 = -0.2234$, $h_q(\triangle r/2) = 1.6192$, $\beta_1 = 0.0610$, $\min h_g(s) = 0.9859$) examples of such solution behavior in the inlet and exit zones are given in Fig.
10.46, 10.47 and 10.48, 10.49, respectively. In the inlet zone the solutions are
obtained for $N = 1,200$ and $\triangle r = 0.015625$ while in the exit zones we used
$N = 480$ and $\triangle r = 0.0078125$. The absolute precision of both inlet and exit
zones solutions was $\varepsilon = 10^{-4}$. To get a better understanding of the solution

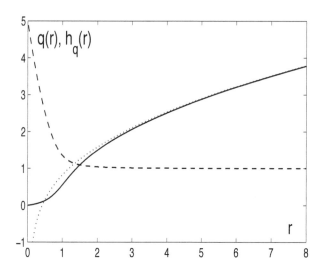

FIGURE 10.46
Main terms of the asymptotic distributions of pressure $q(r)$ (solid curve),
gap $h_q(r)$ (dashed curve), and the Hertzian pressure asymptote $q_a(r)$ (dotted
curve) in the inlet zone of a fully flooded lubricated contact for $A = 0.4$,
$Q_0 = 1$, and $n = 1.25$.

behavior in the exit zone in Fig. 10.50 and 10.51 are presented graphs of pressure $g(s)$ and gap $h_g(s)$ for four series of input parameters: $A = 0.4$, $n = 0.75$,
$Q_0 = 5$, and $Q_0 = 10$ ($\beta_1 = 0.0496$, $\max g(s) = 0.7821$, $\min h_g(s) = 0.8967$,
and $\beta_1 = 0.0397$, $\max g(s) = 0.8537$, $\min h_g(s) = 0.9222$, respectively) and
$A = 0.4$, $n = 1.25$, $Q_0 = 5$, and $Q_0 = 10$ ($\beta_1 = 0.0647$, $\max g(s) =$
0.8850, $\min h_g(s) = 0.8487$, and $\beta_1 = 0.08209$, $\max g(s) = 2.2897$,
$\min h_g(s) = 0.877$, respectively). Obviously, pressure $g(s)$ and gap $h_g(s)$ distributions are smoother for $n = 0.75$ than for $n = 1.25$. The presence of
a pin-like pressure spike for $n = 1.25$ and $Q_0 = 10$ is a manifestation of
mild instability. A similar behavior of pressure $g(s)$ and gap $h_g(s)$ can be
observed for $A = 0.4$, $n = 1.25$, $Q_0 = 20$, $N = 1,200$, $\triangle r = 0.00168$
($\beta_1 = 0.043$, $\max g(s) = 2.236$, $\min h_g(s) = 0.92$) the graphs of which are
depicted in Fig. 10.52.

Qualitatively, there are significant differences between these two series of
solutions. Obviously, for smaller values of Q_0 the sizes of the inlet and exit
zones as well as the inlet gap $h_q(\triangle r/2)$ are much wider than for larger values
of Q_0. It can be clearly seen from the comparison of graphs and effective
zone sizes in Fig. 10.46 and 10.48 in the inlet zones and in Fig. 10.47 and
10.49 in the exit zones, respectively. In the exit zone, $\min h_g(s)$ is smaller

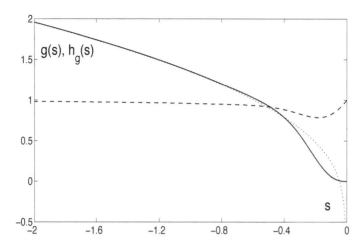

FIGURE 10.47
Main terms of the asymptotic distributions of pressure $g(s)$ (solid curve),
gap $h_g(s)$ (dashed curve), and the Hertzian pressure asymptote $g_a(s)$ (dotted
curve) in the exit zone of a fully flooded lubricated contact for $A = 0.4$,
$Q_0 = 1$, and $n = 1.25$.

and located farther from the exit point for smaller values of Q_0. Moreover,
for smaller values of Q_0 the pressure distribution $g(s)$ in the exit zone is
monotonic while for larger Q_0 it has a spike/local maximum. The value of
this pressure spike increases and it shifts toward the center of the lubricated
contact as Q_0 increases.

In the inlet zone the iteration process converges to a stable solution within
a tried range of the input parameters. In the exit zone the process converges
to a stable solution for, in a sense, relatively small to moderate values of
A, Q_0, and n. To improve convergence of the iteration process, one can modify
calculation of the first several iterations of the methods used in Subsections
10.8.1 and 10.8.2 as follows: $q_k^{i+1} = q_k^i + \gamma v_k$ and $g_k^{i+1} = g_k^i + \gamma v_k$, where
values v_k are determined by solution of systems of linear equations presented
in these subsections. The relaxation parameter γ can, for example, be taken
equal to 0.5. However, the ultimate way to improve convergence and to get
rid of solution instability is to use regularization described in Section 10.11.

Here, we need to reiterate (see Subsection 10.8.4) that to obtain sufficiently
accurate solution the step size $\triangle x$ for numerical solution of the original EHL
problem for lubricants with non-Newtonian rheology should be be determined
as $\epsilon_q \triangle r$, where $\triangle r$ is the step size which provides the adequate precision
for numerical solution of asymptotically valid equations in the inlet and exit
zones.

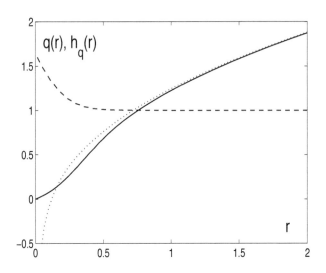

FIGURE 10.48
Main terms of the asymptotic distributions of pressure $q(r)$ (solid curve),
gap $h_q(r)$ (dashed curve), and the Hertzian pressure asymptote $q_a(r)$ (dotted
curve) in the inlet zone of a fully flooded lubricated contact for $A = 0.4$,
$Q_0 = 5$, and $n = 1.25$.

To conclude this subsection it is important to emphasize that the advantages
of the asymptotic approach to lubricated contacts with non-Newtonian fluids
are the same as for the cases with Newtonian fluids (see Subsection 10.8.4).

10.15.2.3   Over-Critical Lubrication Regimes

To develop the asymptotic approach to solution of the EHL problem for non-
Newtonian fluid involved in over-critical lubrication regime, we will closely
follow the procedure of Section 10.6. In particular, we will use the following
system of equation (see equations (10.390), (10.391), and (10.165))

$$\frac{d}{dx}\{M(\mu, p, h, \tfrac{dp}{dx}, V, H_0) - H_0 h\} = 0, \; p(a) = p(c) = \frac{dp(c)}{dx} = 0, \quad (10.443)$$

$$M(\mu, p, h, \tfrac{dp}{dx}, V, H_0) = \frac{H_0}{\mu} \int\limits_{-h/2}^{h/2} zF(H_0 z \tfrac{dp}{dx})dz, \quad (10.444)$$

$$H_0(h - 1) = x^2 - (\tfrac{a_0 + c_0}{2})^2 + \frac{2}{\pi} \int\limits_{a}^{c} p(t) \ln \mid \frac{\frac{a_0 + c_0}{2} - t}{x - t} \mid dt, \quad (10.445)$$

$$\int\limits_{a}^{c} p(t)dt = \frac{\pi}{2}. \quad (10.446)$$

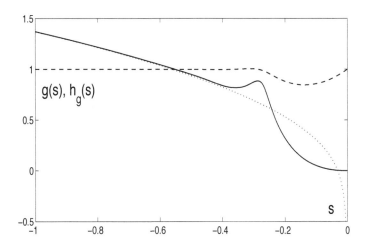

FIGURE 10.49
Main terms of the asymptotic distributions of pressure $g(s)$ (solid curve), gap $h_g(s)$ (dashed curve), and the Hertzian pressure asymptote $g_a(s)$ (dotted curve) in the exit zone of a fully flooded lubricated contact for $A = 0.4$, $Q_0 = 5$, and $n = 1.25$.

to describe the problem.

In Section 10.6 it was shown that in over-critical lubrication regimes the inlet zone is represented by not one but two boundary layers [4] of different sizes. Moreover, the inlet zone has a non-homogeneous structure. It means that in the inlet sub-zone most distant from the Hertzian region (called the inlet ϵ_q-zone) the effects of the viscous fluid flow through a gap are mainly created by the Hertzian pressure, which provides a dominating contribution to the elastic surface displacements. In the inlet sub-zone next to the Hertzian region (called the inlet ϵ_0-zone), the effects of the viscous fluid flow and elastic surface displacements are of the same order of magnitude. The size of the ϵ_0-zone is the critical size of the inlet region ϵ_0 and it is much smaller than ϵ_q - the size of the ϵ_q-zone. The same is true about the two exit ϵ_0- and ϵ_q-zones. In such cases in the ϵ_0-zones we have (see Section 10.6 and equations (10.163) and (10.164), in particular)

$$p(x) = O(\epsilon_0^{1/2}), \ \tfrac{dp(x)}{dx} = O(\epsilon_0^{-1/2}), \ h(x) - 1 = O(\tfrac{\epsilon_0^{3/2}}{H_0}), \ \omega \ll 1, \quad (10.447)$$

while in the ϵ_q-zones we have

$$p = O(\epsilon_0^{1/2}), \ \tfrac{dp}{dx} = O(\epsilon_0^{1/2}\epsilon_q^{-1}), \ h - 1 = O(\tfrac{\epsilon_q^{3/2}}{H_0}), \ \omega \ll 1. \quad (10.448)$$

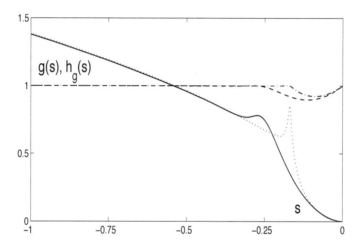

FIGURE 10.50
Main terms of the asymptotic distributions of pressure $g(s)$ (solid curve), gap $h_g(s)$ (dashed curve) for $Q_0 = 5$, and of pressure $g(s)$ (dotted curve), gap $h_g(s)$ (dash-dotted curve) for $Q_0 = 10$, in the exit zone of a fully flooded lubricated contact, $A = 0.4$, $n = 0.75$.

Most of the analysis of Section 10.6 deals with the properties of the solutions of the integral equations related to the equation for gap h. This analysis remains exactly the same for the problem at hand. However, some changes have to be made in the analysis in the inlet and exit ϵ_0- and ϵ_q-zones. To proceed further along the lines of the analysis of over-critical lubrication regimes outlined in Section 10.6 and to obtain formulas for the lubrication film thickness H_0, we need to make two assumptions about the behavior of the rheology function F in the inlet and exit zones. These assumptions can be derived from the assumption (10.416) by replacing in it $\epsilon_q^{-1/2}$ by $\epsilon_0^{1/2}\epsilon_q^{-1}$ in the inlet and exit ϵ_q-zones and by replacing $\epsilon_q^{-1/2}$ by $\epsilon_0^{-1/2}$ in the inlet and exit ϵ_0-zones. As a result of that we get

$$F(H_0\epsilon_0^{1/2}\epsilon_q^{-1}y(t)) = V^{-k}(\omega^{-l}\epsilon_0^{1/2}\epsilon_q^{-1}H_0^{n+1})^{1/m}F_0(y(t)) + \dots,$$

$$\text{(10.449)}$$

$$F_0(y(t)) = O(1) \ for \ y(t) = O(1),$$

$$F(H_0\epsilon_0^{-1/2}y(t)) = V^{-k}(\omega^{-l}\epsilon_0^{-1/2}H_0^{n+1})^{1/m}F_0(y(t)) + \dots,$$

$$\text{(10.450)}$$

$$F_0(y(t)) = O(1) \ for \ y(t) = O(1),$$

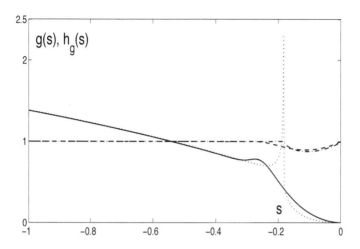

FIGURE 10.51
Main terms of the asymptotic distributions of pressure $g(s)$ (solid curve), gap $h_g(s)$ (dashed curve) for $Q_0 = 5$, and of pressure $g(s)$ (dotted curve), gap $h_g(s)$ (dash-dotted curve) for $Q_0 = 10$, in the exit zone of a fully flooded lubricated contact, $A = 0.4$, $n = 1.25$.

where k, l, m, and n are certain constants. The solution of the problem in the inlet and exit ϵ_q- and ϵ_0-zones will be searched in the form of asymptotic representations (see Section 10.6)

$$p = \epsilon_0^{1/2} q(r) + o(\epsilon_0^{1/2}), \ q(r) = O(1), \ r = \tfrac{x-a_0}{\epsilon_q} = O(1), \ w \ll 1, \quad (10.451)$$

$$p = \epsilon_0^{1/2} g(s) + o(\epsilon_0^{1/2}), \ g(s) = O(1), \ s = \tfrac{x-c_0}{\epsilon_q} = O(1), \ w \ll 1, \quad (10.452)$$

$$p = \epsilon_0^{1/2} q_0 + o(\epsilon_0^{1/2}), \ q_0(r_0) = O(1), \ r_0 = \tfrac{x-a_0}{\epsilon_0} = O(1), \ w \ll 1, \quad (10.453)$$

$$p = \epsilon_0^{1/2} g_0 + o(\epsilon_0^{1/2}), \ g_0(s_0) = O(1), \ s_0 = \tfrac{x-c_0}{\epsilon_0} = O(1), \ w \ll 1, \quad (10.454)$$

where the pair r and s and r_0 and s_0 are the local independent variables in the inlet and exit ϵ_q- and ϵ_0-zones, respectively, $q(r)$ and $g(s)$ are unknown functions representing the major terms of the pressure asymptotic expansions in the inlet and exit ϵ_q-zones while $q_0(r_0)$ and $g_0(s_0)$ are unknown functions representing the major terms of the pressure asymptotic expansions in the inlet and exit ϵ_0-zones. Constants a_0 and c_0 are determined in (10.159) and (10.160), respectively, as follows

$$a_0 = -1 + \alpha_0 \epsilon_0^{1/2} \epsilon_q^{1/2} + \ldots; \ \alpha_0 = O(1), \ w \ll 1, \quad (10.455)$$

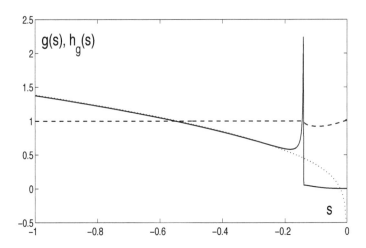

FIGURE 10.52
Main terms of the asymptotic distributions of pressure $g(s)$ (solid curve),
gap $h_g(s)$ (dashed curve), and the Hertzian pressure asymptote $g_a(s)$ (dotted
curve) in the exit zone of a fully flooded lubricated contact for $A = 0.4$,
$Q_0 = 20$, and $n = 1.25$.

$$c_0 = 1 + \beta_0 \epsilon_0^{1/2} \epsilon_q^{1/2} + \ldots; \quad \beta_0 = O(1), \ w \ll 1, \tag{10.456}$$

where constants α_0 and β_0 must be determined from the conditions that the
external solution (see (10.170))

$$p_{ext}(x) = p_0(x) + \epsilon_0^{1/2} \epsilon_q p_1(x) + \ldots, \ x - a_0 \gg \epsilon_0, \ c_0 - x \gg \epsilon_0, \tag{10.457}$$

is finite at points a_0 and c_0 (for details see Section 10.6).

As before (see (10.168)) we assume that in the external region the estimate

$$M(p, h) \ll \epsilon_0^{3/2} \ for \ x - a_0 \gg \epsilon_0, \ c_0 - x \gg \epsilon_0, \ w \ll 1, \tag{10.458}$$

holds. That provides the solutions for $p_0(x)$ and $p_1(x)$ in the external region
in the form of equations (10.173) and (10.182) while constants α_0 and β_0 are
determined by equations (10.184).

After that we estimate the behavior of gap h in the inlet and exit zones (see
Section 10.6). Using (10.449) and estimating the terms of equations (10.443)-
(10.443) in the inlet and exit ϵ_q-zones in the manner similar to Section 10.6
we get the formula for the central film thickness

$$H_0 = A(V^{km} \omega^l \epsilon_0^{-1/2} \epsilon_q^{\frac{3m+2}{2}})^{\frac{1}{m+n+1}}, \ A = O(1), \tag{10.459}$$

and equations for $q(r)$

$$M_0(q, h_q, \mu_q, r) = -\tfrac{4}{3} r \sqrt{-2r} \theta(-r), \ q(\alpha_1) = 0, \tag{10.460}$$

and for $h_q(r)$ for scanty lubrication regimes

$$h_q(r) = 1, \ \epsilon_q \ll (\omega^l \epsilon_0^{-1/2})^{\frac{2}{3n+1}}, \ \omega \ll 1, \tag{10.461}$$

or for $h_q(r)$ for fully flooded lubrication regimes

$$A[h_q(r) - 1] = -\tfrac{4}{3} r \sqrt{-2r} \theta(-r), \ \epsilon_q = (\omega^l \epsilon_0^{-1/2})^{\frac{2}{3n+1}}, \ \omega \ll 1. \tag{10.462}$$

By using the definition of fully flooded lubrication regimes $H_0 = O(\epsilon_q^{3/2})$ and formula (10.459), we can obtain the central film thickness formula and the characteristic size of the inlet and exit ϵ_q-zones for fully flooded lubrication regimes in the form (see Section 10.6)

$$H_0 = A(V^{km} \omega^l \epsilon_0^{-1/2})^{\frac{3}{3n+1}}, \ A = O(1), \ \epsilon_q = (V^{km} \omega^l \epsilon_0^{-1/2})^{\frac{2}{3n+1}}. \tag{10.463}$$

Similarly, we derive equations for $g(s)$

$$M_0(g, h_g, \mu_g, s) = \tfrac{4}{3} s \sqrt{2s} \theta(s) - \tfrac{4}{3} \beta_1 \sqrt{2\beta_1}, \ g(\beta_1) = 0, \tag{10.464}$$

and for $h_g(s)$ for scanty lubrication regimes

$$h_g(s) = 1, \ \epsilon_q \ll (\omega^l \epsilon_0^{-1/2})^{\frac{2}{3n+1}}, \ \omega \ll 1, \tag{10.465}$$

or for $h_g(s)$ for fully flooded lubrication regimes

$$A[h_g(s) - 1] = \tfrac{4}{3} s \sqrt{2s} \theta(s), \ \epsilon_q = (\omega^l \epsilon_0^{-1/2})^{\frac{2}{3n+1}}, \ \omega \ll 1, \tag{10.466}$$

$$M_0(p, h, \mu, x) = \frac{A^{\frac{m+n+1}{m}}}{\mu} \int\limits_{-h/2}^{h/2} F_0(z \tfrac{dp}{dx}) dz, \tag{10.467}$$

where A is a constant independent of ω, ϵ_0, and ϵ_q, functions $h_q(x)$, $\mu_q(q)$ and $h_q(x)$, $\mu_q(q)$ are the major terms of the asymptotic expansions of gap $h(x)$ and viscosity $\mu(p)$ in the inlet and exit ϵ_q-zones, respectively.

Note, that the value of constant A is not determined and cannot be determined by solving the system of equations (10.460)-(10.462) in the inlet ϵ_q-zone as well as the value of β_1 is not determined by equations (10.464)-(10.466) in the exit ϵ_q-zone. These constants are determined by the inlet and exit ϵ_0-zones, respectively. Thus, in the inlet and exit ϵ_q-zones the problem is reduced to integration of nonlinear first-order differential equations.

Obvious mechanical considerations lead to the conclusion that the solution $q(r)$ for $r \geq \alpha_1$ of equations (10.460)-(10.462) is bounded. Moreover, because of monotonicity of the rheological function F, the main term of the asymptotic expansion $q(r)$ is a nondecreasing function of r and $q(r) = q(0)$ for $r > 0$. Therefore, the direct matching of function $q(r) = q(0)$ as $r \to \infty$ without considering the asymptotic in the inlet ϵ_0-zone with the major term of the inner expansion of the external solution is impossible. Therefore, the major term of the two asymptotic expansions that match the solution $\epsilon_0^{1/2} q_0(r_0)$ in the inlet ϵ_0-zone as $r_0 \to \pm\infty$ take the form (for details see Section 10.6)

$$q_a(r_0) = q(0) + \sqrt{2r_0}\theta(r_0). \tag{10.468}$$

A similar asymptotic in the exit zones has the form

$$g_a(s_0) = g(0) + \sqrt{-2s_0}\theta(-s_0). \tag{10.469}$$

The further analysis of the inlet and exit ϵ_0-zones follows exactly the same procedure outlined in Section 10.6.

It is important to remember that from equations (10.232), (10.233), and formula (10.463) for H_0 we obtain that in the inlet and exit ϵ_0-zones

$$h(x) - 1 \ll 1 \ for \ r_0 = O(1) \ and \ s_0 = O(1), \quad \omega \ll 1. \tag{10.470}$$

This is true as long as $\epsilon_0^{3/2} \ll H_0$ (see the formula for H_0 (10.463) and the size of the inlet and exit ϵ_0-zones is much smaller than the size of the inlet and exit ϵ_q-zones for the fully flooded over-critical lubrication regimes from (10.463)), i.e.,

$$\epsilon_0 \ll (\omega^l \epsilon_0^{-1/2})^{\frac{2}{3n+1}}, \ \omega \ll 1. \tag{10.471}$$

The latter inequality represents the condition that discriminates between the pre- and over-critical lubrication regimes. Namely, if

$$\epsilon_0 = O((\omega^l \epsilon_0^{-1/2})^{\frac{2}{3n+1}}) \ or \ \epsilon_0 \gg (\omega^l \epsilon_0^{-1/2})^{\frac{2}{3n+1}}, \ \omega \ll 1, \tag{10.472}$$

then the lubrication regime is pre-critical. Otherwise, the lubrication regime is over-critical.

Therefore, the considered problem for $\omega \ll 1$ is reduced to a set of problems in different zones of the contact area. In particular, in the inlet ϵ_q-zone the asymptotic analogue of the original problem is the system of equations (10.460), (10.467), and (10.461) or (10.462) for function $q(r)$, in the external region the asymptotic solution of the problem is represented by formulas for $p_0(x)$ and $p_1(x)$ in the form of equations (10.173) and (10.182) while constants α_0 and β_0 are determined by equations (10.184), and in the exit ϵ_q-zone the asymptotic analogue of the original problem is the system of equations (10.464), (10.467), and (10.465) or (10.466) for function $g(s)$.

Note, that in the external region the estimate $\mu \gg 1$ leads to the validity of estimates (10.458) and $H_0(h - 1) \ll 1$ in the external region. That validates the entire analysis of the problem. Obviously, the described analysis is valid if parameter of the problem $\omega = Q^{-1} \ll 1$ or $\omega = V \ll 1$. The method is also applicable to the problems with non-Newtonian fluids under non-isothermal conditions. In these cases only solutions in the inlet and exit ϵ_q-zones and function M_0 from (10.242) must be adjusted for thermal conditions or fluid non-Newtonian rheology. The rest of the analysis and formulas remains intact.

Let us consider over-critical lubrication regimes for Examples 1-3 presented earlier when the lubricant viscosity is given by the exponential equation $\mu = \exp(Qp)$, $Q \gg 1$, and the small parameter $\omega = V \ll 1$. Here parameter Q may be a certain function of parameter V. In this case the critical inlet zone size is $\epsilon_0 = Q^{-2}$. Similarly, if $\mu = \exp(Qp^m)$, $Q \gg 1$ then $\epsilon_0 = Q^{-\frac{2}{m}} \ll 1$. Let us consider the first case or the second case for $m = 1$. Also, we can take the small parameter $\omega = V \ll 1$ (see the definitions of k, l, m, and n for different definitions of ω in the preceding subsection of this section).

Example 1. For a Newtonian fluid in equations (10.449) and (10.450), we have $k = 0$, $l = m = n = 1$ for $\omega = V \ll 1$. Therefore, equations (10.459) and (10.463) are reduced to formulas (10.212) and (10.213) for starved and fully flooded lubrication regimes, respectively.

Example 2. For a fluid with the Ostwald-de Waele (power law) rheology described by equations (10.393), we have $k = 0$, $l = m = n$ for $\omega = V \ll 1$. Based on (10.459) and (10.463) we obtain

$$H_0 = A_s(V_n^n Q \epsilon_q^{\frac{3n+2}{2}})^{\frac{1}{2n+1}}, \ A_s = O(1), \ \epsilon_q \ll \epsilon_f = (V_n^n Q)^{\frac{2}{3n+1}}, \quad (10.473)$$

for starved lubrication and

$$H_0 = A_f(V_n^n Q)^{\frac{3}{3n+1}}, \ A_f = O(1), \ \epsilon_q = \epsilon_f, \quad (10.474)$$

for fully flooded lubrication.

Example 3. The case of a lubricant with Reiner-Philippoff-Carreau rheology described by equation (10.395) contains three distinct cases. In both cases when $\frac{V}{12 H_0 G_0} \mid \mu \frac{\partial u}{\partial z} \mid \ll 1$ or $\frac{V}{12 H_0 G_0} \mid \mu \frac{\partial u}{\partial z} \mid = O(1)$ for $\omega \ll 1$, the formulas for the film thickness H_0 coincide with the one for a Newtonian fluid (see Example 1). In case when $V/(12 H_0 G_0) \mid \mu \partial u / \partial z \mid \gg 1$ for $\omega \ll 1$, the rheological function Φ is well approximated by the power law from equations (10.393), where parameter V_n must be replaced by $V_n = V[G_0(1-\eta)]^{\frac{1}{n}}/G_0 \ll 1$. In this case the formulas for the film thickness H_0 coincide with the ones for a fluid with the power law rheology (see Example 2, equations (10.473) and (10.474)). However, the condition $V/(12 H_0 G_0) \mid \mu \partial u / \partial z \mid \gg 1$ for $\omega \ll 1$ can hardly be realized in practice. The reason for that is for pure sliding the shear stress and shear sliding are relatively small and, therefore, the lubrication fluid does not develop non-Newtonian behavior.

10.15.3 Isothermal EHL Problem for Relatively Large Sliding Conditions. Pre- and Over-critical Lubrication Regimes

In general, heavily loaded EHL contact conditions are usually caused by low surface velocities or high load applied to the contact and/or the kind of lubricant viscosity that experiences steep increase with pressure. In any of these cases the EHL problem contains a small parameter $\omega \ll 1$. Depending on what causes the lubricated contact to be heavily loaded (including properties of the lubricant viscosity μ), this small parameter can be defined differently. For example, when the lubricant viscosity μ increases with pressure p in a lubricated contact moderately then the heavily loaded conditions of the contact are caused by the smallness of parameter V. Therefore, in this case the small parameter ω can be taken equal to V, i.e., $\omega = V \ll 1$. In the cases when heavy loading is caused by fast increase of the lubricant viscosity μ with pressure p from 1 to $\mu(1) \gg 1$, the small parameter ω can be taken, for example, equal to the reciprocal of $ln(\mu(1))$, i.e., $\omega = 1/ln(\mu(1)) \ll 1$. In particular, for the exponential law $\mu = \exp(Qp)$ (where $Q = \alpha_p p_H$, α_p is the pressure viscosity coefficient) and the latter definition of ω leads to $\omega = Q^{-1} \ll 1$.

We will analyze the EHL problems with non-Newtonian lubricants in cases of heavily loaded contacts with relatively large slide-to-roll ratios s_0. To make these conditions more precise let us introduce a small function (for comparison see function $\nu(x)$ in (10.328))

$$\nu(x) = \tfrac{H_0 h}{2 f_0} \tfrac{dp}{dx} \ll 1, \ \omega \ll 1, \tag{10.475}$$

representing the ratio of rolling $0.5 H_0 h dp/dx$ and the main term of asymptotic expansion of sliding f frictional stresses. Let us assume that this function is small in the entire contact or just in the inlet zone (see below) of the contact, i.e., $\nu(x) \ll 1$, $\omega \ll 1$, regardless of the particular definition of the parameter ω.

We will try to find the asymptotic representation for $f(x)$ in the form

$$f(x) = f_0(x) + \nu(x)f_1(x) + \nu^2(x)f_2(x) + \nu^3(x)f_3(x)$$
$$+O(\nu^4(x)), \ \nu(x) \ll 1, \tag{10.476}$$

where functions $f_0(x)$, $f_1(x)$, $f_2(x)$, and $f_3(x)$ are the consecutive terms of the asymptotic of $f(x)$, which have to be determined. Originally, this technique was introduces in [26, 52] and then used later by one of the authors in [27, 48] and by other researchers, in particular, in [51].

By substituting the representation for $f(x)$ from equation (10.476) into equation (10.386), expanding it for $\nu(x) \ll 1$, and equating the terms with the same powers of $\nu(x)$, we obtain

$$f_0(x) = \Phi(\tfrac{\mu s_0}{h}), \ f_1(x) = 0, \ f_2(x) = -\tfrac{f_0^2 F''(f_0)}{6 F'(f_0)}, \ f_3(x) = 0, \ \dots \tag{10.477}$$

The obtained solution for $f(x)$ allows for its elimination from the set of the problem unknowns and reducing the problem to determining just two functions: pressure p and gap h. In fact, by substituting equations (10.475)-(10.477) into equation (10.385), we obtain the approximate generalized Reynolds equation

$$\tfrac{d}{dx}\{M(\mu, p, h, \tfrac{dp}{dx}, V, s_0, H_0) - H_0 h\} = 0, \tag{10.478}$$

where the expressions for function M depend on the number of terms retained in equations (10.476) and (10.477). If the first one or two terms of the expansion for $f(x)$ from (10.476) are retained (i.e., $f(x) = f_0(x) + O(\nu^2(x))$, $\nu(x) \ll 1$), then

$$M = \frac{H_0^2 h^3 F'(f_0)}{12\mu} \frac{dp}{dx}, \tag{10.479}$$

while if the first three or four terms of the expansion for $f(x)$ from (10.476) are retained (i.e., $f(x) = f_0(x) + \nu(x) f_2(x) + O(\nu^4(x))$, $\nu(x) \ll 1$), then

$$M = \frac{H_0^2 h^3 F'(f_0)}{12\mu} \frac{dp}{dx} \{ 1 + \frac{H_0^2 h^2}{8} [\frac{F'''(f_0)}{5F'(f_0)} - \frac{1}{3}(\frac{F''(f_0)}{F'(f_0)})^2](\frac{dp}{dx})^2 \}. \tag{10.480}$$

In cases when the lubricant rheology is given by the second equation in (10.382), the derivatives of function $F(f)$ with respect to f involved in equations (10.479) and (10.480) can be expressed the following way:

$$F'(f_0) = \frac{1}{\Phi'(\lambda)}, \quad F''(f_0) = -\frac{\Phi''(\lambda)}{[\Phi'(\lambda)]^3},$$

$$F'''(f_0) = \frac{3[\Phi''(\lambda)]^2 - \Phi'''(\lambda)\Phi'(\lambda)}{[\Phi'(\lambda)]^5}, \quad \lambda = \frac{\mu s_0}{h}, \tag{10.481}$$

where function $\Phi(\lambda)$ is differentiated with respect to λ. Therefore, in these cases the expressions for function M which correspond to equations (10.479) and (10.480) can be represented in the forms

$$M = \frac{H_0^2 h^3}{12\mu\Phi'(\lambda)} \frac{dp}{dx}, \quad \lambda = \frac{\mu s_0}{h}, \tag{10.482}$$

$$M = \frac{H_0^2 h^3}{12\mu\Phi'(\lambda)} \frac{dp}{dx} \{ 1 + \frac{H_0^2 h^2}{120} \frac{4[\Phi''(\lambda)]^2 - 3\Phi'''(\lambda)\Phi'(\lambda)}{[\Phi'(\lambda)]^4} (\frac{dp}{dx})^2 \}, \tag{10.483}$$

respectively.

As a result of this analysis for the regimes for which $\nu(x) \ll 1$ for $a \le x \le c$, we obtain the following simplified approximate formulations of the isothermal EHL problem for the generalized Reynolds equation

$$\frac{d}{dx}\{ M(\mu, p, h, \frac{dp}{dx}, V, s_0, H_0) - H_0 h \} = 0, \tag{10.484}$$

$$p(a) = p(c) = \frac{dp(c)}{dx} = 0, \tag{10.485}$$

$$H_0(h - 1) = x^2 - c^2 + \frac{2}{\pi} \int_a^c p(t) \ln | \frac{c-t}{x-t} | \, dt, \tag{10.486}$$

$$\int_a^c p(t) dt = \frac{\pi}{2}, \tag{10.487}$$

where the expressions for function M are determined by formulas (10.479) and (10.480) or (10.482) and (10.483).

Let us consider some examples of lubricants with Newtonian and non-Newtonian rheologies.

Example 1. For a lubricant of Newtonian rheology, we have (see equations (10.382) and (10.479)):

$$F(x) = \frac{12H_0}{V}x, \quad \Phi(x) = \frac{V}{12H_0}x, \quad M = \frac{H_0^3}{V}\frac{h^3}{\mu}\frac{dp}{dx}. \tag{10.488}$$

It is important to realize that the generalized Reynolds equation (10.484) coincides with the exact Reynolds equation for a lubricant of Newtonian rheology.

Example 2. Let us consider a lubricant with Ostwald-de Waele (power law) rheology [8]. Then we have

$$F(x) = \frac{12H_0}{V_n}\mid x\mid^{(1-n)/n} x, \quad \Phi(x) = (\frac{V_n}{12H_0})^n \mid x \mid^{n-1} x,$$

$$V_n = V(\frac{P}{\pi R'G})^{(n-1)/n}, \tag{10.489}$$

where G is the fluid characteristic shear modulus. It is easy to see that equation (10.479) leads to

$$M = \frac{H_0^2}{12n}(\frac{12H_0}{V_n})^n \frac{h^3}{\mu}(\frac{\mu|s_0|}{h})^{1-n}\frac{dp}{dx}, \tag{10.490}$$

while equation (10.480) leads to the formula

$$M = \frac{H_0^2}{12n}(\frac{12H_0}{V_n})^n \frac{h^3}{\mu}(\frac{\mu|s_0|}{h})^{1-n}\frac{dp}{dx}[1 + \frac{(n-1)(n+2)H_0^2}{120n^2}(\frac{12H_0}{V_n})^{2n}$$

$$\times(\frac{\mu|s_0|}{h})^{-2n}(h\frac{dp}{dx})^2]. \tag{10.491}$$

Interestingly enough, for $n = 1$ the Ostwald-de Waele rheology is reduced to the Newtonian one and equations (10.489)-(10.491) are reduced to equations (10.488) for a lubricant with Newtonian rheology.

Example 3. Let us consider a lubricant with Reiner-Philippoff-Carreau rheology [8] described by the function

$$\Phi(x) = \frac{V}{12H_0}x\{\eta + (1 - \eta)[1 + (\frac{V}{12H_0G_0}\mid x \mid)^m]^{\frac{n-1}{m}}\},$$

$$\eta = \frac{\mu_\infty}{\mu_0}, \quad G_0 = \frac{\pi R'G}{P}, \tag{10.492}$$

where μ_0 and μ_∞ are the lubricant viscosities at the shear rate $\partial u/\partial z = 0$ and $\partial u/\partial z = \infty$, respectively, which are scaled according to formulas (10.383), and G is the characteristic shear stress of the fluid, $G > 0$, n and m are constants, $0 \leq n \leq 1$, $m > 0$. Based on formula (10.482) we conclude that

$$M = \frac{H_0^2 h^3}{12\mu_0 \Phi'(\lambda)}\frac{dp}{dx},$$

$$\Phi'(\lambda) = \frac{V}{12H_0}\{\eta + (1 - \eta)[1+ \mid \lambda \mid^m]^{\frac{n-1}{m}}[1 + (n - 1)\frac{|\lambda|^m}{1+|\lambda|^m}]\}, \tag{10.493}$$

$$\lambda = \frac{V}{12H_0}\frac{\mu_0 s_0}{hG_0}.$$

For the case of formula (10.483) the expression for function M can be represented in a similar form, however, it is much more complex than equation (10.493) and is not presented here. For $n = 1$ expressions (10.493) are reduced to the ones for the case of Newtonian fluids.

Now, let us consider the structure of a heavily loaded lubricated contact. In the central part of a heavily loaded lubricated contact, the Hertzian region, the pressure distribution $p(x)$ is well approximated by the Hertzian one equal to $\sqrt{1 - x^2}\theta(1 - x^2)$ while gap $h(x)$ is practically equal to 1 (see the preceding sections). The behavior of the film thickness H_0 is predominantly determined by the EHL problem solution in the inlet zone that is a small vicinity of the inlet point $x = a$ while the location of the exit point $x = c$ is predominantly determined by the behavior of this solution in the exit zone - small vicinity of $x = c$. In preceding sections it was established that depending on the behavior of the lubricant viscosity and contact operating parameters there are possible two very different mechanisms of lubrication: pre- and over-critical lubrications regimes. These regimes reflect different mechanical processes that cause the contact to be heavily loaded. At the same time these regimes correspond to the different definitions of the small parameter ω discussed earlier (also see [29]). It was established earlier that the sizes of the inlet and exit zones are small in comparison with the size of the Hertzian region, which is approximately equal to 2. Let us assume that the characteristic size of the inlet zone is $\epsilon_q = \epsilon_q(\omega) \ll 1$ for $\omega \ll 1$. Based on the principle of asymptotic matching of the solutions in the Hertzian region and inlet zone, in the inlet zone where $r = O(1)$, $r = (x - a)/\epsilon_q$, $\epsilon_q(\omega) \ll 1$ for $\omega \ll 1$ both the lubricant viscosity μ and gap h are of the order of one while $p(x) = O(\epsilon_q^{1/2})$, $\omega \ll 1$. We will assume that the lubricant viscosity μ is a non-decreasing function of pressure p. To characterize the above–mentioned lubrication regimes, let us introduce a critical size $\epsilon_0(\omega)$ of the inlet zone as the maximum value of $\epsilon_q(\omega)$ for which the following estimate holds (see Section 10.6)

$$\mu(\epsilon_q^{1/2}(\omega)) = O(1), \ \omega \ll 1. \tag{10.494}$$

Therefore, $\mu(\epsilon_q^{1/2}(\omega)) = O(1)$ for $\epsilon_q = O(\epsilon_0(\omega))$ and $\mu(\epsilon_q^{1/2}(\omega)) \gg 1$ for $\epsilon_q \gg \epsilon_0(\omega)$, $\omega \ll 1$. Now, it is possible to define the necessary conditions for pre- and over-critical lubrication regimes. A pre-critical lubrication regime is such a regime that the size of the inlet zone ϵ_q satisfies the estimate

$$\epsilon_q(\omega) = O(\min(1, \epsilon_0(\omega))), \ \omega \ll 1, \tag{10.495}$$

while an over-critical lubrication regime is a regime for which the size of the inlet zone ϵ_q satisfies the estimate

$$\epsilon_0(\omega) \ll \epsilon_q(\omega) \ll 1, \ \omega \ll 1. \tag{10.496}$$

Keeping in mind that $\epsilon_q(\omega) \ll 1$ for $\omega \ll 1$ from estimates (10.494)-(10.496) we obtain that over-critical regimes cannot be realized if $\epsilon_0(\omega) \gg 1$ for $\omega \ll 1$.

Clearly, the existence of the critical size $\epsilon_0(\omega)$ of the inlet zone depends on the particular behavior of the lubricant viscosity μ and the chosen small parameter ω. For example, if $\mu = \exp[Qp/(1+Q_1p)]$, $Q, Q_1 \geq 0$, $Q \geq Q_1$, $Q \gg 1$ and $\omega = Q^{-1} \ll 1$, then $\epsilon_0(\omega) = Q^{-2}$. On the other hand, for the same as in the above example function μ, $Q \gg 1$, and $\omega = V \ll 1$ pre-critical regimes can be realized only for $\epsilon_q(V) = O(Q^{-2})$ while over-critical regimes are possible for $\epsilon_q(V) \gg Q^{-2}$, $Q \gg 1$.

10.15.3.1 Pre-critical Lubrication Regimes

In Section 10.4 it was shown that in a pre-critical lubrication regime the inlet zone is represented by just one boundary layer [4] that has a homogeneous structure. It means that in the entire inlet zone the effects of the viscous fluid flow and elastic displacements are of the same order of magnitude and none of them prevails over another. In such cases in the inlet region, we have (see Section 10.6)

$$p = O(\epsilon_q^{1/2}), \quad \tfrac{dp}{dx} = O(\epsilon_q^{-1/2}), \quad h - 1 = O(\tfrac{\epsilon_q^{3/2}}{H_0}), \quad r = O(1). \qquad (10.497)$$

To proceed further along the lines of the analysis outlined in the preceding sections and to obtain formulas for the lubrication film thickness H_0, we have to make an assumption about the behavior of the rheology function Φ. Therefore, we can assume that the rheology of the lubricant is such that

$$\Phi(\tfrac{\mu s_0}{h}) = V^k \omega^l \mid s_0 \mid^m \, sign(s_0) H_0^{-n} \Phi_0(\tfrac{\mu}{h}),$$

$$\Phi_0(\tfrac{\mu}{h}) = O(1), \quad r = O(1), \quad \omega \ll 1, \qquad (10.498)$$

where $\Phi_0(r)$ is a certain function of r while l, m, k, and n are certain constants. To estimate functions involved in equations (10.476) and (10.477), we obtain the asymptotic behavior of Φ', Φ'', and Φ''' in the inlet zone as follows

$$\Phi'(\tfrac{\mu s_0}{h}) = \tfrac{V^k \omega^l \mid s_0 \mid^{m-1}}{H_0^n} \Phi_1(\tfrac{\mu}{h}), \quad \Phi_1(\tfrac{\mu}{h}) = O(1), \quad r = O(1), \qquad (10.499)$$

$$\Phi''(\tfrac{\mu s_0}{h}) = \tfrac{V^k \omega^l \mid s_0 \mid^{m-2}}{H_0^n} \Phi_2(\tfrac{\mu}{h}), \quad \Phi_2(\tfrac{\mu}{h}) = O(1), \quad r = O(1), \qquad (10.500)$$

$$\Phi'''(\tfrac{\mu s_0}{h}) = \tfrac{V^k \omega^l \mid s_0 \mid^{m-3}}{H_0^n} \Phi_3(\tfrac{\mu}{h}), \quad \Phi_3(\tfrac{\mu}{h}) = O(1), \quad r = O(1), \qquad (10.501)$$

where $\Phi_1(\mu/h)$, $\Phi_2(\mu/h)$, and $\Phi_3(\mu/h)$ are certain functions of r. Similar relationships for Φ, Φ' Φ'', and Φ''' hold in the exit zone. Let us consider the generalized Reynolds equation following from equations (10.479), (10.482), and (10.484). Taking into account equations (10.481), (10.497), and (10.499) and comparing the orders of magnitude of the terms in equation (10.484) in the inlet zone, we obtain a formula for the film thickness

$$H_0 = A(V^k \omega^l \mid s_0 \mid^{m-1} \epsilon_q^2)^{\frac{1}{n+2}}, \quad A = O(1), \quad \omega \ll 1. \qquad (10.502)$$

In the latter equation A is a coefficient of proportionality, which depends only on the specifics of the rheology function F (and/or Φ) and the lubricant viscosity μ, and it is independent of ω, s_0, and ϵ_q. The value of coefficient A can be obtained experimentally or by numerical solution of the system of asymptotically valid in the inlet zone equations similar to the ones derived in Section 10.4 for a Newtonian lubricant. The actual derivation of the asymptotically valid equations in the inlet and exit zones is left as an exercise for the reader. Equation (10.502) is valid for both starved and fully flooded lubrication regimes. The definitions of the starved and fully flooded lubrication regimes are given in (10.124) and (10.125), respectively. Using equation (10.502) and the definitions of starved and fully flooded regimes, we obtain the formulas

$$H_0 = A_s (V^k \omega^l \mid s_0 \mid^{m-1} \epsilon_q^2)^{\frac{1}{n+2}},$$

$$A_s = O(1), \ \epsilon_q \ll \epsilon_f = (V^k \omega^l \mid s_0 \mid^{m-1})^{\frac{2}{3n+2}}, \ \omega \ll 1,$$

(10.503)

for starved lubrication and

$$H_0 = A_f (V^k \omega^l \mid s_0 \mid^{m-1})^{\frac{3}{3n+2}}, \ A_f = O(1), \ \epsilon_q = O(\epsilon_f), \ \omega \ll 1, \quad (10.504)$$

for fully flooded lubrication. The value of ϵ_f (see equation (10.503)) represents the characteristic size of the inlet zone in the case of fully flooded pre-critical lubrication regimes. In equations (10.503) and (10.504) coefficients A_s and A_f differ from each other and can be determined experimentally or numerically. It should be stressed that coefficients A_s and A_f are independent of ω, s_0, V, ϵ_0, and ϵ_q.

Now, let us find the conditions under which the applied perturbation analysis for pre-critical regimes is valid, i.e., let us determine when $\nu(x) \ll 1$, $\omega \ll 1$, in the inlet zone. Using (10.475)-(10.477), and (10.502), we obtain the condition

$$\epsilon_q \ll \epsilon_\nu = (V^k \omega^l \mid s_0 \mid^{m+n+1})^{\frac{2}{3n+2}}, \ \omega \ll 1, \quad (10.505)$$

that guaranties that $\nu(x) \ll 1$, $\omega \ll 1$ in the inlet zone.

By comparing the magnitudes of the values of ϵ_f and ϵ_ν, we easily find that the above analysis is valid for both starved and fully flooded regimes if $\epsilon_f = O(\epsilon_\nu)$, $\omega \ll 1$, which for $n + 2 > 0$ is equivalent to the estimates

$$\mid s_0 \mid \gg 1 \ or \ \mid s_0 \mid = O(1), \ \omega \ll 1. \quad (10.506)$$

Let us consider pre-critical lubrication regimes for Examples 1-3 presented earlier when the small parameter $\omega = V \ll 1$. A similar analysis can be done for the small parameter $\omega = Q \gg 1$.

Example 1. For a Newtonian fluid (see equations (10.488)) in equation (10.498), we have $k = 0$, $l = m = n = 1$. Therefore, equations (10.503) and (10.504) are reduced to (see Section 10.4)

$$H_0 = A_s (V \epsilon_q^2)^{1/3}, \ A_s = O(1), \ \epsilon_q \ll \epsilon_f = V^{2/5}, \ V \ll 1, \quad (10.507)$$

for starved lubrication and

$$H_0 = A_f V^{3/5}, \; A_f = O(1), \; \epsilon_q = O(V^{2/5}), \; V \ll 1, \qquad (10.508)$$

for fully flooded lubrication.

Example 2. For a fluid with the Ostwald-de Waele (power law) rheology described by equations (10.489), we have $k = 0$, $l = m = n$. Therefore, we obtain [48]

$$H_0 = A_s(V_n^n \mid s_0 \mid^{n-1} \epsilon_q^2)^{\frac{1}{n+2}},$$

$$\qquad (10.509)$$

$$A_s = O(1), \; \epsilon_q \ll \epsilon_f = (V_n^n \mid s_0 \mid^{n-1})^{\frac{2}{3n+2}}, \; V_n \ll 1,$$

for starved lubrication and

$$H_0 = A_f(V_n^n \mid s_0 \mid^{n-1})^{\frac{3}{3n+2}}, \; A_f = O(1), \; \epsilon_q = O(\epsilon_f), \; V_n \ll 1, \qquad (10.510)$$

for fully flooded lubrication.

Example 3. The case of a lubricant with Reiner-Philippoff-Carreau rheology described by equation (10.492) contains three distinct cases. In both cases when $V \mid s_0 \mid /(12 H_0 G_0) \ll 1$ or $V \mid s_0 \mid /(12 H_0 G_0) = O(1)$ for $V \ll 1$ the formulas for the film thickness H_0 coincides with the ones for a Newtonian fluid (see Example 1). However, the coefficients of proportionality A_s and A_f in the latter case differ from the Newtonian case and depend on the values of parameters n and m. In case when $V \mid s_0 \mid /(12 H_0 G_0) \gg 1$ for $V \ll 1$ the rheological function Φ is well approximated by the power law represented by formulas (10.489), where parameter V_n must be replaced by $V_n = V[G_0(1 - \eta)]^{\frac{1}{n}}/G_0 \ll 1$. In this case the formulas for the film thickness H_0 coincide with the ones for a fluid with the power law rheology (see Example 2).

Now, let us consider the cases when function M is determined by equations (10.480) or (10.483). The second terms in brackets are supposed to be of the order of magnitude of $\nu^2(x) \ll 1$, $V \ll 1$, in the inlet zone. Using equations (10.483), (10.499)-(10.502) it can be shown that the second terms in brackets of equations (10.480) and (10.483) are small if $\epsilon_q \ll \epsilon_\nu$, $\epsilon_\nu = (V^k \omega^l \mid s_0 \mid^{n+m+1})^{\frac{2}{3n+2}}$, $\omega \ll 1$ (see estimate (10.505)). Therefore, all derived formulas for the lubrication film thickness are valid in the case of the generalized Reynolds equations based on the expressions for the function M from equations (10.480) or (10.483).

10.15.3.2   Over-critical Lubrication Regimes

In Section 10.6 it was shown that in over-critical lubrication regimes the inlet zone is represented by not one but two boundary layers [4] of different sizes. Moreover, the inlet zone has a non-homogeneous structure. It means that in the inlet sub-zone most distant from the Hertzian region (called the inlet ϵ_q-zone) the effects of the viscous fluid flow through a gap mainly created by the

Hertzian pressure dominate over the surface elastic displacements. In the inlet sub-zone next to the Hertzian region (called the inlet ϵ_0-zone) the effects of the viscous fluid flow and surface elastic displacements are of the same order of magnitude. The size of the ϵ_0-zone is the critical size of the inlet region ϵ_0, and it is much smaller than the size ϵ_q of the ϵ_q-zone. In such cases in the ϵ_0-zone we have (see Section 10.6 and equation (10.163), in particular)

$$p = O(\epsilon_0^{1/2}), \quad \tfrac{dp}{dx} = O(\epsilon_0^{-1/2}), \quad h - 1 = O(\tfrac{\epsilon_0^{3/2}}{H_0}), \quad r_0 = O(1), \qquad (10.511)$$

while in the ϵ_q-zone we have

$$p = O(\epsilon_0^{1/2}), \quad \tfrac{dp}{dx} = O(\epsilon_0^{1/2}\epsilon_q^{-1}), \quad h - 1 = O(\tfrac{\epsilon_q^{3/2}}{H_0}), \quad r = O(1). \qquad (10.512)$$

To proceed further along the lines of the analysis for over-critical lubrication regimes outlined in Section 10.6 and to obtain formulas for the lubrication film thickness H_0, we will make the assumption about the behavior of the rheology function Φ given in (10.498). Therefore, estimates (10.499)-(10.501) hold. Let us consider the generalized Reynolds equation following from equations (10.479), (10.482), and (10.484). Taking into account equation (10.481) and estimates (10.511) and (10.512) and comparing the orders of magnitude of the terms in equation (10.484) in the inlet ϵ_q-zone, we obtain a formula for the film thickness

$$H_0 = A(V^k\omega^l \mid s_0 \mid^{m-1} \epsilon_0^{-1/2}\epsilon_q^{5/2})^{\frac{1}{n+2}}, \quad A = O(1), \quad \omega \ll 1. \qquad (10.513)$$

In equation (10.513) A is a coefficient of proportionality that depends only on the specifics of the rheology function F (and/or Φ) and the lubricant viscosity μ and it is independent of ω, s_0, V, ϵ_0, and ϵ_q. Coefficient A can be obtained experimentally or by numerical solution of the system of asymptotically valid in the inlet and exit ϵ_0-zones equations. In the inlet and exit ϵ_q-zones functions $q(r)$, $g(s)$, $h_q(r)$, and $h_g(s)$ satisfy equations (see preceding Section) (10.460), (10.461) or (10.462) and (10.464), (10.465) or (10.466), respectively. In these equations function M_0 can be defined with the help of one of the expressions (10.482) or (10.483) for function M. For example, based on expressions (10.482) and (10.499) the equation for M_0 has the form

$$M_0(p, h, \mu, x) = \tfrac{A^{n+2}}{12} \tfrac{h^3}{\mu\Phi_1(\frac{p}{h})} \tfrac{dp}{dx}. \qquad (10.514)$$

Using the definitions of starved and fully flooded lubrication regimes given in (10.124) and (10.125), respectively, as well as equation (10.513), we derive the formulas

$$H_0 = A_s(V^k\omega^l \mid s_0 \mid^{m-1} \epsilon_0^{-1/2}\epsilon_q^{5/2})^{\frac{1}{n+2}},$$

$$A_s = O(1), \quad \epsilon_q \ll \epsilon_f = (V^k\omega^l \mid s_0 \mid^{m-1} \epsilon_0^{-1/2})^{\frac{2}{3n+1}}, \quad \omega \ll 1, \qquad (10.515)$$

for starved lubrication and

$$H_0 = A_f(V^k\omega^l \mid s_0 \mid^{m-1} \epsilon_0^{-1/2})^{\frac{3}{3n+1}},$$

$$A_f = O(1), \quad \epsilon_q = O(\epsilon_f), \quad \omega \ll 1, \tag{10.516}$$

for fully flooded lubrication. The value of ϵ_f (see equation (10.515)) represents the characteristic size of the ϵ_q-zone of the inlet region in the case of fully flooded over-critical lubrication regimes. In (10.515) and (10.516) coefficients A_s and A_f differ from each other and can be determined experimentally or numerically. It should be stressed that coefficients A_s and A_f are independent of ω, s_0, V, ϵ_0, and ϵ_q.

Now, let us find the conditions under which the applied perturbation analysis for over-critical regimes is valid, i.e., let us determine when $\nu(x) \ll 1$, $\omega \ll 1$, in the ϵ_q-zone of the inlet region. Using equations (10.475)-(10.477), and (10.513) we obtain the condition

$$\epsilon_q \ll \epsilon_{\nu q} = (V^k\omega^l \mid s_0 \mid^{n+m+1} \epsilon_0^{-1/2})^{\frac{2}{3n+1}}, \quad \omega \ll 1, \tag{10.517}$$

that guaranties that $\nu(x) \ll 1$ in the inlet ϵ_q-zone. The condition that guaranties that $\nu(x) \ll 1$ in the ϵ_0-zone is different and has the form

$$\epsilon_q \ll \epsilon_{\nu 0} = (V^k\omega^l \mid s_0 \mid^{n+m+1} \epsilon_0^{\frac{2n+3}{2}})^{\frac{2}{5(n+1)}}, \quad \omega \ll 1. \tag{10.518}$$

We have the relation $\epsilon_{\nu 0} \ll \epsilon_{\nu q}$ if $\epsilon_q \gg \epsilon_0$ that is typical for over-critical regimes. By comparing the magnitudes of the values of ϵ_f and $\epsilon_{\nu q}$, we obtain that the above analysis is valid for both starved and fully flooded regimes if $\epsilon_f = O(\epsilon_{\nu q})$, $\omega \ll 1$, that is represented by estimates (10.506) while the above analysis is valid for just starved lubrication if $\epsilon_{\nu q} \ll \epsilon_f$, $\omega \ll 1$.

Let us consider over-critical lubrication regimes for Examples 1-3 presented earlier when the lubricant viscosity is given by the exponential equation $\mu = \exp(Qp)$, $Q \gg 1$, and the small parameter $\omega = V \ll 1$. Here parameter $Q \gg 1$ may be a certain function of V. In this case the critical inlet zone size is $\epsilon_0 = Q^{-2}$, $Q \gg 1$. The analysis for a different choice of the small parameter ω can be done in a similar fashion.

Example 1. For a Newtonian fluid (see equations (10.488)) in equation (10.498), we have $l = 0$, $m = n = 1$. Therefore, equations (10.515) and (10.516) are reduced to formulas (10.212) and (10.213) for starved lubrication and for fully flooded lubrication regimes, respectively.

Example 2. For a fluid with the Ostwald-de Waele (power law) rheology described by equations (10.489), we have $k = 1$, $l = 0$, $m = n$. We obtain

$$H_0 = A_s(V_n^n Q \mid s_0 \mid^{n-1} \epsilon_q^{5/2})^{\frac{1}{n+2}},$$

$$A_s = O(1), \quad \epsilon_q \ll \epsilon_f = (V_n^n Q \mid s_0 \mid^{n-1})^{\frac{2}{3n+1}}, \quad V \ll 1, \tag{10.519}$$

for starved lubrication and

$$H_0 = A_f(V_n^n Q \mid s_0 \mid^{n-1})^{\frac{3}{3n+1}}, \ A_f = O(1), \ \epsilon_q = O(\epsilon_f), \ V \ll 1, \quad (10.520)$$

for fully flooded lubrication. Here coefficients A_s and A_f are independent of the input parameters V_n, Q, and s_0. Formula (10.520) indicates an interesting property of the film thickness H_0 as a function of the slide-to-roll ratio s_0. Depending on fluid rheology we may have $n < 1$, $n = 1$, or $n > 1$. It follows from formula (10.520) that for $n < 1$ (pseudoplastic fluid) or $n > 1$ (dilatant fluid) the film thickness H_0 is a decreasing or increasing function of s_0, respectively, while for $n = 1$ it is independent of s_0. Therefore, for $n > 1$ and large sliding the film thickness H_0 in a thermal EHL contact may be not as prone to decrease as it usually happens for Newtonian fluids ($n = 1$) and pseudoplastic ones ($n < 1$).

Example 3. The case of a lubricant with Reiner-Philippoff-Carreau rheology described by equation (10.492) contains three distinct cases. In both cases when $\frac{V s_0}{12 H_0 G_0} \mid \mu \frac{\partial u}{\partial z} \mid \ll 1$ or $\frac{V s_0}{12 H_0 G_0} \mid \mu \frac{\partial u}{\partial z} \mid = O(1)$ for $V \ll 1$ the formulas for the film thickness H_0 coincide with the one for a Newtonian fluid (see Example 1). In case when $\frac{V s_0}{12 H_0 G_0} \mid \mu \frac{\partial u}{\partial z} \mid \gg 1$ for $V \ll 1$, the rheological function Φ is well approximated by the power law from equations (10.489), where parameter V_n must be replaced by $V_n = V[G_0(1 - \eta)]^{\frac{1}{n}}/G_0 \ll 1$. In this case the formulas for the film thickness H_0 coincide with the ones for a fluid with the power law rheology (see Example 2, equations (10.519) and (10.520)).

For the case of large sliding the numerical solutions in the inlet and exit zones can be obtained based on an iteration processes practically identical to the ones described in detail and implemented for Newtonian fluids in Subsections 10.8.3, 10.8.4 and Section 10.11 and for non-Newtonian fluids under pure rolling conditions in Subsection 10.15.2. Because of the solution properties are also similar to the ones discussed in these sections, they will not be considered here.

10.15.4 Choosing Pre- and Over-critical Lubrication Regimes, Small Parameter ω, and Pressure Viscosity Coefficient

Nowadays, there are various ways to define the pressure viscosity coefficient of a lubricant that, obviously, lead to different values of the lubrication film thickness if the same formula for the film thickness is used. It is important to establish the correct practical way of determining the pressure viscosity coefficient in exponential like law for lubricant viscosity used for standard calculation of the lubrication film thickness. This standardization would allow for the proper comparison of various lubricants with different pressure-viscosity relationships. However, it is important to remember that the shear stress calculated based on such determined exponential–like viscosity equation may produce a significant error in comparison with the shear stress obtained based on the actual lubricant viscosity relationship $\mu(p)$. We will consider the most frequently used regimes of fully flooded lubrication that allow for obtaining practical and consistent results.

Let us assume that for the known rheology function Φ (and/or F) as well as the function of lubricant viscosity $\mu(p)$ the values of the lubrication film thickness H_0 are obtained numerically or experimentally. The question is how to practically describe these numerically or experimentally obtained values of H_0 by formulas (10.504) and (10.516) derived for the film thickness in fully flooded lubrication regimes that are widely used in engineering practice? We need to find a relatively easy and practical way of determining the unknown components involved in formulas (10.504) and (10.516). These components include coefficient A_f, small parameter ω, and the critical size ϵ_0 of the inlet region.

First, we need to realize that the asymptotic expression (10.498) for the rheological function Φ in the inlet region is independent of any parameters involved in the viscosity dependence $\mu(p)$ because the lubricant rheology is independent of a particular viscosity dependence on pressure and, also, $\mu(p)$ is always of the order of 1 in the inlet region. Second, there are basically only two choices for the problem small parameter: $\omega = V$ or ω equal to some small parameter involved in the relationship for viscosity $\mu(p)$. Third, if the problem small parameter ω is chosen to be equal to V, then the value of power l in (10.498) can be taken equal to zero as parameter V is already reflected in the behavior of the rheological function Φ in (10.498). Therefore, whatever choice of the small parameter ω is made we can consider $l = 0$ and the characteristic size of the inlet zone in fully flooded pre-critical regimes be equal to $\epsilon_f = (\mid s_0 \mid^{m-1} V^k)^{\frac{2}{3n+2}}$ (see the expression for ϵ_f in (10.503)).

Our choice of the proper approximation of the lubricant viscosity relationship is supposed to be based on the lubrication regime a contact is involved in, i.e., either pre- or over-critical fully flooded regime. The actual reason for that is the fact that in these regimes film thickness is determined by different zones of a contact. In particular, for pre-critical lubrication regimes film thickness and coefficient proportionality A_f are determined by the inlet zone of the characteristic size $\epsilon_f = (\mid s_0 \mid^{m-1} V^k)^{\frac{2}{3n+2}}$ while in over-critical lubrication regimes the film thickness and coefficient proportionality A_f are determined by the behavior of pressure in the inlet ϵ_0-zone. Therefore, the type of the lubrication regime should be established first, and then the approximation for viscosity $\mu(p)$ should be chosen.

Now, we need to determine the critical size ϵ_0 of the inlet zone. It is determined based on the solutions of equation (10.494). For practical use we will replace equation (10.494) by

$$\mu(\epsilon_0^{1/2}) = C, \tag{10.521}$$

where C is a constant of the order of magnitude 1. For certainty let us fix

$C = e$. Therefore, the critical size ϵ_0 of the inlet zone is the maximum among all solutions of equation (10.521). If $\mu(p)$ is a monotonically increasing function of p, then ϵ_0 is a unique solution of the above equation.

After that, the value of ϵ_0 must be compared to the characteristic size ϵ_f of the inlet region in a pre-critical lubrication regime. If $\epsilon_0 \gg \epsilon_f$ then the lubrication regime is pre-critical, the problem small parameter w can be taken equal to $V \ll 1$, and for the purpose of lubrication film calculations $\mu(p) = 1$. If ϵ_0 is of the order of ϵ_f, then the problem small parameter w again can be taken equal to $V \ll 1$ or to the parameter involved in the relationship of $\mu(p)$, which causes $\mu(p)$ to increase significantly as pressure p increases from 0 to 1. In this case the lubrication regime is still pre-critical and formula (10.503) should be used for calculating the values of coefficient A_f and film thickness H_0 based on the available numerical or experimental data. In the latter case $\mu(p)$ is no longer can be considered constant in the inlet and exit zones.

If $\epsilon_0 \ll \epsilon_f$, where $\epsilon_f(w)$ is defined by (10.503), then the lubrication regime is over-critical, the small parameter w can be chosen as a specific small parameter involved in the expression for the lubricant viscosity $\mu(p)$ that causes its significant growth from $\mu(p) = 1$ at $p = 0$ to $\mu(p) \gg 1$ at $p = 1$. In addition to that the values of coefficient A_f and film thickness H_0 should be calculated based on formula (10.516) for over-critical lubrication regimes using the available numerical or experimental data (see also the criteria (10.235) and (10.236) derived for the case of Newtonian lubricant).

Now, we are in a position to be able to replace the actual viscosity dependence on pressure by an appropriate exponential like dependence $\mu(p) = \exp(Qp^k)$ and determine its viscosity pressure coefficient Q. To do that we need to expand function $\ln[\mu(p)]$ for small p. Suppose the result of this expansion is

$$\ln[\mu(p)] = \sigma p^\alpha + \dots, \quad p \ll 1, \tag{10.522}$$

where σ and α are nonnegative constants.

Also, suppose $\epsilon_0 \ll \epsilon_f(w) = (| s_0 |^{m-1} V^k)^{\frac{2}{3n+2}}$ then the lubrication regime is over-critical and the problem small parameter w can be chosen as $w = \sigma^{-1} \ll 1$, i.e., dependent on a specific parameter involved in the relationship $\mu(p)$. Therefore, by retaining the same dependence on pressure p as in equation (10.522) and using the just determined ϵ_0, it follows from (10.522) that the actual relationship for the lubricant viscosity $\mu(p)$ can be replaced by the exponential–like relationship

$$\mu(p) = \exp(\epsilon_0^{-\alpha/2} p^\alpha) \tag{10.523}$$

that would provide the film thickness values H_0 practically equal to those obtained for the actual relationship for the lubricant viscosity $\mu(p)$. This is due to the fact that the inlet ϵ_0-zone, where the order of magnitude of pressure p is $\epsilon_0^{1/2}$, is mainly responsible for the formation of the film thickness H_0. Therefore, $Q = \epsilon_0^{-\alpha/2} = \sigma$ can be considered as an appropriate approximation of the viscosity pressure coefficient in an exponential–like viscosity dependence $\mu(p) = \exp(Qp^\alpha)$.

Otherwise, if $\epsilon_0 = O(\epsilon_f(\omega))$ or $\epsilon_0 \gg \epsilon_f(\omega)$, $\epsilon_f(\omega) = (\mid s_0 \mid^{m-1} V^k)^{\frac{2}{3n+2}}$, then the lubrication regime is pre-critical and the problem small parameter ω can be chosen as $\omega = V \ll 1$. In this case formula (10.504) for pre-critical lubrication regimes should be used for calculation of coefficient A_f and film thickness H_0 based on the available numerical or experimental data. Using equation (10.522) for the assumed behavior of the lubricant viscosity $\mu(p)$, we obtain that the actual relationship for the lubricant viscosity $\mu(p)$ can be replaced by the exponential–like relationship of the form

$$\mu(p) = \exp(\epsilon_f^{-\alpha/2} p^\alpha) \; if \; \epsilon_0 = O(\epsilon_f(\omega)) \ll 1, \; \omega \ll 1,$$

$$\mu(p) = 1 \; if \; \epsilon_0 \gg \epsilon_f = (\mid s_0 \mid^{m-1} V^k)^{\frac{2}{3n+2}}, \; \omega \ll 1,$$

$$(10.524)$$

that will provide the film thickness values practically equal to those obtained for the actual relationship for the lubricant viscosity $\mu(p)$. This is due to the fact that the inlet zone, where the order of magnitude of pressure p is $\epsilon_f^{1/2}$, is mainly responsible for the formation of the film thickness H_0. Therefore, in this case $Q = \epsilon_f^{-\alpha/2}(\omega) \gg 1$ if $\epsilon_0 = O(\epsilon_f(\omega)) \ll 1, \omega \ll 1$ and $Q = 0$ if $\epsilon_0 \gg \epsilon_f = (\mid s_0 \mid^{m-1} V^k)^{\frac{2}{3n+2}}, \omega \ll 1$, can be considered as an appropriate approximation of the viscosity pressure coefficient Q in $\mu(p) = \exp(Qp^\alpha)$ as it provides the correct behavior of μ with respect to p for $p \ll 1$ and provides the correct values for $\mu(p)$ in the most important for determining film thickness H_0 part of the contact - the inlet zone.

It is clear that the above definition of the viscosity pressure coefficient Q is not necessarily equal to $d\mu(p)/dp$ at $p = 0$ as it is assumed by many researches and users of the formulas for the lubrication film thickness. In fact, the pressure viscosity coefficient Q is determined by the type of lubrication regime (pre- or over-critical) and behavior of the lubricant viscosity $\mu(p)$ somewhere in the middle of the inlet zone but not necessarily at $p = 0$. The value of constant C in equation (10.521) can be adjusted for the specifics of a particular lubricant viscosity to reflect the particular lubrication conditions better.

It is very important to understand that this viscosity approximation can be used only for the lubrication film calculations. However, an attempt to use this viscosity approximation for friction calculations can lead to significant distortions.

Let us consider couple more examples.

Example 4. Let us consider the case of a constant viscosity $\mu = 1$. Then the maximum solution of equation (10.521) is $\epsilon_0 = \infty$. The above considerations indicate that only pre-critical regimes can be realized in this situation, and the problem small parameter ω can be chosen as $\omega = V \ll 1$. Therefore, the film thickness is described by formulas (10.503) and (10.504).

Example 5. Let us consider the example of a Vinogradov-Malkin viscosity relationship

$$\mu(p) = \exp[\alpha_1 R_0(\tfrac{1}{R-R_0} - \tfrac{1}{1-R_0})],$$

$$R = 1 - \tfrac{1}{\alpha_2}\ln(1+\alpha_3 p), \quad \alpha_3 = \alpha_2\tfrac{p_H}{p_0},$$

(10.525)

where α_1, α_2, α_3, and R_0 are positive dimensionless constants, p_0 is the characteristic pressure. Then for $C = e$ the solution of equation (10.521) has the form

$$\epsilon_0^{1/2} = \tfrac{1}{\alpha_3}\{\exp[\tfrac{\alpha_2(1-R_0)^2}{1-R_0+\alpha_1 R_0}] - 1\}.$$

(10.526)

Note, that for real lubricants the value of $\epsilon_0^{1/2}$ determined from (10.526) is small. By expanding the expression for $\mu(p)$ for small p, we obtain

$$\ln[\mu(p)] = \sigma p + \ldots, \quad \sigma = \tfrac{\alpha_1 \alpha_3}{\alpha_2}\tfrac{R_0}{(1-R_0)^2}, \quad p \ll 1.$$

(10.527)

It is obvious from (10.527) that $d\mu(p)/dp\,|_{p=0} = \sigma$. However, according to the above outlined procedure the adequate pressure viscosity coefficient Q in an exponential–like viscosity relationship is different from σ. Depending on whether one of the following two estimates $\epsilon_0 = O(\epsilon_f(\omega))$ or $\epsilon_0 \gg \epsilon_f(\omega)$ is satisfied or $\epsilon_0 \ll \epsilon_f(\omega)$, where $\epsilon_f(\omega)$ is defined by (10.503), the film thickness H_0 has to be determined from formulas (10.504) or (10.516), which correspond to pre- and over-critical lubrication regimes, respectively. After the determination of the lubrication regime (pre- or over-critical) is done the pressure viscosity coefficient $Q = \epsilon_0^{-1/2}$ (see formula (10.526)) for over-critical regimes and $Q = \epsilon_f^{-1/2}$ (see formula (10.524)) for pre-critical regimes. The small parameter ω we can take as $\omega = \sigma^{-1} \ll 1$.

10.15.5 Non-Newtonian Lubricants and Scale Effects

The estimation or prediction of EHL film thickness requires knowledge of the lubricant properties. Today, in many instances, the properties have been obtained from a measurement of the central film thickness in an optical EHL point contact simulator and the assumption of a classical Newtonian film thickness formula. This technique has the practical advantage of using an effective pressure viscosity coefficient that compensates for shear-thinning. It is shown below that the practice of extrapolating from a laboratory scale measurement of film thickness to the film thickness of an operating contact within a real machine may substantially overestimate the film thickness in the real machine if the machine scale is smaller and the lubricant is shear-thinning within the inlet zone.

The film thickness in EHL concentrated contacts has implications for friction and for the wear and fatigue life of the rollers. The estimation or prediction of EHL film thickness requires knowledge of the lubricant properties. If the lubricant is Newtonian within the pressure-boosting inlet zone, the film thickness may be calculated from a classical film thickness equation such as that offered by Dowson and Higginson [12]. The required liquid properties are the ambient low-shear viscosity μ_0 and the pressure viscosity coefficient α.

The pressure viscosity coefficients that are reported in handbooks and journals come from two sources. Originally, the pressure viscosity coefficient was calculated from measurements of viscosity as a function of pressure at fixed temperature using various definitions of the pressure viscosity coefficient [53]. These are viscosity-derived pressure viscosity coefficients α. Today, in many instances, the reported coefficient has been obtained from a measurement of the central film thickness h_c by an optical EHL point contact simulator and the assumption of a classical Newtonian film thickness formula. Typically, the Newtonian Hamrock and Dowson formula [54] is solved for the value of α which will give the measured h_c. These are film-derived effective pressure viscosity coefficients α_e.

Not surprisingly, it is often found that $\alpha_e < \alpha$ [55] because of the Newtonian assumption implicit in the α_e calculation [56]. The practical advantage of using an effective coefficient that compensates for shear-thinning is obvious. The film-thickness under the same conditions may then be estimated easily using the classical Newtonian formula. The disadvantage stems from the fact that the shear-thinning behavior will change the response of the film thickness to variations in rolling velocity [57], sliding velocity [57], and perhaps to geometry and other material properties.

The material of this section sounds a warning regarding the extrapolation of measurements of film thickness using Newtonian formulas in order to estimate the film thickness between machine elements of a different scale (and perhaps elastic modulus) from the original measurement. Significant over-estimations may result at smaller scales. In a departure from convention, the maximum Hertzian pressure p_H will be used to quantify the contact loading rather than the normal force P since the pressure of an actual machine element contact would be more closely simulated in an experimental measurement than the normal force and, of course, the rheology is dependent on pressure, not normal force. The generalized Newtonian constitutive equation utilized for shear-thinning will be the single-Newtonian Carreau-Yasuda form, which accurately describes the shear dependence of the viscosity that is measured for base oils in viscosimeters [53, 56] - [59].

We will consider the results of an analytical (asymptotic) treatment of the problem, which covers two limiting cases of relatively small and large shear stresses [60]. Then we proceed to a numerical solution, which, in turn, is well suited to covering the intermediate case of moderate shear stresses. A line contact is assumed for simplicity. Extension to the point contact that is usually used in EHL measurements can be accomplished and should not substantially change the conclusions.

10.15.5.1   Results of Perturbation Analysis

Let us consider some results of the perturbation analysis for heavily loaded contacts lubricated with Newtonian and non-Newtonian lubricants. These results were obtained in the preceding sections.

Let us consider a fluid with single-Newtonian Carreau rheology, which, in dimensionless variables in a narrow gap between two heavily loaded elastic solids, is described by equations

$$\tau = \Phi(\mu \tfrac{\partial u}{\partial z}), \ \ \Phi(t) = \tfrac{V}{12H_0} t \{1 + (\tfrac{V}{12H_0 G_0} \mid t \mid)^m\}^{\frac{n-1}{m}}, \tag{10.528}$$

where the dimensionless parameters are introduced based on the Hertzian contact half-width a_H and Hertzian maximum pressure p_H as well as on the ambient low-shear viscosity μ_0, and the central film thickness h_c. The dimensionless parameters in equation (10.528) are defined as follows

$$V = \tfrac{24\mu_0(u_1+u_2)R'^2}{a_H^3 p_H}, \ \ H_0 = \tfrac{2R'h_c}{a_H^2}, \ \ G_0 = \tfrac{\pi R'G}{P}. \tag{10.529}$$

In equation (10.528), parameters m and n are certain positive constants. It is obvious that the rheological equation (10.528) provides for two limiting cases of relatively small and large shear stresses. The case of relatively small shear stresses corresponds to the Newtonian behavior of the lubricant and is described by the relations

$$\tau = \tfrac{V}{12H_0} \mu \tfrac{\partial u}{\partial z} \ if \ \tfrac{V}{12H_0 G} \mid \tfrac{\partial u}{\partial z} \mid \ll 1, \tag{10.530}$$

while the case of relatively large shear stresses leads to the Ostwald-de Waele (power law) rheology as follows

$$\tau = \Phi(\mu \tfrac{\partial u}{\partial z}), \ \ \Phi(t) = \tfrac{V}{12H_0} t (\tfrac{V}{12H_0 G_0} \mid t \mid)^{n-1}, \tag{10.531}$$

which coincides with (10.489) in which V_n is replaced by $V G_0^{\frac{1-n}{n}}$. In case of $n = 1$ equation (10.531) coincides with equation (10.530) describing the Newtonian rheology.

In practice, in the inlet zone of a heavily loaded lubricated contact there is a continuous transition from small to large shear stresses. Therefore, in real situations we can expect that the results (such as film thickness) are somewhere between the results obtained for the two limiting cases. In this subsection we will concentrate on the results of the analytical analysis of the two limiting cases. The value of this analytical analysis is based on the fact that we will be able to derive some conclusions from asymptotically valid (for heavy loading) analytical formulas for the film thickness for the two limiting cases of small and large shear stresses. The case of moderate shear stresses is hardly possible to analyze analytically and it will be done numerically and compared to the analytical results.

Let us consider the case of pure rolling for a heavily loaded contact lubricated by iso-viscous fluids or fluids with moderate pressure dependence of viscosity, i.e., the case of pre-critical lubrication regimes. We will assume that $\mu = \exp(Qp)$, where $Q = \alpha p_H$ is the dimensionless pressure viscosity coefficient. Using the above developed perturbation methods for a Newtonian lubricant under fully flooded conditions, we obtain the formula for the film thickness (see equation (10.128) and (10.436))

$$H_0 = A_f V^{3/5}, \ A_f = O(1), \tag{10.532}$$

where A_f is a constant independent from V. This formula is valid if the following condition are satisfied (see (10.472)

$$Q = O(V^{-1/5}), \ V \ll 1. \tag{10.533}$$

For example, if $V = 0.01$, then the dimensionless pressure viscosity coefficient Q should not be greater than $5 - 10$ because $V^{-1/5} = 2.51$. Condition (10.533) defines lubricants with viscosity moderately dependent on pressure, i.e., pre-critical lubrication regimes. Moreover, if $Q \ll V^{-1/5}$, then constant A_f in formula (10.532) for film thickness H_0 is also independent from Q while for $Q = O(V^{-1/5})$ constant A_f is a function of Q.

Let us consider a case of a power law fluid under the conditions of pure rolling. For heavily loaded contacts under fully flooded conditions, the formula for the film thickness has the form (see equation (10.438))

$$H_0 = A_f(V^n G_0^{1-n})^{\frac{3}{3n+2}}, \ A_f = O(1), \ V \ll 1, \tag{10.534}$$

where A_f is a dimensionless constant dependent on n but independent of V and G_0. In this case the condition (see (10.472)

$$Q = O((V^n G_0^{1-n})^{-\frac{1}{3n+2}}), \ V \ll 1, \tag{10.535}$$

defines lubricants with viscosity moderately dependent on pressure. In case of $Q \ll (V^n G_0^{1-n})^{-\frac{1}{3n+2}}$ constant A_f is also independent from Q. For small slide-to-roll ratios s_0 formula (10.535) remains in force.

Now, let us consider the pure rolling lubrication regimes with high pressure dependence of viscosity, i.e., over-critical lubrication regimes. Using the developed perturbation methods for power law rheology in a contact under fully flooded lubrication conditions, the formula for the film thickness has the form (see equation (10.474))

$$H_0 = A_f(V^n G_0^{1-n} Q)^{\frac{3}{3n+1}}, \ A_f = O(1), \tag{10.536}$$

where A_f is a dimensionless constant dependent on n but independent of V, G_0, and Q. In this case the condition

$$Q \gg (V^n G_0^{1-n})^{-\frac{1}{3n+2}}, \ V \ll 1, \tag{10.537}$$

defines lubricants with viscosity strongly dependent on pressure. For the Newtonian rheology $n = 1$ and the film thickness formula from equation (10.536) assumes the form

$$H_0 = A_f(VQ)^{\frac{3}{4}}, \ A_f = O(1), \tag{10.538}$$

where A_f is a dimensionless constant independent of V and Q. From condition (10.538) for $n = 1$ and, for example, for $V = 0.01$ the dimensionless pressure viscosity coefficient Q should be greater then 10 because $V^{-1/5} = 2.51$. Structurally, formula (10.538) coincides with the Ertel-Grubin formula (see [13, 14]).

It is worth mentioning that for Newtonian lubricants formulas for the film thickness (10.532) as well as formula (10.538) are valid for any slide-to-roll ratio s_0. Also, it is important to realize that introduction of lubricant compressibility does not change the structural formulas for the film thickness derived above.

In case of high slide-to-roll ratio $s_0 \gg 1$ for the limiting case of a lubricant with power rheology the structural formula for the film thickness in a fully flooded lubrication regime for iso-viscous fluids or fluids with viscosity moderately dependent on pressure (pre-critical lubrication regimes) is as follows (see equation (10.510))

$$H_0 = A_f(V^n G_0^{1-n} \mid s_0 \mid^{n-1})^{\frac{3}{3n+2}}, \; A_f = O(1), \; V \ll 1, \qquad (10.539)$$

while for fluids with viscosity strongly dependent on pressure (over-critical lubrication regimes) it is as follows (see equation (10.520))

$$H_0 = A_f(V_n^n Q \mid s_0 \mid^{n-1})^{\frac{3}{3n+1}}, \; A_f = O(1), \; V \ll 1, \qquad (10.540)$$

where constants A_f in (10.539) and (10.540) are different dimensionless constants independent of V, G_0, and s_0 but dependent on n. In the latter case A_f is also independent from Q.

A similar asymptotic analysis of lubricated point contacts leads to exactly the same formulas for the film thickness. The only difference between the formulas for line and point contacts is due to the difference in the relationships for the Hertzian half-width for line contact and radius of point contact a_H respectively, and the maximum Hertzian pressures p_H for line and point contacts.

Notice that due to different lubrication regimes all of the above formulas for the film thickness for the power law fluids and Newtonian fluids are different. However, what they do have in common is for all considered cases of non-Newtonian lubricant behavior the film thickness is a certain nonlinear function of radius R', elastic modulus E', and pressure viscosity coefficient α. The form of this function depends on the nonlinearity of the fluid rheology, i.e., in our case on the value of constant n involved in rheological relationships (10.528) and (10.531). Below we will show that this fact is the source of the scale effect when a lubricating fluid exhibits a non-Newtonian behavior.

10.15.5.2 Application of Analytical Result to an EHL Experiment

For simplicity, we will consider the scaling effect just for one case of pure or near pure rolling in the Ertel-Grubin type of a fully flooded lubrication regime

for a limiting case of high shear stresses for the Carreau-Yasuda fluid rheology
(10.528) represented by the power law rheology (10.531). Solving equations
(10.529) and (10.536) for the dimensional central film thickness h_c, we obtain

$$h_c = 2A_f \left(\frac{R'E'}{p_H}\right)^{\frac{1}{3n+1}} \left\{[3\mu_0(u_1 + u_2)]^n G^{1-n}\alpha\right\}^{\frac{3}{3n+1}}. \qquad (10.541)$$

Suppose we make a measurement of film thickness h_m at a scale of $R' = R_m$
in order to calculate an effective $\alpha = \alpha_e$ for the same values of E', p_H, and
$\mu_0(u_1 + u_2)$ as for some system of machine elements. The parameters G and n
are unknown. Therefore, for the calculation of α_e we assume that $n = 1$ which
makes G to disappear from equation (10.541). Then the effective pressure
viscosity coefficient α_e is determined by the formula

$$\alpha_e = \frac{h_m^{\frac{4}{3}}}{(2A_f)^{\frac{4}{3}}\mu_0(u_1+u_2)} R_m^{-\frac{1}{3}} \left(\frac{p_H}{E'}\right)^{\frac{1}{3}}. \qquad (10.542)$$

Now, let us use this effective pressure viscosity coefficient α_e to calculate
the film thickness h_s of a real system of machine elements of scale $R' = R_s$
assuming that $n = 1$. The calculated film thickness is

$$h_{sc} = 2A_f \left(\frac{R_s E'}{p_H}\right)^{\frac{1}{4}} \left\{3\mu_0(u_1 + u_2)\alpha_e\right\}^{\frac{3}{4}}. \qquad (10.543)$$

Substituting α_e from (10.542) into equation (10.543) gives

$$h_{sc} = h_m \left(\frac{R_s}{R_m}\right)^{\frac{1}{4}}. \qquad (10.544)$$

The actual value of α is equal

$$\alpha = \frac{h_m^{\frac{3n+1}{3}} G^{n-1}}{(2A_f)^{\frac{3n+1}{3}} [\mu_0(u_1+u_2)]^n} R_m^{-\frac{1}{3}} \left(\frac{p_H}{E'}\right)^{\frac{1}{3}}. \qquad (10.545)$$

and the actual value of h_s is

$$h_s = h_m \left(\frac{R_s}{R_m}\right)^{\frac{1}{3n+1}}. \qquad (10.546)$$

The ratio of calculated h_{sc} to actual h_s film thickness is

$$\frac{h_{sc}}{h_s} = \left(\frac{R_s}{R_m}\right)^{\frac{3(n-1)}{4(3n+1)}}. \qquad (10.547)$$

Therefore, if the film thickness of a millimeter–scale system is calculated
from an effective pressure viscosity coefficient obtained in an optical experimental measurement at centimeter scale, then the calculation will overestimate the film thickness by the ratio given in Table 10.18 for the various values
of n. These ratios represent a substantial error in the calculation of film thickness. However, it must be recognized that equation (10.541) results from the
assumption that the asymptotic high-shear behavior of the Carreau-Yasuda
model takes place over the entire inlet zone. In reality, a portion of the inlet

TABLE 10.18
The ratio of calculated to actual film thickness for extrapolation to smaller (1/10) scale for power-law rheology (after Kudish, Kumar, Khonsari, and Bair [60]). Reprinted with permission from the ASME.

n	1	0.8	0.6	0.4	0.3
h_{sc}/h_s	1	1.11	1.28	1.60	1.89

zone will experience nearly linear shear response and the actual error will be less than that given above.

Also, the above described scaling effect occurs in iso-viscous lubricants and lubricants with moderate pressure dependence of viscosity (pre-critical lubrication regimes) under pure or near pure rolling conditions for which the dimensionless film thickness is described by equation (10.534), which leads to

$$h_c = 2A_f \left(\frac{R'^2}{p_H^2 E'}\right)^{\frac{1}{3n+2}} \{[3\mu_0(u_1 + u_2)]^n G^{1-n}\alpha\}^{\frac{3}{3n+2}}. \qquad (10.548)$$

Now, in this regime, the exponent of R' is $2(3n + 1)/(3n + 2)$ times the value of the one in the previous regime and the exponent on the combined elastic modulus E' has gone from a positive value of $1/(3n + 1)$ to a negative value of $-1/(3n + 2)$. In most EHL rigs, glass is used for one element, resulting in $E = 1.2 \cdot 10^{11} Pa$ versus $2.1 \cdot 10^{11}$ for steel on steel. Then extrapolation to a steel/steel system will also depend upon the E' effect. The lower composite modulus of the simulator will partially compensate for the scale effect if the exponent of E' is positive and vice versa.

To further quantify the error for intermediate cases between the two considered limiting regimes that results from extrapolation of optical EHL simulator measurements to smaller scales a numerical solution of the problem is obtained [60].

10.15.5.3 Numerical Solution

Now, instead of considering just the limiting cases of small and large shear stresses we will consider a full EHL solution for a compressible lubricant with the Carreau viscosity from equation (10.528) with $m = 2$ and $n = 0.4$. This numerical analysis will reveal similar scale effects as the asymptotic study of the limiting cases. Using this viscosity model, Jang et al. [61] presented extensive EHL line-contact simulations that closely agreed with published experimental results by Dyson and Wilson [57].

The solution domain in the present simulation ranges from $x = -4$ to $x = 1.5$ with a uniform grid size $\triangle x = 0.0125$. It has been verified that further mesh refinement and extension of the inlet zone cause negligible change in the results. The solution begins by assuming an initial guess for pressure distribution $\{p_i\}$ and offset film thickness. Using this, the film thickness and fluid properties are calculated and substituted in the generalized Reynolds equation

$$\frac{d}{dx}\left\{\rho\left[\frac{1}{\mu}\int_{-h/2}^{h/2} zF(f + H_0z\frac{dp}{dx})dz - h\right]\right\} = 0, \tag{10.549}$$

in which the dimensionless variables are introduced according to equations (10.383), (10.529), $s_0 = 2\frac{u_2-u_1}{u_2+u_1}$ (see equation (10.384)) and the lubricant density ρ is scaled by its value at ambient pressure. The rest of the equations follow from (10.386)-(10.389) and equation (10.388) is rewritten in the form

$$H_0(h - 1) = x^2 + \frac{2}{\pi}\int_a^c p(t) \ln \mid \frac{t}{x-t} \mid dt. \tag{10.550}$$

The lubricant density is determined according to Tait equation [53]

$$\rho = \{1 - \frac{1}{K_0'} \ln[1 + \frac{K_0'}{K_0}p]\}^{-1}, \tag{10.551}$$

where $K_0' = 11.19$ and $K_0 = 0.85\ GPa/p_H$.

The Reynolds equation (10.549) is discretized by using a mixed second order central and first order backward difference scheme. The Reynolds equation is solved along with load balance condition using the Newton-Raphson technique to obtain an improved pressure distribution and offset film thickness. The process of iterations continues until the relative error in pressure and gap distribution decreases below 10^{-5}.

For specific calculations we used the pressure dependence of viscosity in the form $\mu = \mu_0 \exp(\alpha p)$, with $\mu_0 = 0.08 Pa \cdot s$ and $\alpha = 25 GPa^{-1}$, $15 GPa^{-1}$. The average contact pressure is $\frac{\pi}{4}p_H = 0.5 GPa$. The rolling velocity is $0.5(u_1 + u_2) = 1 m/s$ and slide-to-roll ratio $s_0 = 0$. The value of the lubricant critical stress G has been varied from $1 \cdot 10^9 Pa$ to $1 \cdot 10^5 Pa$. The intermediate values of G are selected to place the transition from the lubricant linear to non-linear response within the inlet zone. Therefore, these results are not the high shear behavior of the analysis above but rather should be representative of a real rheological response within the inlet zone. The calculated film thicknesses are given in Table 10.19 for four cases. First, the case of $R' = 0.013 m$, $E' = 110 GPa$, the scale and elasticity of an experimental rig at $\alpha = 25 GPa^{-1}$ is considered. Then, the elastic modulus is increased to $E' = 211 GPa$. This is followed by the case of $R' = 0.0013 m$, $E' = 211 GPa$, the scale and elasticity of, say, a small roller bearing, is investigated while the pressure viscosity coefficient is fixed at $\alpha = 25 GPa^{-1}$. Finally, in order to investigate the sensitivity of pressure viscosity coefficient, the case of $R' = 0.013 m$, $E' = 211 GPa$, is considered with $\alpha = 15 GPa^{-1}$.

Also, given in Table 10.19, are the scale, elasticity and piezo-viscous sensitivities $\triangle \ln h_c / \triangle \ln R'$, $\triangle \ln h_c / \triangle \ln E'$, and $\triangle \ln h_c / \triangle \ln \alpha$, obtained from the examples at each value of G. These sensitivities, as evident from Table 10.19,

TABLE 10.19
The results of a full line contact numerical solution using the Carreau form with $n = 0.4$ (after Kudish, Kumar, Khonsari, and Bair [60]). Reprinted with permission from the ASME.

G $[Pa]$	$1 \cdot 10^9$	$4 \cdot 10^6$	$1 \cdot 10^6$	$3 \cdot 10^5$	$1 \cdot 10^5$
$R' = 0.013m,\ E' = 110GPa,\ \alpha = 25GPa^{-1}$ (the scale and elasticity modulus of an EHL simulator)					
h_c $[nm]$	577	576	561	443	257
$R' = 0.013m,\ E' = 211GPa\ \alpha = 25GPa^{-1}$					
h_c $[nm]$	606	605	588	467	273
$R' = 0.0013m,\ E' = 211GPa,\ \alpha = 25GPa^{-1}$ (the scale and elasticity modulus of a small roller bearing)					
h_c $[nm]$	267	265	235	150	79
$R' = 0.013m,\ E' = 211GPa,\ \alpha = 15GPa^{-1}$					
h_c $[nm]$	457	456	441	336	189
$\frac{\triangle \ln h_c}{\triangle \ln R'}$	0.35	0.36	0.40	0.49	0.54
$\frac{\triangle \ln h_c}{\triangle \ln E'}$	0.074	0.074	0.074	0.082	0.091
$\frac{\triangle \ln h_c}{\triangle \ln \alpha}$	0.55	0.55	0.56	0.65	0.72

increase with decreasing values of G. This occurs because, as G is reduced, the shear-thinning response begins at the lower values of stress, which occur earlier in the inlet zone and more of the inlet experiences shear-thinning.

The trends observed in Table 10.19 are the same as concluded from the perturbation analysis above for which the scale, elasticity and piezo-viscous sensitivities $\triangle \ln h_c / \triangle \ln R'$, $\triangle \ln h_c / \triangle \ln E'$, and $\triangle \ln h_c / \triangle \ln \alpha$ are given in Table 10.20. Shear-thinning increases the sensitivity to changes in scale, and elasticity and pressure-viscosity coefficient. The classical Newtonian formulas for central film thickness from Dowson and Toyoda [62] for line contact and Hamrock and Dowson [63] for circular point contact yield the scale and elasticity sensitivities shown in Table 10.21 for comparison. Notice again that, here, the contact pressure rather than the normal force is used to describe the loading.

10.15.5.4 Application of Numerical Results to an EHL Experiment

A similar analysis of an experimental extrapolation of film thickness to that presented for the perturbation result above can be done for the numerical results of Table 10.18. Suppose we make a measurement of central film thickness h_m in an optical EHL rig simulator, for the same values of p_H and $0.5(u_1 + u_2)$ as for some system of machine elements, but at an experimental scale of $R' = R_m$ in order to calculate an effective $\alpha = \alpha_e$ from some Newtonian film thickness formula. For the calculation of α_e, we assume Newtonian response and use the sensitivity to α from Table 10.20 ($1/0.56 = 1.79$) to calculate $\alpha_e = \alpha(h_m/h_N)^{1.79}$, where h_N is taken to be the film thickness for

TABLE 10.20
The sensitivities of the formulas obtained by the perturbation
analysis (after Kudish, Kumar, Khonsari, and Bair [60]).
Reprinted with permission from the ASME.

	Newtonian rheology, pre-critical regime	Newtonian rheology, over-critical regime	Power law rheology, pre-critical regime	Power law rheology, over-critical regime
$\frac{\triangle \ln h_c}{\triangle \ln R'}$	0.4	0.25	0.625 $\left(\frac{2}{3n+2}\right)$	0.455 $\left(\frac{1}{3n+1}\right)$
$\frac{\triangle \ln h_c}{\triangle \ln E'}$	-0.2	0.25	-0.313 $\left(-\frac{1}{3n+2}\right)$	0.455 $\left(\frac{1}{3n+1}\right)$
$\frac{\triangle \ln h_c}{\triangle \ln \alpha}$	0, A_f depends on α	0.75	0, A_f depends on α	1.364 $\left(\frac{3}{3n+1}\right)$

TABLE 10.21
The sensitivities of the classical Newtonian formulas
(after Kudish, Kumar, Khonsari, and Bair [60]).
Reprinted with permission from the ASME.

	$\frac{\triangle \ln h_c}{\triangle \ln R'}$	$\frac{\triangle \ln h_c}{\triangle \ln E'}$	$\frac{\triangle \ln h_c}{\triangle \ln \alpha}$
Newtonian line contact	0.31	0.070	0.56
Newtonian point contact	0.33	0.061	0.53

the case of $G = 1 \cdot 10^9 Pa$. The effective pressure viscosity coefficients α_e are
listed in Table 10.22 for each of the lower values of G. The film thickness for
the small roller bearing is now extrapolated from the film thicknesses h_m that
would be obtained in the experimental rig by assuming Newtonian response

$$h_{sc} = h_m \left(\frac{R_s}{R_m}\right)^{0.35} \left(\frac{E_s}{E_m}\right)^{0.074} = h_m \left(\frac{0.0013}{0.013}\right)^{0.35} \left(\frac{211}{110}\right)^{0.074}. \qquad (10.552)$$

The extrapolated film thicknesses are listed in Table 10.22 along with the ratio
of calculated to actual film thickness h_{sc}/h_s. The extrapolated film thickness
may be substantially greater than the actual film thickness.

Let us sum up the results of this analysis. We have shown by a perturbation
analysis and by a full EHL numerical solution that the practice of extrapolating from a laboratory scale measurement of film thickness to the film thickness
of an operating contact within a real machine may substantially overestimate
the film thickness in the real machine if the scale is smaller. This observation points to the need for a thorough experimental validation of the classical
Newtonian film thickness calculation using accurate pressure viscosity data
for a wide variety of liquid lubricants so that the limits of applicability of
Newtonian calculations can be established.

TABLE 10.22
The ratio of calculated to actual film thickness
for extrapolation to smaller scale and greater
elastic modulus (after Kudish, Kumar,
Khonsari, and Bair [60]). Reprinted with
permission from the ASME.

$G\ [Pa]$	$4 \cdot 10^6$	$1 \cdot 10^6$	$3 \cdot 10^5$	$1 \cdot 10^5$
$\alpha\ [GPa^{-1}]$	25.0	23.8	15.6	5.90
$h_{sc}[nm]$	270	263	208	121
h_{sc}/h_s	1.02	1.12	1.39	1.53

10.16 TEHL Problems for Lubricants with General Non-Newtonian Rheology

In this section we will analyze both thermal and elastic effects in a contact
lubricated by a fluid with general non-Newtonian rheology including the effect
of heat dissipation in contact solids. The problem will be considered under the
classic assumptions (see Sections 10.2 and 10.3).

10.16.1 TEHL Problem Formulation for Non-Newtonian Fluids

First, let us consider some basic facts related to thermal conductivity of contacting solids. Suppose there is a half-plane that is subjected to the action
of a point heat source with the heat flux q directed into the half-plane. The
heat source is moving along the half-plane boundary with a steady velocity u.
Additionally, the material of the half-plane is characterized by the density ρ,
specific heat c, and thermal diffusivity k. At this point we will ignore thermal
expansion of the solids. We will use the rectangular coordinate system which
is introduced as follows: the x- and z-axes of the coordinate system are directed along the half-plane surface in the direction of the heat source motion
and into the half-plane down, respectively (the y-axis is perpendicular to the
x- and z-axes). According to [64], a moving point source of heat currently
located at the origin of the coordinate system at the point with coordinate x
on the half-plane surface causes the following temperature T of the half-plane
surface

$$T = \tfrac{q}{\pi k \rho c} e^{-\tfrac{ux}{2k}} K_0(|\tfrac{ux}{2k}|)\},\tag{10.553}$$

where K_0 is the modified Bessel functions [65]. Therefore, for a distributed
source of heat using superposition/convolution, we obtain

$$T = \tfrac{1}{\pi k \rho c} \int\limits_{x_i}^{x_e} q(\xi) e^{-\tfrac{u(x-\xi)}{2k}} K_0(|\tfrac{u(x-\xi)}{2k}|) + T_a,\tag{10.554}$$

where T_a is the ambient surface temperature.

Now, let us consider the TEHL problem for two cylinders with parallel axes made of elastic materials with elastic moduli E_i and Poisson's ratios ν_i ($i = 1, 2$) and moving with surface speeds u_1 and u_2. The cylinders are loaded with normal force P. This problem formulation was discussed in detail in previous sections.

To solve the problem we will be using asymptotic methods. We will assume that the shear stress in lubricant generates heat, i.e.,

$$\tau F(\tau) \geq 0. \tag{10.555}$$

In addition, we will assume that the predominant direction of the heat flow generated in the contact is normal to the contact surfaces and is directed into the solids [7]. As in Section 10.10 we will assume that the lubricant viscosity μ satisfies the relationship (10.312).

The derivation of the TEHL problem equations is practically identical to the equation derivations conducted in Sections 10.3 and 10.10. Therefore, based on the assumptions made, the problem can be reduced to the following system of non-linear integro-differential equations [7, 64, 66]

$$\frac{d}{dx}\left\{ \int\limits_{-h/2}^{h/2} \frac{z}{\mu} F(f + z\frac{dp}{dx})dz - \frac{u_1+u_2}{2}h \right\} = 0,$$

$$p(x_i) = p(x_e) = \frac{dp(x_e)}{dx} = 0,$$

$$\int\limits_{-h/2}^{h/2} \frac{1}{\mu} F(f + z\frac{dp}{dx})dz = u_2 - u_1,$$

$$h = h_e + \frac{x^2 - x_e^2}{2R'} + \frac{2}{\pi E'} \int\limits_{x_i}^{x_e} p(t) \ln \mid \frac{x_e-t}{x-t} \mid dt, \tag{10.556}$$

$$\int\limits_{x_i}^{x_e} p(\xi)d\xi = P,$$

$$\frac{\partial^2 T}{\partial z^2} = -\frac{1}{\lambda\mu}(f + z\frac{dp}{dx})F(f + z\frac{dp}{dx}),$$

$$T(x, -\frac{h}{2}) = T_{w1}(x), \ T(x, \frac{h}{2}) = T_{w2}(x),$$

$$q_1 = \lambda\frac{\partial T}{\partial z} \mid_{z=-h/2}, \ q_2 = -\lambda\frac{\partial T}{\partial z} \mid_{z=h/2},$$

$$T_{w1}(x) = \frac{1}{\pi k_1 \rho_1 c_1} \int\limits_{x_i}^{x_e} q_1(\xi) \exp[-\frac{u_1(x-\xi)}{2k_1}]K_0(\mid \frac{u_1(x-\xi)}{2k_1} \mid)d\xi + T_{a1},$$

$$\tag{10.557}$$

$$T_{w2}(x) = \frac{1}{\pi k_2 \rho_2 c_2} \int\limits_{a}^{c} q_2(\xi) \exp[-\frac{u_2(x-\xi)}{2k_2}]K_0(\mid \frac{u_2(x-\xi)}{2k_2} \mid)d\xi + T_{a2},$$

where F is the function describing the lubricant rheology (see equation (10.382)), λ is the coefficient of lubricant heat conductivity, T_{a1} and T_{a2} are the temperatures of the lower and upper contact surfaces far away from the contact region, respectively, λ is the coefficient of lubricant heat conductivity, T_{w1} and T_{w2} are the surface temperatures of the lower and upper cylinders, respectively, and q_1 and q_2 are the heat fluxes directed in the lower and upper cylinders, respectively. The contacting solid materials' densities ρ_1, ρ_2, specific heat parameters c_1, c_2, and the thermal diffusivities k_1 and k_2 for lower (marked with index 1) and upper (marked with index 2) are considered to be known. The rest of the parameters and variables are the same as in Section 10.15.

In addition to equations (10.58), (10.77), (10.305), (10.383), and (10.316), let us introduce some dimensionless variables and parameters as follows

$$T'_{ai} = \frac{T_{ai}}{T_a} - 1 = (-1)^i \Delta_\infty, \; q'_i = \frac{h_e}{\lambda_a T_a} q_i, \; \lambda_i = \frac{u_i a_H}{2k_i}, \; (i = 1, 2),$$

$$T' = \frac{T}{T_a} - 1, \; T_a = \frac{T_{a1} + T_{a2}}{2}, \; \Delta_\infty = \frac{T_{a2} - T_{a1}}{T_{a1} + T_{a2}}, \; s_0 = 2\frac{u_2 - u_1}{u_2 + u_1}, \quad (10.558)$$

$$\delta = \alpha_T T_a, \; \kappa = \frac{(u_1 + u_2)P}{4\pi \lambda_a T_a}\left(\frac{a_H}{R'}\right)^2, \; \eta_0 = \frac{k_1 \rho_1 c_1}{k_2 \rho_2 c_2}, \; \Lambda = \frac{4\lambda_a R'}{\pi k_1 \rho_1 c_1 a_H},$$

where λ_a is the coefficient of lubricant heat conductivity at ambient pressure.

We will postulate that the lubricant viscosity μ can be represented as a product of two parts: one that is independent of the lubricant temperature T and the other that is a function of the lubricant temperature T. In particular, we will assume that in dimensionless variables (10.558), the lubricant viscosity can be represented in the form

$$\mu(p, T) = \mu^0(p)e^{-\delta T}, \; \frac{\partial \mu^0(p)}{\partial T} = 0, \quad (10.559)$$

where $\mu^0(p)$ is the lubricant viscosity at the ambient temperature, δ is the temperature viscosity coefficient, $\delta \geq 0$. Coefficient δ may be a constant or a function of pressure p. With respect to the lubricant heat conductivity λ, we will assume that it may depend only on pressure p and is independent of the lubricant temperature T.

By using the aforementioned dimensionless variables and parameters under the above assumptions the TEHL problem can be reduced to the following system of equations (for simplicity primes at the dimensionless variables are omitted)

$$\frac{d}{dx}\left\{ \int_{-h/2}^{h/2} \frac{z}{\mu}F(f + H_0 z\frac{dp}{dx})dz - h \right\} = 0, \; p(a) = p(c) = \frac{dp(c)}{dx} = 0, \quad (10.560)$$

$$\int_{-h/2}^{h/2} \frac{1}{\mu}F(f + H_0 z\frac{dp}{dx})dz = s_0, \quad (10.561)$$

$$H_0(h-1) = x^2 - c^2 + \frac{2}{\pi} \int_a^c p(t) \ln \left| \frac{c-t}{x-t} \right| \, dt, \tag{10.562}$$

$$\int_a^c p(\xi)d\xi = \frac{\pi}{2}, \tag{10.563}$$

$$\frac{\partial^2 T}{\partial z^2} = -\frac{\kappa H_0}{\lambda \mu}\left(f + H_0 z \frac{dp}{dx}\right)F\left(f + H_0 z \frac{dp}{dx}\right), \tag{10.564}$$

$$T\left(x, -\frac{h}{2}\right) = T_{w1}(x), \;\; T\left(x, \frac{h}{2}\right) = T_{w2}(x).$$

$$q_1 = \lambda \frac{\partial T}{\partial z}\Big|_{z=-h/2}, \;\; q_2 = -\lambda \frac{\partial T}{\partial z}\Big|_{z=h/2}, \tag{10.565}$$

$$T_{w1}(x) = \frac{\Lambda}{2H_0} \int_a^c q_1(\xi)e^{-\lambda_1(x-\xi)}K_0(|\lambda_1(x-\xi)|)d\xi - \Delta_\infty, \tag{10.566}$$

$$T_{w2}(x) = \frac{\Lambda \eta_0}{2H_0} \int_a^c q_2(\xi)e^{-\lambda_2(x-\xi)}K_0(|\lambda_2(x-\xi)|)d\xi + \Delta_\infty.$$

Therefore, for the given values of constants a, Q, s_0, δ, κ, η_0, Λ, λ_1, λ_2, Δ_∞ and functions F, μ^0, and λ we have to find constants H_0 - exit film thickness, c - exit coordinate, and functions $p(x)$ - pressure, $h(x)$ - gap, $f(x)$ - sliding frictional stress, $T(x,z)$ - lubricant temperature, $T_{wi}(x)$ and $q_i(x)$ - surface temperatures and heat fluxes, $(i = 1, 2)$.

10.16.2 Asymptotic Analysis of the Problem for Heavily Loaded Contacts

The whole analysis proposed here is for the case of heavily loaded lubricated contacts. Therefore, it is fair to assume that there is a small parameter ω, which makes these contacts heavily loaded. Our main goal is to reduce equations (10.559)-(10.566) to a problem analogous to the isothermal EHL problem presented in Section 10.15. It means that using some perturbation techniques the temperature $T(x,z)$ and sliding frictional stress $f(x)$ will be determined and solution of equations (10.559)-(10.566) will be reduced to solution of equations for just two functions $p(x)$ and $h(x)$ and two parameters H_0 and c. The reduced equations have to be solved numerically. We will consider only the problem for heavily loaded TEHL contacts with relatively large slide-to-roll ratios s_0. As before we introduce function $\nu(x)$ according to (10.475) which is small in the inlet zone. In case of over-critical regimes we need $\nu(x)$ to be small in the inlet ϵ_0-zone which guarantees that $\nu(x)$ is small in the entire contact. If there is a need to obtain the problem solution in the entire contact region not just the formula for the lubrication film thickness H_0, then $\nu(x)$ would be considered small in the entire contact region. Obviously, it may impose additional limitations on the problem input parameters. The whole procedure of TEHL problem solution consists of a number of steps. On Step 1 we will assume that functions $f(x)$ and $T(x,z)$ can be represented by the expansions

$$f(x) = f_0(x) + \nu(x)f_1(x) + O(\nu^2(x)), \;\; \nu(x) \ll 1, \tag{10.567}$$

$$T(x,z) = T_0(x,z) + \nu(x)T_1(x,z) + O(\nu^2(x)), \quad \nu(x) \ll 1, \qquad (10.568)$$

where functions $f_0(x)$ and $f_1(x)$ are the consecutive terms of the asymptotic of $f(x)$ to be determined from equation (10.561) while functions $T_0(x,z)$ and $T_1(x,z)$ are the consecutive terms of the asymptotic of $T(x,z)$ to be determined from equations (10.559) and (10.564). Based on equations (10.559), (10.567), and (10.568), we can conclude that

$$\frac{1}{\mu(p,T)} = \frac{e^{\delta T_0(x,z)}}{\mu^0(p)}[1 + \nu(x)\delta T_1(x,z) + O(\nu^2(x))], \quad \nu(x) \ll 1, \qquad (10.569)$$

$$F(f + H_0 z \tfrac{dp}{dx}) = F(f_0) + \nu(f_1 + \tfrac{2z}{h}f_0)F'(f_0) + O(\nu^2), \quad \nu \ll 1. \qquad (10.570)$$

Similarly, based on representation (10.568) for the lubricant temperature $T(x,z)$, we can conclude that the surface temperatures $T_{w1}(x)$ and $T_{w2}(x)$ can be searched in the form of the following expansions

$$T_{w1}(x) = T_{w10}(x) + \nu(x)T_{w11}(x) + O(\nu^2(x)),$$
$$T_{w2}(x) = T_{w20}(x) + \nu(x)T_{w21}(x) + O(\nu^2(x)), \quad \nu(x) \ll 1. \qquad (10.571)$$

We have to keep in mind that q_1 and q_2 from equations (10.565) also can be represented as asymptotic expansions in the form

$$q_1(x) = q_{10}(x) + \nu(x)q_{11}(x) + O(\nu^2(x)),$$
$$q_2(x) = q_{20}(x) + \nu(x)q_{21}(x) + O(\nu^2(x)), \quad \nu(x) \ll 1, \qquad (10.572)$$

where q_{10}, q_{20} and q_{11}, q_{21} are used in equations (10.576) and (10.577).

On Step 2 of the solution process we substitute expansions (10.567)-(10.571) into equations (10.561) and (10.564), expand the necessary terms, and equate the terms with the same powers of $\nu(x)$. As a result of that we obtain the following equations

$$f_0(x) = \Phi(\tfrac{\mu^0 s_0}{I_1}), \quad \tfrac{f_1(x)}{f_0(x)} = -\tfrac{1}{I_1}[\tfrac{F(f_0)}{f_0 F'(f_0)}\delta I_{T1} + \tfrac{2}{h}I_z],$$

$$I_1 = \int\limits_{-h/2}^{h/2} e^{\delta T_0}dz, \quad I_{T1} = \int\limits_{-h/2}^{h/2} T_1 e^{\delta T_0}dz, \quad I_z = \int\limits_{-h/2}^{h/2} z e^{\delta T_0}dz, \qquad (10.573)$$

$$\frac{\partial^2 T_0}{\partial z^2} = -\kappa H_0 \frac{f_0 F(f_0)}{\lambda \mu^0}e^{\delta T_0},$$

$$T_0(x,-\tfrac{h}{2}) = T_{w10}(x), \quad T_0(x,\tfrac{h}{2}) = T_{w20}(x), \qquad (10.574)$$

$$\frac{\partial^2 T_1}{\partial z^2} = -\kappa H_0 \frac{f_0 F(f_0)}{\lambda \mu^0}e^{\delta T_0}\{\delta T_1 + [1 + \tfrac{f_0 F'(f_0)}{F(f_0)}](\tfrac{f_1}{f_0} + \tfrac{2z}{h})\},$$

$$T_1(x,-\tfrac{h}{2}) = T_{w11}(x), \quad T_1(x,\tfrac{h}{2}) = T_{w21}(x). \qquad (10.575)$$

$$T_{w10}(x) = \frac{\Lambda}{2H_0} \int\limits_a^c q_{10}(\xi)e^{-\lambda_1(x-\xi)}K_0(|\;\lambda_1(x-\xi)\;|)d\xi - \Delta_\infty,$$

$$\text{(10.576)}$$

$$T_{w20}(x) = \frac{\Lambda\eta_0}{2H_0} \int\limits_a^c q_{20}(\xi)e^{-\lambda_2(x-\xi)}K_0(|\;\lambda_2(x-\xi)\;|)d\xi + \Delta_\infty,$$

$$\nu(x)T_{w11}(x) = \frac{\Lambda}{2H_0} \int\limits_a^c \nu(\xi)q_{11}(\xi)e^{-\lambda_1(x-\xi)}K_0(|\;\lambda_1(x-\xi)\;|)d\xi,$$

$$\text{(10.577)}$$

$$\nu(x)T_{w21}(x) = \frac{\Lambda\eta_0}{2H_0} \int\limits_a^c \nu(\xi)q_{21}(\xi)e^{-\lambda_2(x-\xi)}K_0(|\;\lambda_2(x-\xi)\;|)d\xi.$$

First, let us try to find the exact solution of the nonlinear boundary-value problem (10.574) in the form

$$T_0(x,z) = \frac{T_{w10}+T_{w20}}{2} + \frac{1}{\delta}\ln\frac{R}{\cosh^2[N(z-z_m)]}, \qquad \text{(10.578)}$$

where R, N, and z_m are new unknown functions of x to be determined from solution of boundary-value problem (10.574). By substituting the expression for T_0 from formula (10.578) into equations (10.574), we obtain three equations

$$\frac{2\lambda N^2}{\kappa\delta H_0} = R\frac{f_0 F(f_0)}{\mu^0}\exp[\delta\frac{T_{w10}+T_{w20}}{2}], \qquad \text{(10.579)}$$

$$\frac{R}{\cosh^2[\frac{N}{2}(h+2z_m)]} = e^{-\Delta}, \quad \frac{R}{\cosh^2[\frac{N}{2}(h-2z_m)]} = e^{\Delta}, \quad \Delta = \delta\frac{T_{w20}-T_{w10}}{2}. \quad \text{(10.580)}$$

Based on the fact that the values of κ, δ, and μ^0 are positive from inequality (10.555) and equations (10.580), we get that function R is positive. Therefore, solving equations (10.580) we obtain

$$R = \cosh[\frac{N}{2}(h-2z_m)]\cosh[\frac{N}{2}(h+2z_m)],$$

$$\text{(10.581)}$$

$$z_m = \frac{1}{2N}\ln\{\frac{\sinh[\frac{1}{2}(Nh+\Delta)]}{\sinh[\frac{1}{2}(Nh-\Delta)]}\}.$$

Thus, solution of the boundary-value problem (10.574) is reduced to solution of an algebraic equation (10.579) for N after substituting into it the expressions for functions R and z_m from formulas (10.581). In most cases this equation for N can be solved only numerically. However, in some special cases such as the Newtonian rheology it can be done analytically. As it follows from formula (10.578) function T_0 attains its maximum at $z = z_m$. Obviously, function T_0 reaches its maximum between the contact surfaces if $|\;z_m\;|\leq h/2$ and it is a monotonic function if $|\;z_m\;|\geq h/2$.

The expressions for functions q_{10} and q_{20} are as follows

$$q_{10} = \frac{2\lambda N}{\delta}\tanh[\frac{N}{2}(h+2z_m)], \quad q_{20} = \frac{2\lambda N}{\delta}\tanh[\frac{N}{2}(h-2z_m)], \qquad \text{(10.582)}$$

while for q_{11} and q_{21} we have equations (10.594). Substituting these expressions in equations (10.576) we obtain a nonlinear system of equations for

$T_{w10}(x)$ and $T_{w20}(x)$ in the form

$$T_{w10} = \frac{\Lambda}{H_0} \int_a^c \frac{\lambda N}{\delta} \tanh[N(\tfrac{h}{2} + z_m)] e^{-\lambda_1(x-\xi)} K_0(|\ \lambda_1(x-\xi)\ |) d\xi$$

$$-\Delta_\infty,$$

$$\text{(10.583)}$$

$$T_{w20} = \frac{\Lambda\eta_0}{H_0} \int_a^c \frac{\lambda N}{\delta} \tanh[N(\tfrac{h}{2} - z_m)] e^{-\lambda_2(x-\xi)} K_0(|\ \lambda_2(x-\xi)\ |) d\xi$$

$$+\Delta_\infty,$$

solution of which can be obtained numerically using iterations. As the initial approximations for these iterations, which are described below, can be taken functions $T_{w10}(x) = -\Delta_\infty$, $T_{w20}(x) = \Delta_\infty$.

Obviously, if $\eta_0 = 1$, $\lambda_1 = \lambda_2$, and $z_m = \Delta_\infty = 0$, then the surface temperatures $T_{w10}(x)$ and $T_{w20}(x)$ are equal. On the other hand, based on the fact that function $F(x)$ is an odd function from equation (10.579) we get that function $N(x)$ is independent of the sign of the slide-to-roll ratio s_0, i.e., it is even with respect to s_0. Therefore, we can conclude that in the above case $T_{w2}(x) - T_{w1}(x) = O(\nu(x)) \ll 1$.

Now, let us find the solution of the linear boundary-value problem (10.575) for $T_1(x,z)$ in the form

$$T_1(x,z) = \tfrac{1}{\delta}\{A_1 + B_1(z - z_m) + C_1 \tanh[N(z - z_m)]$$

$$+D_1(z - z_m) \tanh[N(z - z_m)]\},$$

$$\text{(10.584)}$$

where A_1, B_1, C_1, and D_1 are unknown functions of just x that have to be determined from solution of problem (10.575) that takes into account the expressions for f_1 and I_{T1} from equations (10.573) and functions T_{w11} and T_{w21}. After that it is easy to get

$$I_1 = I_A = \exp[\delta \tfrac{T_{w10} + T_{w20}}{2}] \tfrac{\sinh(Nh)}{N},$$

$$I_z = -I_1 \tfrac{R}{N \sinh(Nh)} \{\tfrac{Nh}{2} \tfrac{\sinh(2Nz_m)}{R} - \Delta\},$$

$$I_{T1} = \tfrac{1}{\delta}\{A_1 I_A + B_1 I_B + C_1 I_C + D_1 I_D\}, \quad I_B = I_z - z_m I_1,$$

$$\text{(10.585)}$$

$$I_C = -I_1 \tfrac{\sinh(2Nz_m)}{2R}, \quad I_D = I_1 \tfrac{\sinh(Nh)}{4RN} \{\tfrac{Nh}{2}[1 + \tfrac{\sinh^2(2Nz_m)}{\sinh^2(Nh)}]$$

$$+2Nz_m \tfrac{\sinh(2Nz_m)}{\sinh(Nh)} - \tfrac{2NhR^2}{\sinh^2(Nh)} + \tfrac{2R}{\sinh(Nh)}\},$$

$$\sinh(2Nz_m) = \tfrac{\sinh(Nh)\sinh(\Delta)}{\cosh(Nh) - \cosh(\Delta)}.$$

Equations (10.580) and (10.581) were used to simplify the expressions for I_z and I_{T1} in (10.585).

Taking into account that the boundary-value problem described by equations (10.575) is linear with respect to T_1 it is obvious that the system for functions A_1, B_1, C_1, and D_1 that follows from satisfying (10.575) is a system of linear algebraic equations. Using linear independence of functions 1, $z-z_m$, $\tanh[N(z-z_m)]$, $(z-z_m)\tanh[N(z-z_m)]$ for functions A_1, B_1, C_1, and D_1, we have the system which can be represented in the form

$$B_1 = -\frac{2}{h}[1 + \frac{f_0 F'(f_0)}{F(f_0)}],$$

$$D_1 \frac{2\lambda N}{\kappa \delta H_0} = -\frac{f_0 F(f_0)}{\mu^0} R \exp[\delta \frac{T_{w10}+T_{w20}}{2}]\{A_1$$

$$+[1 + \frac{f_0 F'(f_0)}{F(f_0)}](\frac{f_1}{f_0} + \frac{2z_m}{h})\},$$

(10.586)

$$A_1 + C_1 \tanh[\frac{N}{2}(h - 2z_m)] + D_1(\frac{h}{2} - z_m)\tanh[\frac{N}{2}(h - 2z_m)]$$

$$= -B_1(\frac{h}{2} - z_m) + \delta T_{w21}$$

$$A_1 - C_1 \tanh[\frac{N}{2}(h + 2z_m)] + D_1(\frac{h}{2} + z_m)\tanh[\frac{N}{2}(h + 2z_m)]$$

$$= B_1(\frac{h}{2} + z_m) + \delta T_{w11}.$$

(10.587)

Using the expression for f_1 from (10.573) and some manipulations of the system (10.587) it can be reduced to solution of three systems of linear equations with the same coefficient matrix which have to be solved for three sets of unknowns $\{A_0, C_0, D_0\}$, $\{A_+, C_+, D_+\}$, and $\{A_-, C_-, D_-\}$ and three sets of the right-hand sides so that functions A_1, C_1, and D_1 can be represented in the form

$$\{A_1, C_1, D_1\} = \{A_0, C_0, D_0\} + \{A_+, C_+, D_+\}\frac{\delta(T_{w11}+T_{w21})}{2}$$

$$+\{A_-, C_-, D_-\}\frac{\delta(T_{w21}-T_{w11})}{2}$$

(10.588)

The above systems of linear algebraic equations are as follows

$$A_0 + a_{12}C_0 + a_{13}D_0 = \frac{2}{h}[1 + \frac{f_0 F'(f_0)}{F(f_0)}](\frac{I_z}{I_1} - z_m),$$

$$A_0 + a_{22}C_0 + a_{23}D_0 = -\frac{2z_m}{h}[1 + \frac{f_0 F'(f_0)}{F(f_0)}],$$

$$a_{32}C_0 + a_{33}D_0 = 1 + \frac{f_0 F'(f_0)}{F(f_0)},$$

(10.589)

$$A_+ + a_{12}C_+ + a_{13}D_+ = 0,$$

$$A_+ + a_{22}C_+ + a_{23}D_+ = 1, \tag{10.590}$$

$$a_{32}C_+ + a_{33}D_+ = 0,$$

$$A_- + a_{12}C_- + a_{13}D_- = 0,$$

$$A_- + a_{22}C_- + a_{23}D_- = 0, \tag{10.591}$$

$$a_{32}C_- + a_{33}D_- = 1,$$

respectively, where coefficients are calculated according to the formulas

$$a_{12} = [1 + \tfrac{f_0 F'(f_0)}{F(f_0)}]\tfrac{I_C}{I_1}, \quad a_{13} = [1 + \tfrac{f_0 F'(f_0)}{F(f_0)}]\tfrac{I_D}{I_1},$$

$$a_{22} = -\tfrac{\sinh(2N z_m)}{2R}, \quad a_{23} = \tfrac{h}{4R}[\sinh(Nh) + \tfrac{2z_m}{h}\sinh(2N z_m)], \tag{10.592}$$

$$a_{32} = \tfrac{\sinh(Nh)}{2R}, \quad a_{33} = -\tfrac{h}{4R}[\sinh(2N z_m) + \tfrac{2z_m}{h}\sinh(Nh)].$$

Systems (10.589)-(10.592) have the following solutions

$$A_0 = \tfrac{2}{h}[1 + \tfrac{f_0 F'(f_0)}{F(f_0)}](\tfrac{I_z}{I_1} - z_m) - a_{12}C_0 - a_{13}D_0,$$

$$C_0 = \tfrac{1}{D}[1 + \tfrac{f_0 F'(f_0)}{F(f_0)}]\{\tfrac{2}{h}\tfrac{I_z}{I_1}a_{33} - a_{13} + a_{23}\},$$

$$D_0 = \tfrac{1}{D}[1 + \tfrac{f_0 F'(f_0)}{F(f_0)}]\{a_{12} - a_{22} - \tfrac{2}{h}\tfrac{I_z}{I_1}a_{32}\}, \tag{10.593}$$

$$A_+ = -a_{12}C_+ - a_{13}D_+, \quad C_+ = -\tfrac{a_{33}}{D}, \quad D_+ = \tfrac{a_{32}}{D},$$

$$A_- = -a_{12}C_- - a_{13}D_-, \quad C_- = \tfrac{a_{23} - a_{13}}{D}, \quad D_- = \tfrac{a_{12} - a_{22}}{D},$$

$$D = a_{33}(a_{12} - a_{22}) - a_{32}(a_{13} - a_{23}).$$

It can be shown that for the case of $T_{w10} = T_{w20}$ and $T_{w11} = T_{w22} = 0$ we have $T_0(x, -z) = T_0(x, z)$ and $z_m = I_z = I_C = a_{12} = a_{22} = a_{33} = 0$. That leads to $A_0 = D_0 = 0$ and, therefore, $T_1(x, z) = \tfrac{1}{\delta}[1 + \tfrac{f_0 F'(f_0)}{F(f_0)}]\{\tfrac{\tanh(Nz)}{\tanh(\frac{Nh}{2})} - \tfrac{2z}{h}\}$. For a Newtonian fluid $\tfrac{f_0 F'(f_0)}{F(f_0)} = 1$ and function $T_1(x, z)$ coincides with the one determined for the same conditions by formula (10.337).

Then, based on equations (10.565), (10.572), (10.588), and (10.593), we obtain the expressions for functions $q_{11}(x)$ and $q_{21}(x)$ in the form

$$q_{11}(x) = L_1(x) + P_1(x)T_{w11}(x) + R_1(x)T_{w21}(x),$$

$$q_{21}(x) = L_2(x) + P_2(x)T_{w11}(x) + R_2(x)T_{w21}(x),$$

$$L_1 = \tfrac{\lambda}{\delta}(B_1 + C_0\phi_1 - D_0\theta_1),$$

$$P_1 = \tfrac{\lambda}{2}[(C_+ - C_-)\phi_1 + (D_- - D_+)\theta_1],$$

$$R_1 = \tfrac{\lambda}{2}[(C_+ + C_-)\phi_1 - (D_+ + D_-)\theta_1], \quad \phi_1 = \frac{N}{\cosh^2[\frac{N}{2}(h+2z_m)]},$$

$$L_2 = -\tfrac{\lambda}{\delta}(B_1 + C_0\phi_2 + D_0\theta_2),$$

$$P_2 = \tfrac{\lambda}{2}[(C_- - C_+)\phi_2 + (D_- - D_+)\theta_2], \qquad (10.594)$$

$$R_2 = -\tfrac{\lambda}{2}[(C_+ + C_-)\phi_2 + (D_+ + D_-)\theta_2], \quad \phi_2 = \frac{N}{\cosh^2[\frac{N}{2}(h-2z_m)]},$$

$$\theta_1 = \tanh[\tfrac{N}{2}(h+2z_m)] + \frac{N(h+2z_m)}{2\cosh^2[\frac{N}{2}(h+2z_m)]},$$

$$\theta_2 = \tanh[\tfrac{N}{2}(h-2z_m)] + \frac{N(h-2z_m)}{2\cosh^2[\frac{N}{2}(h-2z_m)]}.$$

By substituting the expressions for functions $q_{11}(x)$ and $q_{21}(x)$ from formulas (10.594) into equations (10.577) and by introducing new unknown functions

$$\Theta_{w1}(x) = \nu(x)T_{w11}(x), \quad \Theta_{w2}(x) = \nu(x)T_{w21}(x), \qquad (10.595)$$

we obtain a system of linear integral equations for functions $\Theta_{w1}(x)$ and $\Theta_{w2}(x)$

$$\Theta_{w1} = \tfrac{\Lambda}{2H_0} \int\limits_a^c [P_1(\xi)\Theta_{w1}(\xi) + R_1(\xi)\Theta_{w2}(\xi)]e^{-\psi_1(x,\xi)}$$

$$\times K_0(|\,\psi_1(x,\xi)\,|)d\xi + \tfrac{\Lambda}{2H_0} \int\limits_a^c \nu(\xi)L_1(\xi)e^{-\psi_1(x,\xi)} \qquad (10.596)$$

$$\times K_0(|\,\psi_1(x,\xi)\,|)d\xi, \quad \psi_1(x,\xi) = \lambda_1(x-\xi),$$

$$\Theta_{w2} = \tfrac{\Lambda\eta_0}{2H_0} \int\limits_a^c [P_2(\xi)\Theta_{w1}(\xi) + R_2(\xi)\Theta_{w2}(\xi)]e^{-\psi_2(x,\xi)}$$

$$\times K_0(|\,\psi_2(x,\xi)\,|)d\xi + \tfrac{\Lambda\eta_0}{2H_0} \int\limits_a^c \nu(\xi)L_2(\xi)e^{-\psi_2(x,\xi)} \qquad (10.597)$$

$$\times K_0(|\,\psi_2(x,\xi)\,|)d\xi, \quad \psi_2(x,\xi) = \lambda_2(x-\xi).$$

System (10.598), (10.597) can be solved numerically for functions Θ_{w1} and Θ_{w2}. Then the two-term asymptotic expansions for lubricant temperatures (see equations (10.571)) in the form

$$T(x,z) = T_0(x,z) + \Theta_1(x,z) + \ldots, \quad \Theta_1(x,z) = \nu(x)T_1(x,z). \qquad (10.598)$$

On the next and last Step 3, we consider the approximation for the generalized Reynolds equation. Using the appropriate expansions of functions involved in equation (10.560) (see above), we arrive at the following approximate generalized Reynolds equation

$$\frac{d}{dx}\left\{\frac{F(f_0)+\nu f_1 F'(f_0)}{\mu^0}I_z + \nu\frac{2}{h}\frac{f_0 F'(f_0)}{\mu^0}I_{z^2} + \nu\delta\frac{F(f_0)}{\mu^0}I_{zT1} - h\right\} = 0,$$

$$I_{z^2} = \int\limits_{-h/2}^{h/2} z^2 e^{\delta T_0}dz, \quad I_{zT1} = \int\limits_{-h/2}^{h/2} zT_1 e^{\delta T_0}dz, \tag{10.599}$$

$$p(a) = p(c) = \frac{dp(c)}{dx} = 0,$$

Substituting the expression for f_1 from formula (10.573) into equation (10.599), we obtain

$$\frac{d}{dx}\left\{\left[\frac{2}{h}\frac{f_0 F'(f_0)}{F(f_0)}\left(I_{z^2}-\frac{I_z^2}{I_1}\right) + \delta\left(I_{zT1}-\frac{I_z I_{T1}}{I_1}\right)\right]\frac{s_0}{I_1}\frac{H_0 h}{2f_0}\frac{dp}{dx} + s_0\frac{I_z}{I_1}\right.$$

$$\left. - h\right\} = 0, \ p(a) = p(c) = \frac{dp(c)}{dx} = 0. \tag{10.600}$$

To bring the generalized Reynolds equation (10.600) to its final form we need the expressions for integrals I_z and I_{T1} from (10.585) which have the form

$$I_{z^2} = \int\limits_{-h/2}^{h/2} z^2 e^{\delta T_0}dz = \frac{R}{N^3}\exp[\delta\frac{T_{w10}+T_{w20}}{2}]\{\frac{N^2 h^2 \sinh(Nh)}{4R}$$

$$\tag{10.601}$$

$$-Nh\ln(R) + 2\int\limits_{-N(h/2+z_m)}^{N(h/2-z_m)}\ln\cosh(y)dy\},$$

$$I_{zT1} = \int\limits_{-h/2}^{h/2} zT_1 e^{\delta T_0}dz = \frac{1}{\delta}\{A_1 I_z + B_1(I_{z^2}-z_m I_z)$$

$$+C_1(I_D + z_m I_C) + D_1(I_E + z_m I_D)\},$$

$$I_E = \int\limits_{-h/2}^{h/2} (z-z_m)^2 \tanh[N(z-z_m)]e^{\delta T_0}dz \tag{10.602}$$

$$= \frac{R}{N^3}\exp[\delta\frac{T_{w10}+T_{w20}}{2}]\{-\frac{N^2}{2R^2}\left(\frac{h^2}{4}+z_m^2\right)\sinh(Nh)\sinh(2Nz_m)$$

$$-\frac{N^2 h z_m}{4R^2}[\sinh^2(Nh)+\sinh^2(2Nz_m)] + N^2 h z_m$$

$$-\frac{Nh}{2R}\sinh(2Nz_m) - \frac{Nz_m}{R}\sinh(Nh) + \Delta\},$$

where the expressions for I_C and I_D follow from (10.585). In formula (10.601) the integral of function $ln[cosh(y)]$ can be calculated using any high precision numerical method or using a proper series (see formulas (10.339) and [26]).

It is easy to see that the contribution of the second group of terms within brackets which depend on $T_1(x,z)$ is of the same order of magnitude as the terms in the first group in Reynolds equation (10.600). There are a number of approaches published aiming at simplifying the Reynolds equation at hand which used just functions $f_0(x)$ and $T_0(x)$ while completely ignoring functions from the next order of approximation, i.e., functions $f_1(x)$ and $T_1(x,z)$. In effect, such approaches neglect the second group of terms within the brackets in (10.600) by assuming that $f_1(x) = T_1(x,z) = 0$. That alone may significantly change the lubrication film thickness obtained from the truncated compared to full Reynolds equations.

With the help of equations (10.585), (10.588), (10.601), and (10.602) equation (10.600) can be rewritten in the form

$$\frac{d}{dx}\left\{\left[\frac{2}{h}\frac{f_0 F'(f_0)}{F(f_0)}(I_{z^2}-\frac{I_z^2}{I_1})+J_0(p,h)\right]\frac{s_0}{I_1}\frac{H_0 h}{2f_0}\frac{dp}{dx}\right.$$

$$+[J_1(p,h)(\Theta_{w11}+\Theta_{w21})+J_2(p,h)(\Theta_{w21}-\Theta_{w11})]\frac{s_0}{I_1}$$

$$\left.+s_0\frac{I_z}{I_1}-h\right\}=0,\ p(a)=p(c)=\frac{dp(c)}{dx}=0,\qquad (10.603)$$

$$J_0(p,h)=B_1[I_{z^2}-\frac{I_z^2}{I_1}-z_m(I_z-I_1)]+C_0(I_D+z_m I_C-\frac{I_z}{I_1}I_C)$$

$$+D_0(I_E+z_m I_D-\frac{I_z}{I_1}I_D),$$

$$J_1(p,h)=\frac{\delta}{2}\{C_+(I_D+z_m I_C-\frac{I_z}{I_1}I_C)$$

$$+D_+(I_E+z_m I_D-\frac{I_z}{I_1}I_D)\},$$

$$\qquad (10.604)$$

$$J_2(p,h)=\frac{\delta}{2}\{C_-(I_D+z_m I_C-\frac{I_z}{I_1}I_C)$$

$$+D_-(I_E+z_m I_D-\frac{I_z}{I_1}I_D)\},$$

where Θ_{w11} and Θ_{w21} are the boundary values (10.595) of function $\Theta_1(x,z)$ from (10.598) at $z=\mp\frac{h}{2}$, respectively.

Above, we presented a relatively simple and well structured way to determine functions f_0, f_1, T_0, T_1, T_{w10}, T_{w20}, T_{w11}, and T_{w21} assuming that functions $p(x)$ and $h(x)$ are known. Therefore, the original problem described by equations (10.560)-(10.566) is reduced to problem of finding functions $p(x)$ and $h(x)$ from equations (10.600) and (10.563) while function $h(x)$ is determined from equation (10.562).

For heavily loaded lubricated contact this TEHL problem can be analyzed asymptotically based on the developed earlier techniques. This analysis can be performed if

$$\left[\frac{2}{h}\frac{f_0 F'(f_0)}{F(f_0)}\left(I_{z^2} - \frac{I_z^2}{I_1}\right) + J_0(p,h)\right]\frac{s_0}{I_1}\frac{H_0^2 h}{2f_0}\frac{dp}{dx}$$

$$+[J_1(p,h)(\Theta_{w11} + \Theta_{w21}) + J_2(p,h)(\Theta_{w21} - \Theta_{w11})]\frac{s_0 H_0}{I_1} \qquad (10.605)$$

$$+s_0 H_0\frac{I_z}{I_1} \ll 1, \; x - a \gg \epsilon_q, \; c - x \gg \epsilon_q, \; \omega \ll 1,$$

where ω is a small parameter involved in the problem while ϵ_q is the characteristic size of the inlet and exit zones of the lubricated contact (see sections on asymptotic analysis of EHL problems). The results of such an asymptotic analysis would be similar to the ones obtained for the corresponding isothermal case of EHL problem for non-Newtonian fluids considered in Subsection 10.15.5.

Interestingly enough, in the special case of an isothermal problem we have $\delta = 0$ and $T_0 = T_1 = 0$. Therefore, $I_z = I_{T1} = I_{zT1} = 0$ and the approximation of the generalized Reynolds equation given by (10.603), (10.604) is reduced to the isothermal case described by equations (10.478) and (10.479), which become exact in the case of Newtonian lubricant.

The above problem can be solved numerically by employing iterative approaches similar to the ones used in Sections 10.9 and 10.12.

The problem analyzed above can be generalized for the cases in which it is necessary to consider normal and tangential surface displacements caused by temperature and tangential stresses. That would require to consider the average speed of the surface $0.5(v_1 + v_2)$ and relative sliding speed $s = v_2 - v_1$ as functions of x. Here velocities v_1 and v_2 can be determined from the formula

$$v_i = u_i\left(1 + \frac{du_{xi}}{dx}\right), \; i = 1, 2,$$

where u_{xi} are the tangential displacements of the lower ($i = 1$) and upper ($i = 2$) surfaces assuming that terms with powers of du_{xi}/dx higher than one can be neglected. A simplifying assumption, which may be appropriate for using formulas from [46, 66] for determining the surface displacements u_{xi} is to replace the actual surface velocities $v_1(x)$ and $v_2(x)$ (which vary from point to point) by their constant rigid counterparts u_1 and u_2, respectively. Most of the problem asymptotic analysis coincides with the one presented above, the analysis of the expressions for $0.5(v_1 + v_2)$, $s = v_2 - v_1$, and $h(x)$ which include tangential stresses and heat fluxes also can be done in a similar manner.

Now, let us make a general conclusion about some properties of the solution of the problem for soft solids considered in Section 10.9, which is related to the one we considered in this section. In a heavily loaded lubricated contact of one hard and another soft elastic solids, we still can expect that outside of the inlet and exit zones the pressure will be close to the Hertzian one $p(x) = \sqrt{1 - x^2}$. Then for high slide-to-roll ratio $s_0 \gg 1$ the surface temperatures $T_{w1}(x)$ and $T_{w2}(x)$ will rise due to heat generation in the contact. Moreover, this rise in

$T_{w1}(x)$ and $T_{w2}(x)$ will be very similar to the increase of functions $T_{w10}(x)$ and $T_{w20}(x)$ (as they differ from $T_{w1}(x)$ and $T_{w2}(x)$ by $O(\nu(x)) \ll 1$). This temperature rise will likely to produce an addition surface displacement which may result in a creation of a "dimple." Now, the question is may such a "dimple" appearance serve as an explanation of the experimentally observed dimple effect [41] which was considered in Section 10.9.

Let us consider this question in more detail. As it follows from equation (10.579) for $N(x)$ and the fact that $F(x)$ is an odd function functions $T_{w10}(x)$ and $T_{w20}(x)$ are even functions with respect to s_0. Therefore, with accuracy of $O(\nu(x)) \ll 1$ functions $T_{w1}(x)$ and $T_{w2}(x)$ are independent of the sign of the slide-to-roll ratio $s_0 \gg 1$. That means that for large s_0 in both cases of $s_0 < 0$ and $s_0 > 0$ we either get or do not get a dimple. Therefore, if we try to explain the development of dimples in heavily loaded lubricated contacts of hard and soft materials using just the thermal effects in lubricated contacts as it was done by Kaneta and Young [43, 44] we will predict that in both cases of $s_0 < 0$ and $s_0 > 0$ we either have a dimple of almost the same size or not have a dimple at all. However, according to the experimental data of Kaneta [41] a dimple was observed in the contact region between hard and soft materials for $s_0 = 2$ and it was not observed for the same materials for $s_0 = -2$. Therefore, for high slide-to-roll ratio s_0 the attempt to explain the development of a dimple for $s_0 = 2$ and its absence for $s_0 = -2$ in a heavily loaded lubricated contact of hard and soft materials by just considering normal thermal displacement of surfaces fails. It means that to explain the dimple phenomenon it is necessary to consider the normal and tangential surface displacements. The tangential displacements cause variations in surface speeds. The influence of the thermal effects in heavily loaded lubricated contacts with $s_0 \gg 1$ would just make the the surface normal displacements larger, i.e., thermal effects would make the development of dimples more pronounced.

10.16.3 Numerical Analysis of the Heat Transfer in the Contact Solids

Based on the solutions and data presented in Sections 10.10 and 10.15 for thermal and isothermal EHL problems for non-Newtonian lubricant rheology it is clear what would be the behavior of most of the contact characteristics (such as H_0, c, $p(x)$, $h(x)$, $T(x, z)$, etc.) if the surfaces temperatures $T_{w1}(x)$ and $T_{w2}(x)$ are given. Therefore, in this subsection we will focus on the analysis of the behavior of surface temperatures $T_{w1}(x)$ and $T_{w2}(x)$. Being concerned only with surface temperatures, we will not determine the actual pressure $p(x)$ and gap $h(x)$ behavior but replace them by the Hertzian ones $p(x) = \sqrt{1 - x^2}$ and $h(x) = 1$ which provide good approximations outside of the inlet and exit zones of heavily loaded lubricated contacts. This choice of $p(x)$ limits our analysis because functions Θ_{w11} and Θ_{w21} are proportional to $\frac{dp}{dx}$ (see (10.595)) which approaches infinity as $x \to \pm 1$. It means that proper analysis of Θ_{w11} and Θ_{w21} requires the actual distributions of $p(x)$ and $h(x)$

realized in a lubricated contact while under the above approximations this analysis would be not reliable. However, these assumptions still allow for adequate analysis of surface temperatures based on the behavior of T_{w10} and T_{w20} (see expansions (10.571)).

For simplicity, let us consider the case of a Newtonian lubricant. Then the rheological function $F(x) = \frac{12H_0}{V}x$ and $\frac{f_0 F'(f_0)}{F(f_0)} = 1$. Therefore, the solution of equations (10.579)-(10.581) is

$$N = \frac{2}{h} \ln(\eta + \sqrt{\eta^2 + 1}),$$

$$\eta = \{\sinh^2(\tfrac{\Delta}{2}) + \tfrac{\kappa \delta V s_0^2}{96\lambda} \mu^0 e^{-\delta \frac{T_{w10}+T_{w20}}{2}}\}^{1/2}, \quad \Delta = \delta \frac{T_{w20}-T_{w10}}{2},$$

(10.606)

while the main term of the sliding frictional stress is

$$f_0 = \frac{V}{12H_0} \frac{\mu^0 s N}{\sinh N h} e^{-\delta \frac{T_{w10}+T_{w20}}{2}}.$$

(10.607)

For $\Delta = 0$, $s = s_0$, $\mu^0 = \mu$, and $T_{w10} = T_{w20} = T_{w0}$ function N coincides with the expression for B obtained independently for the case of Newtonian fluid in Section 10.10 if you take into account that $\kappa_N = \frac{\kappa V}{12}$, which follows from the definitions of parameters κ_N in Section 10.10 (see formulas (10.316)) and κ in this section (see formulas (10.558)). Moreover, for $\delta \to 0$ we have $N \to 0$ and the expression for f_0 approaches to the isothermal one $\frac{\mu^0 s_0}{h}$.

To obtain solutions $T_{w10}(x)$ and $T_{w20}(x)$ of the system of nonlinear equations (10.583), we will use the following iteration process

$$T_{w10}^{n+1} = \frac{\Lambda}{H_0} \int_a^c \frac{\lambda N^{n+1,n}}{\delta} \tanh[N^{n+1,n}(\tfrac{h}{2} + z_m^{n+1,n})] e^{-\lambda_1(x-\xi)}$$

$$\times K_0(|\lambda_1(x-\xi)|)d\xi - \Delta_\infty,$$

(10.608)

$$T_{w20}^{n+1} = \frac{\Lambda \eta_0}{H_0} \int_a^c \frac{\lambda N^{n+1,n+1}}{\delta} \tanh[N^{n+1,n+1}(\tfrac{h}{2} - z_m^{n+1,n+1})]$$

$$\times e^{-\lambda_2(x-\xi)} K_0(|\lambda_2(x-\xi)|)d\xi + \Delta_\infty,$$

(10.609)

where $N^{n+1,n}$ and $z_m^{n+1,n}$ are calculated based on T_{w10}^{n+1} and T_{w20}^n while $N^{n+1,n+1}$ and $z_m^{n+1,n+1}$ are determined based on previously determined T_{w10}^{n+1} and still unknown T_{w20}^{n+1}, (n is the iteration index). The system is solved using Newton's method sequentially, i.e., the iteration cycle is based on solution of decoupled equations and it starts with determining one iteration of equation (10.608) and it continues with determining one iteration of equation (10.610). After that the calculation cycle is repeated until the solution converges to the desired precision. It is assumed here that $a,,\ H_0,\ \delta,\ \lambda,\ \lambda_1,\ \lambda_2,\ \Lambda$, and $p(x)$, $h(x)$ are known. As the initial approximations for these iterations can be taken functions $T_{w10}(x) = -\Delta_\infty$, $T_{w20}(x) = \Delta_\infty$ of the closest available solution of the problem.

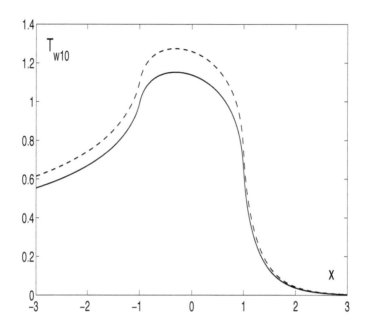

FIGURE 10.53

Dependence of the surface temperatures $T_{w10}(x) = T_{w20}(x)$ on $\frac{\kappa \delta V s_0^2}{96\lambda}$ for $\kappa = 100$ (solid curve) and $\kappa = 200$ (dashed curve), where $Q = 1.2$, $\lambda_1 = \lambda_2 = 1$, $\Delta_\infty = 0$, and $\eta_0 = 1$.

System (10.583) also can be solved by application of Newton's method simultaneously to both equations. However, this approach will require handling solution of algebraic systems of double order compared to the previous approach. This approach might be more appropriate for large pressure viscosity coefficients Q in Barrus viscosity $\mu^0(p) = e^{Qp}$ relationship.

Solution of (10.583) involves the following integral approximations

$$\int_a^c g(\xi) e^{-\lambda_1(x_k-\xi)} K_0(| \lambda_1(x_k - \xi) |) d\xi$$

$$(10.610)$$

$$\approx \sum_{i=1}^{N-1} \frac{g(\xi_i)+g(\xi_{i+1})}{2} e^{-\lambda_2(x_k-\xi_{i+1/2})} K_0(| \lambda_2(x_k - \xi_{i+1/2}) |)\Delta x,$$

where $\{x_k\}$ ($k = 1,\ldots,N_x$) is the set of integer nodes, $x_{k+1} = x_k + \Delta x$, $\{x_{i+1/2}\}$ ($i = 1,\ldots,N_x - 1$) is the set of midpoints $x_{i+1/2} = \frac{x_i+x_{i+1}}{2}$, Δx is the step size.

After the sets of $T_{w10}(x_k)$ and $T_{w20}(x_k)$ ($k = 1,\ldots,N_x$) are determined the surface temperatures outside of the lubricated contact (i.e., for $x < a$ and

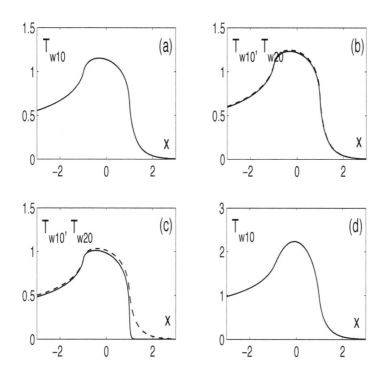

FIGURE 10.54
Dependence of the surface temperatures $T_{w10}(x)$ (solid curves) and $T_{w20}(x)$ (dashed curves) on parameters Q, λ_1, λ_2, and η_0: $Q = 1.2$, $\lambda_1 = \lambda_2 = 1$, $\eta_0 = 1$ - case (a), $Q = 1.2$, $\lambda_1 = \lambda_2 = 1$, $\eta_0 = 5$ - case (b), $Q = 1.2$, $\lambda_1 = 10$, $\lambda_2 = 1$, $\eta_0 = 1$ - case (c), and $Q = 7.5$, $\lambda_1 = \lambda_2 = 1$, $\eta_0 = 1$ - case (d), where $\Delta_\infty = 0$ and $\kappa = 100$.

$x > c$) can be obtained from equations (10.583) by calculating the integrals of the already determined functions.

To illustrate the behavior of the surface temperatures, let us consider some examples. Let us assume that $a = -1$, $c = 1$, $V = 0.15$, $s_0 = 2$, $H_0 = 0.075$, $\delta = 1$, $\frac{\Lambda}{H_0} = 13.333$, $p(x) = \sqrt{1 - x^2}$, and $h(x) = 1$. For simplicity we will assume that λ and δ are constants and $\lambda = 1$. We will consider some numerical results for different combinations of parameters κ, Q, λ_1, λ_2, Δ_∞, and η_0. As Λ increases (H_0 decreases) the surface temperatures $T_{w10}(x)$ and $T_{w20}(x)$ increase as long as $\mid z_m \mid \leq \frac{h}{2}$. That follows from equations (10.583) and the fact that $\tanh[N(\frac{h}{2} \pm z_m)] \geq 0$ for all $\mid x \mid \leq 1$. As the value of $\frac{\kappa \delta V s_0^2}{96\lambda}$ increases the amount of heat generated in the contact increases which causes surface

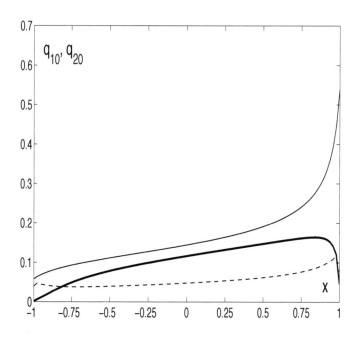

FIGURE 10.55
Dependence of the heat fluxes $q_{10}(x)$ and $q_{20}(x)$ at the contact surfaces obtained for $Q = 1.2$, $\lambda_1 = 10$, $\lambda_2 = 1$ (solid and dashed line, respectively) and for $Q = 7.5$, $\lambda_1 = \lambda_2 = 1$ for which $q_{10}(x) = q_{20}(x)$ (double solid line), where $\Delta_\infty = 0$, $\eta_0 = 1$, and $\kappa = 100$.

temperatures $T_{w10}(x)$ and $T_{w20}(x)$ to increase (see Fig. 10.53).

The numerical results show that if the values of λ_1 and λ_2 are close to each other and Δ_∞ is close to 0 while η_0 is not very far from 1 then the graphs of surface temperatures $T_{w10}(x)$ and $T_{w20}(x)$ are pretty close to each other (see Fig. 10.54). However, the heat fluxes in the contact surfaces $q_{10}(x)$ and $q_{20}(x)$ may differ significantly in these cases. Two examples of the heat fluxes $q_{10}(x)$ and $q_{20}(x)$ and sliding frictional stress f_0 behavior are given in Fig. 10.55 and 10.56. In cases when λ_1 is significantly different from λ_2, Δ_∞ is different from 0 (i.e., surface temperatures of the solids are different far away from the contact) while η_0 is significantly different from 1 we can expect the surface temperatures $T_{w10}(x)$ and $T_{w20}(x)$ of the solids to be different. Some examples illustrating this behavior are shown in Fig. 10.54.

A very interesting dynamics of heat and temperature distributions can be seen in Fig. 10.57 - 10.59 for the case of different surface temperatures far away from the contact (i.e., for $\Delta_\infty = 1$). Obviously, the level of surface temperature

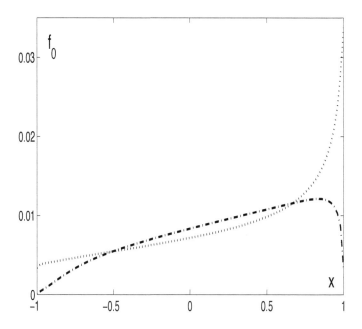

FIGURE 10.56
Two examples of the distribution of the sliding frictional stress $f_0(x)$ obtained for $Q = 1.2$, $\lambda_1 = 10$ (dotted curve) and $Q = 7.5$, $\lambda_1 = 1$ (dashed curve), where $\lambda_2 = 1$, $\Delta_\infty = 0$, $\eta_0 = 1$, and $\kappa = 100$.

rises much more for the originally (far from the contact) cooler surface. In this case the heat fluxes $q_{10}(x)$ and $q_{20}(x)$ only in the center of the contact region directed in both solid surfaces (Fig. 10.58) while at the ends of the contact the heat flows from one contact surface into another (i.e., $q_{10}(x)$ and $q_{20}(x)$ have opposite signs). That finds its reflection in the behavior of functions $N(x)$ and $z_m(x)$ (see Fig. 10.59). It can be seen in the behavior of function $z_m(x)$, which represents the position of the vertex of the lubricant temperature $T(x, z)$ distribution with respect to z. In cases when $\mid z_m \mid \geq \frac{h}{2} = \frac{1}{2}$ the distribution of temperature $T(x, z)$ between the contact surfaces is monotonic and, therefore, the heat flux is directed from one contact surface to another. That, obviously, is represented by the behavior of heat fluxes $q_{10}(x)$ and $q_{20}(x)$ as well as by surface temperatures $T_{w10}(x)$ and $T_{w20}(x)$ (see Fig. 10.57 and 10.58).

If λ_1 is greater than λ_2 (i.e., the lower surface moves faster than the upper one), then the temperature $T_{w10}(x)$ of the lower surface rises slower and to a slightly lower lever than the temperature $T_{w20}(x)$ of the upper surface due to a

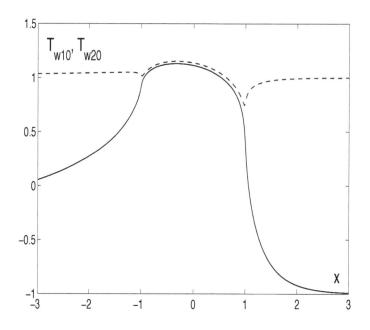

FIGURE 10.57
Distribution of surface temperatures $T_{w10}(x)$ (solid curve) and $T_{w20}(x)$ (dashed curve) obtained for $Q = 1.2$, $\lambda_1 = \lambda_2 = 1$, $\Delta_\infty = 1$, and $\eta_0 = 1$.

lesser exposure to heat generated in the contact. Moreover, for faster–moving surface (assuming that $\lambda_1 > 0$ and $\lambda_2 > 0$, i.e., both surfaces are moving from left to right) the effect of heat generation in the contact extends to a lesser degree ahead of the contact, i.e., for $x > 1$ (see Fig. 10.54, case (c)).

If, for example, the lower surface has larger product $k_1\rho_1c_1$, then the upper surface (i.e., $\eta_0 > 1$) then temperature $T_{w20}(x)$ rises in the contact a little higher than $T_{w10}(x)$ (see Fig. 10.54, case (b)). This is due to slower heat dissipation in the upper solid.

For $\lambda_1 > 0$ and $\lambda_2 > 0$, both solids are moving in the positive direction (from left to right). For small values of the pressure viscosity coefficient Q the maximum of the surface temperatures $T_{w10}(x)$ and $T_{w20}(x)$ occurs to the left of $x = 0$ (i.e., in a lubricated contact the surface temperature maximum is shifted from $x = 0$ toward the contact inlet). As the pressure viscosity coefficient Q increases the temperature distributions lean forward and the temperature maximum point approaches $x = 0$ (see Fig. 10.54, cases (a)-(d)). At the same time the level of the surface temperatures rises due to the rise of lubricant viscosity μ^0 and, therefore, increased heat generation in the contact.

As the value of δ increases the surface temperatures in the contact increase

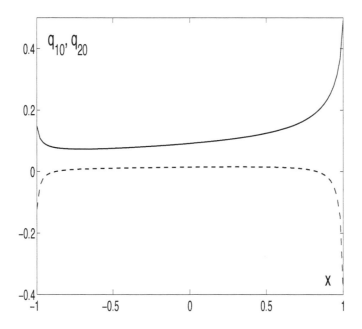

FIGURE 10.58
Distribution of heat fluxes $q_{10}(x)$ (solid curve) and $q_{20}(x)$ (dashed curve) at
contact surfaces obtained for $Q = 1.2$, $\lambda_1 = \lambda_2 = 1$, $\Delta_\infty = 1$, and $\eta_0 = 1$.

to lower levels than for smaller δ This is due to fast decrease of lubricant
viscosity $\mu = \mu^0 e^{-\delta T}$ and heat generation in the contact with increase of
lubricant temperature T.

10.17   Regularization for Heavily Loaded Isothermal EHL Contacts with Non-Newtonian Fluids

There are two major wide classes of non-Newtonian fluids: pseudo-plastic and
dilatant ones. Numerical solution of heavily loaded isothermal EHL problems
for pseudo-plastic lubricant fluids is less and for dilatant lubricant fluids is
more prone to solution instability compared to the case of Newtonian fluids.
Therefore, in many cases there is still need for regularization of such isothermal numerical solutions. In regularization of heavily loaded isothermal EHL
contacts with non-Newtonian fluid, we will follow the same idea as in the case

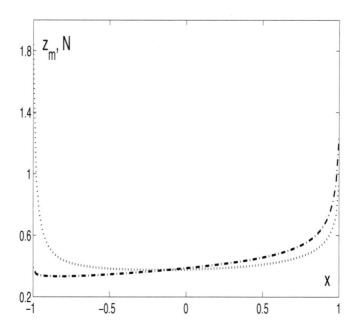

FIGURE 10.59
Distribution of functions $z_m(x)$ (dotted curve) and $N(x)$ (dash-dotted curve) obtained for $Q = 1.2$, $\lambda_1 = \lambda_2 = 1$, $\Delta_\infty = 1$, and $\eta_0 = 1$.

of a Newtonian fluid. However, in this case we need to use the main term of the asymptotic solution for temperature T_0 obtained for a fluid with a particular non-Newtonian rheology. It is easy to do by using the special case of the solution from preceding section. Namely, assuming that the dimensionless $T_{w1} = T_{w2} = 0$ we obtain $z_m = I_z = 0$ (see equations (10.578)-(10.584))

$$T_0(x, z) = \tfrac{1}{\delta} \ln \frac{R}{\cosh^2[N(z - z_m)]}, \quad R = \cosh^2(\tfrac{Nh}{2}), \qquad (10.611)$$

where values of N are determined from solution of the equation

$$\frac{4\lambda}{\kappa \delta H_0 s_0} N \tanh(\tfrac{Nh}{2}) = \Phi(\tfrac{\mu^0 s_0}{h} \frac{Nh}{\sinh(Nh)}). \qquad (10.612)$$

Similarly, from the above conditions and formulas (10.584)-(10.587) follows that

$$T_1(x, z) = \tfrac{1}{\delta}\left[1 + \frac{f_0 F'(f_0)}{F(f_0)}\right]\left\{\frac{\tanh(Nz)}{\tanh(\tfrac{Nh}{2})} - \tfrac{2z}{h}\right\},$$
$$\qquad (10.613)$$

$$f_0 = \Phi(\tfrac{\mu^0 s_0}{h} \frac{Nh}{\sinh(Nh)}).$$

Besides that we obtain $f_1 = I_{T1} = 0$. Therefore, the Reynolds equation is reduced to the equation (see (10.600))

$$\frac{d}{dx}\left\{\left[\frac{2}{h}\frac{f_0 F'(f_0)}{F(f_0)}I_{z^2} + \delta I_{zT1}\right]\frac{N s_0}{\sinh(Nh)}\frac{H_0 h}{2f_0}\frac{dp}{dx} - h\right\} = 0. \tag{10.614}$$

To finalize the generalized Reynolds equation (10.614), we need the particular form of expressions for integrals (see (10.601) and (10.602))

$$I_{z^2} = \frac{\sinh(Nh)}{N^3}\{\frac{N^2 h^2}{4} - \frac{1}{\tanh(\frac{Nh}{2})}[Nh\ln\cosh(\frac{Nh}{2})$$
$$-2\int\limits_0^{Nh/2}\ln\cosh(y)dy]\}, \tag{10.615}$$

$$I_{zT1} = \frac{1}{\delta}[1 + \frac{f_0 F'(f_0)}{F(f_0)}]\{-\frac{2}{h}I_{z^2} + \frac{I_D}{\tanh(\frac{Nh}{2})}\},$$
$$I_D = \frac{\cosh^2(\frac{Nh}{2})}{N^2}[\frac{Nh}{2}\tanh^2(\frac{Nh}{2}) - \frac{Nh}{2} + \tanh(\frac{Nh}{2})], \tag{10.616}$$

where the expression for I_D follows from (10.585).

Finally, recalling the Reynolds equation (10.478) for the isothermal EHL problem with function M from (10.479), we can conclude that the regularized problem is reduced to solution of the following equations

$$\frac{d}{dx}\{M(\mu, p, h, \frac{dp}{dx}, V, s_0, H_0) - H_0 h\} = 0,$$
$$p(a) = p(c) = \frac{dp(c)}{dx} = 0, \tag{10.617}$$

$$M = \frac{H_0^2 h^3 F'(f_0)}{12\mu}\frac{dp}{dx}R_T,$$
$$R_T = \frac{6}{h^2}[\frac{2}{h}I_{z^2} + \frac{F(f_0)}{f_0 F'(f_0)}\delta I_{zT1}], \quad f_0 = \Phi(\frac{\mu s_0}{h}\frac{Nh}{\sinh(Nh)}), \tag{10.618}$$

$$H_0(h - 1) = x^2 - c^2 + \frac{2}{\pi}\int\limits_a^c p(t)\ln\mid\frac{c-t}{x-t}\mid dt, \tag{10.619}$$

$$\int\limits_a^c p(t)dt = \frac{\pi}{2}, \tag{10.620}$$

where I_{z^2} and δI_{zT1} are determined according to (10.615) and (10.616) while function N is obtained by solving equation

$$\frac{V s_0}{24\beta^2 H_0}N\tanh(\frac{Nh}{2}) = \Phi(\frac{\mu s_0}{h}\frac{Nh}{\sinh(Nh)}), \tag{10.621}$$

where β is a fictitious sufficiently small positive constant. This parameter β is chosen in such a way that the numerical solution is stable and, at the same time, the regularized solution is sufficiently close to the non-regularized one. In equations (10.617)-(10.621) it is assumed that $\mu^0 = \mu$, where μ is the isothermal lubricant viscosity at the ambient temperature. To construct a regularized numerical algorithm equations (10.617)-(10.621) have to be resolved for $p(x)$ by inverting equation (10.619) and one has to follow the procedure of numerical calculations outlined in Section 10.12.

For the case of a Newtonian fluid, equations (10.617)-(10.621) coincide with the regularized problem proposed in Section 10.12.

10.18 Friction in Heavily Loaded Lubricated Contacts

After the solution for a heavily loaded EHL contact is obtained, we can determine friction forces at the contact surfaces from the formulas

$$F_T = F_S \pm F_R, \ F_S = \int_a^c f(x)dx, \ F_R = \frac{H_0}{2} \int_a^c h(x)\frac{dp(x)}{dx}dx, \qquad (10.622)$$

where F_S and F_R are the sliding and rolling friction forces, respectively, and $f(x)$ is the sliding frictional stress determined from the problem solution. Formulas for $f(x)$ for various cases can be found in the preceding sections.

For sufficiently hard elastic materials for both pre- and over-critical lubrication regimes in the cental region of the contact the pressure is close to the Hertzian pressure $\sqrt{1-x^2}$ and the gap is practically equal to 1 while outside of the Hertzian region pressure is small and gap is greater than 1 in the inlet zone and it is about 1 in the exit zone. Because of that in these cases calculation of the sliding frictional stress $f(x)$ and sliding friction force F_S pressure $p(x)$ and gap $h(x)$ can be replaced by $\sqrt{1-x^2}$ and 1, respectively. Moreover, due to the symmetry of the Hertzian pressure about $x = 0$ and because of the proximity of $h(x)$ to 1 in the central region of the contact, the main contributions to the rolling friction stress F_R come from the inlet and exit zones.

For example, for Newtonian fluid in an isothermal contact the sliding frictional stress $f(x)$ is given by formula

$$f(x) = \frac{V}{12H_0}\frac{\mu s_0}{h}. \qquad (10.623)$$

Therefore, for $\mu(p) = \exp(Qp)$ and $Q \gg 1$ using Watson Lemma (see [67]) for the dimensionless sliding friction force, we obtain an analytical formula

$$F_S = \int_a^c f(x)dx \approx \frac{V s_0}{12H_0}\sqrt{\frac{2\pi}{Q}}e^Q + \dots \qquad (10.624)$$

Similar, analytical formulas for F_S can be obtained for other lubricant rheologies.

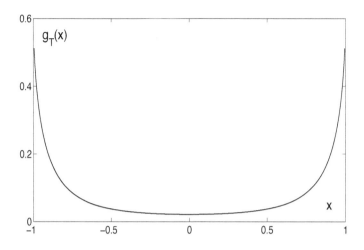

FIGURE 10.60

A typical functional dependence of $g_T(\eta)$ on η in the Hertzian region.

In case of a Newtonian fluid involved in a thermal EHL contact with prevailing sliding frictional stress (i.e., $\nu(x) \ll 1$), the dimensionless viscosity described by $\mu(p, T) = \mu^0(p) \exp(-\delta T)$, and equal surface temperatures $T_{w1} = T_{w2} = T_{w0}$ the sliding frictional stress $f(x)$ is given by the approximate formulas (see Section 10.10)

$$f(x) \approx \frac{V s_0}{12 H_0} \frac{\mu^0 e^{-\delta T_{w0}}}{h} g_T(\eta), \ \ g_T(\eta) = \frac{4(\eta + \sqrt{\eta^2+1})^2 \ln(\eta + \sqrt{\eta^2+1})}{(\eta + \sqrt{\eta^2+1})^4 - 1},$$

$$(10.625)$$

$$\eta = \frac{|s_0|}{2} \sqrt{\frac{\delta \kappa}{2\lambda} \mu^0 e^{-\delta T_{w0}}}.$$

It is important to recognize the fact that the difference in formulas (10.623) and (10.625) for the sliding frictional stress $f(x)$ is in just the multiplier $e^{-\delta T_{w0}} g_T(\eta)$. It can be shown that $g_T(\eta) \to 1$ as $\eta \to 0$. Therefore, for $T_{w0} = 0$ and $\eta \to 0$ the expression for the sliding frictional stress $f(x)$ in a non-isothermal lubricated contact converges to the one in an isothermal contact (see formulas (10.623) and (10.625)). If $\mu^0(p) = \exp(Qp)$, then for sufficiently large values of Q parameter $\eta \gg 1$, which leads to the estimate $g_T(\eta) = \ln(2\eta)/\eta^2$. In turn, in this case it means that the sliding frictional stress $f(x)$ and the sliding friction force F_S are approximately equal to

$$f(x) \approx \frac{VQ}{3H_0} \frac{\lambda}{\delta \kappa s_0} \frac{p}{h}, \ \ F_S \approx \frac{\pi VQ}{6H_0} \frac{\lambda}{\delta \kappa s_0}.$$

$$(10.626)$$

In the expression for F_S it is assumed that λ/δ is a constant. For $T_{w0}(x) = 0$ a typical graph of $g_T(\eta)$ in the Hertzian region $[-1, 1]$ calculated based on the

Hertzian pressure $p = \sqrt{1 - x^2}$ is shown in Fig. 10.60. Obviously, for large values of η the sliding frictional stress $f(x)$ in the most of the Hertzian region and friction force F_S are significantly smaller than the corresponding values in the case of isothermal lubrication.

Formulas (10.625) and (10.626) can be easily generalized for the case of a Newtonian fluid and different surface temperatures $T_{w1} \neq T_{w2}$ (see Subsection 10.16.3 for specific formulas).

In case of pure rolling the slide-to-roll ratio $s_0 = 0$ and $f(x) = 0$. Therefore, under pure rolling conditions $F_S = 0$ and the frictional stresses $\pm \frac{H_0 h}{2} \frac{dp(x)}{dx}$ and friction force F_R are completely determined by the solution behavior in the inlet and exit zones. Usually, under pure rolling conditions friction is low.

Under most lubrication conditions with mixed rolling and sliding, the sliding frictional stress $f(x)$ dominates the rolling one $\pm \frac{H_0 h}{2} \frac{dp(x)}{dx}$, which is equivalent to the case of $\nu(x) \ll 1$ (see preceding sections). Under such conditions the friction force F_T is practically completely represented by its sliding part F_S. For the known lubrication film thickness H_0, the sliding friction force F_S can be determined based on the Hertzian pressure $\sqrt{1 - x^2}$ and, if necessary, the lubricant temperature $T(x, z)$, which, in turn, can be calculated based on the Hertzian pressure. In other words, under such condition the knowledge of the solution behavior in the inlet and exit zones of a lubricated contact is needed only for determining the film thickness H_0. For a particular non-Newtonian lubricant rheology, the specific formulas for the frictional sliding stress $f(x)$ and sliding friction force F_S can be obtained in a similar fashion based on the results of preceding sections.

10.19   Closure

A number of classic and modern EHL problems are considered. In particular, EHL problems for lubricants with Newtonian and non-Newtonian rheology are considered. The analysis is done using regular and matched asymptotic expansions as well as numerical methods. Lubricated contacts under light and heavy loading are considered. Lightly loaded EHL contacts are considered asymptotically and numerically and their results are compared. Asymptotic analysis of lightly loaded EHL contact produced formulas for film thickness and sliding and rolling frictional forces. Asymptotic analysis of heavily loaded contact lubricated by a Newtonian fluid recognizes two different regimes of lubrication: pre- and over-critical lubrication regimes. Overall, the difference between these two regimes is in how strongly the lubricant viscosity depends on pressure and how much lubricant is available at the inlet of the contact. Two different asymptotic approaches to the analysis of the pre- and over-critical regimes are developed while both are based on matched asymptotic expansions. A detailed structure of the inlet and exit zones and the Hertzian region of a lubricated contact is analyzed. A number of asymptotically based formulas for the film thickness in heavily loaded EHL contacts are derived for the cases of Newtonian and non-Newtonian lubricant rheologies. The classic Ertel-Grubin method is analyzed and it's shortcomings are revealed. The numerical methods for EHL problems in the original and asymptotic formulations are proposed and these problems are solved numerically in isothermal and non-isothermal formulations. The issues of numerical precision and stability of the solutions for heavily loaded lubricated contacts of the original and asymptotic problems are discussed. A simple and effective regularization approach to solution of the original and asymptotic problems is proposed. To explain the dimple phenomenon, the elastic tangential displacements of the lubricated contact surfaces are employed. For lubricants with non-Newtonian rheologies, conditions of pure rolling and large sliding are considered. A call for caution in extrapolating the results from experimentally obtained in optical EHL simulators data on pressure viscosity coefficient to the scale of practical applications is made.

10.20   Exercises and Problems

1. List and discuss the assumptions used for the derivation of Reynolds equation for pressure p from Navier-Stokes equations.

2. List and discuss the assumptions used for the derivation of the equation for lubricant temperature T from the energy equation for fluid flow.

3. List and discuss the assumptions used for the derivation of the equation for gap h between the solids in contact.

4. Determine a two-term asymptotic solution of the isothermal EHL problem for a lightly loaded contact (i.e., for $V \gg 1$) lubricated by an incompressible fluid with viscosity determined by $\mu = \exp(Qp)$, where $Q = O(1)$. Graph $p(x)$ and $h(x)$ for $V = 5$ and $V = 10$ and compare this solution with the one determined by formulas (10.53)-(10.57) and data from Table 10.1.

5. (a) What is the difference between pre- and over-critical lubrication regimes in heavily loaded contacts? What are the magnitudes of the contributions of each of the zones in the contact region to the problem solution? Explain the procedure by which pre- and over-critical regimes are determined for the case of lubricant viscosity $\mu = \exp(Qp)$, $Q \gg 1$ (see equations (10.140), (10.141), (10.156), and (10.472)).

(b) Is it possible for an over-critical regime to occur for a lubricant with constant viscosity? Explain why.

6. Describe in your own words the basis for the derivation of the formula for the film thickness H_0 from (10.120) for pre-critical lubrication regimes.

7. For a Newtonian lubricant with constant viscosity, derive formulas (10.149).

8. For a Newtonian lubricant, provide a detailed derivation of formula (10.205) for the film thickness H_0 under over-critical lubrication regimes.

9. Using the numerical data for $q(0)$ and the asymptotic expansion for $p_{ext}(x)$ from (10.209) show that assumption 2 (expressed by estimate (10.246)) laid in the foundation of the Ertel-Grubin method is violated.

10. Provide the reason as to why surface tangential displacements and modified surface linear velocities must be included in the formulation of the EHL problem for soft solids to explain the Kaneta dimple phenomenon.

11. What is the physical meaning of the expression for $\nu(x)$ from formulas (10.328) and (10.475)?

12. Why is it necessary to determine the two-term asymptotic solution for the lubricant temperature T? (Hint: See the expression for function $M(p, h)$ from (10.338).)

13. For starved lubrication regimes for power law fluid derive the formulas of transformation similar to (10.289).

14. Explain the essence of the scale effect described in Subsection 10.15.5, which takes place while extrapolating from measurements made using an optical EHL simulator to practical cases.

BIBLIOGRAPHY

[1] Bhushan, B. 1999. *Principles and Applications of Tribology*. Toronto: John Wiley & Sons, Inc., 586 - 591.

[2] Cheng, H.S. 1980. Fundamentals of Elastohydrodynamic Contact Phenomena. In *Fundamentals of Tribology*, N.P. Suh and N. Saka, eds. Cambridge, MA: The MIT Press, 1009 - 1048.

[3] Lamb, H. 1995. *Hydrodynamics*. New-York: Cambridge University Press, 6th ed.

[4] Van-Dyke, M. 1964. *Perturbation Methods in Fluid Mechanics*. New-York: Academic Press.

[5] Kevorkian, J. and Cole, J.D. 1985. *Perturbation Methods in Applied Mathematics. Applied Mathematics Series, Vol.* 34. New York: Springer-Verlag.

[6] Kapitza, P.L. 1955. Hydrodynamic Theory of Lubrication during Rolling. *Zhurnal Tekhnicheskoy Fiziki* 25:747-762.

[7] Hamrock, B.J. 1994. *Fundamentals of Fluid Film Lubrication*. New York: McGraw-Hill.

[8] Van Wazen, J.R., Lyons, J.W., Kim, K.Y., and Cowell, R.E. 1963. *Viscosity and Flow Measurment. A Laboratory Handbook of Rheology*. New-York-London: Interscience Publishers, John Wiley & Sons.

[9] Kudish, I.I. 1981. Some Problems of the Elastohydrodynamic Theory of Lubrication for a Lightly Loaded Contact. *Mech. of Solids* 16, No. 3:75-88.

[10] Kudish, I.I. and B.G. Yatzko. 1990. Numerical Method for Solution of Problems for Lightly Loaded Contacts with Non-Newtonian Lubricant. *Soviet J. Frict. and Wear* 11, No. 4:594-601.

[11] Fedorenko, R.P. 1978. *Approximate Solution of Optimal Control Problems*. Moscow: Nauka.

[12] Dowson, D. and Higginson, G.R. 1966. *Elastohydrodynamic Lubrication*. London: Pergamon Press.

[13] Ertel, M.A. 1945. Hydrodynamic Calculation of Lubricated Contact for Curvilinear Surfaces. Dissertation, *Proc. of CNIITMASh*, 1-64.

[14] Grubin, A.N. 1949. The Basics of the Hydrodynamic Lubrication Theory for Heavily Loaded Curvilinear Surfaces. *Proc. CNIITMASh* 30:126-184.

[15] Kudish, I.I. and Marchenko, S.M. 1985. Some Problems and Mathematical Methods in the Elastohydrodynamic Lubrication Theory. *Izvestiya Severo—Kavkazskogo Nauchnogo Centra Vysshey Shkoly, Estestwennye Nauki* 3:46-52.

[16] Houpert, L.G. and Hamrock, B.J. 1986. Fast Approach for Calculating Film Thickness and Pressures in Elastohydrodynamically Lubricated Contacts at High Loads. *ASME J. Tribology* 108, No. 3:441-452.

[17] Bissett, E.J. and Glander, D.W. 1988. A Highly Accurate Approach that Resolves the Pressure Spike of Elastohydrodynamic Lubrication. *ASME J. Tribology* 110, No. 2:241-246.

[18] Hamrock, B.J., Ping Pan, and Rong-Tsong Lee. 1988. Pressure Spikes in Elastohydro-dynamically Lubricated Conjunctions. *ASME J. Tribology* 110, No. 2:279-284.

[19] Venner, C.H. and Lubrecht, A.A. 2000. *Multilevel Methods in Lubrication*. Amsterdam: Elsevier.

[20] Evans, H.P. and Hughes, T.G. 2000. Evaluation of Deflection in Semi-Infinite Bodies by a Differential Method. *Proc. Instn. Mech. Engrs.* 214, Part C :563-584.

[21] Archard, J.F., Baglin, K.P. 1986. Elastohydrodynamic Lubrication - Improvements in Analytic Solutions. *Proc. Inst. Mech. Eng.* 200, No. C4 :281-291.

[22] Crook, A.W. 1961. The Lubrication of Rollers II. Film Thickness with Relation to Viscosity and Speed. *Philosophical Trans. Royal Soc. of London, Ser. A, Math., Phys. and Eng. Sci.* 254, No. 1040:223-236.

[23] Kudish, I.I. 1977. Hydrodynamic Lubrication Theory of Rolling Cylindrical Bodies. *Abstracts of the 2nd All — Union Conf. on Elastohydrodynamic Theory of Lubric. and Its Pract. Appl. in Technology*, (1976), Kujbyshev, 11; *Proc. 2nd All — Union Conf. on Elastohydrodynamic Theory of Lubrication and Its Practical Applications in Industry*, Kujbyshev, 33-38.

[24] Kudish, I.I. 1982. Asymptotic Methods of Study for Plane Problems of the Elastohydrodynamic Lubrication Theory in Heavy Loaded Regimes. Part 1. Isothermal Problem. *Proc. Acad. Sci. of Armenia SSR, Mechanics* 35, No. 5:46-64.

[25] Kudish, I.I. 1978. Elastohydrodynamic Problem for a Heavily Loaded Rolling Contact. *Proc. Acad. Sci. of Armenia SSR, Mechanics* 31, No. 1:65-78.

[26] Kudish, I.I. 1978. Asymptotic Analysis of a Plane Non-isothermal Elastohydro-dynamic Problem for a Heavily Loaded Rolling Contact. *Proc. Acad. Sci. of Armenia SSR, Mechanics* 31, No. 6:16-35.

[27] Kudish, I.I. 1983. Asymptotic Method of Study for Plane Problems of the Elastohydrodynamic Lubrication Theory for Heavily Loaded Regimes. Part 2. Non-isothermal Problem. *Proc. Acad. Sci. Armenia SSR, Mechanics* 36, No. 5:47-59.

[28] Vorovich, I.I., Aleksandrov, V.M., and Babeshko, V.A. 1974. *Non – Classical Mixed Problems of Elasticity*. Moscow: Nauka.

[29] Kudish, I.I. 1996. Asymptotic Analysis of a Problem for a Heavily Loaded Lubricated Contact of Elastic Bodies. Pre- and Over-critical Lubrication Regimes for Newtonian Fluids. *Dynamic Systems and Applications*, Dynamic Publishers, Atlanta, 5, No. 3:451-478.

[30] Greenwood, J.A. 1972. An Extension of the Grubin Theory of Elasto-hydrodynamic Lubrication. *Phys. D. Appl. Phys.* 5:2195-2211.

[31] Johnson, K.L. 1970. Regimes of Elastohydrodynamic Lubrication. *J. Mech. Eng. Sci.* 12, No. 1:9-16.

[32] Gakhov, F.D. 1977. *Boundary Value Problems*. 3d ed., Moscow: Nauka.

[33] Bierens De Haan, D. 1867. *Nouvelles Tables, D'Integrales Definies*. New York and London: Hafner Publishing Co.

[34] Pykhteev, G.N. 1980. *Exact Methods of Calculation for the Cochy Type Integrals*. Novosibirsk: Nauka.

[35] Kuznetsov, E.A. 1982. Two-Dimensional Contact Problem with an Additional Load Applied Outside a Stamp Taken into Account. *Soviet Appl. Mech. (English translation of Prikladnaya Mechanika)* 18, No. 5:462-468.

[36] Kostreva, M.M. 1984. Elastohydrodynamic Lubrication: A Nonlinear Complementarity Problem. *Intern. J. Numerical Methods in Fluids* 4:377-397.

[37] Airapetov, E.L., Kudish, I.I., and Panovko, M.Ya. 1992. Numerical Solution of Heavily Loaded Elastohydrodynamic Contact. *Soviet J. Fric. and Wear* 13, No. 6:1-7.

[38] Belotserkovsky, S.M. and Lifanov, I.K. 2000. *Method of Discrete Vortices*. Boca Raton: CRC Press.

[39] Bellman, R.E. and Kalaba, R.E. 1965. *Quasilinearization and Nonlinear Boundary – Value Problems*. New York: Elsevier.

[40] Kaneta M., Nishikawa H., Kameishi K., Sakai T, and Ohno N. 1992. Effects of Elastic Moduli of Contact Surfaces in Elastohydrodynamic Lubrication. *ASME J. Tribology* 114:75-80.

[41] Kaneta M., Nishikawa H., Kanada T., and Matsuda K. 1996. Abnormal Phenomena Appearing in EHL Contacts. *ASME J. Tribology.* 118:886-892.

[42] Cermak J. 1998. Abnormal Phenomena Appearing in EHL Contacts. Discussion on a previously published paper. *ASME J. Tribology* 120:143-144.

[43] Wang, J., Yang, P., Kaneta, M., and Nishikawa, H. 2003. On the Surface Dimple Phenomena in Elliptical TEHL Contacts with Arbitrary Entrainment. *ASME J. Tribology* 125, No. 1:102-109.

[44] Kaneta, M., Shigeta, T., and Young, P. 2006. Film Pressure Distributions in Point Contacts Predicted by Thermal EHL Analysis. *Tribology Intern.* 39:812-819.

[45] Kudish, I.I. 2000. Formulation and Analysis of EHL Problems for Soft Materials. *ASME J. Tribology* 122, No. 3:705-710.

[46] Galin, L.A. 1980. *Contact Problems in the Theory of Elasticity and Viscoelasticity*. Moscow: Nauka.

[47] Petrusevich A.I. 1951. Fundamental Conclusions from the Contact-Hydrodynamic Theory of Lubrication. *Izv. Acad. Nauk, SSSR (OTN)*, 2, 209.

[48] Alexandrov, V.M. and Kudish, I.I. 1980. Problem of Elastohydrodynamic Theory of Lubrication for a Viscous Fluid with Complex Rheology. *J. Mech. Solids* 15, No. 4:57-68.

[49] Kudish, I.I. 1981. On Solution of some Contact and Elastohydrodynamic Problems. *3rd Intern. Tribology Congress EUROTRIB 81, Warsaw, Tribological Processes in Contact Areas of Lubricated Solid Bodies* 2:251-271.

[50] Kudish, I.I. 1983. On Analysis of Plane Contact Problems with a Viscoplastic Lubrication. *Soviet J. Fric. and Wear* 4, No. 3:50-56.

[51] Myllerup, C.M. and Hamrock, B.J. 1994. Perturbation Approach to Hydrodynamic Lubrication Theory. *ASME J. Tribology* 116, No. 1:110-118.

[52] Kudish, I.I. 1979. Elastohydrodynamic Problems for Rough Bodies with non-Newtonian Lubrication. *Dopovidi Akademii Nauk Ukrains'koi RSR*, Seriya A, No. 11:915-920.

[53] Bair, S. 2007 *High − Pressure Rheology for Quantitative Elasto − hydrodynamics*. Amsterdam: Elsevier Science.

[54] Hamrock, B.J. and Dowson, D. 1977. Isothermal Elastohydrodynamic Lubrication of Point Contacts. Part III-Fully Flooded Results. *J. Lubr. Techn.* 99, No. 2:264-276.

[55] Jones, W.R. 1995. Properties of Perfluoropolyethers for Space Applications. *STLE Tribology Trans.* 38, No. 3:557-564.

[56] Bair, S., Vergne, P., and Marchetti, M. 2002. The Effect of Shear-Thinning on Film Thickness for Space Lubricants. *STLE Tribology Trans.* 45, No. 3:330-333.

[57] Dyson, A. and Wilson, A. R. 1965-6. Film Thicknesses in Elastohydrodynamic Lubrication by Silicone Fluids. *Proc. Instn. Mech. Engrs* 180:97-112.

[58] Chapkov, A. D., Bair, S., Cann, P., Lubrecht, A. A. 2007. Film Thickness in Point Contacts under Generalized Newtonian EHL Conditions: Numerical and Experimental Analysis. *Tribology Intern.* 40:1474-1478.

[59] Liu, Y., Wang, Q. J., Bair, S. and Vergne, P. 2007. A Quantitative Solution for the Full Shear-Thinning EHL Point Contact Problem Including Traction. *Tribology Letters* 28, No. 2:171-181.

[60] Kudish, I.I., Kumar, P., Khonsari, M.M., and Bair. S. 2008. Scale Effects in Generalized Newtonian Elastohydrodynamic Films. *ASME J. Tribology*, 130, No. 4:041504-1 - 041504-8.

[61] Jang, J.Y., Khonsari, M.M., and Bair, S. 2007. On the Elastohydrodynamic Analysis of Shear-thinning Fluids. *Proc. Royal Soc.* 463:3271-3290.

[62] Dowson, D. and Toyoda, S. 1979. A Central Film Thickness Formula for Elastohydrodynamic Line Contacts. In Proc. of Leeds-Lyon Symp. on Tribology, 5, September 19-22, 1978, Eds. D. Dowson et al., 60-65.

[63] Hamrock, B.J. and Dowson, D. 1981. *Ball Bearing Lubrication — The Elastohydrodynamics of Elliptical Contacts* New York: Wiley-Interscince.

[64] Carslaw, H.S. and Jaeger, J.C. 1959. *Conduction of Heat in Solids.* 2nd ed. Oxford: Clarendon Press.

[65] *Handbook on Mathematical Functions with Formulas, Graphs and Mathematical Tables*, Eds. Abramowitz, M. and Stegun, I.A., National Bureau of Standards, 55, 1964.

[66] Barber, J. 1984. Thermoelastic Displacements and Stresses due to a Heat Source Moving Over the Surface of a Half-Plane. *ASME J. Appl. Mech.* 51:636-640.

[67] De Bruijn, N.G. 1958. *Asymptotic Methods in Analysis.* Groningen: North-Holland Publishing Co.

11

Lubrication by Greases

11.1 Introduction

Grease is a semi-solid, high viscosity type of lubricant that is well-suited for use in bearings, couplings and open gears where shock loads, high temperatures and/or good adhesion to bearing surfaces are important performance features. Consider the problem of journal bearing lubrication in Fig. 11.1. The outer cylinder (journal) is fixed and is separated from the center rotating shaft by a layer of grease. When the shaft is at rest or turning at low speeds and/or high loads, metal-to-metal surface contact can occur. Wear protection under these conditions can be provided by lubricant decomposition products or surface-active additives which form thin, soft tribo-films which retard metal-to-metal adhesion and reduce friction. As the shaft begins to rotate at higher speeds, it climbs the journal surface in a direction opposite to the direction of rotation. Layers of grease cling to the journal and rotating shaft surfaces, the former remaining stationary and the latter moving in concert with the shaft. Additional grease is carried into the contact zone, and the system enters the hydrodynamic lubrication regime.

The choice between oil and grease lubrication depends upon the ratio of journal speed to viscosity. In general, contacts moving at relatively low relative velocity have higher viscosity requirements than surfaces moving at high velocity. Bearings designated for low speed operation usually are designed with relatively large clearance between the shaft and journal housing to facilitate introduction of high viscosity grease lubricants. High-speed bearings with small clearance are lubricated with lower viscosity oils.

Other bearings commonly lubricated by grease include rolling-element bearings such as ball bearings, cylindrical roller bearings, tapered roller bearings, spherical barrel-shaped roller bearings, and needle bearings. In these types of bearings, extremely high local pressures are formed between the relatively small rotating rolling elements and their raceways (support housings). We learned in Chapter 4 that lubricant viscosity increases rapidly at high pressures that enable the lubricating film to withstand high contact stresses while preventing contact between the rolling surfaces.

Greases are complex formulations consisting of base fluids, thickeners, structural components, and additives designed to meet specific application re-

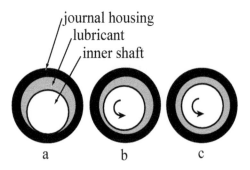

FIGURE 11.1

Development of a hydrodynamic film with increased shaft speed in a plain journal bearing: (a) at rest, (b) low rotational velocity, (c) high rotational velocity.

quirements. Typical base fluids include mineral and synthetic oils. Thickeners and structural components impart shear-thinning non-Newtonian rheological properties to the grease (see Chapter 4, Subsection 4.2.3). At rest, the lubricant is a semi-solid with extremely high apparent viscosity. As shear rate increases, the apparent viscosity falls. Thus, within the gap between a stationary journal and rotating bearing shaft, the lubricating film consists of layers of high viscosity semi-solid at the journal surface and progressively lower viscosity layers closer to the shaft surface.

Thickening agents are prepared in the presence of a base fluid by reacting a fatty acid or ester (from either animal or vegetable sources) with alkali or alkaline earth metal oxides or hydroxides. This so-called saponification process is conducted under controlled conditions of heat, pressure, and agitation. Greases are often referred to by the type of alkali or alkaline earth metal in its thickening system: sodium, calcium, lithium, aluminum, etc. The saponification process produces a three-dimensional microscopic soap fiber structure suspended in the base fluid that serves as an internal scaffolding and defines the rheology and consistency of the grease. Rheology is affected by a number of thickener-related parameters such as soap type and amount, fatty acid/ester chain length, degree of hydrocarbon branching and unsaturation (presence of double bonds) of the fatty acid/ester, presence of other polar substances (such as water), and particle size.

Alkaline Complex Greases are formulated to withstand higher operating temperatures. They are prepared by adding a lower molecular weight organic acid, also called a complexing agent, to the thickening agent mixture. For example, co-reacting 12-hydroxystearic acid (fatty acid) and azelaic acid (complexing agent) with lithium hydroxide (alkaline hydroxide) produces a more intricate lattice structure than that of a simple lithium soap.

TABLE 11.1

Applications of simple and complex greases. Dropping point - the temperature at which grease becomes soft enough to form a drop and fall, ASTM D566 and D2265; EP - Extreme Pressure properties (wear resistance under boundary lubrication); Water resistance - the ability of a grease to resist the adverse effects of water, ASTM D1264, D4049 or D1743.

Thickener Type	Grease Characteristics	Applications
Aluminum	Smooth, gel-like appearance Low dropping point Excellent water resistance Softening/hardening tendencies Highly shear thinning	Low-speed bearings Wet applications
Sodium	Rough, fibrous appearance Moderately high dropping point Poor water resistance Good adhesive (cohesive) properties	Older industrial equipment with frequent re-lubrication Rolling-element bearings
Calcium	Smooth, buttery appearance Low dropping point Good water resistance	Bearings in wet applications Railroad rail lubricants
Lithium	Smooth, buttery to slightly stringy appearance High dropping point Resistant to softening and leakage Moderate water resistance	Automotive chassis and wheel bearings General industrial grease Thread lubricants for oil drilling
Calcium Complex	Smooth, buttery appearance Dropping points above $500°F$ Good water resistance Inherent EP/load-carrying capability	High-temperature industrial and automotive bearings
Aluminum Complex	Smooth, slight gel-like appearance Dropping points below $500°F$ Good water resistance Resistant to softening Shorter life at high temperature	Steel mill roll neck, rolling, and plain bearings
Lithium Complex	Smooth, buttery appearance Dropping points above $500°F$ Resistant to softening and leakage Moderate water resistance	Automotive wheel bearings High-temperature industrial service incl. various rolling-element applications

Other chemical additives are formulated into grease to prevent oxidation, combat corrosion, and reduce friction and wear. Another set of chemically inert ingredients are added to modify viscosity, consistency, or tolerance to water. Examples include viscosity modifiers, pour point depressants, antifoam agents, emulsifiers, and demulsifiers.

Each grease application has its own set of requirements, and each type of grease has its own set of physical and performance properties. Selecting the right grease for a specific end-use is a matter of art, experience, and testing. A summary of industrial and automotive application areas for several families of grease may be found in Table 11.1.

In this chapter we will formulate and study isothermal and non-isothermal problems for EHL contacts in which grease was used as a lubricating medium. We will provide a general qualitative analysis of these problems and determine the conditions under which the problems for grease lubricated contacts can be reduced to the corresponding problems for contact lubricated by fluid lubricants. We will consider an analytical solution of a problem for moving rigid solids separated by grease under certain conditions as well as the asymptotic approach to isothermal and non-isothermal heavily loaded EHL contacts with greases.

11.2   Formulation of the Elastohydrodynamic Lubrication Problem for Greases (Generalized Bingham-Shvedov Viscoplastic Lubricants)

Let us formulate the basic equations pertaining to grease lubricated contacts under steady conditions. We will consider a plane non-isothermal problem about a slow motion of two infinite parallel cylinders separated by a continuous layer of incompressible grease. As in the preceding chapter, we will assume that the cylinders of radii R_1 and R_2 are made of elastic materials with Young's moduli E_i and Poisson's ratios ν_i, $i = 1, 2$. The cylinders' surfaces move with linear velocities u_1 and u_2. A normal force P presses one cylinder into another. Additionally, we will assume that the film thickness is much smaller than the size of the contact region, which, in turn, is much smaller than the cylinders' radii R_1 and R_2. The inertial forces are much smaller than the viscous ones, the heat is generated because of grease viscosity and applied shear stresses while the heat flux is mostly directed across the film thickness. Then in the coordinate system in which the x-axis is directed along the lubrication layer perpendicular to cylinders' axes and the z-axis is directed through cylinders' centers the simplified Navier-Stokes and energy equations of grease motion

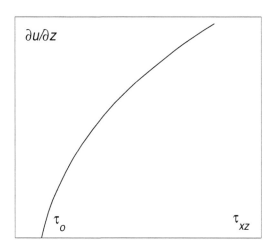

FIGURE 11.2
The general view of a rheological relationship for grease.

take the form (see equations (10.10) and (10.11))

$$\frac{\partial \tau_{xz}}{\partial z} = \frac{\partial p}{\partial x}, \ \frac{\partial p}{\partial z} = 0, \ \frac{\partial u}{\partial x} + \frac{\partial w}{\partial z} = 0, \tag{11.1}$$

$$\frac{\partial}{\partial z}\left(\lambda \frac{\partial T}{\partial z}\right) = -\tau_{xz}\frac{\partial u}{\partial z}, \tag{11.2}$$

where $p(x)$ is pressure, $\tau_{xz}(x, z)$ is the component of the shear stress in grease, $u(x, z)$ and $w(x, z)$ are the components of grease velocity along the lubrication film and across it, $T(x, z)$ is the grease temperature, $\lambda(p, T)$ is the grease heat conductivity coefficient.

Taking into account the above assumptions, the rheological relationships for greases can be represented as follows

$$\tau_{xz} = \tau_0 sign(\tau_{xz}) + \Phi(\mu \frac{\partial u}{\partial z}) \ if \ |\tau_{xz}| \geq \tau_0, \tag{11.3}$$

$$\frac{\partial u}{\partial z} = 0 \ if \ |\tau_{xz}| = 0, \tag{11.4}$$

where τ_0 is the threshold shear (yield) stress beyond which grease flows like a fluid lubricant, $\tau_0 = \tau_0(p, T) \geq 0$, μ is the grease viscosity, $\mu = \mu(p, T)$, Φ is a monotonically increasing sufficiently smooth rheological function, $\Phi(0) = 0$. The general view of such a rheological relationship is presented in Fig. 11.2.

Condition (11.4) describes the so-called core in the grease flow; i.e., it describes the region in the flow in which relative displacement of grease layers is absent. This region of the flow behaves like a rigid solid. Obviously, for

$\tau_0 = 0$, there are no cores in grease flow and the rheological model represented by equations (11.3) and (11.4) gets reduced to the rheological model of generalized non-Newtonian fluid described by equations (10.382).

As an example we can consider the Bingham-Shvedov model of grease [1, 2]

$$\tau_{xz} = \tau_0 sign(\tau_{xz}) + \mu\frac{\partial u}{\partial z} \ if \ \ |\ \tau_{xz}\ | \geq \tau_0; \ \ \frac{\partial u}{\partial z} = 0 \ if \ \ |\ \tau_{xz}\ | = 0. \qquad (11.5)$$

Let us derive the equations governing pressure $p(x)$ and temperature $T(x, z)$ [3, 4]. Integrating the first equation in (11.1), we find

$$\tau_{xz} = f + z\frac{dp}{dx}, \qquad (11.6)$$

where $f = f(x)$ is the sliding frictional stress. Moreover, integrating the continuity equation (last equation in (11.1)) and taking into account the no-slip conditions on solid surfaces $z = \pm\frac{h}{2}$ (see (10.14))

$$u(x, -\tfrac{h}{2}) = u_1, \ u(x, \tfrac{h}{2}) = u_2,$$

$$w(x, (-1)^j\tfrac{h}{2}) = (-1)^j\tfrac{u_i}{2}\tfrac{dh}{dx}, \ j = 1, 2, \qquad (11.7)$$

we obtain the generalized Reynolds equation in the form

$$\frac{d}{dx}\int\limits_{-h/2}^{h/2} u(x, z)dz = 0. \qquad (11.8)$$

To complete the problem formulation, we need to add to equations (11.3), (11.4), (11.6), and (11.8) the conditions of continuity of grease velocity at core boundaries in the flow. From monotonicity of function Φ and equations (11.3) and (11.6) follows that for any fixed x the derivative $\frac{\partial u}{\partial z}$ is equal to zero at no more than two values of z within the lubrication layer $|\ z\ | \leq \frac{h}{2}$. Therefore, at any x in the flow may be encountered at most one core region and all possible types of grease flow in any cross section of the flow are presented in Fig. 11.3 (cores are marked by 1 while flow with shear is marked by 2). We will call these types of flow configurations as flow of types I, IIa, IIb, and III, respectively. The no-slip conditions at the solid surfaces and the continuity conditions at the core boundaries need to be imposed on the grease velocity $u(x, z)$ for each of these flow types. For flow of type I (the core is located in the central region of the flow), we have

$$u(x, -\tfrac{h}{2}) = u_1, \ u(x, \tfrac{h}{2}) = u_2, \ u(x, z_1) = u(x, z_2),$$

$$-\tfrac{h}{2} < z_1 < z_2 < \tfrac{h}{2}, \qquad (11.9)$$

for flow of type IIa (the core is adjacent to the upper solid), we have

$$u(x, -\tfrac{h}{2}) = u_1, \ u(x, z_1) = u_2, \ -\tfrac{h}{2} < z_1 < \tfrac{h}{2} \leq z_2, \qquad (11.10)$$

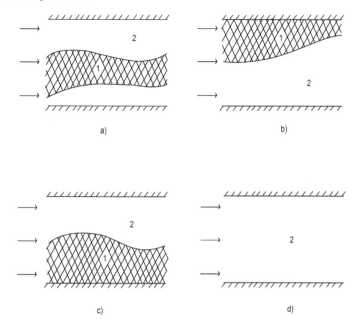

FIGURE 11.3
All possible grease flow configurations in any cross section of the flow: (a) type
I, (b) type IIa, (c) type IIb, and (d) type III (1 marks grease core, 2 marks
shear flow) (after Kudish [3]). Reprinted with permission from Allerton Press.

for flow of type IIb (the core is adjacent to the lower solid), we have

$$u(x, z_2) = u_1, \ u(x, \tfrac{h}{2}) = u_2, \ z_1 \leq -\tfrac{h}{2} < z_2 < \tfrac{h}{2}, \tag{11.11}$$

for flow of type III (the is no core in the flow), we obtain

$$u(x, -\tfrac{h}{2}) = u_1, \ u(x, \tfrac{h}{2}) = u_2,$$

$$z_1 \leq z_2 \leq -\tfrac{h}{2} \text{ or } -\tfrac{h}{2} < z_1 = z_2 < \tfrac{h}{2} \text{ or } \tfrac{h}{2} \leq z_1 \leq z_2, \tag{11.12}$$

where $z_1 = z_1(x)$ and $z_2 = z_2(x)$ are the lower and upper core boundaries,
which are determined by equations (see (11.4))

$$\frac{\partial u(x, z_i)}{\partial z} = 0, \ i = 1, 2. \tag{11.13}$$

Now, let us determine the expressions for functions $u(x, z)$ and $f(x)$. The
grease flow between the solid surfaces we will treat as just a fragment of the
flow in the entire xz-plane. For certainty we will assume that $u_2 - u_1 \geq 0$.

Let us consider all possible types of grease flow in an arbitrary lubrication
film cross–section x and the conditions under which they are realized. First, we

will consider the case when in a chosen cross–section x type I flow is realized. It can be shown that on different sides of the core in this cross section the values of $\frac{\partial u}{\partial z}$ has opposite signs. Let us analyze the case when $\frac{\partial u}{\partial z} > 0$ for $z < z_1$ and $\frac{\partial u}{\partial z} < 0$ for $z > z_2$. Taking into account that $\tau_{xz}\frac{\partial u}{\partial z} \geq 0$ from equations (11.3) and (11.6), we obtain

$$\frac{\partial u}{\partial z} = \frac{1}{\mu}F(f - \tau_0 + z\frac{dp}{dx}), \ z \leq z_1,$$

$$\frac{\partial u}{\partial z} = \frac{1}{\mu}F(f + \tau_0 + z\frac{dp}{dx}), \ z \geq z_2,$$

(11.14)

where F is the rheological function of the liquified grease which is a monotonically increasing function being the inverse to function Φ, $F(0) = 0$.

The core boundaries z_1 and z_2 can be found from equations (11.13) and (11.14)

$$z_1 = \frac{\tau_0 - f}{\frac{\partial p}{\partial x}}, \ z_2 = -\frac{\tau_0 + f}{\frac{\partial p}{\partial x}}.$$

(11.15)

Moreover, using boundary conditions (11.9) and equations (11.14), we obtain the expressions for the grease velocity $u(x, z)$:

$$u(x, z) = u_1 + \int\limits_{-h/2}^{z} \frac{1}{\mu}F(f - \tau_0 + \zeta\frac{dp}{dx})d\zeta, \ z \leq z_1;$$

$$u(x, z) = u(x, z_1), \ z_1 \leq z \leq z_2,$$

(11.16)

$$u(x, z) = u_2 - \int\limits_{z}^{h/2} \frac{1}{\mu}F(f + \tau_0 + \zeta\frac{dp}{dx})d\zeta, \ z \geq z_2.$$

From the third condition in (11.9) and equations (11.15) and (11.16), we derive the equation for $f(x)$:

$$\int\limits_{-h/2}^{z_1} \frac{1}{\mu}F(f - \tau_0 + \zeta\frac{dp}{dx})d\zeta + \int\limits_{z_2}^{h/2} \frac{1}{\mu}F(f + \tau_0 + \zeta\frac{dp}{dx})d\zeta = u_2 - u_1.$$

(11.17)

Employing the expressions for z_1 and z_2 from (11.15), we obtain $\frac{dp}{dx} < 0$. Finally, it follows from (11.9) and (11.15) that the flow of type I can be realized if the following inequality is satisfied

$$\tau_0 + \frac{h}{2}\frac{dp}{dx} < f < -\tau_0 - \frac{h}{2}\frac{dp}{dx}.$$

(11.18)

Now, let us derive the Reynolds and the heat transfer equations. Integrating equations (11.16) for $u(x, z)$ with respect to z over the corresponding intervals and using the generalized Reynolds equation (11.8), we obtain

$$\frac{d}{dx}\left\{ \int\limits_{-h/2}^{z_1} dz \int\limits_{-h/2}^{z} \frac{1}{\mu}F(f - \tau_0 + \zeta\frac{dp}{dx})d\zeta \right.$$

$$- \int\limits_{z_2}^{h/2} dz \int\limits_{z}^{h/2} \tfrac{1}{\mu} F(f + \tau_0 + \zeta \tfrac{dp}{dx}) d\zeta$$

$$\left. + u(x, z_1)(z_2 - z_1) + u_1 z_1 - u_2 z_2 + \tfrac{u_1 + u_2}{2} h \right\} = 0. \tag{11.19}$$

By changing the order of integration in integrals involved in (11.19) and using equation (11.17), we reduce equation (11.19) to the final form

$$\tfrac{d}{dx} \left\{ \int\limits_{-h/2}^{z_1} \tfrac{z}{\mu} F(f - \tau_0 + z \tfrac{dp}{dx}) dz \right.$$

$$\left. + \int\limits_{z_2}^{h/2} \tfrac{z}{\mu} F(f + \tau_0 + z \tfrac{dp}{dx}) dz - \tfrac{u_1 + u_2}{2} h \right\} = 0. \tag{11.20}$$

It is clear that within the core $\tfrac{\partial u}{\partial z} = 0$ and from equation (11.2) follows that within the core heat is not generated and the heat flux is constant. Therefore, from equations (11.2), (11.5), (11.6), and (11.16) for the energy equation, we obtain

$$\tfrac{\partial}{\partial z}(\lambda \tfrac{\partial T}{\partial z}) = -\tfrac{1}{\mu} F[f + \tau_0 sign(z - z_1) + z \tfrac{dp}{dx}](f$$

$$+ z \tfrac{dp}{dx}) \theta[(z - z_1)(z - z_2)], \tag{11.21}$$

where $\theta(x)$ is the Heavyside function such that $\theta(x) = 1$ for $x \geq 0$ and $\theta(x) = 0$ for $x \leq 0$.

In a similar fashion we can analyze the case when $\tfrac{\partial u}{\partial z} < 0$ for $z < z_1$ and $\tfrac{\partial u}{\partial z} > 0$ for $z > z_2$. Under these conditions the equations for z_1, z_2, f, and $u(x, z)$ as well as the generalized Reynolds equation coincide with equations (11.14), (11.17), (11.16), and (11.20) if τ_0 is replaced by $-\tau_0$. It can be shown that in this case $\tfrac{dp}{dx} > 0$. Moreover, the condition for realization of the above case coincide with (11.18) in which $\tfrac{dp}{dx}$ is replaced by $-\tfrac{dp}{dx}$.

Now, let us consider the case when in a chosen cross–section x type IIa flow is realized. It can be shown that $\tfrac{\partial u}{\partial z} > 0$ for $z < z_1$. Then for z_1 and z_2 we obtain formulas (11.15) while for grease velocity $u(x, z)$ we get

$$u(x, z) = u_1 + \int\limits_{-h/2}^{z} \tfrac{1}{\mu} F(f - \tau_0 + \zeta \tfrac{dp}{dx}) d\zeta, \quad -\tfrac{h}{2} \leq z \leq z_1,$$

$$u(x, z) = u_1 2, \quad z_1 \leq z \leq \tfrac{h}{2}. \tag{11.22}$$

Then, from equations (11.10), (11.14), and (11.22), we obtain the equation for $f(x)$:

$$\int\limits_{-h/2}^{z_1} \tfrac{1}{\mu} F(f - \tau_0 + \zeta \tfrac{dp}{dx}) d\zeta = u_2 - u_1. \tag{11.23}$$

From the inequality $z_1 < z_2$ and equations (11.15), we obtain that $\frac{dp}{dx} < 0$. Equations (11.15) together with inequalities in (11.10) lead to the requirement for the realization of type IIa grease flow in the form

$$\tau_0 + \frac{h}{2}\frac{dp}{dx}) < f < \tau_0 - \frac{h}{2}\frac{dp}{dx}, \quad -\tau_0 - \frac{h}{2}\frac{dp}{dx} \leq f. \tag{11.24}$$

Using the generalized Reynolds equation (11.8) and integrating the expressions for the grease velocity $u(x,z)$ from (11.22) with respect to z from $-\frac{h}{2}$ to $\frac{h}{2}$, we derive the equation

$$\frac{d}{dx}\left\{ \int\limits_{-h/2}^{z_1} dz \int\limits_{-h/2}^{z} \frac{1}{\mu}F(f - \tau_0 + \zeta\frac{dp}{dx})d\zeta + (u_1 - u_2)z_1 \right.$$

$$\left. + \frac{u_1+u_2}{2}h \right\} = 0. \tag{11.25}$$

By changing the order of integration in integrals involved in (11.25) and using equation (11.23), we reduce equation (11.25) to the final form

$$\frac{d}{dx}\left\{ \int\limits_{-h/2}^{z_1} \frac{z}{\mu}F(f - \tau_0 + z\frac{dp}{dx})dz - \frac{u_1+u_2}{2}h \right\} = 0. \tag{11.26}$$

Grease flow of type IIb can be analyzed in a similar fashion. In this case $\frac{\partial u}{\partial z} > 0$ for $z > z_2$. Formulas (11.15) are still valid for z_1 and z_2 if τ_0 in them is replaced by $-\tau_0$. It can be shown that in this case $\frac{dp}{dx} > 0$. For the grease velocity $u(x,z)$, we find

$$u(x,z) = u_1, \quad -\frac{h}{2} \leq z \leq z_2,$$

$$u(x,z) = u_1 + \int\limits_{z}^{z_2} \frac{1}{\mu}F(f - \tau_0 + \zeta\frac{dp}{dx})d\zeta, \quad z_2 \leq z \leq \frac{h}{2}. \tag{11.27}$$

Using equations (11.11) and (11.22), we obtain the equation for $f(x)$:

$$\int\limits_{z_2}^{h/2} \frac{1}{\mu}F(f - \tau_0 + \zeta\frac{dp}{dx})d\zeta = u_2 - u_1. \tag{11.28}$$

Besides that, the conditions for the realization of type IIb grease flow follow from conditions (11.24) if $\frac{dp}{dx}$ in them is replaced by $-\frac{dp}{dx}$.

In the fashion described above, we obtain the generalized Reynolds equation

$$\frac{d}{dx}\left\{ \int\limits_{z_2}^{h/2} dz \int\limits_{z_2}^{z} \frac{1}{\mu}F(f - \tau_0 + \zeta\frac{dp}{dx})d\zeta + u_1h \right\} = 0, \tag{11.29}$$

which by changing the order of integration in the integral can be rewritten in the final form

$$\frac{d}{dx}\left\{ \int\limits_{z_2}^{h/2} \frac{z}{\mu}F(f - \tau_0 + z\frac{dp}{dx})dz - \frac{u_1+u_2}{2}h \right\} = 0. \tag{11.30}$$

For types IIa and IIb grease flows, the energy equation is reduced to

$$\tfrac{\partial}{\partial z}(\lambda\tfrac{\partial T}{\partial z}) = -\tfrac{1}{\mu}F[f - \tau_0 sign(z - z_2) + z\tfrac{dp}{dx}](f + z\tfrac{dp}{dx})\theta(z - z_2). \qquad (11.31)$$

Finally, let us consider the case of type III grease flow, i.e., the flow without a core. Then for $u_2 - u_1 \geq 0$ we have $\tfrac{\partial u}{\partial z} \geq 0$ for all $\mid z \mid \leq \tfrac{h}{2}$. The case of $-\tfrac{h}{2} < z_1 = z_2 < \tfrac{h}{2}$ can be realized when either $\tau_0 = 0$ or $\tau_0 > 0$ and $\tfrac{dp}{dx} = 0$. This follows from the fact that z_1 and z_2 satisfy (11.14) for $\tfrac{dp}{dx} > 0$ and the same formulas in which τ_0 is replaced by $-\tau_0$ for $\tfrac{dp}{dx} < 0$. Therefore, for the first and third cases in (11.12) for $u(x, z)$ and $f(x)$, we find

$$u(x, z) = u_1 + \int\limits_{-h/2}^{z} \tfrac{1}{\mu}F(f - \tau_0 + \zeta\tfrac{dp}{dx})d\zeta, \; \mid z \mid \leq \tfrac{h}{2}, \qquad (11.32)$$

$$\int\limits_{-h/2}^{h/2} \tfrac{1}{\mu}F(f - \tau_0 + \zeta\tfrac{dp}{dx})d\zeta = u_2 - u_1. \qquad (11.33)$$

The condition for the realization of this type of flow is

$$\mid \tfrac{dp}{dx} \mid \leq \tfrac{2}{h}(f - \tau_0). \qquad (11.34)$$

The generalized Reynolds equation will assume the form

$$\tfrac{d}{dx}\left\{ \int\limits_{-h/2}^{h/2} dz \int\limits_{-h/2}^{z} \tfrac{1}{\mu}F(f - \tau_0 + \zeta\tfrac{dp}{dx})d\zeta + u_1 h \right\} = 0, \qquad (11.35)$$

which after changing the order of integration in the integral can be rewritten in the final form

$$\tfrac{d}{dx}\left\{ \int\limits_{-h/2}^{h/2} \tfrac{z}{\mu}F(f - \tau_0 + z\tfrac{dp}{dx})dz - \tfrac{u_1+u_2}{2}h \right\} = 0. \qquad (11.36)$$

The energy equation for grease flow of type III has the form

$$\tfrac{\partial}{\partial z}(\lambda\tfrac{\partial T}{\partial z}) = -\tfrac{1}{\mu}F[f - \tau_0 + z\tfrac{dp}{dx}](f + z\tfrac{dp}{dx}). \qquad (11.37)$$

As it was done before (see the preceding chapter) to determine $h(x)$, we need to add to the derived system of equations and inequalities equation (10.19)

$$h(x) = h_e + \tfrac{x^2 - x_e^2}{2R'} + \tfrac{2}{\pi E'} \int\limits_{x_i}^{x_e} p(t) \ln \tfrac{x_e - t}{|x - t|} dt, \qquad (11.38)$$

and the balance condition (10.21):

$$\int\limits_{x_i}^{x_e} p(t)dt = P, \qquad (11.39)$$

where h_e is the lubrication film thickness at the exit point of the contact x_e, i.e., $h_e = h(x_e)$, x_i is the inlet point of the contact, R' and E' are the effective radius and elasticity modulus, $1/R' = 1/R_1 \pm 1/R_2$ (signs $+$ and $-$ are chosen in accordance with cylinders' curvatures) and $1/E' = 1/E_1' + 1/E_2'$, $E_j' = E_j/(1 - \nu_j^2)$, $(j = 1, 2)$, P is the normal force applied to the cylinders.

We also need to add some boundary conditions. Within the lubricated contact, pressure is much higher than the ambient pressure. Moreover, to prevent the lubricant from cavitation at the exit of the contact, we need to require the pressure gradient to be zero (see the preceding chapter). Therefore, we impose the following boundary conditions on pressure (see conditions (10.20)):

$$p(x_i) = p(x_e) = \frac{dp(x_e)}{dx} = 0, \tag{11.40}$$

where the inlet coordinate x_i is considered to be known. The conditions for the temperature at the solid surfaces have the form (see (10.58))

$$T(x, -\tfrac{h}{2}) = T_{w1}(x), \; T(x, \tfrac{h}{2}) = T_{w2}(x), \tag{11.41}$$

where $T_{w1}(x)$ and $T_{w2}(x)$ are the temperatures of the lower and upper solid surfaces, respectively.

However, to solve the EHL problem for grease lubricated contact, some additional conditions are required. In particular, to complete the problem formulation at every point x_c of flow type change and at the core boundaries $z = z_1$ and $z = z_2$, it is necessary to impose some continuity conditions. These conditions follow from the requirement that the functions of pressure, gap, temperature, shear stress, and flow velocity are continuous in the entire contact region. Therefore, at the points with coordinates x_c at which a change of the flow type occurs, we have these conditions in the form

$$[p(x_c)] = [\tfrac{dp(x_c)}{dx}] = 0, \; [T(x_c, z)] = 0, \; \mid z \mid \leq \tfrac{h}{2}. \tag{11.42}$$

The remaining continuity conditions on core boundaries (i.e., on the core/flow interfaces) are obtained from the requirement of continuity of temperature and heat flux and they have the form

$$[T(x, z_j)] = [\tfrac{\partial T(x, z_j)}{\partial z}] = 0, \; j = 1, 2. \tag{11.43}$$

In equations (11.42) and (11.43), the notation $[f(x_0)]$ means the jump of function $f(x)$ at $x = x_0$, i.e., $[f(x_0)] = f(x_0 + 0) - f(x_0 - 0)$.

Therefore, for the known constants x_i, u_1, u_2, R', E', P, and functions F, μ, λ, τ_0, T_{w1}, and T_{w2}, the formulated problem for pressure $p(x)$, gap $h(x)$, frictional sliding stress $f(x)$, grease temperature $T(x, z)$, and two constants: the exit film thickness h_e and exit coordinate x_e is completely formulated. In addition to that, in case of changes of grease flow type, the coordinates of points x_c at which these changes occur are also determined.

The formulated problem is a complex nonlinear problem with a number of free (unknown) boundaries. In different zones of the contact region, different flow types are realized that are described by different equations. Under certain conditions the solution of the problem in this formulation does not exist and requires certain adjustments (see the next section).

After the problem solution is found the friction force F_T, which depends on the sliding F_S and rolling F_R friction forces, can be found from formulas (for comparison see formulas (10.23) valid for $\tau_0 = 0$)

$$F_T = F_S \pm F_R, \quad F_S = \int_{x_i}^{x_e} f(x)dx, \quad F_R = \frac{1}{2} \int_{x_i}^{x_e} h(x)\frac{dp}{dx}dx, \tag{11.44}$$

Finally, let us consider the formulation of the isothermal problem for

$$\frac{\partial \mu}{\partial T} = \frac{\partial \tau_0}{\partial T} = 0, \quad \lambda = \infty, \quad T_{wj} = T_0 = const \ (j = 1, 2). \tag{11.45}$$

In the case of the simplest viscoplastic Binham-Shvedov grease model (11.5), we obtain the following equations for the isothermal EHL problem

$$\frac{d}{dx}\{[-\frac{2}{3\mu}(\tau_0 - \frac{h}{2} \mid \frac{dp}{dx} \mid)^2(2\tau_0 + \frac{h}{2} \mid \frac{dp}{dx} \mid)(\frac{dp}{dx})^{-2}$$

$$+\frac{\mu(u_2-u_1)^2\tau_0}{2}(\tau_0 - \frac{h}{2} \mid \frac{dp}{dx} \mid)^{-2}]s_p - \frac{u_1+u_2}{2}h\} = 0, \quad s_p = sign(\frac{dp}{dx}),$$

$$\tag{11.46}$$

$$f = \frac{\mu(u_2-u_1)}{h}\frac{1}{1-\frac{2\tau_0}{h}(\mid\frac{dp}{dx}\mid)^{-1}},$$

$$\mid \frac{dp}{dx} \mid > \frac{2\tau_0}{h} + \frac{\mu(u_2-u_1)}{h^2} + \frac{1}{h}\sqrt{\frac{\mu(u_2-u_1)}{h}[\frac{\mu(u_2-u_1)}{h} + 4\tau_0]},$$

$$\frac{d}{dx}\{\frac{u_2-u_1}{3}\sqrt{2\mu(u_2 - u_1)}(\mid \frac{dp}{dx} \mid)^{-1}s_p + (\frac{u_1+u_2}{2}s_p + \frac{u_1-u_2}{2})h\} = 0,$$

$$f = \tau_0 - \frac{h}{2} \mid \frac{dp}{dx} \mid + \sqrt{2\mu(u_2 - u_1)} \mid \frac{dp}{dx} \mid,$$

$$\tag{11.47}$$

$$\frac{2\mu(u_2-u_1)}{h^2} < \mid \frac{dp}{dx} \mid \leq \frac{2\tau_0}{h} + \frac{\mu(u_2-u_1)}{h^2}$$

$$+\frac{1}{h}\sqrt{\frac{\mu(u_2-u_1)}{h}[\frac{\mu(u_2-u_1)}{h} + 4\tau_0]},$$

$$\frac{d}{dx}\{\frac{h^3}{12\mu}\frac{dp}{dx} - \frac{u_1+u_2}{2}h\} = 0, \quad f = \tau_0 + \frac{\mu(u_2-u_1)}{h},$$

$$\tag{11.48}$$

$$\mid \frac{dp}{dx} \mid \leq \frac{2\mu(u_2-u_1)}{h^2}, \quad z_1 = -\frac{\tau_0 s_p + f}{\frac{dp}{dx}}, \quad z_2 = \frac{\tau_0 s_p - f}{\frac{dp}{dx}},$$

$$p(x_i) = p(x_e) = \frac{dp(x_e)}{dx} = 0,$$

$$\tag{11.49}$$

$$\int_{x_i}^{x_e} p(t)dt = P,$$

$$[p(x_c)] = [\frac{dp(x_c)}{dx}] = 0, \tag{11.50}$$

$$h(x) = h_e + \frac{x^2-x_e^2}{2R'} + \frac{2}{\pi E'}\int_{x_i}^{x_e} p(t)\ln\frac{x_e-t}{\mid x-t\mid}dt. \tag{11.51}$$

11.3 Some Properties of the EHL Problem Solution for Greases. Qualitative Behavior of the Solution

For $\tau_0 = 0$ the formulated iso- and non-isothermal problems for viscoplastic lubricants are reduced to the EHL problems of the preceding chapter for fluid lubricants. For $\tau_0 \neq 0$ when conditions (11.34) are satisfied equations and solutions of the isothermal EHL problems for viscoplastic and the corresponding fluid lubricants with the same rheological function F differ just by an additive term $\tau_0(p)$ in the expression for the sliding frictional stress $f(x)$ (see equations (11.33) and (11.36)). In the case of thermal EHL problems such a relationship does not hold. This is due to the fact that for conditions satisfying (11.34) function f is involved in equations (11.33), (11.36), and (11.37) as f and $f - \tau_0$.

Moreover, for $\tau_0 \neq 0$, $u_1 = u_2$ in any contact zone where condition (11.34) is not satisfied grease exhibits its viscoplastic properties, which manifest themselves in the presence of a core(s). Cores in a grease flow behave like rigid solids. In certain cases the presence of a core(s) affects the properties of a lubricated contact greatly. Moreover, under certain conditions the formulated problem for grease does not have a solution. In particular, it happens when $u_1 = u_2$ and $\tau_0 > 0$. To alleviate such situations certain adjustments to the problem formulation should be made. They may include taking into account the compressibility of grease and the variations of the surface velocities due to the elastic tangential displacements of the solid surfaces involved in a contact. The latter is usually caused by elevated tangential stresses applied to these surfaces. Such an adjustment can be made in accordance with Section 10.9. A detailed analysis of the flow of viscoplastic medium through pipes and channels with different cross sections is presented in [5, 6].

The analysis of the formulated problem for the Bingham-Shvedov model for $u_1 = u_2 = 0$, $h(x) = h_e = const$, and $\tau_0 > 0$ shows that the nonzero flow of grease through the cross–section $[-\frac{h_e}{2}, \frac{h_e}{2}]$ of the lubrication layer exists only if $| \frac{dp}{dx} |> \frac{2\tau_0}{h}$ (see (11.18)). This condition coincides with a similar one in [6]. Also, the condition which guarantees a nonzero flow of grease through the lubrication film cross section for $u_1 = u_2 > 0$ and $\tau_0 > 0$ is identical to the one in [6]. The same condition is valid for the grease model described by relationships (11.3) and (11.4).

Let us consider the question about possible configurations of a grease flow. Taking into account the continuity of core boundaries it can be shown that the case when type III grease flow is followed by type I grease flow is impossible to realize. In fact, the only possibility of realization of such a case is the flow configuration represented in Fig. 11.4. In such a case at the tip of the core we have $z_1 = z_2$. Based on the equations (11.15) for z_1 and z_2 it is clear that it can happen only when $\tau_0 = 0$. That contradicts the fact of core existence.

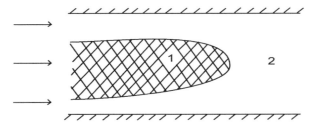

FIGURE 11.4

A grease flow configuration with a center core ending in the middle of the flow (1 marks grease core, 2 marks shear flow) (after Kudish [3]). Reprinted with permission from Allerton Press.

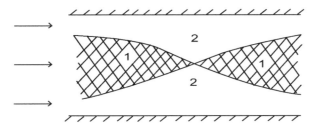

FIGURE 11.5

A grease flow configuration with a center core degenerating into a point in the middle of the flow (1 marks grease core, 2 marks shear flow) (after Kudish [3]). Reprinted with permission from Allerton Press.

This consideration shows that any configuration in which at least at one point a core of type I grease flow degenerates into a point is also impossible (see, for example, Fig. 11.5).

Also, let us show that the grease flow configuration represented in Fig. 11.6 is impossible. These type of configurations are characterized by the existence of two cross sections $x = x_0$ and $x = x_1$ such that for $x_0 < x < x_1$ the flow is of type I while to the left of x_0 and to the right of x_1 the flow is of type IIa and IIb or vice versa. Let us consider three cross sections of the flow. In the aa cross section (see Fig. 11.6), we have $\frac{dp}{dx} < 0$ while in the cc cross section we have $\frac{dp}{dx} > 0$. Assuming that $p(x)$ is a continuous function (which is in agreement with (11.42)), we obtain that between points $x = x_0$ and $x = x_1$ there is a cross section bb in which $\frac{dp}{dx} = 0$ and, therefore, the grease flow does not have a core. However, it contradicts the assumption that for all x from the interval $x_0 < x < x_1$ the flow is of type I, i.e., a grease flow with a central core detached from the solid surfaces.

Obviously, none of the configurations, which include the aforementioned two can be realized.

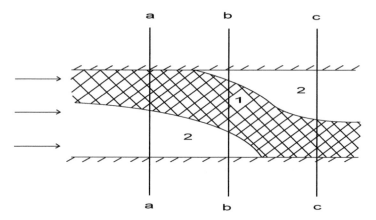

FIGURE 11.6
A grease flow configuration with a continuous core adjacent to the upper and lower contact surfaces (1 marks grease core, 2 marks shear flow) (after Kudish [3]). Reprinted with permission from Allerton Press.

The scaling of the dimensional variables for the regimes of lightly and heavily loaded grease lubricated contacts will be done in accordance with the formulas (10.24), (10.59), (10.60) (where ϑ is a certain parameter that depends on fluid/grease rheology, see Section 10.3) and (10.77), (10.78), (10.58), (10.316), respectively. For both lightly and heavily loaded regimes, the dimensionless variables for τ_0 and z_i are introduced as follows:

$$\tau_0' = \frac{\pi R'}{P}\tau_0, \ \ z_i' = \frac{z_i}{h_e}, \ \ i = 1, 2. \tag{11.52}$$

In further considerations primes at the dimensionless variables are omitted.

11.4   Greases in a Contact of Rigid Solids

In this section we will analyze analytically the simplest isothermal problem for grease lubricated rigid cylinders. The main goal of this study is to get an understanding of the behavior of grease in a contact and to find out the possibility of core existence and location.

11.4.1   Problem Formulation

Let us consider an isothermal problem for grease lubricated contact of two rigid cylinders. We will assume that the grease can be modeled by the Binham-

Shvedov rheological relationships (11.5). In equation (11.51) for the gap function $h(x)$ for rigid cylinders we assume that $E' = \infty$. Then in the dimensionless variables used for lightly loaded contacts (see (11.52) and Section 10.3)) from equations (11.46)-(11.51) we obtain [7]

$$\frac{d}{dx}\{[-\frac{8\gamma^2}{\mu}(\frac{\tau_0}{\gamma} - \frac{h}{2} \mid \frac{dp}{dx} \mid)^2(\frac{2\tau_0}{\gamma} + \frac{h}{2} \mid \frac{dp}{dx} \mid)(\frac{dp}{dx})^{-2}$$

$$+\frac{s_0^2}{24\gamma^2}\mu\tau_0(\frac{\tau_0}{\gamma} - \frac{h}{2} \mid \frac{dp}{dx} \mid)^{-2}]s_p - h\} = 0, \;\; s_p = sign(\frac{dp}{dx}), \tag{11.53}$$

$$f = \frac{s_0^2}{12\gamma^2}\frac{\mu}{h}\{1 - \frac{2\tau_0}{\gamma h}(\mid \frac{dp}{dx} \mid)^{-1}\}^{-1}, \tag{11.54}$$

$$z_1 = -(\frac{\tau_0}{\gamma}s_p + f)(\frac{dp}{dx})^{-1}, \;\; z_2 = (\frac{\tau_0}{\gamma}s_p - f)(\frac{dp}{dx})^{-1}, \tag{11.55}$$

$$\mid \frac{dp}{dx} \mid > \frac{1}{\gamma h}\{2\tau_0 + \frac{s_0}{12\gamma}\frac{\mu}{h} + \sqrt{\frac{s_0}{6\gamma}\frac{\mu}{h}[\frac{s_0}{24\gamma}\frac{\mu}{h} + 2\tau_0]}\}, \tag{11.56}$$

$$\frac{d}{dx}\{-\frac{s_0}{9}\sqrt{\frac{3s_0}{2}}\mu(\mid \frac{dp}{dx} \mid)^{-1}s_p + \gamma(\frac{s_0}{2}s_p - 1)h\} = 0, \tag{11.57}$$

$$f = \frac{\tau_0}{\gamma} - \frac{h}{2} \mid \frac{dp}{dx} \mid + \frac{1}{2\gamma}\sqrt{\frac{2s_0}{3}\mu} \mid \frac{dp}{dx} \mid, \tag{11.58}$$

$$z_1 = (\frac{\tau_0}{\gamma} - f)(\frac{dp}{dx})^{-1}, \;\; z_2 \geq \frac{h}{2} \;\; if \;\; \frac{dp}{dx} < 0,$$

$$z_1 \leq -\frac{h}{2}, \;\; z_2 = (\frac{\tau_0}{\gamma} - f)(\frac{dp}{dx})^{-1} \;\; if \;\; \frac{dp}{dx} > 0, \tag{11.59}$$

$$\frac{s_0}{6\gamma^2}\frac{\mu}{h^2} < \mid \frac{dp}{dx} \mid \leq \frac{1}{\gamma h}\{2\tau_0 + \frac{s_0}{12\gamma}\frac{\mu}{h} + \sqrt{\frac{s_0}{6\gamma}\frac{\mu}{h}[\frac{s_0}{24\gamma}\frac{\mu}{h} + 2\tau_0]}\}, \tag{11.60}$$

$$\frac{d}{dx}\{\frac{\gamma^2h^3}{\mu}\frac{dp}{dx} - h\} = 0, \tag{11.61}$$

$$f = \frac{\tau_0}{\gamma} + \frac{s_0}{12\gamma^2}\frac{\mu}{h}, \tag{11.62}$$

$$\mid z_j \mid \geq \frac{h}{2}, \;\; j = 1, 2, \tag{11.63}$$

$$\mid \frac{dp}{dx} \mid \leq \frac{s_0}{6\gamma^2}\frac{\mu}{h^2}, \tag{11.64}$$

$$p(a) = p(c) = \frac{dp(c)}{dx} = 0, \tag{11.65}$$

$$\int_a^c p(t)dt = \frac{\pi}{2}, \tag{11.66}$$

$$[p(x_{ci})] = [\frac{dp(x_{ci})}{dx}] = 0, \;\; x_{ci} \in X_c, \tag{11.67}$$

$$h = 1 + \frac{x^2 - c^2}{\gamma}, \tag{11.68}$$

where X_c is the set of unknown coordinates x_{ci} of points of grease flow type changes, s_0 is the slide-to-roll ratio, a and c are the dimensionless coordinates of the inlet and exit points, and γ is the dimensionless exit film thickness (see formulas (10.30)).

Therefore, for given functions $\mu(p)$, $\tau_0(p)$, and constants a, s_0, we need to find functions $p(x)$, $h(x)$, $f(x)$, $z_j(x)$ ($j = 1, 2$), and constants c, γ, and x_{ci}. After the solution is found the dimensionless sliding F_S and rolling F_R friction forces are calculated as follows:

$$F_S = \tfrac{2\gamma}{\pi} \int\limits_a^c f(x)dx, \quad F_R = -\tfrac{\gamma}{\pi} \int\limits_a^c p(x)\tfrac{dh}{dx}dx. \tag{11.69}$$

11.4.2 Analysis of Possible Flow Configurations

In spite of the simplification of the problem caused by not taking into account the elastic displacements of the cylinders' surfaces (described by an integral term), the problem is still nonlinear and complex. Let us consider the full range of possible flow configurations in a grease lubricated contact. First, integrating equation (11.61) and taking into account equation (11.68) and boundary conditions (11.65) we obtain

$$\tfrac{dp}{dx} = \tfrac{h-1}{\gamma^3 h^3}\mu. \tag{11.70}$$

From (11.68) and (11.70) it follows that $\tfrac{dp}{dx} = 0$ at two points $x = \pm c$, where at $x = -c$ pressure $p(x)$ reaches its maximum value.

Let us analyze the condition of core absence in the flow given by (11.64). This condition together with equation (11.70) lead to the inequality

$$\mid h - 1 \mid \leq \tfrac{s_0}{6} h. \tag{11.71}$$

From (11.68), (11.70), and (11.71) follows that for $s_0 > 0$ the points with coordinates $x = \pm c$ at which $\tfrac{dp}{dx} = 0$ and $h(x) = 1$ belong to the considered zone of the grease flow. From (11.68) we obtain that

$$h(x) \geq 1 \; for \; x \in [a, -c], \; h(x) \leq 1 \; for \; x \in [-c, c]. \tag{11.72}$$

Now, we consider the number and location of zones in the grease lubricated contact in which relationships (11.57)-(11.60) are valid. To answer this question we need to analyze the solutions of the equation

$$\mid h(x_c) - 1 \mid = \tfrac{s_0}{6} h(x_c), \tag{11.73}$$

i.e., we need to determine the number and the coordinates of the points of grease flow type changes x_c. Using (11.72) and (11.73) we get the expressions for x_{ci}:

$$x_{c1} = -\sqrt{\Delta_1}, \; \Delta_1 = c^2 + \tfrac{\gamma s_0}{6 - s_0} \geq 0, \tag{11.74}$$

$$x_{cj} = (-1)^{j-1}\sqrt{\Delta_2}, \; \Delta_2 = c^2 - \tfrac{\gamma s_0}{6 + s_0} \geq 0, \; j = 2, 3. \tag{11.75}$$

Obviously, under the natural assumptions that $h(0) = 1 - \tfrac{c^2}{\gamma} > 0$ and $\gamma > 0$ inequality $\Delta_1 > 0$ is always valid if $0 < s_0 \leq 6$. Therefore, from formula

(11.74) follows that in the inlet zone of the contact a core is absent as long as $x_{c1} \leq a$ and it is present in the inlet zone if $x_{c1} > a$. Similarly, we obtain that in the central part of the contact a core is absent if $\Delta_2 \leq 0$ and a core is present in this part of the contact if $\Delta_2 > 0$.

Considering all possible locations of cores in the lubrication layer for which relationships (11.57)-(11.60) are valid we can classify these lubrication regimes as follows: flow configuration I (FCI) for $x_{c1} \leq a$ and $\Delta_2 \leq 0$ a core is absent in the flow, flow configuration II (FCII) for $x_{c1} > a$ and $\Delta_2 \leq 0$ there is one adjacent to a solid surface core in the inlet zone of the flow, flow configuration III (FCIII) for $x_{c1} > a$ and $\Delta_2 > 0$ there are two adjacent to solid surfaces cores in the flow, one core is located in the inlet zone of the contact while the second one is in the central part of the contact, and flow configuration IV (FCIV) for $x_{c1} \leq a$ and $\Delta_2 > 0$ there is one adjacent to a solid surface core in the central zone of the flow. In order for this classification to be valid, it is necessary that the inequality (see (11.60) and (11.64)) be satisfied

$$\mid \frac{dp}{dx} \mid \leq \frac{1}{\gamma h}\{2\tau_0 + \frac{s_0}{12\gamma}\frac{\mu}{h} + \sqrt{\frac{s_0}{6\gamma}\frac{\mu}{h}[\frac{s_0}{24\gamma}\frac{\mu}{h} + 2\tau_0]}\}. \qquad (11.76)$$

Integrating (11.57) leads to the equation

$$\frac{s_0}{9}\sqrt{\frac{3s_0}{2}\mu} \mid \frac{dp}{dx} \mid^{-1} s_p = -c_3 + \gamma(\frac{s_0}{2}s_p - 1)h. \qquad (11.77)$$

Constant c_3 coincides with the flux of grease through the gap between the surfaces. Taking into account that for $x = \pm c$ the grease flux is γ and, also, the fact that the lubricant flux through any cross section of the layer is constant from equation (11.77), we obtain

$$c_3 = \gamma. \qquad (11.78)$$

The value of constant c_3 also follows from the absence of a jump in $p(x)$ and $\frac{dp}{dx}$ at points $x = x_{cj}$ $(j = 1, 2, 3)$, i.e., from conditions (11.67). Furthermore, from (11.77) and (11.78), we obtain

$$\frac{dp}{dx} = \frac{s_p s_0^3}{54\gamma^2}\mu[(\frac{s_p s_0}{2} - 1)h + 1]^{-2}. \qquad (11.79)$$

By substituting $\frac{dp}{dx}$ from equation (11.79) into inequality (11.76) after some routine calculations, we will arrive at the inequalities

$$\frac{s_0^3}{54\gamma}\mu h \leq (2\tau_0 + \frac{s_0}{12\gamma}\frac{\mu}{h})[(\frac{s_p s_0}{2} - 1)h + 1]^2,$$

$$\frac{s_0^2}{18\gamma}\mu \mid (\frac{s_p s_0}{2} - 1)h + 1 \mid \geq \mid \frac{s_0^3}{54\gamma}\mu h - 2\tau_0[(\frac{s_p s_0}{2} - 1)h + 1]^2 \mid, \qquad (11.80)$$

which are equivalent to inequality (11.76).

The flow configurations classified earlier (FCI-FCIV) are represented in Fig. 11.7 and marked by (a)-(d), respectively. Therefore, the necessary and

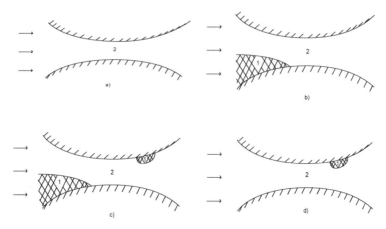

FIGURE 11.7
Grease flow configurations without a core and with cores adjacent to contact surfaces: (a) FCI, (b) FCII, (c) FCIII, (d) FCIV (1 marks grease core, 2 marks shear flow) (after Kudish [7]). Reprinted with permission from Allerton Press.

sufficient condition of realization of the above flow configurations is equivalent to validity of one of the following inequalities:

$$s_0 \geq 6\max(\frac{a^2-c^2}{\gamma+a^2-c^2}, \frac{c^2}{\gamma-c^2}), \tag{11.81}$$

$$\frac{6c^2}{\gamma-c^2} \leq s_0 < 6\frac{a^2-c^2}{\gamma+a^2-c^2}, \tag{11.82}$$

$$0 < s_0 < 6\min(\frac{a^2-c^2}{\gamma+a^2-c^2}, \frac{c^2}{\gamma-c^2}), \tag{11.83}$$

$$6\frac{a^2-c^2}{\gamma+a^2-c^2} \leq s_0 < \frac{6c^2}{\gamma-c^2}. \tag{11.84}$$

Inequalities (11.81)-(11.84) follow from the inequalities for x_{c1} and Δ_2, which characterize each of the considered flow configurations.

The above analysis and continuity of $\frac{dp}{dx}$ indicate that when a core (adjacent to solid surfaces or detached from them within the lubrication layer $-\frac{h}{2} \leq z \leq \frac{h}{2}$) is present in a grease flow the core can be located only within the intervals $[a, x_{c1}]$ and $[x_{c2}, x_{c3}]$.

Now, let us establish the conditions under which grease flows described by equations (11.53)-(11.56) can be realized. From expressions (11.56), (11.76), and (11.80), it follows that the points of flow type changes for flows with cores adjacent to solid surfaces or detached from them can be found from solution of the system

$$p(x_c - 0) = p(x_c + 0), \tag{11.85}$$

$$\frac{s_0^2}{18\gamma}\mu \mid (\frac{s_p s_0}{2} - 1)h(x_c) + 1 \mid$$

$$\geq \mid \frac{s_0^3}{54\gamma}\mu h(x_c) - 2\tau_0[(\frac{s_p s_0}{2} - 1)h(x_c) + 1]^2 \mid. \tag{11.86}$$

In equation (11.85) in the right- and left-hand sides are functions $p(x)$, which satisfy equations (11.53) and (11.57).

It is important to mention that because h is determined by equation (11.68) for $\mu(p) = 1$ and $\tau_0 = const$ it is sufficient to obtain the solution of just equation (11.70).

Without getting into more detail concerning (11.85) and (11.86) let us consider flow configurations I-IV. In these flow configurations there are no points where inequality (11.56) is satisfied. Let us integrate equations (11.70), (11.79), and (11.68). For simplicity we will assume that $\mu(p) = 1$. In agreement with Section 10.3 in the region where inequality (11.64) is satisfied the solution of equation (11.70) has the form

$$p(x) = p_2(x) + c_2$$

$$= c_2 + \frac{1}{8b^3}\left\{\frac{\gamma - 4c^2}{b^2}\left[\arctan\frac{x}{b} + \frac{bx}{b^2 + x^2}\right] - \frac{2\gamma bx}{(b^2 + x^2)^2}\right\},$$

(11.87)

where $b = \sqrt{\gamma - c^2}$. Similarly, by integrating equation (11.79) for $s_0 > 0$, we obtain the following expressions for pressure

$$p(x) = p_4(x) + c_4 = c_4 + \frac{4x}{27\gamma^2} \ for \ s_p = 1, \ s_0 = 2, \tag{11.88}$$

$$p(x) = p_5(x) + c_5 = c_5 + \frac{d}{2d_1^2}\left\{\frac{x}{x^2 + d_1^2} + \frac{1}{d_1}\arctan\frac{x}{d_1}\right\} \ for \ d_1^2 > 0, \tag{11.89}$$

$$p(x) = p_6(x) + c_6 = c_6 - \frac{d}{3x^3} \ for \ d_1 = 0, \tag{11.90}$$

$$p(x) = p_7(x) + c_7 = c_7 + \frac{d}{2d_2^2}\left\{\frac{x}{d_2^2 - x^2} + \frac{1}{2d_2}\ln\left|\frac{x + d_2}{x - d_2}\right|\right\}$$

$$for \ d_2^2 > 0, \tag{11.91}$$

$$d = \frac{s_p s_0^3}{54(1 - \frac{s_p s_0}{2})^2}, \ d_1^2 = \frac{s_p s_0 \gamma}{s_p s_0 - 2} - c^2, \ d_2^2 = c^2 - \frac{s_p s_0 \gamma}{s_p s_0 - 2}. \tag{11.92}$$

Let us assume that $s_p = 1$. As it is clear from the expressions in (11.92)

$$d_1^2 > 0 \ for \ s_0 > 2; \ d_2^2 > 0 \ for \ 0 < s_0 < 2. \tag{11.93}$$

Obviously, for $s_0 > 2$ pressure $p(x)$ in (11.89) is regular in the entire contact region. For $0 < s_0 < 2$ the solution in (11.91) has a singularity at $x = -|d_2|$ (the singularity at $x = |d_2|$ is irrelevant as it is outside of the contact region). Therefore, to satisfy inequalities (11.80) it is necessary to require that functions $p(x)$ and $\frac{dp}{dx}$ from (11.91) are bounded. To satisfy these conditions, it is sufficient to satisfy one of the inequalities: $-|d_2| < a$ or $-|d_2| > x_{c1}$. Therefore, for $0 < s_0 < 2$ we obtain the inequality

$$\frac{2(a^2 - c^2)}{\gamma + a^2 - c^2} < s_0 < 2 \ for \ s_p = 1, \ d_2^2 > 0. \tag{11.94}$$

Now, let $s_p = -1$. Then we will have $d_1^2 > 0$ for $s_0 > \frac{2c^2}{\gamma - c^2}$, $d_1 = 0$ for $s_0 = \frac{2c^2}{\gamma - c^2}$, and $d_2^2 > 0$ for $0 < s_0 < \frac{2c^2}{\gamma - c^2}$. In the considered case, a core can be realized only in the contact zone, which includes point $x = 0$. There-fore, for $d_1 = 0$ and $s_0 = \frac{2c^2}{\gamma - c^2}$ at $x = 0 \in [x_{c2}, x_{c3}]$ functions $p(x)$ and $\frac{dp}{dx}$ from (11.90) have a singularity. Similarly, to satisfy inequality (11.80) in the considered contact zone for $0 < s_0 < \frac{2c^2}{\gamma - c^2}$, it is necessary to require conti-nuity of functions $p(x)$ and $\frac{dp}{dx}$ from (11.91). That is equivalent to inequality $\mid d_2 \mid > x_{c3} = \mid x_{c2} \mid$. However, the analysis of equations (11.75) and (11.92) shows that it is impossible to satisfy the latter inequality. Therefore, in this case the only possibility is

$$s_0 > \frac{2c^2}{\gamma - c^2} \ for \ s_p = -1, \ d_1^2 > 0. \tag{11.95}$$

The condition under which the flow configurations I-IV are realized are in-equality (11.80) and (see (11.94) and (11.95))

$$s_0 > 2 \max\{\tfrac{a^2 - c^2}{\gamma + a^2 - c^2}, \tfrac{c^2}{\gamma - c^2}\}. \tag{11.96}$$

Violation of inequality (11.96) causes the existence of such zones in the contact in which the grease flow is described by relationships (11.53)-(11.56). That causes existence of a core in the flow detached from the solid surfaces.

In the further analysis, we will take advantage of the integrals of functions $p(x)$ from (11.87)-(11.91):

$$I_4(x_1, x_2, c_4) = c_4(x_2 - x_1) + \tfrac{2}{27} \tfrac{x_2^2 - x_1^2}{\gamma^2}, \ s_p = 1, \ s_0 = 2, \tag{11.97}$$

$$I_5(x_1, x_2, c_5) = c_5(x_2 - x_1) + \tfrac{d}{2d_1^3}\{x_2 \arctan \tfrac{x_2}{d_1} - x_1 \arctan \tfrac{x_1}{d_1}\} \tag{11.98}$$

$$for \ d_1^2 > 0,$$

$$I_7(x_1, x_2, c_7) = c_7(x_2 - x_1) + \tfrac{d}{4d_2^3}\{x_2 \ln \mid \tfrac{x_2 + d_2}{x_2 - d_2} \mid$$

$$\tag{11.99}$$

$$-x_1 \ln \mid \tfrac{x_1 + d_2}{x_1 - d_2} \mid\} \ for \ d_2^2 > 0,$$

$$I_2(x_1, x_2, c_2) = c_2(x_2 - x_1) + \tfrac{1}{8b^3}\{\tfrac{\gamma - 4c^2}{b^2}[x_2 \arctan \tfrac{x_2}{b}$$

$$\tag{11.100}$$

$$-x_1 \arctan \tfrac{x_1}{b}] + \gamma b[\tfrac{1}{b^2 + x_2^2} - \tfrac{1}{b^2 + x_1^2}]\}.$$

To complete the determination of pressure in the contact region, it is neces-sary to find the values of constants c, γ, and some of the constants c_2, c_4, c_5, and c_7. To do that we need to satisfy equations (11.65), (11.66), and the first equation in (11.67) while using the relationships (11.74), (11.75), (11.87)-(11.89), (11.91), (11.97)-(11.100). After the solution of this system of transcen-dental equations is obtained it is necessary to check the validity of inequalities (11.80) and (11.96).

Clearly, a similar analysis of the problem can be done in the case of $\mu = \mu(p)$. Moreover, for $\tau_0 \ll 1$ using regular asymptotic expansions the solution of the whole problem (11.53)-(11.68) can be obtained for different combinations of parameters a, τ_0, and s_0. In this case the main term of the asymptotic solution is the solution of the hydrodynamic problem for rigid solids lubricated with Newtonian fluid. This situation allows for realization of cores adjacent to and detached from the solid surfaces. The shape, size, and location of these cores as well as the next terms in the pressure expansion are the solutions of linear algebraic and differential equations. This analysis is based on the presented study of the grease flow with cores adjacent to solid surfaces. Besides that, based on the above analysis and on the method of regular asymptotic expansions [8] the solution of the elastohydrodynamic problem for elastic cylinders in case of lightly loaded contact (when $V \ll 1$) with cores adjacent to solid surfaces can be obtained.

11.4.3 Numerical Results

Let us consider the grease flow configurations defined by inequalities (11.81)-(11.84).

Suppose inequality (11.81) is satisfied. Then in the entire contact region pressure $p(x)$ is determined by equation (11.87) (see also Section 10.3). By satisfying conditions (11.65) and (11.66), we obtain the full problem solution in the absence of cores. For $\mu(p) = 1$ and $s_0 = 6$ the numerical solutions for this case are presented in Table 11.2. It is clear from this data that only the sliding frictional stress $f(x)$ and the sliding friction force F_S depend on the grease threshold stress τ_0. Also, the problem solution is independent from the slide-to-roll ration s_0 as long as inequality (11.81) is satisfied.

TABLE 11.2
Values of γ, c, and F_S versus the inlet coordinate a and threshold stress τ_0 for grease lubricated contact of rigid cylinders obtained for $\mu = 1$ and $s_0 = 6$ (after Kudish [7]). Reprinted with permission from Allerton Press.

a	τ_0	γ	c	F_S
-0.554	0	0.092	0.123	0.133
-0.554	5	0.092	0.123	0.256
-0.954	0	0.123	0.148	0.017
-0.954	5	0.123	0.148	0.038
-5	0	0.157	0.170	0.066
-5	5	0.157	0.170	0.168

Now, let us assume that inequalities (11.82) and (11.96) are satisfied. Then using (11.87) and one of the functions from (11.88), (11.89), and (11.91) and satisfying the first two conditions in (11.65) and the first condition in (11.67) for $x_{ci} = x_{c1}$ from (11.74) with the help of (11.97)-(11.100) satisfying equation (11.66) we obtain three systems of two nonlinear algebraic equations

$$p_5(x_{c1}) - p_5(a) = p_2(x_{c1}) - p_2(c),$$

$$I_5(a, x_{c1}, -p_5(a)) + I_2(x_{c1}, c, -p_2(c)) = \tfrac{\pi}{2}, \ s_0 > 2, \tag{11.101}$$

$$p_4(x_{c1}) - p_4(a) = p_2(x_{c1}) - p_2(c),$$

$$I_4(a, x_{c1}, -p_4(a)) + I_2(x_{c1}, c, -p_2(c)) = \tfrac{\pi}{2}, \ s_0 = 2, \tag{11.102}$$

$$p_7(x_{c1}) - p_7(a) = p_2(x_{c1}) - p_2(c),$$

$$I_7(a, x_{c1}, -p_7(a)) + I_2(x_{c1}, c, -p_2(c)) = \tfrac{\pi}{2}, \ s_0 < 2. \tag{11.103}$$

Suppose inequalities (11.83) and (11.96) are satisfied. Then using (11.87) and one of the functions from (11.88), (11.89), and (11.91) and satisfying the first two conditions in (11.65) and the first condition in (11.67) for $x_{ci} = x_{c1}, \ x_{c2}, \ x_{c3}$ from (11.74) and (11.75) with the help of (11.97)-(11.100) satisfying equation (11.66) we obtain another three systems of two nonlinear algebraic equations

$$p_5(x_{c1}) - p_5(a) = c_{21} + p_2(x_{c1}),$$

$$I_5(a, x_{c1}, -p_5(a)) + g(c_{21}, c_{51}, -p_2(c), \gamma, c) = 0, \ s_0 > 2, \tag{11.104}$$

$$p_4(x_{c1}) - p_4(a) = c_{21} + p_2(x_{c1}),$$

$$I_4(a, x_{c1}, -p_4(a)) + g(c_{21}, c_{51}, -p_2(c), \gamma, c) = 0, \ s_0 = 2, \tag{11.105}$$

$$p_7(x_{c1}) - p_7(a) = c_{21} + p_2(x_{c1}),$$

$$I_7(a, x_{c1}, -p_7(a)) + g(c_{21}, c_{51}, -p_2(c), \gamma, c) = 0, \ s_0 < 2, \tag{11.106}$$

where

$$g(c_{21}, c_{51}, -p_2(c), \gamma, c) = I_2(x_{c1}, x_{c2}, c_{21}) + I_5(x_{c2}, x_{c3}, c_{51})$$

$$+ I_2(x_{c3}, c, -p_2(c)) - \tfrac{\pi}{2}. \tag{11.107}$$

To complete each of the systems of equations (11.104)-(11.107), we need to add to them the following equations

$$c_{21} = c_{51} + p_5(x_{c2}) - p_2(x_{c2}), \ c_{51} = p_2(x_{c3}) - p_2(c) - p_5(x_{c3}). \tag{11.108}$$

Finally, let us assume that inequalities (11.84) and (11.96) are satisfied. Then using (11.87) and (11.89) and satisfying the first two conditions in (11.65) and the first condition in (11.67) for $x_{ci} = x_{c2}$, x_{c3} from (11.75) with the help of (11.98) and (11.100) satisfying equation (11.66) we obtain another system of three nonlinear algebraic equations

$$p_2(x_{c2}) - p_2(a) = p_5(x_{c2}) + c_5, \quad c_5 = p_2(x_{c3}) - p_2(c) - p_5(x_{c3}),$$

$$I_2(a, x_{c2}, -p_2(a)) + I_5(x_{c2}, x_{c3}, c_5) + I_2(x_{c3}, c, -p_2(c)) = \tfrac{\pi}{2}.$$

(11.109)

TABLE 11.3
Grease film thickness γ
and the exit coordinate c
versus the slide-to-roll
ratio s_0 obtained for
$a = -5$ and $\tau_0 = 1$ for a
lubricated contact of
rigid solids in the case of
presence of a core flow
adjacent to a solid sur-
face (after Kudish [13]).
Reprinted with permis-
sion from Allerton Press.

s_0	γ	c
4.808	0.1572	0.1702
4.424	0.1593	0.1715
3.847	0.1645	0.1748
3.482	0.1724	0.1800
3.077	0.1890	0.1906

The results of calculations done for $\mu(p) = 1$, $a = -5$, and various values of s_0 and τ_0 are given in Fig. 11.8-11.11. From Fig. 11.8 we can make a conclusion that for $\tau_0 > 0$ as the slide-to-roll ratio s_0 decreases the dimensionless grease film thickness γ and the coordinate of the exit point c increase significantly in comparison with the values for the corresponding Newtonian fluid while remaining finite. Some data for the inlet coordinate $a = -5$ and threshold shear stress $\tau_0 = 1$ is given in Table 11.3. For $\tau_0 > 0$ the sliding friction force F_S reaches its minimum at approximately $s_0 = 3$ while for a Newtonian fluid ($\tau_0 = 0$) the sliding friction force F_S is a strictly monotonically increasing function of s_0. It is important to notice that the rolling friction force F_R is by about two orders of magnitude smaller than the sliding friction force F_S.

For $\tau_0 = 0.5$ a detached core in the grease flow is present for $s_0 \leq 2.1$ while for $\tau_0 = 5$ a detached core in the grease flow is present for $s_0 \leq 1.9$.

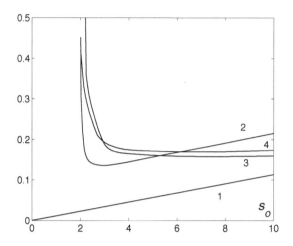

FIGURE 11.8
The dependence of the friction force F_S, film thickness γ, and exit coordinate
c on the slide-to-roll ratio s_0 obtained for $\mu = 1$, $a = -5$: friction force F_S
obtained for $\tau_0 = 0$ - curve 1 and for $\tau_0 = 10$ - curve 2, film thickness γ
obtained for $\tau_0 = 10$ - curve 3, exit coordinate c obtained for $\tau_0 = 10$ - curve
4 (after Kudish [7]). Reprinted with permission from Allerton Press.

TABLE 11.4
Dependence of the critical value of the slide-to-roll
ratio s_* on the inlet coordinate a.

a	-0.031	-0.164	-0.554	-0.954	-5	-10
s_*	0.96	2.64	4.56	5.28	5.96	5.99

For decreasing $\mid a \mid$ (when the inlet coordinate approaches the center of the
contact region), the critical value of s_0 for which a detached core appears in
the grease flow also decreases. The behavior of the critical slide-to-roll ratio
value $s_* = 6\frac{a^2-c^2}{\gamma+a^2-c^2}$ beyond which cores are absent in the grease flow is
similar (see Table 11.4). The core boundaries x_{c1} (marked with 1) and x_{c2}
(marked with 2) as functions of s_0 are presented graphically in Fig. 11.9.
These graphs show that as s_0 decreases the first core appearance occurs at
the slower–moving surface. The graphs of pressure $p(x)$ distribution (see Fig.
11.10) obtained for $\tau_0 = 5$ and $s_0 = 6$ (curve 1), $s_0 = 3$ (curve 2), $s_0 = 2.5$
(curve 3), $s_0 = 2$ (curve 4), show that the pressure in the inlet and exit zones
of the contact increases while in the central part of the contact it decreases
significantly. The curves of the sliding frictional stress $f(x)$ presented in Fig.
11.11 are obtained for a number of different values of τ_0 and s_0 (curve 1 for

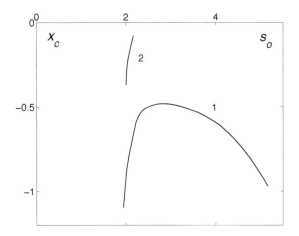

FIGURE 11.9
The dependence of the coordinates of the points of grease flow type changes x_{ci} on the slide-to-roll ratio s_0 obtained for $\mu = 1$, $a = -5$, and $\tau_0 = 10$: $x_{c1} = x_{c1}(s_0)$ - curve 1, $x_{c2} = x_{c2}(s_0)$ - curve 2 (after Kudish [7]). Reprinted with permission from Allerton Press.

$\tau_0 = 0$, $s_0 = 3$, curve 2 for $\tau_0 = 0.5$, $s_0 = 3$, curve 3 for $\tau_0 = 0.5$, $s_0 = 2.5$, curve 4 for $\tau_0 = 2.5$, $s_0 = 2$, and curve 5 for $\tau_0 = 5$, $s_0 = 2$). It follows from these graphs of $f(x)$ that its behavior changes qualitatively and quantitatively as τ_0 increases. In particular, as τ_0 increases the sliding frictional stress $f(x)$ gets distributed more evenly over the contact region and it maximum value decreases.

Based on Fig. 11.9, Tables 11.2 and 11.3 we can point out that the film thickness is greater in a grease lubricated contact compared to the one lubricated with a Newtonian fluid given all other conditions are identical. This theoretical conclusion is confirmed experimentally [9, 10, 11]. Also, in these studies it is found that under isothermal conditions the dependence of the exit film thickness h_e on the average surface velocity $0.5(u_1 + u_2)$ is similar for the grease and the base oil used for grease preparation. The same conclusion follows from the above theoretical analysis done for lightly loaded conditions because from formulas (10.30) and (10.44) for greases described by the Bingham-Shvedov model for the film thickness h_e we get the formula

$$h_e = 6\pi\gamma\frac{\mu_0(u_1+u_2)R'}{P}, \tag{11.110}$$

which coincides with the corresponding formula for h_e for Newtonian lubricant. What is different in this two formulas is the value of the parameter γ, which is greater for grease than for the corresponding fluid lubricant.

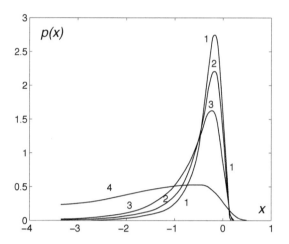

FIGURE 11.10
Pressure distribution $p(x)$ obtained for $\mu = 1$, $a = -5$, and $\tau_0 = 10$: curve 1
for $s_0 = 6$ (Newtonian fluid), curve 2 for $s_0 = 3$, curve 3 for $s_0 = 2.5$, and
curve 4 for $s_0 = 2$ (after Kudish [7]). Reprinted with permission from Allerton
Press.

11.5   Regimes of Grease Lubrication without Cores

In the preceding section we considered some regimes of grease lubrication
of rigid solids with cores adjacent to solid surfaces. In this section we will
consider heavily loaded regimes of grease lubrication without cores.

For greases the rheology of which satisfies equations (11.3) and (11.4) in
the dimensionless variables designed for heavily loaded regimes (see formulas
(10.383), (10.384), and (10.558) in Section 10.15), we obtain the following
system of integro-differential equations

$$\frac{d}{dx}\Big\{ \int\limits_{-h/2}^{z_1} \frac{z}{\mu}F(f + \tau_0 s_p + H_0 z\frac{dp}{dx})dz$$

$$(11.111)$$

$$+ \int\limits_{z_2}^{h/2} \frac{z}{\mu}F(f - \tau_0 s_p + H_0 z\frac{dp}{dx})dz - h\Big\} = 0, \ \ s_p = sign\frac{dp}{dx},$$

$$\int\limits_{-h/2}^{z_1} \frac{1}{\mu}F(f + \tau_0 s_p + H_0 z\frac{dp}{dx})dz$$

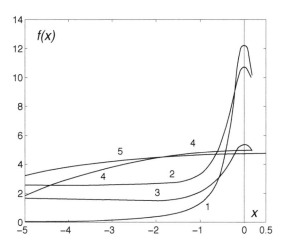

FIGURE 11.11
Sliding frictional stress distribution $f(x)$ obtained for $\mu = 1$, $a = -5$: curve 1 for $s_0 = 6$ and $\tau_0 = 0$ (Newtonian fluid), curve 2 for $s_0 = 3$ and $\tau_0 = 1$, curve 3 for $s_0 = 2.5$ and $\tau_0 = 1$, curve 4 for $s_0 = 2$ and $\tau_0 = 5$, and curve 5 for $s_0 = 2$ and $\tau_0 = 10$ (after Kudish [7]). Reprinted with permission from Allerton Press.

$$+ \int_{z_2}^{h/2} \frac{1}{\mu} F(f - \tau_0 s_p + H_0 z \tfrac{dp}{dx}) dz = s_0, \tag{11.112}$$

$$\tfrac{\partial}{\partial z}(\lambda \tfrac{\partial T}{\partial z}) = -\tfrac{\kappa H_0}{\mu} F[f - \tau_0 s_p sign(z - z_1) + H_0 z \tfrac{dp}{dx}]$$
$$\times (f + H_0 z \tfrac{dp}{dx}) \theta[(z - z_1)(z - z_2)], \tag{11.113}$$

$$z_1 = -\frac{\tau_0 s_p + f}{H_0 \frac{\partial p}{\partial x}}, \ z_2 = \frac{\tau_0 s_p - f}{H_0 \frac{\partial p}{\partial x}}, \tag{11.114}$$

$$\tau_0 - \tfrac{H_0 h}{2} \mid \tfrac{dp}{dx} \mid < f < -\tau_0 + \tfrac{H_0 h}{2} \mid \tfrac{dp}{dx} \mid, \tag{11.115}$$

$$\tfrac{d}{dx} \Big\{ \int_{-h/2}^{z_1} \tfrac{z}{\mu} F(f - \tau_0 + H_0 z \tfrac{dp}{dx}) dz - h \Big\} = 0, \tag{11.116}$$

$$\int_{-h/2}^{z_1} \tfrac{1}{\mu} F(f - \tau_0 + H_0 z \tfrac{dp}{dx}) dz = s_0, \tag{11.117}$$

$$\tfrac{\partial}{\partial z}(\lambda \tfrac{\partial T}{\partial z}) = -\tfrac{\kappa H_0}{\mu} F[f + \tau_0 sign(z - z_1) + H_0 z \tfrac{dp}{dx}](f$$
$$+ H_0 z \tfrac{dp}{dx}) \theta(z_1 - z), \tag{11.118}$$

$$z_1 = \frac{\tau_0 - f}{H_0 \frac{\partial p}{\partial x}}, \quad z_2 = -\frac{\tau_0 + f}{H_0 \frac{\partial p}{\partial x}}, \tag{11.119}$$

$$\tau_0 + \frac{H_0 h}{2} \frac{dp}{dx}) < f < \tau_0 - \frac{H_0 h}{2} \frac{dp}{dx}, \quad -\tau_0 - \frac{H_0 h}{2} \frac{dp}{dx}) \le f, \tag{11.120}$$

$$\frac{d}{dx} \left\{ \int_{z_2}^{h/2} \frac{z}{\mu} F(f - \tau_0 + H_0 z \frac{dp}{dx}) dz - h \right\} = 0, \tag{11.121}$$

$$\int_{z_2}^{h/2} \frac{1}{\mu} F(f - \tau_0 + H_0 z \frac{dp}{dx}) dz = s_0, \tag{11.122}$$

$$\frac{\partial}{\partial z}(\lambda \frac{\partial T}{\partial z}) = -\frac{\kappa H_0}{\mu} F[f - \tau_0 sign(z - z_2) + H_0 z \frac{dp}{dx}](f$$

$$+ H_0 z \frac{dp}{dx}) \theta(z - z_2), \tag{11.123}$$

$$z_1 = -\frac{\tau_0 + f}{H_0 \frac{\partial p}{\partial x}}, \quad z_2 = \frac{\tau_0 - f}{H_0 \frac{\partial p}{\partial x}}, \tag{11.124}$$

$$\tau_0 - \frac{H_0 h}{2} \frac{dp}{dx}) < f < \tau_0 + \frac{H_0 h}{2} \frac{dp}{dx}, \quad -\tau_0 + \frac{H_0 h}{2} \frac{dp}{dx}) \le f, \tag{11.125}$$

$$\frac{d}{dx} \left\{ \int_{-h/2}^{h/2} \frac{z}{\mu} F(f - \tau_0 + H_0 z \frac{dp}{dx}) dz - h \right\} = 0, \tag{11.126}$$

$$\int_{-h/2}^{h/2} \frac{1}{\mu} F(f - \tau_0 + H_0 z \frac{dp}{dx}) dz = s_0, \tag{11.127}$$

$$\frac{\partial}{\partial z}(\lambda \frac{\partial T}{\partial z}) = -\frac{\kappa H_0}{\mu} F[f - \tau_0 + H_0 z \frac{dp}{dx}](f + H_0 z \frac{dp}{dx}), \tag{11.128}$$

$$\mid \frac{dp}{dx} \mid \le \frac{2(f - \tau_0)}{H_0 h}, \tag{11.129}$$

$$p(a) = p(c) = \frac{dp(c)}{dx} = 0, \tag{11.130}$$

$$T(x, -\frac{h}{2}) = T_{w1}(x), \; T(x, \frac{h}{2}) = T_{w2}(x), \tag{11.131}$$

$$\int_{a}^{c} p(t) dt = \frac{\pi}{2}, \tag{11.132}$$

$$H_0(h - 1) = x^2 - c^2 + \frac{2}{\pi} \int_{a}^{c} p(t) \ln \frac{c - t}{|x - t|} dt, \tag{11.133}$$

$$[p(x_{ci})] = [\frac{dp(x_{ci})}{dx}] = 0, \; x_{ci} \in X_c, \tag{11.134}$$

$$[T(x, z_i)] = [\frac{\partial T(x, z_i)}{\partial z}] = 0, \; \mid z_i \mid \le \frac{h}{2}, \tag{11.135}$$

where parameter κ is determined in (10.558).

We will consider regimes of lubrication in which cores in the grease flow do not occur. For such regimes in the case of non-isothermal lubrication, we need to satisfy equations (11.126)-(11.134) while in the case of isothermal lubrication we need to satisfy equations (11.45), (11.126), (11.127), (11.129), (11.130), (11.132)-(11.134). As always, in EHL equations describing heavily loaded contacts, there is a small parameter ω which is usually associated with parameter

$V \ll 1$ (see (10.384)) or dimensionless pressure viscosity coefficient $Q \gg 1$. The latter coefficient usually occurs in the exponential viscosity-pressure relationship, which, in dimensional variables, has the form $\mu = \mu_a \exp(\alpha p)$, where μ_a is the ambient viscosity, α is the dimensional pressure viscosity coefficient and $Q = \alpha p_H$ (p_H is the maximum Hertzian pressure in a dry contact of elastic solids). On detailed discussion of various definitions of the small parameter ω see the preceding chapter.

Let us introduce the analogue of function $\nu(x)$ from (10.475) according to the formula [12]

$$\nu(x) = \frac{H_0 h}{2(f-\tau_0)} \frac{dp}{dx}.$$

(11.136)

Using the definition of $\nu(x)$ from (11.136) inequality (11.129) can be rewritten as $| \nu(x) | \leq 1$. Therefore, when $\nu(x) \ll 1$ or $\nu(x) = O(1)$, $| \nu(x) | \leq 1$ for $\omega \ll 1$, in the inlet zone of the contact the asymptotic approaches developed in Chapter 10 are applicable to the EHL problem for grease. It is clear that the solution of the isothermal EHL problem for grease coincides with the solution of the EHL problem for the corresponding non-Newtonian fluid lubricant ($\tau_0 = 0$) described in detail in Chapter 10. In particular, the formulas for the film thickness H_0 in pre- and over-critical regimes in heavily loaded contacts lubricated by a grease and a non-Newtonian fluid lubricant with rheology identical to the one for the grease with $\tau_0 = 0$ are identical (see Chapter 10). The only difference between these two solutions manifests itself as additive term equal to τ_0 and term proportional to $\int_a^c \tau_0(p)dx$ (see formula (11.69)) in the sliding frictional stress $f(x)$ and force F_S, respectively.

Let us consider in more detail the case of thermal grease lubrication assuming that $\frac{\partial \tau_0}{\partial T} = 0$. The expression for the sliding frictional stress $f(x)$ and lubricant temperature $T(x, z)$ we will search in the form (for comparison see expression (10.476) and (10.567), (10.568)) [12]

$$f(x) = \tau_0(p(x)) + f_0(x) + f_1(x)\nu(x) + O(\nu^2(x)), \quad \omega \ll 1,$$

(11.137)

$$T(x, z) = T_0(x, z) + T_1(x, z)\nu(x) + O(\nu^2(x)), \quad \omega \ll 1,$$

(11.138)

where functions $f_0(x)$, $f_1(x)$, $T_0(x, z)$, and $T_1(x, z)$ are unknown and of the order of unity. Let us assume that the lubricant viscosity μ is determined by the equation (see (10.559))

$$\mu(p, T) = \mu^0(p) \exp(-\delta T),$$

(11.139)

where $\mu^0(p)$ is the lubricant viscosity at $T = 0$ which is dependent only on pressure p, δ is a positive constant or function of p, δ is independent from T. Then expanding the terms of equations (11.127) and (11.128) for $\nu \ll 1$ and following the procedure of Section 10.16 we obtain

$$f_0 = \Phi\left\{\mu^0 s_0\left(\int_{-h/2}^{h/2} e^{\delta T_0} dz\right)^{-1}\right\},$$

(11.140)

$$\lambda \frac{\partial^2 T_0}{\partial z^2} = -\frac{\kappa H_0}{\mu^0} F(f_0)(f_0 + \tau_0) e^{\delta T_0}, \tag{11.141}$$

$$f_1 = -\left\{ \frac{F(f_0)}{F'(f_0)} \delta \int\limits_{-h/2}^{h/2} T_1 e^{\delta T_0} dz + \frac{2f_0}{h} \int\limits_{-h/2}^{h/2} z e^{\delta T_0} dz \right\} \tag{11.142}$$

$$\times \left(\int\limits_{-h/2}^{h/2} e^{\delta T_0} dz \right)^{-1}.$$

$$\lambda \frac{\partial^2 T_1}{\partial z^2} = -\frac{\kappa H_0}{\mu^0} F(f_0)(f_0 + \tau_0) e^{\delta T_0} \{ \delta T_1 + (\frac{f_1}{f_0} + \frac{2z}{h}) \tag{11.143}$$

$$\times [\frac{f_0}{f_0 + \tau_0} + \frac{f_0) F'(f_0)}{F(f_0)}] \}.$$

Let us consider a special case of equal surface temperatures $T_w(x)$ in (11.131)

$$T(x, -\tfrac{h}{2}) = T_w(x), \ T(x, \tfrac{h}{2}) = T_w(x), \tag{11.144}$$

in which solution of equations (11.141) and (11.143) is significantly simplified and it yields (the general case is considered in Section 10.16)

$$e^{\delta T_0} = A_0[1 - \tanh^2 B z], \ A_0 = e^{\delta T_w} \cosh^2 \frac{Bh}{2}, \tag{11.145}$$

$$\Phi(\frac{\mu^0 s_0 e^{-\delta T_w} B}{\sinh Bh}) = \frac{4\lambda}{s_0 \kappa \delta H_0} B \tanh \frac{Bh}{2} - \tau_0, \tag{11.146}$$

$$T_1 = \frac{1}{\delta}[\frac{f_0}{f_0 + \tau_0} + \frac{f_0 F'(f_0)}{F(f_0)}]\{ \frac{\tanh B z}{\tanh \frac{Bh}{2}} - \frac{2z}{h} \}. \tag{11.147}$$

Under the accepted conditions (11.144) we have $f_1 = 0$ and the expression for the first term in the generalized Reynolds equation (11.126) takes the form

$$\int\limits_{-h/2}^{h/2} \frac{z}{\mu} F(f - \tau_0 + H_0 z \frac{dp}{dx}) dz = \frac{H_0 h^3}{12 \mu^0} \frac{dp}{dx} [\Phi'(\frac{\mu^0 s_0}{h})]^{-1} + O(\nu^2), \tag{11.148}$$

where the value of B is determined by equation (11.146).

As an example let us consider the case of Bingham-Shvedov model of grease (11.5). Then in dimensionless variables $\Phi(x) = \frac{V}{12 H_0} x$ and equation (11.146) is reduced to

$$\frac{4\lambda}{s_0 \kappa \delta H_0} B \tanh \frac{Bh}{2} = \frac{V}{12 H_0} \frac{\mu^0 s_0 e^{-\delta T_w} B}{\sinh Bh} + \tau_0, \tag{11.149}$$

where V is determined in (10.384) and the expression for the integral in (11.148) involved in the generalized Reynolds equation (11.126) is reduced to (see (10.338) and (10.339))

$$\int\limits_{-h/2}^{h/2} \frac{z}{\mu} F(f - \tau_0 + H_0 z \frac{dp}{dx}) dz \tag{11.150}$$

$$= \frac{H_0^2}{V} \frac{3h^3 e^{\delta T_w} (1+\beta)}{\mu^0 \ln^2(\sqrt{\beta} + \sqrt{\beta+1})} \left\{ 1 + \ln(1 + \beta) - \sqrt{\frac{1+\beta}{\beta}} \ln(\sqrt{\beta} + \sqrt{\beta+1}) \right\}$$

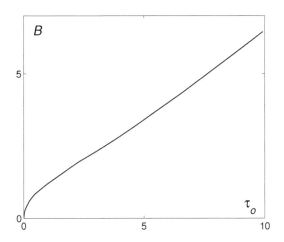

FIGURE 11.12
Dependence of parameter B on τ_0 obtained for $h = \mu^0 = \lambda = 1$, $T_w = 0$, $s_0\kappa\delta H_0 = 2.62$, and $\frac{V s_0}{H_0} = 0.18$.

$$-2\frac{\int\limits_0^{\ln(\sqrt{\beta}+\sqrt{\beta+1})}\ln(\cosh(t))dt}{\ln(\sqrt{\beta}+\sqrt{\beta+1})}\Big\}\frac{dp}{dx} + O(\nu^2), \; \beta = \frac{\kappa s_0^2}{8}\frac{\delta\mu^0}{\lambda}e^{-\delta T_w},$$

$$\omega \ll 1,$$

where

$$\int\limits_0^z \ln(\cosh(t))dt = \frac{z^2}{2} - z\ln 2 + \frac{\pi^2}{24} - \frac{1}{2}\sum_{k=1}^{\infty}\frac{(-1)^{k+1}}{k^2}e^{-2kz},$$

$$(11.151)$$

$$\int\limits_0^z \ln(\cosh(t))dt = \frac{z^3}{6} + O(z^4), \; z \ll 1.$$

Assuming that $B \ll 1$ from equation (11.149), we find

$$B = \{\frac{s_0\kappa\delta H_0}{2\lambda h}(\frac{V}{12H_0}\frac{\mu^0 s_0 e^{-\delta T_w}}{h} + \tau_0)\}^{1/2} \ll 1, \; \omega \ll 1. \qquad (11.152)$$

Notice that it is easy to establish the validity conditions for formula (11.152).

Under these conditions for pre- and over-critical lubrication regimes when in the inlet (ϵ_q- or ϵ_0-) zone $\nu(x) \ll 1$, we get the same formulas for the film thickness H_0 and can conduct the asymptotic analysis of the problem equations similar to the one presented in the preceding chapter.

It can be shown that for the case of $\nu(x) \ll 1$ and $B = O(1)$, $\omega \ll 1$, in the inlet (ϵ_q-inlet) zone formulas for the film thickness H_0 still maintain the same structural form as in the preceding chapter. However, the expression for the integral involved in the generalized Reynolds equation (11.126) and, therefore, the asymptotically valid in the inlet and exit zones equations cannot be effectively simplified.

Assuming that $B \gg 1$ from equation (11.149), we find

$$B = \frac{s_0 \kappa \delta H_0}{4\lambda} \tau_0 \gg 1, \; \omega \ll 1. \tag{11.153}$$

The further analysis of the problem is similar to the one described for the case of $B \ll 1$.

Notice that the same structural formulas for the film thickness H_0 remain in force even in the case when in the inlet (ϵ_q- and ϵ_0-) zone $\mid \nu(x) \mid \leq 1$ and $\nu(x) = O(1)$. Only the coefficient of proportionality of the order of unity changes.

In the preceding chapter it was shown that for fluid lubricants as B increases the film thickness H_0 is monotonically decreasing. For the case of grease the film thickness H_0 behaves the same way. It can be shown that as the threshold shear stress τ_0 increases the value of B decreases and, therefore, the grease film thickness H_0 decreases too. The graph of parameter B as a function of τ_0 obtained for $h = \mu^0 = \lambda = 1$, $T_w = 0$, $s_0 \kappa \delta H_0 = 2.62$, and $\frac{V_{s0}}{H_0} = 0.18$ is given in Fig. 11.12. The value of $B(0)$ corresponds to the value of the parameter B for a fluid lubricant with Newtonian rheology.

Therefore, under thermal contact conditions in a grease flow without cores the grease film thickness may be lower than in a similar fluid lubricant (base oil). In the presence of cores in the grease flow the relationship between the film thicknesses for grease and fluid lubricant can be different.

11.6   Closure

The formulation the problem for a grease lubricated contact is provided. It is shown that the problem is described by a series of different equations of grease flow depending on the presence/absence of cores in the grease flow and their location. Some basic properties of grease flows are established. A detailed solution for the grease lubricated contact of two rigid cylinders is provided for the cases of cores adjacent to cylinders surfaces. It is established that under certain conditions in the provided formulation the solution of the problem may not exist. The changes to the problem formulation, which allow for the problem solution, are suggested. A detailed consideration of isothermal and thermal EHL problems for grease lubricated contacts without cores in the grease flow is provided.

11.7   Exercises and Problems

1. Elaborate on what may happen if the continuity conditions (11.42) and/or (11.43) are not imposed on the problem solution.

2. Provide detailed analysis of the solution properties established in Section 11.3.

3. For rigid cylinders provide the condition that guaranties grease flow without cores. For rigid cylinders compare pressure distributions $p(x)$ for fluid lubricant and grease with cores adjacent to the solid surfaces. What is the main difference between these pressure distributions?

4. Obtain the two-term asymptotic solution for the isothermal grease lubricated contact of rigid cylinders in the case of small stress threshold $\tau_0 \ll 1$.

5. Obtain the two-term asymptotic solution for the isothermal grease lubricated lightly loaded (i.e., $V \gg 1$) contact of elastic cylinders in the case of the possible presence of cores in the grease flow adjacent to solid surfaces (cores detached from the solid surfaces are absent).

6. (a) Show that the value of B from equation (11.149) increases as the stress threshold τ_0 increases. (b) Show that when the value of B increases the film thickness decreases.

BIBLIOGRAPHY

[1] Wilkinson,W.L. 1960 *Non − −Newtonian Fluids : Fluid Mechanics, Mixing and Heat Transfer*. New York: Pergamon Press.

[2] Reiner, M. 1960. *Lectures on Theoretical Rheology*. New York: Interscience Publishers.

[3] Kudish, I.I. 1982. Plane Contact Problems with Viscoplastic Lubrication. *Soviet J. Fric. and Wear* 3, No. 6:1036-1047.

[4] Kudish, I.I. and Semin, V.N. 1983. On Formulation and Analysis of Plane Elastohydrodynamic Problem for Plastic Lubricant. *J. Solid Mech.* 6:107-113.

[5] Volarovich, M.P. and Gutkin, A.M. 1946. Flow of a Viscoplastic Medium Between Two Parallel Plates and in the Gap between Two Coaxial Cylinders. *J. Tech. Phys.* 16, No. 3:321-328.

[6] Mosolov, P.P. and Myasnikov, V.P. 1981. *Mechanics of Rigid − Plastic Media*. Moscow: Nauka.

[7] Kudish, I.I. 1984. Flow of Viscoplastic Medium in a Narrow Gap between Curvilinear Surfaces. *Soviet J. Fric. and Wear* 5, No. 5:841-852.

[8] Van-Dyke, M. 1964. *Perturbation Methods in Fluid Mechanics* New York-London: Academic Press.

[9] Aihara, S. and Dowson, D. 1980. An Experimental Study of Grease Film Thickness under Elastohydrodynamic Conditions. Part 1. General Results. *J. Jpn. Soc. Lubr. Eng.* 25, No. 4:254-260.

[10] Aihara, S. and Dowson, D. 1980. An Experimental Study of Grease Film Thickness under Elastohydrodynamic Conditions. Part 2. Mechanism of Grease Film Formation. *J. Jpn. Soc. Lubr. Eng.* 25, No. 6:379-386.

[11] Jonkisz, W. and Krzeminski-Freda, H. 1982. Wlasnosti Elastohydrodynamic Znego Filmu Smaru Plastycznego. *Archiwum Budowy Maszyn* 39, No. 1:11-25.

[12] Kudish, I.I. 1983. On Analysis of Plane Contact Problems with Viscoplastic Lubricant. *Soviet J. Fric. and Wear* 3, No. 6:449-457.

[13] Kudish, I.I. 1988. About Analysis of Plane Contact Problems in the Presence of Viscoplastic Lubricant. *Soviet J. Fric. and Wear* 4, No. 3:449-457.

12

Elastohydrodynamic Lubrication by Formulated Lubricants That Undergo Stress-Induced Degradation

---

12.1   Introduction

Modern lubricating oils are formulated with a variety of additives designed to (1) provide beneficial rheological characteristics to lubricants, (2) to stabilize their physical and chemical properties, and (3) to protect lubricated equipment against wear, fatigue, and corrosion. Under the influence of chemical and mechanical stresses and elevated temperatures lubricants tend to undergo certain reversible and irreversible changes. The reversible changes are caused by temporary alignment of polymeric additives in the direction of flow, resulting in an apparent drop in viscosity. When the liquid returns to a state of rest, the viscosity returns to its initial value. This is known as non-Newtonian rheology. The irreversible changes are due to a number of ongoing processes such as stress-induced scission of polymeric additives, oxidation, contamination, etc. The latter detrimental processes limit the useful life of lubricants and can lead to costly repairs and down time if a lubricated system is not properly maintained. In this chapter we focus on the combined effects of the lubricants' non-Newtonian rheology and stress-induced polymer molecule scission and on changes in lubricant contact parameters.

It is established experimentally that many lubricants exhibit some reversible non-Newtonian rheological properties (see, for example, Bair and Winer [1] and Hoglund and Jacobson [2]). Some rheological models of lubricant behavior are given in Bair and Winer [1] and in Eyring [3]. Also, there exist a number of theoretical studies of the effect of non-Newtonian behavior on various parameters of lubricated contacts such as lubrication film thickness, frictional stress, etc. Examples include studies by Houpert and Hamrock [4] and Kudish [5].

Several experimental studies of non-reversible stress-induced degradation of lubricants are presented in [6, 8] - [11] while the theoretical studies of lubricant degradation with few exceptions such as [12, 13] dealt only with the processes of homogeneous degradation caused by temperature [14] - [17] and radiation [18] effects. A semi-deterministic attempt of a theoretical study of stress-induced degradation was made by Kudish and Ben-Amotz in [19]. In later papers [20] - [22] and Chapters 6 and 7 of this monograph a kinetic probabilistic approach to stress-induced degradation has been developed. The approach is based on the derivation and usage of a probabilistic kinetic equation(s) describing the process of stress-induced polymer scission. These kinetic equations have been successfully applied to practical cases. In particular, the numerically simulated data are in excellent agreement with experimental data as it is shown in Chapters 6 and 7.

In this chapter we consider the application of the developed kinetics approach to the phenomenon of elastohydrodynamic lubrication of surfaces lubricated with non-Newtonian fluids that undergo some changes caused by lubricant degradation. The problem is reduced to a coupled system of the generalized Reynolds equation for non-Newtonian lubricant flow, the equation for the gap between the surfaces of the elastic solids, the equations for the lubricant flow streamlines, the kinetic equation describing the changes in the polymer molecular weight distribution, and the equations for the lubricant viscosity. The generalized isothermal EHL equations are coupled with the kinetic equation through the lubricant viscosity, which depends not only on lubricant pressure but also on the concentration of polymer molecules and the distribution of their chain lengths. The solution of the problem is obtained using numerical methods similar to the ones described by Kudish and Airapetyan in [23] and [24]. The kinetic equation is solved along the lubricant flow streamlines. The solution of the kinetic equation predicts the density of the probabilistic distribution of the polymer molecule chain lengths. The shear stress and the changes in the distribution of polymer molecular weight caused by lubricant degradation affect local lubricant properties. In particular, the lubricant viscosity experiences reversible and irreversible losses and, in general, is a discontinuous function of spacial variables. The changes in the lubricant viscosity alter virtually all parameters of the lubricated contact such as film thickness, friction stresses, pressure, and gap. Several comparisons of lubricants with Newtonian and non-Newtonian rheologies with and without lubricant degradation are considered.

The material presented in this chapter is mainly based on papers [23] - [26].

12.2 Formulation of the EHL Problems for Lubricants with Newtonian and Non-Newtonian Rheologies That Undergo Stress-Induced Degradation

Let us consider a plane EHL problem. Suppose two infinite parallel cylinders steadily move with linear surface speeds u_1 and u_2. The cylinders have radii R_1 and R_2 and are made of elastic materials with Young's moduli E_1 and

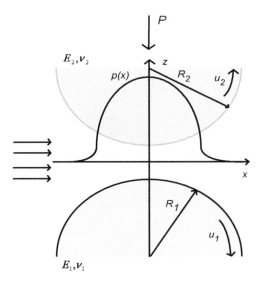

FIGURE 12.1

The general view of a lubricated contact.

E_2 and Poisson's ratios ν_1 and ν_2, respectively. The cylinders are separated by a thin lubrication layer and are loaded by the force P directed along their centers and normal to their axes (see Fig. 12.1).

Let us consider steady isothermal lubrication conditions. We will accept the classic EHL assumptions usually used for non-conformal contacts such as (a) the motion is slow so that the inertia forces can be neglected in comparison with the viscous ones, (b) the lubrication film thickness is much smaller than the size of the contact region, which, in turn, is much smaller than the curvature radii of the contact surfaces, (c) the variation rate of the gap between the cylindrical surfaces in the contact is small, (d) the contact of actual surfaces can be replaced by a contact of two elastic half-planes.

Under the above conditions the lubricant is assumed to be an incompressible fluid with a non-Newtonian rheology described by the equation

$$\frac{\partial u}{\partial z} = \frac{\tau_L}{\mu} G\left(\frac{\tau}{\tau_L}\right), \tag{12.1}$$

where z is the coordinate of a lubricant particle along the coordinate axis directed across the gap between the cylinders, u is the x-component of the lubricant velocity along the x-axis directed along the direction of the cylinders' motion (and the lubrication layer), τ and τ_L are the local and limiting shear stresses in the lubricant, μ is the lubricant viscosity, $\mu = \mu(x, z)$, G is a given function that determines the rheology of the lubricant fluid. The lubricant viscosity may be shear rate-dependent. In pseudoplastic/dilatant fluids increase in shear rate leads to increase/decrease in fluid viscosity. The shear thinning and thickening represent the reversible loss and gain of fluid viscosity, respectively. Most fluids are pseudoplastic. The extreme example of pseudoplastic fluids is the fluids with the limiting shear stress. The viscosity of the latter fluids vanishes as the shear rate exceeds a certain level. Therefore, the particular rheology considered here is given by the formula

$$G(x) = \tanh^{-1} x, \tag{12.2}$$

proposed by Bair and Winer in [1] and based on a series of experimental studies of lubricant rheological behavior.

According to [1, 2] τ_L is usually a linear function of pressure p, i.e.,

$$\tau_L = \tau_{L0} + \tau_{L1} p, \tag{12.3}$$

where parameters τ_{L0} and τ_{L1} are certain functions of the lubricant temperature.

Then the equations describing the process of lubrication are [23, 24, 27]

$$\frac{\partial \tau}{\partial z} = \frac{dp}{dx}, \ p = p(x), \ \tau = \mu \frac{\partial u}{\partial z}, \ \frac{\partial u}{\partial x} + \frac{\partial w}{\partial z} = 0,$$

$$u\left(x, -\frac{h}{2}\right) = \frac{u_1}{\sqrt{1 + \frac{1}{4}\left(\frac{dh}{dx}\right)^2}}, \ u\left(x, \frac{h}{2}\right) = \frac{u_2}{\sqrt{1 + \frac{1}{4}\left(\frac{dh}{dx}\right)^2}}, \tag{12.4}$$

$$w\left(x, -\frac{h}{2}\right) = -u_1 \frac{dh}{dx}, \ w\left(x, \frac{h}{2}\right) = u_2 \frac{dh}{dx},$$

where the last two pairs of conditions in (12.4) represent the boundary conditions imposed on the horizontal u and vertical w components of the lubricant velocity.

Integrating the first relationship in (12.4) with respect to z, we obtain

$$\tau = f + z \frac{dp}{dx}, \tag{12.5}$$

where f is the local unknown frictional stress due to sliding, $f = f(x)$. Consequently, integrating equations (12.1), (12.2) and the third of the relationships in (12.4) with respect to z from $-h/2$ to z we receive the expression for the horizontal component of the lubricant velocity u in the form

$$u(x, z) = \frac{u_1}{\sqrt{1 + \frac{1}{4}\left(\frac{dh}{dx}\right)^2}} + \int_{-h/2}^{z} \frac{\tau_L}{\mu} G\left(\frac{1}{\tau_L}\left(f + \zeta \frac{dp}{dx}\right)\right) d\zeta. \tag{12.6}$$

Setting z equal to $h/2$ in equation (12.5) provides us with the equation for the local sliding frictional stress f as follows [5]

$$\int_{-h/2}^{h/2} \frac{\tau_L}{\mu} G\left(\frac{1}{\tau_L}\left(f + z \frac{dp}{dx}\right)\right) dz = \frac{u_2 - u_1}{\sqrt{1 + \frac{1}{4}\left(\frac{dh}{dx}\right)^2}}. \tag{12.7}$$

Then integrating the continuity equation in (12.4) with respect to z from $-h/2$ to $h/2$ and taking into account the boundary conditions from (12.4) imposed on the components u and w of the lubricant velocity we arrive at the equations governing the isothermal EHL problem based on the generalized Reynolds and elasticity theory equations in the form [5, 24, 27]

$$\int_{-h/2}^{h/2} z\frac{\tau_L}{\mu}G(\frac{1}{\tau_L}(f+z\frac{dp}{dx}))dz = \frac{u_1+u_2}{2}\left[\frac{h}{\sqrt{1+\frac{1}{4}(\frac{dh}{dx})^2}} - \frac{h_e}{\sqrt{1+\frac{1}{4}(\frac{dh(x_e)}{dx})^2}}\right]$$

$$+ \int_{-h_e/2}^{h_e/2} z\frac{\tau_L(x_e)}{\mu(x_e,z)}G(\frac{f(x_e)}{\tau_L(x_e)})dz, \; p(x_i) = p(x_e) = \frac{p(x_e)}{dx} = 0, \qquad (12.8)$$

$$h = h_e + \frac{x^2-x_e^2}{2R'} + \frac{2}{\pi E'}\int_{x_i}^{x_e} p(x')\ln\frac{x_e-x'}{|x-x'|}dx', \; \int_{x_i}^{x_e} p(x')dx' = P,$$

where x_i and x_e are the coordinates of the lubricant inlet and exit points, respectively, $h(x)$ is the gap between the cylinders, h_e is the exit film thickness, $h_e = h(x_e)$, R' is the effective radius of the cylinders, $1/R' = 1/R_1 + 1/R_2$, E' is the effective elasticity modulus, $1/E' = (1-\nu_1^2)/E_1 + (1-\nu_2^2)/E_2$, P is the load per unit length applied to cylinders. The boundary conditions on pressure p at $x = x_i$ and $x = x_e$ reflect the natural assumption that the pressure at the boundary of the contact region is equal to the ambient atmospheric pressure that is much smaller than the pressure inside of the contact region.

The process of polymer additive degradation caused by chain scission, occurs while the additive dissolved in the lubricant moves along the lubricant flow streamlines. Therefore, we need to formulate the equations for the lubricant flow streamlines $z = z(x)$. Similar to [23, 24] these equations can be represented in differential and integral forms as follows:

$$\frac{dz}{dx} = \frac{w}{u},$$

$$\frac{d}{dx}\int_{-h/2}^{z(x)} u(x,\zeta)d\zeta = 0,$$

$$\int_{-h/2}^{z(x)} u(x,\zeta)d\zeta = \frac{h}{2}\frac{u_1}{\sqrt{1+\frac{1}{4}(\frac{dh}{dx})^2}} + zu(x,z) \qquad (12.9)$$

$$- \int_{-h/2}^{z(x)} \zeta\frac{\tau_L}{\mu}G(\frac{1}{\tau_L}(f+\zeta\frac{dp}{dx}))d\zeta.$$

The vertical component of the lubricant velocity $w(x,z)$ theoretically can be determined from the continuity equation and the corresponding boundary conditions [23]. However, the numerical realization of such a process is unstable and, therefore, not recommended for practical calculations.

The stress-induced process of lubricant degradation (i.e., the process of scission of additive polymer molecules with linear structure dissolved in the base stock) is described by the kinetic equation derived and analyzed in Chapter 6 (see also [20, 21]). The kinetic equation of lubricant degradation is written for the probabilistic density distribution $W(x, z, l)$ of polymer molecular weight, where l is the polymer molecule chain length. Function $W(x, z, l)$ is introduced in such a way that $W(x, z, l)\triangle l$ is the polymer weight in a unit fluid volume centered at the point with coordinates (x, z) with the polymer molecule chain lengths in the range from l to $l + \triangle l$. In case of steady two-dimensional lubricant motion, the kinetic equation can be represented as follows (see Chapter 6)

$$\tau_f \{u\frac{\partial W}{\partial x} + w\frac{\partial W}{\partial z}\} = 2l \int_{l}^{\infty} R(x, z, L)p_c(l, L)W(x, z, L)\frac{dL}{L}$$

(12.10)

$$-R(x, z, l)W(x, z, l),$$

where R is the probability of polymer molecule scission, $p_c(l, L)$ is the density of the conditional probability of a polymer molecule with the chain length L to break into two pieces with lengths l and $L - l$, τ_f is the time required for one polymer chain to undergo one act of fragmentation. The expressions for functions R and p_c have been derived in Chapter 6 and have the form

$$R(x, z, l) = 0 \ if \ l \leq L_*$$

$$R(x, z, l) = 1 - (\frac{l}{L_*})^{\frac{2\alpha U_A}{kT}} \exp\left[-\frac{\alpha U_A}{kT}(\frac{l^2}{L_*^2} - 1)\right] \ if \ l \geq L_*,$$

(12.11)

$$p_c(l, L) = ln(2)\frac{4|L-2l|}{L^2} \exp\left[-ln(2)\frac{4l(L-l)}{L^2}\right],$$

where L_* is calculated according to the formulas

$$L_* = \sqrt{\frac{U_A}{Ca_* l_*^2 |\tau|}}, \ U_A = \frac{U}{N_A}.$$

(12.12)

In the above formulas L_* is the characteristic polymer chain length, k is Boltzmann's constant ($1.381 \cdot 10^{-23} \ J/^\circ K$) and N_A is Avogadros number, ($6.022 \cdot 10^{23} \ mole^{-1}$). In equation (12.12) parameter C is the dimensionless shield constant, a_* and l_* are the polymer bead radius and bond length, and U is the bond dissociation energy per mole. The above functions R and p_c were validated by the comparison of numerically simulated and experimentally obtained data in Chapters 6 and 7.

We will assume that the distribution of polymer molecules entering the contact region is uniform across the lubrication layer, i.e.,

$$W(x_i, z, l) = W_a(l) \ if \ u(x_i, z) > 0,$$

(12.13)

$$W(x_e, z, l) = W_a(l) \ if \ u(x_e, z) < 0.$$

It should be noted that at the inlet to the contact region, $x = x_i$, some of the lubricant enters the contact, some may turn around and exit the contact

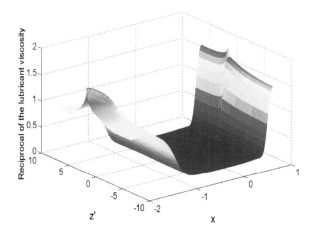

FIGURE 12.2
The reciprocal of the viscosity μ in the lubricants with Newtonian and non-Newtonian rheologies under mixed rolling and sliding conditions and Series I input data ($s_0 = -0.5$). The variable z' is an artificially stretched z-coordinate across the film thickness (namely, $z' = zh(a)/h\,(x)$) to make the relationship more transparent (after Kudish and Airapetyan [24]). Reprinted with permission from the ASME.

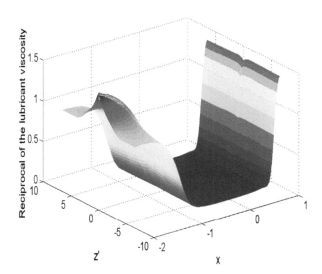

FIGURE 12.11
The reciprocal of the lubricant viscosity μ in Newtonian and non-Newtonian lubrication film under pure rolling conditions ($s_0 = 0$) and Series I input data. The variable z' is an artificially stretched z-coordinate across the film thickness (namely, $z' = zh(a)/h\,(x)$) to make the relationship more transparent (after Kudish and Airapetyan [24]). Reprinted with permission from the ASME.

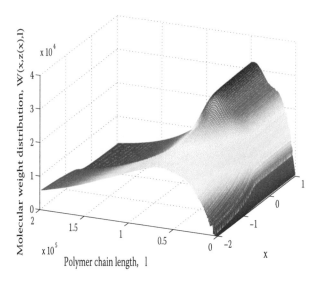

FIGURE 12.12
Molecular weight distribution W of degrading lubricants with Newtonian and non-Newtonian rheologies along the flow streamline closest to $z = 0$ and running through the whole contact below $z = 0$ under pure rolling conditions ($s_0 = 0$) and Series I input data (after Kudish and Airapetyan [24]). Reprinted with permission from the ASME.

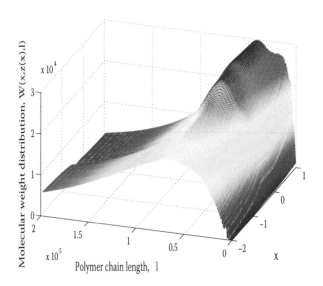

FIGURE 12.15
The molecular weight distribution W of degrading lubricants with Newtonian and non-Newtonian rheologies along the flow streamline z_{31} (x) closest to $z = 0$ and running through the whole contact below $z = 0$ under mixed rolling and sliding conditions ($s_0 = -0.5$) and Series I input data (after Kudish and Airapetyan [24]). Reprinted with permission from the ASME.

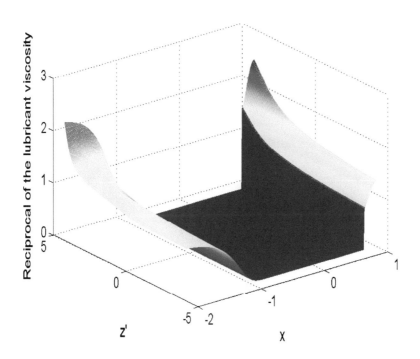

FIGURE 12.17
The reciprocal of the lubricant viscosity μ in a Newtonian lubrication film under pure sliding conditions ($s_0 = -2$) and Series II input data. The variable z' is an artificially stretched z-coordinate across the film thickness (namely, $z' = zh(a)/h(x)$) to make the relationship more transparent (after Kudish and Airapetyan [23]). Reprinted with permission from the ASME.

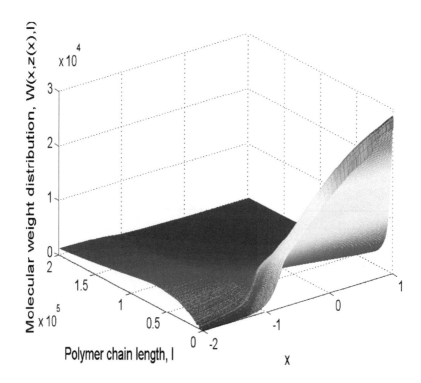

FIGURE 12.18
The molecular weight distribution W of the degrading lubricant with Newtonian rheology along the flow streamline $z_{48}(x)$ running through the entire contact closest to the first turning around streamline under pure sliding conditions ($s_0 = -2$) and Series II input data (after Kudish and Airapetyan [23]). Reprinted with permission from the ASME.

at $x = x_i$. A similar situation is possible at the exit point $x = x_e$ from the contact if the contact surfaces are moving in the opposite directions. In each case it is assumed that when the lubricant enters the contact region (whether through the inlet $x = x_i$ or the exit $x = x_e$ cross sections) the distribution of the polymer molecules is described by $W_a(l)$ from (12.13).

It is important to keep in mind that equations (12.10)-(12.13) need to be solved along the lubricant streamlines $z = z(x)$ described by equations (12.6) and (12.9). Some properties of the kinetic equation (12.10) are established in in Section 6.6. Among these properties is the fact that the mass of the polymer additive is conserved along the lubricant flow streamlines while the number of polymer molecules tends to increase due to polymer molecule scission.

Due to the fact that the lubricant contains some linear polymer additive which may degrade while passing the lubricated contact, the viscosity of the lubricant μ varies as a function of pressure p, concentration c_p, and the density distribution $W(x, z, l)$ of the polymer additive. In particular, we will assume that the lubricant viscosity depends on the polymer additive distribution according to the Huggins and Mark Houwink equations (see Chapter 6 and [6, 7])

$$\mu = \mu_a e^{\alpha_p p} \frac{1 + c_p[\eta] + k_H(c_p[\eta])^2}{1 + c_p[\eta]_a + k_H(c_p[\eta]_a)^2}, \quad [\eta] = k' M_W^\beta,$$

$$M_W = \left\{ \int_0^\infty w_*^\beta W(x, z, w_*) dw_* / \int_0^\infty W(x, z, w_*) dw_* \right\}^{1/\beta},$$

(12.14)

where μ_a is the ambient lubricant viscosity, α_p is the pressure coefficient of viscosity, c_p and $[\eta]$ are the polymer concentration and the intrinsic viscosity, respectively, $[\eta]_a$ is the ambient intrinsic viscosity (at the inlet point x_i), k_H is the Huggins constant ($k_H = 0.25$, see [6, 7]), k' and β are the Mark Houwink constants ($k' = 2.7 \cdot 10^{-4} \ dL/g(g/mole)^{-\beta}$, $\beta = 0.74$, see [6, 7]), w_* is the polymer molecule weight, $w_* = w_m l$, and w_m is the monomer molecular weight.

Therefore, for the given values of x_i, R', E', u_1, u_2, P, μ_a, α_p, α, β, τ_{L0}, τ_{L1}, c_p, w_m, T, U, C, a_*, l_*, k', k_H, and the given ambient distribution of polymer additive $W_a(l)$ the problem is reduced to solution of equations (12.2), (12.3), (12.6)-(12.14) with respect to constants x_e, h_e and functions $p(x)$, $h(x)$, $f(x)$, $\mu(x, z)$, and $W(x, z, l)$.

After the solution of the problem is complete one can determine the frictional stresses τ_1 and τ_2 applied to the contact surfaces according to the formulas

$$\tau_i = (-1)^{i-1} f - \frac{h}{2} \frac{dp}{dx}, \quad (12.15)$$

where index $i = 1$ is for the lower contact surface and $i = 2$ is for the upper contact surface, respectively.

In case of an isothermal problem τ_L may be considered independent of z and τ_L can be pulled out of the integrals in equations (12.6)-(12.9). In addition, in case of no lubricant degradation the latter equations can be simplified further as the lubricant viscosity μ is independent of z. Finally, under the above isothermal conditions, the rheological model described by equations (12.2), (12.3), (12.6)-(12.14) allows for analytical calculation of integrals in equations (12.6)-(12.9) that may simplify and speed up the whole process of numerical solution.

Let us introduce the following dimensionless variables:

$$(x', a, c) = \tfrac{1}{a_H}(x, x_i, x_e), \ (z', h') = \tfrac{1}{h_e}(z, h), \ p' = \tfrac{p}{p_H}, \ \mu' = \tfrac{\mu}{\mu_a},$$

$$u' = \tfrac{2u}{u_1+u_2}, \ w' = \tfrac{2a_H w}{(u_1+u_2)h_e}, \ (f', \tau_i') = \tfrac{2h_e(f, \tau)}{\mu_a(u_1+u_2)}, \ \tau_{L0}' = \tfrac{\tau_{L0}}{p_H}, \qquad (12.16)$$

$$(W', W_a') = \tfrac{1}{W_0}(W, W_a),$$

where a_H and p_H are the half-width of and the maximum pressure in a dry Hertzian contact, W_0 is the characteristic value of the density of molecular weight distribution.

The values of parameters x_i, R', E', u_1, u_2, P, μ_a, α_p, α, β, τ_{L0}, τ_{L1}, c_p, w_m, T, U, C, a_*, l_*, k', and k_H determine the specific values of a number of dimensionless parameters that uniquely identify the solution of the problem. These dimensionless parameters are τ_{L0}/p_H, τ_{L1} and

$$a = \tfrac{x_i}{a_H}, \ V = \tfrac{3\pi^2 \mu_a (u_1+u_2) R' E'}{P^2}, \ Q = \alpha p_H, \ s_0 = \tfrac{2(u_2-u_1)}{u_1+u_2},$$

$$\varepsilon = (\tfrac{a_H}{2R'})^2, \ \gamma = \alpha \tfrac{UA}{kT}, \ \delta = \tfrac{U_A a_H^2}{Ca_* l_*^2 \mu_a (u_1+u_2) R'}, \ \theta = k' c_p w_m^\beta, \qquad (12.17)$$

$$\kappa = \tau_f \tfrac{u_1+u_2}{2a_H}.$$

For simplicity, in the further discussion primes at the dimensionless variables are omitted.

Therefore, the solution of the problem is determined by the values of the dimensionless parameters τ_{L0}, τ_{L1}, the parameters from equations (12.17) and by the function $W_a(l)$. The solution of the problem is represented by the dimensionless functions: pressure $p(x)$, gap $h(x)$, sliding frictional stress $f(x)$, lubricant viscosity $\mu(x, z)$, distribution of molecular weight $W(x, z, l)$ and by two dimensionless constants: the exit coordinate c and the exit film thickness

$$H_0 = \tfrac{2h_e R'}{a_H^2}. \qquad (12.18)$$

In the dimensionless variables, the equations of the problem are

$$\int_{-h/2}^{h/2} \tfrac{\tau_L}{\mu} G(\tfrac{1}{\tau_L}(f + z\tfrac{12H_0^2}{V}\tfrac{dp}{dx}))dz = \frac{s_0}{\sqrt{1+\tfrac{\varepsilon H_0^2}{4}(\tfrac{dh}{dx})^2}},$$

$$\tau_L = \tfrac{12H_0}{V\sqrt{\varepsilon}}(\tau_{L0} + \tau_{L1}p),$$

$$u(x,z) = \frac{1-s_0/2}{\sqrt{1+\frac{\varepsilon H_0^2}{4}(\frac{dh}{dx})^2}} + \int\limits_{-h/2}^{z(x)} \frac{\tau_L}{\mu} G(\frac{1}{\tau_L}(f + \zeta\frac{12H_0^2}{V}\frac{dp}{dx}))d\zeta,$$

$$\frac{d}{dx}\int\limits_{-h/2}^{z(x)} u(x,\zeta)d\zeta = 0, \quad \int\limits_{-h/2}^{z(x)} u(x,\zeta)d\zeta = \frac{h}{2}\frac{1-s_0/2}{\sqrt{1+\frac{\varepsilon H_0^2}{4}(\frac{dh}{dx})^2}} + zu(x,z)$$

$$-\int\limits_{-h/2}^{z(x)} \zeta\frac{\tau_L}{\mu} G(\frac{1}{\tau_L}(f + \zeta\frac{12H_0^2}{V}\frac{dp}{dx}))d\zeta, \quad p(a) = p(c) = \frac{dp(c)}{dx} = 0,$$

(12.19)

$$H_0(h-1) = x^2 - c^2 + \frac{2}{\pi}\int\limits_a^c p(x')\ln\frac{c-x'}{|x-x'|}dx', \quad \int\limits_a^c p(x')dx' = \frac{\pi}{2},$$

$$\tau_i = (-1)^i f + \frac{6H_0^2}{V}h\frac{dp}{dx}, \quad i = 1,2,$$

$$\kappa(u\frac{\partial W}{\partial x} + w\frac{\partial W}{\partial z}) = 2l\int\limits_l^\infty R(x,z,L)p_c(l,L)W(x,z,L)\frac{dL}{L}$$

$$-R(x,z,l)W(x,z,l),$$

(12.20)

$$W(a,z,l) = W_a(l) \; if \; u(a,z) > 0,$$

$$W(c,z,l) = W_a(l) \; if \; u(c,z) < 0,$$

$$R(x,z,l) = 0 \; if \; l \leq L_*$$

$$R(x,z,l) = 1 - (\frac{l}{L_*})^{2\gamma}\exp[-\gamma(\frac{l^2}{L_*^2}-1)] \; if \; l \geq L_*,$$

$$L_* = \sqrt{\frac{H_0\delta}{|\tau|}}, \quad \tau = f + z\frac{12H_0^2}{V}\frac{dp}{dx},$$

$$p_c(l,L) = ln(2)\frac{4|L-2l|}{L^2}\exp[-ln(2)\frac{4l(L-l)}{L^2}],$$

(12.21)

$$\mu(p,W) = e^{Qp}\frac{1+[\eta]+k_H[\eta]^2}{1+[\eta]_a+k_H[\eta]_a^2}, \quad [\eta] = \theta M_W^\beta,$$

$$M_W = \left\{\int\limits_0^\infty l^\beta W(x,z,l)dl / \int\limits_0^\infty W(x,z,l)dl\right\}^{1/\beta}.$$

12.3   Lubricant Flow Topology

Obviously, it is sufficient to only consider cases of non-positive values of the slide-to-roll ratio s_0. Cases with $s_0 > 0$ are identical to the latter ones if the upper and lower solid surfaces are interchanged. Before we consider the numerical method employed for solution of the aforementioned problem, it is necessary to understand the specifics of the topology of the lubricant flow.

In case of pure rolling, $s_0 = 0$ and the flow is symmetric about the x-axis. It has two sets of flow streamlines running through the whole contact and two sets of flow streamlines that enter the contact region through the inlet gap, turn around, and exit the contact region through the same inlet gap. The flow streamlines running through the whole contact region are adjacent to the contact surfaces while the flow streamlines that turn around are adjacent to the x-axis in the inlet zone of the contact. It is important to mention that in the flow there is a special point where both the horizontal u and vertical w components of lubricant velocity are equal to zero. This point is the bifurcation point. The bifurcation point is located in the inlet zone of the lubrication film and outside of the Hertzian region. Because of continuity of the components of the lubricant velocity in the vicinity of the bifurcation point both u and v are small. Moreover, the bifurcation point is the point of intersection of the flow separatrices the curves that separate different zones of the flow, i.e., zones occupied by sets of streamlines running through the whole contact and the ones that turn around and exit through the inlet gap. In the case of pure rolling, there are two symmetric separatrices between the running through the whole contact and turning around streamlines, one separatrix dividing two sets of turning around streamlines, and one separatrix dividing the running through the whole contact streamlines. The latter two coincide with the x-axis.

For cases with $-2 < s_0 < 0$, there are still two pairs of flow streamline sets: two sets of running through the whole contact streamlines and two sets of turning around and exiting through the inlet gap ones. However, they are no longer symmetric about the x-axis. In comparison with the case of pure rolling the entire flow, including the bifurcation point and the separatrices, are shifted toward the upper surface $z = h/2$. For the case of $s_0 = -2$, there is just one set of turning around flow streamlines adjacent to the upper contact surface and one set of flow streamlines running through the whole contact that is adjacent to the lower contact surface everywhere in the contact. Outside of the inlet region the latter set of running through the whole contact flow streamlines is also adjacent to the upper contact surface. For the cases with $-\infty < s_0 < -2$, the lubricant flow is asymmetric. It is represented by two sets of flow streamlines adjacent to the lower and upper contact surfaces and running through the whole contact in opposite directions as well as by two sets of turning around flow streamlines one of which enters the contact region and

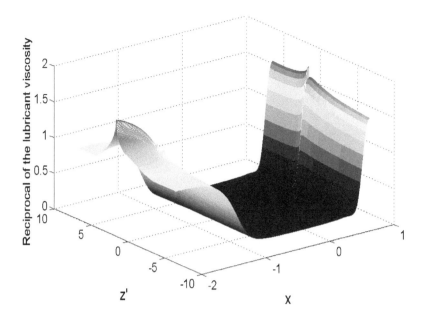

FIGURE 12.2 (See color insert following page 750)

The reciprocal of the viscosity μ in the lubricants with Newtonian and non-Newtonian rheologies under mixed rolling and sliding conditions and Series I input data ($s_0 = -0.5$). The variable z' is an artificially stretched z-coordinate across the film thickness (namely, $z' = zh(a)/h(x)$) to make the relationship more transparent (after Kudish and Airapetyan [24]). Reprinted with permission from the ASME.

exits it through the inlet gap and the other that enters and exits the contact through the exit gap. For $-2 < s_0 \leq 0$, there is only one bifurcation point located between the contact surfaces in the inlet zone while for $s_0 = -2$ the bifurcation point belongs to the upper contact surface in the inlet zone. For $-\infty < s_0 < -2$ there are two bifurcation points between the contact surfaces, one in the inlet and the other in the exit zones of the contact.

Under isothermal conditions in case of no degradation, the lubricant viscosity depends only on pressure p, which, in turn, is a function of x. Therefore, the lubricant viscosity depends only on the distance of a fluid particle from the inlet point $x = a$, i.e., the lubricant viscosity μ depends on x and is independent from z. That makes the function of lubricant viscosity μ a continuous function of x and z because the function of pressure $p(x)$ is a differentiable function.

In case of lubricant degradation, the situation changes completely. Fluid particles that approach the separatrices between two sets of turning around and running through streamlines from different sides follow different flow streamlines and are subjected to different shear stresses over different time periods. Therefore, these fluid particles degrade differently that, in turn, leads to different viscosities of these fluid particles. Thus, we can conclude that degrading lubricants possess a remarkable property. Namely, for all slide-to-roll ratios s_0 (except for $s_0 = 0$ and $s_0 = \pm 2$), the separatrices to which fluid particles approach from both sides in the process of their motion are the curves of the lubricant viscosity $\mu(x, z)$ discontinuity (see Fig. 12.2). In spite of that, the shear stress $\tau(x, z)$ is continuous in the entire lubricated region. This is due to the fact that the sliding frictional stress $f(x)$ and the derivative of pressure dp/dx are continuous functions of x. Taking into account these properties of $\mu(x, z)$ and $\tau(x, z)$ we can conclude that $\frac{\partial u}{\partial z}$ as a function of x and z is also discontinuous across the aforementioned separatrices.

The above–mentioned discontinuities of $\mu(x, z)$ and $\frac{\partial u}{\partial z}$ cause some additional complications in numerical solution of an EHL problem for degrading lubricants with Newtonian and non-Newtonian rheologies. One can expect a much more complex problem formulation for a non-isothermal EHL case due to the fact that conservation of heat flux across of the above separatrices where μ and $\frac{\partial u}{\partial z}$ are discontinuous. It is expected that the lubricant temperature T is also discontinuous across the same separatrices as the lubricant viscosity μ is discontinuous.

12.4 Numerical Method for Solution of Plane EHL Problems for Degrading Lubricants

Certain difficulties in the solution process are caused by the fact that the solution of the above problem is determined within the interval $[a, c]$ where the inlet coordinate a is given while the exit coordinate c is unknown. To simplify the solution process one needs to convert the problem from the interval with a priori unknown boundary $x = c$ into the problem within a region with fixed boundaries, say $[-1, 1]$. That can be achieved by the substitution

$$x = \tfrac{c+a}{2} + \tfrac{c-a}{2} y, \tag{12.22}$$

that makes $-1 \leq y \leq 1$. However, unless it is indicated otherwise, we will discuss the numerical method in the x variable.

The very first step toward the problem solution involves choosing an initial approximation. After that the general iterative process is organized as follows. For the known values of pressure $p(x)$, gap $h(x)$, lubricant viscosity $\mu(x, z)$, film thickness H_0, and exit coordinate c, the numerical procedure for solution of the considered problem consists of several steps in the following order: (a) evaluating the sliding frictional stress $f(x)$, (b) determining the horizontal component of the fluid velocity $u(x, z)$ and fluid flux, (c) calculating the flow streamlines $z(x)$, (d) finding separatrices of the lubricant flow some of which may be the curves of discontinuity of the lubricant viscosity, (e) evaluating the lubricant viscosity $\mu(x, z)$ at every point of the flow by solving the kinetic equation along the flow streamlines, (f) calculating the new approximation of the sliding frictional stress $f(x)$, and (g) solution of the modified Reynolds and gap equations for $p(x)$ and $h(x)$. Each of the above steps of the numerical solution of the problem is considered below in detail. To speed up the solution process, we use the multigrid approach along the x-axis, i.e., we use a series of grids on the interval $[a, c]$ with equidistant nodes x_k, $k = 1, \ldots, N_x$, $x_1 = a$ with increasing sequence of total number of nodes N_x. The grid along the z-axis is introduced by using the convenient basis created by so–called pseudo-streamlines, which can be determined from simple equations $z_j^p(x) = h(x)(2j/N_z^p - 1)/2$, $j = 1, \ldots, N_z^p$, where N_z^p is the number of pseudo-streamlines, including the contact surfaces. Along the l-axis, the axis of polymer molecule chain lengths, the chain nodes l_m, $m = 1, \ldots, N_l$, are introduced in a non-uniform manner in such a way that $l_1 > 0$ and $l_m < l_{m+1}$.

12.4.1   Initial Approximation

The initial approximation of the solution of the formulated EHL problem for degrading lubricant is taken as a solution of the corresponding EHL problem for non-degrading lubricant. To determine this solution we have to perform two tasks: to calculate the sliding frictional stress and to solve the modified Reynolds and gap equations. The latter problems are solved using Newtons method with a damping coefficient. As these procedures are essentially the same as for consequent iterations for the case of degrading lubricant, we will describe them in detail later.

To reach the desired input values of some parameters (such as Q, τ_{L0}, etc.), the process of homotopy is employed. That means that the solution is first obtained for the closest in some sense set of values of the input parameters for which it is known a priori that iterations converge fast. After that the solution is found for a sequence of sets of input parameters (leading to the desired set of values of the input parameters) the values of which are just slightly different from each other. For each new set of values of the input parameters the initial approximation for the problem solution is taken from the solution of the problem for the preceding set of values of the input parameters.

12.4.2   Sliding Frictional Stress $f(x)$

The distribution of the sliding frictional stress $f(x)$ is controlled by the first two equations in (12.19). It is obvious that for any lubricant rheology $f(x) = 0$ for $s_0 = 0$. For $s_0 \neq 0$ the aforementioned equations have an analytical exact solution only in the case of lubricants of Newtonian rheology. Therefore, for $s_0 \neq 0$ the aforementioned equations must be solved numerically. At this stage we assume that the distributions of $p(x)$, $h(x)$, $\mu(x, z)$, and constants H_0 and c are known. To solve the above equations for $f(x_k)$, we use a modified Newtons method:

$$f^{i+1} = f^i - \frac{\alpha_f}{D_f}\left[\int_{-h/2}^{h/2} \frac{\tau_L}{\mu}G\left(\frac{1}{\tau_L}(f^i + z\frac{12H_0^2}{V}\frac{dp}{dx})\right)dz - \frac{s_0}{\sqrt{1+\frac{\varepsilon H_0^2}{4}(\frac{dh}{dx})^2}}\right], \quad (12.23)$$

$$D_f = \frac{V}{12H_0^2}\frac{\tau_L^2}{\mu_{avg}\frac{dp}{dx}}[G(\frac{1}{\tau_L}(f^i + \frac{6H_0^2}{V}h\frac{dp}{dx})) - G(\frac{1}{\tau_L}(f^i - \frac{6H_0^2}{V}h\frac{dp}{dx}))],$$

$$\mu_{avg} = \frac{h}{\int_{-h/2}^{h/2}\frac{dz}{\mu}}. \quad (12.24)$$

In equation (12.23) α_f is an empirically chosen positive coefficient the value of which is adjusted in the process of iterations. Namely, the initial value of α_f is relatively small and it increases with the number of iterations i until it reaches 1. To calculate the above and similar to it integrals, we use the trapezoidal quadrature formula and the grid created by the pseudo-streamlines. To evaluate just one integral with respect to z and to avoid calculating the second similar integral for the derivative of the first one with respect to f, we use an assumption that for the purpose of calculating the above–mentioned derivative of the first integral the lubricant viscosity $\mu(x, z)$ can be approximated and replaced by its average $\mu_{avg}(x)$ across the gap $h(x)$. That allows to calculate the latter integral analytically and, therefore, to simplify and to speed up the process of calculations of the sliding frictional stress $f(x)$. At nodes x_k at which $dp(x_k)/dx = 0$, the expression for D_f can be easily obtained from equations (12.19) by using the limit as $dp/dx \to 0$

$$D_f = \frac{h\tau_L}{\mu_{avg}}G'(\frac{f^i}{\tau_L}) \; if \; \frac{dp(x_k)}{dx} = 0. \quad (12.25)$$

The initial approximation for $f(x)$ we can take as the one-term asymptotic solution of the first equation in (12.19)

$$f = \tau_L G^{-1}\left[\frac{s_0}{\sqrt{1+\frac{\varepsilon H_0^2}{4}(\frac{dh}{dx})^2}}\frac{1}{\tau_L\int_{-h/2}^{h/2}\frac{dz}{\mu}}\right], \quad (12.26)$$

where G^{-1} is the inverse function of the rheological function G. It is obtained under the condition that the sliding frictional stress $f(x)$ is much higher than the rolling frictional stress $6H_0^2/V(hdp/dx)$ (see Chapter 10 and Kudish [5]).

The numerical results show that the iterative method expressed by equations (12.23)-(12.25) converges as well as the modified Newtons method with the same damping coefficient α_f and the value of D_f calculated using the integral of the derivative of the function G. For the present method the required CPU time is almost twice smaller than for the modified Newtons method.

In the process of solution of the Reynolds equation, we will use not only the solution $f(x)$ of the first two equations in (12.19) obtained by using equations

(12.24)-(12.26) but also function $f_{avg}(x)$ that satisfies similar equations in which the local lubricant viscosity $\mu(x, z)$ is replaced by the lubricant viscosity averaged over the film thickness $\mu_{avg}(x)$. The averaged sliding frictional stress $f_{avg}(x)$ is also obtained using equations (12.24)-(12.26) by replacing $\mu(x, z)$ by $\mu_{avg}(x)$ in equation (12.24).

The described numerical method converges very fast for moderate values of the slide-to-roll ratio s_0, pressure coefficient of viscosity Q, speed-load parameter V, and the parameters τ_{L0} and τ_{L1} controlling the limiting shear stress τ_L. However, in cases of the presence of a limiting stress τ_L ($|\tau| \leq \tau_L$) and for high values of s_0, Q, V, and low values of τ_{L0} and τ_{L1} the frictional stress τ may be very close to τ_L (see (12.21)). It means that even very small variations in the sliding frictional stress $f(x)$ may destabilize the iteration process. Such a situation is typical for ill-conditioned problems. Obviously, such variations of $f(x)$ are inevitable during any iterative solution of the first equation in (12.19). One of the ways to deal with such a problem is to dampen the incremental changes of $f(x)$ during iterations based on Newtons method. For higher values of s_0, Q, V, and smaller values of τ_{L0} and τ_{L1}, this dampening must be stronger. Therefore, under the above conditions in cases of strong dampening the iteration process converges relatively slow.

12.4.3 Horizontal Component of the Lubricant Velocity and Flux

After the initial approximation for the solution is chosen or one full iteration of the problem solution is completed, we know the approximate values of pressure $p(x)$, gap $h(x)$, lubricant viscosity $\mu(x, z)$, film thickness H_0, and the exit coordinate c. Based on the above values of $p(x)$, $h(x)$, $\mu(x, z)$, H_0, and c, the formula for $u(x, z)$ from equations (12.19), and using the trapezoidal quadrature formula for integration with respect to ζ on the grid of pseudo-streamlines $z_j^p(x)$, we obtain the distribution of the lubricant velocity $u(x_k, z_j^p(x_k))$, where k is the number of the node x_k. The lubricant flux described by the integral of u with respect to z from equations (12.19) is also calculated on the pseudo-streamlines $z_j^p(x)$ by the trapezoidal quadrature formula for integration with respect to z.

12.4.4 Lubricant Flow Streamlines

Because polymer scission occurs while lubricant particles move along the flow streamlines $z_j(x)$, $j = 1, \ldots, N_z$, there is a necessity to determine the flow streamlines. It seems that an efficient way to do that is to solve a series of initial-value problems

$$\frac{dz_j}{dx} = \frac{w(x, z_j)}{u(x, z_j)}, \quad z_j(x_0) = z_j^p(x_0), \quad j = 1, \ldots, N_z,$$

$$x_0 = a \ if \ u(a, z_j^p(a)) > 0, \quad x_0 = c \ if \ u(c, z_j^p(c)) < 0,$$

$$(12.27)$$

using one of the methods such as the modified Euler method (N_z is the number of flow streamlines). In equations (12.27) $w(x, z)$ is the z-component of the lubricant velocity. The way the x-coordinate of the initial point of the streamline x_0 is determined in (12.27) depends on the topology of the lubricant flow. The only way to find the z-component of the lubricant velocity $w(x, z)$ is to solve the continuity equation

$$\frac{\partial u}{\partial x} + \frac{\partial w}{\partial z} = 0.$$

It can be done by integrating this equation over small rectangular sells with vertices $(x_k, z_j^p(x_k))$, $(x_{k+1}, z_j^p(x_{k+1}))$, $(x_k, z_{j+1}^p(x_k))$, and $(x_{k+1}, z_{j+1}^p(x_{k+1}))$ and using the corresponding boundary conditions [23, 27]. However, our practice showed that the numerical realization of the above process does not provide sufficient precision and stability in determination of the z-component of the lubricant velocity $w(x, z)$ and flow streamlines $z_j(x)$. Therefore, we do not recommend it for practical calculations. In fact, it causes instability in $u(x, z)$.

The calculation of the lubricant flow streamlines is done based on the integrated form of the equation for the x-component of the lubricant flux following from equations (12.19):

$$\int_{-h(x_k)/2}^{z_j(x_k)} u(x_k, z)dz = \int_{-h(x_0)/2}^{z_j(x_0)} u(x_0, z)dz,$$

$$k = 1, \ldots, N_x, \ j = 1, \ldots, N_1,$$

$$x_0 = a \ if \ u(a, z_j^p(a)) > 0, \ x_0 = c \ if \ u(c, z_j^p(c)) < 0,$$

(12.28)

and linear interpolation with respect to z (N_1 is the number of running through and turning around flow streamlines adjacent to the lower contact surface). This is exactly the procedure used to determine the flow streamlines passing through the whole lubricated contact (see Section 12.5). For turning around flow streamlines, the procedure is more complex.

Let us consider the general procedure of the flow streamline computation for the case when $-2 < s_0 < 0$, and there are still two pairs of flow streamline sets: two sets of running through the whole contact streamlines and two sets of turning around and exiting through the inlet gap ones. Prior to calculating the flow streamlines, we determine the so–called zero-curves in the flow, $z_j^0(x)$, $j = 1, 2$, along which the x-component of the lubricant velocity is equal to zero. It is done using the known distribution of the x-component of the lubricant velocity $u(x, z)$ and linear interpolation with respect to z. After the zero-curves $z_j^0(x)$ are determined, we first calculate the sets of lower flow streamlines $z_j(x)$ that are closer to the lower surface $z = -h/2$ (both running through the whole contact and turning around ones) and then the same kind ones that are closer to the upper contact surface $z = h/2$. We determine the initial points of the flow streamlines $z_j(x)$ starting from the lower surface that corresponds to $j =$

1. If $u(a, z_j^p(a)) > 0$, then the point $(a, z_j^p(a))$ is considered to be the starting point for the flow streamline $z_j(x)$, otherwise it is skipped and we go to the next j. Starting from the initial point we calculate the coordinates of the points on the flow streamline based on equation (12.28) until the flow streamline intersects the lower zero-curve or exits through the exit cross section of the contact at $x = c$. If it exits through the exit cross section of the contact at $x = c$, the determination of the flow streamline is complete. Otherwise, using linear interpolation, we find the point $(a, z_j^e(a))$ in the inlet cross section of the contact closest to the initial point $(a, z_j^p(a))$ at which the lubricant flux through the cross section determined by the points $(a, -h(a)/2)$ and $(a, z_j^e(a))$ is equal to the flux through the cross section determined by the points $(a, -h(a)/2)$ and $(a, z_j^p(a))$. Starting from the point $(a, z_j^e(a))$ we solve the equation

$$\int_{-h(x_k)/2}^{z_j(x_k)} u(x_k, z)dz = \int_{-h(x_0)/2}^{z_j(x_0)} u(x_0, z)dz,$$

$$k = 1, \ldots, N_x, \; j = N_{t1}, \ldots, N_1,$$

$$x_0 = a \; if \; u(a, z_j^e(a)) < 0, \; x_0 = c \; if \; u(c, z_j^e(c)) > 0,$$

(12.29)

to find the exiting branches of the turning around flow streamlines (N_{t1} is the number of the turning around flow streamline closest to the lower contact surface). This process continues until the flow streamline in question reaches the lower zero-curve.

For the upper sets of the flow streamlines, the algorithm is similar. However, it is more convenient to solve similar equation instead of solving equations (12.28) and (12.29)

$$\int_{z_j(x_k)}^{h(x_k)/2} u(x_k, z)dz = \int_{z_j(x_0)}^{h(x_0)/2} u(x_0, z)dz,$$

$$k = 1, \ldots, N_x, \; j = N_1 + 1, \ldots, N_z,$$

$$x_0 = a \; if \; u(a, z_j^p(a)) > 0, \; x_0 = c \; if \; u(c, z_j^p(c)) < 0,$$

(12.30)

$$\int_{z_j(x_k)}^{h(x_k)/2} u(x_k, z)dz = \int_{z_j(x_0)}^{h(x_0)/2} u(x_0, z)dz,$$

$$k = 1, \ldots, N_x, \; j = N_1 + 1, \ldots, N_2,$$

$$x_0 = a \; if \; u(a, z_j^e(a)) < 0, \; x_0 = c \; if \; u(c, z_j^e(c)) > 0,$$

where N_2 is the number of the turning around flow streamline closest to the upper contact surface. In all other respects the solution procedure is identical to that described above.

For different flow topologies, the process of calculation of the flow streamlines is similar to that described above but somewhat different. For example, for $s_0 < -2$ or $s_0 > 2$ we still have two sets of both kind of running through the whole contact and turning around flow streamlines. Therefore, after the computation of the flow streamlines that start at the inlet cross section is complete (see equations (12.28), (12.29), or (12.30)) for this topology of the lubricant flow, we employ a similar process of determining of the flow streamlines that start at the exit cross section of the lubricated contact.

In cases of $s_0 = -2$ or $s_0 = 2$, we have just one set of running through the whole contact and turning around flow streamlines and just two sets of equations (12.28), (12.29), or (12.30) are used, respectively.

12.4.5 Separatrices of the Lubricant Flow

As soon as all flow streamlines are found depending on the flow topology we can determine the separatrices in the flow which are needed for the proper determination of the viscosity of a degrading lubricant. The idea behind the approach that allows determining the separatrices is based on conservation of flux. To find the separatrices, we determine the lubricant fluxes U_1, U_2,... through the cross sections between the contact surfaces and the bifurcation points. After that we find curves (separatrices) that represent the boundaries of the flow regions the flux through any cross section of which is preserved, i.e., is equal to U_1, U_2, and so on.

Let us consider the case of $0 \leq |s_0| < 2$. First, we find the bifurcation point as a point of intersection of the zero-curves

$$z_1^0(x) = z_2^0(x). \tag{12.31}$$

Suppose the x-coordinate of the bifurcation point that satisfies equation (12.31) is x_b. To find the separatrices, we need to solve equations

$$\int\limits_{-h(x_k)/2}^{z_1^s(x_k)} u(x_k, z)dz = \int\limits_{-h(x_b)/2}^{z_1^0(x_b)} u(x_b, z)dz, \ z_1^s \leq z_1^0,$$

$$x_k \leq x_b \ if \ s_0 \leq 0 \ or \ x_k \geq x_b \ if \ s_0 \geq 0, \tag{12.32}$$

$$\int\limits_{z_2^s(x_k)}^{h(x_k)/2} u(x_k, z)dz = \int\limits_{z_1^0(x_b)}^{h(x_b)/2} u(x_b, z)dz, \ z_2^s \geq z_2^0,$$

$$x_k \leq x_b \ if \ s_0 \leq 0 \ or \ x_k \geq x_b \ if \ s_0 \geq 0,$$

$$\int\limits_{-h(x_k)/2}^{z_3^s(x_k)} u(x_k, z)dz = \int\limits_{-h(x_b)/2}^{z_1^0(x_b)} u(x_b, z)dz, \ z_1^0 < z_3^s < z_2^0,$$

$$x_k \leq x_b \ if \ s_0 \leq 0 \ or \ x_k \geq x_b \ if \ s_0 \geq 0,$$

$$\int_{-h(x_k)/2}^{z_4^s(x_k)} u(x_k, z)dz = \int_{-h(x_b)/2}^{z_1^0(x_b)} u(x_b, z)dz,$$

$$x_k \geq x_b \ if \ s_0 \leq 0 \ or \ x_k \leq x_b \ if \ s_0 \geq 0.$$

For $| s_0 |= 2$ there is just one bifurcation point either on the upper ($s_0 = -2$) or the lower ($s_0 = 2$) contact surfaces. It is important to remember that for $s_0 = -2$ and $s_0 = 2$ we have $z_2^0(x) = h(x)/2$ and $z_1^0(x) = -h(x)/2$, respectively. Therefore, we need to solve one of the equations

$$\int_{-h(x_k)/2}^{z_1^s(x_k)} u(x_k, z)dz = \int_{-h(x_b)/2}^{z_1^0(x_b)} u(x_b, z)dz,$$

$$x_k \leq x_b, \ z_1^s(x_k) \leq z_1^0(x_k) \ if \ s_0 = -2, \tag{12.33}$$

$$\int_{z_2^s(x_k)}^{h(x_k)/2} u(x_k, z)dz = \int_{z_1^0(x_b)}^{h(x_b)/2} u(x_b, z)dz,$$

$$x_k \leq x_b, \ z_2^s(x_k) \geq z_2^0(x_k) \ if \ s_0 = 2.$$

Now, let us consider the case of $| s_0 |> 2$. First, we need to find the bifurcation points, i.e., the right most point x^L on the zero-curve $z_1^0(x)$ that starts in the inlet zone and the left most point x^R on the zero-curve $z_2^0(x)$ that starts in the exit zone of the contact. After that we need to solve the following equations:

$$\int_{-h(x_k)/2}^{z_1^s(x_k)} u(x_k, z)dz = \int_{-h(x^L)/2}^{z_1^0(x^L)} u(x^L, z)dz,$$

$$x_k \leq x^L, \ z_1^s(x_k) \leq z_1^0(x_k),$$

$$\int_{z_2^s(x_k)}^{h(x_k)/2} u(x_k, z)dz = \int_{z_2^0(x^R)}^{h(x^R)/2} u(x^R, z)dz, \tag{12.34}$$

$$x_k \leq x^R, \ z_2^s(x_k) \geq z_2^0(x_k),$$

$$\int_{-h(x_k)/2}^{z_3^s(x_k)} u(x_k, z)dz = \int_{-h(x^L)/2}^{z_1^0(x^L)} u(x^L, z)dz,$$

$$x^L \leq x_k \leq x^R.$$

Equations (12.32)-(12.34) are easily solved by employing linear interpolation and using the earlier determined lubricant flux.

12.4.6 Solution of the Kinetic Equation and Evaluating the Lubricant Viscosity

First, let us consider the case of $0 \leq |s_0| < 2$ for which there are two zero-curves that start at the inlet cross section and end (intersect) at the bifurcation point in the inlet zone and there is no zero-curve that passes through the whole contact from $x = a$ to $x = c$. To calculate the lubricant viscosity, we have to determine the distribution of the molecular weight $W(x, z, l)$ along the flow streamlines $z = z_j(x)$. That means we have to integrate the kinetic equation (see equations (12.20)) along all flow streamlines $z_j(x)$, $j = 1, \dots, N_z$. Therefore, along each of the streamlines $z_j(x)$, $j = 1, \dots, N_z$, it is convenient to introduce a parameter t (time) according to the formulas

$$dt = \frac{dx}{\kappa u(x, z_j(x))}. \qquad (12.35)$$

By integrating equation (12.35), we can introduce a partition of the t-axis as follows

$$t_{k+1} = t_k + \frac{1}{\kappa} \int_{x_k}^{x_{k+1}} \frac{dx}{u(x, z_j(x))}, \quad k = 1, \dots, N_x, \ t_1 = 0.$$

Discretizing the latter equation gives the set of time nodes t_k

$$t_{k+1} = t_k + \frac{2(x_{k+1} - x_k)}{\kappa[u(x_k, z_j(x_k)) + u(x_{k+1}, z_j(x_{k+1}))]}, \quad k = 1, \dots, N_x, \ t_1 = 0. \qquad (12.36)$$

For simplicity let us introduce the notations

$$R_j(t, l) = R(x_j(t), z_j(t), l), \quad W_j(t, l) = W(x_j(t), z_j(t), l). \qquad (12.37)$$

Then the kinetic equation can be reduced to

$$\frac{dW_j}{dt} = 2l \int_l^\infty R_j(t, L) p_c(l, L) W_j(t, L) \frac{dL}{L} - R_j(t, l) W_j(t, l), \qquad (12.38)$$

$$j = 1, \dots, N_z.$$

Using the trapezoidal quadrature formula for approximation of the integral in equation (12.38), that further is denoted by $I(t_n, l_m)$, we get

$$I(t_k^R, t_k, l_m) = I_1(t_k^R, t_k, l_m) + I_2(t_k^R, t_k, l_m),$$

$$I_1(t_k^R, t_k, l_m) = \frac{l_{m+1} - l_m}{2l_m} = R_j(t_k^R, l_m) p_c(l_m, l_m) W_j(t_k, l_m),$$

$$I_2(t_k^R, t_k, l_m) = \frac{1}{2} \sum_{n=m}^{N_l - 2} \frac{l_{n+2} - l_n}{l_{n+1}} R_j(t_k^R, l_{n+1}) p_c(l_m, l_{n+1})$$

$$(12.39)$$

$$\times W_j(t_k, l_{n+1}) + \frac{l_{N_l} - l_{N_l - 1}}{2l_{N_l}} R_j(t_k^R, l_{N_l}) W_j(t_k, l_{N_l}), \quad m = 1, \dots, N_l,$$

where t_k^R is an approximation of t_k and $t_k^R = t_k$ or $t_k^R = t_{k-1}$.

By employing the forward difference approximation for the derivative of W_j and satisfying equation (12.38) at $t_{k+1/2} = (t_k + t_{k+1})/2$ and $l = l_m$, we obtain

$$\frac{W_j(t_{k+1},l_m) - W_j(t_k,l_m)}{t_{k+1} - t_k} = l_m[I(t_k,t_k,l_m) + I(t_k,t_{k+1},l_m)]$$

$$- \frac{R_j(t_k,l_m)}{2}[W_j(t_k,l_m) + W_j(t_{k+1},l_m)]. \tag{12.40}$$

The specific approximation of the integral in equation (12.38) used in equation (12.40) is caused by the fact that the probability of scission R depends on the lubricant viscosity μ through the sliding frictional stress f (see equations (12.19) and (12.21)) and that the lubricant viscosity $\mu(x_{k+1}, z_j(x_{k+1}))$ (that corresponds to the viscosity μ of the lubricant particle moving along the flow streamline $z_j(x)$ at the time moment t_{k+1}) is unknown while all values of $W_j(t_{k+1}, l_m)$, $m = 1, \ldots, N_l$, are not determined yet. That follows from equations (12.20) and (12.21) for the viscosity μ that represents the integral dependence between μ and $W(t, l)$. Solving equations (12.39) and (12.40) for $W_j(t_{k+1}, l_m)$, we find the final scheme

$$W_j(t_{k+1},l_m) = \frac{2 - R_j(t_k,l_m)\Delta t_k}{2 + R_j(t_k,l_m)\Delta t_k[1-(l_{m+1}-l_m)p_c(l_m,l_m)]}W_j(t_k,l_m)$$

$$+ \frac{2l_m\Delta t_k[I(t_k,t_k,l_m) + I_2(t_k,t_{k+1},l_m)]}{2 + R_j(t_k,l_m)\Delta t_k[1-(l_{m+1}-l_m)p_c(l_m,l_m)]}, \tag{12.41}$$

$$j = 1, 2, \ldots, \quad k = 1, 2, \ldots, \quad m = N_l, \ldots, 1,$$

$$W_j(0,l_m) = W_a(l_m), \quad \Delta t_k = t_{k+1} - t_k.$$

It is important to emphasize that at any given point x_{k+1} (time moment t_{k+1}) on any given flow streamline $z_j(x)$ computation of $W_j(t_{k+1}, l_m)$ with respect to l_m starts with $m = N_l$ and ends with $m = 1$, i.e., it is done in the direction from longer molecular chains to shorter ones.

As soon as all $W_j(t_{k+1}, l_m)$, $m = 1, \ldots, N_l$, for the fixed time moment t_{k+1} are computed from equations (12.20) by simple integration using the trapezoidal rule we can determine the viscosity $\mu_j(t_{k+1}) = \mu(x_{k+1}, z_j(x_{k+1}))$ of a lubricant particle that in the process of its motion underwent degradation. That process gives us the distribution of the lubricant viscosity along the flow streamlines.

In the process of solution of the equations for the sliding frictional stress $f(x)$ and the Reynolds equation, we need to perform integration with respect to z. It is almost impossible to do by using the flow parameters distributed along the flow streamlines and, at the same time, it is easy to do knowing the flow parameters on the grid of pseudo-streamlines $z_j^p(x)$, $j = 1, \ldots, N_z^p$. Therefore, there is a need to consider in more detail the way the lubricant viscosity $\mu(x_{k+1}, z_j(x_{k+1}))$ is evaluated along the flow streamlines $z_j(x)$, $j = 1, \ldots, N_z$, is mapped onto the grid of flow pseudo-streamlines. It is done using linear interpolation with respect to z. In particular, for $s_0 = 0$ the lubricant viscosity is continuous everywhere, and the standard linear interpolation provides its values at any point of interest. In the case of $0 <| s_0 |< 2$, there is a necessity to carefully evaluate the lubricant viscosity above and below the separatrix, the lubricant viscosity discontinuity curve, running from the inlet through the whole contact to the exit. To evaluate the lubricant viscosity in the vicinity of this separatrix, we linearly extrapolate the lubricant viscosity from the two closest adjacent to the separatrix flow streamlines that are located above and below the separatrix while preserving the discontinuity of the lubricant viscosity.

For $| s_0 |= 2$ one of the zero-curves that coincides with one of the contact surfaces, the motionless one, that is at the same time a flow streamline and a pseudo-streamline. To determine the lubricant viscosity along such a motionless contact surface we, obviously, cannot use parameter t introduced by equation (12.35). In this case to evaluate the viscosity of a lubricant particle adjacent to the motionless contact surface, we use linear extrapolation of the lubricant viscosity from the two closest adjacent to the motionless contact surface flow streamlines.

For the case of $| s_0 |> 2$, the only zero-curve in the lubricant flow is a separatrix that separates lubricant particles moving in opposite directions. At the same time, this separatrix is a curve of discontinuity of the lubricant flow. To evaluate the lubricant viscosity in the vicinity of this separatrix, we use the same approach as in the case of $0 <| s_0 |< 2$.

12.4.7   Solution of the Reynolds Equation

To solve the modified Reynolds and gap equations, we use the modified Newton's method that is applied to fourth through eighth equations in (12.19) for $z(x) = h(x)/2$. For this purpose the fourth equation in (12.19) is integrated with respect to z and is represented in the form

$$\int_{-h(x)/2}^{h(x)/2} u(x,z)dz = \int_{-1/2}^{1/2} u(c,z)dz. \qquad (12.42)$$

The modified Reynolds equation from 12.19 is solved on a sequence of grids with decreasing step sizes along the x-axis while the grid created by the pseudo-streamlines along the z-axis remains the same. The further calculations are based on the modified Newton's method. Let us introduce the incremental variations of the major variables: $\triangle p(x_k) = \triangle p_k$, $\triangle f(x_k) = \triangle f_k$, $\triangle h(x_k) = \triangle h_k$, $\triangle \mu(x_k, z)$, $k = 1,\ldots,N_x$, $\triangle H_0$, and $\triangle c$. The unknowns in this iterative process are the incremental variations in pressure $\triangle p(x_k)$, $k = 1,\ldots,N_x$, film thickness $\triangle H_0$, and exit coordinate $\triangle c$ while the variations of $\triangle f(x_k)$ and $\triangle h(x_k)$, $k = 1,\ldots,N_x$, are found in terms of the former ones.

Let us assume that the i-th iterates of all contact parameters are known. The main assumptions used in the following numerical procedure are: (a) the term $\varepsilon H_0^2 (dh/dx)^2$ is small and its variation can be neglected, (b) we will assume that the lubricant viscosity variations $\Delta\mu(x_k, z)$, $k = 1, \ldots, N_x$, are mostly caused by variations of pressure $\Delta p(x_k)$, $k = 1, \ldots, N_x$, and the variations of lubricant viscosity due to variations of the molecular weight W can be neglected, (c) the concept of averaged viscosity introduced in calculation of the sliding frictional stress $f(x)$ will be used for approximation of the Jacobian of the system of linearized equations. Then using equations (12.19) and the two-term Taylor formula, we can linearize the modified Reynolds equation (12.42) in the vicinity of its i-th iterates as well as its boundary conditions for p and the integral condition in the form

$$\frac{\Delta h_k}{\sqrt{1 + \frac{\varepsilon H_0^2}{4}(\frac{dh}{dx})^2}} - \frac{h\Delta h_k}{4}\left[\frac{\tau_L}{\mu(x, h/2)}G(\frac{\tau_+}{\tau_L}) - \frac{\tau_L}{\mu(x, -h/2)}G(\frac{\tau_-}{\tau_L})\right]$$

$$-\Delta p_k \int_{-h/2}^{h/2} \frac{\partial \tau_L}{\partial p}\frac{1}{\mu}G(\frac{\tau}{\tau_L})\zeta d\zeta + \Delta p_k \int_{-h/2}^{h/2} \frac{\tau_L}{\mu^2}\frac{\partial \mu}{\partial p}G(\frac{\tau}{\tau_L})\zeta d\zeta$$

$$- \int_{-h/2}^{h/2} \frac{\tau_L}{\mu}G'(\frac{\tau}{\tau_L})[-\frac{1}{\tau_L^2}\frac{\partial \tau_L}{\partial p}\tau \Delta p_k + \frac{1}{\tau_L}(\Delta f_k + \frac{24 H_0}{V}\zeta \frac{dp}{dx}\Delta H_0$$

$$+\frac{12 H_0^2}{V}\zeta\Delta(\frac{dp}{dx}) - \frac{12 H_0^2}{V}\zeta\frac{dp}{dx}\frac{\Delta c}{c-a})]\zeta d\zeta = \int_{-h/2}^{h/2} \frac{\tau_L}{\mu}G(\frac{\tau}{\tau_L})\zeta d\zeta$$

$$-\frac{h}{\sqrt{1 + \frac{\varepsilon H_0^2}{4}(\frac{dh}{dx})^2}}, \quad k = 2, \ldots, N_x - 1, \quad \Delta p_1 = \Delta p_{N_x} = 0, \qquad (12.43)$$

$$\frac{1}{2}\sum_{k=1}^{N_x-1}[\Delta p_k + \Delta p_{k+1}](x_{k+1} - x_k)$$

$$+\frac{\Delta c}{2(c-a)}\sum_{k=1}^{N_x-1}[\Delta p_k + \Delta p_{k+1}](x_{k+1} - x_k) = \frac{\pi}{2}$$

$$-\frac{1}{2}\sum_{k=1}^{N_x-1}[p_k + p_{k+1}](x_{k+1} - x_k),$$

$$\tau = f + \frac{12 H_0^2}{V}\zeta\frac{dp}{dx}, \quad \tau_+ = f + \frac{6 H_0^2}{V}h\frac{dp}{dx}, \quad \tau_- = f - \frac{6 H_0^2}{V}h\frac{dp}{dx}.$$

To determine the values of $\Delta f(x_k)$ and $\Delta h(x_k)$, $k = 1, \ldots, N_x$, we linearize the equations for $f(x)$ and $h(x)$ from (12.19)

$$\frac{\Delta h_k}{2}\left[\frac{\tau_L}{\mu(x, h/2)}G(\frac{\tau_+}{\tau_L}) + \frac{\tau_L}{\mu(x, -h/2)}G(\frac{\tau_-}{\tau_L})\right]$$

$$+\triangle p_k \left[\int\limits_{-h/2}^{h/2} \frac{\partial \tau_L}{\partial p} \frac{1}{\mu} G(\frac{\tau}{\tau_L}) d\zeta - \int\limits_{-h/2}^{h/2} \frac{\tau_L}{\mu^2} \frac{\partial \mu}{\partial p} G(\frac{\tau}{\tau_L}) d\zeta \right]$$

$$+ \int\limits_{-h/2}^{h/2} \frac{\tau_L}{\mu} G'(\frac{\tau}{\tau_L}) [-\frac{1}{\tau_L^2} \frac{\partial \tau_L}{\partial p} \tau \triangle p_k + \frac{1}{\tau_L}(\triangle f_k + \frac{24 H_0}{V} \zeta \frac{dp}{dx} \triangle H_0$$

$$\tag{12.44}$$

$$+ \frac{12 H_0^2}{V} \zeta \triangle (\frac{dp}{dx}) - \frac{12 H_0^2}{V} \zeta \frac{dp}{dx} \frac{\triangle c}{c-a})] d\zeta = \frac{s_0}{\sqrt{1 + \frac{\varepsilon H_0^2}{4}(\frac{dh}{dx})^2}}$$

$$- \int\limits_{-h/2}^{h/2} \frac{\tau_L}{\mu} G(\frac{\tau}{\tau_L}) d\zeta, \ k = 1, \ldots, N_x,$$

$$H_0 \triangle h_k = \frac{1}{\pi} \sum_{k=1}^{N_x - 1} [\triangle p_k + \triangle p_{k+1}] I(x_k, x_{k+1}, c) - \triangle H_0 (h - 1)$$

$$+ [(x_k - c)(x_k + c - 2a) - H_0(h - 1)] \frac{\triangle c}{c-a}, \tag{12.45}$$

$$I(x_k, x_{k+1}, c) = \int\limits_{x_k}^{x_{k+1}} \ln \frac{|c - x|}{|x_k - x|} dx.$$

In the above equations all variables whose indexes are not shown are calculated at x_k based on the values of ith iterates of the problem solution. Equations (12.43)-(12.45) are obtained by making the transition from the variable x to the variable y according the substitution from equation (12.22), then linearizing equations and, finally, returning back to the variable x.

Taking into account equation (12.44) for $\triangle f(x_k)$ and the fact that $\triangle f(x_k)$ are independent from z, we can eliminate it from equation (12.43). After that with the help of the expression for $\triangle h(x_k)$, we can reduce solution of the modified Reynolds equation to a system of $N_x + 2$ linear algebraic equations for the increments of pressure $\triangle p(x_k)$, $k = 1, \ldots, N_x$, film thickness $\triangle H_0$, and exit coordinate $\triangle c$. Solution of this system is an extremely long process as coefficients of the system depend on the values of a number of integrals in equations (12.43)-(12.45). In equations (12.43) and (12.44), the necessity of numerical evaluation of a number of integrals with respect to z is the main cause of the significant slow down of solution of the above system. It is obvious, that to solve the equation for $f(x)$ and the modified Reynolds and gap equations we must evaluate the integrals in the right-hand sides of equations (12.43) and (12.44). To significantly simplify and to evaluate analytically the rest of the integrals in equations (12.43), (12.44) that control the Jacobian of the system, we will use instead of the local lubricant viscosity $\mu(x, z)$ and the sliding frictional stress $f(x)$ the averaged viscosity $\mu_{avg}(x)$ and the averaged sliding stress $f_{avg}(x)$, respectively. This assumption allows us to do all integrations analytically using integration by parts which significantly speeds up the calculations of the approximate Jacobian of the system. Integrals in equation (12.45) are evaluated analytically.

After the above system is solved for the incremental variations in pressure, film thickness, and exit coordinate the incremental variations of pressure undergo the process of filtering out high frequencies. It is done by applying the Fast Fourier transform to the set of $\triangle p(x_k)$, $k = 1, \ldots, N_x$, cutting off the tail of high frequencies, and inverting the filtered values of $\triangle p(x_k)$, $k = 1, \ldots, N_x$. The degree of filtering decreases with the increase of the iteration number and stops after a certain number of iterations. As soon as filtering is done the new iterates of pressure, film thickness, and exit coordinate are determined by the formulas

$$p_k^{i+1} = p_k^i + \alpha_R \triangle p_k, \ k = 1, \ldots, N_x, \ H_0^{i+1} = H_0^i + \alpha_R \triangle H_0,$$

$$c^{i+1} = c^i + \alpha_R \triangle c, \tag{12.46}$$

where α_R is a positive damping coefficient the value of which is changed depending on the iteration number i. For a certain number of first iterations α_R is kept small and as the iteration process progresses, and the approximate solution gets sufficiently close to the exact one the value of α_R gradually increases.

As we move to a grid with smaller step sizes the Jacobian of the system is calculated on just first few iterations after which it is not changed from iteration to iteration until the solution converges to the desired precision. That significantly accelerates the iterative process without hampering its convergence.

The derivative of the gap h is used in a number of steps of the iteration process. To make the derivative of the gap dh/dx consistent with the Reynolds equation after each iteration dh/dx is determined from the Reynolds equation according to the formula

$$\frac{dh}{dx} = -\sqrt{1 + \frac{\varepsilon H_0^2}{4}\left(\frac{dh}{dx}\right)^2} \int\limits_{-h/2}^{h/2} \frac{\partial u(x,z)}{\partial x} dz. \tag{12.47}$$

The partial derivative $\partial u/\partial x$ is calculated from the expression for the directional derivative du/ds

$$\frac{du}{ds} = \frac{\partial u}{\partial x}\frac{\triangle x}{\sqrt{\triangle x^2 + \varepsilon H_0^2 \triangle z^2}} + \frac{\partial u}{\partial z}\frac{\triangle z}{\sqrt{\triangle x^2 + \varepsilon H_0^2 \triangle z^2}}, \tag{12.48}$$

which is obtained numerically using the expression for $\partial u/\partial z$ that is analytically determined from the equation for $u(x, z)$ (see equations (12.19)).

In case of an isothermal problem τ_L may be considered independent of z, and τ_L can be pulled out of the integrals in equations (12.19). In addition to that, in case of no lubricant degradation the latter equations can be simplified further as the lubricant viscosity μ is independent of z. Finally, under the above isothermal conditions the rheological model described by equation (12.1) allows for analytical calculation of integrals in equations (12.19) that may simplify and speed up the whole process of numerical solution. For a case of non-isothermal lubrication, the numerical method can be design in a similar fashion by averaging not only μ but also τ_L as it may depend on the lubricant temperature and, as a result of that, on variable z.

12.5 Solutions for Lubricants with Newtonian and Non-Newtonian Rheologies without Degradation

In this section we will consider the EHL solutions in cases of lubricants with Newtonian and non-Newtonian rheologies without degradation. That allows determining the reversible loss of lubricant viscosity.

One of the challenges in application of the above approach to practical EHL problems is the necessity to calibrate the kinetics part of the model, i.e., to be able to choose the proper values for constants C and α. In Chapter 6 the values of these parameters were calculated by comparison with experimental data. However, in these experiments the shear stress to which the flow was subjected, the lubricant temperature and velocity were unknown. The lack of knowledge of these parameters was compensated by the proper choice of the value of the shielding constant C that provided the right value for the group $C \mid \tau \mid /T$ that affects the degradation process. Therefore, the actual values of these parameters must be determined from detailed experimental data with all necessary input and output data available. Below we will consider two series of results obtained for different combinations of parameters C and α: Series I for

$$C = 15.5, \ \alpha = 0.008, \tag{12.49}$$

and Series II for

$$C = 0.055, \ \alpha = 0.008. \tag{12.50}$$

In Series II simulation of only nominally Newtonian lubricant will be considered. In both series of simulations, we used the following dimensional

$$U = 347kJ/mole, \ T = 310°K, \ a_* = 0.374nm, \ l_* = 0.154nm,$$

$$w_m = 35.1g/mole, \ a_H = 10^{-3}m, \ p_H = 1.284GPa, \tag{12.51}$$

and dimensionless parameters

$$a = -2, \ V = 0.1, \ \varepsilon = 10^{-4}, \ \gamma = 1.0774, \ \kappa = 1, \tag{12.52}$$

where a_H and p_H are the Hertzian half-length of the contact and the Hertzian maximum pressure, respectively.

For Series I simulations the following additional values of dimensional parameters [28, 29]

$$\mu_a = 0.00125\,Pa\cdot s, \quad c_p = 1.204g/dL \tag{12.53}$$

and dimensionless parameters

$$a = -2, \quad V = 0.1, \quad Q = 5, \quad \tau_{L0} = 0.002, \quad \tau_{L1} = 0.046,$$
$$\delta = 0.671\cdot 10^8, \quad \theta = 4.524\cdot 10^{-3}, \tag{12.54}$$

are used while for Series II simulations the values of additional dimensional parameters [23, 28, 29]

$$\mu_a = 0.00924\,Pa\cdot s, \quad c_p = 0.86g/dL. \tag{12.55}$$

and dimensionless parameters

$$Q = 11, \quad \delta = 0.256\cdot 10^{10}, \quad \theta = 3.231\cdot 10^{-3}, \tag{12.56}$$

are used.

In particular simulations we employed: $M = 3001$ is the number of x_m nodes along the x-axis used for calculation of the flow streamlines and the solution of the kinetic equation, $M_R = 501$ and $M_R = 1501$ are the numbers of nodes used for sequential solution of the Reynolds equation on crude and fine grids, respectively, $J = 641$ is the number of nodes across the gap, the number of nodes with respect to the polymer chain length l was 232. For non-degrading lubricants solution of Reynolds equation was done only on the crude grid with the number of x_m nodes equal to $M_R = 501$. Both grids along and across the film layer were uniform while the grid with respect to the polymer chain length was non-uniform. To speed up the solution process on the fine grid, the Jacobian for Newton's method was calculated just during the first few iterations after which it was kept unchanged until the process converged. The intermediate tolerance levels for Newton's method used for calculation of $f(x)$, $p(x)$, H_0, and c were set equal to 10^{-7}. However, the actual solution precision is controlled by the grid step sizes, and in the discussed cases it was about 10^{-4}. Because the solution depends on three variables (x, z, and l) the numerical solution of the problem requires extremely intensive CPU calculations and high computer memory. For the Newtonian lubricant the predominant fraction of the time required for problem solution is associated with solution of the kinetic equation while for the non-Newtonian lubricant the predominant fraction of the time required for problem solution is associated with solution of the equation for $f(x)$, the Reynolds and gap equations, and the kinetic equation. The initial distribution of the polymer molecular weight $W_a(l)$ is taken from test measurements done by Covitch [11] (see also Chapter 6).

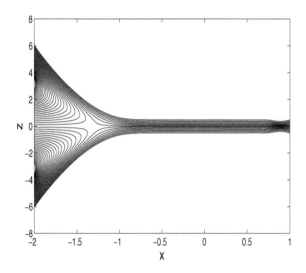

FIGURE 12.3

Flow streamlines $z(x)$ for non-degrading lubricant with Newtonian rheology and Series I input data, $s_0 = 0$ (after Kudish and Airapetyan [24]). Reprinted with permission from the ASME.

First, let us consider the non-Newtonian behavior of the lubricant in the contact. In the inlet zone of a heavily loaded lubricated contact, the pressure is small and the lubricant viscosity is close to one while the gap is large. Therefore, in the inlet zone the local sliding frictional stress $f(x)$ is small. Because of that and the fact that the derivative of pressure dp/dx is relatively small the rheological function G behaves like a linear function. It means that the behavior of G resembles the one of a fluid with Newtonian rheology. Therefore, taking into account that in heavily loaded lubricated contacts the film thickness H_0 is primarily determined by the inlet zone [30] we should expect that the difference between the lubrication film thicknesses in the contacts with the Newtonian and non-Newtonian lubricants is relatively small. The situation is different in case of a lightly loaded lubricated contact (see [31]) where the film thickness is determined by the solution of the problem in the whole contact. On the other hand, in the above cases for high slide-to-roll ratios s_0 the sliding frictional stress $f(x)$ for a lubricant with non-Newtonian rheology $(G(x) = tanh^{-1}x)$ is noticeably lower than that for a lubricant with Newtonian rheology $(G(x) = x)$ given the same ambient lubricant viscosity.

For a non-degrading lubricant with Newtonian rheology, the solution of the isothermal EHL problem is independent of the slide-to-roll ratio s_0. In particular, for the Newtonian lubricant with the input data for Series I we have $H_0 = 0.1966$ and $c = 1.0513$ while for the input data for Series II we have

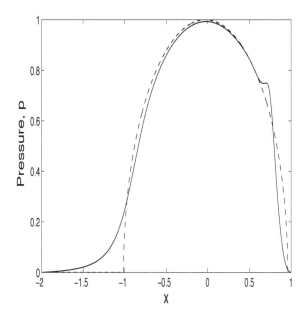

FIGURE 12.4
Pressure distribution $p(x)$ for lubricant with Newtonian (solid line) rheology and Series I input data, $s_0 = 0$. The Hertzian pressure distribution is represented by a dashed line (after Kudish and Airapetyan [24]). Reprinted with permission from the ASME.

$H_0 = 0.339$ and $c = 1.052$. For lubricants with non-Newtonian rheology, the solution of the isothermal EHL problem depends on s_0. For the non-Newtonian lubricant (i.e., for Series I input data), $s_0 = 0$, and τ_{L0} decreasing from 0.01 to 0.002 the film thickness monotonically decreases from $H_0 = 0.1967$ to $H_0 = 0.1963$, respectively, while for $s_0 = -0.5$ the film thickness monotonically decreases from $H_0 = 0.1967$ to $H_0 = 0.1961$, respectively. Within the same range of variations of s_0 and τ_{L0}, the value of c varies by no more than seven units in the fifth place after the decimal point.

The above data indicate that the film thickness H_0 for a lubricant with non-Newtonian rheology is practically identical to the one for the lubricant with the Newtonian rheology. Also, the value of H_0 is almost independent of the slide-to-roll ratio s_0 and depends significantly on the value of the dimensionless pressure coefficient Q. At the same time, the sliding frictional stress $f(x)$ is noticeably affected by the lubricant rheology, pressure coefficient of viscosity Q, speed-load parameter V, limiting stress τ_L, and the slide-to-roll ratio s_0.

Therefore, the behavior of pressure $p(x)$, gap $h(x)$, film thickness H_0, coordinate of the exit point c, and the lubricant flow streamlines $z(x)$ in the

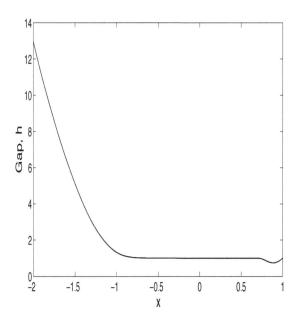

FIGURE 12.5
Gap distribution $h(x)$ for lubricant with Newtonian rheology and Series I input
data, $s_0 = 0$ (after Kudish and Airapetyan [24]). Reprinted with permission
from the ASME.

cases of Newtonian and non-Newtonian lubricants are very similar. For Series
I input data and $s_0 = 0$ the graphs of the flow streamlines $z(x)$, pressure
$p(x)$, and gap $h(x)$ for the lubricant with Newtonian rheology are given in
Figs. 12.3-12.5, respectively. For the lubricant with the non-Newtonian rheology the graphs of p and h are very close to the corresponding ones for the
Newtonian fluid. In case of $s_0 = 0$ for any rheological function G the sliding
frictional stress $f(x) = 0$ because $G(x)$ is an odd function of x. For $s_0 = -0.5$
the graphs of $f(x)$ for the non-degrading Newtonian and non-Newtonian lubricants are presented in Fig. 12.6. The behavior of $f(x)$ is distinctly different
from the one of pressure $p(x)$: the magnitude of the values of $f(x)$ is greater
in the case of the Newtonian lubricant than in the case of the non-Newtonian
one. This effect is due to the reversible viscosity loss of the non-Newtonian
lubricant in comparison with the Newtonian one. A similar behavior of the
surface frictional stresses τ_1 and τ_2 can be seen in Fig. 12.7, where τ_1 and
τ_2 are given for the lubricants with Newtonian and non-Newtonian rheologies
under mixed rolling and sliding conditions ($s_0 = -0.5$) as well as the surface
shear stress $\tau_1 = -\tau_2$ for the Newtonian lubricant (dashed-dotted line) under
pure rolling conditions ($s_0 = 0$). For the case of the pure rolling ($s_0 = 0$)

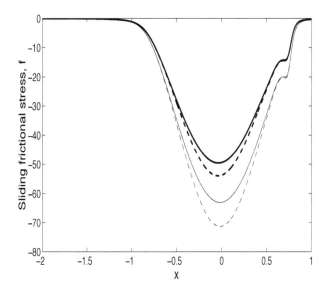

FIGURE 12.6
Sliding frictional stress $f(x)$ distributions for non-degrading lubricants with Newtonian (dashed line) and non-Newtonian (solid line) rheologies and for degrading lubricants with Newtonian (thick dashed line) and non-Newtonian (thick solid line) rheologies for Series I input data, $s_0 = -0.5$ (after Kudish and Airapetyan [24]). Reprinted with permission from the ASME.

for the non-Newtonian lubricant, the surface shear stresses $\tau_1 = -\tau_2$ are very close to the ones for the case of the Newtonian lubricant.

For Series II input data and $s_0 = 0$ the behavior of pressure $p(x)$, gap $h(x)$, and the lubricant flow streamlines $z(x)$ resembles the one shown for the case of Newtonian lubricant for Series I input data in Figs. 12.3-12.5. For Series II input data in the case of pure sliding ($s_0 = -2$), the behavior of pressure $p(x)$ and gap $h(x)$ is still very close to the one for the case of pure rolling ($s_0 = 0$). However, the behavior of the streamlines in case of pure sliding ($s_0 = -2$) is different and it is shown in Fig. 12.8. There are only one set of flow streamlines entering the contact region and turning around and one set of streamlines running through the the entire contact.

The reversible loss of viscosity of the non-Newtonian lubricant can be determined as a ratio of the sliding frictional stresses $f(x)$ for non-Newtonian and Newtonian lubricants in the Hertzian region. It is based on the fact that in the Hertzian region of a heavily loaded lubricated contact $h(x) - 1 \ll 1$ [28]. Therefore, from Fig. 12.6 follows that for $s_0 = -0.5$ the reversible lubricant viscosity loss of the non-Newtonian lubricant reaches 12% of its original value.

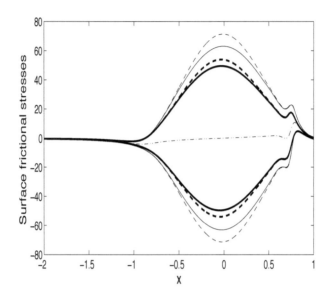

FIGURE 12.7
Frictional stresses τ_1 and τ_2 applied to the contact surfaces for non-degrading
(dashed lines) and degrading (thick dashed lines) Newtonian lubricants and for
non-degrading (solid lines) and degrading (thick solid lines) non-Newtonian
lubricants under mixed rolling and sliding conditions ($s_0 = -0.5$) and surface
frictional stresses $\tau_2 = -\tau_1 = (6H_0^2/V)hdp/dx)$ for non-degrading Newtonian
lubricant (dash-dotted line) under pure rolling conditions ($s_0 = 0$) and Series
I input data (after Kudish and Airapetyan [24]). Reprinted with permission
from the ASME.

12.6   EHL Solutions for Lubricants with Newtonian and Non-Newtonian Rheologies with Degradation

In this section we will compare the EHL solutions for a lubricant with non-
Newtonian rheology experiencing the stress-induced degradation with a solution for a lubricant with a similar rheology but without degradation. That
allows determining the irreversible loss of lubricant viscosity. Moreover, we
will compare the EHL solutions for degrading lubricants with Newtonian and
non-Newtonian rheologies. The simulation input data are taken identical to
those in Section 12.5.

Let us consider some general properties of a solution of the isothermal EHL
problem for a degrading lubricant with non-Newtonian rheology. The behavior
of pressure $p(x)$ and gap $h(x)$ distributions as well as of the film thickness H_0

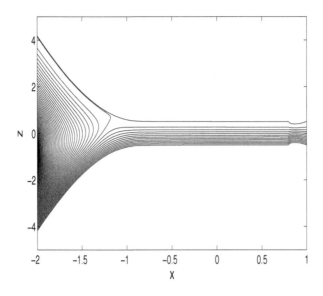

FIGURE 12.8
Flow streamlines $z(x)$ for non-degrading lubricant with Newtonian rheology and Series II input data, $s_0 = -2$ (after Kudish and Airapetyan [23]). Reprinted with permission from the ASME.

and exit coordinate c of the latter problem with respect to parameters a, V, and Q is similar to that described in the preceding chapter for lubricants with Newtonian and non-Newtonian rheologies (also see Kudish [5, 30, 31]). For a lubricant with Newtonian rheology increase in load P leads to increase in the Hertzian pressure p_H, which, in turn, increases Q and causes a relatively slow increase in H_0 and a rapid increase in $f(x)$ and, therefore, in $\tau(x,z)$. That, in turn, leads to a rapid lubricant degradation (see equations (12.20) and (12.21)). For a lubricant with non-Newtonian rheology increase in load P also leads to increase in the sliding frictional stress $f(x)$ and, thus, to increase in the shear stress τ. However, these increases in $f(x)$ and τ are moderated by the lubricant rheology, i.e., these increases are bounded by the limiting stress τ_L. Therefore, for a non-Newtonian lubricant the shear stress τ for high loads P is much lower than for a similar Newtonian lubricant. It means that the degradation process of such a non-Newtonian lubricant runs slower than for a Newtonian counterpart. A usually very small parameter ε almost does not affect the problem solution. For small τ_{L0} and τ_{L1} the shear stress τ in a non-Newtonian lubricant is small, which, in turn, slows down the process of lubricant degradation (see equations (12.21) for R and L_*). As the values of τ_{L0} and τ_{L1} increase, the lubricant shear stress τ and the film thickness H_0

increase. As the slide-to-roll ratio s_0 increases the film thickness H_0 slowly decreases. However, variations in H_0 are insignificant while the increase in τ is directly proportional to the increase in τ_L, i.e., the increase in τ_{L0} and τ_{L1}. The latter causes lubricant degradation to run faster (see equations (12.21) for R and L_*). Moreover, an increase in the value of parameter δ leads to a corresponding increase in the characteristic chain length L_* that causes lubricant degradation to slow down (see equations (12.21) for R and L_*). For smaller values of parameter γ the process of lubricant degradation runs slower than for the larger ones (see equations (12.21) for R). For higher values of θ, the lubricant viscosity responds stronger to changes in the molecular weight distribution, i.e., for higher θ the loss of lubricant viscosity is higher (see equations (12.21) for μ). The value of parameter κ in the left-hand side of the kinetic equation in (12.20) controls the rate of lubricant convection and, therefore, the time a lubricant small volume is present in the contact area. For high values of κ, the time of lubricant presence in the contact area is small and the rate of lubricant degradation is low.

First, let us examine a case of pure rolling ($s_0 = 0$) for Series I input parameters. Under these conditions lubricants with Newtonian and non-Newtonian rheologies degrade to a lesser extent than in cases when $s_0 \neq 0$. The solutions for Newtonian and non-Newtonian lubricant rheologies are qualitatively and quantitatively very close to each other because $|G^{-1}(x)| \leq |x|$ and $G(x) \to x$ as $x \to 0$. The map of flow streamlines for the non-Newtonian lubricant is presented in Fig. 12.9. This map shows that the flow streamlines are symmetric about the line $z = 0$. Because of the symmetry of the problem the lubricant viscosity μ and shear rate $\partial u / \partial z$ are continuous functions of x and z. All of the flow streamlines enter the contact through the inlet cross section of the lubrication film. Some of them that are adjacent to the contact surfaces run through the whole contact. However, the other streamlines turn around and exit the contact region also through the inlet cross section of the film. The pressure and gap distributions are very close to the ones for the non-degrading Newtonian lubricant presented in Figs. 12.4 and 12.5. The maximum difference between the gap distributions for Newtonian and non-Newtonian fluids reaches 8%.

The map of the horizontal component u of the lubricant velocity along the flow streamlines for the non-Newtonian lubricant is given in Fig. 12.10. Because for $s_0 = 0$ the flow is symmetrical about the line $z = 0$, Fig. 12.10 shows the lubricant velocity along four sets of streamlines: two (coinciding) sets of streamlines passing through the whole contact and two (coinciding) sets of streamlines that turn around and exit through the inlet cross section. In this case the exit film thicknesses and exit coordinates for the Newtonian and non-Newtonian degrading lubricants are $H_0 = 0.1829$, $c = 1.0527$ and $H_0 = 0.1834$, $c = 1.0527$, respectively. The comparison of the data for H_0 obtained for degrading and non-degrading lubricants for pure rolling $s_0 = 0$ shows a relatively small affect of lubricant degradation on the film thickness H_0 (only about 8%).

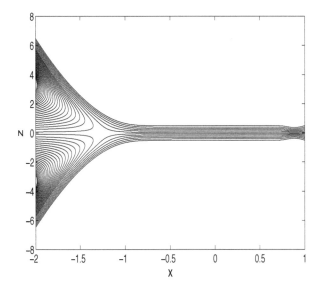

FIGURE 12.9
Flow streamlines for degrading lubricant with non-Newtonian rheology under pure rolling conditions ($s_0 = 0$) and Series I input data (after Kudish and Airapetyan [24]). Reprinted with permission from the ASME.

The distributions of the lubricant viscosity μ and polymer molecular weight W for lubricants with Newtonian and non-Newtonian rheologies are practically identical. The fact that for $s_0 = 0$ lubricants with Newtonian and non-Newtonian rheologies degrade relatively slowly can be also seen from the graphs of the reciprocal of the lubricant viscosity (see Fig. 12.11) and the distribution of the molecular weight W along the flow streamline $z(x)$ that is closest to the line $z = 0$ and running through the whole contact below $z = 0$ (see Figure 12.12). In Fig. 12.5, the variable z' is the artificially stretched z-coordinate across the film thickness (namely, $z' = zh(a)/h(x)$) to make the relationship more transparent. It is clear from the graphs of the reciprocal of the lubricant viscosity (see Fig. 12.11) that for the Newtonian and non-Newtonian lubricants the maximum irreversible viscosity loss reaches about 31%. Lubricant degradation depends on a number of parameters among which the shear stress τ (see equations (12.21)) plays one of the major roles. For $s_0 = 0$ the sliding frictional stress $f(x) = 0$ and, therefore, the shear stress $\tau = (12H_0^2/V)zdp/dx$ reaches its extrema in the inlet and exit zones while in the Hertzian region it is relatively small [30].

That leads to the conclusion that for pure rolling ($s_0 = 0$) practically all lubricant degradation occurs in the inlet zone while the lubricant almost does

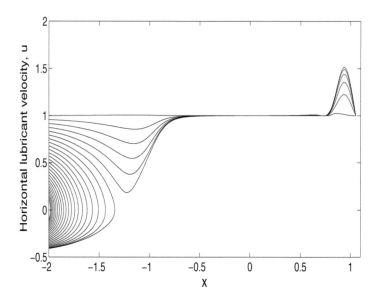

FIGURE 12.10

The horizontal component u of the lubricant velocity along the flow stream-lines for degrading lubricant with non-Newtonian rheology under pure rolling conditions ($s_0 = 0$) and Series I input data (after Kudish and Airapetyan [24]). Reprinted with permission from the ASME.

not degrade in the Hertzian region and the exit zone. The latter also can be clearly seen from the distributions of the molecular weight W in Fig. 12.12.

For the given lubricant flow parameters, the degree of lubricant degradation is higher along the longer streamlines. The length of the flow streamlines adjacent to the contact surfaces is larger and the absolute value of the shear stress along them is slightly higher than for the ones that turn around and exit through the inlet cross section. In spite of that fact lubricant degradation along the turning around streamlines is higher because they are completely located in the inlet of the contact, practically the only contact zone where lubricant degradation takes place. Also, it follows from the expression for the shear stress $\tau = (12H_0^2/V)zdp/dx$ that along the streamlines running through the whole contact the degree of lubricant degradation is higher near the contact surfaces. Among the turning around streamlines the degree of lubricant degradation is higher along the longest streamlines that enter the contact region and are closest to the contact surfaces and exit the contact through the inlet cross section close to $z = 0$. Obviously, for higher viscosity μ, the film thickness H_0 is higher while dp/dx changes insignificantly. Therefore, the shear stress τ is higher that promotes faster lubricant degradation and higher viscosity loss.

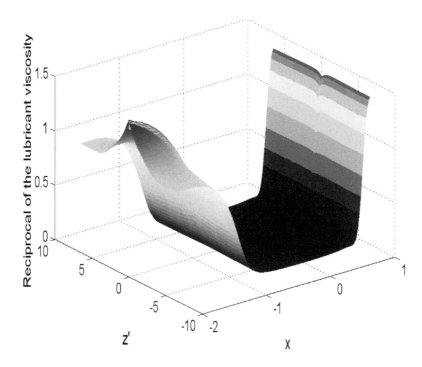

FIGURE 12.11 (See color insert following page 750)

The reciprocal of the lubricant viscosity μ in Newtonian and non-Newtonian lubrication film under pure rolling conditions ($s_0 = 0$) and Series I input data. The variable z' is an artificially stretched z-coordinate across the film thickness (namely, $z' = zh(a)/h(x)$) to make the relationship more transparent (after Kudish and Airapetyan [24]). Reprinted with permission from the ASME.

Similarly, for higher pressure viscosity coefficients Q one can expect faster degradation and higher viscosity loss.

Now, let us examine a case of mixed rolling and sliding with $s_0 = -0.5$ for Series II input data. In this case the solutions of the problem for the lubricants with Newtonian and non-Newtonian rheologies are also very close to each other. The maps of the flow streamlines $z(x)$ and the horizontal component u of the lubricant velocity along the flow streamlines for the degrading Newtonian and non-Newtonian lubricants are given in Figs. 12.13 and 12.14, respectively. The lubricant flows are no longer symmetric about $z = 0$ and the maps of their flow streamlines are completely different from the ones for the pure rolling case. There are still two sets of streamlines with flow reversal as well as two sets of flow streamlines adjacent to the lower and upper contact surfaces and running through the whole contact that can be distinctly seen in Fig.

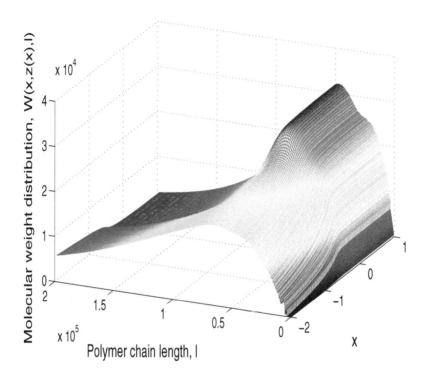

FIGURE 12.12 (See color insert following page 750)

Molecular weight distribution W of degrading lubricants with Newtonian and non-Newtonian rheologies along the flow streamline closest to $z = 0$ and running through the whole contact below $z = 0$ under pure rolling conditions ($s_0 = 0$) and Series I input data (after Kudish and Airapetyan [24]). Reprinted with permission from the ASME.

12.13. The exit lubrication film thickness H_0 and the exit coordinate c exhibit behavior very similar to the case of pure rolling. For the non-degrading and degrading Newtonian lubricants, they are equal to $H_0 = 0.1961$, $c = 1.0513$ and $H_0 = 0.1819$, $c = 1.0531$, respectively. For the non-degrading and degrading non-Newtonian lubricants, the film thickness is equal to $H_0 = 0.1961$ and $H_0 = 0.1813$, respectively. In all other respects the distributions of the pressure p and gap h are very similar to the ones for the case of pure rolling. The numerical results show that in the case of mixed rolling and sliding for both lubricants with Newtonian and non-Newtonian rheologies the sliding frictional stress $f(x)$ (see Fig. 12.6) is smaller than in a similar case without degradation but still large enough to cause lubricant degradation to a much greater extent than in the case of pure rolling. In the case of mixed rolling and sliding, the surface frictional stresses τ_1 and τ_2 are much higher (see Fig.

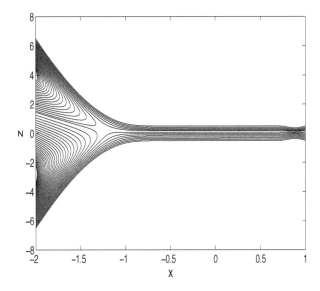

FIGURE 12.13
Flow streamlines $z(x)$ for degrading lubricants with Newtonian and non-Newtonian rheologies under mixed rolling and sliding conditions ($s_0 = -0.5$) and Series I input data (after Kudish and Airapetyan [24]). Reprinted with permission from the ASME.

12.7) than the ones for the case of pure rolling ($\tau_2 = -\tau_1 = (6H_0^2/V)hdp/dx$) and, at the same time, lower than for the case of no degradation. That explains the stronger lubricant degradation in the case of mixed rolling and sliding conditions in comparison with the case of pure rolling conditions (see Figs. 12.11, 12.12, 12.2, and 12.15). For the degrading lubricants, the frictional stresses τ_1 and τ_2 applied to the contact surfaces are on average lower than for the non-degrading lubricant by about 26%. The frictional stress τ in the lubrication film is mostly concentrated near the contact surfaces. Depending on the sign of dp/dx the maximum of the absolute value of the surface frictional stress is reached either on the lower or upper contact surfaces. That leads to faster lubricant degradation near that contact surface and to slower lubricant degradation near the other one. This can be clearly seen from the behavior of the reciprocal of the lubricant viscosity μ in cases of Newtonian and non-Newtonian lubricants (see Fig. 12.2). In the case of $s_0 = -0.5$ the maximum loss of the lubricant viscosity is approximately 41%. That can be seen from the comparison of graphs of the sliding frictional stress $f(x)$ for the non-degrading and degrading lubricants (see Fig. 12.6) as well as from the graph of the distribution of the reciprocal of the lubricant viscosity μ (see

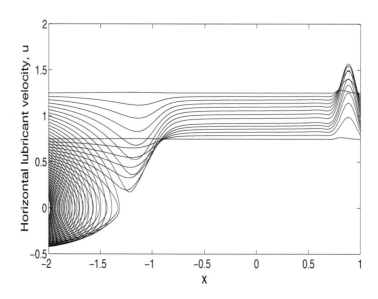

FIGURE 12.14

The horizontal component u of the lubricant velocity along the flow streamlines for degrading lubricants with Newtonian and non-Newtonian rheologies
under mixed rolling and sliding conditions ($s_0 = -0.5$) and Series I input
data (after Kudish and Airapetyan [24]). Reprinted with permission from the
ASME.

Fig. 12.2).

The behavior of the polymer molecule distribution W for the lubricants with
Newtonian and non-Newtonian rheologies along the flow streamline $z_{31}(x)$
running through the whole contact below $z = 0$ and located next to the
first turning around streamline is given in Fig. 12.15. The comparison of the
polymer molecular weight W distributions for the cases of pure rolling and
mixed rolling and sliding (see Fig. 12.12 and 12.15) shows that in the case of
pure rolling the lubricant degrades slower than in the case of mixed rolling
and sliding. In the latter case rapid polymer scission occurs throughout the
entire contact region while in the former case it is mostly concentrated in the
inlet zone of the contact. That can be seen from the values of $W(x, z(x), l)$
for polymer molecules with short chain lengths l at different points along the
flow streamlines. Moreover, from Fig. 12.15, it follows that for Newtonian and
non-Newtonian lubricants under mixed rolling and sliding conditions lubricant
degradation occurs throughout the contact. For non-Newtonian lubricant at
higher values of the slide-to-roll ratio $\mid s_0 \mid$, the relative impact of the rolling
frictional stress $(6H_0^2/V)zdp/dx$ on τ decreases slowly while for the Newtonian

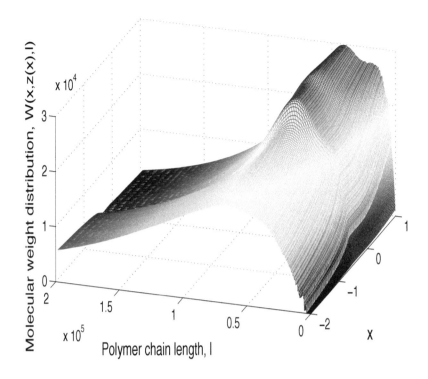

FIGURE 12.15 (See color insert following page 750)

The molecular weight distribution W of degrading lubricants with Newtonian and non-Newtonian rheologies along the flow streamline $z_{31}(x)$ closest to $z = 0$ and running through the whole contact below $z = 0$ under mixed rolling and sliding conditions ($s_0 = -0.5$) and Series I input data (after Kudish and Airapetyan [24]). Reprinted with permission from the ASME.

lubricant it decreases significantly. Moreover, the described differences in the degradation behavior of the lubricants with Newtonian and non-Newtonian rheologies may be the key to understanding why synthetic lubricants, which usually exhibit pseudoplastic behavior, degrade at a much slower pace.

Now, let us turn to the case of Series II input data. Under conditions of pure sliding, the behavior of pressure $p(x)$ and gap $h(x)$ in a lubricated contact is very similar to the one described for Series I input data. Even under conditions of pure sliding ($s_0 = -2$) the behavior of pressure $p(x)$ and gap $h(x)$ in a lubricated contact is very similar to the one described for the cases of pure rolling and mixed rolling and sliding. The reduction in the film thickness H_0 due to polymer degradation is slightly higher than for the cases of $s_0 = 0$ and $s_0 = -0.5$, i.e., $H_0 = 0.295$ and $H_0 = 0.339$ with and without lubricant degradation, respectively. For the case of $s_0 = -2$, there is just one set of

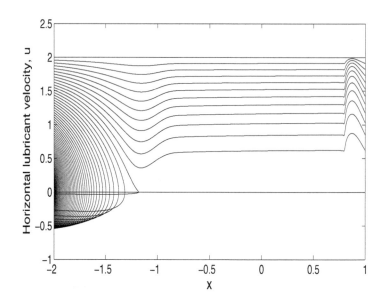

FIGURE 12.16
The horizontal component u of the lubricant velocity along the flow stream-lines for degrading lubricants with Newtonian rheology under pure sliding conditions ($s_0 = -2$) and Series II input data (after Kudish and Airapetyan [23]). Reprinted with permission from the ASME.

turning around flow streamlines adjacent to the motionless upper contact surface and one set of flow streamlines running through the whole contact that is adjacent to the moving lower contact surface everywhere in the contact (see Fig. 12.8). The map of the horizontal component u of the lubricant velocity along these flow streamlines is shown in Fig. 12.16. The frictional stresses τ_1 and τ_2 applied to the surfaces are much higher than in the case of pure rolling $s_0 = 0$. The frictional stress τ in the lubrication layer is mostly concentrated near the motionless upper contact surface ($z = h/2$), and it is relatively low near the moving lower contact surface ($z = -h/2$). That leads to significant lubricant degradation near the upper contact surface and to low lubricant degradation near the moving lower contact surface. The numerical results show that on average in case of pure sliding a lubricant undergoes degradation to a much greater extent than in the case of pure rolling. This can be clearly seen from the behavior of the lubricant viscosity μ in the lubrication layer (see Fig. 12.17). In this case the average loss of the lubricant viscosity reaches about 25%. The local lubricant viscosity losses near the moving lower and motionless upper surfaces reach about 10% and 60%, respectively. For the degrading lubricant, the frictional stresses applied to the contact surfaces are generally

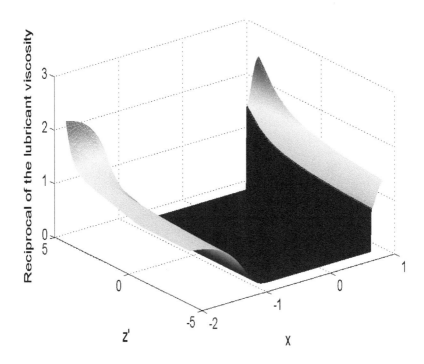

FIGURE 12.17 (See color insert following page 750)
The reciprocal of the lubricant viscosity μ in a Newtonian lubrication film under pure sliding conditions ($s_0 = -2$) and Series II input data. The variable z' is an artificially stretched z-coordinate across the film thickness (namely, $z' = zh(a)/h(x)$) to make the relationship more transparent (after Kudish and Airapetyan [23]). Reprinted with permission from the ASME.

lower than for the non-degrading lubricant by about 20%-30%. The general behavior of the polymer molecule distribution $W(x, z(x), l)$ to a certain extent is similar to the one obtained under pure rolling and mixed rolling and sliding conditions, and it is given along the flow streamline $z_{48}(x)$ running through the entire contact and located next to the first turning around streamline (see Fig. 12.18). Moreover, in case of pure sliding the lubricant degrades along the flow streamlines throughout the entire contact (see Fig. 12.18) while in the considered cases of pure rolling and mixed rolling and sliding the degradation occurs mostly in the inlet and zone (see Fig. 12.12 and 12.15).

It is convenient to analyze the process of polymer degradation based on the behavior of the characteristic polymer chain length L_* (see equations (12.21)), which represents the position of the center of the profile of the probability of polymer scission $R(x, z, l)$. In fact, the probability of polymer scission R is

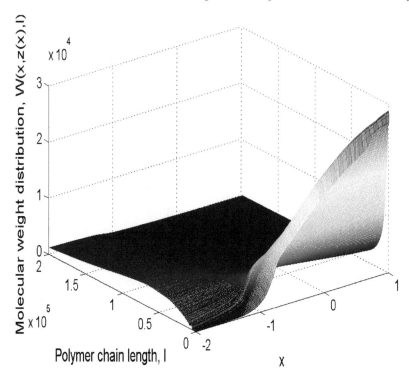

FIGURE 12.18 (See color insert following page 750)
The molecular weight distribution W of the degrading lubricant with Newtonian rheology along the flow streamline $z_{48}(x)$ running through the entire
contact closest to the first turning around streamline under pure sliding conditions ($s_0 = -2$) and Series II input data (after Kudish and Airapetyan [23]).
Reprinted with permission from the ASME.

the main key to the control over the lubricant degradation. As the zone of
the chain lengths l, where $R(x, z, l)$ varies from 0 to 1 is quite narrow, one
can conclude that the polymer molecules with chain lengths l shorter than
L_* practically do not undergo scission while most of the molecules longer
that L_* get degraded. It is important to emphasize that the degree of polymer degradation varies from point to point together with the shear stress τ,
which, in turn, is a certain function of x and z. In the Hertzian region the
shear stress τ is relatively large due to large lubricant viscosity μ while in the
inlet and exit zones the values of the pressure derivative dp/dx and, therefore, the rolling frictional stress $(6H_0^2/V)zdp/dx$, are higher than the ones
in the Hertzian zone. That explains the above–mentioned fact of lubricant
degradation throughout the contact.

The effect of lubricant viscosity loss of up to 70% was observed experimentally and documented by Walker, Sanborn, and Winer [8]. In cases discussed in this chapter, the loss of lubricant viscosity reached about 41%. One has to keep in mind that in reality the value of the pressure coefficient of viscosity Q can be much higher than $Q = 5$, which may lead to a much higher viscosity loss.

It is important to realize that the lubricant viscosity loss caused by degradation is controlled by the contact operating conditions as well as by the lubricant rheology, the nature, concentration, and molecular weight distribution of the polymer additive in the lubricant supplied to the contact. Moreover, significant changes in the lubricant viscosity μ and film thickness H_0 may also lead to noticeable variations in frictional stresses τ_1 and τ_2 applied to the contact surfaces. That can also affect fatigue life of solids involved in a contact with degrading lubricant (see Kudish [32] and Section 12.7). Therefore, for any given operating conditions, it is desirable to select a lubricant with certain rheology and concentration and molecular weight distribution of a polymer additive that will provide sufficiently good lubrication conditions (high lubrication film thickness and low frictional stress) while the viscosity loss is reduced to minimum.

In case of a thermal EHL problem due to heat generation in the lubrication film and elevated lubricant temperature T, the lubricant viscosity (in case of no lubricant degradation) is lower than in a comparable isothermal case. That leads to higher values of not only T but also L_*, which slows down the degradation process. The extent to which the polymer degradation process slows down depends on the lubricant nature, nature and distribution of the molecular weight of polymer molecules entering the contact, and on the contact operating conditions.

The results based on such a model can be used for the evaluation of useful lubricant life and for more realistic estimation of the lubrication film thickness and frictional stresses in lubricated contacts. That may provide a solid basis for maintenance scheduling of lubricated machinery operating under various conditions. Moreover, it may enable researches to create new and more durable lubricant additives for general and specific applications.

To conclude the section we can state that a plane isothermal elastohydrodynamic problem for a line contact lubricated by fluids with Newtonian and non-Newtonian rheologies with degrading polymer additives is modeled. The polymer additive in the lubricant, the viscosity modifier (VM), is considered to be of linear structure. The lubricant degradation is caused by stress-induced scission of polymer molecules while they move through the contact along the flow streamlines. The polymer degradation process is controlled by the probabilistic kinetic equation for the polymer molecular weight distribution. The lubricant flow is described by the generalized Reynolds equation. The lubrication and lubricant degradation processes are coupled through the lubricant viscosity that depends on pressure and temperature as well as on the concentration and distribution of the polymer molecular weight. The problem is solved numerically. In particular, the lubricant flow streamlines, distribution of the polymer molecular weight, reversible and irreversible lubricant viscosity losses, pressure, gap, and film thickness are determined. The stress-induced scission of polymer molecules leads to irreversible changes in the lubricant viscosity - irreversible viscosity loss. Numerical results show that in the cases of pure rolling, mixed rolling and sliding, and pure sliding the viscosity loss reaches about 31%, 41%, and 60%, respectively. For example, in the cases of mixed rolling and sliding, the reversible and irreversible viscosity losses at the exit from the contact can be as high as 12% and 41%, respectively. Moreover, the new property of a flow of degrading lubricant is discovered, i.e., the viscosity of a degrading lubricant generally is a discontinuous function of the position of a fluid particle. The viscosity loss in the inlet zone of the contact causes all parameters of the lubricated contact to change. In particular, in the case of mixed rolling and sliding conditions the lubrication film thickness of the degrading lubricants is about 8% thinner while in the case of pure sliding the lubrication film thickness of the degrading lubricants is about 13% thinner than the one for similar non-degrading lubricants. The effect of lubricant degradation for non-Newtonian lubricants is noticeably lower than for Newtonian lubricants. Reasonably large changes in frictional stresses applied to the contact surfaces (reached up to 40%) may noticeably impact fatigue life.

12.7　Effect of Lubricant Degradation on Contact Fatigue

In practice, most surfaces in relative motion operate in a lubricated environment. One of the often limiting parameters of such contacts is fatigue life. Fatigue life is also used as one of the constraints in design of contacts subjected to cyclic loading. It is well known that contact fatigue is affected by contact pressure, frictional stress, residual stress, initial distribution of material flaws etc. The behavior of contact pressure and, primarily, of frictional stress are determined by lubricant viscous properties. Motor oils are usually formulated by adding to the base oil stock a number of polymeric additives. In spite of the fact that the concentrations of these additives are always low, the presence of the additives significantly modifies and stabilizes lubricant's viscosity and changes lubricant rheology. It is also recognized that in the process of machinery operation lubricant's composition and rheology undergo certain changes. In fact, high operating temperatures and stresses cause degradation of polymeric additives that lead to changes in lubricants viscosity and rheology. Degradation of lubricants causes significant viscosity loss that, in turn, reduces the frictional stress in contacts with sufficiently smooth surfaces operating in elastohydrodynamic regime of lubrication. The reduced frictional stresses raise contact fatigue life. Therefore, to properly predict contact fatigue life it is not sufficient to know the ambient lubricant viscosity and its dependence on pressure in a non-degraded lubricant. To successfully model contact fatigue of lubricated joints with sufficiently smooth surfaces that are completely separated by lubricant, it is necessary to know not only the fatigue resistance and cleanliness of the materials involved in the contact but also to be able to predict the effect of temperature and stress-induced polymer degradation on lubricant viscosity. The latter leads to correct prediction of contact frictional stresses that strongly affect contact fatigue [26, 32] - [36].

The objective of this section is to explore the extent to which lubricant degradation may change contact fatigue life. The analysis is performed numerically based on the developed models of contact fatigue (see Section 9.7 and [26, 32]) and lubricant degradation in an EHL contact (see Section 12.2 and [23]). The results show that for solids completely separated by lubricants that have the same ambient viscosity, material contact fatigue may vary significantly because of the specific way these lubricants were formulated. In particular, contact fatigue is strongly affected by the initial distribution of the molecular weight of the polymeric additive (viscosity modifier) and by contact operating conditions that in some cases promote rapid lubricant degradation caused by high shearing stresses.

12.7.1   Model of Contact Fatigue

The description of the contact fatigue model presented here follows the model from Section 9.7 developed for line contacts. The entire set of assumptions and formulas of the model of contact fatigue are covered in Section 9.7.

12.7.2   Elastohydrodynamic Model for a Lubricant That Undergoes Stress-Induced Scission

In modeling of line contacts with degrading lubricant, we will follow the model from Section 12.2 developed for steady isothermal line elastohydrodynamic contacts. Namely, two cylinders with parallel axes and elastic constants E_1, ν_1 and E_2, ν_2 move with linear velocities u_1 and u_2, respectively, and subjected to a normal compressive force P. The cylinders are completely separated by a lubrication layer. The lubricant is considered to be a Newtonian fluid (base oil) formulated with a polymeric additive of a linear structure. The lubricant viscosity $\mu(p, W)$ depends not only on lubricant pressure p but also on the density of the polymer molecular weight distribution $W(x, z, l)$, where l is the polymer molecule chain length. The density $W(x, z, l)$ is introduced in such a way that $W(x, z, l)dl$ is the molecular weight of the polymer molecules in a unit volume centered at (x, z) with the chain lengths from the interval $[l, l + dl]$.

In the dimensionless form the equations describing the EHL problem for degrading lubricant are presented in Section 12.2 (see equations (12.16)-(12.21)). For a Newtonian lubricating fluid $G(x) = x$ (see (12.1)) in dimensionless variables in (12.16)-(12.18) such equations are given below (see Section 12.2)

$$f = \left\{ \frac{s_0}{\sqrt{1+\frac{\epsilon H_0^2}{4}(\frac{dh}{dx})^2}} - \frac{12H_0^2}{V}\frac{dp}{dx}\int\limits_{-h/2}^{h/2}\frac{\zeta d\zeta}{\mu} \right\} / \left(\int\limits_{-h/2}^{h/2}\frac{d\zeta}{\mu} \right)$$

$$u(x,z) = \frac{1-s_0/2}{\sqrt{1+\frac{\epsilon H_0^2}{4}(\frac{dh}{dx})^2}}\left\{ 1 + \int\limits_{-h/2}^{z}\frac{d\zeta}{\mu} / \int\limits_{-h/2}^{h/2}\frac{d\zeta}{\mu} \right\}$$

$$\hspace{6cm} (12.57)$$

$$+\frac{12H_0^2}{V}\frac{dp}{dx}\left\{ \int\limits_{-h/2}^{z}\frac{\zeta d\zeta}{\mu} - \int\limits_{-h/2}^{h/2}\frac{\zeta d\zeta}{\mu}\int\limits_{-h/2}^{z}\frac{d\zeta}{\mu} / \int\limits_{-h/2}^{h/2}\frac{d\zeta}{\mu} \right\},$$

$$\frac{d}{dx}\int\limits_{-h/2}^{z(x)} u(x,\zeta)d\zeta = 0, \ p(a) = p(c) = \frac{dp(c)}{dx} = 0,$$

$$\int\limits_{-h/2}^{z(x)} u(x,\zeta)d\zeta = \frac{1}{\sqrt{1+\frac{\epsilon H_0^2}{4}(\frac{dh}{dx})^2}}\left\{ \frac{h}{2} - s_0\left[\frac{1}{2} + \int\limits_{-h/2}^{z}\frac{\zeta d\zeta}{\mu} / \int\limits_{-h/2}^{h/2}\frac{d\zeta}{\mu}\right] \right\}$$

$$\hspace{6cm} (12.58)$$

$$+zu(x,z) - \frac{12H_0^2}{V}\frac{dp}{dx}\left\{ \int\limits_{-h/2}^{z}\frac{\zeta^2 d\zeta}{\mu} - \int\limits_{-h/2}^{h/2}\frac{\zeta d\zeta}{\mu}\int\limits_{-h/2}^{z}\frac{\zeta d\zeta}{\mu} / \int\limits_{-h/2}^{h/2}\frac{d\zeta}{\mu} \right\},$$

$$H_0(h-1) = x^2 - c^2 + \frac{2}{\pi}\int\limits_a^c p(x')\ln\frac{c-x'}{|x-x'|}dx', \ \int\limits_a^c p(x')dx' = \frac{\pi}{2},$$

$$\hspace{6cm} (12.59)$$

$$\tau_i = (-1)^i f + \frac{6H_0^2}{V}h\frac{dp}{dx}, \ i = 1,2,$$

$$\kappa(u\frac{\partial W}{\partial x} + w\frac{\partial W}{\partial z}) = 2l\int\limits_l^\infty R(x,z,L)p_c(l,L)W(x,z,L)\frac{dL}{L}$$

$$-R(x,z,l)W(x,z,l),$$

$$W(a,z,l) = W_a(l) \ if \ u(a,z) > 0, \ W(c,z,l) = W_a(l) \ if \ u(c,z) < 0,$$

$$R(x,z,l) = 0 \ if \ l \leq L_*,$$

$$R(x,z,l) = 1 - (\frac{l}{L_*})^{2\gamma}\exp[-\gamma(\frac{l^2}{L_*^2} - 1)] \ if \ l \geq L_*,$$

$$L_* = \sqrt{\frac{H_0\delta}{|\tau|}}, \ \tau = f + y\frac{12H_0^2}{V}\frac{dp}{dx},$$

$$\hspace{6cm} (12.60)$$

$$p_c(l,L) = ln(2)\frac{4|L-2l|}{L^2}\exp[-ln(2)\frac{4l(L-l)}{L^2}],$$

$$\mu(p,W) = e^{Qp}\frac{1+[\eta]+k_H[\eta]^2}{1+[\eta]_a+k_H[\eta]_a^2}, \ [\eta] = \theta M_W^\beta,$$

$$M_W = \left\{ \int\limits_0^\infty l^\beta W(x,z,l)dl \Big/ \int\limits_0^\infty W(x,z,l)dl \right\}^{1/\beta}.$$

Three types of calculations are required for solution of elastohydrodynamic problem for degrading lubricant (see equations (12.57)-(12.60)). The lubricant flow streamlines need to be determined from equations (12.59) and the variations of the density of the molecular weight distribution W along the streamlines have to be calculated from equations (12.60). After the above two calculations are carried out, the EHL problem itself (described by equations (12.59) for $z = h/2$) can be solved. The above process is iterative, and more detail is presented in Section 12.4.

12.7.3 Combined Model for Contact Fatigue and Lubricant That Undergoes Degradation

To combine the two above models, we need to make a number of natural and reasonable assumptions. For heavily loaded smooth contacts in which surfaces are completely separated by a lubricant, the actual pressure p is very close to the Hertzian pressure distribution. The situation with the frictional stress τ applied to the lubricated surfaces is different. There are still no good and reliable methods for determining lubricant rheology and frictional stresses (see [37]). Therefore, the assumptions that the lubricant rheology is Newtonian and the lubricant viscosity changes with pressure according to an exponential law may be crude. Nevertheless, they serve sufficiently well to reveal the main mechanisms of lubricant degradation and provide at least qualitatively correct results (see Chapter 6, and also [11, 32]). A similar analysis for a lubricant with one kind of non-Newtonian rheology was presented in Sections 12.2, 12.5, and 12.6. On the other hand, the results of simulations based on the contact fatigue model show that fatigue life is insensitive to even large local variations in the behavior of the frictional stress applied to the surfaces as long as the frictional force remains the same. It is due to the fact that the maximum of the normal stress intensity factor k_1 at crack tips is reached relatively far from the center of a contact and it is determined by integral expressions given in Chapter 9, which to some extent averages the local variations of frictional stress τ. In particular, at a depth of about half-width of a Hertzian contact beneath the surface, the maximum of the normal stress intensity factor k_1 is reached at a distance of at least $1.5a_H - 2a_H$ behind the contact. The goal of this analysis is mainly to qualitatively demonstrate the effect of lubricant degradation on contact fatigue of completely separated by lubricant smooth surfaces. Therefore, based on the above considerations in fatigue modeling we will accept the frictional stress applied to the surfaces to be determined by Coulomb's law $\tau = -\lambda p$ where λ is the coefficient of friction. For one particular set of the problem, parameters λ will be assigned a reasonable value of the friction coefficient. At the same time, the friction coefficient λ for other cases will be determined as follows:

$$\lambda = \lambda_0 \int_a^c \tau(x)dx / \int_{a_0}^{c_0} \tau_0(x)dx, \qquad (12.61)$$

where λ_0 is the assigned value of the friction coefficient for the frictional stress τ_0 and contact boundaries a_0 and c_0 while λ is the coefficient of friction for the frictional stress τ and contact boundaries a and c. Obviously, the two models of contact fatigue and lubricant degradation are coupled together by equation (12.61). In the future, coupling between the contact fatigue and lubricant degradation parts of the model can be improved as more detailed and accurate information on lubricant rheology and calculation of frictional stresses becomes available.

12.7.4 Numerical Results and Discussion

The fatigue crack resistance parameters of the solid material as well as crack parameters used for particular calculations were as follows $g_0 = 8.863\ MPa^{-n} \cdot m^{1-n/2}$, $n = 6.67$, $\mu_r = 49.62\ \mu m$, and $\sigma_r = 13.9\ \mu m$. The residual stress varied from large compressive $q^0 = -237.9\ MPa$ on the surface to small tensile $q^0 = 0.035\ MPa$ at the depth of 400 μm while the fracture toughness K_f varied between 15 and 95 $MPa \cdot m^{1/2}$. The other parameter values related to lubricant and polymeric additive properties were $U = 347\ kJ/mole$, $T = 310°K$, $a_* = 0.374\ nm$, $l_* = 0.154\ nm$, $w_m = 35.1\ g/mole$, $\mu_a = 0.00125\ Pa \cdot s$, $c_p = 1.204\ g/dL$. The values for the shield constants were chosen as follows $C = 15.5$ and $\alpha = 0.008$. The dimensionless operating parameters $a = -2$, $V = 0.1$, $Q = 5$, $s_0 = -0.5$, $\varepsilon = 10^{-4}$, $\gamma = 1.0774$, $\delta = 0.671 \cdot 10^8$, $\theta = 4.524 \cdot 10^{-3}$, and $\kappa = 1$ correspond to the Hertzian semi-length of the contact $a_H = 10^{-3}m$ and the Hertzian maximum pressure $p_H = 1.284 GPa$.

In this section we compare the global survival probabilities $P_{glob}(N)$ of the lower solid calculated for four cases. The four cases differ from each other only by the initial density of polymer molecular weight distribution $W_a(l)$. It is important to remember that the ambient lubricant viscosity is the same for all four cases. For the base case the initial distribution of the polymer molecular weight $W_{a0}(l)$, $0 \le l \le l_{max}$, $l_{max} > 0$, is taken from the experimental study by Covitch [11]. For the other three cases, the initial densities of polymer molecular weight distributions were

$$W_{a1}(l) = W_{a0}(1.25l),\ 0 \le l \le 1.25 l_{max},$$

$$W_{a2}(l) = W_{a0}(0.5l),\ 0 \le l \le 0.5 l_{max}, \qquad (12.62)$$

$$W_{a3}(l) = W_{a0}(0.25l),\ 0 \le l \le 0.25 l_{max}.$$

The mean polymer chain lengths l_m that correspond to the initial molecular weight distributions from equations (12.62) were as follows

$$l_{m1} = 1.25l_{m0}, \ l_{m2} = 0.5l_{m0}, \ l_{m3} = 0.25l_{m0}, \tag{12.63}$$

where $w_m s_m = w_m l_{m0} = 1.3196 \cdot 10^5$ is the initial mean chain length determined by the initial molecular weight distribution $W_a(l) = W_{a0}(l)$. For the base case a typical value for the friction coefficient in bearings is taken equal to $\lambda_0 = 0.02$ while for the other three cases according to equations (12.61) and (12.62) we obtained $\lambda_1 = 0.01967$, $\lambda_2 = 0.02109$, and $\lambda_3 = 0.02193$ for $s_0 = 0$ and $\lambda_1 = 0.01925$, $\lambda_2 = 0.02231$, and $\lambda_3 = 0.02427$ for $s_0 = -0.5$. To determine the above values of λ, we used $a = a_0 = -2$ and $c = c_0, c_1, c_2, c_3$ obtained from numerical solutions of the elastohydrodynamic lubrication problem.

The main properties of solutions of the isothermal EHL problem for a degrading lubricant with Newtonian and non-Newtonian rheologies are presented in Section 8.6. The behavior of the pressure $p(x)$ and gap $h(x)$ distributions as well as of the film thickness H_0 and exit coordinate c of the EHL problem with respect to parameters a, V, and Q is similar to the one described in Sections 12.5 and 12.6 for non-degrading lubricants with Newtonian and non-Newtonian rheologies. For a lubricant with Newtonian rheology increase in Q causes a relatively slow increase in H_0 and a rapid increase in $f(x)$ and $\tau(x, z)$. That, in turn, leads to rapid lubricant degradation. A usually very small parameter ε almost does not affect the problem solution.

Lubricant degradation depends on a number of parameters among which the shear stress τ (see equations (12.60)) plays the major role. For $s_0 = 0$ the sliding frictional stress $f(x) = 0$ and, therefore, the shear stress $\tau = (12H_0^2/V)zdp/dx$ reaches its extrema in the inlet and exit zones while in the Hertzian region it is relatively small. That leads to the situation when lubricant degradation is relatively slow and practically all lubricant degradation occurs in the inlet zone while the lubricant almost does not degrade in the Hertzian region and the exit zone.

Now, let us examine a case of mixed rolling and sliding with $s_0 = -0.5$. The exit lubrication film thickness H_0, exit coordinate c, the distributions of pressure p and gap h exhibit behavior very similar to the case of pure rolling. From the comparison of the values of the friction coefficient λ (see Tables 12.1 and 12.2) it is obvious that variations in λ due to variations of the initial polymer molecular weight and lubricant degradation are smaller in the case of pure rolling ($s_0 = 0$) than in the case of mixed rolling and sliding ($s_0 = -0.5$). The numerical results from Sections 12.5 and 12.6 show that in the case of mixed rolling and sliding, the surface frictional stresses τ_1 and τ_2 are smaller than in a similar case without degradation but still large enough to cause lubricant degradation to a much greater extent than in the case of pure rolling. On the other hand, in the case of mixed rolling and sliding the surface frictional stresses τ_1 and τ_2 are much higher than the ones for the case of pure rolling ($\tau_2 = -\tau_1 = (6H_0^2/V)hdp/dx$) and, at the same time, lower than for the case of no degradation. That explains the stronger lubricant

TABLE 12.1
Fatigue life N versus the survival probability P_{glob} for
$s_0 = 0$ (after Kudish [26]). Reprinted with permission
from the STLE.

$P_{glob}(N)$	N	λ	l_m
0.9	$0.149 \cdot 10^{10}$		
0.75	$0.1845 \cdot 10^{10}$	$\lambda = 0.02$	$l_m = l_{m0}$
0.5	$0.2345 \cdot 10^{10}$		
0.9	$0.1789 \cdot 10^{10}$		
0.75	$0.2218 \cdot 10^{10}$	$\lambda = 0.01967$	$l_m = 1.25 l_{m0}$
0.5	$0.2619 \cdot 10^{10}$		
0.9	$0.820 \cdot 10^{9}$		
0.75	$0.1018 \cdot 10^{10}$	$\lambda = 0.02109$	$l_m = 0.5 l_{m0}$
0.5	$0.1293 \cdot 10^{10}$		
0.9	$0.527 \cdot 10^{9}$		
0.75	$0.652 \cdot 10^{9}$	$\lambda = 0.02193$	$l_m = 0.25 l_{m0}$
0.5	$0.828 \cdot 10^{9}$		

degradation in the case of mixed rolling and sliding conditions in comparison
with the case of pure rolling conditions (see Sections 12.5 and 12.6). For the
base case of degrading lubricant, the frictional stresses τ_1 and τ_2 applied to
the contact surfaces are on average lower than for the non-degrading lubricant
by about 26% (see Sections 12.5 and 12.6). In the base case for $s_0 = -0.5$, the
maximum loss of the lubricant viscosity reaches about 41% (see Sections 12.5
and 12.6). In many respects the process of lubricant degradation is governed
by the characteristic polymer chain length L_* (see equations (12.60)), which
represents the position of the beginning of the profile of the probability of
polymer scission $R(x, z, l)$. In fact, the probability of polymer scission R is
the main key to achieve control over lubricant degradation. A more detailed
discussion of the role of R in lubricant degradation can be found in Chapter
6 and Sections 12.5 and 12.6.

Now, let us consider the behavior of the global survival probability P_{glob}
(N_{life}) in all four cases for slide-to-roll ratios $s_0 = 0$ and $s_0 = -0.5$. In the
cases when the initial mean polymer chain length l_m varies between $0.25 l_{m0}$
and $1.25 l_{m0}$ the contact fatigue life N_{life} that corresponds to the survival
probability $P_{glob}(N_{life}) = 0.9$ for $s_0 = 0$ changes about 3.4 times, i.e., N_{life}
varies between $0.354 N_{life0}$ and $1.201 N_{life0}$ cycles, respectively (see Table 12.1
and Fig. 12.19) while for $s_0 = -0.5$ the contact fatigue life N_{life} changes more
than 14–fold, i.e., N_{life} varies between $0.108 N_{life0}$ and $1.523 N_{life0}$ cycles,
respectively (see Table 12.2 and Fig. 12.20). Here $N_{life0} = 0.149 \cdot 10^{10}$ for
$s_0 = 0$ and $s_0 = -0.5$ is the contact fatigue life (in cycles) for the base case
when $W_a(l) = W_{a0}(l)$, $l_m = l_{m0}$, and $\lambda = 0.02$. In Tables 12.1 and 12.2 series
of data for the number of cycles N that correspond to the survival probability

TABLE 12.2
Fatigue life N versus the survival probability P_{glob} for
$s_0 = -0.5$ (after Kudish [26]). Reprinted with
permission from the STLE.

$P_{glob}(N)$	N	λ	l_m
0.9	$0.149 \cdot 10^{10}$		
0.75	$0.1845 \cdot 10^{10}$	$\lambda = 0.02$	$l_m = l_{m0}$
0.5	$0.2345 \cdot 10^{10}$		
0.9	$0.227 \cdot 10^{10}$		
0.75	$0.2805 \cdot 10^{10}$	$\lambda = 0.01925$	$l_m = 1.25 l_{m0}$
0.5	$0.357 \cdot 10^{10}$		
0.9	$0.432 \cdot 10^{9}$		
0.75	$0.535 \cdot 10^{9}$	$\lambda = 0.02231$	$l_m = 0.5 l_{m0}$
0.5	$0.680 \cdot 10^{9}$		
0.9	$0.161 \cdot 10^{9}$		
0.75	$0.200 \cdot 10^{9}$	$\lambda = 0.02427$	$l_m = 0.25 l_{m0}$
0.5	$0.254 \cdot 10^{9}$		

$P_{glob}(N)$ equal to 0.9, 0.75, and 0.5 for $s_0 = 0$ and $s_0 = -0.5$ are presented
for the base case and the other three cases from equations (12.61) and (12.62),
respectively.

To understand how polymer concentration c_p affects contact fatigue, we
need to consider the dependence of the lubricant viscosity $\mu(p, W)$ on c_p.
Numerical results show that the distribution of pressure p is almost independent of c_p. Therefore, in dimensional variables the derivative of $\mu(p, W)$ with
respect to c_p can be expressed as follows

$$\frac{\partial \mu}{\partial c_p} \approx e^{Qp}([\eta] - [\eta]_a)\frac{1+2k_H([\eta]+[\eta]_a)+k_H c_p^2[\eta][\eta]_a}{(1+c_p[\eta]_a+k_H c_p^2[\eta]_a^2)^2}$$

$$+e^{Qp}c_p\frac{\partial[\eta]}{\partial c_p}\frac{1+2k_H c_p[\eta]}{1+c_p[\eta]_a+k_H c_p^2[\eta]_a^2}. \tag{12.64}$$

Numerical results show that the first term in equation (12.64) dominates the
second one and as lubricant moves along the contact the intrinsic viscosity
$[\eta]$ monotonically decreases starting with $[\eta]_a$ at the inlet point $x = a$, i.e.,
$[\eta] - [\eta]_a \leq 0$. Therefore, we can expect that the loss of lubricant viscosity
caused by polymer degradation increases with increasing polymer additive
concentration c_p. That, in turn, leads to reduced frictional stresses and higher
fatigue life.

The described above effects of significant increased fatigue life due to reduced lubricant viscosity and frictional stress are valid only for contacts with
sufficiently smooth surfaces so that the direct contact of asperities does not
occur even in case of degraded lubricant. In cases of mixed friction, the dry
friction created by direct contact of asperities usually dominates the friction

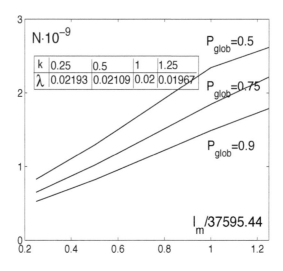

FIGURE 12.19

Graphs of fatigue life N as a function of the average length of polymer additive molecules l_m obtained for pure rolling $s_0 = 0$.

realized in lubricated zones where contact surfaces are completely separated by lubricant. Because the coefficient of dry friction is by about one or two orders of magnitude higher than that of a well–lubricated contact one can expect that for contacts with relatively large portion of direct asperity contact the effect of lubricant viscosity loss may not be that significant. Moreover, in such cases the increase in the ambient viscosity may lead to a reduced overall frictional stress due to increased film thickness that separates asperities better and, thus, reduces the portion of the frictional stress due to direct asperity contact more than it raises the frictional stress due to increased lubricant viscosity. The illustration of this effect can be found in the results of experimental study by Krantz and Kahraman [38].

The effect of lubricant viscosity loss of up to 70% was observed experimentally and documented by Walker, Sanborn, and Winer [8]. Therefore, in real lubricated contacts the viscosity loss may reach high values, which would lead to lower frictional stresses and much wider margins of fatigue life variations.

In case of a thermal EHL problem due to heat generation in the lubrication film and elevated lubricant temperature T, the lubricant viscosity (in case of no lubricant degradation) is lower than in a comparable isothermal case. That leads to higher value of L_*, which slows down the degradation process. The extent to which the polymer degradation process slows down depends on the lubricant nature, on the nature and distribution of the polymer molec-

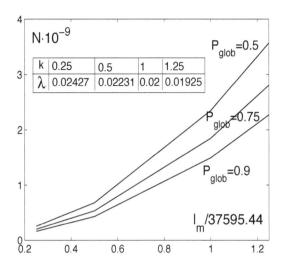

FIGURE 12.20
Graphs of fatigue life N as a function of the average length of polymer additive molecules l_m obtained for mixed rolling and sliding $s_0 = -0.5$.

ular weight entering the contact, and on the contact operating conditions. Therefore, under higher temperature conditions, the margins of fatigue life variations are narrower than for lower temperatures but are still significant.

The results based on such a model can be used for a more realistic estimation of contact fatigue life as a function of not only the parameters of contact materials but also of lubricant parameters such as lubricant viscosity, chemistry, concentration, and details of the polymeric additive initial molecular weight distribution. That may provide a better understanding of contact fatigue performance of lubricated joints working in different applications and under various conditions. Moreover, better understanding of lubricant influence on contact fatigue may lead to better design of bearing and gear testing procedures. In particular, a special attention should be paid to lubricant to keep it identical in different tests used for determination and comparison of fatigue life in a homogeneous sample of tested bearings or gears.

12.8   A Qualitative Model of Lubricant Life

In this section we will develop a model that predicts lubricant life in a simple system of a line contact supplied with lubricant circulating between a lubricant tank and the contact.

Let us consider a steady contact of a roller of length L_r with another cylinder with parallel axes of length greater than L_r lubricated by a lubricant supplied from an oil tank with the volume of lubricant V_T. We will assume that the necessary amount of lubricant is removed from the tank and delivered to the contact and then, after entering and exiting the contact, it is returned back to the oil tank. Let us suppose that the lubricant volume V_T in the oil tank is constant in time and the oil in the tank is continuously mixed. Moreover, we will assume that the lubricant moves from left to right while entering the contact which is true if $\mid s_0 \mid \leq 2$ (see Sections 8.3, 8.5, and 8.6). Based on the assumption that on a time scale equal to the time necessary for lubricant to pass through the contact the flow is quasi-steady, we can derive an equation for the density of the polymer molecular weight distribution $W_T(n, l)$ in the oil tank per unit volume at the time moment n for molecules with chain lengths l from the interval $[l, l + \triangle l]$ (see [25])

$$\frac{dW_T}{dn} = \frac{L_r}{V} \Big\{ - \int\limits_{z_{01}}^{z_{02}} u(x_i, z) W(x_i, z, l) dy$$

$$+ \int\limits_{-h_e/2}^{h_e/2} u(x_e, z) W(x_e, z, l) dz - \Big[\int\limits_{-h_i/2}^{z_{01}} u(x_i, z) dz \qquad (12.65)$$

$$+ \int\limits_{z_{02}}^{h_i/2} u(x_i, z) dz \Big] W_T(n, l) \Big\}, \quad W_T(0, l) = W_{T0}(l),$$

where z_{01} and z_{02} are the lower and upper boundaries in the inlet lubricant flow that separate the lubricant exiting through the inlet gap from the lubricant that enters the contact (see Sections 8.3, 8.5, and 8.6), $h_i = h(x_i)$ and $h_e = h(x_e)$ are the inlet and exit values of gap, respectively, $W(x_i, z, l)$ and $W(x_e, z, l)$ are the solutions of the EHL problem for the degrading lubricant at the inlet and exit film cross sections, respectively, $W_{T0}(l)$ is the initial polymer molecular weight distribution in the oil tank.

The coupling between the kinetic equation in (12.60) describing lubricant degradation in a contact and equation (12.65) occurs naturally through the initial condition for the kinetic equation in (12.60)

$$W(x_i, z, l) = W_T(n, l). \qquad (12.66)$$

It can be shown that in the oil tank the total number of polymer molecules per unit of lubricant volume increases with time n while the total polymer molecular weight per unit of lubricant volume is conserved in time.

The lubricant life can be defined by assuming that the lubricant useful life is determined by the period of time n_{max} during which the lubricant viscosity μ_T in the oil tank is not lower than a certain value μ_{min}, i.e.,

$$\mu_T(n) \geq \mu_{min} \ for \ n \leq n_{max}, \ \mu_T(n_{max}) = \mu_{min}. \tag{12.67}$$

Let us introduce new functions

$$M_\beta = \int\limits_0^\infty W(x,z,l)l^\beta dl / \int\limits_0^\infty W(x,z,l)dl,$$

$$M_{T\beta} = \int\limits_0^\infty W_T(n,l)l^\beta dl / \int\limits_0^\infty W_T(n,l)dl.$$

We will assume that in an incompressible lubricant flow there exists a point (x_*, z_*) such that

$$M_\beta(x_*, z_*) = \varphi M_{T\beta}(n), \ 0 < \varphi < 1,$$

$$-\int\limits_{z_{01}}^{z_{02}} u(x_i, z)M_\beta(x_i, z)dz + \int\limits_{-h_e/2}^{h_e/2} u(x_e, z)M_\beta(x_e, z)dz$$

$$\approx \Big[-\int\limits_{z_{01}}^{z_{02}} u(x_i, z)dz + \int\limits_{-h_e/2}^{h_e/2} u(x_e, z)dz \Big] M_\beta(x_*, z_*),$$

where φ is a certain function that depends on the contact conditions and time n. Function φ varies with time n in such a way that as lubricant degradation goes on the value of φ monotonically increases and, eventually, approaches 1. This approximation may serve as a first crude approach to estimation of lubricant life and it leads to an approximate solution

$$M_{T\beta}(n) = M_{T\beta 0} \exp\big[-\int\limits_0^n \lambda(t)dt \big],$$

$$\lambda(n) = \tfrac{L_r(1-\varphi)}{V_T} \Big[\int\limits_{-h_i/2}^{z_{01}} u(x_i, z)dz + \int\limits_{z_{02}}^{h_e/2} u(x_i, z)dz \Big], \tag{12.68}$$

$$M_W(n) = M_{W0} \exp\Big[-\tfrac{1}{\beta} \int\limits_0^n \lambda(t)dt \Big], \tag{12.69}$$

where M_{W0} is the value of M_W at the time moment $n = 0$. From equations (12.60) and (12.67), we obtain

$$\int\limits_0^n \lambda(t)dt = \ln\Big[\tfrac{2k_H c_p[\eta]_a}{\sqrt{1-4k_H(1-\mu_{min}/\mu_b)}-1} \Big], \tag{12.70}$$

$$[\eta]_a = k'M_{W0}^{\beta}, \quad \mu_b = \frac{\mu_a}{1+c_p[\eta]_a+k_H(c_p[\eta]_a)^2}, \quad (12.71)$$

where μ_b is the viscosity of the base oil without the polymer additive. It follows from equation (12.68) that the rough estimate of the time scale necessary for degrading the tank of oil of volume V_T under given lubricating conditions is

$$n_0 \approx \lambda_0^{-1}, \quad \lambda_0 = \frac{L_r(u_1+u_2)h_e}{2V_T}\left[1 - \frac{M_\beta(x_e,0)}{M_{T\beta}(0)}\right]. \quad (12.72)$$

A more detailed analysis of lubricant life requires extensive computer calculations which are difficult to realize on contemporary computers. These results can be used for evaluation of useful lubricant life and for more realistic estimation of the lubrication film thickness and frictional stresses in lubricated contacts. That may provide a solid basis for maintenance scheduling of lubricated equipment operating under various conditions. Moreover, it may enable researches to create new and more durable lubricant additives for general and specific applications.

12.9   Closure

The effect of stress-induced polymeric additive degradation in a lubricated contact and its effect on contact fatigue are studied. The analysis is performed numerically based on the developed models of lubricant degradation and contact fatigue. The results show that degradation of the polymeric additive in a lubricated contact may be significant and, therefore, leads to reduction of lubricant viscosity and frictional stress. Moreover, the results show that for lubricants with the same ambient viscosity contact fatigue of sufficiently smooth solids completely separated by lubricants may vary significantly due to the specific way these lubricants were formulated. In such cases it is established that contact fatigue is strongly affected by the initial distribution of the molecular weight of the polymeric additive (viscosity modifier) to lubricant and by contact operating conditions that in some cases promote rapid lubricant degradation caused by high lubricant shearing stresses. In particular, for sufficiently smooth surfaces completely separated by lubricant and for lubricants with higher viscosity loss caused by additive degradation, contact fatigue life is higher given same other parameters of the contacts. In cases where the average molecular weight of polymer molecules varied between $3.2989 \cdot 10^4$ and $1.6495 \cdot 10^5$ the contact fatigue life that corresponds to the global survival probability equal to 0.9 for $s_0 = 0$ (pure rolling) varied between $0.179 \cdot 10^{10}$ and $0.527 \cdot 10^9$ and for $s_0 = -0.5$ (mixed rolling and sliding) varied between $0.227 \cdot 10^{10}$ and $0.161 \cdot 10^9$, i.e., contact fatigue life changed more than 3– and 14–fold, respectively. Therefore, to properly determine contact fatigue life it is not sufficient to know the ambient lubricant viscosity and the dependence on pressure of the viscosity of the non-degraded lubricant. Moreover, to success-fully predict contact fatigue of lubricated joints, it is necessary to know not only the fatigue resistance and cleanliness of the contact materials but also to be able to take into account the effect of temperature and stress-induced polymer degradation on lubricant viscosity in joints. Better understanding of lubricant influence on contact fatigue may lead to better design of bearing and gear testing procedures. A qualitative model of lubricant life is proposed.

12.10   Exercises and Problems

1. What are the two general causes of lubricant non-Newtonian behavior in lubricated contacts? List the most important manifestations of lubricant non-Newtonian behavior. What is the usual response of lubricant viscosity to increase in the shear rate? In which cases lubricant viscosity loss is reversible and irreversible?

2. The term $\sqrt{1 + \frac{1}{4}(\frac{dh}{dx})^2}$ is usually very close to 1 and almost never used in formulations of the EHL problems for non-degrading lubricants. What is the primary reason for the inclusion of this term in equations (12.8) for degrading lubricants?

3. Explain why is it necessary to know lubricant flow streamlines to analyze stress-induced lubricant degradation.

4. Describe and discuss the topology of the lubricant flow between the surfaces of contacting solids. What is a possible implication of the topology of flow streamlines in the vicinity of a separatrix running through the entire contact of degrading lubricant for the lubricant viscosity μ in case of some sliding (i.e., $s_0 \neq 0$)?

5. Describe the general approach to the numerical method used for solution of the EHL problem for degrading lubricant. Discuss in detail how: (a) the sliding frictional stress $f(x)$ is determined, (b) the horizontal component $u(x, z)$ of the lubricant velocity and lubricant flux are calculated, (c) the lubricant flow streamlines are obtained, (d) the kinetic equation is solved and lubricant viscosity μ is updated, and (e) the Reynolds equation is solved.

6. Describe the similarities and differences in the behavior of degrading and non-degrading lubricants in heavily loaded contacts. Describe in detail the behavior of the polymer additive molecular weight $W(x, z, l)$ and lubricant viscosity $\mu(x, z)$ in lubricated contacts.

7. Describe the effect the distribution of molecular weight $W(a, z, l)$ or $W(c, z, l)$ of a polymeric additive entering the lubricated contact has on fatigue of a lubricated contact.

BIBLIOGRAPHY

[1] Bair, S. and Winer, W.O. 1979. Shear Strength Measurements of Lubricants at High Pressure. *J. Lubr. Techn.* 101, No. 3:251-257.

[2] Hoglund, E. and Jacobson, B. 1986. Experimental Investigations of the Shear Strength of Lubricants Subjected to High Pressure and Temperature. *ASME J. Tribology* 108, No. 4:571-578.

[3] Eyring, H. 1936. Viscosity, Plasticity, and Diffusion as Examples of Absolute Reaction Rates. *J. Chem. Phys.* 4, No. 4:283-291.

[4] Houpert, L.G. and Hamrock, B.J. 1985. Elastohydrodynamic Lubrication Calculations Used as a Tool to Study Scuffing. In *Mechanisms and Surface Distress : Global Studies of Mechanisms and Local Analyses of Surface Distress Phenomena*, Eds. D. Dowson, et al. Butterworths, England, 146-162.

[5] Kudish, I.I. 1982. Asymptotic Methods for Studying Plane Problems of the Elastohydrodynamic Lubrication Theory in Heavily Loaded Regimes. Part 1. Isothermal Problem. *Izvestija Akademii Nauk Arm. SSR, Mekhanika* 35, No. 5:46-64.

[6] Crail, I.R.H. and Neville, A.L. 1969. The Mechanical Shear Stability of Polymeric VI Improvers. *J. Inst. Petrol.* 55, No. 542:100-108.

[7] Casale, A. and Porter, R.S. 1971. The Mechanochemistry of High Polymers. *J. Rubber Chem. and Techn.* 44, No.2:534-577.

[8] Walker, D.L., Sanborn, D.M., and Winer, W.O. 1975. Molecular Degradation of Lubricants in Sliding Elastohydrodynamic Contacts. *ASME J. Lubr. Techn.* 97, No. 3:390-397.

[9] Yu, J.F.S, Zakin, J.L., and Patterson, G.K. 1979. Mechanical Degradation of High Molecular Weight Polymers in Dilute Solution. *J. Appl. Polymer Sci.* 23:2493-2512.

[10] Odell, J.A., Keller, A., and Rabin, Y. 1988. Flow-Induced Scission of Isolated Macromolecules. *J. Chem. Phys.* 88, No. 6:4022-4028.

[11] Covitch, M.J. 1998. How Polymer Architecture Affects Permanent Viscosity Loss of Multigrade Lubricants. *SAE Technical Paper* No. 982638.

[12] Herbeaux, J.-L., Flamberg, A., Koller, R.D., and Van Arsdale, W.E. 1998. Assesment of Shear Degradation Simulators. *SAE Technical Paper* No. 982637.

[13] Herbeaux, J.-L. 1996. *Mechanochemical Reactions in Polymer Solutions* Ph.D. Dissertation, University of Houston, Houston (and references therein).

[14] Ziff, R.M. and McGrady, E.D. 1985. The Kinetics of Cluster Fragmentation and Depolymerization. *J. Phys. A : Math. Gen.* 18:3027-3037.

[15] Ziff, R.M. and McGrady, E.D. 1986. Kinetics of Polymer Degradation. *AchS, Macromolecules* 19:2513-2519.

[16] McGrady, E.D. and Ziff, R.M. 1988. Analytical Solutions to Fragmentation Equations with Flow. *AIChE* 34, No. 12:2073-2076.

[17] Montroll, E.W. and Simha, R. 1940. Theory of Depolymerization of Long Chain Molecules. *J. Chem. Phys.* 8:721-727.

[18] Saito, O. 1958. On the Effect of High Energy Radiation to Polymers. I, Cross-Linking and Degradation. *J. Phys. Soc. Jpn.* 13:198-206.

[19] Kudish, I.I. and Ben-Amotz, D. 1999. Modeling Polymer Molecule Scission in EHL Contacts. In *The Advancing Frontier of Engineering Tribology, Proc. 1999 STLE/ASME H.S. Cheng Tribology Surveillance*, Eds.: Q. Wang, J. Netzel, and F. Sadeghi, 176-182.

[20] Kudish, I.I., Airapetyan, R.G., and Covitch, M.J., 2002, Modeling of Kinetics of Strain-Induced Degradation of Polymer Additives in Lubricants. *J. Math. Models and Methods Appl. Sci.* 12, No. 6:1-22.

[21] Kudish, I.I., Airapetyan, R.G., and Covitch, M.J. 2003. Modeling of Kinetics of Stress-Induced Degradation of Polymer Additives in Lubricants and Viscosity Loss. *STLE Tribology Trans.* 46, No. 1:1-11.

[22] Kudish, I.I., Airapetyan, R.G., Hayrapetyan, G.R., and Covitch, M.J. 2005. Kinetics Approach to Modeling of Stress Induced Degradation of Lubricants Formulated with Star Polymer Additives. *STLE Tribology Trans.* 48, No. 2:176-189.

[23] Kudish, I.I., and Airapetyan, R.G. 2003. Modeling of Line Contacts with Degrading Lubricant. *ASME J. Tribology* 125, No. 3:513-522.

[24] Kudish, I.I., and Airapetyan, R.G. 2004. Lubricants with Newtonian and Non-Newtonian Rheologies and Their Degradation in Line Contacts. *ASME J. Tribology* 126, No. 1:112-124.

[25] Kudish, I.I., and Airapetyan, R.G. 2003. A New Approach to Modeling of Stress-Induced Degradation of Formulated Lubricants and Their Behavior in Lubricated Contacts. *Proc. 2nd Tribology in Environmental Design Intern. Conf.* September 8-10, Bournemouth, UK.

[26] Kudish, I.I. 2005. Effect of Lubricant Degradation on Contact Fatigue. *STLE Tribology Trans.* 48, No. 1:100-107.

[27] Hamrock, B.J. 1991. *Fundamentals of Fluid Film Lubrication*. Cleveland: NASA Reference Publication 1255.

[28] Billmeyer, F.W., Jr. 1966. *Textbook of Polymer Science*. New York: John Wiley & Sons.

[29] Crespi, G., Valvassori, A., Slisi, U. 1977. Olefin Copolymers. In *The Stereo Rubbers*. Ed. W.M. Saltman. New York: John Wiley & Sons, 365-431.

[30] Kudish, I.I. 1996. Asymptotic Analysis of a Problem for a Heavily Loaded Lubricated Contact of Elastic Bodies. Pre- and Over-critical Lubrication Regimes for Newtonian Fluids. *Dynamic Systems and Applications*, Dynamic Publishers, Atlanta, 5, No. 3:451-476.

[31] Kudish, I.I. 1981. Some Problems of Elastohydrodynamic Theory of Lubrication for a Lightly Loaded Contact. *J. Mech. Solids* 16, No. 3:75-88.

[32] Kudish, I.I. 2000. A New Statistical Model of Contact Fatigue. *STLE Tribology Trans.* 43, No. 4:711-721.

[33] Kudish, I.I. and Burris, K.W. 2000. Modern State of Experimentation and Modeling in Contact Fatigue Phenomenon. Part I. Contact Fatigue versus Normal and Tangential Contact and Residual Stresses. Nonmetallic Inclusions and Lubricant Contamination. Crack Initiation and Crack Propagation. Surface and Subsurface Cracks. *STLE Tribology Trans.* 43, No. 2:187-196.

[34] Tallian, T., Hoeprich, M., and Kudish, I.I. 2001. Author's Closure. *STLE Tribology Trans.* 44, No. 2:153-155.

[35] Kudish, I.I. 2002. Lubricant-Crack Interaction, Origin of Pitting, and Fatigue of Drivers and Followers. *STLE Tribollology Trans.* 45, No. 4:583-594.

[36] Kudish, I.I. 1987. Contact Problems of The Theory of Elasticity for Prestressed Bodies with Cracks. *J. Appl. Mech. and Tech. Phys.* 28, No. 2:295-303.

[37] Jacod, B., Venner, C.H., and Lugt, P.M. 2003. Extension of the Friction Mastercurve to Limiting Shear Stress Models. *ASME J. Tribology* 125, No. 3:739-746.

[38] Krantz, T.L. and Kahraman, A. 2004. An Experimental Investigation of the Influence of the Lubricant Viscosity and Additives on Gear Wear. *STLE Tribology Trans.* 47, No. 1:138-148.

13

Non-steady and Mixed Friction Lubrication Problems

13.1 Introduction

In this chapter we consider several non-steady and mixed lubrication models for infinite journal bearings and non-conformal contacts. Proper modified formulations and some solutions of these problems are presented. The comparison of the solutions of the traditional and modified non-steady problems is provided. A formulation and analysis of a mixed friction problem with zones of dry and lubricant friction are considered. The necessary boundary conditions at the internal contact boundaries separating dry and lubricated zones are considered.

13.2 Properly Formulated Non-steady Plane Lubrication Problems for Contacts of Rigid Solids

This section is dedicated to analysis of two non-steady problems: one for a conformal and the other for non-conformal lubricated contacts. Solutions for lubricated weightless joints involved in a non-steady motion have been studied by Safa and Gohar [1], Ai and Cheng [2], Chang et al. [3], Venner and Lubrecht [4, 5], and Osborn and Sadeghi [6], Cha and Bogy [7], Hashimoto and Mongkolwongrojn [8], Peiran and Shizhu [9, 10]. Non-steady lubrication problems for heavy joints were studied by San Andres and Vance [11], Larsson and Hoglund [12] and Kudish and Panovko [13].

The main purpose of the analysis of this section is to propose the proper formulations of non-steady problems for conformal and non-conformal contacts and to show the difference between the "classic" solutions and the solutions obtained based on the proper problem formulations. In both cases the approaches used and the conclusions made are very similar and follow the material in [14, 15].

13.2.1 A Proper Formulated Non-steady Plane Lubrication Problem for a Non-conformal Contact

We will consider a non-steady line contact and show that the "classic" formulation of the problem has some significant defects that need to be fixed to adequately describe a non-steady regime of lubrication. We will see that a proper use of Newton's second law, which takes into account the inertia of the lubricated contact, removes all mentioned deficiencies.

Let us consider two smooth elastic infinite cylinders with parallel axes and radii R_1 and R_2 slowly rolling one over another and choose a motionless Cartesian coordinate system with the origin at the motionless axis of cylinder 2. In this coordinate system cylinder 1 is acted upon by force P along the z-axis (directed across the lubrication film). The linear velocities of the surface points of the cylinders are u_1 and u_2, respectively.

The contact (see Fig. 13.1) is lubricated by a Newtonian incompressible fluid with viscosity μ. The lubrication process is isothermal. The cylinders are fully separated by the lubrication film. Because of slow relative motion of the contact surfaces the inertia forces in the lubrication layer are small compared to the viscous ones. The lubrication thickness is small relative to the characteristic size of the contact and radii of the cylinders. The lubrication conditions are non-steady. That means that linear velocities u_k $(k = 1, 2)$ and the applied force P as well as the coordinate of the inlet point x_i may vary with time t. The characteristic time of these variations is considered to be small compared to the period of sound waves in elastic solids. The latter requires a non-steady problem formulation only for fluid flow parameters.

Under these assumptions and in the dimensionless variables

$$(x', a, c) = \tfrac{T}{2R'}(x, x_i, x_e), \ t' = \tfrac{(u_{10}+u_{20})T}{4R'}t, \ (p', p'_0) = \tfrac{\pi R'}{TP_0}(p, p_0),$$

$$(h', h'_0, h'^0) = \tfrac{T^2}{2R'}(h, h_0, h^0), \ \mu' = \tfrac{\mu}{\mu_a}, \ u' = \tfrac{u_1+u_2}{u_{10}+u_{20}}, \ P' = \tfrac{P}{P_0}, \quad (13.1)$$

$$T^2 = \tfrac{P_0}{3\pi\mu_a(u_{10}+u_{20})}, \ V = \tfrac{3\pi^2\mu_a(u_{10}+u_{20})R'E'}{P_0^2}$$

the problem can be reduced to the "classic" system of integro-differential equations (for convenience primes are omitted) [16]

$$\tfrac{\partial}{\partial x}\{\tfrac{h^3}{\mu}\tfrac{\partial p}{\partial x}\} = u(x)\tfrac{\partial h}{\partial x} + \tfrac{\partial h}{\partial t}, \quad (13.2)$$

$$h_0(0) = h^0, \ c(0) = c_0, \ p(x, 0) = p_0(x), \quad (13.3)$$

$$p(a, t) = p(c, t) = \tfrac{\partial p(c,t)}{\partial x} = 0, \quad (13.4)$$

$$h(x, t) = h_0 + x^2 + \tfrac{2}{\pi V}\int\limits_a^c p(y, t)\ln|\tfrac{y}{x-y}|\,dy, \quad (13.5)$$

$$\int\limits_a^c p(y, t)dy = \tfrac{\pi}{2}P(t). \quad (13.6)$$

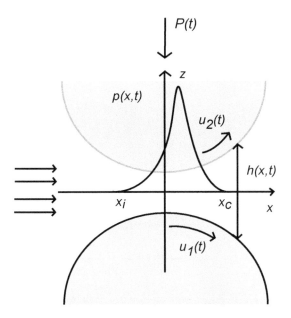

FIGURE 13.1
The general view of the lubricated contact.

In equation (13.1) x_i and x_e are the coordinates of the inlet and exit points, respectively, R' and E' are the effective radius and elastic modulus, respectively, P_0, u_{10}, and u_{20} are the characteristic force applied to the cylinders and their surface velocities, μ_a is the lubricant viscosity at ambient conditions.

In case of a purely squeezed lubrication layer $u(t) = 0$, $c = -a = \infty$ and the boundary conditions in (13.4) must be replaced by the conditions

$$p(\pm\infty, t) = \frac{\partial p(\pm\infty, t)}{\partial x} = 0. \qquad (13.7)$$

The "classic" problem formulation is completely described by equations (13.2)-(13.6) or (13.2), (13.3), (13.5)-(13.7). In addition to these equations, the relations for functions $\mu(p)$, $P(t)$, $u(t)$, $a(t)$ and h^0, c_0, and $p_0(x)$ must be given. The problem solution comprises pressure $p(x,t)$ and gap $h(x,t)$ distributions, exit boundary $c(t)$, and the central film thickness $h_0(t)$.

13.2.1.1 Lubricated Contact of Rigid Solids

Let us consider the case of cylinders made of a rigid material, i.e., $V = \infty$. For simplicity we will assume $\mu(p) = 1$. In this case the problem equations (13.2)-(13.6) can be easily reduced to

$$h_0'(t)\{h_0[\tfrac{1}{h^2(c,t)} - \tfrac{1}{h^2(a,t)}] + c[G(c,t) - G(a,t)]\}$$

$$= u(t)\{2[F(c,t) - F(a,t)] - h(c,t)[G(c,t) - G(a,t)]\}$$

$$h_0(0) = h^0,$$

$$u(t)\{\tfrac{1}{2}\ln\tfrac{h(c,t)}{h(a,t)} + H(a,c,t) - (c-a)F(a,t) \tag{13.8}$$

$$-\tfrac{h(c,t)}{2}[I(a,c,t) - (c-a)G(a,t)]\} - \tfrac{h_0'}{4}\{F(c,t) - F(a,t)$$

$$-\tfrac{2(c-a)h_0}{h^2(a,t)} + 2c[I(a,c,t) - (c-a)G(a,t)]\} = \pi P(t)h_0,$$

where

$$F(x,t) = \tfrac{x}{h(x,t)} + \tfrac{1}{\sqrt{h_0}}\arctan\tfrac{x}{\sqrt{h_0}},$$

$$G(x,t) = \tfrac{x}{h^2(x,t)} + \tfrac{3x}{2h_0 h(x,t)} + \tfrac{3}{2h_0^{3/2}}\arctan\tfrac{x}{\sqrt{h_0}},$$

$$H(a,c,t) = \tfrac{1}{\sqrt{h_0}}\{x\arctan\tfrac{x}{\sqrt{h_0}} = \tfrac{\sqrt{h_0}}{2}\ln h(x,t)\}\,|_a^c, \tag{13.9}$$

$$H(a,c,t) = \tfrac{1}{\sqrt{h_0}}\{x\arctan\tfrac{x}{\sqrt{h_0}} = \tfrac{\sqrt{h_0}}{2}\ln h(x,t)\}\,|_a^c,$$

$$I(a,c,t) = \{-\tfrac{1}{2h(x,t)} + \tfrac{3}{4h_0}\ln h(x,t)\}\,|_a^c + \tfrac{3}{2h_0}H(a,c,t).$$

After this system is solved the pressure distribution $p(x,t)$ is determined according to the formula

$$p(x,t) = \tfrac{u(t)}{2h_0}\{F(x,t) - F(a,t) - \tfrac{h(c,t)}{2}[G(x,t) - G(a,t)]\}$$

$$-\tfrac{h_0'}{4h_0}\{h_0[\tfrac{1}{h^2(x,t)} - \tfrac{1}{h^2(a,t)}] + c[G(x,t) - G(a,t)]\}. \tag{13.10}$$

The structure of system (13.8) and (13.9) allows to make a conclusion that if the external force $P(t)$ and/or the sum of linear velocities $u(t)$ experience an abrupt change (finite discontinuity) at some time moment t then $h_0(t)$, $c(t)$, and $h_0'(t)$ change abruptly as well. Therefore, for $V = \infty$ equations (13.5) and (13.10) make pressure $p(x,t)$ and gap $h(x,t)$ distributions discontinuous functions of time t.

To make this obvious let us consider the case of purely squeezed lubrication film ($u(t) = M(t) = 0$) between two rigid cylinders ($V = \infty$). In this case instead of boundary conditions (13.4), we should use conditions (13.7). The solution of the problem is represented by the formulas

$$p(x,t) = \tfrac{1}{4h^2(x,t)}\tfrac{dh_0(t)}{dt}, \tag{13.11}$$

$$h_0(t) = \frac{h^0}{[1+2\sqrt{h^0}\int_0^t P(v)dv]^2}. \tag{13.12}$$

A simple analysis of the derivative of $h_0(t)$ from (13.12) shows that it is a discontinuous function of t if $P(t)$ is discontinuous. Based on (13.11) we conclude that in this case pressure $p(x,t)$ is a discontinuous function of time t as well.

The next defect of the "classic" problem formulation reveals itself for any external conditions. Because the highest time derivative in the equations of the problem is of the first order for the given initial external conditions, these equations prescribe a certain initial normal speed $h_0'(0)$ of cylinders approach to each other. This prescribed initial normal speed of approach may not correspond the actual initial speed of approach, which generally can be chosen independently of the other parameters of the problem.

The last defect of this classic problem formulation is the fact that unless the external parameters of the problem are described by oscillatory functions this problem does not provide for the possibility to describe such an oscillatory motion. It is also due to the fact that problem is described by equations with the first–order time derivative. In reality, such a lubricated system being slightly perturbed and then left alone (under constant external conditions) may be involved in a damped oscillatory motion [13].

The steady EHL problems are free of all these defects. These defects are typical for all non-steady EHL problems described based on the presented "classic" problem formulation.

13.2.1.2 Lightly Loaded Lubricated Contact of Elastic Solids

Let us consider a lightly loaded regime of lubrication for which the elastic displacements of the contact surfaces are much smaller than the film thickness. Such regimes occur when $V \gg 1$. Assuming that $V \gg 1$ the solution of the problem can be found in the form of regular asymptotic expansions

$$p(x,t) = p_0(x,t) + \tfrac{1}{V}p_1(x,t) + \dots,$$

$$h(x,t) = h^0(x,t) + \tfrac{1}{V}h^1(x,t) + \dots, \tag{13.13}$$

$$h_0(t) = h_0^0(t) + \tfrac{1}{V}h_0^1(t) + \dots, \quad c(t) = c_0(t) + \tfrac{1}{V}c_1(t) + \dots,$$

where $h_0^0(t)$, $c_0(t)$, and $p_0(x,t)$ satisfy equations (13.8) and (13.9) while $h^0(x,t)$ is determined by equation (13.5) for $V = \infty$. The next terms of the expansions such as $h_0^1(t)$, $c_1(t)$, $p_1(x,t)$, and $h^1(x,t)$ are solutions of certain linear differential equations the coefficients of which depend on $h_0^0(t)$, $c_0(t)$, $p_0(x,t)$, and $h^0(x,t)$. Moreover, these coefficients are discontinuous functions of time t if $P(t)$ and/or $u(t)$ are discontinuous. Therefore, for $V \gg 1$ and discontinuous $P(t)$ and/or $u(t)$ the solution of the "classic" problem is discontinuous.

The other way to show discontinuity of $p(x,t)$ and, therefore (see equation (13.5)), discontinuity of $h(x,t)$ is to consider equation (13.6). For any

$V > 0$ if $P(t)$ is a discontinuous function of time t then, obviously, $p(x,t)$ is discontinuous.

Therefore, it is natural to expect that for any $V > 0$ in the "classic" formulation the solution of EHL problem is discontinuous in time t if $P(t)$ and/or $u(t)$ are discontinuous. This situation needs to be fixed because in reality the parameters of a lubricated contact are always continuous functions of time.

13.2.1.3 Modified Problem Formulation for Non-steady EHL Problem

The absence of inertia in the "classic" lubrication system creates the discontinuity effects discussed above. In order to avoid these discontinuities and to "push" them up to the second–order time derivatives (i.e., acceleration) we will "soften" the balance condition in equation (13.6) and replace it by the equation naturally following from Newton's second law [14]

$$m\frac{d^2 h_0}{dt^2} = \int\limits_{x_i}^{x_e} p(x,t)dx - P(t), \tag{13.14}$$

where m is the mass of cylinder 1 per unit length. In equation (13.14) it is assumed that cylinder 2 is either motionless or moves with a constant speed. These conditions for cylinder 2 may be guarantied if cylinder 2 is of infinite mass.

To complete this problem formulation, we need to impose a different set of initial conditions on h_0 and c

$$h_0(0) = h^0, \ h_0'(0) = h^1, \ c(0) = c_0. \tag{13.15}$$

Here the value of h^1 is independent of the system internal parameters and other operating conditions.

This problem formulation, which includes equations (13.14) and (13.15), is the adequate description of the above considered non-steady EHL problem for any loading and operating conditions.

The introduction of the second–order time derivative allows for oscillatory motion, which cannot be observed based on the "classic" problem formulation. It is important to understand that when the abrupt variations in $P(t)$ and $u(t)$ are long gone and all the external parameters of the system such as $P(t)$, $u(t)$, etc., approach constant values, solutions of both the "classic" and modified EHL problem approach the same steady state. The difference is in the manner it occurs, i.e., it is in the solutions transient behavior. After values of $P(t)$, $u(t)$, and $a(t)$ have stabilized the "classic" solution approaches the steady state monotonically. Depending on the lubrication system parameters, the modified solution may approach the steady state in a monotonic or oscillatory manner.

To illustrate the difference in solutions of the problem based on the "classic" and modified formulations, let us consider the case of purely squeezed lubrication film ($u(t) = 0$) between two rigid ($V = \infty$) cylinders. The "classic"

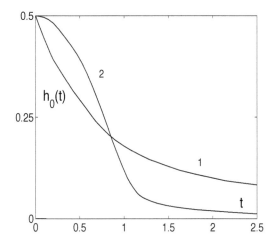

FIGURE 13.2
Graphs of the "classic" (curve marked with 1) and "modified" (curve marked with 2) solutions for $h_0(t)$ for the step load $P(t)$ (after Kudish [14]). Reprinted with permission from the STLE.

problem solution is described by formulas (13.11) and (13.12). With the help of equations (13.14) and (13.15) and the dimensionless variables from (13.1), the modified problem can be reduced to the following initial-value problem

$$\gamma \frac{d^2 h_0(t)}{dt^2} + \frac{1}{4h_0^{3/2}(t)} \frac{dh_0(t)}{dt} = -P(t), \ h_0(0) = h^0, \ h_0'(0) = h^1, \tag{13.16}$$

where $\gamma = \frac{m(u_{10}+u_{20})^2}{8P_0 R'}$ is the dimensionless mass of cylinder 1. In this case the pressure distribution is still described by formula (13.11).

An asymptotic solution of equations (13.16) for $h^1 = 0$ and small t is represented by

$$h_0(t) = h^0 - B \int\limits_0^t P(v)[1 - e^{A(v-t)}]dv + O(t^2),$$
$$\tag{13.17}$$

$$A = \frac{1}{\gamma B}, \ B = 4(h^0)^{3/2}.$$

The comparison of the asymptotic expansions for small t of the "classic" solution from (13.12)

$$h_0(t) = h^0 - B \int\limits_0^t P(v)dv + O(t^2), \ B = 4(h^0)^{3/2}, \tag{13.18}$$

and the modified one (13.17) shows that for small t the modified solution is greater than the "classic" one obtained for the same values of h^0 and h^1.

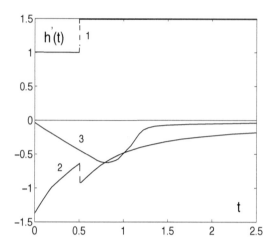

FIGURE 13.3

Graphs of force $P(t)$ (curve marked with 1) and speed $\frac{dh(t)}{dt}$ based on the "classic" (curve marked with 2) and "modified" (curve marked with 3) solutions (after Kudish [14]). Reprinted with permission from the STLE.

The "classic" and modified solutions for $h_0(t)$ are presented in Fig. 13.2. The data for $h_0(t)$ in Fig. 13.2 is obtained with the use of Runge-Kutta method for the case of $\gamma = 1$, $h^0 = 0.5$, $h^1 = 0$, and $P(t) = 1$ for $0 \leq t < 0.5$ and $P(t) = 1.5$ for $t \geq 0.5$. The graphs of the normal velocity $h'_0(t)$ for the "classic" and modified solutions as well as for the applied normal load $P(t)$ are given in Fig. 13.3 for the same initial conditions and load $P(t)$. As it follows from Fig. 13.3, the discontinuity of the normal velocity $h'_0(t)$ at $t = 0.5$ based on the "classic" solution reflects the jump in the load $P(t)$ at the same time moment. For $\gamma = 1$, $h^0 = 0.5$, $h^1 = 0$, and $P(t) = 1$, the graphs of the central film thickness $h_0(t)$ and the normal velocity $h'_0(t)$ based on the "classic" and modified solutions are represented in Figs. 13.4 and 13.5. It follows from Figs. 13.2 and 13.4 that due to the inertia for large time t the film thickness $h_0(t)$ based on the modified solution is smaller than the one based on the "classic" solution.

In a general case of elastic solids ($0 < V < \infty$) and lubricant viscosity dependent on pressure ($\mu(p) \neq 1$), the properties of the "classic" and modified solutions are similar to the ones described above. The modified solution is represented by differentiable functions even if load $P(t)$ and/or the sum of surface speeds $u(t)$ are discontinuous functions of time t.

As a result of this analysis we can conclude that any EHL problem involving non-steady motion such as in cases of non-steady external conditions (applied load, surface speeds, supply of lubricant, etc.), rough surfaces, surfaces with

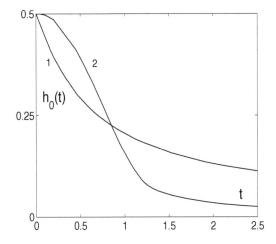

FIGURE 13.4
Graphs of the "classic" (curve marked by 1) and "modified" (curve marked by 2) solutions for $h_0(t)$ for constant load $P(t)$ (after Kudish [14]). Reprinted with permission from the STLE.

dents and/or bumps, surface and/or subsurface cracks should be considered based on a modified problem formulation, which takes into account the system inertia. A similar situation takes places in spatial contacts involved in non-steady motion.

13.2.2 Properly Formulated Non-steady Plane Lubrication Problems for Journal Bearings

Let us consider a non-steady problem for a conformal lubricated contact of two infinite parallel cylindrical solids. We will show that the "classic" formulation of the non-steady EHL problem leads to discontinuous solutions in the cases of abrupt changes in applied load and surface linear velocities. Moreover, we will show the inability of the "classic" solution to accommodate an arbitrary value of the shaft normal initial velocity. We will perform an analytical analysis of the dynamic bearing response to abrupt changes in external load.

Our main goal is to propose a modified formulation of the EHL problem free from the aforementioned defects. The modified problem will be reduced to a system of nonlinear integro-differential Reynolds and integral equations and equations following from Newton's second law.

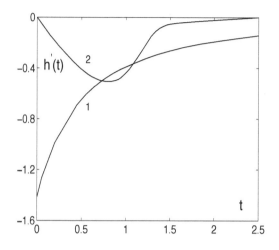

FIGURE 13.5
Graphs of speed $\frac{dh(t)}{dt}$ based on the "classic" (curve marked by 1) and "modified" (curve marked by 2) solutions for constant load $P(t)$ (after Kudish [14]). Reprinted with permission from the STLE.

13.2.2.1 General Assumptions and "Classic" Problem Formulation

Let us consider a shaft with radius R_1 slowly rolling over a busing with radius R_2 (see Fig. 13.6). The shaft and bushing with smooth surfaces are made of the same material with elastic modulus E and Poison's ratio ν. Their axes are parallel and radii are approximately equal, i.e., $\Delta = R_2 - R_1 \ll R_1$. A Cartesian coordinate system (x, y) with the x-axis across the lubrication film and the z-axis perpendicular to it (and to the cylinders axes) is used. The shaft is acted upon by force $P_0 = (X_0, Z_0)$. The angular velocities of the shaft and bushing are ω_1 and ω_2. The joint is lubricated by a Newtonian incompressible fluid with viscosity μ. The shaft and bushing are fully separated by an isothermal lubrication layer. Inertia forces in the lubrication layer can be neglected compared to viscous forces. The lubrication film thickness is small relative to the size of the contact region and radii of the shaft and bushing. EHL conditions are non-steady due to variations in time t of shaft and bushing angular velocities ω_i, applied force P, and/or the inlet boundary α_1. The characteristic time of these variations is small compared to the period of sound waves in elastic material and, therefore, a non-steady problem is required only for the fluid related parameters.

Let us consider the problem in a moving Cartesian coordinate system (x', z') such that the x'-axis passes through the shaft and bushing centers and the z'-axis is tangent to the bushing surface. The angle between the x- and x'-axes

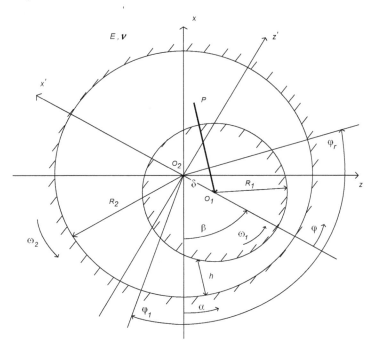

FIGURE 13.6
The general view of the lubricated conjunction (after Kudish [15]). Reprinted
with permission from the ASME.

is measured in the counterclockwise direction and is equal $\beta(t) = \arctan \frac{\delta_z(t)}{\delta_x(t)}$,
where $\delta_x(t)$ and $\delta_z(t)$ are the x- and z-components of the shaft eccentricity
vector. The frictional stresses $\tau_i(\varphi, t)$ and forces Y_{fr}^i, $(i = 1, 2)$ are equal to

$$\tau_1(\varphi, t) = \frac{\mu R(\omega_2 - \omega_1)}{h} - \frac{h}{2R} \frac{\partial p}{\partial \varphi}, \quad \tau_2(\varphi, t) = -\frac{\mu R(\omega_2 - \omega_1)}{h} - \frac{h}{2R} \frac{\partial p}{\partial \varphi},$$

$$Y_{fr}^i(t) = R \int_{\varphi_l}^{\varphi_r} \tau_i(\theta, t) \cos \theta d\theta, \tag{13.19}$$

where φ_l and φ_r are the angles that correspond to the inlet and exit points
of the contact in the moving coordinate system.

Under the stated assumptions and in any stationary Cartesian coordinate
system with the x-axis across the film, the Reynolds equation can be derived
in a manner described in the previous chapters (also see [16]). Taking into
account that $z = R\alpha$ and $R = 0.5(R_1 + R_2) \approx R_1 \approx R_2$ one can get $(\omega(t) =$
$\omega_1(t) + \omega_2(t))$

$$\frac{\partial}{\partial \alpha} \left\{ \frac{h^3}{12\mu} \frac{\partial p}{\partial \alpha} \right\} = R^2 \frac{\omega}{2} \frac{\partial h}{\partial \alpha} + R^2 \frac{\partial h}{\partial t}. \tag{13.20}$$

If $w(t)$ remains positive, then at the contact boundaries $\alpha = \alpha_l(t)$ and $\alpha = \alpha_r(t)$ it is necessary to impose the conditions

$$p(\alpha_l, t) = p(\alpha_r, t) = 0 \tag{13.21}$$

as well as the cavitation boundary condition

$$\frac{\partial p(\alpha_r, t)}{\partial \alpha} = 0 \tag{13.22}$$

and the initial conditions

$$\delta_x(0) = \delta_{x0}, \ \delta_z(0) = \delta_{z0}, \ \alpha_r(0) = \alpha_{r0}. \tag{13.23}$$

If $w(t) = 0$, then contact boundaries $\alpha_r(t) = -\alpha_l(t) = \pi$, and it is necessary to impose the conditions

$$p(\pm\pi, t) = \frac{\partial p(\pm\pi, t)}{\partial \alpha} = 0. \tag{13.24}$$

Now, let us derive the equation for gap $h(\varphi, t)$ in a moving coordinate system. Using the Hertzian assumptions for the radial elastic displacement of the shaft w_{1r}, we get (see [17])

$$w_{1r} + \frac{\partial^2 w_{1r}}{\partial\varphi^2} = \frac{R(1-\kappa)}{4G}p(\varphi, t) + \frac{R(1+\kappa)}{8\pi G}\int_{\varphi_l}^{\varphi_r} \cot\frac{\varphi-\theta}{2}\frac{\partial p(\theta, t)}{\partial\theta}d\theta$$

$$- \frac{1}{2\pi G}(X_0 \cos\varphi + Z_0 \sin\varphi), \tag{13.25}$$

where $G = \frac{E}{2(1+\nu)}$, $\kappa = 3 - 4\nu$ for plane deformation and $\kappa = \frac{3-\nu}{1+\nu}$ for generalized plane stress state. Similarly, for the radial elastic displacement of the bushing w_{2r} we have [17]

$$w_{2r} + \frac{\partial^2 w_{2r}}{\partial\varphi^2} = \frac{R(1-\kappa)}{4G}p(\varphi, t) - \frac{R(1+\kappa)}{8\pi G}\int_{\varphi_l}^{\varphi_r} \cot\frac{\varphi-\theta}{2}\frac{\partial p(\theta, t)}{\partial\theta}d\theta$$

$$+ \frac{R(1+\kappa)}{8\pi G}\int_{\varphi_l}^{\varphi_r} p(\theta, t)d\theta + \frac{\kappa}{2\pi G}(X_0 \cos\varphi + Z_0 \sin\varphi). \tag{13.26}$$

The film thickness equation can be derived based on the fact that the curvature radii

$$\rho_1(\varphi, t) = R_1 + w_{1r}(\varphi, t), \ \rho_2(\varphi, t) = R_2 + w_{2r}(\varphi, t) - h(\varphi, t) \tag{13.27}$$

of the two contact surfaces are equal. Thus, the equation for gap h between the shaft and bushing follows from

$$\Delta = w_{1r} + \frac{\partial^2 w_{1r}}{\partial\varphi^2} - w_{2r} - \frac{\partial^2 w_{2r}}{\partial\varphi^2} + h + \frac{\partial^2 h}{\partial\varphi^2}. \tag{13.28}$$

Assuming that functions w_{1r} and w_{2r} are known the latter equation can be considered as a differential equation for h. To solve it for h two initial conditions must be used

$$h(\pi, t) = \Delta + \sqrt{\delta_x^2 + \delta_z^2}, \quad \frac{\partial h(\pi, t)}{\partial \varphi} = 0. \qquad (13.29)$$

Integrating equations (13.25)-(13.29) we arrive at the equation

$$h(\varphi, t) = \Delta - \sqrt{\delta_x^2 + \delta_z^2} \cos \varphi$$

$$+4R\lambda \int\limits_{\varphi_l}^{\varphi_r} p(\theta, t) \cos(\theta - \varphi) \ln \left| \frac{\cos \frac{\theta}{2}}{\sin \frac{\theta - \varphi}{2}} \right| d\theta$$

$$+R\lambda(1 + \cos \varphi) \int\limits_{\varphi_l}^{\varphi_r} p(\theta, t) d\theta + 2R\lambda \sin \varphi \int\limits_{\varphi_l}^{\varphi_r} p(\theta, t) \sin \theta d\theta \qquad (13.30)$$

$$-4R\lambda \cos \frac{\varphi}{2} \int\limits_{\varphi_l}^{\varphi_r} p(\theta, t) \frac{\cos \frac{\theta - \varphi}{2}}{\cos \frac{\theta}{2}} d\theta, \quad \lambda = \frac{1 + \kappa}{8\pi G}.$$

To complete the problem formulation, the balance conditions

$$\int\limits_{\varphi_l}^{\varphi_r} p(\theta, t) \cos \theta d\theta = \frac{X_0 \cos \beta + Z_0 \sin \beta}{R},$$

$$\qquad (13.31)$$

$$\int\limits_{\varphi_l}^{\varphi_r} p(\theta, t) \sin \theta d\theta = \frac{-X_0 \sin \beta + Z_0 \cos \beta}{R}$$

and the initial conditions for the components $\delta_x(0)$ and $\delta_z(0)$ of the eccentricity vector and the angle $\varphi_r(0)$

$$\delta_x(0) = \delta_{x0}, \quad \delta_z(0) = \delta_{z0}, \quad \varphi_r(0) = \varphi_{r0} \qquad (13.32)$$

must be added to the latter equation. Based on (13.30) and (13.32) and using the given value of $\varphi_l(0)$ the initial condition for h can be calculated.

Equations (13.20)-(13.24) can be rewritten in the moving coordinate system (x', y') taking into account the fact that the line through the shaft and bushing centers is at an angle of $\beta(t)$ (see Fig. 13.6) and

$$\alpha = \varphi + \beta(t), \quad \beta(t) = \arctan \frac{\delta_z(t)}{\delta_x(t)}. \qquad (13.33)$$

Using the following dimensionless variables

$$p' = \frac{p}{p_0}, \quad (h', \delta_x', \delta_z') = \frac{1}{\Delta}(h, \delta_x, \delta_z), \quad \mu' = \frac{\mu}{\mu_a}, \quad w' = \frac{w}{w_0},$$

$$(P_x, P_z) = \frac{1}{X_{00}}(X_0, Z_0), \quad s = \frac{w_2 - w_1}{w_0}, \quad t' = \frac{t}{t_0}, \quad p_0 = \frac{X_{00}}{\pi R},$$

$$\qquad (13.34)$$

$$w_0 = w_{10} + w_{20}, \quad t_0 = \frac{2}{w_0}, \quad V = \frac{6\pi \mu_a w_0 R^3}{\Delta^2 X_{00}}, \quad \lambda_0 = \frac{(1 + \kappa) X_{00}}{8\pi \Delta G},$$

$$\epsilon = \frac{\Delta}{2R}$$

and omitting primes we scale the problem equations

$$\frac{\partial}{\partial \varphi}\left\{\frac{h^3}{12\mu}\frac{\partial p}{\partial \varphi}\right\} = V\left(\omega - \frac{d\beta}{dt}\right)\frac{\partial h}{\partial \varphi} + V\frac{\partial h}{\partial t}, \quad \beta = \arctan\frac{\delta_z}{\delta_x}, \tag{13.35}$$

$$p(\varphi_l, t) = p(\varphi_r, t) = \frac{\partial p(\varphi_r, t)}{\partial \varphi} = 0, \tag{13.36}$$

$$\delta_x(0) = \delta_{x0}, \quad \delta_z(0) = \delta_{z0}, \quad \varphi_r(0) = \varphi_{r0}, \tag{13.37}$$

$$h(\varphi, t) = 1 - \sqrt{\delta_x^2 + \delta_z^2}\,\cos\varphi$$

$$+\frac{4\lambda_0}{\pi}\left\{\int_{\varphi_l}^{\varphi_r} p(\theta, t)\cos(\theta - \varphi)\ln\left|\frac{\cos\frac{\theta}{2}}{\sin\frac{\theta-\varphi}{2}}\right| d\theta\right.$$

$$+\frac{1+\cos\varphi}{4}\int_{\varphi_l}^{\varphi_r} p(\theta, t)d\theta + \frac{\sin\varphi}{2}\int_{\varphi_l}^{\varphi_r} p(\theta, t)\sin\theta d\theta \tag{13.38}$$

$$-\cos\frac{\varphi}{2}\int_{\varphi_l}^{\varphi_r} p(\theta, t)\frac{\cos\frac{\theta-\varphi}{2}}{\cos\frac{\theta}{2}}d\theta\bigg\},$$

$$\int_{\varphi_l}^{\varphi_r} p(\theta, t)\cos\theta d\theta = \pi[P_x(t)\cos\beta(t) + P_z(t)\sin\beta(t)],$$

$$\tag{13.39}$$

$$\int_{\varphi_l}^{\varphi_r} p(\theta, t)\sin\theta d\theta = \pi[-P_x(t)\sin\beta(t) + P_z(t)\cos\beta(t)].$$

For a rolling/sliding contact the problem is completely described by equations (13.35)-(13.39). For a purely squeezed lubrication layer, $\omega(t) = \beta(t) = 0$ and boundary conditions (13.36) must be replaced by

$$p(\pm\pi, t) = \frac{\partial p(\pm\pi, t)}{\partial \varphi} = 0. \tag{13.40}$$

In equations (13.35)-(13.40) functions $\mu(p)$, $\omega(t)$, $P_x(t)$, $P_z(t)$, $\varphi_l(t)$ and constants V, λ_0, δ_{x0}, δ_{z0}, and φ_{r0} are given. The problem solution consists of functions $p(\varphi, t)$, $h(\varphi, t)$, $\delta_x(t)$, $\delta_z(t)$, and $\varphi_r(t)$. The further analysis except for formulas (13.48) is presented in dimensionless variables for $\varphi_l(t) = \varphi_{l0}$.

13.2.2.2   Purely Squeezed Lubrication Layer between Rigid Solids

Suppose the shaft and bushing are made of a rigid material, $\lambda_0 = 0$. For a purely squeezed lubrication layer $P_z(t) = \omega(t) = \beta(t) = \delta_z(t) = 0$. For simplicity let us assume that $\mu(p) = 1$. In this case instead of conditions (13.36), we should use formulas (13.40). Hence, the problem is described by the following equations

$$\frac{d\delta_x(t)}{dt} = \frac{1}{V}P_x(t)[1 - \delta_x^2]^{3/2}, \quad \delta_x(0) = \delta_{x0}. \tag{13.41}$$

The solution of this problem is given by formulas [16]

$$p(\varphi, t) = \frac{V}{2\delta_x(t)} \frac{d\delta_x(t)}{dt} [(1 - \delta_x \cos\varphi)^{-2} - (1 + \delta_x)^{-2}], \tag{13.42}$$

$$\delta_x(t) = \frac{Z(t)}{\sqrt{1 + Z^2(t)}}, \quad \delta_x'(t) = \frac{P_x(t)}{V[1 + Z^2(t)]^{3/2}}, \tag{13.43}$$

$$Z(t) = \frac{\delta_{x0}}{\sqrt{1 - \delta_{x0}^2}} + \frac{1}{V} \int_0^t P_x(u) du.$$

Analysis of $\delta_x'(t)$ from (13.43) shows that it is a discontinuous function of t if $P_x(t)$ is discontinuous. Based on equations (13.42), (13.43), and (13.38) (for $\lambda_0 = 0$), we can conclude that $p(\varphi, t)$ and $\frac{\partial h(\varphi, t)}{\partial t}$ are discontinuous functions in time t.

13.2.2.3 Lubricated Contact of Rolling Rigid Solids

Again, suppose the shaft and bushing are made of a rigid material, $\lambda_0 = 0$. As before we assume that $P_z(t) = \beta(t) = \delta_z(t) = 0$ and $\mu(p) = 1$. In this case instead of conditions (13.36), we should use formulas (13.40). Hence, the problem is described by the following equations (13.35)-(13.39) can be easily reduced to the following system of differential and algebraic equations

$$\delta_x'(t) = -w(t) \frac{r(\delta_x, \varphi_l, \varphi_r)}{q(\delta_x, \varphi_l, \varphi_r)}, \quad \delta_x(0) = \delta_{x0},$$

$$f(\delta_x, \varphi_l, \varphi_r) - g(\delta_x, \varphi_l, \varphi_r) \frac{r(\delta_x, \varphi_l, \varphi_r)}{q(\delta_x, \varphi_l, \varphi_r)} = \frac{\pi P_x(t)}{V w(t)}, \tag{13.44}$$

$$p(\varphi, t) = V w(t) U(\delta_x, \varphi_l, \varphi_r) |_\varphi^{\varphi_l} + V \delta_x'(t) W(\delta_x, \varphi_l, \varphi_r) |_\varphi^{\varphi_l}, \tag{13.45}$$

where f, g, r, q, U, and W are certain continuous functions of δ_x, φ, φ, and φ_l. The structure of system (13.44) and (13.45) allows for a very important conclusion to be drawn: if the external force $P_x(t)$ and/or the sum of angular velocities $w(t)$ experience an abrupt change (i.e., a finite discontinuity) at some moment in time t then $\delta_x(t)$, $\varphi_r(t)$, and $\delta_x'(t)$ change abruptly too. Therefore, formula (13.45) makes pressure $p(\varphi, t)$ a discontinuous function of time t.

13.2.2.4 Lightly Loaded Lubricated Contact of Rolling Elastic Solids

For a lightly loaded lubricated contact of elastic shaft and bushing ($\lambda_0 \ll 1$), we assume that $P_z(t) = w(t) = \beta(t) = \delta_z(t) = 0$ and $\mu(p) = 1$. Assuming that $\lambda_0 \ll 1$ the solution of the problem can be found in the form of a regular perturbation series in powers of λ_0

$$p = p_0(\varphi, t) + \lambda_0 p_1(\varphi, t) + \ldots, \quad h = h_0(\varphi, t) + \lambda_0 h_1(\varphi, t) + \ldots,$$

$$\delta_x = \delta_x^0(t) + \lambda_0 \delta_x^1(t) + \ldots, \quad \varphi_r = \varphi_r^0(t) + \lambda_0 \varphi_r^1(t) + \ldots, \tag{13.46}$$

where $\delta_x^0(t)$, $\varphi_r^0(t)$, and $p_0(\varphi, t)$ satisfy equations (13.44) and (13.45) while $h^0(\varphi, t)$ is determined by equation (13.38) for the case of rigid solids $\lambda_0 = 0$. The next terms of these series such as $\delta_x^1(t)$, $\varphi_r^1(t)$, $p_1(\varphi, t)$ and $h^1(\varphi, t)$ are solutions of certain linear problems described by equations which coefficients depend on $\delta_x^0(t)$, $\varphi_r^0(t)$, $p_0(\varphi, t)$ and $h^0(\varphi, t)$. Taking into account that the latter functions are discontinuous if $P_x(t)$ and/or $\omega(t)$ are discontinuous we can conclude that $p_1(\varphi, t)$ and $h^1(\varphi, t)$ are discontinuous functions of time t as well. Therefore, under such external conditions $h(\varphi, t)$ (see equation (13.38)) is discontinuous in time t. Therefore, in this formulation the problem solution for $\lambda_0 \ll 1$ is discontinuous in time t.

13.2.2.5   Arbitrarily Loaded EHL Contact of Rolling Solids

There is another way to show that the solution of the problem in the "classic" formulation is discontinuous for discontinuous $P_x(t)$. If $P_x(t)$ is discontinuous at some moment t, then it follows from equations (13.39) that the function of pressure $p(\varphi, t)$ is also discontinuous. As a result of that for any $V > 0$ and $\lambda_0 > 0$, the function of gap $h(\varphi, t)$ determined by equation (13.38) is discontinuous in time t as well.

It follows from the Reynolds equation (13.35) that when $\omega(t) - \frac{d\beta(t)}{dt}$ is a discontinuous function of time t functions $p(\varphi, t)$ and $h(\varphi, t)$ are discontinuous in time as well.

13.2.2.6   Other Defects of the "Classic" EHL Problem Formulation

The second defect of the "classic" problem formulation besides just established discontinuity of its solution in some cases can be noticed even for continuously varying input parameters, including $P_x(t)$ and $\omega(t) - \frac{d\beta(t)}{dt}$. The "classic" problem formulation is based on the initial-value problem for the first order differential equation. Therefore, only one initial condition can be imposed on the solution of this equation, namely, the condition of the initial position of the shaft. In other words, the "classic" solution prescribes a certain initial shaft velocity for its particular initial position. It follows from the Reynolds equation (13.35). Moreover, the solution of the problem does not exist if the shaft initial velocity is different from the one prescribed by the Reynolds equation. For example, for the case of a purely squeezed lubrication layer, the prescribed initial shaft velocity is (see equation (13.43))

$$\delta_x'(0) = \tfrac{1}{V} P_x(0)(1 - \delta_{x0}^2)^{3/2}, \tag{13.47}$$

which, apparently, depends on V, the shaft initial position δ_{x0}, and force $P_x(0)$ applied to the shaft at the initial time moment $t = 0$. However, in reality, the shaft initial velocity $\delta_x'(0)$ in most cases is independent of the above parameters and, therefore, is different from the one prescribed by the "classic" solution (13.43). Thus, the solution of the "classic" problem does not exist for an arbitrary initial velocity of the shaft $\delta_x'(0)$. Clearly, this defect of the "classic" problem formulation impairs the very possibility of analysis of such a problem for various realistic initial conditions.

The third defect of the "classic" problem formulation is that the "classic" problem formulation does not allow for oscillatory motion without a presence of an oscillating driving force. For instance, depending on the relationship between the problem parameters a small perturbation in values of external parameters may cause a damped oscillatory behavior of the lubricated system (see the next section and [13]).

Therefore, the "classic" EHL problem solution may be discontinuous if $P_x(t)$ and/or $\omega(t) - \frac{d\beta(t)}{dt}$ change abruptly, it cannot accommodate arbitrary normal initial velocity of the shaft $\delta'_x(0)$, and it does not allow for oscillatory motion. Obviously, such a situation cannot be considered acceptable. Therefore, the "classic" EHL problem is unsatisfactory and it must be modified in such a way that : (a) discontinuity in $p(\varphi, t)$ and $h(\varphi, t)$ would be eliminated, (b) the solution of the problem would exist for any combination of the initial normal shaft position $\delta_x(0)$ and its velocity $\delta'_x(0)$, and (c) the solution would allow for oscillatory motion.

13.2.2.7 Modified Problem Formulation for Non-steady EHL Problem

It is necessary to mention that none of the above–mentioned defects is present in solutions of any steady EHL problem because the acceleration of the system is zero. However, these defects are typical for non-steady EHL problems such as in the cases of non-steady external conditions (applied load, surface velocities, etc.), rough surfaces, surface dents, and/or bumps, etc.

In the considered "classic" problem formulation, it is assumed that the bushing is stationary and the weightless shaft is involved in a non-steady motion. The absence of inertia of a lubricated system causes the described defects. In order to avoid them, we will "soften" the balance conditions (13.39) and replace them by the equations following from Newton's second law

$$m\frac{d^2\delta_x}{dt^2} = X_0(t) - R \int_{\varphi_l}^{\varphi_r} p(\theta, t) \cos(\theta + \beta)d\theta,$$

$$(13.48)$$

$$m\frac{d^2\delta_y}{dt^2} = Z_0(t) - R \int_{\varphi_l}^{\varphi_r} p(\theta, t) \sin(\theta + \beta)d\theta,$$

where m is the mass of the shaft per unit length while the bushing is assumed to be of infinite mass.

The introduction of the second-order time derivatives allows to prevent solution discontinuities, to accommodate an arbitrary normal initial velocity of the shaft, and to consider oscillatory motion.

To illustrate the difference in the problem solutions based on the "classic" and modified formulations, let us consider a purely squeezed lubrication layer

$(P_z(t) = w(t) = \beta(t) = \delta_z(t) = \lambda_0 = 0)$ between two solids made of a rigid material. The "classic" problem is described by equations (13.42) and (13.43). With the help of the first equation from (13.48) and the dimensionless variables from (13.34), the modified problem can be reduced to

$$\gamma \frac{d^2 \delta_x}{dt^2} + \frac{V \delta_x'(t)}{[1-\delta_x^2(t)]^{3/2}} = P_x(t), \quad \delta_x(0) = \delta_{x0}, \quad \delta_x'(0) = \delta_{x1}, \qquad (13.49)$$

where $\gamma = \frac{\pi m \Delta}{2 X_{00}}$ is the dimensionless mass of the shaft and δ_{x1} is the initial shaft normal velocity. In both cases of "classic" and modified problems the pressure distribution is given by formula (13.42).

Now, let us consider an example for $V = 1$, $\gamma = 1$, $\delta_{x0} = 0.2$, and $P_x(t) = 1$ for $t < 0.5$ and $P_x(t) = 1.5$ for $t \geq 0.5$. To obtain the modified problem solution, a Runge-Kutta method is used. The shaft initial velocity $\delta_x'(0)$ dictated by the "classic" solution is 0.9406. Assuming the same value for the shaft initial velocity in the modified solution we obtain that the maximum difference between the "classic" and the modified values of $\delta_x(t)$ is 11%. The shaft velocity based on the modified problem formulation is a smooth function of time t and the one based on the "classic" formulation is discontinuous at $t = 0.5$ when the applied force $P_x(t)$ experiences a discontinuity. At $t = 0.5$ the difference between the shaft velocities in these two case is 38% and the maximum difference is approximately 166%. As it was shown earlier the "classic" solution does not exist for the shaft initial velocity different from the one given by equation (13.47). If the shaft initial velocity is different from the latter, we may see even greater difference between the modified and the "classic" solutions. For example, for $\delta_{x1} = 0$ and the same as earlier other initial and external parameters the graphs of the "classic" and modified shaft displacement $\delta_x(t)$ are given in Fig. 13.7. The graphs of the shaft normal velocity $\delta_x'(t)$ for the "classic" and modified solutions are presented in Fig. 13.8 for the same values of the parameters V, γ, $\gamma P_x(t)$, and the initial conditions. It is obvious from Fig. 13.8 that the discontinuity of the shaft speed $\delta_x'(t)$ at $t = 0.5$ is caused by the jump in the load $P_x(t)$.

Numerical calculations show that for a heavier shaft the modified solution differs from the "classic" one more than for the case of a lighter shaft. Heavy shafts (large γ) accelerate slowly. Therefore, for heavy shafts the modified solution is significantly different from the "classic" one. For example, for the solutions presented in Fig. 13.7 the maximum difference between $\delta_x(t)$ based on the modified and "classic" solutions is 0.308. For heavier shaft with $\gamma = 10$ and all other parameters, the same as before the same value is 0.580. Depending on the initial conditions, the changes in the shaft velocity $\delta_x'(t)$ due to the shaft inertia may be even more significant (see Fig. 13.8).

Therefore, any EHL problem involving non-steady motion such as in the cases of non-steady external conditions (applied load, surface velocities, rough surfaces, surfaces with dents and/or bumps, surface and/or subsurface cracks, etc.) should be analyzed based on the problem formulation similar to the modified one, i.e., taking into account the inertia of the shaft/bushing joint

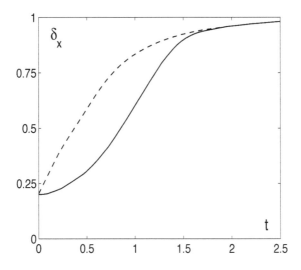

FIGURE 13.7
Graphs of the "classic" (dashed curve) and "modified" (solid curve) solutions for the shaft displacement $\delta_x(t)$ (after Kudish [15]). Reprinted with permission from the ASME.

and Newton's second law. The same situation takes place in spatial EHL problems for conformal and non-conformal lubricated contacts involved in a non-steady motion.

13.3 Non-steady Lubrication Problems for a Journal Bearing with Dents and Bumps

Numerous papers have been dedicated to solution of steady EHL problems for journal bearings (Hamrock [16], Allair and Flack [18], Ghosh et al. [19], Lin and Rylander [20]). Obviously, solutions of non-steady EHL problems are as important as solutions of steady ones. In many practical cases bearing durability and premature failure are to a great extent determined by regimes of starting, stopping, acceleration, deceleration, and riding on a bumpy road. Therefore, it is important to be able to analyze these truly non-steady cases. There have been published very few papers on non-steady motion of journal bearings (see references in the preceding section). Even fewer papers treat the non-steadiness in a proper manner, taking into account the shaft and

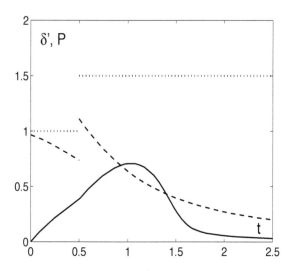

FIGURE 13.8
Graphs of force $P_x(t)$ (dotted curve) and shaft normal speed $d\delta_x(t)dt$ based on the "classic" (dashed curve) and "modified" (solid curve) solutions (after Kudish [15]). Reprinted with permission from the ASME.

bushing masses involved in a non-steady motion. For non-conformal lubricated contacts such problems are considered by Wijnant et al. [21].

The approach to the problem is based on the "modified" problem formulation proposed in the preceding section and used in [22]. The problem solution is free of certain defects such as discontinuity and independence from the initial shaft speed. The problem is reduced to a system of the nonlinear Reynolds equation and integral equations describing lubrication and contact interaction between elastic solids. The additional conditions include initial and boundary conditions and Newton's second law applied to the shaft motion. The main emphasis of the paper is threefold: the analysis of the transient dynamics of the system under constant external conditions, the analysis of the transient dynamics of the system due to abrupt changes in applied load, and the system behavior in the case of a bump/dent presence on the shaft surface. The numerical analysis shows that usually the solution exhibits oscillatory behavior while approaching a steady state. The specific features of the solution behavior depend on the relationship of such parameters as shaft mass, speed, applied load, lubricant viscosity, etc. Under constant external conditions it is observed that in the process of transient motion the radial displacement of the shaft center may vary by no more than 2.5% while the maximum pressure varies by at least 15% and by as much as 350%. Moreover, the variations of pressure are greater for stiffer materials and heavier shafts.

13.3.1 General Assumptions and Problem Formulation

Let us consider a lubricated contact of a smooth shaft with radius R_1 and mass m per unit length slowly moving with angular velocity ω over a motionless bushing with radius R_2. The axes of the shaft and bushing are parallel and the radii are approximately equal, i.e., $\Delta = R_2 - R_1 \ll R_1$. The shaft is acted upon by force $\vec{F}$ and by moment M_0 about the shaft axis.

To simplify the further analysis, let us assume that both the shaft and bushing are made from the same elastic material, for example, steel ($G = 0.5E/(1 + \nu)$, $\kappa = 3 - 4\nu$ for plane deformation and $\kappa = (3 - \nu)/(1 + \nu)$ for generalized plane stress state). It is necessary to point out that in practice the materials of the shaft and bushing are generally different. In case of different materials the problem equations and the analysis are slightly more complicated, however, it does not change significantly the predictions derived from the case of same materials. The joint (see Fig. 13.6) is lubricated by a Newtonian incompressible fluid with viscosity μ. The lubrication process is isothermal. The shaft and bushing are fully separated by the lubrication layer. Because of slow relative motion of the contact surfaces the inertial forces in the lubrication layer are small compared to the viscous forces and can be neglected. The lubrication film thickness is considered to be small relative to the size of the contact region and radii of the shaft and bushing. Under non-steady lubrication conditions, the shaft angular velocity ω, applied force $\vec{F}$, and moment M_0 may vary in time t. The characteristic time of these variations is small compared to the period of sound waves in elastic solids. The latter requires a non-steady problem formulation only for fluid flow parameters.

It is convenient to consider this problem in a moving coordinate system (x', y') shown in Fig. 13.6 with the origin at the bushing center, the x'-axis along the line connecting the centers of the shaft and bushing (across the film thickness), and the y'-axis tangent to the bushing surface. Suppose the center line is moving and is currently at an angle of $\beta = \beta(t)$ with the fixed vertical line shown in Fig. 13.6. The angle φ is measured from the x-axis in the counterclockwise direction. Let us suppose that the total force applied to the shaft $\vec{F}$ is characterized by two components: X_0 - along the x-axis and Y_0 - along the y-axis, i.e., $tan\beta(t) = Y_0/X_0$. Then, assuming that $\omega(t)$ and $\beta(t)$ are given moment $M_0(t)$ applied to the shaft can be determined after the solution is obtained by calculating the moment of friction force applied to the shaft (integral of $p(\varphi, t)$ from (13.56)) and the first derivative of $\omega(t)$ (see Kudish et al. [15]). Let us introduce the dimensionless variables as follows

$$(p', p'_*, \tau'_1, \tau'_2) = \tfrac{1}{p_0}(p, p_*, \tau_1, \tau_2), \quad (h', \delta', \delta'_*) = \tfrac{1}{\Delta}(h, \delta, \delta_*),$$

$$\mu' = \tfrac{\mu}{\mu_a}, \quad \omega' = \tfrac{\omega}{\omega_0}, \quad X = \tfrac{\sqrt{X_0^2 + Y_0^2}}{X_{00}}, \quad Y'_f r^i = \tfrac{Y_f r^i}{X_{00}}, \quad t' = \tfrac{t}{t_0}$$

$$(13.50)$$

$$\delta'_{1*} = \tfrac{t_0}{\Delta}\delta_{1*}, \ \ p_0 = \tfrac{X_{00}}{\pi R}, \ \ \omega_0 = \omega(0), \ \ t_0 = \tfrac{2}{\omega_0},$$

$$V = \tfrac{6\pi\mu_a\omega_0 R^3}{\Delta^2 X_{00}}, \ \ \lambda_0 = \tfrac{(1+\kappa)X_{00}}{8\pi\Delta G}, \ \ \varepsilon = \tfrac{\Delta}{2R}, \ \ \gamma = \varepsilon\tfrac{m\omega_0^2 R}{2X_{00}},$$

$(R = 0.5(R_1 + R_2) \approx R_1 \approx R_2)$, and write down the problem governing equations Kudish et al. [15] in the moving coordinate system in the dimensionless form (primes at dimensionless variables are omitted)

$$\tfrac{\partial}{\partial\varphi}\{\tfrac{h^3}{\mu}\tfrac{\partial p}{\partial\varphi}\} = V(\omega - \tfrac{d\beta}{dt})\tfrac{\partial h}{\partial\varphi} + V\tfrac{\partial h}{\partial t}, \tag{13.51}$$

$$p(\varphi,0) = p_*(\varphi), \ \ \varphi_l(0) = \varphi_{l*}, \ \ \varphi_r(0) = \varphi_{r*}, \tag{13.52}$$

$$p(\varphi_l,t) = p(\varphi_r,t) = 0, \ \ \tfrac{\partial p(\varphi_r,t)}{\partial\varphi} = 0, \tag{13.53}$$

$$h(\varphi,t) = 1 - \delta\cos\varphi + \tfrac{2}{\pi}\lambda_0\int_{\varphi_l}^{\varphi_r} p(\theta,t)L_1(\theta,\varphi)d\theta,$$

$$L_1(\theta,\varphi) = 2cos(\theta - \varphi)\ln\left|\tfrac{\cos\frac{\theta}{2}}{\sin\frac{\theta-\varphi}{2}}\right| - 2\cos\tfrac{\varphi}{2}\tfrac{\cos\frac{\theta-\varphi}{2}}{\cos\frac{\theta}{2}} \tag{13.54}$$

$$+ \tfrac{1+\cos\varphi}{2} + \sin\varphi\sin\theta,$$

$$\gamma\tfrac{d^2\delta}{dt^2} = X - P, \ \ P = \tfrac{1}{\pi}\int_{\varphi_l}^{\varphi_r} p(\theta,t)\cos\theta d\theta, \ \ \delta(0) = \delta_*, \ \ \tfrac{d\delta(0)}{dt} = \delta_{1*}, \tag{13.55}$$

where $\varphi_l(t)$ and $\varphi_r(t)$ are the inlet and exit angular coordinates of the contact region in the moving coordinate system, $\delta(t) = \delta_x(t)$ and $\delta_y(t) = 0$. In equations (13.51)-(13.55) functions $\mu(p)$, $\omega(t)$, $\beta(t)$, $X(t),\varphi_l(t)$, $p_*(\varphi)$ and constants V, λ_0, δ_*, δ_{1*}, φ_{l*}, φ_{r*} are given. The solution for this problem consists of functions $p(\varphi,t)$, $h(\varphi,t)$, $\delta(t)$, and $\varphi_r(t)$. In the further analysis, the inlet angle coordinate $\varphi_l(t)$ is considered to be equal to a constant φ_{l*}.

After the problem is solved, the frictional stresses $\tau_i(\varphi,t)$ and friction forces Y_{fr}^i $(i = 1,2)$ are determined from equations

$$\tau_{1,2}(\varphi,t) = -\tfrac{\varepsilon V}{3}\tfrac{\mu\omega}{h} \mp \varepsilon h\tfrac{\partial p}{\partial\varphi},$$

$$\{Y_{fr}^1, Y_{fr}^2\} = \tfrac{1}{\pi}\int_{\varphi_l}^{\varphi_r}\{\tau_1(\theta,t), \tau_1(\theta,t)\}\cos\theta d\theta. \tag{13.56}$$

13.3.2   Case of Rigid Materials

First, let us consider rigid materials, i.e., $\lambda_0 = 0$. In the case of constant viscosity $\mu = 1$, angular velocity $\omega(t) = 1$, and angle $\beta(t) = 0$, the solution of the problem (13.51)-(13.55) for $p(\varphi,t)$ can be obtained in an analytical form

$$p(\varphi,t) = \tfrac{V\delta}{2(1-\delta^2)}[\cos\varphi_r H_1(\delta,\varphi_l,\varphi) - H_2(\delta,\varphi_l,\varphi)]$$

$$+ \tfrac{V\delta'}{2(1-\delta^2)}[\sin\varphi_r H_1(\delta,\varphi_l,\varphi) + \tfrac{1-\delta^2}{\delta}(\tfrac{1}{h^2} - \tfrac{1}{h_l^2})],$$

$$H_1(\delta, \varphi_l, \varphi) = \delta \left[\frac{\sin \varphi}{h^2} - \frac{\sin \varphi_l}{h_l^2} + \frac{3}{1-\delta^2} \left(\frac{\sin \varphi}{h} - \frac{\sin \varphi_l}{h_l} \right) \right]$$

$$+ \frac{2+\delta^2}{1-\delta^2} H_3(\delta, \varphi_l, \varphi),$$

$$H_2(\delta, \varphi_l, \varphi) = \frac{\sin \varphi}{h^2} - \frac{\sin \varphi_l}{h_l^2} + \frac{1+2\delta^2}{1-\delta^2} \left(\frac{\sin \varphi}{h} - \frac{\sin \varphi_l}{h_l} \right)$$

$$+ \frac{3}{1-\delta^2} H_3(\delta, \varphi_l, \varphi), \quad H_3(\delta, \varphi_l, \varphi) = \frac{2}{1-\delta^2} \arctan \left(\sqrt{\frac{1+\delta}{1-\delta}} \tan \frac{\varphi}{2} \right) \Big|_{\varphi_l}^{\varphi},$$

where $h_l = h(\varphi_l, t)$, $h_r = h(\varphi_r, t)$, and $\delta(t)$ and $\varphi_r(t)$ satisfy the following system of equations (see equations (13.53) and (13.55))

$$\delta' \left[\sin \varphi_r H_1(\delta, \varphi_l, \varphi_r) + \frac{1-\delta^2}{\delta} \left(\frac{1}{h_r^2} - \frac{1}{h_l^2} \right) \right]$$

$$+ \delta \left[\cos \varphi_r H_1(\delta, \varphi_l, \varphi_r) - H_2(\delta, \varphi_l, \varphi_r) \right] = 0,$$

$$\gamma \delta'' = X - P, \quad \delta(0) = \delta_*, \quad \delta'(0) = \delta_{1*},$$

$$P = \frac{V\delta}{2\pi(1-\delta^2)} \left[\cos \varphi_r H_{1i}(\delta, \varphi_l, \varphi_r) - H_{2i}(\delta, \varphi_l, \varphi_r) \right]$$

$$+ \frac{V\delta'}{2\pi(1-\delta^2)} \left\{ \sin \varphi_r H_{1i}(\delta, \varphi_l, \varphi_r) + \frac{1-\delta^2}{\delta} \left[H_i(\delta, \varphi_l, \varphi_r) \right. \right.$$

$$\left. \left. - \frac{\sin \varphi_r - \sin \varphi_l}{h_l^2} \right] \right\},$$

$$H_{1i}(\delta, \varphi_l, \varphi_r) = \delta \left[I_2(\delta, \varphi_l, \varphi_r) + \frac{3}{1-\delta^2} I_1(\delta, \varphi_l, \varphi_r) \right.$$

$$- \frac{\sin \varphi_l}{h_l^2} \left(1 + \frac{3h_1}{1-\delta^2} \right) (\sin \varphi_r - \sin \varphi_l) \right] + \frac{2+\delta^2}{1-\delta^2} I_3(\delta, \varphi_l, \varphi_r),$$

$$H_{2i}(\delta, \varphi_l, \varphi_r) = I_2(\delta, \varphi_l, \varphi_r) + \frac{1+2\delta^2}{1-\delta^2} I_1(\delta, \varphi_l, \varphi_r)$$

$$- \frac{\sin \varphi_l}{h_l^2} \left(1 + \frac{1+2\delta^2}{1-\delta^2} h_1 \right) (\sin \varphi_r - \sin \varphi_l) + \frac{3\delta}{1-\delta^2} I_3(\delta, \varphi_l, \varphi_r),$$

$$H_i(\delta, \varphi_l, \varphi_r) = \frac{1}{1-\delta^2} \left[\frac{\sin \varphi_r}{h_r} - \frac{\sin \varphi_l}{h_l} \right]$$

$$+ \frac{2\delta}{(1-\delta^2)^{3/2}} \arctan \left[\sqrt{\frac{1+\delta}{1-\delta}} \tan \frac{\theta}{2} \right] \Big|_{\varphi_l}^{\varphi_r},$$

$$I_1(\delta, \varphi_l, \varphi_r) = \frac{1}{\delta^2} \left[-h_r + h_l + \ln \frac{h_r}{h_l} \right],$$

$$I_2(\delta, \varphi_l, \varphi_r) = \frac{1}{\delta^2} \left[-\frac{1}{h_r} + \frac{1}{h_l} - \ln \frac{h_r}{h_l} \right],$$

$$I_3(\delta, \varphi_l, \varphi_r) = \sin \varphi_r H_3(\delta, \varphi_l, \varphi_r) - \frac{1}{\delta} \ln \frac{h_r}{h_l}.$$

(13.57)

(13.59)

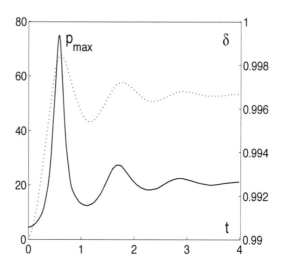

FIGURE 13.9
Distribution of the maximum pressure p_{max} (solid curve) and shaft displacement δ (dotted curve) versus time t for rigid materials, $X = 1$, $\delta_* = 0.99$, $\delta_{1*} = 0.005$, and $\gamma = 10$ (after Kudish [22]). Reprinted with permission from the ASME.

The numerical solution of equations (13.58) and (13.59) can be performed by reducing these equations to a system of first–order differential equations that are discretized using the trapezoidal rule. Let us discuss the numerical results obtained for the case of a constant load $X = 1$, $V = 0.025$, $\varphi_l = -1.309$, $\delta_* = 0.99$, and $\delta_{1*} = 0.005$. Figures 13.9-13.11 illustrate the transient behavior of the maximum pressure p_{max} with time t for the system inertia $\gamma = 10$, 1, 0.1. It is clear from Fig. 13.9 that fluctuations of pressure are greater for greater system inertia γ. Moreover, small variations of the shaft displacement δ of about 0.7% of its steady value ($\delta = 0.99665$) cause variations in the maximum pressure p_{max} of about 350% for $\gamma = 10$ (see Fig. 13.9) and 150% for $\gamma = 1$ (see Fig. 13.10) of its steady value ($p_{max} = 21.26056$), respectively. For $\gamma = 10$, 1, 0.1, the pressure distributions $p(\varphi, t)$ versus φ at the time moments t at which the maximum pressure p_{max} reaches its maximum value are given in Fig. 13.12. In Fig. 13.12, for $\gamma = 0.1$ the pressure distribution $p(\varphi, t)$ versus φ is the steady pressure distribution.

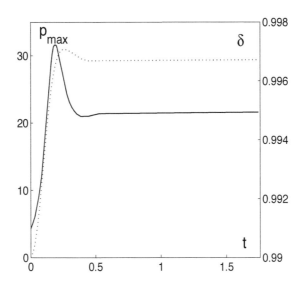

FIGURE 13.10
Distribution of the maximum pressure p_{max} (solid curve) and shaft displacement δ (dotted curve) versus time t for rigid materials, $X = 1$, $\delta_* = 0.99$, $\delta_{1*} = 0.005$, and $\gamma = 1$ (after Kudish [22]). Reprinted with permission from the ASME.

13.3.3 Contact Region Transformation

Equations (13.51)-(13.55) are written for a contact region with one free boundary - the exit coordinate $\varphi_r(t)$. The existence of the free boundary makes the numerical solution of the problem for elastic solids (i.e., for $\lambda_0 > 0$) complicated. To avoid that we introduce a simple transform

$$T = t, \quad \varphi = \frac{\varphi_r + \varphi_l}{2} + \frac{\varphi_r - \varphi_l}{\pi} \psi. \tag{13.60}$$

For $\varphi_l(t) = \varphi_{l*}$ by using transform (13.60) equations (13.51)-(13.55) can be rewritten in the form

$$\frac{\partial}{\partial \psi} \left\{ \frac{h^3}{\mu} \frac{\partial p}{\partial \psi} \right\} = V \frac{\varphi_r - \varphi_l}{\pi} [\omega - \beta' - \frac{\varphi_r' + \varphi_l'}{2} - \frac{\varphi_r' - \varphi_l'}{\pi} \psi] \frac{\partial h}{\partial \psi}$$
$$+ V(\frac{\varphi_r - \varphi_l}{2})^2 \frac{\partial h}{\partial T}, \tag{13.61}$$

$$p(\psi, 0) = p_*(\psi), \quad \varphi_r(0) = \varphi_{r*}, \tag{13.62}$$

$$p(\pm \frac{\pi}{2}, t) = 0, \quad \frac{\partial p(\frac{\pi}{2}, t)}{\partial \psi} = 0, \tag{13.63}$$

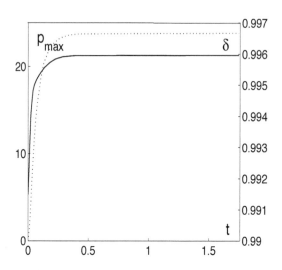

FIGURE 13.11

Distribution of the maximum pressure p_{max} (solid curve) and shaft displacement δ (dotted curve) versus time t for rigid materials, $X = 1$, $\delta_* = 0.99$, $\delta_{1*} = 0.005$, and $\gamma = 0.1$ (after Kudish [22]). Reprinted with permission from the ASME.

$$h = 1 - \delta(T) \cos \Lambda(\psi)$$

$$+ 2\lambda_0 \frac{\varphi_r - \varphi_l}{\pi^2} \int_{-\pi/2}^{\pi/2} p(\Theta, T) L_1(\Theta, \psi, \varphi_l, \varphi_r) d\Theta,$$

$$L_1(\Theta, \psi, \varphi_l, \varphi_r) = 2\cos\left(\frac{\varphi_r - \varphi_l}{\pi}(\Theta - \psi)\right) \ln \left| \frac{\cos \frac{\Lambda(\Theta)}{2}}{\sin[\frac{\varphi_r - \varphi_l}{2\pi}(\Theta - \psi)]} \right| \qquad (13.64)$$

$$- 2\cos\left[\frac{\varphi_r - \varphi_l}{2\pi}(\Theta - \psi)\right] \frac{\cos \frac{\Lambda(\psi)}{2}}{\cos \frac{\Lambda(\Theta)}{2}} + \frac{1}{2}\cos \Lambda(\psi) + \frac{1}{2}$$

$$+ \sin \Lambda(\psi) \sin \Lambda(\Theta), \quad \Lambda(\psi) = \frac{\varphi_r + \varphi_l}{2} + \frac{\varphi_r - \varphi_l}{\pi}\psi,$$

$$\gamma \frac{d^2\delta}{dT^2} = X - P, \ \delta(0) = \delta_*, \ \frac{d\delta(0)}{dT} = \delta_{1*},$$

$$P = \frac{\varphi_r - \varphi_l}{\pi^2} \int\limits_{-\pi/2}^{\pi/2} p(\Theta, T) \cos \Lambda(\Theta) d\Theta, \qquad (13.65)$$

where $\beta'(T) = d\beta(T)/dT$, etc.

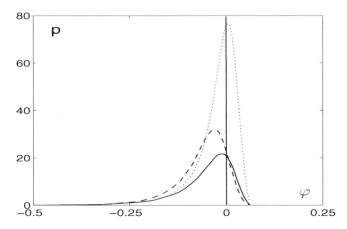

FIGURE 13.12
Pressure distribution $p(\varphi, t)$ for $\gamma = 10$ (dotted curve), (b) $\gamma = 1$ (dashed curve), and (c) $\gamma = 0.1$ (solid curve) for rigid materials, $X = 1$, $\delta_* = 0.99$, $\delta_{1*} = 0.005$ at the time moments t at which the maximum pressure reaches its maximum value (after Kudish [22]). Reprinted with permission from the ASME.

13.3.4 Quadrature Formula and Discretization

Let us introduce two sets of nodes: $\{\psi_k\}$, $k = 1, \ldots, N$, $\psi_1 = -\pi/2$, $\psi_N = \pi/2$, and $\{\psi_{j+1/2}\}$, $\psi_{j+1/2} = 0.5(\psi_j + \psi_{j+1})$, $j = 1, \ldots, N-1$. Further, the low index indicates the node number. We make use of the second–order accuracy trapezoidal quadrature formula for computation of the integrals involved in equations (13.64) and (13.65). Based on that we obtain

$$\int_{-\pi/2}^{\pi/2} L(\Theta, \psi_{j+1/2}) p(\Theta, T) d\Theta = \frac{1}{2} \sum_{k=1}^{N-1} (\psi_{k+1} - \psi_k)$$

$$\times [L(\psi_k, \psi_{j+1/2}) p_k(T) + L(\psi_{k+1}, \psi_{j+1/2}) p_{k+1}(T)], \tag{13.66}$$

where $p_k(T) = p(\psi_k, T)$ and $L(\Theta, \psi)$ is a given kernel.

Let us introduce a set of time moments $\{T_i\}$, $T_0 = 0$, $T_{i+1} = T_i + \Delta T_i$, $i = 0, 1, \ldots$. To provide for better numerical stability, an implicit numerical scheme of the first order of accuracy with respect to time and the second order of accuracy with respect to angle is used. Therefore, when integrating equation (13.61) with respect to time T over the interval $[T_{i-1}, T_i]$, we use the following approximations:

$$\int_{T_{i-1}}^{T_i} f(\psi, T) dT \approx \Delta T_{i-1} f(\psi, T_i),$$

$$\int_{T_{i-1}}^{T_i} \beta'(T)f(\psi,T)dT \approx [\beta(T_i) - \beta(T_{i-1})]f(\psi,T_i).$$

In a similar fashion, an implicit numerical scheme of the second order of accuracy with respect to both time and angle can be obtained by using the trapezoidal rule of integration not only with respect to angle but also with respect to time. However, that scheme will practically double the amount of necessary calculations without much improvement in solution stability.

The second–order differential equation (13.65) for δ is reduced to a system of two first–order differential equations and then discretized using the trapezoidal rule

$$\delta_i = \delta_{i-1} + \eta_{i-1}\triangle T_{i-1}$$

$$+[X(T_i) + X(T_{i-1}) - P(T_i) - P(T_{i-1})]\frac{\triangle T_{i-1}^2}{4\gamma},$$

$$\eta_i = \eta_{i-1} + [X(T_i) + X(T_{i-1}) - P(T_i) - P(T_{i-1})]\frac{\triangle T_{i-1}}{2\gamma}, \tag{13.67}$$

$$i = 1, 2, \ldots,$$

where $\delta_0 = \delta_*$ and $\eta_0 = \delta_{1*}$ are given in (13.65) and $\triangle T_i = T_i - T_{i-1}$.

Now, let us integrate equation (13.61) with respect to T over the interval $[T_{i-1}, T_i]$ $(i = 1, 2, \ldots)$ and, then, integrate the obtained result with respect to ψ over interval $[\psi_{j-1/2}, \psi_{j+1/2}]$ $(j = 2, \ldots, N-1)$. Discretizing in time and space the obtained equation and equations (13.62)-(13.64) give the following formulas:

$$\frac{\triangle T_i}{V}\left\{\frac{h_{j+1/2}^3}{\mu_{j+1/2}}\frac{p_{j+1}-p_j}{\psi_{j+1}-\psi_j} - \frac{h_{j-1/2}^3}{\mu_{j-1/2}}\frac{p_j-p_{j-1}-}{\psi_j-\psi_{j-1}}\right\}$$

$$= \frac{\varphi_r-\varphi_l}{\pi}\left\{[\Omega - \frac{\varphi_r+\varphi_l-\Phi}{2}](h_{j+1/2} - h_{j-1/2})\right.$$

$$\left. - \frac{\varphi_r-\varphi_l-\Psi}{\pi}\left[\psi_{j+1/2}h_{j+1/2}\right.\right. \tag{13.68}$$

$$\left.\left. -\psi_{j-1/2}h_{j-1/2} - \int_{\psi_{j-1/2}}^{\psi_{j+1/2}} hd\psi\right]\right\} + (\frac{\varphi_r-\varphi_l}{2})^2[\int_{\psi_{j-1/2}}^{\psi_{j+1/2}} hd\psi - F_j],$$

$$\Omega = (\omega - \beta')\triangle T_{i-1}, \quad \Phi = \varphi_r + \varphi_l, \quad \Psi = \varphi_r - \varphi_l,$$

$$F_j = \int_{\psi_{j-1/2}}^{\psi_{j+1/2}} hd\psi, \quad \int_{\psi_{j-1/2}}^{\psi_{j+1/2}} hd\psi = [\psi - \frac{\pi\delta}{\varphi_r-\varphi_l}\sin\Lambda(\psi)]\,|_{\psi_{j-1/2}}^{\psi_{j+1/2}} \tag{13.69}$$

$$+\lambda_0\frac{2}{\pi}\int_{-\pi/2}^{\pi/2_{j+1/2}} p(\Theta,T)L_2(\Theta, \psi_{j-1/2}, \psi_{j+1/2}, \varphi_l, \varphi_r)d\Theta,$$

$$L_2(\Theta, \psi_{j-1/2}, \psi_{j+1/2}, \varphi_l, \varphi_r) = \left\{ \sin\psi \ln \frac{\cos^2 \frac{\Lambda(\Theta)}{2}}{|\sin \frac{\psi}{2}|} + \frac{\psi + \sin\psi}{2} \right.$$

$$\left. - \frac{\sin[\psi + \frac{\Lambda(\Theta)}{2}]}{\cos \frac{\Lambda(\Theta)}{2}} \right\} \Big|_{a_{j-1/2}}^{a_{j+1/2}}$$

$$+ \left[\tfrac{1}{2}\sin\Lambda(\psi) - \frac{\varphi_r - \varphi_l}{2\pi}\psi - \cos\Lambda(\psi)\sin\Lambda(\Theta) \right] \Big|_{\psi_{j-1/2}}^{\psi_{j+1/2}},$$

$$a_{j\pm 1/2} = \frac{\varphi_r - \varphi_l}{\pi}(\psi_{j\pm 1/2} - \Theta), \quad j = 2, \ldots, N-1,$$

$$p_j(0) = p_{*j}, \quad j = 2, \ldots, N-1; \quad \delta(0) = \delta_*, \tag{13.70}$$

$$p_1 = p_N = 0, \quad p_{N-2} - 4p_{N-1} = 0, \tag{13.71}$$

$$P(T_i) = \frac{\varphi_r - \varphi_l}{\pi^2} \int_{-\pi/2}^{\pi/2} p(\Theta, T_i) L_3(\Theta, \varphi_l, \varphi_r) d\Theta, \tag{13.72}$$

$$L_3(\Theta, \varphi_l, \varphi_r) = \cos\Lambda(\Theta),$$

$$h_{j+1/2} = 1 - \delta\cos\Lambda(\psi_{j+1/2})$$
$$\tag{13.73}$$
$$+ 2\lambda_0 \frac{\varphi_r - \varphi_l}{\pi^2} \int_{-\pi/2}^{\pi/2} p(\Theta, T) L_1(\Theta, \psi_{j+1/2}, \varphi_l, \varphi_r) d\Theta,$$

where index i, indicating the time step, is omitted and $\varphi_r = \varphi_r(T_i)$, $\Omega = \Omega(T_i)$, $\Phi = \Phi(T_{i-1})$, $\Psi = \Psi(T_{i-1})$, and $X_i = X(T_i)$. In equations (13.69) the value of F_j is calculated based on the solution at the previous time moment T_{i-1}, and the integrals in equations (13.68), (13.69), (13.72), and (13.73) are calculated according to formula (13.66). At any time moment T_i, equations (13.68)-(13.73) represent a system of $N+2$ nonlinear algebraic equations for $N+2$ unknowns: $\{p_j\}$ $(j = 1, \ldots, N)$, δ, and φ_r.

13.3.5   Iterative Numerical Scheme

To solve the system of nonlinear equations (13.68)-(13.73) Newton's method is used. It allows to determine the new iterates of $\{p_j\}$ $(j = 1, \ldots, N)$, δ, and φ_r simultaneously. The system of equations is linearized at each consequent iteration and then solved for the new unknowns:

$$q_j^{l+1} = p_j^{l+1} - p_j^l \ (j = 1, \ldots, N), \quad \Delta\delta = \delta^{l+1} - \delta^l,$$
$$\tag{13.74}$$
$$\Delta\varphi_r = \varphi_r^{l+1} - \varphi_r^l,$$

where l indicates the current solution iteration. After the quantities of $\{q_j^{l+1}\}$ $(j = 1, \ldots, N)$, $\Delta\delta$, and $\Delta\varphi_r$ for the time moment $T = T_i$ are obtained using formulas (13.74), the new iterates p_j^{l+1}, δ^{l+1}, and φ_r^{l+1} can be calculated. The iterations must be continued until a sufficient accuracy is reached. After that the solution for the next time moment $T = T_{i+1}$ can be calculated in the same manner.

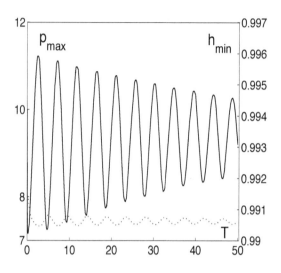

FIGURE 13.13

Distribution of the maximum pressure p_{max} (solid curve) and the minimum film thickness h_{min} (dotted curve) versus time T for $\gamma = 10$, $\delta_{1*} = 0$, and $X = 1$ (after Kudish [22]). Reprinted with permission from the ASME.

An adaptive numerical procedure is employed for choosing the size of the time step $\triangle T_i$. It is done in order to overcome a poor convergence and stability of the numerical solution. In particular, the method allows to decrease or increase the time step $\triangle T_i$, depending on the behavior of the solution iterates at the previous and current time steps.

13.3.6 Analysis of Numerical Results

In this subsection we consider a number of different non-steady problems for a journal bearing such as a transient oscillatory motion under constant external conditions, motion with abruptly changing load, and a shaft motion affected by a bump/dent on its surface. For $\lambda_0 > 0$ all calculations are done with double precision. The number of nodes used is $N = 50$ and the absolute precision used in calculations is 10^{-4}. The CPU time required for one run of the program on a PC is high due to relatively poor convergence and stability of an EHL problem with such a conformal geometry. The CPU time depends on the problem parameters. The longest computation time is required for the oscillatory process for a shaft with large mass ($\gamma = 10$) and certain initial values of δ and δ' to reach the steady state.

Below, all examples of numerical solutions are given for constant viscosity

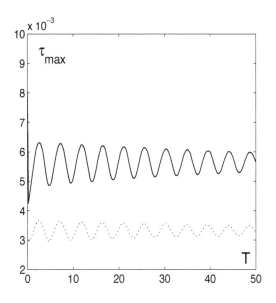

FIGURE 13.14
Distribution of the maximum frictional stresses τ_1 and τ_2 applied to both contact surfaces versus time T for $\gamma = 10$, $\delta_{1*} = 0$, and $X = 1$ (solid curve - lower surface, dotted curve - upper surface) (after Kudish [22]). Reprinted with permission from the ASME.

$\mu = 1$, angular velocity $\omega(T) = 1$ and the following problem parameters: $V = 0.025$, $\lambda_0 = 0.005$, $\varepsilon = 0.003$, $\varphi_l(T) = -1.309$, $\beta(T) = 0$.

13.3.6.1 Transient Oscillatory Motion under Constant External Conditions

Let us consider a transient motion of the shaft subjected to constant external conditions $X(T) = 1$ with the given initial values for parameters $\delta_* = \delta_H + 0.01V^{3/5}/(1 + \lambda_0 V)$, $\varphi_{r*} = 1.16\varphi_H$, and p_{*j} is taken equal to a slightly smoothed out Hertzian pressure distribution $p_H(\varphi(\psi_j))$, $j = 1, \ldots, N$ (where $\varphi(\psi)$ is given by equation (13.57)). The Hertzian values for a dry frictionless elastic contact of a shaft and bushing made from the same material, namely, the shaft radial displacement δ_H, the contact half-width φ_H, and the dimensionless pressure distribution $p_H(\varphi)$ are as follows (see Teplyi [17], pp.

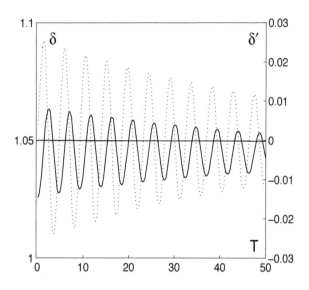

FIGURE 13.15

Distribution of shaft radial displacement δ (solid curve) and its radial speed δ' (dotted curve) versus time T for $\gamma = 10$, $\delta_{1*} = 0$, and $X = 1$ (after Kudish [22]). Reprinted with permission from the ASME.

20-23, 47)

$$p_H(\varphi) = \tfrac{1}{\sqrt{a^2+1}}\{X[\tfrac{\sqrt{a^2+1}-1}{\sqrt{a^2+1}}\chi + \tfrac{2\alpha(\varphi)}{1+\xi^2}] - \tfrac{1}{2}(W + \tfrac{1}{2\lambda_0})\chi\},$$

$$\chi = \ln\frac{\sqrt{a^2+1}-\alpha(\varphi)}{\sqrt{a^2+1}+\alpha(\varphi)}, \tag{13.75}$$

$$W = \frac{2X\{a^2+(\sqrt{a^2+1}-1)[2-\ln(a^2+1)]\}}{\sqrt{a^2+1}[2-\ln(a^2+1)]} + \frac{2(\sqrt{a^2+1}-1)+\ln(a^2+1)}{2\lambda_0[2-\ln(a^2+1)]},$$

$$\frac{2a^4+\ln(a^2+1)-2}{a^4+a^2} = -\frac{1}{\lambda_0 X}, \ \alpha(\varphi) = \sqrt{a^2-\xi^2}, \ \xi = \tan\tfrac{\varphi}{2},$$

$$a = \tan\tfrac{\varphi_H}{2}, \tag{13.76}$$

$$\delta_H = 1 - \lambda_0\frac{2}{\pi}\int\limits_{-\varphi_H}^{\varphi_H} p_H(\Theta)\{1 + 2\cos(\Theta)\ln(\tan\tfrac{|\Theta|}{2})\}d\Theta.$$

To consider the oscillatory motion of the shaft in detail, we analyze solutions for different values of parameters γ and δ_{1*}. First, let us discuss the solutions obtained for $\delta_{1*} = 0$. For $\gamma = 10$ the maximum pressure and minimum film thickness, maximum frictional stresses applied to both contact

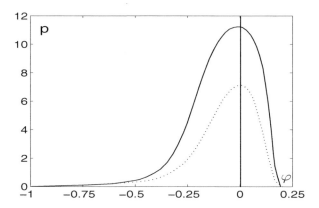

FIGURE 13.16
The maximum pressure distribution at the moments when the reaction force
P applied to the surfaces reaches its minimum ($T = 0.2896$, $p_{max} = 7.0965$
- dotted curve) and maximum ($T = 2.4432$, $p_{max} = 11.2482$ - solid curve),
$\gamma = 10$, $\delta_{1*} = 0$, and $X = 1$ (after Kudish [22]). Reprinted with permission
from the ASME.

surfaces, and shaft displacement and its normal speed versus time, respectively, are presented in Figs. 13.13-13.15. For this case, the pressure and gap
distributions at the moments when the reaction force P reaches its minimum values ($T = 0.2896$, $p_{max} = 7.0965$, $\delta = 1.025937$) and maximum
($T = 2.4432$, $p_{max} = 11.2482$, $\delta = 1.062896$) are given in Figs. 13.16 and
13.17, respectively. In the steady state $p_{max} = 9.3955$ and $\delta = 1.04494$ as it
follows from the numerical results for large times (see also Fig. 13.18). From
Figs. 13.11 and 13.6 it is clear that the maximum contact pressure p_{max} varies
within the range from 75.5% to 119.7% while the shaft displacement δ varies
within a much smaller range from 98.2% to 101.7% of their steady values,
respectively, and p_{max} and δ slowly approach the steady state values. Therefore, small perturbations in the shaft displacement δ cause large variations
in pressure p. It could have been anticipated based on the fact that solution
for pressure p of the equation for gap h is an ill-conditioned problem, i.e., a
problem the solution of which is very sensitive to small variations in the values
of the input data.

Similarly, for $\gamma = 1$ the maximum pressure and minimum film thickness,
maximum frictional stresses applied to both contact surfaces, and shaft displacement and its normal speed versus time, respectively, are presented in
Figs. 13.18 - 13.20. In these cases after a certain time, there is a phase shift of
$\pi/2$ between δ and δ' and a phase shift of π between the maximum pressure
and minimum film thickness; the maximum frictional stresses on the surfaces
are in phase. In particular, it means that pressure reaches its maximum and
minimum values when the shaft normal speed is zero and the shaft displace-

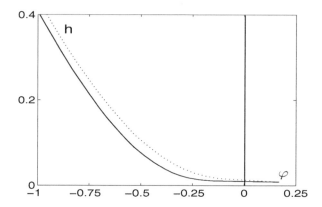

FIGURE 13.17
Gap distributions at the moments when the reaction force P applied to the surfaces reaches its minimum ($T = 0.2896$, $p_{max} = 7.0965$ - dotted curve) and maximum ($T = 2.4432$, $p_{max} = 11.2482$ - solid curve), $\gamma = 10$, $\delta_{1*} = 0$, and $X = 1$ (after Kudish [22]). Reprinted with permission from the ASME.

ment is minimal and maximal, respectively.

It is obvious from Figs. 13.13 - 13.15 and 13.18 - 13.20 that the frequency of oscillations and, especially, damping decrease rapidly as the system inertia increases. After a relatively short period of time, the lubricated system behaves like a free harmonic (linear) oscillator with small amplitude and damping. Therefore, a simplified representation of the system can be done by the well-known equation

$$\delta'' + 2\nu_1\delta' + \nu_0^2\delta = 0, \tag{13.77}$$

where ν_0 is the natural frequency of undamped oscillations, ν_1 is the coefficient of proportionality in the damping term. The natural frequencies and damping coefficient of the lubricated system can be determined by considering small perturbation of a steady solution and by trying to find the characteristics of the oscillatory motion. If $p_0(\varphi)$, $h_0(\varphi)$, δ_0, and φ_{r0} represent the steady solution, then the non-steady solution can be found in the form

$$p(\varphi, T) = p_0(\varphi) + p_1(\varphi)e^{i\nu T} + \dots,$$

$$h(\varphi, T) = h_0(\varphi) + h_1(\varphi)e^{i\nu T} + \dots, \quad \delta(T) = \delta_0 + \delta_1 e^{i\nu T} + \dots, \tag{13.78}$$

$$\varphi_r(T) = \varphi_{r0} + \varphi_{r1}e^{i\nu T} + \dots, \quad \nu = \nu_0 + i\nu_1,$$

where $p_1(\varphi)$, $h_1(\varphi)$, δ_1, and φ_{r1} are small perturbation values, i is the imaginary unit. Then equations (13.61)-(13.65) can be linearized for $p_1(\varphi)$, $h_1(\varphi)$, δ_1, and φ_{r1} in the vicinity of $p_0(\varphi)$, $h_0(\varphi)$, δ_0, and φ_{r0} and after that reduced

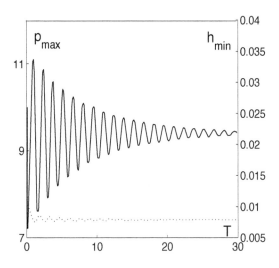

FIGURE 13.18

Distributions of the maximum pressure p_{max} (solid curve) and the minimum film thickness h_{min} (dotted curve) versus time T for $\gamma = 1$, $\delta_{1*} = 0$, and $X = 1$ (after Kudish [22]). Reprinted with permission from the ASME.

to an algebraic equation for ν with complex coefficients. It is important to keep in mind that in this case the initial conditions should be dropped.

The actual derivation of this equation for ν is difficult and not necessary if we have the numerical results presented earlier. It follows from these results that both ν_0 and ν_1 are nonnegative. If $\nu_0 \neq 0$ we have an underdamped motion and if $\nu_0 = 0$ an overdamped one. In particular, using the formula for the frequency of damped oscillations f_{osc} (following from (13.77))

$$f_{osc} = \sqrt{\nu_0^2 - \nu_1^2}$$

and numerical data (see Figs. 13.15 and 13.18) we obtain the values of ν_0 and ν_1 presented in Table 13.1. It follows from Table 13.1 that ν_0 is proportional to $\gamma^{-1/2}$ and ν_1 increases rapidly as γ decreases. Therefore, as the mass of the lubricated system decreases its frequency and damping increase and vice versa. It is important to mention that for a heavy shaft (large γ) damping is a very slow process (small ν_1) and oscillatory motion is sustained for a long period of time. On contrary, for a light shaft (small γ) damping is fast (large ν_1) and oscillations vanish quickly. This trend is in agreement with the physical intuition and can be clearly seen in Figs. 13.13 - 13.15 and 13.18 - 13.20.

The maximum pressure and minimum film thickness for $\gamma = 1$, $\delta_{1*} = -0.06$ and $\delta_{1*} = 0.06$ and the same values of other parameters are presented in

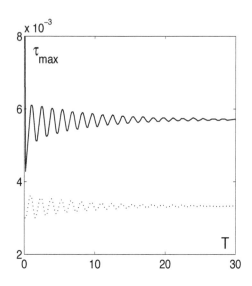

FIGURE 13.19
Distribution of the maximum frictional stresses τ_1 and τ_2 applied to both contact surfaces versus time T for $\gamma = 1$, $\delta_{1*} = 0$, and $X = 1$ (solid curve - lower surface, dotted curve - upper surface) (after Kudish [22]). Reprinted with permission from the ASME.

Figs. 13.21 and 13.22, respectively. For $\delta_{1*} = -0.06$ the pressure maximum p_{max} varies between 5.6101 and 12.0822 and for $\delta_{1*} = 0.06$ it varies between 7.9395 and 10.8289. Therefore, all EHL parameters, including pressure and film thickness, vary significantly in time for different initial values of δ_0 and δ_{1*}.

This analysis indicates that small perturbations of steady conditions caused by instabilities in external load, surface velocities, or surface roughness lead to a relatively long damped oscillatory motion. The analogy presented by equation (13.77) allows to make a conclusion that the potentially most harmful small perturbation (causing large amplitudes of pressure oscillations) is the periodic perturbation with frequency of $(\nu_0^2 - 2\nu_1^2)^{1/2}$. As it was mentioned such a situation can be created not only by external conditions but also by surface waviness and/or roughness developed in the process of work. The data from Table 13.1 indicates that the lighter the shaft the greater effect such small perturbations may cause. The comparison of the solutions for rigid and elastic materials shows that oscillations are damped much faster in the case of rigid materials than the elastic ones (see Figs. 13.11, 13.13, 13.15, 13.18, 13.20 - 13.22. This can be easily understood by taking into account two facts: (1) some of the energy is stored in the elastic materials in the form of the potential

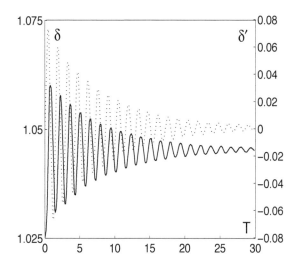

FIGURE 13.20
Distribution of shaft radial displacement δ (solid curve) and its radial speed
δ' (dotted curve) versus time T for $\gamma = 1$, $\delta_{1*} = 0$, and $X = 1$ (after Kudish
[22]). Reprinted with permission from the ASME.

energy of elastic deformation and (2) only the kinetic portion of the system's
total energy is dissipated. The second important conclusion, which follows
from the above comparison, is that in spite of very small variations in the
shaft displacement δ the maximum pressure p_{max} may vary within very large
margins. Therefore, in practice, the fact that the measured shaft displacement
varies insignificantly while the shaft is involved in a non-steady motion does
not mean that the variations of pressure are small as well. In other words, in
the case of a non-steady motion, it is necessary to take into account Newton's
second law (see equation (13.55)) instead of just the balance condition on the
external force X, i.e.,

$$X = \frac{1}{\pi} \int_{\varphi_l}^{\varphi_r} p(\theta, T) \cos \theta d\theta.$$

Finally, the account for elasticity of materials is very important as it not only
changes the dynamics of the system described above but also changes significantly the steady solutions. In fact, for rigid materials the steady parameters
of the lubricated contact are $p_{max} = 21.2606$ and $\delta = 0.99665$ while for the
contact of elastic materials they are $p_{max} = 9.3955$ and $\delta = 1.04494$.

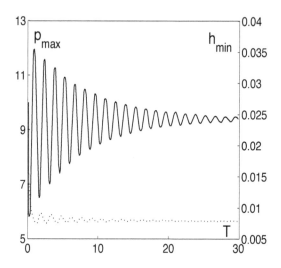

FIGURE 13.21
Distributions of the maximum pressure p_{max} (solid curve) and the minimum film thickness h_{min} (dotted curve) versus time T for $\gamma = 1$, $\delta_{1*} = -0.06$, and $X = 1$ (after Kudish [22]). Reprinted with permission from the ASME.

TABLE 13.1
Dependence of the oscillation frequency and damp- ing
on the system inertia (after
Kudish [22]). Reprin- ted
with permission from the
ASME.

γ	0.1	1	10
ν_0	13.96	4.33	1.37
ν_1	0.41	0.124	0.017

13.3.6.2   Transient Oscillatory Motion under Action of a Step Load

Let us consider a transient motion of the shaft subjected to a step load. The initial parameters of the problem correspond to the solution of the problem for the steady state conditions ($\gamma = 1$, $\delta_{1*} = 0$) described in the preceding subsection. The applied load $X(T)$ for four cases is given by the formulas: case (a) $X(T) = 1$ if $T < 0.1$ and $T > 1.1471794$, and $X(T) = 0.7$ if $0.1 \leq T \leq 1.1471794$; case (b) $X(T) = 1$ if $T < 0.1$ and $T > 2.1743589$, and $X(T) = 1.3$ if $0.1 \leq T \leq 2.1743589$; case (c) $X(T) = 1$ if $T < 0.1$ and $X(T) = 0.7$ if $T \geq 0.1$; and case (d) $X(T) = 1$ if $T < 0.1$ and $X(T) = 1.3$ if $T \geq 0.1$.

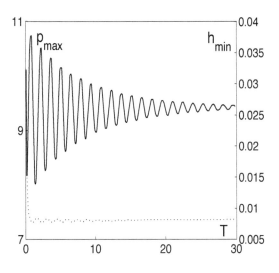

FIGURE 13.22
Distributions of the maximum pressure p_{max} (solid curve) and the minimum film thickness h_{min} (dotted curve) versus time T for $\gamma = 1$, $\delta_{1*} = 0.06$, and $X = 1$ (after Kudish [22]). Reprinted with permission from the ASME.

The numerical solutions for the maximum pressure and shaft displacement obtained for cases (a)-(d) are presented in Figs. 13.23 - 13.26, respectively. The data presented in Figs. 13.23 - 13.26 indicate that after the external load experienced a jump the shaft is involved in oscillatory motion. The maximum pressure varies within the range from 55% to 135% of the steady value. The relative variations of the shaft displacement are not as great, however, the solution of the problem is very sensitive even to small changes in the shaft displacement. As before, the rate of oscillation damping depends primarily on the system inertia, i.e., the greater is inertia, the slower is damping.

13.3.6.3 Oscillatory Motion under Constant External Conditions Due to a Bump/Dent Presence

Suppose there is a bump/dent on the surface of the shaft outside of the contact, which, at a certain time moment, enters the contact region and disturbs the steady process of lubrication. In reality, in the presence of a bump/dent, generally, the initial conditions in the contact region are not steady. In spite of that, the described situation gives a clear understanding as of how in practice a lubricated system would respond to the presence of a bump/dent in the contact. Let us consider a motion of a shaft subjected to a constant force $X(T) = 1$. As the initial contact parameters, let us take those used

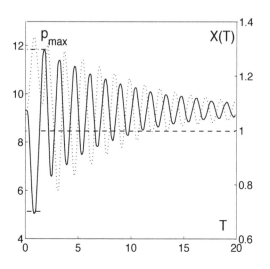

FIGURE 13.23
Distributions of the maximum pressure p_{max} and the external force X versus time T for $\gamma = 1$, $\delta_{1*} = 0$: maximum pressure for case (a) solid curve, external force for case (a) dashed curve, maximum pressure for case (b) dotted curve, external force for case (b) dash-dotted curve (after Kudish [22]). Reprinted with permission from the ASME.

in the preceding two subsections and $\gamma = 1$, $\delta_{1*} = 0$. Let the shape of a bump/dent is described by function $A(\psi)$, where A is a nonnegative function. As a bump/dent passes through the contact its relative position changes. Therefore, the dimensionless gap equation can be presented in the form

$$h(\psi, T) = h_{ideal}(\psi, T) \pm A(\psi - 2T), \tag{13.79}$$

where h_{ideal} is the gap without a bump/dent present (see equation (13.64)) and the upper/lower sign is used in the case of a dent/bump, respectively. In particular calculations function A was defined as follows:

$$A(\psi) = b\{1 + \sin[\omega_b(\psi - \psi_0)]\} \; if \; -\tfrac{\pi}{2} \leq \omega_b(\psi - \psi_0) \leq \tfrac{3\pi}{2},$$

$$A(\psi) = 0 \; otherwise, \tag{13.80}$$

where $2b$ and ψ_0 are the bump/dent amplitude and initial phase shift ($\psi_0 = \psi_l - 1.5\pi/\omega_b - 0.1$), respectively, ω_b is the parameter controlling the bump/dent width, which is equal to $2\pi/\omega_b$. At the initial time moment $T = 0$ the bump/dent is completely outside of the contact region $-\pi/2 \leq \varphi \leq \pi/2$.

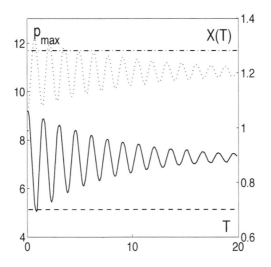

FIGURE 13.24
Distributions of the maximum pressure p_{max} and the external force X versus time T for $\gamma = 1$, $\delta_{1*} = 0$: maximum pressure for case (c) solid curve, external force for case (c) dashed curve, maximum pressure for case (d) dotted curve, external force for case (d) dash-dotted curve (after Kudish [22]). Reprinted with permission from the ASME.

For three different bumps/dents (bump/dent 1: $b = 0.004$, $w_b = 10$; bump/dent 2: $b = 0.008$, $w_b = 10$; bump/dent 3: $b = 0.004$, $w_b = 25$), the graphs of the maximum pressure and shaft normal velocity are given in Figs. 13.27 - 13.30, respectively. At some point in time the bump/dent enters the contact region and perturbs the steady state of a lubricated contact causing shaft displacement and pressure oscillations. These oscillations seem to be greater when the bump/dent is higher/deeper and narrower. For bumps 1-3, the maximum pressure varies within the range from 94.0% to 121.3%, 92.0% to 147.0%, and from 79.8% to 143.2% of the steady maximum pressure $p_{max} = 9.3955$, respectively. Similarly, for dents 1-3 the maximum pressure varies within the range from 83.5% to 110.7%, from 77.9% to 119.2%, and from 94.8% to 118.2% of the steady maximum pressure $p_{max} = 9.3955$, respectively. Obviously, the influence of a bump presence in the contact is stronger than that of a dent with the same shape. The major changes in the maximum pressure take place when the bump/dent center approaches the point of minimum gap. After the bump/dent leaves the contact, the parameter perturbations slowly subside due to viscous damping until the bump/dent enters the contact region the next time. In case when the bump/dent enters the

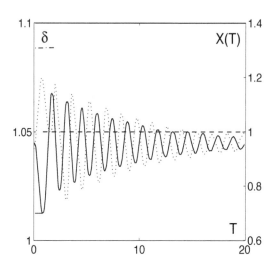

FIGURE 13.25

Distributions of the shaft displacement δ and the external force X versus time T for $\gamma = 1$, $\delta_{1*} = 0$: shaft displacement for case (a) solid curve, external force for case (a) dashed curve, shaft displacement for case (b) dotted curve, external force for case (b) dash-dotted curve (after Kudish [22]). Reprinted with permission from the ASME.

contact multiple times the situation gets more complicated. The behavior of the lubrication parameters depends not only on the bump/dent and contact parameters but also on the value of the angular velocity ω.

Let us summarize the results of the section. A non-steady problem for a conformal EHL contact of two infinite cylindrical surfaces with parallel axes is considered. It takes into account the elasticity of cylinders, lubricant viscosity, contact surface velocities, and applied load. The problem is solved based on the "modified" formulation, which includes Newton's second law of motion. The main emphasis in the section is put on the analysis of the transient dynamics of the system under constant and abruptly changing external load as well as on the system behavior in the case of a bump/dent presence on the shaft surface. A strong dependence of non-steady EHL solutions on the system inertia (mass) and elasticity of materials is established. Generally, the numerical solutions exhibit damped oscillatory behavior. The transient values of pressure are extremely sensitive to small variations in the shaft displacement. This leads to the conclusion that any non-steady motion of a lubricated system must be considered based on Newton's second law.

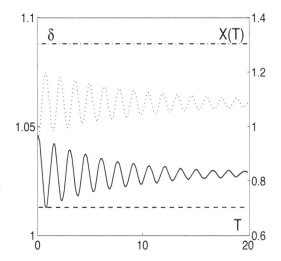

FIGURE 13.26
Distributions of the shaft displacement δ and the external force X versus time T for $\gamma = 1$, $\delta_{1*} = 0$: shaft displacement for case (c) solid curve, external force for case (c) dashed curve, shaft displacement for case (d) dotted curve, external force for case (d) dash-dotted curve (after Kudish [22]). Reprinted with permission from the ASME.

13.4 Starved Lubrication of a Lightly Loaded Spatial EHL Contact. Effect of the Shape of Lubricant Meniscus on Problem Solution

In practice, because of different reasons (starvation, instability of lubrication regime, proximity to critical velocities, influence of lubricant surface tension, etc.) the input oil meniscus is not always far from the center of the contact. Moreover, very often the meniscus is relatively close to it and changes its configuration in time. Usually, this is reflected on graphs of frictional stresses versus time under conditions seemed to be stationary. The oscillations of the friction stress in such a situation may be dramatically large.

Interference methods allow to observe this behavior of EHL film thickness in relation to the configuration (location) of the input oil meniscus. It was registered that in such cases the profile of EHL film thickness (gap between contacting solids) is far from that which may be observed for truly stationary or fully flooded conditions and it depends on the shape of the inlet oil menis-

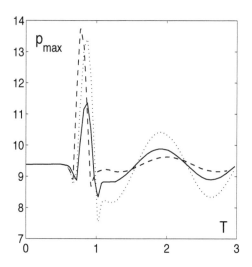

FIGURE 13.27
Distributions of the maximum pressure p_{max} versus time T for bumps 1-3 (bump 1: $a = 0.004$, $\omega_b = 10$ - solid curve; bump 2: $a = 0.008$, $\omega_b = 10$ - dashed curve; bump 3: $a = 0.004$, $\omega_b = 25$ - dotted curve) and $\gamma = 1$, $\delta_{1*} = 0$ (after Kudish [22]). Reprinted with permission from the ASME.

cus. In particular, due to the inlet meniscus shape the gap between lubricated contacting solids may approach or reach zero (direct dry contact of elastic solids) in some zones of the contact area.

In this section we provide a conservative numerical method and some numerical solutions for this problem taking into account a complex shape of the input oil meniscus. The particular physical phenomena (such as instability of lubrication regime, proximity to critical velocities, influence of lubricant surface tension, etc.) which caused the oil meniscus to approach the boundary of the dry contact are not considered here.

The material presented in this section is different from the studies published earlier by Oh [24], Evans and Hughes [25], Hamrock and Dowson [26], Lubrecht and Ten Napel [27] in two ways: (1) the finite-difference equations of the EHL problem are obtained based on the fluid flux conservation considerations. (2) the finite-difference equations of the EHL problem take into account the complex shape of the input oil meniscus. The second provision allows to model some lubrication regimes that are far from fully flooded conditions. The further analysis is based on [23].

Some numerical results for pressure, gap, frictional stress, and subsurface stress distributions in the EHL contact are presented.

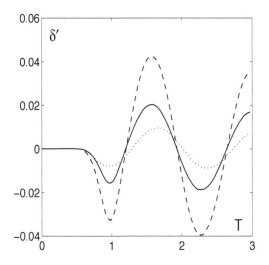

FIGURE 13.28
Distributions of the shaft radial velocity δ' versus time T for bumps 1-3 (bump 1: $a = 0.004$, $\omega_b = 10$ - solid curve; bump 2: $a = 0.008$, $\omega_b = 10$ - dotted curve; bump 3: $a = 0.004$, $\omega_b = 25$ - dashed curve) and $\gamma = 1$, $\delta_{1*} = 0$ (after Kudish [22]). Reprinted with permission from the ASME.

13.4.1 Problem Formulation

Let us assume that two moving (rolling) solids with smooth surfaces made of the same elastic material are lubricated by an incompressible Newtonian viscous fluid under isothermal conditions. The thickness of the lubrication layer is considered to be small compared to characteristic size of the contact area and size of the solids [16]. The solids are in a concentrated (non-conformal) contact and they experience an external compressive force P.

Let us introduce a moving rectangular coordinate system: the z-axis passes through the solids' centers of curvature, and the xy-plane is equidistant from the solids.

It is assumed that the solids' stationary motion is slow and occurs with linear velocities $\vec{u}_1 = (u_{x1}, v_{y1})$ and $\vec{u}_2 = (u_{x2}, v_{y2})$. The inertia forces in the lubrication layer are considered to be small compared with the viscous ones.

In the following dimensionless variables [28]

$$(x', y') = \frac{\vartheta}{2R_x}(x, y), \ (p', \tau') = \frac{8\pi R_x^2}{3P\vartheta^2}(p, \tau), \ h' = \frac{h}{h_0}, \ \mu' = \frac{\mu}{\mu_a},$$

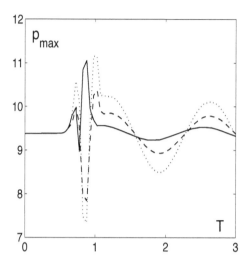

FIGURE 13.29
Distributions of the maximum pressure p_{max} versus time T for dents 1-3 (dent 1: $a = 0.004$, $\omega_b = 10$ - solid curve; dent 2: $a = 0.008$, $\omega_b = 10$ - dashed curve; dent 3: $a = 0.004$, $\omega_b = 25$ - dotted curve) and $\gamma = 1$, $\delta_{1*} = 0$ (after Kudish [22]). Reprinted with permission from the ASME.

$$(\vec{u'}_1, \vec{u'}_2) = \frac{(\vec{u}_1, \vec{u}_2)}{\mid \vec{u}_1 + \vec{u}_2 \mid}, \ \ \gamma = \frac{\vartheta^2}{2R_x} h_0, \ \ \vartheta = \frac{P}{8\pi\mu_a \mid \vec{u}_1 + \vec{u}_2 \mid R_x},$$

which are characteristic for lightly loaded EHL contacts the EHL problem equations are as follows (further, primes are omitted at the dimensionless variables)

$$L(p) = \{\nabla \cdot [\gamma^2 \frac{h^3}{\mu} \nabla p - \vec{u}h]\} = 0, \ \ \nabla = (\frac{\partial}{\partial x}, \frac{\partial}{\partial y}), \ \ \vec{u} = (u_x, v_y), \qquad (13.81)$$

$$h = 1 + \frac{x^2 + \rho y^2}{\gamma} + \frac{2}{\pi V \gamma} \int\int_\Omega \left[\frac{1}{\sqrt{(\xi-x)^2 + (\eta-y)^2}} - \frac{1}{\sqrt{\xi^2 + \eta^2}} \right] p(\xi, \eta) d\xi d\eta,$$

$$\qquad (13.82)$$

$$\int\int_\Omega p(\xi, \eta) d\xi d\eta = \frac{2\pi}{3},$$

$$p \mid_\Gamma = 0,$$

$$\qquad (13.83)$$

$$\gamma = \frac{\theta_0^2}{2R_x} h_0, \ \ V = \frac{8\pi R_x^2 E'}{3P\theta_0^3}, \ \ \rho = \frac{R_x}{R_y}, \ \ \vec{u} = \frac{\vec{u}_1 + \vec{u}_2}{2}, \qquad (13.84)$$

where $p(x, y)$ and $h(x, y)$ are the pressure and gap in the contact, Ω is the contact region, γ is the dimensionless central lubrication film thickness, Γ is the

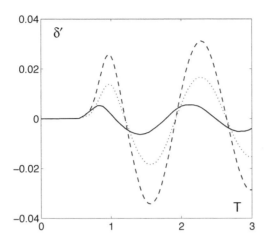

FIGURE 13.30
Distributions of the shaft radial velocity δ' versus time T for dents 1-3 (dent 1: $a = 0.004$, $\omega_b = 10$ - solid curve; dent 2: $a = 0.008$, $\omega_b = 10$ - dashed curve; dent 3: $a = 0.004$, $\omega_b = 25$ - dotted curve) and $\gamma = 1$, $\delta_{1*} = 0$ (after Kudish [22]). Reprinted with permission from the ASME.

boundary of the contact, V and ρ are the given dimensionless parameters, $\vec{u}$ is the average velocity of the solid surfaces in contact. The dimensionless variables are introduced in such a way that for a similar hydrodynamic problem (i.e., for a lubrication problem for rigid solids) $p = O(1)$ and $x^2 + y^2 = O(1)$.

Also, the problem solution must satisfy the boundary conditions for pressure p, which are different at different portions of the boundary of the contact region: the inlet and exit boundaries. Let us assume that Γ_i and Γ_e are the inlet and exit portions of the contact boundary, which are given and unknown in advance, respectively. The whole boundary of the contact Γ is $\Gamma_i \bigcup \Gamma_e$. On the contact inlet boundary, we have the condition:

$$\Gamma_i \text{ is given if } \vec{Q_f} \cdot \vec{n_\Gamma} \mid_{\Gamma_i} < 0, \tag{13.85}$$

representing the fact that lubricant enters the contact through this boundary as well as the boundary condition (13.83) must be satisfied. At the same time, the exit boundary is also defined by condition (13.83) and the cavitation condition

$$\nabla p \cdot \vec{n_\Gamma} \mid_{\Gamma_e} = 0 \text{ if } \vec{Q_f} \cdot \vec{n_\Gamma} \mid_{\Gamma_e} > 0, \tag{13.86}$$

where $\vec{Q_f}$ is the vector of fluid flux given by the expression

$$\vec{Q_f} = h\vec{u} - \gamma^2 \frac{h^3}{\mu} \nabla p, \tag{13.87}$$

and $\vec{n}_\Gamma$ is the external normal vector to the contact boundary Γ.

The location of the contact exit boundary can also be found using complementarity considerations

$$L(p) = 0 \ for \ p > 0, \ i.e., \ within \ the \ contact \ region,$$

$$(13.88)$$

$$L(p) < 0 \ for \ p = 0 \ outside \ the \ contact \ if \ \vec{Q}_f \cdot \vec{n}_\Gamma \ |_{\Gamma_e} \geq 0,$$

which are described in detail by Kostreva [29] and Oh [24]. Oh et al. [30] showed that the complementarity approach leads to the solution of the problem satisfying the cavitation condition on the exit boundary (see equation (13.86)).

In the system of equations and inequalities (13.81)-(13.88), the geometric shape of the input oil meniscus, parameters V (for lightly loaded EHL contact $V \gg 1$), ρ, u_x, v_y, and the function of viscosity $\mu(p)$ are considered to be known. In the presented below numerical results, the exponential relationship for $\mu(p)$ is used

$$\mu(p) = \exp(Qp),$$

$$(13.89)$$

where Q is the dimensionless pressure viscosity coefficient.

The solution of system (13.81)-(13.89) is represented by pressure $p(x, y)$ and gap $h(x, y)$ distributions, dimensionless EHL film thickness γ, and the location of the exit contact boundary Γ_e.

It is necessary to mention that boundary conditions (13.85), (13.87), and (13.88) are of local nature. Thus, some pieces of the contact boundary Γ are known and some are not.

After the solution of the EHL problem is found, the friction stress distributions on the contact surfaces 1 and 2 can be found according to the formulas:

$$\vec{\tau}_{1,2} = \frac{\mu \vec{s}}{12\gamma h} \mp \frac{h}{2}\gamma\nabla p,$$

$$(13.90)$$

where $\vec{s}$ is the dimensionless sliding velocity, $\vec{s} = 2(\vec{u}_2 - \vec{u}_1)/ \mid \vec{u}_2 + \vec{u}_1 \mid$. The first term in (13.90) is the friction stress due to sliding $\vec{\tau}_s$, the second term is the frictional stress due to rolling $\vec{\tau}_r$. The stress components $\sigma_{ij}(x, y, z)$ in the subsurface layers of the contact solids are determined by well–known Boussinesq relations [31]

$$\sigma_{ij}(x, y, z) = \int \int_\Omega G_{ij}(x - \xi, y - \eta, z)p(\xi, \eta)d\xi d\eta,$$

$$(13.91)$$

where $G_{ij}(x, y, z)$ is the Green's tensor obtained from the Boussinesq problem [31]. To characterize the subsurface stress distributions, we will use the octahedral normal and tangential stresses. The latter are determined by the formulas

$$\sigma^{oct} = \frac{\sigma_{11} + \sigma_{22} + \sigma_{33}}{3},$$

$$\tau^{oct} = \frac{1}{3}[(\sigma_{11} - \sigma_{22})^2 + (\sigma_{22} - \sigma_{33})^2 + (\sigma_{11} - \sigma_{33})^2$$

$$(13.92)$$

$$+\tau_{12}^2 + \tau_{23}^2 + \tau_{13}^2]^{1/2}.$$

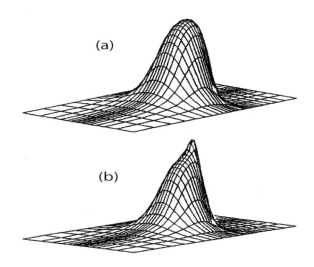

FIGURE 13.31
Pressure distribution for fully flooded lubrication regime: (a) $Q_0 = 0$, $p_{max} = 3.82$, (b) $Q_0 = 0.6$, $p_{max} = 5.53$ (after Kudish and Panovko [13]). Reprinted with permission from the ASME.

13.4.2 Numerical Method

Let us consider solution of the problem equations in a rectangular region $\Omega = \{(x, y) \mid a \leq x \leq b, c \leq y \leq d\}$, where a, b, c, and d are certain constants. Using a transform based on a hyperbolic sine, we can introduce a nonuniform grid with the nodes in the xy-plane with the following coordinates:

$$x_i \ (i = 0, \ldots, i_{max} + 1), \ x_{i-1/2} = (x_i + x_{i-1})/2 \ (i = 1, \ldots, i_{max} + 1),$$

$$y_j \ (j = 0, \ldots, j_{max} + 1), \ y_{j-1/2} = (y_j + y_{j-1})/2 \ (j = 1, \ldots, j_{max} + 1),$$

$$(x_{m0-1/2}, y_{n0-1/2}) = (0, 0).$$

The pressure distribution will be determined at nodes (x_i, y_j) and the gap distribution will be determined at nodes $(x_{i-1/2}, y_{j-1/2})$.

Integrating equation (13.81) over the cell around node (x_i, y_j) gives its integral analogue in the form

$$L_1(p) = \oint_{l_{ij}} \{\gamma^2 \frac{h^3}{\mu} \nabla p \cdot \vec{n} - h\vec{v} \cdot \vec{n}\} dl = 0, \tag{13.93}$$

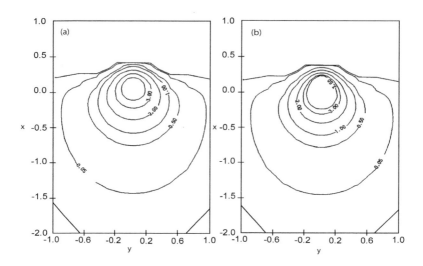

FIGURE 13.32
Isobars for fully flooded lubrication regime: (a) $Q_0 = 0$, (b) $Q_0 = 0.6$ (after Kudish and Panovko [13]). Reprinted with permission from the ASME.

where l_{ij} is the rectangular boundary of the cell surrounding the node (x_i, y_j) and $\vec{n}$ is the external normal vector to the boundary of the cell.

In the complementarity conditions (13.88), operator $L(p)$ should be replaced by operator $L_1(p)$ from equation (13.93).

The iterative process for the EHL problem is based on Newton's method. Equations (13.93), (13.83), and (13.85) linearized near the k-th iterates of the solution $(p_k(x, y), h_k(x, y), \gamma_k, \Gamma_{e,k})$ have the form

$$\oint_{l_{ij}} \{[2\gamma_k \frac{h_k^3}{\mu_k} \frac{\partial p_k}{\partial \vec{n}} - \vec{v} \cdot \vec{n} \frac{\partial h}{\partial \gamma} \mid_k + \gamma_k^2 \frac{3h_k^2}{\mu_k} \frac{\partial p_k}{\partial \vec{n} \frac{\partial h}{\partial \gamma}} \mid_k] \triangle \gamma_{k+1}$$

$$-\gamma_k^2 \frac{h_k^3}{\mu_k^2} \frac{\partial mu}{\partial p} \mid_k \frac{\partial p_k}{\partial \vec{n}} \triangle p_{k+1} + \gamma_k^2 \frac{3h_k^2}{\mu_k} \frac{\partial h}{\partial p} \mid_k \triangle p_{k+1} \partial p_k \partial \vec{n} \qquad (13.94)$$

$$+\gamma_k^2 \frac{h_k^3}{\mu_k} \frac{\partial \triangle p_{k+1}}{\partial \vec{n}} - \vec{u} \cdot \vec{n} \frac{\partial h}{\partial p} \mid_k \triangle p_{k+1}\} = -\oint_{l_{ij}} [\gamma_k^2 \frac{h_k^3}{\mu_k} \frac{\partial p_k}{\partial \vec{n}} - \vec{u} \cdot \vec{n} h_k] dl,$$

$$\iint_{\Omega} \triangle p_{k+1}(\xi, \eta) d\xi d\eta = \frac{2\pi}{3} - \iint_{\Omega} p_k(\xi, \eta) d\xi d\eta, \qquad (13.95)$$

$$\triangle p_{k+1} \mid_{\Gamma_k} = 0, \qquad (13.96)$$

where k is the iteration number, $\triangle p_{k+1} = p_{k+1} - p_k$, $\triangle \gamma_{k+1} = \gamma_{k+1} - \gamma_k$, $\frac{\partial h}{\partial \gamma} \mid_k$ is the partial derivative of h_k with respect to γ_k calculated using equation

(a)

(b)

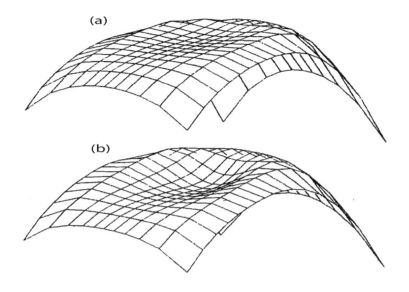

FIGURE 13.33
Gap distribution for fully flooded lubrication regime: (a) $Q_0 = 0$, $h_{min} = 0.726$, (b) $Q_0 = 0.6$, $h_{min} = 0.694$ (after Kudish and Panovko [13]). Reprinted with permission from the ASME.

(13.82), $\frac{\partial \mu}{\partial p}|_k$ is the partial derivative of μ_k with respect to p_k calculated using equation (13.89); $\frac{\partial h}{\partial p}|_k \triangle p_{k+1}$ is the linear operator representing the derivative of h_k with respect to p_k applied to $\triangle p_{k+1}$

$$\frac{\partial h}{\partial p}|_k \triangle p_{k+1} = \frac{2}{\pi V \gamma} \int \int_\Omega \left[\frac{1}{\sqrt{(\xi-x)^2+(\eta-y)^2}} \right.$$

(13.97)

$$\left. - \frac{1}{\sqrt{\xi^2+\eta^2}} \right] \triangle p_{k+1}(\xi, \eta) d\xi d\eta.$$

Integration over the contour l_{ij} is performed along the intervals $[x_{i-1/2}, x_{i+1/2}]$ and $[y_{j-1/2}, y_{j+1/2}]$ taking into account the direction of integration. The following approximations are used for derivatives

$$\frac{\partial p(x_{i\pm1/2}, y_j)}{\partial x} \approx \frac{\pm p(x_{i\pm1}, y_j) \mp p(x_i, y_j)}{\pm x_{i\pm1} \mp x_i},$$

$$\frac{\partial p(x_i, y_{j\pm1/2})}{\partial y} \approx \frac{\pm p(x_i, y_{j\pm1}) \mp p(x_i, y_j)}{\pm y_{j\pm1} \mp y_j}.$$

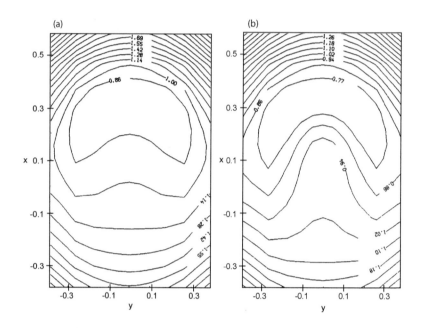

FIGURE 13.34
Level curves of gap distribution for fully flooded lubrication regime: (a) $Q_0 = 0$, (b) $Q_0 = 0.6$ (after Kudish and Panovko [13]). Reprinted with permission from the ASME.

The coefficients of $\triangle p_{k+1}$ and $\triangle \gamma_{k+1}$ are calculated at the points with coordinates $(x_{i\pm1/2}, y_j)$ and $(x_i, y_{j\pm1/2})$. The singular integral in equation (13.93) is calculated according to the following cubature formula (Belotcerkovsky and Lifanov [32]):

$$\iint\limits_{\Omega} \left[\frac{1}{\sqrt{(\xi-x_{m-1/2})^2+(\eta-y_{n-1/2})^2}} - \frac{1}{\sqrt{\xi^2+\eta^2}} \right] p(\xi,\eta)d\xi d\eta$$

$$= \sum_{j=1}^{j_{max}} \sum_{i_1(j)}^{i_2(j)} \left[\frac{1}{\sqrt{(x_i-x_{m-1/2})^2+(y_j-y_{n-1/2})^2}} \right.$$

$$\left. - \frac{1}{\sqrt{(x_i-x_{m0-1/2})^2+(y_j-y_{n0-1/2})^2}} \right] p(x_i, y_j)$$

$$\times (x_{i+1/2} - x_{i-1/2})(y_{j+1/2} - y_{j-1/2}),$$

(13.98)

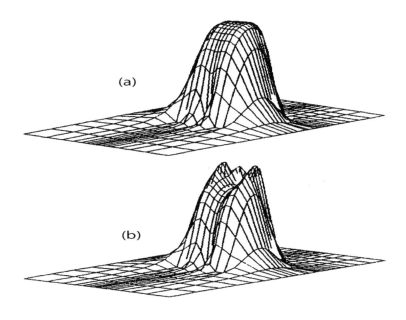

FIGURE 13.35
Pressure distribution for notched input oil meniscus: (a) $Q_0 = 0$, $p_{max} = 3.92$, $p_{min} = 3.915$, (b) $Q_0 = 0.6$, $p_{max} = 4.64$, $p_{min} = 3.28$ (after Kudish and Panovko [13]). Reprinted with permission from the ASME.

where $i_1(j)$ and $i_2(j)$ are the integer arrays describing the location of the inlet and exit boundaries, respectively, $m = 1, \ldots, i_{max} + 1$, $n = 1, \ldots, j_{max} + 1$, node $(x_{m0-1/2}, y_{n0-1/2})$ is the origin of the coordinate system. The array of indexes $i_1(j)$ is known as the inlet boundary Γ_i is given and the array of indexes $i_2(j)$ is found after each iteration from the complementarity conditions (13.88). The values of h at points $(x_{i\pm1/2}, y_j)$ and $(x_i, y_{j\pm1/2})$ are obtained by interpolation. A similar cubature formula is used for calculation of $\frac{\partial h}{\partial p}|_k$ $\triangle p_{k+1}$ (see equation (13.97)).

As an initial approximation for the iterative process, the solution of a similar lubrication problem for rigid solids can be used. The latter problem is solved for the given geometry of the input meniscus and γ_0 by the method of upper relaxation. The aforementioned value of dimensionless film thickness γ_0 is approximately determined from the solution of an equation for the total load applied to the contact $P(\gamma) = 2\pi/3$. The left-hand side of this equation is obtained from the solutions of the lubrication problem for rigid solids for several values of γ. As an initial approximation for the iterative process, the existing solution of a similar lubrication problem for different regime param-

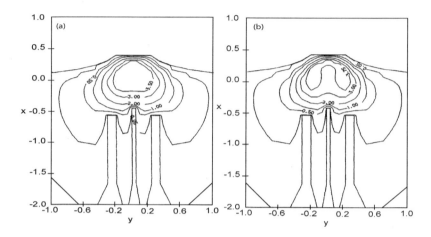

FIGURE 13.36
Isobars for notched input oil meniscus: (a) $Q_0 = 0$, (b) $Q_0 = 0.6$ (after Kudish and Panovko [13]). Reprinted with permission from the ASME.

eters can be used as well.

The first step of the iteration process is represented by the process of solving of the finite-difference analog of equations (13.94)-(13.98) for $\triangle p_{k+1}(x_i, y_j)$ and $\triangle \gamma_{k+1}$ by the Gaussian elimination with partial pivoting. After that, we reconstruct the values of $p_{k+1}(x_i, y_j) = p_k(x_i, y_j) + \triangle p_{k+1}(x_i, y_j)$ and $\gamma_{k+1} = \gamma_k + \triangle \gamma_{k+1}$ and the new location of the exit boundary $\Gamma_{e,k+1}$ is determined according to the complementarity conditions. At this stage the gap distribution $h(x_{i-1/2}, y_{j-1/2})$ is calculated using formula (13.82) and (13.98). The described process of calculations is repeated until the desired relative accuracy of the solution is reached: $\mid p_{k+1}(x_i, y_j)/p_k(x_i, y_j) - 1 \mid < \delta$, etc.

13.4.3   Numerical Results

The described numerical process is used for calculations of a lightly loaded EHL point contact of two spheres with equal radii ($\rho = 1$) rolling along the x-axis ($u_x = 1$, $v_y = 0$) with relative sliding $s_x = 0.01$ and $s_y = 0$. The solutions are obtained for varied geometry of the input oil meniscus and values of parameters V and Q. The calculations are conducted with the relative accuracy $\delta = 10^{-5}$ with more detailed grid at the center of the contact. The entire grid is represented by 26×16 nodes with $\triangle x_{min} = 0.04305$, $\triangle x_{max} = 0.3778$, $\triangle y_{min} = 0.04815$, and $\triangle y_{max} = 0.2647$. All calculations were performed with double precision. The oil meniscus shapes used for calculations

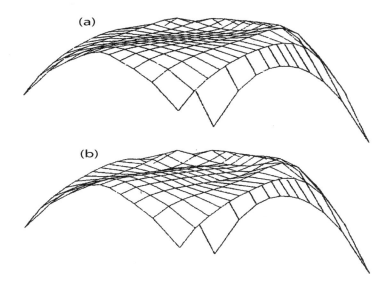

FIGURE 13.37
Gap distribution for notched input oil meniscus: (a) $Q_0 = 0$, $h_{min} = 0.797$, (b) $Q_0 = 0.6$, $h_{min} = 0.896$ (after Kudish and Panovko [13]). Reprinted with permission from the ASME.

had one and three notches. The presented numerical solutions are obtained for $V = 15$, $Q = 0$, and $Q = 0.6$.

For testing purposes series of calculations using two grids 26×16 and 30×30 ($\triangle x_{min} = 0.03158$, $\triangle x_{max} = 0.3698$, $\triangle y_{min} = 0.02419$, and $\triangle y_{max} = 0.1606$) were performed. The obtained results for $V = 15$ and different values of Q are as follows: $\gamma = 0.12$, $h_{min} = 0.726$, $p_{max} = 3.82$, $Q = 0$ and $\gamma = 0.203$, $h_{min} = 0.694$, $p_{max} = 5.53$, $Q = 0.6$ for grid 26×16 and $\gamma = 0.118$, $h_{min} = 0.701$, $p_{max} = 3.805$, $Q = 0$ and $\gamma = 0.200$, $h_{min} = 0.700$, $p_{max} = 5.473$, $Q = 0.6$ for grid 30×30. These results indicate that the relative precision of the solutions obtained for grids 26×16 and 30×30 is in the range of $2\% - 3\%$ for γ, h_{min} and in the range of $0.5\% - 1\%$ for p_{max}. For more detailed grids $N \times M$, the time required for solution of the problem is proportional to $(NM)^3$.

The numerical experiments showed stability of the described numerical method in the following range of the input parameters of the EHL problem for lightly loaded contacts: $15 \leq V < \infty$ and $0 \leq Q \leq 0.6$.

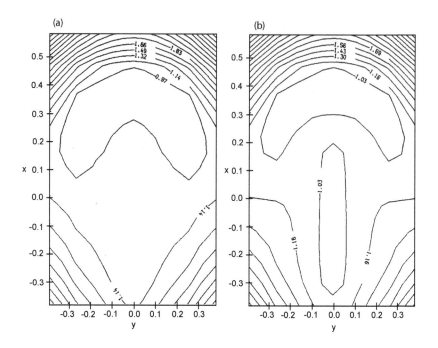

FIGURE 13.38
Level curves of gap distribution for notched input oil meniscus: (a) $Q_0 = 0$, (b) $Q_0 = 0.6$ (after Kudish and Panovko [13]). Reprinted with permission from the ASME.

13.4.3.1 Pressure, Gap, and Frictional Stress Distributions

The distributions of pressure $p(x,y)$ and gap $h(x,y)$ for the case of fully flooded lubrication (the input oil meniscus is far enough from the center of the contact and coincides with the coordinate lines bounding the contact region) are shown in Figs. 13.31 - 13.34. For convenience the distributions of gap $h(x,y)$ are presented only for the portion of the contact region with noticeable pressure $p(x,y)$. The distribution of pressure $p(x,y)$ for $Q = 0.6$ features a sharp peak near the exit boundary, which is absent for the case of $Q = 0$ (see Figs. 13.31 and 13.32). The difference in the pressure maxima for these cases is substantial (see Fig. 13.31). The distribution of gap $h(x,y)$ is more "concave" for the case of $Q = 0.6$ than for $Q = 0$, which is related to the differences in the pressure distributions (see Figs. 13.33 and 13.34). The location of isobars varies with the pressure viscosity coefficient Q. However, the exit boundary remains almost unchanged (see Fig. 13.32). A noticeable

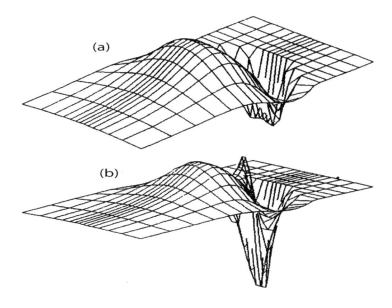

FIGURE 13.39
Rolling frictional stress distribution τ_{rx} for fully flooded lubrication regime:
(a) $Q_0 = 0$, $\tau_{rx,max} = 0.672$, $\tau_{rx,min} = -0.932$, (b) $Q_0 = 0.6$, $\tau_{rx,max} = 0.958$,
$\tau_{rx,min} = -2.61$ (after Kudish and Panovko [13]). Reprinted with permission
from the ASME.

change in the EHL film thickness γ is related to the change in Q: $\gamma = 0.12$ for
$Q = 0$ and $\gamma = 0.203$ for $Q = 0.6$.

The numerical results for the case of an input oil meniscus with complex
geometry are represented in Figs. 13.35 - 13.38. The changes in the geometry of the input oil meniscus employed in calculations (see Fig. 13.36) are
accompanied by a decrease in the fluid flux through the contact that leads to
a decrease in the EHL film thickness: $\gamma = 0.0834$ for $Q = 0$ and $\gamma = 0.107$ for
$Q = 0.6$. The distribution of pressure $p(x, y)$ varies moderately for $Q = 0$ and
much more significantly for $Q = 0.6$ exhibiting several sharp peaks compared
to that for fully flooded lubrication regime (compare Figs. 13.31, 13.32 and
13.35, 13.36). The pressure maximum p_{max} for $Q = 0.6$ decreases substantially
compared to the case of fully flooded regime: $p_{max} = 4.64$ and $p_{max} = 5.53$,
respectively. Similar changes occur with the distributions of gap $h(x, y)$ (compare Figs. 13.33, 13.34 and 13.37, 13.38). These results resemble the behavior
of the pressure distribution in a long and narrow EHL contact with a notched
input meniscus considered in the next section.

The behavior of the x-component of the sliding frictional stress τ_{sx} is some-

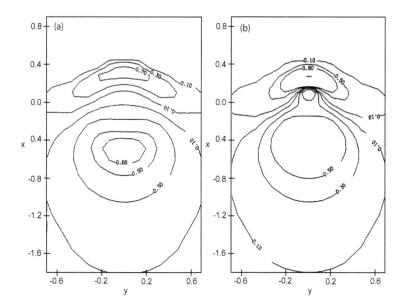

FIGURE 13.40
Level curves of rolling frictional stress distribution τ_{rx} for fully flooded lubrication regime: (a) $Q_0 = 0$, (b) $Q_0 = 0.6$ (after Kudish and Panovko [13]). Reprinted with permission from the ASME.

what similar to the behavior of the pressure distribution in the corresponding cases. The extremum values of τ_{sx} increase as Q increases while condition $s_y = 0$ leads to $\tau_{sy} = 0$.

Figures 13.39 - 13.46 represent the behavior of the components of the rolling frictional stress τ_{rx} and τ_{ry} for the described above cases. The formulas for the components of the rolling frictional stress are as follows:

$$\tau_{rx} = \frac{\gamma h}{2}\frac{\partial p}{\partial x}, \quad \tau_{ry} = \frac{\gamma h}{2}\frac{\partial p}{\partial y}.$$

The behavior of the frictional stress due to rolling $\vec{\tau}_r(x,y)$ is more sophisticated than that for the sliding frictional stress as it is proportional to ∇p. The graphs of $\tau_{rx}(x,y)$ and $\tau_{ry}(x,y)$ are presented in Figs. 13.39 - 13.46. In some portions of the contact, these functions are positive and in others negative. The behavior of $\tau_{rx}(x,y)$ and $\tau_{ry}(x,y)$ is more irregular for the case of $Q = 0.6$ than for $Q = 0$. The shape of these distributions depends on the ge-

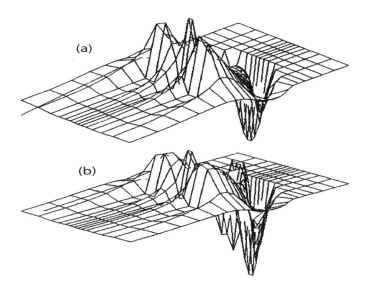

FIGURE 13.41
Rolling frictional stress distribution τ_{rx} for notched input oil meniscus: (a) $Q_0 = 0$, $\tau_{rx,max} = 1.056$, $\tau_{rx,min} = -1.34$, (b) $Q_0 = 0.6$, $\tau_{rx,max} = 1.034$, $\tau_{rx,min} = -2.12$ (after Kudish and Panovko [13]). Reprinted with permission from the ASME.

ometry of the input oil meniscus and the value Q. It may be pointed out that, in general, ranges for $\tau_{rx}(x, y)$ and $\tau_{ry}(x, y)$ increase as Q increases. Besides, the rolling frictional stress distributions are affected by the geometry of the input oil meniscus stronger for $Q = 0$ than for $Q = 0.6$. In particular, the maximum values of the rolling frictional stress may change by 100% or even more (see Figs. 13.39, 13.41 and 13.43, 13.45).

In general, in a lightly loaded EHL contact the rolling frictional stresses prevail over the sliding friction stresses.

13.4.3.2 Octahedral Stresses in the Subsurface Layer

The octahedral subsurface stresses are calculated according to formulas (13.92) using the values for stresses $\sigma_{ij}(x, y, z)$ obtained from formula (13.91). The employed cubature formula is similar to formula (13.98). The calculations are conducted for the typical value of Poisson's ratio $\nu = 0.3$. The calculated results are presented in Figs. 13.47 - 13.52 in the form of level curves in the planes $z = const$, $y = 0$, and $x = const$. The plane $z = const$ ($z = 0.0027$)

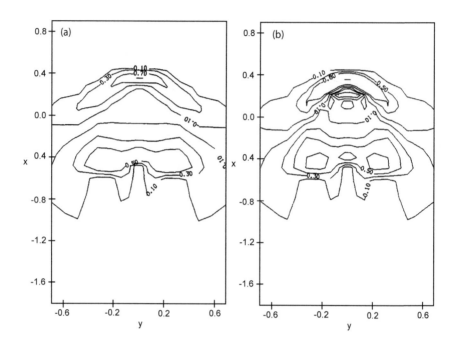

FIGURE 13.42
Level curves of rolling frictional stress distribution τ_{rx} for notched input
oil meniscus: (a) $Q_0 = 0$, (b) $Q_0 = 0.6$ (after Kudish and Panovko [13]).
Reprinted with permission from the ASME.

passes through the node on the z-axis closest to the surface. The location of
the plane $x = const$ ($x = 0.0601$) is related to the location of the pressure
maximum in the contact.

Analysis of the level curves of the normal octahedral stress σ^{oct} leads to
a conclusion that in the subsurface layer σ^{oct} reaches negative extremums.
In case of the fully flooded lubrication regime $\mid \min(\sigma^{oct}) \mid$ increases as
Q increases. For example, $\min(\sigma^{oct}) \approx -2.85$ for $Q = 0$ (reached at $x =
0.0165$, $y = 0$, $z = -0.0447$) and $\min(\sigma^{oct}) \approx -3.54$ for $Q = 0.6$ (reached
at $x = 0.106$, $y = 0$, $z = -0.0447$). In case of the input oil meniscus with
several deep notches that effect is almost negligible: $\min(\sigma^{oct}) \approx -3.01$ for
$Q = 0$ (reached at $x = 0.0165$, $y = 0$, $z = -0.0447$) and $\min(\sigma^{oct}) \approx -2.96$
for $Q = 0.6$ (reached at $x = 0.06$, $y = 0.172$, $z = -0.0447$). However, in the
latter case the level curves of σ^{oct} are deformed in the vicinity of a notch. The
quantitative changes in these distributions are insignificant.

The behavior of the tangential octahedral stress τ^{oct} is of more sophisticated

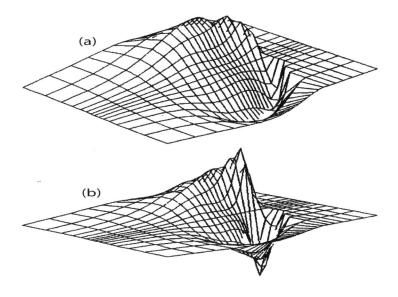

FIGURE 13.43
Rolling frictional stress distribution τ_{ry} for fully flooded lubrication regime:
(a) $Q_0 = 0$, $\tau_{ry,max} = 0.687$, $\tau_{ry,min} = -0.687$, (b) $Q_0 = 0.6$, $\tau_{ry,max} = 1.904$,
$\tau_{ry,min} = -1.904$ (after Kudish and Panovko [13]). Reprinted with permission
from the ASME.

nature (see Figs. 13.47 - 13.52). It can be clearly seen in the z-plane where
several positive local extrema of τ^{oct} exist. Increase in the number of notches in
the input oil meniscus causes an increase in the number of the aforementioned
extrema of τ^{oct} and further deformation of the level curves near notches.
Similar to σ^{oct}, in case of the fully flooded lubrication regime, $\max(\tau^{oct})$
increases as Q increases. For example, $\max(\tau^{oct}) \approx 1.1$ for $Q = 0$ (reached at
$x = 0.106$, $y = 0$, $z = -0.148$) and $\max(\tau^{oct}) \approx 1.47$ for $Q = 0.6$ (reached at
$x = 0.157$, $y = 0$, $z = -0.0825$). The values of the absolute minimum also rise:
$\min(\tau^{oct}) \approx 0.141$ for $Q = 0$ (reached at $x = -0.23$, $y = 0.172$, $z = -0.0447$)
and $\min(\tau^{oct}) \approx 0.18$ for $Q = 0.6$ (reached at $x = -0.3$, $y = 0.172$, $z = -0.0447$).

In case of the input oil meniscus with several deep notches that ef-
fect is almost negligible: $\max(\tau^{oct}) \approx 1.19$ for $Q = 0$ (reached at $x = 0.06$, $y = 0$, $z = -0.26$) and $\max(\tau^{oct}) \approx 1.25$ for $Q = 0.6$ (reached at
$x = 0.157$, $y = -0.172$, $z = -0.148$). However, the absolute minimum of
τ^{oct} continuous to grow as Q increases: $\min(\tau^{oct}) \approx 0.417$ for $Q = 0$ (reached
at $x = -0.0263$, $y = 0$, $z = -0.0228$) and $\min(\tau^{oct}) \approx 0.509$ for $Q = 0.6$

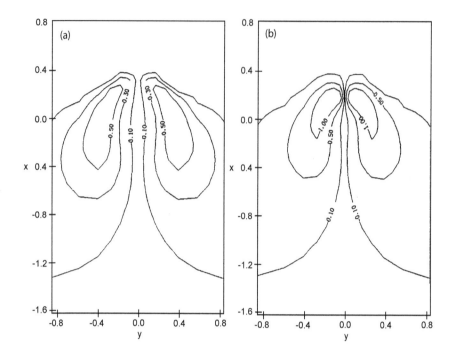

FIGURE 13.44
Level curves of rolling frictional stress distribution τ_{ry} for fully flooded lubrication regime: (a) $Q_0 = 0$, (b) $Q_0 = 0.6$ (after Kudish and Panovko [13]). Reprinted with permission from the ASME.

(reached at $x = 0.157$, $y = 0.103$, $z = -0.0447$).

It should be noted that the locations of the local extrema of functions p, h, τ_{sx}, τ_{sy}, τ_{rx}, τ_{ry}, σ^{oct}, and τ^{oct} substantially depend on the value of the pressure viscosity coefficient Q.

The material presented in this section is only the beginning in understanding of the major mechanisms governing starved lubrication regimes and mixed friction. The numerical results show strong dependence on the geometry of the input oil meniscus and value of the pressure viscosity coefficient Q. For instance, the maximum values of pressure may change by 20% while the rolling frictional stress may change by 100% or even more. Irregularities in the geometry of the input meniscus are reflected in irregularities and complicated structure of pressure, gap, frictional stress, subsurface stresses, and their extrema and may cause significant qualitative and quantitative changes in the distributions of the lubricated contact parameters. Increase in the number of notches in the input oil meniscus causes an increase in the number of extrema

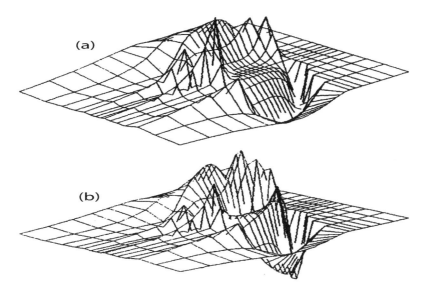

FIGURE 13.45
Rolling frictional stress distribution τ_{ry} for notched input oil meniscus: (a) $Q_0 = 0$, $\tau_{ry,max} = 1.16$, $\tau_{ry,min} = -0.688$, (b) $Q_0 = 0.6$, $\tau_{ry,max} = 1.41$, $\tau_{ry,min} = -1.41$ (after Kudish and Panovko [13]). Reprinted with permission from the ASME.

of τ^{oct}. In practice, this may cause developing of multiple spots of damage (fatigue, wear) on the contact surface and in subsurface layers.

13.5 Formulation and Analysis of a Mixed Lubrication Problem

In practice due to different reasons (starvation, instability of lubrication regime, approaching critical velocities, influence of lubricant surface tension, etc.), the input oil meniscus is not always sufficiently far away from the boundary of purely elastic dry contact. Moreover, very often the meniscus is close to it and changes its configuration in time. Usually, this is reflected on graphs of frictional stresses versus time under conditions seemed to be stationary. The oscillations of frictional stress in such a situation may be dramatically large.

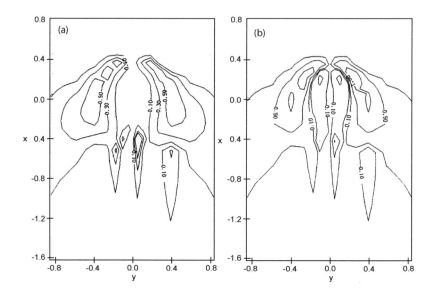

FIGURE 13.46
Level curves of rolling frictional stress distribution τ_{ry} for notched input
oil meniscus: (a) $Q_0 = 0$, (b) $Q_0 = 0.6$ (after Kudish and Panovko [13]).
Reprinted with permission from the ASME.

Optical interference methods allow to observe this behavior of lubrication
film thickness in relation with the configuration (location) of the input oil
meniscus. It was registered that in such cases the profile of the lubrication film
thickness is far from the one which may be observed for truly stationary or
fully flooded conditions and it depends on the shape of the inlet oil meniscus.
In particular, due to the inlet meniscus shape the gap between lubricated
contacting solids may approach or reach zero in some regions of contact area.
    This phenomenon can be easily understood from the consideration of solu-
tions of a plane EHL problem. The general behavior of lubrication film thickness in a plane lubricated contact is well known. If oil meniscus is outside of
the Hertzian dry elastic contact, the lubrication film thickness is greater than
zero and it vanishes as the meniscus approaches the Hertzian elastic contact.
Therefore, if the contact is long and narrow in the directions perpendicular
and parallel to the direction of lubricant flow, respectively, in different sections
of the contact area, the situation is similar to the described one for a plane
EHL problem. It depends on relative location of the input oil meniscus and
boundary of the Hertzian contact. In particular, it can result in the existence

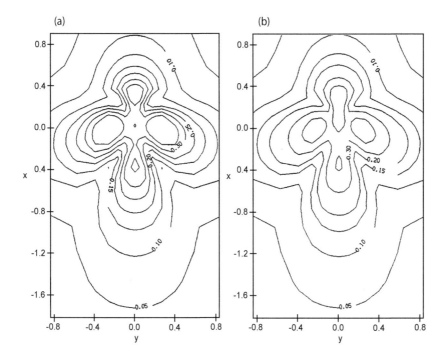

FIGURE 13.47
Level curves of tangential octahedral stress τ^{oct} in plane $z = 0.0027$ for fully flooded lubrication regime: (a) $Q_0 = 0$, (b) $Q_0 = 0.6$ (after Kudish and Panovko [13]). Reprinted with permission from the ASME.

of alternating lubricated and non-lubricated stripes in the contact area.

In this section we provide a mathematical formulation of this problem taking into account the existence of the zones with different types of (dry and fluid) friction. An EHL problem for partially lubricated solids is formulated and analyzed. The consideration of such a problem is unavoidable under mixed friction conditions when the elastic solids are separated by a lubricant film in one part of the contact area while in the other part they are in direct elastic contact without presence of a lubricant. A dry frictional contact occurs with the presence of slippage and adhesion zones. Such severe conditions appear in most practical cases (bearings, gears, etc.). The problem is reduced to a system of alternating nonlinear Reynolds' equation and integral equations and inequalities valid in the contact area. The inlet contact boundary is considered to be known and located close to the boundary of the purely elastic contact subjected to the same conditions. The location of the exit contact boundary must be determined from the problem solution as well as a number of internal

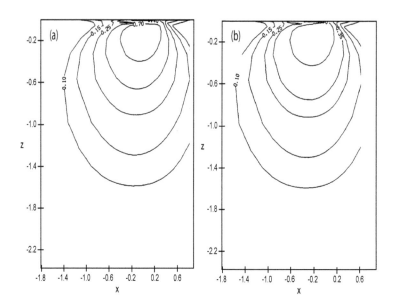

FIGURE 13.48
Level curves of tangential octahedral stress τ^{oct} in plane $y = 0$ for fully flooded
lubrication regime: (a) $Q_0 = 0$, (b) $Q_0 = 0.6$ (after Kudish and Panovko [13]).
Reprinted with permission from the ASME.

contact boundaries separating the zones of dry and lubricated contact. The
conditions of continuity of (dry and fluid) friction stresses on these internal
boundaries are formulated. The special cases of this problem formulation are
the classic contact problem of elasticity with dry friction taking into account
slippage and adhesion zones and the EHL problem for completely lubricated
contact.

The theoretical analysis of a partially lubricated contact is given for the
case of a narrow contact significantly elongated in the direction orthogonal to
the direction of lubricant motion. Under these conditions, in the main part
of the contact region the problem analysis can be reduced to the analysis of
plane elastic and EHL problems. These problems can be successfully analyzed
by asymptotic methods. The specific features of the problem solution, i.e.,
the location and magnitude of dry and lubricated zones, friction stresses,
and pressure are mainly dependent on the configuration of the inlet contact
boundary.

The particular physical phenomena (such as instability of lubrication
regime, approaching critical velocities, influence of lubricant surface tension,
etc.) are not considered here.

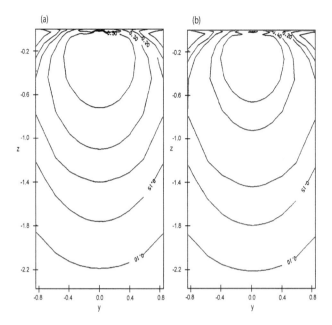

FIGURE 13.49
Level curves of tangential octahedral stress τ^{oct} in plane $x = 0.0601$ for fully flooded lubrication regime: (a) $Q_0 = 0$, (b) $Q_0 = 0.6$ (after Kudish and Panovko [13]). Reprinted with permission from the ASME.

13.5.1 Problem Formulation

Let us assume that two moving (rolling) solids with smooth surfaces made of the same elastic material are lubricated by a Rivlin-type incompressible non-Newtonian1 viscous fluid under isothermal conditions. Consideration of non-Newtonian fluid is essential because it is necessary to take into account transition from fluid to dry friction mechanisms. The lubrication layer is considered to be small compared with the characteristic sizes of the contact region and solids' sizes. The solids form a concentrated (non-conformal) contact and they experience an external compressive force (see Fig. 13.53). Under these assumptions the solids can be replaced by two contacting elastic half-spaces. The further analysis is based on [33].

Let us introduce a moving coordinate system: the z-axis passes through the centers of curvature of the solids and the xy-plane is equidistant from the solid surfaces.

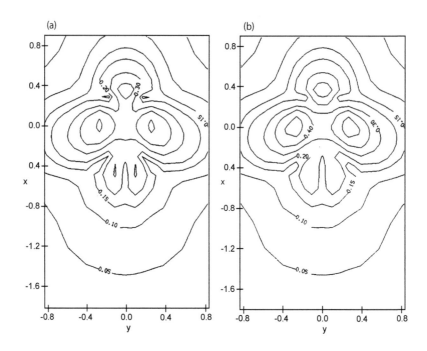

FIGURE 13.50
Level curves of tangential octahedral stress τ^{oct} in plane $z = 0.0027$ for notched input oil meniscus: (a) $Q_0 = 0$, (b) $Q_0 = 0.6$ (after Kudish and Panovko [13]). Reprinted with permission from the ASME.

It is assumed that the solids are involved in steady slow motion with linear velocities $\vec{u}_1 = (u_1, v_1)$ and $\vec{u}_2 = (u_2, v_2)$. The slippage velocity $\vec{s}$ is assumed to be small compared with the rolling velocity $0.5(\vec{u}_1 + \vec{u}_2)$. The inertial forces in lubrication layer are considered to be small compared with viscous ones.

Under these assumptions the tangential stress vector $\vec{\tau}(x, y)$ in the lubrication layer is proportional to the gradient of the lubricant linear velocity $\frac{\partial \vec{u}(x,y)}{\partial z}$

$$\mu \frac{\partial u(x,y)}{\partial z} = F[\tau_{xz}(x, y)], \ \mu \frac{\partial v(x,y)}{\partial z} = F[\tau_{yz}(x, y)] \ or$$

$$\tau_{xz}(x, y) = \Phi[\mu \frac{\partial u(x,y)}{\partial z}], \ \tau_{yz}(x, y) = \Phi[\mu \frac{\partial v(x,y)}{\partial z}],$$

(13.99)

where $\mu = \mu(p)$ is the lubricant viscosity at pressure p, F and Φ are given inverse to each other monotonic smooth enough functions describing the lubricant rheology, $F(0) = \Phi(0) = 0$. The regions of a dry contact are represented

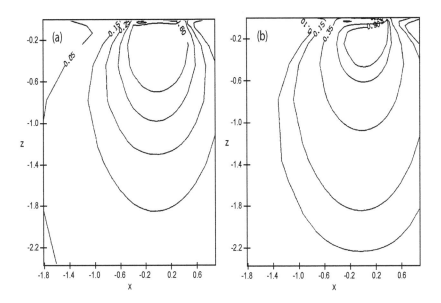

FIGURE 13.51
Level curves of tangential octahedral stress τ^{oct} in plane $y = 0$ for notched
input oil meniscus: (a) $Q_0 = 0$, (b) $Q_0 = 0.6$ (after Kudish and Panovko [13]).
Reprinted with permission from the ASME.

by the adhesion and slippage zones in which the vector of the relative solid
slippage $\vec{s}(x, y)$ is zero and different from zero, respectively. Moreover, the
friction stress and slippage in these zones satisfy Coulomb's law

$$\vec{\tau} = \lambda p \frac{\vec{s}}{|\vec{s}|} \ for \ |\vec{s}| > 0; \ |\vec{\tau}| \le \lambda p \ for \ |\vec{s}| = 0, \tag{13.100}$$

where p is the contact pressure and $\lambda = \lambda(p, |\vec{s}|)$ is the coefficient of dry
friction.

13.5.1.1 Fluid Film Lubrication

First, let us derive the equations governing the considered process in the
lubricated regions. Under conditions of a thin lubricant layer and slow motion
the equations of fluid motion are [16]

$$\frac{\partial \tau_{xz}}{\partial z} = \frac{\partial p}{\partial x}, \ \frac{\partial \tau_{yz}}{\partial z} = \frac{\partial p}{\partial y}, \ \frac{\partial p}{\partial z} = 0, \ \nabla \cdot \vec{w} = 0, \tag{13.101}$$

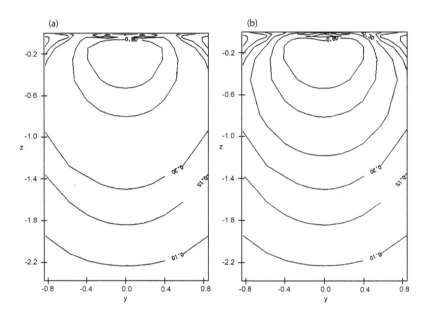

FIGURE 13.52
Level curves of tangential octahedral stress τ^{oct} in plane $x = 0.0601$ for notched input oil meniscus: (a) $Q_0 = 0$, (b) $Q_0 = 0.6$ (after Kudish and Panovko [13]). Reprinted with permission from the ASME.

where $\vec{w} = (u, v, w)$ is the vector of fluid velocity. The latter leads to the relationship $p = p(x, y)$.

Taking into account the assumption of no slippage at the contact surfaces and that the gradient of gap $h(x, y)$ between the contacting solids and $\mid \vec{s} \mid / \mid \vec{u}_1 + \vec{u}_2 \mid$ are small in comparison with 1 the components of the fluid particle velocity $\vec{w} = (u, v, w)$ at the contact surfaces are

$$\vec{u} = \vec{u}_1, \ w = -\tfrac{1}{2}\vec{u}_1 \cdot \nabla h \ for \ z = -\tfrac{h}{2},$$

$$\vec{u} = \vec{u}_2, \ w = \tfrac{1}{2}\vec{u}_2 \cdot \nabla h \ for \ z = \tfrac{h}{2}. \tag{13.102}$$

Integrating the continuity equation $\nabla \cdot \vec{w} = 0$ with respect to z from $-h/2$ to $h/2$ and using boundary conditions (13.102), we obtain (Q_x and Q_y are

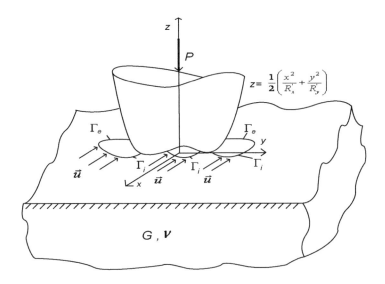

FIGURE 13.53
The general view of the solids involved in a contact with mixed lubrication.

the x- and y-components of the fluid flux)

$$\frac{\partial Q_x}{\partial x} + \frac{\partial Q_y}{\partial y} = 0, \ \vec{Q} = (Q_x, Q_y) = \int\limits_{-h/2}^{h/2} \vec{u}dz. \tag{13.103}$$

Integrating the first two equations in (13.99) with boundary conditions from (13.102), we find

$$u = u_1 + \frac{1}{\mu} \int\limits_{-h/2}^{z} F(f_x + t\frac{\partial p}{\partial x})dt, \ v = v_1 + \frac{1}{\mu} \int\limits_{-h/2}^{z} F(f_y + t\frac{\partial p}{\partial y})dt, \tag{13.104}$$

where $\vec{f} = (f_x, f_y)$ is the vector of the sliding frictional stress the components of which satisfy the equations

$$\frac{1}{\mu} \int\limits_{-h/2}^{h/2} F(f_x + t\frac{\partial p}{\partial x})dt = s_x, \ \frac{1}{\mu} \int\limits_{-h/2}^{h/2} F(f_y + t\frac{\partial p}{\partial y})dt = s_y. \tag{13.105}$$

Finally, using equations (13.103) and (13.104), we derive the generalized Reynolds equation

$$\frac{\partial}{\partial x}\left\{\frac{1}{\mu} \int\limits_{-h/2}^{h/2} zF(f_x + z\frac{\partial p}{\partial x})dz\right\} + \frac{\partial}{\partial y}\left\{\frac{1}{\mu} \int\limits_{-h/2}^{h/2} zF(f_y + z\frac{\partial p}{\partial y})dz\right\} \tag{13.106}$$

$$= \tfrac{1}{2}(\vec{u}_1 + \vec{u}_2) \cdot \nabla h \ if \ h(x, y) > 0,$$

where the fluid flux components can be calculated from

$$Q_x = h\tfrac{u_1 + u_2}{2} - \tfrac{1}{\mu} \int\limits_{-h/2}^{h/2} zF(f_x + z\tfrac{\partial p}{\partial x})dz,$$

(13.107)

$$Q_y = h\tfrac{v_1 + v_2}{2} - \tfrac{1}{\mu} \int\limits_{-h/2}^{h/2} zF(f_y + z\tfrac{\partial p}{\partial y})dz.$$

Obviously, equation (13.105)-(13.107) hold only in the lubricated zones of the contact region.

Now, let us consider the difference in elastic displacements $\triangle \vec{W} = (\triangle U, \triangle V, \triangle W)$ of the surface points of the two (upper and lower) contact solids. The well–known formulas give [34]

$$\triangle U = \tfrac{1}{2\pi G} \int\limits_{\Omega_\tau} \int \{\tfrac{1 - \nu \sin^2 \theta}{R}(\tau_{xz}^+ - \tau_{xz}^-) + \tfrac{\nu \sin \theta \cos \theta}{R}(\tau_{yz}^+ - \tau_{yz}^-)\}d\xi d\eta,$$

$$\triangle V = \tfrac{1}{2\pi G} \int\limits_{\Omega_\tau} \int \{\tfrac{\nu \sin^\theta \cos \theta}{R}(\tau_{xz}^+ - \tau_{xz}^-) + \tfrac{1 - \nu \cos^2 \theta}{R}(\tau_{yz}^+ - \tau_{yz}^-)\}d\xi d\eta,$$

$$\triangle W = \tfrac{1 - 2\nu}{4\pi G} \int\limits_{\Omega_\tau} \int \{\tfrac{\cos \theta}{R}(\tau_{xz}^+ + \tau_{xz}^-) + \tfrac{\sin \theta}{R}(\tau_{yz}^+ + \tau_{yz}^-)\}d\xi d\eta$$

(13.108)

$$+\tfrac{1 - \nu}{\pi G} \int\limits_{\Omega_p} \int \tfrac{p}{R}d\xi d\eta, \ R = \sqrt{(x - \xi)^2 + (y - \eta)^2},$$

$$\sin \theta = \tfrac{y - \eta}{R}, \ \cos \theta = \tfrac{x - \xi}{R},$$

where Ω_p and Ω_τ are the normal and tangential contact area at the boundaries of which pressure p and the frictional stress $\vec{\tau}$ vanish (regions Ω_p and Ω_τ not necessarily coincide), G and ν are the shear elastic modulus and Poisson's ratio of the solid material, $G = \tfrac{E}{2(1+\nu)}$ (E is Young's modulus), superscripts + and − are related to the upper and lower solids, respectively. Moreover, $\tau^+ = (\tau_{xz}^+, \tau_{yz}^+)$ and $\tau^- = (\tau_{xz}^-, \tau_{yz}^-)$ are the tangential stresses acting on the surfaces of the upper and lower solids, which for dry contact zones satisfy conditions (13.100) and for lubricated regions are determined by the equations (see equations (13.99), (13.103), and (13.106))

$$\vec{\tau}^\pm = \mp\vec{f} - \tfrac{h}{2}\nabla p.$$

(13.109)

Using well–known considerations applied to contact problems of elasticity and the expression for difference in vertical displacements $\triangle W$, we get the relation for gap h between the contacting solids

$$h = h_0 + \frac{x^2}{2R_x} + \frac{y^2}{2R_y} + \frac{1-\nu}{\pi G} \int\int_{\Omega_p} \frac{pd\xi d\eta}{R}$$

$$- \frac{1-2\nu}{4\pi G} \int\int_{\Omega_\tau} \{\cos\theta \frac{\partial p}{\partial x} + \sin\theta \frac{\partial p}{\partial y}\} \frac{hd\xi d\eta}{R}, \tag{13.110}$$

where h_0 is an unknown constant, R_x and R_y are the effective radii of the surface curvature for the contact solids in the directions of the x- and y-axes, respectively. Equation (13.110) can be simplified for the case of an incompressible elastic material for which $\nu = 1/2$ and

$$h = h_0 + \frac{x^2}{2R_x} + \frac{y^2}{2R_y} + \frac{1-\nu}{\pi G} \int\int_{\Omega_p} \frac{pd\xi d\eta}{R}. \tag{13.111}$$

13.5.2   Fluid Friction in Lightly and Heavily Loaded Lubricated Contacts

Let us consider the case of a lightly loaded contact of elastic solids lubricated by an incompressible fluid with non-Newtonian rheology. Scaling the problem similar to Sections 10.3 and 13.4 and applying regular perturbation methods would allow to obtain the following structural formula for the dimensionless film thickness $\gamma = \frac{\vartheta^2}{2R_x} h_c$ (compare to formulas (10.38) and (10.43)):

$$\gamma = \gamma_0 + \frac{1}{V}\gamma_1 + \dots, \tag{13.112}$$

where $V = \frac{8\pi R_x^2 E'}{3P\vartheta^3}$ and $\vartheta = \vartheta(\mu_a, u_1, u_2, R', P, \dots)$ is a certain dimensionless parameter that depends on the fluid specific rheology and is independent from the effective elastic modulus E' ($1/E' = 1/E_1' + 1/E_2'$, $E_j' = E_j/(1 - \nu_j^2)$, $j = 1, 2$) while parameters γ_0 and γ_1 are independent of V and may depend on the viscosity pressure coefficient α (assuming that the lubricant viscosity $\mu = \mu_a e^{\alpha p}$, μ_a is the ambient lubricant viscosity), the ratios of velocities $(v_1 + v_2)/(u_1 + u_2)$ and surface radii R_x/R_y, the slide-to-roll ratio $s_0 = 2 \mid \vec{u}_2 - \vec{u}_1 \mid / \mid \vec{u}_2 + \vec{u}_1 \mid$, the lubricant rheology, and the shape of the inlet oil meniscus Γ_i. In case of Newtonian rheology $\vartheta = \frac{P}{8\pi\mu_a|\vec{u}_1+\vec{u}_2|R_x}$.

Let us consider the case of a heavily loaded contact of elastic solids lubricated by an incompressible fluid with Newtonian rheology. By scaling the problem formulated above using the Hertzian semi-axes a_H, b_H, and the maximum pressure p_H we will see that the problem solution depends on the dimensionless parameters V (see (13.130))

$$V = \frac{24\mu_a|\vec{u}_1+\vec{u}_2|R_x^2}{p_H a_H^3}, \quad Q = \alpha p_H,$$

where it is assumed that the lubricant viscosity $\mu = \mu_a e^{\alpha p}$ (μ_a is the ambient lubricant viscosity and α is the viscosity pressure coefficient).

As before, for heavily loaded lubrication regimes we have $V \ll 1$ and/or $Q \gg 1$. Unfortunately, the level of asymptotic analysis, which can be conducted for this problem, is limited due to its complexity. However, using asymptotic estimates of the problem parameter orders of magnitudes, similar to the ones used in line EHL contacts (see Sections 10.4 and 10.6), we still can get some useful information. In particular, for pre-critical starved lubrication regimes (see definitions (10.140), (10.122), and (10.124)) for the dimensionless central film thickness $H_0 = \frac{2R_x h_c}{a_H^2}$ (h_c is the dimensional central film thickness) we get the formula (compare with formula (10.120))

$$H_0 = A(V\epsilon_q^2)^{1/3}, \quad A = O(1), \quad \epsilon_q \ll V^{2/5}, \tag{13.113}$$

while for pre-critical fully flooded lubrication regimes (see definitions (10.123) and (10.125)) the formula for the central film thickness H_0 will get the form (compare with formula (10.128))

$$H_0 = AV^{3/5}, \quad A = O(1), \quad \epsilon_q = V^{2/5}. \tag{13.114}$$

In formulas (13.113) and (13.114), ϵ_q is the characteristic distance between the inlet oil meniscus and the boundary of the Hertzian region, the values of constants A are independent of V but may depend on dimensionless viscosity pressure coefficient Q, the ratios of velocities $(v_1 + v_2)/(u_1 + u_2)$ and surface radii R_x/R_y, and the shape of the inlet meniscus Γ_i of region Ω_p.

For over-critical lubrication regimes (see definition (10.141)), in cases of starved and scanty lubrication (see definition (10.122)) the central film thickness is determined by the formula (compare with formula (10.205))

$$H_0 = A(VQ\epsilon_q^{5/2})^{1/3}, \quad A = O(1), \quad \epsilon_q \ll (VQ)^{1/2}, \tag{13.115}$$

while for fully flooded lubrication regimes (see definition (10.123)) the formula for the central film thickness H_0 will get the form (compare with formula (10.213))

$$H_0 = A(VQ)^{3/4}, \quad A = O(1), \quad \epsilon_q = (VQ)^{1/2}. \tag{13.116}$$

In formulas (13.115) and (13.116), ϵ_q is the characteristic distance between the inlet oil meniscus and the boundary of the Hertzian region, the values of constants A are independent of V and Q but may depend on the ratios of velocities $(v_1 + v_2)/(u_1 + u_2)$ and surface radii R_x/R_y, and the shape of the inlet meniscus Γ_i of region Ω_p.

In formulas (13.113)-(13.116), the particular values of coefficients A (different for different formulas) can be determined numerically or experimentally.

Similar results can be obtained for the cases of lubricants with non-Newtonian rheology (see Section 10.15).

13.5.3   Boundary Friction

Let us derive the equations governing the process in the dry zones. First, we will consider the normal problem. In the dry zones the gap between the contact solids is zero, i.e.,

$$h_0 + \frac{x^2}{2R_x} + \frac{y^2}{2R_y} + \frac{1-\nu}{\pi G} \int\limits_{\Omega_p}\int \frac{pd\xi d\eta}{R}$$

$$-\frac{1-2\nu}{4\pi G} \int\limits_{\Omega_\tau}\int \{\cos\theta \frac{\partial p}{\partial x} + \sin\theta \frac{\partial p}{\partial y}\} \frac{hd\xi d\eta}{R} = 0 \; for \; h(x,y) = 0.$$

(13.117)

Under the conditions of a slow stationary motion and small slippage, the equation for the slippage velocity $\vec{s}$ can be obtained by applying the differential operator $0.5(\vec{u}_1 + \vec{u}_2) \cdot \nabla$ to both sides of the first two equations in (13.108). That operation means differentiation with respect to time. Therefore, the equation for $\vec{s}$ can be expressed in the form

$$\vec{s} = -B(\vec{\tau}) + \vec{u}_2 - \vec{u}_1,$$

$$B(\vec{\tau}) = \frac{u_1+u_2}{2} \int\limits_{\Omega_\tau}\int \mathbf{B}_x \vec{\tau}(\xi,\eta)d\xi d\eta + \frac{v_1+v_2}{2} \int\limits_{\Omega_\tau}\int \mathbf{B}_y \vec{\tau}(\xi,\eta)d\xi d\eta,$$

$$\mathbf{B}_x = \mathbf{D}_x(x-\xi, y-\eta), \quad \mathbf{B}_y = \mathbf{D}_y(x-\xi, y-\eta),$$

(13.118)

$$B_{x11} = -\frac{\cos\theta(3\nu\sin^2\theta-1)}{\pi G R^2}, \quad B_{x12} = B_{x21} = -\frac{\nu\sin\theta(1-3\cos^2\theta)}{\pi G R^2},$$

$$B_{x22} = -\frac{\cos\theta(\nu-1-3\nu\sin^2\theta)}{\pi G R^2}, \quad B_{y11} = -\frac{\nu\sin\theta(\nu-1-3\cos^2\theta)}{\pi G R^2},$$

$$B_{y12} = B_{y21} = -\frac{\nu\cos\theta(1-3\sin^2\theta)}{\pi G R^2}, \quad B_{y22} = -\frac{\sin\theta(3\nu\cos^2\theta-1)}{\pi G R^2},$$

Here $\vec{\tau}$ is the sliding frictional stress that coincides with the frictional stress in the dry zones (determined by equations (13.100)) and is equal to the sliding frictional stress $\vec{f}$ in the lubricated zones (determined by equations (13.105)).

13.5.3.1 Partial Lubrication

When in a contact region in some zones contact takes place directly between smooth surfaces or asperities while in other zones of the contact solid surfaces are separated by fluid film partial lubrication (sometimes referred to as "mixed lubrication") occurs. The behavior of a contact in partial lubrication regime is governed by a combination of boundary friction and fluid film effects.

Everywhere in the contact region contact pressure $p(x,y)$ must be nonnegative

$$p(x,y) \geq 0.$$

(13.119)

Several additional conditions must be imposed on the parameters characterizing a partially lubricated contact. One of them is the static condition

$$\int\limits_{\Omega_p}\int p(x,y)dxdy = P,$$

(13.120)

where P is the external force applied to the contact solids. Everywhere at the contact boundary Γ_p of the region Ω_p

$$p \mid_{\Gamma_p} = 0,$$

$$\text{(13.121)}$$

$$\text{parts of } \Gamma_p \text{ where } h \mid_{\Gamma_p} = 0 \text{ are unknown.}$$

Note that boundary condition (13.121) is the local one. Therefore, some pieces of the contact boundary Γ_p are known and others are unknown.

On the other hand, on the boundary Γ_τ of the region Ω_τ the tangential stress should vanish, i.e.,

$$\vec{\tau} \mid_{\Gamma_\tau} = \vec{0}. \qquad \text{(13.122)}$$

Also, the problem solution must satisfy the boundary conditions for pressure p different at different zones of the contact region boundary belonging whether to lubricated or dry zones, inlet or exit zones of the contact. Suppose that Γ_i and Γ_e are the inlet and exit parts of the contact region Ω_p boundary, which are given and unknown in advance, respectively, $\Gamma_p = \Gamma_i \bigcup \Gamma_e$. The boundary condition for the dry contact zone boundary is given in (13.121). Let us consider the lubricated contact zones. In this case different boundary conditions are required. Here are the conditions that we need to impose on the problem solution:

$$\Gamma_i \text{ is given if } h \mid_{\Gamma_i} > 0 \text{ and } \vec{Q} \cdot \vec{n} \mid_{\Gamma_i} < 0,$$

$$\text{(13.123)}$$

$$\frac{dp}{d\vec{n}} \mid_{\Gamma_e} = 0 \text{ if } h \mid_{\Gamma_e} > 0 \text{ and } \vec{Q} \cdot \vec{n} \mid_{\Gamma_e} \geq 0,$$

where $\vec{n}$ is the vector of the external normal to boundary Γ_p.

The presented boundary conditions (13.121) and (13.123) must be combined with condition (13.119).

Now, the formulation of the boundary conditions imposed on the external boundaries of the contact is complete. In the case of mixed friction conditions, the contact area may contain some internal boundaries between dry and lubricated zones, which, apparently, coincide with flow lines of the lubricant. Therefore, it is necessary to formulate the boundary conditions that must be satisfied on the internal boundaries. As we know the frictional stresses for lubricated and dry conditions are described by equations (13.99) and (13.100) (see also (13.109)). These additional boundary conditions must be imposed on the frictional stress and represent the requirement of the frictional stress continuity on the internal boundaries. Suppose l_i and $\vec{n}_i = (n_{xi}, n_{yi})$ are the i-th internal boundary and a unit normal vector to it, respectively. Therefore, the continuity condition can be expressed in the form

$$\lim_{\epsilon \to 0} \{ \vec{\tau}(x - \epsilon n_{xi}, y - \epsilon n_{yi}) - \vec{\tau}(x + \epsilon n_{xi}, y + \epsilon n_{yi}) \} = \vec{0}$$

$$\text{(13.124)}$$

$$for \ (x, y) \in l_i.$$

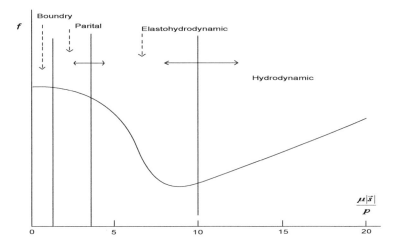

FIGURE 13.54

Variation of friction coefficient with film parameter $\Lambda = \frac{\mu|\vec{s}|}{p}$.

Let us consider the physical nature of the transition from fluid to dry friction. A fundamental experimental study [35] shows that a continuous transition of the frictional stress from fluid friction to dry friction occurs in a small number of fluid molecular layers on a solid surface (see Fig. 13.54).

The specific features of this transition have not been studied extensively, and they depend on the adsorption properties of the lubricant-solid surface interface. We will assume that the equations of the continuous fluid mechanics such as (13.109) can be extended on thin molecular layer of fluid. Let us consider the internal boundary between lubricated and dry regions on which slippage occurs, i.e., $\mid \vec{s} \mid> 0$. Then using equation (13.124) together with (13.100) and (13.109), we obtain

$$\lim_{h \to 0} f(p, h, \mu, \mid \vec{s} \mid) = \lambda p \frac{\vec{s}}{|\vec{s}|} \; for \; \mid \vec{s} \mid> 0. \tag{13.125}$$

It means that due to adsorption effects, the lubricant boundary layers show the properties of structural anisotropic fluids, and in such thin (almost monomolecular) layers lubricant viscosity $\mu = \mu(p, h, \mid \vec{s} \mid)$, i.e., besides pressure p it also depends on film thickness h and sliding speed $\mid \vec{s} \mid$. For a particular case of Newtonian fluid with $F(x) = x$, equation (13.125) becomes more transparent

$$\lim_{h \to 0} \frac{\mu}{h} = \frac{\lambda p}{|\vec{s}|} \; for \; \mid \vec{s} \mid> 0. \tag{13.126}$$

Therefore, we have to assume that the fluid viscosity μ in thin lubrication films is proportional to the film thickness h. That may be related to the property of lubricity of lubricants in thin layers.

Finally, the problem for partially lubricated contact is completely described by equations (13.105), (13.106), (13.109), (13.110) for the lubricated zones, and by equations (13.100), (13.117), (13.118) for the dry zones together with additional conditions (13.119)-(13.124). In addition to these equations and inequalities, the relationships for functions F, Φ, μ, λ, and the inlet boundary Γ_i must be given. It is important that the functions of lubricant viscosity $\mu = \mu(p, h, |\vec{s}|)$ and the coefficient of dry friction $\lambda = \lambda(p, |\vec{s}|)$ satisfy the continuity relationships on the boundaries l_i of dry and lubricated zones. The problem solution consists of functions $p(x, y)$, $h(x, y)$, $\vec{\tau}(x, y)$, $\vec{s}(x, y)$, the exit boundary Γ_e and some parts of the inlet boundary Γ_i of the region Ω_p, which represent the boundaries of the dry zones, the internal boundaries l_i between the lubricated and dry contact zones, the boundary of region Ω_τ, and the central film thickness h_0.

Obviously, if everywhere in the contact region $h(x, y) > 0$, the formulated problem is reduced to the EHL problem, which, for incompressible materials ($\nu = 1/2$), coincides with the EHL problems considered in the preceding chapters. If $h(x, y) = 0$, then the problem is reduced to an analog of a Hertz problem. In this case for incompressible materials ($\nu = 1/2$), the normal problem precisely coincides with the Hertzian one while the tangential problem coincides with the one considered for a dry contact with stick and slip in [36].

In the case of small influence of elastic deformations on slippage, the normal and tangential problems can be decoupled and solved separately. In this case it means that $\vec{s} = \vec{u}_2 - \vec{u}_1 + o(\vec{u}_2 - \vec{u}_1)$, and the normal problem becomes independent of the tangential problem. Thus, first, must be solved the normal problem for $p(x, y)$ and $h(x, y)$, and after that must be solved the tangential problem for $\vec{\tau}(x, y)$ and $\vec{s}(x, y)$.

After the problem is solved, the frictional force $\vec{F}_T^\pm$ applied to the upper/lower contact surfaces can be calculated according to the formulas

$$\vec{F}_T^\pm = \int\int_{\Omega_\tau} \vec{\tau}^\pm d\xi d\eta = \vec{F}_f^\pm + \vec{F}_d^\pm, \quad \vec{F}_f^\pm = \mp \vec{F}_{fs} - \vec{F}_{fr},$$

$$\vec{F}_{fs} = \int\int_{\Omega_{\tau f}} \vec{f} d\xi d\eta, \quad \vec{F}_{fr} = \tfrac{1}{2} \int\int_{\Omega_{\tau f}} h\nabla p d\xi d\eta, \tag{13.127}$$

$$\vec{F}_d^\pm = \int\int_{\Omega_{\tau d}} \vec{\tau}^\pm d\xi d\eta,$$

where $\Omega_{\tau f}$ and $\Omega_{\tau d}$ are the zones of fluid ($h(x, y) > 0$) and dry ($h(x, y) = 0$) friction in the contact region Ω_τ. Here subscripts f and d indicate fluid and dry friction conditions while subscripts s and r indicate sliding and rolling frictional conditions.

As it was mentioned above the shape of the inlet meniscus depends on many different factors which in most cases can not be registered. Therefore, it is appropriate to treat the shape of the inlet meniscus Γ_i as a random function depending on several parameters, for example, $\Gamma_i(a_1, a_2, \ldots, a_n)$. In majority of practical cases it is important to know the average values of such characteristics of the contact as pressure, frictional stress, slippage, and frictional force. Assuming that the probability density function $f(a_1, a_2, \ldots, a_n)$ is known it is easy to do averaging by integrating the function/constant to be averaged over the domain A of the set of parameters $a_1, a_2, \ldots, a_n$.

13.5.4 Partial Lubrication of a Narrow Contact

Let us consider the case of an incompressible elastic material ($\nu = 0.5$) and a long in the direction of the y-axis and narrow in the direction of the x-axis partially lubricated contact. This condition is equivalent to the inequality $R_x/R_y \ll 1$. It means that the eccentricity of the Hertzian (dry) contact ellipse $e = \sqrt{1 - \delta^2} \approx 1$ and it satisfies the equation

$$\frac{\delta^2 D(e)}{K(e) - D(e)} = \frac{R_x}{R_y}, \quad D(e) = \frac{K(e) - E(e)}{e^2}, \tag{13.128}$$

where δ is the relative width of the contact, $\delta = a_H/b_H$ (a_H and b_H are the smaller and larger semi-axes of the Hertzian contact, $a_H = b_H\sqrt{1 - e^2}$, $b_H = \sqrt[3]{\frac{3E(e)}{1-e^2}\frac{PR_x}{\pi E'(1+\rho)}}$, $p_H a_H b_H = \frac{3P}{2\pi}$, p_H is the maximum Hertzian pressure), $K(e)$ and $E(e)$ are the complete elliptic integrals of the first and second kind.

It can be shown that $\delta \ll 1$ for $R_x/R_y \ll 1$. For $\delta \ll 1$ we have $e - 1 \ll 1$. We will use the asymptotic expansions [37, 38] for the integrals over an elongated in the direction of the y-axis narrow contact region $\Omega = \{(x, y) \mid a(y) \leq x \leq c(y), -\delta^{-1} \leq y \leq \delta^{-1}\}$

$$\iint_\Omega \frac{p(\xi,\eta)d\xi d\eta}{\sqrt{(x-\xi)^2+(y-\eta)^2}} = 2\int_a^c p(\xi,y)\ln\frac{1}{|\xi-x|}d\xi$$

$$+F_p(y)\ln[4(\delta^{-2} - y^2)] + \int_{-\delta^{-1}}^{\delta^{-1}} \frac{F_p(\eta)-F_p(y)}{|\eta-y|}d\eta + O(\delta^2\ln\tfrac{1}{\delta}),$$

$$\tag{13.129}$$

$$\iint_\Omega \frac{(x-\xi)p(\xi,\eta)d\xi d\eta}{\sqrt{(x-\xi)^2+(y-\eta)^2}} = \pi\int_a^c p(\xi,y)\text{sign}(x-\xi)d\xi + O(\delta),$$

$$\iint_\Omega \frac{(y-\eta)p(\xi,\eta)d\xi d\eta}{\sqrt{(x-\xi)^2+(y-\eta)^2}} = \int_{-\delta^{-1}}^{\delta^{-1}} \frac{F_p(\eta)d\eta}{y-\eta} + O(\delta), \quad F_p(y) = \int_a^c p(\xi,y)d\xi.$$

Then, for $\delta \ll 1$ let us introduce the following dimensionless variables (further primes are omitted at the dimensionless variables)

$$x' = \frac{x}{a_H}, \ y' = \frac{y}{b_H}, \ z' = \frac{z}{h_e}, \ F' = \frac{2h_e}{\mu_a|\vec{u}_1+\vec{u}_2|}F, \ \mu' = \frac{\mu}{\mu_a},$$

$$\tag{13.130}$$

$$p' = \frac{p}{p_H}, \ (\tau', f') = \frac{a_H^2}{\mu_a|\vec{u}_1+\vec{u}_2|R_x}(\tau, f), \ \vec{s}' = \frac{\vec{s}}{|\vec{u}_1+\vec{u}_2|},$$

$$H_0 = \frac{2R_x h_e}{a_H^2}, \quad V = \frac{24\mu_a|\vec{u}_1+\vec{u}_2|R_x^2}{p_H a_H^3}, \quad \theta = \frac{p_H a_H^2}{\mu_a|\vec{u}_1+\vec{u}_2|R_x}, \quad \eta = \frac{2}{\pi\delta^2 D}\frac{a_H}{R_x},$$

where μ_a is the fluid viscosity at ambient pressure, and h_e is the film thickness at the exit point, $h_e = h_e(y)$.

Using these dimensionless variables and the asymptotic expansions (13.129) for the integrals involved in the problem equations, the normal problem for the main asymptotic terms for $H_0 > 0$ and $H_0 = 0$ can be found in the form

$$\frac{\partial}{\partial x}\left\{\frac{1}{\mu}\int_{-h/2}^{h/2} zF(f_x + zH_0\frac{\partial p}{\partial x})dz - h\right\} = 0, \tag{13.131}$$

$$\frac{1}{\mu}\int_{-h/2}^{h/2} F(f_x + zH_0\frac{\partial p}{\partial x})dz = s_x \ for \ H_0 > 0, \tag{13.132}$$

$$\frac{2}{\pi\delta^2 D}\int_{a_p}^{c_p} p(t,y)\ln\left|\frac{c_p-t}{x-t}\right|dt = c_p^2 - x^2 \ for \ H_0 = 0, \tag{13.133}$$

$$\int_{a_p}^{c_p} p(t,y)dt = \frac{\pi}{2}P_0(y), \tag{13.134}$$

$$p(a_p,y) = p(c_p,y) = 0 \ for \ H_0 \geq 0, \tag{13.135}$$

$$\frac{\partial p(c_p,y)}{\partial x} = 0 \ for \ H_0 > 0, \tag{13.136}$$

$$H_0(h-1) = x^2 - c_p^2 + \frac{2}{\pi\delta^2 D}\int_{a_p}^{c_p} p(t,y)\ln\left|\frac{c_p-t}{x-t}\right|dt \tag{13.137}$$

$$+\frac{H_0\epsilon_x}{\delta^2 D}\int_x^{c_\tau} h(t,y)\frac{\partial p(t,y)}{\partial t}dt \ for \ H_0 > 0,$$

where $\epsilon_x = \frac{1-2\nu}{1-\nu}\frac{a_H}{8R_x}$. In these equations $[a_p, c_p]$ is the interval $p(x,y) \geq 0$ for the particular y, $a_p = a_p(y)$ and $c_p = c_p(y)$ are the inlet and exit points of this interval/contact region, and $P_0(y)$ is a positive function equal to the force applied to the contact region $[a_p, c_p]$. Also, $[a_\tau, c_\tau]$ is the interval where $\vec{\tau}(x,y) \neq \vec{0}$ for the particular y, $a_\tau = a_\tau(y)$ and $c_\tau = c_\tau(y)$ are the inlet and exit points of this interval (see below). It is important to notice that in equations (13.131)-(13.137) constant h_0 is replaced by function $h_e = h_e(y)$ in such a way that $h = h_e$ for $(x,y) \in \Gamma_e$, and by the exit film thickness $H_0 = H_0(y)$. Besides that, for $\delta \ll 1$ the conditions $h(x,y) > 0$ and $h(x,y) = 0$ on the interval $[a_p(y), c_p(y)]$ are equivalent to the conditions $H_0 > 0$ and $H_0 = 0$, respectively.

In a similar fashion, for $\delta \ll 1$ taking into account that s_y, $v_y = O(\delta)$ the tangential problem for the main terms of the asymptotic expansions can be obtained in the form

$$\tau_x = f_x \ for \ H_0 > 0, \tag{13.138}$$

$$\tau_x = \theta \lambda p \frac{s_x}{|s_x|}$$

$$(13.139)$$

$$if \ \ |s_x| > 0 \ and \ |\tau_x| \le \theta \lambda p \ if \ \ |s_x| = 0 \ for \ H_0 = 0,$$

$$s_x = -\eta \int_{a_\tau}^{c_\tau} \frac{\tau_x(t,y)dt}{t-x} + v_x, \ \ \tau_x(a_\tau,y) = \tau_x(c_\tau,y) = 0 \ for \ H_0 \ge 0. \quad (13.140)$$

Therefore, the space problem for mixed friction is reduced to a family of plane problems for lubricated and dry conditions. Here it is important to mention again that if $H_0(y_*) > 0$ for some y_* then the whole segment $[a_p(y_*), c_p(y_*)]$ is lubricated while if $H_0(y_*) = 0$ for some y_* then the whole segment $[a_p(y_*), c_p(y_*)]$ is dry. This follows from the fact that the lubricant volume flow flux is constant in a plane case. The general view of such a partially lubricated contact with one dry strip and two lubricated ones is given in Fig. 13.53.

For $\delta \ll 1$ the systems of equations (13.131)-(13.137) and (13.138)-(13.140) hold outside of the small vicinities, of order of δ, of the points $(x, y) = (x, -d_l)$ and $(x, y) = (x, d_u)$ (which represent the lower and upper tips of the contact boundary Γ) and the points where the radius of inlet boundary Γ_i is of order of δ. In the considered case the contact area is represented by alternating contact stripes (bounded by straight lines $y = const$) in which fluid or dry friction occurs.

In general, systems of equations (13.131)-(13.137) and (13.138)-(13.140) must be solved simultaneously. For incompressible materials ($\nu = 1/2$), the normal and tangential problems get decoupled. In such cases, first the normal problem has to be solved and after that the tangential one. For dry contacts the latter problem was considered in detail in a number of studies such as [31] while for lubricated contact it was analyzed numerically in Chapter 10 lubricated soft elastic materials. Here we will concentrate on the normal problem.

Let us assume that the cross section $y = y_*$ is lubricated, i.e., $H_0(y_*) > 0$. By introducing the following transformation of variables:

$$(x, a_p, c_p) = \sigma(x_1, a_{1p}, c_{1p}), \ p = p_0 p_1, \ H_0 = H_{00} H_1, \ V = V_0 V_1,$$

$$p_0 = \delta\sqrt{P_0 D}, \ \sigma = \frac{1}{\delta}\sqrt{\frac{P_0}{D}}, \ H_{00} = \frac{P_0}{\delta^2 D}, \ P_0 = \frac{V_0}{H_{00}}, \ V_0 = \frac{P_0^2}{\delta^2 D}, \quad (13.141)$$

$$\epsilon_x = \frac{\epsilon_{x1}}{\sigma},$$

equations (13.131)-(13.137) can be reduced to the equations of a plane EHL problem in the form used before. For simplicity, let us consider the case of Newtonian fluid, i.e., the case with $F(x) = x$. Then in the introduced variables equations (13.131), (13.132), (13.134)-(13.137) are reduced to the following ones

$$\frac{\partial}{\partial x_1}\{\frac{H_1^2}{V_1}\frac{h^3}{\mu}\frac{\partial p_1}{\partial x_1} - h\} = 0, \quad (13.142)$$

$$p_1(a_{1p}, y) = p(c_{1p}, y) = 0, \ \frac{p_1(c_{1p}, y)}{\partial x_1} = 0, \quad (13.143)$$

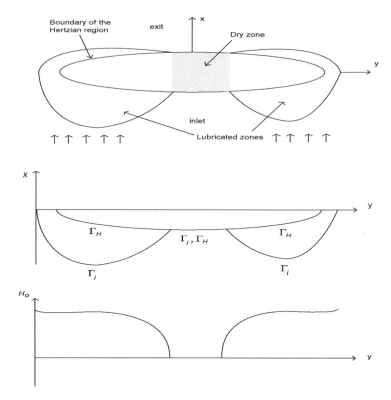

FIGURE 13.55
The view of a partially lubricated contact region with indicated boundary of
the Hertzian (dry) contact Γ_H and the boundary of the partially lubricated
contact region Γ_i (upper and middle sketches) and the film thickness H_0
distribution along the y-axis (bottom sketch).

$$H_1(h-1) = x_1^2 - c_{1p}^2 + \frac{2}{\pi} \int\limits_{a_{1p}}^{c_{1p}} p_1(t,y) \ln\left| \frac{c_{1p}-t}{x_1-t} \right| dt$$

$$(13.144)$$

$$+H_1\epsilon_{x1} \int\limits_{x_1}^{c_{1\tau}} h(t,y)\frac{\partial p_1(t,y)}{\partial t} dt,$$

$$\int\limits_{a_{1p}}^{c_{1p}} p_1(t,y)dt = \frac{\pi}{2}.$$

$$(13.145)$$

Equations (13.142)-(13.145) describe a familiar problem for a line EHL contact. The only difference of these equations from the ones studied in Chapter
10 is the presence of the last term in equation (13.144) proportional to ϵ_{x1},
which represents the influence of the fluid friction on the other parameters of
a lubricated contact. For incompressible solid materials ($\nu = 1/2$), we have

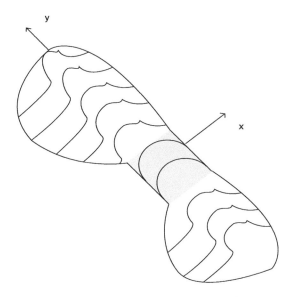

FIGURE 13.56
The schematic pressure distribution in the lubricated and dry zones of the contact region.

$\epsilon_{x1} = 0$ and the lubrication problem described by equations (13.142)-(13.145) is identical to the EHL problem for a Newtonian fluid studied in Chapter 10. Therefore, all conclusions and solutions obtained for this problem in Chapter 10 can be extended to the problem at hand. For compressible solid materials ($\nu < 1/2$), we have $\epsilon_{x1} \neq 0$. In this case the problem still can be analyzed by the same asymptotic methods developed in Chapter 10. Moreover, for starved and fully flooded pre-critical lubrication regimes ($\epsilon_q = O(V_1^{2/5})$, see (10.126) and (10.127)) the analysis (and the results) for the problem described by equations (13.142)-(13.145) is identical to the one used (and the results received) in Chapter 10. In particular, based on formulas (10.149) for $\mu = 1$, we can conclude that for starved lubrication conditions we would have

$$H_0(d) = H_{0*} \mid \tfrac{d}{d_0} \mid^{2/3} \theta(d), \ c_p(d) = \sigma(1 + B \mid \tfrac{d}{d_0} \mid \theta(d)), \qquad (13.146)$$

where H_{0*} is the film thickness when the oil meniscus is located outside of the Hertzian region at a distance of d_0 from the inlet side of the Hertzian contact boundary, d is the distance of the oil meniscus located outside of the Hertzian region from the inlet side of the Hertzian contact boundary ($d > 0$ when the oil meniscus is outside of the Hertzian region), B is a positive constant, $\theta(x)$ is a step function, $\theta(x) = 0$, $x \leq 0$ and $\theta(x) = 1$, $x > 0$. This behavior of the lubrication film thickness H_0 as a function of the distance d of the oil meniscus from the Hertzian contact was clearly observed in optical experimental studies of starved lubrication [39].

Based on (13.146) it is obvious that when the oil meniscus touches the boundary of the Hertzian region the film thickness H_0 becomes equal to zero and the exit coordinate of the contact coincides with the Hertzian one. Qualitatively the behavior of the film thickness H_0 and the exit boundary c_p remains the same for cases of the lubricant viscosity μ varying with pressure p. The described behavior of the inlet boundary and the film thickness H_0 is illustrated in Fig. 13.55. The schematic pressure distribution in the lubricated and dry zones of the contact region is presented in Fig. 13.56.

Usually, under equal conditions dry frictional stress is greater than fluid one. Therefore, for mixed friction the friction force is higher than for purely fluid regime. The increase in friction force depends on a relative portion of the contact region occupied by the dry friction zones that negatively affects contact fatigue life.

To conclude the problem solution function $P_0(y)$ should be determined. We should keep in mind that function $P_0(y)$ for the case of a lubricated contact differs from the one for a dry contact by a small value of $O(\epsilon_q^{3/2}, H_0\epsilon_q^{1/2} \frac{\epsilon_x}{\delta^2 D})$ for pre-critical lubrication regimes and by $O(\epsilon_0^{1/2}\epsilon_q, H_0\epsilon_q^{1/2} \frac{\epsilon_x}{\delta^2 D})$ for over-critical lubrication regimes. Therefore, in most cases for determining function $P_0(y)$ it is sufficient to consider the corresponding dry contacts. That is done in the next section.

The study presented in this section is only the beginning in understanding of major mechanisms governing the phenomenon of mixed friction.

13.6   Dry Narrow Contact of Elastic Solids

Let us consider a narrow (along the x-axis) and significantly stretched along the y-axis contact of a rigid indenter with an elastic half-space. The shape of the indenter surface can be approximated by a paraboloid. In the direction of the x- and y-axes the radii of curvature of the indenter surface are R_x and R_y, respectively. The indenter is impressed in the half-space by a normal force P. In this case $R_x/R_y \ll 1$ and, therefore, $\delta = a_H/b_H \ll 1$ and the eccentricity e of the contact ellipse is close to 1, i.e., $e - 1 \ll 1$ (see equation (13.128)). Using the asymptotic expansions [37, 38] for the integrals over an elongated in the direction of the y-axis narrow contact region in dimensionless variables (13.130) for pressure $p(x, y)$ in any cross section of the contact $y = conts$ at a distance greater than 1 away from the tips of the contact $(0, \pm\delta^{-1})$ we obtain equations (see equations (13.133)-(13.135))

$$\frac{2}{\pi\delta^2 D} \int\limits_{a_p}^{c_p} p(t, y) \ln \left| \frac{c_p - t}{x - t} \right| dt = c_p^2 - x^2, \quad \int\limits_{a_p}^{c_p} p(t, y)dt = \frac{\pi}{2}P_0(y), \qquad (13.147)$$

where $a_p = a_p(y)$ and $c_p = c_p(y)$ are the boundaries of the contact, constants $\delta \ll 1$ and D satisfy equation (13.128), $P_0(y)$ is an unknown function equal to the force applied to this cross section of the contact (see (13.129)).

In this formulation we can consider the cases of the indenter with smooth and sharp edges. The difference between these cases is that the contact boundaries a_p and c_p are unknown and given, respectively.

The solution of this problem can be represented in the form [34]

$$p(x,y) = \delta^2 D q(x,y) + \frac{P_0(y)}{2\sqrt{(x-a_p)(c_p-x)}}, \quad \int_{a_p}^{c_p} q(x,y)dx = 0,$$
$$(13.148)$$

$$q(x,y) = \sqrt{(x-a_p)(c_p-x)} + \frac{2a_p c_p + (c_p-a_p)^2/4 - (c_p+a_p)x}{2\sqrt{(x-a_p)(c_p-x)}}.$$

Outside of the contact in its cross section $y = const$ at a distance of $x = O(\delta^{-1}) \gg 1$, the details of the pressure $p(x,y)$ distribution along the x-axis can be neglected and the contact region can be replaced by a medium line loaded with pressure $p(x,y) = \frac{\pi}{2} P_0(y)\delta(x)$, where $\delta(x)$ is the Dirac's delta-function. Then, at these distances from the contact region the vertical displacement of the indenter w is

$$w(x,y) = -\frac{1}{2\delta^2 D} \int_{-d}^{d} \frac{P_0(\eta)d\eta}{\sqrt{x^2+(y-\eta)^2}} + w_0, \quad (13.149)$$

where $2d = O(\delta^{-1})$ is the diameter of the contact region, i.e., the distance between two most distanced points of the contact, w_0 is the indenter displacement far away from the contact.

Let us determine the asymptotic of the integral in (13.149) for $x = O(1)$. We will have

$$\int_{-d}^{d} \frac{P_0(\eta)d\eta}{\sqrt{x^2+(y-\eta)^2}} = \int_{-d}^{d} \frac{[P_0(\eta)-P_0(y)]d\eta}{\sqrt{x^2+(y-\eta)^2}} + P_0(y) \int_{-d}^{d} \frac{d\eta}{\sqrt{x^2+(y-\eta)^2}}$$
$$(13.150)$$

$$= \int_{-d}^{d} \frac{[P_0(\eta)-P_0(y)]d\eta}{|y-\eta|} + 2P_0(y)[\ln 2 + \ln\sqrt{d^2-y^2} - \ln|x|] + O(1).$$

Therefore, from equations (13.149) and (13.150) for $w(x,y)$, we obtain

$$w(x,y) = -\frac{1}{2\delta^2 D} \int_{-d}^{d} \frac{[P_0(\eta)-P_0(y)]d\eta}{|y-\eta|} - \frac{P_0(y)}{\delta^2 D} \ln \frac{2\sqrt{d^2-y^2}}{|x|} + w_0 + \dots \quad (13.151)$$

From the first equation of (13.147), it is clear that in the region $x = O(1)$ in this cross–section $y = const$ the elastic displacement of the half-space surface $w(x,y)$ has the form

$$w(x,y) = h_w - \frac{2}{\pi\delta^2 D} \int_{a_p}^{c_p} p(t,y) \ln\left|\frac{1}{x-t}\right| dt + \dots \quad (13.152)$$

For $x \gg 1$ from (13.152) and the second equation in (13.147), we obtain the asymptotic representation for $w(x, y)$

$$w(x, y) = h_w - \frac{P_0(y)}{\delta^2 D} \ln \frac{1}{|x|} + \ldots \tag{13.153}$$

Applying the principle of matched asymptotic expansions [40] to the expressions for $w(x, y)$ from (13.151) and (13.153) in the intermediate region $1 \ll |x| \ll \delta^{-1}$, we obtain the equation for function $P_0(y)$

$$\frac{1}{2} \int_{-d}^{d} \frac{[P_0(\eta) - P_0(y)]d\eta}{|y - \eta|} + P_0(y) \ln[2\sqrt{d^2 - y^2}]$$

$$\tag{13.154}$$

$$= \delta^2 D(w_0 - c_p^2 - \rho y^2) - \frac{2}{\pi} \int_{a_p}^{c_p} p(t, y) \ln \left| \frac{1}{c_p - t} \right| dt.$$

This equation involves the unknown constant w_0, which is determined by the force balance equation for the entire contact (see the dimensional analog (13.120))

$$\iint_{\Omega_p} p(\xi, \eta)d\xi d\eta = \frac{2\pi}{3} \delta^{-1}. \tag{13.155}$$

Using the second equation in (13.147) equation (13.155) is easily transformed into

$$\int_{-d}^{d} P_0(y)dy = \frac{4}{3\delta}. \tag{13.156}$$

By substituting (13.148) into (13.154), we obtain

$$\frac{1}{2} \int_{-d}^{d} \frac{[P_0(\eta) - P_0(y)]d\eta}{|y - \eta|} + P_0(y) \ln \frac{8\sqrt{d^2 - y^2}}{c_p - a_p}$$

$$\tag{13.157}$$

$$= \delta^2 D[w_0 - c_p^2 - \rho y^2 - \frac{2}{\pi} \int_{a_p}^{c_p} q(t, y) \ln \frac{1}{c_p - t} dt],$$

where in the case of $q(x, y)$ from (13.148) we have

$$\frac{8}{\pi} \int_{a_p}^{c_p} q(t, y) \ln \frac{1}{c_p - t} dt = -3c_p^2 + 2c_p a_p + a_p^2. \tag{13.158}$$

13.6.1   Examples of Dry Narrow Contacts

First, let us consider the case of an indenter with an ellipsoidal shape such that $c_p = -a_p = \sqrt{1 - \frac{y^2}{d^2}}$ and $d = \delta^{-1} \gg 1$. Then the problem is reduced to equations (see equations (13.156)-(13.158))

$$\frac{1}{2} \int_{-d}^{d} \frac{[P_0(\eta) - P_0(y)]d\eta}{|y - \eta|} + P_0(y) \ln(4d) = \delta^2 D(w_0 - \rho y^2), \tag{13.159}$$

$$\int\limits_{-d}^{d} P_0(y)dy = \tfrac{4}{3\delta},$$

where constant δ satisfies equations (13.128).

For $d \gg 1$ away from points $y = \pm d$ from the first of equations (13.159), we obtain

$$P_0(y) = \tfrac{\delta^2 D(w_0 - \rho y^2)}{\ln(4d)}. \tag{13.160}$$

After that using the second equation in (13.159) and equation (13.160), we find

$$P_0(y) = \tfrac{2}{3} + \rho\tfrac{\delta^2 D(\frac{1}{3}d^2 - y^2)}{\ln(4d)}, \quad w_0 = \tfrac{\rho d^2}{3} + \tfrac{2}{3}\tfrac{\ln(4d)}{\delta^2 D}. \tag{13.161}$$

Now, let us consider the case of an indenter with an ellipsoidal shape or the case of two elliptic elastic solids. In such cases $c_p = -a_p = l(y)$ and $d = \delta^{-1} \gg 1$. Moreover, function $l(y)$ satisfies the equations

$$p(\pm l, y) = 0. \tag{13.162}$$

Using equations (13.148) and (13.162), we find that

$$l(y) = \sqrt{\tfrac{P_0(y)}{\delta^2 D}}, \quad p(x,y) = \delta^2 D\sqrt{l^2(y) - x^2}. \tag{13.163}$$

Substituting the expressions for $l(y)$ and $p(x,y)$ from (13.163) into equations (13.157) and (13.158) for $P_0(y)$, we obtain the equations

$$\tfrac{1}{2}\int\limits_{-d}^{d} \tfrac{[P_0(\eta) - P_0(y)]d\eta}{|y - \eta|} + P_0(y)\ln\tfrac{4\delta D^{1/2}\sqrt{d^2 - y^2}}{P_0^{1/2}(y)} = \delta^2 D(w_0 - \rho y^2),$$
$$\tag{13.164}$$

$$\int\limits_{-d}^{d} P_0(y)dy = \tfrac{4}{3\delta}.$$

The approximate solution of equations (13.164) we will try to find in the form

$$P_0(y) = A^2(d^2 - y^2), \tag{13.165}$$

where A is an unknown constant. Substituting (13.165) into (13.164) and equating the coefficients at y^0 we find $A = d^{-1}$ and

$$P_0(y) = \tfrac{d^2 - y^2}{d}, \quad w_0 = \tfrac{1}{\delta^2 D}[\ln(4D^{1/2}) - \tfrac{1}{2}]. \tag{13.166}$$

It can be shown that the indenter displacement within the contact region is described by a paraboloid of the second degree. Moreover, the error created by this approximation decreases [37] from 15% for $\delta = 0.5$ to 3% for $\delta = 0.2$.

It is worth mentioning that the approximate solution (13.166) for $p(x,y)$ coincides with the exact solution of this problem [41]. As it was mentioned in the preceding section for lubricated contacts function $P_0(y)$ is very close to the one obtained for corresponding dry contacts (see (13.166)).

13.6.2   Optimal Shape of Solids in a Normal Narrow Contact

Now, let us consider the problem of optimal design of contact surfaces in bearings and gears. Solution of this problem is important for increasing fatigue life of ball and roller bearings as well as gears.

Let us consider this problem on an example of a roller bearing working under normal load without skewness. For simplicity we will assume that the surface of a relatively long roller in the direction of the y-axis is described by the equation $z = f_0(y)$. This roller made of an elastic material is normally indented in an elastic half-space. In each cross section $y = const$ the roller radius $R(y)$ is given by $R(y) = R_0 - f_0(y)$, where R_0 is the roller radius in its central cross-section $y = 0$. Then the gap between the non-deformed roller and half-space is described by function

$$f(x, y) = f_0(y) + \frac{x^2}{2[R_0 - f_0(y)]}. \tag{13.167}$$

Let us introduce the dimensionless variables

$$(f', f_0') = \frac{2R_0}{a_H^2}(f, f_0), \quad \epsilon_* = \frac{1}{2}\left(\frac{a_H}{R_0}\right)^2 \tag{13.168}$$

in addition to the ones introduced in (13.130). In (13.168) a_H is the smaller semi-axis of the elliptic contact region of an indenter with radii R_0 and $[f_0''(0)]^{-1}$ along the x- and y-axes, respectively, $e = \sqrt{1 - \delta^2}$ is the eccentricity of the ellipse, which satisfies equation (13.128).

In these dimensionless variables (for simplicity primes are omitted) in cross sections of the contact $y = const$ away from points $(0, \pm d)$, we get equations

$$\frac{x^2 - c_p^2}{1 - \epsilon_* f_0(y)} + \frac{2}{\pi \delta^2 D} \int\limits_{a_p}^{c_p} p(t, y) \ln \left| \frac{c_p - t}{x - t} \right| dt = 0,$$

$$\int\limits_{a_p}^{c_p} p(t, y)dt = \frac{\pi}{2}P_0(y), \tag{13.169}$$

where d is determined by the equations $l(\pm d) = 0$.

The solution of of equations (13.169) has the form

$$p(x, y) = \frac{\delta^2 D}{1 - \epsilon_* f_0(y)}q(x, y) + \frac{P_0(y)}{2\sqrt{(x - a_p)(c_p - x)}}, \quad \int\limits_{a_p}^{c_p} q(x, y)dx = 0, \tag{13.170}$$

$$q(x, y) = \sqrt{(x - a_p)(c_p - x)} + \frac{2a_p c_p + (c_p - a_p)^2/4 - (c_p + a_p)x}{2\sqrt{(x - a_p)(c_p - x)}}.$$

By taking $c_p = -a_p = l(y)$ and satisfying equations (13.162), we find

$$l(y) = \sqrt{[1 - \epsilon_* f_0(y)]\frac{P_0(y)}{\delta^2 D}}, \quad p(x, y) = \frac{\delta^2 D}{1 - \epsilon_* f_0(y)}\sqrt{l^2(y) - x^2}. \tag{13.171}$$

Let us consider the pressure distribution along the roller medium line $x = 0$ at which the maximum of pressure is reached. From (13.171) we have

$$p(0, y) = \sqrt{\tfrac{\delta^2 D P_0(y)}{1 - \epsilon_* f_0(y)}}. \tag{13.172}$$

Because contact fatigue life is inverse proportional to a certain power of the maximum contact pressure (see Section 9.7), the optimal pressure distribution is the one that is constant along the medium roller line. Therefore, the optimal roller configuration we will obtain from the condition $p(0, y)$ is equal to a constant (see (13.172))

$$\tfrac{P_0(y)}{1 - \epsilon_* f_0(y)} = P_0(0), \tag{13.173}$$

where we took into account that $f_0(0) = 0$. From equation (13.173) we find

$$f_0(y) = \tfrac{1}{\epsilon_*} \{1 - \tfrac{P_0(y)}{P_0(0)}\}. \tag{13.174}$$

In the above dimensionless variables using the relationships (13.167) and (13.171), we derive the equation for $P_0(y)$ in the form

$$\tfrac{\epsilon_*}{2} \int\limits_{-d}^{d} \tfrac{[P_0(\eta) - P_0(y)]d\eta}{|y - \eta|} + P_0(y)\{\epsilon_* \ln[\tfrac{4\delta}{P_0(y)} \sqrt{D P_0(0)(d^2 - y^2)}]$$

$$\tag{13.175}$$

$$- \tfrac{\delta^2 D}{P_0(0)}\} = \delta^2 D(\epsilon_* w_0 - 1), \quad \int\limits_{-d}^{d} P_0(y)dy = \tfrac{4}{3\delta}.$$

For $\epsilon_* \ll 1$ for solution of latter equations let us use the regular perturbation method [40]. The solution of equations (13.175) we will try to find in the form

$$P_0(y) = P_{00}(y) + \epsilon_* P_{01}(y) + O(\epsilon_*^2), \quad w_0 = w_{00} + O(\epsilon_*). \tag{13.176}$$

For $\epsilon_* \ll 1$ away from points $(0, \pm d)$, we have $f_0(y) = O(1)$ and, therefore, from equation (13.173) we obtain $P_{00}(y) = P_{00}(0)$. Using that from the integral condition in (13.175), we get

$$P_{00}(y) = \tfrac{2}{3\delta d}. \tag{13.177}$$

By equating the coefficients at ϵ_* in the expansion of equation (13.175) in series of powers of ϵ_*, we derive the equation for $P_{01}(y)$:

$$P_{00} \ln\{4\delta \sqrt{\tfrac{D(d^2 - y^2)}{P_{00}}}\} + \tfrac{\delta^2 D}{P_{00}}[P_{01}(0) - P_{01}(y)] = \delta^2 D w_{00},$$

$$\tag{13.178}$$

$$\int\limits_{-d}^{d} P_{01}(y)dy = 0.$$

The solution of equations (13.178) is

$$P_{01} = \tfrac{4}{9d^2\delta^4 D}\{1 - \ln 2 + \tfrac{1}{2} \ln(1 - \tfrac{y^2}{d^2})\}, \tag{13.179}$$

$$w_{00} = \tfrac{2}{3d\delta^3 D} \ln\{4d\delta \sqrt{\tfrac{3d\delta D}{2}}\}.$$

Therefore, using equations (13.174), (13.176), (13.177), and (13.179), we obtain

$$f_0(y) = -\frac{\ln(1-\frac{y^2}{d^2})}{3d\delta^3 D + 2\epsilon_*(1-\ln 2)} + \dots \qquad (13.180)$$

A simple analysis of problem (13.175) and its solution show that it is a singularly perturbed problem. That is the reason why constant d cannot be determined from the solution of equation $l(d) = 0$ if the solution expansion (13.176), (13.177), and (13.179) is used. Nevertheless, using the terms of this expansion we can estimate the size of the boundary layers adjacent to points $(0, \pm d)$ which is of the order of $\exp(-\frac{3d\delta^3 D}{\epsilon_*})$.

Therefore, if the value of d is known, then the optimal roller shape for which $p(0, y)$ remains constant is given by formula (13.180). In particular, if the contact is realized along the whole length $2L$ of the roller then $d = L$ and the roller shape is completely determined. In spite of the fact that inaccuracy of the obtained approximate optimal shape of the roller increases as we approach the roller end this approximation is acceptable for practical applications.

13.7   Closure

It is shown that the classic formulations of non-steady plane EHL problems for both conformal and non-conformal contacts possess certain serious defects such as a possibility of a discontinuous pressure distribution, etc. The proper formulations of non-steady plane EHL problems, which include the system inertia are proposed. Solutions of some simple problems such a pure squeeze motion demonstrate the advantages of the modified problems formulations. Based on this formulation, which takes into account the system inertia, a non-steady EHL problem for a conformal contact is solved for various conditions that include abruptly varying step load, bumps, and dents on the contact surface, etc. Solution of a spacial EHL problem for starved lightly loaded contact with a complex shape of the inlet meniscus is considered. It is shown that the solution varies considerably depending on the shape of the inlet meniscus. In connection with the latter problem the formulation and analysis of a mixed lubrication problem are considered. The problem formulation takes into account such essential conditions as frictional stress continuity across the boundaries between dry and lubricated zones of the contact. A detailed analysis of the problem is proposed for the case of a contact extended in the direction perpendicular to the motion. The problem is considered asymptotically. Using the fatigue considerations an optimal shape of a roller in a rolling bearing is proposed.

13.8   Exercises and Problems

1. (a) Based on the modified problem formulation obtain a two-term asymptotic solution for a lightly loaded non-steady non-conformal lubricated contact (i.e., for the case of $V \gg 1$) for a lubricant with constant viscosity. Show that all components of the solution such as pressure $p(x,t)$, gap $h(x,t)$, etc., are continuous functions of time t not only for continuous external load $P(t)$ and average surface velocities $u(t)$ but also for finite discontinuous ones.

(b) Repeat the same analysis for a lightly loaded non-steady conformal lubricated contact, i.e., for $\lambda_0 \ll 1$.

2. Solve equation (13.27) and (13.28) to obtain the expression for $h(\varphi, t)$ from (13.29).

3. Describe what influence system inertia has on the behavior of a lubricated contact. What can be expected for relatively small and large system mass?

4. Describe the influence of the shape and distance of the inlet oil meniscus on contact pressure $p(x,y)$, frictional $\tau_{sx}(x,y)$, $\tau_{rx}(x,y)$, and $\tau_{ry}(x,y)$, and octahedral $\sigma^{oct}(x,y,z)$ stresses.

5. Elaborate on why the rheology of fluids in thin films differs from the one of fluids in thick films. To substantiate your analysis use the continuity relationships (13.125) and (13.126).

6. Graph and analyze the optimal shape of a long roller from formula (13.180).

BIBLIOGRAPHY

[1] Safa, M.M.A. and Gohar, R. 1086. Pressure Distribution Under a Ball Impacting a Thin Lubricant Layer. *ASME J. Tribology* 108, No. 3:372-376.

[2] Ai, X. and Cheng, H.S. 1994. A Transient EHL Analysis for Line Contacts with Measured Surface Roughness Using Multigrid Technique. *ASME J. Tribology* 116, No. 3:549-558.

[3] Chang, L., Webster, M.N., and Jackson, A. 1994. A Line-Contact Micro-EHL Model with Three Dimensional Surface Topography. *ASME J. Tribology* 116, No. 1:21-28.

[4] Venner, C.H. and Lubrecht, A.A.. 1994. Transient Analysis of Surface Features in an EHL Line Contact in Case of Sliding. *ASME J. Tribology* 116, No. 2:186-193.

[5] Venner, C.H. and Lubrecht, A.A. 1994. Numerical Simulation of Transverse Ridge in a Circular EHL Contact Under Rolling/Sliding. *ASME J. Tribology* 116, No. 4:751-761.

[6] Osborn, K.F. and Sadeghi, F. 1992. Time Dependent Line EHL Lubrication Using the Multigrid/Multilevel Technique. *ASME J. Tribology* 114, No. 1:68-74.

[7] Cha, E. and Bogy, D.B. 1995. A Numerical Scheme for Static and Dynamic Simulation of Subambient Pressure Shaped Rail Sliders. *ASME J. Tribology* 117, No. 1:36-46.

[8] Hashimoto, H. and Mongkolwongrojn, M. 1994. Adiabatic Approximate Solution for Static and Dynamic Characteristics of Turbulent Partial Journal Bearing with Surface Roughness. *ASME J. Tribology* 116, No. 4:672-680.

[9] Peiran Y. and Shizhu, W. 1991. Pure Squeeze Action in an Isothermal Elastohydrodynamically Lubricated Spherical Conjunction. Part 1. Theory and Dynamic Load Results. *Wear* 142, No. 1:1-16.

[10] Peiran Y. and Shizhu, W. 1991. Pure Squeeze Action in an Isothermal Elastohydrodynamically Lubricated Spherical Conjunction. Part 2. Constant Speed and Constant Load Results. *Wear* 142, No. 1:17-30.

[11] San Andres, L.A. and Vance, J.M. 1987. Effect of Fluid Inertia on Squeeze Film Damper Forces for Small Amplitude Motions about an Off-Center Equilibrium Position. *ASLE Tribology Trans.* 30, No. 1:63-68.

[12] Larsson, R. and Hoglund, E.. 1995. Numerical Simulation of a Ball Impacting and Rebounding a Lubricated Surface. *ASME J. Tribology* 117, No. 1:94-102.

[13] Kudish, I.I. and Panovko, M.Ya. 1992. Oscillations of a Deformable Lubricated Cylinder Rolling Along a Rigid Half-Space. *Soviet J. Fric. and Wear* 13, No. 5:1-11.

[14] Kudish, I.I. 1999. On Formulation of a Non-Steady Lubrication Problem for a Non-Conformal Contact. *STLE Tribology Trans.* 42, No. 1:53-57.

[15] Kudish, I.I., Kelley, F., and Mikrut, D. 1999. Defects of the Classic Formulation of a Non-steady EHL Problem for a Journal Bearing. New Problem Formulation. *ASME J. Tribology* 121, No. 4:995-1000.

[16] Hamrock, B.J. 1991. *Fundamentals of Fluid Film Lubrication*. Cleveland: NASA, Reference Publication 1255.

[17] Teplyi, M.I. 1983. *Contact Problems for Regions with Circular Boundaries*. Lvov: Naukova Dumka.

[18] Allaire, P.E. and Flack, R.D. 1980. Journal Bearing Design for High Speed Turbomachinery. In *Bearing Design – Historical Aspects, Present Technology and Future Problems*, Proc. Intern. Conf. "Century 2 - Emerging Technology," Ed. W.J. Anderson, ASME, New York. 111-160.

[19] Ghosh, M.K., Hamrock, B.J., and Brewe, D. 1985. Hydrodynamic Lubrication of Rigid Non-Conformal Contacts in Combined Rolling and Normal Motion. *ASME J. Tribology* 107:97-103.

[20] Lin, C.R. and Rylander, H.G., Jr. 1991. Performance Characteristics of Compliant Journal Bearings. *ASME J. Tribology* 113, No. 3:639-644.

[21] Wijnant, Y.H., Venner, C.H., Larsson, R., and Erickson, P. 1999. Effects of Structural Vibrations on the Film Thickness in an EHL Circular Contact. *ASME J. Tribology* 121, No. 2:259-264.

[22] Kudish, I.I. 2002. A Conformal Lubricated Contact of Cylindrical Surfaces Involved in a Non-steady Motion. *ASME J. Tribology* 124, No. 1:62-71.

[23] Kudish, I.I. and Panovko, M.Ya. 1997. Influence of an Inlet Oil Meniscus Geometry on Parameters of a Point Elastohydrodynamic Contact. *ASME J. Tribology* 119, No. 1:112-125.

[24] Oh, K.P. 1984. The Numerical Solution of Dynamically Loaded Elastohydrodynamic Contacts as a Nonlinear Complementarity Problem. *ASME J. Tribology* 106, No. 1:88-94.

[25] Evans, H.P. and Hughes, T.G. 2000. Evaluation of Deflection in Semi-infinite Bodies by a Differential Method. *Proc. Instn. Mech. Engrs.* 214, Part C: 563-584.

[26] Hamrock, B.J. and Dowson, D. 1981. *Ball Bearing Lubrication – The Elastohydrodynamics of Elliptical Contacts*. New York: Wiley-Interscience.

[27] Lubrecht, A.A., Ten Napel, W.E., and Bosma, R. 1987. Multigrid Alternative Method of Solution for Two-Dimensional Elastohydrodynamically Lubricated Point Contact Calculations. *ASME J. Tribology* 109, No. 3:437-443.

[28] Kudish, I.I. 1981. Some Problems of Elastohydrodynamic Theory of Lubrication for a Lightly Loaded Contact. *J. Mech. of Solids* 16, No. 3:75-88.

[29] Kostreva, M.M. 1984. Elastohydrodynamic Lubrication: A Nonlinear Complementarity Problem. *Intern. J. Numerical Methods in Fluids* 4:377-397.

[30] Oh, K.P., Li, C.H., and Goenka, P.K. 1987. Elastohydrodynamic Lubrication of Piston Skirts. *ASME J. Tribology* 109, No. 2:356-365.

[31] Johnson, K. 1985. *Contact Mechanics*. Cambridge: Cambridge University Press.

[32] Belotcerkovsky S.M. and Lifanov I.K. 1992. *Method of Discrete Vortices*. Boca Raton: CRC Press.

[33] Kudish, I.I. 1983. On Formulation and Analysis of Spatial Contact Problem for Elastic Solids Under Conditions of Mixed Friction. *J. Appl. Math. and Mech.* 47, No. 6:1006-1014.

[34] Galin, L.A. 1980. *Contact Problems of Elasticity and Viscoelasticity*. Moscow: Nauka.

[35] Akhmatov, A.S. 1963. *Molecular Physics of Boundary Friction*. Moscow: Fizmatgiz.

[36] Goldstein, R.V., Zazovsky, A.F., Spector, A.A., and Fedorenko, R.P. 1979. *Solution of Spatial Rolling Contact Problems with Stick and Slip by a Variational Method*. Reprint of the Inst. for Problems in Mech., USSR Academy of Sciences, Moscow, No. 134.

[37] Kalker, J.J. 1972. On Elastic Line Contact. *ASME Trans., Ser. E, J. Appl. Mech.* 33, No. 4:1125-1132.

[38] Kalker, J.J. 1977. The Surface Displacement of an Elastic Half-Space Loaded in a Slender, Bounded, Curved Surface Region with Application to the Calculation of the Contact Pressure under a Roller. *J. Inst. Maths. Applics.* 19:127-144.

[39] Bakashvili, D.L., Berdenikov, A.I., Imerlishvili, T.V., Manucharov, Yu.S., Mikhailov, I.G., and Shvatsman, V.Sh. 1985. Study of Lubricant Film of Liquids with High Bulk Viscosity in EHL Contact. *Soviet J. Fric. and Wear* 6, No. 2:54-59.

[40] Van-Dyke, M. 1964. *Perturbation Methods in Fluid Mechanics*. New York-London: Academic Press.

[41] Lurye, A.I. 1955. *Spatial Problems in Elasticity*. Moscow: Gostekhizdat.

Index

Alternating contact stripes, 889

Asperities, see also Roughness, 9, 14, 24, 28–30, 32, 35, 38, 73, 93, 100, 243, 401, 416, 417, 454, 455, 479, 480, 883

Asymptotic approximation of stress intensity factors, 394, 422

Barus equation, 115

Bending fatigue, see also Structural fatigue, Torsional fatigue, 247, 464, 465

Block copolymer, 135, 139

Branching, 135

Brinkman equation, 118

Brookfield viscosity, 134

Bump, 817, 825, 826, 838, 847–850, 852, 853

Characteristic size of the inner zone, 37, 91

Chebyshev orthogonal polynomials, 371, 448

Classic problem formulation, 811, 813, 814, 817, 818, 824–826

Classic solution, 809, 815–818, 824, 826–828

Cloud point, 112

Coating, 9, 14, 29, 93

Cold cranking simulator, 130

Concave function, 98, 99

Conditional probability, 151, 152, 154, 157, 159–161, 165, 169, 190, 196, 221, 750

Conjugated dienes, 139

Contact fatigue model, 265, 296, 428, 434, 436, 440, 441, 445, 463, 465, 467, 791, 793

Contact fatigue, see also Pitting, 233–239, 242, 243, 248, 254, 257, 259, 260, 265, 266, 273, 274, 276, 277, 296, 330, 361, 384, 393, 398, 400, 402, 409, 417, 418, 421, 427, 435, 436, 439, 444, 445, 447

Contact problem, 337, 445, 503, 507, 527, 612, 633, 874, 880

    for coated and rough surfaces, 9

    for cracked elastic half-plane, 335, 336

    for smooth surfaces, 10–12, 29, 32, 70

    with fixed and free boundaries, 39

    with fixed boundaries, 22, 98, 101, 341, 344

    with free boundaries, 99, 346

    with friction, 81, 92, 101

    with linear friction, 13

    without friction, 12, 85, 86, 89

Contact Problem 1, 12, 29, 30, 46, 47, 50, 54–59, 83–85, 87

Contact Problem 2, 12, 13, 29, 30, 34, 35, 38, 46, 47, 50, 54, 56, 57, 62, 63, 65, 73, 83, 88–91

Contact Problem 3, 12, 13, 29, 30, 39, 47, 50, 59, 64, 65, 84

Contact Problem 4, 45, 61–63, 65, 66, 72

Contact Problem 5, 62–65, 68, 72, 73

Convex indenter, 28, 93, 98

Copolymer, 135

Core, 713–717, 719, 720, 722–724, 726–731, 733, 734, 736, 738, 742

Crack propagation angle, see also Direction of fatigue crack growth, Plane of fatigue crack propagation, 450, 457

Crack statistics, 428, 459

Damped oscillations, 843

Degree of polymerization, 121

Density, 119

Density of the molecular weight distribution, 165, 793

Density of the number of polymer molecules, 165

Dent, 817, 825, 826, 838, 847–850, 854, 855

Dimple, 576, 577, 582–589, 626, 688

Direction of fatigue crack growth, see also Crack propagation angle, Plane of fatigue crack propagation, 423

Double bond, 140

Driver, 234, 235, 264, 296, 330, 361, 394, 398, 400–402, 409, 411–417

Dry friction, 873, 875, 877, 885, 886, 889, 892

Existence and uniqueness, 17, 26, 28, 94, 151, 162, 165, 187, 205–208, 212

Exit zone, 252, 503, 504, 507–520, 536, 538–541, 543, 546–549, 552, 553, 555, 556, 558, 561, 563, 565–569, 571–576, 579, 588, 596, 600–607, 610, 617, 622, 625, 630, 632–635, 637–643, 645–648, 650, 655–657, 661, 687, 688, 698, 700, 734, 742, 755, 763, 779, 780, 788, 795, 884

External region, see also Hertzian region, Outer region, 41, 42, 44, 60, 67, 85, 87, 90, 507–509, 513, 522, 523, 528, 537, 540, 633, 635, 648, 650

Fatigue life, 233–235, 237, 238, 240, 242, 243, 245, 249–255, 260, 264, 266–270, 272–277, 295, 296, 360, 361, 384, 393, 399–401, 409–411, 413, 414, 416–419, 424, 425, 428, 434, 437, 439–445, 454, 455, 459, 461–465, 467, 665, 789, 793, 796–799, 892, 896, 897

Film thickness, 115, 120, 151, 181, 249, 253, 270, 349, 479, 482–485, 487–491, 493, 495, 496, 501, 503–505, 510–512, 515–518, 521, 533, 540–547, 549, 553, 554, 569, 570, 574, 576, 578, 581, 582, 585, 588, 599–602, 606, 608, 612, 613, 617, 620, 622–626, 628, 630, 633–637, 646, 648, 649, 651, 656, 658, 659, 661–670, 672–675, 678, 700, 712, 720, 725, 734, 735, 739, 741, 742, 746, 749, 752, 755, 756, 759, 766, 768, 772, 773, 776, 778–782, 785, 787, 789, 795, 811, 813, 816, 818, 820, 829, 838, 840, 841, 843, 846, 847, 851, 854, 856, 861, 865, 872, 885, 886, 888, 890–892

Flow streamlines, 152–155, 162, 188, 189, 193, 205, 746, 749, 751, 754, 756, 757, 759–762, 764–766, 771–775, 777–781, 783, 784, 786, 787, 793

Fluid friction, 885, 890

Follower, 234, 235, 264, 296, 330, 361, 394, 398, 400–402, 409, 411–417

Fourier's law, 120

Friction, 13, 81, 82, 86–89, 91–93, 97, 101–103, 157, 189, 228, 233–236, 238, 242, 252, 254, 255, 258, 259, 262–265, 267, 270, 271, 273, 276, 277, 307, 326, 330–344,

346–355, 360, 365, 376, 377, 379, 380, 384, 385, 391–393, 397, 399–401, 404, 406–413, 415, 417, 421–425, 438–441, 443–445, 457, 462, 463, 466, 467, 479–481, 484, 486, 487, 489–491, 493, 496, 497, 499, 553, 555, 556, 574, 583, 586, 588, 590, 592–594, 596, 608, 628, 652, 678, 689, 692, 693, 698–700, 720–722, 726, 731, 733–735, 737, 739, 746, 748, 751, 752, 756–759, 765, 767, 768, 772–777, 782–784, 786, 788, 789, 793, 795–798, 819, 830, 839–841, 844, 851, 856, 865–872, 877, 879, 880, 883–886, 892

Fully flooded lubrication, 511, 512, 516, 556, 559, 560, 562, 568, 569, 571, 635, 636, 638, 657, 658, 660, 661, 668, 857–860, 864–866, 869, 870, 873–875

Functional, 17

Gel permeation chromatography, 137, 142, 169, 220
Grease, 709, 711–713, 715–718
Grease film thickness, 733
Grease rheology, 724
Grease viscosity, 712, 713

Harmonic function, 9, 19, 21, 23
Heat generation, 483, 576, 599, 605, 606, 608, 609, 619, 687, 694, 695, 789, 798
Heat transfer, 626, 688, 716
Heavily loaded lubricated contacts, 252, 253, 336, 421, 506, 515, 619, 678, 688
Hertzian pressure, 57, 61, 63, 252, 266, 267, 271, 341, 348, 376, 421–423, 439, 440, 462, 504, 513, 544, 554, 557, 564, 566, 569–571, 574, 580, 583,

585, 587, 589, 591, 608, 638, 642–645, 648, 659, 666, 669, 698, 700, 739, 773, 777, 793, 887
Hertzian region, see also External region, Outer region, 503, 507, 513, 515–519, 528, 533, 540, 541, 544–547, 552, 553, 574, 575, 578, 579, 583, 584, 589, 600, 630, 631, 633, 637, 645, 655, 658, 659, 698–700, 754, 775, 779, 780, 788, 795, 891, 892
High temperature high shear viscosity, 132
Huggins equation, 121

Inclusions, see also Material defects, 234, 242–250, 257, 264, 265, 267, 273, 417–421, 437, 443
Initial mean of crack half-lengths, 438–440, 443
Initial standard deviation of crack half-lengths, 438–440, 444
Inlet zone, 252, 387, 503, 504, 507–520, 536, 538, 539, 541–549, 552, 553, 555, 556, 558, 562–576, 579, 585, 588, 596, 599–603, 605, 606, 608, 610, 616, 617, 620, 622, 624–626, 630–635, 637–639, 641–646, 648, 651, 652, 655–658, 660–665, 667, 670–673, 678, 687, 688, 698, 700, 727, 734, 739, 742, 754, 755, 763, 764, 772, 779, 784, 788, 795, 884
Inner solution, 33, 34, 37, 38
Inner variable, 32, 37, 70, 90
Inner zone, 32–34, 37, 38, 42–44, 67, 68, 70, 71, 86, 87, 90, 91
Input oil meniscus, 851, 852, 856, 861–865, 867–872, 876–878
Isothermal EHL problem, 504, 515, 599, 604, 609, 610, 617, 619, 626–628, 651, 653, 678, 688,

695, 697, 721, 739, 749, 772, 773, 776, 795

Kinematic viscosity, 111, 130, 132
Kinetic equation, 151–155, 162, 166, 170, 175, 187, 188, 195, 205, 213, 746, 750, 751, 757, 764, 771, 778, 800

Large O, 1
Lightly loaded lubricated contacts, 484, 489, 494, 495, 499, 772, 813, 823, 881
Linear polymer, 135
Logarithmic potential, 98
Lubricated and non-lubricated stripes, 873

Mark-Houwink equation, 122
Matching, 5, 33, 37, 42, 86, 90, 512, 521, 530, 535, 537, 538, 621, 631, 649, 655, 894
Material defects, see also Inclusions, 234, 270, 273, 413, 417, 428, 455, 457
Material fracture toughness, 356, 422, 427, 440, 458, 462
Material local survival probability, 430, 431, 436, 460, 461
Mechanical degradation, 122, 137, 142
Method of matched asymptotic expansions, see also Singular asymptotic method, 5, 31, 36, 90, 503
Method of regular perturbations, see also Regular asymptotic method, 35, 40, 488
Micelle, 139, 141
Mini-rotary viscometer, 131
Mixed friction, 870, 873, 884, 889, 892
Modified problem formulation, 814, 817, 825, 826, 828, 850
Modified solution, 814–818, 826–828

Molecular weight, 114, 115, 117, 121, 122, 129, 135–139, 141, 142, 151, 789
Molecular weight degradation, 117
Mono-grade, 130
Monomer, 121
Multi-grade, 129, 130

Newton's law of viscosity, 110
Newtonian rheology, 772, 777
Non-isothermal conditions, 483, 503, 601, 650
Non-Newtonian lubricant, 113, 480, 496, 515, 590, 626, 652, 665, 669, 688, 700, 746, 771–775, 777–779, 781, 784
Non-Newtonian rheology, 540, 579, 616, 643, 650, 675, 696, 745, 747, 772, 776, 779, 780, 793
Number-average molecular weight, 121
Numerical instability, 556, 605
Numerical stability, 586, 609, 835

Octahedral stress, 868, 873–878
Olefin copolymer, 135
Outer region, see also Hertzian region, External region, 32, 34, 36–38, 43
Outer solution, 32, 34, 37
Over-critical lubrication regime, 504, 512, 517, 518, 538, 541, 542, 545–547, 575, 601, 617, 621, 624, 631, 644–646, 650, 651, 655, 658–660, 663, 665, 668, 669, 698, 741, 882

Paris equation, 418
Partially lubricated contact, 874, 886, 887, 889, 890
Permanent viscosity loss, see also Viscosity loss, 129
Pitting, see also Contact fatigue, 234, 235, 239, 264, 265, 272–274, 276, 277, 296, 394,

397, 398, 406, 411, 415–418, 437, 440, 442–445, 449, 454, 463, 480

Plane of fatigue crack propagation, see also Crack propagation angle, Direction of fatigue crack growth, 457

Polyalkylmethacrylate viscosity modifier, 137

Polymer, 121, 135

Polymer concentration, 136, 138, 152, 161, 179, 204, 227, 751, 797

Pre-critical lubrication regime, 504, 512, 518, 538, 541, 542, 545, 575, 608, 620, 621, 623–625, 631, 632, 636, 655–657, 662–664, 668, 669, 671, 882, 891, 892

Pressure coefficient of viscosity, 116, 348, 505, 582, 759, 773, 789

Principle of maximum, 9, 19, 21

Probability of scission, 152, 153, 156–158, 165, 169, 180, 181, 189, 196, 226, 228, 230, 765, 787, 796

Propagation of fatigue cracks, 235, 404, 424, 425

Random copolymer, 135

Regular asymptotic method, see also Method of regular perturbations, 4, 44, 47, 66, 68, 70, 85, 100, 316, 499, 503, 730, 731

Regularization approach, 480, 604, 609

Residual stress, 234, 236–242, 244, 255, 258, 264, 265, 267, 269, 273–277, 326, 330, 333, 334, 342, 343, 350, 351, 354, 355, 358, 377, 378, 380, 383, 385, 391, 397, 402, 405, 406, 408, 410, 416–418, 421, 423–427, 431, 434, 436, 437, 439–441,

444, 445, 451, 457, 458, 462, 463, 465, 794

Reynolds equation, 714, 716–719, 740, 742, 746, 758, 765–769, 771, 819, 824, 828, 879

Reynolds number, 483

Rheology, 121, 131, 136, 138, 143, 481, 483–485, 488, 490, 491, 496, 498, 499, 503, 504, 571, 579, 627–630, 633, 636, 637, 646, 651, 653, 654, 656–662, 666–671, 674, 677, 680, 688, 700, 736, 739, 742, 745, 747, 748, 757, 772–777, 783, 785, 786, 788, 789, 793, 795, 876

Roughness, see also Asperities, 9, 29, 34, 40, 47, 59, 60, 65–69, 74, 99, 103, 104, 235, 236, 265, 267, 269, 270, 274, 445–447, 452, 454, 455, 479, 546, 844

SAE J300, 130, 133

SAE J306, 134

Separatrix, 754–757, 762, 766

Shape factor, 118

Shear stability, 122, 142, 143

Shear stability index, 122

Single bond, 140

Singular asymptotic method, see also Method of matched asymptotic expansions, 4, 5

Small o, 1

Stable solution, 565, 570, 574, 605, 608, 643

Star polymer, 129, 135, 139, 141, 187–190, 193, 194, 197–205, 213, 214, 218, 220, 222–230

Starved lubrication, 511–514, 516, 556, 559, 560, 562, 564–567, 569, 571, 573, 574, 600, 601, 603, 604, 616, 623, 625, 634, 636–639, 641, 651, 657, 658, 660, 661, 851, 870, 891, 892

Stress intensity factor, 235, 239, 242, 243, 252, 254, 258, 261–264,

271, 276, 300, 308, 318–325,
330–344, 348, 350–358, 361,
365, 369, 371, 374, 377–
380, 382–386, 390–394, 397,
401, 402, 404–406, 408–410,
412–416, 418, 419, 421–428,
432–434, 440, 449, 456, 457,
463, 464, 793

Structural anisotropic fluid, 885

Structural fatigue, see also Bending
fatigue, Torsional fatigue,
267, 455, 465

Styrene diene viscosity modifier, 139

Styrene-alkylmaleate ester viscosity
modifier, 138

Surface tangential displacements, 581,
627

Survival probability of material as a
whole, 431, 434–436, 439

Tapered block copolymer, 135, 141

Thermal conductivity, 120

Thermal EHL problem, 484, 494,
590, 616, 722, 789, 798

Thermal grease lubrication, 739

Threshold shear stress, 713, 733, 742

Topology of the lubricant flow, 760,
762

Torsional fatigue, see also Structural fatigue, Bending fatigue, 463

Transient motion, 828, 839, 846

Uniformly valid asymptotic expansion, 2–4, 6, 31, 34, 36, 41,
44, 55, 57, 59, 63, 69, 72, 85,
87, 89, 101, 513

Unstable solution, 575, 604

Variational inequality, 9, 16–18, 27,
93, 95, 96

Viscosity, 110, 112–116, 120, 121,
130, 131, 152, 156, 158,
161, 165, 169, 174–176, 178,
179, 181, 199, 204, 217, 220,

222, 223, 226, 228, 252, 253,
348, 480, 481, 485, 486, 488,
495, 502, 505, 512, 517, 542,
543, 546, 548, 552, 555, 581,
582, 591, 599, 601, 602, 606,
625, 627–631, 638, 651, 652,
655–657, 659–664, 666, 667,
676, 677, 694, 695, 698, 739,
746, 747, 751, 752, 755–759,
764–769, 778, 779, 781, 783,
786–789, 791, 793, 796, 797,
801, 816, 850, 876, 885, 886,
892

Viscosity grade, 122, 129, 134

Viscosity index, 112

Viscosity loss, see also Permanent
viscosity loss, 122, 123,
179–181, 187, 188, 226, 228,
229, 775, 786, 788, 789, 797,
798

Viscosity modifiers, 109, 117, 124,
129, 130, 137, 142

Walther equation, 111

Weight-average molecular weight,
121

Yield stress, 131

T - #0285 - 071024 - C4 - 234/156/41 - PB - 9780367383794 - Gloss Lamination